STUDENT'S SOLUTIONS MANUAL

GARRET ETGEN

University of Houston

COLLEGE MATHEMATICS FOR BUSINESS, ECONOMICS, LIFE SCIENCES, AND SOCIAL SCIENCES

THIRTEENTH EDITION

Raymond Barnett

Merritt College

Michael Ziegler

Marquette University

Karl Byleen

Marquette University

PEARSON

Boston Columbus Indianapolis New York San Francisco Upper Saddle River
Amsterdam Cape Town Dubai London Madrid Milan Munich Paris Montreal Toronto
Delhi Mexico City São Paulo Sydney Hong Kong Seoul Singapore Taipei Tokyo

D1416284

The author and publisher of this book have used their best efforts in preparing this book. These efforts include the development, research, and testing of the theories and programs to determine their effectiveness. The author and publisher make no warranty of any kind, expressed or implied, with regard to these programs or the documentation contained in this book. The author and publisher shall not be liable in any event for incidental or consequential damages in connection with, or arising out of, the furnishing, performance, or use of these programs.

Reproduced by Pearson from electronic files supplied by the author.

Copyright © 2015, 2011, 2008 Pearson Education, Inc.
Publishing as Pearson, 75 Arlington Street, Boston, MA 02116.

ISBN-13: 978-0-321-94677-5
ISBN-10: 0-321-94677-4

www.pearsonhighered.com

Table of Contents

1   LINEAR EQUATIONS AND GRAPHS

EXERCISE 1-1

Things to remember:

1.   FIRST DEGREE, OR LINEAR, EQUATIONS AND INEQUALITIES

A FIRST DEGREE, or LINEAR, EQUATION in one variable x is an equation that can be written in the form

STANDARD FORM: $ax + b = 0$, $a \neq 0$

If the equality symbol = is replaced by $<, >, \leq$, or $\geq$, then the resulting expression is called a FIRST DEGREE, or LINEAR, INEQUALITY.

2.   SOLUTIONS

A SOLUTION OF AN EQUATION (or inequality) involving a single variable is a number that when substituted for the variable makes the equation (or inequality) true. The set of all solutions is called the SOLUTION SET. To SOLVE AN EQUATION (or inequality) we mean that we find the solution set. Two equations (or inequalities) are EQUIVALENT if they have the same solution set.

3.   EQUALITY PROPERTIES
An equivalent equation will result if:

a) The same quantity is added to or subtracted from each side of a given equation.

b) Each side of a given equation is multiplied by or divided by the same nonzero quantity.

4.   INEQUALITY PROPERTIES
An equivalent inequality will result and the SENSE OR DIRECTION WILL REMAIN THE SAME if each side of the original inequality:

a) Has the same real number added to or subtracted from it.

b) Is multiplied or divided by the same positive number.

An equivalent inequality will result and the SENSE OR DIRECTION WILL REVERSE if each side of the original inequality:

c) Is multiplied or divided by the same negative number.

NOTE: Multiplication and division by 0 is not permitted.

5.   The double inequality $a < x < b$ means that $a < x$ and $x < b$. Other variations, as well as a useful interval notation, are indicated in the following table.

Interval Notation	Inequality Notation	Line Graph
$[a, b]$	$a \leq x \leq b$	
$[a, b)$	$a \leq x < b$	

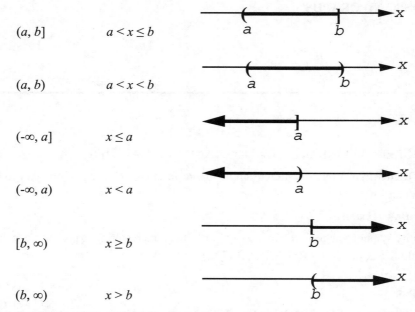

$(a, b]$	$a < x \le b$
(a, b)	$a < x < b$
$(-\infty, a]$	$x \le a$
$(-\infty, a)$	$x < a$
$[b, \infty)$	$x \ge b$
(b, ∞)	$x > b$

[<u>Note</u>: An endpoint on a line graph has a square bracket through it if it is included in the inequality and a parenthesis through it if it is not. An interval of the form $[a, b]$ is a CLOSED INTERVAL, an interval of the form (a, b) is an OPEN INTERVAL. Intervals of the form $(a, b]$ and $[a, b)$ are HALF-OPEN and HALF-CLOSED.]

<u>6</u>.   PROCEDURE FOR SOLVING WORD PROBLEMS

    a.   Read the problem carefully and introduce a variable to represent an unknown quantity in the problem. Often the question asked in a problem will indicate the best way to introduce this variable.

    b.   Identify other quantities in the problem (known or unknown) and, whenever possible, express unknown quantities in terms of the variable you introduced in a.

    c.   Write a verbal statement using the conditions stated in the problem and then write an equivalent mathematical statement (equation or inequality).

    d.   Solve the equation or inequality and answer the questions posed in the problem.

    e.   Check the solution(s) in the original problem.

---

1. $2m + 9 = 5m - 6$

    $2m + 9 - 9 = 5m - 6 - 9$ [using <u>3</u>(a)]

        $2m = 5m - 15$

    $2m - 5m = 5m - 15 - 5m$ [using <u>3</u>(a)]

       $-3m = -15$

       $\dfrac{-3m}{-3} = \dfrac{-15}{-3}$ [using <u>3</u>(b)]

        $m = 5$

3. $2x + 3 < -4$

    $2x + 3 - 3 < -4 - 3$ [using <u>4</u>(a)]

        $2x < -7$

         $x < -\dfrac{7}{2}$ [using <u>4</u>(b)]

5. $-3x \ge -12$

       $\dfrac{-3x}{-3} \le \dfrac{-12}{-3}$ [using <u>4</u>(c)]

        $x \le 4$

7. $-4x - 7 > 5$

$-4x > 5 + 7$

$-4x > 12$

$x < -3$

Graph of $x < -3$ is:

$$\xleftarrow{\hspace{2cm}}\overset{\displaystyle\mathbf{-3}}{)}\xrightarrow{\hspace{3cm}} x$$

9. $2 \le x + 3 \le 5$

$2 - 3 \le x \le 5 - 3$

$-1 \le x \le 2$

Graph of $-1 \le x \le 2$ is:

$$\xleftarrow{\hspace{1.5cm}}\underset{\mathbf{-1}}{[}\rule[0.5ex]{2cm}{1pt}\underset{\mathbf{2}}{]}\xrightarrow{\hspace{1.5cm}} x$$

11. $\dfrac{x}{4} + \dfrac{1}{2} = \dfrac{1}{8}$

Multiply both sides of the equation by 8. We obtain:

$2x + 4 = 1$ [using $\underline{3}$(b)]

$2x = -3, \quad x = -3/2$

13. $\dfrac{y}{-5} > \dfrac{3}{2}$

Multiply both sides of the inequality by 10. We obtain:

$-2y > 15$

$y < -\dfrac{15}{2}$ [using $\underline{4}$(c)]

15. $2u + 4 = 5u + 1 - 7u$

$2u + 4 = -2u + 1$

$4u = -3$

$u = -\dfrac{3}{4}$

17. $10x + 25(x - 3) = 275$

$10x + 25x - 75 = 275$

$35x = 275 + 75$

$35x = 350$

$x = \dfrac{350}{35}$

$x = 10$

19. $3 - y \le 4(y - 3)$

$3 - y \le 4y - 12$

$-5y \le -15$

$y \ge 3$

[$\underline{\text{Note}}$: Division by a negative number, -5.]

21. $\dfrac{x}{5} - \dfrac{x}{6} = \dfrac{6}{5}$

Multiply both sides of the equation by 30. We obtain:

$6x - 5x = 36$

$x = 36$

23. $\dfrac{m}{5} - 3 < \dfrac{3}{5} - \dfrac{m}{2}$

Multiply both sides of the inequality by 10. We obtain:

$2m - 30 < 6 - 5m$ [using $\underline{4}$(b)]

$7m < 36$

$m < \dfrac{36}{7}$

25. $2 \le 3x - 7 < 14$

$7 + 2 \le 3x < 14 + 7$

$9 \le 3x < 21$

$3 \le x < 7$

Graph of $3 \le x < 7$ is:

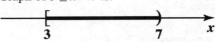

27. $-4 \le \dfrac{9}{5}C + 32 \le 68$

$-36 \le \dfrac{9}{5}C \le 36$

$-36\left(\dfrac{5}{9}\right) \le C \le 36\left(\dfrac{5}{9}\right)$

$-20 \le C \le 20$

Graph of $-20 \le C \le 20$ is:

$$\xleftarrow{\hspace{1cm}}\underset{\mathbf{-20}}{[}\rule[0.5ex]{2cm}{1pt}\underset{\mathbf{20}}{]}\xrightarrow{\hspace{1.5cm}} x$$

29. $3x - 4y = 12$

$3x = 12 + 4y$

$3x - 12 = 4y$

$y = \dfrac{1}{4}(3x - 12)$

$y = \dfrac{3}{4}x - 3$

31. $Ax + By = C$
$By = C - Ax$
$y = \dfrac{C}{B} - \dfrac{Ax}{B}, B \neq 0$

or $\quad y = -\left(\dfrac{A}{B}\right)x + \dfrac{C}{B}$

33. $F = \dfrac{9}{5}C + 32$

$\dfrac{9}{5}C + 32 = F$

$\dfrac{9}{5}C = F - 32$

$C = \dfrac{5}{9}(F - 32)$

35. $-3 \leq 4 - 7x < 18$
$-3 - 4 \leq -7x < 18 - 4$
$-7 \leq -7x < 14.$

Dividing by -7, and recalling $\underline{4}$(c), we have
$1 \geq x > -2$ or $-2 < x \leq 1$
The graph is:

37. (A) $ab > 0$; $a > 0$ $\underline{\text{and}}$ $b > 0$, or $a < 0$ $\underline{\text{and}}$ $b < 0$

(B) $ab < 0$; $a > 0$ $\underline{\text{and}}$ $b < 0$, or $a < 0$ $\underline{\text{and}}$ $b > 0$

(C) $\dfrac{a}{b} > 0$; $a > 0$ $\underline{\text{and}}$ $b > 0$, or $a < 0$ $\underline{\text{and}}$ $b < 0$

(D) $\dfrac{a}{b} < 0$; $a > 0$ $\underline{\text{and}}$ $b < 0$, or $a < 0$ $\underline{\text{and}}$ $b > 0$

39. Let $a, b > 0$. If $\dfrac{b}{a} > 1$, then $b > a$ so $a - b < 0$; $a - b$ is negative.

41. True: Let (a, b) and (c, d) be open intervals such that $(a, b) \cap (c, d) \neq \varnothing$. If $(a, b) \subset (c, d)$, then $(a, b) \cap (c, d) = (a, b)$ an open interval. Similarly, if $(c, d) \subset (a, b)$. If neither interval is contained in the other, then we can assume that $a < c < b < d$, and $(a, b) \cap (c, d) = (c, b)$, an open interval.

43. False: $(0, 1) \cup (2, 3)$ is **not** an open interval.

45. Same as 41.

47. Let x = number of $35 tickets.
Then the number of $55 tickets = $9500 - x$.
Now,

$35x + 55(9500 - x) = 432,500$
$35x + 522,500 - 55x = 432,500$
$-20x = -90,000$
$x = 4,500$

Thus, 4,500 $35 tickets and $9,500 - 4,500 = 5,000$ $55 tickets were sold.

49. Let x = the amount invested in Fund A. Then $500,000 - x$ is the amount invested in Fund B. The annual interest income is
$I = 0.052x + 0.077(500,000 - x)$

Set $I = \$34,000$.

$$0.052x + 0.077(500,000 - x) = 34,000$$
$$52x + 77(500,000 - x) = 34,000,000$$
$$52x - 77x = 34,000,000 - 38,500,000$$
$$-25x = -450,000$$
$$x = 180,000$$

You should invest $180,000 in Fund A and $320,000 in Fund B.

51. Let x be the price of the car in 2012. Then

$$\frac{x}{10,000} = \frac{229.6}{156.9} \quad \text{(refer to Table 2, Example 10)}$$

$$x = 10,000 \cdot \frac{229.6}{156.9} \approx \$14,634 \quad \text{(to the nearest dollar)}$$

53. (A) Wholesale price $300; mark-up $0.4(300) = 120$
Retail price $300 + \$120 = \$420$
(B) Let x = wholesale price. Then
$$x + 0.4x = 77$$
$$1.4x = 77$$
$$x = 55$$

The wholesale price is $55.

55. The difference per round is $8. The number of rounds you must play to recover the cost of the clubs is:
$$\frac{270}{8} = 33.75$$

Since the number of rounds must be a whole number (positive integer), you will recover the cost of the clubs after 34 rounds.

57. The employee must earn $2000 in commission.
If x = sales over $7,000, then
$$0.08x = 2000 \quad \text{and} \quad x = \frac{2000}{0.08} = 25,000$$

Therefore, the employee must sell $32,000 in the month.

59. Let x = number of books produced. Then

Costs: $C = 1.60x + 55,000$

Revenue: $R = 11x$

To find the break-even point, set $R = C$:

$$11x = 1.60x + 55,000$$
$$9.40x = 55,000$$
$$x = 5851.06383$$

Thus, 5851 books will have to be sold for the publisher to break even.

61. Let x = number of books produced.

   Costs: $C = 55,000 + 2.10x$

Revenue: $R = 11x$

(A)   The obvious strategy is to raise the sales price of the book.

(B)   To find the break-even point, set $R(x) = C(x)$:

$$11x = 55,000 + 2.10x$$
$$8.90x = 55,000$$
$$x = 6179.78$$

The company must sell more than 6180 books.

(C)   From Problem 59, the production level at the break-even point is: 5,851 books. At this production level, the costs are:

$$C = 55,000 + 2.10(5,851) = \$67,287.10.$$

If p is the new price of the book, then we need

$$5851p = 67,287.10 \text{ and } p \approx 11.50.$$

The company should increase the price at least $0.50 (50 cents).

63. Let x = the number of rainbow trout in the lake. Then,

$$\frac{x}{200} = \frac{200}{8} \quad \text{(since proportions are the same)}$$

$$x = \frac{200}{8}(200)$$

$$x = 5,000$$

65. $\text{IQ} = \dfrac{\text{Mental age}}{\text{Chronological age}}(100)$

$$\frac{\text{Mental age}}{9}(100) = 140$$

$$\text{Mental age} = \frac{140}{100}(9)$$

$$= 12.6 \text{ years}$$

EXERCISE 1-2

Things to remember:

1.   CARTESIAN (RECTANGULAR) COORDINATE SYSTEM

The Cartesian coordinate system is formed by the intersection of a horizontal real number line, called the **x-axis**, and a vertical real number line, called the **y-axis**. The two number lines intersect at their origins. The axes, called the COORDINATE AXES, divide the plane into four parts called QUADRANTS, which are numbered counterclockwise from I to IV. (See the figure.)

COORDINATES are assigned to each point in the plane as follows. Given a point P in the plane, pass a horizontal and a vertical line through P (see the figure). The vertical line will intersect the x-axis at a point with coordinate a; the horizontal line will intersect the y-axis at a point with coordinate b. The ORDERED PAIR (a, b) are the coordinates of P. The first coordinate, a, is called the ABSCISSA OF P; the second coordinate, b, is called the

ORDINATE OF P. The point with coordinates $(0, 0)$ is called the ORIGIN. There is a one-to-one correspondence between the points in the plane and the set of all ordered pairs of real numbers.

2.   LINEAR EQUATIONS IN TWO VARIABLES

A LINEAR EQUATION IN TWO VARIABLES is an equation that can be written in the STANDARD FORM

$$Ax + By = C$$

where A, B, and C are constants and A and B are not both zero.

3.   GRAPH OF A LINEAR EQUATION IN TWO VARIABLES

The graph of any equation of the form

$$Ax + By = C \quad (A \text{ and } B \text{ not both } 0)$$

is a line, and every line in a Cartesian coordinate system is the graph of a linear equation in two variables. The graph of the equation $x = a$ is a VERTICAL LINE; the graph of the equation $y = b$ is a HORIZONTAL LINE.

4.   INTERCEPTS

If a line crosses the x-axis at a point with x coordinate a, then a is called an x intercept of the line; if it crosses the y-axis at a point with y coordinate b, then b is called the y intercept.

5.   SLOPE OF A LINE

If a line passes through two distinct points $P_1(x_1, y_1)$ and $P_2(x_2, y_2)$, then its slope is given by the formula

$$m = \frac{y_2 - y_1}{x_2 - x_1} = \frac{\text{vertical change (rise)}}{\text{horizontal change (run)}}$$

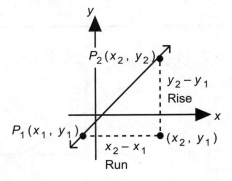

6. GEOMETRIC INTERPRETATION OF SLOPE

Line	Slope	Example
Rising as x moves from left to right	Positive	
Falling as x moves from left to right	Negative	
Horizontal	0	
Vertical	Not defined	

7. EQUATIONS OF A LINE; SPECIAL FORMS

Standard form	$Ax + By = C$	A and B not both 0
Slope-intercept form	$y = mx + b$	Slope: m; y-intercept: b
Point-slope form	$y - y_1 = m(x - x_1)$	Slope: m; point: (x_1, y_1)
Horizontal line	$y = b$	Slope: 0
Vertical line	$x = a$	Slope: Undefined

1. (D)

3. (C); The slope is 0.

5. $y = 2x - 3$

x	y
0	−3
1	−1
4	5

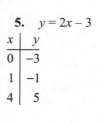

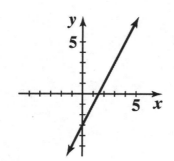

7. $2x + 3y = 12$

x	y
0	4
6	0
9	−2

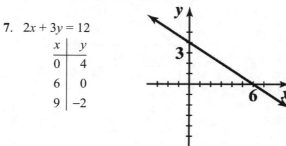

9. $y = 5x - 7$; slope: $m = 5$; y intercept: $b = -7$

11. $y = -\dfrac{5}{2}x - 9$; slope: $m = -\dfrac{5}{2}$; y intercept: $b = -9$

13. $y = \dfrac{x}{4} + \dfrac{2}{3} = \dfrac{1}{4}x + \dfrac{2}{3}$; slope: $m = \dfrac{1}{4}$; y intercept: $b = \dfrac{2}{3}$

15. $m = 2, b = 1$; using $\underline{7}$, equation: $y = 2x + 1$

17. $m = -\dfrac{1}{3}, \; b = 6$; using $\underline{7}$, equation: $y = -\dfrac{1}{3}x + 6$

19. x intercept: -1 [or $(-1, 0)$]

y intercept: -2 [or $(0, -2)$]

slope: $m = \dfrac{-2 - 0}{0 - (-1)} = \dfrac{-2}{1} = -2$

equation: $y = -2x - 2$

21. x intercept: -3 [or $(-3, 0)$]

y intercept: 1 [or $(0, 1)$]

slope: $m = \dfrac{1 - 0}{0 - (-3)} = \dfrac{1}{3}$

equation: $y = \dfrac{1}{3}x + 1$

23. $y = -\dfrac{2}{3}x - 2$

$m = -\dfrac{2}{3}, b = -2$

x	y
0	-2
3	-4
-3	0

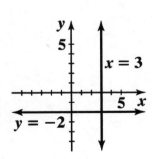

25. $3x - 2y = 10$

x	y
0	-5
10	10
4	1

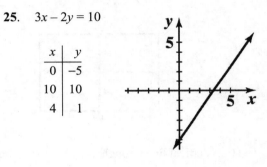

27.

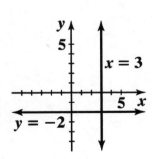

29. $4x + y = 3$

$\quad y = -4x + 3$; slope: $m = -4$

31. $3x + 5y = 15$

$\quad 5y = -3x + 15$

$\quad y = -\dfrac{3}{5}x + 3$; slope: $m = -\dfrac{3}{5}$

33. $-4x + 2y = 9$

$\quad 2y = 4x + 9$

$\quad y = 2x + \dfrac{9}{2}$; slope: $m = 2$

35.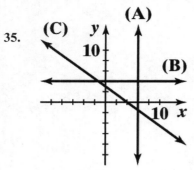

37.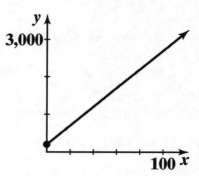

39. (A)

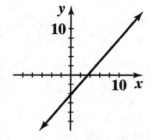

(B) x intercept — set $y = 0$:

$$1.2x - 4.2 = 0$$
$$x = 3.5$$

y intercept — set $x = 0$:

$$y = -4.2$$

(C)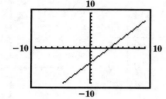

(D) x-intercept: 3.5;
    y-intercept: –4.2

41.   Vertical line through $(4, -3)$: $x = 4$; horizontal line through $(4, -3)$: $y = -3$.

43.   Vertical line through $(-1.5, -3.5)$: $x = -1.5$; horizontal line through $(-1.5, -3.5)$: $y = -3.5$.

45.   Slope: $m = 5$; point: $(3, 0)$. Using the point-slope form:

$$y - 0 = 5(x - 3)$$
$$y = 5x - 15$$

47.   Slope: $m = -2$; point: $(-1, 9)$. Using the point-slope form:

$$y - 9 = -2[x - (-1)]$$
$$y - 9 = -2(x + 1)$$
$$y = -2x + 7$$

49.   Slope: $m = 1 / 3$; point: $(-4, -8)$. Using the point-slope form:

$$y - (-8) = \frac{1}{3}[x - (-4)]$$

$$y + 8 = \frac{1}{3}(x + 4)$$

$$y = \frac{1}{3}x + \frac{4}{3} - 8$$

$$y = \frac{1}{3}x - \frac{20}{3}$$

51. Slope: $m = -3.2$; point: $(5.8, 12.3)$.   Using the point-slope form:

$$y - 12.3 = -3.2(x - 5.8)$$

$$y - 12.3 = -3.2x + 18.56$$

$$y = -3.2x + 30.86$$

53. Points $(2, 5), (5, 7)$

(A) Using 5, slope: $m = \dfrac{7 - 5}{5 - 2} = \dfrac{2}{3}$

(B) Using the point-slope form:

$$y - 5 = \frac{2}{3}(x - 2)$$

Simplifying: $3y - 15 = 2x - 4$ and $-2x + 3y = 11$

From (B), $y = \dfrac{2}{3}x + \dfrac{11}{3}$

55. Points $(-2, -1), (2, -6)$

(A) Using 5, slope: $m = \dfrac{-6 - (-1)}{2 - (-2)} = -\dfrac{5}{4}$

(B) Using the point-slope form:

$$y - (-1) = -\frac{5}{4}[x - (-2)]$$

$$y + 1 = -\frac{5}{4}(x + 2)$$

Simplifying: $4y + 4 = -5x - 10$ and $5x + 4y = -14$

(C) From (B), $y = -\dfrac{5}{4}x - \dfrac{7}{2}$

57. Points $(5, 3), (5, -3)$

(A) Using 5, slope: $m = m = \dfrac{-3 - 3}{5 - 5} = \dfrac{-6}{0}$, not defined

(B) The line is vertical; equation $x = 5$.

(C) The slope is not defined; no slope-intercept form.

59. Points $(-2, 5), (3, 5)$

(A) Using 5, slope: $m = \dfrac{5-5}{3-(-2)} = \dfrac{0}{5} = 0$

(B) The line is horizontal; equation $y = 5$.

(C) $y = 0x + 5$ or $y = 5$

61. The graphs of $y = mx + 2$, m any real number, all have the same y-intercept $(0, 2)$; for each real number m, $y = mx + 2$ is a non-vertical line that passes through the point $(0, 2)$.

63. Fixed cost: $124; variable cost $0.12 per doughnut.

Total daily cost for producing x doughnuts

$C = 0.12x + 124$

At a total daily cost of $250,

$0.12x + 124 = 250$

$0.12x = 126$

$x = 1050$

Thus, 1050 doughnuts can be produced at a total daily cost of $250.

65. (A) Since daily cost and production are linearly related,

$$C = mx + b$$

From the given information, the points $(80, 7647)$ and $(100, 9147)$ satisfy this equation. Therefore:

$$\text{slope } m = \frac{9147 - 7647}{100 - 80} = \frac{1500}{20} = 75$$

Using the point-slope form with $(x_1, C_1) = (80, 7647)$:

$C - 7467 = 75(x - 80)$

$C = 75x - 6000 + 7647$

$C = 75x + 1647$

(B)

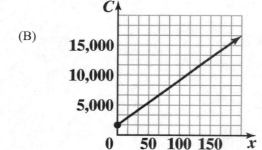

(C) The y-intercept, $1,647, is the fixed cost and the slope, $75, is the cost per club.

67. (A) $R = mC + b$

From the given information, the points $(85, 112)$ and $(175, 238)$ satisfy this equation. Therefore,

$$\text{slope } m = \frac{238 - 112}{175 - 85} = \frac{126}{90} = 1.4$$

Using the point-slope form with $(C_1, R_1) = (85, 112)$

$R - 112 = 1.4(C - 85)$

$R = 1.4C - 119 + 112$

$R = 1.4C - 7$

(B) Set $R = 185$. Then

$$185 = 1.4C - 7$$
$$1.4C = 192$$
$$C = 137.14$$

To the nearest dollar, the store pays $137.

69. (A) $V = mt + 157,000$. At $t = 10$,

$$V = 10m + 157,000 = 82,000$$
$$10m = 82,000 - 157,000 = -75,000$$
$$m = -7,500$$
$$V = -7,500t + 157,000$$

(B) At $t = 6$, $V = -7,500(6) + 157,000 = 112,000$

The value of the tractor after 6 years is $112,000.

(C) Solve $-7,500t + 157,000 < 70,000$ for t:

$$-7,500t < 70,000 - 157,000 = -87,000$$
$$t > 11.6$$

The value of the tractor will fall below $70,000 in the 12th year.

(D)

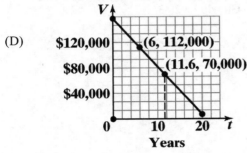

71. (A) $T = mx + b$

At $x = 0$, $T = 212$. Therefore, $b = 212$ and $T = mx + 212$.

At $x = 10$ (thousand)

$$193.6 = 10m + 212$$
$$10m = 193.6 - 212 = -18.4$$
$$m = -1.84x + 212$$

Therefore, $T = -1.84x + 212$

(B) At $x = 3.5$, $T = -1.84(3.5) + 212 = 205.56$

The boiling point at 3,500 feet is 205.56°F.

(C) Solve $200 = -1.84x + 212$ for x:

$$-1.84x = 200 - 212 = -12$$
$$x \approx 6.5217$$

The boiling point is 200°F at 6,522 feet.

(D)

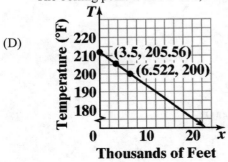

73. (A) $T = -3.6A + b$

At $A = 0$, $T = 70$. Therefore, $T = -3.6A + 70$.

(B) Solve $34 = -3.6A + 70$ for A:

$-3.6A = 34 - 70 = -36$

$A = 10$

The altitude of the aircraft is 10,000 feet.

75. (A) $N = mt + b$

At $t = 0$, $N = 2.76$. Therefore, $N = mt + 2.76$.

At $t = 32$,

$2.55 = 32m + 2.76$

$32m = -0.21$

$m = -0.0066$

Therefore, $N = -0.0066t + 2.76$.

(B) At $t = 50$, $N = -0.0066(50) + 2.76 = 2.43$

The average number of persons per household in 2030 will be 2.43.

77. (A) $f = mt + b$

At $t = 0$, $f = 21$. Therefore, $f = mt + 21$.

At $t = 10$,

$17.3 = 10m + 21$

$10m = -3.7$

$m = -0.37$

Therefore, $f = -0.37t + 21$.

(B) Solve $-0.37t + 21 < 12$ for t:

$-0.37t < -9$

$t > 24.32$

The percentage of female smokers will fall below 12% in 2024.

79. (A) $p = mx + b$

At $x = 7,500$, $p = 2.28$; at $x = 7,900$, $p = 2.37$.

Therefore, slope $m = \dfrac{2.37 - 2.28}{7,900 - 7,500} = \dfrac{0.09}{400} = 0.000225$

Using the point-slope form with $(x_1, p_1) = (7,500, 2.28)$:

$p - 2.28 = 0.000225(x - 7,500) = 0.000225x + 0.5925$

$p = 0.000225x + 0.5925$ Price-supply equation

(B) $p = mx + b$

At $x = 7,900$, $p = 2.28$; at $x = 7,800$, $p = 2.37$.

Therefore, slope $m = \dfrac{2.37 - 2.28}{7,800 - 7,900} = -\dfrac{0.09}{100} = -0.0009$

Using the point-slope form with $(x_1, p_1) = (7,900, 2.37)$:

$$p - 2.28 = -0.0009(x - 7,900) = -0.0009x + 7.11$$
$$p = -0.0009x + 9.39 \quad \text{Price-demand equation}$$

(C) To find the equilibrium point, solve
$$0.000225x + 0.5925 = -0.0009x + 9.39$$
$$0.001125x = 9.39 - 0.5925 = 8.7975$$
$$x = 7,820$$
At $x = 7,820$, $p = 0.000225(7,820) + 0.5925 = 2.352$
The equilibrium point is $(7,820, 2.352)$.

(D)

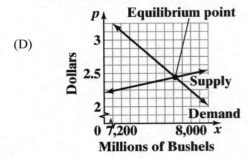

81. (A) $s = mw + b$

At $w = 0$, $s = 0$. Therefore, $b = 0$ and $s = mw$.

At $w = 5$, $s = 2$. Therefore,

$$2 = 5m \text{ and } m = \frac{2}{5}$$

Thus, $s = \frac{2}{5}w$.

(B) At $w = 20$, $s = \frac{2}{5}(20) = 8$

The spring stretches 8 inches.

(C) If $s = 3.6$,

$$3.6 = \frac{2}{5}w \text{ and } w = \frac{3.6(5)}{2} = 9$$

A 9 pound weight will stretch the spring 3.6 inches.

EXERCISE 1-3

Things to remember:

1.   LINEAR RELATION

If the variables x and y are related by the equation

$y = mx + b$ where m and b are constants with $m \neq 0$, then x and y are LINEARLY RELATED. The slope m is the rate of change of y with respect to x. If $P_1(x_1, y_1)$ and $P_2(x_2, y_2)$ are two distinct points on the line $y = mx + b$, then

$$m = \frac{y_2 - y_1}{x_2 - x_1} = \frac{\text{change in } y}{\text{change in } x}$$

This ratio is called the RATE OF CHANGE OF y with respect to x.

2.   LINEAR REGRESSION

REGRESSION ANALYSIS is a process for finding a function that models a given data set. Finding a linear model for a data set is called LINEAR REGRESSION and the line is called the REGRESSION LINE. A graph of the set of points in a data set is called a SCATTER PLOT. A regression model can be used to INTERPOLATE between points in a data set or to EXTRAPOLATE or predict points outside the data set.

---

1. (A) Let h = height in inches over 5 ft.
         w = weight in kilograms
    The linear model for Robinson's estimate is
    $w = 1.7h + 49$.

    (B) The rate of change of weight with respect to height is 1.7 kilograms per inch.

    (C) At $h = 4$, $w = 1.7(4) + 49 = 55.8$ kg.

    (D) Solve $60 = 1.7h + 49$ for h:
        $1.7h = 60 - 49 = 11$
        $h \approx 6.5$ inches
    The woman's height is approximately 5' 6.5".

3. (A) $p = md + b$
    At $d = 0$, $p = 14.7$. Therefore, $p = md + 14.7$.
    At $d = 33$, $p = 29.4$. Therefore,
    $$\text{slope } m = \frac{29.4 - 14.7}{33 - 0} = \frac{14.7}{33} = 0.4\overline{45}$$
    and $p = 0.4\overline{45}\,d + 14.7$.

    (B) The rate of change of pressure with respect to depth is $0.4\overline{45}$ lb/in^2 per foot.

    (C) At $d = 50$, $p = 0.4\overline{45}(50) + 14.7 \approx 37$ lb/in^2.

    (D) 4 atmospheres $= 14.7(4) = 58.8$ lb/in^2.
    Solve $58.8 = 0.4\overline{45}\,d + 14.7$ for d:
    $0.4\overline{45}\,d = 58.8 - 14.7 = 44.1$
        $d \approx 99$ feet

5. (A) $a = mt + b$
         At time $t = 0$, $a = 2880$. Therefore, $b = 2880$.
         At time $t = 120$, $a = 0$. The rate of change of altitude with respect to time is:
         $$m = \frac{0 - 2880}{120 - 0} = -24$$
        Therefore, $a = -24t + 2880$.

    (B) The rate of descent is -24 ft/sec.

    (C) The speed at landing is 24 ft/sec.

7. $s = mt + b$
    At temperature $t = 0$, $s = 331$. Therefore, $b = 331$.
    At temperature $t = 20$, $s = 343$. Therefore,

slope $m = \dfrac{343 - 331}{20 - 0} = \dfrac{12}{20} = 0.6$

Therefore, $s = 0.6t + 331$; the rate of change of the speed of sound with respect to temperature is 0.6 m/sec per °C.

9. $y = -0.3x + 84.4$

(A)

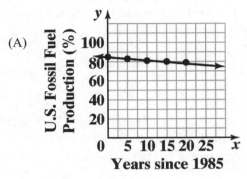

(B)   The rate of change of the percentage of fossil fuel with respect to time is − 0.3% per year.

(C)   At $x = 40$, $y \approx -0.3(40) + 84.4 = 72.4$; fossil fuel production as a percentage of total production will be 72.4%.

(D) Solve $-0.3x + 84.4 < 70$ for x:

$-0.3x < 70 - 84.4 = -14.4$

$x > 48$

Fossil fuel production will be less than 70% of total energy production in 2034.

11. $f = -0.39t + 21.93$

(A)

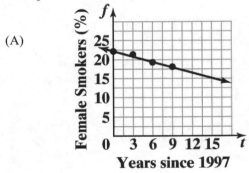

(B)   Solve $-0.39t + 21.93 < 10$ for t:

$-0.39t < 10 - 21.93 = -11.93$

$t > 30.59$

The first year in which the percentage of female smokers will be less than 10% is 2028.

13. $y = 0.23x + 9.56$

(A)

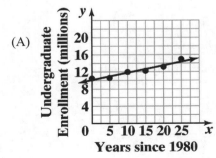

(B)   At $x = 45$, $y = 0.23(45) + 9.56 = 19.91$. The undergraduate enrollment on 2025 to the nearest hundred thousand will be 19,900,000.

(C)   Undergraduate enrollment is increasing at the rate of 230,000 students per year.

15. $y = 0.75x$

(A)

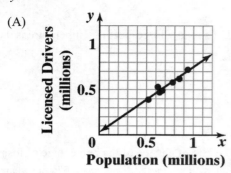

(B)   At $x = 1.6$, $y = 0.75(1.6) = 1.2$. There were approximately 1,200,000 licensed drivers in Idaho in 2010.

(C)   Solve $0.75 = 0.75x$ for x: $x = 1$
The population of Rhode Island in 2010 was approximately 1,000,000.

17. $S = 15.8t + 251$

(A)

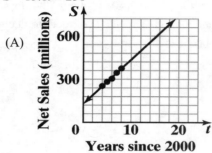

(B)   At $t = 22$, $S = 15.8(22) + 251 = 598.6 \approx 599$. Wal-Mart's net sales for 2022 will be approximately \$599 billion.

19. $E = -0.55T + 31$

(A)

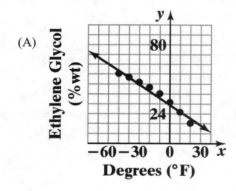

(B)   Solve $30 = -0.55T + 31$ for T:
$-0.55T = 30 - 31 = -1$
$T \approx 1.\overline{81}$
The freezing temperature of the solution is approximately 2°F.

(C)   at 15°F, $E = -0.55(15) + 31 = 22.75$.
A solution that freezes at 15°F is 22.75% ethylene glycol.

21. $y = 1.37x - 2.58$

(A) The rate of change of height with respect to Dbh is 1.37 ft/in.

(B) A 1 in. increase in Dbh produces a 1.37 foot increase in height.

(C) At $x = 15$, $y = 1.37(15) - 2.58 = 17.97$. The spruce is approximately 18 feet tall.

(D) Solve $25 = 1.37x - 2.58$ for x:
$1.37x = 25 + 2.58 = 27.58$
$x \approx 20.13$
The Dbh is approximately 20 inches.

23. $y = 1.70x + 30.90$

(A) The average monthly price is increasing at the rate of \$1.70 per year.

(B) At $x = 24$, $y = 1.70(24) + 30.90 = 71.70$. The average monthly price in 2024 will be $71.70.

25. Male enrollment: $y = 0.07x + 5.0$;
female enrollment: $y = 0.18x + 3.8$

(A) The male enrollment is increasing at the rate of $0.07(1,000,000) = 70,000$ students per year; the female enrollment is increasing at the rate of $0.18(1,000,000) = 180,000$ students per year.

(B) Male enrollment in 2025 ($x = 55$):
$y = 0.07(55) + 5.0 = 8.9$ or 8.9 million;
female enrollment in 2025:
$y = 0.18(55) + 3.8 \approx 13.7$ or 13.7 million

(C) Solve $0.18x + 3.8 > 0.07x + 5.0 + 5$ for x:.
$$0.11x > 6.2$$
$$x > 56.36$$

Female enrollment will exceed male enrollment by at least 5 million in 2027.

27. Linear regression $y = a + bx$

Men: the regression formula is: Women: the regression formula is:
$y = -0.087x + 49.033$ $y = -0.088x + 54.708$

Yes; the women's times are decreasing at a faster rate than the men's times.

29. Linear regression $y = a + bx$
Supply: Demand:

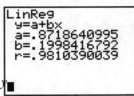   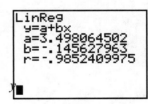

To find the equilibrium price,
solve $0.87 + 0.20x = 3.5 - 0.15x$ for x:
$$0.87 + 0.20x = 3.5 - 0.15x$$
$$0.35x = 2.63$$
$$x \approx 7.51$$
At $x = 7.51$, $y = 0.87 + 0.20(7.51) = 2.37$. The equilibrium price is $2.37.

CHAPTER 1 REVIEW

1. $2x + 3 = 7x - 11$
$-5x = -14$
$x = \dfrac{14}{5} = 2.8$ (1-1)

2. $\dfrac{x}{12} - \dfrac{x-3}{3} = \dfrac{1}{2}$
Multiply each term by 12:
$x - 4(x - 3) = 6$
$x - 4x + 12 = 6$
$-3x = 6 - 12$
$-3x = -6$
$x = 2$ (1-1)

3. $2x + 5y = 9$

$5y = 9 - 2x$

$y = \dfrac{9}{5} - \dfrac{2}{5}x = 1.8 - 0.4x$ (1-1)

4. $3x - 4y = 7$

$3x = 7 + 4y$

$x = \dfrac{7}{3} + \dfrac{4}{3}y$ (1-1)

5. $4y - 3 < 10$

$4y < 13$

$y < \dfrac{13}{4}$ or $\left(-\infty, \dfrac{13}{4}\right)$

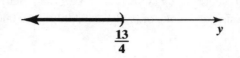

(1-1)

6. $-1 < -2x + 5 \le 3$

$-6 < -2x \le -2$ (Divide the inequalities by -2 and reverse the direction.)

$3 > x \ge 1$

or $[1, 3)$

(1-1)

7. $1 - \dfrac{x-3}{3} \le \dfrac{1}{2}$

Multiply both sides of the inequality by 6. We do not reverse the direction of the inequality, since $6 > 0$.

$6 - 2(x - 3) \le 3$

$6 - 2x + 6 \le 3$

$-2x \le 3 - 12$

$-2x \le -9$

Divide both sides by -2 and reverse the direction of the inequality, since $-2 < 0$.

$x \ge \dfrac{9}{2}$ or $\left[\dfrac{9}{2}, \infty\right)$

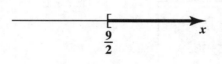

(1-1)

8. $3x + 2y = 9$

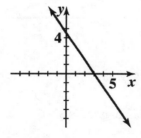

(1-2)

9. The line passes through $(6, 0)$ and $(0, 4)$

slope $m = \dfrac{4 - 0}{0 - 6} = -\dfrac{2}{3}$

From the slope-intercept form: $y = -\dfrac{2}{3}x + 4$; multiplying by 3 gives: $3y = -2x + 12$, so

$2x + 3y = 12$ (1-2)

Graph:

10. x-intercept: $2x = 18$, $x = 9$;
y-intercept: $-3y = 18$, $x = -6$;
slope-intercept form:

$$y = \frac{2}{3}x - 6; \text{ slope} = \frac{2}{3}$$

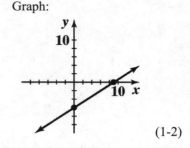

(1-2)

11. $y = -\frac{2}{3}x + 6$ (1-2)

12. Vertical line: $x = -6$; horizontal line: $y = 5$ (1-2)

13. Use the point-slope form:

(A) $y - 2 = -\frac{2}{3}[x - (-3)]$

$y - 2 = -\frac{2}{3}(x + 3)$

$y = -\frac{2}{3}x$

(B) $y - 3 = 0(x - 3)$

$y = 3$

(1-2)

14. (A) Slope: $\dfrac{-1-5}{1-(-3)} = -\dfrac{3}{2}$

$y - 5 = -\dfrac{3}{2}(x + 3)$

$3x + 2y = 1$

(B) Slope: $\dfrac{5-5}{4-(-1)} = 0$

$y - 5 = 0(x - 1)$

$y = 5$

(C) Slope: $\dfrac{-2-7}{-2-(-2)}$ not defined since $2 - (-2) = 0$

$x = -2$

(1-2)

15. $3x + 25 = 5x$

$-2x = -25$

$x = \dfrac{25}{2}$ (1-1)

16. $\dfrac{u}{5} = \dfrac{u}{6} + \dfrac{6}{5}$ (multiply by 30)

$6u = 5u + 36$

$u = 36$ (1-1)

17. $\dfrac{5x}{3} - \dfrac{4+x}{2} = \dfrac{x-2}{4} + 1$ (multiply by 12)

$20x - 6(4 + x) = 3(x - 2) + 12$

$20x - 24 - 6x = 3x - 6 + 12$

$11x = 30$

$x = \dfrac{30}{11}$ (1-1)

18. $0.05x + 0.25(30 - x) = 3.3$

$0.05x + 7.5 - 0.25x = 3.3$

$-0.20x = -4.2$

$x = \dfrac{-4.2}{-0.20} = 21$ (1-1)

19. $0.2(x - 3) + 0.05x = 0.4$

$0.2x - 0.6 + 0.05x = 0.4$

$0.25x = 1$

$x = 4$ (1-1)

20. $2(x + 4) > 5x - 4$

$2x + 8 > 5x - 4$

$2x - 5x > -4 - 8$

$-3x > -12$ (Divide both sides by –3 and reverse the inequality)

$x < 4$ or $(-\infty, 4)$

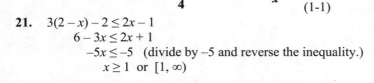

(1-1)

21. $3(2 - x) - 2 \leq 2x - 1$

$6 - 3x \leq 2x + 1$

$-5x \leq -5$ (divide by –5 and reverse the inequality.)

$x \geq 1$ or $[1, \infty)$

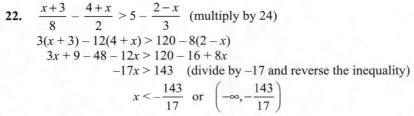

(1-1)

22. $\dfrac{x+3}{8} - \dfrac{4+x}{2} > 5 - \dfrac{2-x}{3}$ (multiply by 24)

$3(x + 3) - 12(4 + x) > 120 - 8(2 - x)$

$3x + 9 - 48 - 12x > 120 - 16 + 8x$

$-17x > 143$ (divide by –17 and reverse the inequality)

$x < -\dfrac{143}{17}$ or $\left(-\infty, -\dfrac{143}{17}\right)$

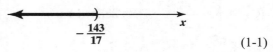

(1-1)

23. $-5 \leq 3 - 2x < 1$

$-8 \leq -2x < -2$ (divide by –2 and reverse the directions of the inequalities.)

$4 \geq x > 1$ which is the same as $1 < x \leq 4$ or $(1, 4]$

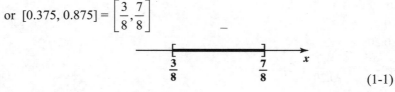

(1-1)

24. $-1.5 \leq 2 - 4x \leq 0.5$

$-3.5 \leq -4x \leq -1.5$ (divide by –4 and reverse the directions of the inequalities.)

$0.875 \geq x \geq 0.375$ which is the same as $0.375 \leq x \leq 0.875$

or $[0.375, 0.875] = \left[\dfrac{3}{8}, \dfrac{7}{8}\right]$

(1-1)

25.

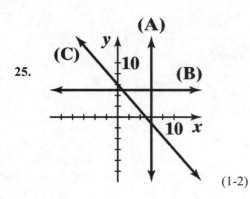

(1-2)

26. The graph of $x = -3$ is a vertical line 3 units to the *left* of the y-axis; $y = 2$ is a horizontal line 2 units *above* the x-axis.

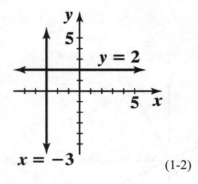

(1-2)

27. (A) $3x + 4y = 0$; $\quad y = -\dfrac{3}{4}x \quad$ an oblique line through the origin with slope $-\dfrac{3}{4}$

(B) $3x + 4 = 0$; $\quad x = -\dfrac{4}{3} \quad$ a vertical line with x intercept $-\dfrac{4}{3}$

(C) $4y = 0$; $\quad y = 0 \quad$ the x-axis

(D) $3x + 4y = 36 \quad$ an oblique line with x intercept 12 and y-intercept 9. $\quad$ (1-2)

28. $A = \dfrac{1}{2}(a + b)h$; solve for a; assume $h \neq 0$

$A = \dfrac{1}{2}(ah + bh) = \dfrac{1}{2}ah + \dfrac{1}{2}bh$

$A - \dfrac{1}{2}bh = \dfrac{1}{2}ah \quad$ (multiply by $\frac{2}{h}$)

$\dfrac{2A}{h} - b = a \quad$ and $\quad a = \dfrac{2A - bh}{h} \quad$ (1-1)

29. $S = \dfrac{P}{1 - dt}$; solve for d; assume $dt \neq 0$ or 1

$S(1 - dt) = P \quad$ (multiply by $1 - dt$)

$\quad S - Sdt = P$

$\quad\quad -Sdt = P - S \quad$ (divide by $-St$)

$\quad\quad\quad d = \dfrac{P - S}{-St} = \dfrac{S - P}{St} \quad\quad$ (1-1)

30. $a + b < b - a$

$\quad\quad 2a < 0$

$\quad\quad\ \ a < 0$

The inequality is true for $a < 0$ and b any number.   (1-1)

31. $b < a < 0$ (divide by b; reverse the direction of the inequalities since $b < 0$).

$1 > \dfrac{a}{b} > 0$

Thus, $0 < \dfrac{a}{b} < 1$; $\dfrac{a}{b} < 1$; $\dfrac{a}{b}$ is less than 1.   (1-1)

32.   The graphs of the pairs $\{y = 2x, y = -\dfrac{1}{2}x\}$ and

$\{y = \dfrac{2}{3}x + 2, y = -\dfrac{3}{2}x + 2\}$ are shown below:

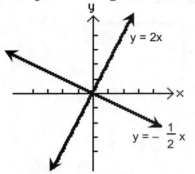
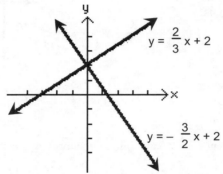

In each case, the graphs appear to be perpendicular to each other. It can be shown that two slant lines are perpendicular if and only if their slopes are negative reciprocals.   (1-2)

33.   Let x = amount invested at 5%.

Then $300,000 - x$ = amount invested at 9%.

Yield = $300,000(0.08) = 24,000$

Solve $x(0.05) + (300,000 - x)(0.09) = 24,000$ for x:

$\quad\quad\quad\quad 0.05x + 27,000 - 0.09x = 24,000$

$\quad\quad\quad\quad\quad\quad\quad\quad -0.04x = -3,000$

$\quad\quad\quad\quad\quad\quad\quad\quad\quad\quad x = 75,000$

Invest \$75,000 at 5%, \$225,000 at 9%.   (1-1)

34.   Let x = the number of DVD's.

Cost: $C(x) = 90,000 + 5.10x$

Revenue: $R(x) = 14.70x$

Break-even point: $C(x) = R(x)$

$\quad\quad\quad 90,000 + 5.10x = 14.70x$

$\quad\quad\quad\quad\quad\quad 90,000 = 9.60x$

$\quad\quad\quad\quad\quad\quad\quad\quad\quad x = 9,375$

9,375 DVDs must be sold to break even.   (1-1)

35. Let x = person's age in years.
  (A) Minimum heart rate: $m = (220 - x)(0.6) = 132 - 0.6x$
  (B) Maximum heart rate: $M = (220 - x)(0.85) = 187 - 0.85x$

  (C) At $x = 20$, $m = 132 - 0.6(20) = 120$
  $\qquad\qquad\quad M = 187 - 0.85(20) = 170$
  range – between 120 and 170 beats per minute.

  (D) At $x = 50$, $m = 132 - 0.6(50) = 102$
  $\qquad\qquad\quad M = 187 - 0.85(50) = 144.5$
  range – between 102 and 144.5 beats per minute.   (1-3)

36. $V = mt + b$
  (A) *At* $t = 0$, $V = 224,000$; at $t = 8$, $V = 100,000$
  $$\text{slope } m = \frac{100,000 - 224,000}{8 - 0} = -\frac{124,000}{8} = -15,500$$
  $V = -15,500t + 224,000$

  (B) At $t = 12$, $V = -15,500(12) + 224,000 = 38,000.$
  The bulldozer will be worth \$38,000 after 12 years.   (1-2)

37. $R = mC + b$
  (A) From the given information, the points (50, 80) and (130, 208) satisfy this equation. Therefore:
  $$\text{slope } m = \frac{208 - 80}{130 - 50} = \frac{128}{80} = 1.6$$
  Using the point-slope form with $(C_1, R_1) = (50, 80)$:

  $R - 80 = 1.6(C - 50) = 1.6C - 80$
  $R = 1.6C$

  (B) At $C = 120$, $R = 1.6(120) = 192$; \$192.
  (C) At $R = 176$, $176 = 1.6C$
  $$C = \frac{176}{1.6} = 110; \$110$$

  (D) Slope $m = 1.6$. The slope is the rate of change of retail price with respect to cost.   (1-2)

38. Let x = weekly sales
  $E = 400 + 0.10(x - 6,000)$ for $x \geq 6,000$
  At $x = 4000$, $E = 400$; \$400
  At $x = 10,000$, $E = 400 + 0.10(10,000 - 6,000) = 400 + 0.10(4,000) = 800$; \$800.   (1-1)

39. $p = mx + b$
  From the given information, the points (1,160, 3.79) and
  (1,320, 3.59) satisfy this equation. Therefore,
  $$\text{slope } m = \frac{3.59 - 3.79}{1,320 - 1,160} = -\frac{0.20}{160} = -0.00125$$
  Using the point-slope form with $(x_1, p_1) = (1,160, 3.79)$

  $p - 3.79 = -0.00125(x - 1,160) = -0.00125x + 1.45$
  $\qquad\quad p = -0.00125x + 5.24$

If $p = 3.29$, solve $3.29 = -0.00125x + 5.24$ for x:

$$-0.00125x = 3.29 - 5.24 = -1.95$$
$$x = 1,560$$

The stores would sell 1,560 bottles.   (1-2)

40. $T = 40 - 2M$

(A)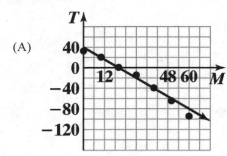

(B)   At $M = 35$, $T = 40 - 2(35) = -30$; $-30°$F.

(C)   At $T = -50$,
$$-50 = 40 - 2M$$
$$-2M = -90$$
$$M = 45;\ 45\% \quad (1\text{-}3)$$

41. $r = -0.198t + 14.2$

(A) The dropout rate is decreasing at the rate of 0.198 percentage points per year.

(B)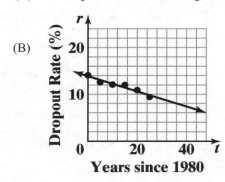

(C)   Solve $5 = -0.198t + 14.2$ for t:
$$-0.198t = -9.2$$
$$t = 46.\overline{46}$$

The dropout rate will be less than 5% in 2027.

(1-3)

42. $y = 4.75x + 171$

(A) The CPI is increasing at the rate of 4.75 units per year.

(B) At $x = 24$, $y = 4.75(24) + 171 = 285$; the CPI in 2024 will be 285.00.   (1-3)

43. $y = 0.74x + 2.83$

(A) The rate of change of tree height with respect to Dbh is 0.74.

(B) Tree height increases by 0.74 feet.

(C) At $x = 25$, $y = 0.74(25) + 2.83 = 21.33$. To the nearest foot, the tree is 21 feet high.

(D) Solve $15 = 0.74x + 2.83$ for x:
$$0.74x = 15 - 2.83 = 12.17$$
$$x = 16.446$$

To the nearest inch, the Dbh is 16 inches.   (1-3)

2 FUNCTIONS AND GRAPHS

EXERCISE 2-1

Things to remember:

1. POINT-BY-POINT PLOTTING

    To sketch the graph of an equation in two variables, plot enough points from its solution set in a rectangular coordinate system so that the total graph is apparent and then connect these points with a smooth curve.

2. A FUNCTION is a correspondence between one set of elements, called the DOMAIN, and a second set of elements, called the RANGE, such that to each element in the domain there corresponds one and only one element in the range.

3. EQUATIONS AND FUNCTIONS

    Given an equation in two variables. If there corresponds exactly one value of the dependent variable (output) to each value of the independent variable (input), then the equation specifies a function. If there is more than one output for at least one input, then the equation does not specify a function.

4. VERTICAL LINE TEST FOR A FUNCTION

    An equation specifies a function if each vertical line in the coordinate system passes through at most one point on the graph of the equation. If any vertical line passes through two or more points on the graph of an equation, then the equation does not specify a function.

5. AGREEMENT ON DOMAINS AND RANGES

    If a function is specified by an equation and the domain is not given explicitly, then assume that the domain is the set of all real number replacements of the independent variable (inputs) that produce real values for the dependent variable (outputs). The range is the set of all outputs corresponding to input values.

    In many applied problems, the domain is determined by practical considerations within the problem.

6. FUNCTION NOTATION — THE SYMBOL $f(x)$

    For any element x in the domain of the function f, the symbol $f(x)$ represents the element in the range of f corresponding to x in the domain of f. If x is an input value, then $f(x)$ is the corresponding output value. If x is an element which is not in the domain of f, then f is NOT DEFINED at x and $f(x)$ DOES NOT EXIST.

1. $y = x + 1$:

x	-4	-2	0	2	4
y	-3	-1	1	3	5

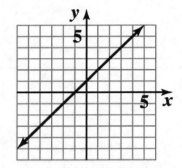

3. $x = y^2$:

x	0	1	4	9	16
y	0	±1	±2	±3	±4

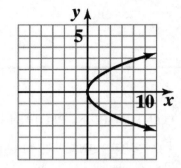

5. $y = x^3$:

x	-2	-1	0	1	2
y	-8	-1	0	1	8

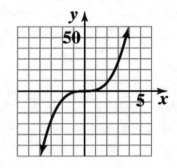

7. $xy = -6$:

x	-6	-3	-1	1	3	6
y	1	2	6	-6	-2	-1

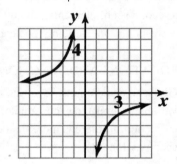

9. The table specifies a function, since for each domain value there corresponds one and only one range value.

11. The table does not specify a function, since more than one range value corresponds to a given domain value. (Range values 5, 6 correspond to domain value 3; range values 6, 7 correspond to domain value 4.)

13. This is a function.

15. The graph specifies a function; each vertical line in the plane intersects the graph in at most one point.

17. The graph does not specify a function. There are vertical lines which intersect the graph in more than one point. For example, the y-axis intersects the graph in three points.

19. The graph specifies a function.

21. $y - 2x = 7$ or $y = 2x + 7$; a linear function.

23. $xy - 4 = 0$ or $y = \dfrac{4}{x}$; neither a linear nor a constant function.

25. $y = 5x + \dfrac{1}{2}(7 - 10x) = 5x + \dfrac{7}{2} - 5x = \dfrac{7}{2}$; a constant function.

27. $3x + 4y = 5$ or $y = -\dfrac{3}{4}x + \dfrac{5}{4}$; a linear function

29. $f(x) = 1 - x$: Since f is a linear function, we only need to plot two points.

x	$f(x)$
-2	3
2	-1

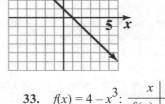

31. $f(x) = x^2 - 1$:

x	-3	-2	-1	0	1	2	3
$f(x)$	8	3	0	-1	0	3	8

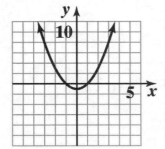

33. $f(x) = 4 - x^3$:

x	-2	-1	0	1	2
$f(x)$	12	5	4	3	-4

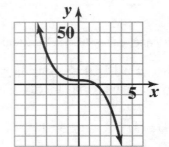

35. $f(x) = \dfrac{8}{x}$

x	-8	-4	-2	-1	1	2	4	8
$f(x)$	-1	-2	-4	-8	8	4	2	1

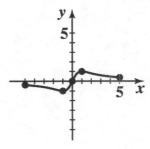

37. The graph of f is:

39. $y = f(-5) = 0$

41. $y = f(5) = 4$

43. $f(x) = 0$ at $x = -5, 0, 4$

45. $f(x) = -4$ at $x = -6$

47. domain: all real numbers or $(-\infty, \infty)$

49. domain: all real numbers except -4

51. domain: $x \leq 7$

53. Given $2x + 5y = 10$. Solving for y, we have: $5y = 10 - 2x$ and $y = 2 - \dfrac{2}{5}x$.

This equation specifies a function. The domain is R, the set of real numbers.

55.   Given $y(x+y)=4$. Solving for y, we have:

$$y^2+xy=4 \quad \text{or} \quad y^2+xy-4=0. \quad \text{So} \quad y=\frac{-x\pm\sqrt{x^2+16}}{2}$$

This equation does not specify y as a function x. For example, when $x=0$, $y=\pm2$, when $x=3$, $y=-4,1$.

57.   Given $x^{-3}+y^3=27$. Solving for y, we have:

$$y^3=27-\frac{1}{x^3}=\frac{27x^3-1}{x^3} \quad \text{and} \quad y=\frac{\sqrt[3]{27x^3-1}}{x}.$$

This equation specifies a function. The domain is all real numbers except $x=0$.

59.   Given $x^3-y^2=0$. Solving for y, we have: $y=\pm\sqrt{x^3}$.

This equation does not specify y as a function x. For example, when $x=1$, $y=\pm1$.

61. $f(4)=(4)^2-4=16-4=12$

63. $f(x+1)=(x+1)^2-4=x^2+2x+1-4=x^2+2x-3$ **65.** $f(-6x)=(-6x)^2-4=36x^2-4$

67. $f(x^3)=(x^3)^2-4=x^6-4$

69. $f(2)+f(h)=[(2)^2-4]+[(h)^2-4]=[4-4]+[h^2-4]=h^2-4$

71. $f(2+h)=(2+h)^2-4=4+4h+h^2-4=4h+h^2$

73. $f(2+h)-f(2)=[(2+h)^2-4]-[(2)^2-4]=[4+4h+h^2-4]-0=4h+h^2$

75. $f(x)=4x-3$

   (A) $f(x+h)=4(x+h)-3=4x+4h-3$

   (B) $f(x+h)-f(x)=4x+4h-3-(4x-3)=4h$

   (C) $\dfrac{f(x+h)-f(x)}{h}=\dfrac{4h}{h}=4$

77. $f(x)=4x^2-7x+6$

   (A) $f(x+h)=4(x+h)^2-7(x+h)+6$

   $\qquad =4(x^2+2xh+h^2)-7x-7h+6$

   $\qquad =4x^2+8xh+4h^2-7x-7h+6$

   (B) $f(x+h)-f(x)=4x^2+8xh+4h^2-7x-7h+6-(4x^2-7x+6)$

   $\qquad\quad =8xh+4h^2-7h$

   (C) $\dfrac{f(x+h)-f(x)}{h}=\dfrac{8xh+4h^2-7h}{h}=\dfrac{h(8x+4h-7)}{h}=8x+4h-7$

79. $f(x) = x(20 - x) = 20x - x^2$

(A) $f(x + h) = 20(x + h) - (x + h)^2 = 20x + 20h - x^2 - 2xh - h^2$

(B) $f(x + h) - f(x) = 20x + 20h - x^2 - 2xh - h^2 - (20x - x^2)$

$= 20h - 2xh - h^2$

(C) $\dfrac{f(x+h) - f(x)}{h} = \dfrac{20h - 2xh - h^2}{h} = \dfrac{h(20 - 2x - h)}{h} = 20 - 2x - h$

81. Given $A = \ell\, w = 25.$

Thus, $\ell = \dfrac{25}{w}$. Now $P = 2\,\ell + 2w$

$= 2\left(\dfrac{25}{w}\right) + 2w = \dfrac{50}{w} + 2w,\ \ w > 0$

The domain is $w > 0.$

83. Given $P = 2\,\ell + 2w = 100$ or $\ell + w = 50$ and $w = 50 - \ell.$

Now $A = \ell\, w = \ell\,(50 - \ell)$ and $A = 50\,\ell - \ell^2.$

The domain is $0 < \ell < 50.$ [Note: $\ell < 50$ since $\ell \geq 50$ implies $w \leq 0.$]

85.

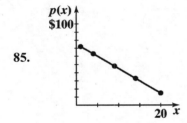

(B) $p(x) = 75 - 3x$
$p(7) = 75 - 3(7) = 75 - 21 = 54;$
$p(11) = 75 - 3(11) = 75 - 33 = 42$
Estimated price per chip for a demand of 7 million chips: \$54;
for a demand of 11 million chips: \$42

87. (A) $R(x) = x \cdot p(x) = x(75 - 3x) = 75x - 3x^2, 1 \leq x \leq 20$

(B)

x	R(x)
1	72
4	252
8	408
12	468
16	432
20	300

(C)

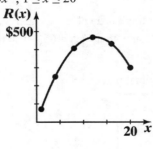

89. (A) Profit: $P(x) = R(x) - C(x) = 75x - 3x^2 - [125 + 16x] = 59x - 3x^2 - 125,\ 1 \le x \le 20$

(B)

x	$P(x)$
1	−69
4	63
8	155
12	151
16	51
20	−145

(C)

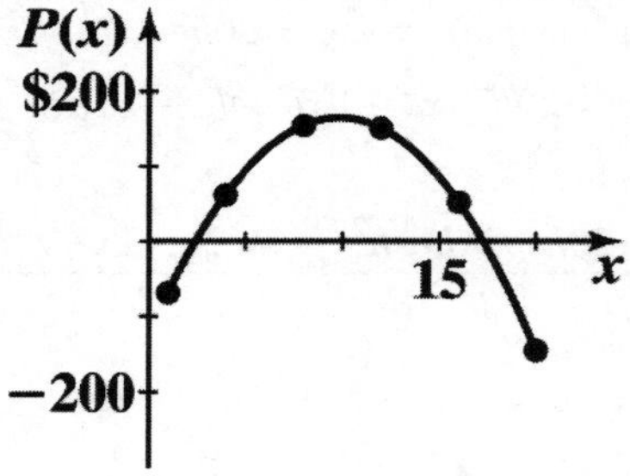

91.

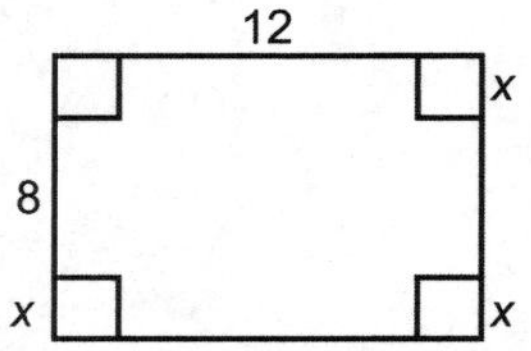

(A) $V = \text{(length)(width)(height)}$

$V(x) = (12 - 2x)(8 - 2x)x$

$= x(8 - 2x)(12 - 2x)$

(B) Domain: $0 < x < 4$

(C) $V(1) = (12 - 2)(8 - 2)(1)$

$= (10)(6)(1) = 60$

$V(2) = (12 - 4)(8 - 4)(2)$

$= (8)(4)(2) = 64$

$V(3) = (12 - 6)(8 - 6)(3)$

$= (6)(2)(3) = 36$

Thus,

Volume

x	$V(x)$
1	60
2	64
3	36

(D)

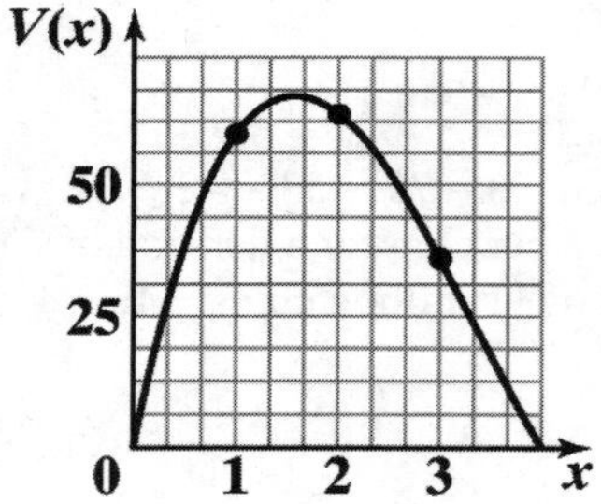

93. (A) The graph indicates that there is a value of x near 2, and slightly less than 2, such that $V(x) = 65$. The table is shown at the right.

Thus, $x = 1.9$ to one decimal place.

(B)

x	Y_1
1.7	67.252
1.8	66.528
1.9	65.436
2.0	64.000

(C)

X	Y1
1.9	65.436
1.91	65.307
1.92	65.176
1.93	65.04
1.94	64.902
1.95	64.76
1.96	64.614

X=1.93

$x = 1.93$ to two decimal places.

95. Given $(w + a)(v + b) = c$. Let $a = 15$, $b = 1$, and $c = 90$. Then: $(w + 15)(v + 1) = 90$
Solving for v, we have

$$v + 1 = \frac{90}{w+15} \quad \text{and} \quad v = \frac{90}{w+15} - 1 = \frac{90 - (w+15)}{w+15} = \frac{75 - w}{w+15}.$$

If $w = 16$, then $v = \frac{75 - 16}{16 + 15} = \frac{59}{31} \approx 1.9032$ cm/sec.

EXERCISE 2-2

Things to remember:

1.   LIBRARY OF ELEMENTARY FUNCTIONS

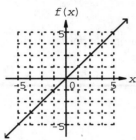

Identity Function

$f(x) = x$
Domain: All real numbers
Range: All real numbers
     (a)

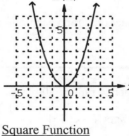

Square Function

$h(x) = x^2$
Domain: All real numbers
Range: $[0, \infty)$
     (b)

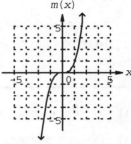

Cube Function

$m(x) = x^3$
Domain: All real numbers
Range: All real numbers
     (c)

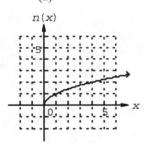

Square-Root Function

$n(x) = \sqrt{x}$
Domain: $[0, \infty)$
Range: $[0, \infty)$
     (d)

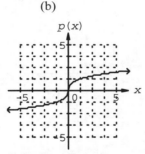

Cube-Root Function

$p(x) = \sqrt[3]{x}$
Domain: All real numbers
Range: All real numbers
     (e)

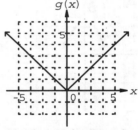

Absolute Value Function

$g(x) = |x|$
Domain: All real numbers
Range: $[0, \infty)$
     (f)

NOTE: Letters used to designate the above functions may vary from context to context.

2.   GRAPH TRANSFORMATIONS SUMMARY

Vertical Translation:

$$y = f(x) + k \qquad \begin{cases} k > 0 & \text{Shift graph of } y = f(x) \text{ up } k \text{ units} \\ k < 0 & \text{Shift graph of } y = f(x) \text{ down } |k| \text{ units} \end{cases}$$

Horizontal Translation:

$$y = f(x + h) \qquad \begin{cases} h > 0 & \text{Shift graph of } y = f(x) \text{ left } h \text{ units} \\ h < 0 & \text{Shift graph of } y = f(x) \text{ right } |h| \text{ units} \end{cases}$$

Reflection:
$y = -f(x)$ Reflect the graph of $y = f(x)$ in the x axis

Vertical Stretch and Shrink:

$$y = Af(x) \begin{cases} A > 1 & \text{Stretch graph of } y = f(x) \text{ vertically by multiplying each} \\ & \text{ordinate value by } A \\ 0 < A < 1 & \text{Shrink graph of } y = f(x) \text{ vertically by multiplying each} \\ & \text{ordinate value by } A \end{cases}$$

<u>3.</u>   PIECEWISE-DEFINED FUNCTIONS

Functions whose definitions involve more than one rule are called PIECEWISE-DEFINED FUNCTIONS.

For example,

$$f(x) = |x| = \begin{cases} -x & \text{if } x < 0 \\ x & \text{if } x \geq 0 \end{cases}$$

is a piecewise-defined function.

To graph a piecewise-defined function, graph each rule over the appropriate portion of the domain.

---

1. $f(x) = 5x - 10$;   domain: all real numbers;   range: all real numbers

3. $f(x) = 15 - \sqrt{x}$;   domain: $[0, \infty)$;   range: $(-\infty, 15]$

5. $f(x) = 2|x| + 7$;   domain: all real numbers;   range: $[7, \infty)$

7. $f(x) = -\sqrt[3]{x} + 100$;   domain: all real numbers;   range: all real numbers

9.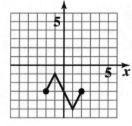

11.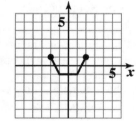

13.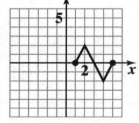

15.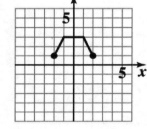

17.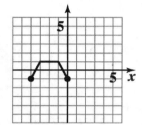

19.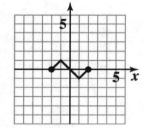

21.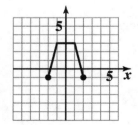

23.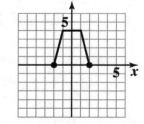

25. The graph of $g(x) = -|x + 3|$ is the graph of $y = |x|$ reflected in the x axis and shifted 3 units to the left.

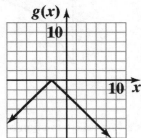

27. The graph of $f(x) = (x - 4)^2 - 3$ is the graph of $y = x^2$ shifted 4 units to the right and 3 units down.

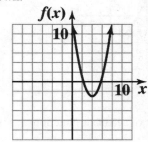

29. The graph of $f(x) = 7 - \sqrt{x}$ is the graph of $y = \sqrt{x}$ reflected in the x axis and shifted 7 units up.

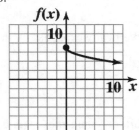

31. The graph of $h(x) = -3|x|$ is the graph of $y = |x|$ reflected in the x axis and vertically expanded by a factor of 3.

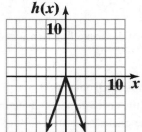

33. The graph of the basic function $y = x^2$ is shifted 2 units to the left and 3 units down.
Equation: $y = (x + 2)^2 - 3$.

35. The graph of the basic function $y = x^2$ is reflected in the x axis, shifted 3 units to the right and 2 units up.
Equation: $y = 2 - (x - 3)^2$.

37. The graph of the basic function $y = \sqrt{x}$ is reflected in the x axis and shifted 4 units up.
Equation: $y = 4 - \sqrt{x}$.

39. The graph of the basic function $y = x^3$ is shifted 2 units to the left and 1 unit down.
Equation: $y = (x + 2)^3 - 1$.

41. $g(x) = \sqrt{x - 2} - 3$

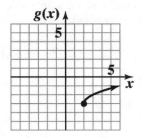

43. $g(x) = -|x + 3|$

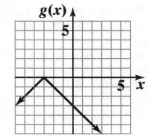

45. $g(x) = -(x - 2)^3 - 1$

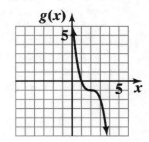

47.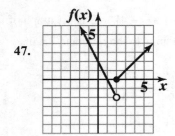

49.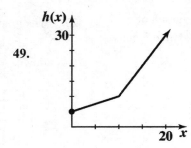

51.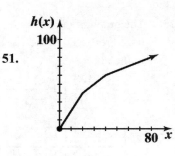

53. The graph of the basic function: $y = |x|$ is reflected in the x axis and has a vertical contraction by the factor 0.5.
    Equation: $y = -0.5|x|$.

55. The graph of the basic function $y = x^2$ is reflected in the x axis and is vertically expanded by the factor 2.
    Equation: $y = -2x^2$.

57. The graph of the basic function $y = \sqrt[3]{x}$ is reflected in the x axis and is vertically expanded by the factor 3.
    Equation: $y = -3\sqrt[3]{x}$.

59. Vertical shift, horizontal shift.
    Reversing the order does not change the result. Consider a point
    (a, b) in the plane. A vertical shift of k units followed by a horizontal shift of h units moves (a, b) to $(a, b + k)$ and then to
    $(a + h, b + k)$.

    In the reverse order, a horizontal shift of h units followed by a vertical shift of k units moves (a, b) to $(a + h, b)$ and then to
    $(a + h, b + k)$. The results are the same.

61. Vertical shift, reflection in the x axis.
    Reversing the order can change the result. For example, let (a, b) be a point in the plane with $b > 0$. A vertical shift of k units, $k \neq 0$, followed by a reflection in the x axis moves (a, b) to $(a, b + k)$ and then to $(a, -[b + k]) = (a, -b - k)$.
    In the reverse order, a reflection in the x axis followed by the vertical shift of k units moves (a, b) to $(a, -b)$ and then to
    $(a, -b + k)$; $(a, -b - k) \neq (a, -b + k)$ when $k \neq 0$.

63. Horizontal shift, reflection in y axis.
    Reversing the order can change the result. For example, let (a, b) be a point in the plane with $a > 0$. A horizontal shift of h units followed by a reflection in the y axis moves (a, b) to the point
    $(a + h, b)$ and then to $(-[a + h], b) = (-a - h, b)$.
    In the reverse order, a reflection in the y axis followed by the horizontal sift of h units moves (a, b) to $(-a, b)$ and then the
    $(-a + h, b)$, $(-a - h, b) \neq (-a + h, b)$ when $h \neq 0$.

65. (A) The graph of the basic function $y = \sqrt{x}$ is reflected in the x axis, vertically expanded by a factor of 4, and shifted up 115 units.

(B)

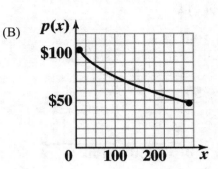

67. (A) The graph of the basic function $y = x^3$ is vertically contracted by a factor of 0.00048 and shifted right 500 units and up 60,000 units.

(B)

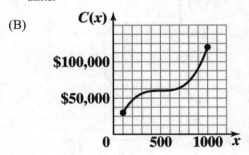

69. (A) $S(x) = \begin{cases} 8.50 + 0.0650x & \text{if } 0 \le x \le 700 \\ 8.50 + 0.0650(700) + 0.09(x - 700) & \text{if } x > 700 \end{cases}$

$= \begin{cases} 8.50 + 0.0650x & \text{if } 0 \le x \le 700 \\ -9 + 0.09x & \text{if } x > 700 \end{cases}$

(B)

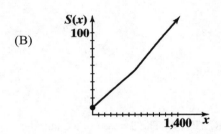

71. (A) If $0 \le x \le 30{,}000$, $T(x) = 0.035x$, and $T(30{,}000) = 1{,}050$.

If $30{,}000 < x \le 60{,}000$, $T(x) = 1{,}050 + 0.0625(x - 30{,}000) = 0.0625x - 825$, and $T(60{,}000) = 2{,}925$.

If $x > 60{,}000$, $T(x) = 2{,}925 + 0.0645(x - 60{,}000) = 0.0645x - 945$.

Thus, $T(x) = \begin{cases} 0.035x & \text{if } 0 \le x \le 30{,}000 \\ 0.0625x - 825 & \text{if } 30{,}000 < x \le 60{,}000 \\ 0.0645x - 945 & \text{if } x > 60{,}000 \end{cases}$

(B)

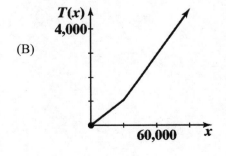

(C) $T(40{,}000) = 0.0625(40{,}000) - 825$
$= 1{,}675;$
$\$1{,}675$

$T(70{,}000) = 0.0645(70{,}000) - 945$
$= 3{,}570; \quad \$3{,}570$

73. (A)   The graph of the basic function $y = x$ is vertically expanded by a factor of 5.5 and shifted down 220 units.

(B)

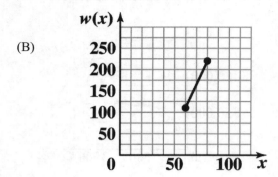

75. (A)   The graph of the basic function $y = \sqrt{x}$ is vertically expanded by a factor of 7.08.

(B)

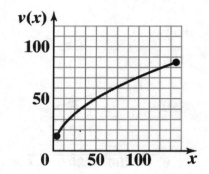

EXERCISE 2-3

Things to remember:

<u>1.</u>   QUADRATIC FUNCTION

If a, b, and c are real numbers with $a \neq 0$, then the function

$$f(x) = ax^2 + bx + c \qquad \text{STANDARD FORM}$$

is a QUADRATIC FUNCTION and its graph is a PARABOLA. The domain of a quadratic function is the set of all real numbers.

<u>2.</u>   PROPERTIES OF A QUADRATIC FUNCTION AND ITS GRAPH

Given a quadratic function

$$f(x) = ax^2 + bx + c, \quad a \neq 0,$$

and the VERTEX FORM obtained by completing the square

$$f(x) = a(x-h)^2 + k.$$

The general properties of f are as follows:

a. The graph of f is a parabola:

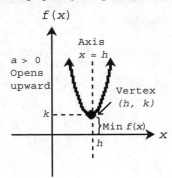

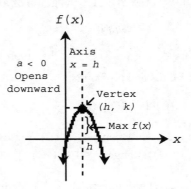

b.   Vertex: (h, k) [parabola increases on one side of the vertex and decreases on the other]

c.   Axis (of symmetry): $x = h$ (parallel to y axis)

d. $f(h) = k$ is the minimum if $a > 0$ and the maximum if $a < 0$

e.   Domain: All real numbers

Range: $(-\infty, k]$ if $a < 0$ or $[k, \infty)$ if $a > 0$

f.   The graph of f is the graph of $g(x) = ax^2$ translated horizontally h units and vertically k units.

---

1. $f(x) = x^2 - 10x$ (standard form)

$= x^2 - 10x + 25 - 25$ (completing the square)

$= (x - 5)^2 - 25$ (vertex form)

3. $f(x) = x^2 + 20x + 50$ (standard form)

$= (x^2 + 20x) + 50$

$= (x^2 + 20x + 100) + 50 - 100$ (completing the square)

$= (x + 10)^2 - 50$ (vertex form)

5. $f(x) = -2x^2 + 4x - 5$ (standard form)

$= -2(x^2 - 2x) - 5$

$= -2(x^2 - 2x + 1 - 1) - 5$ (completing the square)

$= -2(x - 1)^2 - 5 + 2$

$= -2(x - 1)^2 - 3$ (vertex form)

7. $f(x) = 2x^2 + 2x + 1$ (standard form)

$= 2(x^2 + x) + 1$

$= 2(x^2 + x + \frac{1}{4} - \frac{1}{4}) + 1$ (completing the square)

$= 2(x + \frac{1}{2})^2 + 1 - \frac{1}{4}$

$= 2(x + \frac{1}{2})^2 + \frac{1}{2}$ (vertex form)

9. $f(x) = x^2 - 4x + 3 = (x - 2)^2 - 1$:   the graph of $f(x)$ is the graph of $y = x^2$ shifted right 2 units and down 1 unit.

11. $m(x) = -x^2 + 6x - 4 = -(x - 3)^2 + 5$:   the graph of $m(x)$ is the graph of $y = x^2$ reflected in the x axis, then shifted right 3 units and up 5 units.

13.   (A) m   (B) g   (C) f   (D) n

15.   (A) x-intercepts: 1, 3; y-intercept: -3 (B) Vertex: (2, 1)
     (C) Maximum: 1   (D) Range: $y \le 1$ or $(-\infty, 1]$

17.   (A) x-intercepts: $-3, -1$; y-intercept: 3   (B) Vertex: $(-2, -1)$
     (C) Minimum: -1 (D) Range: $y \ge -1$ or $[-1, \infty)$

19. $f(x) = -(x - 3)^2 + 2$

    (A) y-intercept: $f(0) = -(0 - 3)^2 + 2 = -7$
          x-intercepts: $f(x) = 0$
$$-(x - 3)^2 + 2 = 0$$
$$(x - 3)^2 = 2$$
$$x - 3 = \pm\sqrt{2}$$
$$x = 3 \pm \sqrt{2}$$

    (B) Vertex: (3, 2)   (C) Maximum: 2   (D) Range: $y \le 2$ or $(-\infty, 2]$

21. $m(x) = (x + 1)^2 - 2$

    (A) y-intercept: $m(0) = (0 + 1)^2 - 2 = 1 - 2 = -1$
          x-intercepts: $m(x) = 0$
$$(x + 1)^2 - 2 = 0$$
$$(x + 1)^2 = 2$$
$$x + 1 = \pm\sqrt{2}$$
$$x = -1 \pm \sqrt{2}$$

    (B) Vertex: $(-1, -2)$ (C) Minimum: -2 (D) Range: $y \ge -2$ or $[-2, \infty)$

23. $y = -[x - (-2)]^2 + 5 = -(x + 2)^2 + 5$

25. $y = (x - 1)^2 - 3$

27. $f(x) = x^2 - 8x + 12 = (x^2 - 8x) + 12$
$$= (x^2 - 8x + 16) + 12 - 16$$
$$= (x - 4)^2 - 4 \quad \text{(vertex form)}$$

    (A) y-intercept: $f(0) = 0^2 - 8(0) + 12 = 12$
          x-intercepts: $f(x) = 0$
$$(x - 4)^2 - 4 = 0$$
$$(x - 4)^2 = 4$$
$$x - 4 = \pm 2$$
$$x = 2, 6$$

    (B) Vertex: $(4, -4)$ (C) Minimum: -4 (D) Range: $y \ge -4$ or $[-4, \infty)$

29. $r(x) = -4x^2 + 16x - 15 = -4(x^2 - 4x) - 15$
$$= -4(x^2 - 4x + 4) - 15 + 16$$
$$= -4(x - 2)^2 + 1 \quad \text{(vertex form)}$$

(A) y-intercept: $r(0) = -4(0)^2 + 16(0) - 15 = -15$
   x-intercepts: $r(x) = 0$
$$-4(x - 2)^2 + 1 = 0$$
$$(x - 2)^2 = \frac{1}{4}$$
$$x - 2 = \pm \frac{1}{2}$$
$$x = \frac{3}{2}, \frac{5}{2}$$

(B) Vertex: $(2, 1)$ (C) Maximum: 1   (D) Range: $y \le 1$ or $(-\infty, 1]$

31. $u(x) = 0.5x^2 - 2x + 5 = 0.5(x^2 - 4x) + 5$
$$= 0.5(x^2 - 4x + 4) + 3$$
$$= 0.5(x - 2)^2 + 3 \quad \text{(vertex form)}$$

(A) y-intercept: $u(0) = 0.5(0)^2 - 2(0) + 5 = 5$
   x-intercepts: $u(x) = 0$
$$0.5(x - 2)^2 + 3 = 0$$
$$(x - 2)^2 = -6; \text{ no solutions.}$$

There are no x-intercepts.

(B) Vertex: $(2, 3)$ (C) Minimum: 3   (D) Range: $y \ge 3$ or $[3, \infty)$

33. $f(x) = 0.3x^2 - x - 8$

(A) $f(x) = 4$: $0.3x^2 - x - 8 = 4$
$$0.3x^2 - x - 12 = 0$$

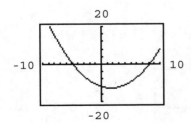

$x = -4.87, 8.21$

(B) $f(x) = -1$: $0.3x^2 - x - 8 = -1$
$$0.3x^2 - x - 7 = 0$$

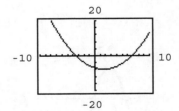

$x = -3.44, 6.78$

(C) $f(x) = -9$: $0.3x^2 - x - 8 = -9$
$$0.3x^2 - x + 1 = 0$$

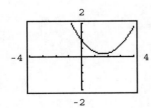

No solutions.

35.

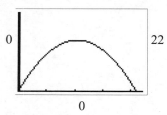

maximum value $f(10.41667) = 651.0417$

37. $g(x) = 0.25x^2 - 1.5x - 7 = 0.25(x^2 - 6x + 9) - 2.25 - 7 = 0.25(x - 3)^2 - 9.25$

(A) x-intercepts: $0.25(x - 3)^2 - 9.25 = 0$

$$(x - 3)^2 = 37$$
$$x - 3 = \pm\sqrt{37}$$
$$x = 3 + \sqrt{37} \approx 9.0828, \ 3 - \sqrt{37} \approx -3.0828$$

y-intercept: -7

(B) Vertex: $(3, -9.25)$ (C) Minimum: -9.25 (D) Range: $y \geq -9.25$ or $[-9.25, \infty)$

39. $f(x) = -0.12x^2 + 0.96x + 1.2$

$$= -0.12(x^2 - 8x + 16) + 1.92 + 1.2$$
$$= -0.12(x - 4)^2 + 3.12$$

(A) x-intercepts: $-0.12(x - 4)^2 + 3.12 = 0$

$$(x - 4)^2 = 26$$
$$x - 4 = \pm\sqrt{26}$$
$$x = 4 + \sqrt{26} \approx 9.0990, \ 4 - \sqrt{26} \approx -1.0990$$

y-intercept: 1.2

(B) Vertex: $(4, 3.12)$ (C) Maximum: 3.12 (D) Range: $y \leq 3.12$ or $(-\infty, 3.12]$

41.

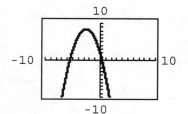

$x = -5.37, 0.37$

43.

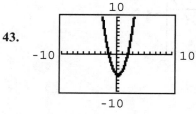

$-1.37 < x < 2.16$

45.

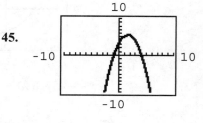

$x \leq -0.74$ or $x \geq 4.19$

47. f is a quadratic function and min $f(x) = f(2) = 4$

Axis: $x = 2$

Vertex: $(2, 4)$

Range: $y \geq 4$ or $[4, \infty)$

x intercepts: None

49. (A)

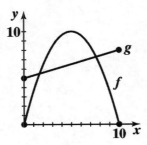

(B) $f(x) = g(x)$

$$-0.4x(x - 10) = 0.3x + 5$$
$$-0.4x^2 + 4x = 0.3x + 5$$
$$-0.4x^2 + 3.7x = 5$$
$$-0.4x^2 + 3.7x - 5 = 0$$

$$x = \frac{-3.7 \pm \sqrt{3.7^2 - 4(-0.4)(-5)}}{2(-0.4)}$$

$$x = \frac{-3.7 \pm \sqrt{5.69}}{-0.8} \approx 1.64, 7.61$$

(C) $f(x) > g(x)$ for $1.64 < x < 7.61$

(D) $f(x) < g(x)$ for $0 \le x < 1.64$ or $7.61 < x \le 10$

51. (A)

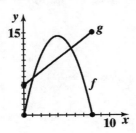

(B) $f(x) = g(x)$

$$-0.9x^2 + 7.2x = 1.2x + 5.5$$
$$-0.9x^2 + 6x = 5.5$$
$$-0.9x^2 + 6x - 5.5 = 0$$

$$x = \frac{-6 \pm \sqrt{36 - 4(-0.9)(-5.5)}}{2(-0.9)}$$

$$x = \frac{-6 \pm \sqrt{16.2}}{-1.8} \approx 1.1, 5.57$$

(C) $f(x) > g(x)$ for $1.10 < x < 5.57$

(D) $f(x) < g(x)$ for $0 \le x < 1.10$ or $5.57 < x \le 8$

53. A quadratic function has exactly one real zero if its graph is tangent to the x–axis.

55. A quadratic function has two real zeros if $b^2 - 4ac > 0$.

57. If a and k have the same sign, then the quadratic function has no real zeros.

59.

$$ax^2 + bx + c = a\left(x^2 + \frac{b}{a}x + \frac{c}{a}\right)$$

$$= a\left(x^2 + \frac{b}{a}x + \frac{b^2}{4a^2} - \frac{b^2}{4a^2} + \frac{c}{a}\right)$$

$$= a\left(\left[x^2 + \frac{b}{a}x + \frac{b^2}{4a^2}\right] + \frac{4ac - b^2}{4a^2}\right)$$

$$= a\left(x + \frac{b}{2a}\right)^2 + \frac{4ac - b^2}{4a}$$

Therefore, $h = -\dfrac{b}{2a}$

61.   Mathematical model: $f(x) = -0.518x^2 + 33.3x - 481$

(A)

x	28	30	32	34	36
Mileage	45	52	55	51	47
$f(x)$	45.3	51.8	54.2	52.4	46.5

(B)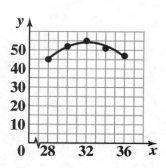

(C) $x = 31: f(31) = -0.518(31)^2 + 33.3(31) - 481 = 53.502$
    $f(31) \approx 53.50$ thousand miles

   $x = 35: f(35) = -0.518(35)^2 + 33(35) - 481 \approx 49.95$ thousand miles

(D)   The maximum mileage is achieved at 32 lb/in^2 pressure. Increasing the pressure or decreasing the pressure reduces the mileage.

63.   Quadratic regression using the data in Problem 61:
    $f(x) = -0.518x^2 + 33.3x - 481$

65. $p(x) = 75 - 3x$; $R(x) = xp(x)$; $1 \le x \le 20$
    $R(x) = x(75 - 3x) = 75x - 3x^2$

$$= -3(x^2 - 25x)$$

$$= -3\left(x^2 - 25x + \frac{625}{4}\right) + \frac{1875}{4}$$

$$= -3\left(x - \frac{25}{2}\right)^2 + \frac{1875}{4}$$

$$= -3(x - 12.5)^2 + 468.75$$

(A)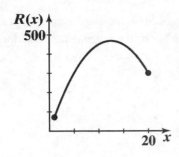

(B) Output for maximum revenue: $x = 12.5$ (12,500,000 chips); maximum revenue: $468,750,000

(C) Wholesale price per chip at maximum revenue:
$p(12.5) = 75 - 3(12.5) = 37.5$ or $37.50

67. Revenue function: $R(x) = x(75 - 3x) = 75 - 3x^2$
Cost function: $C(x) = 125 + 16x$

(A)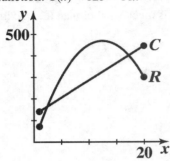

(B) Break-even points: $R(x) = C(x)$

$$75x - 3x^2 = 125 + 16x$$
$$-3x^2 + 59x - 125 = 0 \quad \text{or}$$
$$3x^2 - 59x + 125 = 0$$

$$x = \frac{59 \pm \sqrt{(59)^2 - 4(3)(125)}}{2(3)}$$

$$= \frac{59 \pm \sqrt{1981}}{6} = \frac{59 \pm 44.508}{6}$$

$x = 2.415$ or 17.251
The company breaks even at $x = 2,451,000$ chips and $17,251,000$ chips.

(C) Using the results from (A) and (B):
Loss: $1 \leq x < 2.415$ or $17.251 < x \leq 20$
Profit: $2.415 < x < 17.251$

69. Revenue function: $R(x) = x(75 - 3x)$
Cost function: $C(x) = 125 + 16x$

(A) Profit function: $P(x) = x(75 - 3x) - (125 + 16x)$
$$= 75x - 3x^2 - 125 - 16x;$$
$$P(x) = 59x - 3x^2 - 125$$

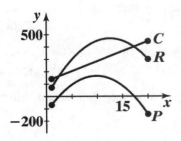

(B) The x coordinates of the intersection points of R and C are the same as the x-intercepts of P.

(C) x-intercepts of P: $-3x^2 + 59x - 125$

$$x = \frac{-59 \pm \sqrt{(59)^2 - 4(-3)(-125)}}{-6}$$

$x = 2.415$ or 17.251 (See Problem 67)
The break–even points are: 2,415,000 chips and 17,251,000 chips.

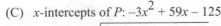

(D)   The maximum profit and the maximum revenue do not occur at the same output level and they are not equal. The profit function involves both revenue and cost; the revenue function does not involve the production costs.

(E) $P(x) = -3x^2 + 59x - 125$

$$= -3\left(x^2 - \frac{59}{3}x\right) - 125$$

$$= -3\left[x - \left(\frac{59}{6}\right)\right]^2 - 125 + 3\left(\frac{59}{6}\right)^2$$

$$= -3(x - 9.833)^2 + 165.083$$

The maximum profit is $165,083,000; it occurs at an output level of 9,833,000 chips. From Problem 67(B), the maximum revenue is $468,750,000. The maximum profit is much smaller than the maximum revenue.

71.   Solve: $f(x) = 1,000(0.04 - x^2) = 20$

$$40 - 1000x^2 = 20$$

$$1000x^2 = 20$$

$$x^2 = 0.02$$

$$x = 0.14 \text{ or } -0.14$$

Since we are measuring distance, we take the positive solution:

$x = 0.14$ cm

73. Quadratic regression model

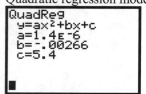

$y \approx 0.0000014(3100)^2 - 0.00266(3100) + 5.4 \approx 10.6$ mph.

EXERCISE 2-4

Things to remember:

<u>1.</u>   POLYNOMIAL FUNCTION

A POLYNOMIAL FUNCTION is a function that can be written in the form

$$f(x) = a_n x^n + a_{n-1} x^{n-1} + \ldots + a_1 x + a_0$$

for n a nonnegative integer, called the DEGREE of the polynomial. The coefficients $a_0, a_1, \ldots, a_n$ are real numbers with $a_n \neq 0$, a_n is called the LEADING COEFFICIENT of f. The DOMAIN of a polynomial function is the set of all real numbers. The graph of a polynomial function is continuous, with no holes or breaks.

2.   A RATIONAL FUNCTION is any function that can be written in the form

$$f(x) = \frac{n(x)}{d(x)}, \quad d(x) \neq 0$$

where $n(x)$ and $d(x)$ are polynomials. The DOMAIN is the set of all real numbers such that $d(x) \neq 0$.

3.   PROCEDURE: VERTICAL AND HORIZONTAL ASYMPTOTES OF RATIONAL FUNCTIONS
Consider the rational function

$$f(x) = \frac{n(x)}{d(x)}$$

where $n(x)$ and $d(x)$ are polynomials.

VERTICAL ASYMPTOTES:

Case 1. Suppose $n(x)$ and $d(x)$ have no real zeros in common. If c is a number such that $d(c) = 0$, then the line $x = c$ is a vertical asymptote of the graph of f.

Case 2. If $n(x)$ and $d(x)$ have one or more real zeros in common, cancel the common linear factors and apply Case 1 to the reduced function. (The reduced function has the same asymptotes as f.)

HORIZONTAL ASYMPTOTES:

Case 1. If degree $n(x) <$ degree $d(x)$, then $y = 0$ is the horizontal asymptote.

Case 2. If degree $n(x) =$ degree $d(x)$, then $y = a/b$ is the horizontal asymptote, where a is the leading coefficient of $n(x)$ and b is the leading coefficient of $d(x)$.

Case 3. If degree $n(x) >$ degree $d(x)$, then there is no horizontal asymptote.

1. $f(x) = 50 - 5x$
   (A) degree 1
   (B) x-intercept: $f(x) = 0$
$$50 - 5x = 0$$
$$x = 10$$

   (C) y-intercept: $f(0) = 50$

3. $f(x) = x^4(x-1)$
   (A) degree: 5
   (B) x-intercepts: $f(x) = 0$
$$x^4(x-1) = 0$$
$$x = 0, 1$$

   (C) y-intercept: $f(0) = 0$

5. $f(x) = x^2 + 3x + 2 = (x+2)(x+1)$

   (A) degree: 2

   (B) x-intercepts: $f(x) = 0$
$$x = -2, -1$$

   (C) y-intercept: $f(0) = 2$

7. $f(x) = (x^2 - 1)(x^2 - 9)$
   (A) degree: 4
   (B) x-intercepts: $f(x) = 0$
$$x = -1, 1, -3, 3$$
   (C) y-intercept: $f(0) = 9$

9. $f(x) = (2x+3)^4(x-5)^5$

  (A) degree: 9

  (B) x-intercepts: $f(x) = 0$
$$x = -\tfrac{3}{2}, \ 5$$

  (C) y-intercept: $f(0) = 3^4(-5)^5 = -253,125$

11.   (A) 4   (B) negative

13.   (A) 5   (B) negative

15. (A) 1   (B) negative   **17.** (A) 6   (B) positive

19. 10

21. 1; polynomials of odd degree cross the x-axis at least once.

23. $f(x) = \dfrac{x+2}{x-2}$

   (A) *Intercepts:*

   x-intercepts: $f(x) = 0$ only if $x + 2 = 0$ or $x = -2$.

   The x intercept is -2.

   y-intercept: $f(0) = \dfrac{0+2}{0-2} = -1$

   The y intercept is -1.

   (B) *Domain:* The denominator is 0 at $x = 2$. Thus, the domain is the set of all real numbers except 2.

   (C) *Asymptotes:*

   Vertical asymptotes: $f(x) = \dfrac{x+2}{x-2}$ The denominator is 0 at $x = 2$. Therefore, the line $x = 2$ is

   a vertical asymptote.

   Horizontal asymptotes: $f(x) = \dfrac{x+2}{x-2} = \dfrac{1+\dfrac{2}{x}}{1-\dfrac{2}{x}}$

   As x increases or decreases without bound, the numerator tends to 1 and the denominator tends to 1. Therefore, the line $y = 1$ is a horizontal asymptote.

   (D) 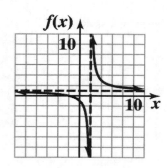   (E)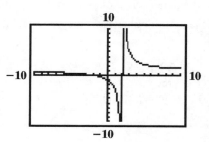

25. $f(x) = \dfrac{3x}{x+2}$

   (A) *Intercepts:*

   x-intercepts: $f(x) = 0$ only if $3x = 0$ or $x = 0$.

   The x-intercept is 0.

   y-intercept: $f(0) = \dfrac{3 \cdot 0}{0+2} = 0$

   The y-intercept is 0.

   (B) *Domain:*   The denominator is 0 at $x = -2$. Thus, the domain is the set of all real numbers except -2.

   (C) *Asymptotes:*

   Vertical asymptotes: $f(x) = \dfrac{3x}{x+2}$

The denominator is 0 at $x = -2$. Therefore, the line $x = -2$ is a vertical asymptote.

Horizontal asymptotes: $$f(x) = \frac{3x}{x+2} = \frac{3}{1 + \frac{2}{x}}$$

As x increases or decreases without bound, the numerator is 3 and the denominator tends to 1. Therefore, the line $y = 3$ is a horizontal asymptote.

(D)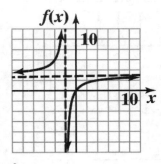

(E)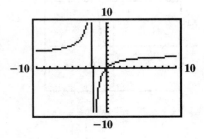

27. $f(x) = \dfrac{4 - 2x}{x - 4}$

(A) *Intercepts:*

x-intercepts: $f(x) = 0$ only if $4 - 2x = 0$ or $x = 2$.
            The x-intercept is 2.

y-intercept: $f(0) = \dfrac{4 - 2 \cdot 0}{0 - 4} = -1$
            The y-intercept is -1.

(B) *Domain:* The denominator is 0 at $x = 4$. Thus, the domain is the set of all real numbers except 4.

(C) *Asymptotes:*

Vertical asymptotes: $f(x) = \dfrac{4 - 2x}{x - 4}$

The denominator is 0 at $x = 4$. Therefore, the line $x = 4$ is a vertical asymptote.

Horizontal asymptotes: $f(x) = \dfrac{4 - 2x}{x - 4} = \dfrac{\dfrac{4}{x} - 2}{1 - \dfrac{4}{x}}$

As x increases or decreases without bound, the numerator tends to -2 and the denominator tends to 1. Therefore, the line $y = -2$ is a horizontal asymptote.

(D)

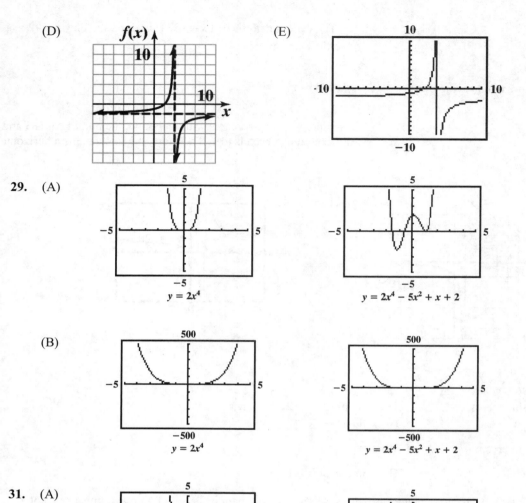

(E)

29. (A)

$y = 2x^4$

$y = 2x^4 - 5x^2 + x + 2$

(B)

$y = 2x^4$

$y = 2x^4 - 5x^2 + x + 2$

31. (A)

$y = -x^5$

$y = -x^5 + 4x^3 - 4x + 1$

(B)

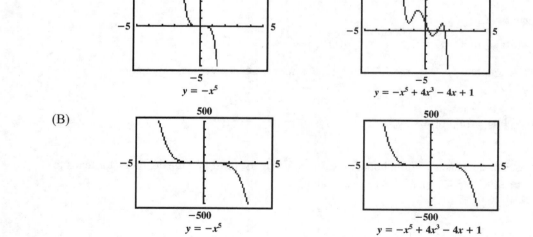

$y = -x^5$

$y = -x^5 + 4x^3 - 4x + 1$

33. $f(x) = \dfrac{n(x)}{d(x)} = \dfrac{5x^3 + 2x - 3}{6x^3 - 7x + 1}$.   Since degree $n(x) = 3 =$ degree $d(x)$, $y = \dfrac{5}{6}$ is the horizontal asymptote.

35. $f(x) = \dfrac{n(x)}{d(x)} = \dfrac{1 - 5x + x^2}{2 + 3x + 4x^2}$. Since degree $n(x) = 2 = $ degree $d(x)$, $y = \dfrac{1}{4}$ is the horizontal asymptote.

37. $f(x) = \dfrac{n(x)}{d(x)} = \dfrac{x^4 + 2x^2 + 1}{1 - x^5}$. Since degree $n(x) = 4 < 5 = $ degree $d(x)$, $y = 0$ is the horizontal asymptote.

39. $f(x) = \dfrac{n(x)}{d(x)} = \dfrac{x^2 + 6x + 1}{x - 5}$. Since degree $n(x) = 2 > 1 = $ degree $d(x)$, there is no horizontal asymptote.

41. $f(x) = \dfrac{n(x)}{d(x)} = \dfrac{x^2 + 1}{(x^2 - 1)(x^2 - 9)} = \dfrac{x^2 + 1}{(x - 1)(x + 1)(x - 3)(x + 3)}$. Since $n(x) = x^2 + 1$ has no real zeros and

$d(1) = d(-1) = d(3) = d(-3) = 0$, $x = 1$, $x = -1$, $x = 3$, $x = -3$ are the vertical asymptotes of the graph of f.

43. $f(x) = \dfrac{n(x)}{d(x)} = \dfrac{x^2 - x - 6}{x^2 - 3x - 10} = \dfrac{(x - 3)(x + 2)}{(x - 5)(x + 2)} = \dfrac{x - 3}{x - 5}$, $x \neq -2$, $x = 5$ is a vertical asymptote of the graph of

f.

45. $f(x) = \dfrac{n(x)}{d(x)} = \dfrac{x^2 + 3x}{x^3 - 36x} = \dfrac{x(x + 3)}{x(x^2 - 36)} = \dfrac{x + 3}{(x - 6)(x + 6)}$, $x \neq 0$. $x = 6$, $x = -6$ are the vertical asymptotes

of the graph of f.

47. $f(x) = \dfrac{2x^2}{x^2 - x - 6}$

   (A) *Intercepts:*

         x-intercepts: $f(x) = 0$ only if $2x^2 = 0$ or $x = 0$.
                                         The x-intercept is 0.

         y-intercept: $f(0) = \dfrac{2 \cdot 0^2}{0^2 - 0 - 6} = 0$
                                         The y-intercept is 0.

   (B) *Asymptotes:*

         Vertical asymptotes: $f(x) = \dfrac{2x^2}{x^2 - x - 6} = \dfrac{2x^2}{(x - 3)(x + 2)}$

         The denominator is 0 at $x = -2$ and $x = 3$. Thus, the lines $x = -2$
                                     and $x = 3$ are vertical asymptotes.

         Horizontal asymptotes: $f(x) = \dfrac{2x^2}{x^2 - x - 6} = \dfrac{2}{1 - \dfrac{1}{x} - \dfrac{6}{x^2}}$

As x increases or decreases without bound, the numerator is 2 and the denominator tends to 1. Therefore, the line $y = 2$ is a horizontal asymptote.

   (C)   (D)

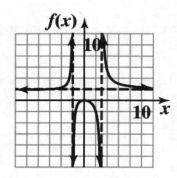

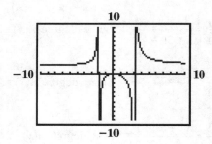

49. $f(x) = \dfrac{6 - 2x^2}{x^2 - 9}$

(A)   *Intercepts:*

x-intercepts: $f(x) = 0$ only if $6 - 2x^2 = 0$

$$2x^2 = 6$$
$$x^2 = 3$$
$$x = \pm\sqrt{3}$$

The x-intercepts are $\pm\sqrt{3}$.

y-intercept: $f(0) = \dfrac{6 - 2 \cdot 0^2}{0^2 - 9} = -\dfrac{2}{3}$

The y-intercept is $-\dfrac{2}{3}$.

(B)   *Asymptotes:*

Vertical asymptotes: $f(x) = \dfrac{6 - 2x^2}{x^2 - 9} = \dfrac{6 - 2x^2}{(x - 3)(x + 3)}$

The denominator is 0 at $x = -3$ and $x = 3$. Thus, the lines
$x = -3$ and $x = 3$ are vertical asymptotes.

Horizontal asymptotes: $f(x) = \dfrac{6 - 2x^2}{x^2 - 9} = \dfrac{\dfrac{6}{x^2} - 2}{1 - \dfrac{9}{x^2}}$

As x increases or decreases without bound, the numerator tends to -2 and the denominator tends to 1.
Therefore, the line $y = -2$ is a horizontal asymptote.

(C)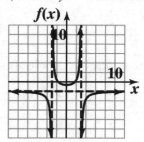

(D)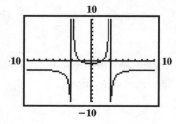

51. $f(x) = \dfrac{-4x + 24}{x^2 + x - 6}$

(A)   *Intercepts:*

x-intercepts: $f(x) = 0$ only if $-4x + 24 = 0$ or $x = 6$.

The x-intercept is 6.

y-intercept: $f(0) = \dfrac{-4(0) + 24}{0^2 + 0 - 6} = -4$

The y-intercept is –4.

(B)   *Asymptotes:*

Vertical asymptotes: $f(x) = \dfrac{-4x + 24}{x^2 + x - 6} = \dfrac{-4x + 24}{(x+3)(x-2)}$

The denominator is 0 at $x = -3$ and $x = 2$. Thus, the lines

$x = -3$ and $x = 2$ are vertical asymptotes.

Horizontal asymptotes: $f(x) = \dfrac{-4x + 24}{x^2 + x - 6} = \dfrac{-\dfrac{4}{x} + \dfrac{24}{x^2}}{1 + \dfrac{1}{x} + \dfrac{6}{x^2}}$

As x increases or decreases without bound, the numerator tends to 0 and the denominator tends to 1. Therefore, the line $y = 0$ (the x axis) is a horizontal asymptote.

(C)   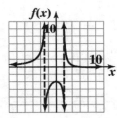   (D)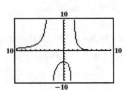

53.   The graph has 1 turning point which implies degree $n = 2$. The x-intercepts are $x = -1$ and $x = 2$.
Thus, $f(x) = (x + 1)(x - 2) = x^2 - x - 2$.

55.   The graph has 2 turning points which implies degree $n = 3$. The x-intercepts are $x = -2$, $x = 0$, and $x = 2$.
The direction of the graph indicates that leading coefficient is negative

$f(x) = -(x + 2)(x)(x - 2) = 4x - x^3$.

57.   (A)   Since $C(x)$ is a linear function of x, it can be written in the form

$C(x) = mx + b$

Since the fixed costs are \$200, $b = 200$.

Also, $C(20) = 3800$, so

$3800 = m(20) + 200$

$20m = 3600$

$m = 180$

Therefore, $C(x) = 180x + 200$

(B) $\overline{C}(x) = \dfrac{C(x)}{x} = \dfrac{180x + 200}{x}$

(C)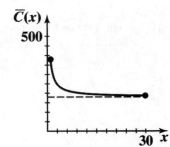

(D) $\overline{C}(x) = \dfrac{180x + 200}{x} = \dfrac{180 + \dfrac{200}{x}}{1}$

As x increases, the numerator tends to 180 and the denominator is 1.
Therefore, $\overline{C}(x)$ tends to 180 or \$180 per board.

59.   (A) $\overline{C}(n) = \dfrac{2500 + 175n + 25n^2}{n}$

(B)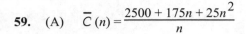

(C) Using the graph, we calculate

$\overline{C}(8) = \dfrac{2500 + 175(8) + 25(8)^2}{8} = 687.50$

$\overline{C}(9) = \dfrac{2500 + 175(9) + 25(9)^2}{9} = 677.78$

$\overline{C}(10) = \dfrac{2500 + 175(10) + 25(10)^2}{10} = 675.00$

$\overline{C}(11) = \dfrac{2500 + 175(11) + 25(11)^2}{11} = 677.27$

$\overline{C}(12) = \dfrac{2500 + 175(12) + 25(12)^2}{12} = 683.33$

Thus, it appears that the average cost per year is a minimum at $n = 10$ years; at 10 years, the average minimum cost is \$675.00 per year.

(D)   10 years; \$675.00 per year

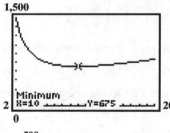

61.   (A) $\overline{C}(x) = \dfrac{0.00048(x - 500)^3 + 60,000}{x}$

(B)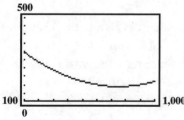

(C)   The caseload which yields the minimum average cost per case is 750 cases per month. At 750 cases per month, the average cost per case is \$90.

63. (A)   Cubic regression model

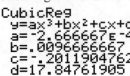

CubicReg
y=ax³+bx²+cx+d
a=-2.666667E-4
b=.0096666667
c=-.2011904762
d=17.84761905

(B)   Per capita consumption of ice cream in 2025:
$y(45) \approx 4.1$ lbs.

65. (A) $v(x) = \dfrac{26 + 0.06x}{x} = \dfrac{\dfrac{26}{x} + 0.06}{1}$

As x increases, the numerator tends to 0.06 and the denominator is 1. Therefore, $v(x)$ approaches 0.06 centimeters per second as x increases.

(B)

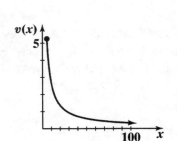

67. (A)   Cubic regression model

CubicReg
y=ax³+bx²+cx+d
a=8.7037037E-5
b=-.0108492063
c=.2907407407
d=8.546031746

(B)   The marriage rate per 1000 population for 2025 will be 5.5.

EXERCISE 2-5

Things to remember:

1.   EXPONENTIAL FUNCTION

The equation

$f(x) = b^x, b > 0, b \neq 1$

defines an EXPONENTIAL FUNCTION for each different constant b, called the BASE. The DOMAIN of f is all real numbers, and the RANGE of f is the set of positive real numbers.

2.   BASIC PROPERTIES OF THE GRAPH OF $f(x) = b^x, b > 0, b \neq 1$

a.   All graphs pass through $(0,1)$; $b^0 = 1$ for any base b.

b.   All graphs are continuous curves; there are no holes or jumps.

c.   The x-axis is a horizontal asymptote.

d.   If $b > 1$, then b^x increases as x increases.

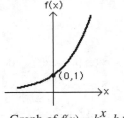

Graph of $f(x) = b^x, b > 1$

e. If $0 < b < 1$, then b^x decreases as x increases.

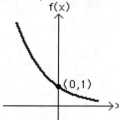

Graph of $f(x) = b^x$, $0 < b < 1$

3. PROPERTIES OF EXPONENTIAL FUNCTIONS

For $a, b > 0$, $a \neq 1$, $b \neq 1$, and x, y real numbers:

a. EXPONENT LAWS

(i) $a^x a^y = a^{x+y}$ (iv) $(ab)^x = a^x b^x$

(ii) $\dfrac{a^x}{a^y} = a^{x-y}$ (v) $\left(\dfrac{a}{b}\right)^x = \dfrac{a^x}{b^x}$

(iii) $(a^x)^y = a^{xy}$

b. $a^x = a^y$ if and only if $x = y$.

c. For $x \neq 0$, $a^x = b^x$ if and only if $a = b$.

4. EXPONENTIAL FUNCTION WITH BASE $e = 2.71828\ldots$

Exponential functions with base e and base $1/e$ are respectively defined by $y = e^x$ and $y = e^{-x}$.

Domain: $(-\infty, \infty)$

Range: $(0, \infty)$

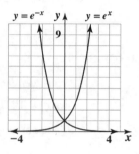

5. Functions of the form $y = ce^{kt}$, where c and k are constants and the independent variable t represents time, are often used to model population growth and radioactive decay. Since $y(0) = c$, c represents the initial population or initial amount. The constant k represents the growth or decay rate; $k > 0$ in the case of population growth, $k < 0$ in the case of radioactive decay.

6. COMPOUND INTEREST

If a principal P (present value) is invested at an annual rate r (expressed as a decimal) compounded m times per year, then the amount A (future value) in the account at the end of t years is given by:

$$A = P\left(1 + \frac{r}{m}\right)^{mt}.$$

If a principal P is invested at an annual rate r (expressed as a decimal) compounded continuously, then the amount in the account at the end of t years is given by

$$A = Pe^{rt}$$

where $e \approx 2.71828$ is the base of the exponential function.

1.   (A) k (B) g (C) h (D) f

3. $y = 5^x$, $-2 \le x \le 2$

x	y
-2	$\frac{1}{25}$
-1	$\frac{1}{5}$
0	1
1	5
2	25

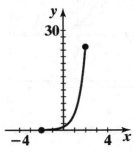

5. $y = \left(\dfrac{1}{5}\right)^x = 5^{-x}$, $-2 \le x \le 2$

x	y
-2	25
-1	5
0	1
1	$\frac{1}{5}$
2	$\frac{1}{25}$

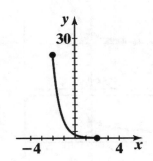

7. $f(x) = -5^x$, $-2 \le x \le 2$

x	y
-2	$-\frac{1}{25}$
-1	$-\frac{1}{5}$
0	-1
1	-5
2	-25

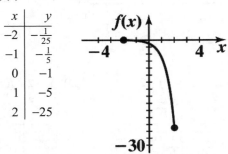

9. $y = -e^{-x}$, $-3 \le x \le 3$

x	y
-3	≈ -20
-2	≈ -7.4
-1	≈ -2.7
0	-1
1	≈ -0.4
2	≈ -0.1
3	≈ 0.05

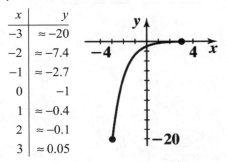

11. $g(x) = -f(x)$; the graph of g is the graph of f reflected in the x axis.

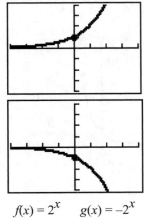

$f(x) = 2^x$ $g(x) = -2^x$

13. $g(x) = f(x + 1)$; the graph of g is the graph of f shifted one unit to the left.

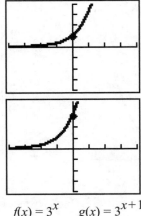

$f(x) = 3^x$ $g(x) = 3^{x+1}$

15. $g(x) = f(x) + 1$; the graph of g is the graph of f shifted one unit up.

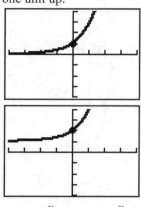

$$f(x) = e^x \qquad g(x) = e^x + 1$$

17. $g(x) = 2f(x + 2)$; the graph of g is the graph of f vertically expanded by a factor of 2 and shifted to the left 2 units.

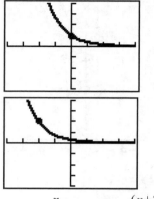

$$f(x) = e^{-x} \qquad g(x) = 2e^{-(x+2)}$$

19. (A) $y = f(x) - 1$

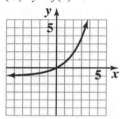

(B) $y = f(x + 2)$

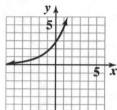

(C) $y = 3f(x) - 2$

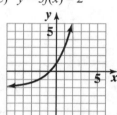

(D) $y = 2 - f(x - 3)$

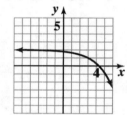

21. $f(t) = 2^{t/10}$, $-30 \le t \le 30$

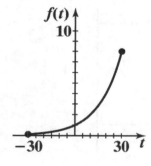

x	y
-4	≈ -3
-2	≈ -2.6
-1	-2
0	≈ 0.3
1	≈ 4.4
2	≈ 17.1

23. $y = -3 + e^{1+x}$, $-4 \le x \le 2$

25. $y = e^{|x|}$, $-3 \le x \le 3$

x	y
-3	≈ 20.1
-1	≈ 2.7
0	1
1	≈ 2.7
3	≈ 20.1

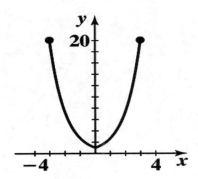

27. Solve

$$a^2 = a^{-2}$$
$$a^2 = \frac{1}{a^2}$$
$$a^4 = 1$$
$$a^4 - 1 = 0$$
$$(a^2 - 1)(a^2 + 1) = 0$$

$a^2 - 1 = 0$ implies $a = 1, -1$

$a^2 + 1 = 0$ has no real solutions

The exponential function property: $a^x = a^y$ if and only if $x = y$ assumes $a > 0$ and $a \ne 1$. Our solutions are $a = 1, -1$; $1^x = 1^y$ for all real numbers x, y, $(-1)^x = (-1)^y$ for all even integers x, y.

29. $10^{2-3x} = 10^{5x-6}$ implies (see 3b)
$$2 - 3x = 5x - 6$$
$$-8x = -8$$
$$x = 1$$

31. $4^{5x-x^2} = 4^{-6}$ implies
$$5x - x^2 = -6$$
or $-x^2 + 5x + 6 = 0$
$$x^2 - 5x - 6 = 0$$
$$(x - 6)(x + 1) = 0$$
$$x = 6, -1$$

33. $5^3 = (x + 2)^3$ implies (by property 3c)
$$5 = x + 2$$
Thus, $x = 3$.

35. $xe^{-x} + 7e^{-x} = 0$
$$e^{-x}(x + 7) = 0$$
$$x + 7 = 0 \quad (\text{since } e^{-x} \ne 0)$$
$$x = -7$$

37. $2x^2 e^x - 8e^x = 0$
$$2e^x(x^2 - 4) = 0$$
$$x^2 - 4 = 0 \quad (\text{since } e^x \ne 0)$$
$$x = -2, 2$$

39. $e^{4x} - e = 0$
$$e^{4x} = e^1$$
$$4x = 1$$
$$x = \tfrac{1}{4}$$

41. $e^{3x-1} + e > 0$ for all x. Therefore, $e^{3x-1} + e = 0$ has no solutions.

43. $h(x) = x2^x$, $-5 \le x \le 0$

x	$h(x)$
-5	$-\frac{5}{32}$
-4	$-\frac{1}{4}$
-3	$-\frac{3}{8}$
-2	$-\frac{1}{2}$
-1	$-\frac{1}{2}$
0	0

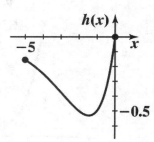

45. $N = \dfrac{100}{1 + e^{-t}}$, $0 \le t \le 5$

t	N
0	50
1	≈ 73.1
2	≈ 88.1
3	≈ 95.3
5	≈ 99.3

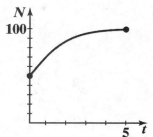

47.   Use $A = Pe^{rt}$, $P = 10,000$, $r = 0.0395$, and $t = 12$.

$A = 10,000e^{0.0395(12)} \approx 10,000e^{0.474} \approx \$16,064.07$.

49. $A = P\left(1 + \dfrac{r}{m}\right)^{mt}$, we have:

(A) $P = 2,500$, $r = 0.07$, $m = 4$, $t = \dfrac{3}{4}$

$A = 2,500\left(1 + \dfrac{0.07}{4}\right)^{4 \cdot 3/4} = 2,500(1 + 0.0175)^3 = 2,633.56$

Thus, $A = \$2,633.56$.

(B) $A = 2,500\left(1 + \dfrac{0.07}{4}\right)^{4 \cdot 15} = 2,500(1 + 0.0175)^{60} = 7079.54$

Thus, $A = \$7,079.54$.

51. $A = P\left(1 + \dfrac{r}{m}\right)^{mt}$. With $A = 15,000$, $r = 0.0675$, and $m = 52$, we have:

$15,000 = P\left(1 + \dfrac{0.0675}{52}\right)^{52(5)} = P(1.4011)$.

Therefore, $P = \dfrac{15,000}{1.4011} \approx \$10,706$.

53. $A = P\left(1 + \dfrac{r}{m}\right)^{mt}$. $P = 10,000, \ t = 1$.

   (A)   Stonebridge Bank: $r = 0.0095, \ m = 12$

$$A = 10,000\left(1 + \frac{0.0095}{12}\right)^{12(1)} = \$10,095.41$$

   (B)   Deep Green Bank: $r = 0.0080, \ m = 365$

$$A = 10,000\left(1 + \frac{0.0080}{365}\right)^{365(1)} = \$10,080.32$$

   (C)   Provident Bank: $r = 0.0085, \ m = 4$

$$A = 10,000\left(1 + \frac{0.0085}{4}\right)^{4(1)} = \$10,085.27$$

55.   Given $N = 2(1 - e^{-0.037t}), \ 0 \le t \le 50$

t	N
0	0
10	≈ 0.62
30	≈ 1.34
50	≈ 1.69

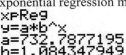

N approaches 2 as t increases without bound.

57.   (A)   Exponential regression model

$y = ab^x$; 2022 is year 32; $y(32) \approx \$9,781,000$

   (B)   According to the model, $y(10) \approx \$1,647,000$.

59.   Given $I = I_0 e^{-0.23d}$

   (A) $I = I_0 e^{-0.23(10)} = I_0 e^{-2.3} \approx I_0(0.10)$

Thus, about 10% of the surface light will reach a depth of 10 feet.

   (B) $I = I_0 e^{-0.23(20)} = I_0 e^{-4.6} \approx I_0(0.010)$

Thus, about 1% of the surface light will reach a depth of 20 feet.

61.   (A)   Model: $P(t) = 7.1.8e^{0.011t}$

   (B)   In the year 2025, $t = 12$; $P(12) = 7.1e^{0.011(12)} = 7.1e^{0.132} \approx 8,100,000,000$ (nearest 100 million)

   In the year 2035, $t = 22$; $P(22) = 7.1e^{0.011(22)} = 7.1e^{0.242} \approx 9,000,000$ or 9 billion

63.   (A)   Exponential regression model

From Problem 61, the world population in 2022 is projected to be 7,8000,000. The model implies that the number of internet hosts will be almost 6 times larger than the estimated world population in 2022.

2022: $t = 28$, $y(28) \approx 42,772,000,000$

EXERCISE 2-6

Things to remember:

1. ONE-TO-ONE FUNCTIONS

    A function f is said to be ONE-TO-ONE if each range value corresponds to exactly one domain value.

2. INVERSE OF A FUNCTION

    If f is a one-to-one function, then the INVERSE of f is the function formed by interchanging the independent and dependent variables for f. Thus, if (a, b) is a point on the graph of f, then (b, a) is a point on the graph of the inverse of f.

    Note: If f is not one-to-one, then f DOES NOT HAVE AN INVERSE.

3. LOGARITHMIC FUNCTIONS

    The inverse of an exponential function is called a LOGARITHMIC FUNCTION.
    For $b > 0$ and $b \neq 1$,

Logarithmic form		Exponential form
$y = \log_b x$	is equivalent to	$x = b^y$

    The LOG TO THE BASE b OF x is the exponent to which b must be raised to obtain x. [Remember: A logarithm is an exponent.] The DOMAIN of the logarithmic function is the range of the corresponding exponential function, and the RANGE of the logarithmic function is the domain of the corresponding exponential function. Typical graphs of an exponential function and its inverse, a logarithmic function, for $b > 1$, are shown in the figure below:

    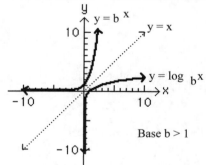

    Base b > 1

4. PROPERTIES OF LOGARITHMIC FUNCTIONS

    If b, M, and N are positive real numbers, $b \neq 1$, and p and x are real numbers, then:

a. $\log_b 1 = 0$

b. $\log_b b = 1$

c. $\log_b b^x = x$

d. $b^{\log_b x} = x, \; x > 0$

e. $\log_b MN = \log_b M + \log_b N$

f. $\log_b \dfrac{M}{N} = \log_b M - \log_b N$

g. $\log_b M^p = p \log_b M$

h. $\log_b M = \log_b N$ if and only if $M = N$

5. LOGARITHMIC NOTATION; LOGARITHMIC-EXPONENTIAL RELATIONSHIPS

Common logarithm: $\log x$ means $\log_{10} x$

Natural logarithm: $\ln x$ means $\log_e x$

$\log x = y$ is equivalent to $x = 10^y$

$\ln x = y$ is equivalent to $x = e^y$

---

1. $27 = 3^3$ (using 3)

3. $1 = 10^0$

5. $8 = 4^{3/2}$

7. $\log_7 49 = 2$

9. $\log_4 8 = \dfrac{3}{2}$

11. $\log_b A = u$

13. $\log_{10} 100 = \log_{10} 10^2 = 2$

15. $\log_2 16 = \log_2 2^4 = 4$

17. $\log_5 \dfrac{1}{25} = \log_5 5^{-2} = -2$

19. $\ln \dfrac{1}{e^4} = \ln e^{-4} = -4$

(using 2a)

21. $\log_b \dfrac{P}{Q} = \log_b P - \log_b Q$

23. $\log_b L^5 = 5 \log_b L$

25. $3^{p \log_3 q} = 3^{\log_3 q^p} = q^p$ (using 4g and 4d)

27. $\log_3 x = 2$

$x = 3^2$

$x = 9$

29. $\log_7 49 = y$

$\log_7 7^2 = y$

$2 = y$

Thus, $y = 2$.

31. $\log_b 10^{-4} = -4$

$10^{-4} = b^{-4}$

This equality implies $b = 10$ (since the exponents are the same).

33. $\log_4 x = \dfrac{1}{2}; \quad x = 4^{1/2}; \quad x = 2$

35. False; counter-example: $y = x^2$.

37. True; if g is the inverse of f, then f is the inverse of g so g must be one-to-one.

39. True; if $y = 2x$, then $x = 2y$ implies $y = \dfrac{x}{2}$.

41. False; $f(x) = \ln x$ is one-to-one; domain of $f = (0, \infty)$, range of $f = (-\infty, \infty)$.

43. $\log_b x = \dfrac{2}{3} \log_b 8 + \dfrac{1}{2} \log_b 9 - \log_b 6 = \log_b 8^{2/3} + \log_b 9^{1/2} - \log_b 6$

$= \log_b 4 + \log_b 3 - \log_b 6 = \log_b \dfrac{4 \cdot 3}{6}$

$$\log_b x = \log_b 2$$
$$x = 2$$

45. $\log_b x = \dfrac{3}{2}\,\log_b 4 - \dfrac{2}{3}\,\log_b 8 + 2\log_b 2 = \log_b 4^{3/2} - \log_b 8^{2/3} + \log_b 2^2$

$$= \log_b 8 - \log_b 4 + \log_b 4 = \log_b 8$$

$$\log_b x = \log_b 8$$

$$x = 8$$

47. $\log_b x + \log_b(x - 4) = \log_b 21$

$$\log_b x(x - 4) = \log_b 21$$

Therefore, $x(x - 4) = 21$

$$x^2 - 4x - 21 = 0$$
$$(x - 7)(x + 3) = 0$$

Thus, $\qquad x = 7$.

[Note: $x = -3$ is not a solution

since $\log_b(-3)$ is not defined.]

49. $\log_{10}(x - 1) - \log_{10}(x + 1) = 1$

$$\log_{10}\left(\frac{x-1}{x+1}\right) = 1$$

Therefore, $\dfrac{x-1}{x+1} = 10^1 = 10$

$$x - 1 = 10(x + 1)$$
$$x - 1 = 10x + 10$$
$$-9x = 11$$
$$x = -\frac{11}{9}$$

There is *no solution*, since

$$\log_{10}\left(-\frac{11}{9} - 1\right) = \log_{10}\left(-\frac{20}{9}\right)$$

is not defined. Similarly,

$$\log_{10}\left(-\frac{11}{9} + 1\right) = \log_{10}\left(-\frac{2}{9}\right)$$

is not defined.

51. $y = \log_2(x - 2)$

$$x - 2 = 2^y$$
$$x = 2^y + 2$$

x	y
$\frac{9}{4}$	-2
$\frac{5}{2}$	-1
3	0
4	1
6	2
18	4

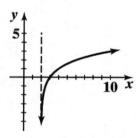

53. The graph of $y = \log_2(x - 2)$ is the graph of $y = \log_2 x$ shifted to the right 2 units.

55. Since logarithmic functions are defined only for positive "inputs", we must have $x + 1 > 0$ or $x > -1$; domain: $(-1, \infty)$. The range of $y = 1 + \ln(x + 1)$ is the set of all real numbers.

57. (A) 3.54743

(B) −2.16032

(C) 5.62629

(D) −3.19704

59. (A) $\log x = 1.1285$
$$x = 10^{1.1285} \approx 13.4431$$

(B) $\log x = -2.0497$
$$x = 10^{-2.0497} \approx 0.0089$$

(C) $\ln x = 2.7763$
$$x = e^{2.7763} \approx 16.0595$$

$$\text{(D)} \quad \ln x = -1.8879$$
$$x = e^{-1.8879} \approx 0.1514$$

61. $10^x = 12$ (Take common logarithms of both sides)

$\log 10^x = \log 12 \approx 1.0792$

$\quad x \approx 1.0792$ ($\log 10^x = x \quad \log 10 = x$; $\log 10 = 1$)

63. $e^x = 4.304$ (Take natural logarithms of both sides)

$\ln e^x = \ln 4.304 \approx 1.4595$

$\quad x \approx 1.4595$ ($\ln e^x = x \quad \ln e = x$; $\ln e = 1$)

65. $1.005^{12t} = 3$ (Take either common or natural logarithms of both sides; here we'll use natural logarithms.)

$\ln 1.005^{12t} = \ln 3$

$\quad 12t = \dfrac{\ln 3}{\ln 1.005} \approx 220.2713$

$\quad\quad t = 18.3559$

67. $y = \ln x, \ x > 0$

x	y
0.5	≈ -0.69
1	0
2	≈ 0.69
4	≈ 1.39
5	≈ 1.61

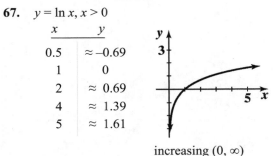

increasing $(0, \infty)$

69. $y = |\ln x|, \ x > 0$

x	y
0.5	≈ 0.69
1	0
2	≈ 0.69
4	≈ 1.39
5	≈ 1.6

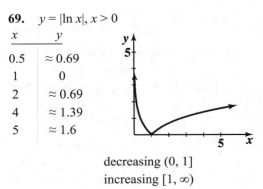

decreasing $(0, 1]$
increasing $[1, \infty)$

71. $y = 2 \ln(x + 2), \ x > -2$

x	y
-1.5	≈ -1.39
-1	0
0	≈ 1.39
1	≈ 2.2
5	≈ 3.89
10	≈ 4.97

increasing $(-2, \infty)$

73. $y = 4 \ln x - 3, \ x > 0$

x	y
0.5	≈ -5.77
1	-3
5	≈ 3.44
10	≈ 6.21

increasing $(0, \infty)$

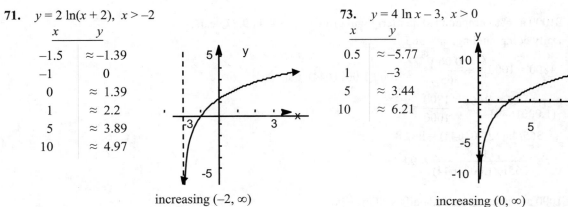

75. For any number b, $b > 0$, $b \neq 1$, $\log_b 1 = y$ is equivalent to $b^y = 1$ which implies $y = 0$. Thus, $\log_b 1 = 0$ for any permissible base b.

77. 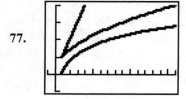 A function f is "larger than" a function g on an interval $[a, b]$ if $f(x) > g(x)$ for $a \leq x \leq b$.

$r(x) > q(x) > p(x)$ for $1 < x \leq 16$, that is $x > \sqrt{x} > \ln x$ for $1 < x \leq 16$

79. From the compound interest formula $A = P(1 + r)^t$, we have:

$2P = P(1 + 0.2136)^t$ or $(1.2136)^t = 2$

Take the natural log of both sides of this equation:

$\ln(1.2136)^t = \ln 2$ [Note: the common log could have been used instead of the natural log.]

$t \ln(1.2136) = \ln 2$

$t = \dfrac{\ln 2}{\ln 1.2136} \approx \dfrac{0.69135}{0.19359} = 3.58 \approx 4$ years

81. $A = A = P\left(1 + \dfrac{r}{m}\right)^{mt}$, $r = 0.06$, $P = 1000$, $A = 1800$.

Quarterly compounding: $m = 4$

$1800 = 1000\left(1 + \dfrac{0.06}{4}\right)^{4t} = 1000(1.015)^{4t}$

$(1.015)^{4t} = \dfrac{1800}{1000} = 1.8$

$4t \ln(1.015) = \ln(1.8)$

$t = \dfrac{\ln(1.8)}{4\ln(1.015)} \approx 9.87$

$1000 at 6% compounded quarterly will grow to \$1800 in 9.87 years.

Daily compounding: $m = 365$

$1800 = 1000\left(1 + \dfrac{0.06}{365}\right)^{365t} = 1000(1.0001644)365t$

$(1.0001644)^{365t} = \dfrac{1800}{1000} = 1.8$

$365t \ln(1.0001644) = \ln 1.8$

$t = \dfrac{\ln 1.8}{365 \ln(1.0001644)} \approx 9.80$

$1000 at 6% compounded daily will grow to $1800 in 9.80 years.

83. $A = Pe^{rt}$; $r = 0.0475$, $P = 35,000$, $A = 50,000$

$$50{,}000 = 35{,}000e^{0.0475t}$$

$$e^{0.0475t} = \frac{50{,}000}{35{,}000} = \frac{10}{7}$$

$$0.0475t = \ln\left(\frac{10}{7}\right)$$

$$t = \frac{\ln(10/7)}{0.0475} \approx 7.51 \quad \text{years}$$

85. (A)   Logarithmic regression model,

Table 1:

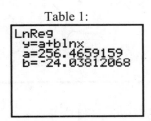

```
LnReg
 y=a+blnx
 a=256.4659159
 b=-24.03812068
```

To estimate the demand at a price level of $50, we solve
$$a + b \ln x = 50$$
for x. The result is $x \approx 5{,}373$ screwdrivers per month

(B) Logarithmic regression model,

Table 2:

```
LnReg
 y=a+blnx
 a=-127.8085281
 b=20.01315349
```

To estimate the supply at a price level of $50, we solve
$$a + b \ln x = 50$$
for x. The result is $x \approx 7{,}220$ screwdrivers per month.

(C) The condition is not stable, the price is likely to decrease since the demand at a price level of $50 is much lower than the supply at this level.

87. $I = I_0 10^{N/10}$

Take the common log of both sides of this equation. Then:

$$\log I = \log(I_0 10^{N/10}) = \log I_0 + \log 10^{N/10}$$

$$= \log I_0 + \frac{N}{10}\log \ 10 = \log I_0 + \frac{N}{10} \ (\text{since } \log 10 = 1)$$

So, $\dfrac{N}{10} = \log I - \log I_0 = \log\left(\dfrac{I}{I_0}\right)$ and $N = 10\log\left(\dfrac{I}{I_0}\right)$.

89.   Logarithmic regression model

```
LnReg
 y=a+blnx
 a=-551.2132518
 b=149.1505358

■
```

The yield in 2024: $t = 124$, $y(124) \approx 168$ bushels/acre.

91. Assuming that the current population is 7.1 billion and that the growth rate is 1.1% compounded continuously, the population after t years will be

$$P(t) = 7.1e^{0.011t}$$

Given that there are $1.68 \times 10^{14} = 168,000$ billion square yards of land, solve

$$7.1e^{0.011t} = 168,000$$

for t:

$$e^{0.011t} = \frac{168,000}{7.1} \approx 23,662$$

$$0.011t = \ln 23,662$$

$$t = \frac{\ln 23,662}{0.011} \approx 916$$

It will take approximately 916 years.

CHAPTER 2 REVIEW

1.

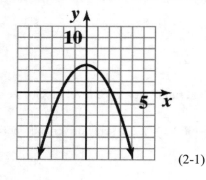

(2-1)

2. $x^2 = y^2$:

x	-3	-2	-1	0	1	2	3
y	±3	±2	±1	0	±1	±2	±3

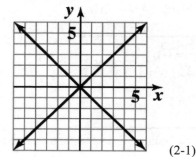

(2-1)

3. $y^2 = 4x^2$:

x	-3	-2	-1	0	1	2	3
y	±6	±4	±2	0	±2	±4	±6

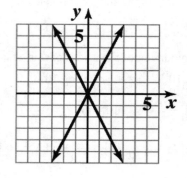

(2-1)

4. (A) Not a function; fails vertical line test

(B) A function

(C) A function

(D) Not a function; fails vertical line test   (2-1)

5. $f(x) = 2x - 1$, $g(x) = x^2 - 2x$

(A) $f(-2) + g(-1) = 2(-2) - 1 + (-1)^2 - 2(-1) = -2$

(B) $f(0) \cdot g(4) = (2 \cdot 0 - 1)(4^2 - 2 \cdot 4) = -8$

(C) $\dfrac{g(2)}{f(3)} = \dfrac{2^2 - 2\cdot 2}{2\cdot 3 - 1} = 0$

(D) $\dfrac{f(3)}{g(2)}$ not defined because $g(2) = 0$ (2-1)

6. $u = e^y$
$v = \ln u$ (2-6)

7. $x = 10^y$
$y = \log x$ (2-6)

8. $\ln M = N$
$M = e^N$ (2-6)

9. $\log u = v$
$u = 10^v$ (2-6)

10. $\log_3 x = 2$
$x = 3^2 = 9$ (2-6)

11. $\log_x 36 = 2$
$x^2 = 36$
$x = 6$
(2-6)

12 $\log_2 16 = x$
$2^x = 16$
$x = 4$
(2-6)

13. $10^x = 143.7$
$x = \log 143.7$
$x \approx 2.157$
(2-6)

14. $e^x = 503,000$
$x = \ln 503,000 \approx 13.128$ (2-6)

15. $\log x = 3.105$
$x = 10^{3.105} \approx 1273.503$ (2-6)

16. $\ln x = -1.147$
$x = e^{-1.147} \approx 0.318$ (2-6)

17. (A) $y = 4$ (B) $x = 0$ (C) $y = 1$ (D) $x = -1$ or 1
(E) $y = -2$ (F) $x = -5$ or 5 (2-1)

18. (A)

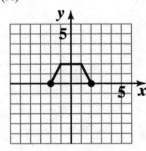

(B)

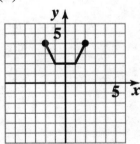

(C)

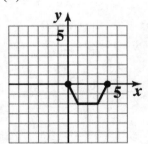

(D)

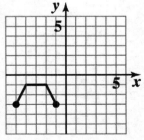

(2-2)

19. $f(x) = -x^2 + 4x = -(x^2 - 4x)$
$$= -(x^2 - 4x + 4) + 4$$
$$= -(x - 2)^2 + 4 \quad \text{(vertex form)}$$

The graph of $f(x)$ is the graph of $y = x^2$ reflected in the x axis, then shifted right 2 units and up 4 units. (2-3)

20. (A) g (B) m (C) n (D) f (2-2, 2-3)

21. $y = f(x) = (x + 2)^2 - 4$

(A) x intercepts: $(x + 2)^2 - 4 = 0$; y intercept: 0
$$(x + 2)^2 = 4$$
$$x + 2 = -2 \text{ or } 2$$
$$x = -4, 0$$

(B) Vertex: $(-2, -4)$ (C) Minimum: -4 (D) Range: $y \geq -4$ or $[-4, \infty)$ (2-3)

22. $y = 4 - x + 3x^2 = 3x^2 - x + 4$; quadratic function.   (2-3)

23. $y = \dfrac{1 + 5x}{6} = \dfrac{5}{6}x + \dfrac{1}{6}$; linear function.   (2-1, 2-3)

24. $y = \dfrac{7 - 4x}{2x} = \dfrac{7}{2x} - 2$; none of these.   (2-1), (2-3)

25. $y = 8x + 2(10 - 4x) = 8x + 20 - 8x = 20$; constant function   (2-1)

26. $\log(x + 5) = \log(2x - 3)$
$$x + 5 = 2x - 3$$
$$-x = -8$$
$$x = 8 \qquad \text{(2-6)}$$

27. $2\ln(x - 1) = \ln(x^2 - 5)$
$$\ln(x - 1)^2 = \ln(x^2 - 5)$$
$$(x - 1)^2 = x^2 - 5$$
$$x^2 - 2x + 1 = x^2 - 5$$
$$-2x = -6$$
$$x = 3 \qquad \text{(2-6)}$$

28. $9^{x-1} = 3^{1+x}$
$$(3^2)^{x-1} = 3^{1+x}$$
$$3^{2x-2} = 3^{1+x}$$
$$2x - 2 = 1 + x$$
$$x = 3 \qquad \text{(2-5)}$$

29. $e^{2x} = e^{x^2 - 3}$
$$2x = x^2 - 3$$
$$x^2 - 2x - 3 = 0$$
$$(x - 3)(x + 1) = 0$$
$$x = 3, -1 \qquad \text{(2-5)}$$

30. $2x^2 e^x = 3xe^x$

$2x^2 = 3x$

$2x^2 - 3x = 0$

$x(2x - 3) = 0$

$x = 0, \ 3/2$ (2-5)

31. $\log_{1/3} 9 = x$

$\left(\dfrac{1}{3}\right)^x = 9$

$\dfrac{1}{3^x} = 9$

$3^x = \dfrac{1}{9}$

$x = -2$ (2-6)

32. $\log_x 8 = -3$

$x^{-3} = 8$

$\dfrac{1}{x^3} = 8$

$x^3 = \dfrac{1}{8}$

$x = \dfrac{1}{2}$ (2-6)

33. $\log_9 x = \dfrac{3}{2}$

$9^{3/2} = x$

$x = 27$ (2-6)

34. $x = 3(e^{1.49}) \approx 13.3113$

35. $x = 230(10^{-0.161}) \approx 158.7552$ (2-5)

36. $\log x = -2.0144$

$x \approx 10^{-2.0144} \approx 0.0097$ (2-6)

37. $\ln x = 0.3618$

$x = e^{0.3618} \approx 1.4359$ (2-6)

38. $35 = 7(3^x)$

$3^x = 5$

$\ln 3^x = \ln 5$

$x \ln 3 = \ln \ 5$

$x = \dfrac{\ln 5}{\ln 3} \approx 1.4650$ (2-6)

39. $0.01 = e^{-0.05x}$

$\ln(0.01) = \ln(e^{-0.05x}) = -0.05x$

Thus, $x = \dfrac{\ln(0.01)}{-0.05} \approx 92.1034$

40. $8{,}000 = 4{,}000(1.08)^x$

$(1.08)^x = 2$

$\ln(1.08)^x = \ln \ 2$

$x \ln 1.08 = \ln \ 2$

$x = \dfrac{\ln 2}{\ln 1.08} \approx 9.0065$ (2-6)

41. $5^{2x-3} = 7.08$

$\ln(5^{2x-3}) = \ln 7.08$

$(2x - 3) \ln 5 = \ln 7.08$

$2x \ln 5 - 3 \ln 5 = \ln 7.08$

$x = \dfrac{\ln 7.08 + 3 \ln 5}{2 \ln 5}$

$x \approx 2.1081$ (2-6)

42. (A) $x^2 - x - 6 = 0$ at $x = -2, 3$

Domain: all real numbers except $x = -2, 3$

(B) $5 - x > 0$ for $x < 5$

Domain: $x < 5$ or $(-\infty, 5)$ (2-1)

43. $f(x) = 4x^2 + 4x - 3 = 4(x^2 + x) - 3$

$$= 4\left(x^2 + x + \frac{1}{4}\right) - 3 - 1$$

$$= 4\left(x + \frac{1}{2}\right)^2 - 4 \quad \text{(vertex form)}$$

Intercepts:

y intercept: $f(0) = 4(0)^2 + 4(0) - 3 = -3$
x intercepts: $f(x) = 0$

$$4\left(x + \frac{1}{2}\right)^2 - 4 = 0$$

$$\left(x + \frac{1}{2}\right)^2 = 1$$

$$x + \frac{1}{2} = \pm 1$$

$$x = -\frac{1}{2} \pm 1 = -\frac{3}{2}, \frac{1}{2}$$

Vertex: $\left(-\frac{1}{2}, -4\right)$;   minimum: -4;   range: $y \geq -4$ or $[-4, \infty)$ (2-3)

44. $f(x) = e^x - 1,\ g(x) = \ln(x + 2)$

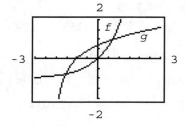

Points of intersection:
$(-1.54, -0.79), (0.69, 0.99)$

(2-5, 2-6)

45. $f(x) = \dfrac{50}{x^2 + 1}$:

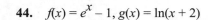

x	-3	-2	-1	0	1	2	3
$f(x)$	5	10	25	50	25	10	5

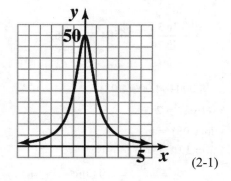

(2-1)

46. $f(x) = \dfrac{-66}{2 + x^2}$:

x	-3	-2	-1	0	1	2	3
$f(x)$	-6	-11	-22	-66	-22	-11	-6

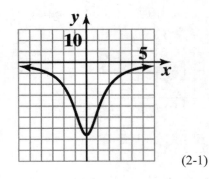

(2-1)

For Problems 47 – 50, $f(x) = 5x + 1$.

47. $f(f(0)) = f(5(0) + 1) = f(1) = 5(1) + 1 = 6$ (2-1)

48. $f(f(-1)) = f(5(-1) + 1) = f(-4) = 5(-4) + 1 = -19$ (2-1)

49. $f(2x - 1) = 5(2x - 1) + 1 = 10x - 4$ (2-1)

50. $f(4 - x) = 5(4 - x) + 1 = 20 - 5x + 1 = 21 - 5x$ (2-1)

51. $f(x) = 3 - 2x$
   (A) $f(2) = 3 - 2(2) = 3 - 4 = -1$
   (B) $f(2 + h) = 3 - 2(2 + h) = 3 - 4 - 2h = -1 - 2h$
   (C) $f(2 + h) - f(2) = -1 - 2h - (-1) = -2h$
   (D) $\dfrac{f(2 + h) - f(2)}{h} = -\dfrac{2h}{h} = -2$ (2-1)

52. $f(x) = x^2 - 3x + 1$
   (A) $f(a) = a^2 - 3a + 1$
   (B) $f(a + h) = (a + h)^2 - 3(a + h) + 1 = a^2 + 2ah + h^2 - 3a - 3h + 1$
   (C) $f(a + h) - f(a) = a^2 + 2ah + h^2 - 3a - 3h + 1 - (a^2 - 3a + 1)$
        $= 2ah + h^2 - 3h$
   (D) $\dfrac{f(a + h) - f(a)}{h} = \dfrac{2ah + h^2 - 3h}{h} = \dfrac{h(2a + h - 3)}{h} = 2a + h - 3$ (2-1)

53. The graph of m is the graph of $y = |x|$ reflected in the x axis and shifted 4 units to the right.   (2-2)

54. The graph of g is the graph of $y = x^3$ vertically contracted by a factor of 0.3 and shifted up 3 units.   (2-2)

55. The graph of $y = x^2$ is vertically expanded by a factor of 2, reflected in the x axis and shifted to the left 3 units.
Equation: $y = -2(x + 3)^2$ (2-2)

56.   Equation: $f(x) = 2\sqrt{x+3} - 1$

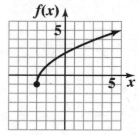

(2-2)

57. $f(x) = \dfrac{n(x)}{d(x)} = \dfrac{5x+4}{x^2 - 3x + 1}$.   Since degree $n(x) = 1 < 2 = $ degree $d(x)$, $y = 0$ is the horizontal asymptote.

(2-4)

58. $f(x) = \dfrac{n(x)}{d(x)} = \dfrac{3x^2 + 2x - 1}{4x^2 - 5x + 3}$.   Since degree $n(x) = 2 = $ degree $d(x)$, $y = \dfrac{3}{4}$ is the horizontal asymptote.

(2-4)

59. $f(x) = \dfrac{n(x)}{d(x)} = \dfrac{x^2 + 4}{100x + 1}$.   Since degree $n(x) = 2 > 1 = $ degree $d(x)$, there is no horizontal asymptote   (2-4)

60. $f(x) = \dfrac{n(x)}{d(x)} = \dfrac{x^2 + 100}{x^2 - 100} = \dfrac{x^2 + 100}{(x-10)(x+10)}$.   Since $n(x) = x^2 + 100$ has no real zeros and

$d(10) = d(-10) = 0$, $x = 10$ and $x = -10$ are the vertical asymptotes of the graph of f.   (2-4)

61. $f(x) = \dfrac{n(x)}{d(x)} = \dfrac{x^2 + 3x}{x^2 + 2x} = \dfrac{x(x+3)}{x(x+2)} = \dfrac{x+3}{x+2}$, $x \neq 0$. $x = -2$ is a vertical asymptote of the graph of f.   (2-4)

62.   True; $p(x) = \dfrac{p(x)}{1}$ is a rational function for every polynomial p.   (2-4)

63.   False; $f(x) = \dfrac{1}{x} = x^{-1}$ is not a polynomial function.   (2-4)

64.   False; $f(x) = \dfrac{1}{x^2 + 1}$ has no vertical asymptotes.   (2-4)

65.   True: let $f(x) = b^x$, $(b > 0, b \neq 1)$, then the positive x-axis is a horizontal asymptote if $0 < b < 1$, and the negative x-axis is a horizontal asymptote if $b > 1$.   (2-5)

66.   True: let $f(x) = \log_b x$ $(b > 0, b \neq 1)$. If $0 < b < 1$, then the positive y-axis is a vertical asymptote; if $b > 1$, then the negative y-axis is a vertical asymptote.   (2-6)

67.   True; $f(x) = \dfrac{x}{x-1}$ has vertical asymptote $x = 1$ and horizontal asymptote $y = 1$.   (2-4)

68.

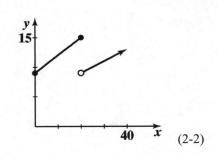

(2-2)

69.

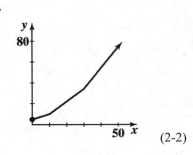

(2-2)

70. $y = -(x-4)^2 + 3$ (2-2, 2-3)

71. $f(x) = -0.4x^2 + 3.2x + 1.2 = -0.4(x^2 - 8x + 16) + 7.6$

$$= -0.4(x-4)^2 + 7.6$$

(A) y intercept: 1.2

x intercepts: $-0.4(x-4)^2 + 7.6 = 0$

$$(x-4)^2 = 19$$

$$x = 4 + \sqrt{19} \approx 8.4,\ 4 - \sqrt{19} \approx -0.4$$

(B) Vertex: (4.0, 7.6)   (C) Maximum: 7.6   (D) Range: $y \le 7.6$ or $(-\infty, 7.6]$ (2-3)

72.

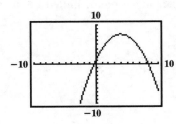

(A) y intercept: 1.2

x intercepts: -0.4, 8.4

(B) Vertex: (4.0, 7.6)

(C) Maximum: 7.6

(D) Range: $y \le 7.6$ or $(-\infty, 7.6]$ (2-3)

73. $\log 10^\pi = \pi \log 10 = \pi$ (see logarithm properties $\underline{4}$.b & g, Section 2-5)

$10^{\log \sqrt{2}} = y$ is equivalent to $\log y = \log \sqrt{2}$

which implies $y = \sqrt{2}$

Similarly, $\ln e^\pi = \pi \ln e = \pi$ (Section 2-5, $\underline{4}$.b & g) and $e^{\ln \sqrt{2}} = y$ implies $\ln y = \ln \sqrt{2}$ and

$y = \sqrt{2}$.   (2-6)

74. $\log x - \log 3 = \log 4 - \log (x + 4)$

$$\log \frac{x}{3} = \log \frac{4}{x+4}$$

$$\frac{x}{3} = \frac{4}{x+4}$$

$$x(x+4) = 12$$

$$x^2 + 4x - 12 = 0$$

$$(x+6)(x-2) = 0$$

$$x = -6,\ 2$$

Since $\log(-6)$ is not defined, -6 is not a solution. Therefore, the solution is $x = 2$.   (2-6)

75. $\ln(2x-2) - \ln(x-1) = \ln x$

$$\ln\left(\frac{2x-2}{x-1}\right) = \ln x$$

$$\ln\left[\frac{2(x-1)}{x-1}\right] = \ln x$$

$$\ln 2 = \ln x$$

$$x = 2 \qquad (2\text{-}6)$$

76. $\ln(x+3) - \ln x = 2\ln 2$

$$\ln\left(\frac{x+3}{x}\right) = \ln(2^2)$$

$$\frac{x+3}{x} = 4$$

$$x + 3 = 4x$$

$$3x = 3$$

$$x = 1 \qquad (2\text{-}6)$$

77.
$$\log 3x^2 = 2 + \log 9x$$

$$\log\ 3x^2 - \log\ 9x = 2$$

$$\log\left(\frac{3x^2}{9x}\right) = 2$$

$$\log\left(\frac{x}{3}\right) = 2$$

$$\frac{x}{3} = 10^2 = 100$$

$$x = 300 \qquad (2\text{-}6)$$

78.
$$\ln\ y = -5t + \ln\ c$$

$$\ln\ y - \ln\ c = -5t$$

$$\ln\frac{y}{c} = -5t$$

$$\frac{y}{c} = e^{-5t}$$

$$y = ce^{-5t} \qquad (2\text{-}6)$$

79.   Let x be *any* positive real number and suppose $\log_1 x = y$. Then $1^y = x$. But, $1^y = 1$, so $x = 1$, i.e., $x = 1$ for all positive real numbers x. This is clearly impossible.   (2-6)

80.   The graph of $y = \sqrt[3]{x}$ is vertically expanded by a factor of 2, reflected in the x axis, shifted 1 unit to the left and 1 unit down.
Equation: $y = -2\sqrt[3]{x+1} - 1$ (2-2)

81. $G(x) = 0.3x^2 + 1.2x - 6.9 = 0.3(x^2 + 4x + 4) - 8.1$
$$= 0.3(x+2)^2 - 8.1$$

(A) y intercept: -6.9

    x intercepts: $0.3(x+2)^2 - 8.1 = 0$
$$(x+2)^2 = 27$$
$$x = -2 + \sqrt{27} \approx 3.2, -2 - \sqrt{27} \approx -7.2$$

(B) Vertex: $(-2, -8.1)$ (C) Minimum: -8.1 (D) Range: $y \geq -8.1$ or $[-8.1, \infty)$ (2-3)

82.

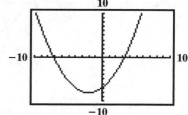

(A) y intercept: -6.9
    x intercept: $-7.2, 3.2$

(B) Vertex: $(-2, -8.1)$

(C) Minimum: -8.1

(D) Range: $y \geq -8.1$ or $[-8.1, \infty)$ (2-3)

83.   (A) $S(x) = 3$ if $0 \leq x \leq 20$;

        $S(x) = 3 + 0.057(x - 20)$

           $= 0.057x + 1.86$ if $20 < x \leq 200$;

        $S(200) = 13.26$

         $S(x) = 13.26 + 0.0346(x - 200)$

$= 0.0346x + 6.34$ if $200 < x \le 1000$;

$S(1000) = 40.94$

$S(x) = 40.94 + 0.0217(x - 1000)$

$= 0.0217x + 19.24$ if $x > 1000$

Therefore, $S(x) = \begin{cases} 3 & \text{if} \quad 0 \le x \le 20 \\ 0.057x + 1.86 & \text{if} \quad 20 < x \le 200 \\ 0.0346x + 6.34 & \text{if} \quad 200 < x \le 1000 \\ 0.0217x + 19.24 & \text{if} \quad x > 1000 \end{cases}$

(B)

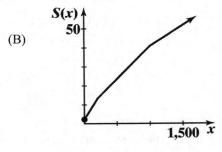

(2-2)

84. $A = P\left(1 + \dfrac{r}{m}\right)^{mt}$; $P = 5{,}000$, $r = 0.0125$, $m = 4$, $t = 5$.

$A = 5000\left(1 + \dfrac{0.0125}{4}\right)^{4(5)} = 5000\left(1 + \dfrac{0.0125}{4}\right)^{20} \approx 5321.95$

After 5 years, the CD will be worth \$5,321.95   (2-5)

85. $A = P\left(1 + \dfrac{r}{m}\right)^{mt}$; $P = 5{,}000$, $r = 0.0105$, $m = 365$, $t = 5$

$A = 5000\left(1 + \dfrac{0.0105}{365}\right)^{365(5)} = 5000\left(1 + \dfrac{0.0105}{365}\right)^{1825} \approx 5269.51$

After 5 years, the CD will be worth \$5,269.51.   (2-5)

86. $A = P\left(1 + \dfrac{r}{m}\right)^{mt}$, $r = 0.0659$, $m = 12$

Solve $P\left(1 + \dfrac{0.0659}{12}\right)^{12t} = 3P$ or $(1.005492)^{12t} = 3$

for t:

$12t \ln(1.005492) = \ln 3$

$t = \dfrac{\ln 3}{12 \ln(1.005492)} \approx 16.7$ year.   (2-5)

87. $A = Pe^{rt}$, $r = 0.0739$. Solve $2P = Pe^{0.0739t}$ for t.

$2P = Pe^{0.0739t}$

$e^{0.0739t} = 2$

$0.0739t = \ln 2$

$t = \dfrac{\ln 2}{0.0739} \approx 9.38$ years.

(2-5)

88. $p(x) = 50 - 1.25x$ Price-demand function

$C(x) = 160 + 10x$ Cost function

$R(x) = xp(x)$

$\quad = x(50 - 1.25x)$ Revenue function

(A)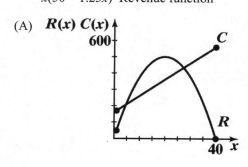

(B) $R = C$

$$x(50 - 1.25x) = 160 + 10x$$
$$-1.25x^2 + 50x = 160 + 10x$$
$$-1.25x^2 + 40x = 160$$
$$-1.25(x^2 - 32x + 256) = 160 - 320$$
$$-1.25(x - 16)^2 = -160$$
$$(x - 16)^2 = 128$$
$$x = 16 + \sqrt{128} \approx 27.314,$$
$$16 - \sqrt{128} \approx 4.686$$

$R = C$ at $x = 4.686$ thousand units

(4,686 units) and $x = 27.314$ thousand units (27,314 units)

$R < C$ for $1 \le x < 4.686$ or $27.314 < x \le 40$

$R > C$ for $4.686 < x < 27.314$

(C) Max Rev: $50x - 1.25x^2 = R$

$$-1.25(x^2 - 40x + 400) + 500 = R$$
$$-1.25(x - 20)^2 + 500 = R$$

Vertex at (20, 500)

Max. Rev. = 500 thousand ($500,000) occurs when <u>output</u> is 20 thousand (20,000 units)

<u>Wholesale price</u> at this output: $p(x) = 50 - 1.25x$

$$p(20) = 50 - 1.25(20) = \$25 \qquad (2\text{-}3)$$

89. (A) $P(x) = R(x) - C(x) = x(50 - 1.25x) - (160 + 10x)$

$$= -1.25x^2 + 40x - 160$$

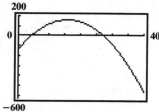

(B) $P = 0$ for $x = 4.686$ thousand units (4,686 units) and $x = 27.314$ thousand units (27,314 units)

$P < 0$ for $1 \le x < 4.686$ or $27.314 < x \le 40$

$P > 0$ for $4.686 < x < 27.314$

(C) Maximum profit is 160 thousand dollars ($160,000), and this occurs at $x = 16$ thousand units (16,000 units). The wholesale price at this output is $p(16) = 50 - 1.25(16) = \$30$, which is $5 greater than the $25 found in 88(C). $\qquad (2\text{-}3)$

90. (A) The area enclosed by the storage areas is given by

$$A = (2y)x$$

Now, $3x + 4y = 840$

so $y = 210 - \dfrac{3}{4}x$

Thus $A(x) = 2\left(210 - \dfrac{3}{4}x\right)x$

$$= 420x - \dfrac{3}{2}x^2$$

(B) Clearly x and y must be nonnegative; the fact that $y \geq 0$ implies

$$210 - \dfrac{3}{4}x \geq 0$$

and $210 \geq \dfrac{3}{4}x$

$$840 \geq 3x$$

$$280 \geq x$$

Thus, domain A: $0 \leq x \leq 280$

(C)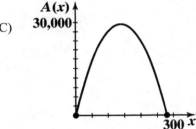

(D) Graph $A(x) = 420x - \dfrac{3}{2}x^2$ and $y = 25{,}000$ together.

There are two values of x that will produce storage areas with a combined area of 25,000 square feet, one near $x = 90$ and the other near $x = 190$.

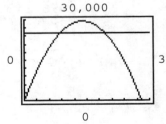

(E) $x = 86$, $x = 194$

(F) $A(x) = 420x - \dfrac{3}{2}x^2 = -\dfrac{3}{2}(x^2 - 280x)$

Completing the square, we have

$$A(x) = -\dfrac{3}{2}(x^2 - 280x + 19{,}600 - 19{,}600)$$

$$= -\dfrac{3}{2}[(x - 140)^2 - 19{,}600]$$

$$= -\dfrac{3}{2}(x - 140)^2 + 29{,}400$$

The dimensions that will produce the maximum combined area are: $x = 140$ ft, $y = 105$ ft. The maximum area is 29,400 sq. ft.   (2-3)

91. (A) Quadratic regression model,

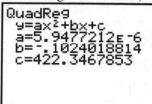

To estimate the demand at price level of $180, we solve the equation
$$ax^2 + bx + c = 180$$
for x. The result is $x \approx 2{,}833$ sets.

(B) Linear regression model,
Table 2:

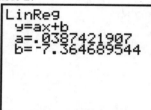

To estimate the supply at a price level of $180, we solve the equation
$$ax + b = 180$$
for x. The result is $x \approx 4{,}836$ sets.

(C) The condition is not stable; the price is likely to decrease since the supply at the price level of $180 exceeds the demand at this level.

(D) Equilibrium price: $131.59
Equilibrium quantity: 3,587 cookware set. (2-3)

92. (A) Cubic Regression

```
CubicReg
 y=ax³+bx²+cx+d
 a=.5813376005
 b=-21.33512555
 c=96.11031183
 d=5571.799951
```

(B) $y(35) = 7725$ (?) (2-5)

93. (A) $N(0) = 1$

$$N\left(\frac{1}{2}\right) = 2$$

$$N(1) = 4 = 2^2$$

$$N\left(\frac{3}{2}\right) = 8 = 2^3$$

$$N(2) = 16 = 2^4$$

$$\vdots$$

Thus, we conclude that
$N(t) = 2^{2t}$ or $N = 4^t$.

(B) We need to solve:

$$2^{2t} = 10^9$$

$$\log 2^{2t} = \log 10^9 = 9$$

$$2t \log 2 = 9$$

$$t = \frac{9}{2\log 2} \approx 14.95$$

Thus, the mouse will die in 15 days.

(2-6)

94. Given $I = I_0 e^{-kd}$. When $d = 73.6$, $I = \frac{1}{2} I_0$. Thus, we have:

$$\frac{1}{2} I_0 = I_0 e^{-k(73.6)}$$

$$e^{-k(73.6)} = \frac{1}{2}$$

$$-k(73.6) = \ln \frac{1}{2}$$

$$k = \frac{\ln(0.5)}{-73.6} \approx 0.00942$$

Thus, $k \approx 0.00942$.

To find the depth at which 1% of the surface light remains, we set $I = 0.01 I_0$ and solve

$$0.01 I_0 = I_0 e^{-0.00942d} \quad \text{for } d:$$

$$0.01 = e^{-0.00942d}$$

$$-0.00942d = \ln 0.01$$

$$d = \frac{\ln 0.01}{-0.00942} \approx 488.87$$

Thus, 1% of the surface light remains at approximately 489 feet.   (2-6)

95.   (A) Logarithmic regression model:

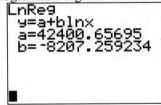

Year 2023 corresponds to $x = 83$; $y(83) \approx 6,134,000$ cows.

   (B)   ln (0) is not defined.   (2-6)

96. Using the continuous compounding model, we have:

$$2P_0 = P_0 e^{0.03t}$$

$$2 = e^{0.03t}$$

$$0.03t = \ln 2$$

$$t = \frac{\ln 2}{0.03} \approx 23.1$$

Thus, the model predicts that the population will double in approximately 23.1 years.   (2-5)

97. (A) Exponential regression model

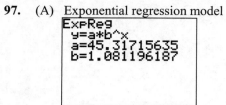

Year 2022 corresponds to $x = 42$; $y(42) \approx \$1{,}203$ billion.

(B) To find when the expenditures will reach two trillion, solve $ab^x = 2{,}000$ for x. The result is $x \approx 48.51$ years; that is, in 2028.   (2-5)

3   MATHEMATICS OF FINANCE

EXERCISE 3-1

Things to remember:

<u>1.</u>   SIMPLE INTEREST

$$I = Prt$$

where I = interest
P = principal
r = annual simple interest rate expressed as a decimal
t = time in years

<u>2.</u>   AMOUNT AT SIMPLE INTEREST

$$A = P + Prt = P(1 + rt)$$

where A = amount or *future value*
P = principal or *present value*
r = annual simple interest rate expressed as a decimal
t = time in years

1.   Tax = Amount x rate: $A = 449.99$, $r = 0.0565$; $T = A\text{r} = 449.99(0.0565) \approx 25.42$. The tax will be \$25.42.

3.   21 games, 16 wins: Winning percentage: $\frac{16}{21} \approx 0.761920$, 76.2% to the nearest percentage point.

5. $y = 12{,}000 + 120x$; slope: $= 120$, y-intercept: $= 1200$.

7. $y = 2000(1 + 0.025x) = 2000 + 50x$; slope: $= 50$, y-intercept: $= 2000$.

9. $r = 1.5\%$; decimal $d = \frac{1.5}{100} = 0.015$.

11.   Decimal $d = 0.006$; rate $r = 100(0.006) = 0.6$ %.

13. $r = 0.4\%$; decimal $d = \frac{0.4}{100} = 0.004$.

15.   Decimal $d = 0.2499$; rate $r = 100(0.2499) = 24.99$ %.

17.   4 months $= \frac{4}{12} = \frac{1}{3}$ year

19.   240 days $= \frac{240}{360} = \frac{2}{3}$ year

21.   12 weeks $= \frac{12}{52} = \frac{3}{13}$ year

23.   2 quarters $= \frac{2}{4} = \frac{1}{2}$ year

25. $I = Prt$; $P = 300$, $r = 0.07$, $t = 2$; $I = 300(0.07)2 = \$42$

27. $I = Prt$; $I = 36$, $r = 0.04$, $t = 6$ months $= \frac{1}{2}$ year; $P = \frac{I}{rt} = \frac{36}{(0.04)(1/2)} = \$1{,}800$

29. $I = Prt$; $I = 48$, $P = 600$, $t = 240$ days $= \frac{2}{3}$ year; $r = \frac{I}{Pt} = \frac{48}{600(2/3)} = 0.12$; 12%

31. $I = Prt$; $I = 60$, $P = 2,400$, $r = 0.05$; $t = \dfrac{I}{Pr} = \dfrac{60}{2400(0.05)} = \dfrac{1}{2}$ year

33. $A = P(1 + rt)$; $P = 4,500$, $r = 0.10$, $t = \dfrac{1}{4}$ year

$A = 4,500(1 + (0.1)(0.25)) = 4500(1.025) = \$4612.50$

35. $A = P(1 + rt)$; $A = 910$, $r = 0.16$, $t = 13$ weeks $= \dfrac{1}{4}$ year

$910 = P(1 + 0.16[0.25]) = P(1.04)$; $P = \dfrac{910}{1.04} = \$875$

37. $A = P(1 + rt)$; $A = 14,560$, $P = 13,000$, $t = 4$ months $= \dfrac{1}{3}$ year

$14,560 = 13,000(1 + r/3) = 13,000 + \dfrac{13,000r}{3}$,

$\dfrac{13,000r}{3} = 1,560$; $r = \dfrac{3(1,560)}{13,000} = 0.36$ or 36%

39. $A = P(1 + rt)$; $A = 736$, $P = 640$, $r = 0.15$
$736 = 640(1 + 0.15t) = 640 + 96t$,
$96t = 96$, $t = 1$ year

41. $I = Prt$
Divide both sides by Pt.

$\dfrac{I}{Pt} = \dfrac{Prt}{Pt}$

$\dfrac{I}{Pt} = r$ or $r = \dfrac{I}{Pt}$

43. $A = P + Prt = P(1 + rt)$
Divide both sides by $(1 + rt)$.

$\dfrac{A}{1+rt} = \dfrac{P(1+rt)}{1+rt}$

$\dfrac{A}{1+rt} = P$ or $P = \dfrac{A}{1+rt}$

45. $A = P(1 + rt) = P + Prt$, $Prt = A - P$, $t = \dfrac{A - P}{rP}$

47. Each of the graphs is a straight line; the y intercept in each case is 1000 and their slopes are 40, 80, and 120.

49. $P = \$3000$, $r = 4.5\% = 0.045$, $t = 4$ months $= \dfrac{1}{3}$ year; $I = Prt = 3000(0.045)\left(\dfrac{1}{3}\right) = \$45$.

51. $P = \$554$, $r = 20\% = 0.2$, $t = 1$ month $= \dfrac{1}{12}$ year; $I = Prt = 554(0.2)\left(\dfrac{1}{12}\right) = \$9.23$

53. $P = \$7260$, $r = 8\% = 0.08$, $t = 8$ months $= \dfrac{2}{3}$ year

$A = 7260\left[1 + 0.08\left(\dfrac{2}{3}\right)\right] = 7260[1.05\overline{33}] = \$7647.20$

55. $P = \$4000$, $A = \$4270$, $t = 10$ months $= \dfrac{5}{6}$ year. The interest on the loan is $I = A - P = \$270$.

$r = \dfrac{I}{Pt} = \dfrac{270}{4000\left(\frac{5}{6}\right)} = 0.081$. Thus, $r = 8.1\%$.

57. $P = \$1000$, $I = \$30$, $t = 60$ days $= \dfrac{1}{6}$ year. $r = \dfrac{I}{Pt} = \dfrac{30}{1000\left(\frac{1}{6}\right)} = 0.18$. Thus, $r = 18\%$.

59. $P = \$1500$. The amount of interest paid is $I = (0.29)(3)(120) = \$104.40$. Thus, the total amount repaid is
$\$1500 + \$104.40 = \$1604.40$. To find the annual interest rate, we let $t = 120$ days $= \dfrac{1}{3}$ year. Then

$r = \dfrac{I}{Pt} = \dfrac{104.40}{1500\left(\frac{1}{3}\right)} = 0.20880$. Thus, $r = 20.880\%$.

61. Use Formula 2: $A = P(1+rt)$ with $A = 1{,}000$, $P = 989.37$ and $t = \dfrac{13}{52} = 0.25$.

$$1{,}000 = 989.37(1 + 0.25r)$$
$$= 989.37 + 247.3425r$$
$$247.3425r = 10.63$$
$$r = \dfrac{10.63}{247.3425} \approx 0.04298 \text{ or } 4.298\%$$

63. Use Formula 2: $A = P(1 + rt)$ with $A = 1{,}000$, $r = 5.53\% = 0.0553$ and $t = \dfrac{50}{360} = 0.1389$.

$$1{,}000 = P[1 + (0.0553)(0.1389)]$$
$$= P[1.00768]$$
$$P = \$992.38$$

65. 2% of the unpaid balance: $1{,}215.45(0.02) = 24.309$ or $\$24.31$; minimum payment $\$24.31$

67. Interest owed: $I = \Pr t = 869.89(.1699)\left(\frac{1}{12}\right) \approx 12.32$.
Minimum payment: the larger of $869.89(0.02) = \$17.40$ and $\$20.00$; the minimum payment is $\$20.00$
Minimum payment – interest: $20.00 - 12.32 = \$7.68$.

69. Principal plus interest on the original note:

$$A = P(1 + rt) = \$5500\left[1 + 0.08\left(\dfrac{90}{360}\right)\right] = \$5610$$

The third party pays $\$5{,}560$ and will receive $\$5610$ in 60 days. We want to find r given that $A = 5610$,
$P = 5{,}560$ and $t = \dfrac{60}{360} = \dfrac{1}{6}$

$$A = P + Prt$$
$$r = \dfrac{A - P}{Pt}$$
$$r = \dfrac{5610 - 5560}{5560\left(\frac{1}{6}\right)} \approx 0.05396 \text{ or } 5.396\%$$

71. The principal P is the cost of the stock plus the firm's commission:
  Cost: $200(14.20) = 2840$
  P: $2840 + 25 + 0.018(2840) = 2916.12$
The investor sells the stock for $200(15.75) = 3150$, and the firm's commission is:
  $37 + 0.014(3150) = 81.10$
Thus, the investor has $3150 - 81.10 = \$3068.90$ after selling the stock; the investor has earned
$3068.90 - 2916.12 = \$152.78$.

Now, using formula 1, with $P = 2916.12$, $I = 152.78$, and $t = 39$ weeks $= \dfrac{3}{4}$ year, we have

$$r = \dfrac{152.78}{2916.12\left(\frac{3}{4}\right)} = 0.06986 \text{ or } 6.986\%$$

73.   The principal P is the cost of the stock plus the firm's commission:

Cost: $215(45.75) = 9836.25$

P: $9836.25 + 56 + 0.01(9836.25) = 9990.61$

The investor sells the stock for $215(51.90) = 11{,}158.50$, and the firm's commission is:

$106 + 0.005(11{,}158.50) = 161.79$

Thus, the investor has $11{,}158.50 - 161.79 = 10{,}996.71$ after selling the stock; the investor has earned $10{,}996.71 - 9990.61 = \$1006.10$.

Now, using formula $\underline{1}$ with $P = 9990.61$, $I = 1006.10$ and $t = \dfrac{300}{360} = \dfrac{5}{6}$ year,

we have $r = \dfrac{1006.10}{9990.61\left(\frac{5}{6}\right)} \approx 0.12085$ or 12.085%

75. $P = \$475$, $I = \$29$, $t = \dfrac{20}{360} = \dfrac{1}{18}$ year

$r = \dfrac{I}{Pt} = \dfrac{29}{475\left(\frac{1}{18}\right)} \approx 1.09895$ or 109.895%

77. $P = \$1{,}900$, $I = \$69$, $t = \dfrac{15}{360} = \dfrac{1}{24}$ year.

$r = \dfrac{I}{Pt} = \dfrac{69}{1900\left(\frac{1}{24}\right)}$ • 0.87158 or 87.158%

79.   Days $1 - 29$: $3{,}000$; Day 30: $1{,}500$. Average daily balance: $\dfrac{29(3000) + 1(1500)}{30} = \dfrac{88{,}500}{30} = 2{,}950.$

Interest: $2950(0.1999)\left(\dfrac{30}{360}\right) = \$49.14.$

81.   Days $1 - 11$: 523.18;   days $12 - 16$: 671.16;   days $17 - 25$: 471.16;   days $26 - 28$: 507.43.

Average daily balance: $\dfrac{11(523.18) + 5(671.16) + 9(471.16) + 3(507.43)}{28} = 531.20.$

Interest: $531.20(0.1999)\dfrac{28}{360} = \$8.26$

EXERCISE 3-2

Things to remember:

1.   AMOUNT—COMPOUND INTEREST

$A = P(1 + i)^{n}$,   where $i = \dfrac{r}{m}$ and

A = amount (future value) at the end of n periods
P = principal (present value)
r = annual (quoted) rate
m = number of compounding periods per year
n = total number of compounding periods

$i = \dfrac{r}{m}$ = rate per compounding period

2.   CONTINUOUS COMPOUND INTEREST FORMULA
     If a principal P is invested at an annual rate r (expressed as a decimal) compounded continuously, then the amount A in the account at the end of t years is given by

$$A = Pe^{rt}.$$

3.   ANNUAL PERCENTAGE YIELD
     If a principal is invested at the annual (nominal) rate r compounded m times per year, then the annual percentage yield is

$$APY = \left(1 + \frac{r}{m}\right)^m - 1$$

If a prinicipal is invested at the annual (nominal) rate r compounded continuously, then the annual percentage yield is

$$APY = e^r - 1$$

The annual percentage yield is also referred to as the EFFECTIVE RATE or the TRUE INTEREST RATE.

---

1. $1{,}641.6 = P(1.2)^3$
     $1641.6 = P(1.728)$
     $P = 950$

3. $12x^3 = 58{,}956$
     $x^3 = 4913$
     $x = \sqrt[3]{4913} = 17$

5. $6.75 = 3(1 + i)^2$
     $(1 + i)^2 = 2.25$
     $1 + i = \sqrt{2.25} = 1.5$
     $i = 0.5$

7. $14{,}641 = 10{,}000(1.1)^n$
     $(1.1)^n = 1.4641$
     $\ln(1.1)^n = n\ln(1.1) = \ln(1.4641)$
     $n = \dfrac{\ln(1.4641)}{\ln(1.1)} = 4$

9. $P = \$5{,}000$; $I = 0.005$; $n = 36$. Find A.
     Using 1, $A = (P(1 + i)^n = 5{,}000(1 + 0.005)^{36} = 5{,}000(1.005)^{36} = \$5{,}983.40$.

11. $A = 8{,}000$; $i = 0.02$; $n = 32$. Find P. Using 1, $P = \dfrac{A}{(1 + i)^n} = \dfrac{8{,}000}{(1 + 0.02)^{32}} = \dfrac{8{,}000}{(1.02)^{32}} = \$4{,}245.07$.

13. $A = Pe^{rt}$; $P = 2{,}450$, $r = 0.0812$, $t = 3$
     $A = 2{,}450e^{0.0812(3)} = 2{,}450e^{0.2436} \approx \$3{,}125.79$

15. $A = Pe^{rt}$; $A = 6{,}300$, $r = 0.0945$, $t = 8$
     $6{,}300 = Pe^{0.0945(8)} = Pe^{0.756} \approx P(2.12974)$
     $P = \dfrac{6{,}300}{2.12974} \approx \$2958.11$

17. $A = Pe^{rt}$; $A = 88{,}000$, $P = 71{,}153$, $r = 0.085$
     $88{,}000 = 71{,}153e^{0.085t}$
     $e^{0.085t} = \dfrac{88{,}000}{71{,}153}$

$$0.085t = \ln\left(\frac{88,000}{71,153}\right) \approx 0.212504$$

$$t = \frac{0.212504}{0.085} \approx 2.5 \text{ years}$$

19. $A = Pe^{rt}$; $A = 15,875$, $P = 12,100$, $t = 48 \text{ months} = 4 \text{ years}$

$$15,875 = 12,100e^{r(4)}$$

$$e^{4r} = \frac{15,875}{12,100} \approx 1.31198$$

$$4r = \ln(1.31198)$$

$$r = \frac{\ln(1.31198)}{4} \approx 0.0679 \text{ or } 6.79\%$$

21. $r = 9\% = 0.09$, $m = 12$; $i = \dfrac{r}{m} = \dfrac{0.09}{12} = 0.0075 = 0.75\%$ per month.

23. $r = 14.6\% = 0.146$, $m = 360$; $i = \dfrac{r}{m} = \dfrac{0.146}{360} \approx 0.00040 = 0.04\%$ per day.

25. $r = 4.8\% = 0.048$; $m = 4$; $i = \dfrac{r}{m} = \dfrac{0.048}{4} = 0.012 = 1.2\%$ per quarter.

27. $i = 0.395\% = 0.00395$; $m = 12$; $r = im = (0.00395)12 = 0.0474 = 4.74\%$

29. $i = 0.9\% = 0.009$; $m = 4$; $r = im = (0.009)4 = 0.036 = 3.6\%$

31. $I = 2.1\% = 0.021$; $m = 2$; $r = im = (0.021)2 = 0.42 = 4.2\%$

33. $P = \$100$, $r = 6\% = 0.06$

(A) $m = 1$, $i = 0.06$, $n = 4$

$$A = (1 + i)^n$$
$$= 100(1 + 0.06)^4$$
$$= 100(1.06)^4 = \$126.25$$

Interest $= 126.25 - 100 = \$26.25$

(B) $m = 4$, $i = \dfrac{0.06}{4} = 0.015$

$$n = 4(4) = 16$$
$$A = 100(1 + 0.015)^{16}$$
$$= 100(1.015)^{16} = \$126.90$$

Interest $= 126.90 - 100 = \$26.90$

(C) $m = 12$, $i = \dfrac{0.06}{12} = 0.005$, $n = 4(12) = 48$

$$A = 100(1 + 0.005)^{48} = 100(1.005)^{48} = \$127.05$$

Interest $= 127.05 - 100 = \$27.05$

35. $P = \$5000$, $r = 5\%$, $m = 12$

(A) $n = 2(12) = 24$

$$i = \frac{0.05}{12} = 0.0042$$

$$A = 5000\left(1 + \frac{0.05}{12}\right)^{24} = \$5,524.71$$

(B) $n = 4(12) = 48$

$$i = \frac{0.05}{12} = 0.0042$$

$$A = 5000\left(1 + \frac{0.05}{12}\right)^{48} = \$6,104.48$$

37. $A = Pe^{rt}$; $P = 8,000$, $r = 0.07$, $t = 6$

$A = 8,000e^{0.07(6)} = 8,000e^{0.42} \approx \$12,175.69$

39. Each of the graphs is increasing, curves upward and has y intercept 1000. The greater the interest rate, the greater the increase. The amounts at the end of 8 years are:

At 4%: $A = 1000\left(1 + \dfrac{0.04}{12}\right)^{96} = \$1376.40$

At 8%: $A = 1000\left(1 + \dfrac{0.08}{12}\right)^{96} = \$1892.46$

At 12%: $A = 1000\left(1 + \dfrac{0.12}{12}\right)^{96} = \$2599.27$

41. $P = 1,000$, $r = 9.75\% = 0.0975 = i$ since the interest is compounded annually.

1st year: $A = P(1 + i)^n = 1000(1 + 0.0975)^1 = \$1,097.50$; interest \$97.50

2nd year: $A = 1000(1 + 0.0975)^2 = \$1,204.51$; interest: $1,204.51 - 1,097.50 = \$107.01$

3rd year: $A = 1000(1 + 0.0975)^3 = \$1,321.95$; interest: $1,321.95 - 1,204.51 = \$117.44$

and so on. The results are:

Period	Interest	Amount
0		\$1,000.00
1	\$97.50	\$1,097.50
2	\$107.01	\$1,204.51
3	\$117.44	\$1,321.95
4	\$128.89	\$1,450.84
5	\$141.46	\$1,592.29
6	\$155.25	\$1,747.54

43. $A = \$10,000$, $r = 6\% = 0.06$, $i = \dfrac{0.06}{2} = 0.03$

(A) $n = 2(5) = 10$

$$A = P(1 + i)^n$$
$$10,000 = P(1 + 0.03)^{10}$$
$$= P(1.03)^{10}$$
$$P = \frac{10,000}{(1.03)^{10}} = \$7440.94$$

(B) $n = 2(10) = 20$

$$P = \frac{A}{(1+i)^n} = \frac{10,000}{(1+0.03)^{20}}$$
$$= \frac{10,000}{(1.03)^{20}}$$
$$= \$5536.76$$

45. $A = Pe^{rt}$

(A) $A = 25,000$, $r = 9\% = 0.09$, $t = 3$

$25,000 = Pe^{0.09(3)} = Pe^{0.27}$

$$P = \frac{25,000}{e^{0.27}} \approx \$19,084.49$$

(B) $25,000 = Pe^{0.09(9)} = Pe^{0.81}$

$$P = \frac{25,000}{e^{0.81}} \approx \$11,121.45$$

47. $\text{APY} = \left(1 + \dfrac{r}{m}\right)^{m} - 1$

(A) $r = 3.9\% = 0.039, \ m = 12$

$\text{APY} = \left(1 + \dfrac{0.039}{12}\right)^{12} - 1 = (1.00325)^{12} - 1 \approx 0.0397 = 3.97\%$

(B) $r = 2.3\% = 0.023, \ m = 4$

$\text{APY} = \left(1 + \dfrac{0.023}{4}\right)^{4} - 1 = (1.00575)^{4} - 1 \approx 0.0232 = 2.32\%$

49. $\text{APY} = e^{r} - 1$

(A) $r = 5.15\% = 0.0515$

$\text{APY} = e^{0.0515} - 1 \approx 0.0528 = 5.28\%$

(B) $\text{APY} = \left(1 + \dfrac{r}{m}\right)^{m} - 1; \ r = 5.20\% = 0.0520; \ m = 2$

$\text{APY} = \left(1 + \dfrac{0.0520}{2}\right)^{2} - 1 = (1.026)^{2} - 1 \approx 0.0527 = 5.27\%$

51.   We have $P = \$4000, \ A = \$9000, \ r = 7\% = 0.07, \ m = 12,$ and

$i = \dfrac{0.07}{12} = 0.0058.$ Since $A = P(1 + i)^{n}$, we have:

$9000 = 4000(1 + 0.0058)^{n}$ or $(1.0058)^{n} = 2.25$

We solve for n by taking logarithms of both sides:

$\ln(1.0058)^{n} = \ln(2.25)$
$n \ln(1.0058) = \ln(2.25)$

$n = \dfrac{\ln(2.25)}{\ln(1.0058)} \approx \dfrac{0.8109}{0.0058} \approx 140.22$

Thus, $n = 140$ months or 11 years and 8 months.

53. $A = Pe^{rt}, \ A = 8{,}600, \ P = 6{,}000, \ r = 9.6\% = 0.096$

$8{,}600 = 6{,}000e^{0.096t}$

$e^{0.096t} = \dfrac{8{,}600}{6{,}000} = \dfrac{43}{30}; \ 0.096t = \ln(43/30), \ t \approx 3.75$ years

55. $A = 2P, \ i = 0.06$

$A = P(1 + i)^{n}$

$2P = P(1 + 0.06)^{n}$

$(1.06)^{n} = 2$

$\ln(1.06)^{n} = \ln 2$
$n \ln(1.06) = \ln 2$

$n = \dfrac{\ln 2}{\ln 1.06} \approx \dfrac{0.6931}{0.0583} \approx 11.9 \approx 12$

57.   We have $A = P(1 + i)^n$. To find the doubling time, set $A = 2P$.
This yields:

$2P = P(1 + i)^n$ or $(1 + i)^n = 2$

Taking the natural logarithm of both sides, we obtain:

$\ln(1 + i)^n = \ln 2$

$n \ln(1 + i) = \ln 2$

$$n = \frac{\ln 2}{\ln(1+i)}$$

(A) $r = 10\% = 0.1$, $m = 4$. Thus,

$i = \dfrac{0.1}{4} = 0.025$ and $n = \dfrac{\ln 2}{\ln(1.025)} \approx 28.07$ quarters or $7\frac{1}{4}$ years.

(B) $r = 12\% = 0.12$, $m = 4$. Thus,

$i = \dfrac{0.12}{4} = 0.03$ and $n = \dfrac{\ln 2}{\ln(1.03)} \approx 23.44$ quarters.

That is, 24 quarters or 6 years.

59.   Doubling time T: $2P = Pe^{rT}$, $e^{rT} = 2$, $rT = \ln 2$, $T = \dfrac{\ln 2}{r}$

(A) $T = \dfrac{\ln 2}{0.09} \approx 7.7$ years

(B) $T = \dfrac{\ln 2}{0.11} \approx 6.3$ years

61. $P = 20{,}000$, $r = 7\% = 0.07$, $m = 4$, $i = \dfrac{0.07}{4} = 0.0175$,

$n = 17(4) = 68$

$A = P(1 + i)^n = 20{,}000(1.0175)^{68} \approx \$65{,}068.44$

63. $P = 210{,}000$, $r = 3\% = 0.03$, $m = 1$, $i = \dfrac{0.03}{1} = 0.03$, $n = 10$

$A = P(1 + i)^n = 210{,}000(1.03)^{10} \approx \$282{,}222.44$

65. $A = 25$, $r = 4.8\% = 0.048$, $m = 1$, $i = \dfrac{0.048}{1} = 0.048$, $n = 5$

$P = \dfrac{A}{(1+i)^n} = \dfrac{25}{(1.048)^5} \approx \$19.78$ per sq ft/mo.

67.   From Problem 59, the doubling time is

$T = \dfrac{\ln 2}{r} = \dfrac{\ln 2}{0.017} \approx 41$ years

69. (A) $A = P\left(1 + \dfrac{r}{m}\right)^{mt}$; $P = 100$, $r = 0.03$, $m = 4$, $t = 250(4) = 1{,}000$

$$A = 100\left(1 + \frac{0.03}{4}\right)^{1000} = 175{,}814.5499; \quad \$175{,}814.55$$

(B) Monthly: $A = 100\left(1 + \dfrac{0.03}{12}\right)^{(12)250} = 179{,}119.9161; \quad \$179{,}119.92$

Daily: $A = 100\left(1 + \dfrac{0.03}{365}\right)^{91250} = 180{,}748.5169; \quad \$180{,}748.52$

Continuously: $A = 100e^{0.03(250)} = 180{,}804.2414; \quad \$180{,}804.24$

71. $A = Pe^{rt}$; $A = 50{,}000$, $P = 28{,}000$, $t = 6$

$50{,}000 = 28{,}000\, e^{6r}$

$$e^{6r} = \frac{50{,}000}{28{,}000} = \frac{25}{14}$$

$$6r = \ln\left(\frac{25}{14}\right), \quad r = \frac{\ln(25/14)}{6} \approx 0.9664 \text{ or } 9.664\%$$

73. $P = \$7000$, $A = \$9000$, $r = 9\% = 0.09$, $m = 12$, $i = \dfrac{0.09}{12} = 0.0075$

Since $A = P(1 + i)^n$, we have:

$9000 = 7000(1 + 0.0075)^n$ or $(1.0075)^n = \dfrac{9}{7}$

Therefore, $\ln(1.0075)^n = \ln\left(\dfrac{9}{7}\right)$

$$n \ln(1.0075) = \ln\left(\frac{9}{7}\right)$$

$$n = \frac{\ln\left(\frac{9}{7}\right)}{\ln(1.0075)} \approx 33.6$$

Thus, it will take 34 months or 2 years and 10 months.

75. $P = \$20{,}000$, $r = 6\% = 0.06$, $m = 365$, $i = \dfrac{0.06}{365}$,

$n = (365)35 = 12{,}775$

Since $A = P(1 + i)^n$, we have:

$$A = 20{,}000\left(1 + \frac{0.06}{365}\right)^{12{,}775} \approx \$163{,}295.21$$

77. From Problem 57, the doubling time is:

$$n = \frac{\ln 2}{\ln(1+i)}$$

At 7% = 0.07 compounded daily, $i = \frac{0.07}{365} \approx 0.000191781$ and

$$n = \frac{\ln 2}{\ln(1.000191781)} \approx 3615 \text{ days or } 9.904 \text{ years}$$

From Problem 59, the doubling time is

$$T = \frac{\ln 2}{r}$$

At 8.2% = 0.082 compounded continuously

$$T = \frac{\ln 2}{0.082} \approx 8.453 \text{ years}$$

79. If an investment of P dollars doubles in n years at an interest rate r compounded annually, then r satisfies the equation:

$$2P = P(1 + r)^n$$

so $1 + r = 2^{1/n}$

$$r = 2^{1/n} - 1$$

This is the exact value of r.
The approximate value of r given by

$$r = \frac{72}{n}$$

is called the Rule of 72.

Setting $n = 6, 7, 8, 9, 10, 11, 12$ in these formulas gives:

Years	Exact rate	Rule of 72
6	12.2	12.0
7	10.4	10.3
8	9.1	9.0
9	8.0	8.0
10	7.2	7.2
11	6.5	6.5
12	5.9	6.0

81. $2,400 at 13% compounded quarterly:

Value after n quarters: $A_1 = 2400\left(1 + \frac{0.13}{4}\right)^n = 2400(1.0325)^n$

$3,000 at 6% compounded quarterly:

Value after n quarters: $A_2 = 3000\left(1 + \frac{0.06}{4}\right)^n = 3000(1.015)^n$

Graph A_1 and A_2:

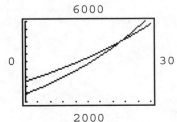

The graphs intersect at (13.05, 3643.56)
$A_2(n) > A_1(n)$ for $n < 13.05$, $A_1(n) > A_2(n)$ for $n > 13.05$.

Thus it will take 14 quarters for the $2,400 investment to be worth more than the $3,000 investment.

83.   The value of P dollars at 10% simple interest after t years is given by

$A_S = P(1 + 0.1t)$

The value of P dollars at 7% interest compounded annually after

t years is given by

$A_C = P(1.07)^t$

Let $P = \$1$ and graph

$A_S = 1 + 0.1t, \quad A_C = (1.07)^t.$

The graphs intersect at the point where $t = 10.89 \approx 11$ years.

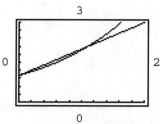

For investments of less than 11 years, simple interest at 10% is better; for investments of more than 11 years, interest compounded annually at 7% is better.

85.   The relationship between the annual percentage yield, APY, and the annual nominal rate r is:

$$\text{APY} = \left(1 + \frac{r}{m}\right)^m - 1$$

In this case, APY = 0.068 and $m = 365$. Thus, we must solve

$$0.068 = \left(1 + \frac{r}{365}\right)^{365} - 1$$

for r:

$$\left(1 + \frac{r}{365}\right)^{365} = 1.068$$

$$1 + \frac{r}{365} = \sqrt[365]{1.068} = (1.068)^{1/365}$$

$$r = 365[(1.068)^{1/365} - 1] \approx 0.0658; \quad r = 6.58\%$$

87.   The APY corresponding to 7% = 0.07 compounded continuously is:

$\text{APY} = e^r - 1 = e^{0.07} - 1 \approx 0.07251$ or 7.251%

To find the annual nominal rate compounded monthly for the

APY = 7.251%, we solve

$$0.07251 = \left(1 + \frac{r}{12}\right)^{12} - 1$$

for r:

$$\left(1 + \frac{r}{12}\right)^{12} = 1.07251$$

$$1 + \frac{r}{12} = \sqrt[12]{1.07251} = (1.07251)^{1/12} \approx 1.005851$$

$$\frac{r}{12} = 0.005851$$

$$r = 0.0702 \text{ or } 7.02\%$$

89. $A = \$30,000,\ r = i = 4.348\% = 0.04348,\ n = 15$

From $A = P(1 + r)^n$, we have

$$P = \frac{A}{(1+r)^n} = \frac{30,000}{(1.04348)^{15}} \approx \$15,843.80$$

91. $A = \$10,000,\ P = \$4,126,\ n = 20$

Using $A = P(1 + r)^n$, we have

$$10,000 = 4,126(1 + r)^{20}$$

$$(1 + r)^{20} = \frac{10,000}{4,126} \approx 2.4237$$

Therefore

$$1 + r = \sqrt[20]{2.4237} = (2.4237)^{1/20} \approx 1.04526$$

and $r \approx 0.04526$ or $r = 4.526\%$

93. $r = 1.02\%$ compounded daily; $m = 365$. $APY = \left(1 + \frac{0.0102}{365}\right)^{365} - 1 = 0.0102520;\ 1.025\%$

95.   The principal P is the cost of the stock plus the firm's commission:

Cost: $100(65) = 6500$,

$P = 6500 + 75 + 0.003(6500) = 6594.50$

The investor sells the stock for $100(125) = 12,500$ and the firm's commission is:

$75 + 0.003(12,500) = 112.50$

Thus, the investor has $12,500 - 112.50 = 12,387.50$ after selling the stock. To find the annual compound rate of return, use

$A = P(1 + r)^n$ with $A = 12,387.50,\ P = 6594.50,\ n = 5,$ and solve for r:

$$12,387.50 = 6594.50(1 + r)^5$$

$$(1 + r)^5 = \frac{12,387.50}{6594.50} \approx 1.8785$$

$$1 + r = \sqrt[5]{1.8785} = (1.8785)^{1/5} \approx 1.1344$$

$$r \approx 0.1344 = 13.44\%$$

97.   The principal P is the cost of the stock plus the firm's commission:

Cost: $200(28) = 5600$

$P = 5600 + 57 + 0.006(5600) = 5690.60$

The investor sells the stock for $200(55) = 11,000$ and the firm's commission is:

$75 + 0.003(11,000) = 108.$

Thus, the investor has $11,000 - 108 = 10,892$ after selling the stock. To find the annual compound rate of return, use

$A = P(1 + r)^n$ with $A = 10,892,\ P = 5690.60,\ n = 4,$ and solve for r:

$$10,892 = 5690.60(1 + r)^4$$

$$(1 + r)^4 = \frac{10,892}{5690.60} \approx 1.9140$$

$$1 + r = \sqrt[4]{1.9140} = (1.9140)^{1/4} \approx 1.1762$$

$$r \approx 0.1762 = 17.62\%$$

EXERCISE 3-3

Things to remember:

<u>1</u>.   FUTURE VALUE OF AN ORDINARY ANNUITY

$$FV = PMT \left(\frac{(1+i)^n - 1}{i} \right)$$

where

       FV = future value (amount)

    PMT = periodic payment

        i = rate per period

       n = number of payments (periods)

(Payments are made at the end of each period.)

<u>2</u>.   SINKING FUND PAYMENT

$$PMT = FV \frac{i}{(1+i)^n - 1}$$

<u>3</u>.   APPROXIMATING INTEREST RATES

Algebra can be used to solve the future value formula in <u>1</u> for *PMT* or *n*, or the payment formula in <u>2</u> for *FV* or *n*. Graphical techniques or equation solvers can be used to approximate *i* to as many places as desired in each of these formulas.

---

1. $S = \dfrac{a(r^n - 1)}{r - 1}$; $a = 1$, $r = 2$, $n = 10$; $S = \dfrac{1(2^{10} - 1)}{2 - 1} = 2^{10} - 1 = 1024 - 1 = 1023$.

3. $a = 30$, $r = 1$, $n = 100$; $S = 100(30) = 3,000$.

5. $a = 10$, $r = 3$, $n = 15$; $S = \dfrac{10(3^{15} - 1)}{3 - 1} = \dfrac{10(14,348,906)}{2} = 71,744,530$.

7. $r = 8\% = 0.08$, $m = 4$, $i = \dfrac{0.08}{4} = 0.02$

     $n = 20(4) = 80$

9. $r = 7.5\% = 0.075$, $m = 2$, $i = \dfrac{0.075}{2} = 0.0375$

     $n = 12(2) = 24$

11. $r = 9\% = 0.09$, $m = 12$, $i = \dfrac{0.09}{12} = 0.0075$

     $n = 4(12) = 48$

13. $r = 5.95\% = 0.0595$, $m = 1$, $i = 0.0595$

     $n = 12(1) = 12$

15. $n = 20,\ i = 0.03,\ PMT = \$500$

$$FV = PMT\left(\frac{(1+i)^n - 1}{i}\right) = 500\frac{(1+0.03)^{20} - 1}{0.03} = 500(26.87037449) = \$13,435.19.$$

17. $FV = \$5,000,\ n = 15,\ i = 0.01$

$$PMT = FV\frac{i}{(1+i)^n - 1} = \frac{FV}{s_{\overline{n}|i}} = 5000\frac{0.01}{(1+0.01)^{15} - 1}$$

$$= \frac{5000}{16.09689554} \approx \$310.62$$

19. $FV = \$4000,\ i = 0.02,\ PMT = 200,\ n = ?$

$$FV = PMT\frac{(1+i)^n - 1}{i}$$

$$\frac{iFV}{PMT} = (1+i)^n - 1$$

$$(1+i)^n = \frac{iFV}{PMT} + 1$$

$$\ln(1+i)^n = \ln\left[\frac{iFV}{PMT} + 1\right]$$

$$n\ln(1+i) = \ln\left[\frac{iFV}{PMT} + 1\right]$$

$$n = \frac{\ln\left[\dfrac{iFV}{PMT} + 1\right]}{\ln(1+i)} = \frac{\ln\left[\dfrac{(0.02)4000}{200} + 1\right]}{\ln(1.02)} = \frac{\ln(1.4)}{\ln(1.02)} \approx \frac{0.3365}{0.01980} = 16.99 \text{ or } 17 \text{ periods.}$$

21. $FV = \$7,600;\ PMT = \$500;\ n = 10,\ i = ?$

$$FV = PMT\frac{(1+i)^n - 1}{i}$$

Substituting the given values into this formula gives

$$7600 = 500\frac{(1+i)^{10} - 1}{i}$$

and $\dfrac{(1+i)^{10} - 1}{i} = 15.2$

Graph $Y_1 = \dfrac{(1+x)^{10} - 1}{x}$,

$Y_2 = 15.2.$

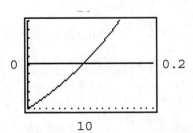

The graphs intersect at the point where $x \approx 0.09$. Thus, $i = 0.09$.

23. An ordinary annuity is an annuity in which the payments are made at the <u>end</u> of each time interval.

25. $FV = PMT \dfrac{(1+i)^n - 1}{i}$

$$PMT\left[(1+i)^n - 1\right] = i\,FV$$

$$(1+i)^n - 1 = \frac{i\,FV}{PMT}$$

$$(1+i)^n = 1 + \frac{i\,FV}{PMT}$$

$$n\,\ln(1+i) = \ln\left[1 + \frac{i\,FV}{PMT}\right]$$

$$n = \frac{\ln\left[1 + \dfrac{i\,FV}{PMT}\right]}{\ln(1+i)}$$

27. $PMT = \$500,\ n = 10(12) = 120,\ i = \dfrac{0.0665}{12}$

$$FV = 500\,\frac{\left(1 + \frac{0.0665}{12}\right)^{120} - 1}{\frac{0.0665}{12}} = 500\,s_{\overline{120}|\,0.0665} = 500(169.7908065) = \$84{,}895.40$$

Total deposits: $500(120) = \$60,000$

Interest: $FV - 60{,}000 = 84{,}895.40 - 60{,}000 = \$24{,}895.40$

29. $PMT = \$300,\ i = \dfrac{0.06}{12} = 0.005,\ n = 5(12) = 60$

$$FV = 300\,\frac{(1 + 0.005)^{60} - 1}{0.005} = 300\,s_{\overline{60}|\,0.005} \qquad \text{(using 1)}$$

$$= 300(69.77003051) = \$20{,}931.01$$

After five years, \$20,931.01 will be in the account.

31. $FV = \$200{,}000,\ i = \dfrac{0.0635}{12},\ n = 15(12) = 180$

$$PMT = FV\,\frac{i}{(1+i)^{180} - 1} = 200{,}000(0.003337158) = \$667.43 \text{ per month}$$

33. $FV = \$100{,}000,\ i = \dfrac{0.075}{12} = 0.00625,\ n = 8(12) = 96$

$$PMT = FV\,\frac{i}{(1+i)^{96} - 1} = 100{,}000(0.00763387) = \$763.39 \text{ per month}$$

35. $PMT = \$1{,}000,\ i = \dfrac{0.0832}{1} = 0.0832,\ n = 5$

$$FV = PMT\,\frac{(1+i)^n - 1}{i} = 1000\,\frac{(1.0832)^n - 1}{0.0832}$$

$n = 1:\ FV = \$1{,}000$

$n = 2:\ FV = 1000\,\dfrac{(1.0832)^2 - 1}{0.0832} = \$2{,}083.20$

Interest: $2083.20 - 2000 = \$83.20$

$n = 3:\ FV = 1000\,\dfrac{(1.0832)^3 - 1}{0.0832} = \$3{,}256.52$

Interest: 3256.52 − 2083.20 = $173.32

and so on.

Balance Sheet

Period	Amount	Interest	Balance
1	$1,000.00	$0.00	$1,000.00
2	$1,000.00	$83.20	$2,083.20
3	$1,000.00	$173.32	$3,256.52
4	$1,000.00	$270.94	$4,527.46
5	$1,000.00	$376.69	$5,904.15

37. $FV = PMT \dfrac{(1+i)^n - 1}{i} = 100 \dfrac{(1+0.0050)^{12} - 1}{0.0050}$ (after one year)

$\quad = 100 \dfrac{(1.0050)^{12} - 1}{0.0050} \left[\text{Note:} PMT = \$100, i = \dfrac{0.06}{12} = 0.0050, n = 12 \right]$

$\quad = \$1233.56$ \hfill (1)

Total deposits in one year = 12(100) = $1200.

Interest earned in first year = $FV - 1200 = 1233.56 - 1200 = \$33.56$.

At the end of the second year:

$FV = 100 \dfrac{(1+0.0050)^{24} - 1}{0.0050}$ [Note: $n = 24$]

$\quad = 100 \dfrac{(1.0050)^{24} - 1}{0.0050} = \$2543.20$ \hfill (2)

Total deposits plus interest in the second year = (2) − (1)

$\qquad\qquad\qquad\qquad\qquad\qquad = 2543.20 - 1233.56 = \$1309.64$ \hfill (3)

Interest earned in the second year = (3) − 1200

$\qquad\qquad\qquad\qquad\qquad = 1309.64 - 1200 = \$109.64$ \hfill (4)

At the end of the third year,

$FV = 100 \dfrac{(1+0.0050)^{36} - 1}{0.0050}$ [Note: $n = 36$]

$\quad = 100 \dfrac{(1.0050)^{36} - 1}{0.0050} = \$3933.61$

Total deposits plus interest in the third year = (4) − (2)

$\qquad\qquad\qquad\qquad\qquad\qquad = 3933.61 - 2543.20 = \$1390.41$ \hfill (5)

Interest earned in the third year = (5) − 1200

$\qquad\qquad\qquad\qquad\qquad = 1390.41 - 1200 = \$190.41$

Thus

Year	Interest earned
1	$ 33.56
2	$109.64
3	$190.41

39. $PMT = \$1000$, $i = 0.064$, $n = 12$

$$FV = 1000\frac{(1+0.064)^{12} - 1}{0.064} = 1000(17.26921764) = 17{,}269.21764$$

Thus, Bob will have \$17,269.22 in his IRA account on his 35th birthday. On his 65th birthday, he will have

$A = 17{,}269.22(1 + 0.064)^{30} = 17{,}269.22(6.430561) = \$111{,}050.77$

41. $FV = \$111{,}050.77$, $i = 0.064$, $n = 30$

$$PMT = 111{,}050.77\frac{0.064}{(1+0.064)^{30} - 1}$$

$$= 111{,}050.77(0.011785155) = \$1308.75 \text{ per year}$$

43. (A) From Section 3.2, $\text{APY} = \left(1 + \frac{r}{m}\right)^m - 1$. Set $\text{APY} = 0.01551$, $m = 12$, and solve for r:

$$0.01551 = \left(1 + \frac{r}{12}\right)^{12} - 1$$

$$\left(1 + \frac{r}{12}\right)^{12} - 1 = 0.01551$$

$$\left(1 + \frac{r}{12}\right)^{12} = 1.01551$$

$$1 + \frac{r}{12} = (1.01551)^{1/12}$$

$$r = 12\left[(1.01551)^{1/12} - 1\right] \approx 0.01540$$

The annual nominal rate is 1.540%.

(B) $PMT = FV\dfrac{i}{(1+i)^n - 1}$; $FV = 10{,}000$, $i = \dfrac{0.01540}{12} = 0.001283$, $n = 4(12) = 48$.

$$PMT = 10{,}000\frac{0.001283}{(1+0.001283)^{48} - 1} \approx 202.1177; \text{ monthly payment: } \$202.12.$$

45. $PMT = \$200$, $FV = \$7{,}000$, $i = \dfrac{0.057}{12}$

From Problem 19:

$$n = \frac{\ln\left[\dfrac{FV(i)}{PMT} + 1\right]}{\ln(1+i)} = \frac{\ln\left[\dfrac{7000\left(\frac{0.057}{12}\right)}{200} + 1\right]}{\ln\left(1 + \frac{0.057}{12}\right)} = \frac{0.153793473}{0.004738754} \approx 32.45$$

Thus, $n = 33$ months.

47. This problem was done with a graphics calculator.

Start with the equation $\dfrac{(1+i)^n - 1}{i} - \dfrac{FV}{PMT} = 0$

where $FV = 5840$, $PMT = 1000$, $n = 5$ and $i = \dfrac{r}{1} = r$,

where r is the nominal annual rate. With these values, the equation is:

$$\frac{(1+r)^5 - 1}{r} - \frac{5840}{1000} = 0$$

or $(1 + r)^5 - 1 - 5.840r = 0$

Set $y = (1 + r)^5 - 1 - 5.840r$ and use your calculator to find the zero r of the function y, where $0 < r < 1$. The result is $r = 0.077$ or 7.77% to two decimal places.

49. Start with the equation $\dfrac{(1+i)^n - 1}{i} - \dfrac{FV}{PMT} = 0$

where $FV = 620$, $PMT = 50$, $n = 12$ and $i = \dfrac{r}{12}$,

where r is the annual nominal rate. With these values, the equation becomes

$$\dfrac{(1+i)^{12} - 1}{i} - \dfrac{620}{50} = 0$$

or $(1+i)^{12} - 1 - 12.4i = 0$

Set $y = (1+i)^{12} - 1 - 12.4i$ and use your calculator to find the zero i of the function y, where $0 < i < 1$. The result is $i = 0.005941$. Thus $r = 12(0.005941) = 0.0713$ or $r = 7.13\%$ to two decimal places.

51. Annuity: $PMT = 500$, $i = \dfrac{0.06}{4} = 0.015$

$$Y_1 = 500 \dfrac{[(1 + 0.015)^{4x} - 1]}{0.015}$$

Simple interest: $P = 5000$, $r = 0.04$

$$Y_2 = 5000(1 + 0.04x)$$

The graphs of Y_1 and Y_2 are shown in the figure at the right.

intersection: $x = 2.57$; $y = 5514$

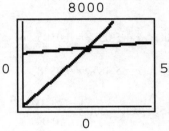

The annuity will be worth more after 2.57 years, or 11 quarterly payments.

EXERCISE 3-4

Things to remember:

1. PRESENT VALUE OF AN ORDINARY ANNUITY

$$PV = PMT \dfrac{1 - (1+i)^{-n}}{i}$$

where

$PV =$ present value of all payments

$PMT =$ periodic payment

$i =$ rate per period

$n =$ number of periods

(Payments are made at the end of each period.)

2. AMORTIZATION FORMULA

$$PMT = PV \dfrac{i}{1 - (1+i)^{-n}}$$

---

1. $S = \dfrac{a(r^n - 1)}{r - 1}$; $a = 1$, $r = \frac{1}{2}$, $n - 1 = 8$, $n = 9$; $S = \dfrac{1\left(\dfrac{1}{2^9} - 1\right)}{\frac{1}{2} - 1} = \dfrac{2(2^9 - 1)}{2^9} = \dfrac{511}{256}$.

3. $a = 30$, $r = \frac{1}{10}$, $n - 1 = 6$, $n = 7$; $S = \dfrac{30\left(\dfrac{1}{10^7} - 1\right)}{\frac{1}{10} - 1} = \dfrac{30(10^7 - 1)}{10^7} \cdot \dfrac{10}{9} = \dfrac{10^7 - 1}{3(10^5)} = \dfrac{3,333,333}{100,000}$.

5. $a = 1$, $r = -\frac{1}{2}$, $n - 1 = 8$, $n = 9$; $S = \dfrac{1\left[\left(-\frac{1}{2}\right)^9 - 1\right]}{-\frac{1}{2} - 1} = \dfrac{2^9 + 1}{2^9} \cdot \dfrac{2}{3} = \dfrac{171}{256}$

7. $r = 7.2\% = 0.072$, $m = 12$, $i = \dfrac{0.072}{12} = 0.006$, $n = 4(12) = 48$

9. $r = 9.9\% = 0.099$, $m = 4$, $i = \dfrac{0.099}{4} = 0.02475$, $n = 10(4) = 40$

11. $r = 5.05\% = 0.0505$, $m = 2$, $i = \dfrac{0.0505}{2} = 0.02525$, $n = 16(2) = 32$

13. $r = 5.48\% = 0.0548$, $m = 1$, $i = 0.0548$, $n = 9(1) = 9$

15. $PV = 200\dfrac{1 - (1 + .04)^{-30}}{0.04} = 200(17.29203330) = \$3458.41$

17. $PMT = 40,000\dfrac{0.0075}{1 - (1 + 0.0075)^{-96}} = \dfrac{40,000}{68.25843856} = \$586.01$

19. $PV = \$5000$, $i = 0.01$, $PMT = 200$

   We have, $PV = PMT\,\dfrac{1 - (1 + i)^{-n}}{i}$

   $5000 = 200\dfrac{1 - (1 + 0.01)^{-n}}{0.01} = 20,000\left[1 - (1.01)^{-n}\right]$

   $\dfrac{1}{4} = 1 - (1.01)^{-n}$

   $(1.01)^{-n} = \dfrac{3}{4} = 0.75$

   $\ln(1.01)^{-n} = \ln(0.75)$

   $-n\ln(1.01) = \ln(0.75)$

   $n = \dfrac{-\ln(0.75)}{\ln(1.01)} \approx 29$

21. $PV = \$9,000, \ PMT = \$600, \ n = 20, \ i = ?$

$$PV = PMT \ \frac{1-(1+i)^{-n}}{i}$$

Substituting the given values into this formula gives

$$9000 = 600 \ \frac{1-(1+i)^{-20}}{i}$$

$$15i = 1 - (1+i)^{-20} = 1 - \frac{1}{(1+i)^{20}}$$

$$15i + \frac{1}{(1+i)^{20}} = 1$$

Graph $Y_1 = 15x + \dfrac{1}{(1+x)^{20}}$, $Y_2 = 1$.

The curves intersect at $x = 0$ and
$x \approx 0.029$. Thus $i = 0.029$.

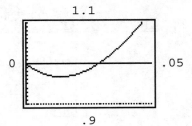

23. The present value of an annuity is the current value of a series of equal payments made at equally spaced time intervals in the future.

25. You are going to make monthly payments to a bank to pay off a loan of $\$P$ at an interest rate r. This is an annuity with present value $\$P$.

27. $PMT = \$5,000, \ i = 0.0665, \ n = 10$

$$PV = PMT \ \frac{1-(1+i)^{-n}}{i} = 5000 \ \frac{1-(1.0665)^{-10}}{0.0665}$$

$$= 5000(7.138636854) = \$35,693.18$$

29. $PMT = 350, \ i = \dfrac{0.0756}{12} = 0.0063, \ n = 36$

$$PV = PMT \ \frac{1-(1+i)^{-n}}{i} = 350 \frac{1-(1.0063)^{-36}}{0.0063} = 350(32.11945179) = 11,241.81$$

You can borrow $11,241.81. The total amount paid on this loan is: $350(36) = \$12,600$. Thus, the total interest you will pay is

$$12,600 - 11,241.81 = \$1,358.19$$

31. $PV = 2,500, \ i = 0.0125, \ n = 48$

$$PMT = PV \ \frac{i}{1-(1+i)^{-n}} = 2500 \frac{0.0125}{1-(1.0125)^{-48}} = 2500(0.027830748) = \$69.58$$

Total amount paid: $69.58(48) = \$3339.84$
Total interest paid: $3335.04 - 2500 = \$839.84$

33. $PV = PMT \dfrac{1-(1+i)^{-n}}{i};\quad PV = 2,487.56,\ PMT = 100,\ i = \dfrac{0.1699}{12}.$

Solve $\quad 2,487.56 = 100\ \dfrac{1-\left(1+\dfrac{0.1699}{12}\right)^{-n}}{\dfrac{0.1699}{12}}\quad$ for n.

$$\dfrac{0.1699}{12}(2487.56) = 100 - 100\left(1+\dfrac{0.1699}{12}\right)^{-n}$$

$$\left(1+\dfrac{0.1699}{12}\right)^{-n} = \dfrac{\dfrac{0.1699}{12}(2487.56)-100}{-100}$$

$$-n = \dfrac{\ln\left[\dfrac{\dfrac{0.1699}{12}(2487.56)-100}{-100}\right]}{\ln\left(1+\dfrac{0.1699}{12}\right)} = -30.8818;\quad n = 31\ \text{months}$$

35. $PV = PMT \dfrac{1-(1+i)^{-n}}{i};\quad PV = 937.14,\ PMT = 20,\ i = \dfrac{0.1499}{12}.$

Solve $\quad 937.14 = 20\ \dfrac{1-\left(1+\dfrac{0.1499}{12}\right)^{-n}}{\dfrac{0.1499}{12}}\quad$ for n.

$$\dfrac{0.1499}{12}(937.14) = 20 - 20\left(1+\dfrac{0.1499}{12}\right)^{-n}$$

$$\left(1+\dfrac{0.1499}{12}\right)^{-n} = \dfrac{\dfrac{0.1499}{12}(937.14)-20}{-20}$$

$$-n = \dfrac{\ln\left[\dfrac{20-\dfrac{0.1499}{12}(937.14)}{20}\right]}{\ln\left(1+\dfrac{0.1499}{12}\right)} \approx -70.9064;\quad n = 71\ \text{months}$$

37. 0% financing for 72 months implies that the monthly payment should be $\$\dfrac{9330}{72} = \$129.58$ not \$179. The interest rate actually being charged can be found by solving

$$9330 = 179\ \dfrac{1-(1+i)^{-72}}{i}$$

for i. The result is: $i \approx 0.009412$. The interest rate $r = 12i \approx 0.1129$ or 11.29% compounded monthly.

39. If you choose 0% financing, your monthly payment will be:

$$PMT_1 = \dfrac{27,300}{60} = \$455$$

If you choose the \$5000 rebate and borrow \$22,300 at 6.3% compounded monthly for 60 months, your monthly payment will be

$$PMT_2 = PV\frac{i}{1-(1+i)^{-n}}$$

$$= 22{,}300 \cdot \frac{0.00525}{1-(1.00525)^{-60}} \quad (i = \frac{0.063}{12} = 0.00525)$$

$$= \$434.24$$

You should choose the \$5000 rebate. You will save \$20.76 per month or $60(20.76) = \$1245.60$ over the life of the loan.

41. Amortized amount: $35{,}000 - 0.20(35{,}000) = \$28{,}000$

We now have: $PV = 28{,}000, \quad i = \frac{0.0875}{12} \approx 0.007292, \quad n = 12(12) = 144$

$$PMT = PV\frac{i}{1-(1+i)^{-n}} = 28{,}000\frac{0.007292}{1-(1.007292)^{-144}}$$

$$= 28{,}000(0.011240) = \$314.72$$

Total amount paid: $144(314.72) = \$45{,}319.68$
Total interest paid: $45{,}319.68 - 28{,}000 = \$17{,}319.68$

43. First, we compute the quarterly payment for $PV = \$5000, \ i = 0.028$ and $n = 8$:

$$PMT = PV\frac{i}{1-(1+i)^{-n}} = 5000\frac{0.028}{1-(1.028)^{-8}} = 5000(0.14125701) = \$706.29$$

The amortization schedule is as follows:

Payment number	Payment	Interest	Unpaid balance reduction	Unpaid balance
0				$5,000.00
1	$706.29	$140.00	$566.29	4,433.71
2	706.29	124.14	582.15	3,851.56
3	706.29	107.84	598.45	3,253.11
4	706.29	91.09	615.20	2,637.91
5	706.29	73.86	632.43	2,005.48
6	706.29	56.15	650.14	1,355.34
7	706.29	37.95	668.34	687.00
8	706.24	19.24	687.00	0.00
Total	$5,650.27	$650.27	$5,000.00	

45. First, we compute the monthly payment for

$$PV = \$6000, \ i = \frac{0.09}{12} = 0.0075, \ n = 3(12) = 36:$$

$$PMT = PV\frac{i}{1-(1+i)^{-n}} = 6000\frac{0.0075}{1-(1.0075)^{-36}}$$

$$= 6000(0.031799732) = \$190.80$$

Now, compute the unpaid balance after 12 payments by considering 24 unpaid payments:

$PMT = 190.80$, $i = 0.0075$, $n = 24$

$$PV = PMT \frac{1-(1+i)^{-n}}{i} = 190.80 \frac{1-(1.0075)^{-24}}{0.0075}$$
$$= 190.80(21.88914614) = \$4176.45$$

The amount of loan paid in 12 months is:
  $6000 - 4176.45 = \$1823.55$

The total amount paid during the first year is:
  $190.80(12) = \$2289.60$

Thus, the interest paid during the first year is:
  $2289.60 - 1823.55 = \$466.05$

Next, we compute the unpaid balance after 24 months by considering 12 unpaid payments:

$PMT = \$190.80$, $i = 0.0075$, $n = 12$

$$PV = PMT \frac{1-(1+i)^{-n}}{i} = 190.80 \frac{1-(1.0075)^{-12}}{0.0075}$$
$$= 190.80(11.43491267) = \$2181.78$$

The amount of the loan paid in 24 months is:
  $6000 - 2181.78 = \$3818.22$

and the amount of the loan paid during the second year is:
  $3818.22 - 1823.55 = \$1994.67$

The total amount paid during the second year is $190.80(12) = \$2289.60$.
Therefore, the interest paid during the second year is:
  $2289.60 - 1994.67 = \$294.93$

The total amount paid in 36 months is:
  $190.80(36) = \$6868.80$

and the total interest paid is: $6868.80 - 6000 = \$868.80$
Therefore, the interest paid during the third year is:

  $868.80 - [466.05 + 294.93] = \$107.82$

47. $PMT = \$525$, $i = \dfrac{0.078}{12} = 0.0065$, $n = 30(12) = 360$

The present value of these payments is:

$$PV = PMT \frac{1-(1+i)^{-n}}{i} = 525 \frac{1-(1.0065)^{-360}}{0.0065} = 525(138.9138739) = \$72,929.78$$

Selling price = loan + down payment
          $= 72,929.78 + 25,000 = \$97,929.78$

Total amount paid in 30 years: $525(360) = \$189,000$
Total interest paid: $189,000 - 72,929.78 = \$116,070.22$

49. $P = \$6000$, $n = 2(12) = 24$, $i = \dfrac{0.035}{12} \approx 0.0029167$

The total amount owed at the end of the two years is:
$A = P(1+i)^n = 6000(1 + 0.0029167)^{24} = 6000(1.0029167)^{24} \approx 6434.39$
Now, the monthly payment is:

$$PMT = PV \frac{i}{1-(1+i)^{-n}}$$

where $n = 4(12) = 48$, $PV = \$6434.39$, $i = \frac{0.035}{12} \approx 0.0029167$. Thus,

$$PMT = 6434.39 \frac{0.0029167}{1-(1+0.0029167)^{-48}} = \$143.85 \text{ per month}$$

The total amount paid in 48 payments is $143.85(48) = \$6904.80$. Thus, the interest paid is $6904.80 - 6000 = \$904.80$.

51. Monthly payment: $PV = \$150,000$, $i = \frac{0.061}{12} = 0.005083\overline{3}$, $n = 30(12) = 360$

$$PMT = PV \frac{i}{1-(1+i)^{-n}} = 150,000 \frac{0.005083\overline{3}}{1-(1.005083\overline{3})^{-360}} = \$908.99$$

(A) To compute the balance after 10 years (with the balance of the loan to be paid in 20 years), use
$PMT = \$908.99$, $i = 0.005803$, $n = 20(12) = 240$

$$\text{Balance after 10 years} = PMT \frac{1-(1+i)^{-240}}{i} = PMT \frac{1-(1.005083\overline{3})^{-240}}{0.005083\overline{3}} = 908.99 \frac{1-(1.005083\overline{3})^{-240}}{0.005083\overline{3}}$$

$$= \$125,862$$

(B) To compute the balance after 20 years, use
$PMT = \$908.99$, $i = 0.005803$, $n = 10(12) = 120$

$$\text{Balance after 20 years} = PMT \frac{1-(1+i)^{-120}}{i} = PMT \frac{1-(1.005083\overline{3})^{-120}}{0.005083\overline{3}} = 908.99 \frac{1-(1.005083\overline{3})^{-120}}{0.005083\overline{3}}$$

$$= \$81,507$$

(C) To compute the balance after 25 years, use
$PMT = \$908.99$, $i = 0.005803$, $n = 5(12) = 60$

$$\text{Balance after 25 years} = PMT \frac{1-(1+i)^{-60}}{i} = PMT \frac{1-(1.005083\overline{3})^{-60}}{0.005083\overline{3}} = 908.99 \frac{1-(1.005083\overline{3})^{-60}}{0.005083\overline{3}}$$

$$= \$46,905$$

53. (A) $PV = \$129,000$, $i = \frac{0.072}{12} = 0.006$, $n = 20(12) = 240$.

Monthly payment: $PMT = PV \frac{i}{1-(1+i)^{-n}} = 129,000 \frac{0.006}{1-(1.006)^{-240}} = \$1,015.68$

Total amount paid in 240 payments: $1,015.68(240) = \$243,763.34$.
Total interest paid: $243,763.34 - 129,000 = \$114,763.34$.

(B) New payment: $PMT = 1,015.68 + 102.41 = \$1118.09$

$PV = \$129,000, \quad i = \dfrac{0.072}{12} = 0.006$ We solve $\quad PMT = PV \dfrac{i}{1-(1+i)^{-n}} \quad$ for n:

$$1,118.09 = 129,000 \dfrac{0.006}{1-(1.006)^{-n}} = \dfrac{774}{1-(1.006)^{-n}}$$

$$1-(1.006)^{-n} = \dfrac{774}{1,118.09}$$

$$(1.006)^{-n} = 1 - \dfrac{774}{1,118.09}$$

$$-n\ln(1.006) = \ln\left(1 - \dfrac{774}{1,118.09}\right)$$

$$n = \dfrac{-\ln\left(1 - \dfrac{774}{1,118.09}\right)}{\ln(1.006)} \approx 197$$

The mortgage will be paid off in 197 months.

Total amount paid in 197 payments @ \$1,118.09: $1,118.09\,(197) = \$220,263.73$.

Total interest paid: $220,263.73 - 129,000 = \$91,263.73$.

Savings on interest: $114,763 - 91,264 = \$23,499$.

55. $PV = \$500,000, \quad i = \dfrac{0.075}{12} = 0.00625$

$$500,000 = PMT \dfrac{1-(1.00625)^{-n}}{0.00625}$$

$$\dfrac{500,000(0.00625)}{PMT} = 1-(1.00625)^{-n}$$

$$(1.00625)^{-n} = 1 - \dfrac{500,000(0.00625)}{PMT}$$

(A) $PMT = \$5,000$

$$(1.00625)^{-n} = 1 - \dfrac{500,000(0.00625)}{5000} = 0.375$$

$$-n = \dfrac{\ln(0.375)}{\ln(1.00625)} = -157; \quad \text{and} \quad n = 157$$

Thus, 157 withdrawals.

(B) $PMT = \$4,000$

$$(1.00625)^{-n} = 1 - \dfrac{500,000(0.00625)}{4000} = 0.21875$$

$$-n = \dfrac{\ln(0.21875)}{\ln(1.00625)} = -243; \quad \text{and} \quad n = 243$$

Thus, 243 withdrawals.

(C) The interest per month on \$500,000 at 7.5% compounded monthly is greater than \$3,000. For example, the interest in the first month is

$$500,000\left(\dfrac{0.075}{12}\right) = \$3,125.$$

Thus, the owner can withdraw \$3,000 per month forever.

57. (A) First, calculate the future value of the ordinary annuity:

$$PMT = \$100, \quad i = \frac{0.0744}{12} = 0.0062, \quad n = 30(12) = 360$$

$$FV = PMT\,\frac{(1+i)^n - 1}{i} = 100\,\frac{(1.0062)^{360} - 1}{0.0062} = \$133,137$$

The interest earned during the 30 year period is:

$$133,137 - 360(100) = 133,137 - 36,000 = \$97,137$$

Next, using \$133,137 as the present value, determine the monthly payment:

$$PV = \$133,137, \quad i = 0.0062, \quad n = 15(12) = 180$$

$$PMT = PV\,\frac{i}{1 - (1+i)^{-n}} = 133,137\,\frac{0.0062}{1 - (1.0062)^{-180}} = 1229.66$$

The monthly withdrawals are \$1229.66. Interest earned during the 15 year period is: $(1229.66)(180) - 133,137 = \$88,201.80$

Total interest: $97,137 + 88,201.80 = \$185,338.80$

(B) First find the present value of the ordinary annuity:

$$PMT = \$2,000, \quad i = 0.0062, \quad n = 180$$

$$PV = 2000\,\frac{1 - (1.0062)^{-180}}{0.0062} = \$216,542.54$$

Next, using \$216,542.54, as the future value of an ordinary annuity, calculate the monthly payment.

$$FV = \$216,542.54, \quad i = 0.0062, \quad n = 360$$

$$PMT = 216,542.54\,\frac{0.0062}{(1.0062)^{360} - 1} = 162.65$$

The monthly deposit is \$162.65.

59. $PV = (\$179,000)(0.80) = \$143,200, \quad i = \dfrac{0.084}{12} = 0.007, \quad n = 12(30) = 360.$

Monthly payment: $PMT = PV\,\dfrac{i}{1 - (1+i)^{-n}} = 143,200\,\dfrac{0.007}{1 - (1 + 0.007)^{-360}} = \dfrac{1002.4}{1 - (1.007)^{-360}} = \$1,090.95$

Next, we find the present value of a \$1,090.95 per month, 18-year annuity.

$PMT = \$1,090.95, \quad i = 0.007, \quad$ and $\quad n = 12(18) = 216.$

$$PV = PMT\,\frac{1 - (1+i)^{-n}}{i} = 1,090.95\,\frac{1 - (1.007)^{-216}}{0.007} = \$121,308.50$$

Finally, equity = (current market value) − (unpaid loan balance)

$$= 215,000 - 121,308.50 = \$93,691.50$$

The couple can borrow $(\$93,691.50)(0.70) = \$65,584.$

61. Original mortgage: $145,000 - 0.20(145,000) = \$116,000.$

With $PV = \$116,000, \quad i = \dfrac{0.079}{12} = 0.006583\overline{3}, \quad$ and $\quad n = 30(12) = 360,$

the monthly payment is:

$$PMT = PV\,\frac{i}{1 - (1+i)^{-n}} = 116,000\,\frac{0.006583\overline{3}}{1 - (1.006853\overline{3})^{-360}} = \$843.10.$$

To compute the unpaid balance after 10 years, use

$PMT = \$843.10$, $i = 0.006583\overline{3}$, $n = 20(12) = 240$.

Unpaid balance: $PMT \dfrac{1-(1+i)^{-240}}{i} = 843.10\dfrac{1-(1.006583)^{-240}}{0.006583\overline{3}} = \$101,550.51$.

Now, refinancing this amount at the new interest rate for 20 years, we find the new monthly payment:

$PV = \$101,550.51$, $i = \dfrac{0.055}{12} = 0.004583\overline{3}$, $n = 20(12) = 240$

$PMT = PV \dfrac{i}{1-(1+i)^{-n}} = 101,550.51\dfrac{0.004583\overline{3}}{1-(1.004583\overline{3})^{-240}} = \$698.55$

If the original mortgage had been completed, the interest paid would have been

   $843.10(360) - 116,000 = 303,516 - 116,000 = \$187,516$.

The amount of the mortgage paid during the first 10 years is:

   $116,000 - 101,550.51 = \$14,449.49$.

Thus, the interest paid during the first 10 years is:

   $843.10(120) - 14,449.49 = 101,172 - 14,449.49 = \$86,722.51$.

The interest paid on the new 20-year mortgage will be

   $698.55(240) - 101,550.51 = 167,652 - 101,550.51 = \$66,101.49$.

Therefore, the total interest on the new mortgage will be

   $86,722.51 + 66,101.49 = \$152,824$.

The savings on interest is

   $187,516 - 152,824 = \$34,692$.

63.   The graphs are decreasing, curve downward, and have x intercept 30. The unpaid balances are always in the ratio 2:3:4. The monthly payments and total interest in each case are:

   (use $PMT = PV \dfrac{i}{1-(1+i)^{-n}}$ where $i = \dfrac{0.09}{12} = 0.0075$, $n = 360$)

<u>$50,000 mortgage</u>

   $PMT = 50,000\dfrac{0.0075}{1-(1.0075)^{-360}}$

     $= 50,000(0.0080462)$

     $= \$402.31$ per month

Total interest paid $= 360(402.31) - 50,000 = \$94,831.60$

<u>$75,000 mortgage</u>

   $PMT = 75,000\dfrac{0.0075}{1-(1.0075)^{-360}}$

     $= 75,000(0.0080462)$

     $= \$603.47$ per month

Total interest paid $= 360(603.47) - 75,000 = \$142,249.20$

<u>$100,000 mortgage</u>

   $PMT = 100,000\dfrac{0.0075}{1-(1.0075)^{-360}}$

$= 100,000(0.0080462)$

$= \$804.62$ per month

Total interest paid $= 360(804.62) - 100,000 = \$189,663.20$

Problems 65 thru 67 start from the equation

(*) $\dfrac{1-(1+i)^{-n}}{i} - \dfrac{PV}{PMT} = 0$

A graphics calculator was used to solve these problems

65. $PV = 1000$, $PMT = 90$, $n = 12$, $i = \dfrac{r}{12}$ where r is the annual nominal rate.

With these values, the equation (*) becomes

$$\dfrac{1-(1+i)^{-12}}{i} - \dfrac{1000}{90} = 0$$

or $1 - (1+i)^{-12} - 11.11i = 0$

Put $y = 1 - (1+i)^{-12} - 11.11i$ and use your calculator to find the zero i of y, where $0 < i < 1$. The result is $i \approx 0.01204$ and $r = 12(0.01204) = 0.14448$. Thus, $r = 14.45\%$ (two decimal places).

67. $PV = 90,000$, $PMT = 1200$, $n = 12(10) = 120$, $i = \dfrac{r}{12}$ where r is the annual nominal rate. With these values,

the equation (*) becomes

$$\dfrac{1-(1+i)^{-120}}{i} - \dfrac{90,000}{1200} = 0$$

or $1 - (1+i)^{-120} - 75i = 0$

This equation can be written

$$(1+i)^{120} - (1 - 75i)^{-1} = 0$$

Put $y = (1+x)^{120} - (1 - 75x)^{-1}$ and use your calculator to find the zero x of y where $0 < x < 1$. The result is $x \approx 0.00851$ and $r = 12(0.00851) = 0.10212$. Thus $r = 10.21\%$ (two decimal places).

CHAPTER 3 REVIEW

1. $A = 100\left(1 + \dfrac{0.09}{2}\right)$

$= 100(1.045) = \$104.50$ (3-1)

2. $808 = P\left(1 + \dfrac{0.12}{12}\right)$

$P = \dfrac{808}{1.01} = \$800$ (3-1)

3. $212 = 200(1 + 0.08 \cdot t)$

$$1 + 0.08t = \frac{212}{200}$$

$$0.08t = \frac{212}{200} - 1 = \frac{12}{200}$$

$$t = \frac{0.06}{0.08} = 0.75 \text{ yr. or 9 mos.}$$

$(3\text{-}1)$

4. $4120 = 4000\left(1 + r\frac{1}{2}\right)$

$$1 + \frac{r}{2} = \frac{4120}{4000}$$

$$\frac{r}{2} = \frac{4120}{4000} - 1 = \frac{120}{4000} = 0.03$$

$$r = 0.06 \text{ or } 6\% \qquad (3\text{-}1)$$

5. $A = 1200(1 + 0.005)^{30}$

$\quad = 1200(1.005)^{30} = \$1393.68$

$(3\text{-}2)$

6. $P = \dfrac{A}{(1+i)^n} = \dfrac{5000}{(1+0.0075)^{60}} = \dfrac{5000}{(1.0075)^{60}}$

$\quad = \$3193.50$

$(3\text{-}2)$

7. $A = Pe^{rt}$; $P = 4{,}750$, $r = 6.8\% = 0.068$, $t = 3$

$A = 4{,}750e^{0.068(3)} \approx 4{,}750e^{0.204} \approx \$5{,}824.92 \qquad (3\text{-}2)$

8. $A = Pe^{rt}$; $A = 36{,}000$, $r = 9.3\% = 0.093$, $t = 60$ months $= 5$ years

$36{,}000 = Pe^{0.093(5)}$, $P = \dfrac{36{,}000}{e^{0.465}} \approx \$22{,}612.86 \qquad (3\text{-}2)$

9. $FV = 1000\left[\dfrac{(1+0.005)^{60} - 1}{0.005}\right]$

$\quad = 1000 \cdot 69.77003051$

$\quad = \$69{,}770.03 \qquad (3\text{-}3)$

10. $PMT = \dfrac{(0.015)8{,}000}{(1.015)^{48} - 1}$

$\quad \approx \$115.00 \qquad (3\text{-}3)$

11. $PV = 2500\left[\dfrac{1 - (1+0.02)^{-16}}{0.02}\right]$

$\quad = 2500(13.57770931)$

$\quad \approx \$33{,}944.27 \qquad (3\text{-}4)$

12. $PMT = \dfrac{(0.0075)8{,}000}{1 - (1+0.0075)^{-60}}$

$\quad = \dfrac{60}{0.3613003014} = \$166.07 \qquad (3\text{-}4)$

13. (A) $\qquad 2500 = 1000(1.06)^n$

$\qquad (1.06)^n = 2.5$

$\qquad \ln(1.06)^n = \ln 2.5$

$\qquad n \ln 1.06 = \ln 2.5$

$\qquad n = \dfrac{\ln 2.5}{\ln 1.06} = 15.73 \approx 16$

(B)   We find the intersection of
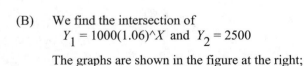
$Y_1 = 1000(1.06)\char`^X$ and $Y_2 = 2500$

The graphs are shown in the figure at the right; intersection: $x = 15.73$; $y = 2500$

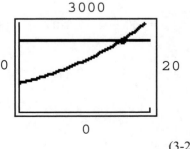

$(3\text{-}2)$

14.   (A) $$5000 = 100 \frac{(1.01)^n - 1}{0.01}$$

$$5000 = 10{,}000[(1.01)^n - 1]$$

$$0.5 = (1.01)^n - 1$$

$$(1.01)^n = 1.5$$

$$n \ln(1.01) = \ln(1.5)$$

$$n = \frac{\ln(1.5)}{\ln(1.01)} = 40.75 \approx 41$$

(B)   We find the intersection of

$$Y_1 = 100 \frac{(1.01)^x - 1}{0.01} = 10{,}000[(1.01)^\wedge X - 1] \text{ and } Y_2 = 5000$$

The graphs are shown in the figure at the right.
intersection: $x = 40.75$; $y = 5000$

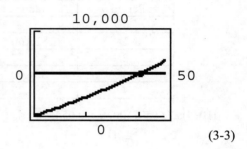

(3-3)

15. $P = \$3000$, $r = 0.14$, $t = \dfrac{10}{12}$

$$A = 3000 \left(1 + 0.14 \frac{10}{12} \right) \quad [\text{using } A = P(1 + rt)]$$

$$= \$3350$$

Interest $= 3350 - 3000 = \$350$ (3-1)

16. $P = \$6{,}000$, $r = 7\% = 0.07$, $i = \dfrac{0.07}{12} = 0.00583\overline{3}$, $n = 17(12) = 204$

$$A = P(1 + i)^n = 6000(1 + 0.00583\overline{3})^{204} = 6000(1.00583\overline{3})^{204} = \$19{,}654.42 \text{ or } \$19{,}654 \text{ rounded to the nearest dollar.}$$ (3-2)

17. $A = \$25{,}000$, $r = 6.6\% = 0.066$, $i = \dfrac{0.066}{12} = 0.0055$, $n = 10(12) = 120$

$$P = \frac{A}{(1 + i)^n} = \frac{25{,}000}{(1.0055)^{120}} = \$12{,}944.67$$ (3-2)

18.   (A) $P = \$400$, $i = 0.054$

$$A = P(1 + i)^n$$

Year 1: $A = 400(1.054)^1 = \$421.60$
Interest: \$21.60

Year 2: $A = 400(1.054)^2 = \$444.37$
Interest: $444.37 - 421.60 = \$22.77$

Year 3: $A = 400(1.054)^3 = \$468.36$
Interest: $468.36 - 444.37 = \$23.99$

Year 4: $A = 400(1.054)^4 = \$493.65$
Interest: $493.95 - 468.36 = \$25.29$

Period	Interest	Amount
0		$400.00
1	$21.60	$421.60
2	$22.77	$444.37
3	$24.00	$468.36
4	$25.29	$493.65

(B) $PMT = \$100$, $i = 0.054$, $n = 4$

$$FV = PMT\,\frac{(1+i)^n - 1}{i}$$

Year 1: $FV = \$100$

Year 2: $FV = 100\,\dfrac{(1.054)^2 - 1}{0.054} = \$205.40$

Interest: $205.40 - 200.00 = \$5.40$

Year 3: $FV = 100\,\dfrac{(1.054)^3 - 1}{0.054} = \$316.49$

Interest: $316.49 - 205.40 = \$11.09$

Year 4: $FV = 100\,\dfrac{(1.054)^4 - 1}{0.054} = \$433.58$

Interest: $433.58 - 316.49 = \$17.09$

Period	Interest	Payment	Balance
1		$100.00	$100.00
2	$5.40	$100.00	$205.40
3	$11.09	$100.00	$316.49
4	$17.09	$100.00	$433.58

(3-2, 3-3)

19. The value of $1 at 13% simple interest after t years is:
$$A_s = 1(1 + 0.13t) = 1 + 0.13t$$

The value of $1 at 9% interest compounded annually for t years is:
$$A_c = 1(1 + 0.09)^t = (1.09)^t$$

Graph $Y_1 = 1 + 0.13x$, $Y_2 = (1.09)^x$

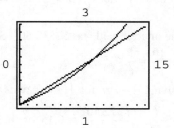

The graphs intersect at the point where $x \approx 9$. For investments lasting less than 9 years, simple interest at 13% is better; for investments lasting more than 9 years, interest compounded annually at 9% is better.   (3-2)

20. $P = \$10,000$, $r = 7\% = 0.07$, $m = 365$, $i = \dfrac{0.07}{365}$, and $n = 40(365) = 14{,}600$

$$A = P(1 + i)^n = 10{,}000\left(1 + \frac{0.07}{365}\right)^{14{,}600} \approx \$164{,}402 \qquad (3\text{-}2)$$

21. $A = Pe^{rt}$; $A = 40{,}000$, $P = 25{,}000$, $t = 6$
$$40{,}000 = 25{,}000e^{6r}$$
$$e^{6r} = \frac{40{,}000}{25{,}000} = \frac{8}{5}$$
$$6r = \ln(8/5), \quad r = \frac{\ln(8/5)}{6} \approx 0.7833 \text{ or } 7.833\% \qquad (3\text{-}2)$$

22. The effective rate for 9% compounded quarterly is:
$$\text{APY} = \left(1 + \frac{r}{m}\right)^m - 1, \quad r = 0.09, \quad m = 4$$
$$= \left(1 + \frac{0.09}{4}\right)^4 - 1 = (1.0225)^4 - 1 \approx 0.0931 \text{ or } 9.31\%$$

The effective rate for 9.25% compounded annually is 9.25%.
Thus, 9% compounded quarterly is the better investment.   (3-2)

23. $PMT = \$200$, $r = 7.2\% = 0.072$, $i = \dfrac{0.072}{12} = 0.006$, $n = 8(12) = 96$

$$FV = PMT\frac{(1+i)^n - 1}{i} = 200\frac{(1.006)^{96} - 1}{0.006} = 200(129.308244) = \$25{,}861.65$$

The total amount invested is: $200(96) = \$19{,}200$.
Thus, the interest earned with this annuity is:
$$I = 25{,}861.65 - 19{,}200 = \$6{,}661.65. \qquad (3\text{-}3)$$

24. $P = \$635$, $r = 22\% = 0.22$, $t = \dfrac{1}{12}$, $i = Prt = 635(0.22)\dfrac{1}{12} = \$11.64$ \qquad (3-1)

25. $A = P(1 + i)^n$; $P = \$23{,}000$, $r = 5\% = 0.05$, $m = 1$, $i = \dfrac{r}{m} = 0.05$, $n = 5$

$$A = 23{,}000(1.05)^5 \approx \$29{,}354 \qquad (3\text{-}2)$$

26. $A = P(1 + i)^n$; $A = \$23{,}000$, $r = 5\% = 0.05$, $m = 1$, $i = \dfrac{r}{m} = 0.05$, $n = 5$

$23{,}000 = P(1.05)^5$; $P = \dfrac{23{,}000}{(1.05)^5} = \$18{,}021$ (3-2)

27. The interest paid was $\$2812.50 - \$2500 = \$312.50$. $P = \$2500$,

$t = \dfrac{10}{12} = \dfrac{5}{6}$

Solving $I = Prt$ for r, we have:

$r = \dfrac{I}{Pt} = \dfrac{312.50}{2500\left(\frac{5}{6}\right)} = 0.15$ or 15% (3-1)

28. If you choose 0% financing, your monthly payment will be

$PMT_1 = \dfrac{21{,}600}{48} = \$450$

If you choose the \$3000 rebate and borrow \$18,600 at 4.8% compounded monthly for 48 months, your monthly payment will be:

$$PMT_2 = PV\dfrac{i}{1 - (1+i)^{-n}}$$

$$= 18{,}600\dfrac{0.004}{1 - (1.004)^{-48}} \qquad (i = \dfrac{0.048}{12} = 0.004)$$

$$= \$426.66$$

You should choose the \$3000 rebate. You will save \$23.34 per month or $48(23.34) = \$1120.32$ over the life of the loan.   (3-4)

29. (A) $r = 6.25\% = 0.0625$, $m = 12$,

$$\text{APY} = \left(1 + \dfrac{0.0625}{12}\right)^{12} - 1 = 0.06432 \text{ or } 6.432\%$$

(B) $r = 0.0625$ compounded continuously

$$\text{APY} = e^{0.0625} - 1 \approx 0.06449 \text{ or } 6.449\%$$ (3-2)

30. $P = \$5{,}000$, $r = 9\% = 0.09$, $m = 4$, $i = \dfrac{0.09}{4} = 0.0225$, $A = \$6000$

$$A = P(1+i)^n$$

$$6000 = 5000(1.0225)^n$$

$$(1.0225)^n = \dfrac{6000}{5000} = \dfrac{6}{5}$$

$$n\ln(1.0225) = \ln(6/5)$$

$$n = \dfrac{\ln(6/5)}{\ln(1.0225)} = 8.19$$

Thus, it will take 9 quarters, or 2 years and 3 months.   (3-2)

CHAPTER 3 REVIEW   **3-35**

31. (A) $r = 6\% = 0.06$, $m = 12$, $i = \dfrac{0.06}{12} = 0.005$

If we invest P dollars, then we want to know how long it will take to have $2P$ dollars:

$$A = P(1 + i)^n$$
$$2P = P(1.005)^n$$
$$(1.005)^n = 2$$
$$\ln(1.005)^n = \ln 2$$
$$n \ln (1.005) = \ln 2$$
$$n = \frac{\ln 2}{\ln(1.005)} \approx 138.98 \text{ or } 139 \text{ months}$$

Thus, it will take 139 months, or 11 years and 6 months, for an investment to double at 6% interest compounded monthly.

(B) $r = 9\% = 0.09$, $m = 12$, $i = \dfrac{0.09}{12} = 0.0075$

$$2P = P(1.0075)^n$$
$$(1.0075)^n = 2$$
$$\ln(1.0075)^n = \ln 2$$
$$n = \frac{\ln 2}{\ln(1.0075)} \approx 92.77 \text{ or } 93 \text{ months}$$

Thus, it will take 93 months, or 7 years and 9 months, for an investment to double at 9% compounded monthly.   (3-2)

32. (A) $PMT = \$2000$, $m = 1$, $r = i = 7\% = 0.07$, $n = 45$

$$FV = PMT \, \frac{(1+i)^n - 1}{i}$$
$$= 2000 \, \frac{(1 + 0.07)^{45} - 1}{0.07} = 2000 \, \frac{(1.07)^{45} - 1}{0.07} \approx \$571,499$$

(B) $PMT = \$2000$, $m = 1$, $r = i = 11\% = 0.11$, $n = 45$

$$FV = PMT \, \frac{(1+i)^n - 1}{i}$$
$$= 2000 \, \frac{(1 + 0.11)^{45} - 1}{0.11} = 2000 \, \frac{(1.11)^{45} - 1}{0.11} \approx \$1,973,277 \qquad (3\text{-}3)$$

33. $A = \$17,388.17$, $P = \$12,903.28$, $m = 1$, $r = i$, $n = 3$

$$A = P(1 + i)^n$$
$$17,388.17 = 12,903.28(1 + i)^3$$
$$(1 + i)^3 = \frac{17,388.17}{12,903.28} \approx 1.3475775$$
$$1 + i = \sqrt[3]{\frac{17,388.17}{12,903.28}} \approx 1.104547979$$
$$i \approx 0.1045 \text{ or } 10.45\% \qquad (3\text{-}2)$$

Copyright © 2015 Pearson Education, Inc.

34. (A) $P = \$400$, $t = \dfrac{15}{360}$, $I = \$29.00$

$I = Prt$; $r = \dfrac{I}{Pt} = \dfrac{29}{400\left(\frac{15}{360}\right)} = 1.74$

The annual rate of interest is 174%.

(B) $P = \$1,800$, $t = \dfrac{21}{360}$, $I = \$69.00$

$r = \dfrac{I}{Pt} = \dfrac{69}{1,800\left(\frac{21}{360}\right)} \approx 0.6571$

The annual rate of interest is 65.71%.　　(3-1)

35. $FV = \$50,000$, $i = \dfrac{0.055}{12} = 0.004583\overline{3}$, $n = 5(12) = 60$

$PMT = FV\ \dfrac{i}{(1+i)^n - 1} = 50,000\ \dfrac{0.004583\overline{3}}{(1.004583\overline{3})^{60} - 1}$

$= 50,000(0.014517828) = \$725.89$

The monthly deposit should be $725.89.　　(3-3)

36. (A) The present value of an annuity which provides for quarterly withdrawals of $5,000 for 10 years at 7.32% interest compounded quarterly is given by:

$PV = PMT\ \dfrac{1-(1+i)^{-n}}{i}$ with $PMT = \$5,000$, $i = \dfrac{0.0732}{4} = 0.0183$, and $n = 10(4) = 40$

$= 5,000\ \dfrac{1-(1.0183)^{-40}}{0.0183} = 5,000(28.18911469) = \$140,945.57$

This amount will have to be in the account when he retires.

(B) To determine the quarterly deposit to accumulate the amount in part (A), we use the formula:

$PMT = FV\ \dfrac{i}{(1+i)^n - 1}$ where $FV = \$140,945.57$, $i = 0.0183$ and $n = 4(20) = 80$

$= \$140,945.57\ \dfrac{0.0183}{(1.0183)^{80} - 1} = 140,945.57(0.005602533)$

$= 789.65$ (quarterly payment)

(C) The amount collected during the 10-year period is
($5000)40 = $200,000.
The amount deposited during the 20-year period is:
(789.65)80 ≈ $63,172.
Thus, the interest earned during the 30-year period is:
$200,000 − 63,172 = $136,828.　　(3-3, 3-4)

37. $PV = \$4,000$, $i = 0.009$, $n = 48$

$PMT = PV\ \dfrac{i}{1-(1+i)^{-n}} = 4,000\ \dfrac{0.009}{1-(1.009)^{-48}} = 4,000(0.025748504) = \$102.99$

The monthly payment is $102.99.

The total amount paid is: $102.99(48) = $4,943.52.

Therefore, the interest paid is 4,943.52 – 4,000 = $943.52.   (3-4)

38. $FV = \$50,000$, $r = 6.12\% = 0.0612$, $m = 12$, $i = \dfrac{0.0612}{12} = 0.0051$, $n = 12(6) = 72$

$$PMT = FV \frac{i}{(1+i)^n - 1} = 50,000 \frac{0.0051}{(1.0051)^{72} - 1} = \$576.48 \text{ per month.}   \text{(3-3)}$$

39. To determine how long it will take money to double, we need to solve the equation $2P = P(1 + i)^n$ for n. From this equation, we obtain:

$$(1 + i)^n = 2$$
$$\ln(1 + i)^n = \ln 2$$
$$n \ln(1 + i) = \ln 2$$
$$n = \frac{\ln 2}{\ln(1+i)}$$

(A) $i = \dfrac{0.075}{365} = 0.000205479$

Thus, $n = \dfrac{\ln 2}{\ln(1.000205479)} \approx 3373.67$ days or 9.24 years

(B) $i = 0.075$

Thus, $n = \dfrac{\ln 2}{\ln(1.075)} \approx 9.58$ years or 10 years to the nearest year.   (3-2)

40. First, we must calculate the future value of $8000 at 5.5% interest compounded monthly for 2.5 years.

$A = P(1 + i)^n$ where $P = \$8000$, $i = \dfrac{0.055}{12}$ and $n = 30$

$$= 8000\left(1 + \frac{0.055}{12}\right)^{30} = \$9176.33$$

Now, we calculate the monthly payment to amortize this debt at 5.5% interest compounded monthly over 5 years.

$PMT = PV \dfrac{i}{1 + (1+i)^{-n}}$ where $PV = \$9176.33$, $i = \dfrac{0.055}{12} \approx 0.004583\overline{3}$, and $n = 12(5) = 60$

$$= 9176.33 \frac{0.004583\overline{3}}{1 - (1 + 0.004583\overline{3})^{-60}} = \frac{42.058179}{1 - (1.004583\overline{3})^{-60}} \approx \$175.28$$

The total amount paid on the loan is:

$175.28(60) = $10,516.80.

Thus, the interest paid is:

$I = \$10,516.80 - \$8000 = \$2516.80.$ (3-4)

41. $A = Pe^{rt}$; $P = 5,650$, $r = 8.65\% = 0.0865$, $t = 10$

$A = 5,650e^{0.0865(10)} = 5,650e^{0.865} \approx \$13,418.78$ (3-2)

42.   Use $FV = PMT \dfrac{(1+i)^n - 1}{i}$ where $PMT = 1200$ and $i = \dfrac{0.06}{12} = 0.005$.

The graphs of

$Y_1 = 1200 \dfrac{(1.005)^x - 1}{0.005} = 240{,}000[(1.005)^x - 1]$

$Y_2 = 100{,}000$

are shown in the figure at the right.

Intersection: $x \approx 70$; $y = 100{,}000$

The fund will be worth \$100,000 after 70 payments, that is, after 5 years, 10 months.   (3-3)

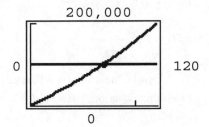

43.   We first find the monthly payment: $PV = \$50{,}000$, $i = \dfrac{0.09}{12} = 0.0075$, $n = 12(20) = 240$

$PMT = PV \dfrac{i}{1-(1+i)^{-n}}$

$\qquad = 50{,}000 \dfrac{0.0075}{1-(1.0075)^{-240}} = \$449.86$ per month

The present value of the \$449.86 per month, 20 year annuity at 9%, after x years, is given by

$y = 449.86 \dfrac{1-(1.0075)^{-12(20-x)}}{0.0075}$

$\qquad = 59{,}981.33 \left[1 - (1.0075)^{-12(20-x)} \right]$

The graphs of

$Y_1 = 59{,}981.33 \left[1 - (1.0075)^{-12(20-x)} \right]$

$Y_2 = 10{,}000$

are shown in the figure at the right.

Intersection: $x \approx 18$; $y = 10{,}000$

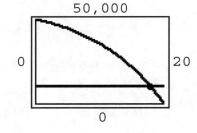

The unpaid balance will be below \$10,000 after 18 years.   (3-4)

44. $P = \$100$, $I = \$0.08$, $t = \dfrac{1}{360}$

$r = \dfrac{I}{Pt} = \dfrac{0.08}{100\left(\frac{1}{360}\right)} = 0.288$ or 28.8%.   (3-1)

45. $PV = \$1000, \ i = 0.025, \ n = 4$

The quarterly payment is:

$$PMT = PV \ \frac{i}{1-(1+i)^{-n}}$$

$$= 1000 \ \frac{0.025}{1-(1+0.025)^{-4}} = \frac{25}{1-(1.025)^{-4}} \approx \$265.82$$

Payment number	Payment	Interest	Unpaid balance reduction	Unpaid balance	
0					$1000.00
1	$265.82	$25.00	$240.82	759.18	
2	265.82	18.98	246.84	512.34	
3	265.82	12.81	253.01	259.33	
4	265.81	6.48	259.33	0.00	
Totals	$1063.27	$63.27	$1000.00		

(3-4)

46. $PMT = \$300, \ FV = \$9,000, \ i = \dfrac{0.0798}{12} = 0.00665; \quad FV = PMT \dfrac{(1+i)^n - 1}{i}$

$$9000 = 300 \frac{(1.00665)^n - 1}{0.00665}$$

$$(1.00665)^n - 1 = \frac{9000(0.00665)}{300} = 0.1995$$

$$(1.00665)^n = 1.1995$$

$$n \ln(1.00665) = \ln(1.1995)$$

$$n = \frac{\ln(1.1995)}{\ln(1.00665)} \approx 27.44$$

It will take 28 months.   (3-3)

47. $FV = \$850,000, \ r = 8.76\% = 0.0876, \ m = 2, \ i = \dfrac{0.0876}{2} = 0.0438, \ n = 2(6) = 12$

$$PMT = FV \ \frac{i}{(1+i)^n - 1}$$

$$= 850,000 \ \frac{0.0438}{(1+0.0438)^{12} - 1} = \frac{37,230}{(1.0438)^{12} - 1} \approx \$55,347.48$$

The total amount invested is:

$12(55,347.48) = \$664,169.76.$

Thus, the interest earned with this annuity is:

$I = \$850,000 - \$664,169.76 = \$185,830.24.$ (3-3)

48. $APY = 2.50\% = 0.025,\ m = 12;\quad APY = \left(1+\dfrac{r}{m}\right)^m - 1.$ Solve $0.025 = \left(1+\dfrac{r}{12}\right)^{12} - 1$ for r:

$$\left(1+\frac{r}{12}\right)^{12} = 1.025$$

$$1+\frac{r}{12} = (1.025)^{1/12}$$

$$\frac{r}{12} = (1.025)^{1/12} - 1$$

$$r = 12\left[(1.025)^{1/12} - 1\right] \approx 0.0247180$$

The annual nominal rate is 2.47%   (3-2)

49. The interest earned is $I = \$5{,}000 - \$4{,}922.15 = \$77.85.$

So $P = \$4{,}922.15,\ t = \dfrac{13}{52} = 0.25$ and $I = Prt.$ Solving for r, we have

$$r = \frac{I}{Pt} = \frac{77.85}{4{,}922.15(0.25)} \approx 0.0633 \text{ or } 6.33\%. \quad (3\text{-}1)$$

50. Using the sinking fund formula

$$PMT = FV\ \frac{i}{(1+i)^n - 1}$$

with $PMT = \$200.00,\ FV = \$10{,}000,$ and $i = \dfrac{0.0702}{12} = 0.00585,$ we have

$$200 = 10{,}000\,\frac{0.00585}{(1.00585)^n - 1} = \frac{58.5}{(1.00585)^n - 1}$$

Therefore, $\quad (1.00585)^n - 1 = \dfrac{58.5}{200} = 0.2925$

$$(1.00585)^n = 1.2925$$

$$\ln(1.00585)^n = \ln(1.2925)$$

$$n = \frac{\ln(1.2925)}{\ln(1.00585)} \approx 43.99$$

The couple will have to make 44 deposits.   (3-3)

51. $PV = \$80{,}000,\ i = \dfrac{0.0942}{12} = 0.00785,\ n = 8(12) = 96$

(A) $\quad PMT = PV\ \dfrac{i}{1-(1+i)^{-n}}$

$$= 80{,}000\,\frac{0.00785}{1-(1.00785)^{-96}} = 80{,}000(0.014869003) = \$1{,}189.52 \text{ monthly payment}$$

(B) Now use $PMT = \$1{,}189.52,\ i = 0.00785,$ and $n = 96 - 12 = 84$ to calculate the unpaid balance.

$$PV = PMT\ \frac{1-(1+i)^{-n}}{i}$$

$$= 1{,}189.52 \frac{1-(1.00785)^{-84}}{0.00785} = 1{,}189.52(61.33824469)$$

$$= \$72{,}963.07 \text{ unpaid balance after the first year}$$

(C) Amount of loan paid during the first year:
$80{,}000 - 72{,}963.07 = \$7{,}036.93$.
Amount of payments during the first year:
$12(\$1{,}189.52) = \$14{,}274.24$.
Thus, the interest paid during the first year is:
$\$14{,}274.24 - 7{,}036.93 = \$7{,}237.31$.   (3-4)

52. <u>Certificate of Deposit</u>: \$10,000 at 7% = 0.07 compounded monthly for $360 - 72 = 288$ months:

$$P = \$10{,}000, \quad i = \frac{0.07}{12} = 0.005\overline{83}, \quad n = 288$$

$$A = P(1 + i)^n$$
$$= 10{,}000(1.005\overline{83})^{288} = 10{,}000(5.339429847) = \$53{,}394.30$$

<u>Reduce the Principal</u>:

<u>Step 1</u>: Find the monthly payment on the mortgage:

$$PV = \$60{,}000, \quad i = \frac{0.082}{12} = 0.006\overline{83}, \quad n = 30(12) = 360$$

$$PMT = PV \frac{i}{1 - (1 + i)^{-n}}$$

$$= 60{,}000 \frac{0.006\overline{83}}{1 - (1.006\overline{83})^{-360}} = 60{,}000(0.007477544) = \$448.65 \text{ per month}$$

<u>Step 2</u>: Find the unpaid balance after 72 payments, that is, find the present value of a \$448.65 per month annuity at 8.2% for 288 payments:

$$PMT = \$448.65, \quad i = \frac{0.082}{12} = 0.006\overline{83}, \quad n = 288$$

$$PV = 448.65 \frac{1 - (1.006\overline{83})^{-288}}{0.006\overline{83}} = 448.65(125.7549493) = \$56{,}419.96$$

<u>Step 3</u>: Reduce the principal by \$10,000 and determine the time to pay off the loan, that is find out how long it will take to pay off $56{,}419.96 - 10{,}000 = \$46{,}419.96$ with monthly payments of \$448.65:

$$46{,}419.96 = 448.65 \frac{1 - (1.006\overline{83})^{-n}}{0.006\overline{83}}$$

$$1 - \left(1.006\overline{83}\right)^{-n} = \frac{46{,}419.96(0.006\overline{83})}{448.65} = 0.707016394$$

$$\left(1.006\overline{83}\right)^{-n} = 0.292983605$$

$$-n \ln(1.006\overline{83}) = \ln(0.292983605)$$

$$n = \frac{-\ln(0.292983605)}{\ln(1.006\overline{83})} \approx 180$$

The loan will be paid off after 180 payments. Thus, by reducing the principal after 72 payments, the entire mortgage will be paid after $72 + 180 = 252$ payments.

Step 4: Calculate the future value of a $448.65 per month annuity at 7% for $360 - 252 = 108$ months.

$$PMT = \$448.65, \ i = \frac{0.07}{12} = 0.0058\overline{3}, \ n = 108$$

$$FV = PMT \frac{(1+i)^n - 1}{i} = 448.65 \frac{(1.0058\overline{3})^{108} - 1}{0.0058\overline{3}}$$
$$= 448.65(149.8589179) = \$67,234.20$$

Conclusion: Use the $10,000 to reduce the principal and invest the monthly payment at 7% for 108 months. (3-2, 3-3, 3-4)

53.   We find the monthly payment and the total interest for each of the options. The monthly payment is given by:

$$PMT = PV \frac{i}{1 - (1+i)^{-n}}, \quad PV = \$75.000, \ n = 12(30) = 360$$

<u>7.54% mortgage:</u> $i = \dfrac{0.0754}{12} = 0.00628\,\overline{3}$

$$PMT = 75,000 \frac{0.00628\overline{3}}{1 - (1.00628\overline{3})^{-360}}$$
$$= 75,000(0.007019555) = \$526.47 \text{ per month}$$

Total interest paid $= 360(526.47) - 75,000 = \$114,529.20$

<u>6.87% mortgage:</u> $i = \dfrac{0.0687}{12} = 0.005725$

$$PMT = 75,000 \frac{0.005725}{1 - (1.005725)^{-360}}$$
$$= 75,000(0.006565947) = \$492.45 \text{ per month}$$

Total interest paid $= 360(492.45) - 75,000 = \$102,282.00$
The lower rate would save over $12,247.20 in interest.   (3-4)

54. $A = \$5000, \ r = i = 5.6\% = 0.056, \ n = 5$

$$P = \frac{A}{(1+i)^n} = \frac{5000}{(1.056)^5} = \$3,807.59. \quad (3-2)$$

55. $P = \$5,695, \ A = \$10,000, \ n = 10, \ m = 1, \ r = i$

$$A = P(1 + i)^n$$
$$10,000 = 5,695(1 + i)^{10}$$

$$(1 + i)^{10} = \frac{10,000}{5,695} = 1.755926251$$

$$10 \ln(1 + i) = \ln(1.755926251) = 0.562996496$$

$$\ln(1 + i) = 0.0562996496$$
$$1 + i = e^{0.0562996496} \approx 1.057914639$$

$$I \approx 0.0579 \text{ or } 5.79\% \quad (3-2)$$

56. $A = \$5000,\ r = 6.4\% = 0.064,\ t = \dfrac{26}{52} = 0.5$

$P = \dfrac{A}{1+rt} = \dfrac{5000}{1+(0.064)(0.5)} = \dfrac{5{,}000}{1.032} = \$4{,}844.96$ (3-1)

57. We first compute the monthly payment using $PV = \$10{,}000,\ i = \dfrac{0.12}{12} = 0.01,$ and $n = 5(12) = 60.$

$PMT = PV\ \dfrac{i}{1-(1+i)^{-n}}$

$\qquad = 10{,}000\ \dfrac{0.01}{1-(1+0.01)^{-60}} = \dfrac{100}{1-(1.01)^{-60}} = \$222.44$ per month

Now, we calculate the unpaid balance after 24 payments by using
$PMT = \$222.44,\ i = 0.01,$ and $n = 60 - 24 = 36.$

$PV\ \ = PMT\ \dfrac{1-(1+i)^{-n}}{i}$

$\qquad = 222.44\ \dfrac{1-(1+0.01)^{-36}}{0.01} = 22{,}244[1-(1.01)^{-36}] = \$6697.11$

Thus, the unpaid balance after 2 years is $6697.11.   (3-4)

58. First find the annual percentage yield for 7.28% compounded quarterly:
$r = 7.28\% = 0.0728,\ m = 4$

$\text{APY} = \left(1+\dfrac{r}{m}\right)^{m} - 1 = \left(1+\dfrac{0.0728}{4}\right)^{4} - 1 = (1.0182)^{4} - 1 = 0.074812$ or 7.4812%

Now find the rate r compounded monthly that has the APY of 7.4812%:
$\text{APY} = 0.074812,\ m = 12$

$0.074812 = \left(1+\dfrac{r}{12}\right)^{12} - 1$

$\left(1+\dfrac{r}{12}\right)^{12} = 1.074812$

$1+\dfrac{r}{12} = \sqrt[12]{1.074812} = (1.074812)^{1/12} \approx 1.00603$

$\qquad r = 12(0.00603) \approx 0.0724;\ \ r = 7.24\%$ (3-2)

59. **(A)**   We first calculate the future value of an annuity of $2000 at 8% compounded annually for 9 years.

$FV = PMT\ \dfrac{(1+i)^{n}-1}{i}$ where $PMT = \$2000,\ i = 0.08,$ and $n = 9$

$\qquad = 2000\ \dfrac{(1+0.08)^{9}-1}{0.08} = 25{,}000[(1.08)^{9} - 1] \approx \$24{,}975.12$

Now, we calculate the future value of this amount at 8% compounded annually for 36 years.

$A = P(1+i)^{n},$ where $P = \$24{,}975.12,\ i = 0.08,$ and $n = 36$

$\qquad = 24{,}975.12(1+0.08)^{36} = 24{,}975.12(1.08)^{36} \approx \$398{,}807$

(B)   This is the future value of a $2000 annuity at 8% compounded annually for 36 years.

$$FV = PMT \, \frac{(1+i)^n - 1}{i} \text{ where } PMT = \$2000, \; i = 0.08, \text{ and } n = 36$$

$$= 2000 \, \frac{(1+0.08)^{36} - 1}{0.08} = 25{,}000[(1.08)^{36} - 1] \approx \$374{,}204 \qquad (3\text{-}3)$$

60. $A = Pe^{rt}$; $A = 27{,}000$, $r = 5.5\% = 0.055$, $t = 10$

$27{,}000 = Pe^{0.055(10)} = Pe^{0.550}$

$$P = \frac{27{,}000}{e^{0.550}} \approx \$15{,}577.64 \qquad (3\text{-}2)$$

61.   The amount of the loan is $0.8(100{,}000) = \$80{,}000$, and

$$PMT = PV \, \frac{i}{1 - (1+i)^{-n}} \, .$$

(A)   First let $i = \dfrac{0.0768}{12} = 0.0064$, $n = 30(12) = 360$. Then

$$PMT = 80{,}000 \, \frac{0.0064}{1 - (1.0064)^{-360}} = 80{,}000(0.00711581) = 569.26$$

The monthly payment on the 30 year mortgage is: $569.26.

Next, let $i = \dfrac{0.0768}{12} = 0.0064$, $n = 15(12) = 180$. Then

$$PMT = 80{,}000 \, \frac{0.0064}{1 - (1.0064)^{-180}} = 80{,}000(0.00937271) = 749.82$$

The monthly payment on the 15 year mortgage is: $749.82.

(B)   To find the unpaid balance after 10 years, we use

$$PV = PMT \, \frac{1 - (1+i)^{-n}}{i} \, .$$

For the 30 year mortgage, there are 20 years remaining at $569.26 per month.

$PMT = \$569.26$, $i = 0.0064$, $n = 20(12) = 240$

$$PV = 569.26 \, \frac{1 - (1.0064)^{-240}}{0.0064} = 569.26(122.4536903) = 69{,}707.99$$

The unpaid balance for the 30 year mortgage is $69,707.99.

For the 15 year mortgage, there are 5 years remaining at $749.82 per month:

$PMT = \$749.82$, $i = 0.0064$, $n = 5(12) = 60$

$$PV = 749.82 \, \frac{1 - (1.0064)^{-60}}{0.0064} = 749.82(49.69291289) = 37{,}260.74$$

The unpaid balance for the 15 year mortgage is: $37,260.74.   (3-4)

62. The amount of the mortgage is: $0.8(83,000) = \$66,400$.
The monthly payment is given by:

$$PMT = PV\,\frac{i}{1-(1+i)^{-n}} \quad \text{where } PV = \$66,400,\; i = \frac{0.084}{12} = 0.007,\; n = 30(12) = 360$$

$$PMT = 66,400\,\frac{0.007}{1-(1.007)^{-360}} = 66,400(0.007618376) = 505.86$$

The monthly payment is: $505.86.

Next, we find the present value of a $505.86 per month,
22 year annuity at 8.4%.

$$PMT = \$505.86,\; i = \frac{0.084}{12} = 0.007,\; \text{and } n = 22(12) = 264$$

$$PV = PMT\,\frac{1-(1+i)^{-n}}{i} = 505.86\,\frac{1-(1.007)^{-264}}{0.007}$$
$$= 505.86(120.204351) = 60,806.57$$

The unpaid loan balance is $60,806.57.

Finally, equity = (current market value) − (unpaid loan balance)
$$= 95,000 - 60,806.57 = \$34,193.43.$$

The family can borrow up to $0.6(34,193.43) = \$20,516$.   (3-4)

63. $PV = \$600$, $PMT = 110$, $n = 6$

Solve $600 = 110\,\dfrac{1-(1+i)^{-6}}{i}$ for i and multiply the

result by 12 to find r.

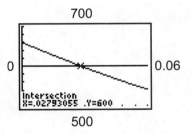

$i \approx 0.02793055$
$r \approx 0.33517$ or $r = 33.52\%$
(3-4)

64. (A) $FV = \$220,000$, $PMT = \$2,000$, $n = 25$.

Solve $220,000 = 2,000\,\dfrac{(1+i)^{25}-1}{i}$ for i:

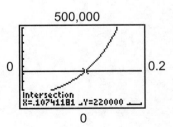

$i \approx 0.10741$ or $i = 10.74\%$

(B)   Withdrawals at $30,000 per year:

Solve $220{,}000 = 30{,}000 \dfrac{1-(1.10741)^{-n}}{0.10741}$ for n

$n \approx 15$ years

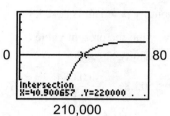

Withdrawals $24,000 per year:

Solve $220{,}000 = 24{,}000 \dfrac{1-(1.10741)^{-n}}{0.10741}$

$n \approx 40$ years   (3-3, 3-4)

4   SYSTEMS OF LINEAR EQUATIONS; MATRICES

EXERCISE 4-1

Things to remember:

1.   SYSTEMS OF TWO EQUATIONS IN TWO VARIABLES

Given the LINEAR SYSTEM
$$ax + by = h$$
$$cx + dy = k$$
where a, b, c, d, h, and k are real constants, a pair of numbers $x = x_0$ and $y = y_0$ [also written as an ordered pair (x_0, y_0)] is a SOLUTION to this system if each equation is satisfied by the pair. The set of all such ordered pairs is called the SOLUTION SET for the system. To SOLVE a system is to find its solution set.

2.   SYSTEMS OF LINEAR EQUATIONS: BASIC TERMS

A system of linear equations is CONSISTENT if it has one or more solutions and INCONSISTENT if no solutions exist. Furthermore, a consistent system is said to be INDEPENDENT if it has exactly one solution (often referred to as the UNIQUE SOLUTION) and DEPENDENT if it has more than one solution. Two systems of equations are EQUIVALENT if they have the same solution set.

3.   The system of two linear equations in two variables
$$ax + by = h$$
$$cx + dy = k$$
can be solved by:
(a)   graphing;
(b)   substitution;
(c)   elimination by addition.

4.   POSSIBLE SOLUTIONS TO A LINEAR SYSTEM

The linear system
$$ax + by = h$$
$$cx + dy = k$$
must have:

(a)   exactly one solution (consistent and independent); or

(b)   no solution (inconsistent); or

(c)   infinitely many solutions (consistent and dependent).

<u>5</u>. OPERATIONS THAT PRODUCE EQUIVALENT SYSTEMS

A system of linear equations is transformed into an equivalent system if:

(a) two equations are interchanged;

(b) an equation is multiplied by a nonzero constant;

(c) a constant multiple of one equation is added to another equation.

---

1. Set $x = 0$. Then $y = 7$; $(0, 7)$

3. Set $y = 0$. Then $x = 24$; $(24, 0)$

5. Set $x = 5$. Then $6(5) - 5y = 120$, $-5y = 90$, $y = -18$; $(5, -18)$

7. Slope $= \dfrac{-5 - 7}{4 - 2} = -6$; $y - 7 = -6(x - 2)$, $y = -6x + 19$,

9. (B); no solution

11. (A); $x = -3$, $y = 1$

13. $3x - y = 2$

$x + 2y = 10$

Point of intersection: (2, 4)

Solution: $x = 2$; $y = 4$

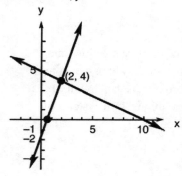

15. $m + 2n = 4$

$2m + 4n = -8$

Since the graphs of the given equations are parallel lines, there is no solution.

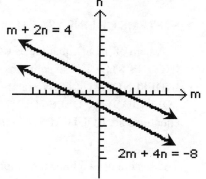

17. $y = 2x - 3$ (1)

$x + 2y = 14$ (2)

By substituting y from (1) into (2), we get:

$x + 2(2x - 3) = 14$

$x + 4x - 6 = 14$

$5x = 20$

$x = 4$

Now, substituting $x = 4$ into (1), we have:

$y = 2(4) - 3$

$y = 5$

Solution: $x = 4$, $y = 5$

19. $2x + y = 6$ (1)

$x - y = -3$ (2)

Solve (2) for y to obtain the system:

$2x + y = 6$ (3)

$y = x + 3$ (4)

Substitute y from (4) into (3):

$2x + x + 3 = 6$

$3x = 3$

$x = 1$

Now, substituting $x = 1$ into (4), we get:

$y = 1 + 3$

$y = 4$

Solution: $x = 1$, $y = 4$

21. $3u - 2v = 12$ (1)
$7u + 2v = 8$ (2)

Add (1) and (2):

$10u = 20$
$u = 2$

Substituting $u = 2$ into (2), we get:

$7(2) + 2v = 8$
$2v = -6$
$v = -3$

Solution: $u = 2$, $v = -3$

23. $2m - n = 10$ (1)
$m - 2n = -4$ (2)

Multiply (1) by -2 and add to (2) to obtain:

$-3m = -24$
$m = 8$

Substituting $m = 8$ into (2), we get:

$8 - 2n = -4$
$-2n = -12$
$n = 6$

Solution: $m = 8$, $n = 6$

25. $9x - 3y = 24$ (1)
$11x + 2y = 1$ (2)

Solve (1) for y to obtain:

$y = 3x - 8$ (3)

and substitute into (2):

$11x + 2(3x - 8) = 1$
$11x + 6x - 16 = 1$
$17x = 17$
$x = 1$

Now, substitute $x = 1$ into (3):

$y = 3(1) - 8$
$y = -5$

Solution: $x = 1$, $y = -5$

27. $2x - 3y = -2$ (1)
$-4x + 6y = 7$ (2)

Multiply (1) by 2 and add to (2) to get:

$0 = 3$

This implies that the system is inconsistent, and thus there is no solution.

29. $3x + 8y = 4$ (1)
$15x + 10y = -10$ (2)

Multiply (1) by -5 and add to (2) to get:

$-30y = -30$
$y = 1$

Substituting $y = 1$ into (1), we get:

$3x + 8(1) = 4$
$3x = -4$
$x = -\dfrac{4}{3}$

Solution: $x = -\dfrac{4}{3}$, $y = 1$

31. $-6x + 10y = -30$ (1)
$3x - 5y = 15$ (2)

Multiply (2) by 2 and add to (1). This yields:

$0 = 0$

which implies that (1) and (2) are equivalent equations and there are infinitely many solutions. Geometrically, the two lines are coincident. The system is dependent.

33. $x + y = 1$ (1)
$0.3x - 0.4y = 0$ (2)

Multiply equation (2) by 10 to remove the decimals

$x + y = 1$ (1)
$3x - 4y = 0$ (3)

Multiply (1) by 4 and add to (2) to get

$7x = 4$

$x = \dfrac{4}{7}$

Now substitute $x = \dfrac{4}{7}$ in (1):

$\dfrac{4}{7} + y = 1$

$y = 1 - \dfrac{4}{7} = \dfrac{3}{7}$

Solution: $x = \dfrac{4}{7}$, $y = \dfrac{3}{7}$

35. $x + 0y = 7$ implies $x = 7$
$0x + y = 3$ implies $y = 3$

37. $5x + 0y = 4$ implies $x = 4/5$
$0x + 3y = -2$ implies $y = -2/3$

39. $x + y = 0$ (1)
$x - y = 0$ (2)

Adding (1) and (2) gives $2x = 0$, so $x = 0$, which implies $y = 0$.

41. $x - 2y = 4$ (1)
$0x + y = 5$ implies $y = 5$

Substituting $y = 5$ into (1) gives $x - 2(5) = 4$, which implies $x = 14$.

43.   The price will tend to go down.

45.   If $m \neq n$, then the system has a unique solution independent of the values of b and c.

47.   If the system has infinitely many solutions, then there is a non-zero number k such that $n = km$ and $c = kb$.

49.   Solution: $x = -14$, $y = -37$; the lines intersect at the point

(−14, −37).

51.   No solution; the lines are parallel.

53.   First solve each equation for y:

$y = \dfrac{3}{2}x - \dfrac{15}{2}$

$y = -\dfrac{4}{3}x + \dfrac{13}{3}$

The graphs of the equations are shown at the right.

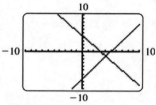

intersection: $x = 4.176$, $y = -1.235$
(4.176, −1.235)

55. Multiply each equation by 10 and then solve for y:

$$y = \frac{24}{35}x + \frac{1}{35}$$

$$y = \frac{17}{26}x - \frac{1}{13}$$

The graphs of these equations are almost indistinguishable.

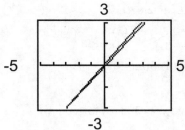

intersection: $x = -3.310$, $y = -2.241$

$(-3.310, -2.241)$

57. $x - 2y = -6$ (L_1)

$2x + y = 8$ (L_2)

$x + 2y = -2$ (L_3)

(A) L_1 and L_2 intersect:

$x - 2y = -6$ (1)

$2x + y = 8$ (2)

Multiply (2) by 2 and add to (1):

$5x = 10$

$x = 2$

Substitute $x = 2$ in (1) to get

$2 - 2y = -6$

$-2y = -8$

$y = 4$

Solution: $x = 2$, $y = 4$

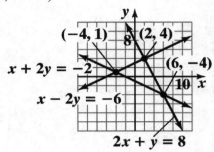

(B) L_1 and L_3 intersect:

$x - 2y = -6$ (3)

$x + 2y = -2$ (4)

Add (3) and (4):

$2x = -8$

$x = -4$

Substitute $x = -4$ in (3)   to get

$-4 - 2y = -6$

$-2y = -2$

$y = 1$

Solution: $x = -4$, $y = 1$

(C) L_2 and L_3 intersect:

$2x + y = 8$ (5)

$x + 2y = -2$ (6)

Multiply (6) by -2 and add   to (5)

$-3y = 12$

$y = -4$

Substitute $y = -4$ in (5) to get

$2x - 4 = 8$

$2x = 12$

$x = 6$

Solution: $x = 6$, $y = -4$

59. $x + y = 1$ (L_1)

$x - 2y = -8$ (L_2)

$3x + y = -3$ (L_3)

(A) L_1 and L_2 intersect

$x + y = 1$ (1)

$x - 2y = -8$ (2)

Subtract (2) from (1):

$3y = 9$

$y = 3$

Substitute $y = 3$ in (1) to get $x + 3 = 1$, $x = -2$

Solution: $x = -2$, $y = 3$

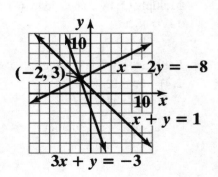

(B) L_1 and L_3 intersect:

$x + y = 1$ (3)

$3x + y = -3$ (4)

Subtract (4) from (3):

$-2x = 4$

$x = -2$

Substitute $x = -2$ in (3) to get

$-2 + y = 1$

$y = 3$

Solution: $x = -2, \ y = 3$

(C)   It follows from (A) and (B), that L_2 and L_3 intersect at

$x = -2, \ y = 3.$

61. $4x - 3y = -24$ (L_1)

$2x + 3y = 12$ (L_2)

$8x - 6y = 24$ (L_3)

(A) L_1 and L_2 intersect:

$4x - 3y = -24$ (1)

$2x + 3y = 12$ (2)

Add (1) and (2):

$6x = -12$

$x = -2$

Substitute $x = -2$ in (2) to get

$2(-2) + 3y = 12$

$3y = 16$

$y = \dfrac{16}{3}$

Solution: $x = -2, \ y = \dfrac{16}{3}$

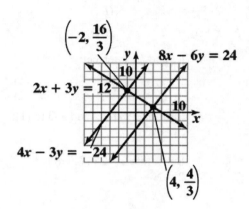

(B)   In slope-intercept form, (L_1) and (L_3) have equations

$y = \dfrac{4}{3}x + 8$ (L_1)

$y = \dfrac{4}{3}x - 4$ (L_2)

Thus, (L_1) and (L_3) have the same slope and different

y-intercepts; (L_1) and (L_3) are parallel; they do not intersect.

(C) L_2 and L_3 intersect:

$2x + 3y = 12$ (3)

$8x - 6y = 24$ (4)

Multiply (3) by 2 and add to (4):

$12x = 48$

$x = 4$

Substitute $x = 4$ in (3) to get

$2(4) + 3y = 12$

$3y = 4$

$y = \dfrac{4}{3};$ solution: $x = 4, \ y = \dfrac{4}{3}$

63.   (A) $5x + 4y = 4$ Multiply the top equation by 9 and the bottom
            $11x + 9y = 4$ equation by –4.

            $45x + 36y = 36$ Add the equations.
            $\underline{-44x - 36y = -16}$

                  $x = 20$

            $5(20) + 4y = 4$ Substitute $x = 20$ in the first equation.
                  $4y = -96$
                  $y = -24$

            Solution: $(20, -24)$

       (B) $5x + 4y = 4$ Multiply the top equation by 8 and the bottom
            $11x + 8y = 4$ equation by –4.

            $40x + 32y = 32$ Add the equations.
            $\underline{-44x - 32y = -16}$

                 $-4x = 16$
                  $x = -4$

            $5(-4) + 4y = 4$ Substitute $x = 4$ in the first equation.
                  $4y = 24$
                   $y = 6$

            Solution: $(-4, 6)$

       (C) $5x + 4y = 4$ Multiply the top equation by 8 and the bottom
            $10x + 8y = 4$ equation by –4.

            $40x + 32y = 32$ Add the equations.
            $\underline{-40x - 32y = -16}$

                   $0 = -16$ This system has no solutions.

65. $p = 0.7q + 3$ Supply equation
        $p = -1.7q + 15$ Demand equation

       (A) $p = \$4$

            Supply: $4 = 0.7q + 3$
                  $0.7q = 1$
                  $q = \dfrac{1}{0.7} \approx 1.429$

            Thus, the supply will be 143 T-shirts.

            Demand: $4 = -1.7q + 15$
                  $1.7q = 11$
                  $q = \dfrac{11}{1.7} \approx 6.471$

            Thus, the demand will be 647 T-shirts. At this price level, the
            demand exceeds the supply; the price will rise.

(B) $p = \$9$

Supply: $9 = 0.7q + 3$

$\qquad 0.7q = 6$

$\qquad\quad q = \dfrac{6}{0.7} \approx 8.571$

Thus, the supply will be 857 T-shirts.

Demand: $9 = -1.7q + 15$

$\qquad\quad 1.7q = 6$

$\qquad\qquad q = \dfrac{6}{1.7} \approx 3.529$

Thus, the demand will be 353 T-shirts. At this price level, the supply exceeds the demand; the price will fall.

(C)   Solve the pair of equations to find the equilibrium price and the equilibrium quantity.

$\qquad 0.7q + 3 = -1.7q + 15$

$\qquad\quad 2.4q = 12$

$\qquad\qquad q = \dfrac{12}{2.4} = 5$

The equilibrium quantity is 500 T-shirts. Substitute $q = 5$ in either of the two equations to find p.

$\quad p = (0.7)5 + 3 = 3.5 + 3$

$\quad p = 6.50$

The equilibrium price is $6.50.

(D)

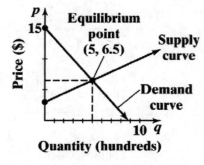

67.   (A)   Supply equation: $p = aq + b$

At $p = 4.80$, $q = 1.9$: $4.80 = 1.9a + b$

At $p = 5.10$, $q = 2.1$: $5.10 = 2.1a + b$

Solve

$\quad 1.9a + b = 4.80$ (1)

$\quad 2.1a + b = 5.10$ (2)

Subtract equation (1) from equation (2):

$0.2a = 0.3$

$\quad a = \dfrac{3}{2} = 1.5$

Substitute $a = 1.5$ into equation (1) (or equation (2)) to find b:

$1.9(1.5) + b = 4.80$

$\qquad\qquad b = 4.80 - 2.85 = 1.95$

The supply equation is:

$\quad p = 1.5q + 1.95$

(B)   Demand equation: $p = aq + b$

At 4.80, $q = 2.0$: $4.80 = 2.0a + b$

At 5.10, $q = 1.8$: $5.10 = 1.8a + b$

Solve $2.0a + b = 4.80$ (1)

$\qquad\quad 1.8a + b = 5.10$ (2)

Subtract equation (1) from equation (2):

$-0.2a = 0.3$

$$a = -\frac{3}{2} = -1.5$$

Substitute $a = -1.5$ into equation (1) (or into equation (2)) to find b:

$2.0(-1.5) + b = 4.80$

$b = 4.80 + 2.0(1.5) = 7.80$

The demand equation is:

$p = -1.5q + 7.80$

(C) Equilibrium price and quantity

Solve $1.5q + 1.95 = -1.5q + 7.80$ for q:

$3q = 7.80 - 1.95 = 5.85$

$q = 1.95$

Substitute $q = 1.95$ into either the supply equation or the demand equation:

$p = 1.5(1.95) + 1.95 = 4.875$

The equilibrium price is $4.875 or $4.88; the equilibrium quantity is 1.95 billion bushels.

(D)

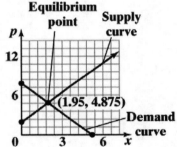

69. (A) The company breaks even when:

Cost = Revenue

$48,000 + 1400x = 1800x$

$48,000 = 1800x - 1400x$

or $400x = 48,000$

$$x = \frac{48,000}{400}$$

$x = 120$

Thus, 120 units must be manufactured and sold to break even.

Cost = $48,000 + 1400(120)$

$= \$216,000 = $ Revenue

(B)

y↑
$400,000
Revenue line
Profit
Cost line
Loss   (120, 216,000)
Break-even point
200 x

71. Let x = number of DVD's marketed per month

(A) Revenue: $R = 19.95x$

Cost: $C = 7.45x + 24,000$

(B) At the break-even point Revenue = Cost, that is

$19.95x = 7.45x + 24,000$

$12.50x = 24,000$

$x = 1920$

Thus, 1920 DVD's must be sold per month to break even.

Cost = Revenue = $38,304 at the break-even point.

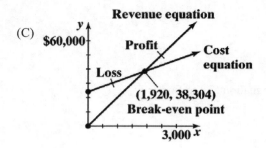

73. Let x = base price

    y = surcharge

    5 pound package: $x + 4y = 27.75$
    20 pound package: $x + 19y = 64.50$
    Solve the two equations:

    $x + 4y = 27.75$ (1)
    $x + 19y = 64.50$ (2)

    Multiply (1) by (–1) and add to (2):

    $15y = 36.75$

    $y = 2.45$

    Substitute $y = 2.45$ into (1):

    $x + 4(2.45) = 27.75$

    $x = 27.75 - 9.80 = 17.95$

    Thus, the base price is \$17.95; the surcharge is \$2.45 per pound.

75. Let x = number of pounds of robust blend

    y = number of pounds of mild blend

    Total amount of Columbian beans: $132 \times 50 = 6,600$ lbs.

    Total amount of Brazilian beans: $132 \times 40 = 5,280$ lbs.

    Columbian beans needed: $\dfrac{12}{16}x + \dfrac{6}{16}y$ or $\dfrac{3}{4}x + \dfrac{3}{8}y$

    Brazilian beans needed: $\dfrac{4}{16}x + \dfrac{10}{16}y$ or $\dfrac{1}{4}x + \dfrac{5}{8}y$

    Thus, we need to solve:

    $\dfrac{3}{4}x + \dfrac{3}{8}y = 6,600$ (1)

    $\dfrac{1}{4}x + \dfrac{5}{8}y = 5,280$ (2)

    Multiply (2) by –3 and add to (1):

    $-\dfrac{12}{8}y = -9,240$

    $y = 6,160$

    Substitute $y = 6,160$ into (1):

    $\dfrac{3}{4}x + \dfrac{3}{8}(6,160) = 6,600$

    $\dfrac{3}{4}x = 6,600 - 2,310 = 4,290$

    $x = 5,720$

    Therefore, the manufacturer should produce 5,720 pounds of the robust blend and 6,160 pounds of the mild blend

77. Let x = amount of mix A, and

y = amount of mix B.

We want to solve the following system of equations:

$0.1x + 0.2y = 20$ (1)
$0.06x + 0.02y = 6$ (2)

Clear the decimals from (1) and (2) by multiplying both sides of (1) by 10 and both sides of (2) by 100.

$x + 2y = 200$ (3)
$6x + 2y = 600$ (4)

Multiply (3) by –1 and add to (4):

$5x = 400$
$x = 80$

Now substitute $x = 80$ into (3):

$80 + 2y = 200$
$2y = 120$
$y = 60$

Solution: x = mix A = 80 grams; y = mix B = 60 grams

79. Let x = number of hours of Mexico plant
y = number of hours of Taiwan plant

Then we need to solve

$40x + 20y = 4000$ (1)
$32x + 32y = 4000$ (2)

Multiply (1) by 32, and (2) by (–20) and add the resulting equations:

$640x = 48{,}000$
$x = 75$

Substitute $x = 75$ into (1) (or (2))

$40(75) + 20y = 4000$
$20y = 4000 - 3000 = 1000$
$y = 50$

The plant in Mexico should be operated 75 hours, the plant in Taiwan should be operated 50 hours.

81. $s = a + bt^2$

(A) At $t = 1, s = 180:$ $a + b = 180$
At $t = 2, s = 132:$ $a + 4b = 132$

Solve the equations

$a + b = 180$ (1)
$a + 4b = 132$ (2)

Subtract (1) from (2):

$3b = -48$
$b = -16$

Substitute $b = -16$ in (1):

$a - 16 = 180$
$a = 196$

Thus, $s = 196 - 16t^2$

(B) At $t = 0, s = 196;$ the building is 196 feet high

(C) At $s = 0,$ $196 - 16t^2 = 0$

$16t^2 = 196$
$t^2 = 12.25$
$t = 3.5$

The object falls for 3.5 seconds.

83. Let d = the distance to the earthquake. Then it took $d/5$ seconds for the primary wave to reach the station and $d/3$ seconds for the secondary wave to reach the station. We are given that

$$\frac{d}{3} - \frac{d}{5} = 16$$
$$5d - 3d = 240$$
$$2d = 240$$
$$d = 120$$

The earthquake was 120 miles from the station. The primary wave traveled $\frac{120}{5}$ = 24 seconds; the secondary

wave traveled $\frac{120}{3}$ = 40 seconds.

85. $p = -\dfrac{1}{5}d + 70$ [Approach equation]

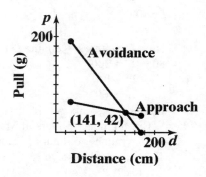

$p = -\dfrac{4}{3}d + 230$ [Avoidance equation]

(A) The figure shows the graphs of the two equations.

(B) Setting the two equations equal to each other, we have

$$-\frac{1}{5}d + 70 = -\frac{4}{3}d + 230$$

$$-\frac{1}{5}d + \frac{4}{3}d = 230 - 70$$

$$\frac{17}{15}d = 160$$

$$d = 141 \text{ cm (approx.)}$$

(C) The rat would be very confused (!); it would vacillate.

EXERCISE 4-2

Things to remember:

1. MATRICES

   A MATRIX is a rectangular array of numbers written within brackets. Each number in a matrix is called an ELEMENT. If a matrix has m rows and n columns, it is called an $m \times n$ MATRIX; $m \times n$ is the SIZE; m and n are the DIMENSIONS. A matrix with n rows and n columns is a SQUARE MATRIX OF ORDER n. A matrix with only one column is a COLUMN MATRIX; a matrix with only one row is a ROW MATRIX. The element in the ith row and jth column of a matrix A is denoted a_{ij}. The PRINCIPAL DIAGONAL of a

   matrix A consists of the elements
   $a_{11}, a_{22}, a_{33}, \ldots$.

2. Associated with the linear system
   $$a_1 x_1 + b_1 x_2 = k_1$$
   $$a_2 x_1 + b_2 x_2 = k_2 \qquad \text{(I)}$$

   is the AUGMENTED MATRIX of the system

   $$\begin{bmatrix} a_1 & b_1 & | & k_1 \\ a_2 & b_2 & | & k_2 \end{bmatrix}. \qquad \text{(II)}$$

3.   OPERATIONS THAT PRODUCE ROW-EQUIVALENT MATRICES

An augmented matrix is transformed into a row-equivalent matrix if:

(a)   two rows are interchanged $(R_i \leftrightarrow R_j)$;

(b)   a row is multiplied by a nonzero constant $(kR_i \rightarrow R_i)$;

(c)   a constant multiple of one row is added to another row
$(kR_j + R_i \rightarrow R_i)$.

(Note: The arrow $\rightarrow$ means "replaces.")

4.   Given the system of linear equations (I) and its associated augmented matrix (II). If (II) is row equivalent to a matrix of the form:

(1) $\begin{bmatrix} 1 & 0 & | & m \\ 0 & 1 & | & n \end{bmatrix}$,   then (I) has a unique solution (consistent and independent);

(2) $\begin{bmatrix} 1 & m & | & n \\ 0 & 0 & | & 0 \end{bmatrix}$,   then (I) has infinitely many solutions (consistent and dependent);

(3) $\begin{bmatrix} 1 & m & | & n \\ 0 & 0 & | & p \end{bmatrix}$, $p \neq 0$, then (I) has no solution (inconsistent).

---

1.   Elements in A: 6; in C: 3

3. B: 3×3; D: 2×1

5. D is a column matrix

7. B is a square matrix

9.   2, 1

11.   2, 8, 0

13. $c_{11} + c_{12} + c_{13} = 2 + -3 + 0 = -1$

15.   Coefficient matrix: $\begin{bmatrix} 3 & 5 \\ 2 & -4 \end{bmatrix}$; augmented matrix: $\begin{bmatrix} 3 & 5 & | & 8 \\ 2 & -4 & | & -7 \end{bmatrix}$

17.   Coefficient matrix: $\begin{bmatrix} 1 & 4 \\ 6 & 0 \end{bmatrix}$; augmented matrix: $\begin{bmatrix} 1 & 4 & | & 15 \\ 6 & 0 & | & 18 \end{bmatrix}$

19. $\begin{aligned} 2x_1 + 5x_2 &= 7 \\ x_1 + 4x_2 &= 9 \end{aligned}$

21. $\begin{aligned} 4x_1 &= -10 \\ 8x_2 &= 40 \end{aligned}$

23.   Interchange row 1 and row 2.
$\begin{bmatrix} 4 & -6 & | & -8 \\ 1 & -3 & | & 2 \end{bmatrix}$

25.   Multiply row 1 by -4.
$\begin{bmatrix} -4 & 12 & | & -8 \\ 4 & -6 & | & -8 \end{bmatrix}$

27.   Multiply row 2 by 2.
$\begin{bmatrix} 1 & -3 & | & 2 \\ 8 & -12 & | & -16 \end{bmatrix}$

29.   Replace row 2 by the sum of row 2 and -4 times row 1.
$\begin{bmatrix} 1 & -3 & | & 2 \\ 0 & 6 & | & -16 \end{bmatrix}$

31.   Replace row 2 by the sum of row 2 and -2 times row 1.
$\begin{bmatrix} 1 & -3 & | & 2 \\ 2 & 0 & | & -12 \end{bmatrix}$

33.   Replace row 2 by the sum of row 2 and -1 times row 1.
$\begin{bmatrix} 1 & -3 & | & 2 \\ 3 & -3 & | & -10 \end{bmatrix}$

35. $\begin{bmatrix} -1 & 2 & | & -3 \\ 6 & -3 & | & 12 \end{bmatrix} \dfrac{1}{3}R_2 \rightarrow R_2 \sim \begin{bmatrix} -1 & 2 & | & -3 \\ 2 & -1 & | & 4 \end{bmatrix}$

37. $\begin{bmatrix} -1 & 2 & | & -3 \\ 6 & -3 & | & 12 \end{bmatrix} 6R_1 + R_2 \rightarrow R_2 \sim \begin{bmatrix} -1 & 2 & | & 3 \\ 0 & 9 & | & -6 \end{bmatrix}$

39. $\begin{bmatrix} -1 & 2 & | & -3 \\ 6 & -3 & | & 12 \end{bmatrix} \dfrac{1}{3}R_2 + R_1 \rightarrow R_1 \sim \begin{bmatrix} 1 & 1 & | & 1 \\ 6 & -3 & | & 12 \end{bmatrix}$

41. $\begin{bmatrix} -1 & 2 & | & -3 \\ 6 & -3 & | & 12 \end{bmatrix} R_1 \leftrightarrow R_2 \sim \begin{bmatrix} 6 & -3 & | & 12 \\ -1 & 2 & | & -3 \end{bmatrix}$

43. $\begin{bmatrix} 3 & -2 & | & 6 \\ 4 & -3 & | & 6 \end{bmatrix} \sim \begin{bmatrix} 1 & -1 & | & 0 \\ 4 & -3 & | & 6 \end{bmatrix} \sim$

$R_2 - R_1 \rightarrow R_1 , \quad -4R_1 + R_2 \rightarrow R_2$

$\begin{bmatrix} 1 & -1 & | & 0 \\ 0 & 1 & | & 6 \end{bmatrix} \sim \begin{bmatrix} 1 & 0 & | & 6 \\ 0 & 1 & | & 6 \end{bmatrix}$

$R_2 - R_1 \rightarrow R_1 , \quad -4R_1 + R_2 \rightarrow R_2 ,$

Thus, $x_1 = 6$ and $x_2 = 6$.

The graph of each equation in the
system passes through the point (6, 6).

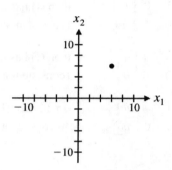

45. $\begin{bmatrix} 3 & -2 & | & -3 \\ -6 & 4 & | & 6 \end{bmatrix} \sim \begin{bmatrix} 3 & -2 & | & -3 \\ 0 & 0 & | & 0 \end{bmatrix} \sim \begin{bmatrix} 1 & -\frac{2}{3} & | & -1 \\ 0 & 0 & | & 0 \end{bmatrix}$

$2R_1 + R_2 \rightarrow R_2 \qquad \dfrac{1}{3}R_1 \rightarrow R_1$

From $\underline{4}$, Form (2), the system has infinitely many solutions (consistent
and dependent).

If $x_2 = t$, then $x_1 = \dfrac{2}{3}t - 1$;

solution set: $\{\frac{2}{3}t - 1, t \mid t$ any real number$\}$

The graph of the solution set is the same as the graph of each equation in
the system.

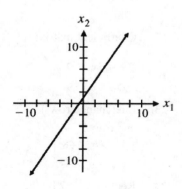

47.   System   Augmented matrix   Graphs:

$$x_1 + x_2 = 5$$
$$x_1 - x_2 = 1$$

$$\begin{bmatrix} 1 & 1 & | & 5 \\ 1 & -1 & | & 1 \end{bmatrix}$$

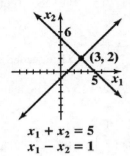

$x_1 + x_2 = 5$
$x_1 - x_2 = 1$

$$\begin{bmatrix} 1 & 1 & | & 5 \\ 1 & -1 & | & 1 \end{bmatrix} (-1)R_1 + R_2 \rightarrow R_2 \begin{bmatrix} 1 & 1 & | & 5 \\ 0 & -2 & | & -4 \end{bmatrix}$$

$$x_1 + x_2 = 5$$
$$-2x_2 = -4$$

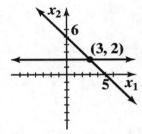

$x_1 + x_2 = 5$
$-2x_2 = -4$

$$\begin{bmatrix} 1 & 1 & | & 5 \\ 0 & -2 & | & -4 \end{bmatrix} -\frac{1}{2}R_2 \rightarrow R_2 \begin{bmatrix} 1 & 1 & | & 5 \\ 0 & 1 & | & 2 \end{bmatrix}$$

$$x_1 + x_2 = 5$$
$$x_2 = 2$$

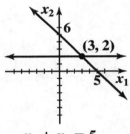

$x_1 + x_2 = 5$
$x_2 = 2$

$$\begin{bmatrix} 1 & 1 & | & 5 \\ 0 & 1 & | & 2 \end{bmatrix} (-1)R_2 + R_1 \rightarrow R_1 \begin{bmatrix} 1 & 0 & | & 3 \\ 0 & 1 & | & 2 \end{bmatrix}$$

$$x_1 = 3$$
$$x_2 = 2$$

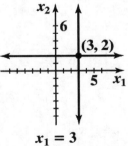

$x_1 = 3$
$x_2 = 2$

Solution: $x_1 = 3$, $x_2 = 2$. Each pair of lines has the same intersection point.

49. $\begin{bmatrix} 1 & 0 & | & -4 \\ 0 & 1 & | & 6 \end{bmatrix}$ implies $x_1 = -4$, $x_2 = 6$.

51. $\begin{bmatrix} 1 & 3 & | & 2 \\ 0 & 0 & | & 4 \end{bmatrix}$ The second equation is $0x_1 + 0x_2 = 4$ which has no solution.

53. $\begin{bmatrix} 1 & -2 & | & 15 \\ 0 & 0 & | & 0 \end{bmatrix}$ The second equation is $0x_1 + 0x_2 = 0$ which has infinitely many solutions. Set $x_2 = t$.
Then $x_1 = 2t + 15$ for any real number t.

55. $\begin{bmatrix} 1 & -2 & | & 1 \\ 2 & -1 & | & 5 \end{bmatrix} \sim \begin{bmatrix} 1 & -2 & | & 1 \\ 0 & 3 & | & 3 \end{bmatrix} \sim \begin{bmatrix} 1 & -2 & | & 1 \\ 0 & 1 & | & 1 \end{bmatrix} \sim \begin{bmatrix} 1 & 0 & | & 3 \\ 0 & 1 & | & 1 \end{bmatrix}$

$(-2)R_1 + R_2 \rightarrow R_2 \quad \frac{1}{3}R_2 \rightarrow R_2 \quad 2R_2 + R_1 \rightarrow R_1$

Thus, $x_1 = 3$ and $x_2 = 1$.

57. $\begin{bmatrix} 1 & -4 & | & -2 \\ -2 & 1 & | & -3 \end{bmatrix} \sim \begin{bmatrix} 1 & -4 & | & -2 \\ 0 & -7 & | & -7 \end{bmatrix} \sim \begin{bmatrix} 1 & -4 & | & -2 \\ 0 & 1 & | & 1 \end{bmatrix} \sim \begin{bmatrix} 1 & 0 & | & 2 \\ 0 & 1 & | & 1 \end{bmatrix}$

$2R_1 + R_2 \rightarrow R_2 \quad \left(-\frac{1}{7}\right)R_2 \rightarrow R_2 \quad 4R_2 + R_1 \rightarrow R_1$

Thus, $x_1 = 2$ and $x_2 = 1$.

59. $\begin{bmatrix} 3 & -1 & | & 2 \\ 1 & 2 & | & 10 \end{bmatrix} \sim \begin{bmatrix} 1 & 2 & | & 10 \\ 3 & -1 & | & 2 \end{bmatrix} \sim \begin{bmatrix} 1 & 2 & | & 10 \\ 0 & -7 & | & -28 \end{bmatrix} \sim \begin{bmatrix} 1 & 2 & | & 10 \\ 0 & 1 & | & 4 \end{bmatrix} \sim \begin{bmatrix} 1 & 0 & | & 2 \\ 0 & 1 & | & 4 \end{bmatrix}$

$R_1 \leftrightarrow R_2 \quad (-3)R_1 + R_2 \rightarrow R_2 \quad \left(-\frac{1}{7}\right)R_2 \rightarrow R_2 \quad (-2)R_2 + R_1 \rightarrow R_1$

Thus, $x_1 = 2$ and $x_2 = 4$.

61. $\begin{bmatrix} 1 & 2 & | & 4 \\ 2 & 4 & | & -8 \end{bmatrix} \sim \begin{bmatrix} 1 & 2 & | & 4 \\ 0 & 0 & | & -16 \end{bmatrix}$ From 4, Form (3), the system is inconsistent; there is no solution.

$(-2)R_1 + R_2 \rightarrow \square R_2$

63. $\begin{bmatrix} 2 & 1 & | & 6 \\ 1 & -1 & | & -3 \end{bmatrix} \sim \begin{bmatrix} 1 & -1 & | & -3 \\ 2 & 1 & | & 6 \end{bmatrix} \sim \begin{bmatrix} 1 & -1 & | & -3 \\ 0 & 3 & | & 12 \end{bmatrix} \sim \begin{bmatrix} 1 & -1 & | & -3 \\ 0 & 1 & | & 4 \end{bmatrix} \sim \begin{bmatrix} 1 & 0 & | & 1 \\ 0 & 1 & | & 4 \end{bmatrix}$

$R_1 \leftrightarrow R_2 \quad (-2)R_1 + R_2 \rightarrow R_2 \quad \frac{1}{3}R_2 \rightarrow R_2 \quad R_2 + R_1 \rightarrow R_1$

Thus, $x_1 = 1$ and $x_2 = 4$.

65. $\begin{bmatrix} 3 & -6 & | & -9 \\ -2 & 4 & | & 6 \end{bmatrix} \sim \begin{bmatrix} 1 & -2 & | & -3 \\ -2 & 4 & | & 6 \end{bmatrix} \sim \begin{bmatrix} 1 & -2 & | & -3 \\ 0 & 0 & | & 0 \end{bmatrix}$

$\frac{1}{3}R_1 \rightarrow R_1 \quad 2R_1 + R_2 \rightarrow R_2$

From 4, Form (2), the system has infinitely many solutions (consistent and dependent). If $x_2 = s$, then
$x_1 - 2s = -3$ or $x_1 = 2s - 3$.
Thus, $x_2 = s$, $x_1 = 2s - 3$, for any real number s, are the solutions.

67. $\begin{bmatrix} 4 & -2 & | & 2 \\ -6 & 3 & | & -3 \end{bmatrix} \sim \begin{bmatrix} 1 & -\frac{1}{2} & | & \frac{1}{2} \\ -6 & 3 & | & -3 \end{bmatrix} \sim \begin{bmatrix} 1 & -\frac{1}{2} & | & \frac{1}{2} \\ 0 & 0 & | & 0 \end{bmatrix}$

$\frac{1}{4}R_1 \rightarrow R_1 \qquad 6R_1 + R_2 \rightarrow R_2$

Thus, the system has infinitely many solutions (consistent and dependent). Let $x_2 = s$. Then

$x_1 - \frac{1}{2}s = \frac{1}{2}$ or $x_1 = \frac{1}{2}s + \frac{1}{2}$.

The set of solutions is $x_2 = s$, $x_1 = \frac{1}{2}s + \frac{1}{2}$ for any real number s.

69. $\begin{bmatrix} 2 & 1 & | & 1 \\ 4 & -1 & | & -7 \end{bmatrix} \sim \begin{bmatrix} 1 & \frac{1}{2} & | & \frac{1}{2} \\ 4 & -1 & | & -7 \end{bmatrix} \sim \begin{bmatrix} 1 & \frac{1}{2} & | & \frac{1}{2} \\ 0 & -3 & | & -9 \end{bmatrix} \sim \begin{bmatrix} 1 & \frac{1}{2} & | & \frac{1}{2} \\ 0 & 1 & | & 3 \end{bmatrix} \sim \begin{bmatrix} 1 & 0 & | & -1 \\ 0 & 1 & | & 3 \end{bmatrix}$

$\frac{1}{2}R_1 \rightarrow R_1 \quad (-4)R_1 + R_2 \rightarrow R_2 \quad \left(-\frac{1}{3}\right)R_2 \rightarrow R_2 \quad \left(-\frac{1}{2}\right)R_2 + R_1 \rightarrow R_1$

Thus, $x_1 = -1$ and $x_2 = 3$.

71. $\begin{bmatrix} 4 & -6 & | & 8 \\ -6 & 9 & | & -10 \end{bmatrix} \sim \begin{bmatrix} 1 & -\frac{3}{2} & | & 2 \\ -6 & 9 & | & -10 \end{bmatrix} \sim \begin{bmatrix} 1 & -\frac{3}{2} & | & 2 \\ 0 & 0 & | & 2 \end{bmatrix}$

$\frac{1}{4}R_1 \rightarrow R_1 \qquad 6R_1 + R_2 \rightarrow R_2$

The second row of the final augmented matrix corresponds to the equation

$0x_1 + 0x_2 = 2$ which has no solution. Thus, the system has no solution; it is inconsistent.

73. $\begin{bmatrix} -4 & 6 & | & -8 \\ 6 & -9 & | & 12 \end{bmatrix} \sim \begin{bmatrix} 1 & -\frac{3}{2} & | & 2 \\ 6 & -9 & | & 12 \end{bmatrix} \sim \begin{bmatrix} 1 & -\frac{3}{2} & | & 2 \\ 0 & 0 & | & 0 \end{bmatrix}$

$\left(-\frac{1}{4}\right)R_1 \rightarrow R_1 \quad (-6)R_1 + R_2 \rightarrow R_2$

The system has infinitely many solutions (consistent and dependent). If $x_2 = t$, then

$x_1 - \frac{3}{2}t = 2$ or $x_1 = \frac{3}{2}t + 2$

Thus, the set of solutions is $x_2 = t$, $x_1 = \frac{3}{2}t + 2$ for any real number t.

75. $\begin{bmatrix} 3 & -1 & | & 7 \\ 2 & 3 & | & 1 \end{bmatrix} \sim \begin{bmatrix} 1 & -\frac{1}{3} & | & \frac{7}{3} \\ 2 & 3 & | & 1 \end{bmatrix} \sim \begin{bmatrix} 1 & -\frac{1}{3} & | & \frac{7}{3} \\ 0 & \frac{11}{3} & | & -\frac{11}{3} \end{bmatrix} \sim \begin{bmatrix} 1 & -\frac{1}{3} & | & \frac{7}{3} \\ 0 & 1 & | & -1 \end{bmatrix} \sim \begin{bmatrix} 1 & 0 & | & 2 \\ 0 & 1 & | & -1 \end{bmatrix}$

$\frac{1}{3}R_1 \rightarrow R_1 \quad (-2)R_1 + R_2 \rightarrow R_2 \quad \frac{3}{11}R_2 \rightarrow R_2 \quad \frac{1}{3}R_2 + R_1 \rightarrow R_1$

Thus $x_1 = 2$, $x_2 = -1$.

77. $\begin{bmatrix} 3 & 2 & | & 4 \\ 2 & -1 & | & 5 \end{bmatrix} \sim \begin{bmatrix} 1 & \frac{2}{3} & | & \frac{4}{3} \\ 2 & -1 & | & 5 \end{bmatrix} \sim \begin{bmatrix} 1 & \frac{2}{3} & | & \frac{4}{3} \\ 0 & -\frac{7}{3} & | & \frac{7}{3} \end{bmatrix} \sim \begin{bmatrix} 1 & \frac{2}{3} & | & \frac{4}{3} \\ 0 & 1 & | & -1 \end{bmatrix} \sim \begin{bmatrix} 1 & 0 & | & 2 \\ 0 & 1 & | & -1 \end{bmatrix}$

$\frac{1}{3}R_1 \rightarrow R_1 \quad (-2)R_1 + R_2 \rightarrow R_2 \quad \left(-\frac{3}{7}\right)R_2 \rightarrow R_2 \quad \left(-\frac{2}{3}\right)R_2 + R_1 \rightarrow R_1$

Thus $x_1 = 2$, $x_2 = -1$.

79. $\begin{bmatrix} 0.2 & -0.5 & | & 0.07 \\ 0.8 & -0.3 & | & 0.79 \end{bmatrix} \sim \begin{bmatrix} 1 & -2.5 & | & 0.35 \\ 0.8 & -0.3 & | & 0.79 \end{bmatrix} \sim \begin{bmatrix} 1 & -2.5 & | & 0.35 \\ 0 & 1.7 & | & 0.51 \end{bmatrix}$

$\dfrac{1}{0.2} R_1 \rightarrow R_1 \qquad (-0.8)R_1 + R_2 \rightarrow R_2 \qquad \dfrac{1}{1.7} R_2 \rightarrow R_2$

$\sim \begin{bmatrix} 1 & -2.5 & | & 0.35 \\ 0 & 1 & | & 0.3 \end{bmatrix} \sim \begin{bmatrix} 1 & 0 & | & 1.1 \\ 0 & 1 & | & 0.3 \end{bmatrix}$ Thus $x_1 = 1.1, \quad x_2 = 0.3.$

$2.5R_2 + R_1 \rightarrow R_1$

81. $0.8x_1 + 2.88x_2 = 4$

$1.25x_1 + 4.34x_2 = 5$

$\begin{bmatrix} 0.8 & 2.88 & | & 4 \\ 1.25 & 4.34 & | & 5 \end{bmatrix} \sim \begin{bmatrix} 1 & 3.6 & | & 5 \\ 1.25 & 4.34 & | & 5 \end{bmatrix} \sim \begin{bmatrix} 1 & 3.6 & | & 5 \\ 0 & -0.16 & | & -1.25 \end{bmatrix}$

$(1.25)R_1 \rightarrow R_1 \qquad (-1.25)R_1 + R_2 \rightarrow R_2 \qquad (-6.25)R_2 \rightarrow R_2$

$\sim \begin{bmatrix} 1 & 3.6 & | & 5 \\ 0 & 1 & | & 7.8125 \end{bmatrix} \sim \begin{bmatrix} 1 & 0 & | & -23.125 \\ 0 & 1 & | & 7.8125 \end{bmatrix}$

$(-3.6)R_2 + R_1 \rightarrow R_1$

Solution: $x_1 = -23.125, \quad x_2 = 7.8125$

83. $4.8x_1 - 40.32x_2 = 295.2$

$-3.75x_1 + 28.7x_2 = -211.2$

$\begin{bmatrix} 4.8 & -40.32 & | & 295.2 \\ -3.75 & 28.7 & | & -211.2 \end{bmatrix} \sim \begin{bmatrix} 1 & -8.4 & | & 61.5 \\ -3.75 & 28.7 & | & -211.2 \end{bmatrix} \sim \begin{bmatrix} 1 & -8.4 & | & 61.5 \\ 0 & -2.8 & | & 19.425 \end{bmatrix}$

$(1/4.8)R_1 \rightarrow R_1 \qquad (3.75)R_1 + R_2 \rightarrow R_2 \qquad (1/-2.8)R_2 \rightarrow R_2$

$\sim \begin{bmatrix} 1 & -8.4 & | & 61.5 \\ 0 & 1 & | & -6.9375 \end{bmatrix} \sim \begin{bmatrix} 1 & 0 & | & 3.225 \\ 0 & 1 & | & -6.9375 \end{bmatrix}$

$(8.4)R_2 + R_1 \rightarrow R_1$

Solution: $x_1 = 3.225, \quad x_2 = -6.9375$

EXERCISE 4-3

Things to remember:

1. A matrix is said to be in REDUCED ROW ECHELON FORM or, more simply, in REDUCED FORM if

    (a) each row consisting entirely of zeros is below any row having at least one nonzero element;

    (b) the left-most nonzero element in each row is 1;

    (c) all other elements in the column containing the left-most 1 of a given row are zeros;

    (d) the left-most 1 in any row is to the right of the left-most 1 in any row above.

2.   GAUSS-JORDAN ELIMINATION

Step 1. Choose the leftmost nonzero column and use appropriate row operations to get a 1 at the top.

Step 2. Use multiples of the row containing the 1 from step 1 to get zeros in all remaining places in the column containing this 1.

Step 3. Repeat step 1 with the SUBMATRIX formed by (mentally) deleting the row used in step 2 and all rows above this row.

Step 4. Repeat step 2 with the ENTIRE MATRIX, including the mentally deleted rows. Continue this process until the entire matrix is in reduced form.

[*Note:* If at any point in this process we obtain a row with all zeros to the left of the vertical line and a nonzero number to the right, we can stop before we find the reduced form, since we will have a contradiction: $0 = n$, $n \neq 0$. We can then conclude that the system has no solution.]

---

1. Augmented matrix: $\begin{bmatrix} 1 & 2 & 3 & | & 12 \\ 1 & 7 & 5 & | & 15 \end{bmatrix}$

3. Augmented matrix: $\begin{bmatrix} 1 & 0 & 6 & | & 2 \\ 0 & 1 & -1 & | & 5 \\ 1 & 3 & 0 & | & 7 \end{bmatrix}$

5. System of equations: $\begin{aligned} x_1 - 3x_2 &= 4 \\ 3x_1 + 2x_2 &= 5 \\ -x_1 + 6x_2 &= 3 \end{aligned}$

7. System of equations: $5x_1 - 2x_2 + 8x_4 = 4$

9. $\begin{bmatrix} 1 & 0 & | & 2 \\ 0 & 1 & | & -1 \end{bmatrix}$

is in reduced form

11. $\begin{bmatrix} 1 & 0 & 2 & | & 3 \\ 0 & 0 & 0 & | & 0 \\ 0 & 1 & -1 & | & 4 \end{bmatrix}$

is not in reduced form: condition (a) is violated; the second row should be at the bottom.

$R_2 \leftrightarrow R_3$

13. $\begin{bmatrix} 0 & 1 & 0 & | & 2 \\ 0 & 0 & 3 & | & -1 \\ 0 & 0 & 0 & | & 0 \end{bmatrix}$

is not in reduced form: The first non-zero element in the second row is not 1; condition (b) is violated

$\dfrac{1}{3} R_2 \to R_2$

15. $\begin{bmatrix} 1 & 1 & 0 & | & 1 \\ 0 & 0 & 1 & | & 1 \\ 0 & 0 & 0 & | & 0 \end{bmatrix}$

is in reduced form.

17. $\begin{bmatrix} 1 & 0 & -2 & 0 & | & 1 \\ 0 & 0 & 1 & 1 & | & 0 \end{bmatrix}$

$2R_2 + R_1 \to R_1$

not in reduced form: The column containing the left-most 1 in row 2 has a nonzero element; condition (c) is violated

19. $x_1 = -2$

$x_2 = 3$

$x_3 = 0$

21. $x_1 \quad - 2x_3 = 3$ (1)

$x_2 + x_3 = -5$ (2)

Let $x_3 = t$. From (2), $x_2 = -5 - t$.

From (1), $x_1 = 3 + 2t$. Thus, the solution is

$x_1 = 2t + 3$

$x_2 = -t - 5$

$x_3 = t$

t any real number.

23. $x_1 = 0$

$x_2 = 0$

$0 = 1$

Inconsistent; no solution.

25. $x_1 \quad - 3x_3 = 5$

$x_2 + 2x_3 = -7$

Let $x_3 = t$. Then $x_1 = 3t + 5$,

$x_2 = -2t - 7$, t any real number.

27. $x_1 - 2x_2 \quad - 3x_4 = -5$

$x_3 + 3x_4 = 2$

Let $x_2 = s$ and $x_4 = t$. Then

$x_1 = 2s + 3t - 5$, $\quad x_2 = s$, $\quad x_3 = -3t + 2$, $\quad x_4 = t$, $\quad s$ and t any real numbers.

29. Problem 19.

31. Problems 21, 25, 27.

33. False. Counterexample: Problem 23

35. True. For example, the reduced form of the augmented matrix in the 3x3 case will be

$$\begin{bmatrix} 1 & 0 & 0 & | & r_1 \\ 0 & 1 & 0 & | & r_2 \\ 0 & 0 & 1 & | & r_3 \end{bmatrix}$$ which has exactly one solution.

37. False. Counterexample:

$x_1 + x_2 = 1$

$x_1 - x_2 = 2$

$2x_1 + 2x_2 = 3$

39. $\begin{bmatrix} 1 & 2 & | & -1 \\ 0 & 1 & | & 3 \end{bmatrix} \sim \begin{bmatrix} 1 & 0 & | & -7 \\ 0 & 1 & | & 3 \end{bmatrix}$

$(-2)R_2 + R_1 \to R_1$

41. $\begin{bmatrix} 1 & 1 & 1 & | & 16 \\ 2 & 3 & 4 & | & 25 \end{bmatrix} \sim \begin{bmatrix} 1 & 1 & 1 & | & 16 \\ 0 & 1 & 2 & | & -7 \end{bmatrix} \sim \begin{bmatrix} 1 & 0 & -1 & | & 23 \\ 0 & 1 & 2 & | & -7 \end{bmatrix}$

$(-2)R_1 + R_2 \to R_2 \quad (-1)R_2 + R_1 \to R_1$

43. $\begin{bmatrix} 1 & 0 & -3 & | & 1 \\ 0 & 1 & 2 & | & 0 \\ 0 & 0 & 3 & | & -6 \end{bmatrix} \sim \begin{bmatrix} 1 & 0 & -3 & | & 1 \\ 0 & 1 & 2 & | & 0 \\ 0 & 0 & 1 & | & -2 \end{bmatrix} \sim \begin{bmatrix} 1 & 0 & 0 & | & -5 \\ 0 & 1 & 0 & | & 4 \\ 0 & 0 & 1 & | & -2 \end{bmatrix}$

$\dfrac{1}{3}R_3 \to R_3 \qquad\qquad 3R_3 + R_1 \to R_1$

$(-2)R_3 + R_2 \to R_2$

45. $\begin{bmatrix} 1 & 2 & -2 & | & -1 \\ 0 & 3 & -6 & | & 1 \\ 0 & -1 & 2 & | & -\frac{1}{3} \end{bmatrix} \sim \begin{bmatrix} 1 & 2 & -2 & | & -1 \\ 0 & 1 & -2 & | & \frac{1}{3} \\ 0 & -1 & 2 & | & -\frac{1}{3} \end{bmatrix} \sim \begin{bmatrix} 1 & 2 & -2 & | & -1 \\ 0 & 1 & -2 & | & \frac{1}{3} \\ 0 & 0 & 0 & | & 0 \end{bmatrix} \sim \begin{bmatrix} 1 & 0 & 2 & | & -\frac{5}{3} \\ 0 & 1 & -2 & | & \frac{1}{3} \\ 0 & 0 & 0 & | & 0 \end{bmatrix}$

$\dfrac{1}{3}R_2 \to R_2 \qquad R_2 + R_3 \to R_3 \qquad (-2)R_2 + R_1 \to R_1$

47. The corresponding augmented matrix is:

$\begin{bmatrix} 2 & 4 & -10 & | & -2 \\ 3 & 9 & -21 & | & 0 \\ 1 & 5 & -12 & | & 1 \end{bmatrix} \sim \begin{bmatrix} 1 & 2 & -5 & | & -1 \\ 3 & 9 & -21 & | & 0 \\ 1 & 5 & -12 & | & 1 \end{bmatrix} \sim \begin{bmatrix} 1 & 2 & -5 & | & -1 \\ 0 & 3 & -6 & | & 3 \\ 0 & 3 & -7 & | & 2 \end{bmatrix}$

$(1/2)R_1 \to R_1 \qquad\quad (-3)R_1 + R_2 \to R_2 \qquad (1/3R_2 \to R_2$

$(-1)R_1 + R_3 + \to R_3$

$\sim \begin{bmatrix} 1 & 2 & -5 & | & -1 \\ 0 & 1 & -2 & | & 1 \\ 0 & 3 & -7 & | & 2 \end{bmatrix} \sim \begin{bmatrix} 1 & 0 & -1 & | & -3 \\ 0 & 1 & -2 & | & 1 \\ 0 & 0 & -1 & | & -1 \end{bmatrix} \sim \begin{bmatrix} 1 & 0 & -1 & | & -3 \\ 0 & 1 & -2 & | & 1 \\ 0 & 0 & 1 & | & 1 \end{bmatrix} \sim \begin{bmatrix} 1 & 0 & 0 & | & -2 \\ 0 & 1 & 0 & | & 3 \\ 0 & 0 & 1 & | & 1 \end{bmatrix}$

$(-3)R_2 + R_3 \to R_3 \qquad (-1)R_3 \to R_3 \qquad 2R_3 + R_2 \to R_2$

$(-2)R_2 + R_1 \to R_1 \qquad\qquad\qquad\qquad R_3 + R_1 \to R_1$

Thus, $x_1 = -2; \ x_2 = 3; \ x_3 = 1.$

49. The corresponding augmented matrix is:

$\begin{bmatrix} 3 & 8 & -1 & | & -18 \\ 2 & 1 & 5 & | & 8 \\ 2 & 4 & 2 & | & -4 \end{bmatrix} \sim \begin{bmatrix} 2 & 4 & 2 & | & -4 \\ 2 & 1 & 5 & | & 8 \\ 3 & 8 & -1 & | & -18 \end{bmatrix} \sim \begin{bmatrix} 1 & 2 & 1 & | & -2 \\ 2 & 1 & 5 & | & 8 \\ 3 & 8 & -1 & | & -18 \end{bmatrix}$

$R_1 \leftrightarrow R_3 \qquad\qquad \dfrac{1}{2}R_1 \to R_1 \qquad (-2)R_1 + R_2 \to R_2$

$(-3)R_1 + R_3 \to R_3$

$\sim \begin{bmatrix} 1 & 2 & 1 & | & -2 \\ 0 & -3 & 3 & | & 12 \\ 0 & 2 & -4 & | & -12 \end{bmatrix} \sim \begin{bmatrix} 1 & 2 & 1 & | & -2 \\ 0 & 1 & -1 & | & -4 \\ 0 & 2 & -4 & | & -12 \end{bmatrix} \sim \begin{bmatrix} 1 & 0 & 3 & | & 6 \\ 0 & 1 & -1 & | & -4 \\ 0 & 0 & -2 & | & -4 \end{bmatrix}$

$$\left(-\frac{1}{3}\right)R_2 \to R_2 \qquad (-2)R_2 + R_3 \to R_3 \qquad \left(-\frac{1}{2}\right)R_3 \to R_3$$
$$(-2)R_2 + R_1 \to R_1$$

$$\sim \begin{bmatrix} 1 & 0 & 3 & | & 6 \\ 0 & 1 & -1 & | & -4 \\ 0 & 0 & 1 & | & 2 \end{bmatrix} \sim \begin{bmatrix} 1 & 0 & 0 & | & 0 \\ 0 & 1 & 0 & | & -2 \\ 0 & 0 & 1 & | & 2 \end{bmatrix} \qquad \text{Thus} \quad x_1 = 0, \ x_2 = -2, \ x_3 = 2.$$
$$(-3)R_3 + R_1 \to R_1$$
$$R_3 + R_2 \to R_2$$

51. $\begin{bmatrix} 2 & -1 & -3 & | & 8 \\ 1 & -2 & 0 & | & 7 \end{bmatrix} \sim \begin{bmatrix} 1 & -2 & 0 & | & 7 \\ 2 & -1 & -3 & | & 8 \end{bmatrix} \sim \begin{bmatrix} 1 & -2 & 0 & | & 7 \\ 0 & 3 & -3 & | & -6 \end{bmatrix}$

$\qquad R_1 \leftrightarrow R_2 \qquad\quad (-2)R_1 + R_2 \to R_2 \qquad \frac{1}{3}R_2 \ \to \ R_2$

$\sim \begin{bmatrix} 1 & -2 & 0 & | & 7 \\ 0 & 1 & -1 & | & -2 \end{bmatrix} \sim \begin{bmatrix} 1 & 0 & -2 & | & 3 \\ 0 & 1 & -1 & | & -2 \end{bmatrix}$
$\qquad\qquad 2R_2 + R_1 \to R_1$

Thus, $\begin{aligned} x_1 \quad - 2x_3 &= 3 \quad (1) \\ x_2 - \ x_3 &= -2 \quad (2) \end{aligned}$

Let $x_3 = t$, where t is any real number. Then: $x_1 = 2t + 3, \ x_2 = t - 2, \ x_3 = t$

53. $\begin{bmatrix} 2 & -1 & | & 0 \\ 3 & 2 & | & 7 \\ 1 & -1 & | & -1 \end{bmatrix} \sim \begin{bmatrix} 1 & -1 & | & -1 \\ 3 & 2 & | & 7 \\ 2 & -1 & | & 0 \end{bmatrix} \sim \begin{bmatrix} 1 & -1 & | & -1 \\ 0 & 5 & | & 10 \\ 0 & 1 & | & 2 \end{bmatrix} \sim \begin{bmatrix} 1 & -1 & | & -1 \\ 0 & 1 & | & 2 \\ 0 & 5 & | & 10 \end{bmatrix}$

$\quad R_1 \leftrightarrow R_3 \qquad (-3)R_1 + R_2 \to R_2 \quad R_2 \leftrightarrow R_3 \qquad R_2 + R_1 \to R_1$
$\qquad\qquad\qquad (-2)R_1 + R_3 \to R_3 \qquad\qquad\qquad (-5)R_2 + R_3 \to R_3$

$\sim \begin{bmatrix} 1 & 0 & | & 1 \\ 0 & 1 & | & 2 \\ 0 & 0 & | & 0 \end{bmatrix} \qquad \text{Thus} \quad x_1 = 1, \ x_2 = 2.$

55. $\begin{bmatrix} 3 & -4 & -1 & | & 1 \\ 2 & -3 & 1 & | & 1 \\ 1 & -2 & 3 & | & 2 \end{bmatrix} \sim \begin{bmatrix} 1 & -2 & 3 & | & 2 \\ 2 & -3 & 1 & | & 1 \\ 3 & -4 & -1 & | & 1 \end{bmatrix} \sim \begin{bmatrix} 1 & -2 & 3 & | & 2 \\ 0 & 1 & -5 & | & -3 \\ 0 & 2 & -10 & | & -5 \end{bmatrix} \sim \begin{bmatrix} 1 & -2 & 3 & | & 2 \\ 0 & 1 & -5 & | & -3 \\ 0 & 0 & 0 & | & 1 \end{bmatrix}$

$\quad R_1 \leftrightarrow R_3 \qquad\qquad (-2)R_1 + R_2 \to R_2 \qquad (-2)R_2 + R_3 \to R_3$
$\qquad\qquad\qquad\qquad (-3)R_1 + R_3 \to R_3$

From the last row, we conclude that there is no solution; the system is inconsistent.

57. $\begin{bmatrix} 3 & -2 & 1 & | & -7 \\ 2 & 1 & -4 & | & 0 \\ 1 & 1 & -3 & | & 1 \end{bmatrix} \sim \begin{bmatrix} 1 & 1 & -3 & | & 1 \\ 2 & 1 & -4 & | & 0 \\ 3 & -2 & 1 & | & -7 \end{bmatrix} \sim \begin{bmatrix} 1 & 1 & -3 & | & 1 \\ 0 & -1 & 2 & | & -2 \\ 0 & -5 & 10 & | & -10 \end{bmatrix} \sim \begin{bmatrix} 1 & 1 & -3 & | & 1 \\ 0 & 1 & -2 & | & 2 \\ 0 & -5 & 10 & | & -10 \end{bmatrix}$

$\quad R_1 \leftrightarrow R_3 \qquad\qquad (-2)R_1 + R_2 \to R_2 \qquad (-1)R_2 \to R_2 \qquad\qquad (-1)R_2 + R_1 \to R_1$
$\qquad\qquad\qquad\qquad (-3)R_1 + R_3 \to R_3 \qquad\qquad\qquad\qquad\qquad 5R_2 + R_3 \to R_3$

$$\sim \begin{bmatrix} 1 & 0 & -1 & | & -1 \\ 0 & 1 & -2 & | & 2 \\ 0 & 0 & 0 & | & 0 \end{bmatrix}$$

From this matrix, $x_1 - x_3 = -1$ and $x_2 - 2x_3 = 2$. Let $x_3 = t$ be any real number, then $x_1 = t - 1$, $x_2 = 2t + 2$, and $x_3 = t$.

59. $\begin{bmatrix} 2 & 4 & -2 & | & 2 \\ -3 & -6 & 3 & | & -3 \end{bmatrix} \sim \begin{bmatrix} 1 & 2 & -1 & | & 1 \\ -3 & -6 & 3 & | & -3 \end{bmatrix} \sim \begin{bmatrix} 1 & 2 & -1 & | & 1 \\ 0 & 0 & 0 & | & 0 \end{bmatrix}$

$\quad\quad \frac{1}{2}R_1 \rightarrow R_1 \quad\quad\quad 3R_1 + R_2 \rightarrow R_2$

From this matrix, $x_1 + 2x_2 - x_3 = 1$. Let $x_2 = s$ and $x_3 = t$. Then
$x_1 = -2s + t + 1$, $x_2 = s$, and $x_3 = t$, s and t any real numbers.

61. $\begin{bmatrix} 4 & -1 & 2 & | & 3 \\ -4 & 1 & -3 & | & -10 \\ 8 & -2 & 9 & | & -1 \end{bmatrix} \begin{array}{c} \frac{1}{4}R_1 \rightarrow R_1 \end{array} \sim \begin{bmatrix} 1 & -\frac{1}{4} & \frac{1}{2} & | & \frac{3}{4} \\ -4 & 1 & -3 & | & -10 \\ 8 & -2 & 9 & | & -1 \end{bmatrix} \begin{array}{c} 4R_1 + R_2 \rightarrow R_2 \\ (-8)R_1 + R_3 \rightarrow R_3 \end{array}$

$\sim \begin{bmatrix} 1 & -\frac{1}{4} & \frac{1}{2} & | & \frac{3}{4} \\ 0 & 0 & -1 & | & -7 \\ 0 & 0 & 5 & | & -7 \end{bmatrix} (-1)R_2 \rightarrow R_2 \sim \begin{bmatrix} 1 & -\frac{1}{4} & \frac{1}{2} & | & \frac{3}{4} \\ 0 & 0 & 1 & | & 7 \\ 0 & 0 & 5 & | & -7 \end{bmatrix} (-5)R_2 + R_3 \rightarrow R_3$

$\sim \begin{bmatrix} 1 & -\frac{1}{4} & \frac{1}{2} & | & \frac{3}{4} \\ 0 & 0 & 1 & | & 7 \\ 0 & 0 & 0 & | & -42 \end{bmatrix}$ No solution.

63. (A) The system is dependent with two parameters and an infinite number of solutions.

(B) The system is dependent with one parameter and an infinite number of solutions.

(C) The system is independent with a unique solution.

(D) Impossible

65. $\begin{bmatrix} 1 & 2 & -4 & -1 & | & 7 \\ 2 & 5 & -9 & -4 & | & 16 \\ 1 & 5 & -7 & -7 & | & 13 \end{bmatrix} \sim \begin{bmatrix} 1 & 2 & -4 & -1 & | & 7 \\ 0 & 1 & -1 & -2 & | & 2 \\ 0 & 3 & -3 & -6 & | & 6 \end{bmatrix} \sim \begin{bmatrix} 1 & 0 & -2 & 3 & | & 3 \\ 0 & 1 & -1 & -2 & | & 2 \\ 0 & 0 & 0 & 0 & | & 0 \end{bmatrix}$

$(-2)R_1 + R_2 \rightarrow R_2 \quad (-2)R_2 + R_1 \rightarrow R_1$
$(-1)R_1 + R_3 \rightarrow R_3 \quad (-3)R_2 + R_3 \rightarrow R_3$

Thus, $x_1 - 2x_3 + 3x_4 = 3$ and $x_2 - x_3 - 2x_4 = 2$. Let $x_3 = s$ and $x_4 = t$. Then $x_1 = 2s - 3t + 3$, $x_2 = s + 2t + 2$, $x_3 = s$, $x_4 = t$, where s, t are any real numbers.

67. $\begin{bmatrix} 1 & -1 & 3 & -2 & | & 1 \\ -2 & 4 & -3 & 1 & | & 0.5 \\ 3 & -1 & 10 & -4 & | & 2.9 \\ 4 & -3 & 8 & -2 & | & 0.6 \end{bmatrix} \begin{array}{c} 2R_1 + R_2 \rightarrow R_2 \\ (-3)R_1 + R_3 \rightarrow R_3 \\ (-4)R_1 + R_4 \rightarrow R_4 \end{array} \sim \begin{bmatrix} 1 & -1 & 3 & -2 & | & 1 \\ 0 & 2 & 3 & -3 & | & 2.5 \\ 0 & 2 & 1 & 2 & | & -0.1 \\ 0 & 1 & -4 & 6 & | & -3.4 \end{bmatrix} R_2 \leftrightarrow R_4$

$$\sim \begin{bmatrix} 1 & -1 & 3 & -2 & | & 1 \\ 0 & 1 & -4 & 6 & | & -3.4 \\ 0 & 2 & 1 & 2 & | & -0.1 \\ 0 & 2 & 3 & -3 & | & 2.5 \end{bmatrix} \quad \begin{array}{l} R_2 + R_1 \to R_1 \\ (-2)R_2 + R_3 \to R_3 \\ (-2)R_2 + R_4 \to R_4 \end{array}$$

$$\sim \begin{bmatrix} 1 & 0 & -1 & 4 & | & -2.4 \\ 0 & 1 & -4 & 6 & | & -3.4 \\ 0 & 0 & 9 & -10 & | & 6.7 \\ 0 & 0 & 11 & -15 & | & 9.3 \end{bmatrix} \quad (-1)R_4 + R_3 \to R_3 \text{ (to simplify arithmetic)}$$

$$\sim \begin{bmatrix} 1 & 0 & -1 & 4 & | & -2.4 \\ 0 & 1 & -4 & 6 & | & -3.4 \\ 0 & 0 & -2 & 5 & | & -2.6 \\ 0 & 0 & 11 & -15 & | & 9.3 \end{bmatrix} \quad \left(-\dfrac{1}{2}\right) R_3 \to R_3$$

$$\sim \begin{bmatrix} 1 & 0 & -1 & 4 & | & -2.4 \\ 0 & 1 & -4 & 6 & | & -3.4 \\ 0 & 0 & 1 & -2.5 & | & 1.3 \\ 0 & 0 & 11 & -15 & | & 9.3 \end{bmatrix} \quad \begin{array}{l} R_3 + R_1 \to R_1 \\ 4R_3 + R_2 \to R_2 \\ (-11)R_3 + R_4 \to R_4 \end{array}$$

$$\sim \begin{bmatrix} 1 & 0 & 0 & 1.5 & | & -1.1 \\ 0 & 1 & 0 & -4 & | & 1.8 \\ 0 & 0 & 1 & -2.5 & | & 1.3 \\ 0 & 0 & 0 & 12.5 & | & -5 \end{bmatrix} \quad \dfrac{1}{12.5} \, R_4 \to R_4$$

$$\sim \begin{bmatrix} 1 & 0 & 0 & 1.5 & | & -1.1 \\ 0 & 1 & 0 & -4 & | & 1.8 \\ 0 & 0 & 1 & -2.5 & | & 1.3 \\ 0 & 0 & 0 & 1 & | & -0.4 \end{bmatrix} \quad \begin{array}{l} (-1.5)R_4 + R_1 \to R_1 \\ 4R_4 + R_2 \to R_2 \\ 2.5R_4 + R_3 \to R_3 \end{array}$$

$$\sim \begin{bmatrix} 1 & 0 & 0 & 0 & | & -0.5 \\ 0 & 1 & 0 & 0 & | & 0.2 \\ 0 & 0 & 1 & 0 & | & 0.3 \\ 0 & 0 & 0 & 1 & | & -0.4 \end{bmatrix} \quad \text{Solution: } x_1 = -0.5, \ x_2 = 0.2, \ x_3 = 0.3, \ x_2 = -0.4$$

69.
$$\begin{bmatrix} 1 & -2 & 1 & 1 & 2 & | & 2 \\ -2 & 4 & 2 & 2 & -2 & | & 0 \\ 3 & -6 & 1 & 1 & 5 & | & 4 \\ -1 & 2 & 3 & 1 & 1 & | & 3 \end{bmatrix} \quad \begin{array}{l} 2R_1 + R_2 \to R_2 \\ (-3)R_1 + R_3 \to R_3 \\ R_1 + R_4 \to R_4 \end{array}$$

$$\sim \begin{bmatrix} 1 & -2 & 1 & 1 & 2 & | & 2 \\ 0 & 0 & 4 & 4 & 2 & | & 4 \\ 0 & 0 & -2 & -2 & -1 & | & -2 \\ 0 & 0 & 4 & 2 & 3 & | & 5 \end{bmatrix} \quad \dfrac{1}{4} R_2 \to R_2$$

$$\sim \begin{bmatrix} 1 & -2 & 1 & 1 & 2 & | & 2 \\ 0 & 0 & 1 & 1 & \frac{1}{2} & | & 1 \\ 0 & 0 & -2 & -2 & -1 & | & -2 \\ 0 & 0 & 4 & 2 & 3 & | & 5 \end{bmatrix} \quad \begin{array}{l} (-1)R_2 + R_1 \to R_1 \\ 2R_2 + R_3 \to R_3 \\ (-4)R_2 + R_4 \to R_4 \end{array}$$

$$\sim \begin{bmatrix} 1 & -2 & 0 & 0 & \frac{3}{2} & 1 \\ 0 & 0 & 1 & 1 & \frac{1}{2} & 1 \\ 0 & 0 & 0 & 0 & 0 & 0 \\ 0 & 0 & 0 & -2 & 1 & 1 \end{bmatrix} \begin{matrix} \\ \\ R_3 \leftrightarrow R_4 \\ \\ \end{matrix} \sim \begin{bmatrix} 1 & -2 & 0 & 0 & \frac{3}{2} & 1 \\ 0 & 0 & 1 & 1 & \frac{1}{2} & 1 \\ 0 & 0 & 0 & -2 & 1 & 1 \\ 0 & 0 & 0 & 0 & 0 & 0 \end{bmatrix} \left(-\frac{1}{2}\right)R_3 \to R_3$$

$$\sim \begin{bmatrix} 1 & -2 & 0 & 0 & \frac{3}{2} & 1 \\ 0 & 0 & 1 & 1 & \frac{1}{2} & 1 \\ 0 & 0 & 0 & 1 & -\frac{1}{2} & -\frac{1}{2} \\ 0 & 0 & 0 & 0 & 0 & 0 \end{bmatrix} \begin{matrix} \\ \\ (-1)R_3 + R_2 \to R_2 \\ \\ \end{matrix} \sim \begin{bmatrix} 1 & -2 & 0 & 0 & \frac{3}{2} & 1 \\ 0 & 0 & 1 & 0 & 1 & \frac{3}{2} \\ 0 & 0 & 0 & 1 & -\frac{1}{2} & -\frac{1}{2} \\ 0 & 0 & 0 & 0 & 0 & 0 \end{bmatrix}$$

The system of equations is:

$$x_1 - 2x_2 \qquad\qquad\quad + \frac{3}{2}x_5 = 1$$

$$x_3 \qquad\quad + \; x_5 = \frac{3}{2}$$

$$x_4 \; - \frac{1}{2}x_5 = -\frac{1}{2}$$

or $\; x_1 = 2x_2 - \dfrac{3}{2}x_5 + 1, \quad x_3 = -x_5 + \dfrac{3}{2}, \quad x_4 = \dfrac{1}{2}x_5 - \dfrac{1}{2}$

Let $x_2 = s$ and $x_5 = t$. Then $x_1 = 2s - \dfrac{3}{2}t + 1, \; x_2 = s, x_3 = -t + \dfrac{3}{2}, \quad x_4 = \dfrac{1}{2}t - \dfrac{1}{2}, x_5 = t$ for any real

numbers s and t.

71. The coordinates of the points must satisfy the quadratic equation: $\; y = ax^2 + bx + c$.

$(-2, 9)$: $\quad 9 = a(-2)^2 + b(-2) + c \quad$ or $\qquad 4a - 2b + c = 9$

$(1, -9)$: $\; -9 = a(1)^2 + b(1) + c \qquad$ or $\qquad a + b + c = -9$

$\;(4, 9)$: $\quad 9 = a(4)^2 + b(4) + c \qquad$ or $\qquad 16a + 4b + c = 9$

The system of equations is:

$$4a - 2b + c = 9$$
$$a + \; b + c = -9$$
$$16a + 4b + c = 9$$

The augmented matrix is:

$$\begin{bmatrix} 4 & -2 & 1 & 9 \\ 1 & 1 & 1 & -9 \\ 16 & 4 & 1 & 9 \end{bmatrix} \sim \begin{bmatrix} 1 & 1 & 1 & -9 \\ 4 & -2 & 1 & 9 \\ 16 & 4 & 1 & 9 \end{bmatrix} \sim \begin{bmatrix} 1 & 1 & 1 & -9 \\ 0 & -6 & -3 & 45 \\ 0 & -12 & -15 & 153 \end{bmatrix}$$

$$\begin{matrix} R_1 \to R_2 \qquad\qquad\quad -4R_1 + R_2 \to R_2 \quad \left(-\frac{1}{6}\right)R_2 \to R_2 \\ -16R_1 + R_3 \to R_3 \end{matrix}$$

$$\sim \begin{bmatrix} 1 & 1 & 1 & -9 \\ 0 & 1 & \frac{1}{2} & -\frac{15}{2} \\ 0 & -12 & -15 & 153 \end{bmatrix} \sim \begin{bmatrix} 1 & 1 & 1 & -9 \\ 0 & 1 & \frac{1}{2} & -\frac{15}{2} \\ 0 & 0 & -9 & 63 \end{bmatrix} \sim \begin{bmatrix} 1 & 1 & 1 & -9 \\ 0 & 1 & \frac{1}{2} & -\frac{15}{2} \\ 0 & 0 & 1 & -7 \end{bmatrix}$$

$$\begin{matrix} 12R_2 + R_3 \to R_3 \qquad \left(-\frac{1}{9}\right)R_3 \to R_3 \qquad \left(-\frac{1}{2}\right)R_3 + R_2 \to R_2 \\ (-1)R_3 + R_1 \to R_1 \end{matrix}$$

$$\sim \begin{bmatrix} 1 & 1 & 0 & | & -2 \\ 0 & 1 & 0 & | & -4 \\ 0 & 0 & 1 & | & -7 \end{bmatrix} \sim \begin{bmatrix} 1 & 0 & 0 & | & 2 \\ 0 & 1 & 0 & | & -4 \\ 0 & 0 & 1 & | & -7 \end{bmatrix} \quad \text{Thus} \quad a = 2, \quad b = -4, \quad c = -7.$$

$$(-1)R_2 + R_1 \to R_1$$

73. Let x_1 = Number of one-person boats

x_2 = Number of two-person boats

x_3 = Number of four-person boats.

(A) The mathematical model is:

$$0.5x_1 + x_2 + 1.5x_3 = 380$$
$$0.6x_1 + 0.9x_2 + 1.2x_3 = 330$$
$$0.2x_1 + 0.3x_2 + 0.5x_3 = 120$$

$$\begin{bmatrix} 0.5 & 1 & 1.5 & | & 380 \\ 0.6 & 0.9 & 1.2 & | & 330 \\ 0.2 & 0.3 & 0.5 & | & 120 \end{bmatrix} \sim \begin{bmatrix} 1 & 2 & 3 & | & 760 \\ 0.6 & 0.9 & 1.2 & | & 330 \\ 0.2 & 0.3 & 0.5 & | & 120 \end{bmatrix} \sim \begin{bmatrix} 1 & 2 & 3 & | & 760 \\ 0 & -0.3 & -0.6 & | & -126 \\ 0 & -0.1 & -0.1 & | & -32 \end{bmatrix}$$

$$2R_1 \to R_1 \qquad\qquad (-0.6)R_1 + R_2 \to R_2 \qquad \left(-\frac{1}{0.3}\right)R_2 \to R_2$$

$$(-0.2)R_1 + R_3 \to R_3$$

$$\sim \begin{bmatrix} 1 & 2 & 3 & | & 760 \\ 0 & 1 & 2 & | & 420 \\ 0 & -0.1 & -0.1 & | & -32 \end{bmatrix} \sim \begin{bmatrix} 1 & 0 & -1 & | & -80 \\ 0 & 1 & 2 & | & 420 \\ 0 & 0 & 0.1 & | & 10 \end{bmatrix} \sim \begin{bmatrix} 1 & 0 & -1 & | & -80 \\ 0 & 1 & 2 & | & 420 \\ 0 & 0 & 1 & | & 100 \end{bmatrix}$$

$$(0.1)R_2 + R_3 \to R_3 \qquad 10R_3 \to R_3 \qquad\qquad R_3 + R_1 \to R_1$$

$$(-2)R_2 + R_1 \to R_1 \qquad\qquad\qquad (-2)R_3 + R_2 \to R_2$$

$$\sim \begin{bmatrix} 1 & 0 & 0 & | & 20 \\ 0 & 1 & 0 & | & 220 \\ 0 & 0 & 1 & | & 100 \end{bmatrix}$$

Thus, $x_1 = 20$, $x_2 = 220$, and $x_3 = 100$, or 20 one-person boats, 220 two-person boats, and 100 four-person boats.

(B) The mathematical model is:

$$0.5x_1 + x_2 + 1.5x_3 = 380$$
$$0.6x_1 + 0.9x_2 + 1.2x_3 = 330$$

$$\begin{bmatrix} 0.5 & 1 & 1.5 & | & 380 \\ 0.6 & 0.9 & 1.2 & | & 330 \end{bmatrix} \sim \begin{bmatrix} 1 & 2 & 3 & | & 760 \\ 0.6 & 0.9 & 1.2 & | & 330 \end{bmatrix} \sim \begin{bmatrix} 1 & 2 & 3 & | & 760 \\ 0 & -0.3 & -0.6 & | & -126 \end{bmatrix}$$

$$2R_1 \to R_1 \qquad\qquad (-0.6)R_1 + R_2 \to R_2 \qquad\qquad \left(-\frac{1}{0.3}\right)R_2 \to R_2$$

$$\sim \begin{bmatrix} 1 & 2 & 3 & | & 760 \\ 0 & 1 & 2 & | & 420 \end{bmatrix} \sim \begin{bmatrix} 1 & 0 & -1 & | & -80 \\ 0 & 1 & 2 & | & 420 \end{bmatrix}$$

$$(-2)R_2 + R_1 \to R_1$$

Thus, $x_1 \quad - \quad x_3 = -80 \qquad (1)$

$\qquad\qquad x_2 + 2x_3 = 420 \qquad (2)$

Let $x_3 = t$ (t any real number). Then, $x_2 = 420 - 2t$ [from (2)] and $x_1 = t - 80$ [from (1)].

In order to keep x_1 and x_2 positive, $t \leq 210$ and $t \geq 80$.

Thus, $x_1 = t - 80$ (one-person boats)

$\qquad x_2 = 420 - 2t$ (two-person boats)

$\qquad x_3 = t \qquad$ (four-person boats)

where $80 \leq t \leq 210$ and t is an integer.

(C)   The mathematical model is:

$\qquad 0.5x_1 + \quad x_2 = 380$

$\qquad 0.6x_1 + 0.9x_2 = 330$

$\qquad 0.2x_1 + 0.3x_2 = 120$

$$\begin{bmatrix} 0.5 & 1 & \vrule & 380 \\ 0.6 & 0.9 & \vrule & 330 \\ 0.2 & 0.3 & \vrule & 120 \end{bmatrix} \sim \begin{bmatrix} 1 & 2 & \vrule & 760 \\ 0.6 & 0.9 & \vrule & 330 \\ 0.2 & 0.3 & \vrule & 120 \end{bmatrix} \sim \begin{bmatrix} 1 & 1 & \vrule & 760 \\ 0 & -0.3 & \vrule & -126 \\ 0 & -0.1 & \vrule & -32 \end{bmatrix} \sim \begin{bmatrix} 1 & 2 & \vrule & 760 \\ 0 & 1 & \vrule & 420 \\ 0 & -0.1 & \vrule & -32 \end{bmatrix}$$

$\quad 2R_1 \to R_1 \qquad (-0.6R_1) + R_2 \to R_2 \quad \left(-\dfrac{1}{0.3}\right)R_2 \to R_2 \quad 0.1R_2 + R_3 \to R_3$

$\qquad\qquad\qquad (-0.2R_1) + R_3 \to R_3$

$\sim \begin{bmatrix} 1 & 2 & \vrule & 760 \\ 0 & 1 & \vrule & 420 \\ 0 & 0 & \vrule & 10 \end{bmatrix}$ From this matrix, we conclude that there is no solution; there is no production schedule that will use all the labor-hours in all departments.

75.   Let x_1 = number of 8,000 gallon tank cars

$\qquad x_2$ = number of 16,000 gallon tank cars

$\qquad x_3$ = number of 24,000 gallon tank cars

Then, the mathematical model is:

$\qquad\quad x_1 + \qquad x_2 + \qquad x_3 = 24$

$\quad 8{,}000x_1 + 16{,}000x_2 + 24{,}000x_3 = 520{,}000$

Dividing the second equation by 8,000, we get the equivalent system:

$\quad x_1 + \ x_2 + \ x_3 = 24$

$\quad x_1 + 2x_2 + 3x_3 = 65$

The augmented matrix corresponding to this system is

$$\begin{bmatrix} 1 & 1 & 1 & \vrule & 24 \\ 1 & 2 & 3 & \vrule & 65 \end{bmatrix} \sim \begin{bmatrix} 1 & 1 & 1 & \vrule & 24 \\ 0 & 1 & 2 & \vrule & 41 \end{bmatrix} \sim \begin{bmatrix} 1 & 0 & -1 & \vrule & -17 \\ 0 & 1 & 2 & \vrule & 41 \end{bmatrix}$$

$(-1)R_1 + R_2 \to R_2 \quad (-1)R_2 + R_1 \to R_1$

Thus, $x_1 \quad - \quad x_3 = -17$

$\qquad\qquad x_2 + 2x_3 = 41$

Let $x_3 = t$. Then $x_1 = t - 17$ and $x_2 = 41 - 2t$. Thus, $(t - 17)$ 8,000-gallon tank cars, $(41 - 2t)$ 16,000-gallon tank cars and (t) 24,000-gallon tank cars should be purchased. Also, since t, $41 - 2t$ and $t - 17$ must each be non-negative integers, it follows that $t = 17, 18, 19,$ or 20.

77. The cost $C(t)$ of leasing (t) 24,000 gallon tank cars, $(41 - 2t)$ 16,000-gallon tank cars, and $(t - 17)$ 8,000-gallon tank cars is:

$C(t) = 450(t - 17) + 650(41 - 2t) + 1,150t$ dollars per month.

$C(17) = 650(41 - 34) + 1,150(17) = 650(7) + 1,150(17) = \$24,100$
$C(18) = 450(1) + 650(5) + 1,150(18) = \$24,400$
$C(19) = 450(2) + 650(3) + 1,150(19) = \$24,700$
$C(20) = 450(3) + 650(1) + 1,150(20) = \$25,000$

The minimum monthly cost is \$24,100 when 17 24,000-gallon tank cars and 7 16,000-gallon tank cars are leased.

79. Let x_1 = federal income tax

   x_2 = state income tax

   x_3 = local income tax

Then, the mathematical model is:

$x_1 = 0.50[7,650,000 - (x_2 + x_3)]$

$x_2 = 0.20[7,650,000 - (x_1 + x_3)]$

$x_3 = 0.10[7,650,000 - (x_1 + x_2)]$

and

$\begin{aligned} x_1 + 0.5x_2 + 0.5x_3 &= 3,825,000 \\ 0.2x_1 + x_2 + 0.2x_3 &= 1,530,000 \\ 0.1x_1 + 0.1x_2 + x_3 &= 765,000 \end{aligned}$

The corresponding augmented matrix is:

$$\left[\begin{array}{ccc|c} 1 & 0.5 & 0.5 & 3,825,000 \\ 0.2 & 1 & 0.2 & 1,530,000 \\ 0.1 & 0.1 & 1 & 765,000 \end{array}\right] \quad \begin{array}{l} (-0.2)R_1 + R_2 \to R_2 \\ (-0.1)R_1 + R_3 \to R_3 \end{array}$$

$$\sim \left[\begin{array}{ccc|c} 1 & 0.5 & 0.5 & 3,825,000 \\ 0 & 0.9 & 0.1 & 765,000 \\ 0 & 0.05 & 0.95 & 382,500 \end{array}\right] \quad 20R_3 \to R_3 \text{ (simplify arithmetic)}$$

$$\sim \left[\begin{array}{ccc|c} 1 & 0.5 & 0.5 & 3,825,000 \\ 0 & 0.9 & 0.1 & 765,000 \\ 0 & 1 & 19 & 7,650,000 \end{array}\right] \quad R_2 \leftrightarrow R_3$$

$$\sim \left[\begin{array}{ccc|c} 1 & 0.5 & 0.5 & 3,825,000 \\ 0 & 1 & 19 & 7,650,000 \\ 0 & 0.9 & 0.1 & 765,000 \end{array}\right] \quad \begin{array}{l} (-0.5)R_2 + R_1 \to R_1 \\ (-0.9)R_2 + R_3 \to R_3 \end{array}$$

$$\sim \left[\begin{array}{ccc|c} 1 & 0 & -9 & 0 \\ 0 & 1 & 19 & 7,650,000 \\ 0 & 0 & -17 & -6,120,000 \end{array}\right] \quad \left(-\dfrac{1}{17}\right)R_3 \to R_3$$

$$\sim \left[\begin{array}{ccc|c} 1 & 0 & -9 & 0 \\ 0 & 1 & 19 & 7,650,000 \\ 0 & 0 & 1 & 360,000 \end{array}\right] \quad \begin{array}{l} 9R_3 + R_1 \to R_1 \\ (-19)R_3 + R_2 \to R_2 \end{array}$$

$$\sim \begin{bmatrix} 1 & 0 & 0 & | & 3,240,000 \\ 0 & 1 & 0 & | & 810,000 \\ 0 & 0 & 1 & | & 360,000 \end{bmatrix}$$

Thus, $x_1 = \$3,240,000$, $x_2 = \$810,000$, $x_3 = \$360,000$. The total tax liability is

$x_1 + x_2 + x_3 = \$4,410,000$ which is 57.65% of the taxable income $\left(\dfrac{4,410,000}{7,650,000} = 0.5765 \text{ or } 57.65\% \right)$.

81. Let x_1 = Taxable income of company A

x_2 = Taxable income of company B

x_3 = Taxable income of company C

x_4 = Taxable income of company D

The taxable income of each company is given by the system of equations:

$x_1 = 0.71(3.2) + 0.08x_2 + 0.03x_3 + 0.07x_4$

$x_2 = 0.12x_1 + 0.81(2.6) + 0.11x_3 + 0.13x_4$

$x_3 = 0.11x_1 + 0.09x_2 + 0.72(3.8) + 0.08x_4$

$x_4 = 0.06x_1 + 0.02x_2 + 0.14x_3 + 0.72(4.4)$

which is the same as:

$x_1 - 0.08x_2 - 0.03x_3 - 0.07x_4 = 2.272$

$-0.12x_1 + x_2 - 0.11x_3 - 0.13x_4 = 2.106$

$-0.11x_1 - 0.09x_2 + x_3 - 0.08x_4 = 2.736$

$-0.06x_1 - 0.02x_2 - 0.14x_3 + x_4 = 3.168$

The corresponding augmented coefficient matrix is

$$\begin{bmatrix} 1 & -0.08 & -0.03 & -0.07 & | & 2.272 \\ -0.12 & 1 & -0.11 & -0.13 & | & 2.106 \\ -0.11 & -0.09 & 1 & -0.08 & | & 2.736 \\ -0.06 & -0.02 & -0.14 & 1 & | & 3.168 \end{bmatrix}$$

and the reduced form (from a graphing utility) is

$$\begin{bmatrix} 1 & 0 & 0 & 0 & | & 2.927 \\ 0 & 1 & 0 & 0 & | & 3.372 \\ 0 & 0 & 1 & 0 & | & 3.675 \\ 0 & 0 & 0 & 1 & | & 3.926 \end{bmatrix}$$

The taxable incomes are: Company A - \$2,927,000, Company B - \$3,372,000, Company C - \$3,675,000, Company D - \$3,926,000

83. Let x_1 = number of ounces of food A,

x_2 = number of ounces of food B,

x_3 = number of ounces of food C.

(A)   The mathematical model is:

$30x_1 + 10x_2 + 20x_3 = 340$

$10x_1 + 10x_2 + 20x_3 = 180$

$10x_1 + 30x_2 + 20x_3 = 220$

$$\begin{bmatrix} 30 & 10 & 20 & | & 340 \\ 10 & 10 & 20 & | & 180 \\ 10 & 30 & 20 & | & 220 \end{bmatrix} \qquad \sim \begin{bmatrix} 10 & 10 & 20 & | & 180 \\ 30 & 10 & 20 & | & 340 \\ 10 & 30 & 20 & | & 220 \end{bmatrix} \qquad \sim \begin{bmatrix} 1 & 1 & 2 & | & 18 \\ 3 & 1 & 2 & | & 34 \\ 1 & 3 & 2 & | & 22 \end{bmatrix}$$

$$R_1 \leftrightarrow R_2 \qquad\qquad\qquad \frac{1}{10}R_1 \to R_1 \qquad\qquad\qquad (-3)R_1 + R_2 \to R_2$$
$$\qquad\qquad\qquad\qquad\qquad \frac{1}{10}R_2 \to R_2 \qquad\qquad\qquad (-1)R_1 + R_3 \to R_3$$
$$\qquad\qquad\qquad\qquad\qquad \frac{1}{10}R_3 \to R_3$$

$$\sim \begin{bmatrix} 1 & 1 & 2 & | & 18 \\ 0 & -2 & -4 & | & -20 \\ 0 & 2 & 0 & | & 4 \end{bmatrix} \qquad \sim \begin{bmatrix} 1 & 1 & 2 & | & 18 \\ 0 & 1 & 2 & | & 10 \\ 0 & 2 & 0 & | & 4 \end{bmatrix} \qquad \sim \begin{bmatrix} 1 & 0 & 0 & | & 8 \\ 0 & 1 & 2 & | & 10 \\ 0 & 0 & -4 & | & -16 \end{bmatrix}$$

$$-\frac{1}{2}R_2 \to R_2 \qquad\qquad (-1)R_2 + R_1 \to R_1 \qquad\qquad -\frac{1}{4}R_3 \to R_3$$
$$\qquad\qquad\qquad\qquad (-2)R_2 + R_3 \to R_3$$

$$\sim \begin{bmatrix} 1 & 0 & 0 & | & 8 \\ 0 & 1 & 2 & | & 10 \\ 0 & 0 & 1 & | & 4 \end{bmatrix} \sim \begin{bmatrix} 1 & 0 & 0 & | & 8 \\ 0 & 1 & 0 & | & 2 \\ 0 & 0 & 1 & | & 4 \end{bmatrix}$$

$$(-2)R_3 + R_2 \to R_2$$

Thus, $x_1 = 8$, $x_2 = 2$, $x_3 = 4$, or 8 ounces of food A, 2 ounces of food B, and 4 ounces of food C.

(B)   The mathematical model is:
$$30x_1 + 10x_2 = 340$$
$$10x_1 + 10x_2 = 180$$
$$10x_1 + 30x_2 = 220$$

$$\begin{bmatrix} 30 & 10 & | & 340 \\ 10 & 10 & | & 180 \\ 10 & 30 & | & 220 \end{bmatrix} \sim \begin{bmatrix} 3 & 1 & | & 34 \\ 1 & 1 & | & 18 \\ 1 & 3 & | & 22 \end{bmatrix} \sim \begin{bmatrix} 1 & 1 & | & 18 \\ 3 & 1 & | & 34 \\ 1 & 3 & | & 22 \end{bmatrix} \sim \begin{bmatrix} 1 & 1 & | & 18 \\ 0 & -2 & | & -20 \\ 0 & 2 & | & 4 \end{bmatrix}$$

$$\frac{1}{10}R_1 \to R_1 \qquad R_1 \leftrightarrow R_2 \qquad (-3)R_1 + R_2 \to R_2 \qquad \left(-\frac{1}{2}\right)R_2 \to R_2$$
$$\frac{1}{10}R_2 \to R_2 \qquad\qquad\qquad (-1)R_1 + R_3 \to R_3$$
$$\frac{1}{10}R_3 \to R_3$$

$$\sim \begin{bmatrix} 1 & 1 & | & 18 \\ 0 & 1 & | & 10 \\ 0 & 2 & | & 4 \end{bmatrix} \sim \begin{bmatrix} 1 & 1 & | & 18 \\ 0 & 1 & | & 10 \\ 0 & 0 & | & -16 \end{bmatrix}$$

$$(-2)R_2 + R_3 \to R_3$$

From this matrix, we conclude that there is no solution.

(C)   The mathematical model is:

$$30x_1 + 10x_2 + 20x_3 = 340$$
$$10x_1 + 10x_2 + 20x_3 = 180$$

$$\begin{bmatrix} 30 & 10 & 20 & | & 340 \\ 10 & 10 & 20 & | & 180 \end{bmatrix} \sim \begin{bmatrix} 10 & 10 & 20 & | & 180 \\ 30 & 10 & 20 & | & 340 \end{bmatrix} \sim \begin{bmatrix} 1 & 1 & 2 & | & 18 \\ 3 & 1 & 2 & | & 34 \end{bmatrix} \sim \begin{bmatrix} 1 & 1 & 2 & | & 18 \\ 0 & -2 & -4 & | & -20 \end{bmatrix}$$

$$R_1 \leftrightarrow R_2 \qquad \frac{1}{10}R_1 \to R_1 \qquad (-3)R_1 + R_2 \to R_2 \qquad \left(-\frac{1}{2}\right)R_2 \to R_2$$

$$\frac{1}{10}R_2 \to \square R_2$$

$$\sim \begin{bmatrix} 1 & 1 & 2 & | & 18 \\ 0 & 1 & 2 & | & 10 \end{bmatrix} \sim \begin{bmatrix} 1 & 0 & 0 & | & 8 \\ 0 & 1 & 2 & | & 10 \end{bmatrix} \qquad \text{Thus,} \quad x_1 \qquad\quad = 8$$
$$(-1)R_2 + R_1 \to R_1 \qquad\qquad\qquad\qquad x_2 + 2x_3 = 10$$

Let $x_3 = t$ (t any real number). Then, $x_2 = 10 - 2t$, $0 \le t \le 5$, for x_2 to be positive.

The solution is: $x_1 = 8$ ounces of food A; $x_2 = 10 - 2t$ ounces of food B; $x_3 = t$ ounces of food C, $0 \le t \le 5$.

85.   Let x_1 = number of barrels of mix A,

   x_2 = number of barrels of mix B,

   x_3 = number of barrels of mix C,

   x_4 = number of barrels of mix D,

The mathematical model is:

$$30x_1 + 30x_2 + 30x_3 + 60x_4 = 900 \quad (1)$$
$$50x_1 + 75x_2 + 25x_3 + 25x_4 = 750 \quad (2)$$
$$30x_1 + 20x_2 + 20x_3 + 50x_4 = 700 \quad (3)$$

Divide each side of equation (1) by 30, each side of equation (2) by 25, and each side of equation (3) by 10. This yields the system of linear equations:

$$x_1 + x_2 + x_3 + 2x_4 = 30$$
$$2x_1 + 3x_2 + x_3 + x_4 = 30$$
$$3x_1 + 2x_2 + 2x_3 + 5x_4 = 70$$

$$\begin{bmatrix} 1 & 1 & 1 & 2 & | & 30 \\ 2 & 3 & 1 & 1 & | & 30 \\ 3 & 2 & 2 & 5 & | & 70 \end{bmatrix} \sim \begin{bmatrix} 1 & 1 & 1 & 2 & | & 30 \\ 0 & 1 & -1 & -3 & | & -30 \\ 0 & -1 & -1 & -1 & | & -20 \end{bmatrix} \sim \begin{bmatrix} 1 & 0 & 2 & 5 & | & 60 \\ 0 & 1 & -1 & -3 & | & -30 \\ 0 & 0 & -2 & -4 & | & -50 \end{bmatrix}$$

$$(-2)R_1 + R_2 \to R_2 \qquad\qquad R_2 + R_3 \to R_3 \qquad\qquad\qquad \left(-\frac{1}{2}\right)R_3 \to R_3$$
$$(-3)R_1 + R_3 \to R_3 \qquad\qquad (-1)R_2 + R_1 \to R_1$$

$$\sim \begin{bmatrix} 1 & 0 & 2 & 5 & | & 60 \\ 0 & 1 & -1 & -3 & | & -30 \\ 0 & 0 & 1 & 2 & | & 25 \end{bmatrix} \sim \begin{bmatrix} 1 & 0 & 0 & 1 & | & 10 \\ 0 & 1 & 0 & -1 & | & -5 \\ 0 & 0 & 1 & 2 & | & 25 \end{bmatrix}$$

$R_3 + R_2 \rightarrow R_2$

$(-2)R_3 + R_1 \rightarrow R_1$

Thus, $x_1 \quad + \quad x_4 = 10$

$\qquad x_2 \quad - \quad x_4 = -5$

$\qquad\qquad x_3 + 2x_4 = 25$

Let $x_4 = t =$ number of barrels of mix D. Then $x_1 = 10 - t =$ number of barrels of mix A, $x_2 = t - 5 =$ number of barrels of mix B, and $x_3 = 25 - 2t =$ number of barrels of mix C. Since the number of barrels of each mix must be nonnegative, $5 \le t \le 10$. Also, t is an integer.

87. From Problem 85, the cost of the four mixes is

$C(t) = 46(10 - t) + 72(t - 5) + 57(25 - 2t) + 63t$

$\qquad = -25t + 1525, \ 5 \le t \le 10.$

The graph of C is a straight line with negative slope. Therefore, the minimum occurs at $t = 10$. The minimizing solution is: 0 barrels of mix A, 5 barrels of mix B, 5 barrels of mix C, and 10 barrels of mix D.

89. $y = ax^2 + bx + c$

Let 1900 correspond to $x = 0$. Then $P(0) = 75 = a(0)^2 + b(0) + c$.

Thus, $c = 75$.

1950: $a(50)^2 + b(50) + 75 = 150 \quad$ or $\quad 50a + b = \dfrac{3}{2}$

2000: $a(100)^2 + b(100) + 75 = 275 \quad$ or $\quad 100a + b = 2$

Substitute $b = 2 - 100a$ into $50a + b = \dfrac{3}{2}$. This gives

$$50a + (2 - 100a) = \frac{3}{2}$$

$$-50a = -\frac{1}{2}$$

$$a = \frac{1}{100} = 0.01$$

Therefore, $b = 2 - 100\left(\dfrac{1}{100}\right) = 2 - 1 = 1$ and $P = 0.01x^2 + x + 75.$

The year 2050 corresponds to $t = 150$: $P(150) = 0.01(150)^2 + 150 + 75 = 450$. The model estimates a population of 450 million in 2050.

91. $L = ax^2 + bx + c$

Let $1980 - 1985$ correspond to $t = 0$. Then

$L(0) = c = 77.6.$

1985–1990: $L(5) = a(5)^2 + b(5) + 77.6 = 78 \quad$ or $\quad 25a + 5b = 0.4$

1990–1995: $L(10) = a(10)^2 + b(10) + 77.6 = 78.6 \quad$ or $\quad 100a + 10b = 1$

The augmented matrix for the equations is:

$$\begin{pmatrix} 25 & 5 & | & 0.4 \\ 100 & 10 & | & 1 \end{pmatrix} \sim \begin{pmatrix} 1 & \frac{1}{5} & | & 0.016 \\ 100 & 10 & | & 1 \end{pmatrix} \sim \begin{pmatrix} 1 & \frac{1}{5} & | & 0.016 \\ 0 & -10 & | & -0.6 \end{pmatrix}$$

$$\left(\frac{1}{25}\right)R_1 \rightarrow R_1 \quad (-100)R_1 + R_2 \rightarrow R_2 \quad \left(-\frac{1}{10}\right)R_2 \rightarrow R_2$$

$$\begin{pmatrix} 1 & \frac{1}{5} & | & 0.016 \\ 0 & 1 & | & 0.06 \end{pmatrix} \sim \begin{pmatrix} 1 & 0 & | & 0.004 \\ 0 & 1 & | & 0.06 \end{pmatrix}$$

$$\left(-\frac{1}{5}\right)R_2 + R_1 \rightarrow R_1$$

Therefore, $a = 0.004$, $b = 0.06$, and $L = 0.004x^2 + 0.06x + 77.6$.
Life expectancy 1995–2000

$$L(15) = 0.004(15)^2 + 0.06(15) + 77.6 \approx 79.4 \text{ years}$$

Life expectancy 2000 – 2005

$$L(20) = 0.004(20)^2 + 0.06(20) + 77.6 \approx 80.4 \text{ years}$$

93. Quadratic Regression

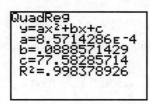 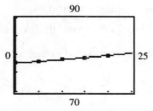

95. Let x_1 = number of hours for Company A,
and x_2 = number of hours for Company B.

The mathematical model is: $30x_1 + 20x_2 = 600$
$$10x_1 + 20x_2 = 400$$

Divide each side of each equation by 10. This yields the system of linear equations:

$3x_1 + 2x_2 = 60$
$x_1 + 2x_2 = 40$

$$\begin{bmatrix} 3 & 2 & | & 60 \\ 1 & 2 & | & 40 \end{bmatrix} \sim \begin{bmatrix} 1 & 2 & | & 40 \\ 3 & 2 & | & 60 \end{bmatrix} \sim \begin{bmatrix} 1 & 2 & | & 40 \\ 0 & -4 & | & -60 \end{bmatrix} \sim \begin{bmatrix} 1 & 2 & | & 40 \\ 0 & 1 & | & 15 \end{bmatrix} \sim \begin{bmatrix} 1 & 0 & | & 10 \\ 0 & 1 & | & 15 \end{bmatrix}$$

$$R_1 \leftrightarrow R_2 \quad (-3)R_1 + R_2 \rightarrow R_2 \quad \left(-\frac{1}{4}\right)R_2 \rightarrow R_2 \quad (-2)R_2 + R_1 \rightarrow R_1$$

Thus, $x_1 = 10$ and $x_2 = 15$, or 10 hours for Company A and 15 hours for Company B.

97. (A)   6th and Washington Ave.: $x_1 + x_2 = 1200$
6th and Lincoln Ave.: $x_2 + x_3 = 1000$
5th and Lincoln Ave.: $x_3 + x_4 = 1300$

(B)   The system of equations is:

$$
\begin{aligned}
x_1 \qquad\quad + x_4 &= 1500 \\
x_1 + x_2 \qquad\quad &= 1200 \\
x_2 + x_3 \qquad &= 1000 \\
x_3 + x_4 &= 1300
\end{aligned}
$$

$$
\begin{bmatrix}
1 & 0 & 0 & 1 & | & 1500 \\
1 & 1 & 0 & 0 & | & 1200 \\
0 & 1 & 1 & 0 & | & 1000 \\
0 & 0 & 1 & 1 & | & 1300
\end{bmatrix}
\sim
\begin{bmatrix}
1 & 0 & 0 & 1 & | & 1500 \\
0 & 1 & 0 & -1 & | & -300 \\
0 & 1 & 1 & 0 & | & 1000 \\
0 & 0 & 1 & 1 & | & 1300
\end{bmatrix}
\sim
\begin{bmatrix}
1 & 0 & 0 & 1 & | & 1500 \\
0 & 1 & 0 & -1 & | & -300 \\
0 & 0 & 1 & 1 & | & 1300 \\
0 & 0 & 1 & 1 & | & 1300
\end{bmatrix}
$$

$(-1)R_1 + R_2 \rightarrow R_2$ $\qquad$ $(-1)R_2 + R_3 \rightarrow R_3$ $\qquad$ $(-1)R_3 + R_4 \rightarrow R_4$

$$
\sim
\begin{bmatrix}
1 & 0 & 0 & 1 & | & 1500 \\
0 & 1 & 0 & -1 & | & -300 \\
0 & 0 & 1 & 1 & | & 1300 \\
0 & 0 & 0 & 0 & | & 0
\end{bmatrix}
$$

Thus
$$
\begin{aligned}
x_1 \qquad\quad + x_4 &= 1500 \\
x_2 \qquad - x_4 &= -300 \\
x_3 + x_4 &= 1300
\end{aligned}
$$

Let $x_4 = t$. Then $x_1 = 1500 - t$, $x_2 = t - 300$ and $x_3 = 1300 - t$. Since x_1, x_2, x_3, and x_4 must be nonnegative integers, we have $300 \le t \le 1300$.

(C)   The flow from Washington Ave. to Lincoln Ave. on 5th Street is given by $x_4 = t$. As shown in part (B), $300 \le t \le 1300$, that is, the maximum number of vehicles is 1300 and the minimum number is 300.

(D)   If $x_4 = t = 1000$, then   Washington Ave.: $x_1 = 1500 - 1000 = 500$,   6th St.: $x_2 = 1000 - 300 = 700$,   Lincoln Ave.: $x_3 = 1300 - 1000 = 300$.

EXERCISE 4-4

Things to remember:

1.   A matrix with m rows and n columns is said to have SIZE $m \times n$. If a matrix has the same number of rows and columns, then it is called a SQUARE MATRIX. A matrix with only one column is a COLUMN MATRIX, and a matrix with only one row is a ROW MATRIX.

2.   Two matrices are EQUAL if they have the same size and their corresponding elements are equal.

3.   The SUM of two matrices of the same size, $m \times n$, is an $m \times n$ matrix whose elements are the sum of the corresponding elements of the two given matrices. Addition is not defined for matrices with different sizes. Matrix addition is commutative: $A + B = B + A$, and associative: $(A + B) + C = A + (B + C)$.

4.   A matrix with all elements equal to zero is called a ZERO MATRIX.

5.   The NEGATIVE OF A MATRIX M, denoted by $-M$, is the matrix whose elements are the negatives of the elements of M.

6. If A and B are matrices of the same size, then subtraction is defined by $A - B = A + (-B)$. Thus, to subtract B from A, simply subtract corresponding elements.

7. If M is a matrix and k is a number, then kM is the matrix formed by multiplying each element of M by k.

8. PRODUCT OF A ROW MATRIX AND A COLUMN MATRIX

The product of a $1 \times n$ row matrix and an $n \times 1$ column matrix is the 1×1 matrix given by

$$
\underset{\left[a_1 \ a_2 \cdots a_n\right]}{\overset{1 \times n}{}} \ \overset{n \times 1}{\begin{bmatrix} b_1 \\ b_2 \\ \vdots \\ b_n \end{bmatrix}} = \left[a_1 b_1 + a_2 b_2 + \cdots a_n b_n \right]
$$

Note that the number of elements in the row matrix and the number of elements in the column matrix must be the same for the product to be defined.

9. Let A be an $m \times p$ matrix and B be a $p \times n$ matrix. The MATRIX PRODUCT of A and B, denoted AB, is the $m \times n$ matrix whose element in the ith row and the jth column is the real number obtained from the product of the ith row of A and the jth column of B. If the number of columns in A does not equal the number of rows in B, then the matrix product AB is not defined.

NOTE: Matrix multiplication is *not* commutative. That is AB does not always equal BA, even when both multiplications are defined.

---

1. $\begin{bmatrix} 1 & 5 \end{bmatrix} + \begin{bmatrix} 3 & 10 \end{bmatrix} = \begin{bmatrix} 4 & 15 \end{bmatrix}$

3. $\begin{bmatrix} 2 & 0 \\ -3 & 6 \end{bmatrix} - \begin{bmatrix} 0 & -4 \\ -1 & 0 \end{bmatrix} = \begin{bmatrix} 2 & 4 \\ -2 & 6 \end{bmatrix}$

5. $\begin{bmatrix} 3 \\ 6 \end{bmatrix} + \begin{bmatrix} -1 & 9 \end{bmatrix}$ Addition not defined; the matrices have different sizes.

7. $7\begin{bmatrix} 3 & -5 & 9 & 4 \end{bmatrix} = \begin{bmatrix} 7(3) & 7(-5) & 7(9) & 7(4) \end{bmatrix} = \begin{bmatrix} 21 & -35 & 63 & 28 \end{bmatrix}$

9. $\begin{bmatrix} 3 & 4 \\ -1 & -2 \end{bmatrix}\begin{bmatrix} -1 \\ 2 \end{bmatrix} = \begin{bmatrix} \begin{bmatrix} 3 & 4 \end{bmatrix}\begin{bmatrix} -1 \\ 2 \end{bmatrix} \\ \begin{bmatrix} -1 & -2 \end{bmatrix}\begin{bmatrix} -1 \\ 2 \end{bmatrix} \end{bmatrix} = \begin{bmatrix} -3+8 \\ 1-4 \end{bmatrix} = \begin{bmatrix} 5 \\ -3 \end{bmatrix}$

11. $\begin{bmatrix} 2 & -3 \\ 1 & 2 \end{bmatrix}\begin{bmatrix} 1 & -1 \\ 0 & -2 \end{bmatrix} = \begin{bmatrix} \begin{bmatrix} 2 & -3 \end{bmatrix}\begin{bmatrix} 1 \\ 0 \end{bmatrix} & \begin{bmatrix} 2 & -3 \end{bmatrix}\begin{bmatrix} -1 \\ -2 \end{bmatrix} \\ \begin{bmatrix} 1 & 2 \end{bmatrix}\begin{bmatrix} 1 \\ 0 \end{bmatrix} & \begin{bmatrix} 1 & 2 \end{bmatrix}\begin{bmatrix} -1 \\ -2 \end{bmatrix} \end{bmatrix} = \begin{bmatrix} 2+0 & -2+6 \\ 1+0 & -1-4 \end{bmatrix} = \begin{bmatrix} 2 & 4 \\ 1 & -5 \end{bmatrix}$

13. $\begin{bmatrix} 1 & -1 \\ 0 & -2 \end{bmatrix} \begin{bmatrix} 2 & -3 \\ 1 & 2 \end{bmatrix}$

$$= \begin{bmatrix} [1 \ -1]\begin{bmatrix} 2 \\ 1 \end{bmatrix} & [1 \ -1]\begin{bmatrix} -3 \\ 2 \end{bmatrix} \\ [0 \ -2]\begin{bmatrix} 2 \\ 1 \end{bmatrix} & [0 \ -2]\begin{bmatrix} -3 \\ 2 \end{bmatrix} \end{bmatrix}$$

$$= \begin{bmatrix} 2-1 & -3-2 \\ 0-2 & 0-4 \end{bmatrix} = \begin{bmatrix} 1 & -5 \\ -2 & -4 \end{bmatrix}$$

15. $\begin{bmatrix} 5 & 0 \\ 0 & 5 \end{bmatrix} \begin{bmatrix} 1 & 3 \\ 2 & 4 \end{bmatrix} = \begin{bmatrix} 5 & 15 \\ 10 & 20 \end{bmatrix}$

17. $\begin{bmatrix} 1 & 3 \\ 2 & 4 \end{bmatrix} \begin{bmatrix} 5 & 0 \\ 0 & 5 \end{bmatrix} = \begin{bmatrix} 5 & 15 \\ 10 & 20 \end{bmatrix}$

19. $\begin{bmatrix} 0 & 1 \\ 0 & 0 \end{bmatrix} \begin{bmatrix} 3 & 7 \\ 5 & 9 \end{bmatrix} = \begin{bmatrix} 5 & 9 \\ 0 & 0 \end{bmatrix}$

21. $\begin{bmatrix} 3 & 7 \\ 5 & 9 \end{bmatrix} \begin{bmatrix} 0 & 1 \\ 0 & 0 \end{bmatrix} = \begin{bmatrix} 0 & 3 \\ 0 & 5 \end{bmatrix}$

23. $[5 \ -2]\begin{bmatrix} -3 \\ -4 \end{bmatrix} = [-15+8] = [-7]$

25. $\begin{bmatrix} -3 \\ -4 \end{bmatrix}[5 \ -2] = \begin{bmatrix} (-3)(5) & (-3)(-2) \\ (-4)(5) & (-4)(-2) \end{bmatrix} = \begin{bmatrix} -15 & 6 \\ -20 & 8 \end{bmatrix}$

27. $[3 \ -2 \ -4]\begin{bmatrix} 1 \\ 2 \\ -3 \end{bmatrix} = [(3-4+12)] = [11]$

29. $\begin{bmatrix} 1 \\ 2 \\ -3 \end{bmatrix}[3 \ -2 \ -4] = \begin{bmatrix} (1)(3) & (1)(-2) & (1)(-4) \\ (2)(3) & (2)(-2) & (2)(-4) \\ (-3)(3) & (-3)(-2) & (-3)(-4) \end{bmatrix} = \begin{bmatrix} 3 & -2 & -4 \\ 6 & -4 & -8 \\ -9 & 6 & 12 \end{bmatrix}$

31. $AC = \begin{bmatrix} 2 & -1 & 3 \\ 0 & 4 & -2 \end{bmatrix} \begin{bmatrix} -1 & 0 & 2 \\ 4 & -3 & 1 \\ -2 & 3 & 5 \end{bmatrix} = \begin{bmatrix} -12 & 12 & 18 \\ 20 & -18 & -6 \end{bmatrix}$

33. AB is not defined; the number of columns of A (3) does not equal the number of rows of B (2).

35. $B^2 = BB = \begin{bmatrix} -3 & 1 \\ 2 & 5 \end{bmatrix} \begin{bmatrix} -3 & 1 \\ 2 & 5 \end{bmatrix} = \begin{bmatrix} 11 & 2 \\ 4 & 27 \end{bmatrix}$

37. $B + AD = \begin{bmatrix} -3 & 1 \\ 2 & 5 \end{bmatrix} + \begin{bmatrix} 2 & -1 & 3 \\ 0 & 4 & -2 \end{bmatrix} \begin{bmatrix} 3 & -2 \\ 0 & -1 \\ 1 & 2 \end{bmatrix}$

$$= \begin{bmatrix} -3 & 1 \\ 2 & 5 \end{bmatrix} + \begin{bmatrix} 9 & 3 \\ -2 & -8 \end{bmatrix} = \begin{bmatrix} 6 & 4 \\ 0 & -3 \end{bmatrix}$$

39. $0.1DB = 0.1\begin{bmatrix} 3 & -2 \\ 0 & -1 \\ 1 & 2 \end{bmatrix} \begin{bmatrix} -3 & 1 \\ 2 & 5 \end{bmatrix} = 0.1\begin{bmatrix} -13 & -7 \\ -2 & -5 \\ 1 & 11 \end{bmatrix} = \begin{bmatrix} -1.3 & -0.7 \\ -0.2 & -0.5 \\ 0.1 & 1.1 \end{bmatrix}$

41. $3BA + 4AC = 3\begin{bmatrix} -3 & 1 \\ 2 & 5 \end{bmatrix}\begin{bmatrix} 2 & -1 & 3 \\ 0 & 4 & -2 \end{bmatrix} + 4\begin{bmatrix} 2 & -1 & 3 \\ 0 & 4 & -2 \end{bmatrix}\begin{bmatrix} -1 & 0 & 2 \\ 4 & -3 & 1 \\ -2 & 3 & 5 \end{bmatrix}$

$\qquad\qquad = 3\begin{bmatrix} -6 & 7 & -11 \\ 4 & 18 & -4 \end{bmatrix} + 4\begin{bmatrix} -12 & 12 & 18 \\ 20 & -18 & -6 \end{bmatrix}$

$\qquad\qquad = \begin{bmatrix} -18 & 21 & -33 \\ 12 & 54 & -12 \end{bmatrix} + \begin{bmatrix} -48 & 48 & 72 \\ 80 & -72 & -24 \end{bmatrix} = \begin{bmatrix} -66 & 69 & 39 \\ 92 & -18 & -36 \end{bmatrix}$

43. $(-2)BA + 6CD$ is not defined; $-2BA$ is 2×3, $6CD$ is 3×2.

45. $ACD = A(CD) = \begin{bmatrix} 2 & -1 & 3 \\ 0 & 4 & -2 \end{bmatrix}\left(\begin{bmatrix} -1 & 0 & 2 \\ 4 & -3 & 1 \\ -2 & 3 & 5 \end{bmatrix}\begin{bmatrix} 3 & -2 \\ 0 & -1 \\ 1 & 2 \end{bmatrix}\right)$

$\qquad\qquad = \begin{bmatrix} 2 & -1 & 3 \\ 0 & 4 & -2 \end{bmatrix}\begin{bmatrix} -1 & 6 \\ 13 & -3 \\ -1 & 11 \end{bmatrix} = \begin{bmatrix} -18 & 48 \\ 54 & -34 \end{bmatrix}$

47. $DBA = D(BA) = \begin{bmatrix} 3 & -2 \\ 0 & -1 \\ 1 & 2 \end{bmatrix}\left(\begin{bmatrix} -3 & 1 \\ 2 & 5 \end{bmatrix}\begin{bmatrix} 2 & -1 & 3 \\ 0 & 4 & -2 \end{bmatrix}\right)$

$\qquad\qquad = \begin{bmatrix} 3 & -2 \\ 0 & -1 \\ 1 & 2 \end{bmatrix}\begin{bmatrix} -6 & 7 & -11 \\ 4 & 18 & -4 \end{bmatrix} = \begin{bmatrix} -26 & -15 & -25 \\ -4 & -18 & 4 \\ 2 & 43 & -19 \end{bmatrix}$

49. $AB = \begin{bmatrix} a & a \\ b & b \end{bmatrix}\begin{bmatrix} a & a \\ -a & -a \end{bmatrix} = \begin{bmatrix} a^2 - a^2 & a^2 - a^2 \\ ab - ab & ab - ab \end{bmatrix} = \begin{bmatrix} 0 & 0 \\ 0 & 0 \end{bmatrix}$

$\qquad BA = \begin{bmatrix} a & a \\ -a & -a \end{bmatrix}\begin{bmatrix} a & a \\ b & b \end{bmatrix} = \begin{bmatrix} a^2 + ab & a^2 + ab \\ -a^2 - ab & -a^2 - ab \end{bmatrix}$

51. $A^2 = \begin{bmatrix} ab & b^2 \\ -a^2 & -ab \end{bmatrix}\begin{bmatrix} ab & b^2 \\ -a^2 & -ab \end{bmatrix} = \begin{bmatrix} a^2b^2 - a^2b^2 & ab^3 - ab^3 \\ -a^3b + a^3b & -a^2b^2 + a^2b^2 \end{bmatrix} = \begin{bmatrix} 0 & 0 \\ 0 & 0 \end{bmatrix}$

53. $A = [0.3 \quad 0.7], B = \begin{bmatrix} 0.4 & 0.6 \\ 0.2 & 0.8 \end{bmatrix}$

$\qquad B^2 = \begin{bmatrix} 0.28 & 0.72 \\ 0.24 & 0.76 \end{bmatrix}, \quad B^3 = \begin{bmatrix} 0.256 & 0.744 \\ 0.248 & 0.752 \end{bmatrix} \dots,$

$\qquad B^8 = \begin{bmatrix} 0.250 & 0.74999 \\ 0.24999 & 0.75000 \end{bmatrix}, \quad B^n \to \begin{bmatrix} 0.25 & 0.75 \\ 0.25 & 0.75 \end{bmatrix}$

$\qquad AB = [0.26 \quad 0.74], AB^2 = [0.252 \quad 0.748],$

$\qquad AB^3 = [0.2504 \quad 0.7496] \dots,$

$\qquad AB^{10} = [0.2500\dots \quad 0.7499\dots], \quad AB^n \to [0.25 \quad 0.75]$

55. $\begin{bmatrix} a & b \\ c & d \end{bmatrix} + \begin{bmatrix} 2 & -3 \\ 0 & 1 \end{bmatrix} = \begin{bmatrix} a+2 & b-3 \\ c+0 & d+1 \end{bmatrix} = \begin{bmatrix} 1 & -2 \\ 3 & -4 \end{bmatrix}$

Thus, $\begin{aligned} a + 2 &= 1, & a &= -1 \\ b - 3 &= -2, & b &= 1 \\ c + 0 &= 3, & c &= 3 \\ d + 1 &= -4, & d &= -5 \end{aligned}$

57. $\begin{bmatrix} 1 & -2 \\ 2 & -3 \end{bmatrix} \begin{bmatrix} a & b \\ c & d \end{bmatrix} = \begin{bmatrix} 1 & 0 \\ 3 & 2 \end{bmatrix}$

$\begin{bmatrix} a-2c & b-2d \\ 2a-3c & 2b-3d \end{bmatrix} = \begin{bmatrix} 1 & 0 \\ 3 & 2 \end{bmatrix}$

implies $\quad \begin{aligned} a - 2c &= 1 & b - 2d &= 0 \\ 2a - 3c &= 3 & 2b - 3d &= 2 \end{aligned}$

The augmented matrix for the first system is:

$\begin{bmatrix} 1 & -2 & | & 1 \\ 2 & -3 & | & 3 \end{bmatrix} (-2)R_1 + R_2 \to R_2 \sim \begin{bmatrix} 1 & -2 & | & 1 \\ 0 & 1 & | & 1 \end{bmatrix} 2R_2 + R_1 \to R_1$

$\sim \begin{bmatrix} 1 & 0 & | & 3 \\ 0 & 1 & | & 1 \end{bmatrix}$ Thus, $a = 3, \; c = 1.$

For the second system, substitute $b = 2d$ from the first equation into the second equation:

$\begin{aligned} 2(2d) - 3d &= 2 \\ d &= 2 \\ b &= 4 \end{aligned}$

Solution: $\; a = 3, \; b = 4, \; c = 1, \; d = 2$

59. False. A 1x1 matrix is equivalent to a real number; multiplication of real numbers is commutative.

61. True. For example, $A = \begin{bmatrix} 0 & 0 \\ 1 & 2 \end{bmatrix}$, $B = \begin{bmatrix} 2 & 0 \\ -1 & 0 \end{bmatrix}$; $AB = \begin{bmatrix} 0 & 0 \\ 0 & 0 \end{bmatrix}$

63. Let $A = \begin{bmatrix} a_1 & 0 \\ 0 & a_2 \end{bmatrix}$ and $B = \begin{bmatrix} b_1 & 0 \\ 0 & b_2 \end{bmatrix}$

(A) Always true:

$A + B = \begin{bmatrix} a_1 & 0 \\ 0 & a_2 \end{bmatrix} + \begin{bmatrix} b_1 & 0 \\ 0 & b_2 \end{bmatrix} = \begin{bmatrix} a_1+b_1 & 0 \\ 0 & a_2+b_2 \end{bmatrix}$

(B) Always true: matrix addition is commutative, $A + B = B + A$ for *any* pair of matrices of the same size.

(C) Always true:

$AB = \begin{bmatrix} a_1 & 0 \\ 0 & a_2 \end{bmatrix} \begin{bmatrix} b_1 & 0 \\ 0 & b_2 \end{bmatrix} = \begin{bmatrix} a_1 b_1 & 0 \\ 0 & a_2 b_2 \end{bmatrix}$

(D) Always true:

$BA = \begin{bmatrix} b_1 & 0 \\ 0 & b_2 \end{bmatrix} \begin{bmatrix} a_1 & 0 \\ 0 & a_2 \end{bmatrix} = \begin{bmatrix} b_1 a_1 & 0 \\ 0 & b_2 a_2 \end{bmatrix} = \begin{bmatrix} a_1 b_1 & 0 \\ 0 & a_2 b_2 \end{bmatrix} = AB$

65. $A + B = \begin{bmatrix} \$47 & \$39 \\ \$90 & \$125 \end{bmatrix} + \begin{bmatrix} \$56 & \$42 \\ \$84 & \$115 \end{bmatrix} = \begin{bmatrix} \$103 & \$81 \\ \$174 & \$240 \end{bmatrix}$

$$\frac{1}{2}(A + B) = \frac{1}{2} \begin{matrix} \text{Guitar} & \text{Banjo} \\ \begin{bmatrix} \$103 & \$81 \\ \$174 & \$240 \end{bmatrix} \end{matrix} = \begin{bmatrix} \$51.50 & \$40.50 \\ \$87.00 & \$120.00 \end{bmatrix} \begin{matrix} \text{Materials} \\ \text{Labor} \end{matrix}$$

67. The dealer is increasing the retail prices by 10%. Thus, the new retail price matrix M' is:

$$M' = M + 0.1M = (1.1)M = (1.1) \begin{bmatrix} 35,075 & 2560 & 1070 & 640 \\ 39,045 & 1840 & 770 & 460 \\ 45,535 & 3400 & 1415 & 850 \end{bmatrix}$$

$$= \begin{bmatrix} 38,582.50 & 2816 & 1177 & 704 \\ 42,949.50 & 2024 & 847 & 506 \\ 50,088.50 & 3740 & 1556.50 & 935 \end{bmatrix}$$

$$N' = N + 0.15N = (1.15)N = (1.15) \begin{bmatrix} 30,996 & 2050 & 850 & 510 \\ 34,857 & 1585 & 660 & 395 \\ 41,667 & 2890 & 1200 & 725 \end{bmatrix}$$

$$= \begin{bmatrix} 35,645.40 & 2357.50 & 977.50 & 586.50 \\ 40,085.55 & 1822.75 & 759 & 454.25 \\ 47,917.05 & 3323.50 & 1380 & 833.75 \end{bmatrix}$$

The new markup (to the nearest dollar) is:

$$M' - N' = \begin{matrix} & \text{Basic Car} & \text{Air Cond} & \text{AM/FM} & \text{Cruise} \\ \text{Model } A \\ \text{Model } B \\ \text{Model } C \end{matrix} \begin{bmatrix} \$2937 & \$459 & \$200 & \$118 \\ \$2864 & \$201 & \$88 & \$52 \\ \$2171 & \$417 & \$177 & \$101 \end{bmatrix}$$

69. (A) $[0.6 \quad 0.6 \quad 0.2] \begin{bmatrix} 17.30 \\ 12.22 \\ 10.63 \end{bmatrix} = \$19.84$

The labor cost per boat for one-person boats at the Massachusetts plant is: $19.84.

(B) $[1.5 \quad 1.2 \quad 0.4] \begin{bmatrix} 14.65 \\ 10.29 \\ 9.66 \end{bmatrix} = \$38.19$

The labor cost per boat for four-person boats at the Virginia plant is: $38.19.

(C) MN gives the labor cost per boat at each plant; NM is not defined.

(D) $MN = \begin{matrix} \text{MA} & \text{VA} \\ \begin{bmatrix} 19.84 & 16.90 \\ 31.49 & 26.81 \\ 44.87 & 38.19 \end{bmatrix} \end{matrix} \begin{matrix} \text{One–person boat} \\ \text{Two–person boat} \\ \text{Four–person boat} \end{matrix}$

71. (A) $[4 \; 2] \begin{bmatrix} 15 \\ 5 \end{bmatrix} = 70$ There are 70 g of protein in Mix X.

(B) $[3 \; 1] \begin{bmatrix} 5 \\ 15 \end{bmatrix} = 30$ There are 30 g of fat in Mix Z.

(C) *MN* gives the amount (in grams) of protein, carbohydrates and fat in 20 ounces of each mix. The product *NM* has no meaningful interpretation.

(D)
$$MN = \begin{bmatrix} 4 & 2 \\ 20 & 16 \\ 3 & 1 \end{bmatrix} \begin{bmatrix} 15 & 10 & 5 \\ 5 & 10 & 15 \end{bmatrix} = \begin{array}{c} \\ \\ \end{array} \overset{\text{Mix X Mix Y Mix Z}}{\begin{bmatrix} 70 & 60 & 50 \\ 380 & 360 & 340 \\ 50 & 40 & 30 \end{bmatrix}} \begin{array}{l} \text{Protein} \\ \text{Carbohydrate} \\ \text{Fat} \end{array}$$

73. (A) $[1000 \; 500 \; 5000] \begin{bmatrix} 1.20 \\ 3.00 \\ 1.45 \end{bmatrix} = 9,950;$ total amount spent in Berkeley = \$9,950.

(B) $[2000 \; 800 \; 8000] \begin{bmatrix} 1.20 \\ 3.00 \\ 1.45 \end{bmatrix} = 16,400;$ total amount spent in Oakland = \$16,400.

(C) *MN* is not defined; *NM* gives the total cost for each city.

(D)
$$NM = \begin{bmatrix} 1000 & 500 & 5000 \\ 2000 & 800 & 8000 \end{bmatrix} \begin{bmatrix} 1.20 \\ 3.00 \\ 1.45 \end{bmatrix} = \overset{\text{Cost/City}}{\begin{bmatrix} \$9,950 \\ \$16,400 \end{bmatrix}} \begin{array}{l} \text{Berkeley} \\ \text{Oakland} \end{array}$$

(E)
$$[1 \; 1] \cdot N = [1 \; 1] \begin{bmatrix} 1000 & 500 & 5000 \\ 2000 & 800 & 8000 \end{bmatrix}$$

$$\overset{\text{Telephone \quad House \quad letters}}{= [3,000 \quad 1,300 \quad 13,000]}$$

(F)
$$N \cdot \begin{bmatrix} 1 \\ 1 \\ 1 \end{bmatrix} = \begin{bmatrix} 1,000 & 500 & 5,000 \\ 2,000 & 800 & 8,000 \end{bmatrix} \begin{bmatrix} 1 \\ 1 \\ 1 \end{bmatrix} = \overset{\text{Total contacts}}{\begin{bmatrix} 6,500 \\ 10,800 \end{bmatrix}} \begin{array}{l} \text{Berkeley} \\ \text{Oakland} \end{array}$$

EXERCISE 4-5

Things to remember:

1. The IDENTITY element for multiplication for the set of square matrices of order *n* (dimension $n \times n$) is the square matrix *I* of order *n* which has 1's on the principal diagonal (upper left corner to lower right corner) and 0's elsewhere. The identity matrices of order 2 and 3, respectively, are

$$I = \begin{bmatrix} 1 & 0 \\ 0 & 1 \end{bmatrix} \text{ and } I = \begin{bmatrix} 1 & 0 & 0 \\ 0 & 1 & 0 \\ 0 & 0 & 1 \end{bmatrix}.$$

2. If *M* is any square matrix of order *n* and *I* is the identity matrix of order *n*, then

$$IM = MI = M.$$

3.   INVERSE OF A SQUARE MATRIX
     Let M be a square matrix of order n and I be the identity matrix of order n. If there exists a matrix M^{-1} such that
     $$MM^{-1} = M^{-1}M = I$$
     then M^{-1} is called the MULTIPLICATIVE INVERSE OF M or, more simply, the INVERSE OF M. M^{-1} is read "M inverse."

4.   If the augmented matrix $[M\,|\,I]$ is transformed by row operations into $[I\,|\,B]$, then the resulting matrix B is M^{-1}. However, if all zeros are obtained in one or more rows to the left of the vertical line during the row transformation procedure, then M^{-1} does not exist. Matrices that do not have inverses are called SINGULAR MATRICES.

---

1.   (A) 4: Additive inverse: -4; multiplicative inverse: $\dfrac{1}{4}$

     (B) –3: Additive inverse: 3; multiplicative inverse: $-\dfrac{1}{3}$

     (C) 0: Additive inverse: 0; multiplicative inverse: not defined

3.   (A) $\dfrac{2}{3}$: Additive inverse: $-\dfrac{2}{3}$; multiplicative inverse: $\dfrac{3}{2}$

     (B) $-\dfrac{1}{7}$: Additive inverse: $\dfrac{1}{7}$; multiplicative inverse: -7

     (C) 1.6: Additive inverse: -1.6; multiplicative inverse: $\dfrac{1}{1.6} = 0.625$

5. $\begin{bmatrix} 2 & 5 \end{bmatrix}$ does not have an inverse; it's not a square matrix.

7. $\begin{bmatrix} 0 & 0 \\ 0 & 0 \end{bmatrix}$ A matrix with one or more rows of zeros does not have an inverse.

9.   (a) $\begin{bmatrix} 1 & 0 \\ 0 & 0 \end{bmatrix}\begin{bmatrix} 2 & -3 \\ 4 & 5 \end{bmatrix} = \begin{bmatrix} 2 & -3 \\ 0 & 0 \end{bmatrix}$ (b) $\begin{bmatrix} 2 & -3 \\ 4 & 5 \end{bmatrix}\begin{bmatrix} 1 & 0 \\ 0 & 0 \end{bmatrix} = \begin{bmatrix} 2 & 0 \\ 4 & 0 \end{bmatrix}$

11. (a) $\begin{bmatrix} 0 & 0 \\ 0 & 1 \end{bmatrix}\begin{bmatrix} 2 & -3 \\ 4 & 5 \end{bmatrix} = \begin{bmatrix} 0 & 0 \\ 4 & 5 \end{bmatrix}$ (b) $\begin{bmatrix} 2 & -3 \\ 4 & 5 \end{bmatrix}\begin{bmatrix} 0 & 0 \\ 0 & 1 \end{bmatrix} = \begin{bmatrix} 0 & -3 \\ 0 & 5 \end{bmatrix}$

13. (a) $\begin{bmatrix} 1 & 0 \\ 0 & 1 \end{bmatrix}\begin{bmatrix} 2 & -3 \\ 4 & 5 \end{bmatrix} = \begin{bmatrix} 2 & -3 \\ 4 & 5 \end{bmatrix}$ (b) $\begin{bmatrix} 2 & -3 \\ 4 & 5 \end{bmatrix}\begin{bmatrix} 1 & 0 \\ 0 & 1 \end{bmatrix} = \begin{bmatrix} 2 & -3 \\ 4 & 5 \end{bmatrix}$

15. $\begin{bmatrix} 1 & 0 & 0 \\ 0 & 1 & 0 \\ 0 & 0 & 1 \end{bmatrix} \begin{bmatrix} -2 & 1 & 3 \\ 2 & 4 & -2 \\ 5 & 1 & 0 \end{bmatrix}$

$= \begin{bmatrix} 1(-2)+0\cdot2+0\cdot5 & 1\cdot1+0\cdot4+0\cdot1 & 1\cdot3+0(-2)+0\cdot0 \\ 0(-2)+1\cdot2+0\cdot5 & 0\cdot1+1\cdot4+0\cdot1 & 0\cdot3+1(-2)+0\cdot0 \\ 0(-2)+0\cdot2+1\cdot5 & 0\cdot1+0\cdot4+1\cdot1 & 0\cdot3+0(-2)+1\cdot0 \end{bmatrix} = \begin{bmatrix} -2 & 1 & 3 \\ 2 & 4 & -2 \\ 5 & 1 & 0 \end{bmatrix}$

17. $\begin{bmatrix} -2 & 1 & 3 \\ 2 & 4 & -2 \\ 5 & 1 & 0 \end{bmatrix} \begin{bmatrix} 1 & 0 & 0 \\ 0 & 1 & 0 \\ 0 & 0 & 1 \end{bmatrix}$

$= \begin{bmatrix} (-2)\cdot1+1\cdot0+3\cdot0 & (-2)0+1\cdot1+3\cdot0 & (-2)0+1\cdot0+3\cdot1 \\ 2\cdot1+4\cdot0+(-2)0 & 2\cdot0+4\cdot1+(-2)0 & 2\cdot0+4\cdot0+(-2)1 \\ 5\cdot1+1\cdot0+0\cdot0 & 5\cdot0+1\cdot1+0\cdot0 & 5\cdot0+1\cdot0+0\cdot1 \end{bmatrix} = \begin{bmatrix} -2 & 1 & 3 \\ 2 & 4 & -2 \\ 5 & 1 & 0 \end{bmatrix}$

19. $\begin{bmatrix} 3 & -4 \\ -2 & 3 \end{bmatrix} \begin{bmatrix} 3 & 4 \\ 2 & 3 \end{bmatrix} = \begin{bmatrix} 1 & 0 \\ 0 & 1 \end{bmatrix}$ Yes

21. $\begin{bmatrix} 2 & 2 \\ -1 & -1 \end{bmatrix} \begin{bmatrix} 1 & 1 \\ -1 & -1 \end{bmatrix} = \begin{bmatrix} 0 & 0 \\ 0 & 0 \end{bmatrix}$ No

23. $\begin{bmatrix} -5 & 2 \\ -8 & 3 \end{bmatrix} \begin{bmatrix} 3 & -2 \\ 8 & -5 \end{bmatrix} = \begin{bmatrix} 1 & 0 \\ 0 & 1 \end{bmatrix}$ Yes

25. No. The second matrix has a column of zeros; it does not have an inverse.

27. $\begin{bmatrix} 1 & -1 & 1 \\ 0 & 2 & -1 \\ 2 & 3 & 0 \end{bmatrix} \begin{bmatrix} 3 & 3 & -1 \\ -2 & -2 & 1 \\ -4 & -5 & 2 \end{bmatrix} = \begin{bmatrix} 1 & 0 & 0 \\ 0 & 1 & 0 \\ 0 & 0 & 1 \end{bmatrix}$ Yes

29. The matrix is not square.

31. The matrix is not square.

33. The matrix has a row of zeros.

35. The matrix has a column of zeros.

37. $D = 1 \cdot 6 - 2 \cdot 3 = 0$

39. $\begin{bmatrix} -1 & 0 & | & 1 & 0 \\ -3 & 1 & | & 0 & 1 \end{bmatrix} \sim \begin{bmatrix} 1 & 0 & | & -1 & 0 \\ -3 & 1 & | & 0 & 1 \end{bmatrix} \sim \begin{bmatrix} 1 & 0 & | & -1 & 0 \\ 0 & 1 & | & -3 & 1 \end{bmatrix}$

$\quad (-1)R_1 \rightarrow R_1 \quad 3R_1 + R_2 \rightarrow R_2$

Thus, $M^{-1} = \begin{bmatrix} -1 & 0 \\ -3 & 1 \end{bmatrix}$

Check:

$M^{-1}M = \begin{bmatrix} -1 & 0 \\ -3 & 1 \end{bmatrix} \begin{bmatrix} -1 & 0 \\ -3 & 1 \end{bmatrix} = \begin{bmatrix} 1 & 0 \\ 0 & 1 \end{bmatrix}$

41. $\begin{bmatrix} 1 & 2 & | & 1 & 0 \\ 1 & 3 & | & 0 & 1 \end{bmatrix} \sim \begin{bmatrix} 1 & 2 & | & 1 & 0 \\ 0 & 1 & | & -1 & 1 \end{bmatrix} \sim \begin{bmatrix} 1 & 0 & | & 3 & -2 \\ 0 & 1 & | & -1 & 1 \end{bmatrix}$

$\quad (-1)R_1 + R_2 \rightarrow R_2 \quad (-2)R_2 + R_1 \rightarrow R_1$

Thus, $M^{-1} = \begin{bmatrix} 3 & -2 \\ -1 & 1 \end{bmatrix}$.

Check:

$$M^{-1}M = \begin{bmatrix} 3 & -2 \\ -1 & 1 \end{bmatrix}\begin{bmatrix} 1 & 2 \\ 1 & 3 \end{bmatrix} = \begin{bmatrix} 1 & 0 \\ 0 & 1 \end{bmatrix}$$

43. $\begin{bmatrix} 1 & 3 & | & 1 & 0 \\ 2 & 7 & | & 0 & 1 \end{bmatrix} \sim \begin{bmatrix} 1 & 3 & | & 1 & 0 \\ 0 & 1 & | & -2 & 1 \end{bmatrix} \sim \begin{bmatrix} 1 & 0 & | & 7 & -3 \\ 0 & 1 & | & -2 & 1 \end{bmatrix}$

$(-2)R_1 + R_2 \to R_2 \quad (-3)R_2 + R_1 \to R_1$

Thus, $M^{-1} = \begin{bmatrix} 7 & -3 \\ -2 & 1 \end{bmatrix}$.

Check:

$$M^{-1}M = \begin{bmatrix} 7 & -3 \\ -2 & 1 \end{bmatrix}\begin{bmatrix} 1 & 3 \\ 2 & 7 \end{bmatrix} = \begin{bmatrix} 1 & 0 \\ 0 & 1 \end{bmatrix}$$

45. $\begin{bmatrix} 1 & -3 & 0 & | & 1 & 0 & 0 \\ 0 & 1 & 1 & | & 0 & 1 & 0 \\ 2 & -1 & 4 & | & 0 & 0 & 1 \end{bmatrix} (-2)R_1 + R_3 \to R_3 \sim \begin{bmatrix} 1 & -3 & 0 & | & 1 & 0 & 0 \\ 0 & 1 & 1 & | & 0 & 1 & 0 \\ 0 & 5 & 4 & | & -2 & 0 & 1 \end{bmatrix} \begin{matrix} 3R_2 + R_1 \to R_1 \\ (-5)R_2 + R_3 \to R_3 \end{matrix}$

$\sim \begin{bmatrix} 1 & 0 & 3 & | & 1 & 3 & 0 \\ 0 & 1 & 1 & | & 0 & 1 & 0 \\ 0 & 0 & -1 & | & -2 & -5 & 1 \end{bmatrix} (-1)R_3 \to R_3 \sim \begin{bmatrix} 1 & 0 & 3 & | & 1 & 3 & 0 \\ 0 & 1 & 1 & | & 0 & 1 & 0 \\ 0 & 0 & 1 & | & 2 & 5 & -1 \end{bmatrix} \begin{matrix} (-3)R_3 + R_1 \to R_1 \\ (-1)R_3 + R_2 \to R_2 \end{matrix}$

$\sim \begin{bmatrix} 1 & 0 & 0 & | & -5 & -12 & 3 \\ 0 & 1 & 0 & | & -2 & -4 & 1 \\ 0 & 0 & 1 & | & 2 & 5 & -1 \end{bmatrix}; \quad M^{-1} = \begin{bmatrix} -5 & -12 & 3 \\ -2 & -4 & 1 \\ 2 & 5 & -1 \end{bmatrix}$

$$M^{-1}M = \begin{bmatrix} -5 & -12 & 3 \\ -2 & -4 & 1 \\ 2 & 5 & -1 \end{bmatrix}\begin{bmatrix} 1 & -3 & 0 \\ 0 & 1 & 1 \\ 2 & -1 & 4 \end{bmatrix} = \begin{bmatrix} 1 & 0 & 0 \\ 0 & 1 & 0 \\ 0 & 0 & 1 \end{bmatrix} = \begin{bmatrix} 1 & 0 & 0 \\ 0 & 1 & 0 \\ 0 & 0 & 1 \end{bmatrix}$$

47. $\begin{bmatrix} 1 & 1 & 0 & | & 1 & 0 & 0 \\ 2 & 3 & -1 & | & 0 & 1 & 0 \\ 1 & 0 & 2 & | & 0 & 0 & 1 \end{bmatrix} \begin{matrix} (-2)R_1 + R_2 \to R_2 \\ (-1)R_1 + R_3 \to R_3 \end{matrix} \sim \begin{bmatrix} 1 & 1 & 0 & | & 1 & 0 & 0 \\ 0 & 1 & -1 & | & -2 & 1 & 0 \\ 0 & -1 & 2 & | & -1 & 0 & 1 \end{bmatrix} \begin{matrix} (-1)R_2 + R_1 \to R_1 \\ R_2 + R_3 \to R_3 \end{matrix}$

$\sim \begin{bmatrix} 1 & 0 & 1 & | & 3 & -1 & 0 \\ 0 & 1 & -1 & | & -2 & 1 & 0 \\ 0 & 0 & 1 & | & -3 & 1 & 1 \end{bmatrix} \begin{matrix} (-1)R_3 + R_1 \to R_1 \\ R_3 + R_2 \to R_2 \end{matrix} \sim \begin{bmatrix} 1 & 0 & 0 & | & 6 & -2 & -1 \\ 0 & 1 & 0 & | & -5 & 2 & 1 \\ 0 & 0 & 1 & | & -3 & 1 & 1 \end{bmatrix}; \quad M^{-1} = \begin{bmatrix} 6 & -2 & -1 \\ -5 & 2 & 1 \\ -3 & 1 & 1 \end{bmatrix}$

$$M^{-1}M = \begin{bmatrix} 6 & -2 & -1 \\ -5 & 2 & 1 \\ -3 & 1 & 1 \end{bmatrix}\begin{bmatrix} 1 & 1 & 0 \\ 2 & 3 & -1 \\ 1 & 0 & 2 \end{bmatrix} = \begin{bmatrix} 1 & 0 & 0 \\ 0 & 1 & 0 \\ 0 & 0 & 1 \end{bmatrix}$$

49. $\begin{bmatrix} 4 & 3 & | & 1 & 0 \\ -3 & -2 & | & 0 & 1 \end{bmatrix} R_2 + R_1 \rightarrow R_1 \sim \begin{bmatrix} 1 & 1 & | & 1 & 1 \\ -3 & -2 & | & 0 & 1 \end{bmatrix} 3R_1 + R_2 \rightarrow R_2$

$\sim \begin{bmatrix} 1 & 1 & | & 1 & 1 \\ 0 & 1 & | & 3 & 4 \end{bmatrix} (-1)R_2 + R_1 \rightarrow R_1 \sim \begin{bmatrix} 1 & 0 & | & -2 & -3 \\ 0 & 1 & | & 3 & 4 \end{bmatrix} (-1)R_2 + R_1 \rightarrow R_1$

$M^{-1} = \begin{bmatrix} -2 & -3 \\ 3 & 4 \end{bmatrix}$

51. $\begin{bmatrix} 2 & 6 & | & 1 & 0 \\ 3 & 9 & | & 0 & 1 \end{bmatrix} \frac{1}{2}R_1 \rightarrow R_1 \sim \begin{bmatrix} 1 & 3 & | & \frac{1}{2} & 0 \\ 3 & 9 & | & 0 & 1 \end{bmatrix} (-3)R_1 + R_2 \rightarrow R_2 \sim \begin{bmatrix} 1 & 3 & | & \frac{1}{2} & 0 \\ 0 & 0 & | & -\frac{3}{2} & 1 \end{bmatrix}$

The inverse does not exist.

53. $\begin{bmatrix} 2 & 1 & | & 1 & 0 \\ 4 & 3 & | & 0 & 1 \end{bmatrix} \frac{1}{2}R_1 \rightarrow R_1 \sim \begin{bmatrix} 1 & \frac{1}{2} & | & \frac{1}{2} & 0 \\ 4 & 3 & | & 0 & 1 \end{bmatrix} (-4)R_1 + R_2 \rightarrow R_2$

$\sim \begin{bmatrix} 1 & \frac{1}{2} & | & \frac{1}{2} & 0 \\ 0 & 1 & | & -2 & 1 \end{bmatrix} \left(-\frac{1}{2}\right)R_2 + R_1 \rightarrow R_1 \sim \begin{bmatrix} 1 & 0 & | & \frac{3}{2} & -\frac{1}{2} \\ 0 & 1 & | & -2 & 1 \end{bmatrix}$

$M^{-1} = \begin{bmatrix} \frac{3}{2} & -\frac{1}{2} \\ -2 & 1 \end{bmatrix} = \begin{bmatrix} 1.5 & -0.5 \\ -2 & 1 \end{bmatrix}$

55. $\begin{bmatrix} 2 & 0 \\ 0 & 2 \end{bmatrix}^{-1} = \begin{bmatrix} \frac{1}{2} & 0 \\ 0 & \frac{1}{2} \end{bmatrix}$

57. $\begin{bmatrix} 3 & 0 \\ 0 & -5 \end{bmatrix}^{-1} = \begin{bmatrix} \frac{1}{3} & 0 \\ 0 & -\frac{1}{5} \end{bmatrix}$

59. $\begin{bmatrix} 4 & 0 & 0 \\ 0 & 2 & 0 \\ 0 & 0 & -8 \end{bmatrix}^{-1} = \begin{bmatrix} \frac{1}{4} & 0 & 0 \\ 0 & \frac{1}{2} & 0 \\ 0 & 0 & -\frac{1}{8} \end{bmatrix}$

61. $\begin{bmatrix} -5 & -2 & -2 & | & 1 & 0 & 0 \\ 2 & 1 & 0 & | & 0 & 1 & 0 \\ 1 & 0 & 1 & | & 0 & 0 & 1 \end{bmatrix} \sim \begin{bmatrix} 1 & 0 & 1 & | & 0 & 0 & 1 \\ 2 & 1 & 0 & | & 0 & 1 & 0 \\ -5 & -2 & -2 & | & 1 & 0 & 0 \end{bmatrix} \sim \begin{bmatrix} 1 & 0 & 1 & | & 0 & 0 & 1 \\ 0 & 1 & -2 & | & 0 & 1 & -2 \\ 0 & -2 & 3 & | & 1 & 0 & 5 \end{bmatrix}$

$\qquad\qquad R_1 \leftrightarrow R_3 \qquad\qquad\qquad (-2)R_1 + R_2 \rightarrow R_2 \qquad\qquad 2R_2 + R_3 \rightarrow R_3$
$\qquad\qquad\qquad\qquad\qquad\qquad 5R_1 + R_3 \rightarrow R_3$

$\sim \begin{bmatrix} 1 & 0 & 1 & | & 0 & 0 & 1 \\ 0 & 1 & -2 & | & 0 & 1 & -2 \\ 0 & 0 & -1 & | & 1 & 2 & 1 \end{bmatrix} \sim \begin{bmatrix} 1 & 0 & 1 & | & 0 & 0 & 1 \\ 0 & 1 & -2 & | & 0 & 1 & -2 \\ 0 & 0 & 1 & | & -1 & -2 & -1 \end{bmatrix} \sim \begin{bmatrix} 1 & 0 & 0 & | & 1 & 2 & 2 \\ 0 & 1 & 0 & | & -2 & -3 & -4 \\ 0 & 0 & 1 & | & -1 & -2 & -1 \end{bmatrix}$

$\qquad\qquad (-1)R_3 \rightarrow R_3 \qquad\qquad\qquad 2R_3 + R_2 \rightarrow R_2$
$\qquad\qquad\qquad\qquad\qquad\qquad (-1)R_3 + R_1 \rightarrow R_1$

Thus, the inverse is $\begin{bmatrix} 1 & 2 & 2 \\ -2 & -3 & -4 \\ -1 & -2 & -1 \end{bmatrix}$.

63.
$$\begin{bmatrix} 2 & 1 & 1 & | & 1 & 0 & 0 \\ 1 & 1 & 0 & | & 0 & 1 & 0 \\ -1 & -1 & 0 & | & 0 & 0 & 1 \end{bmatrix} \rightarrow \begin{bmatrix} 1 & 1 & 0 & | & 0 & 1 & 0 \\ 2 & 1 & 1 & | & 1 & 0 & 0 \\ -1 & -1 & 0 & | & 0 & 0 & 1 \end{bmatrix} \rightarrow \begin{bmatrix} 1 & 1 & 0 & | & 0 & 1 & 0 \\ 0 & -1 & 1 & | & 1 & -2 & 0 \\ 0 & 0 & 0 & | & 0 & 1 & 1 \end{bmatrix}$$

$\qquad R_1 \leftrightarrow R_2 \qquad\qquad (-2)R_1 + R_2 \rightarrow R_2$
$\qquad\qquad\qquad\qquad R_1 + R_3 \rightarrow R_3$

From this matrix, we conclude that the inverse does not exist.

65.
$$\begin{bmatrix} -1 & -2 & 2 & | & 1 & 0 & 0 \\ 4 & 3 & 0 & | & 0 & 1 & 0 \\ 4 & 0 & 4 & | & 0 & 0 & 1 \end{bmatrix} (-1)R_1 \rightarrow R_1 \sim \begin{bmatrix} 1 & 2 & -2 & | & -1 & 0 & 0 \\ 4 & 3 & 0 & | & 0 & 1 & 0 \\ 4 & 0 & 4 & | & 0 & 0 & 1 \end{bmatrix} \begin{matrix} (-4)R_1 + R_2 \rightarrow R_2 \\ (-4)R_1 + R_3 \rightarrow R_3 \end{matrix}$$

$$\sim \begin{bmatrix} 1 & 2 & -2 & | & -1 & 0 & 0 \\ 0 & -5 & 8 & | & 4 & 1 & 0 \\ 0 & -8 & 12 & | & 4 & 0 & 1 \end{bmatrix} \left(-\frac{1}{5}\right)R_2 \rightarrow R_2 \sim \begin{bmatrix} 1 & 2 & -2 & | & -1 & 0 & 0 \\ 0 & 1 & -\frac{8}{5} & | & -\frac{4}{5} & -\frac{1}{5} & 0 \\ 0 & -8 & 12 & | & 4 & 0 & 1 \end{bmatrix} \begin{matrix} (-2)R_2 + R_1 \rightarrow R_1 \\ 8R_2 + R_3 \rightarrow R_3 \end{matrix}$$

$$\sim \begin{bmatrix} 1 & 0 & \frac{6}{5} & | & \frac{3}{5} & \frac{2}{5} & 0 \\ 0 & 1 & -\frac{8}{5} & | & -\frac{4}{5} & -\frac{1}{5} & 0 \\ 0 & 0 & -\frac{4}{5} & | & -\frac{12}{5} & -\frac{8}{5} & 1 \end{bmatrix} \left(-\frac{5}{4}\right)R_3 \rightarrow R_3$$

$$\sim \begin{bmatrix} 1 & 0 & \frac{6}{5} & | & \frac{3}{5} & \frac{2}{5} & 0 \\ 0 & 1 & -\frac{8}{5} & | & -\frac{4}{5} & -\frac{1}{5} & 0 \\ 0 & 0 & 1 & | & 3 & 2 & -\frac{5}{4} \end{bmatrix} \begin{matrix} \left(-\frac{6}{5}\right)R_3 + R_1 \rightarrow R_1 \\ \left(\frac{8}{5}\right)R_3 + R_2 \rightarrow R_2 \end{matrix}$$

$$\sim \begin{bmatrix} 1 & 0 & 0 & | & -3 & -2 & \frac{3}{2} \\ 0 & 1 & 0 & | & 4 & 3 & -2 \\ 0 & 0 & 1 & | & 3 & 2 & -\frac{5}{4} \end{bmatrix}$$

$$M^{-1} = \begin{bmatrix} -3 & -2 & \frac{3}{2} \\ 4 & 3 & -2 \\ 3 & 2 & -\frac{5}{4} \end{bmatrix} = \begin{bmatrix} -3 & -2 & 1.5 \\ 4 & 3 & -2 \\ 3 & 2 & -1.25 \end{bmatrix}$$

67.
$$\begin{bmatrix} 2 & -1 & -2 & | & 1 & 0 & 0 \\ -4 & 2 & 8 & | & 0 & 1 & 0 \\ 6 & -2 & -1 & | & 0 & 0 & 1 \end{bmatrix} \frac{1}{2}R_1 \rightarrow R_1$$

$$\sim \begin{bmatrix} 1 & -\frac{1}{2} & -1 & | & \frac{1}{2} & 0 & 0 \\ -4 & 2 & 8 & | & 0 & 1 & 0 \\ 6 & -2 & -1 & | & 0 & 0 & 1 \end{bmatrix} \begin{matrix} 4R_1 + R_2 \rightarrow R_2 \\ (-6)R_1 + R_3 \rightarrow R_3 \end{matrix}$$

$$\sim \begin{bmatrix} 1 & -\frac{1}{2} & -1 & | & \frac{1}{2} & 0 & 0 \\ 0 & 0 & 4 & | & 2 & 1 & 0 \\ 0 & 1 & 5 & | & -3 & 0 & 1 \end{bmatrix} R_2 \leftrightarrow R_3$$

$$\sim \begin{bmatrix} 1 & -\frac{1}{2} & -1 & | & \frac{1}{2} & 0 & 0 \\ 0 & 1 & 5 & | & -3 & 0 & 1 \\ 0 & 0 & 4 & | & 2 & 1 & 0 \end{bmatrix} \begin{matrix} \frac{1}{2}R_2 + R_1 \rightarrow R_1 \\ \frac{1}{4}R_3 \rightarrow R_3 \end{matrix}$$

$$\sim \begin{bmatrix} 1 & 0 & \frac{3}{2} & -1 & 0 & \frac{1}{2} \\ 0 & 1 & 5 & -3 & 0 & 1 \\ 0 & 0 & 1 & \frac{1}{2} & \frac{1}{4} & 0 \end{bmatrix} \begin{matrix} \left(-\dfrac{3}{2}\right)R_3 + R_1 \to R_1 \\ \\ (-5)R_3 + R_2 \to R_2 \end{matrix}$$

$$\sim \begin{bmatrix} 1 & 0 & 0 & -\frac{7}{4} & -\frac{3}{8} & \frac{1}{2} \\ 0 & 1 & 0 & -\frac{11}{2} & -\frac{5}{4} & 1 \\ 0 & 0 & 1 & \frac{1}{2} & \frac{1}{4} & 0 \end{bmatrix}$$

$$M^{-1} = \begin{bmatrix} -\frac{7}{4} & -\frac{3}{8} & \frac{1}{2} \\ -\frac{11}{2} & -\frac{5}{4} & 1 \\ \frac{1}{2} & \frac{1}{4} & 0 \end{bmatrix} = \begin{bmatrix} -1.75 & -0.375 & 0.5 \\ -5.5 & -1.25 & 1 \\ 0.5 & 0.25 & 0 \end{bmatrix}$$

69. $A = \begin{bmatrix} 4 & 3 \\ 3 & 2 \end{bmatrix};$ $\begin{bmatrix} 4 & 3 & 1 & 0 \\ 3 & 2 & 0 & 1 \end{bmatrix}(-1)R_2 + R_1 \to R_1$

$\sim \begin{bmatrix} 1 & 1 & 1 & -1 \\ 3 & 2 & 0 & 1 \end{bmatrix}(-3)R_1 + R_2 \to R_2 \sim \begin{bmatrix} 1 & 1 & 1 & -1 \\ 0 & -1 & -3 & 4 \end{bmatrix}(-1)R_2 \to R_2$

$\sim \begin{bmatrix} 1 & 1 & 1 & -1 \\ 0 & 1 & 3 & -4 \end{bmatrix}(-1)R_2 + R_1 \to R_1 \sim \begin{bmatrix} 1 & 0 & -2 & 3 \\ 0 & 1 & 3 & -4 \end{bmatrix}$

$A^{-1} = \begin{bmatrix} -2 & 3 \\ 3 & -4 \end{bmatrix};$ $\begin{bmatrix} -2 & 3 & 1 & 0 \\ 3 & -4 & 0 & 1 \end{bmatrix}R_2 + R_1 \to R_1$

$\sim \begin{bmatrix} 1 & -1 & 1 & 1 \\ 3 & -4 & 0 & 1 \end{bmatrix}(-3)R_1 + R_2 \to R_2 \sim \begin{bmatrix} 1 & -1 & 1 & 1 \\ 0 & -1 & -3 & -2 \end{bmatrix}(-1)R_2 \to R_2$

$\sim \begin{bmatrix} 1 & -1 & 1 & 1 \\ 0 & 1 & 3 & 2 \end{bmatrix}R_2 + R_1 \to R_1 \sim \begin{bmatrix} 1 & 0 & 4 & 3 \\ 0 & 1 & 3 & 2 \end{bmatrix}$

Thus, $(A^{-1})^{-1} = \begin{bmatrix} 4 & 3 \\ 3 & 2 \end{bmatrix} = A$

71. $\begin{bmatrix} a & 0 & 1 & 0 \\ 0 & d & 0 & 1 \end{bmatrix}\begin{matrix} \frac{1}{a}R_1 \to R_1, \text{ provided } a \neq 0 \\ \frac{1}{d}R_2 \to R_2, \text{ provided } d \neq 0 \end{matrix}$

$\begin{bmatrix} 1 & 0 & \frac{1}{a} & 0 \\ 0 & 1 & 0 & \frac{1}{d} \end{bmatrix}$. M^{-1} exists and equals $\begin{bmatrix} \frac{1}{a} & 0 \\ 0 & \frac{1}{d} \end{bmatrix}$ if and only if $a \neq 0$, $d \neq 0$. In general, the inverse of a

diagonal matrix exists if and only if each of the diagonal elements is non-zero.

73. $A = \begin{bmatrix} 3 & 2 \\ -4 & -3 \end{bmatrix}$

$A^{-1}:\begin{bmatrix} 3 & 2 & 1 & 0 \\ -4 & -3 & 0 & 1 \end{bmatrix} \sim \begin{bmatrix} -1 & -1 & 1 & 1 \\ -4 & -3 & 0 & 1 \end{bmatrix} \sim \begin{bmatrix} 1 & 1 & -1 & -1 \\ -4 & -3 & 0 & 1 \end{bmatrix}$

$\qquad\qquad R_1 + R_2 \to R_1 \qquad (-1)R_1 \to R_1 \qquad 4R_1 + R_2 \to R_2$

$\sim \begin{bmatrix} 1 & 1 & -1 & -1 \\ 0 & 1 & -4 & -3 \end{bmatrix} \sim \begin{bmatrix} 1 & 0 & 3 & 2 \\ 0 & 1 & -4 & -3 \end{bmatrix}$

$\quad (-1)R_2 + R_1 \to R_1$

$A^{-1} = \begin{bmatrix} 3 & 2 \\ -4 & -3 \end{bmatrix} = A;$ $A^2 = AA = AA^{-1} = \begin{bmatrix} 1 & 0 \\ 0 & 1 \end{bmatrix} = I$

75. $A = \begin{bmatrix} 4 & 3 \\ -5 & -4 \end{bmatrix}$

$A^{-1}:$ $\begin{bmatrix} 4 & 3 & | & 1 & 0 \\ -5 & -4 & | & 0 & 1 \end{bmatrix} \sim \begin{bmatrix} -1 & -1 & | & 1 & 1 \\ -5 & -4 & | & 0 & 1 \end{bmatrix} \sim \begin{bmatrix} 1 & 1 & | & -1 & -1 \\ -5 & -4 & | & 0 & 1 \end{bmatrix}$

$\qquad R_2 + R_1 \rightarrow R_1 \qquad (-1)R_1 \rightarrow R_1 \qquad 5R_1 + R_2 \rightarrow R_2$

$\sim \begin{bmatrix} 1 & 1 & | & -1 & -1 \\ 0 & 1 & | & -5 & -4 \end{bmatrix} \sim \begin{bmatrix} 1 & 0 & | & 4 & 3 \\ 0 & 1 & | & -5 & -4 \end{bmatrix}$

$\qquad (-1)R_2 + R_1 \rightarrow R_1$

$A^{-1} = \begin{bmatrix} 4 & 3 \\ -5 & -4 \end{bmatrix} = A; \quad A^2 = AA = AA^{-1} = I$

77. $A = \begin{bmatrix} 1 & 2 \\ 1 & 3 \end{bmatrix}$ The message "WINGARDIUM LEVIOSA" corresponds to the sequence

23 9 14 7 1 18 4 9 21 13 0 12 5 22 9 15 19 1

The matrix corresponding to this message is: $B = \begin{bmatrix} 23 & 14 & 1 & 4 & 21 & 0 & 5 & 9 & 19 \\ 9 & 7 & 18 & 9 & 13 & 12 & 22 & 15 & 1 \end{bmatrix}$

$AB = \begin{bmatrix} 41 & 28 & 37 & 22 & 47 & 24 & 49 & 39 & 21 \\ 50 & 35 & 55 & 31 & 60 & 36 & 71 & 54 & 22 \end{bmatrix}$. The coded message is:

41 50 28 35 37 55 22 31 47 60 24 36 49 71 39 54 21 22.

79. First we must find the inverse of $A = \begin{bmatrix} 1 & 2 \\ 1 & 3 \end{bmatrix}$

$\begin{bmatrix} 1 & 2 & | & 1 & 0 \\ 1 & 3 & | & 0 & 1 \end{bmatrix} \quad \sim \quad \begin{bmatrix} 1 & 2 & | & 1 & 0 \\ 0 & 1 & | & -1 & 1 \end{bmatrix} \sim \begin{bmatrix} 1 & 0 & | & 3 & -2 \\ 0 & 1 & | & -1 & 1 \end{bmatrix}$

$(-1)R_1 + R_2 \rightarrow R_2 \qquad (-2)R_2 + R_1 \rightarrow R_1$

Thus, $A^{-1} = \begin{bmatrix} 3 & -2 \\ -1 & 1 \end{bmatrix}$

Now $\begin{bmatrix} 3 & -2 \\ -1 & 1 \end{bmatrix} \begin{bmatrix} 52 & 17 & 5 & 29 & 4 & 52 & 25 & 29 & 15 & 5 \\ 70 & 21 & 5 & 43 & 4 & 70 & 35 & 33 & 18 & 5 \end{bmatrix} = \begin{bmatrix} 16 & 9 & 5 & -14 & 4 & 16 & 5 & 21 & 9 & 5 \\ 18 & 4 & 0 & 19 & 0 & 18 & 10 & 4 & 3 & 0 \end{bmatrix}$

Thus, the decoded message is
16 18 9 4 5 0 1 14 4 0 16 18 5 10 21 4 9 3 5 0

which corresponds to PRIDE AND PREJUDICE

81. The matrix corresponding to DEPART ISTANBUL ORIENT EXPRESS is:

$$C = \begin{bmatrix} 4 & 18 & 19 & 2 & 15 & 14 & 24 & 19 \\ 5 & 20 & 20 & 21 & 18 & 20 & 16 & 19 \\ 16 & 0 & 1 & 12 & 9 & 0 & 18 & 0 \\ 1 & 9 & 14 & 0 & 5 & 5 & 5 & 0 \end{bmatrix}, \quad BC = \begin{bmatrix} 37 & 103 & 121 & 58 & 90 & 83 & 113 & 76 \\ 47 & 67 & 75 & 68 & 74 & 59 & 97 & 57 \\ 10 & 47 & 53 & 23 & 38 & 39 & 45 & 38 \\ 58 & 123 & 142 & 91 & 117 & 103 & 147 & 95 \end{bmatrix}$$

The encoded message is 37 47 10 58 103 67 47 123 121 75 53 142 58 68 23 91 90 74 38 117 83 59 39 103 113 97 45 147 76 57 38 95

83. First, find the inverse of B: $B^{-1} = \begin{bmatrix} 2 & 2 & 1 & -3 \\ -2 & -1 & 1 & 2 \\ 1 & 1 & -1 & -1 \\ 0 & -1 & -1 & 1 \end{bmatrix}$. The matrix for the encoded message is:

$$C = \begin{bmatrix} 85 & 31 & 139 & 61 & 69 & 18 \\ 74 & 27 & 73 & 70 & 59 & 13 \\ 27 & 13 & 58 & 18 & 23 & 9 \\ 109 & 40 & 154 & 93 & 87 & 22 \end{bmatrix}. \quad B^{-1}C = \begin{bmatrix} 18 & 9 & 20 & 1 & 18 & 5 \\ 1 & 4 & 15 & 12 & 0 & 4 \\ 23 & 5 & 0 & 20 & 18 & 0 \\ 8 & 0 & 23 & 5 & 5 & 0 \end{bmatrix}$$

The message is RAWHIDE TO WALTER REED.

85. The matrix corresponding to THE EAGLE HAS LANDED is:

$$D = \begin{bmatrix} 20 & 1 & 8 & 1 \\ 8 & 7 & 1 & 14 \\ 5 & 12 & 19 & 4 \\ 0 & 5 & 0 & 5 \\ 5 & 0 & 12 & 4 \end{bmatrix}, \quad CD = \begin{bmatrix} 30 & 13 & 39 & 9 \\ 28 & 19 & 56 & 30 \\ 58 & 26 & 48 & 29 \\ 15 & 12 & 43 & 12 \\ 38 & 30 & 40 & 33 \end{bmatrix}$$

The encoded message is 30 28 58 15 38 13 19 26 12 30 39 56 48 43 40 9 30 29 12 33

87. First find the inverse of C: $C^{-1} = \begin{bmatrix} -2 & -1 & 2 & 2 & -1 \\ 3 & 2 & -2 & -4 & 1 \\ 6 & 2 & -4 & -5 & 2 \\ -2 & -1 & 1 & 2 & 0 \\ -3 & -1 & 2 & 3 & -1 \end{bmatrix}$. The matrix for the encoded message is:

$$D = \begin{bmatrix} 37 & 30 & 27 & 28 & 32 & 22 \\ 72 & 67 & 77 & 24 & 58 & 38 \\ 58 & 50 & 41 & 52 & 70 & 70 \\ 45 & 46 & 45 & 14 & 36 & 12 \\ 56 & 60 & 39 & 37 & 76 & 67 \end{bmatrix}. \quad C^{-1}D = \begin{bmatrix} 4 & 5 & 2 & 15 & 14 & 15 \\ 15 & 0 & 12 & 9 & 4 & 21 \\ 21 & 4 & 5 & 12 & 0 & 2 \\ 2 & 15 & 0 & 0 & 20 & 12 \\ 12 & 21 & 20 & 1 & 18 & 5 \end{bmatrix}$$

The message is: DOUBLE DOUBLE TOIL AND TROUBLE

EXERCISE 4-6

Things to remember:

1. BASIC PROPERTIES OF MATRICES

    Assuming all products and sums are defined for the indicated matrices *A, B, C, I,* and *O*, then

    ADDITION PROPERTIES

Associative: :	$(A + B) + C = A + (B + C)$
Commutative:	$A + B = B + A$
Additive Identity:	$A + 0 = 0 + A = A$
Additive Inverse:	$A + (-A) = (-A) + A = 0$

    MULTIPLICATION PROPERTIES

Associative Property:	$A(BC) = (AB)C$
Multiplicative Identity:	$AI = IA = A$
Multiplicative Inverse:	If A is a square matrix and A^{-1} exists, then $AA^{-1} = A^{-1}A = I$.

    COMBINED PROPERTIES

Left Distributive:	$A(B + C) = AB + AC$
Right Distributive:	$(B + C)A = BA + CA$

    EQUALITY

Addition:	If $A = B$, then $A + C = B + C$.
Left Multiplication:	If $A = B$, then $CA = CB$.
Right Multiplication:	If $A = B$, then $AC = BC$.

2. USING INVERSE METHODS TO SOLVE SYSTEMS OF EQUATIONS

    If the number of equations in a system equals the number of variables and the coefficient matrix has an inverse, then the system will always have a unique solution that can be found by using the inverse of the coefficient matrix to solve the corresponding matrix equation.

Matrix Equation	Solution
$AX = B$	$X = A^{-1}B$

---

1. $5x = -3;\ x = -3/5$

3. $4x = 8x + 7,\ -4x = 7,\ x = -7/4$

5. $6x + 8 = -2x + 17,\ 8x = 9,\ x = 9/8$

7. $10 - 3x = 7x + 9,\ -10x = -1,\ x = 1/10$

9. $\begin{bmatrix} 3 & 1 \\ 2 & -1 \end{bmatrix} \begin{bmatrix} x_1 \\ x_2 \end{bmatrix} = \begin{bmatrix} 5 \\ -4 \end{bmatrix}$

$\begin{bmatrix} 3x_1 + x_2 \\ 2x_1 - x_2 \end{bmatrix} = \begin{bmatrix} 5 \\ -4 \end{bmatrix}$

Thus, $3x_1 + x_2 = 5$
$\qquad 2x_1 - x_2 = -4$

11. $\begin{bmatrix} -3 & 1 & 0 \\ 2 & 0 & 1 \\ -1 & 3 & -2 \end{bmatrix} \begin{bmatrix} x_1 \\ x_2 \\ x_3 \end{bmatrix} = \begin{bmatrix} 3 \\ -4 \\ 2 \end{bmatrix}$

$\begin{bmatrix} -3x_1 + x_2 \\ 2x_1 + x_3 \\ -x_1 + 3x_2 - 2x_3 \end{bmatrix} = \begin{bmatrix} 3 \\ -4 \\ 2 \end{bmatrix}$

Thus, $\quad -3x_1 + x_2 \qquad = 3$
$\qquad\qquad 2x_1 \qquad + x_3 = -4$
$\qquad\qquad -x_1 + 3x_2 - 2x_3 = 2$

13. $3x_1 - 4x_2 = 1$
$\quad 2x_1 + \ x_2 = 5$

$\begin{bmatrix} 3x_1 - 4x_2 \\ 2x_1 + x_2 \end{bmatrix} = \begin{bmatrix} 1 \\ 5 \end{bmatrix}$ and $\begin{bmatrix} 3 & -4 \\ 2 & 1 \end{bmatrix} \begin{bmatrix} x_1 \\ x_2 \end{bmatrix} = \begin{bmatrix} 1 \\ 5 \end{bmatrix}$

15. $\quad x_1 - 3x_2 + 2x_3 = -3$
$\ -2x_1 + 3x_2 \qquad = 1$
$\quad x_1 + \ x_2 + 4x_3 = -2$

$\begin{bmatrix} x_1 & -3x_2 & +2x_3 \\ -2x_1 & +3x_2 & \\ x_1 & +x_2 & +4x_3 \end{bmatrix} = \begin{bmatrix} -3 \\ 1 \\ -2 \end{bmatrix}$ and $\begin{bmatrix} 1 & -3 & 2 \\ -2 & 3 & 0 \\ 1 & 1 & 4 \end{bmatrix} \begin{bmatrix} x_1 \\ x_2 \\ x_3 \end{bmatrix} = \begin{bmatrix} -3 \\ 1 \\ -2 \end{bmatrix}$

17. $\begin{bmatrix} x_1 \\ x_2 \end{bmatrix} = \begin{bmatrix} 3 & -2 \\ 1 & 4 \end{bmatrix} \begin{bmatrix} -2 \\ 1 \end{bmatrix} = \begin{bmatrix} 3(-2) + (-2)1 \\ 1(-2) + 4 \cdot 1 \end{bmatrix} = \begin{bmatrix} -8 \\ 2 \end{bmatrix}$ Thus, $x_1 = -8$
$\qquad\qquad\qquad\qquad\qquad\qquad\qquad\qquad\qquad\qquad\qquad\qquad\qquad$ and $x_2 = 2$

19. $\begin{bmatrix} x_1 \\ x_2 \end{bmatrix} = \begin{bmatrix} -2 & 3 \\ 2 & -1 \end{bmatrix} \begin{bmatrix} 3 \\ 2 \end{bmatrix} = \begin{bmatrix} (-2)3 + 3 \cdot 2 \\ 2 \cdot 3 + (-1)2 \end{bmatrix} = \begin{bmatrix} 0 \\ 4 \end{bmatrix}$ Thus, $x_1 = 0$
$\qquad\qquad\qquad\qquad\qquad\qquad\qquad\qquad\qquad\qquad\qquad\qquad\qquad$ and $x_2 = 4$

21. $\begin{bmatrix} 1 & -1 \\ 1 & -2 \end{bmatrix} \begin{bmatrix} x_1 \\ x_2 \end{bmatrix} = \begin{bmatrix} 5 \\ 7 \end{bmatrix}$

If $A = \begin{bmatrix} 1 & -1 \\ 1 & -2 \end{bmatrix}$ has an inverse, then $\begin{bmatrix} x_1 \\ x_2 \end{bmatrix} = A^{-1} \begin{bmatrix} 5 \\ 7 \end{bmatrix}$.

$\begin{bmatrix} 1 & -1 & | & 1 & 0 \\ 1 & -2 & | & 0 & 1 \end{bmatrix} \sim \begin{bmatrix} 1 & -1 & | & 1 & 0 \\ 0 & -1 & | & -1 & 1 \end{bmatrix} \sim \begin{bmatrix} 1 & -1 & | & 1 & 0 \\ 0 & 1 & | & 1 & -1 \end{bmatrix}$

$(-1)R_1 + R_2 \to R_2 \qquad (-1)R_2 \to R_2 \qquad\qquad R_2 + R_1 \to R_1$

$\sim \begin{bmatrix} 1 & 0 & | & 2 & -1 \\ 0 & 1 & | & 1 & -1 \end{bmatrix}$; $A^{-1} = \begin{bmatrix} 2 & -1 \\ 1 & -1 \end{bmatrix}$, $\begin{bmatrix} x_1 \\ x_2 \end{bmatrix} = \begin{bmatrix} 2 & -1 \\ 1 & -1 \end{bmatrix} \begin{bmatrix} 5 \\ 7 \end{bmatrix} = \begin{bmatrix} 3 \\ -2 \end{bmatrix}$

Therefore, $x_1 = 3$, $x_2 = -2$.

23. $\begin{bmatrix} 1 & 1 \\ 2 & -3 \end{bmatrix} \begin{bmatrix} x_1 \\ x_2 \end{bmatrix} = \begin{bmatrix} 15 \\ 10 \end{bmatrix}$

If $A = \begin{bmatrix} 1 & 1 \\ 2 & -3 \end{bmatrix}$ has an inverse, then $\begin{bmatrix} x_1 \\ x_2 \end{bmatrix} = A^{-1} \begin{bmatrix} 15 \\ 10 \end{bmatrix}$

$\left[\begin{array}{cc|cc} 1 & 1 & 1 & 0 \\ 2 & -3 & 0 & 1 \end{array} \right] \sim \left[\begin{array}{cc|cc} 1 & 1 & 1 & 0 \\ 0 & -5 & -2 & 1 \end{array} \right] \sim \left[\begin{array}{cc|cc} 1 & 1 & 1 & 0 \\ 0 & 1 & \frac{2}{5} & -\frac{1}{5} \end{array} \right]$

$(-2)R_1 + R_2 \rightarrow R_2 \qquad \left(-\frac{1}{5}\right)R_2 \rightarrow R_2 \qquad (-1)R_2 + R_1 \rightarrow R_1$

$\sim \left[\begin{array}{cc|cc} 1 & 0 & \frac{3}{5} & \frac{1}{5} \\ 0 & 1 & \frac{2}{5} & -\frac{1}{5} \end{array} \right]; \quad A^{-1} = \begin{bmatrix} \frac{3}{5} & \frac{1}{5} \\ \frac{2}{5} & -\frac{1}{5} \end{bmatrix}, \quad \begin{bmatrix} x_1 \\ x_2 \end{bmatrix} = \begin{bmatrix} \frac{3}{5} & \frac{1}{5} \\ \frac{2}{5} & -\frac{1}{5} \end{bmatrix} \begin{bmatrix} 15 \\ 10 \end{bmatrix} = \begin{bmatrix} 11 \\ 4 \end{bmatrix}$

Therefore, $x_1 = 11$, $x_2 = 4$.

25. $\begin{bmatrix} 1 & 2 \\ 1 & 1 \end{bmatrix} \begin{bmatrix} x_1 \\ x_2 \end{bmatrix} + \begin{bmatrix} 3 \\ 4 \end{bmatrix} = \begin{bmatrix} 9 \\ 9 \end{bmatrix}$

$\begin{bmatrix} 1 & 2 \\ 1 & 1 \end{bmatrix} \begin{bmatrix} x_1 \\ x_2 \end{bmatrix} = \begin{bmatrix} 9 \\ 9 \end{bmatrix} - \begin{bmatrix} 3 \\ 4 \end{bmatrix} = \begin{bmatrix} 6 \\ 5 \end{bmatrix}$

If $A = \begin{bmatrix} 1 & 2 \\ 1 & 1 \end{bmatrix}$ has an inverse, then $\begin{bmatrix} x_1 \\ x_2 \end{bmatrix} = A^{-1} \begin{bmatrix} 6 \\ 5 \end{bmatrix}$

$\left[\begin{array}{cc|cc} 1 & 2 & 1 & 0 \\ 1 & 1 & 0 & 1 \end{array} \right] \sim \left[\begin{array}{cc|cc} 1 & 2 & 1 & 0 \\ 0 & -1 & -1 & 1 \end{array} \right] \sim \left[\begin{array}{cc|cc} 1 & 2 & 1 & 0 \\ 0 & 1 & 1 & -1 \end{array} \right] \sim \left[\begin{array}{cc|cc} 1 & 0 & -1 & 2 \\ 0 & 1 & 1 & -1 \end{array} \right]$

$(-1)R_1 + R_2 \rightarrow R_2 \quad (-1)R_2 \rightarrow R_2 \qquad (-2)R_2 + R_1 \rightarrow R_1$

$A^{-1} = \begin{bmatrix} -1 & 2 \\ 1 & -1 \end{bmatrix}; \quad \begin{bmatrix} x_1 \\ x_2 \end{bmatrix} = \begin{bmatrix} -1 & 2 \\ 1 & -1 \end{bmatrix} \begin{bmatrix} 6 \\ 5 \end{bmatrix} = \begin{bmatrix} 4 \\ 1 \end{bmatrix}.$

Therefore, $x_1 = 4$, $x_2 = 1$.

27. $\begin{bmatrix} 2 & 2 \\ 2 & 3 \end{bmatrix} \begin{bmatrix} x_1 \\ x_2 \end{bmatrix} + \begin{bmatrix} -5 \\ 2 \end{bmatrix} = \begin{bmatrix} 2 \\ 3 \end{bmatrix}$

$\begin{bmatrix} 2 & 2 \\ 2 & 3 \end{bmatrix} \begin{bmatrix} x_1 \\ x_2 \end{bmatrix} = \begin{bmatrix} 2 \\ 3 \end{bmatrix} - \begin{bmatrix} -5 \\ 2 \end{bmatrix} = \begin{bmatrix} 7 \\ 1 \end{bmatrix}$

If $A = \begin{bmatrix} 2 & 2 \\ 2 & 3 \end{bmatrix}$ has an inverse, then $\begin{bmatrix} x_1 \\ x_2 \end{bmatrix} = A^{-1} \begin{bmatrix} 7 \\ 1 \end{bmatrix}$

$\left[\begin{array}{cc|cc} 2 & 2 & 1 & 0 \\ 2 & 3 & 0 & 1 \end{array} \right] \sim \left[\begin{array}{cc|cc} 2 & 2 & 0 & 1 \\ 0 & 1 & -1 & 1 \end{array} \right] \sim \left[\begin{array}{cc|cc} 2 & 0 & 3 & -2 \\ 0 & 1 & -1 & 1 \end{array} \right] \sim \left[\begin{array}{cc|cc} 1 & 0 & \frac{3}{2} & -1 \\ 0 & 1 & -1 & 1 \end{array} \right]$

$(-1)R_1 + R_2 \rightarrow R_2 \quad (-2)R_2 + R_1 \rightarrow R_1 \qquad \left(\frac{1}{2}\right)R_1 \rightarrow R_1$

$A^{-1} = \begin{bmatrix} \frac{3}{2} & -1 \\ -1 & 1 \end{bmatrix}; \quad \begin{bmatrix} x_1 \\ x_2 \end{bmatrix} = \begin{bmatrix} \frac{3}{2} & -1 \\ -1 & 1 \end{bmatrix} \begin{bmatrix} 7 \\ 1 \end{bmatrix} = \begin{bmatrix} \frac{19}{2} \\ -6 \end{bmatrix}.$

Therefore, $x_1 = \dfrac{19}{2}$, $x_2 = -6$.

29.
$$\begin{bmatrix} 3 & -2 \\ 6 & -4 \end{bmatrix}\begin{bmatrix} x_1 \\ x_2 \end{bmatrix} + \begin{bmatrix} 0 \\ -3 \end{bmatrix} = \begin{bmatrix} 1 \\ -2 \end{bmatrix}$$

$$\begin{bmatrix} 3 & -2 \\ 6 & -4 \end{bmatrix}\begin{bmatrix} x_1 \\ x_2 \end{bmatrix} = \begin{bmatrix} 1 \\ -2 \end{bmatrix} - \begin{bmatrix} 0 \\ -3 \end{bmatrix} = \begin{bmatrix} 1 \\ 1 \end{bmatrix}$$

If $A = \begin{bmatrix} 3 & -2 \\ 6 & -4 \end{bmatrix}$ has an inverse, then $\begin{bmatrix} x_1 \\ x_2 \end{bmatrix} = A^{-1}\begin{bmatrix} 1 \\ 1 \end{bmatrix}$.

$$\begin{bmatrix} 3 & -2 & | & 1 & 0 \\ 6 & -4 & | & 0 & 1 \end{bmatrix} \sim \begin{bmatrix} 3 & -2 & | & 1 & 0 \\ 0 & 0 & | & -2 & 1 \end{bmatrix};\ A \text{ does not have an inverse.}$$

$(-2)R_1 + R_2 \to R_2$

We conclude that the equation has either no solution or infinitely many solutions. We row reduce the augmented matrix:

$$\begin{bmatrix} 3 & -2 & | & 1 \\ 6 & -4 & | & 1 \end{bmatrix} \sim \begin{bmatrix} 3 & -2 & | & 1 \\ 0 & 0 & | & -1 \end{bmatrix};\quad \text{there is no solution.}$$

$(-2)R_1 + R_2 \to R_2$

31. The matrix equation for the given system is:
$$\begin{bmatrix} 1 & 2 \\ 1 & 3 \end{bmatrix}\begin{bmatrix} x_1 \\ x_2 \end{bmatrix} = \begin{bmatrix} k_1 \\ k_2 \end{bmatrix}$$

From Exercise 4-5, Problem 41, $\begin{bmatrix} 1 & 2 \\ 1 & 3 \end{bmatrix}^{-1} = \begin{bmatrix} 3 & -2 \\ -1 & 1 \end{bmatrix}$

Thus, $\begin{bmatrix} x_1 \\ x_2 \end{bmatrix} = \begin{bmatrix} 3 & -2 \\ -1 & 1 \end{bmatrix}\begin{bmatrix} k_1 \\ k_2 \end{bmatrix}$

(A) $\begin{bmatrix} x_1 \\ x_2 \end{bmatrix} = \begin{bmatrix} 3 & -2 \\ -1 & 1 \end{bmatrix}\begin{bmatrix} 1 \\ 3 \end{bmatrix} = \begin{bmatrix} -3 \\ 2 \end{bmatrix}$ Thus, $x_1 = -3$ and $x_2 = 2$

(B) $\begin{bmatrix} x_1 \\ x_2 \end{bmatrix} = \begin{bmatrix} 3 & -2 \\ -1 & 1 \end{bmatrix}\begin{bmatrix} 3 \\ 5 \end{bmatrix} = \begin{bmatrix} -1 \\ 2 \end{bmatrix}$ Thus, $x_1 = -1$ and $x_2 = 2$

(C) $\begin{bmatrix} x_1 \\ x_2 \end{bmatrix} = \begin{bmatrix} 3 & -2 \\ -1 & 1 \end{bmatrix}\begin{bmatrix} -2 \\ 1 \end{bmatrix} = \begin{bmatrix} -8 \\ 3 \end{bmatrix}$ Thus, $x_1 = -8$ and $x_2 = 3$

33. The matrix equation for the given system is:
$$\begin{bmatrix} 1 & 3 \\ 2 & 7 \end{bmatrix}\begin{bmatrix} x_1 \\ x_2 \end{bmatrix} = \begin{bmatrix} k_1 \\ k_2 \end{bmatrix}$$

From Exercise 4-5, Problem 43, $\begin{bmatrix} 1 & 3 \\ 2 & 7 \end{bmatrix}^{-1} = \begin{bmatrix} 7 & -3 \\ -2 & 1 \end{bmatrix}$

Thus, $\begin{bmatrix} x_1 \\ x_2 \end{bmatrix} = \begin{bmatrix} 7 & -3 \\ -2 & 1 \end{bmatrix}\begin{bmatrix} k_1 \\ k_2 \end{bmatrix}$

(A) $\begin{bmatrix} x_1 \\ x_2 \end{bmatrix} = \begin{bmatrix} 7 & -3 \\ -2 & 1 \end{bmatrix}\begin{bmatrix} 2 \\ -1 \end{bmatrix} = \begin{bmatrix} 17 \\ -5 \end{bmatrix}$ Thus, $x_1 = 17$ and $x_2 = -5$

(B) $\begin{bmatrix} x_1 \\ x_2 \end{bmatrix} = \begin{bmatrix} 7 & -3 \\ -2 & 1 \end{bmatrix}\begin{bmatrix} 1 \\ 0 \end{bmatrix} = \begin{bmatrix} 7 \\ -2 \end{bmatrix}$ Thus, $x_1 = 7$ and $x_2 = -2$

(C) $\begin{bmatrix} x_1 \\ x_2 \end{bmatrix} = \begin{bmatrix} 7 & -3 \\ -2 & 1 \end{bmatrix}\begin{bmatrix} 3 \\ -1 \end{bmatrix} = \begin{bmatrix} 24 \\ -7 \end{bmatrix}$ Thus, $x_1 = 24$
and $x_2 = -7$

35. The matrix equation for the given system is:

$$\begin{bmatrix} 1 & -3 & 0 \\ 0 & 1 & 1 \\ 2 & -1 & 4 \end{bmatrix}\begin{bmatrix} x_1 \\ x_2 \\ x_3 \end{bmatrix} = \begin{bmatrix} k_1 \\ k_2 \\ k_3 \end{bmatrix}$$

From Exercise 4-5, Problem 45, $\begin{bmatrix} 1 & -3 & 0 \\ 0 & 1 & 1 \\ 2 & -1 & 4 \end{bmatrix}^{-1} = \begin{bmatrix} -5 & -12 & 3 \\ -2 & -4 & 1 \\ 2 & 5 & -1 \end{bmatrix}$

Thus,

$$\begin{bmatrix} x_1 \\ x_2 \\ x_3 \end{bmatrix} = \begin{bmatrix} -5 & -12 & 3 \\ -2 & -4 & 1 \\ 2 & 5 & -1 \end{bmatrix}\begin{bmatrix} k_1 \\ k_2 \\ k_3 \end{bmatrix}$$

(A) $\begin{bmatrix} x_1 \\ x_2 \\ x_3 \end{bmatrix} = \begin{bmatrix} -5 & -12 & 3 \\ -2 & -4 & 1 \\ 2 & 5 & -1 \end{bmatrix}\begin{bmatrix} 1 \\ 0 \\ 2 \end{bmatrix} = \begin{bmatrix} 1 \\ 0 \\ 0 \end{bmatrix}$; $x_1 = 1$, $x_2 = 0$, $x_3 = 0$

(B) $\begin{bmatrix} x_1 \\ x_2 \\ x_3 \end{bmatrix} = \begin{bmatrix} -5 & -12 & 3 \\ -2 & -4 & 1 \\ 2 & 5 & -1 \end{bmatrix}\begin{bmatrix} -1 \\ 1 \\ 0 \end{bmatrix} = \begin{bmatrix} -7 \\ -2 \\ 3 \end{bmatrix}$; $x_1 = -7$, $x_2 = -2$, $x_3 = 3$

(C) $\begin{bmatrix} x_1 \\ x_2 \\ x_3 \end{bmatrix} = \begin{bmatrix} -5 & -12 & 3 \\ -2 & -4 & 1 \\ 2 & 5 & -1 \end{bmatrix}\begin{bmatrix} 2 \\ -2 \\ 1 \end{bmatrix} = \begin{bmatrix} 17 \\ 5 \\ -7 \end{bmatrix}$; $x_1 = 17$, $x_2 = 5$, $x_3 = -7$

37. The matrix equation for the given system is:

$$\begin{bmatrix} 1 & 1 & 0 \\ 2 & 3 & -1 \\ 1 & 0 & 2 \end{bmatrix}\begin{bmatrix} x_1 \\ x_2 \\ x_3 \end{bmatrix} = \begin{bmatrix} k_1 \\ k_2 \\ k_3 \end{bmatrix}$$

From Exercise 4-5, Problem 47, $\begin{bmatrix} 1 & 1 & 0 \\ 2 & 3 & -1 \\ 1 & 0 & 2 \end{bmatrix}^{-1} = \begin{bmatrix} 6 & -2 & -1 \\ -5 & 2 & 1 \\ -3 & 1 & 1 \end{bmatrix}$

Thus,

$$\begin{bmatrix} x_1 \\ x_2 \\ x_3 \end{bmatrix} = \begin{bmatrix} 6 & -2 & -1 \\ -5 & 2 & 1 \\ -3 & 1 & 1 \end{bmatrix}\begin{bmatrix} k_1 \\ k_2 \\ k_3 \end{bmatrix}$$

(A) $\begin{bmatrix} x_1 \\ x_2 \\ x_3 \end{bmatrix} = \begin{bmatrix} 6 & -2 & -1 \\ -5 & 2 & 1 \\ -3 & 1 & 1 \end{bmatrix} \begin{bmatrix} 2 \\ 0 \\ 4 \end{bmatrix} = \begin{bmatrix} 8 \\ -6 \\ -2 \end{bmatrix}$; $x_1 = 8,\ x_2 = -6,\ x_3 = -2$

(B) $\begin{bmatrix} x_1 \\ x_2 \\ x_3 \end{bmatrix} = \begin{bmatrix} 6 & -2 & -1 \\ -5 & 2 & 1 \\ -3 & 1 & 1 \end{bmatrix} \begin{bmatrix} 0 \\ 4 \\ -2 \end{bmatrix} = \begin{bmatrix} -6 \\ 6 \\ 2 \end{bmatrix}$; $x_1 = -6,\ x_2 = 6,\ x_3 = 2$

(C) $\begin{bmatrix} x_1 \\ x_2 \\ x_3 \end{bmatrix} = \begin{bmatrix} 6 & -2 & -1 \\ -5 & 2 & 1 \\ -3 & 1 & 1 \end{bmatrix} \begin{bmatrix} 4 \\ 2 \\ 0 \end{bmatrix} = \begin{bmatrix} 20 \\ -16 \\ -10 \end{bmatrix}$; $x_1 = 20,\ x_2 = -16,\ x_3 = -10$

39. $AX = B$, $X = \dfrac{B}{A}$, the expression $\dfrac{B}{A}$, a quotient of matrices is not defined.

41. $XA = B$, $X = A^{-1}B$

Solution:

$(XA)A^{-1} = BA^{-1}$ multiply on the right by A^{-1}

$X(AA^{-1}) = BA^{-1}$ associative property

$\quad XI = BA^{-1}$ $AA^{-1} = I$, the identity matrix

$\quad X = BA^{-1}$

$X \ne A^{-1}B$ because matrix multiplication is not commutative; $A^{-1}B \ne BA^{-1}$ in general.

43. $AX = B$, $X = A^{-1}BA$, $X = B$

Solution:

$A^{-1}(AX) = A^{-1}(BA)$ multiply on the left by A^{-1}

$(A^{-1}A)X = A^{-1}BA$ associative property

$\quad IX = A^{-1}BA$ $AA^{-1} = I$, the identity matrix

$\quad X = A^{-1}BA$

$A^{-1}(BA) \ne (BA)A^{-1} = B(AA^{-1}) = B$ because matrix multiplication is not commutative.

45. $-2x_1 + 4x_2 = 5$

$6x_1 - 12x_2 = 15$

The second equation is a multiple (-3) of the first. Therefore, the system has infinitely many solutions. The solutions are:

$$x_1 = 2t + \frac{5}{2},\ \ x_2 = t,\ \ t \text{ any real number.}$$

47. $x_1 - 3x_2 - 2x_3 = -1$

$-2x_1 + 6x_2 + 4x_3 = 3$

The system is not "square" – 2 equations in 3 unknowns. The matrix of coefficients is 2×3; it does not have an inverse.

Solve the system by Gauss-Jordan elimination:

$$\begin{pmatrix} 1 & -3 & -2 & | & -1 \\ -2 & 6 & 4 & | & 3 \end{pmatrix} \sim \begin{pmatrix} 1 & -3 & -2 & | & -1 \\ 0 & 0 & 0 & | & 1 \end{pmatrix}$$

$2R_1 + R_2 \rightarrow R_2$

The system has no solution.

49. e_1: $\quad x_1 - 2x_2 + 3x_3 = 1$

e_2: $\quad 2x_1 - 3x_2 - 2x_3 = 3$

e_3: $\quad x_1 - x_2 - 5x_3 = 2$

Note that $e_3 = (-1)e_1 + e_2$. This implies that the system has infinitely many solutions; the coefficient

matrix does not have an inverse. Solve the system by Gauss-Jordan elimination:

$$\begin{bmatrix} 1 & -2 & 3 & | & 1 \\ 2 & -3 & -2 & | & 3 \\ 1 & -1 & -5 & | & 2 \end{bmatrix} \sim \begin{bmatrix} 1 & -2 & 3 & | & 1 \\ 0 & 1 & -8 & | & 1 \\ 0 & 1 & -8 & | & 1 \end{bmatrix} \sim \begin{bmatrix} 1 & -2 & 3 & | & 1 \\ 0 & 1 & -8 & | & 1 \\ 0 & 0 & 0 & | & 0 \end{bmatrix} \sim \begin{bmatrix} 1 & 0 & -13 & | & 3 \\ 0 & 1 & -8 & | & 1 \\ 0 & 0 & 0 & | & 0 \end{bmatrix}$$

$(-2)R_1 + R_2 \rightarrow R_2 \quad (-1)R_2 + R_3 \rightarrow R_3 \quad 2R_2 + R_1 \rightarrow R_1$

$(-1)R_1 + R_3 \rightarrow R_3$

Solutions: $x_1 = 3 + 13t$, $x_2 = 1 + 8t$, $x_3 = t$, t any real number.

51. $AX - BX = C$

$(A - B)X = C$

$\quad X = (A - B)^{-1}C$

53. $\quad AX + X = C$

$(A + I)X = C$, where I is the identity matrix of order n

$\quad X = (A + I)^{-1}C$

55. $\quad AX - C = D - BX$

$AX + BX = C + D$

$(A + B)X = C + D$

$\quad X = (A + B)^{-1}(C + D)$

57. The matrix equation for the given system is:

$$\begin{bmatrix} 1 & 2.001 \\ 1 & 2 \end{bmatrix} \begin{bmatrix} x_1 \\ x_2 \end{bmatrix} = \begin{bmatrix} k_1 \\ k_2 \end{bmatrix}$$

First we compute the inverse of $\begin{bmatrix} 1 & 2.001 \\ 1 & 2 \end{bmatrix}$

$$\begin{bmatrix} 1 & 2.001 & | & 1 & 0 \\ 1 & 2 & | & 0 & 1 \end{bmatrix} \sim \begin{bmatrix} 1 & 2.001 & | & 1 & 0 \\ 0 & -0.001 & | & -1 & 1 \end{bmatrix} \sim \begin{bmatrix} 1 & 2.001 & | & 1 & 0 \\ 0 & 1 & | & 1000 & -1000 \end{bmatrix}$$

$(-1)R_1 + R_2 \rightarrow R_2 \qquad (-1000)R_2 \rightarrow R_2 \qquad (-2.001)R_2 + R_1 \rightarrow R_1$

$$\sim \begin{bmatrix} 1 & 0 & | & -2000 & 2001 \\ 0 & 1 & | & 1000 & -1000 \end{bmatrix}$$

Thus, $\begin{bmatrix} 1 & 2.001 \\ 1 & 2 \end{bmatrix}^{-1} = \begin{bmatrix} -2000 & 2001 \\ 1000 & -1000 \end{bmatrix}$ and $\begin{bmatrix} x_1 \\ x_2 \end{bmatrix} = \begin{bmatrix} -2000 & 2001 \\ 1000 & -1000 \end{bmatrix} \begin{bmatrix} k_1 \\ k_2 \end{bmatrix}$

(A) $\begin{bmatrix} x_1 \\ x_2 \end{bmatrix} = \begin{bmatrix} -2000 & 2001 \\ 1000 & -1000 \end{bmatrix} \begin{bmatrix} 1 \\ 1 \end{bmatrix} = \begin{bmatrix} 1 \\ 0 \end{bmatrix}$; $x_1 = 1,\ x_2 = 0$

(B) $\begin{bmatrix} x_1 \\ x_2 \end{bmatrix} = \begin{bmatrix} -2000 & 2001 \\ 1000 & -1000 \end{bmatrix} \begin{bmatrix} 1 \\ 0 \end{bmatrix} = \begin{bmatrix} -2000 \\ 1000 \end{bmatrix}$; $x_1 = -2,000,\ x_2 = 1,000$

(C) $\begin{bmatrix} x_1 \\ x_2 \end{bmatrix} = \begin{bmatrix} -2000 & 2001 \\ 1000 & -1000 \end{bmatrix} \begin{bmatrix} 0 \\ 1 \end{bmatrix} = \begin{bmatrix} 2001 \\ -1000 \end{bmatrix}$; $x_1 = 2,001,\ x_2 = -1,000$

59. The matrix equation for the given system is:

$$\begin{bmatrix} 1 & 8 & 7 \\ 6 & 6 & 8 \\ 3 & 4 & 6 \end{bmatrix} \begin{bmatrix} x_1 \\ x_2 \\ x_3 \end{bmatrix} = \begin{bmatrix} 135 \\ 155 \\ 75 \end{bmatrix}$$

Thus, $\begin{bmatrix} x_1 \\ x_2 \\ x_3 \end{bmatrix} = \begin{bmatrix} 1 & 8 & 7 \\ 6 & 6 & 8 \\ 3 & 4 & 6 \end{bmatrix}^{-1} \begin{bmatrix} 135 \\ 155 \\ 75 \end{bmatrix} = \begin{bmatrix} -0.08 & 0.4 & -0.44 \\ 0.24 & 0.3 & -0.68 \\ -0.12 & -0.4 & 0.84 \end{bmatrix} \begin{bmatrix} 135 \\ 155 \\ 75 \end{bmatrix} = \begin{bmatrix} 18.2 \\ 27.9 \\ -15.2 \end{bmatrix}$ and

$x_1 = 18.2,\ x_2 = 27.9,\ x_3 = -15.2$

61. The matrix equation for the given system is:

$$\begin{bmatrix} 6 & 9 & 7 & 5 \\ 6 & 4 & 7 & 3 \\ 4 & 5 & 3 & 2 \\ 4 & 3 & 8 & 2 \end{bmatrix} \begin{bmatrix} x_1 \\ x_2 \\ x_3 \\ x_4 \end{bmatrix} = \begin{bmatrix} 250 \\ 195 \\ 145 \\ 125 \end{bmatrix}$$

Thus

$$\begin{bmatrix} x_1 \\ x_2 \\ x_3 \\ x_4 \end{bmatrix} = \begin{bmatrix} 6 & 9 & 7 & 5 \\ 6 & 4 & 7 & 3 \\ 4 & 5 & 3 & 2 \\ 4 & 3 & 8 & 2 \end{bmatrix}^{-1} \begin{bmatrix} 250 \\ 195 \\ 145 \\ 125 \end{bmatrix} = \begin{bmatrix} -0.25 & 0.37 & 0.28 & -0.21 \\ 0 & -0.4 & 0.4 & 0.2 \\ 0 & -0.16 & -0.04 & 0.28 \\ 0.5 & 0.5 & -1 & -0.5 \end{bmatrix} \begin{bmatrix} 250 \\ 195 \\ 145 \\ 125 \end{bmatrix} = \begin{bmatrix} 24 \\ 5 \\ -2 \\ 15 \end{bmatrix}$$

and $x_1 = 24,\ x_2 = 5,\ x_3 = -2,\ x_4 = 15$.

63. (A) Let x_1 = Number of \$25 tickets sold

$\qquad\quad$ x_2 = Number of \$35 tickets sold

The mathematical model is:

$$x_1 + x_2 = 10,000$$
$$25x_1 + 35x_2 = k\ ,\ \ k = 275,000,\ 300,000,\ 325,000$$

The corresponding matrix equation is:

$$\begin{bmatrix} 1 & 1 \\ 25 & 35 \end{bmatrix} = \begin{bmatrix} 10,000 \\ k \end{bmatrix}$$

Compute the inverse of the coefficient matrix A.

$$\begin{bmatrix} 1 & 1 & | & 1 & 0 \\ 25 & 35 & | & 0 & 1 \end{bmatrix} \sim \begin{bmatrix} 1 & 1 & | & 1 & 0 \\ 0 & 10 & | & -25 & 1 \end{bmatrix} \sim \begin{bmatrix} 1 & 1 & | & 1 & 0 \\ 0 & 1 & | & -\frac{5}{2} & \frac{1}{10} \end{bmatrix}$$

$(-25)R_1 + R_2 \to R_2 \quad \frac{1}{10}R_2 \to R_2 \quad (-1)R_2 + R_1 \to R_1$

$$\sim \begin{bmatrix} 1 & 0 & | & \frac{7}{2} & -\frac{1}{10} \\ 0 & 1 & | & -\frac{5}{2} & \frac{1}{10} \end{bmatrix} \quad \text{Thus,} \quad A^{-1} = \begin{bmatrix} \frac{7}{2} & -\frac{1}{10} \\ -\frac{5}{2} & \frac{1}{10} \end{bmatrix}$$

<u>Concert 1</u>: Return $275,000

$$\begin{bmatrix} x_1 \\ x_2 \end{bmatrix} = \begin{bmatrix} \frac{7}{2} & -\frac{1}{10} \\ -\frac{5}{2} & \frac{1}{10} \end{bmatrix} \begin{bmatrix} 10,000 \\ 275,000 \end{bmatrix} = \begin{bmatrix} 7,500 \\ 2,500 \end{bmatrix}$$

Thus, 7,500 $25 tickets and 2,500 $35 tickets must be sold.

<u>Concert 2</u>: Return $300,000

$$\begin{bmatrix} x_1 \\ x_2 \end{bmatrix} = \begin{bmatrix} \frac{7}{2} & -\frac{1}{10} \\ -\frac{5}{2} & \frac{1}{10} \end{bmatrix} \begin{bmatrix} 10,000 \\ 300,000 \end{bmatrix} = \begin{bmatrix} 5,000 \\ 5,000 \end{bmatrix}$$

Thus, 5,000 $25 tickets and 5,000 $35 tickets must be sold.

<u>Concert 3</u>: Return $325,000

$$\begin{bmatrix} x_1 \\ x_2 \end{bmatrix} = \begin{bmatrix} \frac{7}{2} & -\frac{1}{10} \\ -\frac{5}{2} & \frac{1}{10} \end{bmatrix} \begin{bmatrix} 10,000 \\ 325,000 \end{bmatrix} = \begin{bmatrix} 2,500 \\ 7,500 \end{bmatrix}$$

Thus, 2,500 $25 tickets and 7,500 $35 tickets must be sold.

(B) $200,000 Return?

$$\begin{bmatrix} x_1 \\ x_2 \end{bmatrix} \stackrel{?}{=} \begin{bmatrix} \frac{7}{2} & -\frac{1}{10} \\ -\frac{5}{2} & \frac{1}{10} \end{bmatrix} \begin{bmatrix} 10,000 \\ 200,000 \end{bmatrix} = \begin{bmatrix} 15,000 \\ -5,000 \end{bmatrix}$$

This is not possible; x_2 cannot be negative.

$400,000 Return?

$$\begin{bmatrix} x_1 \\ x_2 \end{bmatrix} \stackrel{?}{=} \begin{bmatrix} \frac{7}{2} & -\frac{1}{10} \\ -\frac{5}{2} & \frac{1}{10} \end{bmatrix} \begin{bmatrix} 10,000 \\ 400,000 \end{bmatrix} = \begin{bmatrix} -5,000 \\ 15,000 \end{bmatrix}$$

This is not possible; x_1 cannot be negative.

(C)   Fix a return k. Then

$$\begin{bmatrix} x_1 \\ x_2 \end{bmatrix} = \begin{bmatrix} \frac{7}{2} & -\frac{1}{10} \\ -\frac{5}{2} & \frac{1}{10} \end{bmatrix} \begin{bmatrix} 10,000 \\ k \end{bmatrix} = \begin{bmatrix} 35,000 - \frac{k}{10} \\ -25,000 + \frac{k}{10} \end{bmatrix}$$

Thus, $x_1 = 35,000 - \dfrac{k}{10}$ and $x_2 = -25,000 + \dfrac{k}{10}$.

Since $x_1 \geq 0$, $35,000 - \dfrac{k}{10} \geq 0$, $-\dfrac{k}{10} \geq -35,000$, $k \leq 350,000$

Since $x_2 \geq 0$, $-25,000 + \dfrac{k}{10} \geq 0$, $\dfrac{k}{10} \geq 25,000$, $k \geq 250,000$.

Thus, $250,000 \leq k \leq 350,000$; any number between (and including) $250,000 and $350,000 is a possible return.

65. Let x_1 = number of hours at Plant A

and x_2 = number of hours at Plant B

Then, the mathematical model is:

$10x_1 + 8x_2 = k_1$ (number of car frames)

$5x_1 + 8x_2 = k_2$ (number of truck frames)

The corresponding matrix equation is:

$$\begin{bmatrix} 10 & 8 \\ 5 & 8 \end{bmatrix}\begin{bmatrix} x_1 \\ x_2 \end{bmatrix} = \begin{bmatrix} k_1 \\ k_2 \end{bmatrix}$$

First we compute the inverse of $\begin{bmatrix} 10 & 8 \\ 5 & 8 \end{bmatrix}$

$$\begin{bmatrix} 10 & 8 & | & 1 & 0 \\ 5 & 8 & | & 0 & 1 \end{bmatrix} \sim \begin{bmatrix} 1 & \frac{4}{5} & | & \frac{1}{10} & 0 \\ 5 & 8 & | & 0 & 1 \end{bmatrix} \sim \begin{bmatrix} 1 & \frac{4}{5} & | & \frac{1}{10} & 0 \\ 0 & 4 & | & -\frac{1}{2} & 1 \end{bmatrix} \sim \begin{bmatrix} 1 & \frac{4}{5} & | & \frac{1}{10} & 0 \\ 0 & 1 & | & -\frac{1}{8} & \frac{1}{4} \end{bmatrix}$$

$$\frac{1}{10}R_1 \to R_1 \qquad (-5)R_1 + R_2 \to R_2 \qquad \frac{1}{4}R_2 \to R_2 \qquad \left(-\frac{4}{5}\right)R_2 + R_1 \to R_1$$

$$\sim \begin{bmatrix} 1 & 0 & | & \frac{1}{5} & -\frac{1}{5} \\ 0 & 1 & | & -\frac{1}{8} & \frac{1}{4} \end{bmatrix}$$

Thus $\begin{bmatrix} 10 & 8 \\ 5 & 8 \end{bmatrix}^{-1} = \begin{bmatrix} \frac{1}{5} & -\frac{1}{5} \\ -\frac{1}{8} & \frac{1}{4} \end{bmatrix}$ and $\begin{bmatrix} x_1 \\ x_2 \end{bmatrix} = \begin{bmatrix} \frac{1}{5} & -\frac{1}{5} \\ -\frac{1}{8} & \frac{1}{4} \end{bmatrix}\begin{bmatrix} k_1 \\ k_2 \end{bmatrix}$

Now, for order 1:

$$\begin{bmatrix} x_1 \\ x_2 \end{bmatrix} = \begin{bmatrix} \frac{1}{5} & -\frac{1}{5} \\ -\frac{1}{8} & \frac{1}{4} \end{bmatrix}\begin{bmatrix} 3000 \\ 1600 \end{bmatrix} = \begin{bmatrix} 280 \\ 25 \end{bmatrix}$$ and $\begin{array}{l} x_1 = 280 \quad \text{hours at Plant A} \\ x_2 = 25 \quad \text{hours at Plant B} \end{array}$

For order 2:

$$\begin{bmatrix} x_1 \\ x_2 \end{bmatrix} = \begin{bmatrix} \frac{1}{5} & -\frac{1}{5} \\ -\frac{1}{8} & \frac{1}{4} \end{bmatrix}\begin{bmatrix} 2800 \\ 2000 \end{bmatrix} = \begin{bmatrix} 160 \\ 150 \end{bmatrix}$$ and $\begin{array}{l} x_1 = 160 \quad \text{hours at Plant A} \\ x_2 = 150 \quad \text{hours at Plant B} \end{array}$

For order 3:

$$\begin{bmatrix} x_1 \\ x_2 \end{bmatrix} = \begin{bmatrix} \frac{1}{5} & -\frac{1}{5} \\ -\frac{1}{8} & \frac{1}{4} \end{bmatrix}\begin{bmatrix} 2600 \\ 2200 \end{bmatrix} = \begin{bmatrix} 80 \\ 225 \end{bmatrix}$$ and $\begin{array}{l} x_1 = 80 \quad \text{hours at Plant A} \\ x_2 = 225 \quad \text{hours at Plant B} \end{array}$

67. Let x_1 = President's bonus

x_2 = Executive Vice President's bonus

x_3 = Associate Vice President's bonus

x_4 = Assistant Vice President's bonus

Then, the mathematical model is:

$x_1 = 0.03(2,000,000 - x_2 - x_3 - x_4)$

$x_2 = 0.025(2,000,000 - x_1 - x_3 - x_4)$

$x_3 = 0.02(2,000,000 - x_1 - x_2 - x_4)$

$x_4 = 0.015(2,000,000 - x_1 - x_2 - x_3)$

or

$$x_1 + 0.03x_2 + 0.03x_3 + 0.03x_4 = 60{,}000$$
$$0.025x_1 + x_2 + 0.025x_3 + 0.025x_4 = 50{,}000$$
$$0.02x_1 + 0.02x_2 + x_3 + 0.02x_4 = 40{,}000$$
$$0.015x_1 + 0.015x_2 + 0.015x_3 + x_4 = 30{,}000$$

and
$$\begin{bmatrix} 1 & 0.03 & 0.03 & 0.03 \\ 0.025 & 1 & 0.025 & 0.025 \\ 0.02 & 0.02 & 1 & 0.02 \\ 0.015 & 0.015 & 0.015 & 1 \end{bmatrix} \begin{bmatrix} x_1 \\ x_2 \\ x_3 \\ x_4 \end{bmatrix} = \begin{bmatrix} 60{,}000 \\ 50{,}000 \\ 40{,}000 \\ 30{,}000 \end{bmatrix}$$

Thus
$$\begin{bmatrix} x_1 \\ x_2 \\ x_3 \\ x_4 \end{bmatrix} = \begin{bmatrix} 1 & 0.03 & 0.03 & 0.03 \\ 0.025 & 1 & 0.025 & 0.025 \\ 0.02 & 0.02 & 1 & 0.02 \\ 0.015 & 0.015 & 0.015 & 1 \end{bmatrix}^{-1} \begin{bmatrix} 60{,}000 \\ 50{,}000 \\ 40{,}000 \\ 30{,}000 \end{bmatrix} \approx \begin{bmatrix} 56{,}600 \\ 47{,}000 \\ 37{,}400 \\ 27{,}900 \end{bmatrix}$$

or $x_1 = \$56{,}600$, $x_2 = \$47{,}000$, $x_3 = \$37{,}400$, $x_4 = \$27{,}900$ to the nearest hundred dollars.

69.   Let x_1 = number of ounces of mix A.

x_2 = number of ounces of mix B.

The mathematical model is:
$$0.2x_1 + 0.14x_2 = k_1 \text{ (protein)}$$
$$0.04x_1 + 0.03x_2 = k_2 \text{ (fat)}$$

The corresponding matrix equation is:
$$\begin{bmatrix} 0.2 & 0.14 \\ 0.04 & 0.03 \end{bmatrix} \begin{bmatrix} x_1 \\ x_2 \end{bmatrix} = \begin{bmatrix} k_1 \\ k_2 \end{bmatrix}$$

Next, compute the inverse of the coefficient matrix A:
$$\begin{bmatrix} 0.2 & 0.14 & | & 1 & 0 \\ 0.04 & 0.03 & | & 0 & 1 \end{bmatrix} \sim \begin{bmatrix} 1 & 0.7 & | & 5 & 0 \\ 0.04 & 0.03 & | & 0 & 1 \end{bmatrix} \sim \begin{bmatrix} 1 & 0.7 & | & 5 & 0 \\ 0 & 0.002 & | & -0.2 & 1 \end{bmatrix}$$

$5R_1 \rightarrow R_1$ $(-0.04)R_1 + R_2 \rightarrow R_2$ $500R_2 \rightarrow R_2$

$$\sim \begin{bmatrix} 1 & 0.7 & | & 5 & 0 \\ 0 & 1 & | & -100 & 500 \end{bmatrix} \sim \begin{bmatrix} 1 & 0 & | & 75 & -350 \\ 0 & 1 & | & -100 & 500 \end{bmatrix}$$

$(-0.7)R_2 + R_1 \rightarrow R_1$

Thus, $A^{-1} = \begin{bmatrix} 75 & -350 \\ -100 & 500 \end{bmatrix}$

(A)   Diet 1: Protein - 80 oz., Fat - 17 oz.
$$\begin{bmatrix} x_1 \\ x_2 \end{bmatrix} = \begin{bmatrix} 75 & -350 \\ -100 & 500 \end{bmatrix} \begin{bmatrix} 80 \\ 17 \end{bmatrix} = \begin{bmatrix} 50 \\ 500 \end{bmatrix}$$

Thus, 50 ounces of mix A, 500 ounces of mix B.

Diet 2: Protein - 90 oz., Fat - 18 oz.
$$\begin{bmatrix} x_1 \\ x_2 \end{bmatrix} = \begin{bmatrix} 75 & -350 \\ -100 & 500 \end{bmatrix} \begin{bmatrix} 90 \\ 18 \end{bmatrix} = \begin{bmatrix} 450 \\ 0 \end{bmatrix}$$

Thus, 450 ounces of mix A, 0 ounces of mix B.

<u>Diet 3</u>: Protein - 100 oz., Fat - 21 oz.

$$\begin{bmatrix} x_1 \\ x_2 \end{bmatrix} = \begin{bmatrix} 75 & -350 \\ -100 & 500 \end{bmatrix} \begin{bmatrix} 100 \\ 21 \end{bmatrix} = \begin{bmatrix} 150 \\ 500 \end{bmatrix}$$

Thus, 150 ounces of mix A, 500 ounces of mix B.

(B) Protein - 100 oz., Fat - 22 oz.

$$\begin{bmatrix} x_1 \\ x_2 \end{bmatrix} = \begin{bmatrix} 75 & -350 \\ -100 & 500 \end{bmatrix} \begin{bmatrix} 100 \\ 22 \end{bmatrix} = \begin{bmatrix} -200 \\ -1000 \end{bmatrix}$$

This is not possible; x_1 and x_2 must both be non-negative.

Protein - 80 oz., Fat - 15 oz.

$$\begin{bmatrix} x_1 \\ x_2 \end{bmatrix} = \begin{bmatrix} 75 & -350 \\ -100 & 500 \end{bmatrix} \begin{bmatrix} 80 \\ 15 \end{bmatrix} = \begin{bmatrix} 750 \\ -500 \end{bmatrix}$$

This is not possible; x_1 and x_2 must both be non-negative.

EXERCISE 4-7

Things to remember:

<u>1.</u> Given two industries C_1 and C_2, with

$$M = \begin{matrix} & C_1 & C_2 \\ \begin{matrix} C_1 \\ C_2 \end{matrix} & \begin{bmatrix} a_{11} & a_{12} \\ a_{21} & a_{22} \end{bmatrix} \end{matrix}, \quad X = \begin{bmatrix} x_1 \\ x_2 \end{bmatrix}, \quad D = \begin{bmatrix} d_1 \\ d_2 \end{bmatrix},$$

| Technology | Output | Final Demand |
| Matrix | Matrix | Matrix |

where a_{ij} is the input required from C_i to produce a dollar's worth of output for C_j. The solution to the input-output matrix equation

$$X = MX + D \quad \text{is} \quad X = (I - M)^{-1}D,$$

where I is the identity matrix, assuming $I - M$ has an inverse.

1. $x = 3x + 6, \quad -2x = 6, \quad x = -3$ **3.** $x = 0.9x + 10, \quad 0.1x = 10, \quad x = 100$

5. $x = 0.2x + 3.2,\quad 0.8x = 3.2,\quad x = 4$ **7.** $x = 0.68x + 2.56,\quad 0.32x = 2.56,\quad x = 8$

9.   40¢ from A and 20¢ from E are required to produce a dollar's worth of output for A.

11. $I - M = \begin{bmatrix} 1 & 0 \\ 0 & 1 \end{bmatrix} - \begin{bmatrix} 0.4 & 0.2 \\ 0.2 & 0.1 \end{bmatrix} = \begin{bmatrix} 0.6 & -0.2 \\ -0.2 & 0.9 \end{bmatrix}$

Converting the decimals to fractions to calculate the inverse, we have:

$\begin{bmatrix} \frac{3}{5} & -\frac{1}{5} & 1 & 0 \\ \frac{1}{5} & \frac{9}{10} & 0 & 1 \end{bmatrix} \sim \begin{bmatrix} 1 & -\frac{1}{3} & \frac{5}{3} & 0 \\ -\frac{1}{5} & \frac{9}{10} & 0 & 1 \end{bmatrix} \sim \begin{bmatrix} 1 & -\frac{1}{3} & \frac{5}{3} & 0 \\ 0 & \frac{5}{6} & \frac{1}{3} & 1 \end{bmatrix} \sim \begin{bmatrix} 1 & -\frac{1}{3} & \frac{5}{3} & 0 \\ 0 & 1 & \frac{2}{5} & \frac{6}{5} \end{bmatrix}$

$\quad\quad \frac{5}{3}R_1 \to R_1 \quad\quad\quad \frac{1}{5}R_1 + R_2 \to R_2 \quad\quad \frac{6}{5}R_2 \to R_2 \quad\quad \frac{1}{3}R_2 + R_1 \to R_1$

$\sim \begin{bmatrix} 1 & 0 & \frac{9}{5} & \frac{2}{5} \\ 0 & 1 & \frac{2}{5} & \frac{6}{5} \end{bmatrix}$ Thus, $I - M = \begin{bmatrix} 0.6 & -0.2 \\ -0.2 & 0.9 \end{bmatrix}$ and $(I - M)^{-1} = \begin{bmatrix} 1.8 & 0.4 \\ 0.4 & 1.2 \end{bmatrix}$.

13. $X = (I - M)^{-1}D_2 = \begin{bmatrix} 1.8 & 0.4 \\ 0.4 & 1.2 \end{bmatrix}\begin{bmatrix} 8 \\ 5 \end{bmatrix}$ Thus, $\begin{bmatrix} x_1 \\ x_2 \end{bmatrix} = \begin{bmatrix} 16.4 \\ 9.2 \end{bmatrix}$ and $x_1 = 16.4$, $x_2 = 9.2$.

15.   20¢ from A, 10¢ from B, and 10¢ from E are required to produce a dollar's worth of output for B.

17. $\begin{bmatrix} 1 & 0 & 0 \\ 0 & 1 & 0 \\ 0 & 0 & 1 \end{bmatrix} - \begin{bmatrix} 0.3 & 0.2 & 0.2 \\ 0.1 & 0.1 & 0.1 \\ 0.2 & 0.1 & 0.1 \end{bmatrix} = \begin{bmatrix} 0.7 & -0.2 & -0.2 \\ -0.1 & 0.9 & -0.1 \\ -0.2 & -0.1 & -0.9 \end{bmatrix}$

19. $X = (I - M)^{-1}D_1$

Therefore, $\begin{bmatrix} x_1 \\ x_2 \\ x_3 \end{bmatrix} = \begin{bmatrix} 1.6 & 0.4 & 0.4 \\ 0.22 & 1.18 & 0.18 \\ 0.38 & 0.22 & 1.22 \end{bmatrix}\begin{bmatrix} 5 \\ 10 \\ 15 \end{bmatrix} = \begin{bmatrix} 18 \\ 15.6 \\ 22.4 \end{bmatrix}$

Thus, agriculture, \$18 billion; building, \$15.6 billion; and energy, \$22.4 billion.

21. $I - M = \begin{bmatrix} 1 & 0 \\ 0 & 1 \end{bmatrix} - \begin{bmatrix} 0.2 & 0.2 \\ 0.3 & 0.3 \end{bmatrix} = \begin{bmatrix} 0.8 & -0.2 \\ -0.3 & 0.7 \end{bmatrix} = \begin{bmatrix} \frac{4}{5} & -\frac{1}{5} \\ -\frac{3}{10} & \frac{7}{10} \end{bmatrix}$,

converting the decimals to fractions.

$\begin{bmatrix} \frac{4}{5} & -\frac{1}{5} & 1 & 0 \\ -\frac{3}{10} & \frac{7}{10} & 0 & 1 \end{bmatrix} \sim \begin{bmatrix} 1 & -\frac{1}{4} & \frac{5}{4} & 0 \\ -\frac{3}{10} & \frac{7}{10} & 0 & 1 \end{bmatrix} \sim \begin{bmatrix} 1 & -\frac{1}{4} & \frac{5}{4} & 0 \\ 0 & \frac{5}{8} & \frac{3}{8} & 1 \end{bmatrix}$

$\quad\quad \frac{5}{4}R_1 \to R_1 \quad\quad\quad \frac{3}{10}R_1 + R_2 \to R_2 \quad\quad \frac{8}{5}R_2 \to R_2$

$$\sim \begin{bmatrix} 1 & -\frac{1}{4} & \Big| & \frac{5}{4} & 0 \\ 0 & 1 & \Big| & \frac{3}{5} & \frac{8}{5} \end{bmatrix} \sim \begin{bmatrix} 1 & 0 & \Big| & \frac{7}{5} & \frac{2}{5} \\ 0 & 1 & \Big| & \frac{3}{5} & \frac{8}{5} \end{bmatrix} \quad \text{Thus,} \quad (I-M)^{-1} = \begin{bmatrix} 1.4 & 0.4 \\ 0.6 & 1.6 \end{bmatrix}.$$

$$\frac{1}{4}R_2 + R_1 \to R_1$$

Now, $X = (I-M)^{-1}D = \begin{bmatrix} 1.4 & 0.4 \\ 0.6 & 1.6 \end{bmatrix}\begin{bmatrix} 10 \\ 25 \end{bmatrix} = \begin{bmatrix} 24 \\ 46 \end{bmatrix}.$

23. $I-M = \begin{bmatrix} 1 & 0 \\ 0 & 1 \end{bmatrix} - \begin{bmatrix} 0.7 & 0.8 \\ 0.3 & 0.2 \end{bmatrix} = \begin{bmatrix} 0.3 & -0.8 \\ -0.3 & 0.8 \end{bmatrix}$

$I-M$ is singular $(R_2 = (-1)R_1)$; X does not exist.

25. $I-M = \begin{bmatrix} 1 & 0 & 0 \\ 0 & 1 & 0 \\ 0 & 0 & 1 \end{bmatrix} - \begin{bmatrix} 0.3 & 0.1 & 0.3 \\ 0.2 & 0.1 & 0.2 \\ 0.1 & 0.1 & 0.1 \end{bmatrix} = \begin{bmatrix} 0.7 & -0.1 & -0.3 \\ -0.2 & 0.9 & -0.2 \\ -0.1 & -0.1 & 0.9 \end{bmatrix}$

$$\begin{bmatrix} 0.7 & -0.1 & -0.3 & \Big| & 1 & 0 & 0 \\ -0.2 & 0.9 & -0.2 & \Big| & 0 & 1 & 0 \\ -0.1 & -0.1 & 0.9 & \Big| & 0 & 0 & 1 \end{bmatrix} \sim \begin{bmatrix} 7 & -1 & -3 & \Big| & 10 & 0 & 0 \\ -2 & 9 & -2 & \Big| & 0 & 10 & 0 \\ 1 & 1 & -9 & \Big| & 0 & 0 & -10 \end{bmatrix}$$

$\qquad 10R_1 \to R_1 \qquad\qquad\qquad\qquad R_1 \leftrightarrow R_3$

$\qquad 10R_2 \to R_2$

$\qquad -10R_3 \to R_3$

$$\sim \begin{bmatrix} 1 & 1 & -9 & \Big| & 0 & 0 & -10 \\ -2 & 9 & -2 & \Big| & 0 & 10 & 0 \\ 7 & -1 & -3 & \Big| & 10 & 0 & 0 \end{bmatrix} \sim \begin{bmatrix} 1 & 1 & -9 & \Big| & 0 & 0 & -10 \\ 0 & 11 & -20 & \Big| & 0 & 10 & -20 \\ 0 & -8 & 60 & \Big| & 10 & 0 & 70 \end{bmatrix}$$

$\qquad\quad 2R_1 + R_2 \to R_2 \qquad\qquad\qquad \frac{1}{11}R_2 \to R_2$

$\qquad\quad (-7)R_1 + R_3 \to R_3$

$$\sim \begin{bmatrix} 1 & 1 & -9 & \Big| & 0 & 0 & -10 \\ 0 & 1 & -1.82 & \Big| & 0 & 0.91 & -1.82 \\ 0 & -8 & 60 & \Big| & 10 & 0 & 70 \end{bmatrix} \sim \begin{bmatrix} 1 & 0 & -7.18 & \Big| & 0 & -0.91 & -8.18 \\ 0 & 1 & -1.82 & \Big| & 0 & 0.91 & -1.82 \\ 0 & 0 & 45.44 & \Big| & 10 & 7.28 & 55.44 \end{bmatrix}$$

$\qquad\quad 8R_2 + R_3 \to R_3 \qquad\qquad\qquad \frac{1}{45.44}R_3 \to R_3$

$$\sim \begin{bmatrix} 1 & 0 & -7.18 & \Big| & 0 & -0.91 & -8.18 \\ 0 & 1 & -1.82 & \Big| & 0 & 0.91 & -1.82 \\ 0 & 0 & 1 & \Big| & 0.22 & 0.16 & 1.22 \end{bmatrix} \sim \begin{bmatrix} 1 & 0 & 0 & \Big| & 1.58 & 0.24 & 0.58 \\ 0 & 1 & 0 & \Big| & 0.4 & 1.2 & 0.4 \\ 0 & 0 & 1 & \Big| & 0.22 & 0.16 & 1.22 \end{bmatrix}$$

$\qquad 1.82R_3 + R_2 \to R_2$

$\qquad 7.18R_3 + R_1 \to R_1$

Thus, $(I-M)^{-1} = \begin{bmatrix} 1.58 & 0.24 & 0.58 \\ 0.4 & 1.2 & 0.4 \\ 0.22 & 0.16 & 1.22 \end{bmatrix}$,

and $X = (I-M)^{-1}D = \begin{bmatrix} 1.58 & 0.24 & 0.58 \\ 0.4 & 1.2 & 0.4 \\ 0.22 & 0.16 & 1.22 \end{bmatrix}\begin{bmatrix} 20 \\ 5 \\ 10 \end{bmatrix} = \begin{bmatrix} 38.6 \\ 18 \\ 17.4 \end{bmatrix}.$

27. **(A)** The technology matrix $M = \begin{bmatrix} 0.3 & 0.25 \\ 0.1 & 0.25 \end{bmatrix}$ and the final demand matrix

$D = \begin{bmatrix} 40 \\ 40 \end{bmatrix}$. The input-output matrix equation is $X = MX + D$ or

$X = \begin{bmatrix} 0.3 & 0.25 \\ 0.1 & 0.25 \end{bmatrix} X + \begin{bmatrix} 40 \\ 40 \end{bmatrix}$, where $X = \begin{bmatrix} x_1 \\ x_2 \end{bmatrix}$.

The solution is $X = (I - M)^{-1}D$, provided $I - M$ has an inverse. Now,

$I - M = \begin{bmatrix} 1 & 0 \\ 0 & 1 \end{bmatrix} - \begin{bmatrix} 0.3 & 0.25 \\ 0.1 & 0.25 \end{bmatrix} = \begin{bmatrix} 0.7 & -0.25 \\ -0.1 & 0.75 \end{bmatrix}$

$(I - M)^{-1}: \begin{bmatrix} 0.7 & -0.25 & | & 1 & 0 \\ -0.1 & 0.75 & | & 0 & 1 \end{bmatrix} -10R_2 \rightarrow R_2 \sim \begin{bmatrix} 0.7 & -0.25 & | & 1 & 0 \\ 1 & -7.5 & | & 0 & -10 \end{bmatrix} R_1 \leftrightarrow R_2$

$\sim \begin{bmatrix} 1 & -7.5 & | & 0 & -10 \\ 0.7 & -0.25 & | & 1 & 0 \end{bmatrix} (-0.7)R_1 + R_2 \rightarrow R_2 \sim \begin{bmatrix} 1 & -7.5 & | & 0 & -10 \\ 0 & 5 & | & 1 & 7 \end{bmatrix} (0.2)R_2 \rightarrow R_2$

$\sim \begin{bmatrix} 1 & -7.5 & | & 0 & -10 \\ 0 & 1 & | & 0.2 & 1.4 \end{bmatrix} (7.5)R_2 + R_1 \rightarrow R_1 \sim \begin{bmatrix} 1 & 0 & | & 1.5 & 0.5 \\ 0 & 1 & | & 0.2 & 1.4 \end{bmatrix}$

Thus, $(I - M)^{-1} = \begin{bmatrix} 1.5 & 0.5 \\ 0.2 & 1.4 \end{bmatrix}$ and $X = \begin{bmatrix} 1.5 & 0.5 \\ 0.2 & 1.4 \end{bmatrix} \begin{bmatrix} 40 \\ 40 \end{bmatrix} = \begin{bmatrix} 80 \\ 64 \end{bmatrix}$

Thus, the output for each sector is:
Agriculture: $80 million; Manufacturing: $64 million

(B) If the agricultural output is increased by $20 million and the manufacturing output remains at $64 million, then the final demand D is given by

$D = (I - M)X = \begin{bmatrix} 0.7 & -0.25 \\ -0.1 & 0.75 \end{bmatrix} \begin{bmatrix} 100 \\ 64 \end{bmatrix} = \begin{bmatrix} 54 \\ 38 \end{bmatrix}$

The final demand for agriculture increases to $54 million and the final demand for manufacturing decreases to $38 million.

29. From Problem 27, the technology matrix T in this case is:

$T = \begin{bmatrix} 0.25 & 0.1 \\ 0.25 & 0.3 \end{bmatrix}$.

The final demand matrix $D = \begin{bmatrix} 40 \\ 40 \end{bmatrix}$.

The input-output matrix is $X = TX + D$ and the solution is
$X = (I - T)^{-1}D$ provided

$I - T = \begin{bmatrix} 1 & 0 \\ 0 & 1 \end{bmatrix} - \begin{bmatrix} 0.25 & 0.1 \\ 0.25 & 0.3 \end{bmatrix} = \begin{bmatrix} 0.75 & -0.1 \\ -0.25 & 0.7 \end{bmatrix}$

has an inverse

$\begin{bmatrix} 0.75 & -0.1 & | & 1 & 0 \\ -0.25 & 0.7 & | & 0 & 1 \end{bmatrix} \sim \begin{bmatrix} 1 & -0.8 & | & 1 & -1 \\ -0.25 & 0.7 & | & 0 & 1 \end{bmatrix} \sim \begin{bmatrix} 1 & -0.8 & | & 1 & -1 \\ 0 & 0.5 & | & 0.25 & 0.75 \end{bmatrix}$

$\quad (-1)R_2 + R_1 \rightarrow R_1 \qquad\qquad (0.25)R_1 + R_2 \rightarrow R_2 \qquad 2R_2 \rightarrow R_2$

$$\sim \begin{bmatrix} 1 & -0.8 & | & 1 & -1 \\ 0 & 1 & | & 0.5 & 1.5 \end{bmatrix} \sim \begin{bmatrix} 1 & 0 & | & 1.4 & 0.2 \\ 0 & 1 & | & 0.5 & 1.5 \end{bmatrix}$$
$$(0.8)R_2 + R_1 \rightarrow R_1$$

Thus $(I - T)^{-1} = \begin{bmatrix} 1.4 & 0.2 \\ 0.5 & 1.5 \end{bmatrix}$ and $X = \begin{bmatrix} 1.4 & 0.2 \\ 0.5 & 1.5 \end{bmatrix} \begin{bmatrix} 40 \\ 40 \end{bmatrix} = \begin{bmatrix} 64 \\ 80 \end{bmatrix}$

The output for each sector is: manufacturing: \$64 million; agriculture: \$80 million.

31. Let x_1 = total output of energy

x_2 = total output of mining

Then the final demand matrix $D = \begin{pmatrix} 0.4x_1 \\ 0.4x_2 \end{pmatrix}$ and the input-output matrix equation is:

$$\begin{bmatrix} x_1 \\ x_2 \end{bmatrix} = \begin{bmatrix} 0.2 & 0.3 \\ 0.4 & 0.3 \end{bmatrix} \begin{bmatrix} x_1 \\ x_2 \end{bmatrix} + \begin{bmatrix} 0.4x_1 \\ 0.4x_2 \end{bmatrix}$$

This yields the dependent system of equations
$$0.6x_1 = 0.2x_1 + 0.3x_2$$
$$0.6x_2 = 0.4x_1 + 0.3x_2$$

which is equivalent to
$$0.4x_1 - 0.3x_2 = 0$$

or $\qquad\qquad x_1 = \dfrac{3}{4}x_2$

Thus, the total output of the energy sector should be 75% of the total output of the mining sector.

33. Each element of a technology matrix represents the input needed from C_i to produce \$1 dollar's worth of output for C_j. Hence, each element must be a number between 0 and 1, inclusive.

35. The technology matrix $M = \begin{bmatrix} 0.1 & 0.2 \\ 0.2 & 0.4 \end{bmatrix}$ and the final demand matrix $D = \begin{bmatrix} 20 \\ 10 \end{bmatrix}$.

The input-output matrix equation is $X = MX + D$ or
$$X = \begin{bmatrix} 0.1 & 0.2 \\ 0.2 & 0.4 \end{bmatrix} X + \begin{bmatrix} 20 \\ 10 \end{bmatrix} \quad \text{where} \quad X = \begin{bmatrix} x_1 \\ x_2 \end{bmatrix}.$$

The solution is $X = (I - M)^{-1}D$, provided $(I - M)$ has an inverse. Now,

$$I - M = \begin{bmatrix} 1 & 0 \\ 0 & 1 \end{bmatrix} - \begin{bmatrix} 0.1 & 0.2 \\ 0.2 & 0.4 \end{bmatrix} = \begin{bmatrix} 0.9 & -0.2 \\ -0.2 & 0.6 \end{bmatrix} = \begin{bmatrix} \frac{9}{10} & -\frac{1}{5} \\ -\frac{1}{5} & \frac{3}{5} \end{bmatrix}$$

$$\begin{bmatrix} \frac{9}{10} & -\frac{1}{5} & | & 1 & 0 \\ -\frac{1}{5} & \frac{3}{5} & | & 0 & 1 \end{bmatrix} \sim \begin{bmatrix} 1 & -\frac{2}{9} & | & \frac{10}{9} & 0 \\ -\frac{1}{5} & \frac{3}{5} & | & 0 & 1 \end{bmatrix} \sim \begin{bmatrix} 1 & -\frac{2}{9} & | & \frac{10}{9} & 0 \\ 0 & \frac{5}{9} & | & \frac{2}{9} & 1 \end{bmatrix} \sim \begin{bmatrix} 1 & -\frac{2}{9} & | & \frac{10}{9} & 0 \\ 0 & 1 & | & \frac{2}{5} & \frac{9}{5} \end{bmatrix}$$

$$\dfrac{10}{9}R_1 \rightarrow R_1 \qquad\qquad \dfrac{1}{5}R_1 + R_2 \rightarrow R_2 \qquad \dfrac{9}{5}R_2 \rightarrow R_2 \qquad \dfrac{2}{9}R_2 + R_1 \rightarrow R_1$$

$$\sim \begin{bmatrix} 1 & 0 & | & \frac{6}{5} & \frac{2}{5} \\ 0 & 1 & | & \frac{2}{5} & \frac{9}{5} \end{bmatrix} \text{Thus,} \quad (I-M)^{-1} = \begin{bmatrix} \frac{6}{5} & \frac{2}{5} \\ \frac{2}{5} & \frac{9}{5} \end{bmatrix} = \begin{bmatrix} 1.2 & 0.4 \\ 0.4 & 1.8 \end{bmatrix}, \text{ and}$$

$$X = \begin{bmatrix} 1.2 & 0.4 \\ 0.4 & 1.8 \end{bmatrix} \begin{bmatrix} 20 \\ 10 \end{bmatrix} = \begin{bmatrix} 28 \\ 26 \end{bmatrix}.$$

Therefore, the output for each sector is: coal, \$28 billion;
steel, \$26 billion.

37. The technology matrix $M = \begin{bmatrix} 0.20 & 0.40 \\ 0.15 & 0.30 \end{bmatrix} = \begin{bmatrix} \frac{1}{5} & \frac{2}{5} \\ \frac{3}{20} & \frac{3}{10} \end{bmatrix}$ and the final demand matrix $D = \begin{bmatrix} 60 \\ 80 \end{bmatrix}$. The

input-output matrix equation is $X = MX + D$ or

$X = \begin{bmatrix} \frac{1}{5} & \frac{2}{5} \\ \frac{3}{20} & \frac{3}{10} \end{bmatrix} X + \begin{bmatrix} 60 \\ 80 \end{bmatrix}$ where $X = \begin{bmatrix} x_1 \\ x_2 \end{bmatrix}$. The solution is $X = (I - M)^{-1} D$, provided $(I - M)$ has an

inverse. Now

$I - M = \begin{bmatrix} 1 & 0 \\ 0 & 1 \end{bmatrix} - \begin{bmatrix} \frac{1}{5} & \frac{2}{5} \\ \frac{3}{20} & \frac{3}{10} \end{bmatrix} = \begin{bmatrix} \frac{4}{5} & -\frac{2}{5} \\ -\frac{3}{20} & \frac{7}{10} \end{bmatrix}.$

$\begin{bmatrix} \frac{4}{5} & -\frac{2}{5} & | & 1 & 0 \\ -\frac{3}{20} & \frac{7}{10} & | & 0 & 1 \end{bmatrix} \sim \begin{bmatrix} 1 & -\frac{1}{2} & | & \frac{5}{4} & 0 \\ -\frac{3}{20} & \frac{7}{10} & | & 0 & 1 \end{bmatrix} \sim \begin{bmatrix} 1 & -\frac{1}{2} & | & \frac{5}{4} & 0 \\ 0 & \frac{5}{8} & | & \frac{3}{16} & 1 \end{bmatrix}$

$\quad\; \frac{5}{4} R_1 \to R_1 \qquad\qquad \frac{3}{20} R_1 + R_2 \to R_2 \qquad \frac{8}{5} R_2 \to R_2$

$\sim \begin{bmatrix} 1 & -\frac{1}{2} & | & \frac{5}{4} & 0 \\ 0 & 1 & | & \frac{3}{10} & \frac{8}{5} \end{bmatrix} \sim \begin{bmatrix} 1 & 0 & | & \frac{7}{5} & \frac{4}{5} \\ 0 & 1 & | & \frac{3}{10} & \frac{8}{5} \end{bmatrix}$

$\quad \frac{1}{2} R_2 + R_1 \to R_1$

Thus $(I - M)^{-1} = \begin{bmatrix} \frac{7}{5} & \frac{4}{5} \\ \frac{3}{10} & \frac{8}{5} \end{bmatrix}$ and $X = \begin{bmatrix} \frac{7}{5} & \frac{4}{5} \\ \frac{3}{10} & \frac{8}{5} \end{bmatrix} \begin{bmatrix} 60 \\ 80 \end{bmatrix} = \begin{bmatrix} 148 \\ 146 \end{bmatrix}$

Therefore, the output for each sector is: Agriculture—$148 million; Tourism—$146 million

39. The technology matrix $M = \begin{bmatrix} 0.2 & 0.4 & 0.3 \\ 0.2 & 0.1 & 0.1 \\ 0.2 & 0.1 & 0.1 \end{bmatrix}$ and the final demand matrix $D = \begin{bmatrix} 10 \\ 15 \\ 20 \end{bmatrix}$.

The input-output matrix equation is $X = MX + D$ or

$X = \begin{bmatrix} 0.2 & 0.4 & 0.3 \\ 0.2 & 0.1 & 0.1 \\ 0.2 & 0.1 & 0.1 \end{bmatrix} X + \begin{bmatrix} 10 \\ 15 \\ 20 \end{bmatrix}.$

The solution is $X = (I - M)^{-1} D$, provided $I - M$ has an inverse. Now,

$I - M = \begin{bmatrix} 1 & 0 & 0 \\ 0 & 1 & 0 \\ 0 & 0 & 1 \end{bmatrix} - \begin{bmatrix} 0.2 & 0.4 & 0.3 \\ 0.2 & 0.1 & 0.1 \\ 0.2 & 0.1 & 0.1 \end{bmatrix} = \begin{bmatrix} 0.8 & -0.4 & -0.3 \\ -0.2 & 0.9 & -0.1 \\ -0.2 & -0.1 & 0.9 \end{bmatrix}.$

$\begin{bmatrix} 0.8 & -0.4 & -0.3 & | & 1 & 0 & 0 \\ -0.2 & 0.9 & -0.1 & | & 0 & 1 & 0 \\ -0.2 & -0.1 & 0.9 & | & 0 & 0 & 1 \end{bmatrix} \sim \begin{bmatrix} 8 & -4 & -3 & | & 10 & 0 & 0 \\ -2 & 9 & -1 & | & 0 & 10 & 0 \\ -2 & -1 & 9 & | & 0 & 0 & 10 \end{bmatrix}$

$\qquad 10 R_1 \to R_1 \qquad\qquad\qquad \left(-\frac{1}{2}\right) R_2 \to R_2$

$$10R_2 \to R_2$$
$$10R_3 \to R_3$$

$$\sim \begin{bmatrix} 8 & -4 & -3 & | & 10 & 0 & 0 \\ 1 & -\frac{9}{2} & \frac{1}{2} & | & 0 & -5 & 0 \\ -2 & -1 & 9 & | & 0 & 0 & 10 \end{bmatrix} \sim \begin{bmatrix} 1 & -\frac{9}{2} & \frac{1}{2} & | & 0 & -5 & 0 \\ 8 & -4 & -3 & | & 10 & 0 & 0 \\ -2 & -1 & 9 & | & 0 & 0 & 10 \end{bmatrix}$$

$$R_1 \leftrightarrow R_2 \qquad\qquad (-8)R_1 + R_2 \to R_2$$
$$2R_1 + R_3 \to R_3$$

$$\sim \begin{bmatrix} 1 & -\frac{9}{2} & \frac{1}{2} & | & 0 & -5 & 0 \\ 0 & 32 & -7 & | & 10 & 40 & 0 \\ 0 & -10 & 10 & | & 0 & -10 & 10 \end{bmatrix} \sim \begin{bmatrix} 1 & -\frac{9}{2} & \frac{1}{2} & | & 0 & -5 & 0 \\ 0 & 32 & -7 & | & 10 & 40 & 0 \\ 0 & 1 & -1 & | & 0 & 1 & -1 \end{bmatrix}$$

$$\left(-\frac{1}{10}\right)R_3 \to R_3 \qquad\qquad R_2 \leftrightarrow R_3$$

$$\sim \begin{bmatrix} 1 & -\frac{9}{2} & \frac{1}{2} & | & 0 & -5 & 0 \\ 0 & 1 & -1 & | & 0 & 1 & -1 \\ 0 & 32 & -7 & | & 10 & 40 & 0 \end{bmatrix} \sim \begin{bmatrix} 1 & 0 & -4 & | & 0 & -\frac{1}{2} & -\frac{9}{2} \\ 0 & 1 & -1 & | & 0 & 1 & -1 \\ 0 & 0 & 25 & | & 10 & 8 & 32 \end{bmatrix}$$

$$(-32)R_2 + R_3 \to R_3 \qquad\qquad \frac{1}{25}R_3 \to R_3$$

$$\frac{9}{2}R_2 + R_1 \to R_1$$

$$\sim$$

$$\begin{bmatrix} 1 & 0 & -4 & | & 0 & -\frac{1}{2} & -\frac{9}{2} \\ 0 & 1 & -1 & | & 0 & 1 & -1 \\ 0 & 0 & 1 & | & 0.4 & 0.32 & 1.28 \end{bmatrix} \sim \begin{bmatrix} 1 & 0 & 0 & | & 1.6 & 0.78 & 0.62 \\ 0 & 1 & 0 & | & 0.4 & 1.32 & 0.28 \\ 0 & 0 & 1 & | & 0.4 & 0.32 & 1.28 \end{bmatrix}$$

$$R_3 + R_2 \to R_2$$
$$4R_3 + R_1 \to R_1$$

Thus, $(I - M)^{-1} = \begin{bmatrix} 1.6 & 0.78 & 0.62 \\ 0.4 & 1.32 & 0.28 \\ 0.4 & 0.32 & 1.28 \end{bmatrix}$, and $X = (I - M)^{-1}D = \begin{bmatrix} 1.6 & 0.78 & 0.62 \\ 0.4 & 1.32 & 0.28 \\ 0.4 & 0.32 & 1.28 \end{bmatrix}\begin{bmatrix} 10 \\ 15 \\ 20 \end{bmatrix} = \begin{bmatrix} 40.1 \\ 29.4 \\ 34.4 \end{bmatrix}$

Therefore, agriculture, \$40.1 billion; manufacturing, \$29.4 billion; and energy, \$34.4 billion.

41. The technology matrix is $M = \begin{bmatrix} 0.05 & 0.17 & 0.23 & 0.09 \\ 0.07 & 0.12 & 0.15 & 0.19 \\ 0.25 & 0.08 & 0.03 & 0.32 \\ 0.11 & 0.19 & 0.28 & 0.16 \end{bmatrix}$.

The input-output matrix equation is $X = MX + D$

where $X = \begin{bmatrix} A \\ E \\ L \\ M \end{bmatrix}$ and D is the final demand matrix. Thus $X = (I - M)^{-1}D$, where

$$I - M = \begin{bmatrix} 0.95 & -0.17 & -0.23 & -0.09 \\ -0.07 & 0.88 & -0.15 & -0.19 \\ -0.25 & -0.08 & 0.97 & -0.32 \\ -0.11 & -0.19 & -0.28 & 0.84 \end{bmatrix}$$

Now, $X = \begin{bmatrix} 1.25 & 0.37 & 0.47 & 0.40 \\ 0.26 & 1.33 & 0.41 & 0.48 \\ 0.47 & 0.36 & 1.39 & 0.66 \\ 0.38 & 0.47 & 0.62 & 1.57 \end{bmatrix} D$

Year 1: $D = \begin{bmatrix} 23 \\ 41 \\ 18 \\ 31 \end{bmatrix}$ and $(I - M)^{-1}D \approx \begin{bmatrix} 65 \\ 83 \\ 71 \\ 88 \end{bmatrix}$

Agriculture: \$65 billion; Energy: \$83 billion; Labor: \$71 billion; Manufacturing: \$88 billion

Year 2: $D = \begin{bmatrix} 32 \\ 48 \\ 21 \\ 33 \end{bmatrix}$ and $(I - M)^{-1}D \approx \begin{bmatrix} 81 \\ 97 \\ 83 \\ 99 \end{bmatrix}$

Agriculture: \$81 billion; Energy: \$97 billion; Labor: \$83 billion; Manufacturing: \$99 billion

Year 3: $D = \begin{bmatrix} 55 \\ 62 \\ 25 \\ 35 \end{bmatrix}$ and $(I - M)^{-1}D \approx \begin{bmatrix} 117 \\ 124 \\ 106 \\ 120 \end{bmatrix}$

Agriculture: \$117 billion; Energy: \$124 billion; Labor: \$106 billion; Manufacturing: \$120 billion

CHAPTER 4 REVIEW

1. $y = 2x - 4$

(1)

$y = \dfrac{1}{2}x + 2$

(2)

The point of intersection is the solution. This is $x = 4$, $y = 4$.   (4-1)

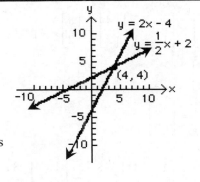

2. Substitute equation (1) into (2):

$2x - 4 = \dfrac{1}{2}x + 2$

$\dfrac{3}{2}x = 6$

$x = 4$

Substitute $x = 4$ into (1):

$y = 2 \cdot 4 - 4 = 4$

Solution: $x = 4$, $y = 4$

(4-1)

3. (A) $\begin{bmatrix} 0 & 1 & | & 2 \\ 1 & 0 & | & 3 \end{bmatrix}$ is not in reduced form; the left-most 1 in the second row is not to the right of the left-most 1 in the first row. [condition (d)] $R_1 \leftrightarrow R_2$

(B) $\begin{bmatrix} 1 & 0 & | & 2 \\ 0 & 3 & | & 3 \end{bmatrix}$ is not in reduced form; the left-most nonzero element in row 2 is not 1. [condition (b)]

$$\frac{1}{3} R_2 \to R_2$$

(C) $\begin{bmatrix} 1 & 0 & 1 & 2 \\ 0 & 1 & 1 & 3 \end{bmatrix}$ is in reduced form.

(D) $\begin{bmatrix} 1 & 1 & 0 & 2 \\ 0 & 1 & 1 & 3 \end{bmatrix}$ is not in reduced form; the left-most 1 in the second row is not the only non-zero element in its column. [condition (c)]

$$(-1)R_2 + R_1 \to R_1 \qquad (4\text{-}3)$$

4. $A = \begin{bmatrix} 5 & 3 & -1 & 0 & 2 \\ -4 & 8 & 1 & 3 & 0 \end{bmatrix},\ B = \begin{bmatrix} -3 & 2 \\ 0 & 4 \\ -1 & 7 \end{bmatrix}$

(A) A is 2×5, B is 3×2 (B) $a_{24} = 3,\ a_{15} = 2,\ b_{31} = -1,\ b_{22} = 4$

(C) AB is not defined; the number of columns of $A \neq$ the number of rows of B. BA is defined. (4-2, 4-4)

5. (A) $\begin{bmatrix} 1 & -2 \\ 1 & -3 \end{bmatrix}\begin{bmatrix} x_1 \\ x_2 \end{bmatrix} = \begin{bmatrix} 4 \\ 2 \end{bmatrix}$

$$\begin{bmatrix} 1 & -2 & | & 4 \\ 1 & -3 & | & 2 \end{bmatrix} \sim \begin{bmatrix} 1 & -2 & | & 4 \\ 0 & -1 & | & -2 \end{bmatrix} \sim \begin{bmatrix} 1 & -2 & | & 4 \\ 0 & 1 & | & 2 \end{bmatrix}$$
$$(-1)R_1 + R_2 \to R_2 \qquad (-1)R_2 \to R_2 \qquad (2)R_2 + R_1 \to R_1$$

$\sim \begin{bmatrix} 1 & 0 & | & 8 \\ 0 & 1 & | & 2 \end{bmatrix}$ Therefore, $x_1 = 8,\ x_2 = 2$.

or calculate the inverse of the coefficient matrix A:

$$A^{-1} = \begin{bmatrix} 3 & -2 \\ 1 & -1 \end{bmatrix}\ \text{ and }\ \begin{bmatrix} x_1 \\ x_2 \end{bmatrix} = \begin{bmatrix} 3 & -2 \\ 1 & -1 \end{bmatrix}\begin{bmatrix} 4 \\ 2 \end{bmatrix} = \begin{bmatrix} 8 \\ 2 \end{bmatrix};\ x_1 = 8,\ x_2 = 2.$$

(B) $\begin{bmatrix} 5 & 3 \\ 1 & 1 \end{bmatrix}\begin{bmatrix} x_1 \\ x_2 \end{bmatrix} + \begin{bmatrix} 25 \\ 14 \end{bmatrix} = \begin{bmatrix} 18 \\ 22 \end{bmatrix}$

$\begin{bmatrix} 5 & 3 \\ 1 & 1 \end{bmatrix}\begin{bmatrix} x_1 \\ x_2 \end{bmatrix} = \begin{bmatrix} 18 \\ 22 \end{bmatrix} - \begin{bmatrix} 25 \\ 14 \end{bmatrix} = \begin{bmatrix} -7 \\ 8 \end{bmatrix}$

Augmented matrix

$$\begin{bmatrix} 5 & 3 & | & -7 \\ 1 & 1 & | & 8 \end{bmatrix} \sim \begin{bmatrix} 1 & 1 & | & 8 \\ 5 & 3 & | & -7 \end{bmatrix} \sim \begin{bmatrix} 1 & 1 & | & 8 \\ 0 & -2 & | & -47 \end{bmatrix} \sim \begin{bmatrix} 1 & 1 & | & 8 \\ 0 & 1 & | & \frac{47}{2} \end{bmatrix}$$

$$R_1 \leftrightarrow R_2 \quad (-5)R_1 + R_2 \to R_2 \quad \left(-\frac{1}{2}\right)R_2 \to R_2 \quad (-1)R_2 + R_1 \to R_1$$

$$\sim \begin{bmatrix} 1 & 0 & \Big| & -\frac{31}{2} \\ 0 & 1 & \Big| & \frac{47}{2} \end{bmatrix}. \text{ Therefore, } x_1 = -\frac{31}{2}, \quad x_2 = \frac{47}{2}.$$

Inverse matrix

$$A^{-1} = \begin{bmatrix} 0.5 & -1.5 \\ -0.5 & 2.5 \end{bmatrix} \text{ and } \begin{bmatrix} x_1 \\ x_2 \end{bmatrix} = \begin{bmatrix} 0.5 & -1.5 \\ -0.5 & 2.5 \end{bmatrix} \begin{bmatrix} -7 \\ 8 \end{bmatrix} = \begin{bmatrix} -15.5 \\ 23.5 \end{bmatrix};$$

$$x_1 = -15.5 = -\frac{31}{2}, \quad x_2 = 23.5 = \frac{47}{2}. \qquad (4\text{-}2, 4\text{-}6)$$

6. $A + B = \begin{bmatrix} 1+2 & 2+1 \\ 3+1 & 1+1 \end{bmatrix} = \begin{bmatrix} 3 & 3 \\ 4 & 2 \end{bmatrix}$

$(4\text{-}4)$

7. $B + D = \begin{bmatrix} 2 & 1 \\ 1 & 1 \end{bmatrix} + \begin{bmatrix} 1 \\ 2 \end{bmatrix}$

The matrices B and D cannot be added because their dimensions are different. $(4\text{-}4)$

8. $A - 2B = \begin{bmatrix} 1 & 2 \\ 3 & 1 \end{bmatrix} - 2\begin{bmatrix} 2 & 1 \\ 1 & 1 \end{bmatrix} = \begin{bmatrix} 1 & 2 \\ 3 & 1 \end{bmatrix} + \begin{bmatrix} -4 & -2 \\ -2 & -2 \end{bmatrix} = \begin{bmatrix} -3 & 0 \\ 1 & -1 \end{bmatrix}$ $(4\text{-}4)$

9. $AB = \begin{bmatrix} 1 & 2 \\ 3 & 1 \end{bmatrix}\begin{bmatrix} 2 & 1 \\ 1 & 1 \end{bmatrix} = \begin{bmatrix} [1 \ 2]\begin{bmatrix}2\\1\end{bmatrix} & [1 \ 2]\begin{bmatrix}1\\1\end{bmatrix} \\ [3 \ 1]\begin{bmatrix}2\\1\end{bmatrix} & [3 \ 1]\begin{bmatrix}1\\1\end{bmatrix} \end{bmatrix} = \begin{bmatrix} 4 & 3 \\ 7 & 4 \end{bmatrix}$ $(4\text{-}4)$

10. AC is *not defined* because the dimension of A is 2×2 and the dimension of C is 1×2. So, the number of columns in A is not equal to the number of rows in C. $(4\text{-}4)$

11. $AD = \begin{bmatrix} 1 & 2 \\ 3 & 1 \end{bmatrix}\begin{bmatrix} 1 \\ 2 \end{bmatrix} = \begin{bmatrix} [1 \ 2]\begin{bmatrix}1\\2\end{bmatrix} \\ [3 \ 1]\begin{bmatrix}1\\2\end{bmatrix} \end{bmatrix} = \begin{bmatrix} 5 \\ 5 \end{bmatrix}$ $(4\text{-}4)$

12. $DC = \begin{bmatrix} 1 \\ 2 \end{bmatrix}[2 \ \ 3] = \begin{bmatrix} (1)\cdot(2) & (1)\cdot(3) \\ (2)\cdot(2) & (2)\cdot(3) \end{bmatrix} = \begin{bmatrix} 2 & 3 \\ 4 & 6 \end{bmatrix}$ $(4\text{-}4)$

13. $CD = [2 \ \ 3]\begin{bmatrix} 1 \\ 2 \end{bmatrix} = [2 + 6] = [8]$ $(4\text{-}4)$

14. $C + D = [2 \ \ 3] + \begin{bmatrix} 1 \\ 2 \end{bmatrix}$

Not defined because the dimensions of C and D are different. $(4\text{-}4)$

15. $\begin{bmatrix} 4 & 3 & \Big| & 1 & 0 \\ 3 & 2 & \Big| & 0 & 1 \end{bmatrix} (-1)R_2 + R_1 \to R_1 \ \sim \ \begin{bmatrix} 1 & 1 & \Big| & 1 & -1 \\ 3 & 2 & \Big| & 0 & 1 \end{bmatrix} (-3)R_1 + R_2 \to R_2$

$$\sim \begin{bmatrix} 1 & 1 & | & 1 & -1 \\ 0 & -1 & | & -3 & 4 \end{bmatrix} (-1)R_2 \rightarrow R_2 \sim \begin{bmatrix} 1 & 1 & | & 1 & -1 \\ 0 & 1 & | & 3 & -4 \end{bmatrix} (-1)R_2 + R_1 \rightarrow R_1$$

$$\sim \begin{bmatrix} 1 & 0 & | & -2 & 3 \\ 0 & 1 & | & 3 & -4 \end{bmatrix}$$

Thus, $A^{-1} = \begin{bmatrix} -2 & 3 \\ 3 & -4 \end{bmatrix}$ and $A^{-1}A = \begin{bmatrix} -2 & 3 \\ 3 & -4 \end{bmatrix} \begin{bmatrix} 4 & 3 \\ 3 & 2 \end{bmatrix} = \begin{bmatrix} 1 & 0 \\ 0 & 1 \end{bmatrix}$ (4-5)

16. $4x_1 + 3x_2 = 3$ (1)   Multiply (1) by 2 and (2) by –3.

$3x_1 + 2x_2 = 5$ (2)

$8x_1 + 6x_2 = 6$ Add the two equations.

$-9x_1 - 6x_2 = -15$

$-x_1 = -9$

$x_1 = 9$ Substitute $x_1 = 9$ into either (1) or (2);
                                       we choose (2).

$3(9) + 2x_2 = 5$

$27 + 2x_2 = 5$

$2x_2 = -22$

$x_2 = -11$

Solution: $x_1 = 9$, $x_2 = -11$.   (4-1)

17.   The augmented matrix of the system is:

$$\begin{bmatrix} 4 & 3 & | & 3 \\ 3 & 2 & | & 5 \end{bmatrix} (-1)R_2 + R_1 \rightarrow R_1 \sim \begin{bmatrix} 1 & 1 & | & -2 \\ 3 & 2 & | & 5 \end{bmatrix} (-3)R_1 + R_2 \rightarrow R_2 \sim \begin{bmatrix} 1 & 1 & | & -2 \\ 0 & -1 & | & 11 \end{bmatrix} (-1)R_2 \rightarrow R_2$$

$$\sim \begin{bmatrix} 1 & 1 & | & -2 \\ 0 & 1 & | & -11 \end{bmatrix} (-1)R_2 + R_1 \rightarrow R_1$$

$$\sim \begin{bmatrix} 1 & 0 & | & 9 \\ 0 & 1 & | & -11 \end{bmatrix}$$

The system of equations is:
$x_1 = 9$
$x_2 = -11$

Solution: $x_1 = 9$, $x_2 = -11$.   (4-2)

18.   The system of equations in matrix form is:

$$\begin{bmatrix} 4 & 3 \\ 3 & 2 \end{bmatrix} \begin{bmatrix} x_1 \\ x_2 \end{bmatrix} = \begin{bmatrix} 3 \\ 5 \end{bmatrix}$$

Thus, $\begin{bmatrix} x_1 \\ x_2 \end{bmatrix} = \begin{bmatrix} 4 & 3 \\ 3 & 2 \end{bmatrix}^{-1} \begin{bmatrix} 3 \\ 5 \end{bmatrix} = \begin{bmatrix} -2 & 3 \\ 3 & -4 \end{bmatrix} \begin{bmatrix} 3 \\ 5 \end{bmatrix}$ (by Problem 15) $= \begin{bmatrix} 9 \\ -11 \end{bmatrix}$

Solution: $x_1 = 9$, $x_2 = -11$.

Replacing the constants 3, 5 by 7, 10, respectively:

$$\begin{bmatrix} x_1 \\ x_2 \end{bmatrix} = \begin{bmatrix} -2 & 3 \\ 3 & -4 \end{bmatrix} \begin{bmatrix} 7 \\ 10 \end{bmatrix} = \begin{bmatrix} 16 \\ -19 \end{bmatrix}$$

Solution: $x_1 = 16$, $x_2 = -19$

Replacing the constants 3, 5 by 4, 2, respectively:

$$\begin{bmatrix} x_1 \\ x_2 \end{bmatrix} = \begin{bmatrix} -2 & 3 \\ 3 & -4 \end{bmatrix} \begin{bmatrix} 4 \\ 2 \end{bmatrix} = \begin{bmatrix} -2 \\ 4 \end{bmatrix}$$

Solution: $x_1 = -2$, $x_2 = 4$.   (4-6)

19. $A + D = \begin{bmatrix} 2 & -2 \\ 1 & 0 \\ 3 & 2 \end{bmatrix} + \begin{bmatrix} 3 & -2 & 1 \\ -1 & 1 & 2 \end{bmatrix}$ Not defined because the dimensions of A and D are different.

(4-4)

20. $E + DA = \begin{bmatrix} 3 & -4 \\ -1 & 0 \end{bmatrix} + \begin{bmatrix} 3 & -2 & 1 \\ -1 & 1 & 2 \end{bmatrix} \begin{bmatrix} 2 & -2 \\ 1 & 0 \\ 3 & 2 \end{bmatrix} = \begin{bmatrix} 3 & -4 \\ -1 & 0 \end{bmatrix} + \begin{bmatrix} 7 & -4 \\ 5 & 6 \end{bmatrix} = \begin{bmatrix} 10 & -8 \\ 4 & 6 \end{bmatrix}$ (4-4)

21. From Problem 20, $DA = \begin{bmatrix} 7 & -4 \\ 5 & 6 \end{bmatrix}$. Thus,

$$DA - 3E = \begin{bmatrix} 7 & -4 \\ 5 & 6 \end{bmatrix} - 3 \begin{bmatrix} 3 & -4 \\ -1 & 0 \end{bmatrix} = \begin{bmatrix} 7 & -4 \\ 5 & 6 \end{bmatrix} + \begin{bmatrix} -9 & 12 \\ 3 & 0 \end{bmatrix} = \begin{bmatrix} -2 & 8 \\ 8 & 6 \end{bmatrix}$$ (4-4)

22. $BC = \begin{bmatrix} -1 \\ 2 \\ 3 \end{bmatrix} \begin{bmatrix} 2 & 1 & 3 \end{bmatrix} = \begin{bmatrix} -2 & -1 & -3 \\ 4 & 2 & 6 \\ 6 & 3 & 9 \end{bmatrix}$ (4-4)

23. $CB = \begin{bmatrix} 2 & 1 & 3 \end{bmatrix} \begin{bmatrix} -1 \\ 2 \\ 3 \end{bmatrix} = [-2 + 2 + 9] = [9]$ (a 1×1 matrix)   (4-4)

24. $AD - BC$

$$AD = \begin{bmatrix} 2 & -2 \\ 1 & 0 \\ 3 & 2 \end{bmatrix} \begin{bmatrix} 3 & -2 & 1 \\ -1 & 1 & 2 \end{bmatrix} = \begin{bmatrix} 8 & -6 & -2 \\ 3 & -2 & 1 \\ 7 & -4 & 7 \end{bmatrix}$$

$$BC = \begin{bmatrix} -1 \\ 2 \\ 3 \end{bmatrix} \begin{bmatrix} 2 & 1 & 3 \end{bmatrix} = \begin{bmatrix} -2 & -1 & -3 \\ 4 & 2 & 6 \\ 6 & 3 & 9 \end{bmatrix}$$

$$AD - BC = \begin{bmatrix} 8 & -6 & -2 \\ 3 & -2 & 1 \\ 7 & -4 & 7 \end{bmatrix} - \begin{bmatrix} -2 & -1 & -3 \\ 4 & 2 & 6 \\ 6 & 3 & 9 \end{bmatrix} = \begin{bmatrix} 8-(-2) & -6-(-1) & -2-(-3) \\ 3-4 & -2-2 & 1-6 \\ 7-6 & -4-3 & 7-9 \end{bmatrix}$$

$$= \begin{bmatrix} 10 & -5 & 1 \\ -1 & -4 & -5 \\ 1 & -7 & -2 \end{bmatrix}$$ (4-4)

25. $\begin{bmatrix} 1 & 2 & 3 & | & 1 & 0 & 0 \\ 2 & 3 & 4 & | & 0 & 1 & 0 \\ 1 & 2 & 1 & | & 0 & 0 & 1 \end{bmatrix} \sim \begin{bmatrix} 1 & 2 & 3 & | & 1 & 0 & 0 \\ 0 & -1 & -2 & | & -2 & 1 & 0 \\ 0 & 0 & -2 & | & -1 & 0 & 1 \end{bmatrix} \sim \begin{bmatrix} 1 & 2 & 3 & | & 1 & 0 & 0 \\ 0 & 1 & 2 & | & 2 & -1 & 0 \\ 0 & 0 & -2 & | & -1 & 0 & 1 \end{bmatrix}$

$(-2)R_1 + R_2 \rightarrow R_2 \qquad\qquad (-1)R_2 \rightarrow R_2 \qquad\qquad (-2)R_2 + R_1 \rightarrow R_1$

$(-1)R_1 + R_3 \rightarrow R_3 \qquad\qquad\qquad\qquad\qquad\qquad \left(-\dfrac{1}{2}\right)R_3 \rightarrow R_3$

$\sim \begin{bmatrix} 1 & 0 & -1 & | & -3 & 2 & 0 \\ 0 & 1 & 2 & | & 2 & -1 & 0 \\ 0 & 0 & 1 & | & \frac{1}{2} & 0 & -\frac{1}{2} \end{bmatrix} \sim \begin{bmatrix} 1 & 0 & 0 & | & -\frac{5}{2} & 2 & -\frac{1}{2} \\ 0 & 1 & 0 & | & 1 & -1 & 1 \\ 0 & 0 & 1 & | & \frac{1}{2} & 0 & -\frac{1}{2} \end{bmatrix}; \quad A^{-1} = \begin{bmatrix} -\frac{5}{2} & 2 & -\frac{1}{2} \\ 1 & -1 & 1 \\ \frac{1}{2} & 0 & -\frac{1}{2} \end{bmatrix}$

$R_3 + R_1 \rightarrow R_1$

$(-2)R_3 + R_2 \rightarrow R_2$

Check:

$A^{-1}A = \begin{bmatrix} -\frac{5}{2} & 2 & -\frac{1}{2} \\ 1 & -1 & 1 \\ \frac{1}{2} & 0 & -\frac{1}{2} \end{bmatrix} \begin{bmatrix} 1 & 2 & 3 \\ 2 & 3 & 4 \\ 1 & 2 & 1 \end{bmatrix} = \begin{bmatrix} 1 & 0 & 0 \\ 0 & 1 & 0 \\ 0 & 0 & 1 \end{bmatrix}$ (4-5)

26. (A) The augmented matrix corresponding to the given system is:

$\begin{bmatrix} 1 & 2 & 3 & | & 1 \\ 2 & 3 & 4 & | & 3 \\ 1 & 2 & 1 & | & 3 \end{bmatrix} \sim \begin{bmatrix} 1 & 2 & 3 & | & 1 \\ 0 & -1 & -2 & | & 1 \\ 0 & 0 & -2 & | & 2 \end{bmatrix} \sim \begin{bmatrix} 1 & 2 & 3 & | & 1 \\ 0 & 1 & 2 & | & -1 \\ 0 & 0 & -2 & | & 2 \end{bmatrix}$

$(-2)R_1 + R_2 \rightarrow R_2 \qquad (-1)R_2 \rightarrow R_2 \qquad (-2)R_2 + R_1 \rightarrow R_1$

$(-1)R_1 + R_3 \rightarrow R_3$

$\sim \begin{bmatrix} 1 & 0 & -1 & | & 3 \\ 0 & 1 & 2 & | & -1 \\ 0 & 0 & -2 & | & 2 \end{bmatrix} \sim \begin{bmatrix} 1 & 0 & -1 & | & 3 \\ 0 & 1 & 2 & | & -1 \\ 0 & 0 & 1 & | & -1 \end{bmatrix} \sim \begin{bmatrix} 1 & 0 & 0 & | & 2 \\ 0 & 1 & 0 & | & 1 \\ 0 & 0 & 1 & | & -1 \end{bmatrix}$

$\left(-\dfrac{1}{2}\right)R_3 \rightarrow R_3 \qquad\qquad \begin{array}{l}(-2)R_3 + R_2 \rightarrow R_2 \\ R_3 + R_1 \rightarrow \square R_1\end{array}$

Thus, the solution is: $x_1 = 2, \ x_2 = 1, \ x_3 = -1$.

(B) The augmented matrix corresponding to the given system is:

$\begin{bmatrix} 1 & 2 & -1 & | & 2 \\ 2 & 3 & 1 & | & -3 \\ 3 & 5 & 0 & | & -1 \end{bmatrix} \sim \begin{bmatrix} 1 & 2 & -1 & | & 2 \\ 0 & -1 & 3 & | & -7 \\ 0 & -1 & 3 & | & -7 \end{bmatrix} \sim \begin{bmatrix} 1 & 2 & -1 & | & 2 \\ 0 & 1 & -3 & | & 7 \\ 0 & -1 & 3 & | & -7 \end{bmatrix}$

$(-2)R_1 + R_2 \rightarrow R_2 \qquad (-1)R_2 \rightarrow R_2 \qquad R_2 + R_3 \rightarrow R_3$

$(-3)R_1 + R_3 \rightarrow R_3 \qquad\qquad\qquad\qquad (-2)R_2 + R_1 \rightarrow R_1$

$\sim \begin{bmatrix} 1 & 0 & 5 & | & -12 \\ 0 & 1 & -3 & | & 7 \\ 0 & 0 & 0 & | & 0 \end{bmatrix}$

Thus, $x_1 \qquad\quad + 5x_3 = -12$ (1)

            $x_2 - 3x_3 = 7$ (2)

Let $x_3 = t$ (t any real number). Then, from (1), $x_1 = -5t - 12$ and, from (2), $x_2 = 3t + 7$.

Thus, the solution is $x_1 = -5t - 12$, $x_2 = 3t + 7$, $x_3 = t$.

(C)   The augmented matrix corresponding to the given system is:

$$\begin{bmatrix} 1 & 1 & 1 & | & 8 \\ 3 & 2 & 4 & | & 21 \end{bmatrix} \sim \begin{bmatrix} 1 & 1 & 1 & | & 8 \\ 0 & -1 & 1 & | & -3 \end{bmatrix} \sim \begin{bmatrix} 1 & 1 & 1 & | & 8 \\ 0 & 1 & -1 & | & 3 \end{bmatrix} \sim \begin{bmatrix} 1 & 0 & 2 & | & 5 \\ 0 & 1 & -1 & | & 3 \end{bmatrix}$$

$(-3)R_1 + R_2 \rightarrow R_2$ $(-1)R_2 \rightarrow R_2$ $(-1)R_2 + R_1 \rightarrow R_1$

Thus, $x_1 + 2x_3 = 5$

$x_2 - x_3 = 3$

Let $x_3 = t$. Then, $x_1 = 5 - 2t$, $x_2 = 3 + t$ and the solution is: $x_1 = 5 - 2t$, $x_2 = 3 + t$,
$x_3 = t$, t any real number   (4-3)

27. (A)   The matrix equation for the given system is:

$$\begin{bmatrix} 1 & 2 & 3 \\ 2 & 3 & 4 \\ 1 & 2 & 1 \end{bmatrix} \begin{bmatrix} x_1 \\ x_2 \\ x_3 \end{bmatrix} = \begin{bmatrix} 1 \\ 3 \\ 3 \end{bmatrix}$$

The inverse matrix of the coefficient matrix of the system, from Problem 25, is:

$$\begin{bmatrix} -\frac{5}{2} & 2 & -\frac{1}{2} \\ 1 & -1 & 1 \\ \frac{1}{2} & 0 & -\frac{1}{2} \end{bmatrix} \quad \text{Thus, } \begin{bmatrix} x_1 \\ x_2 \\ x_3 \end{bmatrix} = \begin{bmatrix} -\frac{5}{2} & 2 & -\frac{1}{2} \\ 1 & -1 & 1 \\ \frac{1}{2} & 0 & -\frac{1}{2} \end{bmatrix} \begin{bmatrix} 1 \\ 3 \\ 3 \end{bmatrix} = \begin{bmatrix} 2 \\ 1 \\ -1 \end{bmatrix}$$

Solution: $x_1 = 2$, $x_2 = 1$, $x_3 = -1$.

(B) $\begin{bmatrix} x_1 \\ x_2 \\ x_3 \end{bmatrix} = \begin{bmatrix} -\frac{5}{2} & 2 & -\frac{1}{2} \\ 1 & -1 & 1 \\ \frac{1}{2} & 0 & -\frac{1}{2} \end{bmatrix} \begin{bmatrix} 0 \\ 0 \\ -2 \end{bmatrix} = \begin{bmatrix} 1 \\ -2 \\ 1 \end{bmatrix}$ Solution: $x_1 = 1$, $x_2 = -2$, $x_3 = 1$.

(C) $\begin{bmatrix} x_1 \\ x_2 \\ x_3 \end{bmatrix} = \begin{bmatrix} -\frac{5}{2} & 2 & -\frac{1}{2} \\ 1 & -1 & 1 \\ \frac{1}{2} & 0 & -\frac{1}{2} \end{bmatrix} \begin{bmatrix} -3 \\ -4 \\ 1 \end{bmatrix} = \begin{bmatrix} -1 \\ 2 \\ -2 \end{bmatrix}$ Solution: $x_1 = -1$, $x_2 = 2$, $x_3 = -2$.   (4-6)

28. $2x_1 - 6x_2 = 4$
$-x_1 + kx_2 = -2$
If $k = 3$, then the first equation is a multiple (-2) of the second equation which implies there are infinitely many solutions. If $k \neq 3$, then the system has a unique solution.   (4-3)

29. $M = \begin{bmatrix} 0.2 & 0.15 \\ 0.4 & 0.3 \end{bmatrix}$; $I - M = \begin{bmatrix} 0.8 & -0.15 \\ -0.4 & 0.7 \end{bmatrix} = \begin{bmatrix} \frac{4}{5} & -\frac{3}{20} \\ -\frac{2}{5} & \frac{7}{10} \end{bmatrix}$

$(I - M)^{-1}$: $\begin{bmatrix} \frac{4}{5} & -\frac{3}{20} & | & 1 & 0 \\ -\frac{2}{5} & \frac{7}{10} & | & 0 & 1 \end{bmatrix} \sim \begin{bmatrix} \frac{4}{5} & -\frac{3}{20} & | & 1 & 0 \\ 0 & \frac{5}{8} & | & \frac{1}{2} & 1 \end{bmatrix}$

$\left(\frac{1}{2}\right)R_1 + R_2 \rightarrow R_2$ $\left(\frac{5}{4}\right)R_1 \rightarrow R_1$, $\left(\frac{8}{5}\right)R_2 \rightarrow R_2$

$$\sim \begin{bmatrix} 1 & -\frac{3}{16} & \frac{5}{4} & 0 \\ 0 & 1 & \frac{4}{5} & \frac{8}{5} \end{bmatrix} \sim \begin{bmatrix} 1 & 0 & \frac{7}{5} & \frac{3}{10} \\ 0 & 1 & \frac{4}{5} & \frac{8}{5} \end{bmatrix}$$

$$\left(\frac{3}{16}\right)R_2 + R_1 \rightarrow R_1$$

Thus, $(I - M)^{-1} = \begin{bmatrix} \frac{7}{5} & \frac{3}{10} \\ \frac{4}{5} & \frac{8}{5} \end{bmatrix} = \begin{bmatrix} 1.4 & 0.3 \\ 0.8 & 1.6 \end{bmatrix}$

The output matrix X is given by

$$X = (I - M)^{-1}D = \begin{bmatrix} 1.4 & 0.3 \\ 0.8 & 1.6 \end{bmatrix} \begin{bmatrix} 30 \\ 20 \end{bmatrix} = \begin{matrix} A \\ E \end{matrix} \begin{bmatrix} 48 \\ 56 \end{bmatrix}$$

Agriculture: \$48 billion; Energy: \$56 billion.   (4-7)

30. $T = \begin{bmatrix} 0.3 & 0.4 \\ 0.15 & 0.2 \end{bmatrix}$, $D = \begin{bmatrix} 20 \\ 30 \end{bmatrix}$

$$I - T = \begin{bmatrix} 0.7 & -0.4 \\ -0.15 & 0.8 \end{bmatrix} = \begin{bmatrix} \frac{7}{10} & -\frac{2}{5} \\ -\frac{3}{20} & \frac{4}{5} \end{bmatrix}$$

$(I - T)^{-1}$: $\begin{bmatrix} \frac{7}{10} & -\frac{2}{5} & 1 & 0 \\ -\frac{3}{20} & \frac{4}{5} & 0 & 1 \end{bmatrix} \sim \begin{bmatrix} 1 & -\frac{4}{7} & \frac{10}{7} & 0 \\ -\frac{3}{20} & \frac{4}{5} & 0 & 1 \end{bmatrix} \sim \begin{bmatrix} 1 & -\frac{4}{7} & \frac{10}{7} & 0 \\ 0 & \frac{5}{7} & \frac{3}{14} & 1 \end{bmatrix}$

$$\left(\frac{10}{7}\right)R_1 \rightarrow R_1 \qquad \left(\frac{3}{20}\right)R_1 + R_2 \rightarrow R_2 \qquad \left(\frac{7}{5}\right)R_2 \rightarrow R_2$$

$$\sim \begin{bmatrix} 1 & -\frac{4}{7} & \frac{10}{7} & 0 \\ 0 & 1 & \frac{3}{10} & \frac{7}{5} \end{bmatrix} \sim \begin{bmatrix} 1 & 0 & \frac{8}{5} & \frac{4}{5} \\ 0 & 1 & \frac{3}{10} & \frac{7}{5} \end{bmatrix}$$

$$\left(\frac{4}{7}\right)R_2 + R_1 \rightarrow R_1$$

Thus, $(I - T)^{-1} = \begin{bmatrix} \frac{8}{5} & \frac{4}{5} \\ \frac{3}{10} & \frac{7}{5} \end{bmatrix} = \begin{bmatrix} 1.6 & 0.8 \\ 0.3 & 1.4 \end{bmatrix}$.

The output matrix X is given by

$$X = (I - T)^{-1}D = \begin{bmatrix} 1.6 & 0.8 \\ 0.3 & 1.4 \end{bmatrix} \begin{bmatrix} 20 \\ 30 \end{bmatrix} = \begin{matrix} E \\ A \end{matrix} \begin{bmatrix} 56 \\ 48 \end{bmatrix};$$ Energy: \$56 billion; Agriculture: \$48 billion.   (4-7)

31. $M = \begin{bmatrix} 0.45 & 0.65 \\ 0.55 & 0.35 \end{bmatrix}$; $I - M = \begin{bmatrix} 0.55 & -0.65 \\ -0.55 & 0.65 \end{bmatrix}$

Since $R_2 = (-1)R_1$, $(I - M)$ is singular and X does not exist.   (4-5)

32.   The graphs of the two equations are:

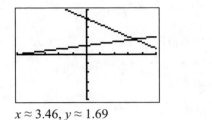

$x \approx 3.46, y \approx 1.69$ (4-1)

33. $\begin{bmatrix} 4 & 5 & 6 & | & 1 & 0 & 0 \\ 4 & 5 & -4 & | & 0 & 1 & 0 \\ 1 & 1 & 1 & | & 0 & 0 & 1 \end{bmatrix} R_1 \leftrightarrow R_3 \sim \begin{bmatrix} 1 & 1 & 1 & | & 0 & 0 & 1 \\ 4 & 5 & -4 & | & 0 & 1 & 0 \\ 4 & 5 & 6 & | & 1 & 0 & 0 \end{bmatrix} \begin{array}{l} (-4)R_1 + R_2 \to R_2 \\ (-4)R_1 + R_3 \to R_3 \end{array}$

$\sim \begin{bmatrix} 1 & 1 & 1 & | & 0 & 0 & 1 \\ 0 & 1 & -8 & | & 0 & 1 & -4 \\ 0 & 1 & 2 & | & 1 & 0 & -4 \end{bmatrix} \begin{array}{l} (-1)R_2 + R_1 \to R_1 \\ (-1)R_2 + R_3 \to R_3 \end{array} \sim \begin{bmatrix} 1 & 0 & 9 & | & 0 & -1 & 5 \\ 0 & 1 & -8 & | & 0 & 1 & -4 \\ 0 & 0 & 10 & | & 1 & -1 & 0 \end{bmatrix} \frac{1}{10} R_3 \to R_3$

$\sim \begin{bmatrix} 1 & 0 & 9 & | & 0 & -1 & 5 \\ 0 & 1 & -8 & | & 0 & 1 & -4 \\ 0 & 0 & 1 & | & \frac{1}{10} & -\frac{1}{10} & 0 \end{bmatrix} \begin{array}{l} (-9)R_3 + R_1 \to R_1 \\ 8R_3 + R_2 \to R_2 \end{array} \sim \begin{bmatrix} 1 & 0 & 0 & | & -\frac{9}{10} & -\frac{1}{10} & 5 \\ 0 & 1 & 0 & | & \frac{8}{10} & \frac{2}{10} & -4 \\ 0 & 0 & 1 & | & \frac{1}{10} & -\frac{1}{10} & 0 \end{bmatrix}$

Thus, $A^{-1} = \begin{bmatrix} -0.9 & -0.1 & 5 \\ 0.8 & 0.2 & -4 \\ 0.1 & -0.1 & 0 \end{bmatrix}$;

$A^{-1}A = \begin{bmatrix} -0.9 & -0.1 & 5 \\ 0.8 & 0.2 & -4 \\ 0.1 & -0.1 & 0 \end{bmatrix} \begin{bmatrix} 4 & 5 & 6 \\ 4 & 5 & -4 \\ 1 & 1 & 1 \end{bmatrix} = \begin{bmatrix} 1 & 0 & 0 \\ 0 & 1 & 0 \\ 0 & 0 & 1 \end{bmatrix}$ (4-5)

34.   The given system is equivalent to:
$$4x_1 + 5x_2 + 6x_3 = 36{,}000$$
$$4x_1 + 5x_2 - 4x_3 = 12{,}000$$
$$x_1 + x_2 + x_3 = 7{,}000$$

In matrix form, this system is:
$$\begin{bmatrix} 4 & 5 & 6 \\ 4 & 5 & -4 \\ 1 & 1 & 1 \end{bmatrix} \begin{bmatrix} x_1 \\ x_2 \\ x_3 \end{bmatrix} = \begin{bmatrix} 36{,}000 \\ 12{,}000 \\ 7{,}000 \end{bmatrix}$$

Thus,
$$\begin{bmatrix} x_1 \\ x_2 \\ x_3 \end{bmatrix} = \begin{bmatrix} 4 & 5 & 6 \\ 4 & 5 & -4 \\ 1 & 1 & 1 \end{bmatrix}^{-1} \begin{bmatrix} 36{,}000 \\ 12{,}000 \\ 7{,}000 \end{bmatrix} = \begin{bmatrix} -0.9 & -0.1 & 5 \\ 0.8 & 0.2 & -4 \\ 0.1 & -0.1 & 0 \end{bmatrix} \begin{bmatrix} 36{,}000 \\ 12{,}000 \\ 7{,}000 \end{bmatrix} = \begin{bmatrix} 1{,}400 \\ 3{,}200 \\ 2{,}400 \end{bmatrix}$$

Solution: $x_1 = 1{,}400$, $x_2 = 3{,}200$, $x_3 = 2{,}400$.   (4-6)

35.   First, multiply the first two equations of the system by 100. Then the augmented matrix of the resulting system is:

$$\begin{bmatrix} 4 & 5 & 6 & | & 36,000 \\ 4 & 5 & -4 & | & 12,000 \\ 1 & 1 & 1 & | & 7,000 \end{bmatrix} R_1 \leftrightarrow R_3 \sim \begin{bmatrix} 1 & 1 & 1 & | & 7,000 \\ 4 & 5 & -4 & | & 12,000 \\ 4 & 5 & 6 & | & 36,000 \end{bmatrix} \begin{array}{l} (-4)R_1 + R_2 \to R_2 \\ \\ (-4)R_1 + R_3 \to R_3 \end{array}$$

$$\sim \begin{bmatrix} 1 & 1 & 1 & | & 7,000 \\ 0 & 1 & -8 & | & -16,000 \\ 0 & 1 & 2 & | & 8,000 \end{bmatrix} \begin{array}{l} (-1)R_2 + R_1 \to R_1 \\ \\ (-1)R_2 + R_3 \to R_3 \end{array} \sim \begin{bmatrix} 1 & 0 & 9 & | & 23,000 \\ 0 & 1 & -8 & | & -16,000 \\ 0 & 0 & 10 & | & 24,000 \end{bmatrix} \frac{1}{10} R_3 \to R_3$$

$$\sim \begin{bmatrix} 1 & 0 & 9 & | & 23,000 \\ 0 & 1 & -8 & | & -16,000 \\ 0 & 0 & 1 & | & 2,400 \end{bmatrix} \begin{array}{l} (-9)R_3 + R_1 \to R_1 \\ \\ 8R_3 + R_2 \to R_2 \end{array} \sim \begin{bmatrix} 1 & 0 & 0 & | & 1,400 \\ 0 & 1 & 0 & | & 3,200 \\ 0 & 0 & 1 & | & 2,400 \end{bmatrix} \begin{array}{l} x_1 = 1,400 \\ x_2 = 3,200 \\ x_3 = 2,400 \end{array}$$

Solution: $x_1 = 1,400$, $x_2 = 3,200$, $x_3 = 2,400$.   (4-3)

36. $M = \begin{bmatrix} 0.2 & 0 & 0.4 \\ 0.1 & 0.3 & 0.1 \\ 0 & 0.4 & 0.2 \end{bmatrix} = \begin{bmatrix} \frac{1}{5} & 0 & \frac{2}{5} \\ \frac{1}{10} & \frac{3}{10} & \frac{1}{10} \\ 0 & \frac{2}{5} & \frac{1}{5} \end{bmatrix}$ and $D = \begin{bmatrix} 40 \\ 20 \\ 30 \end{bmatrix}$

$$I - M = \begin{bmatrix} 1 & 0 & 0 \\ 0 & 1 & 0 \\ 0 & 0 & 1 \end{bmatrix} - \begin{bmatrix} \frac{1}{5} & 0 & \frac{2}{5} \\ \frac{1}{10} & \frac{3}{10} & \frac{1}{10} \\ 0 & \frac{2}{5} & \frac{1}{5} \end{bmatrix} = \begin{bmatrix} \frac{4}{5} & 0 & -\frac{2}{5} \\ -\frac{1}{10} & \frac{7}{10} & -\frac{1}{10} \\ 0 & -\frac{2}{5} & \frac{4}{5} \end{bmatrix}.$$

$$\begin{bmatrix} \frac{4}{5} & 0 & -\frac{2}{5} & | & 1 & 0 & 0 \\ -\frac{1}{10} & \frac{7}{10} & -\frac{1}{10} & | & 0 & 1 & 0 \\ 0 & -\frac{2}{5} & \frac{4}{5} & | & 0 & 0 & 1 \end{bmatrix} \sim \begin{bmatrix} 1 & 0 & -\frac{1}{2} & | & \frac{5}{4} & 0 & 0 \\ -\frac{1}{10} & \frac{7}{10} & -\frac{1}{10} & | & 0 & 1 & 0 \\ 0 & -\frac{2}{5} & \frac{4}{5} & | & 0 & 0 & 1 \end{bmatrix}$$

$$\frac{5}{4} R_1 \to R_1 \qquad\qquad \frac{1}{10} R_1 + R_2 \to R_2$$

$$\sim \begin{bmatrix} 1 & 0 & -\frac{1}{2} & | & \frac{5}{4} & 0 & 0 \\ 0 & \frac{7}{10} & -\frac{3}{20} & | & \frac{1}{8} & 1 & 0 \\ 0 & -\frac{2}{5} & \frac{4}{5} & | & 0 & 0 & 1 \end{bmatrix} \sim \begin{bmatrix} 1 & 0 & -\frac{1}{2} & | & \frac{5}{4} & 0 & 0 \\ 0 & -\frac{2}{5} & \frac{4}{5} & | & 0 & 0 & 1 \\ 0 & \frac{7}{10} & -\frac{3}{20} & | & \frac{1}{8} & 1 & 0 \end{bmatrix}$$

$$R_2 \leftrightarrow R_3 \qquad\qquad \left(-\frac{5}{2}\right) R_2 \to R_2$$

$$\sim \begin{bmatrix} 1 & 0 & -\frac{1}{2} & | & \frac{5}{4} & 0 & 0 \\ 0 & 1 & -2 & | & 0 & 0 & -\frac{5}{2} \\ 0 & \frac{7}{10} & -\frac{3}{20} & | & \frac{1}{8} & 1 & 0 \end{bmatrix} \sim \begin{bmatrix} 1 & 0 & -\frac{1}{2} & | & \frac{5}{4} & 0 & 0 \\ 0 & 1 & -2 & | & 0 & 0 & -\frac{5}{2} \\ 0 & 0 & \frac{5}{4} & | & \frac{1}{8} & 1 & \frac{7}{4} \end{bmatrix}$$

$$\left(-\frac{7}{10}\right) R_2 + R_3 \to R_3 \qquad\qquad \frac{4}{5} R_3 \to R_3$$

$$\sim \begin{bmatrix} 1 & 0 & -\frac{1}{2} & | & \frac{5}{4} & 0 & 0 \\ 0 & 1 & -2 & | & 0 & 0 & -\frac{5}{2} \\ 0 & 0 & 1 & | & \frac{1}{10} & \frac{4}{5} & \frac{7}{5} \end{bmatrix} \sim \begin{bmatrix} 1 & 0 & 0 & | & \frac{13}{10} & \frac{2}{5} & \frac{7}{10} \\ 0 & 1 & 0 & | & \frac{1}{5} & \frac{8}{5} & \frac{3}{10} \\ 0 & 0 & 1 & | & \frac{1}{10} & \frac{4}{5} & \frac{7}{5} \end{bmatrix}$$

$$2R_3 + R_2 \to R_2, \frac{1}{2} R_3 + R_1 \to R_1$$

Thus $(I-M)^{-1} = \begin{bmatrix} \frac{13}{10} & \frac{2}{5} & \frac{7}{10} \\ \frac{1}{5} & \frac{8}{5} & \frac{3}{10} \\ \frac{1}{10} & \frac{4}{5} & \frac{7}{5} \end{bmatrix} = \begin{bmatrix} 1.3 & 0.4 & 0.7 \\ 0.2 & 1.6 & 0.3 \\ 0.1 & 0.8 & 1.4 \end{bmatrix}$ and

$X=(I-M)^{-1}D = \begin{bmatrix} 1.3 & 0.4 & 0.7 \\ 0.2 & 1.6 & 0.3 \\ 0.1 & 0.8 & 1.4 \end{bmatrix} \begin{bmatrix} 40 \\ 20 \\ 30 \end{bmatrix} = \begin{bmatrix} 81 \\ 49 \\ 62 \end{bmatrix}$ (4-7)

37. (A)   The system has a unique solution.

  (B)   The system either has no solutions or infinitely many solutions.   (4-6)

38. (A)   The system has a unique solution.

  (B)   The system has no solutions.

  (C)   The system has infinitely many solutions.   (4-3)

39.   The third step in (A) is incorrect:
    $X-MX=(I-M)X$ **not** $X(I-M)$
Each step in (B) is correct.   (4-6)

40.   Let x = the number of machines produced.

  (A) $C(x) = 243{,}000 + 22.45x$
    $R(x) = 59.95x$

  (B)   Set $C(x) = R(x)$:
      $59.95x = 243{,}000 + 22.45x$
       $37.5x = 243{,}000$
          $x = 6{,}480$

    If 6,480 machines are produced, $C = R = \$388{,}476$; break-even point (6,480, 388,476).

  (C)   A profit occurs if $x > 6{,}480$; a loss occurs if $x < 6{,}480$

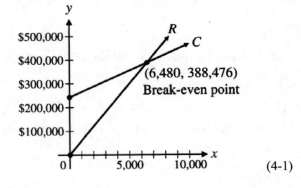

(4-1)

41.   Let x_1 = Number of tons of Voisey's Bay ore
       x_2 = number of tons of Hawk Ridge ore

    Then, we have the following system of equations
    $0.02x_1 + 0.03x_2 = 6$
    $0.04x_1 + 0.02x_2 = 8$

Multiply each equation by 100. This yields

$$2x_1 + 3x_2 = 600$$
$$4x_1 + 2x_2 = 800$$

The augmented matrix corresponding to this system is:

$$\begin{bmatrix} 2 & 3 & | & 600 \\ 4 & 2 & | & 800 \end{bmatrix} \sim \begin{bmatrix} 2 & 3 & | & 600 \\ 0 & -4 & | & -400 \end{bmatrix} \sim \begin{bmatrix} 2 & 3 & | & 600 \\ 0 & 1 & | & 100 \end{bmatrix} \sim \begin{bmatrix} 2 & 0 & | & 300 \\ 0 & 1 & | & 100 \end{bmatrix} \sim \begin{bmatrix} 1 & 0 & | & 150 \\ 0 & 1 & | & 100 \end{bmatrix}$$

$$(-2)R_1 + R_2 \to R_2 \quad \left(-\frac{1}{4}\right)R_2 \to R_2 \quad (-3)R_2 + R_1 \to R_1 \quad \left(\frac{1}{2}\right)R_1 \to R_1$$

Thus, the solution is: $x_1 = 150$ tons of Voisey's Bay ore, $x_2 = 100$ tons of Hawk Ridge ore.   (4-3)

42. (A)   The matrix equation for Problem 41 is:

$$\begin{bmatrix} 0.02 & 0.03 \\ 0.04 & 0.02 \end{bmatrix} \begin{bmatrix} x_1 \\ x_2 \end{bmatrix} = \begin{bmatrix} 6 \\ 8 \end{bmatrix}$$

First, we compute the inverse of $\begin{bmatrix} 0.02 & 0.03 \\ 0.04 & 0.02 \end{bmatrix}$;

$$\begin{bmatrix} 0.02 & 0.03 & | & 1 & 0 \\ 0.04 & 0.02 & | & 0 & 1 \end{bmatrix} \sim \begin{bmatrix} 1 & 1.5 & | & 50 & 0 \\ 0.04 & 0.02 & | & 0 & 1 \end{bmatrix} \sim \begin{bmatrix} 1 & 1.5 & | & 50 & 0 \\ 0 & -0.04 & | & -2 & 1 \end{bmatrix}$$

$$50R_1 \to R_1 \qquad (-0.04)R_1 + R_2 \to R_2 \qquad \left(-\frac{1}{0.04}\right)R_2 \to R_2$$

$$\sim \begin{bmatrix} 1 & 1.5 & | & 50 & 0 \\ 0 & 1 & | & 50 & -25 \end{bmatrix} \sim \begin{bmatrix} 1 & 0 & | & -25 & 37.5 \\ 0 & 1 & | & 50 & -25 \end{bmatrix}$$

$$(-1.5)R_2 + R_1 \to R_1$$

Thus, the inverse of the coefficient matrix is:

$$\begin{bmatrix} -25 & 37.5 \\ 50 & -25 \end{bmatrix}$$

Now, $\begin{bmatrix} x_1 \\ x_2 \end{bmatrix} = \begin{bmatrix} -25 & 37.5 \\ 50 & -25 \end{bmatrix} \begin{bmatrix} 6 \\ 8 \end{bmatrix} = \begin{bmatrix} 150 \\ 100 \end{bmatrix}$

Again, the solution is: $x_1 = 150$ tons of Voisey's Bay ore, $x_2 = 100$ tons of Hawk Ridge ore.

(B) $\begin{bmatrix} x_1 \\ x_2 \end{bmatrix} = \begin{bmatrix} -25 & 37.5 \\ 50 & -25 \end{bmatrix} \begin{bmatrix} 7.5 \\ 7 \end{bmatrix} = \begin{bmatrix} 75 \\ 200 \end{bmatrix}$

Now the solution is: $x_1 = 75$ tons of Voisey's Bay ore, $x_2 = 200$ tons of Hawk Ridge ore.

(4-6)

43. (A)   Let $x_1 =$ number of 3,000 cubic foot hoppers

$x_2 =$ number of 4,500 cubic foot hoppers

$x_3 =$ number of 6,000 cubic foot hoppers

Then

$$x_1 + x_2 + x_3 = 20$$
$$3,000x_1 + 4,500x_2 + 6,000x_3 = 108,000$$

or

$$x_1 + x_2 + x_3 = 20$$
$$3x_1 + 4.5x_2 + 6x_3 = 108$$

The augmented matrix for this system is:

$$\begin{bmatrix} 1 & 1 & 1 & | & 20 \\ 3 & 4.5 & 6 & | & 108 \end{bmatrix}$$

Now

$$\begin{bmatrix} 1 & 1 & 1 & | & 20 \\ 3 & 4.5 & 6 & | & 108 \end{bmatrix} \sim \begin{bmatrix} 1 & 1 & 1 & | & 20 \\ 0 & 1.5 & 3 & | & 48 \end{bmatrix} \sim \begin{bmatrix} 1 & 1 & 1 & | & 20 \\ 0 & 1 & 2 & | & 32 \end{bmatrix}$$

$$(-3)R_1 + R_2 \to R_2 \qquad \tfrac{2}{3}R_2 \to R_2 \qquad (-1)R_2 + R_1 \to R_1$$

$$\sim \begin{bmatrix} 1 & 0 & -1 & | & -12 \\ 0 & 1 & 2 & | & 32 \end{bmatrix}$$

The corresponding system of equations is:

$$x_1 \qquad - x_3 = -12$$
$$x_2 + 2x_3 = 32$$

and the solutions are: $x_1 = t - 12$ 3,000 cubic foot hoppers, $x_2 = 32 - 2t$ 4,500 cubic foot hoppers, $x_3 = t$ 6,000 cubic foot hoppers.

Since $x_1, x_2,$ and x_3 are non-negative integers, it follows that $t = 12, 13, 14, 15,$ or 16.

(B) Cost: $C(t) = 180(t - 12) + 225(32 - 2t) + 325(t)$
$C(12) = 225(8) + 325(12) = \$5,700$
$C(13) = 180(1) + 225(6) + 325(13) = \$5,755$
$C(14) = 180(2) + 225(4) + 325(14) = \$5,810$
$C(15) = 180(3) + 225(2) + 325(15) = \$5,865$
$C(16) = 180(4) + 325(16) = \$5,920$

The minimum monthly cost is $5,700 when 8 4,500 cubic foot hoppers and 12 6,000 cubic foot hoppers are leased.   (4-3)

44. (A) The elements of *MN* give the cost of materials for each alloy from each supplier. The product *NM* is also defined, but does not have an interpretation in the context of this problem.

(B) $MN = \begin{bmatrix} 4,800 & 600 & 300 \\ 6,000 & 1,400 & 700 \end{bmatrix} \begin{bmatrix} 0.75 & 0.70 \\ 6.50 & 6.70 \\ 0.40 & 0.50 \end{bmatrix}$

$\qquad$ Supplier A Supplier B
$= \begin{bmatrix} \$7,620 & \$7,530 \\ \$13,880 & \$13,930 \end{bmatrix} \begin{matrix} \text{Alloy 1} \\ \text{Alloy 2} \end{matrix}$

(C) The total costs of materials from Supplier *A* is:
$\$7,620 + \$13,880 = \$21,500$

The total costs of materials from Supplier *B* is:
$\$7,530 + \$13,930 = \$21,460$

These values can be obtained from the matrix product

$$[1 \quad 1]\begin{bmatrix} 7,620 & 7,530 \\ 13,880 & 13,930 \end{bmatrix}$$

Supplier B will provide the materials at lower cost.   (4-4)

45. (A) $[0.25 \quad 0.20 \quad 0.05]\begin{bmatrix} 12 \\ 15 \\ 7 \end{bmatrix} = 6.35$

The labor cost for one model B calculator at the California plant
is $6.35.

(B)   The elements of MN give the total labor costs for each calculator at each plant. The product NM is
also defined, but does not have an interpretation in the context of this problem.

(C) $MN = \begin{bmatrix} 0.15 & 0.10 & 0.05 \\ 0.25 & 0.20 & 0.05 \end{bmatrix}\begin{bmatrix} 12 & 10 \\ 15 & 12 \\ 7 & 6 \end{bmatrix} = \begin{array}{cc} CA & TX \\ \begin{bmatrix} \$3.65 & \$3.00 \\ \$6.35 & \$5.20 \end{bmatrix} & \begin{array}{l} Model\ A \\ Model\ B \end{array} \end{array}$

46.   Let $x_1 = $ amount invested at 5%

and $x_2 = $ amount invested at 10%.

Then, $x_1 + x_2 = 5000$

$0.05x_1 + 0.1x_2 = 400$

The augmented matrix for the system given above is:

$$\begin{bmatrix} 1 & 1 & | & 5000 \\ 0.05 & 0.1 & | & 400 \end{bmatrix} \sim \begin{bmatrix} 1 & 1 & | & 5000 \\ 0 & 0.05 & | & 150 \end{bmatrix} \sim \begin{bmatrix} 1 & 1 & | & 5000 \\ 0 & 1 & | & 3000 \end{bmatrix} \sim \begin{bmatrix} 1 & 0 & | & 2000 \\ 0 & 1 & | & 3000 \end{bmatrix}$$

$(-0.05)R_1 + R_2 \rightarrow R_1 \qquad \dfrac{1}{0.05}R_2 \rightarrow R_2 \qquad (-1)R_2 + R_1 \rightarrow R_1$

Hence, $x_1 = \$2000$ at 5%, $x_2 = \$3000$ at 10%.   (4-3)

47.   The matrix equation corresponding to the system in Problem 46 is:

$$\begin{bmatrix} 1 & 1 \\ 0.05 & 0.1 \end{bmatrix}\begin{bmatrix} x_1 \\ x_2 \end{bmatrix} = \begin{bmatrix} 5000 \\ 400 \end{bmatrix}$$

Now we compute the inverse matrix of $\begin{bmatrix} 1 & 1 \\ 0.05 & 0.1 \end{bmatrix}$.

$$\begin{bmatrix} 1 & 1 & | & 1 & 0 \\ 0.05 & 0.1 & | & 0 & 1 \end{bmatrix} \sim \begin{bmatrix} 1 & 1 & | & 1 & 0 \\ 0 & 0.05 & | & -0.05 & 1 \end{bmatrix} \sim \begin{bmatrix} 1 & 1 & | & 1 & 0 \\ 0 & 1 & | & -1 & 20 \end{bmatrix}$$

$(-0.05)R_1 + R_2 \rightarrow R_1 \qquad \dfrac{1}{0.05}R_2 \rightarrow R_2 \qquad (-1)R_2 + R_1 \rightarrow R_1$

$$\sim \begin{bmatrix} 1 & 0 & | & 2 & -20 \\ 0 & 1 & | & -1 & 20 \end{bmatrix}$$

Thus, the inverse of the coefficient matrix is $\begin{bmatrix} 2 & -20 \\ -1 & 20 \end{bmatrix}$, and

$$\begin{bmatrix} x_1 \\ x_2 \end{bmatrix} = \begin{bmatrix} 2 & -20 \\ -1 & 20 \end{bmatrix}\begin{bmatrix} 5000 \\ 400 \end{bmatrix} = \begin{bmatrix} 10,000 - 8,000 \\ -5,000 + 8,000 \end{bmatrix} = \begin{bmatrix} 2000 \\ 3000 \end{bmatrix}.$$ So, $x_1 = \$2000$ at 5%,   (4-6)
$x_2 = \$3000$ at 10%.

48. From Problem 46, the system of equations is:
$$x_1 + x_2 = 5000$$
$$0.05x_1 + 0.1x_2 = k \text{ (annual return)}$$

From Problem 47, the inverse of the coefficient matrix A is: $A^{-1} = \begin{bmatrix} 2 & -20 \\ -1 & 20 \end{bmatrix}$

$k = \$200?$
$$\begin{bmatrix} x_1 \\ x_2 \end{bmatrix} = \begin{bmatrix} 2 & -20 \\ -1 & 20 \end{bmatrix}\begin{bmatrix} 5000 \\ 200 \end{bmatrix} = \begin{bmatrix} 6000 \\ -1000 \end{bmatrix};\ \ x_1 = \$6,000,\ \ x_2 = -\$1,000$$
This is not possible, x_2 cannot be negative.

$k = \$600?$
$$\begin{bmatrix} x_1 \\ x_2 \end{bmatrix} = \begin{bmatrix} 2 & -20 \\ -1 & 20 \end{bmatrix}\begin{bmatrix} 5000 \\ 600 \end{bmatrix} = \begin{bmatrix} -2000 \\ 7000 \end{bmatrix};\ \ x_1 = -\$2,000,\ \ x_2 = \$7,000$$
This is not possible, x_1 cannot be negative.

Fix a return k. Then
$$\begin{bmatrix} x_1 \\ x_2 \end{bmatrix} = \begin{bmatrix} 2 & -20 \\ -1 & 20 \end{bmatrix}\begin{bmatrix} 5000 \\ k \end{bmatrix} = \begin{bmatrix} 10,000 - 20k \\ -5,000 + 20k \end{bmatrix}$$
so $x_1 = 10,000 - 20k$, $x_2 = -5,000 + 20k$

Since $x_1 \geq 0$, we have $10,000 - 20k \geq 0$
$$20k \leq 10,000$$
$$k \leq 500$$
Since $x_2 \geq 0$, we have $-5,000 + 20k \geq 0$
$$20k \geq 5,000$$
$$k \geq 250$$

The possible annual yields must satisfy $250 \leq k \leq 500$.   (4-6)

49. Let x_1 = number of \$8 tickets
x_2 = number of \$12 tickets
x_3 = number of \$20 tickets

Since the number of \$8 tickets must equal the number of \$20 tickets, we have
$$x_1 = x_3 \text{ or } x_1 - x_3 = 0$$
Also, since all seats are sold
$$x_1 + x_2 + x_3 = 25,000$$
Finally, the return is
$$8x_1 + 12x_2 + 20x_3 = R \text{ (where } R \text{ is the return required).}$$
Thus, the system of equations is:

$$\begin{aligned} x_1 \quad\quad\; - \; x_3 &= 0 \\ x_1 + \; x_2 + \; x_3 &= 25{,}000 \\ 8x_1 + 12x_2 + 20x_3 &= R \end{aligned} \quad \text{or} \quad \begin{bmatrix} 1 & 0 & -1 \\ 1 & 1 & 1 \\ 8 & 12 & 20 \end{bmatrix} \begin{bmatrix} x_1 \\ x_2 \\ x_3 \end{bmatrix} = \begin{bmatrix} 0 \\ 25{,}000 \\ R \end{bmatrix}$$

First, we compute the inverse of the coefficient matrix

$$\left[\begin{array}{ccc|ccc} 1 & 0 & -1 & 1 & 0 & 0 \\ 1 & 1 & 1 & 0 & 1 & 0 \\ 8 & 12 & 20 & 0 & 0 & 1 \end{array}\right] \sim \left[\begin{array}{ccc|ccc} 1 & 0 & -1 & 1 & 0 & 0 \\ 0 & 1 & 2 & -1 & 1 & 0 \\ 0 & 12 & 28 & -8 & 0 & 1 \end{array}\right]$$

$$(-1)R_1 + R_2 \to R_2 \quad\quad (-12)R_2 + R_3 \to R_3$$
$$(-8)R_1 + R_3 \to R_3$$

$$\sim \left[\begin{array}{ccc|ccc} 1 & 0 & -1 & 1 & 0 & 0 \\ 0 & 1 & 2 & -1 & 1 & 0 \\ 0 & 0 & 4 & 4 & -12 & 1 \end{array}\right] \sim \left[\begin{array}{ccc|ccc} 1 & 0 & -1 & 1 & 0 & 0 \\ 0 & 1 & 2 & -1 & 1 & 0 \\ 0 & 0 & 1 & 1 & -3 & \frac{1}{4} \end{array}\right]$$

$$\tfrac{1}{4}R_3 \to R_3 \quad\quad (-2)R_3 + R_2 \to R_2$$
$$R_3 + R_1 \to R_1$$

$$\sim \left[\begin{array}{ccc|ccc} 1 & 0 & 0 & 2 & -3 & \frac{1}{4} \\ 0 & 1 & 0 & -3 & 7 & -\frac{1}{2} \\ 0 & 0 & 1 & 1 & -3 & \frac{1}{4} \end{array}\right]. \quad \text{Thus, the inverse is} \quad \begin{bmatrix} 2 & -3 & \frac{1}{4} \\ -3 & 7 & -\frac{1}{2} \\ 1 & -3 & \frac{1}{4} \end{bmatrix}.$$

Concert 1:

$$\begin{aligned} x_1 \quad\quad\; - \; x_3 &= 0 \\ x_1 + \; x_2 + \; x_3 &= 25{,}000 \\ 8x_1 + 12x_2 + 20x_3 &= 320{,}000 \end{aligned} \quad \text{or} \quad \begin{bmatrix} 1 & 0 & -1 \\ 1 & 1 & 1 \\ 8 & 12 & 20 \end{bmatrix} \begin{bmatrix} x_1 \\ x_2 \\ x_3 \end{bmatrix} = \begin{bmatrix} 0 \\ 25{,}000 \\ 320{,}000 \end{bmatrix}$$

Thus $\begin{bmatrix} x_1 \\ x_2 \\ x_3 \end{bmatrix} = \begin{bmatrix} 2 & -3 & \frac{1}{4} \\ -3 & 7 & -\frac{1}{2} \\ 1 & -3 & \frac{1}{4} \end{bmatrix} \begin{bmatrix} 0 \\ 25{,}000 \\ 320{,}000 \end{bmatrix} = \begin{bmatrix} 5{,}000 \\ 15{,}000 \\ 5{,}000 \end{bmatrix}$

and $x_1 = 5{,}000$ $8 tickets
  $x_2 = 15{,}000$ $12 tickets
  $x_3 = 5{,}000$ $20 tickets

Concert 2:

$$\begin{aligned} x_1 \quad\quad\; - \; x_3 &= 0 \\ x_1 + \; x_2 + \; x_3 &= 25{,}000 \\ 8x_1 + 12x_2 + 20x_3 &= 330{,}000 \end{aligned} \quad \text{or} \quad \begin{bmatrix} 1 & 0 & -1 \\ 1 & 1 & 1 \\ 8 & 12 & 20 \end{bmatrix} \begin{bmatrix} x_1 \\ x_2 \\ x_3 \end{bmatrix} = \begin{bmatrix} 0 \\ 25{,}000 \\ 330{,}000 \end{bmatrix}$$

Thus $\begin{bmatrix} x_1 \\ x_2 \\ x_3 \end{bmatrix} = \begin{bmatrix} 2 & -3 & \frac{1}{4} \\ -3 & 7 & -\frac{1}{2} \\ 1 & -3 & \frac{1}{4} \end{bmatrix} \begin{bmatrix} 0 \\ 25{,}000 \\ 330{,}000 \end{bmatrix} = \begin{bmatrix} 7{,}500 \\ 10{,}000 \\ 7{,}500 \end{bmatrix}$

and $x_1 = 7{,}500$ \$8 tickets

$\quad x_2 = 10{,}000$ \$12 tickets

$\quad x_3 = 7{,}500$ \$20 tickets

Concert 3:

$$\begin{aligned}
x_1 \quad - \quad x_3 &= 0 \\
x_1 + x_2 + x_3 &= 25{,}000 \\
8x_1 + 12x_2 + 20x_3 &= 340{,}000
\end{aligned}
\qquad \text{or} \qquad
\begin{bmatrix} 1 & 0 & -1 \\ 1 & 1 & 1 \\ 8 & 12 & 20 \end{bmatrix}
\begin{bmatrix} x_1 \\ x_2 \\ x_3 \end{bmatrix} =
\begin{bmatrix} 0 \\ 25{,}000 \\ 340{,}000 \end{bmatrix}$$

Thus
$$\begin{bmatrix} x_1 \\ x_2 \\ x_3 \end{bmatrix} =
\begin{bmatrix} 2 & -3 & \frac{1}{4} \\ -3 & 7 & -\frac{1}{2} \\ 1 & -3 & \frac{1}{4} \end{bmatrix}
\begin{bmatrix} 0 \\ 25{,}000 \\ 340{,}000 \end{bmatrix} =
\begin{bmatrix} 10{,}000 \\ 5{,}000 \\ 10{,}000 \end{bmatrix}$$

and $x_1 = 10{,}000$ \$8 tickets

$\quad x_2 = 5{,}000$ \$12 tickets

$\quad x_3 = 10{,}000$ \$20 tickets   (4-6)

50. From Problem 49, if it is not required to have an equal number of \$8 tickets and \$12 tickets, then the new mathematical model is:

$$\begin{aligned}
x_1 + x_2 + x_3 &= 25{,}000 \\
8x_1 + 12x_2 + 20x_3 &= k \quad \text{(return requested)}
\end{aligned}$$

The augmented matrix is:

$$\begin{bmatrix} 1 & 1 & 1 & | & 25{,}000 \\ 8 & 12 & 20 & | & k \end{bmatrix} \sim
\begin{bmatrix} 1 & 1 & 1 & | & 25{,}000 \\ 0 & 4 & 12 & | & k - 200{,}000 \end{bmatrix}$$

$(-8)R_1 + R_2 \rightarrow R_2 \qquad \frac{1}{4}R_2 \rightarrow R_2$

$$\sim \begin{bmatrix} 1 & 1 & 1 & | & 25{,}000 \\ 0 & 1 & 3 & | & \frac{k}{4} - 50{,}000 \end{bmatrix} \sim
\begin{bmatrix} 1 & 0 & -2 & | & -\frac{k}{4} + 75{,}000 \\ 0 & 1 & 3 & | & \frac{k}{4} - 50{,}000 \end{bmatrix}$$

$(-1)R_2 + R_1 \rightarrow R_1$

Concert 1: $k = \$320{,}000$; $\frac{1}{4}k = 80{,}000$

$$\begin{bmatrix} 1 & 0 & -2 & | & -5{,}000 \\ 0 & 1 & 3 & | & 30{,}000 \end{bmatrix};$$

$x_1 = 2t - 5{,}000$ \$8 tickets

$x_2 = 30{,}000 - 3t$ \$12 tickets

$x_3 = t$ \$20 tickets, t an integer

Since $x_1, x_2 \geq 0$, t must satisfy $2{,}500 \leq t \leq 10{,}000$.

Concert 2: $k = \$330{,}000$; $\frac{1}{4}k = 82{,}500$

$$\begin{bmatrix} 1 & 0 & -2 & | & -7{,}500 \\ 0 & 1 & 3 & | & 32{,}500 \end{bmatrix};$$

$x_1 = 2t - 7{,}500$ \$8 tickets

$x_2 = 32{,}500 - 3t$ \$12 tickets

$x_3 = t$ \$20 tickets, t an integer

Since $x_1, x_2 \geq 0$, t must satisfy $3{,}750 \leq t \leq 10{,}833$.

Concert 3: $k = \$340{,}000$; $\frac{1}{4}k = 85{,}000$

$$\begin{bmatrix} 1 & 0 & -2 & | & -10{,}000 \\ 0 & 1 & 3 & | & 35{,}000 \end{bmatrix};$$

$$x_1 = 2t - 10{,}000 \quad \$8 \text{ tickets}$$
$$x_2 = 35{,}000 - 3t \quad \$12 \text{ tickets}$$
$$x_3 = t \quad \$20 \text{ tickets, } t \text{ an integer}$$

Since $x_1, x_2 \geq 0$, t must satisfy $5{,}000 \leq t \leq 11{,}666$.   (4-3)

51.   The technology matrix is

$$M = \begin{matrix} \\ \text{Agriculture} \\ \text{Fabrication} \end{matrix} \overset{\text{Agriculture Fabrication}}{\begin{bmatrix} 0.30 & 0.10 \\ 0.20 & 0.40 \end{bmatrix}}$$

Now $I - M = \begin{bmatrix} 1 & 0 \\ 0 & 1 \end{bmatrix} - \begin{bmatrix} 0.30 & 0.10 \\ 0.20 & 0.40 \end{bmatrix} = \begin{bmatrix} 0.70 & -0.10 \\ -0.20 & 0.60 \end{bmatrix} = \begin{bmatrix} \frac{7}{10} & -\frac{1}{10} \\ -\frac{1}{5} & \frac{3}{5} \end{bmatrix}$

Next, we calculate the inverse of $I - M$

$$\begin{bmatrix} \frac{7}{10} & -\frac{1}{10} & | & 1 & 0 \\ -\frac{1}{5} & \frac{3}{5} & | & 0 & 1 \end{bmatrix} \sim \begin{bmatrix} 1 & -3 & | & 0 & -5 \\ \frac{7}{10} & -\frac{1}{10} & | & 1 & 0 \end{bmatrix} \sim \begin{bmatrix} 1 & -3 & | & 0 & -5 \\ 0 & 2 & | & 1 & \frac{7}{2} \end{bmatrix} \sim \begin{bmatrix} 1 & -3 & | & 0 & -5 \\ 0 & 1 & | & \frac{1}{2} & \frac{7}{4} \end{bmatrix}$$

$$-5R_2 \to R_2 \qquad \left(-\frac{7}{10}\right)R_1 + R_2 \to R_2 \qquad \frac{1}{2}R_2 \to R_2 \qquad 3R_2 + R_1 \to R_1$$
$$R_1 \leftrightarrow R_2$$

$$\sim \begin{bmatrix} 1 & 0 & | & \frac{3}{2} & \frac{1}{4} \\ 0 & 1 & | & \frac{1}{2} & \frac{7}{4} \end{bmatrix} \quad \text{Thus, } (I-M)^{-1} = \begin{bmatrix} \frac{3}{2} & \frac{1}{4} \\ \frac{1}{2} & \frac{7}{4} \end{bmatrix}.$$

Let x_1 = output for agriculture and x_2 = output for fabrication.

Then the output $X = \begin{bmatrix} x_1 \\ x_2 \end{bmatrix}$ needed to satisfy a final demand $D = \begin{bmatrix} d_1 \\ d_2 \end{bmatrix}$ for agriculture and fabrication is given by $X = (I-M)^{-1}D$.

(A)   Let $D = \begin{bmatrix} 50 \\ 20 \end{bmatrix}$. Then $X = \begin{bmatrix} \frac{3}{2} & \frac{1}{4} \\ \frac{1}{2} & \frac{7}{4} \end{bmatrix}\begin{bmatrix} 50 \\ 20 \end{bmatrix} = \begin{bmatrix} 75 + 5 \\ 25 + 35 \end{bmatrix} = \begin{bmatrix} 80 \\ 60 \end{bmatrix}$

Thus, the total output for agriculture is $80 billion; the total output for fabrication is $60 billion.

(B)   Let $D = \begin{bmatrix} 80 \\ 60 \end{bmatrix}$. Then $X = \begin{bmatrix} \frac{3}{2} & \frac{1}{4} \\ \frac{1}{2} & \frac{7}{4} \end{bmatrix}\begin{bmatrix} 80 \\ 60 \end{bmatrix} = \begin{bmatrix} 120 + 15 \\ 40 + 105 \end{bmatrix} = \begin{bmatrix} 135 \\ 145 \end{bmatrix}$

Thus, the total output for agriculture is $135 billion; the total output for fabrication is $145 billion.
(4-7)

52.   First we find the inverse of $B = \begin{bmatrix} 1 & 1 & 0 \\ 1 & 0 & 1 \\ 1 & 1 & 1 \end{bmatrix}$.

$$\begin{bmatrix} 1 & 1 & 0 & | & 1 & 0 & 0 \\ 1 & 0 & 1 & | & 0 & 1 & 0 \\ 1 & 1 & 1 & | & 0 & 0 & 1 \end{bmatrix} \sim \begin{bmatrix} 1 & 1 & 0 & | & 1 & 0 & 0 \\ 0 & -1 & 1 & | & -1 & 1 & 0 \\ 0 & 0 & 1 & | & -1 & 0 & 1 \end{bmatrix} \sim \begin{bmatrix} 1 & 1 & 0 & | & 1 & 0 & 0 \\ 0 & 1 & -1 & | & 1 & -1 & 0 \\ 0 & 0 & 1 & | & -1 & 0 & 1 \end{bmatrix}$$

$$(-1)R_1 + R_2 \to R_2 \qquad (-1)R_2 \to R_2 \qquad (-1)R_2 + R_1 \to R_1$$
$$(-1)R_1 + R_3 \to R_3$$

$$\sim \begin{bmatrix} 1 & 0 & 1 & 0 & 1 & 0 \\ 0 & 1 & -1 & 1 & -1 & 0 \\ 0 & 0 & 1 & -1 & 0 & 1 \end{bmatrix} \sim \begin{bmatrix} 1 & 0 & 1 & 1 & 1 & -1 \\ 0 & 1 & 0 & 0 & -1 & 1 \\ 0 & 0 & 1 & -1 & 0 & 1 \end{bmatrix} \quad \text{Thus, } B^{-1} = \begin{bmatrix} 1 & 1 & -1 \\ 0 & -1 & 1 \\ -1 & 0 & 1 \end{bmatrix}.$$

$$(-1)R_3 + R_1 \rightarrow R_1$$
$$R_3 + R_2 \rightarrow R_2$$

$$\text{Now, } \begin{bmatrix} 1 & 1 & -1 \\ 0 & -1 & 1 \\ -1 & 0 & 1 \end{bmatrix} \begin{bmatrix} 7 & 19 & 20 & 5 & 9 & 15 & 13 & 21 \\ 25 & 6 & 8 & 14 & 23 & 6 & 1 & 26 \\ 30 & 24 & 28 & 14 & 28 & 21 & 14 & 29 \end{bmatrix} = \begin{bmatrix} 2 & 1 & 0 & 5 & 4 & 0 & 0 & 18 \\ 5 & 18 & 20 & 0 & 5 & 15 & 13 & 3 \\ 23 & 5 & 8 & 9 & 19 & 6 & 1 & 8 \end{bmatrix}$$

Thus, the decoded message is
2 5 23 1 18 5 0 20 8 5 0 9 4 5 19 0 15 6 0 13 1 18 3 8
which corresponds to BEWARE THE IDES OF MARCH..   (4-5)

53. (A) 1st & Elm: $x_1 + x_4 = 1300$

2nd & Elm: $x_1 - x_2 = 400$

2nd & Oak: $x_2 + x_3 = 700$

1st & Oak: $x_3 - x_4 = -200$

(B) The augmented matrix for the system in part (A) is:

$$\begin{bmatrix} 1 & 0 & 0 & 1 & 1300 \\ 1 & -1 & 0 & 0 & 400 \\ 0 & 1 & 1 & 0 & 700 \\ 0 & 0 & 1 & -1 & -200 \end{bmatrix} \sim \begin{bmatrix} 1 & 0 & 0 & 1 & 1300 \\ 0 & -1 & 0 & -1 & -900 \\ 0 & 1 & 1 & 0 & 700 \\ 0 & 0 & 1 & -1 & -200 \end{bmatrix}$$

$$(-1)R_1 + R_2 \rightarrow R_2 \qquad (-1)R_2 \rightarrow R_2$$

$$\sim \begin{bmatrix} 1 & 0 & 0 & 1 & 1300 \\ 0 & 1 & 0 & 1 & 900 \\ 0 & 1 & 1 & 0 & 700 \\ 0 & 0 & 1 & -1 & -200 \end{bmatrix} \sim \begin{bmatrix} 1 & 0 & 0 & 1 & 1300 \\ 0 & 1 & 0 & 1 & 900 \\ 0 & 0 & 1 & -1 & -200 \\ 0 & 0 & 1 & -1 & -200 \end{bmatrix} \sim \begin{bmatrix} 1 & 0 & 0 & 1 & 1300 \\ 0 & 1 & 0 & 1 & 900 \\ 0 & 0 & 1 & -1 & -200 \\ 0 & 0 & 0 & 0 & 0 \end{bmatrix}$$

$$(-1)R_2 + R_3 \rightarrow R_3 \qquad (-1)R_3 + R_4 \rightarrow R_4$$

The corresponding system of equations is

$$x_1 \qquad\quad + x_4 = 1300$$
$$x_2 \;\; + x_4 = 900$$
$$x_3 - x_4 = -200$$

Let $x_4 = t$. Then, $x_1 = 1300 - t$, $x_2 = 900 - t$, $x_3 = t - 200$, $x_4 = t$ where $200 \le t \le 900$.

*($t \ge 200$ so that x_3 is non-negative; $t \le 900$ so that x_2 is non-negative)

(C) maximum: 900, minimum: 200

(D) Elm St.: $x_1 = 800$; 2nd St.: $x_2 = 400$; Oak St.: $x_3 = 300$.   (4-3)

5   LINEAR INEQUALITIES AND LINEAR PROGRAMMING

EXERCISE 5-1

Things to remember:

1. A line divides the plane into two sets called HALF-PLANES. A vertical line divides the plane into LEFT and RIGHT HALF-PLANES; a nonvertical line divides the plane into UPPER and LOWER HALF-PLANES. In either case, the dividing line is called the BOUNDARY LINE of each half-plane.

2. The graph of the linear inequality
    $$Ax + By < C \text{ or } Ax + By > C$$
    with $B \neq 0$ is either the upper half-plane or the lower half-plane (but not both) determined by the line $Ax + By = C$.
    If $B = 0$, the graph of
    $$Ax < C \text{ or } Ax > C$$
    is either the right half-plane or the left half-plane (but not both) determined by the vertical line $Ax = C$.

3. For strict inequalities ("<" or ">"), the line is not included in the graph. For weak inequalities ("≤" or "≥"), the line is included in the graph.

4. PROCEDURE FOR GRAPHING LINEAR INEQUALITIES
    Step 1. First graph $Ax + By = C$ as a dashed line if equality is not included in the original statement or as a solid line if equality is included.
    Step 2. Choose a test point anywhere in the plane not on the line [the origin (0, 0) often requires the least computation] and substitute the coordinates into the inequality.
    Step 3. Does the test point satisfy the original inequality? If so, shade the half-plane that contains the test point. If not, shade the opposite half-plane.

---

1. $5 \neq 2(3) + 1 = 7$;   No

3. $5 \leq 2(3) + 1 = 7$;   Yes

5. $13(10) - 11(12) = 130 - 132 = -2$;   No

7. $13(10) - 11(12) = 130 - 132 = -2 < 2$;   No

9. $y \leq x - 1$

Graph $y = x - 1$ as a solid line.

Test point (0, 0):

x	y
0	−1
1	0

$0 \leq 0 - 1$
$0 \leq -1$

The inequality is false. Thus, the graph is the half-plane below the line $y = x - 1$, including the line.

11. $3x - 2y > 6$
Graph $3x - 2y = 6$ as a dashed line.

Test point (0, 0):

$3 \cdot 0 - 2 \cdot 0 > 6$
$0 > 6$

The inequality is false. Thus, the graph is the half-plane below the line $3x - 2y = 6$, not including the line.

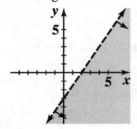

13. $x \geq -4$
Graph $x = -4$ [the vertical line through (−4, 0)] as a solid line.

Test point (0, 0):

$0 \geq -4$

The inequality is true. Thus, the graph is the half-plane to the right of the line $x = -4$, including the line.

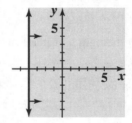

15. $6x + 4y \geq 24$

Graph the line $6x + 4y = 24$ as a solid line.

Test point (0, 0):

$6 \cdot 0 + 4 \cdot 0 \geq 24$
$0 \geq 24$

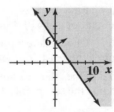

The inequality is false. Thus, the graph is the half-plane above the line, including the line.

17. $5x \leq -2y$ or $5x + 2y \leq 0$

Graph the line $5x + 2y = 0$ as a solid line. Since the line passes through the origin
(0, 0), we use (1, 0) as a test point:

$5 \cdot 1 + 2 \cdot 0 \leq 0$
$5 \leq 0$

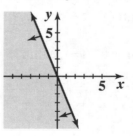

This inequality is false. Thus, the graph

is the half-plane below the line $5x + 2y = 0$, including the line.

19. (A) Graph $2x + 3y = 18$ as a dashed line

Test point $(0, 0)$:
$$2 \cdot 0 + 3 \cdot 0 < 18$$
$$0 < 18$$

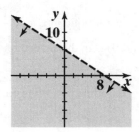

The inequality is true. Thus, the graph is the half-plane below the line $2x + 3y = 18$, not including the line.

(B) The set of points that do not satisfy the inequality is the half-plane above the line, including the line.

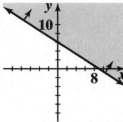

21. (A) Graph $5x - 2y = 20$ as a solid line

Test point $(0, 0)$:
$$5 \cdot 0 - 2 \cdot 0 \geq 20$$
$$0 \geq 20$$

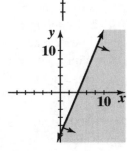

This inequality is false. Thus, the graph is the half-plane below the line $5x - 2y = 20$, including the line.

(B) The set of points that do not satisfy the inequality is the half-plane above the line, not including the line.

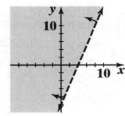

23. Let h = number of overtime hours; $h < 20$.

25. Let s = annual salary; $s \geq \$65,000$.

27. Let a = number of freshmen admitted; $a \leq 1,700$.

29. The boundary line passes through the points $(-3, 0)$ and $(0, -2)$.

slope: $m = \dfrac{-2-0}{0-(-3)} = -\dfrac{2}{3}$; y intercept: $b = -2$

Boundary line equation: $y = -\dfrac{2}{3}x - 2$ or $2x + 3y = -6$

Since $(0, 0)$ is in the shaded region and the boundary line is solid, the graph is the graph of $2x + 3y \geq -6$.

31. The equation of the boundary line is $y = 3$. Since $(0, 0)$ is in the shaded region and the boundary line is dashed, the graph is the graph of $y < 3$.

33. The boundary line passes through the origin $(0, 0)$ and the point $(5, 4)$.

Slope $m = \dfrac{4}{5}$

Boundary line equation: $y = \dfrac{4}{5}x$ or $4x - 5y = 0$.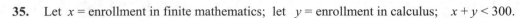

Since $(1, 0)$ is in the shaded region and the boundary line is solid, the graph is the graph of $4x - 5y \geq 0$.

35. Let $x =$ enrollment in finite mathematics; let $y =$ enrollment in calculus; $x + y < 300$.

37. Let $x =$ revenue; let $y =$ cost; $x + 20{,}000 \leq y$ or $x \leq y - 20{,}000$.

39. Let $x =$ number of grams of saturated fat; let $y =$ number of grams of unsaturated fat;

$x > 3y$.

41. $25x + 40y \leq 3{,}000,\ x \geq 0,\ y \geq 0$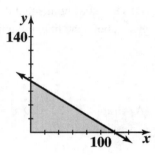

Step 1.   Graph the line $25x + 40y = 3000$ as
               a solid line.

Step 2.   Substituting $x = 0, y = 0$ in the inequality produces a
               true statement, so $(0, 0)$ is in the solution set.

Step 3.   With $x \geq 0, y \geq 0$, the graph is shown
               at the right.

43. $15x - 50y < 1{,}500,\ x \geq 0,\ y \geq 0$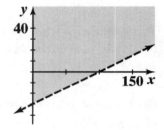

Step 1. Graph the line $15x - 50y = 1{,}500$
               as a dashed line.

Step 2.   Substituting $x = 0, y = 0$ in the
               inequality produces a true statement,
               so $(0, 0)$ is in the solution set.

Step 3.   With $x \geq 0, y \geq 0$, the graph is shown
               at the right.

45. $-18x + 30y \geq 2{,}700,\ x \geq 0,\ y \geq 0$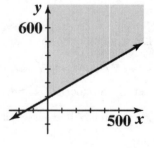

Step 1.   Graph the line $-18x + 30y = 2{,}700$
               as a solid line.

Step 2.   Substituting $x = 0, y = 0$ in the
               inequality produces a false statement,
               so $(0, 0)$ is not in the solution set.

Step 3.   With $x \geq 0, y \geq 0$, the graph is shown
               at the right.

47. $40x - 55y > 0, x \geq 0, y \geq 0$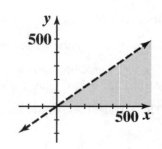

Step 1.   Graph the line $40x - 55y = 0$
               as a dashed line.

Step 2.   Since the line passes through the
               origin choose $(1, 0)$ as a test point;
               substituting $x = 1, y = 0$ in the
               inequality produces a true statement,
               so $(1, 0)$ is in the solution set.

Step 3.   With $x \geq 0, y \geq 0$, the graph is shown at the right.

49. $25x + 75y < -600, x \geq 0, y \geq 0$
Step 1.   Graph the line $25x + 75y = -600$ as a dashed line.
Step 2.   Substituting $x = 0, y = 0$ into the inequality produces a false statement, so $(0, 0)$ is not in the solution set.
Step 3.   With $x \geq 0, y \geq 0$, the solution set is empty and has no graph.

51.   Let x = number of acres for corn
   y = number of acres for soybeans
Then $40x + 32y \leq 5,000$.
Dividing by 8, we get the inequality
   $5x + 4y \leq 625$.
We must also have $x \geq 0, y \geq 0$.
Graph the line $5x + 4y = 625$ as a solid line. Since $x = 0, y = 0$
satisfies the inequality, $(0, 0)$, is in the solution set. The graph is
shown at the right.

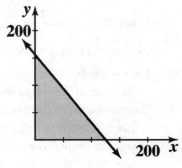

53.   Let x = number of pounds of Brand A;
   y = number of pounds of Brand B.
(A)   Number of pounds of nitrogen: $0.26x + 0.16y$.
   He wants
      $0.26x + 0.16y \geq 120, x \geq 0, y \geq 0$
   Multiplying the inequality by 100, we get
   $26x + 16y \geq 12,000, x \geq 0, y \geq 0$
   or $13x + 8y \geq 6,000, x \geq 0, y \geq 0$
Graph the line $13x + 8y = 6,000$ as a solid line.
Since $x = 0, y = 0$ does not satisfy the inequality $(0, 0)$ is not in the
solution set.
The graph is shown at the right.

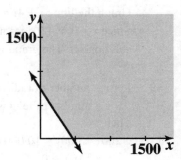

(B)   Number of pounds of phosphate: $0.03x + 0.08y$.
   She wants
      $0.03x + 0.08y \leq 28, x \geq 0, y \geq 0$
   Multiplying by 100, we get
      $3x + 8y \leq 2800, x \geq 0, y \geq 0$
Graph the line $3x + 8y = 2800$ as a solid line.
Since $x = 0, y = 0$ satisfies the inequality,
$(0, 0)$ is in the solution set. The graph is shown at the right.

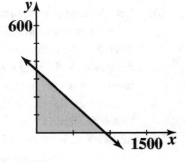

55.   Let x = number of pounds of standard blend
   y = number of pounds of deluxe blend
The mill wants to produce a fabric that is at least 20% acrylic.

Therefore,

$$0.30x + 0.09y \geq 0.20(x+y), \quad x \geq 0, \quad y \geq 0.$$

Multiplying by 100, we get

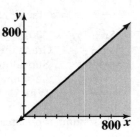

$$30x + 9y \geq 20(x+y) \quad \text{which implies} \quad y \leq \frac{10}{11}x$$

Graph the line $y = \frac{10}{11}x$ as a solid line. Since $x = 1, y = 0$ satisfies the

inequality, $(1, 0)$ is in the solution set. The graph is shown at the right.

57. Let x = number of weeks to operate plant A,

$\quad\quad y$ = number of weeks to operate plant B.

To produce at least 400 sedans, we must have

$\quad 10x + 8y \geq 400, x \geq 0, y \geq 0$

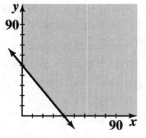

Divide the inequality by 2:

$\quad 5x + 4y \geq 200, x \geq 0, y \geq 0$

Graph the line $5x + 4y = 200$ as a solid line.

Since $x = 0, y = 0$ does not satisfy the inequality, $(0, 0)$ is not in the

solution set. The graph is shown at the right.

59. Let x = number of radio ads,

$\quad\quad y$ = number of television ads.

Then

$\quad 200x + 800y \leq 10,000$ or $x + 4y \leq 50,$

$\quad\quad x \geq 0, y \geq 0$

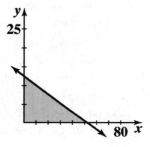

Graph the line $x + 4y = 50$ as a solid line.

Since $x = 0, y = 0$ satisfies the inequality

$(0, 0)$ is in the solution set.

61. Let x = number of regular mattresses,

$\quad\quad y$ = number of king size mattresses

Then $5x + 6y$ is the number of minutes required to cut x regular

mattresses and y king size mattresses. 50 labor hours = 50(60) = 3000

labor minutes are available. Therefore,

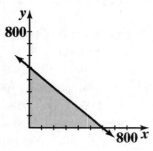

$\quad 5x + 6y \leq 3,000, x \geq 0, y \geq 0$

Graph the line $5x + 6y = 3,000$ as a solid line.

Since $x = 0, y = 0$ satisfies the inequality

$(0, 0)$ is in the solution set.

EXERCISE 5-2

Things to remember:

1. To solve a system of linear inequalities graphically, graph each inequality in the system and then take the intersection of all the graphs. The resulting graph is called the SOLUTION REGION, or FEASIBLE REGION.

2. A CORNER POINT of a solution region is a point in the solution region that is the intersection of two boundary lines.

3. The solution region of a system of linear inequalities is BOUNDED if it can be enclosed within a circle; if it cannot be enclosed within a circle, then it is UNBOUNDED.

---

1. $4(3) + 5 = 17 \leq 20$, $3(3) + 5(5) = 34 \leq 37$, $3 \geq 0$, $5 \geq 0$;   Yes

3. $4(3) + 6 = 18 \leq 20$, $3(3) + 5(6) = 39 > 37$;   No

5. $5(4) + 3 = 23 \leq 32$, $7(4) + 4(3) = 40 < 45$;   No

7. $5(6) + 2 = 32 \leq 32$, $7(6) + 4(2) = 50 \geq 45$, $6 \geq 0$, $2 \geq 0$;   Yes

9. The graph of $x + 2y \leq 8$ is the half-plane below the line $x + 2y = 8$ [e.g., (0, 0) satisfies the inequality]. The graph of $3x - 2y \geq 0$ is the half-plane below the line $3x - 2y = 0$ [e.g., (1, 0) satisfies the inequality]. The intersection of these two regions is region IV.

11. The graph of $x + 2y \geq 8$ is the half-plane above the line $x + 2y = 8$ [e.g., (0, 0) does not satisfy the inequality]. The graph of $3x - 2y \geq 0$ is the half-plane below the line $3x - 2y = 0$ [e.g., (1, 0) satisfies the inequality]. The intersection of these two regions is region I.

13. The graphs of the inequalities $3x + y \geq 6$ and $x \leq 4$ are:

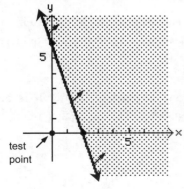
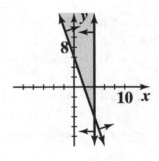

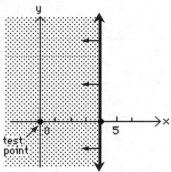

The intersection of these regions (drawn on the same coordinate plane) is shown in the graph at the right.

15.   The graphs of the inequalities $x - 2y \le 12$ and $2x + y \ge 4$ are:

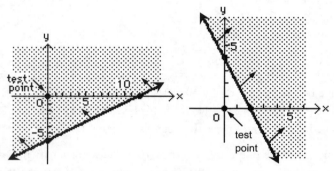

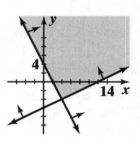

The intersection of these regions (drawn on the same coordinate plane) is shown in the graph at the right.

17.   The graph of $x + 3y \le 18$ is the region below the line $x + 3y = 18$ and the graph of $2x + y \ge 16$ is the region above the line $2x + y = 16$. The graph of $x \ge 0, y \ge 0$ is the first quadrant. The intersection of these regions is region IV. The corner points are $(8, 0)$, $(18, 0)$, and $(6, 4)$.

19.   The graph of $x + 3y \ge 18$ is the region above the line $x + 3y = 18$ and the graph of $2x + y \ge 16$ is the region above the line $2x + y = 16$. The graph of $x \ge 0, y \ge 0$ is the first quadrant. The intersection of these regions is region I. The corner points are $(0, 16)$, $(6, 4)$, and $(18, 0)$.

21.   The graphs of the inequalities are shown at the right. The solution region is indicated by the shaded region. The solution region is *bounded*.
The corner points of the solution region are:
$(0, 0)$, the intersection of $x = 0, y = 0$;
$(0, 4)$, the intersection of $x = 0$,
        $2x + 3y = 12$;
$(6, 0)$, the intersection of $y = 0$,
        $2x + 3y = 12$.

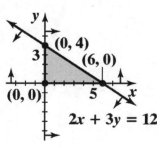

23. The graphs of the inequalities are shown at the right. The solution region is shaded. The solution region is *bounded*.

The corner points of the solution region are:

(0, 0), the intersection of $x = 0$, $y = 0$;
(0, 4), the intersection of $x = 0$, $x + 2y = 8$;
(4, 2), the intersection of $x + 2y = 8$,
        $2x + y = 10$;
(5, 0), the intersection of $y = 0$,
        $2x + y = 10$.

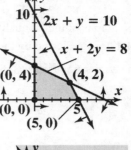

25. The graphs of the inequalities are shown at the right. The solution region is shaded. The solution region is *unbounded*.

The corner points of the solution region are:

(0, 10), the intersection of $x = 0$,
        $2x + y = 10$;
(4, 2), the intersection of $x + 2y = 8$,
        $2x + y = 10$;
(8, 0), the intersection of $y = 0$, $x + 2y = 8$.

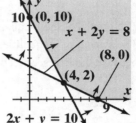

27. The graphs of the inequalities are shown at the right. The solution is indicated by the shaded region. The solution region is *bounded*.
The corner points of the solution region are:
(0, 0), the intersection of $x = 0$, $y = 0$,
(0, 6), the intersection of $x = 0$,
        $x + 2y = 12$;
(2, 5), the intersection of $x + 2y = 12$,
        $x + y = 7$;
(3, 4), the intersection of $x + y = 7$,
        $2x + y = 10$;
(5, 0), the intersection of $y = 0$,
        $2x + y = 10$.

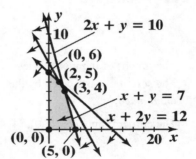

Note that the point of intersection of the lines $2x + y = 10$, $x + 2y = 12$ is not a corner point because it is not in the solution region.

29. The graphs of the inequalities are shown at the right. The solution is indicated by the shaded region, which is *unbounded*.

The corner points are:

(0, 16), the intersection of $x = 0$,
        $2x + y = 16$;
(4, 8), the intersection of $2x + y = 16$,
        $x + y = 12$;
(10, 2), the intersection of $x + y = 12$,
        $x + 2y = 14$;
(14, 0), the intersection of $y = 0$,
        $x + 2y = 14$.

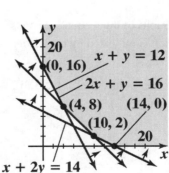

The intersection of $x + 2y = 14$, $2x + y = 16$ is not a corner point because it is not in the solution region.

31. The graphs of the inequalities are shown at the right. The solution is indicated by the shaded region, which is *bounded*.

The corner points are (8, 6), (4, 7), and (9, 3).

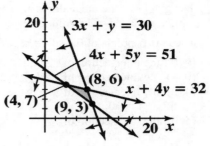

33. The graphs of the inequalities are shown at the right. The system of inequalities does not have a solution because the intersection of the graphs is empty.

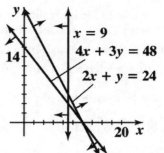

35. The graphs of the inequalities are shown at the right. The solution is indicated by the shaded region, which is *unbounded*.

The corner points are (0, 0), (4, 4), and (8, 12).

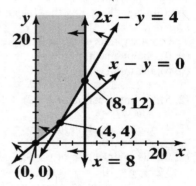

37. The graphs of the inequalities are shown at the right. The solution is indicated by the shaded region, which is *bounded*.

The corner points are (2, 1), (3, 6), (5, 4), and (5, 2).

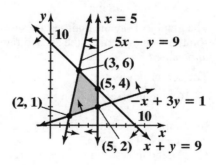

39. The graphs of the inequalities are shown at the right. The solution is indicated by the shaded region, which is *bounded*. The corner points are (1.24, 5.32), (2.17, 6.56), and (6.2, 1.6).

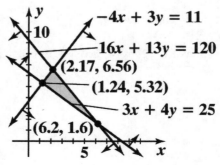

41.   (A) $3x + 4y = 36$

$\underline{3x + 2y = 30}$ subtract

$2y = 6$

$y = 3$

$x = 8$

intersection point: (8, 3)

$3x + 4y = 36$

$\underline{y = 0}$

$3x = 36$

$x = 12$

intersection point: (12, 0)

$3x + 2y = 30$

$\underline{y = 0}$

$3x = 30$

$x = 10$

intersection point: (10, 0)

$3x + 4y = 36$

$\underline{x = 0}$

$4y = 36$

$y = 9$

intersection point: (0, 9)

$3x + 2y = 30$

$\underline{x = 0}$

$2y = 30$

$y = 15$

intersection point: (0, 15)

$x = 0$

$y = 0$

intersection point: (0, 0)

(B)   The corner points are: (8, 3), (0, 9), (10, 0), (0, 0);
(0, 15) does not satisfy $3x + 4y \le 36$,
(12, 0) does not satisfy $3x + 2y \le 30$.

43.   Let x = the number of trick skis and y = the number of slalom skis produced per day. The information is summarized in the following table.

	Hours per ski		Maximum labor-hours per day available
	Trick ski	Slalom ski	
Fabrication	6 hrs	4 hrs	108 hrs
Finishing	1 hr	1 hr	24 hrs

We have the following inequalities:

$6x + 4y \le 108$ for fabrication

$x + y \le 24$ for finishing

Also, $x \ge 0$ and $y \ge 0$.

The graphs of these inequalities are shown at the right. The shaded region indicates the set of feasible solutions.

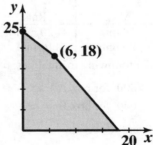

45.   (A)   If x is the number of trick skis and y is the number of slalom skis per day, then the profit per day is given by

$P(x, y) = 50x + 60y$

All the production schedules in the feasible region that lie on the graph of the line

$50x + 60y = 1{,}100$

will provide a profit of \$1,100.

(B)   There are many possible choices. For example, producing 5 trick
      skis and 15 slalom skis per day will produce a profit of

   $$P(5, 15) = 50(5) + 60(15) = 1,150$$

   All the production schedules in the feasible region that lie on the
   graph of $50x + 60y = 1,150$ will provide a profit of $1,150.

47.   Let x = the number of cubic yards of mix A and y = the number of cubic yards of mix B. The information
      is summarized in the following table:

	Amount of substance per cubic yard		Minimum Monthly requirement
	Mix A	Mix B	
Phosphoric acid	20 lbs	10 lbs	460 lbs
Nitrogen	30 lbs	30 lbs	960 lbs
Potash	5 lbs	10 lbs	220 lbs

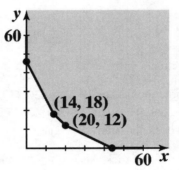

We have the following inequalities:

$20x + 10y \geq 460$

$30x + 30y \geq 960$

$5x + 10y \geq 220$

Also, $x \geq 0$ and $y \geq 0$.

The graphs of these inequalities are
shown at the right. The shaded
region indicates the set of feasible
solutions.

49.   Let x = the number of mice used and y = the number of rats used.
      The information is summarized in the following table.

	Mice	Rats	Maximum time available per day
Box A	10 min	20 min	800 min
Box B	20 min	10 min	640 min

We have the following inequalities:

$10x + 20y \leq 800$ for box A

$20x + 10y \leq 640$ for box B

Also, $x \geq 0$ and $y \geq 0$.

The graphs of these inequalities are shown at
the right. The shaded region indicates the set
of feasible solutions.

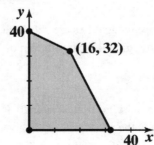

EXERCISE 5-3

Things to remember:

1. A LINEAR PROGRAMMING PROBLEM is a problem that is concerned with finding the OPTIMAL VALUE (maximum or minimum value) of a linear OBJECTIVE FUNCTION of the form

    $$z = ax + by,$$

    where the DECISION VARIABLES x, y are subject to PROBLEM CONSTRAINTS in the form of linear inequalities and equations. In addition, the decision variables must satisfy the NONNEGATIVE CONSTRAINTS x, $y \geq 0$. The set of points satisfying both the problem constraints and the nonnegative constraints is called the FEASIBLE REGION for the problem. Any point in the feasible region that produces the optimal value of the objective function over the feasible region is called an OPTIMAL SOLUTION.

2. CONSTRUCTING A MATHEMATICAL MODEL FOR AN APPLIED LINEAR PROGRAMMING PROBLEM

    Step 1. Introduce decision variables

    Step 2. Summarize relevant material in table form, relating columns to the decision variables, if possible.

    Step 3. Determine the objective and write a linear objective function.

    Step 4. Write problem constraints using linear equations and/or inequalities.

    Step 5. Write non-negative constraints.

3. FUNDAMENTAL THEOREM OF LINEAR PROGRAMMING

    If the optimal value of the objective function in a linear programming problem exists, then that value must occur at one (or more) of the corner points of the feasible region.

4. EXISTENCE OF SOLUTIONS

    (A)   If the feasible region for a linear programming problem is bounded, then both the maximum value and the minimum value of the objective function always exist.

    (B)   If the feasible region is unbounded, and the coefficients of the objective function are positive, then the minimum value of the objective function exists, but the maximum value does not.

    (C)   If the feasible region is empty (that is, there are no points that satisfy all the constraints), then both the maximum value and the minimum value of the objective function do not exist.

5. GEOMETRIC METHOD FOR SOLVING A LINEAR PROGRAMMING PROBLEM WITH TWO DECISION VARIABLES.

    Step 1.   Graph the feasible region. Then, if according to 4 an optimal solution exists, find the coordinates of each corner point.

    Step 2.   Construct a CORNER POINT TABLE listing the value of the objective function at each corner point.

    Step 3.   Determine the optimal solution(s) from the table in Step (2).

    Step 4.   For an applied problem, interpret the optimal solution(s) in terms of the original problem.

1. Evaluate Q at each corner point:

Corner point	$Q = 7x + 14y$
$(0, 0)$	$Q = 7(0) + 14(0) = 0$
$(12, 0)$	$Q = 7(12) + 14(0) = 84$
$(0, 5)$	$Q = 7(0) + 14(5) = 70$
$(12, 5)$	$Q = 7(12) + 14(5) = 154$

Max $Q = 154$, min $Q = 0$

3. Evaluate Q at each corner point:

Corner point	$Q = 10x - 12y$
$(0, 0)$	$Q = 10(0) - 12(0) = 0$
$(12, 0)$	$Q = 10(12) - 12(0) = 120$
$(0, 5)$	$Q = 10(0) - 12(5) = -60$
$(12, 5)$	$Q = 10(12) - 12(5) = 60$

Max $Q = 120$, min $Q = -60$

5. Evaluate Q at each corner point:

Corner point	$Q = -4x - 3y$
$(0, 0)$	$-4(0) - 3(0) = 0$
$(8, 0)$	$-4(8) - 3(0) = -32$
$(0, 10)$	$-4(0) - 3(10) = -30$

Max $Q = 0$, min $Q = -32$

7. Evaluate Q at each corner point:

Corner point	$Q = -6x + 4y$
$(0, 0)$	$-6(0) + 4(0) = 0$
$(8, 0)$	$-6(8) + 4(0) = -48$
$(0, 10)$	$-6(0) + 4(10) = 40$

Max $Q = 40$, min $Q = -48$

9.

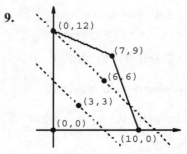

From the figure:

maximum profit $P = 16$ at $x = 7$, $y = 9$.

Step (2):   Evaluate the objective function at each corner point.

Corner Point	$P = x + y$
(0, 0)	0
(0, 12)	12
(7, 9)	16
(10, 0)	10

Step (3):   Determine the optimal solution from Step (2).
The maximum value of P is 16 at (7, 9).

11.

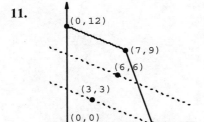

From the figure:
   maximum profit $P = 84$ at $x = 0$, $y = 12$; at $x = 7$, $y = 9$;
   and at every point on the line segment joining (0, 12) and
   (7, 9).

Step (2):   Evaluate the objective function at each corner point.

corner point	$P = 3x + 7y$
(0,0)	0
(0,12)	84
(7,9)	84
(10,0)	30

Step (3):   Determine the optimal solution from Step (2).
The maximum value of P is 84 at (0, 12) *and* (7, 9). This is a multiple optimal solution.

13. $C = 7x + 4y$

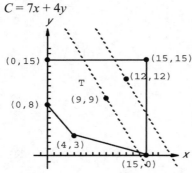

From the figure: minimum cost $C = 32$ at $x = 0$, $y = 8$.

Step (2): Evaluate the objective function at each corner point.

Corner Point	$C = 7x + 4y$
(15, 15)	165
(0, 15)	60
(0, 8)	32
(4, 3)	40
(15, 0)	105

Step (3): Determine the optimal solution from Step (2).
The minimum value of C is 32 at $x = 0$, $y = 8$.

15. $C = 3x + 8y$

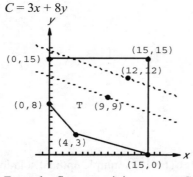

From the figure: minimum cost $C = 36$ at $x = 4$, $y = 3$.

Step (2): Evaluate the objective function at each corner point.

Corner Point	$C = 3x + 8y$
(15, 15)	165
(0, 15)	120
(0, 8)	64
(4, 3)	36
(15, 0)	45

Step (3): Determine the optimal solution from Step (2).
The minimum value of C is 36 at $x = 4$, $y = 3$.

17. Step (1):   Graph the feasible region and find the corner points.

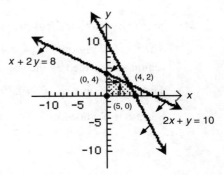

The feasible region S is the solution set of the given inequalities. This region is indicated by the shading in the graph at the right.

The corner points are $(0, 0)$, $(0, 4)$, $(4, 2)$, and $(5, 0)$.

Since S is bounded, it follows from $\underline{4}$(a) that P has a maximum value

Step (2):   Evaluate the objective function at each corner point.

The value of P at each corner point is given in the following table.

Corner Point	$P = 5x + 5y$
$(0,0)$	$P = 5(0) + 5(0) = 0$
$(0,4)$	$P = 5(0) + 5(4) = 20$
$(4,2)$	$P = 5(4) + 5(2) = 30$
$(5,0)$	$P = 5(5) + 5(0) = 25$

Step (3):   Determine the optimal solution.

The maximum value of P is 30 at $x = 4$, $y = 2$.

19. Step (1): Graph the feasible region and find the corner points.

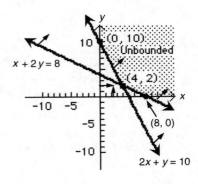

The feasible region S is the solution set of the given inequalities. This region is indicated by the shading in the graph at the right.

The corner points are $(0, 10)$, $(4, 2)$, and $(8, 0)$.

Since S is unbounded and $a = 2 > 0$, $b = 3 > 0$, it follows from $\underline{4}$(b) that z has a minimum value but not a maximum value.

Step (2):   Evaluate the objective function at each corner point.

The value of z at each corner point is given in the following table:

Corner Point	$z = 2x + 3y$
$(0,10)$	$z = 2(0) + 3(10) = 30$
$(4,2)$	$z = 2(4) + 3(2) = 14$
$(8,0)$	$z = 2(8) + 3(0) = 16$

Step (3):   Determine the optimal solutions.

The minimum occurs at $x = 4$, $y = 2$, and the minimum value is $z = 14$; z does not have a maximum value.

21.   <u>Step (1)</u>: Graph the feasible region and find the corner points.

The feasible region S is the solution set of the given inequalities. This region is indicated by the shading in the graph at the right.

The corner points are $(0, 0)$, $(0, 6)$, $(2, 5)$, $(3, 4)$, and $(5, 0)$.

Since S is bounded, it follows from <u>4</u>(a) that P has a maximum value.

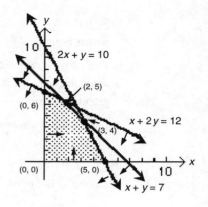

<u>Step (2)</u>:   Evaluate the objective function at each corner point.

The value of P at each corner point is:

Corner Point	$P = 30x + 40y$
$(0,0)$	$P = 30(0) + 40(0) = 0$
$(0,6)$	$P = 30(0) + 40(6) = 240$
$(2,5)$	$P = 30(2) + 40(5) = 260$
$(3,4)$	$P = 30(3) + 40(4) = 250$
$(5,0)$	$P = 30(5) + 40(0) = 150$

<u>Step (3)</u>:   Determine the optimal solution.
The maximum occurs at $x = 2$, $y = 5$, and the maximum value is $P = 260$.

23.   <u>Step (1)</u>:   Graph the feasible region and find the corner points.

The feasible region S is the solution set of the given inequalities. This region is indicated by the shading in the graph at the right.

The corner points are $(0, 16)$, $(4, 8)$, $(10, 2)$, and $(14, 0)$.

Since S is unbounded and $a = 10 > 0$, $b = 30 > 0$, it follows from <u>4</u>(b) that z has a minimum value but not a maximum value.

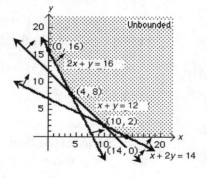

<u>Step (2)</u>:   Evaluate the objective function at each corner point.
The value of z at each corner point is:

Corner Point	$z = 10x + 30y$
$(0,16)$	$z = 10(0) + 30(16) = 480$
$(4,8)$	$z = 10(4) + 30(8) = 280$
$(10,2)$	$z = 10(10) + 30(2) = 160$
$(14,0)$	$z = 10(14) + 30(0) = 140$

<u>Step (3)</u>:   Determine the optimal solution.

The minimum occurs at $x = 14$, $y = 0$, and the minimum value is $z = 140$; z does not have a maximum value.

25. Step (1): Graph the feasible region and find the corner points.

The feasible region S is the solution set of the given inequalities, and is indicated by the shading in the graph at the right.

The corner points are $(0, 2)$, $(0, 9)$, $(2, 6)$, $(5, 0)$, and $(2, 0)$.

Since S is bounded, it follows from $\underline{4}$(a) that P has a maximum value and a minimum value.

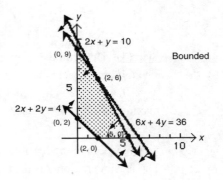

Step (2): Evaluate the objective function at each corner point.
The value of P at each corner point is given in the following table:

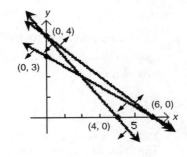

Corner Point	$P = 30x + 10y$
$(0, 2)$	$P = 30(0) + 10(2) = 20$
$(0, 9)$	$P = 30(0) + 10(9) = 90$
$(2, 6)$	$P = 30(2) + 10(6) = 120$
$(5, 0)$	$P = 30(5) + 10(0) = 150$
$(2, 0)$	$P = 30(2) + 10(0) = 60$

Step (3): Determine the optimal solutions.
The maximum occurs at $x = 5$, $y = 0$, and the maximum value is $P = 150$; the minimum occurs at $x = 0$, $y = 2$, and the minimum value is $P = 20$.

27. Step (1): Graph the feasible region and find the corner points.

The feasible region S is the solution set of the given inequalities. As indicated, the feasible region is empty. Thus, by $\underline{4}$(c), there are no optimal solutions.

29. Step (1): Graph the feasible region and find the corner points.

The feasible region S is the solution set of the given inequalities, and is indicated by the shading in the graph at the right.

The corner points are $(3, 8)$, $(8, 10)$, and $(12, 2)$.

Since S is bounded, it follows from $\underline{4}$(a) that P has a maximum value and a minimum value.

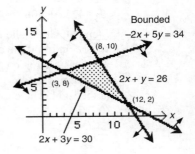

Step (2): Evaluate the objective function at each corner point.
The value of P at each corner point is:

Corner Point	$P = 20x + 10y$
$(3, 8)$	$P = 20(3) + 10(8) = 140$
$(8, 10)$	$P = 20(8) + 10(10) = 260$
$(12, 2)$	$P = 20(12) + 10(2) = 260$

Step (3): Determine the optimal solutions.

The minimum occurs at $x = 3$, $y = 8$, and the minimum value   is $P = 140$; the maximum occurs at $x = 8$, $y = 10$, at $x = 12$, $y = 2$, and at any point along the line segment joining $(8, 10)$ and $(12, 2)$. The maximum value is $P = 260$.

31. Step (1): Graph the feasible region and find the corner points. The feasible region S is the set of solutions of the given inequalities, and is indicated by the shading in the graph at the right.

The corner points are $(0, 0)$, $(0, 800)$, $(400, 600)$, $(600, 450)$, and $(900, 0)$. Since S is bounded, it follows from $\underline{4}$(a) that P has a maximum value.

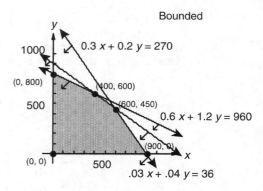

Step (2): Evaluate the objective function at each corner point.
The value of P at each corner point is:

Corner Point	$P = 20x + 30y$
$(0, 0)$	$P = 20(0) + 30(0) = 0$
$(0, 800)$	$P = 20(0) + 30(800) = 24,000$
$(400, 600)$	$P = 20(400) + 30(600) = 26,000$
$(600, 450)$	$P = 20(600) + 30(450) = 25,500$
$(900, 0)$	$P = 20(900) + 30(0) = 18,000$

Step (3): Determine the optimal solution.
The maximum occurs at $x = 400$, $y = 600$, and the maximum value is $P = 26,000$.

33. $\ell_1 : 275x + 322y = 3,381$

$\ell_2 : 350x + 340y = 3,762$

$\ell_3 : 425x + 306y = 4,114$.

Step (1): Graph the feasible region and find the corner points.
The feasible region S is the solution set of the given inequalities, and is indicated by the shading in the graph at the right. The corner points are $(0, 0)$, $(0, 10.5)$, $(3.22, 7.75)$, $(6.62, 4.25)$, $(9.68, 0)$.

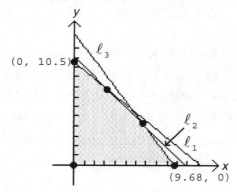

<u>Step (2)</u>: Evaluate the objective function at each corner point.

The value of P at each corner point is

Corner Point	$P = 525x + 478y$
$(0, 0)$	$P = 525(0) + 478(0) = 0$
$(0, 10.5)$	$P = 525(0) + 478(10.5) = 5{,}019$
$(3.22, 7.75)$	$P = 525(3.22) + 478(7.75) = 5{,}395$
$(6.62, 4.25)$	$P = 525(6.62) + 478(4.25) = 5{,}507$
$(9.68, 0)$	$P = 525(9.68) + 478(0) = 5{,}082$

<u>Step (3)</u>: Determine the optimal solution.

The maximum occurs at $x = 6.62$, $y = 4.25$, and the maximum value is $P = 5{,}507$.

35.   Minimize and maximize $z = x - y$

Subject to

$$x - 2y \leq 0$$
$$2x - y \leq 6$$
$$x, y \geq 0$$

The feasible region and several values of the objective function are shown in the figure.

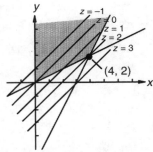

The points $(0, 0)$ and $(4, 2)$ are the corner points; $z = x - y$ does not have a minimum value. Its maximum value is 2 at $(4, 2)$.

37.   The value of $P = ax + by$, $a > 0$, $b > 0$, at each corner point is:

Corner Point	$P = ax + by$
$O: (0, 0)$	$P = a(0) + b(0) = 0$
$A: (0, 5)$	$P = a(0) + b(5) = 5b$
$B: (4, 3)$	$P = a(4) + b(3) = 4a + 3b$
$C: (5, 0)$	$P = a(5) + b(0) = 5a$

(A)   For the maximum value of P to occur at A only, we must have $5b > 4a + 3b$ and $5b > 5a$. Solving the first inequality, we get $2b > 4a$ or $b > 2a$; from the second inequality, we get $b > a$. Therefore, we must have $b > 2a$ or $2a < b$ in order for P to have its maximum value at A only.

(B)   For the maximum value of P to occur at B only, we must have $4a + 3b > 5b$ and $4a + 3b > 5a$.

Solving this pair of inequalities, we get $4a > 2b$ and $3b > a$, which is the same as $\dfrac{a}{3} < b < 2a$.

(C)   For the maximum value of P to occur at C only, we must have $5a > 4a + 3b$ and $5a > 5b$. This pair of inequalities implies that $a > 3b$ or $b < \dfrac{a}{3}$.

(D)   For the maximum value of P to occur at both A and B, we must have $5b = 4a + 3b$ or $b = 2a$.

(E)   For the maximum value of P to occur at both B and C, we must have $4a + 3b = 5a$ or $b = \dfrac{a}{3}$.

39.   (A)   Construct the mathematical model

a.   Decision variables:

Let x = Number of trick skis

y = Number of slalom skis

b. Relevant material in table form:

	Trick ski	Slalom ski	Labor-hours available
Fabricating	6	4	108
Finishing	1	1	24
Profit	$40/ski	$30/ski	

c. Objective function:

Maximize profit $P = 40x + 30y$

d. Problem constraints:

$6x + 4y \leq 108$ [Fabricating constraint]

$x + y \leq 24$ [Finishing constraint]

e. Non-negativity constraints

$x \geq 0, y \geq 0$

The mathematical model for this problem is:

$$\text{Maximize } P = 40x + 30y$$
$$\text{Subject to: } 6x + 4y \leq 108$$
$$x + y \leq 24$$
$$x \geq 0, y \geq 0$$

Step (1):   Graph the feasible region and find the corner points.

The feasible region S is the solution set of the given system of inequalities, and is indicated by the shading in the graph below.

The corner points are (0, 0), (0, 24), (6, 18), and (18, 0).

Since S is bounded, P has a maximum value by 4(a).

Step (2): Evaluate the objective function at each corner point.
The value of P at each corner point is:

Corner Point	$P = 40x + 30y$
(0,0)	$P = 40(0) + 30(0) = 0$
(0,24)	$P = 40(0) + 30(24) = 720$
(6,18)	$P = 40(6) + 30(18) = 780$
(18,0)	$P = 40(18) + 30(0) = 720$

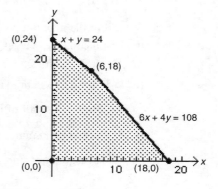

Step (3): Determine the optimal solution.
The maximum occurs when $x = 6$ (trick skis) and

$y = 18$ (slalom skis) are produced. The maximum profit is $P = \$780$.

Corner Point	$P = 40x + 25y$
$(0,0)$	$P = 40(0) + 25(0) = 0$
(B) $(0,24)$	$P = 40(0) + 25(24) = 600$
$(6,18)$	$P = 40(6) + 25(18) = 690$
$(18,0)$	$P = 40(18) + 25(0) = 720$

The maximum profit decreases to \$720 when 18 trick skis and no slalom skis are produced.

Corner Point	$P = 40x + 45y$
$(0,0)$	$P = 40(0) + 45(0) = 0$
(C) $(0,24)$	$P = 40(0) + 45(24) = 1080$
$(6,18)$	$P = 40(6) + 45(18) = 1050$
$(18,0)$	$P = 40(18) + 45(0) = 720$

The maximum profit increases to \$1,080 when no trick skis and 24 slalom skis are produced.

41. (A) Construct the mathematical model

a. Decision variables:

Let x = Number of days to operate Plant A

y = Number of days to operate Plant B

b. Relevant material in table form:

	Plant A	Plant B	Amount required
Tables	20	25	200
Chairs	60	50	500
Cost/day	\$1000	\$900	

c. Objective function:

Minimize the cost $C = 1000x + 900y$

d. Problem constraints:

$20x + 25y \geq 200$ [Table constraint]

$60x + 50y \geq 500$ [Chair constraint]

e. Non-negativity constraints

$x \geq 0, y \geq 0$

The mathematical model for this problem is:

Minimize $C = 1000x + 900y$

Subject to: $20x + 25y \geq 200$

$60x + 50y \geq 500$

$x \geq 0, y \geq 0$

Step (1): Graph the feasible region and find the corner points.

The feasible region S is the solution set of the system of inequalities, and is indicated by the shading in the graph shown below.

The corner points are (0, 10), (5, 4), and (10, 0).

Since S is unbounded and $a = 1000 > 0$, $b = 900 > 0$, C has a minimum value by 4(b).

Step (2): Evaluate the objective function at each corner point.

The value of C at each corner point is:

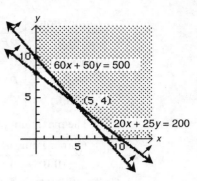

Corner Point	$C = 1000x + 900y$
$(0, 10)$	$C = 1000(0) + 900(10) = 9,000$
$(5, 4)$	$C = 1000(5) + 900(4) = 8,600$
$(10, 0)$	$C = 1000(10) + 900(0) = 10,000$

Step (3): Determine the optimal solution.

The minimum occurs when $x = 5$ and $y = 4$.

That is, Plant A should be operated five days and Plant B should be operated four days. The minimum cost is $C = \$8600$.

(B)

Corner Point	$C = 600x + 900y$
$(0, 10)$	$C = 600(0) + 900(10) = 9,000$
$(5, 4)$	$C = 600(5) + 900(4) = 6,600$
$(10, 0)$	$C = 600(10) + 900(0) = 6,000$

The minimum cost decreases to $6,000 per day when Plant A is operated 10 days and Plant B is operated 0 days.

(C)

Corner Point	$C = 1000x + 800y$
$(0, 10)$	$C = 1000(0) + 800(10) = 8,000$
$(5, 4)$	$C = 1000(5) + 800(4) = 8,200$
$(10, 0)$	$C = 1000(10) + 800(0) = 10,000$

The minimum cost decreases to $8,000 per day when Plant A is operated 0 days and Plant B is operated 10 days.

43. Construct the mathematical model:
   a.   Decision variables:
        Let x = Number of buses
        y = Number of vans
   b.   Relevant material in table form:

	Buses	Vans	Number to accommodate
Students	40	8	400
Chaperones	3	1	36
Rental cost	$1200/bus	$100/van	

   c.   Objective function:
        Minimize the cost $C = 1200x + 100y$
   d.   Problem constraints:
        $40x + 8y \geq 400$ [Student constraint]
        $3x + y \leq 36$ [Chaperone constraint]
   e.   Non-negative constraints
        $x \geq 0, y \geq 0$

   The mathematical model for this problem is:

   Minimize $C = 1200x + 100y$
   Subject to: $40x + 8y \geq 400$
   $3x + y \leq 36$
   $x \geq 0, y \geq 0$

Step (1): Graph the feasible region and find the corner points.
The feasible region S is the solution set of the system of
inequalities, and is indicated by the shading in the graph at the
right.
The corner points are (10, 0), (7, 15), and (12, 0).
Since S is bounded, C has a minimum value by <u>4</u>(a).

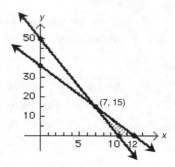

Step (2): Evaluate the objective function at each corner point.
The value of C at each corner point is:

Corner Point	$C = 1200x + 100y$
(10,0)	$C = 1200(10) + 100(0) = 12,000$
(7,15)	$C = 1200(7) + 100(15) = 9,900$
(12,0)	$C = 1200(12) + 100(0) = 14,400$

Step (3): Determine the optimal solution.
The minimum occurs when $x = 7$ and $y = 15$. That is, the officers should rent 7 buses and 15 vans at
the minimum cost of $9900.

45. **(A)**   Construct the mathematical model
a.   Decision variables:
     Let x = Amount invested in the CD
         y = Amount invested in the mutual fund
c.   Objective function:
     Maximize the return $P = 0.05x + 0.09y$
d.   Problem constraints:
     $x + y \leq 60,000$ [Amount available constraint]
     $y \geq 10,000$ [Mutual fund constraint]
     $x \geq 2y$ [Investor constraint]
e.   Non-negative constraints
     $x \geq 0, y \geq 0$

The mathematical model for this problem is
     Maximize $P = 0.05x + 0.09y$
     Subject to: $x + y \leq 60,000$
                  $y \geq 10,000$
                  $x \geq 2y$
                  $x, y \geq 0$

The feasible region S is the solution set of the system of inequalities and is
indicated by the shading in the graph.

The corner points are (20,000, 10,000), (40,000, 20,000) and (50,000, 10,000).

Since S is bounded, P has a maximum value by 4<u>a</u>.

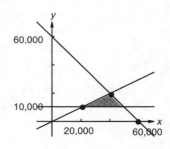

The value of P at each corner point is given in the table below.

Corner Point	$P = 0.05x + 0.09y$
(20,000,10,000)	$P = 0.05(20,000) + 0.09(10,000) = 1900$
(40,000,20,000)	$P = 0.05(40,000) + 0.09(20,000) = 3800$
(50,000,10,000)	$P = 0.05(50,000) + 0.09(10,000) = 3400$

Thus, the maximum return is $3,800 when $40,000 is invested in the CD and $20,000 is invested in the mutual fund.

47.   Construct the mathematical model

    a.   Decision variables:

        Let x = Number of gallons produced by old process

          y = Number of gallons produced by new process

    b.   Relevant material in table form:

	Grams/gallon old process	Grams/gallon new process	Maximum allowed
Sulfur dioxide	20	5	16,000
Particulate	40	20	30,000
Profit	60¢/gal	20¢/gal	

    c.   Objective function:

        Maximize the profit function $P = 60x + 20y$

    d.   Problem constraints:

        $20x + 5y \le 16{,}000$ [Sulfur dioxide constraint]

        $40x + 20y \le 30{,}000$ [Particulate constraint]

    e.   Non-negative constraints

        $x \ge 0, y \ge 0$

    The mathematical model for this problem is:

$$\text{Maximize } P = 60x + 20y$$
$$\text{Subject to: } 20x + 5y \le 16{,}000$$
$$40x + 20y \le 30{,}000$$
$$x \ge 0, y \ge 0$$

<u>Step (1)</u>: Graph the feasible region and find the corner points.

The feasible region S is the solution set of the given inequalities, and is indicated by the shading in the graph at the right.

The corner points are (0, 0), (0, 1500), and (750, 0).

    Since S is bounded, P has a maximum value by <u>4</u>(a).

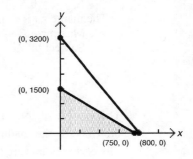

<u>Step (2)</u>: Evaluate the objective function at each corner point.

The value of P at each corner point is:

Corner Point	$P = 60x + 20y$
(0,0)	$P = 60(0) + 20(0) = 0$ (cents)
(0,1500)	$P = 60(0) + 20(1500) = 30{,}000$ (cents)
(750,0)	$P = 60(750) + 20(0) = 45{,}000$ (cents)

<u>Step (3)</u>: The maximum profit is $450 when 750 gallons are produced using the old process exclusively.

(B)   The mathematical model for this problem is:

Maximize $P = 60x + 20y$

Subject to: $20x + 5y \le 11{,}500$

$40x + 20y \le 30{,}000$

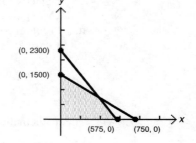

The feasible region S for this problem is indicated by the shading in the graph at the right.

The corner points are $(0, 0)$, $(0, 1500)$, $(400, 700)$, and $(575, 0)$.

The value P at each corner point is:

Corner Point	$P = 60x + 20y$
$(0,0)$	$P = 60(0) + 20(0) = 0$
$(0,1500)$	$P = 60(0) + 20(1{,}500) = 30{,}000$
$(400,700)$	$P = 60(400) + 20(700) = 38{,}000$
$(575,0)$	$P = 60(575) + 20(0) = 34{,}500$

The maximum profit is $380 when 400 gallons are produced using the old process and 700 gallons using the new process.

(C)   The mathematical model for this problem is:

Maximize $P = 60x + 20y$

Subject to: $20x + 5y \le 7{,}200$

$40x + 20y \le 30{,}000$

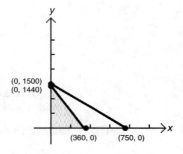

The feasible region S for this problem is indicated by the shading in the graph at the right. The corner points are $(0, 0)$, $(0, 1440)$, and $(360, 0)$.

The value of P at each corner point is:

Corner Point	$P = 60x + 20y$
$(0,0)$	$P = 60(0) + 20(0) = 0$
$(0,1440)$	$P = 60(0) + 20(1{,}440) = 28{,}800$
$(360,0)$	$P = 60(360) + 20(0) = 21{,}600$

The maximum profit is $288 when 1,440 gallons are produced by the new process exclusively.

49.   Construct the mathematical model

a.   Decision variables:

Let x = Number of bags of Brand A

y = Number of bags of Brand B

b.   Relevant material in table form:

Amounts	Brand A	Brand B	
Phosphoric acid	4	4	1000
Chloride	2	1	400
Nitrogen	8 lbs.	3 lbs.	

c.   Objective function:
 Maximize the amount of nitrogen $N = 8x + 3y$

d.   Problem constraints:
 $4x + 4y \geq 1,000$ [Phosphoric acid constraint]
 $2x + y \leq 400$ [Chloride constraint]

e.   Non-negative constraints
 $x \geq 0, y \geq 0$

(A)   The mathematical model for this problem is:

Maximize $N = 8x + 3y$

Subject to: $4x + 4y \geq 1000$
 $2x + y \leq 400$
 $x \geq 0, y \geq 0$

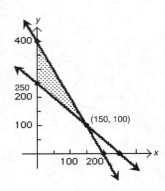

is   The feasible region S is the solution set of the system of inequalities, and indicated by the shading in the graph at the right. The corner points are $(0, 250)$, $(0, 400)$, and $(150, 100)$.

Since S is bounded, N has a maximum value by <u>4</u>(a).

The value of N at each corner point is given in the table below:

Corner Point	$N = 8x + 3y$
$(0, 250)$	$N = 8(0) + 3(250) = 750$
$(150, 100)$	$N = 8(150) + 3(100) = 1500$
$(0, 400)$	$N = 8(0) + 3(400) = 1200$

Thus, the maximum occurs when $x = 150$ and $y = 100$. That is, the grower should use 150 bags of Brand A and 100 bags of Brand B. The maximum number of pounds of nitrogen is 1500.

(B)   The mathematical model for this problem is:

Minimize $N = 8x + 3y$

Subject to: $4x + 4y \geq 1000$
 $2x + y \leq 400$
 $x \geq 0, y \geq 0$

The feasible region S and the corner points are the same as in part (A). Thus, the minimum occurs when $x = 0$ and $y = 250$. That is, the grower should use 0 bags of Brand A and 250 bags of Brand B. The minimum number of pounds of nitrogen is 750.

51.   Construct the mathematical model

a.   Decision variables:
 Let x = Number of cubic yards of mix A
 y = Number of cubic yards of mix B

b.   Relevant material in table form:

	Amount per Cubic Yard (in pounds)		Minimum monthly requirement
	Mix A	Mix B	
Phosphoric acid	20	10	460
Nitrogen	30	30	960
Potash	5	10	220
Cost/cubic yd.	$30	$35	

c. Objective function:
    Minimize the cost $C = 30x + 35y$

d. Problem constraints:
    $20x + 10y \geq 460$ [Phosphoric acid constraint]
    $30x + 30y \geq 960$ [Nitrogen constraint]
    $5x + 10y \geq 220$ [Potash constraint]

e. Non-negative constraints
    $x \geq 0, y \geq 0$

The mathematical model for this problem is:

Minimize $C = 30x + 35y$
Subject to: $20x + 10y \geq 460$
$\phantom{\text{Subject to: }}30x + 30y \geq 960$
$\phantom{\text{Subject to: }}5x + 10y \geq 220$
$\phantom{\text{Subject to: }}x \geq 0, y \geq 0$

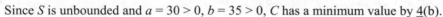

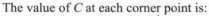

The feasible region S is the solution set of the given inequalities and is indicated by the shading in the graph at the right.

The corner points are $(0, 46)$, $(14, 18)$, $(20, 12)$, and $(44, 0)$.

Since S is unbounded and $a = 30 > 0$, $b = 35 > 0$, C has a minimum value by <u>4</u>(b).

The value of C at each corner point is:

Corner Point	$C = 30x + 35y$
(0, 46)	$C = 30(0) + 35(46) = 1610$
(14, 18)	$C = 30(14) + 35(18) = 1050$
(20, 12)	$C = 30(20) + 35(12) = 1020$
(44, 0)	$C = 30(44) + 35(0) = 1320$

Thus, the minimum occurs when the amount of mix A used is 20 cubic yards and the amount of mix B used is 12 cubic yards. The minimum cost is $C = \$1020$.

53. Construct the mathematical model

    a.   Decision variables:

        Let x = Number of mice used

          y = Number of rats used

    c.   Objective function:

        Maximize the number of mice and rats used $P = x + y$

    d.   Problem constraints:

        $10x + 20y \le 800$ [Box A constraint]

        $20x + 10y \le 640$ [Box B constraint]

    e.   Non-negative constraints

        $x \ge 0, y \ge 0$

The mathematical model for this problem is:

Maximize $P = x + y$

Subject to: $10x + 20y \le 800$

        $20x + 10y \le 640$

        $x \ge 0, y \ge 0$

The feasible region S is the solution set of the given inequalities, and is indicated by the shading in the graph at the right.

The corner points are (0, 0), (0, 40), (16, 32), and (32, 0).

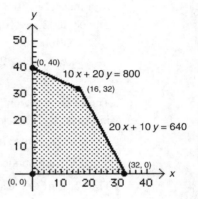

Since S is bounded, P has a maximum value by <u>4</u>(a).

The value of P at each corner point is:

Corner Point	$P = x + y$
(0, 0)	$P = 0 + 0 = 0$
(0, 40)	$P = 0 + 40 = 40$
(16, 32)	$P = 16 + 32 = 48$
(32, 0)	$P = 32 + 0 = 32$

Thus, the maximum occurs when the number of mice used is 16 and the number of rats used is 32. The maximum number of mice and rats that can be used is 48.

CHAPTER 5 REVIEW

1. $x > 2y - 3$ or $x - 2y > -3$

    Graph the line $x - 2y = -3$ as a dashed line. Substituting $x = 0$, $y = 0$ in the inequality produces a true statement, so (0, 0) is in the solution set.

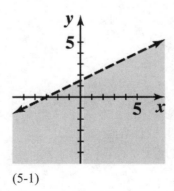

(5-1)

2. $3y - 5x \leq 30$

Graph the line $3y - 5x = 30$ as a solid line. Substituting $x = 0$, $y = 0$ into the inequality produces a true statement, so $(0, 0)$ is in the solution set.

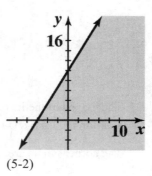

(5-2)

3. $5x + 9y \leq 90,\ x \geq 0,\ y \geq 0$

The graph of $5x + 9y \leq 90$ is the half-plane below the line $5x + 9y = 90$, including the line. With $x \geq 0$, $y \geq 0$, the graph of the system is the shaded region. The solution region is bounded. The coordinates of the corner points are: $(0, 0)$, $(18, 0)$, $(0, 10)$.

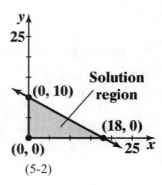

(5-2)

4. $15x + 16y \geq 1{,}200,\ x \geq 0,\ y \geq 0$

The graph of $15x + 16y \geq 1200$ is the half-plane above the line $15x + 16y = 1{,}200$, including the line. With $x \geq 0$, $y \geq 0$, the graph of the system is the shaded region. The solution region is unbounded. The coordinates of the corner points are $(80, 0)$, $(0, 75)$.

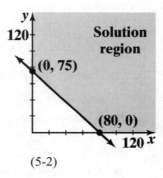

(5-2)

5. $2x + y \leq 8$
   $3x + 9y \leq 27$
   $x, y \geq 0$

The graphs of the inequalities are shown at the right. The solution region is shaded; it is *bounded*.

The corner points are: $(0, 0)$, $(0, 3)$, $(3, 2)$, $(4, 0)$

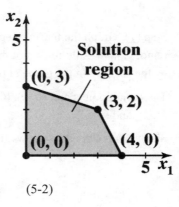

(5-2)

6. $3x + y \geq 9$
    $2x + 4y \geq 16$
    $x, y \geq 0$

    The graphs of the inequalities are shown at the right. The solution region is shaded; it is *unbounded*.

    The corner points are: (0, 9), (2, 3), (8, 0)

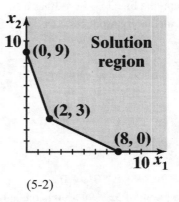

(5-2)

7. The boundary line passes through (6, 0) and (0, –4).

    slope: $m = \dfrac{0-(-4)}{6-0} = \dfrac{2}{3}$

    y intercept: $b = -4$

    Boundary line equation: $y = \dfrac{2}{3}x - 4$

    $3y = 2x - 12$
    $2x - 3y = 12$

    Since (0, 0) is in the shaded region and the boundary line is solid, the graph is the graph of $2x - 3y \leq 12$.
    (5-1)

8. The boundary line passes through (2, 0) and (0, 8).

    slope: $m = \dfrac{0-8}{2-0} = -4$

    y intercept: $b = 8$

    Boundary line equation: $y = -4x + 8$

    $4x + y = 8$

    Since (0, 0) is not in the shaded region and the boundary line is solid, the graph is the graph of $4x + y \geq 8$.   (5-1)

9. Step (1): Graph the feasible region and find the corner points. The feasible region S is the solution set of the given inequalities. This region is indicated by the shading in the graph at the right.

    The corner points are: (0, 0), (0, 4), (4, 2), and (5, 0).

    Since S is bounded it follows from 4(a) that P has a maximum value.

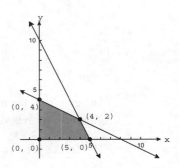

Step (2): Evaluate the objective function at each corner point. The value of P at each corner point is given in the following table.

Corner Point	$P = 2x + 6y$
$(0, 0)$	$P = 2(0) + 6(0) = 0$
$(0, 4)$	$P = 2(0) + 6(4) = 24$
$(4, 2)$	$P = 2(4) + 6(2) = 20$
$(5, 0)$	$P = 2(5) + 6(0) = 10$

Step (3): Determine the optimal solution.
The maximum value of P is 24 at $x = 0$, $y = 4$.   (5-3)

10. Step (1): Graph the feasible region and find the corner points. The feasible region S is the solution set of the given inequalities. This region is indicated by the shading in the graph at the right.

The corner points are: (0, 20), (9, 2), and (15, 0).

Since S is unbounded and the coefficients of C are positive it follows from 4(b) that C has a minimum value.

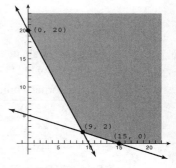

Step (2): Evaluate the objective function at each corner point. The value of C at each corner point is given in the following table.

Corner Point	$C = 5x + 2y$
$(0, 20)$	$C = 5(0) + 2(20) = 40$
$(9, 2)$	$C = 5(9) + 2(2) = 49$
$(15, 0)$	$C = 5(15) + 2(0) = 75$

Step (3): Determine the optimal solution.
The minimum value of C is 40 at $x = 0$, $y = 20$.   (5-3)

11. Step (1): Graph the feasible region and find the corner points. The feasible region S is the solution set of the given inequalities. This region is indicated by the shading in the graph at the right.
The corner points are: (0, 0), (0, 6), (2, 5), (3, 4) and (5, 0).
Since S is bounded it follows from 4(a) that P has a maximum value.

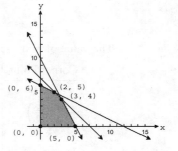

Step (2): Evaluate the objective function at each corner point. The value of P at each corner point is given in the following table.

Corner Point	$P = 3x + 4y$
$(0, 0)$	$P = 3(0) + 4(0) = 0$
$(0, 6)$	$P = 3(0) + 4(6) = 24$
$(2, 5)$	$P = 3(2) + 4(5) = 26$
$(3, 4)$	$P = 3(3) + 4(4) = 25$
$(5, 0)$	$P = 3(5) + 4(0) = 15$

Step (3): Determine the optimal solution.
The maximum value of P is 26 at $x = 2$, $y = 5$.   (5-3)

12. Step (1): Graph the feasible region and find the corner points. The feasible region S is the solution set of the given inequalities. This region is indicated by the shading in the graph at the right.

The corner points are: (3, 9), (5, 5), and (10, 0).

Since S is unbounded and the coefficients of C are positive, it follows from 4(b) that C has a minimum value.

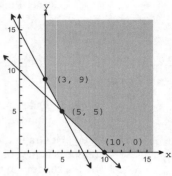

Step (2): Evaluate the objective function at each corner point. The value of C at each corner point is given in the following table.

Corner Point	$C = 8x + 3y$
$(3,9)$	$C = 8(3) + 3(9) = 51$
$(5,5)$	$C = 8(5) + 3(5) = 55$
$(10,0)$	$C = 8(10) + 3(0) = 80$

Step (3): Determine the optimal solution.
The minimum value of C is 51 at $x = 3$, $y = 9$.   (5-3)

13. Step (1): Graph the feasible region and find the corner points. The feasible region S is the solution set of the given inequalities. This region is indicated by the shading in the graph at the right.

The corner points are: (0, 0), $\left(0, \frac{26}{3}\right)$, (8,6), (10, 2), and (10, 0).

Since S is bounded it follows from 4(a) that P has a maximum value.

Step (2): Evaluate the objective function at each corner point.

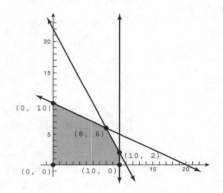

The value of P at each corner point is given in the following table.

Corner Point	$P = 3x + 2y$
$(0,0)$	$P = 3(0) + 2(0) = 0$
$\left(0, \frac{26}{3}\right)$	$P = 3(0) + 2\left(\frac{26}{3}\right) = \frac{52}{3} = 17\frac{1}{3}$
$(8,6)$	$P = 3(8) + 2(6) = 36$
$(10,2)$	$P = 3(10) + 2(2) = 34$
$(10,0)$	$P = 3(10) + 2(0) = 30$

Step (3): Determine the optimal solution. The maximum value of P is 36 at $x = 8$, $y = 6$.   (5-3)

14. Let x = number of calculator boards
        y = number of toaster boards

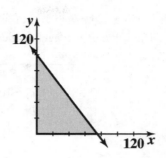

   (A) 5 hours = 5(60) = 300 minutes
        The oven is available for 300
        minutes. Therefore,
        $4x + 3y \le 300, x \ge 0, y \ge 0.$

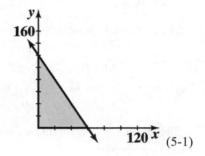

   (B) 2 hours = 2(60) = 120 minutes
        The wave machine is available for 120 minutes.
        Therefore,
        $2x + y \le 120, x \ge 0, y \ge 0$

                                                              (5-1)

15. (A) Let x = the number of regular sails
            and y = the number of competition sails.
            The mathematical model for this problem is:
                 Maximize $P = 100x + 200y$

                 Subject to: $2x + 3y \le 150$
                               $4x + 10y \le 380$
                                  $x, y \ge 0$

        The feasible region is indicated by the shading in the graph below.

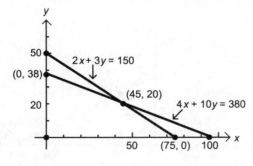

The corner points are (0, 0), (0, 38),
(45, 20), (75, 0).

The value P at each corner point is:

Corner point	$P = 100x + 200y$
$(0,0)$	$P = 100(0) + 200(0) = 0$
$(0,38)$	$P = 100(0) + 200(38) = 7,600$
$(45,20)$	$P = 100(45) + 200(20) = 8,500$
$(75,0)$	$P = 100(75) + 200(0) = 7,500$

        Optimal solution: max P = \$8,500 when 45 regular and
        20 competition sails are produced.

   (B) The mathematical model for this problem is:
                 Maximize $P = 100x + 260y$
                 Subject to: $2x + 3y \le 150$
                               $4x + 10y \le 380$
                                  $x, y \ge 0$

The feasible region and the corner points are the same as in part (A). The value of P at each corner point is:

Corner point	$P = 100x + 260y$
$(0,0)$	$P = 100(0) + 260(0) = 0$
$(0,38)$	$P = 100(0) + 260(38) = 9,880$
$(45,20)$	$P = 100(45) + 260(20) = 9,700$
$(75,0)$	$P = 100(75) + 260(0) = 7,500$

The maximum profit increases to \$9,880 when 38 competition and 0 regular sails are produced.

(C)   The mathematical model for this problem is:

Maximize $P = 100x + 140y$

Subject to: $2x + 3y \le 150$

$4x + 10y \le 380$

$x, y \ge 0$

The feasible region and the corner points are the same as in parts (A) and (B). The value of P at each corner point is:

Corner point	$P = 100x + 140y$
$(0,0)$	$P = 100(0) + 140(0) = 0$
$(0,38)$	$P = 100(0) + 140(38) = 5,320$
$(45,20)$	$P = 100(45) + 140(20) = 7,300$
$(75,0)$	$P = 100(75) + 140(0) = 7,500$

The maximum profit decreases to \$7,500 when 0 competition and 75 regular sails are produced.

(5-3)

16.   Let x = number of grams of mix A

y = number of grams of mix B

The constraints are:

vitamins: $2x + 5y \ge 850$

minerals: $2x + 4y \ge 800$

calories: $4x + 5y \ge 1,150$

$x, y \ge 0$

The feasible region is indicated by the shading in the graph at the right. The corner points are:

$(0, 230)$, $(100, 150)$, $(300, 50)$, $(425, 0)$.

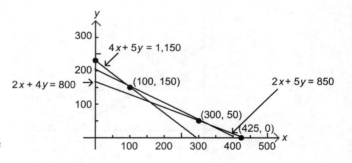

(A)   The mathematical model for this problem is:

minimize $C = 0.04x + 0.09y$ subject to the constraints given above.

The value of C at each corner point is:

Corner point	$C = 0.04x + 0.09y$
$(0, 230)$	$C = 0.04(0) + 0.09(230) = 20.70$
$(100, 150)$	$C = 0.04(100) + 0.09(150) = 17.50$
$(300, 50)$	$C = 0.04(300) + 0.09(50) = 16.50$
$(425, 0)$	$C = 0.04(425) + 0.09(0) = 17.00$

The minimum cost is $16.50 when 300 grams of mix A and 50 grams of mix B are used.

(B)   The mathematical model for this problem is:

minimize $C = 0.04x + 0.06y$ subject to the constraints given above.

The value of C at each corner point is:

Corner point	$C = 0.04x + 0.06y$
$(0, 230)$	$C = 0.04(0) + 0.06(230) = 13.80$
$(100, 150)$	$C = 0.04(100) + 0.06(150) = 13.00$
$(300, 50)$	$C = 0.04(300) + 0.06(50) = 15.00$
$(425, 0)$	$C = 0.04(425) + 0.06(0) = 17.00$

The minimum cost decreases to $13.00 when 100 grams of mix A and 150 grams of mix B are used.

(C)   The mathematical model for this problem is:

minimize $C = 0.04x + 0.12y$ subject to the constraints given above.

The value of C at each corner point is:

Corner point	$C = 0.04x + 0.12y$
$(0, 230)$	$C = 0.04(0) + 0.12(230) = 27.60$
$(100, 150)$	$C = 0.04(100) + 0.12(150) = 22.00$
$(300, 50)$	$C = 0.04(300) + 0.12(50) = 18.00$
$(425, 0)$	$C = 0.04(425) + 0.12(0) = 17.00$

The minimum cost increases to $17.00 when 425 grams of mix A and 0 grams of mix B are used.
(5-3)

6   LINEAR PROGRAMMING: THE SIMPLEX METHOD

EXERCISE 6-1

Things to remember:

<u>1</u>.   STANDARD MAXIMIZATION PROBLEM IN STANDARD FORM

A linear programming problem is said to be a STANDARD MAXIMIZATION PROBLEM IN STANDARD FORM if its mathematical model is:

Maximize $P = c_1 x_1 + c_2 x_2 + \cdots + c_n x_n$

Subject to problem constraints of the form:

$a_1 x_1 + a_2 x_2 + \cdots + a_n x_n \leq b, \ b \geq 0$

with nonnegative constraints:

$x_1, x_2, ..., x_n \geq 0.$

[<u>Note</u>: The coefficients of the objective function can be any real numbers.]

<u>2</u>.   SLACK VARIABLES

Given a linear programming problem. SLACK VARIABLES are nonnegative quantities that are introduced to convert problem constraint inequalities into equations.

<u>3</u>.   THE TABLE METHOD (Two Decision Variables)

Assume that a standard maximization problem in standard form has two decision variables , x_1 and x_2, and m problem constraints.

<u>Step 1</u>: Use slack variables $s_1, s_2, ..., s_m$ to convert the i-system to an e-system.

<u>Step 2</u>: Form a table with $(m+2)(m+1)/2$ rows and $m+2$ columns labeled

$x_1, x_2, s_1, s_2, ..., s_m$. In the first row, assign 0 to x_1 and x_2. In the second row, assign 0 to x_1 and s_1. Continue until all the rows contain al possible combinations of assigning two 0's to the variables.

<u>Step 3</u>: Complete each row to a solution of the e-system, if possible. Because two of the variables have the value 0, this involves solving a system of m linear equation in m variables. If the system has no solution of infinitely many solutions, do not complete the row.

<u>Step 4</u>: Solve the linear programming problem by finding the maximum value of P over thos completed rows that have no negative values.

<u>4</u>.   FUNDAMENTAL THEOREM OF LINEAR PROGRAMMING

If the optimal value of the objective function in a linear programming problem exists, then that value must occur at one (or more) of the basic feasible solutions.

---

1. $_5C_2 = \dfrac{5!}{2! \cdot 3!} = \dfrac{5 \cdot 4}{2} = 10$

3. $_6C_2 = \dfrac{6!}{2! \cdot 4!} = \dfrac{6 \cdot 5}{2} = 15$

5. Set $x_1 = s_1 = 0$. Then $\begin{aligned} 5x_2 &= 10 \\ 3x_2 + s_2 &= 8 \end{aligned}$ $x_2 = s_2 = 2$. Solution: $(0, 2, 0, 2)$

7. Set $x_2 = s_2 = 0$. Then $\begin{aligned} 2x_1 + s_1 &= 10 \\ x_1 \quad\;\; &= 8 \end{aligned}$ $x_1 = 8$, $s_1 = -6$. Solution: $(8, 0, -6, 0)$.

9. e-system: $\begin{aligned} 2x_1 + 3x_2 + s_1 \quad\;\; &= 9 \\ 6x_1 + 7x_2 \quad\;\; + s_2 &= 13 \end{aligned}$

11. e-system: $\begin{aligned} 12x_1 - 14x_2 + s_1 \qquad\qquad &= 55 \\ 19x_1 + 5x_2 \qquad + s_2 \qquad &= 40 \\ -8x_1 + 11x_2 \qquad\qquad + s_3 &= 64 \end{aligned}$

13. e-system: $6x_1 + 5x_2 + s_1 = 18$

15. e-system

$$\begin{aligned} 4x_1 - 3x_2 + s_1 \qquad\qquad\qquad &= 12 \\ 5x_1 + 2x_2 \quad + s_2 \qquad\qquad &= 25 \\ -3x_1 + 7x_2 \qquad + s_3 \qquad &= 32 \\ 2x_1 + x_2 \qquad\qquad + s_4 &= 9 \end{aligned}$$

17. s_1, s_2 are basic variables.

19. x_1, s_2 are nonbasic variables.

21. x_2, s_2 are nonbasic variables.

23. (A), (B), (E), (F) are feasible since all variables are nonnegative.

25. $P = 2x_1 + 5x_2$:

(A) $P = 2(0) + 5(0) = 0$, (B) $P = 2(0) + 5(8) = 40$, (E) $P = 2(9) + 5(0) = 18$, (F) $P = 2(6) + 5(4) = 32$
Max $P = 32$ at $x_1 = 6$, $x_2 = 4$.

27. x_2, s_1, s_3 are basic variables.

29. x_2, s_3 are nonbasic variables.

31. (C), (D), (E), (F) are not feasible; each has at least one negative variable.

33. Set $x_2 = s_3 = 0$. Then $\begin{aligned} x_1 + s_1 \qquad\quad &= 24 \\ 2x_1 \quad + s_2 &= 30 \\ 4x_1 \qquad\quad &= 48 \end{aligned}$ Therefore, $x_1 = 12$, $s_1 = 12$, $s_2 = 6$ and

$(x_1, x_2, s_1, s_2, s_3) = (12, 0, 12, 6, 0)$.

35. Set $s_1 = s_3 = 0$. Then $\begin{aligned} x_1 + x_2 \qquad\quad &= 24 \\ 2x_1 + x_2 + s_2 &= 30 \\ 4x_1 + x_2 \qquad\quad &= 48 \end{aligned}$ Therefore, $x_1 = 8$, $x_2 = 16$, $s_2 = -2$ and

$(x_1, x_2, s_1, s_2, s_3) = (8, 16, 0, -2, 0)$.

37.

e-system: $4x_1 + 5x_2 + s_1 = 20$

x_1	x_2	s_1	feasible
0	0	20	yes
0	4	0	yes
5	0	0	yes

39.

e-system:
$$x_1 + x_2 + s_1 \qquad = 6$$
$$x_1 + 4x_2 \qquad + s_2 = 12$$

x_1	x_2	s_1	s_2	feasible
0	0	6	12	yes
0	6	0	−12	no
0	3	3	0	yes
6	0	0	6	yes
12	0	−6	0	no
4	2	0	0	yes

41.

e-system:
$$2x_1 + 5x_2 + s_1 \qquad = 20$$
$$x_1 + 2x_2 \qquad + s_2 = 9$$

x_1	x_2	s_1	s_2	feasible
0	0	20	9	yes
0	4	0	1	yes
0	9/2	−5/2	0	no
10	0	0	−1	no
9	0	2	0	yes
5	2	0	0	yes

43.

e-system:
$$x_1 + 2x_2 + s_1 \qquad = 24$$
$$x_1 + x_2 \qquad + s_2 \qquad = 15$$
$$2x_1 + x_2 \qquad + s_3 = 24$$

x_1	x_2	s_1	s_2	s_3	feasible
0	0	24	15	24	yes
0	12	0	3	12	yes
0	15	−6	0	9	no
0	24	−24	−9	0	no
24	0	0	−9	−24	no
15	0	9	0	−6	no
12	0	12	3	0	yes
6	9	0	0	3	yes
8	8	0	−1	0	no
9	6	3	0	0	yes

45. Corner points: (0,0), (0,4), (5,0)

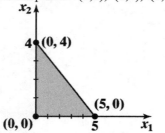

47. Corner points (0,0), (0,3), (6,0), (4,2)

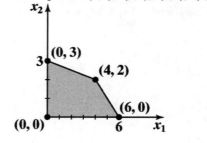

49. Corner points: (0,0), (0,4), (9,0), (5,2)

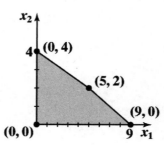

51.

x_1	x_2	s_1	$P = 10x_1 + 9x_2$
0	0	20	0
0	4	0	36
5	0	0	50

Max $P = 50$ at $x_1 = 5$, $x_2 = 0$.

53.

x_1	x_2	s_1	s_2	$P = 15x_1 + 20x_2$
0	0	6	12	0
0	6	0	−12	−
0	3	3	0	60
6	0	0	6	90
12	0	−6	0	−
4	2	0	0	100

Max $P = 100$ at $x_1 = 4$, $x_2 = 2$.

55.

x_1	x_2	s_1	s_2	$P = 25x_1 + 10x_2$
0	0	20	9	0
0	4	0	1	40
0	9/2	−5/2	0	−
10	0	0	−1	−
9	0	2	0	225
5	2	0	0	145

Max $P = 225$ at $x_1 = 9$, $x_2 = 0$.

57.

x_1	x_2	s_1	s_2	s_3	$P = 30x_1 + 40x_2$
0	0	24	15	24	0
0	12	0	3	12	480
0	15	−6	0	9	−
0	24	−24	−9	0	−
24	0	0	−9	−24	−
15	0	9	0	−6	−
12	0	12	3	0	360
6	9	0	0	3	540
8	8	0	−1	0	−
9	6	3	0	0	510

Max $P = 540$ at $x_1 = 6$, $x_2 = 9$.

59. $_{10}C_4 = \dfrac{10!}{4! \cdot (10-6)!} = \dfrac{10!}{4! \cdot 6!} = 210$

61. $_{72}C_{30} = \dfrac{72!}{30! \cdot (72-30)!} = \dfrac{72!}{30! \cdot 42!} = 1.64 \times 10^{20}$

EXERCISE 6-2

Things to remember:

<u>1.</u>   PROCEDURE: SELECTING BASIC AND NONBASIC VARIABLES FOR THE SIMPLEX PROCESS
Given a simplex tableau,

Step 1. NUMBERS OF VARIABLES: Determine the number of basic and the number of nonbasic variables. These numbers do not change during the simplex process.

Step 2. SELECTING BASIC VARIABLES: A variable can be selected as a basic variable only if it corresponds to a column in the tableau that has exactly one nonzero element (usually 1) and the nonzero element in the column is not in the same row as the nonzero element in the column of another basic variable. This procedure always selects P as a basic variable, since the P column never changes during the simplex process.

Step 3. SELECTING NONBASIC VARIABLES: After the basic variables are selected in Step 2, the remaining variables are selected as the nonbasic variables. The tableau columns under the nonbasic variables will usually contain more than one nonzero element.

<u>2.</u>   PROCEDURE: SELECTING THE PIVOT ELEMENT

Step 1. Locate the most negative indicator in the bottom row of the tableau to the left of the P column (the negative number with the largest absolute value). The column containing this element is the PIVOT COLUMN. If there is a tie for the most negative indicator, choose either column.

Step 2. Divide each POSITIVE element in the pivot column above the dashed line into the corresponding element in the last column. The PIVOT ROW is the row corresponding to the smallest quotient. If there is a tie for the smallest quotient, choose either row. If the pivot column above the dashed line has no positive elements, then there is no solution and we stop.

Step 3. The PIVOT (or PIVOT ELEMENT) is the element in the intersection of the pivot column and pivot row.

[Note: The pivot element is always positive and never appears in the bottom row.]

[Remember: The entering variable is at the top of the pivot column and the exiting variable is at the left of the pivot row.]

3.   PROCEDURE: PERFORMING THE PIVOT OPERATION

A PIVOT OPERATION or PIVOTING consists of performing row operations as follows:

Step 1. Multiply the pivot row by the reciprocal of the pivot element to transform the pivot element into a 1. (If the pivot element is already a 1, omit this step.)

Step 2. Add multiples of the pivot row to other rows in the tableau to transform all other nonzero elements in the pivot column into 0's.

[Note: **In a pivot operation, you can never interchange two rows.**]

4.   SIMPLEX ALGORITHM FOR STANDARD MAXIMIZATION PROBLEMS
Problem constraints are of the ≤ form with nonnegative constants on the right hand side. The coefficients of the objective function can be any real numbers.

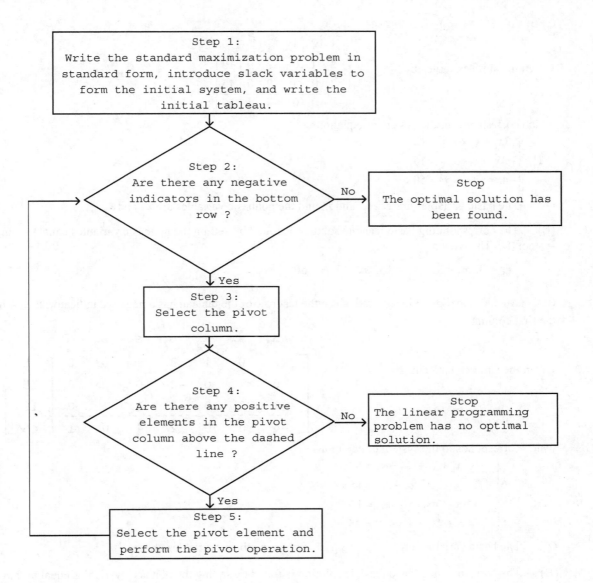

1. Given the simplex tableau:

$$\begin{array}{c} \begin{array}{ccccc} x_1 & x_2 & s_1 & s_2 & P \end{array} \\ \left[\begin{array}{ccccc|c} 2 & 1 & 0 & 3 & 0 & 12 \\ 3 & 0 & 1 & -2 & 0 & 15 \\ \hline -4 & 0 & 0 & 4 & 1 & 50 \end{array}\right] \end{array}$$

which corresponds to the system of equations:

$$(I)\ \begin{cases} 2x_1 + x_2 + 3s_2 = 12 \\ 3x_1 + s_1 - 2s_2 = 15 \\ -4x_1 + 4s_2 + P = 50 \end{cases}$$

(A)   The basic variables are x_2, s_1, and P, and the nonbasic variables are x_1 and s_2.

(B)   The corresponding basic feasible solution is found by setting the nonbasic variables equal to 0 in system (I). This yields:

$$x_1 = 0,\ x_2 = 12,\ s_1 = 15,\ s_2 = 0, P = 50$$

(C)   An additional pivot is required, since the last row of the tableau has a negative indicator, the –4 in the first column.

3. Given the simplex tableau:

$$\begin{array}{c} \begin{array}{ccccccc} x_1 & x_2 & x_3 & s_1 & s_2 & s_3 & P \end{array} \\ \left[\begin{array}{ccccccc|c} -2 & 0 & 1 & 3 & 1 & 0 & 0 & 5 \\ 0 & 1 & 0 & -2 & 0 & 0 & 0 & 15 \\ -1 & 0 & 0 & 4 & 1 & 1 & 0 & 12 \\ \hline -4 & 0 & 0 & 2 & 4 & 0 & 1 & 45 \end{array}\right] \end{array}$$

which corresponds to the system of equations:

$$(I)\ \begin{cases} -2x_1 + x_3 + 3s_1 + s_2 = 5 \\ x_2 - 2s_1 = 15 \\ -x_1 + 4s_1 + s_2 + s_3 = 12 \\ -4x_1 + 2s_1 + 4s_2 + P = 45 \end{cases}$$

(A)   The basic variables are x_2, x_3, s_3, P, and the nonbasic variables are x_1, s_1, s_2.

(B)   The corresponding basic feasible solution is found by setting the nonbasic variables equal to 0 in system (I). This yields:

$$x_1 = 0,\ x_2 = 15,\ x_3 = 5,\ s_1 = 0,\ s_2 = 0,\ s_3 = 12,\ \ P = 45$$

(C)   Since the last row of the tableau has a negative indicator, the –4 in the first column, an additional pivot should be required. However, since there are no positive elements in the pivot column (the first column), the problem has *no optimal solution*.

5. Given the simplex tableau:

$$\begin{array}{c} \begin{array}{ccccc} x_1 & x_2 & s_1 & s_2 & P \end{array} \\ \left[\begin{array}{ccccc|c} 1 & 4 & 1 & 0 & 0 & 4 \\ 3 & 5 & 0 & 1 & 0 & 24 \\ \hline -8 & -5 & 0 & 0 & 1 & 0 \end{array}\right] \end{array}$$

The most negative indicator is –8 in the first column. Thus, the first column is the pivot column. Now, $\dfrac{4}{1}$ = 4 and $\dfrac{24}{3}$ = 8. Thus, the first row is the pivot row and the pivot element is the element in the first row, first column. These are indicated in the following tableau.

Enter

$$\begin{array}{c} \\ \text{Exit } s_1 \\ s_2 \\ \\ P \end{array} \begin{array}{cccccc} x_1 & x_2 & s_1 & s_2 & P & \\ \left[\begin{array}{ccccc|c} ① & 4 & 1 & 0 & 0 & 4 \\ 3 & 5 & 0 & 1 & 0 & 24 \\ \hline -8 & -5 & 0 & 0 & 1 & 0 \end{array}\right] \end{array}$$

$\dfrac{4}{1} = 4$ (minimum)

$\dfrac{24}{3} = 8$

$$\left[\begin{array}{ccccc|c} ① & 4 & 1 & 0 & 0 & 4 \\ 3 & 5 & 0 & 1 & 0 & 24 \\ \hline -8 & -5 & 0 & 0 & 1 & 0 \end{array}\right] \sim \left[\begin{array}{ccccc|c} 1 & 4 & 1 & 0 & 0 & 4 \\ 0 & -7 & -3 & 1 & 0 & 12 \\ \hline 0 & 27 & 8 & 0 & 1 & 32 \end{array}\right]$$

$(-3)R_1 + R_2 \rightarrow R_2$
$8R_1 + R_3 \rightarrow R_3$

7.   Given the simplex tableau:

$$\begin{array}{cccccc} x_1 & x_2 & s_1 & s_2 & s_3 & P \\ \left[\begin{array}{cccccc|c} 2 & 1 & 1 & 0 & 0 & 0 & 4 \\ 3 & 0 & 1 & 1 & 0 & 0 & 8 \\ 0 & 0 & 2 & 0 & 1 & 0 & 2 \\ \hline -4 & 0 & -3 & 0 & 0 & 1 & 5 \end{array}\right] \end{array}$$

The most negative indicator is –4. Thus, the first column is the pivot column. Now, $\dfrac{4}{2} = 2$, $\dfrac{8}{3} = 2\dfrac{2}{3}$.

Thus, the first row is the pivot row, and the pivot element is the element in the first row, first column. These are indicated in the tableau.

Enter

$$\begin{array}{c} \\ \text{Exit } x_2 \\ s_2 \\ s_3 \\ P \end{array} \begin{array}{ccccccc} x_1 & x_2 & s_1 & s_2 & s_3 & P & \\ \left[\begin{array}{cccccc|c} ② & 1 & 1 & 0 & 0 & 0 & 4 \\ 3 & 0 & 1 & 1 & 0 & 0 & 8 \\ 0 & 0 & 2 & 0 & 1 & 0 & 2 \\ \hline -4 & 0 & -3 & 0 & 0 & 1 & 5 \end{array}\right] \end{array}$$

$\dfrac{4}{2} = 2$ (minimum)

$\dfrac{8}{3} = 2\dfrac{2}{3}$

$$\left[\begin{array}{cccccc|c} ② & 1 & 1 & 0 & 0 & 0 & 4 \\ 3 & 0 & 1 & 1 & 0 & 0 & 8 \\ 0 & 0 & 2 & 0 & 1 & 0 & 2 \\ \hline -4 & 0 & -3 & 0 & 0 & 1 & 5 \end{array}\right] \sim$$

$$\begin{array}{cccccc} x_1 & x_2 & s_1 & s_2 & s_3 & P \\ \left[\begin{array}{cccccc|c} ① & \frac{1}{2} & \frac{1}{2} & 0 & 0 & 0 & 2 \\ 3 & 0 & 1 & 1 & 0 & 0 & 8 \\ 0 & 0 & 2 & 0 & 1 & 0 & 2 \\ \hline -4 & 0 & -3 & 0 & 0 & 1 & 5 \end{array}\right] \end{array}$$

$\dfrac{1}{2}R_1 \rightarrow R_1$

$(-3)R_1 + R_2 \rightarrow R_2,$ $4R_1 + R_4 \rightarrow R_4$

$$\sim \begin{bmatrix} 1 & \frac{1}{2} & \frac{1}{2} & 0 & 0 & 0 & 2 \\ 0 & -\frac{3}{2} & -\frac{1}{2} & 1 & 0 & 0 & 2 \\ 0 & 0 & 2 & 0 & 1 & 0 & 2 \\ \hdashline 0 & 2 & -1 & 0 & 0 & 1 & 13 \end{bmatrix}$$

9. **(A)** Introduce slack variables s_1 and s_2 to obtain:

Maximize $P = 15x_1 + 10x_2$

Subject to: $2x_1 + x_2 + s_1 \qquad = 10$

$\qquad\qquad x_1 + 3x_2 \qquad + s_2 = 10$

$\qquad\qquad x_1,\ x_2,\ s_1,\ s_2 \geq 0$

This system can be written in initial form:

$$2x_1 + x_2 + s_1 \qquad\qquad = 10$$
$$x_1 + 3x_2 \qquad + s_2 \qquad = 10$$
$$-15x_1 - 10x_2 \qquad\qquad + P = 0$$
$$x_1,\ x_2,\ s_1,\ s_2 \geq 0$$

(B) The simplex tableau for this problem is:

$$\begin{array}{c} \text{Enter} \\ \begin{array}{cc} & \begin{array}{ccccc} x_1 & x_2 & s_1 & s_2 & P \end{array} \\ \begin{array}{c} \text{Exit} \ s_1 \\ s_2 \\ P \end{array} & \left[\begin{array}{ccccc|c} ② & 1 & 1 & 0 & 0 & 10 \\ 1 & 3 & 0 & 1 & 0 & 10 \\ \hline -15 & -10 & 0 & 0 & 1 & 0 \end{array}\right] \end{array} \end{array} \quad \begin{array}{c} \dfrac{10}{2} = 5 \text{ (minimum)} \\[2mm] \dfrac{10}{1} = 10 \end{array}$$

Column 1 is the pivot column (-15 is the most negative indicator). Row 1 is the pivot row (5 is the smallest positive quotient). Thus, the pivot element is the circled 2.

(C) We use the simplex method as outlined above. The pivot elements are circled.

$$\begin{array}{c} \begin{array}{ccccc} x_1 & x_2 & s_1 & s_2 & P \end{array} \\ \begin{array}{c} s_1 \\ s_2 \\ P \end{array} \left[\begin{array}{ccccc|c} ② & 1 & 1 & 0 & 0 & 10 \\ 1 & 3 & 0 & 1 & 0 & 10 \\ \hline -15 & -10 & 0 & 0 & 1 & 0 \end{array}\right] \end{array} \sim \left[\begin{array}{ccccc|c} ① & \frac{1}{2} & \frac{1}{2} & 0 & 0 & 5 \\ 1 & 3 & 0 & 1 & 0 & 10 \\ \hline -15 & -10 & 0 & 0 & 1 & 0 \end{array}\right] \sim$$

$$\frac{1}{2}R_1 \to R_1 \qquad\qquad\qquad (-1)R_1 + R_2 \to R_2$$
$$15R_1 + R_3 \to R_3$$

$$\sim \begin{bmatrix} 1 & \frac{1}{2} & \frac{1}{2} & 0 & 0 & | & 5 \\ 0 & \boxed{\frac{5}{2}} & -\frac{1}{2} & 1 & 0 & | & 5 \\ \hline 0 & -\frac{5}{2} & \frac{15}{2} & 0 & 1 & | & 75 \end{bmatrix} \quad \sim \begin{bmatrix} 1 & \frac{1}{2} & \frac{1}{2} & 0 & 0 & | & 5 \\ 0 & \boxed{1} & -\frac{1}{5} & \frac{2}{5} & 0 & | & 2 \\ \hline 0 & -\frac{5}{2} & \frac{15}{2} & 0 & 1 & | & 75 \end{bmatrix}$$

$$\frac{2}{5}R_2 \rightarrow R_2 \qquad\qquad \left(-\frac{1}{2}\right)R_2 + R_1 \rightarrow R_1$$

$$\left(\frac{5}{2}\right)R_2 + R_3 \rightarrow R_3$$

$$\sim \begin{array}{c} \\ x_1 \\ x_2 \\ P \end{array} \begin{array}{cc} \begin{array}{ccccc} x_1 & x_2 & s_1 & s_2 & P \end{array} \\ \begin{bmatrix} 1 & 0 & \frac{3}{5} & -\frac{1}{5} & 0 & | & 4 \\ 0 & 1 & -\frac{1}{5} & \frac{2}{5} & 0 & | & 2 \\ \hline 0 & 0 & 7 & 1 & 1 & | & 80 \end{bmatrix} \end{array}$$

All elements in the last row are nonnegative. Thus, max $P = 80$ at $x_1 = 4$, $x_2 = 2$, $s_1 = 0$, $s_2 = 0$.

11. **(A)**   Introduce slack variables s_1 and s_2 to obtain:

Maximize $P = 30x_1 + x_2$
Subject to: $2x_1 + x_2 + s_1 \qquad = 10$
$\qquad\qquad x_1 + 3x_2 \qquad + s_2 = 10$
$\qquad\qquad x_1, x_2, s_1, s_2 \geq 0$

This system can be written in the initial form:

$$2x_1 + x_2 + s_1 \qquad\qquad = 10$$
$$x_1 + 3x_2 \qquad + s_2 \qquad = 10$$
$$-30x_1 - x_2 \qquad\qquad + P = 0$$

(B)   The simplex tableau for this problem is

$$\begin{array}{c} \\ \text{Exit}\quad s_1 \\ s_2 \\ P \end{array} \begin{array}{c} \overset{\text{Enter}}{\begin{array}{ccccc} x_1 & x_2 & s_1 & s_2 & P \end{array}} \\ \begin{bmatrix} \boxed{2} & 1 & 1 & 0 & 0 & | & 10 \\ 1 & 3 & 0 & 1 & 0 & | & 10 \\ \hline -30 & -1 & 0 & 0 & 1 & | & 0 \end{bmatrix} \end{array} \begin{array}{l} \frac{10}{2} = 5 \text{ (minimum)} \\ \\ \frac{10}{1} = 10 \end{array}$$

$$\underset{\substack{\uparrow \\ \text{pivot} \\ \text{column}}}{}$$

$$\begin{array}{c} \\ s_1 \\ s_2 \\ P \end{array} \begin{array}{c} \begin{array}{ccccc} x_1 & x_2 & s_1 & s_2 & P \end{array} \\ \begin{bmatrix} \boxed{2} & 1 & 1 & 0 & 0 & | & 10 \\ 1 & 3 & 0 & 1 & 0 & | & 10 \\ \hline -30 & -1 & 0 & 0 & 1 & | & 0 \end{bmatrix} \end{array} \sim \begin{bmatrix} \boxed{1} & \frac{1}{2} & \frac{1}{2} & 0 & 0 & | & 5 \\ 1 & 3 & 0 & 1 & 0 & | & 10 \\ \hline -30 & -1 & 0 & 0 & 1 & | & 0 \end{bmatrix}$$

$$\frac{1}{2}R_1 \rightarrow R_1 \qquad\qquad (-1)R_1 + R_2 \rightarrow R_2$$

$$30R_1 + R_3 \rightarrow R_3$$

$$\sim \begin{array}{c} x_1 \\ s_2 \\ P \end{array} \left[\begin{array}{ccccc|c} x_1 & x_2 & s_1 & s_2 & P & \\ 1 & \frac{1}{2} & \frac{1}{2} & 0 & 0 & 5 \\ 0 & \frac{5}{2} & -\frac{1}{2} & 1 & 0 & 5 \\ \hline 0 & 14 & 15 & 0 & 1 & 150 \end{array} \right]$$

All the elements in the last row are nonnegative. Thus, max $P = 150$ at $x_1 = 5$, $x_2 = 0$, $s_1 = 0$, $s_2 = 5$.

13. The simplex tableau for this problem is:

Enter

$$\begin{array}{c} s_1 \\ s_2 \\ \text{pivot} \to s_3 \\ \text{row} \\ \text{Exit} \end{array} \left[\begin{array}{cccccc|c} x_1 & x_2 & s_1 & s_2 & s_3 & P & \\ 2 & 1 & 1 & 0 & 0 & 0 & 10 \\ 1 & 1 & 0 & 1 & 0 & 0 & 7 \\ 1 & ② & 0 & 0 & 1 & 0 & 12 \\ \hline -30 & -40 & 0 & 0 & 0 & 1 & 0 \end{array} \right] \begin{array}{l} 10 \\ 7 \\ \frac{12}{2} = 6 \text{ (minimum)} \end{array}$$

[<u>Note</u>: The pivot elements have been circled.]

pivot column $\frac{1}{2}R_3 \to R_3$

$$\sim \left[\begin{array}{cccccc|c} 2 & 1 & 1 & 0 & 0 & 0 & 10 \\ 1 & 1 & 0 & 1 & 0 & 0 & 7 \\ \frac{1}{2} & ① & 0 & 0 & \frac{1}{2} & 0 & 6 \\ \hline -30 & -40 & 0 & 0 & 0 & 1 & 0 \end{array} \right]$$

$(-1)R_3 + R_1 \to R_1$, $(-1)R_3 + R_2 \to R_2$, and $40R_3 + R_4 \to R_4$

$$\begin{array}{c} \\ \text{pivot} \to \\ \text{row} \sim \\ \\ \end{array} \left[\begin{array}{cccccc|c} \frac{3}{2} & 0 & 1 & 0 & -\frac{1}{2} & 0 & 4 \\ ⓵ & 0 & 0 & 1 & -\frac{1}{2} & 0 & 1 \\ \frac{1}{2} & 1 & 0 & 0 & \frac{1}{2} & 0 & 6 \\ \hline -10 & 0 & 0 & 0 & 20 & 1 & 240 \end{array} \right] \begin{array}{l} \frac{4}{3/2} = \frac{8}{3} \\ \frac{1}{1/2} = 2 \text{ (minimum)} \\ \frac{6}{1/2} = 12 \end{array}$$

pivot column $2R_2 \to R_2$

$$\sim \left[\begin{array}{cccccc|c} \frac{3}{2} & 0 & 1 & 0 & -\frac{1}{2} & 0 & 4 \\ ① & 0 & 0 & 2 & -1 & 0 & 2 \\ \frac{1}{2} & 1 & 0 & 0 & \frac{1}{2} & 0 & 6 \\ \hline -10 & 0 & 0 & 0 & 20 & 1 & 240 \end{array} \right]$$

$\left(-\frac{3}{2} \right) R_2 + R_1 \to R_1$, $\left(-\frac{1}{2} \right) R_2 + R_3 \to R_3$, and $10R_2 + R_4 \to R_4$

$$\begin{array}{c} \\ s_1 \\ \sim \quad x_1 \\ x_2 \\ \\ \end{array} \begin{array}{c} \begin{array}{cccccc} x_1 & x_2 & s_1 & s_2 & s_3 & P \end{array} \\ \left[\begin{array}{cccccc|c} 0 & 0 & 1 & -3 & 1 & 0 & 1 \\ 1 & 0 & 0 & 2 & -1 & 0 & 2 \\ 0 & 1 & 0 & -1 & 1 & 0 & 5 \\ \hdashline 0 & 0 & 0 & 20 & 10 & 1 & 260 \end{array}\right] \end{array}$$

Optimal solution: max $P = 260$ at $x_1 = 2$, $x_2 = 5$, $s_1 = 1$, $s_2 = 0$, $s_3 = 0$.

15. The simplex tableau for this problem is:

$$\begin{array}{c}\\ \text{Exit} \\ \downarrow \\ \text{pivot} \rightarrow s_1 \\ \text{row} \quad \quad s_2 \\ s_3 \\ P \end{array} \begin{array}{c} \quad\quad\text{Enter} \\ \begin{array}{cccccc} x_1 & x_2 & s_1 & s_2 & s_3 & P \end{array} \\ \left[\begin{array}{cccccc|c} -2 & ① & 1 & 0 & 0 & 0 & 2 \\ -1 & 1 & 0 & 1 & 0 & 0 & 5 \\ 0 & 1 & 0 & 0 & 1 & 0 & 6 \\ \hdashline -2 & -3 & 0 & 0 & 0 & 1 & 0 \end{array}\right] \end{array} \begin{array}{l} \frac{2}{1} = 2 \text{ (minimum)} \\ \frac{5}{1} = 5 \\ \frac{6}{1} = 6 \end{array}$$

pivot column $(-1)R_1 + R_2 \rightarrow R_2$, $(-1)R_1 + R_3 \rightarrow R_3$, and $3R_1 + R_4 \rightarrow R_4$

$$\begin{array}{c}\\ \\ \text{pivot} \rightarrow \\ \sim \quad \text{row} \\ \\ \end{array} \left[\begin{array}{cccccc|c} -2 & 1 & 1 & 0 & 0 & 0 & 2 \\ 1 & 0 & -1 & 1 & 0 & 0 & 3 \\ ② & 0 & -1 & 0 & 1 & 0 & 4 \\ \hdashline -8 & 0 & 3 & 0 & 0 & 1 & 6 \end{array}\right] \begin{array}{l} \\ \frac{3}{1} = 3 \\ \frac{4}{2} = 2 \text{ (minimum)} \end{array}$$

pivot column $\frac{1}{2}R_3 \rightarrow R_3$

$$\sim \left[\begin{array}{cccccc|c} -2 & 1 & 1 & 0 & 0 & 0 & 2 \\ 1 & 0 & -1 & 1 & 0 & 0 & 3 \\ ① & 0 & -\frac{1}{2} & 0 & \frac{1}{2} & 0 & 2 \\ \hdashline -8 & 0 & 3 & 0 & 0 & 1 & 6 \end{array}\right] \sim \begin{array}{c} \begin{array}{cccccc} x_1 & x_2 & s_1 & s_2 & s_3 & P \end{array} \\ \left[\begin{array}{cccccc|c} 0 & 1 & 0 & 0 & 1 & 0 & 6 \\ 0 & 0 & -\frac{1}{2} & 1 & -\frac{1}{2} & 0 & 1 \\ 1 & 0 & -\frac{1}{2} & 0 & \frac{1}{2} & 0 & 2 \\ \hdashline 0 & 0 & -1 & 0 & 4 & 1 & 22 \end{array}\right] \end{array}$$

$2R_3 + R_1 \rightarrow R_1$, $(-1)R_3 + R_2 \rightarrow R_2$, and $8R_3 + R_4 \rightarrow R_4$

pivot column

Since there are no positive elements in the pivot column (above the dashed line), we conclude that there is no optimal solution.

17. The simplex tableau for this problem is:

$$\begin{array}{c} \\ \text{pivot} \rightarrow s_1 \\ \text{row} \\ s_2 \\ \\ s_3 \\ \\ P \end{array} \begin{array}{cccccc} x_1 & x_2 & s_1 & s_2 & s_3 & P \\ \left[\begin{array}{cccccc|c} -1 & \textcircled{1} & 1 & 0 & 0 & 0 & 2 \\ -1 & 3 & 0 & 1 & 0 & 0 & 12 \\ 1 & -4 & 0 & 0 & 1 & 0 & 4 \\ \hline 1 & -2 & 0 & 0 & 0 & 1 & 0 \end{array}\right] \end{array} \begin{array}{l} \frac{2}{1} = 2 \text{ (minimum)} \\ \\ \frac{12}{3} = 4 \end{array}$$

$\uparrow$
pivot column

pivot $(-3)R_1 + R_2 \rightarrow R_2$, $4R_1 + R_3 \rightarrow R_3$, and $2R_1 + R_4 \rightarrow R_4$

$$\begin{array}{c} \\ \text{pivot} \rightarrow \\ \text{row} \\ \sim \\ \\ \\ \end{array} \left[\begin{array}{cccccc|c} -1 & 1 & 1 & 0 & 0 & 0 & 2 \\ \textcircled{2} & 0 & -3 & 1 & 0 & 0 & 6 \\ -3 & 0 & 4 & 0 & 1 & 0 & 12 \\ \hline -1 & 0 & 2 & 0 & 0 & 1 & 4 \end{array}\right] \begin{array}{l} \\ \frac{6}{2} = 3 \leftarrow \text{pivot row} \\ [\underline{\text{Note}}: \text{We only use the} \\ \textit{positive} \text{ elements above} \\ \text{the dashed line in the} \\ \text{pivot column.}] \end{array}$$

$\uparrow$
pivot column $\frac{1}{2}R_2 \rightarrow R_2$

$$\sim \left[\begin{array}{cccccc|c} -1 & 1 & 1 & 0 & 0 & 0 & 2 \\ \textcircled{1} & 0 & -\frac{3}{2} & \frac{1}{2} & 0 & 0 & 3 \\ -3 & 0 & 4 & 0 & 1 & 0 & 12 \\ \hline -1 & 0 & 2 & 0 & 0 & 1 & 4 \end{array}\right]$$

$R_2 + R_1 \rightarrow R_1$, $3R_2 + R_3 \rightarrow R_3$, and $R_2 + R_4 \rightarrow R_4$

$$\begin{array}{c} x_2 \\ x_1 \\ s_3 \\ \\ \end{array} \sim \begin{array}{cccccc} x_1 & x_2 & s_1 & s_2 & s_3 & P \\ \left[\begin{array}{cccccc|c} 0 & 1 & -\frac{1}{2} & \frac{1}{2} & 0 & 0 & 5 \\ 1 & 0 & -\frac{3}{2} & \frac{1}{2} & 0 & 0 & 3 \\ 0 & 0 & -\frac{1}{2} & \frac{3}{2} & 1 & 0 & 21 \\ \hline 0 & 0 & \frac{1}{2} & \frac{1}{2} & 0 & 1 & 7 \end{array}\right] \end{array}$$

Optimal solution: max $P = 7$ at $x_1 = 3$, $x_2 = 5$, $s_1 = 0$, $s_2 = 0$, $s_3 = 21$.

19. The simplex tableau for this problem is:

$$\begin{array}{c} \\ \text{pivot} \rightarrow s_1 \\ \text{row} \\ s_2 \\ \\ P \end{array} \begin{array}{cccccc} x_1 & x_2 & x_3 & s_1 & s_2 & P \\ \left[\begin{array}{cccccc|c} \textcircled{1} & 1 & -1 & 1 & 0 & 0 & 10 \\ 2 & 4 & 3 & 0 & 1 & 0 & 30 \\ \hline -5 & -2 & 1 & 0 & 0 & 1 & 0 \end{array}\right] \end{array} \begin{array}{l} \frac{10}{1} = 10 \text{ (minimum)} \\ \\ \frac{30}{2} = 15 \end{array}$$

$\uparrow$
pivot column $(-2)R_1 + R_2 \rightarrow R_2$, $5R_1 + R_3 \rightarrow R_3$

$$\sim \left[\begin{array}{cccccc|c} 1 & 1 & -1 & 1 & 0 & 0 & 10 \\ 0 & 2 & \textcircled{5} & -2 & 1 & 0 & 10 \\ \hline 0 & 3 & -4 & 5 & 0 & 1 & 50 \end{array}\right] \sim \left[\begin{array}{cccccc|c} 1 & 1 & -1 & 1 & 0 & 0 & 10 \\ 0 & \frac{2}{5} & \textcircled{1} & -\frac{2}{5} & \frac{1}{5} & 0 & 2 \\ \hline 0 & 3 & -4 & 5 & 0 & 1 & 50 \end{array}\right]$$

$\frac{1}{5}R_2 \rightarrow R_2$ $R_2 + R_1 \rightarrow R_1$, $4R_2 + R_3 \rightarrow R_3$

$$\sim \begin{array}{c} \\ x_1 \\ x_3 \\ P \end{array} \begin{array}{cccccc} x_1 & x_2 & x_3 & s_1 & s_2 & P \\ \begin{bmatrix} 1 & \frac{7}{5} & 0 & \frac{3}{5} & \frac{1}{5} & 0 \\ 0 & \frac{2}{5} & 1 & -\frac{2}{5} & \frac{1}{5} & 0 \\ \hdashline 0 & \frac{23}{5} & 0 & \frac{17}{5} & \frac{4}{5} & 1 \end{bmatrix} & \begin{matrix} 12 \\ 2 \\ \\ 58 \end{matrix} \end{array}$$

Optimal solution: max $P = 58$
$x_1 = 12$, $x_2 = 0$, $x_3 = 2$, $s_1 = s_2 = 0$.

21. The simplex tableau for this problem is:

$$\begin{array}{c} \\ \\ s_1 \\ \text{pivot} \rightarrow s_2 \\ \text{row} \\ P \end{array} \begin{array}{cccccc} x_1 & x_2 & x_3 & s_1 & s_2 & P \\ \begin{bmatrix} 1 & 0 & 1 & 1 & 0 & 0 \\ 0 & 1 & ① & 0 & 1 & 0 \\ \hdashline -2 & -3 & -4 & 0 & 0 & 1 \end{bmatrix} & \begin{matrix} 4 \\ 3 \\ \\ 0 \end{matrix} \end{array} \quad \begin{matrix} \frac{4}{1} = 4 \\ \frac{3}{1} = 3 \text{ (minimum)} \end{matrix}$$

$$\underset{\substack{\text{pivot} \\ \text{column}}}{\uparrow} \quad (-1)R_2 + R_1 \rightarrow R_1, \ 4R_2 + R_3 \rightarrow R_3$$

$$\sim \begin{bmatrix} ① & -1 & 0 & 1 & -1 & 0 \\ 0 & 1 & 1 & 0 & 1 & 0 \\ \hdashline -2 & 1 & 0 & 0 & 4 & 1 \end{bmatrix} \begin{matrix} 1 \\ 3 \\ \\ 12 \end{matrix} \quad \sim \begin{bmatrix} 1 & -1 & 0 & 1 & -1 & 0 \\ 0 & ① & 1 & 0 & 1 & 0 \\ \hdashline 0 & -1 & 0 & 2 & 2 & 1 \end{bmatrix} \begin{matrix} 1 \\ 3 \\ \\ 14 \end{matrix}$$

$$2R_1 + R_3 \rightarrow R_3 \qquad\qquad\qquad R_2 + R_1 \rightarrow R_1 \text{ and } R_2 + R_3 \rightarrow R_3$$

$$\sim \begin{array}{cccccc} x_1 & x_2 & x_3 & s_1 & s_2 & P \\ \begin{bmatrix} 1 & 0 & 1 & 1 & 0 & 0 \\ 0 & 1 & 1 & 0 & 1 & 0 \\ \hdashline 0 & 0 & 1 & 2 & 3 & 1 \end{bmatrix} & \begin{matrix} 4 \\ 3 \\ \\ 17 \end{matrix} \end{array}$$

Optimal solution: max $P = 17$ at
$x_1 = 4$, $x_2 = 3$, $x_3 = 0$, $s_1 = 0$, $s_2 = 0$.

23. The simplex tableau for this problem is:

$$\begin{array}{c} \\ \\ s_1 \\ \text{pivot} \rightarrow s_2 \\ \text{row} \\ s_3 \\ \\ \end{array} \begin{array}{ccccccc} x_1 & x_2 & x_3 & s_1 & s_2 & s_3 & P \\ \begin{bmatrix} 3 & 2 & 5 & 1 & 0 & 0 & 0 \\ ② & 1 & 1 & 0 & 1 & 0 & 0 \\ 1 & 1 & 2 & 0 & 0 & 1 & 0 \\ \hdashline -4 & -3 & -2 & 0 & 0 & 0 & 1 \end{bmatrix} & \begin{matrix} 23 \\ 8 \\ 7 \\ \\ 0 \end{matrix} \end{array} \quad \begin{matrix} \frac{23}{3} = 7\frac{2}{3} \\ \frac{8}{2} = 4 \text{ (minimum)} \\ \frac{7}{1} = 7 \end{matrix}$$

$$\underset{\substack{\text{pivot} \\ \text{column}}}{\uparrow} \quad \frac{1}{2}R_2 \rightarrow R_2$$

$$\sim \begin{bmatrix} 3 & 2 & 5 & 1 & 0 & 0 & 0 & | & 23 \\ \boxed{1} & \frac{1}{2} & \frac{1}{2} & 0 & \frac{1}{2} & 0 & 0 & | & 4 \\ 1 & 1 & 2 & 0 & 0 & 1 & 0 & | & 7 \\ \hline -4 & -3 & -2 & 0 & 0 & 0 & 1 & | & 0 \end{bmatrix} \sim \begin{bmatrix} 0 & \frac{1}{2} & \frac{7}{2} & 1 & -\frac{3}{2} & 0 & 0 & | & 11 \\ 1 & \frac{1}{2} & \frac{1}{2} & 0 & \frac{1}{2} & 0 & 0 & | & 4 \\ 0 & \boxed{\frac{1}{2}} & \frac{3}{2} & 0 & -\frac{1}{2} & 1 & 0 & | & 3 \\ \hline 0 & -1 & 0 & 0 & 2 & 0 & 1 & | & 16 \end{bmatrix}$$

$(-3)R_2 + R_1 \rightarrow R_1,\ (-1)\ R_2 + R_3 \rightarrow R_3,\qquad 2R_3 \rightarrow R_3$

$4R_2 + R_4 \rightarrow R_4$

$$\sim \begin{bmatrix} 0 & \frac{1}{2} & \frac{7}{2} & 1 & -\frac{3}{2} & 0 & 0 & | & 11 \\ 1 & \frac{1}{2} & \frac{1}{2} & 0 & \frac{1}{2} & 0 & 0 & | & 4 \\ 0 & \boxed{1} & 3 & 0 & -1 & 2 & 0 & | & 6 \\ \hline 0 & -1 & 0 & 0 & 2 & 0 & 1 & | & 16 \end{bmatrix} \sim$$

$\left(-\dfrac{1}{2}\right)R_3 + R_1 \rightarrow R_1,\ \left(-\dfrac{1}{2}\right)R_3 + R_2 \rightarrow R_2,\ \text{and}$

$R_3 + R_4 \rightarrow R_4$

$$\begin{array}{c} \\ s_1 \\ x_1 \\ x_2 \\ P \end{array} \begin{array}{c} \begin{matrix} x_1 & x_2 & x_3 & s_1 & s_2 & s_3 & P \end{matrix} \\ \begin{bmatrix} 0 & 0 & 2 & 1 & -1 & -1 & 0 & | & 8 \\ 1 & 0 & -1 & 0 & 1 & -1 & 0 & | & 1 \\ 0 & 1 & 3 & 0 & -1 & 2 & 0 & | & 6 \\ \hline 0 & 0 & 3 & 0 & 1 & 2 & 1 & | & 22 \end{bmatrix} \end{array}$$

Optimal solution: max $P = 22$ at $x_1 = 1$, $x_2 = 6$, $x_3 = 0$, $s_1 = 8$, $s_2 = 0$, $s_3 = 0$.

25. Multiply the first problem constraint by $\dfrac{10}{6}$, the second by 100, and the third by 10 to clear the fractions.

Then, the simplex tableau for this problem is:

$$\begin{array}{c} \\ s_1 \\ s_2 \\ s_3 \\ P \end{array} \begin{array}{c} \begin{matrix} x_1 & x_2 & s_1 & s_2 & s_3 & P \end{matrix} \\ \begin{bmatrix} 1 & \boxed{2} & 1 & 0 & 0 & 0 & | & 1{,}600 \\ 3 & 4 & 0 & 1 & 0 & 0 & | & 3{,}600 \\ 3 & 2 & 0 & 0 & 1 & 0 & | & 2{,}700 \\ \hline -20 & -30 & 0 & 0 & 0 & 1 & | & 0 \end{bmatrix} \end{array} \begin{array}{l} \dfrac{1{,}600}{2} = 800 \\[8pt] \dfrac{3{,}600}{4} = 900 \\[8pt] \dfrac{2{,}700}{2} = 1{,}350 \end{array}$$

$\dfrac{1}{2}R_1 \rightarrow R_1$

$$\sim \begin{bmatrix} \frac{1}{2} & ① & \frac{1}{2} & 0 & 0 & 0 & | & 800 \\ 3 & 4 & 0 & 1 & 0 & 0 & | & 3,600 \\ 3 & 2 & 0 & 0 & 1 & 0 & | & 2,700 \\ \hline -20 & -30 & 0 & 0 & 0 & 1 & | & 0 \end{bmatrix}$$

$(-4)R_1 + R_2 \to R_2,\ (-2)R_1 + R_3 \to R_3,$ and $30R_1 + R_4 \to R_4$

$$\sim \begin{bmatrix} \frac{1}{2} & 1 & \frac{1}{2} & 0 & 0 & 0 & | & 800 \\ ① & 0 & -2 & 1 & 0 & 0 & | & 400 \\ 2 & 0 & -1 & 0 & 1 & 0 & | & 1,100 \\ \hline -5 & 0 & 15 & 0 & 0 & 1 & | & 24,000 \end{bmatrix} \quad \begin{array}{l} \frac{800}{1/2} = 1,600 \\ \frac{400}{1} = 400 \\ \frac{1,100}{2} = 550 \end{array}$$

$\left(-\frac{1}{2}\right)R_2 + R_1 \to R_1,\ (-2)R_2 + R_3 \to R_3,$ and $5R_2 + R_4 \to R_4$

$$\sim \begin{array}{c} \\ x_2 \\ x_1 \\ s_3 \\ P \end{array} \begin{array}{c} \begin{array}{cccccc} x_1 & x_2 & s_1 & s_2 & s_3 & P \end{array} \\ \begin{bmatrix} 0 & 1 & \frac{3}{2} & -\frac{1}{2} & 0 & 0 & | & 600 \\ 1 & 0 & -2 & 1 & 0 & 0 & | & 400 \\ 0 & 0 & 3 & -2 & 1 & 0 & | & 300 \\ \hline 0 & 0 & 5 & 5 & 0 & 1 & | & 26,000 \end{bmatrix} \end{array}$$

Optimal solution: max $P = 26,000$ at $x_1 = 400$, $x_2 = 600$, $s_1 = 0$, $s_2 = 0$, $s_3 = 300$.

27. The simplex tableau for this problem is:

$$\begin{array}{c} \\ s_1 \\ s_2 \\ s_3 \\ P \end{array} \begin{array}{c} \begin{array}{ccccccc} x_1 & x_2 & x_3 & s_1 & s_2 & s_3 & P \end{array} \\ \begin{bmatrix} 2 & 2 & ⑧ & 1 & 0 & 0 & 0 & | & 600 \\ 1 & 3 & 2 & 0 & 1 & 0 & 0 & | & 600 \\ 3 & 2 & 1 & 0 & 0 & 1 & 0 & | & 400 \\ \hline -1 & -2 & -3 & 0 & 0 & 0 & 1 & | & 0 \end{bmatrix} \end{array} \quad \begin{array}{l} \frac{600}{8} = 75 \\ \frac{600}{2} = 300 \\ \frac{400}{1} = 400 \end{array}$$

$\frac{1}{8}R_1 \to R_1$

$$\sim \begin{bmatrix} \frac{1}{4} & \frac{1}{4} & ① & \frac{1}{8} & 0 & 0 & 0 & | & 75 \\ 1 & 3 & 2 & 0 & 1 & 0 & 0 & | & 600 \\ 3 & 2 & 1 & 0 & 0 & 1 & 0 & | & 400 \\ \hline -1 & -2 & -3 & 0 & 0 & 0 & 1 & | & 0 \end{bmatrix}$$

$(-2)R_1 + R_2 \to R_2,\ (-1)R_1 + R_3 \to R_3,$ and $3R_1 + R_4 \to R_4$

$$\sim \begin{bmatrix} \frac{1}{4} & \frac{1}{4} & 1 & \frac{1}{8} & 0 & 0 & 0 & 75 \\ \frac{1}{2} & \boxed{\tfrac{5}{2}} & 0 & -\frac{1}{4} & 1 & 0 & 0 & 450 \\ \frac{11}{4} & \frac{7}{4} & 0 & -\frac{1}{8} & 0 & 1 & 0 & 325 \\ \hdashline -\frac{1}{4} & -\frac{5}{4} & 0 & \frac{3}{8} & 0 & 0 & 1 & 225 \end{bmatrix} \begin{matrix} \frac{75}{1/4} = 300 \\[4pt] \frac{450}{5/2} = 180 \\[4pt] \frac{325}{7/4} = 185.71 \\[4pt] \\ \end{matrix}$$

$$\frac{2}{5} R_2 \rightarrow R_2$$

$$\sim \begin{bmatrix} \frac{1}{4} & \frac{1}{4} & 1 & \frac{1}{8} & 0 & 0 & 0 & 75 \\ \frac{1}{5} & \boxed{1} & 0 & -\frac{1}{10} & \frac{2}{5} & 0 & 0 & 180 \\ \frac{11}{4} & \frac{7}{4} & 0 & -\frac{1}{8} & 0 & 1 & 0 & 325 \\ \hdashline -\frac{1}{4} & -\frac{5}{4} & 0 & \frac{3}{8} & 0 & 0 & 1 & 225 \end{bmatrix}$$

$$\left(-\frac{1}{4}\right) R_2 + R_1 \rightarrow R_1, \ \left(-\frac{7}{4}\right) R_2 + R_3 \rightarrow R_3, \ \text{and} \ \frac{5}{4} R_2 + R_4 \rightarrow R_4$$

$$\sim \begin{array}{c} \\ x_3 \\ x_2 \\ s_3 \\ P \end{array} \begin{array}{cccccccc} x_1 & x_2 & x_3 & s_1 & s_2 & s_3 & P & \\ \begin{bmatrix} \frac{1}{5} & 0 & 1 & \frac{3}{20} & -\frac{1}{10} & 0 & 0 & 30 \\ \frac{1}{5} & 1 & 0 & -\frac{1}{10} & \frac{2}{5} & 0 & 0 & 180 \\ \frac{12}{5} & 0 & 0 & \frac{1}{20} & -\frac{7}{10} & 1 & 0 & 10 \\ \hdashline 0 & 0 & 0 & \frac{1}{4} & \frac{1}{2} & 0 & 1 & 450 \end{bmatrix} \end{array}$$

Optimal solution: max $P = 450$ at $x_1 = 0$, $x_2 = 180$, $x_3 = 30$, $s_1 = 0$, $s_2 = 0$, $s_3 = 10$.

29. The simplex tableau for this problem is:

$$\begin{array}{c} \\ s_1 \\ s_2 \\ s_3 \\ s_4 \\ P \end{array} \begin{array}{cccccccc} x_1 & x_2 & s_1 & s_2 & s_3 & s_4 & P & \\ \begin{bmatrix} 1 & 2 & 1 & 0 & 0 & 0 & 0 & 40 \\ 1 & 3 & 0 & 1 & 0 & 0 & 0 & 48 \\ 1 & 4 & 0 & 0 & 1 & 0 & 0 & 60 \\ 0 & \boxed{1} & 0 & 0 & 0 & 1 & 0 & 14 \\ \hdashline -2 & -5 & 0 & 0 & 0 & 0 & 1 & 0 \end{bmatrix} \end{array} \begin{matrix} \frac{40}{2} = 20 \\[4pt] \frac{48}{3} = 16 \\[4pt] \frac{60}{4} = 15 \\[4pt] \frac{14}{1} = 14 \\[4pt] \\ \end{matrix}$$

$$(-2) R_4 + R_1 \rightarrow R_1, \ (-3) R_4 + R_2 \rightarrow R_2, \ (-4) R_4 + R_3 \rightarrow R_3,$$
$$\text{and } 5 R_4 + R_5 \rightarrow R_5$$

$$\sim \begin{bmatrix} 1 & 0 & 1 & 0 & 0 & -2 & 0 & | & 12 \\ 1 & 0 & 0 & 1 & 0 & -3 & 0 & | & 6 \\ ① & 0 & 0 & 0 & 1 & -4 & 0 & | & 4 \\ 0 & 1 & 0 & 0 & 0 & 1 & 0 & | & 14 \\ \hdashline -2 & 0 & 0 & 0 & 0 & 5 & 1 & | & 70 \end{bmatrix} \begin{matrix} \frac{12}{1} = 12 \\ \frac{6}{1} = 6 \\ \frac{4}{1} = 4 \\ \\ \\ \end{matrix}$$

$(-1)R_3 + R_1 \rightarrow R_1$, $(-1)R_3 + R_2 \rightarrow R_2$, and $2R_3 + R_5 \rightarrow R_5$

$$\sim \begin{bmatrix} 0 & 0 & 1 & 0 & -1 & 2 & 0 & | & 8 \\ 0 & 0 & 0 & 1 & -1 & ① & 0 & | & 2 \\ 1 & 0 & 0 & 0 & 1 & -4 & 0 & | & 4 \\ 0 & 1 & 0 & 0 & 0 & 1 & 0 & | & 14 \\ \hdashline 0 & 0 & 0 & 0 & 2 & -3 & 1 & | & 78 \end{bmatrix} \begin{matrix} \frac{8}{2} = 4 \\ \frac{2}{1} = 2 \\ \\ \frac{14}{1} = 14 \\ \end{matrix}$$

$(-2)R_2 + R_1 \rightarrow R_1$, $4R_2 + R_3 \rightarrow R_3$, $(-1)R_2 + R_4 \rightarrow R_4$,
and $3R_2 + R_5 \rightarrow R_5$

$$\sim \begin{bmatrix} 0 & 0 & 1 & -2 & ① & 0 & 0 & | & 4 \\ 0 & 0 & 0 & 1 & -1 & 1 & 0 & | & 2 \\ 1 & 0 & 0 & 4 & -3 & 0 & 0 & | & 12 \\ 0 & 1 & 0 & -1 & 1 & 0 & 0 & | & 12 \\ \hdashline 0 & 0 & 0 & 3 & -1 & 0 & 1 & | & 84 \end{bmatrix} \begin{matrix} \frac{4}{1} = 4 \\ \\ \\ \frac{12}{1} = 12 \\ \end{matrix}$$

$R_1 + R_2 \rightarrow R_2$, $3R_1 + R_3 \rightarrow R_3$, $(-1)R_1 + R_4 \rightarrow R_4$, and $R_1 + R_5 \rightarrow R_5$

$$\sim \begin{matrix} \\ s_3 \\ s_4 \\ x_1 \\ x_2 \\ \\ P \end{matrix} \begin{bmatrix} x_1 & x_2 & s_1 & s_2 & s_3 & s_4 & P & \\ 0 & 0 & 1 & -2 & 1 & 0 & 0 & | & 4 \\ 0 & 0 & 1 & -1 & 0 & 1 & 0 & | & 6 \\ 1 & 0 & 3 & -2 & 0 & 0 & 0 & | & 24 \\ 0 & 1 & -1 & 1 & 0 & 0 & 0 & | & 8 \\ \hdashline 0 & 0 & 1 & 1 & 0 & 0 & 1 & | & 88 \end{bmatrix}$$

Optimal solution: max $P = 88$ at $x_1 = 24$, $x_2 = 8$, $s_1 = 0$, $s_2 = 0$, $s_3 = 4$, $s_4 = 6$.

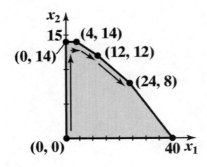

31. Simplex Method:

The simplex tableau for this problem is:

$$
\begin{array}{c}
 \\
s_1 \\
s_2
\end{array}
\begin{array}{cccccc}
x_1 & x_2 & s_1 & s_2 & P & \\
\end{array}
\left[
\begin{array}{ccccc|c}
-2 & \boxed{1} & 1 & 0 & 0 & 4 \\
0 & 1 & 0 & 1 & 0 & 10 \\
\hline
-2 & -3 & 0 & 0 & 1 & 0
\end{array}
\right]
\begin{array}{l}
\frac{4}{1} = 4 \\
\frac{10}{1} = 10
\end{array}
$$

$$(-1)R_1 + R_2 \to R_2, \quad 3R_1 + R_3 \to R_3$$

$$
\sim
\left[
\begin{array}{ccccc|c}
-2 & 1 & 1 & 0 & 0 & 4 \\
\boxed{2} & 0 & -1 & 1 & 0 & 6 \\
\hline
-8 & 0 & 3 & 0 & 1 & 12
\end{array}
\right]
\quad \frac{6}{2} = 3
$$

$$\frac{1}{2}R_2 \to R_2$$

$$
\sim
\left[
\begin{array}{ccccc|c}
-2 & 1 & 0 & 0 & 0 & 4 \\
\boxed{1} & 0 & -\frac{1}{2} & \frac{1}{2} & 0 & 3 \\
\hline
-8 & 0 & 3 & 0 & 1 & 12
\end{array}
\right]
$$

$$2R_2 + R_1 \to R_1, \quad 8R_2 + R_3 \to R_3$$

$$
\sim
\left[
\begin{array}{ccccc|c}
0 & 1 & 0 & 1 & 0 & 10 \\
1 & 0 & -\frac{1}{2} & \frac{1}{2} & 0 & 3 \\
\hline
0 & 0 & -1 & 4 & 1 & 36
\end{array}
\right]
$$

No positive elements in the pivot column; no optimal solution exists.

Geometric Method:

<u>Step (1)</u>: Graph the feasible region and find the corner points. The feasible region S is the solution set of the inequalities. This region is indicated by the shading in the graph at the right.

The corner points are $(0, 0)$, $(0, 4)$, and $(3, 10)$.

Since S is unbounded and the coefficients of the objective function are positive, P does not have a maximum value.

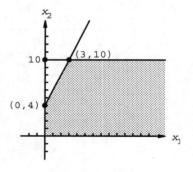

33. The simplex tableau for this problem is:

$$
\begin{array}{c}
s_1 \\
\\
s_2 \\
\\
s_3 \\
P
\end{array}
\left[
\begin{array}{cccccc|c}
x_1 & x_2 & s_1 & s_2 & s_3 & P & \\
2 & 1 & 1 & 0 & 0 & 0 & 16 \\
1 & 0 & 0 & 1 & 0 & 0 & 6 \\
0 & 1 & 0 & 0 & 1 & 0 & 10 \\
\hline
-1 & -1 & 0 & 0 & 0 & 1 & 0
\end{array}
\right]
$$

(A)   Solution using the first column as the pivot column

$$
\begin{array}{cccccc|c}
x_1 & x_2 & s_1 & s_2 & s_3 & P & \\
2 & 1 & 1 & 0 & 0 & 0 & 16 \\
① & 0 & 0 & 1 & 0 & 0 & 6 \\
0 & 1 & 0 & 0 & 1 & 0 & 10 \\
\hline
-1 & -1 & 0 & 0 & 0 & 1 & 0
\end{array}
\quad
\begin{array}{l}
\dfrac{16}{2} = 8 \\[4pt]
\dfrac{6}{1} = 6
\end{array}
$$

$$(-2)R_2 + R_1 \rightarrow R_1, \quad R_2 + R_4 \rightarrow R_4$$

$$
\sim
\begin{array}{cccccc|c}
0 & ① & 1 & -2 & 0 & 0 & 4 \\
1 & 0 & 0 & 1 & 0 & 0 & 6 \\
0 & 1 & 0 & 0 & 1 & 0 & 10 \\
\hline
0 & -1 & 0 & 1 & 0 & 1 & 6
\end{array}
\quad
\begin{array}{l}
\dfrac{4}{1} = 4 \\[8pt]
\dfrac{10}{1} = 10
\end{array}
$$

$$(-1)R_1 + R_3 \rightarrow R_3, \quad R_1 + R_4 \rightarrow R_4$$

$$
\sim
\begin{array}{cccccc|c}
0 & 1 & 1 & -2 & 0 & 0 & 4 \\
1 & 0 & 0 & 1 & 0 & 0 & 6 \\
0 & 0 & -1 & ② & 1 & 0 & 6 \\
\hline
0 & 0 & 1 & -1 & 0 & 1 & 10
\end{array}
\quad
\begin{array}{l}
\dfrac{6}{1} = 6 \\[4pt]
\dfrac{6}{2} = 3
\end{array}
$$

$$\dfrac{1}{2}R_3 \rightarrow R_3$$

$$
\sim
\begin{array}{cccccc|c}
0 & 1 & 1 & -2 & 0 & 0 & 4 \\
1 & 0 & 0 & 1 & 0 & 0 & 6 \\
0 & 0 & -\frac{1}{2} & 1 & \frac{1}{2} & 0 & 3 \\
\hline
0 & 0 & 1 & -1 & 0 & 1 & 10
\end{array}
$$

$$2R_3 + R_1 \rightarrow R_1, \quad (-1)R_3 + R_2 \rightarrow R_2, \quad R_3 + R_4 \rightarrow R_4$$

$$
\sim
\begin{array}{l}
\\ x_2 \\ x_1 \\ s_2 \\ P
\end{array}
\begin{array}{cccccc|c}
x_1 & x_2 & s_1 & s_2 & s_3 & P & \\
0 & 1 & 0 & 0 & 1 & 0 & 10 \\
1 & 0 & \frac{1}{2} & 0 & -\frac{1}{2} & 0 & 3 \\
0 & 0 & -\frac{1}{2} & 1 & \frac{1}{2} & 0 & 3 \\
\hline
0 & 0 & \frac{1}{2} & 0 & \frac{1}{2} & 1 & 13
\end{array}
$$

Optimal solution: max $P = 13$ at $x_1 = 3$, $x_2 = 10$, $s_1 = 0$, $s_2 = 3$, $s_3 = 0$

(B)   Solution using the second column as the pivot column

$$
\begin{array}{c}
\begin{array}{ccccccc} x_1 & x_2 & s_1 & s_2 & s_3 & P & \end{array}\\
\begin{array}{c} s_1 \\ s_2 \\ s_3 \\ P \end{array}
\left[
\begin{array}{cccccc|c}
2 & 1 & 1 & 0 & 0 & 0 & 16 \\
1 & 0 & 0 & 1 & 0 & 0 & 6 \\
0 & ① & 0 & 0 & 1 & 0 & 10 \\
\hline
-1 & -1 & 0 & 0 & 0 & 1 & 0
\end{array}
\right]
\begin{array}{l} \frac{16}{1} = 16 \\[6pt] \\[6pt] \frac{10}{1} = 10 \end{array}
\end{array}
$$

$$(-1)R_3 + R_1 \to R_1, \quad R_3 + R_4 \to R_4$$

$$
\sim
\left[
\begin{array}{cccccc|c}
② & 0 & 1 & 0 & -1 & 0 & 6 \\
1 & 0 & 0 & 1 & 0 & 0 & 6 \\
0 & 1 & 0 & 0 & 1 & 0 & 10 \\
\hline
-1 & 0 & 0 & 0 & 1 & 1 & 10
\end{array}
\right]
\begin{array}{l} \frac{6}{2} = 3 \\[6pt] \frac{6}{1} = 6 \end{array}
$$

$$\tfrac{1}{2}R_1 \to R_1$$

$$
\sim
\left[
\begin{array}{cccccc|c}
① & 0 & \frac{1}{2} & 0 & -\frac{1}{2} & 0 & 3 \\
1 & 0 & 0 & 1 & 0 & 0 & 6 \\
0 & 1 & 0 & 0 & 1 & 0 & 10 \\
\hline
-1 & 0 & 0 & 0 & 1 & 1 & 10
\end{array}
\right]
$$

$$(-1)R_1 + R_2 \to R_2, \quad R_1 + R_4 \to R_4$$

$$
\begin{array}{c}
\begin{array}{ccccccc} x_1 & x_2 & s_1 & s_2 & s_3 & P & \end{array}\\
\begin{array}{c} x_1 \\ s_2 \\ x_2 \\ P \end{array}
\sim
\left[
\begin{array}{cccccc|c}
1 & 0 & \frac{1}{2} & 0 & -\frac{1}{2} & 0 & 3 \\
0 & 0 & -\frac{1}{2} & 1 & \frac{1}{2} & 0 & 3 \\
0 & ① & 0 & 0 & 1 & 0 & 10 \\
\hline
0 & 0 & \frac{1}{2} & 0 & \frac{1}{2} & 1 & 13
\end{array}
\right]
\end{array}
$$

```
Choosing either solution produces
the same optimal solution.
```

```
Optimal solution: max P = 13 at x₁ = 3, x₂ = 10, s₁ = 0,
s₂ = 3, s₃ = 0
```

Optimal solution: max $P = 13$ at $x_1 = 3$, $x_2 = 10$, $s_1 = 0$, $s_2 = 3$, $s_3 = 0$

35.   The simplex tableau for this problem is:

$$
\begin{array}{c}
\begin{array}{ccccccc} x_1 & x_2 & x_3 & s_1 & s_2 & P & \end{array}\\
\begin{array}{c} s_1 \\ s_2 \\ P \end{array}
\left[
\begin{array}{cccccc|c}
1 & 1 & 2 & 1 & 0 & 0 & 20 \\
2 & 1 & 4 & 0 & 1 & 0 & 32 \\
\hline
-3 & -3 & -2 & 0 & 0 & 1 & 0
\end{array}
\right]
\end{array}
$$

(A)   Solution using the first column as the pivot column

$$\begin{array}{ccccccc} x_1 & x_2 & x_3 & s_1 & s_2 & P & \end{array}$$

$$\left[\begin{array}{cccccc|c} 1 & 1 & 2 & 1 & 0 & 0 & 20 \\ \textcircled{2} & 1 & 4 & 0 & 1 & 0 & 32 \\ \hdashline -3 & -3 & -2 & 0 & 0 & 1 & 0 \end{array}\right] \begin{array}{l} \frac{20}{1} = 20 \\[4pt] \frac{32}{2} = 16 \end{array}$$

$$\tfrac{1}{2}R_2 \;\to\; R_2$$

$$\sim \left[\begin{array}{cccccc|c} 1 & 1 & 2 & 1 & 0 & 0 & 20 \\ 1 & \tfrac{1}{2} & 2 & 0 & \tfrac{1}{2} & 0 & 16 \\ \hdashline -3 & -3 & -2 & 0 & 0 & 1 & 0 \end{array}\right]$$

$$(-1)R_2 + R_1 \;\to\; R_1, \quad 3R_2 + R_3 \;\to\; R_3$$

$$\sim \left[\begin{array}{cccccc|c} 0 & \textcircled{$\tfrac{1}{2}$} & 0 & 1 & -\tfrac{1}{2} & 0 & 4 \\ 1 & \tfrac{1}{2} & 2 & 0 & \tfrac{1}{2} & 0 & 16 \\ \hdashline 0 & -\tfrac{3}{2} & 4 & 0 & \tfrac{3}{2} & 1 & 48 \end{array}\right] \begin{array}{l} \frac{4}{1/2} = 8 \\[4pt] \frac{16}{1/2} = 32 \end{array}$$

$$2R_1 \;\to\; R_1$$

$$\sim \left[\begin{array}{cccccc|c} 0 & 1 & 0 & 2 & -1 & 0 & 8 \\ 1 & \tfrac{1}{2} & 2 & 0 & \tfrac{1}{2} & 0 & 16 \\ \hdashline 0 & -\tfrac{3}{2} & 4 & 0 & \tfrac{3}{2} & 1 & 48 \end{array}\right]$$

$$\left(-\tfrac{1}{2}\right)R_1 + R_2 \;\to\; R_2, \quad \tfrac{3}{2}R_1 + R_3 \;\to\; R_3$$

$$\begin{array}{ccccccc} & x_1 & x_2 & x_3 & s_1 & s_2 & P \end{array}$$

$$\sim \begin{array}{c} x_2 \\ x_1 \\ P \end{array}\left[\begin{array}{cccccc|c} 0 & 1 & 0 & 2 & -1 & 0 & 8 \\ 1 & 0 & 2 & -1 & 1 & 0 & 12 \\ \hdashline 0 & 0 & 4 & 3 & 0 & 1 & 60 \end{array}\right]$$

Optimal solution: max $P = 60$ at $x_1 = 12$, $x_2 = 8$, $x_3 = 0$, $s_1 = 0$, $s_2 = 0$

(B)   Solution using the second column as the pivot column

$$\begin{array}{ccccccc} & x_1 & x_2 & x_3 & s_1 & s_2 & P \end{array}$$

$$\begin{array}{c} s_1 \\ s_2 \\ P \end{array}\left[\begin{array}{cccccc|c} 1 & \textcircled{1} & 2 & 1 & 0 & 0 & 20 \\ 2 & 1 & 4 & 0 & 1 & 0 & 32 \\ \hdashline -3 & -3 & -2 & 0 & 0 & 1 & 0 \end{array}\right] \begin{array}{l} \frac{20}{1} = 20 \\[4pt] \frac{32}{1} = 32 \end{array}$$

$$(-1)R_1 + R_2 \;\to\; R_2, \quad 3R_1 + R_3 \;\to\; R_3$$

$$\sim \begin{array}{c} x_2 \\ s_2 \\ P \end{array} \left[\begin{array}{cccccc|c} x_1 & x_2 & x_3 & s_1 & s_2 & P & \\ 1 & 1 & 2 & 1 & 0 & 0 & 20 \\ 1 & 0 & 2 & -1 & 1 & 0 & 12 \\ \hline 0 & 0 & 4 & 3 & 0 & 1 & 60 \end{array} \right]$$

Optimal solution: max $P = 60$
at $x_1 = 0$, $x_2 = 20$, $x_3 = 0$,
$s_1 = 0$, $s_2 = 12$

The maximum value of P is 60. Since the optimal solution is obtained at two corner points, $(12, 8, 0)$ and $(0, 20, 0)$, every point on the line segment connecting these points is also an optimal solution.

37. Let x_1 = the number of A components
x_2 = the number of B components
x_3 = the number of C components
The mathematical model for this problem is:
Maximize $P = 7x_1 + 8x_2 + 10x_3$
Subject to $2x_1 + 3x_2 + 2x_3 \le 1000$
$x_1 + x_2 + 2x_3 \le 800$
$x_1, x_2, x_3 \ge 0$

We introduce slack variables s_1, s_2 to obtain the equivalent form:

$$\begin{aligned} 2x_1 + 3x_2 + 2x_3 + s_1 \qquad\qquad &= 1000 \\ x_1 + x_2 + 2x_3 \qquad + s_2 \quad &= 800 \\ -7x_1 - 8x_2 - 10x_3 \qquad\qquad + P &= 0 \end{aligned}$$

The simplex tableau for this problem is:

$$\begin{array}{c} s_1 \\ s_2 \\ P \end{array} \left[\begin{array}{cccccc|c} x_1 & x_2 & x_3 & s_1 & s_2 & P & \\ 2 & 3 & 2 & 1 & 0 & 0 & 1000 \\ 1 & 1 & ② & 0 & 1 & 0 & 800 \\ \hline -7 & -8 & -10 & 0 & 0 & 1 & 0 \end{array} \right] \begin{array}{l} \dfrac{1000}{2} = 500 \\[2mm] \dfrac{800}{2} = 400 \end{array}$$

$$\tfrac{1}{2}R_2 \to R_2$$

$$\sim \left[\begin{array}{cccccc|c} 2 & 3 & 2 & 1 & 0 & 0 & 1000 \\ \frac{1}{2} & \frac{1}{2} & ① & 0 & \frac{1}{2} & 0 & 400 \\ \hline -7 & -8 & -10 & 0 & 0 & 1 & 0 \end{array} \right]$$

$$(-2)R_2 + R_1 \to R_1, \quad 10R_2 + R_3 \to R_3$$

$$\sim \left[\begin{array}{cccccc|c} 1 & ② & 0 & 1 & -1 & 0 & 200 \\ \frac{1}{2} & \frac{1}{2} & 1 & 0 & \frac{1}{2} & 0 & 400 \\ \hline -2 & -3 & 0 & 0 & 5 & 1 & 4000 \end{array} \right] \begin{array}{l} \dfrac{200}{2} = 100 \\[2mm] \dfrac{400}{1/2} = 800 \end{array}$$

$$\tfrac{1}{2}R_1 \to R_1$$

$$\sim \begin{bmatrix} \frac{1}{2} & ① & 0 & \frac{1}{2} & -\frac{1}{2} & 0 & 100 \\ \frac{1}{2} & \frac{1}{2} & 1 & 0 & \frac{1}{2} & 0 & 400 \\ \hline -2 & -3 & 0 & 0 & 5 & 1 & 4000 \end{bmatrix}$$

$$\left(-\frac{1}{2}\right)R_1 + R_2 \to R_2, \quad 3R_1 + R_3 \to R_3$$

$$\sim \begin{bmatrix} ⓐ\frac{1}{2} & 1 & 0 & \frac{1}{2} & -\frac{1}{2} & 0 & 100 \\ \frac{1}{4} & 0 & 1 & -\frac{1}{4} & \frac{3}{4} & 0 & 350 \\ \hline -\frac{1}{2} & 0 & 0 & \frac{3}{2} & \frac{7}{2} & 1 & 4300 \end{bmatrix} \begin{matrix} \frac{100}{1/2} = 200 \\ \frac{350}{1/4} = 1400 \\ \\ \end{matrix}$$

$$2R_1 \to R_1$$

$$\sim \begin{bmatrix} 1 & 2 & 0 & 1 & -1 & 0 & 200 \\ \frac{1}{4} & 0 & 1 & -\frac{1}{4} & \frac{3}{4} & 0 & 350 \\ \hline -\frac{1}{2} & 0 & 0 & \frac{3}{2} & \frac{7}{2} & 1 & 4300 \end{bmatrix}$$

$$\left(-\frac{1}{4}\right)R_1 + R_2 \to R_2, \quad \frac{1}{2}R_1 + R_3 \to R_3$$

$$\sim \begin{array}{c} \\ x_1 \\ x_3 \\ P \end{array} \begin{bmatrix} x_1 & x_2 & x_3 & s_1 & s_2 & P & \\ 1 & 2 & 0 & 1 & -1 & 0 & 200 \\ 0 & -\frac{1}{2} & 1 & -\frac{1}{2} & 1 & 0 & 300 \\ \hline 0 & 1 & 0 & 2 & 3 & 1 & 4400 \end{bmatrix}$$

Optimal solution: the maximum profit is $4400 when 200 *A* components, 0 *B* components and 300 *C* components are manufactured.

39. Let x_1 = the amount invested in government bonds,

x_2 = the amount invested in mutual funds,

x_3 = the amount invested in money market funds.

The mathematical model for this problem is:

Maximize $P = .08x_1 + .13x_2 + .15x_3$

Subject to: $x_1 + x_2 + x_3 \le 100{,}000$

$$x_2 + x_3 \le x_1$$

$$x_1, \ x_2, \ x_3 \ge 0$$

We introduce slack variables s_1 and s_2 to obtain the equivalent form:

$$\begin{aligned} x_1 + \quad x_2 + \quad x_3 + s_1 \qquad\quad &= 100{,}000 \\ -x_1 + \quad x_2 + \quad x_3 \qquad + s_2 \quad &= 0 \\ -.08x_1 - .13x_2 - .15x_3 \qquad\qquad + P &= 0 \end{aligned}$$

The simplex tableau for this problem is:

$$
\begin{array}{c}
\\
s_1 \\
s_2 \\
\\
P
\end{array}
\begin{array}{c}
\begin{array}{cccccc}
x_1 & x_2 & x_3 & s_1 & s_2 & P
\end{array} \\
\left[
\begin{array}{cccccc|c}
1 & 1 & 1 & 1 & 0 & 0 & 100{,}000 \\
-1 & 1 & ① & 0 & 1 & 0 & 0 \\
\hline
-.08 & -.13 & -.15 & 0 & 0 & 1 & 0
\end{array}
\right]
\begin{array}{l}
\dfrac{100{,}000}{1} = 100{,}000 \\[4pt]
\dfrac{0}{1} = 0
\end{array}
\end{array}
$$

$(-1)R_2 + R_1 \rightarrow R_1$ and $.15R_2 + R_3 \rightarrow R_3$

$$
\sim
\left[
\begin{array}{cccccc|c}
② & 0 & 0 & 1 & -1 & 0 & 100{,}000 \\
-1 & 1 & 1 & 0 & 1 & 0 & 0 \\
\hline
-.23 & .02 & 0 & 0 & .15 & 1 & 0
\end{array}
\right]
\sim
\left[
\begin{array}{cccccc|c}
① & 0 & 0 & \frac{1}{2} & -\frac{1}{2} & 0 & 50{,}000 \\
-1 & 1 & 1 & 0 & 1 & 0 & 0 \\
\hline
-.23 & .02 & 0 & 0 & .15 & 1 & 0
\end{array}
\right]
$$

$$\frac{1}{2}R_1 \rightarrow R_1 \qquad\qquad\qquad\qquad R_1 + R_2 \rightarrow R_2 \text{ and } .23R_1 + R_3 \rightarrow R_3$$

$$
\begin{array}{c}
\\
x_1 \\
x_2 \\
\\
P
\end{array}
\begin{array}{c}
\begin{array}{cccccc}
x_1 & x_2 & x_3 & s_1 & s_2 & P
\end{array} \\
\sim
\left[
\begin{array}{cccccc|c}
1 & 0 & 0 & \frac{1}{2} & -\frac{1}{2} & 0 & 50{,}000 \\
0 & 1 & 1 & \frac{1}{2} & \frac{1}{2} & 0 & 50{,}000 \\
\hline
0 & .02 & 0 & .115 & .035 & 1 & 11{,}500
\end{array}
\right]
\end{array}
$$

Optimal solution: the maximum return is $11,500 when $x_1 = $50,000$ is invested in government bonds, $x_2 = \$0$ is invested in mutual funds, and $x_3 = \$50,000$ is invested in money market funds.

41.   Let x_1 = the number of daytime ads,

x_2 = the number of prime-time ads,

x_3 = the number of late-night ads.

The mathematical model for this problem is:

Maximize $P = 14{,}000x_1 + 24{,}000x_2 + 18{,}000x_3$

Subject to: $1000x_1 + 2000x_2 + 1500x_3 \le 20{,}000$

$\qquad\qquad\quad x_1 + \quad x_2 + \quad x_3 \le 15$

$\qquad\qquad\qquad\quad x_1,\ x_2,\ x_3 \ge 0$

We introduce slack variables to obtain the following initial form:

$$1000x_1 + \quad 2000x_2 + \quad 1500x_3 + s_1 \qquad\qquad = 20{,}000$$

$$x_1 + \qquad x_2 + \qquad x_3 \qquad + s_2 \qquad = 15$$

$$-14{,}000x_1 - 24{,}000x_2 - 18{,}000x_3 \qquad\qquad + P = 0$$

The simplex tableau for this problem is:

$$
\begin{array}{c}
\\
s_1 \\
s_2 \\
\\
P
\end{array}
\begin{array}{c}
\begin{array}{cccccc}
x_1 & x_2 & x_3 & s_1 & s_2 & P
\end{array} \\
\left[
\begin{array}{cccccc|c}
1000 & ⦿2000⦿ & 1500 & 1 & 0 & 0 & 20{,}000 \\
1 & 1 & 1 & 0 & 1 & 0 & 15 \\
\hline
-14{,}000 & -24{,}000 & -18{,}000 & 0 & 0 & 1 & 0
\end{array}
\right]
\begin{array}{l}
\dfrac{20{,}000}{2000} = 10 \\[4pt]
\dfrac{15}{1} = 15
\end{array}
\end{array}
$$

$$\frac{1}{2000}\,R_1 \rightarrow R_1$$

$$\sim \begin{bmatrix} \frac{1}{2} & \textcircled{1} & \frac{3}{4} & \frac{1}{2000} & 0 & 0 & | & 10 \\ 1 & 1 & 1 & 0 & 1 & 0 & | & 15 \\ \hline -14{,}000 & -24{,}000 & -18{,}000 & 0 & 0 & 1 & | & 0 \end{bmatrix}$$

$(-1)R_1 + R_2 \rightarrow R_2,\ 24{,}000R_1 + R_3 \rightarrow R_3$

$$\sim \begin{bmatrix} \frac{1}{2} & 1 & \frac{3}{4} & \frac{1}{2000} & 0 & 0 & | & 10 \\ \textcircled{$\frac{1}{2}$} & 0 & \frac{1}{4} & -\frac{1}{2000} & 1 & 0 & | & 5 \\ \hline -2000 & 0 & 0 & 12 & 0 & 1 & | & 240{,}000 \end{bmatrix}$$

$2R_2 \rightarrow R_2$

$$\sim \begin{bmatrix} \frac{1}{2} & 1 & \frac{3}{4} & \frac{1}{2000} & 0 & 0 & | & 10 \\ \textcircled{1} & 0 & \frac{1}{2} & -\frac{1}{1000} & 2 & 0 & | & 10 \\ \hline -2000 & 0 & 0 & 12 & 0 & 1 & | & 240{,}000 \end{bmatrix}$$

$\left(-\frac{1}{2}\right)R_2 + R_1 \rightarrow R_1,\ 2000R_2 + R_3 \rightarrow R_3$

$$\begin{array}{c} \\ x_2 \\ \sim\ x_1 \\ P \end{array} \begin{array}{cccccc} x_1 & x_2 & x_3 & s_1 & s_2 & P \\ \end{array}$$

$$\begin{array}{c} x_2 \\ \sim\ x_1 \\ P \end{array} \begin{bmatrix} 0 & 1 & \frac{1}{2} & \frac{1}{1000} & -1 & 0 & | & 5 \\ 1 & 0 & \frac{1}{2} & -\frac{1}{1000} & 2 & 0 & | & 10 \\ \hline 0 & 0 & 1000 & 10 & 4000 & 1 & | & 260{,}000 \end{bmatrix}$$

Optimal solution: maximum number of potential customers is 260,000 when $x_1 = 10$ daytime ads, $x_2 = 5$ prime-time ads, and $x_3 = 0$ late-night ads are placed.

43. Let x_1 = the number of colonial houses,

$\quad\quad x_2$ = the number of split-level houses,

$\quad\quad x_3$ = the number of ranch-style houses.

(A)   The mathematical model for this problem is:

Maximize $P = 20{,}000x_1 + 18{,}000x_2 + 24{,}000x_3$

Subject to:
$$\frac{1}{2}x_1 + \frac{1}{2}x_2 + \quad\quad x_3 \le 30$$
$$60{,}000x_1 + 60{,}000x_2 + 80{,}000x_3 \le 3{,}200{,}000$$
$$4{,}000x_1 + 3{,}000x_2 + 4{,}000x_3 \le 180{,}000$$
$$x_1,\ x_2,\ x_3 \ge 0$$

We simplify the inequalities and then introduce slack variables to obtain the initial form:

$$
\begin{aligned}
\tfrac{1}{2}x_1 + \tfrac{1}{2}x_2 + x_3 + s_1 &= 30 \\
6x_1 + 6x_2 + 8x_3 + s_2 &= 320 \\
4x_1 + 3x_2 + 4x_3 + s_3 &= 180 \\
-20{,}000x_1 - 18{,}000x_2 - 24{,}000x_3 + P &= 0
\end{aligned}
$$

[Note: This simplification will change the interpretation of the slack variables.]
The simplex tableau for this problem is:

$$
\begin{array}{c}
\begin{array}{c}
\\ s_1 \\ s_2 \\ s_3 \\ \\ P
\end{array}
\left[
\begin{array}{ccccccc|c}
x_1 & x_2 & x_3 & s_1 & s_2 & s_3 & P & \\
\tfrac{1}{2} & \tfrac{1}{2} & \textcircled{1} & 1 & 0 & 0 & 0 & 30 \\
6 & 6 & 8 & 0 & 1 & 0 & 0 & 320 \\
4 & 3 & 4 & 0 & 0 & 1 & 0 & 180 \\
\hline
-20{,}000 & -18{,}000 & -24{,}000 & 0 & 0 & 0 & 1 & 0
\end{array}
\right]
\begin{array}{l}
\tfrac{30}{1} = 30 \\[4pt]
\tfrac{320}{8} = 40 \\[4pt]
\tfrac{180}{4} = 45
\end{array}
\end{array}
$$

$(-8)R_1 + R_2 \to R_2,\ (-4)R_1 + R_3 \to R_3,\ 24{,}000R_1 + R_4 \to R_4$

$$
\sim
\left[
\begin{array}{ccccccc|c}
\tfrac{1}{2} & \tfrac{1}{2} & 1 & 1 & 0 & 0 & 0 & 30 \\
2 & 2 & 0 & -8 & 1 & 0 & 0 & 80 \\
\textcircled{2} & 1 & 0 & -4 & 0 & 1 & 0 & 60 \\
\hline
-8000 & -6000 & 0 & 24{,}000 & 0 & 0 & 1 & 720{,}000
\end{array}
\right]
$$

$\tfrac{1}{2}R_3 \to R_3$

$$
\sim
\left[
\begin{array}{ccccccc|c}
\tfrac{1}{2} & \tfrac{1}{2} & 1 & 1 & 0 & 0 & 0 & 30 \\
2 & 2 & 0 & -8 & 1 & 0 & 0 & 80 \\
\textcircled{1} & \tfrac{1}{2} & 0 & -2 & 0 & \tfrac{1}{2} & 0 & 30 \\
\hline
-8000 & -6000 & 0 & 24{,}000 & 0 & 0 & 1 & 720{,}000
\end{array}
\right]
$$

$\left(-\tfrac{1}{2}\right)R_3 + R_1 \to R_1,\ (-2)R_3 + R_2 \to R_2,\ 8000R_3 + R_4 \to R_4$

$$
\sim
\left[
\begin{array}{ccccccc|c}
0 & \tfrac{1}{4} & 1 & 2 & 0 & -\tfrac{1}{4} & 0 & 15 \\
0 & \textcircled{1} & 0 & -4 & 1 & -1 & 0 & 20 \\
1 & \tfrac{1}{2} & 0 & -2 & 0 & \tfrac{1}{2} & 0 & 30 \\
\hline
0 & -2000 & 0 & 8000 & 0 & 4000 & 1 & 960{,}000
\end{array}
\right]
$$

$\left(-\tfrac{1}{4}\right)R_2 + R_1 \to R_1,\ \left(-\tfrac{1}{2}\right)R_2 + R_3 \to R_3,\ 2000R_2 + R_4 \to R_4$

$$
\begin{array}{c}
\begin{array}{ccccccc}
\;x_1 & x_2 & x_3 & s_1 & s_2 & s_3 & P
\end{array}\\
\sim
\begin{array}{c}
x_3\\x_2\\x_1\\P
\end{array}
\left[
\begin{array}{ccccccc|c}
0 & 0 & 1 & 3 & -\frac{1}{4} & 0 & 0 & 10\\
0 & 1 & 0 & -4 & 1 & -1 & 0 & 20\\
1 & 0 & 0 & 0 & -\frac{1}{2} & 1 & 0 & 20\\
\hline
0 & 0 & 0 & 0 & 2000 & 2000 & 1 & 1{,}000{,}000
\end{array}
\right]
\end{array}
$$

Optimal solution: maximum profit is $1,000,000 when $x_1 = 20$
colonial houses, $x_2 = 20$ split-level houses, and $x_3 = 10$ ranch-
style houses are built.

45. Refer to Problem 43. The mathematical model for this problem is:

Maximize $P = 17{,}000x_1 + 18{,}000x_2 + 24{,}000x_3$

Subject to:
$$\frac{1}{2}x_1 + \frac{1}{2}x_2 + x_3 \le 30$$
$$60{,}000x_1 + 60{,}000x_2 + 80{,}000x_3 \le 3{,}200{,}000$$
$$4{,}000x_1 + 3{,}000x_2 + 4{,}000x_3 \le 180{,}000$$

Following the solution in Problem 43, we obtain the simplex tableau:

$$
\begin{array}{c}
\begin{array}{ccccccc}
x_1 & x_2 & x_3 & s_1 & s_2 & s_3 & P
\end{array}\\
\begin{array}{c}
s_1\\s_2\\s_3\\P
\end{array}
\left[
\begin{array}{ccccccc|c}
\frac{1}{2} & \frac{1}{2} & ① & 1 & 0 & 0 & 0 & 30\\
6 & 6 & 8 & 0 & 1 & 0 & 0 & 320\\
4 & 3 & 4 & 0 & 0 & 1 & 0 & 180\\
\hline
-17{,}000 & -18{,}000 & -24{,}000 & 0 & 0 & 0 & 1 & 0
\end{array}
\right]
\begin{array}{l}
\frac{30}{1}=30\\[4pt]
\frac{320}{8}=40\\[4pt]
\frac{180}{4}=45
\end{array}
\end{array}
$$

$(-8)R_1 + R_2 \rightarrow R_2,\;\; (-4)R_1 + R_3 \rightarrow R_3,\;\; 24{,}000R_1 + R_4 \rightarrow R_4$

$$
\sim
\left[
\begin{array}{ccccccc|c}
\frac{1}{2} & \frac{1}{2} & 1 & 1 & 0 & 0 & 0 & 30\\
2 & ② & 0 & -8 & 1 & 0 & 0 & 80\\
2 & 1 & 0 & -4 & 0 & 1 & 0 & 60\\
\hline
-5000 & -6000 & 0 & 24{,}000 & 0 & 0 & 1 & 720{,}000
\end{array}
\right]
\begin{array}{l}
\frac{30}{1/2}=60\\[4pt]
\frac{80}{2}=40\\[4pt]
\frac{60}{1}=60
\end{array}
$$

$\frac{1}{2}R_2 \rightarrow R_2$

$$
\sim
\left[
\begin{array}{ccccccc|c}
\frac{1}{2} & \frac{1}{2} & 1 & 1 & 0 & 0 & 0 & 30\\
1 & ① & 0 & -4 & \frac{1}{2} & 0 & 0 & 40\\
2 & 1 & 0 & -4 & 0 & 1 & 0 & 60\\
\hline
-5000 & -6000 & 0 & 24{,}000 & 0 & 0 & 1 & 720{,}000
\end{array}
\right]
$$

$\left(-\dfrac{1}{2}\right)R_2 + R_1 \rightarrow R_1,\;\; (-1)R_2 + R_3 \rightarrow R_3,\;\; 6{,}000R_2 + R_4 \rightarrow R_4$

$$
\begin{array}{c}
\quad\ \ x_1\ \ x_2\ \ x_3\ \ s_1\ \ \ \ s_2\ \ \ s_3\ \ \ P \\
\sim
\begin{array}{c}
x_3 \\
x_2 \\
s_3 \\
\\
P
\end{array}
\left[
\begin{array}{ccccccc|c}
0 & 0 & 1 & 3 & -\frac{1}{4} & 0 & 0 & 10 \\
1 & 1 & 0 & -4 & \frac{1}{2} & 0 & 0 & 40 \\
1 & 0 & 0 & 0 & -\frac{1}{2} & 1 & 0 & 20 \\
\hline
1000 & 0 & 0 & 0 & 3000 & 0 & 1 & 960{,}000
\end{array}
\right]
\end{array}
$$

Optimal solution: maximum profit is $960,000 when x_1 = 0 colonial houses, x_2 = 40 split level houses and x_3 = 10 ranch houses are built. In this case, s_3 = 20 (thousand) labor hours are not used.

47. Refer to Problem 43. The mathematical model for this problem is:

Maximize $P = 25{,}000x_1 + 18{,}000x_2 + 24{,}000x_3$

Subject to:
$$\frac{1}{2}x_1 + \frac{1}{2}x_2 + x_3 \le 30$$
$$60{,}000x_1 + 60{,}000x_2 + 80{,}000x_3 \le 3{,}200{,}000$$
$$4{,}000x_1 + 3{,}000x_2 + 4{,}000x_3 \le 180{,}000$$

Following the solutions in Problems 43 and 45, we obtain the simplex tableau:

$$
\begin{array}{c}
\quad\quad x_1\quad\quad\ x_2\quad\ \ x_3\ \ s_1\ s_2\ s_3\ \ P \\
\begin{array}{c}
s_1 \\
s_2 \\
s_3 \\
\\
P
\end{array}
\left[
\begin{array}{ccccccc|c}
\frac{1}{2} & \frac{1}{2} & 1 & 1 & 0 & 0 & 0 & 30 \\
6 & 6 & 8 & 0 & 1 & 0 & 0 & 320 \\
④ & 3 & 4 & 0 & 0 & 1 & 0 & 180 \\
\hline
-25{,}000 & -18{,}000 & -24{,}000 & 0 & 0 & 0 & 1 & 0
\end{array}
\right]
\begin{array}{l}
\frac{30}{1/2}=60 \\
\frac{320}{6}=53.33 \\
\frac{180}{4}=45
\end{array}
\end{array}
$$

$$\frac{1}{4}R_3 \to R_3$$

$$
\sim
\left[
\begin{array}{ccccccc|c}
\frac{1}{2} & \frac{1}{2} & 1 & 1 & 0 & 0 & 0 & 30 \\
6 & 6 & 8 & 0 & 1 & 0 & 0 & 320 \\
① & \frac{3}{4} & 1 & 0 & 0 & \frac{1}{4} & 0 & 45 \\
\hline
-25{,}000 & -18{,}000 & -24{,}000 & 0 & 0 & 0 & 1 & 0
\end{array}
\right]
$$

$$\left(-\frac{1}{2}\right)R_3 + R_1 \to R_1,\quad -6R_3 + R_2 \to R_2,\quad 25{,}000R_3 + R_4 \to R_4$$

$$
\begin{array}{c}
\quad\quad x_1\ \ x_2\ \ \ x_3\ \ s_1\ s_2\ \ \ s_3\ \ \ P \\
\sim
\begin{array}{c}
s_1 \\
s_2 \\
x_1 \\
\\
P
\end{array}
\left[
\begin{array}{ccccccc|c}
0 & \frac{1}{8} & \frac{1}{2} & 1 & 0 & -\frac{1}{8} & 0 & 7.5 \\
0 & \frac{3}{2} & 2 & 0 & 1 & \frac{3}{2} & 0 & 50 \\
1 & \frac{3}{4} & 1 & 0 & 0 & \frac{1}{4} & 0 & 45 \\
\hline
0 & 750 & 1000 & 0 & 0 & 6250 & 1 & 1{,}125{,}000
\end{array}
\right]
\end{array}
$$

Optimal solution: maximum profit is $1,125,000 when x_1 = 45 colonial houses, x_2 = 0 split level houses and x_3 = 0 ranch houses are built. In this case, s_1 = 7.5 acres of land, and s_2 = 50(10,000) = $500,000 of capital are not used.

49. Let x_1 = the number of grams of food A,

x_2 = the number of grams of food B,

x_3 = the number of grams of food C.

The mathematical model for this problem is:

Maximize $P = 3x_1 + 4x_2 + 5x_3$

Subject to:
$$x_1 + 3x_2 + 2x_3 \leq 30$$
$$2x_1 + x_2 + 2x_3 \leq 24$$
$$x_1, \ x_2, \ x_3 \geq 0$$

We introduce slack variables s_1 and s_2 to obtain the initial form:

$$x_1 + 3x_2 + 2x_3 + s_1 \qquad\qquad = 30$$
$$2x_1 + x_2 + 2x_3 \qquad + s_2 \qquad = 24$$
$$-3x_1 - 4x_2 - 5x_3 \qquad\qquad + P = 0$$

The simplex tableau for this problem is:

$$
\begin{array}{c}
\\ s_1 \\ s_2 \\ P
\end{array}
\begin{bmatrix}
x_1 & x_2 & x_3 & s_1 & s_2 & P & \\
1 & 3 & 2 & 1 & 0 & 0 & 30 \\
2 & 1 & ② & 0 & 1 & 0 & 24 \\
\hline
-3 & -4 & -5 & 0 & 0 & 1 & 0
\end{bmatrix}
\begin{array}{l}
\\ \frac{30}{2} = 15 \\ \frac{24}{2} = 12 \\
\end{array}
$$

$$\frac{1}{2}R_2 \rightarrow R_2$$

$$
\sim
\begin{bmatrix}
1 & 3 & 2 & 1 & 0 & 0 & 30 \\
1 & \frac{1}{2} & ① & 0 & \frac{1}{2} & 0 & 12 \\
\hline
-3 & -4 & -5 & 0 & 0 & 1 & 0
\end{bmatrix}
\sim
\begin{bmatrix}
-1 & ② & 0 & 1 & -1 & 0 & 6 \\
1 & \frac{1}{2} & 1 & 0 & \frac{1}{2} & 0 & 12 \\
\hline
2 & -\frac{3}{2} & 0 & 0 & \frac{5}{2} & 1 & 60
\end{bmatrix}
\begin{array}{l}
\frac{6}{2} = 3 \\ \frac{12}{1/2} = 24 \\
\end{array}
$$

$$(-2)R_2 + R_1 \rightarrow R_1, \ 5R_2 + R_3 \rightarrow R_3 \qquad\qquad \frac{1}{2}R_1 \rightarrow R_1$$

$$
\sim
\begin{bmatrix}
-\frac{1}{2} & ① & 0 & \frac{1}{2} & -\frac{1}{2} & 0 & 3 \\
1 & \frac{1}{2} & 1 & 0 & \frac{1}{2} & 0 & 12 \\
\hline
2 & -\frac{3}{2} & 0 & 0 & \frac{5}{2} & 1 & 60
\end{bmatrix}
\sim
\begin{array}{c}
\\ x_2 \\ x_3 \\ P
\end{array}
\begin{bmatrix}
x_1 & x_2 & x_3 & s_1 & s_2 & P & \\
-\frac{1}{2} & 1 & 0 & \frac{1}{2} & -\frac{1}{2} & 0 & 3 \\
\frac{5}{4} & 0 & 1 & -\frac{1}{4} & \frac{3}{4} & 0 & \frac{21}{2} \\
\hline
\frac{5}{4} & 0 & 0 & \frac{3}{4} & \frac{7}{4} & 1 & \frac{129}{2}
\end{bmatrix}
$$

$$\left(-\frac{1}{2}\right)R_1 + R_2 \rightarrow R_2, \ \frac{3}{2}R_1 + R_3 \rightarrow R_3$$

```
Optimal solution: the maximum amount of protein is 64.5 units when
x₁ = 0 grams of food A, x₂ = 3 grams of food B and x₃ = 10.5 grams of
food C are used.
```

51. Let x_1 = the number of undergraduate students,

   x_2 = the number of graduate students,

   x_3 = the number of faculty members.

The mathematical model for this problem is:

Maximize $P = 18x_1 + 25x_2 + 30x_3$

Subject to:
$$x_1 + x_2 + x_3 \le 20$$
$$100x_1 + 150x_2 + 200x_3 \le 3200$$
$$x_1,\ x_2,\ x_3 \ge 0$$

Divide the second inequality by 50 to simplify the arithmetic. Then introduce slack variables s_1 and s_2 to obtain the initial form.

$$x_1 + x_2 + x_3 + s_1 \qquad\qquad = 20$$
$$2x_1 + 3x_2 + 4x_3 \qquad + s_2 \qquad = 64$$
$$-18x_1 - 25x_2 - 30x_3 \qquad\qquad + P = 0$$

The simplex tableau for this problem is:

$$
\begin{array}{c}
\begin{array}{cccccc}
x_1 & x_2 & x_3 & s_1 & s_2 & P
\end{array}\\
\begin{array}{c}s_1\\s_2\\P\end{array}
\left[\begin{array}{cccccc|c}
1 & 1 & 1 & 1 & 0 & 0 & 20\\
2 & 3 & ④ & 0 & 1 & 0 & 64\\
\hline
-18 & -25 & -30 & 0 & 0 & 1 & 0
\end{array}\right]
\begin{array}{l}\frac{20}{1}=20\\ \frac{64}{4}=16\end{array}
\end{array}
$$

$\frac{1}{4}R_2 \to R_2$

$$
\sim
\left[\begin{array}{cccccc|c}
1 & 1 & 1 & 1 & 0 & 0 & 20\\
\frac{1}{2} & \frac{3}{4} & ① & 0 & \frac{1}{4} & 0 & 16\\
\hline
-18 & -25 & -30 & 0 & 0 & 1 & 0
\end{array}\right]
\sim
\left[\begin{array}{cccccc|c}
⑫ & \frac{1}{4} & 0 & 1 & -\frac{1}{4} & 0 & 4\\
\frac{1}{2} & \frac{3}{4} & 1 & 0 & \frac{1}{4} & 0 & 16\\
\hline
-3 & -\frac{5}{2} & 0 & 0 & \frac{15}{2} & 1 & 480
\end{array}\right]
\begin{array}{l}\frac{4}{1/2}=8\\ \frac{16}{1/2}=32\end{array}
$$

$(-1)R_2 + R_1 \to R_1,\ \ 30R_2 + R_3 \to R_3$ $\qquad\qquad\qquad 2R_1 \to R_1$

$$
\sim
\left[\begin{array}{cccccc|c}
① & \frac{1}{2} & 0 & 2 & -\frac{1}{2} & 0 & 8\\
\frac{1}{2} & \frac{3}{4} & 1 & 0 & \frac{1}{4} & 0 & 16\\
\hline
-3 & -\frac{5}{2} & 0 & 0 & \frac{15}{2} & 1 & 480
\end{array}\right]
\sim
\left[\begin{array}{cccccc|c}
1 & ⑫ & 0 & 2 & -\frac{1}{2} & 0 & 8\\
0 & \frac{1}{2} & 1 & -1 & \frac{1}{2} & 0 & 12\\
\hline
0 & -1 & 0 & 6 & 6 & 1 & 504
\end{array}\right]
\begin{array}{l}\frac{8}{1/2}=16\\ \frac{12}{1/2}=24\end{array}
$$

$\left(-\frac{1}{2}\right)R_1 + R_2 \to R_2,\ \ 3R_1 + R_3 \to R_3$ $\qquad\qquad 2R_1 \to R_1$

$$
\begin{array}{c}
\\
\sim
\left[\begin{array}{cccccc|c}
2 & 1 & 0 & 4 & -1 & 0 & 16\\
0 & \frac{1}{2} & 1 & -1 & \frac{1}{2} & 0 & 12\\
\hline
0 & -1 & 0 & 6 & 6 & 1 & 504
\end{array}\right]
\end{array}
\begin{array}{c}
\begin{array}{cccccc}
x_1 & x_2 & x_3 & s_1 & s_2 & P
\end{array}\\
\begin{array}{c}x_2\\x_3\\P\end{array}
\sim
\left[\begin{array}{cccccc|c}
2 & 1 & 0 & 4 & -1 & 0 & 16\\
-1 & 0 & 1 & -3 & 1 & 0 & 4\\
\hline
2 & 0 & 0 & 10 & 5 & 1 & 520
\end{array}\right]
\end{array}
$$

$\left(-\frac{1}{2}\right)R_1 + R_2 \to R_2,\ \ R_1 + R_3 \to R_3$

Optimal solution: the maximum number of interviews is 520 when $x_1 = 0$ undergraduates, $x_2 = 16$ graduate students, and $x_3 = 4$ faculty members are hired.

EXERCISE 6-3

Things to remember:

1. Given a matrix A. The transpose of A, denoted A^T, is the matrix formed by interchanging the rows and corresponding columns of A (first row with first column, second row with second column, and so on.)

2. FORMATION OF THE DUAL PROBLEM

   Given a minimization problem with $\geq$ problem constraints:

   Step 1. Use the coefficients and constants in the problem constraints and the objective function to form a matrix A with the coefficients of the objective function in the last row.

   Step 2. Interchange the rows and columns of matrix A to form the matrix A^T, the transpose of A.

   Step 3. Use the rows of A^T to form a maximization problem with $\leq$ problem constraints.

3. THE FUNDAMENTAL PRINCIPLE OF DUALITY

   A minimization problem has a solution if and only if its dual problem has a solution. If a solution exists, then the optimal value of the minimization problem is the same as the optimal value of the dual problem.

4. SOLUTION OF A MINIMIZATION PROBLEM

   Given a minimization problem with nonnegative coefficients in the objective function:

   Step 1. Write all problem constraints as $\geq$ inequalities. (This may introduce negative numbers on the right side of the problem constraints.)

   Step 2. Form the dual problem.

   Step 3. Write the initial system of the dual problem, using the variables from the minimization problem as the slack variables.

   Step 4. Use the simplex method to solve the dual problem.

   Step 5. Read the solution of the minimization problem from the bottom row of the final simplex tableau in Step 4.

   [Note: If the dual problem has no solution, then the minimization problem has no solution.]

1. $A = [-5 \ \ 0 \ \ 3 \ \ -1 \ \ 8]; A^T = \begin{bmatrix} -5 \\ 0 \\ 3 \\ -1 \\ 8 \end{bmatrix}$

3. $A = \begin{bmatrix} 1 \\ -2 \\ 0 \\ 4 \end{bmatrix}; A^T = [1 \ \ -2 \ \ 0 \ \ 4]$

5. $A = \begin{bmatrix} 2 & 1 & -6 & 0 & -1 \\ 5 & 2 & 0 & 1 & 3 \end{bmatrix}$; $A^T = \begin{bmatrix} 2 & 5 \\ 1 & 2 \\ -6 & 0 \\ 0 & 1 \\ -1 & 3 \end{bmatrix}$

7. $A = \begin{bmatrix} 1 & 2 & -1 \\ 0 & 2 & -7 \\ 8 & 0 & 1 \\ 4 & -1 & 3 \end{bmatrix}$; $A^T = \begin{bmatrix} 1 & 0 & 8 & 4 \\ 2 & 2 & 0 & -1 \\ -1 & -7 & 1 & 3 \end{bmatrix}$

9. **(A)**   Given the minimization problem:

Minimize $C = 8x_1 + 9x_2$

Subject to: $x_1 + 3x_2 \geq 4$

$2x_1 + x_2 \geq 5$

$x_1, x_2 \geq 0$

The matrix A corresponding to this problem is: $A = \begin{bmatrix} 1 & 3 & | & 4 \\ 2 & 1 & | & 5 \\ 8 & 9 & | & 1 \end{bmatrix}$

The matrix A^T corresponding to the dual problem has the rows of A as its columns. Thus:

$A^T = \begin{bmatrix} 1 & 2 & | & 8 \\ 3 & 1 & | & 9 \\ 4 & 5 & | & 1 \end{bmatrix}$

The dual problem is: Maximize $P = 4y_1 + 5y_2$

Subject to: $y_1 + 2y_2 \leq 8$

$3y_1 + y_2 \leq 9$

$y_1, y_2 \geq 0$

(B)   Letting x_1 and x_2 be slack variables, the initial system for the dual problem is:

$y_1 + 2y_2 + x_1 \qquad = 8$

$3y_1 + y_2 \qquad + x_2 \qquad = 9$

$-4y_1 - 5y_2 \qquad + P = 0$

(C)   The simplex tableau for this problem is:

	y_1	y_2	x_1	x_2	P	
x_1	1	2	1	0	0	8
x_2	3	1	0	1	0	9
P	-4	-5	0	0	1	0

\ **11.**   From the final simplex table

$$
\begin{array}{c}
 \begin{array}{ccccc} y_1 & y_2 & x_1 & x_2 & P \end{array} \\
\begin{array}{c} y_2 \\ y_1 \\ P \end{array}
\left[\begin{array}{ccccc|c}
0 & 1 & 5 & -2 & 0 & 5 \\
1 & 0 & -7 & 3 & 0 & 3 \\
\hline
0 & 0 & 1 & 2 & 1 & 121
\end{array}\right]
\end{array}
$$

(A) the optimal solution of the dual problem is:
     maximum value of P = 121 at y_1 = 3 and y_2 = 5;

(B) the optimal solution of the minimization problem is:
     minimum value of C = 121 at x_1 = 1, x_2 = 2.

13.   (A)   The matrix corresponding to the given problem is: $A = \begin{bmatrix} 4 & 1 & 13 \\ 3 & 1 & 12 \\ \hline 9 & 2 & 1 \end{bmatrix}$

The matrix A^T corresponding to the dual problem has the rows of
A as its columns, that is:

$$A^T = \begin{bmatrix} 4 & 3 & 9 \\ 1 & 1 & 2 \\ \hline 13 & 12 & 1 \end{bmatrix}$$

Thus, the dual problem is: Maximize $P = 13y_1 + 12y_2$

Subject to: $4y_1 + 3y_2 \le 9$
$y_1 + y_2 \le 2$
$y_1, y_2 \ge 0$

(B)   We introduce slack variables x_1 and x_2 to obtain the initial
system for the dual problem:

$$
\begin{aligned}
4y_1 + 3y_2 + x_1 \quad\quad &= 9 \\
y_1 + y_2 \quad + x_2 \quad &= 2 \\
-13y_1 - 12y_2 \quad\quad\quad + P &= 0
\end{aligned}
$$

The simplex tableau for this problem is:

$$
\begin{array}{c}
 \begin{array}{ccccc} y_1 & y_2 & x_1 & x_2 & P \end{array} \\
\begin{array}{c} x_1 \\ x_2 \\ P \end{array}
\left[\begin{array}{ccccc|c}
4 & 3 & 1 & 0 & 0 & 9 \\
① & 1 & 0 & 1 & 0 & 2 \\
\hline
-13 & -12 & 0 & 0 & 1 & 0
\end{array}\right]
\end{array}
\begin{array}{l}
\frac{9}{4} = 2.25 \\
\frac{2}{1} = 2 \\

\end{array}
\sim
\begin{array}{c}
 \begin{array}{ccccc} y_1 & y_2 & x_1 & x_2 & P \end{array} \\
\begin{array}{c} x_1 \\ y_1 \\ P \end{array}
\left[\begin{array}{ccccc|c}
0 & -1 & 1 & -4 & 0 & 1 \\
1 & 1 & 0 & 1 & 0 & 2 \\
\hline
0 & 1 & 0 & 13 & 1 & 26
\end{array}\right]
\end{array}
$$

$(-4)R_2 + R_1 \rightarrow R_1$ and $13R_2 + R_3 \rightarrow R_3$

Optimal solution: min C = 26 at x_1 = 0, x_2 = 13.

15. (A) The matrix corresponding to the given problem is: $A = \begin{bmatrix} 2 & 3 & 15 \\ 1 & 2 & 8 \\ \hline 7 & 12 & 1 \end{bmatrix}$

The matrix A^T corresponding to the dual problem has the rows of A as its columns, that is:

$$A^T = \begin{bmatrix} 2 & 1 & 7 \\ 3 & 2 & 12 \\ \hline 15 & 8 & 1 \end{bmatrix}$$

Thus, the dual problem is: Maximize $P = 15y_1 + 8y_2$

Subject to: $2y_1 + y_2 \le 7$
$3y_1 + 2y_2 \le 12$
$y_1, y_2 \ge 0$

(B) We introduce slack variables x_1 and x_2 to obtain the initial system for the dual problem:

$$2y_1 + y_2 + x_1 \qquad = 7$$
$$3y_1 + 2y_2 \qquad + x_2 \quad = 12$$
$$-15y_1 - 8y_2 \qquad + P = 0$$

The simplex tableau for this problem is:

$$
\begin{array}{c}
x_1 \\ x_2 \\ P
\end{array}
\begin{array}{ccccc}
y_1 & y_2 & x_1 & x_2 & P \\
\end{array}
\left[\begin{array}{ccccc|c}
\textcircled{2} & 1 & 1 & 0 & 0 & 7 \\
3 & 2 & 0 & 1 & 0 & 12 \\
\hline
-15 & -8 & 0 & 0 & 1 & 0
\end{array}\right]
\begin{array}{c}
\frac{7}{2} = 3.5 \\
\frac{12}{3} = 4 \\

\end{array}
\sim
\left[\begin{array}{ccccc|c}
\textcircled{1} & \frac{1}{2} & \frac{1}{2} & 0 & 0 & \frac{7}{2} \\
3 & 2 & 0 & 1 & 0 & 12 \\
\hline
-15 & -8 & 0 & 0 & 1 & 0
\end{array}\right]
$$

$$\frac{1}{2}R_1 \to R_1 \qquad\qquad\qquad (-3)R_1 + R_2 \to R_2 \text{ and } 15R_1 + R_3 \to R_3$$

$$
\sim
\left[\begin{array}{ccccc|c}
1 & \frac{1}{2} & \frac{1}{2} & 0 & 0 & \frac{7}{2} \\
0 & \textcircled{\tfrac{1}{2}} & -\frac{3}{2} & 1 & 0 & \frac{3}{2} \\
\hline
0 & -\frac{1}{2} & \frac{15}{2} & 0 & 1 & \frac{105}{2}
\end{array}\right]
\begin{array}{c}
\frac{7/2}{1/2} = 7 \\
\frac{3/2}{1/2} = 3 \\

\end{array}
\sim
\left[\begin{array}{ccccc|c}
1 & \frac{1}{2} & \frac{1}{2} & 0 & 0 & \frac{7}{2} \\
0 & \textcircled{1} & -3 & 2 & 0 & 3 \\
\hline
0 & -\frac{1}{2} & \frac{15}{2} & 0 & 1 & \frac{105}{2}
\end{array}\right]
$$

$$2R_2 \to R_2 \qquad\qquad\qquad \left(-\frac{1}{2}\right)R_2 + R_1 \to R_1 \text{ and } \frac{1}{2}R_2 + R_3 \to R_3$$

$$
\begin{array}{c}
y_1 \\ y_2 \\ P
\end{array}
\begin{array}{ccccc}
y_1 & y_2 & x_1 & x_2 & P \\
\end{array}
\sim
\left[\begin{array}{ccccc|c}
1 & 0 & 2 & -1 & 0 & 2 \\
0 & 1 & -3 & 2 & 0 & 3 \\
\hline
0 & 0 & 6 & 1 & 1 & 54
\end{array}\right]
$$

Optimal solution: min $C = 54$ at $x_1 = 6$, $x_2 = 1$.

17. (A) The matrices corresponding to the given problem and to the dual problem are:

$$A = \begin{bmatrix} 2 & 1 & | & 8 \\ -2 & 3 & | & 4 \\ \hline 11 & 4 & | & 1 \end{bmatrix} \quad \text{and} \quad A^T = \begin{bmatrix} 2 & -2 & | & 11 \\ 1 & 3 & | & 4 \\ \hline 8 & 4 & | & 1 \end{bmatrix} \text{ respectively.}$$

Thus, the dual problem is: Maximize $P = 8y_1 + 4y_2$

Subject to: $2y_1 - 2y_2 \le 11$
$$y_1 + 3y_2 \le 4$$
$$y_1, \ y_2 \ge 0$$

(B) We introduce slack variables x_1 and x_2 to obtain the initial system for the dual problem:

$$2y_1 - 2y_2 + x_1 \qquad = 11$$
$$y_1 + 3y_2 \qquad + x_2 \qquad = 4$$
$$-8y_1 - 4y_2 \qquad \qquad + P = 0$$

The simplex tableau for this problem is:

$$
\begin{array}{c}
\begin{array}{ccccc}
y_1 & y_2 & x_1 & x_2 & P
\end{array} \\
\begin{array}{c} x_1 \\ \sim x_2 \\ P \end{array}
\left[
\begin{array}{ccccc|c}
2 & -2 & 1 & 0 & 0 & 11 \\
① & 3 & 0 & 1 & 0 & 4 \\
\hline
-8 & -4 & 0 & 0 & 1 & 0
\end{array}
\right]
\end{array}
\quad
\begin{array}{c}
\frac{11}{2} = 5.5 \\
\frac{4}{1} = 4
\end{array}
\quad
\begin{array}{c}
\begin{array}{ccccc}
y_1 & y_2 & x_1 & x_2 & P
\end{array} \\
\begin{array}{c} x_1 \\ \sim y_1 \\ P \end{array}
\left[
\begin{array}{ccccc|c}
0 & -8 & 1 & -2 & 0 & 3 \\
1 & 3 & 0 & 1 & 0 & 4 \\
\hline
0 & 20 & 0 & 8 & 1 & 32
\end{array}
\right]
\end{array}
$$

$$(-2)R_2 + R_1 \to R_1 \text{ and } 8R_2 + R_3 \to R_3$$

```
Optimal solution: min C = 32 at x₁ = 0, x₂ = 8.
```

19. (A) The matrices corresponding to the given problem and the dual problem are:

$$A = \begin{bmatrix} -3 & 1 & | & 6 \\ 1 & -2 & | & 4 \\ \hline 7 & 9 & | & 1 \end{bmatrix} \quad \text{and} \quad A^T = \begin{bmatrix} -3 & 1 & | & 7 \\ 1 & -2 & | & 9 \\ \hline 6 & 4 & | & 1 \end{bmatrix}$$

respectively.

Thus, the dual problem is:

Maximize $P = 6y_1 + 4y_2$

Subject to: $-3y_1 + y_2 \le 7$
$$y_1 - 2y_2 \le 9$$
$$y_1, \ y_2 \ge 0$$

(B) We introduce slack variables x_1 and x_2 to obtain the initial system for the dual problem:

$$
\begin{aligned}
-3y_1 + y_2 + x_1 &= 7 \\
y_1 - 2y_2 + x_2 &= 9 \\
-6y_1 - 4y_2 + P &= 0
\end{aligned}
$$

The simplex tableau for this problem is:

$$
\begin{array}{c}
\quad\; y_1 \;\; y_2 \;\; x_1 \;\; x_2 \;\; P \\
\begin{array}{c}
x_1 \\
x_2 \\
P
\end{array}
\left[
\begin{array}{ccccc|c}
-3 & 1 & 1 & 0 & 0 & 7 \\
\textcircled{1} & -2 & 0 & 1 & 0 & 9 \\
\hline
-6 & -4 & 0 & 0 & 1 & 0
\end{array}
\right]
\end{array}
\sim
\left[
\begin{array}{ccccc|c}
0 & -5 & 1 & 3 & 0 & 34 \\
1 & -2 & 0 & 1 & 0 & 9 \\
\hline
0 & -16 & 0 & 6 & 1 & 54
\end{array}
\right]
$$

$3R_2 + R_1 \rightarrow R_1$ and $6R_2 + R_3 \rightarrow R_3$

```
The negative elements in the second
column above the dashed line
indicate that the problem does not
have a solution.
```

21. The matrices corresponding to the given problem and the dual problem are:

$$
A = \left[
\begin{array}{cc|c}
2 & 1 & 8 \\
1 & 2 & 8 \\
\hline
3 & 9 & 1
\end{array}
\right]
\quad \text{and} \quad
A^T = \left[
\begin{array}{cc|c}
2 & 1 & 3 \\
1 & 2 & 9 \\
\hline
8 & 8 & 1
\end{array}
\right]
$$

respectively. Thus, the dual problem is:

$\quad$ Maximize $P = 8y_1 + 8y_2$

$\quad$ Subject to: $2y_1 + y_2 \le 3$

$\qquad\qquad\quad y_1 + 2y_2 \le 9$

$\qquad\qquad\quad y_1, \, y_2 \ge 0$

We introduce slack variables x_1 and x_2 to obtain the initial system:

$$
\begin{aligned}
2y_1 + y_2 + x_1 &= 3 \\
y_1 + 2y_2 + x_2 &= 9 \\
-8y_1 - 8y_2 + P &= 0
\end{aligned}
$$

The simplex tableau for this problem is:

$$
\begin{array}{c}
\quad\; y_1 \;\; y_2 \;\; x_1 \;\; x_2 \;\; P \\
\begin{array}{c}
x_1 \\
x_2 \\
P
\end{array}
\left[
\begin{array}{ccccc|c}
2 & \textcircled{1} & 1 & 0 & 0 & 3 \\
1 & 2 & 0 & 1 & 0 & 9 \\
\hline
-8 & -8 & 0 & 0 & 1 & 0
\end{array}
\right]
\begin{array}{l}
\frac{3}{1} = 3 \\[2pt]
\frac{9}{2} = 4.5
\end{array}
\end{array}
\sim
\begin{array}{c}
\begin{array}{c}
y_2 \\
x_2 \\
P
\end{array}
\left[
\begin{array}{ccccc|c}
2 & 1 & 1 & 0 & 0 & 3 \\
-3 & 0 & -2 & 1 & 0 & 3 \\
\hline
8 & 0 & 8 & 0 & 1 & 24
\end{array}
\right]
\end{array}
$$

$(-2)R_1 + R_2 \rightarrow R_2$ and $8R_1 + R_3 \rightarrow R_3$

```
Optimal solution: min C = 24
at x₁ = 8, x₂ = 0.
```

```
[Note: We could use either column 1 or column 2 as the pivot column.
Column 2 involves slightly simpler calculations.]
```

23. The matrices corresponding to the given problem and the dual problem are:

$$A = \begin{bmatrix} 1 & 1 & | & 4 \\ 1 & -2 & | & -8 \\ -2 & 1 & | & -8 \\ \hline 7 & 5 & | & 1 \end{bmatrix} \quad \text{and} \quad A^T = \begin{bmatrix} 1 & 1 & -2 & | & 7 \\ 1 & -2 & 1 & | & 5 \\ \hline 4 & -8 & -8 & | & 1 \end{bmatrix} \text{ respectively.}$$

Thus, the dual problem is: Maximize $P = 4y_1 - 8y_2 - 8y_3$

Subject to: $y_1 + y_2 - 2y_3 \le 7$

$y_1 - 2y_2 + y_3 \le 5$

$y_1, \ y_2, \ y_3 \ge 0$

We introduce slack variables x_1 and x_2 to obtain the initial system:

$$y_1 + y_2 - 2y_3 + x_1 \qquad = 7$$
$$y_1 - 2y_2 + y_3 \qquad + x_2 \quad = 5$$
$$-4y_1 + 8y_2 + 8y_3 \qquad + P = 0$$

The simplex tableau for this problem is:

$$
\begin{array}{c}

\end{array}
\begin{array}{c}
y_1 \quad y_2 \quad y_3 \quad x_1 \quad x_2 \quad P
\end{array}
$$

$$
\begin{array}{c} x_1 \\ x_2 \\ P \end{array}
\begin{bmatrix}
1 & 1 & -2 & 1 & 0 & 0 & | & 7 \\
① & -2 & 1 & 0 & 1 & 0 & | & 5 \\
\hline
-4 & 8 & 8 & 0 & 0 & 1 & | & 0
\end{bmatrix}
\begin{array}{l} \frac{7}{1} = 7 \\ \frac{5}{1} = 5 \end{array}
$$

$$
\begin{array}{c} x_1 \\ \sim \ y_1 \\ P \end{array}
\begin{bmatrix}
0 & 3 & -3 & 1 & -1 & 0 & | & 2 \\
1 & -2 & 1 & 0 & 1 & 0 & | & 5 \\
\hline
0 & 0 & 12 & 0 & 4 & 1 & | & 20
\end{bmatrix}
$$

$(-1)R_2 + R_1 \rightarrow R_1$ and $4R_2 + R_3 \rightarrow R_3$

Optimal solution: min $C = 20$
at $x_1 = 0$, $x_2 = 4$.

25. The matrices corresponding to the given problem and the dual problem are:

$$A = \begin{bmatrix} 2 & 1 & | & 16 \\ 1 & 1 & | & 12 \\ 1 & 2 & | & 14 \\ \hline 10 & 30 & | & 1 \end{bmatrix} \quad \text{and} \quad A^T = \begin{bmatrix} 2 & 1 & 1 & | & 10 \\ 1 & 1 & 2 & | & 30 \\ \hline 16 & 12 & 14 & | & 1 \end{bmatrix} \text{ respectively.}$$

Thus, the dual problem is: Maximize $P = 16y_1 + 12y_2 + 14y_3$

Subject to: $2y_1 + y_2 + y_3 \le 10$

$y_1 + y_2 + 2y_3 \le 30$

$y_1, \ y_2, \ y_3 \ge 0$

We introduce slack variables x_1 and x_2 to obtain the initial system:

$$2y_1 + y_2 + y_3 + x_1 \qquad = 10$$
$$y_1 + y_2 + 2y_3 \qquad + x_2 = 30$$
$$-16y_1 - 12y_2 - 14y_3 \qquad + P = 0$$

The simplex tableau for this problem is:

$$
\begin{array}{c}
 \\
x_1 \\
x_2 \\
P
\end{array}
\begin{array}{c}
y_1 \quad y_2 \quad y_3 \quad x_1 \quad x_2 \quad P \\
\left[
\begin{array}{cccccc|c}
② & 1 & 1 & 1 & 0 & 0 & 10 \\
1 & 1 & 2 & 0 & 1 & 0 & 30 \\
\hline
-16 & -12 & -14 & 0 & 0 & 1 & 0
\end{array}
\right]
\end{array}
\begin{array}{c}
\frac{10}{2} = 5 \\
\frac{30}{1} = 30 \\

\end{array}
\sim
\left[
\begin{array}{cccccc|c}
① & \frac{1}{2} & \frac{1}{2} & \frac{1}{2} & 0 & 0 & 5 \\
1 & 1 & 2 & 0 & 1 & 0 & 30 \\
\hline
-16 & -12 & -14 & 0 & 0 & 1 & 0
\end{array}
\right]
$$

$$\tfrac{1}{2}R_1 \to R_1 \qquad\qquad\qquad\qquad (-1)R_1 + R_2 \to R_2 \text{ and } 16R_1 + R_3 \to R_3$$

$$
\sim
\left[
\begin{array}{cccccc|c}
1 & \frac{1}{2} & ⓵\hspace{-0.6em}\frac{1}{2} & \frac{1}{2} & 0 & 0 & 5 \\
0 & \frac{1}{2} & \frac{3}{2} & -\frac{1}{2} & 1 & 0 & 25 \\
\hline
0 & -4 & -6 & 8 & 0 & 1 & 80
\end{array}
\right]
\begin{array}{c}
\frac{5}{1/2} = 10 \\
\frac{25}{3/2} = 16.66 \\

\end{array}
\sim
\left[
\begin{array}{cccccc|c}
2 & 1 & ① & 1 & 0 & 0 & 10 \\
0 & \frac{1}{2} & \frac{3}{2} & -\frac{1}{2} & 1 & 0 & 25 \\
\hline
0 & -4 & -6 & 8 & 0 & 1 & 80
\end{array}
\right]
$$

$$2R_1 \to R_1 \qquad\qquad\qquad\qquad \left(-\tfrac{3}{2}\right)R_1 + R_2 \to R_2 \text{ and } 6R_1 + R_3 \to R_3$$

$$
\sim
\begin{array}{c}
y_3 \\
x_2 \\
P
\end{array}
\left[
\begin{array}{cccccc|c}
2 & 1 & 1 & 1 & 0 & 0 & 10 \\
-3 & -1 & 0 & -2 & 1 & 0 & 10 \\
\hline
12 & 2 & 0 & 14 & 0 & 1 & 140
\end{array}
\right]
$$

Optimal solution: min $C = 140$ at $x_1 = 14$, $x_2 = 0$.

27. The matrices corresponding to the given problem and the dual problem are:

$$
A = \left[
\begin{array}{cc|c}
1 & 0 & 4 \\
1 & 1 & 8 \\
1 & 2 & 10 \\
\hline
5 & 7 & 1
\end{array}
\right]
\quad \text{and} \quad
A^T = \left[
\begin{array}{ccc|c}
1 & 1 & 1 & 5 \\
0 & 1 & 2 & 7 \\
\hline
4 & 8 & 10 & 1
\end{array}
\right]
\text{ respectively.}
$$

Thus, the dual problem is: Maximize $P = 4y_1 + 8y_2 + 10y_3$

Subject to: $y_1 + y_2 + y_3 \le 5$
$$y_2 + 2y_3 \le 7$$
$$y_1, \, y_2, \, y_3 \ge 0$$

We introduce slack variables x_1 and x_2 to obtain the initial system:

$$
\begin{aligned}
y_1 + y_2 + y_3 + x_1 &= 5 \\
y_2 + 2y_3 + x_2 &= 7 \\
-4y_1 - 8y_2 - 10y_3 + P &= 0
\end{aligned}
$$

The simplex tableau for this problem is:

$$\begin{array}{c} \\ x_1 \\ x_2 \\ P \end{array} \begin{array}{cccccc} y_1 & y_2 & y_3 & x_1 & x_2 & P \\ \end{array} \left[\begin{array}{cccccc|c} 1 & 1 & 1 & 1 & 0 & 0 & 5 \\ 0 & 1 & ② & 0 & 1 & 0 & 7 \\ \hline -4 & -8 & -10 & 0 & 0 & 1 & 0 \end{array} \right] \begin{array}{l} \frac{5}{1} = 5 \\ \frac{7}{2} = 3.5 \end{array} \sim \left[\begin{array}{cccccc|c} 1 & 1 & 1 & 1 & 0 & 0 & 5 \\ 0 & \frac{1}{2} & ① & 0 & \frac{1}{2} & 0 & \frac{7}{2} \\ \hline -4 & -8 & -10 & 0 & 0 & 1 & 0 \end{array} \right]$$

$$\frac{1}{2}R_2 \to R_2 \qquad\qquad\qquad (-1)R_2 + R_1 \to R_1 \ \text{ and } \ 10R_2 + R_3 \to R_3$$

$$\sim \left[\begin{array}{cccccc|c} ① & \frac{1}{2} & 0 & 1 & -\frac{1}{2} & 0 & \frac{3}{2} \\ 0 & \frac{1}{2} & 1 & 0 & \frac{1}{2} & 0 & \frac{7}{2} \\ \hline -4 & -3 & 0 & 0 & 5 & 1 & 35 \end{array} \right] \sim \left[\begin{array}{cccccc|c} 1 & ①\hspace{-0.2em}\frac{1}{2} & 0 & 1 & -\frac{1}{2} & 0 & \frac{3}{2} \\ 0 & \frac{1}{2} & 1 & 0 & \frac{1}{2} & 0 & \frac{7}{2} \\ \hline 0 & -1 & 0 & 4 & 3 & 1 & 41 \end{array} \right] \begin{array}{l} \frac{3/2}{1/2} = 3 \\ \frac{7/2}{1/2} = 7 \end{array}$$

$$4R_1 + R_3 \to R_3 \qquad\qquad\qquad\qquad 2R_1 \to R_1$$

$$\sim \left[\begin{array}{cccccc|c} 2 & ① & 0 & 2 & -1 & 0 & 3 \\ 0 & \frac{1}{2} & 1 & 0 & \frac{1}{2} & 0 & \frac{7}{2} \\ \hline 0 & -1 & 0 & 4 & 3 & 1 & 41 \end{array} \right] \begin{array}{c} \\ \\ \\ \end{array} \begin{array}{c} y_2 \\ \sim y_3 \\ P \end{array} \begin{array}{cccccc} y_1 & y_2 & y_3 & x_1 & x_2 & P \\ \end{array} \left[\begin{array}{cccccc|c} 2 & 1 & 0 & 2 & -1 & 0 & 3 \\ -1 & 0 & 1 & -1 & 1 & 0 & 2 \\ \hline 2 & 0 & 0 & 6 & 2 & 1 & 44 \end{array} \right]$$

$$\left(-\frac{1}{2}\right)R_1 + R_2 \to R_2 \qquad\qquad \text{Optimal solution: min } C = 44 \text{ at}$$
$$\text{and } R_3 + R_1 \to R_3 \qquad\qquad\qquad x_1 = 6, \ x_2 = 2.$$

29. The matrices corresponding to the given problem and the dual problem are:

$$A = \left[\begin{array}{ccc|c} 1 & 1 & 2 & 7 \\ 2 & 1 & 1 & 4 \\ \hline 10 & 7 & 12 & 1 \end{array} \right] \quad \text{and} \quad A^T = \left[\begin{array}{cc|c} 1 & 2 & 10 \\ 1 & 1 & 7 \\ 2 & 1 & 12 \\ \hline 7 & 4 & 1 \end{array} \right] \quad \text{respectively.}$$

Thus, the dual problem is: Maximize $P = 7y_1 + 4y_2$

$$\begin{aligned} \text{Subject to: } \ & y_1 + 2y_2 \le 10 \\ & y_1 + y_2 \le 7 \\ & 2y_1 + y_2 \le 12 \\ & y_1, \ y_2 \ge 0 \end{aligned}$$

We introduce slack variables $x_1, x_2,$ and x_3 to obtain the initial system:

$$\begin{aligned} y_1 + 2y_2 + x_1 \qquad\qquad\quad &= 10 \\ y_1 + y_2 \quad + x_2 \qquad\quad &= 7 \\ 2y_1 + y_2 \qquad\quad + x_3 \quad &= 12 \\ -7y_1 - 4y_2 \qquad\qquad\quad + P &= 0 \end{aligned}$$

The simplex tableau for this problem is:

$$
\begin{array}{c}
 \\
x_1 \\
x_2 \\
x_3 \\
P
\end{array}
\begin{array}{c}
y_1 \quad y_2 \quad x_1 \quad x_2 \quad x_3 \quad P \\
\left[\begin{array}{cccccc|c}
1 & 2 & 1 & 0 & 0 & 0 & 10 \\
1 & 1 & 0 & 1 & 0 & 0 & 7 \\
② & 1 & 0 & 0 & 1 & 0 & 12 \\
\hdashline
-7 & -4 & 0 & 0 & 0 & 1 & 0
\end{array}\right]
\end{array}
\begin{array}{c}
\frac{10}{1}=10 \\
\frac{7}{1}=7 \\
\frac{12}{2}=6 \\
\;
\end{array}
\sim
\left[\begin{array}{cccccc|c}
1 & 2 & 1 & 0 & 0 & 0 & 10 \\
1 & 1 & 0 & 1 & 0 & 0 & 7 \\
① & \frac{1}{2} & 0 & 0 & \frac{1}{2} & 0 & 6 \\
\hdashline
-7 & -4 & 0 & 0 & 0 & 1 & 0
\end{array}\right]
$$

$$\frac{1}{2}R_3 \to R_3 \qquad\qquad\qquad (-1)R_3 + R_1 \to R_1,\ (-1)\,R_3 + R_2 \to R_2,$$
$$\text{and } 7R_3 + R_4 \to R_4$$

$$
\sim
\left[\begin{array}{cccccc|c}
0 & \frac{3}{2} & 1 & 0 & -\frac{1}{2} & 0 & 4 \\
0 & ⓵ & 0 & 1 & -\frac{1}{2} & 0 & 1 \\
1 & \frac{1}{2} & 0 & 0 & \frac{1}{2} & 0 & 6 \\
\hdashline
0 & -\frac{1}{2} & 0 & 0 & \frac{7}{2} & 1 & 42
\end{array}\right]
\begin{array}{c}
\frac{4}{3/2}=\frac{8}{3} \\
\frac{1}{1/2}=2 \\
\frac{6}{1/2}=12 \\
\;
\end{array}
\sim
\left[\begin{array}{cccccc|c}
0 & \frac{3}{2} & 1 & 0 & -\frac{1}{2} & 0 & 4 \\
0 & ① & 0 & 2 & -1 & 0 & 2 \\
1 & \frac{1}{2} & 0 & 0 & \frac{1}{2} & 0 & 6 \\
\hdashline
0 & -\frac{1}{2} & 0 & 0 & \frac{7}{2} & 1 & 42
\end{array}\right]
$$

$$2R_2 \to R_2 \qquad\qquad \left(-\frac{3}{2}\right)R_2 + R_1 \to R_1,\ \left(-\frac{1}{2}\right)R_2 + R_3 \to R_3,$$
$$\text{and } \frac{1}{2}R_2 + R_4 \to R_4$$

$$
\begin{array}{c}
 \\
x_1 \\
y_2 \\
y_1 \\
P
\end{array}
\begin{array}{c}
y_1 \quad y_2 \quad x_1 \quad x_2 \quad x_3 \quad P \\
\sim
\left[\begin{array}{cccccc|c}
0 & 0 & 1 & -3 & 1 & 0 & 1 \\
0 & 1 & 0 & 2 & -1 & 0 & 2 \\
1 & 0 & 0 & -1 & 1 & 0 & 5 \\
\hdashline
0 & 0 & 0 & 1 & 3 & 1 & 43
\end{array}\right]
\end{array}
$$

Optimal solution: min $C = 43$ at $x_1 = 0,\ x_2 = 1,\ x_3 = 3$.

31. The matrices corresponding to the given problem and the dual problem are:

$$
A = \left[\begin{array}{ccc|c}
1 & -4 & 1 & 6 \\
-1 & 1 & -2 & 4 \\
\hline
5 & 2 & 2 & 1
\end{array}\right]
\quad \text{and} \quad
A^T = \left[\begin{array}{cc|c}
1 & -1 & 5 \\
-4 & 1 & 2 \\
1 & -2 & 2 \\
\hline
6 & 4 & 1
\end{array}\right]
$$

Thus, the dual problem is: Maximize $P = 6y_1 + 4y_2$

$$
\begin{aligned}
\text{Subject to:} \quad y_1 - y_2 &\le 5 \\
-4y_1 + y_2 &\le 2 \\
y_1 - 2y_2 &\le 2 \\
y_1, \ y_2 &\ge 0
\end{aligned}
$$

We introduce slack variables $x_1, x_2,$ and x_3 to obtain the initial system:

$$
\begin{aligned}
y_1 - y_2 + x_1 &= 5 \\
-4y_1 + y_2 + x_2 &= 2 \\
y_1 - 2y_2 + x_3 &= 2 \\
-6y_1 - 4y_2 + P &= 0
\end{aligned}
$$

The simplex tableau for this problem is:

$$
\begin{array}{c}
\begin{array}{ccccccc}
 & y_1 & y_2 & x_1 & x_2 & x_3 & P \\
\end{array} \\
\begin{array}{c}
x_1 \\
x_2 \\
x_3 \\
P
\end{array}
\left[
\begin{array}{cccccc|c}
1 & -1 & 1 & 0 & 0 & 0 & 5 \\
-4 & 1 & 0 & 1 & 0 & 0 & 2 \\
\boxed{1} & -2 & 0 & 0 & 1 & 0 & 2 \\
\hline
-6 & -4 & 0 & 0 & 0 & 1 & 0
\end{array}
\right]
\begin{array}{l}
\frac{5}{1} = 5 \\[6pt]
\frac{2}{1} = 2 \\[6pt]
\end{array}
\end{array}
$$

$$
\sim
\left[
\begin{array}{cccccc|c}
0 & \boxed{1} & 1 & 0 & -1 & 0 & 3 \\
0 & -7 & 0 & 1 & 4 & 0 & 10 \\
1 & -2 & 0 & 0 & 1 & 0 & 2 \\
\hline
0 & -16 & 0 & 0 & 6 & 1 & 12
\end{array}
\right]
$$

$(-1)R_3 + R_1 \rightarrow R_1, \quad 4R_3 + R_2 \rightarrow R_2,$

and $6R_3 + R_4 \rightarrow R_4$

$7R_1 + R_2 \rightarrow R_2, \quad 2R_1 + R_3 \rightarrow R_3,$

and $16R_1 + R_4 \rightarrow R_4$

$$
\begin{array}{c}
\begin{array}{ccccccc}
 & y_1 & y_2 & x_1 & x_2 & x_3 & P \\
\end{array} \\
\sim
\begin{array}{c}
y_2 \\
x_2 \\
y_1 \\
P
\end{array}
\left[
\begin{array}{cccccc|c}
0 & 1 & 1 & 0 & -1 & 0 & 3 \\
0 & 0 & 7 & 1 & -3 & 0 & 31 \\
1 & 0 & 2 & 0 & -1 & 0 & 8 \\
\hline
0 & 0 & 16 & 0 & -10 & 1 & 60
\end{array}
\right]
\end{array}
$$

Since all the entries above the dashed line in the pivot column, the x_3 column, are negative, the problem does not have an optimal solution.

33. The dual problem has 2 variables and 4 problem constraints.

35. The original problem must have two problem constraints, and any number of variables.

37. No. The dual problem will not be a standard maximization problem (one of the elements in the last column will be negative.)

39. Yes. Multiply both sides of the inequality by –1.

41. The matrices corresponding to the given problem and the dual problem are:

$$
A = \left[
\begin{array}{ccc|c}
3 & 2 & 2 & 16 \\
4 & 3 & 1 & 14 \\
5 & 3 & 1 & 12 \\
\hline
16 & 8 & 4 & 1
\end{array}
\right]
\quad \text{and} \quad
A^T = \left[
\begin{array}{ccc|c}
3 & 4 & 5 & 16 \\
2 & 3 & 3 & 8 \\
2 & 1 & 1 & 4 \\
\hline
16 & 14 & 12 & 1
\end{array}
\right]
\text{ respectively.}
$$

Thus, the dual problem is: Maximize $P = 16y_1 + 14y_2 + 12y_3$

Subject to: $3y_1 + 4y_2 + 5y_3 \leq 16$

$2y_1 + 3y_2 + 3y_3 \leq 8$

$2y_1 + y_2 + y_3 \leq 4$

$y_1, y_2, y_3 \geq 0$

We introduce slack variables x_1, x_2, and x_3 to obtain the initial system:

$$3y_1 + 4y_2 + 5y_3 + x_1 \qquad\qquad = 16$$
$$2y_1 + 3y_2 + 3y_3 \qquad + x_2 \qquad = 8$$
$$2y_1 + y_2 + y_3 \qquad\qquad + x_3 \quad = 4$$
$$-16y_1 - 14y_2 - 12y_3 \qquad\qquad\quad + P = 0$$

The simplex tableau for this problem is:

$$
\begin{array}{c}
\begin{array}{ccccccc} y_1 & y_2 & y_3 & x_1 & x_2 & x_3 & P \end{array} \\
\begin{array}{c} x_1 \\ x_2 \\ x_3 \\ P \end{array}
\left[\begin{array}{ccccccc|c}
3 & 4 & 5 & 1 & 0 & 0 & 0 & 16 \\
2 & 3 & 3 & 0 & 1 & 0 & 0 & 8 \\
②\;2 & 1 & 1 & 0 & 0 & 1 & 0 & 4 \\
\hdashline
-16 & -14 & -12 & 0 & 0 & 0 & 1 & 0
\end{array}\right]
\begin{array}{l}
\frac{16}{3} = 5.33 \\
\frac{8}{2} = 4 \\
\frac{4}{2} = 2
\end{array}
\end{array}
$$

$$\tfrac{1}{2}R_3 \;\to\; R_3$$

$$
\sim
\left[\begin{array}{ccccccc|c}
3 & 4 & 5 & 1 & 0 & 0 & 0 & 16 \\
2 & 3 & 3 & 0 & 1 & 0 & 0 & 8 \\
① & \frac{1}{2} & \frac{1}{2} & 0 & 0 & \frac{1}{2} & 0 & 2 \\
\hdashline
-16 & -14 & -12 & 0 & 0 & 0 & 1 & 0
\end{array}\right]
$$

$$(-3)R_3 + R_1 \to R_1, \quad (-2)R_3 + R_2 \to R_2, \quad \text{and } 16R_3 + R_4 \to R_4$$

$$
\sim
\left[\begin{array}{ccccccc|c}
0 & \frac{5}{2} & \frac{7}{2} & 1 & 0 & -\frac{3}{2} & 0 & 10 \\
0 & ②\;2 & 2 & 0 & 1 & -1 & 0 & 4 \\
1 & \frac{1}{2} & \frac{1}{2} & 0 & 0 & \frac{1}{2} & 0 & 2 \\
\hdashline
0 & -6 & -4 & 0 & 0 & 8 & 1 & 32
\end{array}\right]
\begin{array}{l}
\frac{10}{5/2} = 4 \\
\frac{4}{2} = 2 \\
\frac{2}{1/2} = 4
\end{array}
\qquad
\sim
\left[\begin{array}{ccccccc|c}
0 & \frac{5}{2} & \frac{7}{2} & 1 & 0 & -\frac{3}{2} & 0 & 10 \\
0 & ① & 1 & 0 & \frac{1}{2} & -\frac{1}{2} & 0 & 2 \\
1 & \frac{1}{2} & \frac{1}{2} & 0 & 0 & \frac{1}{2} & 0 & 2 \\
\hdashline
0 & -6 & -4 & 0 & 0 & 8 & 1 & 32
\end{array}\right]
$$

$$\tfrac{1}{2}R_2 \;\to\; R_2 \qquad\qquad \left(-\tfrac{5}{2}\right)R_2 + R_1 \to R_1, \quad \left(-\tfrac{1}{2}\right)R_2 + R_3 \to R_3,$$
$$\text{and } 6R_2 + R_4 \to R_4$$

$$
\begin{array}{c}
\begin{array}{ccccccc} y_1 & y_2 & y_3 & x_1 & x_2 & x_3 & P \end{array} \\
\begin{array}{c} y_3 \\ y_2 \\ y_1 \\ P \end{array}
\sim
\left[\begin{array}{ccccccc|c}
0 & 0 & 1 & 1 & -\frac{5}{4} & -\frac{1}{4} & 0 & 5 \\
0 & 1 & 1 & 0 & \frac{1}{2} & -\frac{1}{2} & 0 & 2 \\
1 & 0 & 0 & 0 & -\frac{1}{4} & \frac{3}{4} & 0 & 1 \\
\hdashline
0 & 0 & 2 & 0 & 3 & 5 & 1 & 44
\end{array}\right]
\end{array}
$$

```
Optimal solution: min C = 44
at x₁ = 0, x₂ = 3, x₃ = 5.
```

43.   The first and second inequalities must be rewritten before forming the dual.

Minimize $C = 5x_1 + 4x_2 + 5x_3 + 6x_4$

Subject to:
$$
\begin{aligned}
-x_1 - x_2 \quad\qquad\quad &\geq -12 \\
-x_3 - x_4 &\geq -25 \\
x_1 \qquad + x_3 \quad\;\; &\geq 20 \\
x_2 \qquad + x_4 &\geq 15 \\
x_1,\ x_2,\ x_3,\ x_4 &\geq 0
\end{aligned}
$$

The matrices corresponding to the given problem and the dual problem are:

$$A = \begin{bmatrix} -1 & -1 & 0 & 0 & | & -12 \\ 0 & 0 & -1 & -1 & | & -25 \\ 1 & 0 & 1 & 0 & | & 20 \\ 0 & 1 & 0 & 1 & | & 15 \\ \hline 5 & 4 & 5 & 6 & | & 1 \end{bmatrix} \quad \text{and} \quad A^T = \begin{bmatrix} -1 & 0 & 1 & 0 & | & 5 \\ -1 & 0 & 0 & 1 & | & 4 \\ 0 & -1 & 1 & 0 & | & 5 \\ 0 & -1 & 0 & 1 & | & 6 \\ \hline -12 & -25 & 20 & 15 & | & 1 \end{bmatrix}$$

The dual problem is: Maximize $P = -12y_1 - 25y_2 + 20y_3 + 15y_4$

$$\begin{aligned}
\text{Subject to: } -y_1 \quad\quad + y_3 \quad\quad &\le 5 \\
-y_1 \quad\quad\quad\quad + y_4 &\le 4 \\
-y_2 + y_3 \quad\quad &\le 5 \\
-y_2 \quad\quad + y_4 &\le 6 \\
y_1, y_2, y_3, y_4 &\ge 0
\end{aligned}$$

We introduce the slack variables $x_1, x_2, x_3,$ and x_4 to obtain the initial system:

$$\begin{aligned}
-y_1 \quad + y_3 \quad + x_1 \quad\quad\quad\quad\quad\quad\quad &= 5 \\
-y_1 \quad\quad\quad + y_4 \quad + x_2 \quad\quad\quad\quad\quad &= 4 \\
-y_2 + y_3 \quad\quad\quad\quad + x_3 \quad\quad\quad &= 5 \\
-y_2 \quad\quad + y_4 \quad\quad\quad\quad + x_4 \quad &= 6 \\
+12y_1 + 25y_2 - 20y_3 - 15y_4 \quad\quad\quad\quad + P &= 0
\end{aligned}$$

The simplex tableau for this problem is:

	y_1	y_2	y_3	y_4	x_1	x_2	x_3	x_4	P		
x_1	-1	0	1	0	1	0	0	0	0	5	$\frac{5}{1} = 5$
x_2	-1	0	0	1	0	1	0	0	0	4	
x_3	0	-1	①	0	0	0	1	0	0	5	$\frac{5}{1} = 5$
x_4	0	-1	0	1	0	0	0	1	0	6	
P	12	25	-20	-15	0	0	0	0	1	0	

$(-1)R_3 + R_1 \rightarrow R_1$ and $20R_3 + R_5 \rightarrow R_5$

$$\sim \begin{bmatrix} -1 & 1 & 0 & 0 & 1 & 0 & -1 & 0 & 0 & | & 0 \\ -1 & 0 & 0 & ① & 0 & 1 & 0 & 0 & 0 & | & 4 \\ 0 & -1 & 1 & 0 & 0 & 0 & 1 & 0 & 0 & | & 5 \\ 0 & -1 & 0 & 1 & 0 & 0 & 0 & 1 & 0 & | & 6 \\ \hline 12 & 5 & 0 & -15 & 0 & 0 & 20 & 0 & 1 & | & 100 \end{bmatrix} \begin{matrix} \\ \frac{4}{1} = 4 \\ \\ \frac{6}{1} = 6 \\ \\ \end{matrix}$$

$(-1)R_2 + R_4 \rightarrow R_4$ and $15R_2 + R_5 \rightarrow R_5$

$$\sim \begin{bmatrix} -1 & 1 & 0 & 0 & 1 & 0 & -1 & 0 & 0 & | & 0 \\ -1 & 0 & 0 & 1 & 0 & 1 & 0 & 0 & 0 & | & 4 \\ 0 & -1 & 1 & 0 & 0 & 0 & 1 & 0 & 0 & | & 5 \\ ① & -1 & 0 & 0 & 0 & -1 & 0 & 1 & 0 & | & 2 \\ \hdashline -3 & 5 & 0 & 0 & 0 & 15 & 20 & 0 & 1 & | & 160 \end{bmatrix}$$

$$3R_4 + R_5 \rightarrow R_5, \ R_4 + R_1 \rightarrow R_1, \ \text{and} \ R_4 + R_2 \rightarrow R_2$$

$$\sim \begin{array}{c} \\ x_1 \\ y_4 \\ y_3 \\ y_1 \\ P \end{array} \begin{array}{cccccccccc} y_1 & y_2 & y_3 & y_4 & x_1 & x_2 & x_3 & x_4 & P & \\ \begin{bmatrix} 0 & 0 & 0 & 0 & 1 & -1 & -1 & 1 & 0 & | & 2 \\ 0 & -1 & 0 & 1 & 0 & 0 & 0 & 1 & 0 & | & 6 \\ 0 & -1 & 1 & 0 & 0 & 0 & 1 & 0 & 0 & | & 5 \\ 1 & -1 & 0 & 0 & 0 & -1 & 0 & 1 & 0 & | & 2 \\ \hdashline 0 & 2 & 0 & 0 & 0 & 12 & 20 & 3 & 1 & | & 166 \end{bmatrix} \end{array}$$

Optimal solution:
min $C = 166$ at $x_1 = 0$,
$x_2 = 12$, $x_3 = 20$,
$x_4 = 3$.

45. Let x_1 = the number of hours the Cedarburg plant is operated,

x_2 = the number of hours the Grafton plant is operated,

x_3 = the number of hours the West Bend plant is operated.

The mathematical model for this problem is:

Minimize $C = 70x_1 + 75x_2 + 90x_3$
Subject to: $20x_1 + 10x_2 + 20x_3 \geq 300$
$10x_1 + 20x_2 + 20x_3 \geq 200$
$x_1, x_2, x_3 \geq 0$

Divide each of the problem constraint inequalities by 10 to simplify the calculations. The matrices corresponding to the given problem and the dual problem are:

$$A = \begin{bmatrix} 2 & 1 & 2 & | & 30 \\ 1 & 2 & 2 & | & 20 \\ \hline 70 & 75 & 90 & | & 1 \end{bmatrix} \quad \text{and} \quad A^T = \begin{bmatrix} 2 & 1 & | & 70 \\ 1 & 2 & | & 75 \\ 2 & 2 & | & 90 \\ \hline 30 & 20 & | & 1 \end{bmatrix} \quad \text{respectively.}$$

Thus, the dual problem is:

Maximize $P = 30y_1 + 20y_2$
Subject to: $2y_1 + y_2 \leq 70$
$y_1 + 2y_2 \leq 75$
$2y_1 + 2y_2 \leq 90$
$y_1, y_2 \geq 0$

We introduce slack variables x_1, x_2, and x_3 to obtain the initial system:

$$
\begin{aligned}
2y_1 + y_2 + x_1 &= 70 \\
y_1 + 2y_2 + x_2 &= 75 \\
2y_1 + 2y_2 + x_3 &= 90 \\
-30y_1 - 20y_2 + P &= 0
\end{aligned}
$$

The simplex tableau for this problem is:

$$
\begin{array}{c}
\begin{array}{cccccc}
y_1 & y_2 & x_1 & x_2 & x_3 & P
\end{array} \\
\begin{array}{c}
x_1 \\ x_2 \\ x_3 \\ P
\end{array}
\left[
\begin{array}{cccccc|c}
② & 1 & 1 & 0 & 0 & 0 & 70 \\
1 & 2 & 0 & 1 & 0 & 0 & 75 \\
2 & 2 & 0 & 0 & 1 & 0 & 90 \\
\hline
-30 & -20 & 0 & 0 & 0 & 1 & 0
\end{array}
\right]
\end{array}
\quad
\begin{array}{l}
\frac{70}{2} = 35 \\[4pt]
\frac{75}{1} = 75 \\[4pt]
\frac{90}{2} = 45
\end{array}
$$

$$\frac{1}{2}R_1 \rightarrow R_1$$

$$
\sim
\left[
\begin{array}{cccccc|c}
① & \frac{1}{2} & \frac{1}{2} & 0 & 0 & 0 & 35 \\
1 & 2 & 0 & 1 & 0 & 0 & 75 \\
2 & 2 & 0 & 0 & 1 & 0 & 90 \\
\hline
-30 & -20 & 0 & 0 & 0 & 1 & 0
\end{array}
\right]
$$

$$(-1)R_1 + R_2 \rightarrow R_2, \quad (-2)R_1 + R_3 \rightarrow R_3,$$
$$\text{and } 30R_1 + R_4 \rightarrow R_4$$

$$
\sim
\left[
\begin{array}{cccccc|c}
1 & \frac{1}{2} & \frac{1}{2} & 0 & 0 & 0 & 35 \\
0 & \frac{3}{2} & -\frac{1}{2} & 1 & 0 & 0 & 40 \\
0 & ① & -1 & 0 & 1 & 0 & 20 \\
\hline
0 & -5 & 15 & 0 & 0 & 1 & 1050
\end{array}
\right]
\quad
\begin{array}{l}
\frac{35}{1/2} = 70 \\[4pt]
\frac{40}{3/2} = \frac{80}{3} \approx 26.67 \\[4pt]
\frac{20}{1} = 20
\end{array}
$$

$$\left(-\frac{1}{2}\right)R_3 + R_1 \rightarrow R_1, \quad \left(-\frac{3}{2}\right)R_3 + R_2 \rightarrow R_2, \quad \text{and } 5R_3 + R_4 \rightarrow R_4$$

$$
\sim
\begin{array}{c}
\begin{array}{cccccc}
y_1 & y_2 & x_1 & x_2 & x_3 & P
\end{array} \\
\begin{array}{c}
y_1 \\ x_2 \\ y_2 \\ P
\end{array}
\left[
\begin{array}{cccccc|c}
1 & 0 & 1 & 0 & -\frac{1}{2} & 0 & 25 \\
0 & 0 & 1 & 1 & -\frac{3}{2} & 0 & 10 \\
0 & 1 & -1 & 0 & 1 & 0 & 20 \\
\hline
0 & 0 & 10 & 0 & 5 & 1 & 1150
\end{array}
\right]
\end{array}
$$

The minimal production cost is \$1150 when the Cedarburg plant is operated 10 hours per day, the West Bend plant is operated 5 hours per day, and the Grafton plant is not used.

47. Refer to Problem 45. If the demand for deluxe ice cream increases to 300 gallons per day and all other data remains the same, then the matrices for this problem and the dual problem are:

$$
A = \left[
\begin{array}{ccc|c}
2 & 1 & 2 & 30 \\
1 & 2 & 2 & 30 \\
\hline
70 & 75 & 90 & 1
\end{array}
\right]
\quad \text{and} \quad
A^T = \left[
\begin{array}{cc|c}
2 & 1 & 70 \\
1 & 2 & 75 \\
2 & 2 & 90 \\
\hline
30 & 30 & 1
\end{array}
\right]
\quad \text{respectively.}
$$

Thus, the dual problem is:

Maximize $P = 30y_1 + 30y_2$

Subject to: $2y_1 + y_2 \le 70$

$y_1 + 2y_2 \le 75$

$$2y_1 + 2y_2 \le 90$$
$$y_1,\ y_2 \ge 0$$

We introduce slack variables $x_1, x_2,$ and x_3 to obtain the initial system:

$$
\begin{aligned}
2y_1 + y_2 + x_1 &= 70 \\
y_1 + 2y_2 + x_2 &= 75 \\
2y_1 + 2y_2 + x_3 &= 90 \\
-30y_1 - 30y_2 + P &= 0
\end{aligned}
$$

The simplex tableau for this problem is:

$$
\begin{array}{c}
\quad\ \ y_1\ \ y_2\ \ x_1\ \ x_2\ \ x_3\ \ P \\
\begin{array}{c} x_1 \\ x_2 \\ x_3 \\ P \end{array}
\left[
\begin{array}{cccccc|c}
② & 1 & 1 & 0 & 0 & 0 & 70 \\
1 & 2 & 0 & 1 & 0 & 0 & 75 \\
2 & 2 & 0 & 0 & 1 & 0 & 90 \\
\hline
-30 & -30 & 0 & 0 & 0 & 1 & 0
\end{array}
\right]
\begin{array}{l}
\frac{70}{2} = 35 \\[4pt]
\frac{75}{1} = 75 \\[4pt]
\frac{90}{2} = 45 \\
\end{array}
\end{array}
$$

$$\tfrac{1}{2}R_1 \to R_1$$

```
Note: Either column 1 or column 2 can be used as the pivot column.
      We chose column 1.
```

$$
\sim
\left[
\begin{array}{cccccc|c}
① & \frac{1}{2} & \frac{1}{2} & 0 & 0 & 0 & 35 \\
1 & 2 & 0 & 1 & 0 & 0 & 75 \\
2 & 2 & 0 & 0 & 1 & 0 & 90 \\
\hline
-30 & -30 & 0 & 0 & 0 & 1 & 0
\end{array}
\right]
\quad \sim
\left[
\begin{array}{cccccc|c}
1 & \frac{1}{2} & \frac{1}{2} & 0 & 0 & 0 & 35 \\
0 & \frac{3}{2} & -\frac{1}{2} & 1 & 0 & 0 & 40 \\
0 & ① & -1 & 0 & 1 & 0 & 20 \\
\hline
0 & -15 & 15 & 0 & 0 & 1 & 1050
\end{array}
\right]
\begin{array}{l}
\frac{35}{1/2} = 70 \\[4pt]
\frac{40}{3/2} = \frac{80}{3} \approx 26.67 \\[4pt]
\frac{20}{1} = 20
\end{array}
$$

$$(-1)R_1 + R_2 \to R_2, \qquad\qquad \left(-\tfrac{1}{2}\right)R_3 + R_1 \to R_1,$$

$$(-2)R_1 + R_3 \to R_3, \qquad\qquad \left(-\tfrac{3}{2}\right)R_3 + R_2 \to R_2,$$

$$30R_1 + R_4 \to R_4 \qquad\qquad 15R_3 + R_4 \to R_4$$

$$
\begin{array}{c}
\quad\ \ y_1\ \ y_2\ \ x_1\ \ x_2\ \ x_3\ \ P \\
\begin{array}{c} y_1 \\ x_2 \\ y_2 \\ P \end{array}
\sim
\left[
\begin{array}{cccccc|c}
1 & 0 & 1 & 0 & -\frac{1}{2} & 0 & 25 \\
0 & 0 & 1 & 1 & -\frac{3}{2} & 0 & 10 \\
0 & 1 & -1 & 0 & 1 & 0 & 20 \\
\hline
0 & 0 & 0 & 0 & 15 & 1 & 1350
\end{array}
\right]
\end{array}
$$

The minimal production cost is $1350 when the West Bend plant is operated 15 hours per day, and the Cedarburg and Grafton plants are not used.

49. Refer to Problem 45. If the demand for deluxe ice cream increases to 400 gallons per day and all other data remains the same, then the matrices for this problem and the dual are:

$$
A =
\left[
\begin{array}{ccc|c}
2 & 1 & 2 & 30 \\
1 & 2 & 2 & 40 \\
\hline
70 & 75 & 90 & 1
\end{array}
\right]
\quad \text{and} \quad
A^T =
\left[
\begin{array}{cc|c}
2 & 1 & 70 \\
1 & 2 & 75 \\
2 & 2 & 90 \\
\hline
30 & 40 & 1
\end{array}
\right]
\quad \text{respectively.}
$$

Thus, the dual problem is:

Maximize $P = 30y_1 + 40y_2$

Subject to: $2y_1 + y_2 \le 70$

$$y_1 + 2y_2 \le 75$$

$$2y_1 + 2y_2 \leq 90$$
$$y_1, \ y_2 \geq 0$$

We introduce slack variables $x_1, x_2,$ and x_3 to obtain the initial system:

$$2y_1 + \ y_2 + x_1 \qquad\qquad = 70$$
$$y_1 + 2y_2 \qquad + x_2 \qquad = 75$$
$$2y_1 + 2y_2 \qquad\qquad + x_3 \quad = 90$$
$$-30y_1 - 40y_2 \qquad\qquad\quad + P = 0$$

The simplex tableau for this problem is:

$$
\begin{array}{c}
\\ x_1 \\ x_2 \\ x_3 \\ P
\end{array}
\begin{array}{c}
y_1 \ \ y_2 \ \ x_1 \ \ x_2 \ \ x_3 \ \ P \\
\left[\begin{array}{cccccc|c}
2 & 1 & 1 & 0 & 0 & 0 & 70 \\
1 & ② & 0 & 1 & 0 & 0 & 75 \\
2 & 2 & 0 & 0 & 1 & 0 & 90 \\
\hline
-30 & -40 & 0 & 0 & 0 & 1 & 0
\end{array}\right]
\end{array}
\begin{array}{l}
\frac{70}{1} = 70 \\[4pt]
\frac{75}{2} = 37.5 \\[4pt]
\frac{90}{2} = 45
\end{array}
$$

$$\tfrac{1}{2}R_2 \to R_2$$

$$
\sim
\left[\begin{array}{cccccc|c}
2 & 1 & 1 & 0 & 0 & 0 & 70 \\
\frac{1}{2} & ① & 0 & \frac{1}{2} & 0 & 0 & \frac{75}{2} \\
2 & 2 & 0 & 0 & 1 & 0 & 90 \\
\hline
-30 & -40 & 0 & 0 & 0 & 1 & 0
\end{array}\right]
\sim
\left[\begin{array}{cccccc|c}
\frac{3}{2} & 0 & 1 & -\frac{1}{2} & 0 & 0 & \frac{65}{2} \\
\frac{1}{2} & 1 & 0 & \frac{1}{2} & 0 & 0 & \frac{75}{2} \\
① & 0 & 0 & -1 & 1 & 0 & 15 \\
\hline
-10 & 0 & 0 & 20 & 0 & 1 & 1500
\end{array}\right]
\begin{array}{l}
\frac{65/2}{3/2} \approx 21.67 \\[4pt]
\frac{75/2}{1/2} = 75 \\[4pt]
\frac{15}{1} = 15
\end{array}
$$

$$(-1)R_2 + R_1 \to R_1, \qquad\qquad \left(-\tfrac{3}{2}\right)R_3 + R_1 \to R_1,$$

$$(-2)R_2 + R_3 \to R_3, \qquad\qquad \left(-\tfrac{1}{2}\right)R_3 + R_2 \to R_2,$$

$$40R_2 + R_4 \to R_4 \qquad\qquad\quad 10R_3 + R_4 \to R_4$$

$$
\sim
\begin{array}{c}
\\ x_1 \\ y_2 \\ y_1 \\ P
\end{array}
\begin{array}{c}
y_1 \ \ y_2 \ \ x_1 \ \ x_2 \ \ x_3 \ \ P \\
\left[\begin{array}{cccccc|c}
0 & 0 & 1 & 1 & -\frac{3}{2} & 0 & 10 \\
0 & 1 & 0 & 1 & -\frac{1}{2} & 0 & 30 \\
1 & 0 & 0 & -1 & 1 & 0 & 15 \\
\hline
0 & 0 & 0 & 10 & 10 & 1 & 1650
\end{array}\right]
\end{array}
$$

The minimal production cost is $1650 when the Grafton plant and West Bend plant are each operated 10 hours per day, and the Cedarburg plant is not used.

51. Let x_1 = the number of ounces of food L,

x_2 = the number of ounces of food M,

x_3 = the number of ounces of food N.

Mathematical model: Minimize $C = 20x_1 + 24x_2 + 18x_3$

Subject to: $20x_1 + 10x_2 + 10x_3 \geq 300$

$10x_1 + 10x_2 + 10x_3 \geq 200$

$10x_1 + 15x_2 + 10x_3 \geq 240$

$x_1, \ x_2, \ x_3 \geq 0$

Divide the first two problem constraints by 10 and the third by 5. This will simplify the calculations.

$$A = \begin{bmatrix} 2 & 1 & 1 & | & 30 \\ 1 & 1 & 1 & | & 20 \\ 2 & 3 & 2 & | & 48 \\ \hline 20 & 24 & 18 & | & 1 \end{bmatrix} \quad \text{and} \quad A^T = \begin{bmatrix} 2 & 1 & 2 & | & 20 \\ 1 & 1 & 3 & | & 24 \\ 1 & 1 & 2 & | & 18 \\ \hline 30 & 20 & 48 & | & 1 \end{bmatrix}$$

The dual problem is: Maximize $P = 30y_1 + 20y_2 + 48y_3$

$$\text{Subject to: } 2y_1 + y_2 + 2y_3 \le 20$$
$$y_1 + y_2 + 3y_3 \le 24$$
$$y_1 + y_2 + 2y_3 \le 18$$
$$y_1, \; y_2, \; y_3 \ge 0$$

We introduce slack variables x_1, x_2, and x_3 to obtain the initial system:

$$2y_1 + y_2 + 2y_3 + x_1 \qquad\qquad = 20$$
$$y_1 + y_2 + 3y_3 \qquad + x_2 \qquad = 24$$
$$y_1 + y_2 + 2y_3 \qquad\qquad + x_3 \quad = 18$$
$$-30y_1 - 20y_2 - 48y_3 \qquad\qquad + P = 0$$

The simplex tableau for this problem is:

$$
\begin{array}{c}
\begin{array}{ccccccc}
y_1 & y_2 & y_3 & x_1 & x_2 & x_3 & P
\end{array} \\
\begin{array}{c} x_1 \\ x_2 \\ x_3 \\ P \end{array}
\left[\begin{array}{ccccccc|c}
2 & 1 & 2 & 1 & 0 & 0 & 0 & 20 \\
1 & 1 & \text{③} & 0 & 1 & 0 & 0 & 24 \\
1 & 1 & 2 & 0 & 0 & 1 & 0 & 18 \\
\hline
-30 & -20 & -48 & 0 & 0 & 0 & 1 & 0
\end{array}\right]
\begin{array}{l}
\frac{20}{2} = 10 \\[4pt]
\frac{24}{3} = 8 \\[4pt]
\frac{18}{2} = 9 \\[4pt]
\;
\end{array}
\end{array}
$$

$\frac{1}{3}R_2 \rightarrow R_2$

$$
\sim \left[\begin{array}{ccccccc|c}
2 & 1 & 2 & 1 & 0 & 0 & 0 & 20 \\
\frac{1}{3} & \frac{1}{3} & \text{①} & 0 & \frac{1}{3} & 0 & 0 & 8 \\
1 & 1 & 2 & 0 & 0 & 1 & 0 & 18 \\
\hline
-30 & -20 & -48 & 0 & 0 & 0 & 1 & 0
\end{array}\right]
$$

$(-2)R_2 + R_1 \rightarrow R_1$, $(-2)R_2 + R_3 \rightarrow R_3$, and $48R_2 + R_4 \rightarrow R_4$

$$
\sim \left[\begin{array}{ccccccc|c}
\text{④/③} & \frac{1}{3} & 0 & 1 & -\frac{2}{3} & 0 & 0 & 4 \\
\frac{1}{3} & \frac{1}{3} & 1 & 0 & \frac{1}{3} & 0 & 0 & 8 \\
\frac{1}{3} & \frac{1}{3} & 0 & 0 & -\frac{2}{3} & 1 & 0 & 2 \\
\hline
-14 & -4 & 0 & 0 & 16 & 0 & 1 & 384
\end{array}\right]
\begin{array}{l}
\frac{4}{4/3} = 3 \\[4pt]
\frac{8}{1/3} = 24 \\[4pt]
\frac{2}{1/3} = 6 \\[4pt]
\;
\end{array}
$$

$\frac{3}{4}R_1 \rightarrow R_1$

$$\sim \begin{bmatrix} ① & \frac{1}{4} & 0 & \frac{3}{4} & -\frac{1}{2} & 0 & 0 & 3 \\ \frac{1}{3} & \frac{1}{3} & 1 & 0 & \frac{1}{3} & 0 & 0 & 8 \\ \frac{1}{3} & \frac{1}{3} & 0 & 0 & -\frac{2}{3} & 1 & 0 & 2 \\ \hdashline -14 & -4 & 0 & 0 & 16 & 0 & 1 & 384 \end{bmatrix}$$

$$\left(-\frac{1}{3}\right)R_1 + R_2 \rightarrow R_2, \quad \left(-\frac{1}{3}\right)R_1 + R_3 \rightarrow R_3, \quad \text{and } 14R_1 + R_4 \rightarrow R_4$$

$$\sim \begin{bmatrix} 1 & \frac{1}{4} & 0 & \frac{3}{4} & -\frac{1}{2} & 0 & 0 & 3 \\ 0 & \frac{1}{4} & 1 & -\frac{1}{4} & \frac{1}{2} & 0 & 0 & 7 \\ 0 & ⓵\!\!\!\frac{1}{4} & 0 & -\frac{1}{4} & -\frac{1}{2} & 1 & 0 & 1 \\ \hdashline 0 & -\frac{1}{2} & 0 & \frac{21}{2} & 9 & 0 & 1 & 426 \end{bmatrix} \quad \begin{array}{l} \frac{3}{1/4} = 12 \\ \frac{7}{1/4} = 28 \\ \frac{1}{1/4} = 4 \end{array}$$

$$4R_3 \rightarrow R_3$$

$$\sim \begin{bmatrix} 1 & \frac{1}{4} & 0 & \frac{3}{4} & -\frac{1}{2} & 0 & 0 & 3 \\ 0 & \frac{1}{4} & 1 & -\frac{1}{4} & \frac{1}{2} & 0 & 0 & 7 \\ 0 & ① & 0 & -1 & -2 & 4 & 0 & 4 \\ \hdashline 0 & -\frac{1}{2} & 0 & \frac{21}{2} & 9 & 0 & 1 & 426 \end{bmatrix}$$

$$\begin{array}{cccccccc} & y_1 & y_2 & y_3 & x_1 & x_2 & x_3 & P \\ \sim \begin{matrix} y_1 \\ y_3 \\ y_2 \\ P \end{matrix} & \begin{bmatrix} 1 & 0 & 0 & 1 & 0 & -1 & 0 & 2 \\ 0 & 0 & 1 & 0 & 1 & -1 & 0 & 6 \\ 0 & 1 & 0 & -1 & -2 & 4 & 0 & 4 \\ \hdashline 0 & 0 & 0 & 10 & 8 & 2 & 1 & 428 \end{bmatrix} \end{array}$$

$$\left(-\frac{1}{4}\right)R_3 + R_1 \rightarrow R_1, \quad \left(-\frac{1}{4}\right)R_3 + R_2 \rightarrow R_2,$$

$$\text{and } \frac{1}{2}R_3 + R_4 \rightarrow R_4$$

The minimal cholesterol intake is 428 units when 10 ounces of food L, 8 ounces of food M, and 2 ounces of food N are used.

53. Let x_1 = the number of students bused from North Division to Central,

x_2 = the number of students bused from North Division to Washington,

x_3 = the number of students bused from South Division to Central,

x_4 = the number of students bused from South Division to Washington.

The mathematical model for this problem is:

Minimize $C = 5x_1 + 2x_2 + 3x_3 + 4x_4$

Subject to: $x_1 + x_2 \qquad\qquad \geq 300$

$\qquad\qquad\quad x_3 + x_4 \geq 500$

$\qquad x_1 \quad + x_3 \qquad \leq 400$

$\qquad\qquad x_2 \qquad + x_4 \leq 500$

$\qquad x_1, \; x_2, \; x_3, \; x_4 \geq 0$

We multiply the last two problem constraints by –1 so that all the constraints are of the $\geq$ type. The model becomes:

Minimize $C = 5x_1 + 2x_2 + 3x_3 + 4x_4$

$$\text{Subject to: } x_1 + x_2 \qquad \geq 300$$
$$x_3 + x_4 \geq 500$$
$$-x_1 \quad -x_3 \qquad \geq -400$$
$$-x_2 \quad -x_4 \geq -500$$
$$x_1,\ x_2,\ x_3,\ x_4 \geq 0$$

The matrices for this problem and the dual problem are:

$$A = \begin{bmatrix} 1 & 1 & 0 & 0 & | & 300 \\ 0 & 0 & 1 & 1 & | & 500 \\ -1 & 0 & -1 & 0 & | & -400 \\ 0 & -1 & 0 & -1 & | & -500 \\ \hline 5 & 2 & 3 & 4 & | & 1 \end{bmatrix} \quad \text{and} \quad A^T = \begin{bmatrix} 1 & 0 & -1 & 0 & | & 5 \\ 1 & 0 & 0 & -1 & | & 2 \\ 0 & 1 & -1 & 0 & | & 3 \\ 0 & 1 & 0 & -1 & | & 4 \\ \hline 300 & 500 & -400 & -500 & | & 1 \end{bmatrix}$$

The dual problem is: Maximize $P = 300y_1 + 500y_2 - 400y_3 - 500y_4$

$$\text{Subject to: } y_1 \qquad - y_3 \qquad \leq 5$$
$$y_1 \qquad\qquad -y_4 \leq 2$$
$$y_2 - y_3 \qquad \leq 3$$
$$y_2 \qquad -y_4 \leq 4$$
$$y_1,\ y_2,\ y_3,\ y_4 \geq 0$$

We introduce slack variables x_1, x_2, x_3, and x_4 to obtain the initial system:

$$y_1 \qquad - \quad y_3 \qquad + x_1 \qquad\qquad\qquad = 5$$
$$y_1 \qquad\qquad - \quad y_4 \qquad + x_2 \qquad\qquad = 2$$
$$y_2 - \quad y_3 \qquad\qquad + x_3 \qquad = 3$$
$$y_2 \qquad - \quad y_4 \qquad\qquad + x_4 = 4$$
$$-300y_1 - 500y_2 + 400y_3 + 500y_4 \qquad\qquad + P = 0$$

The simplex tableau for this problem is:

	y_1	y_2	y_3	y_4	x_1	x_2	x_3	x_4	P	
x_1	1	0	-1	0	1	0	0	0	0	5
x_2	1	0	0	-1	0	1	0	0	0	2
x_3	0	①	-1	0	0	0	1	0	0	3
x_4	0	1	0	-1	0	0	0	1	0	4
P	-300	-500	400	500	0	0	0	0	1	0

$\dfrac{3}{1} = 3$

$\dfrac{4}{1} = 4$

$(-1)R_3 + R_4 \rightarrow R_4$ and $500R_3 + R_5 \rightarrow R_5$

$$
\begin{bmatrix}
1 & 0 & -1 & 0 & 1 & 0 & 0 & 0 & 0 & 5 \\
\textcircled{1} & 0 & 0 & -1 & 0 & 1 & 0 & 0 & 0 & 2 \\
0 & 1 & -1 & 0 & 0 & 0 & 1 & 0 & 0 & 3 \\
0 & 0 & 1 & -1 & 0 & 0 & -1 & 1 & 0 & 1 \\
\hline
-300 & 0 & -100 & 500 & 0 & 0 & 500 & 0 & 1 & 1500
\end{bmatrix}
\begin{array}{l} \frac{5}{1} = 5 \\[4pt] \frac{2}{1} = 2 \end{array}
$$

$(-1)R_2 + R_1 \rightarrow R_1$ and $300R_2 + R_5 \rightarrow R_5$

$$
\sim
\begin{bmatrix}
0 & 0 & -1 & 1 & 1 & -1 & 0 & 0 & 0 & 3 \\
1 & 0 & 0 & -1 & 0 & 1 & 0 & 0 & 0 & 2 \\
0 & 1 & -1 & 0 & 0 & 0 & 1 & 0 & 0 & 3 \\
0 & 0 & \textcircled{1} & -1 & 0 & 0 & -1 & 1 & 0 & 1 \\
\hline
0 & 0 & -100 & 200 & 0 & 300 & 500 & 0 & 1 & 2100
\end{bmatrix}
$$

$R_4 + R_1 \rightarrow R_1$, $R_4 + R_3 \rightarrow R_3$, and $100R_4 + R_5 \rightarrow R_5$

	y_1	y_2	y_3	y_4	x_1	x_2	x_3	x_4	P	
x_1	0	0	0	0	1	-1	-1	1	0	4
y_1	1	0	0	-1	0	1	0	0	0	2
$\sim y_2$	0	1	0	-1	0	0	0	1	0	4
y_3	0	0	1	-1	0	0	-1	1	0	1
P	0	0	0	100	0	300	400	100	1	2200

The minimal cost is $2200 when 300 students are bused from North Division to Washington, 400 students are bused from South Division to Central, and 100 students are bused from South Division to Washington. No students are bused from North Division to Central.

EXERCISE 6-4

Things to remember:

The BIG M method is a solution method for a linear programming problem with an objective function to be maximized and problem constraints that are a combination of $\geq$ inequalities, $\leq$ inequalities, and equations.

1. THE BIG M METHOD—INTRODUCING SLACK, SURPLUS AND ARTIFICIAL VARIABLES TO FORM THE MODIFIED PROBLEM

   Step 1. If any problem constraints have negative constants on the right-hand side, . multiply both sides by –1 to obtain a constraint with a nonnegative constant. [If the constraint is an inequality, this will reverse the direction of the inequality.]

   Step 2. Introduce a SLACK VARIABLE in each $\leq$ constraint.

   Step 3. Introduce a SURPLUS VARIABLE and an ARTIFICIAL VARIABLE in each $\geq$ constraint.

   Step 4. Introduce an artificial variable in each = constraint.

   Step 5. For each artificial variable a_i, add $-Ma_i$ to the objective function. Use the same constant M for all artificial variables.

2. THE BIG M METHOD—SOLVING THE PROBLEM

   Step 1. Form the preliminary simplex tableau for the modified problem.

   Step 2. Use row operations to eliminate the M's in the bottom row of the preliminary simplex tableau in the columns corresponding to the artificial variables. The resulting tableau is the initial simplex tableau.

   Step 3. Solve the modified problem by applying the simplex method to the initial simplex tableau found in Step 2.

   Step4. Relate the optimal solution of the modified problem to the original problem.

   (a) If the modified problem has no optimal solution, then the original problem has no optimal solution.

   (b) If all artificial variables are zero in the solution to the modified problem, then delete the artificial variables to find an optimal solution to the original problem.

   (c) If any artificial variables are nonzero in the optimal solution to the modified problem, then the original problem has no optimal solution.

---

1. (A) We introduce a slack variable s_1 to convert the first inequality ($\leq$) into an equation, and we use a surplus variable s_2 and an artificial variable a_1 to convert the second inequality ($\geq$) into an equation.

   The modified problem is: Maximize $P = 5x_1 + 2x_2 - Ma_1$

   $$\text{Subject to: } \begin{aligned} x_1 + 2x_2 + s_1 \quad\quad\quad &= 12 \\ x_1 + x_2 \quad -s_2 + a_1 &= 4 \\ x_1,\ x_2,\ s_1,\ s_2,\ a_1 &\geq 0 \end{aligned}$$

(B)   The preliminary simplex tableau for the modified problem is:

$$
\begin{array}{cccccc}
x_1 & x_2 & s_1 & s_2 & a_1 & P \\
\end{array}
$$

$$
\left[\begin{array}{cccccc|c}
1 & 2 & 1 & 0 & 0 & 0 & 12 \\
1 & 1 & 0 & -1 & 1 & 0 & 4 \\
\hline
-5 & -2 & 0 & 0 & M & 1 & 0
\end{array}\right]
\sim
\left[\begin{array}{cccccc|c}
1 & 2 & 1 & 0 & 0 & 0 & 12 \\
1 & 1 & 0 & -1 & 1 & 0 & 4 \\
\hline
-M-5 & -M-2 & 0 & M & 0 & 1 & -4M
\end{array}\right]
$$

$(-M)R_2 + R_3 \rightarrow R_3$

Thus, the initial simplex tableau is:

$$
\begin{array}{cccccc}
x_1 & x_2 & s_1 & s_2 & a_1 & P \\
\end{array}
$$

$$
\left[\begin{array}{cccccc|c}
1 & 2 & 1 & 0 & 0 & 0 & 12 \\
1 & 1 & 0 & -1 & 1 & 0 & 4 \\
\hline
-M-5 & -M-2 & 0 & M & 0 & 1 & -4M
\end{array}\right]
$$

(C)   We use the simplex method to solve the modified problem.

$$
\begin{array}{c}
 \\
s_1 \\
a_1 \\
P
\end{array}
\left[\begin{array}{cccccc|c}
x_1 & x_2 & s_1 & s_2 & a_1 & & \\
1 & 2 & 1 & 0 & 0 & 0 & 12 \\
\boxed{1} & 1 & 0 & -1 & 1 & 0 & 4 \\
\hline
-M-5 & -M-2 & 0 & M & 0 & 1 & -4M
\end{array}\right]
\begin{array}{l}
\frac{12}{1} = 12 \\
\frac{4}{1} = 4
\end{array}
$$

$(-1)R_2 + R_1 \rightarrow R_1$ and $(M+5)R_2 + R_3 \rightarrow R_3$

$$
\begin{array}{cccccc}
 & & & & & \\
\end{array}
$$

$$
\sim
\left[\begin{array}{cccccc|c}
0 & 1 & 1 & \boxed{1} & -1 & 0 & 8 \\
1 & 1 & 0 & -1 & 1 & 0 & 4 \\
\hline
0 & 3 & 0 & -5 & M+5 & 1 & 20
\end{array}\right]
\begin{array}{c}
s_2 \\
\sim x_1 \\
P
\end{array}
\left[\begin{array}{cccccc|c}
x_1 & x_2 & s_1 & s_2 & a_1 & P & \\
0 & 1 & 1 & 1 & -1 & 0 & 8 \\
1 & 2 & 1 & 0 & 0 & 0 & 12 \\
\hline
0 & 8 & 5 & 0 & M & 1 & 60
\end{array}\right]
$$

$R_1 + R_2 \rightarrow R_2$ and $5R_1 + R_3 \rightarrow R_3$

Thus, the optimal solution of the modified problem is: max $P = 60$ at $x_1 = 12$, $x_2 = 0$, $s_1 = 0$, $s_2 = 8$, $a_1 = 0$.

(D)   The optimal solution of the original problem is: max $P = 60$ at $x_1 = 12$, $x_2 = 0$.

3.   (A)   We introduce the slack variable s_1 and the artificial variable a_1 to obtain the modified problem:

Maximize $P = 3x_1 + 5x_2 - Ma_1$

Subject to: $2x_1 + x_2 + s_1 \quad = 8$

$\qquad\quad x_1 + x_2 \quad\quad + a_1 = 6$

$\qquad\quad x_1, \, x_2, \, s_1, \, a_1 \geq 0$

(B)   The preliminary simplex tableau for the modified problem is:

$$
\begin{array}{ccccc}
x_1 & x_2 & s_1 & a_1 & P
\end{array}
$$

$$
\left[\begin{array}{ccccc|c}
2 & 1 & 1 & 0 & 0 & 8 \\
1 & 1 & 0 & 1 & 0 & 6 \\
\hline
-3 & -5 & 0 & M & 1 & 0
\end{array}\right]
\sim
\left[\begin{array}{ccccc|c}
2 & 1 & 1 & 0 & 0 & 8 \\
1 & 1 & 0 & 1 & 0 & 6 \\
\hline
-M-3 & -M-5 & 0 & 0 & 1 & -6M
\end{array}\right]
$$

$(-M)R_2 + R_3 \rightarrow R_3$

Thus, the initial simplex tableau is:

$$
\begin{array}{cccccc}
 & x_1 & x_2 & s_1 & a_1 & P
\end{array}
$$

$$
\begin{array}{c}
s_1 \\
a_1 \\
\\
P
\end{array}
\left[\begin{array}{ccccc|c}
2 & 1 & 1 & 0 & 0 & 8 \\
1 & 1 & 0 & 1 & 0 & 6 \\
\hline
-M-3 & -M-5 & 0 & 0 & 1 & -6M
\end{array}\right]
$$

(C)   We use the simplex method to solve the modified problem.

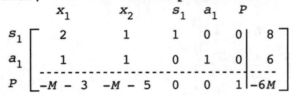

$(-1)R_2 + R_1 \rightarrow R_1$ and $(M+5)R_2 + R_3 \rightarrow R_3$

Thus, the optimal solution of the modified problem is max P = 30 at x_1 = 0, x_2 = 6, s_1 = 2, a_1 = 0.

(D)   The optimal solution of the original problem is: max $P = 30$ at $x_1 = 0$, $x_2 = 6$.

5.   (A)   We introduce slack, surplus, and artificial variables to obtain the modified problem:
Maximize $P = 4x_1 + 3x_2 - Ma_1$

Subject to:
$$-x_1 + 2x_2 + s_1 \qquad = 2$$
$$x_1 + x_2 \qquad - s_2 + a_1 = 4$$
$$x_1,\ x_2,\ s_1,\ s_2,\ a_1 \geq 0$$

(B)   The preliminary simplex tableau for the modified problem is:

$$
\begin{array}{cccccc}
x_1 & x_2 & s_1 & s_2 & a_1 & P \\
\end{array}
$$

$$
\left[\begin{array}{cccccc|c}
-1 & 2 & 1 & 0 & 0 & 0 & 2 \\
1 & 1 & 0 & -1 & 1 & 0 & 4 \\
\hline
-4 & -3 & 0 & 0 & M & 1 & 0
\end{array}\right]
\sim
\left[\begin{array}{cccccc|c}
-1 & 2 & 1 & 0 & 0 & 0 & 2 \\
1 & 1 & 0 & -1 & 1 & 0 & 4 \\
\hline
-M-4 & -M-3 & 0 & M & 0 & 1 & -4M
\end{array}\right]
$$

$(-M)R_2 + R_3 \rightarrow R_3$

Thus, the initial simplex tableau is:

$$
\begin{array}{ccccccc}
 & x_1 & x_2 & s_1 & s_2 & a_1 & P \\
\end{array}
$$

$$
\begin{array}{c}
s_1 \\
a_1 \\
P
\end{array}
\left[\begin{array}{cccccc|c}
-1 & 2 & 1 & 0 & 0 & 0 & 2 \\
1 & 1 & 0 & -1 & 1 & 0 & 4 \\
\hline
-M-4 & -M-3 & 0 & M & 0 & 1 & -4M
\end{array}\right]
$$

(C)   We use the simplex method to solve the modified problem:

$$
\begin{array}{cccccc}
x_1 & x_2 & s_1 & s_2 & a_1 & P \\
\end{array}
$$

$$
\left[\begin{array}{cccccc|c}
-1 & 2 & 1 & 0 & 0 & 0 & 2 \\
①& 1 & 0 & -1 & 1 & 0 & 4 \\
\hline
-M-4 & -M-3 & 0 & M & 0 & 1 & -4M
\end{array}\right]
\sim
\left[\begin{array}{cccccc|c}
0 & 3 & 1 & -1 & 1 & 0 & 6 \\
1 & 1 & 0 & -1 & 1 & 0 & 4 \\
\hline
0 & 1 & 0 & -4 & M+4 & 1 & 16
\end{array}\right]
$$

$R_2 + R_1 \rightarrow R_1, \ (M+4)R_2 + R_3 \rightarrow R_3$

No optimal solution exists because the elements in the pivot column (the s_2 column) above the dashed line are negative.

(D)   No optimal solution exists.

7. (A)   We introduce slack, surplus, and artificial variables to obtain the modified problem:

Maximize $P = 5x_1 + 10x_2 - Ma_1$

Subject to: $x_1 + x_2 + s_1 \qquad\qquad = 3$

$\qquad\qquad 2x_1 + 3x_2 \qquad - s_2 + a_1 = 12$

$\qquad\qquad x_1,\ x_2,\ s_1,\ s_2,\ a_1 \geq 0$

(B)   The preliminary simplex tableau for the modified problem is:

$$
\begin{array}{c} \\ s_1 \\ a_1 \\ \\ P \end{array}
\begin{array}{cccccc}
x_1 & x_2 & s_1 & s_2 & a_1 & P \\
\end{array}
\left[\begin{array}{cccccc|c}
1 & 1 & 1 & 0 & 0 & 0 & 3 \\
2 & 3 & 0 & -1 & 1 & 0 & 12 \\
\hline
-5 & -10 & 0 & 0 & M & 1 & 0 \\
\end{array}\right]
\sim
\left[\begin{array}{cccccc|c}
1 & 1 & 1 & 0 & 0 & 0 & 3 \\
2 & 3 & 0 & -1 & 1 & 0 & 12 \\
\hline
-2M-5 & -3M-10 & 0 & M & 0 & 1 & -12M \\
\end{array}\right]
$$

$(-M)R_2 + R_3 \rightarrow R_3$

Thus, the initial simplex tableau is:

$$
\begin{array}{c} \\ s_1 \\ a_1 \\ \\ P \end{array}
\begin{array}{cccccc}
x_1 & x_2 & s_1 & s_2 & a_1 & P \\
\end{array}
\left[\begin{array}{cccccc|c}
1 & 1 & 1 & 0 & 0 & 0 & 3 \\
2 & 3 & 0 & -1 & 1 & 0 & 12 \\
\hline
-2M-5 & -3M-10 & 0 & M & 0 & 1 & -12M \\
\end{array}\right]
$$

(C)   Applying the simplex method to the initial tableau, we have:

$$
\left[\begin{array}{cccccc|c}
1 & ① & 1 & 0 & 0 & 0 & 3 \\
2 & 3 & 0 & -1 & 1 & 0 & 12 \\
\hline
-2M-5 & -3M-10 & 0 & M & 0 & 1 & -12M \\
\end{array}\right]
$$

$(-3)R_1 + R_2 \rightarrow R_2,\ (3M+10)R_1 + R_3 \rightarrow R_3$

$$
\begin{array}{c} \\ x_2 \\ a_1 \\ \\ P \end{array}
\begin{array}{cccccc}
x_1 & x_2 & s_1 & s_2 & a_1 & P \\
\end{array}
\sim\left[\begin{array}{cccccc|c}
1 & 1 & 1 & 0 & 0 & 0 & 3 \\
-1 & 0 & -3 & -1 & 1 & 0 & 3 \\
\hline
M+5 & 0 & 3M+10 & M & 0 & 1 & -3M+30 \\
\end{array}\right]
$$

The optimal solution of the modified problem is: max $P = -3M + 30$ at $x_1 = 0$, $x_2 = 3$, $s_1 = 0$, $s_2 = 0$, and $a_1 = 3$.

(D)   The original problem does not have an optimal solution, since the artificial variable a_1 in the solution of the modified problem has a nonzero value.

9.   To minimize $P = 2x_1 - x_2$, we maximize $T = -P = -2x_1 + x_2$. Introducing slack, surplus, and artificial variables, we obtain the modified problem:

Maximize $T = -2x_1 + x_2 - Ma_1$

Subject to: $x_1 + x_2 + s_1 \qquad\qquad = 8$

$\qquad\qquad 5x_1 + 3x_2 \qquad - s_2 + a_1 = 30$

$\qquad\qquad\qquad x_1,\ x_2,\ s_1,\ s_2,\ a_1 \geq 0$

The preliminary simplex tableau for this problem is:

$$
\begin{array}{cccccc}
x_1 & x_2 & s_1 & s_2 & a_1 & T \\
\end{array}
$$

$$
\left[
\begin{array}{cccccc|c}
1 & 1 & 1 & 0 & 0 & 0 & 8 \\
5 & 3 & 0 & -1 & 1 & 0 & 30 \\
\hline
2 & -1 & 0 & 0 & M & 1 & 0 \\
\end{array}
\right]
$$

$$(-M)R_2 + R_3 \rightarrow R_3$$

$$
\begin{array}{c}
s_1 \\
\sim \quad a_1 \\
T
\end{array}
\left[
\begin{array}{cccccc|c}
x_1 & x_2 & s_1 & s_2 & a_1 & T & \\
1 & 1 & 1 & 0 & 0 & 0 & 8 \\
⑤ & 3 & 0 & -1 & 1 & 0 & 30 \\
\hline
-5M + 2 & -3M - 1 & 0 & M & 0 & 1 & -30M \\
\end{array}
\right]
\begin{array}{l}
\frac{8}{1} = 8 \\
\frac{30}{5} = 6 \\
\end{array}
$$

(This is the initial simplex tableau.) $\dfrac{1}{5}R_2 \rightarrow R_2$

$$
\sim
\left[
\begin{array}{cccccc|c}
1 & 1 & 1 & 0 & 0 & 0 & 8 \\
① & \frac{3}{5} & 0 & -\frac{1}{5} & \frac{1}{5} & 0 & 6 \\
\hline
-5M + 2 & -3M - 1 & 0 & M & 0 & 1 & -30M \\
\end{array}
\right]
$$

$$(-1)R_2 + R_1 \rightarrow R_1, \quad (5M - 2)R_2 + R_3 \rightarrow R_3$$

$$
\sim
\left[
\begin{array}{cccccc|c}
0 & ②⁄₅ & 1 & \frac{1}{5} & -\frac{1}{5} & 0 & 2 \\
1 & \frac{3}{5} & 0 & -\frac{1}{5} & \frac{1}{5} & 0 & 6 \\
\hline
0 & -\frac{11}{5} & 0 & \frac{2}{5} & M - \frac{2}{5} & 1 & -12 \\
\end{array}
\right]
\begin{array}{l}
\frac{2}{2/5} = 5 \\
\frac{6}{3/5} = 10 \\
\end{array}
$$

$$\dfrac{5}{2}R_1 \rightarrow R_1$$

$$
\sim
\left[
\begin{array}{cccccc|c}
0 & ① & \frac{5}{2} & \frac{1}{2} & -\frac{1}{2} & 0 & 5 \\
1 & \frac{3}{5} & 0 & -\frac{1}{5} & \frac{1}{5} & 0 & 6 \\
\hline
0 & -\frac{11}{5} & 0 & \frac{2}{5} & M - \frac{2}{5} & 1 & -12 \\
\end{array}
\right]
\begin{array}{c}
x_2 \\
\sim \quad x_1 \\
T
\end{array}
\left[
\begin{array}{cccccc|c}
x_1 & x_2 & s_1 & s_2 & a_1 & T & \\
0 & 1 & \frac{5}{2} & \frac{1}{2} & -\frac{1}{2} & 0 & 5 \\
1 & 0 & -\frac{3}{2} & -\frac{1}{2} & \frac{1}{2} & 0 & 3 \\
\hline
0 & 0 & \frac{11}{2} & \frac{3}{2} & M - \frac{3}{2} & 1 & -1 \\
\end{array}
\right]
$$

$$\left(-\dfrac{3}{5}\right)R_1 + R_2 \rightarrow R_2, \quad \left(\dfrac{11}{5}\right)R_1 + R_3 \rightarrow R_3$$

Thus, the optimal solution is: max $T = -1$ at $x_1 = 3$, $x_2 = 5$, and min $P = -\max T = 1$.

The modified problem for maximizing $P = 2x_1 - x_2$ subject to the given constraints is: Maximize $P = 2x_1 - x_2 - Ma_1$

$$
\begin{array}{rcl}
\text{Subject to:} \quad x_1 + x_2 + s_1 & = & 8 \\
5x_1 + 3x_2 \quad\quad - s_2 + a_1 & = & 30 \\
x_1, \, x_2, \, s_1, \, s_2, \, a_1 & \geq & 0 \\
\end{array}
$$

The preliminary simplex tableau for the modified problem is:

$$
\begin{array}{c}
\begin{array}{cccccc} x_1 & x_2 & s_1 & s_2 & a_1 & P \end{array}\\
\left[\begin{array}{cccccc|c}
1 & 1 & 1 & 0 & 0 & 0 & 8\\
5 & 3 & 0 & -1 & 1 & 0 & 30\\
\hline
-2 & 1 & 0 & 0 & M & 1 & 0
\end{array}\right]
\end{array}
$$

$$(-M)R_2 + R_3 \rightarrow R_3$$

$$
\begin{array}{c}
\begin{array}{cccccc} x_1 & x_2 & s_1 & s_2 & a_1 & P \end{array}\\
\sim\begin{array}{c} s_1\\ a_1\\ \\ P \end{array}
\left[\begin{array}{cccccc|c}
1 & 1 & 1 & 0 & 0 & 0 & 8\\
\circled{5} & 3 & 0 & -1 & 1 & 0 & 30\\
\hline
-5M-2 & -3M+1 & 0 & M & 0 & 1 & -30M
\end{array}\right]
\begin{array}{l}\frac{8}{1}=8\\ \frac{30}{5}=6\end{array}
\end{array}
$$

(This is the initial simplex tableau.) $\frac{1}{5}R_2 \rightarrow R_2$

$$
\sim\left[\begin{array}{cccccc|c}
1 & 1 & 1 & 0 & 0 & 0 & 8\\
\circled{1} & \frac{3}{5} & 0 & -\frac{1}{5} & \frac{1}{5} & 0 & 6\\
\hline
-5M-2 & -3M+1 & 0 & M & 0 & 1 & -30M
\end{array}\right]
\sim\left[\begin{array}{cccccc|c}
0 & \frac{2}{5} & 1 & \circled{\frac{1}{5}} & -\frac{1}{5} & 0 & 2\\
1 & \frac{3}{5} & 0 & -\frac{1}{5} & \frac{1}{5} & 0 & 6\\
\hline
0 & \frac{11}{5} & 0 & -\frac{2}{5} & M+\frac{2}{5} & 1 & 12
\end{array}\right]
$$

$$(-1)R_2 + R_1 \rightarrow R_1,\ (5M+2)R_2 + R_3 \rightarrow R_3 \qquad\qquad 5R_1 \rightarrow R_1$$

$$
\sim\left[\begin{array}{cccccc|c}
0 & 2 & 5 & 1 & -1 & 0 & 10\\
1 & \frac{3}{5} & 0 & -\frac{1}{5} & \frac{1}{5} & 0 & 6\\
\hline
0 & \frac{11}{5} & 0 & -\frac{2}{5} & M+\frac{2}{5} & 1 & 12
\end{array}\right]
\sim
\begin{array}{c}
\begin{array}{cccccc} x_1 & x_2 & s_1 & s_2 & a_1 & P \end{array}\\
\begin{array}{c} s_2\\ x_2\\ \\ P \end{array}
\left[\begin{array}{cccccc|c}
0 & 2 & 5 & 1 & -1 & 0 & 10\\
1 & 1 & 1 & 0 & 0 & 0 & 8\\
\hline
0 & 3 & 2 & 0 & M & 1 & 16
\end{array}\right]
\end{array}
$$

$$\frac{1}{5}R_1 + R_2 \rightarrow R_2 \text{ and } \frac{2}{5}R_1 + R_3 \rightarrow R_3$$

Thus, the optimal solution is: max $P = 16$ at $x_1 = 8$, $x_2 = 0$.

11. We introduce slack, surplus, and artificial variables to obtain the modified problem:
Maximize $P = 2x_1 + 5x_2 - Ma_1$

Subject to:
$$
\begin{aligned}
x_1 + 2x_2 + s_1 &\qquad\qquad\qquad = 18\\
2x_1 + x_2 &\quad + s_2 \qquad\qquad = 21\\
x_1 + x_2 &\qquad\qquad -s_3 + a_1 = 10\\
x_1,\ x_2,\ s_1,\ s_2,\ s_3,\ a_1 &\geq 0
\end{aligned}
$$

The preliminary simplex tableau for this problem is:

$$
\begin{array}{ccccccc}
x_1 & x_2 & s_1 & s_2 & s_3 & a_1 & P \\
\end{array}
$$

$$
\begin{bmatrix}
1 & 2 & 1 & 0 & 0 & 0 & 0 & | & 18 \\
2 & 1 & 0 & 1 & 0 & 0 & 0 & | & 21 \\
1 & 1 & 0 & 0 & -1 & 1 & 0 & | & 10 \\
\hline
-2 & -5 & 0 & 0 & 0 & M & 1 & | & 0
\end{bmatrix}
$$

$$(-M)R_3 + R_4 \rightarrow R_4$$

$$
\begin{array}{c}
\\ s_1 \\ s_2 \\ a_1 \\ \\ P
\end{array}
\begin{bmatrix}
x_1 & x_2 & s_1 & s_2 & s_3 & a_1 & P & & \\
1 & ② & 1 & 0 & 0 & 0 & 0 & | & 18 \\
2 & 1 & 0 & 1 & 0 & 0 & 0 & | & 21 \\
1 & 1 & 0 & 0 & -1 & 1 & 0 & | & 10 \\
\hline
-M-2 & -M-5 & 0 & 0 & M & 0 & 1 & | & -10M
\end{bmatrix}
\begin{array}{l}
\frac{18}{2} = 9 \\
\frac{21}{1} = 21 \text{ (This is the initial} \\
\qquad\quad\text{simplex tableau.)} \\
\frac{10}{1} = 10
\end{array}
$$

$$\frac{1}{2}R_1 \rightarrow R_1$$

$$
\begin{bmatrix}
\frac{1}{2} & ① & \frac{1}{2} & 0 & 0 & 0 & 0 & | & 9 \\
2 & 1 & 0 & 1 & 0 & 0 & 0 & | & 21 \\
1 & 1 & 0 & 0 & -1 & 1 & 0 & | & 10 \\
\hline
-M-2 & -M-5 & 0 & 0 & M & 0 & 1 & | & -10M
\end{bmatrix}
$$

$$(-1)R_1 + R_2 \rightarrow R_2, \ (-1)R_1 + R_3 \rightarrow R_3, \text{ and } (M+5)R_1 + R_4 \rightarrow R_4$$

$$
\begin{bmatrix}
\frac{1}{2} & 1 & \frac{1}{2} & 0 & 0 & 0 & 0 & | & 9 \\
\frac{3}{2} & 0 & -\frac{1}{2} & 1 & 0 & 0 & 0 & | & 12 \\
① & 0 & -\frac{1}{2} & 0 & -1 & 1 & 0 & | & 1 \\
\hline
-\frac{1}{2}M+\frac{1}{2} & 0 & \frac{1}{2}M+\frac{5}{2} & 0 & M & 0 & 1 & | & -M+45
\end{bmatrix}
\begin{array}{l}
\frac{9}{1/2} = 18 \\
\frac{12}{3/2} = 8 \\
\frac{1}{1/2} = 2
\end{array}
$$

$$2R_3 \rightarrow R_3$$

$$
\begin{bmatrix}
\frac{1}{2} & 1 & \frac{1}{2} & 0 & 0 & 0 & 0 & | & 9 \\
\frac{3}{2} & 0 & -\frac{1}{2} & 1 & 0 & 0 & 0 & | & 12 \\
① & 0 & -1 & 0 & -2 & 2 & 0 & | & 2 \\
\hline
-\frac{1}{2}M+\frac{1}{2} & 0 & \frac{1}{2}M+\frac{5}{2} & 0 & M & 0 & 1 & | & -M+45
\end{bmatrix}
$$

$$\left(-\frac{1}{2}\right)R_3 + R_1 \rightarrow R_1, \ \left(-\frac{3}{2}\right)R_3 + R_2 \rightarrow R_2, \text{ and } \left(\frac{1}{2}M - \frac{1}{2}\right)R_3 + R_4 \rightarrow R_4$$

$$
\begin{array}{c}
\\ x_2 \\ s_2 \\ x_1 \\ \\ P
\end{array}
\begin{bmatrix}
x_1 & x_2 & s_1 & s_2 & s_3 & a_1 & P & & \\
0 & 1 & 1 & 0 & 1 & -1 & 0 & | & 8 \\
0 & 0 & 1 & 1 & 3 & -3 & 0 & | & 9 \\
1 & 0 & -1 & 0 & -2 & 2 & 0 & | & 2 \\
\hline
0 & 0 & 3 & 0 & 1 & M-1 & 1 & | & 44
\end{bmatrix}
$$

Optimal solution: max $P = 44$
at $x_1 = 2$, $x_2 = 8$.

13.   We introduce surplus and artificial variables to obtain the modified problem:

Maximize $P = 10x_1 + 12x_2 + 20x_3 - Ma_1 - Ma_2$

Subject to: $3x_1 + x_2 + 2x_3 - s_1 + a_1 \quad\quad = 12$

$\quad\quad x_1 - x_2 + 2x_3 \quad\quad\quad\quad + a_2 = 6$

$\quad\quad\quad x_1,\ x_2,\ x_3,\ s_1,\ a_1,\ a_2 \ge 0$

The preliminary simplex tableau for the modified problem is:

$$
\begin{array}{ccccccc}
x_1 & x_2 & x_3 & s_1 & a_1 & a_2 & P \\
\end{array}
$$

$$
\left[\begin{array}{ccccccc|c}
3 & 1 & 2 & -1 & 1 & 0 & 0 & 12 \\
1 & -1 & 2 & 0 & 0 & 1 & 0 & 6 \\
\hline
-10 & -12 & -20 & 0 & M & M & 1 & 0
\end{array}\right]
$$

$(-M)R_1 + R_3 \rightarrow R_3$

$$
\sim \left[\begin{array}{ccccccc|c}
3 & 1 & 2 & -1 & 1 & 0 & 0 & 12 \\
1 & -1 & 2 & 0 & 0 & 1 & 0 & 6 \\
\hline
-3M-10 & -M-12 & -2M-20 & M & 0 & M & 1 & -12M
\end{array}\right]
$$

$(-M)R_2 + R_3 \rightarrow R_3$

$$
\begin{array}{ccccccc}
x_1 & x_2 & x_3 & s_1 & a_1 & a_2 & P \\
\end{array}
$$

$$
\sim \begin{array}{c}
a_1 \\
a_2 \\
P
\end{array}
\left[\begin{array}{ccccccc|c}
3 & 1 & 2 & -1 & 1 & 0 & 0 & 12 \\
1 & -1 & ② & 0 & 0 & 1 & 0 & 6 \\
\hline
-4M-10 & -12 & -4M-20 & M & 0 & 0 & 1 & -18M
\end{array}\right]
\begin{array}{l}
\frac{12}{2}=6 \\
\frac{6}{2}=3
\end{array}
$$

$\frac{1}{2}R_2 \rightarrow R_2$

$$
\sim \left[\begin{array}{ccccccc|c}
3 & 1 & 2 & -1 & 1 & 0 & 0 & 12 \\
\frac{1}{2} & -\frac{1}{2} & ① & 0 & 0 & \frac{1}{2} & 0 & 3 \\
\hline
-4M-10 & -12 & -4M-20 & M & 0 & 0 & 1 & -18M
\end{array}\right]
$$

$(-2)R_2 + R_1 \rightarrow R_1$ and $(4M+20)R_2 + R_3 \rightarrow R_3$

$$
\sim \left[\begin{array}{ccccccc|c}
2 & ② & 0 & -1 & 1 & -1 & 0 & 6 \\
\frac{1}{2} & -\frac{1}{2} & 1 & 0 & 0 & \frac{1}{2} & 0 & 3 \\
\hline
-2M & -2M-22 & 0 & M & 0 & 2M+10 & 1 & -6M+60
\end{array}\right]
$$

$\frac{1}{2}R_1 \rightarrow R_1$

$$
\sim \left[\begin{array}{ccccccc|c}
1 & ① & 0 & -\frac{1}{2} & \frac{1}{2} & -\frac{1}{2} & 0 & 3 \\
\frac{1}{2} & -\frac{1}{2} & 1 & 0 & 0 & \frac{1}{2} & 0 & 3 \\
\hline
-2M & -2M-22 & 0 & M & 0 & 2M+10 & 1 & -6M+60
\end{array}\right]
$$

$\frac{1}{2}R_1 + R_2 \rightarrow R_2,\ (2M+22)R_1 + R_3 \rightarrow R_3$

$$
\begin{array}{cccccccc}
x_1 & x_2 & x_3 & s_1 & a_1 & & a_2 & P \\
\end{array}
$$

$$
\sim \begin{array}{c}
x_2 \\
x_3 \\
P
\end{array}
\left[\begin{array}{ccccccc|c}
1 & 1 & 0 & -\frac{1}{2} & \frac{1}{2} & -\frac{1}{2} & 0 & 3 \\
1 & 0 & 1 & -\frac{1}{4} & \frac{1}{4} & \frac{1}{4} & 0 & \frac{9}{2} \\
\hline
22 & 0 & 0 & -11 & M+11 & M-1 & 1 & 126
\end{array}\right]
$$

```
No optimal solution exists because there are no positive numbers in
the pivot column.
```

15. We will maximize $P = -C = 5x_1 + 12x_2 - 16x_3$ subject to the given constraints. Introduce slack, surplus, and artificial variables to obtain the modified problem:

Maximize $P = 5x_1 + 12x_2 - 16x_3 - Ma_1 - Ma_2$

Subject to:
$$x_1 + 2x_2 + x_3 + s_1 \qquad\qquad = 10$$
$$2x_1 + 3x_2 + x_3 \qquad - s_2 + a_1 \qquad = 6$$
$$2x_1 + x_2 - x_3 \qquad\qquad\quad + a_2 = 1$$
$$x_1, x_2, x_3, s_1, s_2, a_1, a_2 \geq 0$$

The preliminary simplex tableau for the modified problem is:

x_1	x_2	x_3	s_1	s_2	a_1	a_2	P	
1	2	1	1	0	0	0	0	10
2	3	1	0	-1	1	0	0	6
2	1	-1	0	0	0	1	0	1
-5	-12	16	0	0	M	M	1	0

$(-M)R_2 + R_4 \rightarrow R_4$

1	2	1	1	0	0	0	0	10
2	3	1	0	-1	1	0	0	6
2	1	-1	0	0	0	1	0	1
$-2M-5$	$-3M-12$	$-M+16$	0	M	0	M	1	$-6M$

$(-M)R_3 + R_4 \rightarrow R_4$

	x_1	x_2	x_3	s_1	s_2	a_1	a_2	P		
s_1	1	2	1	1	0	0	0	0	10	$\frac{10}{2} = 5$
a_1	2	3	1	0	-1	1	0	0	6	$\frac{6}{3} = 2$
a_2	2	①	-1	0	0	0	1	0	1	$\frac{1}{1} = 1$
P	$-4M-5$	$-4M-12$	16	0	M	0	0	1	$-7M$	

$(-2)R_3 + R_1 \rightarrow R_1$, $(-3)R_3 + R_2 \rightarrow R_2$, and $(4M+12)R_3 + R_4 \rightarrow R_4$

-3	0	3	1	0	0	-2	0	8	$\frac{8}{3} \approx 2.67$
-4	0	④	0	-1	1	-3	0	3	$\frac{3}{4} = .75$
2	1	-1	0	0	0	1	0	1	
$4M+19$	0	$-4M+4$	0	M	0	$4M+12$	1	$-3M+12$	

$\frac{1}{4}R_2 \rightarrow R_2$

-3	0	3	1	0	0	-2	0	8
-1	0	①	0	$-\frac{1}{4}$	$\frac{1}{4}$	$-\frac{3}{4}$	0	$\frac{3}{4}$
2	1	-1	0	0	0	1	0	1
$4M+19$	0	$-4M+4$	0	M	0	$4M+12$	1	$-3M+12$

$(-3)R_2 + R_1 \rightarrow R_1$, $R_2 + R_3 \rightarrow R_3$, and $(4M-4)R_2 + R_4 \rightarrow R_4$

	x_1	x_2	x_3	s_1	s_2	a_1	a_2	P	
s_1	0	0	0	1	$\frac{3}{4}$	$-\frac{3}{4}$	$\frac{1}{4}$	0	$\frac{23}{4}$
x_3	-1	0	1	0	$-\frac{1}{4}$	$\frac{1}{4}$	$-\frac{3}{4}$	0	$\frac{3}{4}$
x_2	1	1	0	0	$-\frac{1}{4}$	$\frac{1}{4}$	$\frac{1}{4}$	0	$\frac{7}{4}$
P	23	0	0	0	1	$M-1$	$M+15$	1	9

Optimal solution:
min $C = -9$ at
$x_1 = 0$, $x_2 = \dfrac{7}{4}$, and
$x_3 = \dfrac{3}{4}$.

17. We introduce a slack and an artificial variable to obtain the modified problem:

Maximize $P = 3x_1 + 5x_2 + 6x_3 - Ma_1$

Subject to: $2x_1 + x_2 + 2x_3 + s_1 \qquad = 8$

$\qquad\qquad 2x_1 + x_2 - 2x_3 \qquad + a_1 = 0$

$\qquad\qquad x_1,\ x_2,\ x_3,\ s_1,\ a_1 \geq 0$

The preliminary simplex tableau for the modified problem is:

$$
\begin{array}{c}
\begin{array}{cccccc} x_1 & x_2 & x_3 & s_1 & a_1 & P \end{array} \\
\begin{array}{c} s_1 \\ a_1 \\ P \end{array}
\left[\begin{array}{cccccc|c}
2 & 1 & 2 & 1 & 0 & 0 & 8 \\
2 & 1 & -2 & 0 & 1 & 0 & 0 \\
\hline
-3 & -5 & -6 & 0 & M & 1 & 0
\end{array} \right]
\end{array}
\sim
\left[\begin{array}{cccccc|c}
2 & 1 & 2 & 1 & 0 & 0 & 8 \\
\textcircled{2} & 1 & -2 & 0 & 1 & 0 & 0 \\
\hline
-3-2M & -5-M & -6+2M & 0 & 0 & 1 & 0
\end{array} \right]
$$

$(-M)R_2 + R_3 \rightarrow R_3$ $\qquad\qquad \frac{1}{2}R_2 \rightarrow R_2$

$$
\sim
\left[\begin{array}{cccccc|c}
2 & 1 & 2 & 1 & 0 & 0 & 8 \\
\textcircled{1} & \frac{1}{2} & -1 & 0 & \frac{1}{2} & 0 & 0 \\
\hline
-3-2M & -5-M & -6+2M & 0 & 0 & 1 & 0
\end{array} \right]
\sim
\left[\begin{array}{cccccc|c}
0 & 0 & \textcircled{4} & 1 & -1 & 0 & 8 \\
1 & \frac{1}{2} & -1 & 0 & \frac{1}{2} & 0 & 0 \\
\hline
0 & -\frac{7}{2} & -9 & 0 & \frac{3}{2}+M & 1 & 0
\end{array} \right]
$$

$(-2)R_2 + R_1 \rightarrow R_1,\ (3+2M)R_2 + R_3 \rightarrow R_3$ $\qquad \frac{1}{4}R_1 \rightarrow R_1$

$$
\sim
\left[\begin{array}{cccccc|c}
0 & 0 & \textcircled{1} & \frac{1}{4} & -\frac{1}{4} & 0 & 2 \\
1 & \frac{1}{2} & -1 & 0 & \frac{1}{2} & 0 & 0 \\
\hline
0 & -\frac{7}{2} & -9 & 0 & \frac{3}{2}+M & 1 & 0
\end{array} \right]
\sim
\left[\begin{array}{cccccc|c}
0 & 0 & 1 & \frac{1}{4} & -\frac{1}{4} & 0 & 2 \\
1 & \textcircled{$\frac{1}{2}$} & 0 & \frac{1}{4} & \frac{1}{4} & 0 & 2 \\
\hline
0 & -\frac{7}{2} & 0 & \frac{9}{4} & -\frac{3}{4}+M & 1 & 18
\end{array} \right]
$$

$R_1 + R_2 \rightarrow R_2,\ 9R_1 + R_3 \rightarrow R_3$ $\qquad\qquad 2R_2 \rightarrow R_2$

$$
\sim
\left[\begin{array}{cccccc|c}
0 & 0 & 1 & \frac{1}{4} & -\frac{1}{4} & 0 & 2 \\
2 & \textcircled{1} & 0 & \frac{1}{2} & \frac{1}{2} & 0 & 4 \\
\hline
0 & -\frac{7}{2} & 0 & \frac{9}{4} & -\frac{3}{4}+M & 1 & 18
\end{array} \right]
\sim
\begin{array}{c}
\begin{array}{cccccc} x_1 & x_2 & x_3 & s_1 & a_1 & P \end{array} \\
\begin{array}{c} x_3 \\ x_2 \\ P \end{array}
\left[\begin{array}{cccccc|c}
0 & 0 & 1 & \frac{1}{4} & -\frac{1}{4} & 0 & 2 \\
2 & 1 & 0 & \frac{1}{2} & \frac{1}{2} & 0 & 4 \\
\hline
7 & 0 & 0 & 4 & 1+M & 1 & 32
\end{array} \right]
\end{array}
$$

$\frac{7}{2}R_2 + R_3 \rightarrow R_3$

Optimal solution: max $P = 32$ at
$x_1 = 0,\ x_2 = 4,\ x_3 = 2.$

19. We introduce slack, surplus, and artificial variables to obtain the modified problem:

Maximize $P = 2x_1 + 3x_2 + 4x_3 - Ma_1$

Subject to: $\quad x_1 + 2x_2 + x_3 + s_1 \qquad\qquad\qquad = 25$

$\qquad\qquad x_1 + x_2 + 2x_3 \qquad + s_2 \qquad\qquad = 60$

$\qquad\qquad x_1 + 2x_2 - x_3 \qquad\qquad - s_3 + a_1 = 10$

$\qquad\qquad x_1,\ x_2,\ x_3,\ s_1,\ s_2,\ s_3,\ a_1 \geq 0$

The preliminary simplex tableau for the modified problem is:

$$
\begin{array}{cccccccc|c}
x_1 & x_2 & x_3 & s_1 & s_2 & s_3 & a_1 & P & \\
1 & 2 & 1 & 1 & 0 & 0 & 0 & 0 & 25 \\
2 & 1 & 2 & 0 & 1 & 0 & 0 & 0 & 60 \\
1 & 2 & -1 & 0 & 0 & -1 & 1 & 0 & 10 \\
\hline
-2 & -3 & -4 & 0 & 0 & 0 & M & 1 & 0
\end{array}
$$

$(-M)R_3 + R_4 \rightarrow R_4$

$$
\begin{array}{c}
s_1 \\ s_2 \\ a_1 \\ P
\end{array}
\begin{array}{cccccccc|c}
x_1 & x_2 & x_3 & s_1 & s_2 & s_3 & a_1 & P & \\
1 & 2 & 1 & 1 & 0 & 0 & 0 & 0 & 25 \\
2 & 1 & 2 & 0 & 1 & 0 & 0 & 0 & 60 \\
1 & ② & -1 & 0 & 0 & -1 & 1 & 0 & 10 \\
\hline
-2-M & -3-2M & -4+M & 0 & 0 & M & 0 & 1 & -10M
\end{array}
$$

$\frac{1}{2}R_3 \rightarrow R_3$

$$
\begin{array}{cccccccc|c}
1 & 2 & 1 & 1 & 0 & 0 & 0 & 0 & 25 \\
2 & 1 & 2 & 0 & 1 & 0 & 0 & 0 & 60 \\
\frac{1}{2} & ① & -\frac{1}{2} & 0 & 0 & -\frac{1}{2} & \frac{1}{2} & 0 & 5 \\
\hline
-2-M & -3-2M & -4+M & 0 & 0 & M & 0 & 1 & -10M
\end{array}
$$

$(-2)R_3 + R_1 \rightarrow R_1, \quad (-1)R_3 + R_2 \rightarrow R_2, \quad (3+2M)R_3 + R_4 \rightarrow R_4$

$$
\begin{array}{cccccccc|c}
0 & 0 & ② & 1 & 0 & 1 & -1 & 0 & 15 \\
\frac{3}{2} & 0 & \frac{5}{2} & 0 & 1 & \frac{1}{2} & -\frac{1}{2} & 0 & 55 \\
\frac{1}{2} & 1 & -\frac{1}{2} & 0 & 0 & -\frac{1}{2} & \frac{1}{2} & 0 & 5 \\
\hline
-\frac{1}{2} & 0 & -\frac{11}{2} & 0 & 0 & -\frac{3}{2} & \frac{3}{2}+M & 1 & 15
\end{array}
$$

$\frac{1}{2}R_1 \rightarrow R_1$

$$
\begin{array}{cccccccc|c}
0 & 0 & ① & \frac{1}{2} & 0 & \frac{1}{2} & -\frac{1}{2} & 0 & \frac{15}{2} \\
\frac{3}{2} & 0 & \frac{5}{2} & 0 & 1 & \frac{1}{2} & -\frac{1}{2} & 0 & 55 \\
\frac{1}{2} & 1 & -\frac{1}{2} & 0 & 0 & -\frac{1}{2} & \frac{1}{2} & 0 & 5 \\
\hline
-\frac{1}{2} & 0 & -\frac{11}{2} & 0 & 0 & -\frac{3}{2} & \frac{3}{2}+M & 1 & 15
\end{array}
$$

$\left(-\frac{5}{2}\right)R_1 + R_2 \rightarrow R_2, \quad \frac{1}{2}R_1 + R_3 \rightarrow R_3, \quad \frac{11}{2}R_1 + R_4 \rightarrow R_4$

$$
\begin{array}{cccccccc|c}
0 & 0 & 1 & \frac{1}{2} & 0 & \frac{1}{2} & -\frac{1}{2} & 0 & \frac{15}{2} \\
\frac{3}{2} & 0 & 0 & -\frac{5}{4} & 1 & -\frac{3}{4} & \frac{3}{4} & 0 & \frac{145}{4} \\
\left(\frac{1}{2}\right) & 1 & 0 & \frac{1}{4} & 0 & -\frac{1}{4} & \frac{1}{4} & 0 & \frac{35}{4} \\
\hline
-\frac{1}{2} & 0 & 0 & \frac{11}{4} & 0 & \frac{5}{4} & -\frac{5}{4}+M & 1 & \frac{225}{4}
\end{array}
$$

$2R_3 \rightarrow R_3$

$$\sim \begin{bmatrix} 0 & 0 & 1 & \frac{1}{2} & 0 & \frac{1}{2} & -\frac{1}{2} & 0 & \frac{15}{2} \\ \frac{3}{2} & 0 & 0 & -\frac{5}{4} & 1 & -\frac{3}{4} & \frac{3}{4} & 0 & \frac{145}{4} \\ ① & 2 & 0 & \frac{1}{2} & 0 & -\frac{1}{2} & \frac{1}{2} & 0 & \frac{35}{2} \\ \hline -\frac{1}{2} & 0 & 0 & \frac{11}{4} & 0 & \frac{5}{4} & -\frac{5}{4}+M & 1 & \frac{225}{4} \end{bmatrix}$$

$$\left(-\frac{3}{2}\right)R_3 + R_2 \to R_2, \quad \frac{1}{2}R_3 + R_4 \to R_4$$

$$\begin{array}{c} \\ x_3 \\ s_2 \\ \sim \;\; x_1 \\ \\ P \end{array}
\begin{array}{cccccccc} x_1 & x_2 & x_3 & s_1 & s_2 & s_3 & a_1 & P \end{array}
\begin{bmatrix} 0 & 0 & 1 & \frac{1}{2} & 0 & \frac{1}{2} & -\frac{1}{2} & 0 & \frac{15}{2} \\ 0 & -3 & 0 & -2 & 1 & 0 & 0 & 0 & 10 \\ 1 & 2 & 0 & \frac{1}{2} & 0 & -\frac{1}{2} & \frac{1}{2} & 0 & \frac{35}{2} \\ \hline 0 & 1 & 0 & 3 & 0 & 1 & -1+M & 1 & 65 \end{bmatrix}$$

Optimal solution: max $P = 65$ at $x_1 = \frac{35}{2}$, $x_2 = 0$, $x_3 = \frac{15}{2}$.

21. We introduce slack, surplus, and artificial variables to obtain the modified problem:

Maximize $P = x_1 + 2x_2 + 5x_3 - Ma_1 - Ma_2$

Subject to:
$$x_1 + 3x_2 + 2x_3 + s_1 \qquad\qquad\qquad = 60$$
$$2x_1 + 5x_2 + 2x_3 \qquad - s_2 + a_1 \qquad\qquad = 50$$
$$x_1 - 2x_2 + x_3 \qquad\qquad\qquad - s_3 + a_2 = 40$$
$$x_1,\ x_2,\ x_3,\ s_1,\ s_2,\ s_3,\ a_1,\ a_2 \ge 0$$

The preliminary simplex tableau for the modified problem is:

$$\begin{array}{ccccccccc} x_1 & x_2 & x_3 & s_1 & s_2 & a_1 & s_3 & a_2 & P \end{array}$$
$$\begin{bmatrix} 1 & 3 & 2 & 1 & 0 & 0 & 0 & 0 & 0 & 60 \\ 2 & 5 & 2 & 0 & -1 & 1 & 0 & 0 & 0 & 50 \\ 1 & -2 & 1 & 0 & 0 & 0 & -1 & 1 & 0 & 40 \\ \hline -1 & -2 & -5 & 0 & 0 & M & 0 & M & 1 & 0 \end{bmatrix}$$

$$(-M)R_2 + R_4 \to R_4$$

$$\sim \begin{bmatrix} 1 & 3 & 2 & 1 & 0 & 0 & 0 & 0 & 0 & 60 \\ 2 & 5 & 2 & 0 & -1 & 1 & 0 & 0 & 0 & 50 \\ 1 & -2 & 1 & 0 & 0 & 0 & -1 & 1 & 0 & 40 \\ \hline -2M-1 & -5M-2 & -2M-5 & 0 & M & 0 & 0 & M & 1 & -50M \end{bmatrix}$$

$$(-M)R_3 + R_4 \to R_4$$

$$\sim \begin{bmatrix} 1 & 3 & 2 & 1 & 0 & 0 & 0 & 0 & 0 & 60 \\ 2 & 5 & ② & 0 & -1 & 1 & 0 & 0 & 0 & 50 \\ 1 & -2 & 1 & 0 & 0 & 0 & -1 & 1 & 0 & 40 \\ \hline -3M-1 & -3M-2 & -3M-5 & 0 & M & 0 & M & 0 & 1 & -90M \end{bmatrix}$$

$$\frac{1}{2}R_2 \to R_2$$

$$\sim \begin{bmatrix} 1 & 3 & 2 & 1 & 0 & 0 & 0 & 0 & 0 & 60 \\ 1 & \frac{5}{2} & \textcircled{1} & 0 & -\frac{1}{2} & \frac{1}{2} & 0 & 0 & 0 & 25 \\ 1 & -2 & 1 & 0 & 0 & 0 & -1 & 1 & 0 & 40 \\ \hline -3M-1 & -3M-2 & -3M-5 & 0 & M & 0 & M & 0 & 1 & -90M \end{bmatrix}$$

$(-2)R_2 + R_1 \to R_1,\ (-1)R_2 + R_3 \to R_3,\ (3M+5)R_2 + R_4 \to R_4$

$$\sim \begin{bmatrix} -1 & -2 & 0 & 1 & \textcircled{1} & -1 & 0 & 0 & 0 & 10 \\ 1 & \frac{5}{2} & 1 & 0 & -\frac{1}{2} & \frac{1}{2} & 0 & 0 & 0 & 25 \\ 0 & -\frac{9}{2} & 0 & 0 & \frac{1}{2} & -\frac{1}{2} & -1 & 1 & 0 & 15 \\ \hline 4 & \frac{9}{2}M + \frac{21}{2} & 0 & 0 & -\frac{1}{2}M - \frac{5}{2} & \frac{3}{2}M + \frac{5}{2} & M & 0 & 1 & -15M + 125 \end{bmatrix}$$

$\frac{1}{2}R_1 + R_2 \to R_2,\ \left(-\frac{1}{2}\right)R_1 + R_3 \to R_3,\ \left(\frac{1}{2}M + \frac{5}{2}\right)R_1 + R_4 \to R_4$

$$\sim \begin{bmatrix} -1 & -2 & 0 & 1 & 1 & -1 & 0 & 0 & 0 & 10 \\ \frac{1}{2} & \frac{3}{2} & 1 & \frac{1}{2} & 0 & 0 & 0 & 0 & 0 & 30 \\ \textcircled{$\frac{1}{2}$} & -\frac{7}{2} & 0 & -\frac{1}{2} & 0 & 0 & -1 & 1 & 0 & 10 \\ \hline -\frac{1}{2}M + \frac{3}{2} & \frac{7}{2}M + \frac{11}{2} & 0 & \frac{1}{2}M + \frac{5}{2} & 0 & M & M & 0 & 1 & -10M + 150 \end{bmatrix}$$

$2R_3 + R_1 \to R_1,\ (-1)R_3 + R_2 \to R_2,\ (M-3)R_3 + R_4 \to R_4,\ 2R_3 \to R_3$

$$\sim \begin{array}{c} \\ s_2 \\ x_3 \\ x_1 \\ \\ \end{array} \begin{bmatrix} x_1 & x_2 & x_3 & s_1 & s_2 & a_1 & s_3 & a_2 & P & \\ 0 & -9 & 0 & 0 & 1 & -1 & -2 & 2 & 0 & 30 \\ 0 & 5 & 1 & 1 & 0 & 0 & 1 & -1 & 0 & 20 \\ 1 & -7 & 0 & -1 & 0 & 0 & -2 & 2 & 0 & 20 \\ \hline 0 & 16 & 0 & 4 & 0 & M & 3 & M-3 & 1 & 120 \end{bmatrix}$$

Optimal solution: max $P = 120$ at $x_1 = 20$, $x_2 = 0$, $x_3 = 20$.

23. **(A)** Refer to Problem 5.
The graph of the feasible region is shown at the right. Since it is unbounded, $P = 4x_1 + 3x_2$ does not have a maximum value by Theorem 2(B) in Section 5.3.

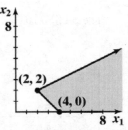

(B) Refer to Problem 7.
The graph of the feasible region is empty. Therefore, $P = 5x_1 + 10x_2$ does not have a maximum value, by Theorem 2(C) in Section 5.3.

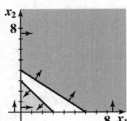

25. We will maximize $P = -C = -10x_1 + 40x_2 + 5x_3$

Subject to: $x_1 + 3x_2 \quad + s_1 \quad = 6$

$\qquad\qquad 4x_2 + x_3 \quad + s_2 = 3$

$\qquad\qquad x_1,\ x_2,\ x_3,\ s_1,\ s_2 \geq 0$

where s_1, s_2 are slack variables. The simplex tableau for this problem is:

$$
\begin{array}{c}
\begin{array}{ccccccc}
x_1 & x_2 & x_3 & s_1 & s_2 & P & \\
\end{array}\\
\begin{array}{c}
s_1 \\ s_2 \\ P
\end{array}
\left[
\begin{array}{cccccc|c}
1 & 3 & 0 & 1 & 0 & 0 & 6 \\
0 & ④ & 1 & 0 & 1 & 0 & 3 \\
\hline
10 & -40 & -5 & 0 & 0 & 1 & 0
\end{array}
\right]
\begin{array}{l}
\frac{6}{3} = 2 \\[4pt]
\frac{3}{4} = .75
\end{array}
\end{array}
$$

$\frac{1}{4}R_2 \rightarrow R_2$

$$
\sim
\left[
\begin{array}{cccccc|c}
1 & 3 & 0 & 1 & 0 & 0 & 6 \\
0 & ① & \frac{1}{4} & 0 & \frac{1}{4} & 0 & \frac{3}{4} \\
\hline
10 & -40 & -5 & 0 & 0 & 1 & 0
\end{array}
\right]
$$

$(-3)R_2 + R_1 \rightarrow R_1$ and $40R_2 + R_3 \rightarrow R_3$

$$
\begin{array}{c}
\begin{array}{ccccccc}
x_1 & x_2 & x_3 & s_1 & s_2 & P & \\
\end{array}\\
\sim
\begin{array}{c}
x_1 \\ x_2 \\ P
\end{array}
\left[
\begin{array}{cccccc|c}
1 & 0 & -\frac{3}{4} & 1 & -\frac{3}{4} & 0 & \frac{15}{4} \\
0 & 1 & \frac{1}{4} & 0 & \frac{1}{4} & 0 & \frac{3}{4} \\
\hline
10 & 0 & 5 & 0 & 10 & 1 & 30
\end{array}
\right]
\end{array}
$$

Optimal solution: min $C = -30$ at $x_1 = 0$, $x_2 = \frac{3}{4}$, $x_3 = 0$.

27. Introduce slack, surplus, and artificial variables to obtain the modified problem:

Maximize $P = -5x_1 + 10x_2 + 15x_3 - Ma_1$

Subject to: $2x_1 + 3x_2 + x_3 + s_1 \qquad = 24$

$\qquad\qquad x_1 - 2x_2 - 2x_3 \qquad - s_2 + a_1 = 1$

$\qquad\qquad x_1,\ x_2,\ x_3,\ s_1,\ s_2,\ a_1 \geq 0$

The preliminary simplex tableau for the modified problem is:

$$
\begin{array}{ccccccc}
x_1 & x_2 & x_3 & s_1 & s_2 & a_1 & P
\end{array}
$$

$$
\left[
\begin{array}{ccccccc|c}
2 & 3 & 1 & 1 & 0 & 0 & 0 & 24 \\
1 & -2 & -2 & 0 & -1 & 1 & 0 & 1 \\
\hline
5 & -10 & -15 & 0 & 0 & M & 1 & 0
\end{array}
\right]
$$

$(-M)R_2 + R_3 \rightarrow R_3$

$$
\begin{array}{c}
\begin{array}{ccccccc}
x_1 \quad & x_2 \quad & x_3 \quad & s_1 & s_2 & a_1 & P
\end{array}\\
\sim
\begin{array}{c}
s_1 \\ a_1 \\ P
\end{array}
\left[
\begin{array}{ccccccc|c}
2 & 3 & 1 & 1 & 0 & 0 & 0 & 24 \\
① & -2 & -2 & 0 & -1 & 1 & 0 & 1 \\
\hline
-M+5 & 2M-10 & 2M-15 & 0 & M & 0 & 1 & -M
\end{array}
\right]
\begin{array}{l}
\frac{24}{2} = 12 \\[4pt]
\frac{1}{1} = 1
\end{array}
\end{array}
$$

$(-2)R_2 + R_1 \rightarrow R_1$ and $(M-5)R_2 + R_3 \rightarrow R_3$

$$\sim \begin{bmatrix} 0 & 7 & \circled{5} & 1 & 2 & -2 & 0 & 22 \\ 1 & -2 & -2 & 0 & -1 & 1 & 0 & 1 \\ \hline 0 & 0 & -5 & 0 & 5 & M-5 & 1 & -5 \end{bmatrix} \sim \begin{bmatrix} 0 & \frac{7}{5} & \circled{1} & \frac{1}{5} & \frac{2}{5} & -\frac{2}{5} & 0 & \frac{22}{5} \\ 1 & -2 & -2 & 0 & -1 & 1 & 0 & 1 \\ \hline 0 & 0 & -5 & 0 & 5 & M-5 & 1 & -5 \end{bmatrix}$$

$$\frac{1}{5}R_1 \rightarrow R_1 \qquad\qquad\qquad 2R_1 + R_2 \rightarrow R_2 \text{ and } 5R_1 + R_3 \rightarrow R_3$$

$$\sim \begin{array}{c} x_3 \\ x_1 \\ P \end{array} \begin{array}{c} \end{array} \begin{bmatrix} \begin{array}{ccccccc} x_1 & x_2 & x_3 & s_1 & s_2 & a_1 & P \\ 0 & \frac{7}{5} & 1 & \frac{1}{5} & \frac{2}{5} & -\frac{2}{5} & 0 \\ 1 & \frac{4}{5} & 0 & \frac{2}{5} & -\frac{1}{5} & \frac{1}{5} & 0 \\ \hline 0 & 7 & 0 & 1 & 7 & M-7 & 1 \end{array} \end{bmatrix} \begin{array}{c} \frac{22}{5} \\ \frac{49}{5} \\ 17 \end{array}$$

Optimal solution: max $P = 17$ at $x_1 = \dfrac{49}{5}$, $x_2 = 0$, $x_3 = \dfrac{22}{5}$.

29. The matrices corresponding to the given problem and the dual problem are:

$$A = \begin{bmatrix} 1 & 3 & 0 & 6 \\ 0 & 4 & 1 & 3 \\ \hline 10 & 40 & 5 & 1 \end{bmatrix} \quad \text{and} \quad A^T = \begin{bmatrix} 1 & 0 & 10 \\ 3 & 4 & 40 \\ 0 & 1 & 5 \\ \hline 6 & 3 & 1 \end{bmatrix}$$

Thus, the dual problem is: Maximize $P = 6y_1 + 3y_2$

$$\begin{aligned} \text{Subject to:} \quad y_1 &\leq 10 \\ 3y_1 + 4y_2 &\leq 40 \\ y_2 &\leq 5 \\ y_1, y_2 &\geq 0 \end{aligned}$$

We introduce the slack variables x_1, x_2, and x_3 to obtain the initial system:

$$\begin{aligned} y_1 + x_1 &= 10 \\ 3y_1 + 4y_2 + x_2 &= 40 \\ y_2 + x_3 &= 5 \\ -6y_1 - 3y_2 + P &= 0 \end{aligned}$$

The simplex tableau for this problem is:

$$\begin{array}{c} x_1 \\ x_2 \\ x_3 \\ P \end{array} \begin{bmatrix} \begin{array}{cccccc} y_1 & y_2 & x_1 & x_2 & x_3 & P \\ \circled{1} & 0 & 1 & 0 & 0 & 0 & 10 \\ 3 & 4 & 0 & 1 & 0 & 0 & 40 \\ 0 & 1 & 0 & 0 & 1 & 0 & 5 \\ \hline -6 & -3 & 0 & 0 & 0 & 1 & 0 \end{array} \end{bmatrix} \begin{array}{l} \frac{10}{1} = 10 \\ \frac{40}{3} \approx 13.33 \\ \\ \end{array}$$

$$(-3)R_1 + R_2 \rightarrow R_2 \text{ and } 6R_1 + R_4 \rightarrow R_4$$

$$\sim \begin{bmatrix} 1 & 0 & 1 & 0 & 0 & 0 & 10 \\ 0 & ④ & -3 & 1 & 0 & 0 & 10 \\ 0 & 1 & 0 & 0 & 1 & 0 & 5 \\ \hline 0 & -3 & 6 & 0 & 0 & 1 & 60 \end{bmatrix} \begin{matrix} \frac{10}{4} = 2.5 \\ \\ \frac{5}{1} = 5 \end{matrix} \sim \begin{bmatrix} 1 & 0 & 1 & 0 & 0 & 0 & 10 \\ 0 & ① & -\frac{3}{4} & \frac{1}{4} & 0 & 0 & \frac{5}{2} \\ 0 & 1 & 0 & 0 & 1 & 0 & 5 \\ \hline 0 & -3 & 6 & 0 & 0 & 1 & 60 \end{bmatrix}$$

$$\frac{1}{4}R_2 \to R_2 \qquad\qquad\qquad (-1)R_2 + R_3 \to R_3 \text{ and } 3R_2 + R_4 \to R_4$$

$$\begin{matrix} & y_1 & y_2 & x_1 & x_2 & x_3 & P \\ \begin{matrix} y_1 \\ y_2 \\ x_3 \\ \\ P \end{matrix} & \sim \begin{bmatrix} 1 & 0 & 1 & 0 & 0 & 0 & 10 \\ 0 & 1 & -\frac{3}{4} & \frac{1}{4} & 0 & 0 & \frac{5}{2} \\ 0 & 0 & \frac{3}{4} & -\frac{1}{4} & 1 & 0 & \frac{5}{2} \\ \hline 0 & 0 & \frac{15}{4} & \frac{3}{4} & 0 & 1 & \frac{135}{2} \end{bmatrix} \end{matrix}$$

Optimal solution: min $C = \dfrac{135}{2}$,

$$x_1 = \frac{15}{4}, \quad x_2 = \frac{3}{4}, \quad x_3 = 0.$$

31. We introduce the slack variables s_1 and s_2 to obtain the initial system:

$$\begin{aligned} x_1 + 3x_2 + x_3 + s_1 &= 40 \\ 2x_1 + x_2 + 3x_3 + s_2 &= 60 \\ -12x_1 - 9x_2 - 5x_3 + P &= 0 \end{aligned}$$

The simplex tableau for this problem is:

$$\begin{matrix} & x_1 & x_2 & x_3 & s_1 & s_2 & P \\ \begin{matrix} s_1 \\ s_2 \\ \\ \end{matrix} & \begin{bmatrix} 1 & 3 & 1 & 1 & 0 & 0 & 40 \\ ② & 1 & 3 & 0 & 1 & 0 & 60 \\ \hline -12 & -9 & -5 & 0 & 0 & 1 & 0 \end{bmatrix} \end{matrix} \begin{matrix} \frac{40}{1} = 40 \\ \frac{60}{2} = 30 \\ \\ \end{matrix} \sim \begin{bmatrix} 1 & 3 & 1 & 1 & 0 & 0 & 40 \\ ① & \frac{1}{2} & \frac{3}{2} & 0 & \frac{1}{2} & 0 & 30 \\ \hline -12 & -9 & -5 & 0 & 0 & 1 & 0 \end{bmatrix}$$

$$\frac{1}{2}R_2 \to R_2 \qquad\qquad (-1)R_2 + R_1 \to R_1 \text{ and } 12R_2 + R_3 \to R_3$$

$$\sim \begin{bmatrix} 0 & ⑤⁄₂ & -\frac{1}{2} & 1 & -\frac{1}{2} & 0 & 10 \\ 1 & \frac{1}{2} & \frac{3}{2} & 0 & \frac{1}{2} & 0 & 30 \\ \hline 0 & -3 & 13 & 0 & 6 & 1 & 360 \end{bmatrix} \begin{matrix} \frac{10}{5/2} = 4 \\ \frac{30}{1/2} = 60 \\ \\ \end{matrix} \sim \begin{bmatrix} 0 & ① & -\frac{1}{5} & \frac{2}{5} & -\frac{1}{5} & 0 & 4 \\ 1 & \frac{1}{2} & \frac{3}{2} & 0 & \frac{1}{2} & 0 & 30 \\ \hline 0 & -3 & 13 & 0 & 6 & 1 & 360 \end{bmatrix}$$

$$\frac{2}{5}R_1 \to R_1 \qquad\qquad \left(-\frac{1}{2}\right)R_1 + R_2 \to R_2 \text{ and } 3R_1 + R_3 \to R_3$$

$$\begin{matrix} & x_1 & x_2 & x_3 & s_1 & s_2 & P \\ \begin{matrix} x_2 \\ x_1 \\ \\ P \end{matrix} & \sim \begin{bmatrix} 0 & 1 & -\frac{1}{5} & \frac{2}{5} & -\frac{1}{5} & 0 & 4 \\ 1 & 0 & \frac{8}{5} & -\frac{1}{5} & \frac{3}{5} & 0 & 28 \\ \hline 0 & 0 & \frac{62}{5} & \frac{6}{5} & \frac{27}{5} & 1 & 372 \end{bmatrix} \end{matrix}$$

Optimal solution: max $P = 372$ at $x_1 = 28$, $x_2 = 4$, $x_3 = 0$.

33. Let x_1 = the number of ads placed in the *Sentinel*,

x_2 = the number of ads placed in the *Journal*,

x_3 = the number of ads placed in the *Tribune*.

The mathematical model is: Minimize $C = 200x_1 + 200x_2 + 100x_3$

$$\begin{aligned} \text{Subject to:} \quad x_1 + x_2 + x_3 &\le 10 \\ 2000x_1 + 500x_2 + 1500x_3 &\ge 16{,}000 \\ x_1, x_2, x_3 &\ge 0 \end{aligned}$$

Divide the second constraint inequality by 100 to simplify the calculations, and introduce slack, surplus, and artificial variables to obtain the equivalent form:

Maximize $P = -C = -200x_1 - 200x_2 - 100x_3 - Ma_1$

Subject to:
$$x_1 + x_2 + x_3 + s_1 \qquad = 10$$
$$20x_1 + 5x_2 + 15x_3 \qquad - s_2 + a_1 = 160$$
$$x_1,\ x_2,\ x_3,\ s_1,\ s_2,\ a_1 \geq 0$$

The simplex tableau for the modified problem is:

$$
\begin{array}{ccccccc|c}
x_1 & x_2 & x_3 & s_1 & s_2 & a_1 & P & \\
1 & 1 & 1 & 1 & 0 & 0 & 0 & 10 \\
20 & 5 & 15 & 0 & -1 & 1 & 0 & 160 \\
\hline
200 & 200 & 100 & 0 & 0 & M & 1 & 0
\end{array}
$$

$(-M)R_2 + R_3 \rightarrow R_3$

$$
\begin{array}{c}
s_1 \\
\sim\ a_1 \\
P
\end{array}
\begin{array}{ccccccc|c}
x_1 & x_2 & x_3 & s_1 & s_2 & a_1 & P & \\
1 & 1 & 1 & 1 & 0 & 0 & 0 & 10 \\
\textcircled{20} & 5 & 15 & 0 & -1 & 1 & 0 & 160 \\
\hline
-20M+200 & -5M+200 & -15M+100 & 0 & M & 0 & 1 & -160M
\end{array}
$$

$\dfrac{10}{1} = 10$

$\dfrac{160}{20} = 8$

$\dfrac{1}{20}R_2 \rightarrow R_2$

$$
\begin{array}{ccccccc|c}
1 & 1 & 1 & 1 & 0 & 0 & 0 & 10 \\
\textcircled{1} & \frac{1}{4} & \frac{3}{4} & 0 & -\frac{1}{20} & \frac{1}{20} & 0 & 8 \\
\hline
-20M+200 & -5M+200 & -15M+100 & 0 & M & 0 & 1 & -160M
\end{array}
$$

$(-1)R_2 + R_1 \rightarrow R_1$ and $(20M - 200)R_2 + R_3 \rightarrow R_3$

$$
\begin{array}{ccccccc|c}
0 & \frac{3}{4} & \textcircled{$\frac{1}{4}$} & 1 & \frac{1}{20} & -\frac{1}{20} & 0 & 2 \\
1 & \frac{1}{4} & \frac{3}{4} & 0 & -\frac{1}{20} & \frac{1}{20} & 0 & 8 \\
\hline
0 & 150 & -50 & 0 & 10 & M-10 & 1 & -1600
\end{array}
$$

$\dfrac{2}{1/4} = 8$

$\dfrac{8}{3/4} = \dfrac{32}{3} \approx 10.67$

$4R_1 \rightarrow R_1$

$$
\begin{array}{ccccccc|c}
0 & 3 & \textcircled{1} & 4 & \frac{1}{5} & -\frac{1}{5} & 0 & 8 \\
1 & \frac{1}{4} & \frac{3}{4} & 0 & -\frac{1}{20} & \frac{1}{20} & 0 & 8 \\
\hline
0 & 150 & -50 & 0 & 10 & M-10 & 1 & -1600
\end{array}
$$

$\left(-\dfrac{3}{4}\right)R_1 + R_2 \rightarrow R_2$ and $50R_1 + R_3 \rightarrow R_3$

$$
\begin{array}{c}
x_3 \\
\sim\ x_1 \\
P
\end{array}
\begin{array}{ccccccc|c}
x_1 & x_2 & x_3 & s_1 & s_2 & a_1 & P & \\
0 & 3 & 1 & 4 & \frac{1}{5} & -\frac{1}{5} & 0 & 8 \\
1 & -2 & 0 & -3 & -\frac{1}{5} & \frac{1}{5} & 0 & 2 \\
\hline
0 & 300 & 0 & 200 & 20 & M-20 & 1 & -1200
\end{array}
$$

The minimal cost is \$1200 when two ads are placed in the *Sentinel*, no ads are placed in the *Journal*, and eight ads are placed in the *Tribune*.

35. Let x_1 = the number of bottles of brand A

x_2 = the number of bottles of brand B,

x_3 = the number of bottles of brand C.

The mathematical model is: Minimize $C = 0.6x_1 + 0.4x_2 + 0.9x_3$

Subject to: $10x_1 + 10x_2 + 20x_3 \geq 100$

$2x_1 + 3x_2 + 4x_3 \leq 24$

$x_1,\ x_2,\ x_3 \geq 0$

Divide the first inequality by 10, and introduce slack, surplus, and artificial variables to obtain the equivalent form:

Maximize $P = -10C = -6x_1 - 4x_2 - 9x_3 - Ma_1$

Subject to: $x_1 + x_2 + 2x_3 - s_1 + a_1 = 10$

$2x_1 + 3x_2 + 4x_3 + s_2 = 24$

$x_1,\ x_2,\ x_3,\ s_1,\ s_2,\ a_1 \geq 0$

The simplex tableau for the modified problem is:

$$
\begin{array}{c}
\begin{array}{ccccccc}
x_1 & x_2 & x_3 & s_1 & a_1 & s_2 & P
\end{array}\\
\left[\begin{array}{ccccccc|c}
1 & 1 & 2 & -1 & 1 & 0 & 0 & 10 \\
2 & 3 & 4 & 0 & 0 & 1 & 0 & 24 \\
\hline
6 & 4 & 9 & 0 & M & 0 & 1 & 0
\end{array}\right]
\end{array}
$$

$(-M)R_1 + R_3 \rightarrow R_3$

$$
\begin{array}{c}
\begin{array}{ccccccc}
\ x_1 & x_2 & x_3 & s_1 & a_1 & s_2 & P
\end{array}\\
\begin{array}{c}
a_1 \\ {\sim}\ s_2 \\ \\
\end{array}
\left[\begin{array}{ccccccc|c}
1 & 1 & ② & -1 & 1 & 0 & 0 & 10 \\
2 & 3 & 4 & 0 & 0 & 1 & 0 & 24 \\
\hline
-M+6 & -M+4 & -2M+9 & M & 0 & 0 & 1 & -10M
\end{array}\right]
\begin{array}{l}
\frac{10}{2} = 5 \\ \frac{24}{4} = 6 \\ \\
\end{array}
\end{array}
$$

$\dfrac{1}{2}R_1 \rightarrow R_1$

$$
{\sim}\ \left[\begin{array}{ccccccc|c}
\frac{1}{2} & \frac{1}{2} & ① & -\frac{1}{2} & \frac{1}{2} & 0 & 0 & 5 \\
2 & 3 & 4 & 0 & 0 & 1 & 0 & 24 \\
\hline
-M+6 & -M+4 & -2M+9 & M & 0 & 0 & 1 & -10M
\end{array}\right]
$$

$(-4)R_1 + R_2 \rightarrow R_2$ and $(2M - 9)R_1 + R_3 \rightarrow R_3$

$$
{\sim}\ \left[\begin{array}{ccccccc|c}
\frac{1}{2} & \frac{1}{2} & 1 & -\frac{1}{2} & \frac{1}{2} & 0 & 0 & 5 \\
0 & ① & 0 & 2 & -2 & 1 & 0 & 4 \\
\hline
-\frac{3}{2} & -\frac{1}{2} & 0 & \frac{9}{2} & M - \frac{9}{2} & 0 & 1 & -45
\end{array}\right]
\begin{array}{l}
\frac{5}{1/2} = 10 \\ \frac{4}{1} = 4 \\ \\
\end{array}
$$

$\left(-\dfrac{1}{2}\right)R_2 + R_1 \rightarrow R_1$ and $\dfrac{1}{2}R_2 + R_3 \rightarrow R_3$

$$
\begin{array}{c}
\begin{array}{ccccccc}
x_1 & x_2 & x_3 & s_1 & a_1 & s_2 & P
\end{array}\\
\begin{array}{c}
x_3 \\ {\sim}\ x_2 \\ P
\end{array}
\left[\begin{array}{ccccccc|c}
\frac{1}{2} & 0 & 1 & -\frac{3}{2} & \frac{3}{2} & -\frac{1}{2} & 0 & 3 \\
0 & 1 & 0 & 2 & -2 & 1 & 0 & 4 \\
\hline
\frac{3}{2} & 0 & 0 & \frac{11}{2} & M - \frac{11}{2} & \frac{1}{2} & 1 & -43
\end{array}\right]
\end{array}
$$

The minimal cost is $4.30 when 0 bottles of brand A, 4 bottles of brand B and 3 bottles of brand C are consumed.

37. Let x_1 = the number of cubic yards of mix A,

x_2 = the number of cubic yards of mix B,

x_3 = the number of cubic yards of mix C.

The mathematical model is: Maximize $P = 12x_1 + 16x_2 + 8x_3$

Subject to: $16x_1 + 8x_2 + 16x_3 \geq 800$

$$12x_1 + 8x_2 + 16x_3 \leq 700$$

$$x_1, \ x_2, \ x_3 \geq 0$$

We simplify the inequalities, and introduce slack, surplus, and artificial variables to obtain the modified problem:

Maximize $P = 12x_1 + 16x_2 + 8x_3 - Ma_1$

Subject to: $4x_1 + 2x_2 + 4x_3 - s_1 + a_1 \quad = 200$

$$3x_1 + 2x_2 + 4x_3 \qquad + s_2 = 175$$

$$x_1, \ x_2, \ x_3, \ s_1, \ s_2, \ a_1 \geq 0$$

The simplex tableau for the modified problem is:

$$
\begin{array}{ccccccc|c}
x_1 & x_2 & x_3 & s_1 & a_1 & s_2 & P & \\
4 & 2 & 4 & -1 & 1 & 0 & 0 & 200 \\
3 & 2 & 4 & 0 & 0 & 1 & 0 & 175 \\
\hline
-12 & -16 & -8 & 0 & M & 0 & 1 & 0
\end{array}
$$

$(-M)R_1 + R_3 \rightarrow R_3$

$$
\begin{array}{c}
a_1 \\
\sim \ s_2 \\
\\
\end{array}
\begin{array}{ccccccc|c}
x_1 & x_2 & x_3 & s_1 & a_1 & s_2 & P & \\
④ & 2 & 4 & -1 & 1 & 0 & 0 & 200 \\
3 & 2 & 4 & 0 & 0 & 1 & 0 & 175 \\
\hline
-4M-12 & -2M-16 & -4M-8 & M & 0 & 0 & 1 & -200M
\end{array}
\quad
\begin{array}{l}
\frac{200}{4} = 50 \\
\frac{175}{3} \approx 58.33
\end{array}
$$

$\frac{1}{4}R_1 \rightarrow R_1$

$$
\sim
\begin{array}{ccccccc|c}
① & \frac{1}{2} & 1 & -\frac{1}{4} & \frac{1}{4} & 0 & 0 & 50 \\
3 & 2 & 4 & 0 & 0 & 1 & 0 & 175 \\
\hline
-4M-12 & -2M-16 & -4M-8 & M & 0 & 0 & 1 & -200M
\end{array}
$$

$(-3)R_1 + R_2 \rightarrow R_2$ and $(4M+12)R_1 + R_3 \rightarrow R_3$

$$
\sim
\begin{array}{ccccccc|c}
1 & \frac{1}{2} & 1 & -\frac{1}{4} & \frac{1}{4} & 0 & 0 & 50 \\
0 & ⓵\frac{1}{2} & 1 & \frac{3}{4} & -\frac{3}{4} & 1 & 0 & 25 \\
\hline
0 & -10 & 4 & -3 & M+3 & 0 & 1 & 600
\end{array}
\quad
\begin{array}{l}
\frac{50}{1/2} = 100 \\
\frac{25}{1/2} = 50
\end{array}
$$

$2R_2 \rightarrow R_2$

$$
\sim
\begin{array}{ccccccc|c}
1 & \frac{1}{2} & 1 & -\frac{1}{4} & \frac{1}{4} & 0 & 0 & 50 \\
0 & ① & 2 & \frac{3}{2} & -\frac{3}{2} & 2 & 0 & 50 \\
\hline
0 & -10 & 4 & -3 & M+3 & 0 & 1 & 600
\end{array}
$$

$\left(-\frac{1}{2}\right)R_2 + R_1 \rightarrow R_1$ and $10R_2 + R_3 \rightarrow R_3$

$$
\begin{array}{c}
 \\
x_1 \\
\sim x_2 \\

\end{array}
\begin{array}{c}
\begin{array}{ccccccc}
x_1 & x_2 & x_3 & s_1 & a_1 & s_2 & P \\
\end{array} \\
\left[
\begin{array}{ccccccc|c}
1 & 0 & 0 & -1 & 1 & -1 & 0 & 25 \\
0 & 1 & 2 & \frac{3}{2} & -\frac{3}{2} & 2 & 0 & 50 \\
\hline
0 & 0 & 24 & 12 & M-12 & 20 & 1 & 1100 \\
\end{array}
\right]
\end{array}
$$

The maximum amount of nitrogen is 1100 pounds when 25 cubic yards of mix A, 50 cubic yards of mix B, and 0 cubic yards of mix C are used.

39. Let x_1 = the number of car frames produced in Milwaukee,

x_2 = the number of truck frames produced in Milwaukee,

x_3 = the number of car frames produced in Racine,

x_4 = the number of truck frames produced in Racine.

The mathematical model for this problem is:

Maximize $P = 50x_1 + 70x_2 + 50x_3 + 70x_4$

Subject to:
$$
\begin{aligned}
x_1 \quad\quad + x_3 \quad\quad &\le 250 \\
x_2 \quad\quad + x_4 &\le 350 \\
x_1 + x_2 \quad\quad\quad &\le 300 \\
x_3 + x_4 &\le 200 \\
150x_1 + 200x_2 \quad\quad\quad &\le 50{,}000 \\
135x_3 + 180x_4 &\le 35{,}000 \\
x_1,\ x_2,\ x_3,\ x_4 &\ge 0
\end{aligned}
$$

41. Let x_1 = the number of barrels of A used in regular gasoline,

x_2 = the number of barrels of A used in premium gasoline,

x_3 = the number of barrels of B used in regular gasoline,

x_4 = the number of barrels of B used in premium gasoline,

x_5 = the number of barrels of C used in regular gasoline,

x_6 = the number of barrels of C used in premium gasoline.

Cost $C = 28(x_1 + x_2) + 30(x_3 + x_4) + 34(x_5 + x_6)$

Revenue $R = 38(x_1 + x_3 + x_5) + 46(x_2 + x_4 + x_6)$

Profit $P = R - C = 10x_1 + 18x_2 + 8x_3 + 16x_4 + 4x_5 + 12x_6$

Thus, the mathematical model for this problem is:

Maximize $P = 10x_1 + 18x_2 + 8x_3 + 16x_4 + 4x_5 + 12x_6$

Subject to:
$$
\begin{aligned}
x_1 + x_2 \quad\quad\quad\quad\quad &\le 40{,}000 \\
x_3 + x_4 \quad\quad\quad &\le 25{,}000 \\
x_5 + x_6 &\le 15{,}000 \\
x_1 \quad\quad + x_3 \quad\quad + x_5 \quad\quad &\ge 30{,}000 \\
x_2 \quad\quad + x_4 \quad\quad + x_6 &\ge 25{,}000 \\
-5x_1 \quad\quad + 5x_3 \quad\quad + 15x_5 \quad\quad &\ge 0 \\
-15x_2 \quad\quad - 5x_4 \quad\quad + 5x_6 &\ge 0 \\
x_1,\ x_2,\ x_3,\ x_4,\ x_5,\ x_6 &\ge 0
\end{aligned}
$$

43. Let x_1 = percentage of funds invested in high-tech funds

x_2 = percentage of funds invested in global funds

x_3 = percentage of funds invested in corporate bonds

x_4 = percentage of funds invested in municipal bonds

x_5 = percentage of funds invested in CD's

Risk levels:

High-tech funds: $2.7x_1$

Global funds: $1.8x_2$

Corporate Bonds: $1.2x_3$

Municipal Bonds: $0.5x_4$

CD's: $0x_5$

Total risk level: $2.7x_1 + 1.8x_2 + 1.2x_3 + 0.5x_4$

Return: $0.11x_1 + 0.1x_2 + 0.09x_3 + 0.08x_4 + 0.05x_5$

The mathematical model for this problem is:

Maximize $P = 0.11x_1 + 0.1x_2 + 0.09x_3 + 0.08x_4 + 0.05x_5$

$$\begin{aligned}
\text{Subject to:} \quad & x_1 + x_2 + x_3 + x_4 + x_5 = 1 \\
& 2.7x_1 + 1.8x_2 + 1.2x_3 + 0.5x_4 \leq 1.8 \\
& x_5 \geq 0.2 \\
& x_1, \ x_2, \ x_3, \ x_4, \ x_5 \geq 0
\end{aligned}$$

45. Let x_1 = the number of ounces of food L,

x_2 = the number of ounces of food M,

x_3 = the number of ounces of food N.

The mathematical model for this problem is:

Minimize $C = 0.4x_1 + 0.6x_2 + 0.8x_3$

$$\begin{aligned}
\text{Subject to:} \quad & 30x_1 + 10x_2 + 30x_3 \geq 400 \\
& 10x_1 + 10x_2 + 10x_3 \geq 200 \\
& 10x_1 + 30x_2 + 20x_3 \geq 300 \\
& 8x_1 + 4x_2 + 6x_3 \leq 150 \\
& 60x_1 + 40x_2 + 50x_3 \leq 900 \\
& x_1, \ x_2, \ x_3 \geq 0
\end{aligned}$$

47. Let x_1 = the number of students from A enrolled in school I,

x_2 = the number of students from A enrolled in school II,

x_3 = the number of students from B enrolled in school I,

x_4 = the number of students from B enrolled in school II,

x_5 = the number of students from C enrolled in school I,

x_6 = the number of students from C enrolled in school II.

The mathematical model for this problem is:

Minimize $C = 4x_1 + 8x_2 + 6x_3 + 4x_4 + 3x_5 + 9x_6$

Subject to:
$$
\begin{aligned}
x_1 + x_2 &= 500 \\
x_3 + x_4 &= 1200 \\
x_5 + x_6 &= 1800 \\
x_1 + x_3 + x_5 &\geq 1400 \\
x_2 + x_4 + x_6 &\geq 1400 \\
x_1 + x_3 + x_5 &\leq 2000 \\
x_2 + x_4 + x_6 &\leq 2000 \\
x_1 &\leq 300 \\
x_2 &\leq 300 \\
x_3 &\leq 720 \\
x_4 &\leq 720 \\
x_5 &\leq 1080 \\
x_6 &\leq 1080 \\
x_1, x_2, x_3, x_4, x_5, x_6 &\geq 0
\end{aligned}
$$

1. Given the linear programming problem
    Maximize $P = 6x_1 + 2x_2$

    Subject to: $2x_1 + x_2 \le 8$
    $$x_1 + 2x_2 \le 10$$
    $$x_1, \ x_2 \ge 0$$

    We introduce the slack variables s_1 and s_2 to obtain the system of equations:

    $$2x_1 + x_2 + s_1 \quad\ = 8$$
    $$x_1 + 2x_2 \quad\ + s_2 = 10 \qquad (6\text{-}1)$$

2. There are 2 basic and 2 nonbasic variables.   (6-1)

3. The basic solutions are given in the following table.

x_1	x_2	s_1	s_2	Intersection Point	Feasible?
0	0	8	10	O	Yes
0	8	0	–6	B	No
0	5	3	0	A	Yes
4	0	0	6	D	Yes
10	0	–12	0	E	No
2	4	0	0	C	Yes

    (6-1)

4. The simplex tableau for Problem 1 is:

    $$
    \begin{array}{c}
    \text{Enter} \\
    \begin{array}{cccccc}
     & x_1 & x_2 & s_1 & s_2 & P \\
    \text{Exit}\ s_1 & \boxed{2} & 1 & 1 & 0 & 0 \\
    s_2 & 1 & 2 & 0 & 1 & 0 \\
    P & -6 & -2 & 0 & 0 & 1
    \end{array}
    \end{array}
    \begin{array}{c|l}
    8 & \frac{8}{2} = 4 \\
    10 & \frac{10}{1} = 10 \\
    0 &
    \end{array}
    $$

    (6-2)

5.

$$
\begin{array}{c}
s_1 \\
\sim\ s_2 \\
P
\end{array}
\begin{array}{cccccc}
x_1 & x_2 & s_1 & s_2 & P & \\
\end{array}
\left[\begin{array}{ccccc|c}
② & 1 & 1 & 0 & 0 & 8 \\
1 & 2 & 0 & 1 & 0 & 10 \\
\hline
-6 & -2 & 0 & 0 & 1 & 0
\end{array}\right]
\sim
\left[\begin{array}{ccccc|c}
① & \frac{1}{2} & \frac{1}{2} & 0 & 0 & 4 \\
1 & 2 & 0 & 1 & 0 & 10 \\
\hline
-6 & -2 & 0 & 0 & 1 & 0
\end{array}\right]
$$

$\frac{1}{2}R_1 \to R_1$ $(-1)R_1 + R_2 \to R_2$ and $6R_1 + R_3 \to R_3$

$$
\begin{array}{c}
x_1 \\
\sim\ s_2 \\
P
\end{array}
\begin{array}{cccccc}
x_1 & x_2 & s_1 & s_2 & P & \\
\end{array}
\left[\begin{array}{ccccc|c}
1 & \frac{1}{2} & \frac{1}{2} & 0 & 0 & 4 \\
0 & \frac{3}{2} & -\frac{1}{2} & 1 & 0 & 6 \\
\hline
0 & 1 & 3 & 0 & 1 & 24
\end{array}\right]
$$

Optimal solution: max $P = 24$ at $x_1 = 4$, $x_2 = 0$.   (6-2)

6.

Enter

$$
\begin{array}{c}
x_2 \\
s_2 \\
\text{Exit}\ s_3 \\
P
\end{array}
\begin{array}{cccccccc}
x_1 & x_2 & x_3 & s_1 & s_2 & s_3 & P & \\
\end{array}
\left[\begin{array}{ccccccc|c}
2 & 1 & 3 & -1 & 0 & 0 & 0 & 20 \\
3 & 0 & 4 & 1 & 1 & 0 & 0 & 30 \\
② & 0 & 5 & 2 & 0 & 1 & 0 & 10 \\
\hline
-8 & 0 & -5 & 3 & 0 & 0 & 1 & 50
\end{array}\right]
\begin{array}{l}
\frac{20}{2} = 10 \\
\frac{30}{3} = 10 \\
\frac{10}{2} = 5 \\
\end{array}
$$

The basic variables are x_2, s_2, s_3, and P, and the nonbasic variables are x_1, x_3, and s_1.

The first column is the pivot column and the third row is the pivot row. The pivot element is circled.

$$
\sim
\left[\begin{array}{ccccccc|c}
2 & 1 & 3 & -1 & 0 & 0 & 0 & 20 \\
3 & 0 & 4 & 1 & 1 & 0 & 0 & 30 \\
① & 0 & \frac{5}{2} & 1 & 0 & \frac{1}{2} & 0 & 5 \\
\hline
-8 & 0 & -5 & 3 & 0 & 0 & 1 & 50
\end{array}\right]
$$

$(-2)R_3 + R_1 \to R_1$, $(-3)R_3 + R_2 \to R_2$, $8R_3 + R_4 \to R_4$

$$
\begin{array}{c}
x_2 \\
s_2 \\
\sim\ x_1 \\
P
\end{array}
\begin{array}{cccccccc}
x_1 & x_2 & x_3 & s_1 & s_2 & s_3 & P & \\
\end{array}
\left[\begin{array}{ccccccc|c}
0 & 1 & -2 & -3 & 0 & -1 & 0 & 10 \\
0 & 0 & -\frac{7}{2} & -2 & 1 & -\frac{3}{2} & 0 & 15 \\
1 & 0 & \frac{5}{2} & 1 & 0 & \frac{1}{2} & 0 & 5 \\
\hline
0 & 0 & 15 & 11 & 0 & 4 & 1 & 90
\end{array}\right]
$$

(6-2)

7.   (A)   The basic feasible solution is: $x_1 = 0$, $x_2 = 2$, $s_1 = 0$, $s_2 = 5$, $P = 12$. Additional pivoting is required because the last row contains a negative indicator.

(B)   The basic feasible solution is: $x_1 = 0$, $x_2 = 0$, $s_1 = 0$, $s_2 = 7$, $P = 22$. There is no optimal solution because there are no positive elements above the dashed line in the pivot column, column 1.

(C)   The basic feasible solution is: $x_1 = 6$, $x_2 = 0$, $s_1 = 15$, $s_2 = 0$, $P = 10$. This is the optimal solution.   (6-2)

8.   The matrices corresponding to the given problem and the dual problem are:

$$A = \begin{bmatrix} 1 & 3 & | & 15 \\ 2 & 1 & | & 20 \\ 5 & 2 & | & 1 \end{bmatrix} \quad \text{and} \quad A^T = \begin{bmatrix} 1 & 2 & | & 5 \\ 3 & 1 & | & 2 \\ 15 & 20 & | & 1 \end{bmatrix}$$

Thus, the dual problem is: Maximize $P = 15y_1 + 20y_2$

$$\text{Subject to:} \quad y_1 + 2y_2 \le 5$$
$$3y_1 + y_2 \le 2$$
$$y_1, y_2 \ge 0 \qquad (6\text{-}3)$$

9.   Introduce the slack variables x_1 and x_2 to obtain the initial system:

$$y_1 + 2y_2 + x_1 \qquad = 5$$
$$3y_1 + y_2 \qquad + x_2 \qquad = 2$$
$$-15y_1 - 20y_2 \qquad + P = 0 \qquad (6\text{-}3)$$

10.   The first simplex tableau for the dual problem, Problem 8, is:

$$\begin{array}{c}
 \\
x_1 \\
x_2 \\
P
\end{array}
\begin{bmatrix}
y_1 & y_2 & x_1 & x_2 & P & \\
1 & 2 & 1 & 0 & 0 & 5 \\
3 & 1 & 0 & 1 & 0 & 2 \\
\hdashline
-15 & -20 & 0 & 0 & 1 & 0
\end{bmatrix} \qquad (6\text{-}3)$$

11.   Using the simplex method, we have:

$$\begin{bmatrix}
1 & 2 & 1 & 0 & 0 & 5 \\
3 & ① & 0 & 1 & 0 & 2 \\
\hdashline
-15 & -20 & 0 & 0 & 1 & 0
\end{bmatrix}
\begin{array}{l}
\frac{5}{2} = 2.5 \\
\frac{2}{1} = 2
\end{array}
\quad \sim \quad
\begin{array}{c}
 \\
x_1 \\
y_2 \\
P
\end{array}
\begin{bmatrix}
y_1 & y_2 & x_1 & x_2 & P & \\
-5 & 0 & 1 & -2 & 0 & 1 \\
3 & 1 & 0 & 1 & 0 & 2 \\
\hdashline
45 & 0 & 0 & 20 & 1 & 40
\end{bmatrix}$$

$$(-2)R_2 + R_1 \to R_1, \quad 20R_2 + R_3 \to R_3$$

Optimal solution: max $P = 40$ at $y_1 = 0$ and $y_2 = 2$.   (6-2)

12.   Minimum $C = 40$ at $x_1 = 0$ and $x_2 = 20$.   (6-3)

13.   Maximize $P = 3x_1 + 4x_2$

Subject to: $2x_1 + 4x_2 \le 24$

$3x_1 + 3x_2 \le 21$

$4x_1 + 2x_2 \le 20$

$x_1, \; x_2 \ge 0$

We simplify the inequalities and introduce the slack variables $s_1, s_2,$ and s_3 to obtain the equivalent form:

$$x_1 + 2x_2 + s_1 \qquad\qquad = 12$$
$$x_1 + x_2 + s_2 \qquad\quad = 7$$
$$2x_1 + x_2 + s_3 \quad = 10$$
$$-3x_1 - 4x_2 + P = 0$$

The simplex tableau for this problem is:

$$
\begin{array}{c}
\text{Exit } s_1 \\ s_2 \\ s_3 \\ P
\end{array}
\left[
\begin{array}{cccccc|c}
x_1 & x_2 & s_1 & s_2 & s_3 & P & \\
1 & ② & 1 & 0 & 0 & 0 & 12 \\
1 & 1 & 0 & 1 & 0 & 0 & 7 \\
2 & 1 & 0 & 0 & 1 & 0 & 10 \\
\hline
-3 & -4 & 0 & 0 & 0 & 1 & 0
\end{array}
\right]
\begin{array}{l}
\frac{12}{2} = 6 \\[4pt]
\frac{7}{1} = 7 \\[4pt]
\frac{10}{1} = 10 \\[4pt]
\end{array}
$$

(Enter x_2)

$$\tfrac{1}{2} R_1 \rightarrow R_1$$

$$
\sim
\left[
\begin{array}{cccccc|c}
\tfrac{1}{2} & ① & \tfrac{1}{2} & 0 & 0 & 0 & 6 \\
1 & 1 & 0 & 1 & 0 & 0 & 7 \\
2 & 1 & 0 & 0 & 1 & 0 & 10 \\
\hline
-3 & -4 & 0 & 0 & 0 & 1 & 0
\end{array}
\right]
\sim
\left[
\begin{array}{cccccc|c}
\tfrac{1}{2} & 1 & \tfrac{1}{2} & 0 & 0 & 0 & 6 \\
① & 0 & -\tfrac{1}{2} & 1 & 0 & 0 & 1 \\
\tfrac{3}{2} & 0 & -\tfrac{1}{2} & 0 & 1 & 0 & 4 \\
\hline
-1 & 0 & 2 & 0 & 0 & 1 & 24
\end{array}
\right]
\begin{array}{l}
\frac{6}{1/2} = 12 \\[4pt]
\frac{1}{1/2} = 2 \\[4pt]
\frac{4}{3/2} \approx 2.67
\end{array}
$$

$(-1)R_1 + R_2 \rightarrow R_2, \;\; (-1)R_1 + R_3 \rightarrow R_3,$
and $4R_1 + R_4 \rightarrow R_4$

$2R_2 \rightarrow R_2$

$$
\sim
\left[
\begin{array}{cccccc|c}
\tfrac{1}{2} & 1 & \tfrac{1}{2} & 0 & 0 & 0 & 6 \\
① & 0 & -1 & 2 & 0 & 0 & 2 \\
\tfrac{3}{2} & 0 & -\tfrac{1}{2} & 0 & 1 & 0 & 4 \\
\hline
-1 & 0 & 2 & 0 & 0 & 1 & 24
\end{array}
\right]
\begin{array}{c}
x_2 \\ x_1 \\ s_3 \\ P
\end{array}
\sim
\left[
\begin{array}{cccccc|c}
x_1 & x_2 & s_1 & s_2 & s_3 & P & \\
0 & 1 & 1 & -1 & 0 & 0 & 5 \\
1 & 0 & -1 & 2 & 0 & 0 & 2 \\
0 & 0 & 1 & -3 & 1 & 0 & 1 \\
\hline
0 & 0 & 1 & 2 & 0 & 1 & 26
\end{array}
\right]
$$

$\left(-\tfrac{1}{2}\right)R_2 + R_1 \rightarrow R_1, \;\; \left(-\tfrac{3}{2}\right)R_2 + R_3 \rightarrow R_3,$
and $R_2 + R_4 \rightarrow R_4$

Optimal solution: max $P = 26$ at
$x_1 = 2, \; x_2 = 5.$

(6-2)

14.   Minimize $C = 3x_1 + 8x_2$

Subject to: $x_1 + x_2 \ge 10$

$x_1 + 2x_2 \ge 15$

$x_2 \ge 3$

$x_1, \; x_2 \ge 0$

The matrices corresponding to the given problem and the dual problem are:

$$A = \begin{bmatrix} 1 & 1 & | & 10 \\ 1 & 2 & | & 15 \\ 0 & 1 & | & 3 \\ \hline 3 & 8 & | & 1 \end{bmatrix} \quad \text{and} \quad A^T = \begin{bmatrix} 1 & 1 & 0 & | & 3 \\ 1 & 2 & 1 & | & 8 \\ \hline 10 & 15 & 3 & | & 1 \end{bmatrix}$$

Thus, the dual problem is: Maximize $P = 10y_1 + 15y_2 + 3y_3$

$$\begin{aligned}
\text{Subject to:} \quad y_1 + y_2 &\le 3 \\
y_1 + 2y_2 + y_3 &\le 8 \\
y_1,\ y_2,\ y_3 &\ge 0 \qquad (6\text{-}3)
\end{aligned}$$

15. Introduce the slack variables x_1 and x_2 to obtain the initial system:

$$\begin{aligned}
y_1 + y_2 + x_1 &= 3 \\
y_1 + 2y_2 + y_3 + x_2 &= 8 \\
-10y_1 - 15y_2 - 3y_3 + P &= 0
\end{aligned}$$

The simplex tableau for this problem is:

$$\begin{array}{c}
\begin{array}{cccccc}
y_1 & y_2 & y_3 & x_1 & x_2 & P
\end{array} \\
\begin{array}{c} x_1 \\ x_2 \\ P \end{array}
\left[\begin{array}{cccccc|c}
1 & \textcircled{1} & 0 & 1 & 0 & 0 & 3 \\
1 & 2 & 1 & 0 & 1 & 0 & 8 \\
\hline
-10 & -15 & -3 & 0 & 0 & 1 & 0
\end{array}\right]
\begin{array}{l} \frac{3}{1} = 3 \\ \frac{8}{2} = 4 \end{array}
\end{array}
\sim
\left[\begin{array}{cccccc|c}
1 & 1 & 0 & 1 & 0 & 0 & 3 \\
-1 & 0 & \textcircled{1} & -2 & 1 & 0 & 2 \\
\hline
5 & 0 & -3 & 15 & 0 & 1 & 45
\end{array}\right]$$

$(-2)R_1 + R_2 \to R_2$ and $15R_1 + R_3 \to R_3$ $\qquad$ $3R_2 + R_3 \to R_3$

$$\begin{array}{c}
\begin{array}{cccccc}
y_1 & y_2 & y_3 & x_1 & x_2 & P
\end{array} \\
\begin{array}{c} y_2 \\ \sim \ y_3 \\ {} \end{array}
\left[\begin{array}{cccccc|c}
1 & 1 & 0 & 1 & 0 & 0 & 3 \\
-1 & 0 & 1 & -2 & 1 & 0 & 2 \\
\hline
2 & 0 & 0 & 9 & 3 & 1 & 51
\end{array}\right]
\end{array}$$

Optimal solution: min $C = 51$ at $x_1 = 9$, $x_2 = 3$. $\qquad\qquad$ (6-3)

16. Introduce slack variables s_1 and s_2 to obtain the equivalent form:

$$\begin{aligned}
x_1 - x_2 - 2x_3 + s_1 &= 3 \\
2x_1 + 2x_2 - 5x_3 + s_2 &= 10 \\
-5x_1 - 3x_2 + 3x_3 + P &= 0
\end{aligned}$$

The simplex tableau for this problem is:

Enter

$$
\begin{array}{c}
\quad\quad\ x_1\ \ x_2\ \ x_3\ \ s_1\ \ s_2\ \ P \\
\begin{array}{c}
\text{Exit } s_1 \\ s_2 \\ P
\end{array}
\left[\begin{array}{cccccc|c}
\textcircled{1} & -1 & -2 & 1 & 0 & 0 & 3 \\
2 & 2 & -5 & 0 & 1 & 0 & 10 \\
\hline
-5 & -3 & 3 & 0 & 0 & 1 & 0
\end{array}\right]
\begin{array}{c}
\frac{3}{1}=3 \\ \frac{10}{2}=5 \\ \
\end{array}
\sim
\left[\begin{array}{cccccc|c}
1 & -1 & -2 & 1 & 0 & 0 & 3 \\
0 & \textcircled{4} & -1 & -2 & 1 & 0 & 4 \\
\hline
0 & -8 & -7 & 5 & 0 & 1 & 15
\end{array}\right]
\end{array}
$$

$(-2)R_1 + R_2 \to R_2$ and $5R_1 + R_3 \to R_3$ $\qquad\qquad$ $\frac{1}{4}R_2 \to R_2$

$$
\sim
\left[\begin{array}{cccccc|c}
1 & -1 & -2 & 1 & 0 & 0 & 3 \\
0 & \textcircled{1} & -\frac{1}{4} & -\frac{1}{2} & \frac{1}{4} & 0 & 1 \\
\hline
0 & -8 & -7 & 5 & 0 & 1 & 15
\end{array}\right]
\sim
\begin{array}{c}
\quad x_1\ \ x_2\ \ x_3\ \ s_1\ \ s_2\ \ P \\
\left[\begin{array}{cccccc|c}
1 & 0 & -\frac{9}{4} & \frac{1}{2} & \frac{1}{4} & 0 & 4 \\
0 & 1 & -\frac{1}{4} & -\frac{1}{2} & \frac{1}{4} & 0 & 1 \\
\hline
0 & 0 & -9 & 1 & 2 & 1 & 23
\end{array}\right]
\end{array}
$$

$R_2 + R_1 \to R_1$ and $8R_2 + R_3 \to R_3$

No optimal solution exists; the elements in the pivot column (the x_3 column) above the dashed line are negative.   (6-2)

17. Introduce slack variables s_1 and s_2 to obtain the equivalent form:

$$
\begin{aligned}
x_1 - x_2 - 2x_3 + s_1 \qquad\quad &= 3 \\
x_1 + x_2 \qquad\quad + s_2 \quad &= 5 \\
-5x_1 - 3x_2 + 3x_3 \qquad + P &= 0
\end{aligned}
$$

The simplex tableau for this problem is:

Enter

$$
\begin{array}{c}
\quad\quad\ x_1\ \ x_2\ \ x_3\ \ s_1\ \ s_2\ \ P \\
\begin{array}{c}
\text{Exit } s_1 \\ s_2 \\ P
\end{array}
\left[\begin{array}{cccccc|c}
\textcircled{1} & -1 & -2 & 1 & 0 & 0 & 3 \\
1 & 1 & 0 & 0 & 1 & 0 & 5 \\
\hline
-5 & -3 & 3 & 0 & 0 & 1 & 0
\end{array}\right]
\begin{array}{c}
\frac{3}{1}=3 \\ \frac{5}{1}=5 \\ \
\end{array}
\sim
\left[\begin{array}{cccccc|c}
1 & -1 & -2 & 1 & 0 & 0 & 3 \\
0 & \textcircled{2} & 2 & -1 & 1 & 0 & 2 \\
\hline
0 & -8 & -7 & 5 & 0 & 1 & 15
\end{array}\right]
\end{array}
$$

$(-1)R_1 + R_2 \to R_2$ and $5R_1 + R_3 \to R_3$ $\qquad\qquad$ $\frac{1}{2}R_2 \to R_2$

$$
\sim
\left[\begin{array}{cccccc|c}
1 & -1 & -2 & 1 & 0 & 0 & 3 \\
0 & \textcircled{1} & 1 & -\frac{1}{2} & \frac{1}{2} & 0 & 1 \\
\hline
0 & -8 & -7 & 5 & 0 & 1 & 15
\end{array}\right]
\sim
\begin{array}{c}
\quad\ \ x_1\ \ x_2\ \ x_3\ \ s_1\ \ s_2\ \ P \\
\begin{array}{c}
x_1 \\ x_2 \\ P
\end{array}
\left[\begin{array}{cccccc|c}
1 & 0 & -1 & \frac{1}{2} & \frac{1}{2} & 0 & 4 \\
0 & 1 & 1 & -\frac{1}{2} & \frac{1}{2} & 0 & 1 \\
\hline
0 & 0 & 1 & 1 & 4 & 1 & 23
\end{array}\right]
\end{array}
$$

$R_2 + R_1 \to R_1$ and $8R_2 + R_3 \to R_3$ Optimal solution: max $P = 23$ at $x_1 = 4,\ x_2 = 1,\ x_3 = 0$.   (6-2)

18. (A)   We introduce a surplus variable s_1 and an artificial variable a_1 to convert the first inequality ($\geq$) into an equation; we introduce a slack variable s_2 to convert the second inequality ($\leq$) into an equation.

The modified problem is: Maximize $P = x_1 + 3x_2 - Ma_1$

$$
\begin{aligned}
\text{Subject to: } x_1 + x_2 - s_1 + a_1 \qquad\quad &= 6 \\
x_1 + 2x_2 \qquad\quad + s_2 &= 8 \\
x_1,\ x_2,\ s_1,\ s_2,\ a_1 &\geq 0
\end{aligned}
$$

(B)   The preliminary simplex tableau is:

$$
\begin{array}{cccccc}
x_1 & x_2 & s_1 & a_1 & s_2 & P \\
\end{array}
$$

$$
\left[
\begin{array}{cccccc|c}
1 & 1 & -1 & 1 & 0 & 0 & 6 \\
1 & 2 & 0 & 0 & 1 & 0 & 8 \\
\hline
-1 & -3 & 0 & M & 0 & 1 & 0
\end{array}
\right]
$$

Now

$$
\begin{array}{c}
a_1 \\
s_2 \\
P
\end{array}
\left[
\begin{array}{cccccc|c}
1 & 1 & -1 & 1 & 0 & 0 & 6 \\
1 & 2 & 0 & 0 & 1 & 0 & 8 \\
\hline
-1 & -3 & 0 & M & 0 & 1 & 0
\end{array}
\right]
\sim
\left[
\begin{array}{cccccc|c}
1 & 1 & -1 & 1 & 0 & 0 & 6 \\
1 & 2 & 0 & 0 & 1 & 0 & 8 \\
\hline
-M-1 & -M-3 & M & 0 & 0 & 1 & -6M
\end{array}
\right]
$$

$$(-M)R_1 + R_3 \to R_3$$

(C)   Thus, the initial simplex tableau is:

$$
\begin{array}{c}
a_1 \\
\sim \; s_1 \\
P
\end{array}
\left[
\begin{array}{cccccc|c}
1 & 1 & -1 & 1 & 0 & 0 & 6 \\
1 & 2 & 0 & 0 & 1 & 0 & 8 \\
\hline
-M-1 & -M-3 & M & 0 & 0 & 1 & -6M
\end{array}
\right]
$$

$$
\begin{array}{c}
a_1 \\
\sim \; s_2 \\
P
\end{array}
\left[
\begin{array}{cccccc|c}
1 & 1 & -1 & 1 & 0 & 0 & 6 \\
1 & ② & 0 & 0 & 1 & 0 & 8 \\
\hline
-M-1 & -M-3 & M & 0 & 0 & 1 & -6M
\end{array}
\right]
\begin{array}{l}
\frac{6}{1} = 6 \\
\frac{8}{2} = 4
\end{array}
$$

$$\frac{1}{2}R_2 \to R_2$$

$$
\sim
\left[
\begin{array}{cccccc|c}
1 & 1 & -1 & 1 & 0 & 0 & 6 \\
\frac{1}{2} & ① & 0 & 0 & \frac{1}{2} & 0 & 4 \\
\hline
-M-1 & -M-3 & M & 0 & 0 & 1 & -6M
\end{array}
\right]
$$

$$(-1)R_2 + R_1 \to R_1, \quad (M+3)R_2 + R_3 \to R_3$$

$$\sim \begin{bmatrix} \boxed{\tfrac{1}{2}} & 0 & -1 & 1 & -\tfrac{1}{2} & 0 & 2 \\ \tfrac{1}{2} & 1 & 0 & 0 & \tfrac{1}{2} & 0 & 4 \\ \hdashline -\tfrac{1}{2}M+\tfrac{1}{2} & 0 & M & 0 & \tfrac{1}{2}M+\tfrac{3}{2} & 1 & -2M+12 \end{bmatrix} \begin{matrix} \tfrac{2}{1/2}=4 \\ \tfrac{4}{1/2}=8 \\ \\ \end{matrix}$$

$$2R_1 \rightarrow R_1$$

$$\sim \begin{bmatrix} \boxed{1} & 0 & -2 & 2 & -1 & 0 & 4 \\ \tfrac{1}{2} & 1 & 0 & 0 & \tfrac{1}{2} & 0 & 4 \\ \hdashline -\tfrac{1}{2}M+\tfrac{1}{2} & 0 & M & 0 & \tfrac{1}{2}M+\tfrac{3}{2} & 1 & -2M+12 \end{bmatrix}$$

$$\left(-\tfrac{1}{2}\right)R_1 + R_2 \rightarrow R_2, \quad \left(\tfrac{1}{2}M - \tfrac{1}{2}\right)R_1 + R_3 \rightarrow R_3$$

$$\sim \begin{matrix} x_1 \\ x_2 \\ P \end{matrix} \begin{matrix} x_1 & x_2 & s_1 & a_1 & s_2 & P \\ \begin{bmatrix} 1 & 0 & -2 & 2 & -1 & 0 & 4 \\ 0 & 1 & 1 & -1 & 1 & 0 & 2 \\ \hdashline 0 & 0 & 1 & M-1 & 2 & 1 & 10 \end{bmatrix} \end{matrix}$$

```
The optimal solution to the modified problem is: Maximum P = 10
at x₁ = 4, x₂ = 2, s₁ = 0, a₁ = 0, s₂ = 0.
```

(D)   Since $a_1 = 0$, the solution to the original problem is: Maximum $P = 10$ at $x_1 = 4$, $x_2 = 2$.   (6-4)

19.   (A)   We introduce a surplus variable s_1 and an artificial variable a_1 to convert the first inequality ($\geq$) into an equation; we introduce a slack variable s_2 to convert the second inequality ($\leq$) into an equation.

The modified problem is: Maximize $P = x_1 + x_2 - Ma_1$

$$\begin{aligned} \text{Subject to: } & x_1 + x_2 - s_1 + a_1 && = 5 \\ & x_1 + 2x_2 && + s_2 = 4 \\ & x_1, \ x_2, \ s_1, \ s_2, \ a_1 \geq 0 \end{aligned}$$

(B)   The preliminary simplex tableau is:

$$\begin{matrix} x_1 & x_2 & s_1 & a_1 & s_2 & P \\ \begin{bmatrix} 1 & 1 & -1 & 1 & 0 & 0 & 5 \\ 1 & 2 & 0 & 0 & 1 & 0 & 4 \\ \hdashline -1 & -1 & 0 & M & 0 & 1 & 0 \end{bmatrix} \end{matrix}$$

Now

$$\begin{bmatrix} 1 & 1 & -1 & 1 & 0 & 0 & 5 \\ 1 & 2 & 0 & 0 & 1 & 0 & 4 \\ \hdashline -1 & -1 & 0 & M & 0 & 1 & 0 \end{bmatrix} \sim \begin{matrix} a_1 \\ s_2 \\ P \end{matrix} \begin{matrix} x_1 & x_2 & s_1 & a_1 & s_2 & P \\ \begin{bmatrix} 1 & 1 & -1 & 1 & 0 & 0 & 5 \\ 1 & 2 & 0 & 0 & 1 & 0 & 4 \\ \hdashline -M-1 & -M-1 & M & 0 & 0 & 1 & -5M \end{bmatrix} \end{matrix}$$

$$(-M)R_1 + R_3 \rightarrow R_3 \qquad\qquad\qquad \text{Initial simplex tableau}$$

(C)

$$\begin{bmatrix} 1 & 1 & -1 & 1 & 0 & 0 & | & 5 \\ ① & 2 & 0 & 0 & 1 & 0 & | & 4 \\ \hline -M-1 & -M-1 & M & 0 & 0 & 1 & | & -5M \end{bmatrix}$$

$(-1)R_2 + R_1 \rightarrow R_1, \ (M+1)R_2 + R_3 \rightarrow R_3$

$$\begin{array}{c} \\ a_1 \\ \sim \ x_1 \\ P \end{array} \begin{array}{cccccc} x_1 & x_2 & s_1 & a_1 & s_2 & P \\ \begin{bmatrix} 0 & -1 & -1 & 1 & -1 & 0 & | & 1 \\ 1 & 2 & 0 & 0 & 1 & 0 & | & 4 \\ \hline 0 & M+1 & M & 0 & M+1 & 1 & | & -M+4 \end{bmatrix} \end{array}$$

The optimal solution to the modified problem is: $x_1 = 4$, $x_2 = 0$, $s_1 = 0$, $a_1 = 1$, $s_2 = 0$, $P = -M + 4$.

(D)   Since $a_1 \neq 0$, the original problem does not have a solution.   (6-4)

20. Multiply the second inequality by −1 to obtain a positive number on the right-hand side. This yields the problem: Maximize $P = 2x_1 + 3x_2 + x_3$

Subject to: $x_1 - 3x_2 + x_3 \leq 7$

$x_1 + x_2 - 2x_3 \geq 2$ (Note: Direction of inequality is reversed.)

$3x_1 + 2x_2 - x_3 = 4$

$x_1, \ x_2, \ x_3 \geq 0$

Now, introduce a slack variable s_1 to convert the first inequality ($\leq$) into an equation; introduce a surplus variable s_2 and an artificial variable a_1 to convert the second inequality (≥ 0) into an equation; introduce an artificial variable a_2 into the equation.

The modified problem is:

Maximize $P = 2x_1 + 3x_2 + x_3 - Ma_1 - Ma_2$

Subject to: $x_1 - 3x_2 + x_3 + s_1 \qquad\qquad = 7$

$x_1 + x_2 - 2x_3 \qquad - s_2 + a_1 \qquad = 2$

$3x_1 + 2x_2 - x_3 \qquad\qquad\quad + a_2 = 4$

$x_1, \ x_2, \ x_3, \ s_1, \ s_2, \ a_1, \ a_2 \geq 0$ (6-4)

21. The basic simplex method with slack variables solves standard maximization problems involving $\leq$ constraints with nonnegative constants on the right side.   (6-2)

22. The dual method solves minimization problems with positive coefficients in the objective function.   (6-3)

23. The big M method solves any linear programming problem.   (6-4)

24. Introduce slack variables s_1, s_2, s_3, and s_4 to obtain:

Maximize $P = 2x_1 + 3x_2$

Subject to: $x_1 + 2x_2 + s_1 \qquad\qquad = 22$

$$3x_1 + x_2 \qquad + s_2 \qquad\qquad = 26$$
$$x_1 \qquad\qquad\qquad + s_3 \qquad = 8$$
$$x_2 \qquad\qquad\qquad\qquad + s_4 = 10$$
$$x_1,\ x_2,\ s_1,\ s_2,\ s_3,\ s_4 \geq 0$$

This system can be written in the initial form:

$$x_1 + 2x_2 + s_1 \qquad\qquad\qquad = 22$$
$$3x_1 + x_2 \qquad + s_2 \qquad\qquad = 26$$
$$x_1 \qquad\qquad\qquad + s_3 \qquad = 8$$
$$x_2 \qquad\qquad\qquad\qquad + s_4 \qquad = 10$$
$$-2x_1 - 3x_2 \qquad\qquad\qquad\qquad + P = 0$$

The simplex tableau for this problem is:

$$
\begin{array}{c}
\\
s_1 \\
s_2 \\
s_3 \\
s_4 \\
P
\end{array}
\begin{array}{c}
\begin{array}{ccccccc}
x_1 & x_2 & s_1 & s_2 & s_3 & s_4 & P
\end{array} \\
\left[
\begin{array}{ccccccc|c}
1 & 2 & 1 & 0 & 0 & 0 & 0 & 22 \\
3 & 1 & 0 & 1 & 0 & 0 & 0 & 26 \\
1 & 0 & 0 & 0 & 1 & 0 & 0 & 8 \\
0 & ① & 0 & 0 & 0 & 1 & 0 & 10 \\
\hdashline
-2 & -3 & 0 & 0 & 0 & 0 & 1 & 0
\end{array}
\right]
\end{array}
\quad
\begin{array}{l}
\frac{22}{2} = 11 \\[6pt]
\frac{26}{1} = 26 \\[6pt]
\\
\frac{10}{1} = 10
\end{array}
$$

$(-2)R_4 + R_1 \rightarrow R_1,\ (-1)R_4 + R_2 \rightarrow R_2,$
$\quad 3R_4 + R_5 \rightarrow R_5$

	Basic Solution						Corner Point
x_1	x_2	s_1	s_2	s_3	s_4		(0, 0)
0	0	22	26	8	10		

$$
\begin{array}{c}
\\
s_1 \\
s_2 \\
\sim\ s_3 \\
x_2 \\
\\
\end{array}
\begin{array}{c}
\begin{array}{ccccccc}
x_1 & x_2 & s_1 & s_2 & s_3 & s_4 & P
\end{array} \\
\left[
\begin{array}{ccccccc|c}
① & 0 & 1 & 0 & 0 & -2 & 0 & 2 \\
3 & 0 & 0 & 1 & 0 & -1 & 0 & 16 \\
1 & 0 & 0 & 0 & 1 & 0 & 0 & 8 \\
0 & 1 & 0 & 0 & 0 & 1 & 0 & 10 \\
\hdashline
-2 & 0 & 0 & 0 & 0 & 3 & 1 & 30
\end{array}
\right]
\end{array}
\quad
\begin{array}{l}
\frac{2}{1} = 2 \\[6pt]
\frac{16}{3} = 5.33 \\[6pt]
\frac{8}{1} = 8 \\
\end{array}
$$

$(-3)R_1 + R_2 \rightarrow R_2,\ (-1)R_1 + R_3 \rightarrow R_3,$
$\quad 2R_1 + R_5 \rightarrow R_5$

	Basic Solution						Corner Point
x_1	x_2	s_1	s_2	s_3	s_4		(0, 10)
0	10	2	16	8	0		

$$
\begin{array}{c}
\begin{array}{ccccccc}
x_1 & x_2 & s_1 & s_2 & s_3 & s_4 & P
\end{array}\\
\begin{array}{c}
x_1\\ s_2\\ \sim \ s_3\\ x_2\\ \\ P
\end{array}
\left[\begin{array}{ccccccc|c}
1 & 0 & 1 & 0 & 0 & -2 & 0 & 2\\
0 & 0 & -3 & 1 & 0 & ⑤ & 0 & 10\\
0 & 0 & -1 & 0 & 1 & 2 & 0 & 6\\
0 & 1 & 0 & 0 & 0 & 1 & 0 & 10\\
\hline
0 & 0 & 2 & 0 & 0 & -1 & 1 & 34
\end{array}\right]
\begin{array}{l}
\frac{10}{5}=2\\[4pt]
\frac{6}{2}=3\\[4pt]
\frac{10}{1}=10
\end{array}
\end{array}
$$

$$\tfrac{1}{5}R_2 \rightarrow R_2$$

Basic Solution | Corner Point

x_1	x_2	s_1	s_2	s_3	s_4	(2, 10)
2	10	0	10	6	0	

$$
\sim
\left[\begin{array}{ccccccc|c}
1 & 0 & 1 & 0 & 0 & -2 & 0 & 2\\
0 & 0 & -\frac{3}{5} & \frac{1}{5} & 0 & ① & 0 & 2\\
0 & 0 & -1 & 0 & 1 & 2 & 0 & 6\\
0 & 1 & 0 & 0 & 0 & 1 & 0 & 10\\
\hline
0 & 0 & 2 & 0 & 0 & -1 & 1 & 34
\end{array}\right]
$$

$$2R_2 + R_1 \rightarrow R_1, \ (-2)R_2 + R_3 \rightarrow R_3, \ (-1)R_2 + R_4 \rightarrow R_4, \ R_2 + R_5 \rightarrow R_5$$

$$
\begin{array}{c}
\begin{array}{ccccccc}
x_1 & x_2 & s_1 & s_2 & s_3 & s_4 & P
\end{array}\\
\begin{array}{c}
x_1\\ s_4\\ \sim \ s_3\\ x_2\\ \\ P
\end{array}
\left[\begin{array}{ccccccc|c}
1 & 0 & -\frac{1}{5} & \frac{2}{5} & 0 & 0 & 0 & 6\\
0 & 0 & -\frac{3}{5} & \frac{1}{5} & 0 & 1 & 0 & 2\\
0 & 0 & \frac{1}{5} & -\frac{2}{5} & 1 & 0 & 0 & 2\\
0 & 1 & \frac{3}{5} & -\frac{1}{5} & 0 & 0 & 0 & 8\\
\hline
0 & 0 & \frac{7}{5} & \frac{1}{5} & 0 & 0 & 1 & 36
\end{array}\right]
\end{array}
$$

Basic Solution | Corner Point

x_1	x_2	s_1	s_2	s_3	s_4	(6, 8)
6	8	0	0	2	2	

Optimal solution:

max $P = 36$ at $x_1 = 6$, $x_2 = 8$.

The graph of the feasible region and the path to the optimal solution is shown at the right

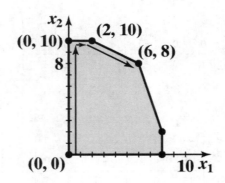

25. Multiply the first constraint inequality by -1 to transform it into a $\geq$ inequality. Now the problem is:
Minimize $C = 3x_1 + 2x_2$

Subject to: $-2x_1 - x_2 \geq -20$
$$2x_1 + x_2 \geq 9$$
$$x_1 + x_2 \geq 6$$
$$x_1, \ x_2 \geq 0$$

The matrices corresponding to this problem and its dual are respectively:

$$A = \begin{bmatrix} -2 & -1 & -20 \\ 2 & 1 & 9 \\ 1 & 1 & 6 \\ \hline 3 & 2 & 1 \end{bmatrix} \quad \text{and} \quad A^T = \begin{bmatrix} -2 & 2 & 1 & 3 \\ -1 & 1 & 1 & 2 \\ \hline -20 & 9 & 6 & 1 \end{bmatrix}$$

Thus, the dual problem is: Maximize $P = -20y_1 + 9y_2 + 6y_3$

Subject to: $-2y_1 + 2y_2 + y_3 \leq 3$
$$-y_1 + y_2 + y_3 \leq 2$$
$$y_1, \ y_2, \ y_3 \geq 0$$

We introduce slack variables x_1 and x_2 to obtain the initial system for the dual problem:

$$-2y_1 + 2y_2 + y_3 + x_1 \qquad\quad = 3$$
$$-y_1 + y_2 + y_3 \qquad + x_2 \qquad = 2$$
$$20y_1 - 9y_2 - 6y_3 \qquad\qquad + P = 0$$

The simplex tableau for this problem is:

$$
\begin{array}{c}
\begin{array}{c} \quad\; y_1 \quad y_2 \;\; y_3 \;\; x_1 \;\; x_2 \quad P \end{array} \\
\begin{array}{c} x_1 \\ x_2 \\ P \end{array}
\left[\begin{array}{cccccc|c}
-2 & ② & 1 & 1 & 0 & 0 & 3 \\
-1 & 1 & 1 & 0 & 1 & 0 & 2 \\ \hline
20 & -9 & -6 & 0 & 0 & 1 & 0
\end{array} \right]
\begin{array}{c} \frac{3}{2} = 1.5 \\ \frac{2}{1} = 2 \\ \; \end{array}
\sim
\left[\begin{array}{cccccc|c}
-1 & ① & \frac{1}{2} & \frac{1}{2} & 0 & 0 & \frac{3}{2} \\
-1 & 1 & 1 & 0 & 1 & 0 & 2 \\ \hline
20 & -9 & -6 & 0 & 0 & 1 & 0
\end{array} \right]
\end{array}
$$

$$\frac{1}{2}R_1 \rightarrow R_1 \qquad\qquad\qquad\qquad (-1)R_1 + R_2 \rightarrow R_2, \; 9R_1 + R_3 \rightarrow R_3$$

$$
\sim
\left[\begin{array}{cccccc|c}
-1 & 1 & \frac{1}{2} & \frac{1}{2} & 0 & 0 & \frac{3}{2} \\
0 & 0 & ⑴\tfrac{1}{2} & -\frac{1}{2} & 1 & 0 & \frac{1}{2} \\ \hline
11 & 0 & -\frac{3}{2} & \frac{9}{2} & 0 & 1 & \frac{27}{2}
\end{array} \right]
\begin{array}{c} \frac{3/2}{1/2} = 3 \\ \frac{1/2}{1/2} = 1 \\ \; \end{array}
\sim
\left[\begin{array}{cccccc|c}
-1 & 1 & \frac{1}{2} & \frac{1}{2} & 0 & 0 & \frac{3}{2} \\
0 & 0 & ① & -1 & 2 & 0 & 1 \\ \hline
11 & 0 & -\frac{3}{2} & \frac{9}{2} & 0 & 1 & \frac{27}{2}
\end{array} \right]
$$

$$2R_2 \rightarrow R_2 \qquad\qquad\qquad\qquad \left(-\frac{1}{2}\right)R_2 + R_1 \rightarrow R_1, \; \frac{3}{2}R_2 + R_3 \rightarrow R_3$$

$$
\begin{array}{c}
\begin{array}{c} \quad\; y_1 \;\; y_2 \;\; y_3 \;\; x_1 \;\; x_2 \quad P \end{array} \\
\sim \begin{array}{c} y_2 \\ y_3 \\ P \end{array}
\left[\begin{array}{cccccc|c}
-1 & 1 & 0 & 1 & -1 & 0 & 1 \\
0 & 0 & 1 & -1 & 2 & 0 & 1 \\ \hline
11 & 0 & 0 & 3 & 3 & 1 & 15
\end{array} \right]
\end{array}
$$

Optimal solution: min $C = 15$ at $x_1 = 3$ and $x_2 = 3$. $\qquad\qquad$ (6-3)

26. First convert the problem to a maximization problem by seeking the maximum of $P = -C = -3x_1 - 2x_2$.
Next, introduce a slack variable s_1 into the first inequality to convert it into an equation; introduce surplus
variables s_2 and s_3 and artificial variables a_1 and a_2 into the second and third inequalities to convert them
into equations.

The modified problem is: Maximize $P = -3x_1 - 2x_2 - Ma_1 - Ma_2$

$$
\begin{aligned}
\text{Subject to: } 2x_1 + x_2 + s_1 \quad\quad\quad\quad &= 20 \\
2x_1 + x_2 \quad\quad - s_2 + a_1 \quad\quad &= 9 \\
x_1 + x_2 \quad\quad\quad\quad - s_3 + a_2 &= 6 \\
x_1,\ x_2,\ s_1,\ s_2,\ s_3,\ a_1,\ a_2 &\geq 0
\end{aligned}
$$

The preliminary simplex tableau is:

$$
\begin{array}{cccccccc|c}
x_1 & x_2 & s_1 & s_2 & a_1 & s_3 & a_2 & P & \\
2 & 1 & 1 & 0 & 0 & 0 & 0 & 0 & 20 \\
2 & 1 & 0 & -1 & 1 & 0 & 0 & 0 & 9 \\
1 & 1 & 0 & 0 & 0 & -1 & 1 & 0 & 6 \\
\hline
3 & 2 & 0 & 0 & M & 0 & M & 1 & 0
\end{array}
$$

$(-M)R_2 + R_4 \rightarrow R_4$

$$
\sim
\begin{bmatrix}
2 & 1 & 1 & 0 & 0 & 0 & 0 & 0 & 20 \\
2 & 1 & 0 & -1 & 1 & 0 & 0 & 0 & 9 \\
1 & 1 & 0 & 0 & 0 & -1 & 1 & 0 & 6 \\
\hline
-2M+3 & -M+2 & 0 & M & 0 & 0 & M & 1 & -9M
\end{bmatrix}
$$

$(-M)R_3 + R_4 \rightarrow R_4$

$$
\begin{array}{c}
\\ s_1 \\ a_1 \\ a_2 \\ P
\end{array}
\sim
\begin{array}{cccccccc|c}
x_1 & x_2 & s_1 & s_2 & a_1 & s_3 & a_2 & P & \\
2 & 1 & 1 & 0 & 0 & 0 & 0 & 0 & 20 \\
② & 1 & 0 & -1 & 1 & 0 & 0 & 0 & 9 \\
1 & 1 & 0 & 0 & 0 & -1 & 1 & 0 & 6 \\
\hline
-3M+3 & -2M+2 & 0 & M & 0 & M & 0 & 1 & -15M
\end{array}
\quad
\begin{array}{l}
\frac{20}{2} = 10 \\
\frac{9}{2} = 4.5 \\
\frac{6}{1} = 6
\end{array}
$$

$\dfrac{1}{2}R_2 \rightarrow R_2$

$$
\sim
\begin{bmatrix}
2 & 1 & 1 & 0 & 0 & 0 & 0 & 0 & 20 \\
① & \frac{1}{2} & 0 & -\frac{1}{2} & \frac{1}{2} & 0 & 0 & 0 & \frac{9}{2} \\
1 & 1 & 0 & 0 & 0 & -1 & 1 & 0 & 6 \\
\hline
-3M+3 & -2M+2 & 0 & M & 0 & M & 0 & 1 & -15M
\end{bmatrix}
$$

$(-2)R_2 + R_1 \rightarrow R_1,\ (-1)R_2 + R_3 \rightarrow R_3,\ (3M-3)R_2 + R_4 \rightarrow R_4$

$$\sim \begin{bmatrix} 0 & 0 & 1 & 1 & -1 & 0 & 0 & 0 & 11 \\ 1 & \frac{1}{2} & 0 & -\frac{1}{2} & \frac{1}{2} & 0 & 0 & 0 & \frac{9}{2} \\ 0 & \boxed{\tfrac{1}{2}} & 0 & \frac{1}{2} & -\frac{1}{2} & -1 & 1 & 0 & \frac{3}{2} \\ \hdashline 0 & -\frac{1}{2}M + \frac{1}{2} & 0 & -\frac{1}{2}M + \frac{3}{2} & \frac{3}{2}M - \frac{3}{2} & M & 0 & 1 & -\frac{3}{2}M - \frac{27}{2} \end{bmatrix} \begin{matrix} \\ \dfrac{9}{1/2} = 18 \\ \dfrac{3/2}{1/2} = 3 \\ \\ \end{matrix}$$

$$2R_3 \to R_3$$

$$\sim \begin{bmatrix} 0 & 0 & 1 & 1 & -1 & 0 & 0 & 0 & 11 \\ 1 & \frac{1}{2} & 0 & -\frac{1}{2} & \frac{1}{2} & 0 & 0 & 0 & \frac{9}{2} \\ 0 & \circled{1} & 0 & 1 & -1 & -2 & 2 & 0 & 3 \\ \hdashline 0 & -\frac{1}{2}M + \frac{1}{2} & 0 & -\frac{1}{2}M + \frac{3}{2} & \frac{3}{2}M - \frac{3}{2} & M & 0 & 1 & -\frac{3}{2}M - \frac{27}{2} \end{bmatrix}$$

$$\left(-\frac{1}{2}\right)R_3 + R_2 \to R_2, \quad \left(\frac{1}{2}M - \frac{1}{2}\right)R_3 + R_4 \to R_4$$

$$\sim \begin{array}{c} \\ s_1 \\ x_1 \\ x_2 \\ P \end{array} \begin{array}{cccccccc} x_1 & x_2 & s_1 & s_2 & a_1 & s_3 & a_2 & P \\ \end{array}$$
$$\sim \begin{array}{c} s_1 \\ x_1 \\ x_2 \\ P \end{array} \begin{bmatrix} 0 & 0 & 1 & 1 & -1 & 0 & 0 & 0 & 11 \\ 1 & 0 & 0 & -1 & 1 & 1 & -1 & 0 & 3 \\ 0 & 1 & 0 & 1 & -1 & -2 & 2 & 0 & 3 \\ \hdashline 0 & 0 & 0 & 1 & M-1 & 1 & M-1 & 1 & -15 \end{bmatrix}$$

Optimal solution: Max $P = -15$ at $x_1 = 3$, $x_2 = 3$. Thus, the optimal solution of the original problem is: Min $C = 15$ at $x_1 = 3$, $x_2 = 3$. (6-4)

27. Multiply the first two constraint inequalities by -1 to transform them into $\geq$ inequalities. The problem now is:

$$\text{Minimize } C = 15x_1 + 12x_2 + 15x_3 + 18x_4$$
$$\begin{aligned} \text{Subject to: } -x_1 - x_2 \qquad\qquad &\geq -240 \\ -x_3 - x_4 &\geq -500 \\ x_1 \quad + x_3 \qquad &\geq 400 \\ x_2 \quad + x_4 &\geq 300 \\ x_1,\ x_2,\ x_3,\ x_4 &\geq 0 \end{aligned}$$

The matrices corresponding to this problem and its dual are, respectively:

$$A = \begin{bmatrix} -1 & -1 & 0 & 0 & -240 \\ 0 & 0 & -1 & -1 & -500 \\ 1 & 0 & 1 & 0 & 400 \\ 0 & 1 & 0 & 1 & 300 \\ \hline 15 & 12 & 15 & 18 & 1 \end{bmatrix} \quad \text{and} \quad A^T = \begin{bmatrix} -1 & 0 & 1 & 0 & 15 \\ -1 & 0 & 0 & 1 & 12 \\ 0 & -1 & 1 & 0 & 15 \\ 0 & -1 & 0 & 1 & 18 \\ \hline -240 & -500 & 400 & 300 & 1 \end{bmatrix}$$

Thus, the dual problem is: Maximize $P = -240y_1 - 500y_2 + 400y_3 + 300y_4$
$$\begin{aligned} \text{Subject to: } -y_1 \qquad + y_3 \qquad &\leq 15 \\ -y_1 \qquad\qquad + y_4 &\leq 12 \\ -y_2 + y_3 \qquad &\leq 15 \\ -y_2 \qquad + y_4 &\leq 18 \\ y_1,\ y_2,\ y_3,\ y_4 &\geq 0 \end{aligned}$$

We introduce the slack variables x_1, x_2, x_3, and x_4 to obtain the initial system for the dual problem:

$$
\begin{aligned}
-y_1 \quad\quad + \quad y_3 \quad\quad + x_1 \quad\quad\quad\quad\quad &= 15 \\
-y_1 \quad\quad\quad\quad\quad + \quad y_4 \quad\quad + x_2 \quad\quad\quad &= 12 \\
- \quad y_2 + \quad y_3 \quad\quad\quad\quad\quad + x_3 \quad\quad &= 15 \\
- \quad y_2 \quad\quad\quad + \quad y_4 \quad\quad\quad\quad\quad + x_4 &= 18 \\
240y_1 + 500y_2 - 400y_3 - 300y_4 \quad\quad\quad\quad\quad\quad + P &= 0
\end{aligned}
$$

The simplex tableau for this problem is:

	y_1	y_2	y_3	y_4	x_1	x_2	x_3	x_4	P	
x_1	-1	0	①	0	1	0	0	0	0	15
x_2	-1	0	0	1	0	1	0	0	0	12
x_3	0	-1	1	0	0	0	1	0	0	15
x_4	0	-1	0	1	0	0	0	1	0	18
P	240	500	-400	-300	0	0	0	0	1	0

$\dfrac{15}{1} = 15$

[Note: Either element can be chosen as the pivot; we choose the element in the first row.]

$(-1)R_1 + R_3 \to R_3, \quad 400R_1 + R_5 \to R_5$

$$
\sim
\begin{bmatrix}
-1 & 0 & 1 & 0 & 1 & 0 & 0 & 0 & 0 & 15 \\
-1 & 0 & 0 & ① & 0 & 1 & 0 & 0 & 0 & 12 \\
1 & -1 & 0 & 0 & -1 & 0 & 1 & 0 & 0 & 0 \\
0 & -1 & 0 & 1 & 0 & 0 & 0 & 1 & 0 & 18 \\
\hline
-160 & 500 & 0 & -300 & 400 & 0 & 0 & 0 & 1 & 6000
\end{bmatrix}
$$

$\dfrac{15}{1} = 12$

$\dfrac{18}{1} = 18$

$(-1)R_2 + R_4 \to R_4, \quad 300R_2 + R_5 \to R_5$

$$
\sim
\begin{bmatrix}
-1 & 0 & 1 & 0 & 1 & 0 & 0 & 0 & 0 & 15 \\
-1 & 0 & 0 & 1 & 0 & 1 & 0 & 0 & 0 & 12 \\
① & -1 & 0 & 0 & -1 & 0 & 1 & 0 & 0 & 0 \\
1 & -1 & 0 & 0 & 0 & -1 & 0 & 1 & 0 & 6 \\
\hline
-460 & 500 & 0 & 0 & 400 & 300 & 0 & 0 & 1 & 9600
\end{bmatrix}
$$

$\dfrac{0}{1} = 0$

$\dfrac{6}{1} = 6$

$R_3 + R_1 \to R_1, \quad R_3 + R_2 \to R_2, \quad (-1)R_3 + R_4 \to R_4, \quad 460R_3 + R_5 \to R_5$

$$
\sim
\begin{bmatrix}
0 & -1 & 1 & 0 & 0 & 0 & 1 & 0 & 0 & 15 \\
0 & -1 & 0 & 1 & -1 & 1 & 1 & 0 & 0 & 12 \\
1 & -1 & 0 & 0 & -1 & 0 & 1 & 0 & 0 & 0 \\
0 & 0 & 0 & 0 & ① & -1 & -1 & 1 & 0 & 6 \\
\hline
0 & 40 & 0 & 0 & -60 & 300 & 460 & 0 & 1 & 9600
\end{bmatrix}
$$

$60R_4 + R_5 \to R_5, \quad R_4 + R_2 \to R_2, \quad R_4 + R_3 \to R_3$

	y_1	y_2	y_3	y_4	x_1	x_2	x_3	x_4	P	
y_3	0	-1	1	0	0	0	1	0	0	15
y_4	0	-1	0	1	0	1	1	0	0	18
~ y_1	1	-1	0	0	0	0	1	0	0	6
x_4	0	0	0	0	1	-1	-1	1	0	6
P	0	40	0	0	0	240	400	60	1	9960

Optimal solution: min C = 9960 at x_1 = 0, x_2 = 240, x_3 = 400, x_4 = 60.   (6-3)

28.   (A)   Let x_1 = amount invested in oil stock

x_2 = amount invested in steel stock

x_3 = amount invested in government bonds

The mathematical model for this problem is:

Maximize $P = 0.12x_1 + 0.09x_2 + 0.05x_3$

Subject to: $x_1 + x_2 + x_3 \le 150{,}000$

$x_1 \qquad\qquad \le 50{,}000$

$x_1 + x_2 - x_3 \le 25{,}000$

$x_1,\ x_2,\ x_3 \ge 0$

Introduce slack variables s_1, s_2, s_3 to obtain the initial system:

$x_1 + \quad x_2 + \quad x_3 + s_1 \qquad\qquad\quad = 150{,}000$

$x_1 \qquad\qquad\qquad\qquad + s_2 \qquad\quad = 50{,}000$

$x_1 + \quad x_2 - \quad x_3 \qquad\qquad + s_3 \quad\ = 25{,}000$

$-0.12x_1 - 0.09x_2 - 0.05x_3 \qquad\qquad\qquad + P = 0$

$x_1,\ x_2,\ x_3,\ s_1,\ s_2,\ s_3 \ge 0$

The simplex tableau for this problem is:

	x_1	x_2	x_3	s_1	s_2	s_3	P	
s_1	1	1	1	1	0	0	0	150,000
s_2	1	0	0	0	1	0	0	50,000
s_3	①	1	-1	0	0	1	0	25,000
P	-0.12	-0.09	-0.05	0	0	0	1	0

$(-1)R_3 + R_1 \rightarrow R_1,\ (-1)R_3 + R_2 \rightarrow R_1,\ 0.12R_3 + R_4 \rightarrow R_4$

	x_1	x_2	x_3	s_1	s_2	s_3	P		
	0	0	2	1	0	-1	0	125,000	$\dfrac{125{,}000}{2}$ = 62,500
~	0	-1	①	0	1	-1	0	25,000	$\dfrac{25{,}000}{1}$ = 25,000
	1	1	-1	0	0	1	0	25,000	
	0	0.03	-0.17	0	0	0.12	1	3,000	

$(-2)R_2 + R_1 \rightarrow R_1,\ R_2 + R_3 \rightarrow R_3,\ (0.17)R_2 + R_4 \rightarrow R_4$

$$\sim \begin{bmatrix} 0 & ② & 0 & 1 & -2 & 1 & 0 & 75{,}000 \\ 0 & -1 & 1 & 0 & 1 & -1 & 0 & 25{,}000 \\ 1 & 0 & 0 & 0 & 1 & 0 & 0 & 50{,}000 \\ \hline 0 & -0.14 & 0 & 0 & 0.17 & -0.05 & 1 & 7{,}250 \end{bmatrix}$$

$$\frac{1}{2}R_1 \to R_1$$

$$\sim \begin{bmatrix} 0 & ① & 0 & \frac{1}{2} & -1 & \frac{1}{2} & 0 & 37{,}500 \\ 0 & -1 & 1 & 0 & 1 & -1 & 0 & 25{,}000 \\ 1 & 0 & 0 & 0 & 1 & 0 & 0 & 50{,}000 \\ \hline 0 & -0.14 & 0 & 0 & 0.17 & -0.05 & 1 & 7{,}250 \end{bmatrix}$$

$$R_1 + R_2 \to R_2,\ 0.14R_1 + R_4 \to R_4$$

$$\begin{array}{c}\\ x_2 \\ x_3 \\ x_1 \\ \\ P \end{array}\sim \begin{bmatrix} x_1 & x_2 & x_3 & s_1 & s_2 & s_3 & P & \\ 0 & 1 & 0 & \frac{1}{2} & -1 & \frac{1}{2} & 0 & 37{,}500 \\ 0 & 0 & 1 & \frac{1}{2} & 0 & -\frac{1}{2} & 0 & 62{,}500 \\ 1 & 0 & 0 & 0 & 1 & 0 & 0 & 50{,}000 \\ \hline 0 & 0 & 0 & 0.07 & 0.03 & 0.02 & 1 & 12{,}500 \end{bmatrix}$$

The maximum return is \$12,500 when \$50,000 is invested in oil stock, \$37,500 in steel stock, and \$62,500 in government bonds.

(B)   The mathematical model for this problem is:

$$\text{Maximize } P = 0.09x_1 + 0.12x_2 + 0.05x_3$$

$$\text{Subject to: } x_1 + x_2 + x_3 \le 150{,}000$$
$$x_1 \qquad\qquad \le 50{,}000$$
$$x_1 + x_2 - x_3 \le 25{,}000$$
$$x_1,\ x_2,\ x_3 \ge 0$$

Introduce slack variables s_1, s_2, s_3 to obtain the initial system:

$$x_1 + x_2 + x_3 + s_1 \qquad\qquad = 150{,}000$$
$$x_1 \qquad\qquad + s_2 \qquad = 50{,}000$$
$$x_1 + x_2 - x_3 \qquad\quad + s_3 = 25{,}000$$
$$-0.09x_1 - 0.12x_2 - 0.05x_3 \qquad\qquad + P = 0$$

The simplex tableau for this problem is:

$$\begin{array}{c} \\ s_1 \\ s_2 \\ s_3 \\ \\ P \end{array}\begin{bmatrix} x_1 & x_2 & x_3 & s_1 & s_2 & s_3 & P & \\ 1 & 1 & 1 & 1 & 0 & 0 & 0 & 150{,}000 \\ 1 & 0 & 0 & 0 & 1 & 0 & 0 & 50{,}000 \\ 1 & ① & -1 & 0 & 0 & 1 & 0 & 25{,}000 \\ \hline -0.09 & -0.12 & -0.05 & 0 & 0 & 0 & 1 & 0 \end{bmatrix}$$

$$(-1)R_3 + R_1 \to R_1,\ 0.12R_3 + R_4 \to R_4$$

$$\sim \begin{bmatrix} 0 & 0 & ② & 1 & 0 & -1 & 0 & | & 125{,}000 \\ 1 & 0 & 0 & 0 & 1 & 0 & 0 & | & 50{,}000 \\ 1 & 1 & -1 & 0 & 0 & 1 & 0 & | & 25{,}000 \\ \hline 0.03 & 0 & -0.17 & 0 & 0 & 0.12 & 1 & | & 3{,}000 \end{bmatrix}$$

$$\tfrac{1}{2}R_1 \;\to\; R_1$$

$$\sim \begin{bmatrix} 0 & 0 & ① & \tfrac{1}{2} & 0 & -\tfrac{1}{2} & 0 & | & 62{,}500 \\ 1 & 0 & 0 & 0 & 1 & 0 & 0 & | & 50{,}000 \\ 1 & 1 & -1 & 0 & 0 & 1 & 0 & | & 25{,}000 \\ \hline 0.03 & 0 & -0.17 & 0 & 0 & 0.12 & 1 & | & 3{,}000 \end{bmatrix}$$

$$R_1 + R_3 \;\to\; R_3, \quad 0.17R_1 + R_4 \;\to\; R_4$$

	x_1	x_2	x_3	s_1	s_2	s_3	P	
x_3	0	0	1	$\tfrac{1}{2}$	0	$-\tfrac{1}{2}$	0	62,500
s_2	1	0	0	0	1	0	0	50,000
x_2	1	1	0	$\tfrac{1}{2}$	0	$\tfrac{1}{2}$	0	87,500
P	0.03	0	0	0.085	0	0.35	1	13,625

The maximum return is $13,625 when $0 is invested in oil stock, $87,500 is invested in steel stock, and $62,500 in invested in government bonds.   (6-2)

29.   Let x_1 = the number of regular chairs

   x_2 = the number of rocking chairs

   x_3 = the number of chaise lounges

(A) The mathematical model for this problem is:

   Maximize $P = 17x_1 + 24x_2 + 31x_3$

   Subject to: $x_1 + 2x_2 + 3x_3 \le 2{,}500$

   $2x_1 + 2x_2 + 4x_3 \le 3{,}000$

   $3x_1 + 3x_2 + 2x_3 \le 3{,}500$

   $x_1,\ x_2,\ x_3 \ge 0$

   We introduce slack variables s_1, s_2, s_3 to obtain the equivalent form

   $$\begin{aligned} x_1 + \ 2x_2 + \ 3x_3 + s_1 \qquad\qquad\quad &= 2{,}500 \\ 2x_1 + \ 2x_2 + \ 4x_3 \qquad + s_2 \qquad\quad &= 3{,}000 \\ 3x_1 + \ 3x_2 + \ 2x_3 \qquad\qquad + s_3 \quad &= 3{,}500 \\ -17x_1 - 24x_2 - 31x_3 \qquad\qquad\qquad + P &= 0 \end{aligned}$$

The simplex tableau for this problem is:

$$
\begin{array}{c}
\\ s_1 \\ s_2 \\ s_3 \\ P
\end{array}
\begin{array}{ccccccc|c}
x_1 & x_2 & x_3 & s_1 & s_2 & s_3 & P & \\
1 & 2 & 3 & 1 & 0 & 0 & 0 & 2,500 \\
2 & 2 & ④ & 0 & 1 & 0 & 0 & 3,000 \\
3 & 3 & 2 & 0 & 0 & 1 & 0 & 3,500 \\
\hline
-17 & -24 & -31 & 0 & 0 & 0 & 1 & 0
\end{array}
\begin{array}{l}
\frac{2500}{3}=833.33 \\
\frac{3000}{4}=750 \\
\frac{3500}{2}=1750
\end{array}
$$

$$\tfrac{1}{4}R_1 \rightarrow R_2$$

$$
\sim
\begin{bmatrix}
1 & 2 & 3 & 1 & 0 & 0 & 0 & 2,500 \\
\frac{1}{2} & \frac{1}{2} & ① & 0 & \frac{1}{4} & 0 & 0 & 750 \\
3 & 3 & 2 & 0 & 0 & 1 & 0 & 3,500 \\
-17 & -24 & -31 & 0 & 0 & 0 & 1 & 0
\end{bmatrix}
$$

$$(-3)R_2+R_1\rightarrow R_1,\quad (-2)R_2+R_3\rightarrow R_3,\quad (31)R_1+R_4\rightarrow R_4$$

$$
\sim
\begin{bmatrix}
-\frac{1}{2} & ⓵\!\frac{1}{2} & 0 & 1 & -\frac{3}{4} & 0 & 0 & 250 \\
\frac{1}{2} & \frac{1}{2} & 1 & 0 & \frac{1}{4} & 0 & 0 & 750 \\
2 & 2 & 0 & 0 & -\frac{1}{2} & 1 & 0 & 2,000 \\
-\frac{3}{2} & -\frac{17}{2} & 0 & 0 & \frac{31}{4} & 0 & 1 & 23,250
\end{bmatrix}
\begin{array}{l}
\frac{250}{1/2}=500 \\
\frac{750}{1/2}=1,500 \\
\frac{2000}{2}=1,000
\end{array}
$$

$$2R_1\rightarrow R_1$$

$$
\sim
\begin{bmatrix}
-1 & ① & 0 & 2 & -\frac{3}{2} & 0 & 0 & 500 \\
\frac{1}{2} & \frac{1}{2} & 1 & 0 & \frac{1}{4} & 0 & 0 & 750 \\
2 & 2 & 0 & 0 & -\frac{1}{2} & 1 & 0 & 2,000 \\
-\frac{3}{2} & -\frac{17}{2} & 0 & 0 & \frac{31}{4} & 0 & 1 & 23,250
\end{bmatrix}
$$

$$\left(-\tfrac{1}{2}\right)R_1+R_2\rightarrow R_2,\quad -2R_1+R_3\rightarrow R_3,\quad \tfrac{17}{2}R_1+R_4\rightarrow R_4$$

$$
\sim
\begin{bmatrix}
-1 & 1 & 0 & 2 & -\frac{3}{2} & 0 & 0 & 500 \\
1 & 0 & 1 & -1 & 1 & 0 & 0 & 500 \\
④ & 0 & 0 & -4 & \frac{5}{2} & 1 & 0 & 1,000 \\
-10 & 0 & 0 & 17 & -5 & 0 & 1 & 27,500
\end{bmatrix}
\begin{array}{l}
\frac{500}{1}=500 \\
\frac{1000}{4}=250
\end{array}
$$

$$\left(\tfrac{1}{4}\right)R_3\rightarrow R_3$$

$$
\sim
\begin{bmatrix}
-1 & 1 & 0 & 2 & -\frac{3}{2} & 0 & 0 & 500 \\
1 & 0 & 1 & -1 & 1 & 0 & 0 & 500 \\
① & 0 & 0 & -1 & \frac{5}{8} & \frac{1}{4} & 0 & 250 \\
-10 & 0 & 0 & 17 & -5 & 0 & 1 & 27,500
\end{bmatrix}
$$

$$R_3+R_1\rightarrow R_1,\quad (-\ )R_3+R_2\rightarrow R_2,\quad 10R_3+R_4\rightarrow R_4$$

$$\begin{array}{c} \\ x_2 \\ x_3 \\ x_1 \\ \\ \end{array} \begin{array}{ccccccc} x_1 & x_2 & x_3 & s_1 & s_2 & s_3 & P \\ \left[\begin{array}{ccccccc|c} 0 & 1 & 0 & 1 & -\frac{7}{8} & \frac{1}{4} & 0 & 750 \\ 0 & 0 & 1 & 0 & \frac{3}{8} & -\frac{1}{4} & 0 & 250 \\ 1 & 0 & 0 & -1 & \frac{5}{8} & \frac{1}{4} & 0 & 250 \\ \hline 0 & 0 & 0 & 7 & \frac{5}{4} & \frac{5}{2} & 1 & 30{,}000 \end{array}\right] \end{array}$$

The maximum profit is $30,000 when 250 regular chairs, 750 rocking chairs and 250 chaise lounges are produced.

(B)   If the profit on a regular chair is increased to $25, the model becomes:

Maximize $P = 25x_1 + 24x_2 + 31x_3$

Subject to: $x_1 + 2x_2 + 3x_3 \le 2{,}500$

$2x_1 + 2x_2 + 4x_3 \le 3{,}000$

$3x_1 + 3x_2 + 2x_3 \le 3{,}500$

$x_1,\ x_2,\ x_3 \ge 0$

Introducing slack variables s_1, s_2, s_3, we obtain the simplex tableau

$$\begin{array}{ccccccc} x_1 & x_2 & x_3 & s_1 & s_2 & s_3 & P \\ \left[\begin{array}{ccccccc|c} 1 & 2 & 3 & 1 & 0 & 0 & 0 & 2{,}500 \\ 2 & 2 & ④ & 0 & 1 & 0 & 0 & 3{,}000 \\ 3 & 3 & 2 & 0 & 0 & 1 & 0 & 3{,}500 \\ \hline -25 & -24 & -31 & 0 & 0 & 0 & 1 & 0 \end{array}\right] \end{array} \begin{array}{l} \frac{2500}{3} = 833.33 \\ \frac{3000}{4} = 750 \\ \frac{3500}{2} = 1{,}750 \end{array}$$

$\frac{1}{4}R_2 \to R_2$

$$\left[\begin{array}{ccccccc|c} 1 & 2 & 3 & 1 & 0 & 0 & 0 & 2{,}500 \\ \frac{1}{2} & \frac{1}{2} & ① & 0 & \frac{1}{4} & 0 & 0 & 750 \\ 3 & 3 & 2 & 0 & 0 & 1 & 0 & 3{,}500 \\ \hline -25 & -24 & -31 & 0 & 0 & 0 & 1 & 0 \end{array}\right]$$

$-3R_2 + R_1 \to R_1,\ (-2)R_2 + R_3 \to R_3,\ 31R_2 + R_4 \to R_4$

$$\left[\begin{array}{ccccccc|c} -\frac{1}{2} & \frac{1}{2} & 0 & 1 & -\frac{3}{4} & 0 & 0 & 250 \\ \frac{1}{2} & \frac{1}{2} & 1 & 0 & \frac{1}{4} & 0 & 0 & 750 \\ ② & 2 & 0 & 0 & -\frac{1}{2} & 1 & 0 & 2{,}000 \\ \hline -\frac{19}{2} & -\frac{17}{2} & 0 & 0 & \frac{31}{4} & 0 & 1 & 23{,}250 \end{array}\right] \begin{array}{l} \\ \frac{750}{1/2} = 1{,}500 \\ \frac{2000}{2} = 1{,}000 \\ \\ \end{array}$$

$\frac{1}{2}R_3 \to R_3$

$$\left[\begin{array}{ccccccc|c} -\frac{1}{2} & \frac{1}{2} & 0 & 1 & -\frac{3}{4} & 0 & 0 & 250 \\ \frac{1}{2} & \frac{1}{2} & 1 & 0 & \frac{1}{4} & 0 & 0 & 750 \\ ① & 1 & 0 & 0 & -\frac{1}{4} & \frac{1}{2} & 0 & 1{,}000 \\ \hline -\frac{19}{2} & -\frac{17}{2} & 0 & 0 & \frac{31}{4} & 0 & 1 & 23{,}250 \end{array}\right]$$

$\frac{1}{2}R_3 + R_1 \to R_1,\ \left(-\frac{1}{2}\right)R_3 + R_2 \to R_2,\ \frac{19}{2}R_3 + R_4 \to R_4$

$$
\begin{array}{c}
\quad\ \ x_1\ x_2\ x_3\ s_1\quad s_2\quad s_3\ P \\
\begin{array}{c} x_2 \\ x_3 \\ x_1 \\ \ \end{array}
\left[\begin{array}{cccccccc|c}
0 & 1 & 0 & 1 & -\frac{7}{8} & -\frac{1}{4} & 0 & -250 \\
0 & 0 & 1 & 0 & \frac{3}{8} & -\frac{1}{4} & 0 & 250 \\
1 & 1 & 0 & 0 & -\frac{1}{4} & \frac{1}{2} & 0 & 1{,}000 \\
\hline
0 & 1 & 0 & 0 & \frac{43}{8} & 0 & 1 & 32{,}750
\end{array}\right]
\end{array}
$$

The maximum profit is \$32,750 when 1,000 regular chairs, 0 rocking chairs, and 250 chaise lounges are produced.

(C) The available hours in the finishing department are reduced to 3000 hours, the initial simplex tableau becomes

$$
\left[\begin{array}{ccccccc|c}
1 & 2 & 3 & 1 & 0 & 0 & 0 & 2{,}500 \\
2 & 2 & ④ & 0 & 1 & 0 & 0 & 3{,}000 \\
3 & 3 & 2 & 0 & 0 & 1 & 0 & 3{,}000 \\
\hline
-17 & -24 & -31 & 0 & 0 & 0 & 1 & 0
\end{array}\right]
\begin{array}{l}
\frac{2500}{3} = 833.33 \\
\frac{3000}{4} = 750 \\
\frac{3000}{2} = 1{,}500
\end{array}
$$

$$\frac{1}{4}R_2 \rightarrow R_2$$

$$
\left[\begin{array}{ccccccc|c}
1 & 2 & 3 & 1 & 0 & 0 & 0 & 2{,}500 \\
\frac{1}{2} & \frac{1}{2} & ① & 0 & \frac{1}{4} & 0 & 0 & 750 \\
3 & 3 & 2 & 0 & 0 & 1 & 0 & 3{,}000 \\
\hline
-17 & -24 & -31 & 0 & 0 & 0 & 1 & 0
\end{array}\right]
$$

$$-3R_2 + R_1 \rightarrow R_1,\ -2R_2 + R_3 \rightarrow R_3,\ 31R_2 + R_4 \rightarrow R_4$$

$$
\left[\begin{array}{ccccccc|c}
-\frac{1}{2} & \left(\frac{1}{2}\right) & 0 & 1 & -\frac{3}{4} & 0 & 0 & 250 \\
\frac{1}{2} & \frac{1}{2} & 1 & 0 & \frac{1}{4} & 0 & 0 & 750 \\
2 & 2 & 0 & 0 & -\frac{1}{2} & 1 & 0 & 1{,}500 \\
\hline
-\frac{3}{2} & -\frac{17}{2} & 0 & 0 & \frac{31}{4} & 0 & 1 & 23{,}250
\end{array}\right]
\begin{array}{l}
\frac{250}{1/2} = 500 \\
\frac{750}{1/2} = 1{,}500 \\
\frac{1500}{2} = 750
\end{array}
$$

$$
\left[\begin{array}{ccccccc|c}
-1 & ① & 0 & 2 & -\frac{3}{2} & 0 & 0 & 500 \\
\frac{1}{2} & \frac{1}{2} & 1 & 0 & \frac{1}{4} & 0 & 0 & 750 \\
2 & 2 & 0 & 0 & -\frac{1}{2} & 1 & 0 & 1{,}500 \\
\hline
-\frac{3}{2} & -\frac{17}{2} & 0 & 0 & \frac{31}{4} & 0 & 1 & 23{,}250
\end{array}\right]
$$

$$-\frac{1}{2}R_1 + R_2 \rightarrow R_2,\ -2R_1 + R_3 \rightarrow R_3,\ \frac{17}{2}R_1 + R_4 \rightarrow R_4$$

$$
\left[\begin{array}{ccccccc|c}
-1 & 1 & 0 & 2 & -\frac{3}{2} & 0 & 0 & 500 \\
1 & 0 & 1 & -1 & 1 & 0 & 0 & 500 \\
④ & 0 & 0 & -4 & \frac{5}{2} & 1 & 0 & 500 \\
\hline
-10 & 0 & 0 & 17 & -5 & 0 & 1 & 27{,}500
\end{array}\right]
\begin{array}{l}
\frac{500}{1} = 500 \\
\frac{500}{4} = 125
\end{array}
$$

$$\frac{1}{4}R_3 \rightarrow R_3$$

$$\left[\begin{array}{ccccccc|c} -1 & 1 & 0 & 2 & -\frac{3}{2} & 0 & 0 & 500 \\ 1 & 0 & 1 & -1 & 1 & 0 & 0 & 500 \\ ① & 0 & 0 & -1 & \frac{5}{8} & \frac{1}{4} & 0 & 125 \\ \hline -10 & 0 & 0 & 17 & -5 & 0 & 1 & 27,500 \end{array}\right]$$

$$R_3 + R_1 \rightarrow R_1, \quad -R_3 + R_2 \rightarrow R_2, \quad 10R_3 + R_4 \rightarrow R_4$$

$$\begin{array}{c} \\ x_2 \\ x_3 \\ x_1 \\ \\ \end{array} \begin{array}{cccccc} x_1 & x_2 & x_3 & s_1 & s_2 & s_3 \quad P \\ \end{array}$$

$$\begin{array}{c} \\ x_2 \\ x_3 \\ x_1 \\ \\ \end{array}\left[\begin{array}{ccccccc|c} 0 & 1 & 0 & 1 & -\frac{7}{8} & \frac{1}{4} & 0 & 625 \\ 0 & 0 & 1 & 0 & \frac{3}{8} & -\frac{1}{4} & 0 & 375 \\ 1 & 0 & 0 & -1 & \frac{5}{8} & \frac{1}{4} & 0 & 125 \\ \hline 0 & 0 & 0 & 7 & \frac{5}{4} & \frac{5}{2} & 1 & 28,750 \end{array}\right]$$

In this case, the maximum profit is \$28,750 when 125 regular chairs, 625 rocking chairs and 375 chaise lounges are produced. Reducing the available hours in the finishing department reduces the profit by \$1,250.　　　　　　　　　　　　　　　　　　　　　　　(6-3)

30. Let x_1 = the number of motors from A to X,

x_2 = the number of motors from A to Y,

x_3 = the number of motors from B to X,

x_4 = the number of motors from B to Y.

The mathematical model for this problem is:

Minimize $C = 5x_1 + 8x_2 + 9x_3 + 7x_4$

Subject to: $\quad x_1 + x_2 \qquad\qquad\ \le 1,500$

$$x_3 + x_4 \le 1,000$$
$$x_1 \qquad + x_3 \qquad\ \ge\ \ 900$$
$$x_2 \qquad + x_4 \ge 1,200$$
$$x_1, x_2, x_3, x_4 \ge 0$$

Multiply the first two inequalities by –1 to obtain $\ge$ inequalities. The model then becomes:

Minimize $C = 5x_1 + 8x_2 + 9x_3 + 7x_4$

Subject to: $\ -x_1 - x_2 \qquad\qquad \ge -1,500$

$$-x_3 - x_4 \ge -1,000$$
$$x_1 \qquad + x_3 \qquad\ \ge 900$$
$$x_2 \qquad + x_4 \ge 1,200$$
$$x_1, x_2, x_3, x_4 \ge 0$$

The matrices for this problem and the dual problem are:

$$A = \left[\begin{array}{cccc|c} -1 & -1 & 0 & 0 & -1,500 \\ 0 & 0 & -1 & -1 & -1,000 \\ 1 & 0 & 1 & 0 & 900 \\ 0 & 1 & 0 & 1 & 1,200 \\ \hline 5 & 8 & 9 & 7 & 1 \end{array}\right] \text{ and } A^T = \left[\begin{array}{cccc|c} -1 & 0 & 1 & 0 & 5 \\ -1 & 0 & 0 & 1 & 8 \\ 0 & -1 & 1 & 0 & 9 \\ 0 & -1 & 0 & 1 & 7 \\ \hline -1,500 & -1,000 & 900 & 1,200 & 1 \end{array}\right]$$

The dual problem is:

Maximize $P = -1{,}500y_1 - 1{,}000y_2 + 900y_3 + 1{,}200y_4$

Subject to:
$$-y_1 \qquad + y_3 \qquad \le 5$$
$$-y_1 \qquad\qquad + y_4 \le 8$$
$$-y_2 + y_3 \qquad \le 9$$
$$-y_2 \qquad + y_4 \le 7$$
$$y_1,\ y_2,\ y_3,\ y_4 \ge 0$$

Introduce slack variables x_1, x_2, x_3, x_4 to obtain the initial system:

$$-y_1 \qquad\quad + \quad y_3 \qquad\quad + x_1 \qquad\qquad\qquad = 5$$
$$-y_1 \qquad\qquad\qquad + \quad y_4 \quad + x_2 \qquad\qquad = 8$$
$$- \quad y_2 + \quad y_3 \qquad\qquad\qquad + x_3 \qquad = 9$$
$$- \quad y_2 \qquad + \quad y_4 \qquad\qquad + x_4 = 7$$
$$1{,}500y_1 + 1{,}000y_2 - 900y_3 - 1{,}200y_4 \qquad\qquad\qquad + P = 0$$

$$
\begin{array}{c}
\begin{array}{ccccccccc}
y_1 & y_2 & y_3 & y_4 & x_1 & x_2 & x_3 & x_4 & P
\end{array}\\
\begin{array}{c}
x_1 \\ x_2 \\ x_3 \\ x_4 \\ P
\end{array}
\left[
\begin{array}{ccccccccc|c}
-1 & 0 & 1 & 0 & 1 & 0 & 0 & 0 & 0 & 5 \\
-1 & 0 & 0 & 1 & 0 & 1 & 0 & 0 & 0 & 8 \\
0 & -1 & 1 & 0 & 0 & 0 & 1 & 0 & 0 & 9 \\
0 & -1 & 0 & ① & 0 & 0 & 0 & 1 & 0 & 7 \\
1{,}500 & 1{,}000 & -900 & -1{,}200 & 0 & 0 & 0 & 0 & 1 & 0
\end{array}
\right]
\end{array}
$$

$(-1)R_4 + R_2 \rightarrow R_2,\ 1{,}200R_4 + R_5 \rightarrow R_5$

$$
\sim
\left[
\begin{array}{ccccccccc|c}
-1 & 0 & ① & 0 & 1 & 0 & 0 & 0 & 0 & 5 \\
-1 & 1 & 0 & 0 & 0 & 1 & 0 & -1 & 0 & 1 \\
0 & -1 & 1 & 0 & 0 & 0 & 1 & 0 & 0 & 9 \\
0 & -1 & 0 & 1 & 0 & 0 & 0 & 1 & 0 & 7 \\
1{,}500 & -200 & -900 & 0 & 0 & 0 & 0 & 1{,}200 & 1 & 8{,}400
\end{array}
\right]
\begin{array}{l}
\frac{5}{1} = 5 \\[2pt] \\
\frac{9}{1} = 9 \\
\\
\end{array}
$$

$(-1)R_1 + R_3 \rightarrow R_3,\ 900R_1 + R_5 \rightarrow R_5$

$$
\sim
\left[
\begin{array}{ccccccccc|c}
-1 & 0 & 1 & 0 & 1 & 0 & 0 & 0 & 0 & 5 \\
-1 & ① & 0 & 0 & 0 & 1 & 0 & -1 & 0 & 1 \\
1 & -1 & 0 & 0 & -1 & 0 & 1 & 0 & 0 & 4 \\
0 & -1 & 0 & 1 & 0 & 0 & 0 & 1 & 0 & 7 \\
600 & -200 & 0 & 0 & 900 & 0 & 0 & 1{,}200 & 1 & 12{,}900
\end{array}
\right]
$$

$R_2 + R_3 \rightarrow R_3,\ R_2 + R_4 \rightarrow R_4,\ 200R_2 + R_5 \rightarrow R_5$

$$
\begin{array}{c}
\begin{array}{ccccccccc}
y_1 & y_2 & y_3 & y_4 & x_1 & x_2 & x_3 & x_4 & P
\end{array}\\
\begin{array}{c}
y_3 \\ y_2 \\ x_3 \\ y_4 \\ P
\end{array}
\sim
\left[
\begin{array}{ccccccccc|c}
-1 & 0 & 1 & 0 & 1 & 0 & 0 & 0 & 0 & 5 \\
-1 & 1 & 0 & 0 & 0 & 1 & 0 & -1 & 0 & 1 \\
0 & 0 & 0 & 0 & -1 & 1 & 1 & -1 & 0 & 5 \\
-1 & 0 & 0 & 1 & 0 & 1 & 0 & 0 & 0 & 8 \\
400 & 0 & 0 & 0 & 900 & 200 & 0 & 1{,}000 & 1 & 13{,}100
\end{array}
\right]
\end{array}
$$

Optimal solution: min $C = \$13{,}100$ when 900 motors are shipped from factory A to plant X, 200 motors are shipped from factory A to plant Y, 0 motors are shipped from factory B to plant X, and 1,000 motors are shipped from factory B to plant Y.   (6-3)

31.   Let x_1 = number of pounds of long grain rice in Brand A

x_2 = number of pounds of long grain rice in Brand B

x_3 = number of pounds of wild rice in Brand A

x_4 = number of pounds of wild rice in Brand B

The mathematical model for this problem is:

Maximize $P = 0.8x_1 + 0.5x_2 - 1.9x_3 - 2.2x_4$

Subject to:
$$x_3 \geq 0.1(x_1 + x_3)$$
$$x_4 \geq 0.05(x_2 + x_4)$$
$$x_1 + x_2 \leq 8,000$$
$$x_3 + x_4 \leq 500$$
$$x_1, \ x_2, \ x_3, \ x_4 \geq 0$$

which is the same as:

Maximize $P = 0.8x_1 + 0.5x_2 - 1.9x_3 - 2.2x_4$

Subject to:
$$0.1x_1 - 0.9x_3 \leq 0$$
$$0.05x_2 - 0.95x_4 \leq 0$$
$$x_1 + x_2 \leq 8,000$$
$$x_3 + x_4 \leq 500$$
$$x_1, \ x_2, \ x_3, \ x_4 \geq 0$$

Introduce slack variables $s_1, s_2, s_3,$ and s_4 to obtain the initial system:

$$0.1x_1 - 0.9x_3 + s_1 = 0$$
$$0.05x_2 - 0.95x_4 + s_2 = 0$$
$$x_1 + x_2 + s_3 = 0$$
$$x_3 + x_4 + s_4 = 0$$

	x_1	x_2	x_3	x_4	s_1	s_2	s_3	s_4	P	
y_3	⓪.1	0	-0.9	0	1	0	0	0	0	0
y_2	0	0.05	0	-0.95	0	1	0	0	0	0
x_3	1	1	0	0	0	0	1	0	0	8,000
y_4	0	0	1	1	0	0	0	1	0	500
P	-0.8	-0.5	1.9	2.2	0	0	0	0	1	0

$\frac{0}{0.1} = 0$

$\frac{8000}{1} = 8,000$

$$10R_1 \to R_1$$

①	0	-9	0	10	0	0	0	0	0
0	0.05	0	-0.95	0	1	0	0	0	0
1	1	0	0	0	0	1	0	0	8,000
0	0	1	1	0	0	0	1	0	500
-0.8	-0.5	1.9	2.2	0	0	0	0	1	0

$$(-1)R_1 + R_3 \to R_3, \quad 0.8R_1 + R_5 \to R_5$$

1	0	-9	0	10	0	0	0	0	0
0	0.05	0	-0.95	0	1	0	0	0	0
0	1	9	0	-10	0	1	0	0	8,000
0	0	①	1	0	0	0	1	0	500
0	-0.5	-5.3	2.2	8	0	0	0	1	0

$\frac{8000}{9} \approx 888.9$

$\frac{500}{1} = 500$

$$9R_4 + R_1 \to R_1, \quad (-9)R_4 + R_3 \to R_3, \quad 5.3R_4 + R_5 \to R_5$$

$$\sim \begin{bmatrix} 1 & 0 & 0 & 9 & 10 & 0 & 0 & 9 & 0 & | & 4,500 \\ 0 & \boxed{0.05} & 0 & -0.95 & 0 & 1 & 0 & 0 & 0 & | & 0 \\ 0 & 1 & 0 & -9 & -10 & 0 & 1 & -9 & 0 & | & 3,500 \\ 0 & 0 & 1 & 1 & 0 & 0 & 0 & 1 & 0 & | & 500 \\ 0 & -0.5 & 0 & 7.5 & 8 & 0 & 0 & 5.3 & 1 & | & 2,650 \end{bmatrix} \quad \begin{array}{l} \frac{0}{0.05} = 0 \\[6pt] \frac{3,500}{1} = 3,500 \end{array}$$

$$20R_2 \rightarrow R_2$$

$$\sim \begin{bmatrix} 1 & 0 & 0 & 9 & 10 & 0 & 0 & 9 & 0 & | & 4,500 \\ 0 & \boxed{1} & 0 & -19 & 0 & 20 & 0 & 0 & 0 & | & 0 \\ 0 & 1 & 0 & -9 & -10 & 0 & 1 & -9 & 0 & | & 3,500 \\ 0 & 0 & 1 & 1 & 0 & 0 & 0 & 1 & 0 & | & 500 \\ 0 & -0.5 & 0 & 7.5 & 8 & 0 & 0 & 5.3 & 1 & | & 2,650 \end{bmatrix}$$

$$(-1)R_2 + R_3 \rightarrow R_3, \quad (0.5)R_2 + R_5 \rightarrow R_5$$

$$\sim \begin{bmatrix} 1 & 0 & 0 & 9 & 10 & 0 & 0 & 9 & 0 & | & 4,500 \\ 0 & 1 & 0 & -19 & 0 & 20 & 0 & 0 & 0 & | & 0 \\ 0 & 0 & 0 & \boxed{10} & -10 & -20 & 1 & -9 & 0 & | & 3,500 \\ 0 & 0 & 1 & 1 & 0 & 0 & 0 & 1 & 0 & | & 500 \\ 0 & 0 & 0 & -2 & 8 & 10 & 0 & 5.3 & 1 & | & 2,650 \end{bmatrix} \quad \begin{array}{l} \frac{4,500}{9} = 500 \\[6pt] \frac{3,500}{10} = 350 \\[6pt] \frac{500}{1} = 500 \end{array}$$

$$\frac{1}{10}R_3 \rightarrow R_3$$

$$\sim \begin{bmatrix} 1 & 0 & 0 & 9 & 10 & 0 & 0 & 9 & 0 & | & 4,500 \\ 0 & 1 & 0 & -19 & 0 & 20 & 0 & 0 & 0 & | & 0 \\ 0 & 0 & 0 & \boxed{1} & -1 & -2 & \frac{1}{10} & -\frac{9}{10} & 0 & | & 350 \\ 0 & 0 & 1 & 1 & 0 & 0 & 0 & 1 & 0 & | & 500 \\ 0 & 0 & 0 & -2 & 8 & 10 & 0 & 5.3 & 1 & | & 2,650 \end{bmatrix}$$

$$(-9)R_3 + R_1 \rightarrow R_1, \quad 19R_3 + R_2 \rightarrow R_2, \quad (-1)R_3 + R_4 \rightarrow R_4, \quad 2R_3 + R_5 \rightarrow R_5$$

	x_1	x_2	x_3	x_4	s_1	s_2	s_{31}	s_4	P	
x_1	1	0	0	0	19	18	$\frac{9}{10}$	0	$\frac{171}{10}$	1,350
x_2	0	1	0	0	-19	-18	$-\frac{19}{10}$	0	$\frac{171}{10}$	6,650
x_4	0	0	0	1	-1	-2	$-\frac{1}{10}$	0	$\frac{9}{10}$	350
x_3	0	0	1	0	1	2	$\frac{1}{10}$	0	$\frac{19}{10}$	150
P	0	0	0	0	6	6	$\frac{1}{5}$	1	$\frac{7}{2}$	3,350

The maximum profit is \$3,350 when 1,350 pounds of long grain rice and 150 pounds of wild rice are used to produce 1,500 pounds of brand *A*, and 6,650 pounds of long grain rice and 350 pounds of wild rice are used to produce 7,000 pounds of brand *B*. (6-3)

7 LOGIC, SETS, AND COUNTING

EXERCISE 7-1

Things to remember:

1. A PROPOSITION is a statement (not a question or command) that is either true or false.

2. NEGATION
    If p is a proposition, then the proposition $\neg p$, read NOT p, or the NEGATION of p, is false if p is true and true if p is false.

p	$\neg p$
T	F
F	T

3. DISJUNCTION
    If p and q are propositions, then the proposition $p \vee q$, read p OR q, or the DISJUNCTION of p and q, is true if p is true, or if q is true, or if both are true, and is false otherwise.

p	q	$p \vee q$
T	T	T
T	F	T
F	T	T
F	F	F

4. CONJUNCTION
    If p and q are propositions, then the proposition $p \wedge q$, read p AND q, or the CONJUNCTION of p and q, is true if both p and q are true, and is false otherwise.

p	q	$p \wedge q$
T	T	T
T	F	F
F	T	F
F	F	F

5. CONDITIONAL
    If p and q are propositions, then the proposition $p \rightarrow q$, read IF p THEN q, or the CONDITIONAL WITH HYPOTHESIS p AND CONCLUSION q, is false if p is true and q is false, but is true otherwise.

p	q	$p \rightarrow q$
T	T	T
T	F	F
F	T	T
F	F	T

6.   CONVERSE AND CONTRAPOSITIVE

Let $p \rightarrow q$ be a conditional proposition. The proposition $q \rightarrow p$ is called the CONVERSE of $p \rightarrow q$. The proposition $\neg q \rightarrow \neg p$ is called the CONTRAPOSITIVE of $p \rightarrow q$.

7.   TAUTOLOGY, CONTRADICTION, CONTINGENCY

A proposition is a TAUTOLOGY if each entry in its column of the truth table is T, a CONTRADICTION if each entry is F, and a CONTINGENCY if at least one entry is T and at least one entry is F.

8.   LOGICAL IMPLICATION

Consider the rows of the truth tables for the compound propositions P and Q. If whenever P is true, Q is also true, we say that P LOGICALLY IMPLIES Q, and write $P \Rightarrow Q$. We call $P \Rightarrow Q$ a LOGICAL IMPLICATION.

9.   LOGICAL EQUIVALENCE

If the compound propositions P and Q have identical truth tables, we say that P and Q are LOGICALLY EQUIVALENT, and write $P \equiv Q$. We call $P \equiv Q$ a LOGICAL EQUIVALENCE.

---

1.   1, 2, 4, 5, 10, 20   **3.** 11, 22, 33, 44, 55   **5.** 23, 29

7.   An odd integer has the form $2k+1$. If $2k+1$ and $2n+1$ are odd integers, then $(2k+1)+(2n+1) = 2k+2n+2$ which is even.

9.   91 is not odd.   False

11.   91 is prime and 91 is odd.   True

13.   If 91 is odd, then 91 is prime.   False

15.   If $2^9 + 2^8 + 2^7 > 987$, then $9 \cdot 10^2 + 8 \cdot 10 + 7 = 987$; true

17. $2^9 + 2^8 + 2^7 > 987$ or $9 \cdot 10^2 + 8 \cdot 10 + 7 = 987$; true

19.   If $9 \cdot 10^2 + 8 \cdot 10 + 7$ is not equal to 987, then $2^9 + 2^8 + 2^7$ is not greater than 987; true

21. $-3 < 0$ or $-3 > 0$; disjunction; true.

23   If $-3 < 0$, then $(-3)^2 < 0$; conditional; false.

25.   11 is not prime; negation; false.

27.   7 is odd and 7 is prime; conjunction; true.

29.   Converse: If triangle ABC is equiangular, then triangle ABC is equilateral. Contrapositive: If triangle ABC is not equiangular, then triangle ABC is not equilateral.

31.   Converse: If f(x) is an increasing function, then $f(x)$ is a linear function with positive slope. Contrapositive: If $f(x)$ is not an increasing function, then $f(x)$ is not a linear function with positive slope.

33. Converse: If n is an integer that is a multiple of 2 and a multiple of 4, then n is an integer that is a multiple of 8. Contrapositive: If n is an integer that is not a multiple of 2 or not a multiple of 4, then n is an integer that is not a multiple of 8.

35.

p	q	$\neg p$	$\neg p \wedge q$
T	T	F	F
T	F	F	F
F	T	T	T
F	F	T	F

Contingency; the fourth column has both T and F entries.

37.

p	q	$\neg p$	$\neg p \rightarrow q$
T	T	F	T
T	F	F	T
F	T	T	T
F	F	T	F

Contingency; the fourth column has both T and F entries.

39.

p	q	$p \vee q$	$q \wedge (p \vee q)$
T	T	T	T
T	F	T	F
F	T	T	T
F	F	F	F

Contingency; the fourth column has both T and F entries.

41.

p	q	$p \rightarrow q$	$p \vee (p \rightarrow q)$
T	T	T	T
T	F	F	T
F	T	T	T
F	F	T	T

Tautology; each entry in the fourth column is T.

43.

p	q	$p \wedge q$	$p \rightarrow (p \wedge q)$
T	T	T	T
T	F	F	F
F	T	F	T
F	F	F	T

Contingency; the fourth column has both T and F entries.

45.

p	q	$p \rightarrow q$	$\neg p$	$(p \rightarrow q) \rightarrow \neg p$
T	T	T	F	F
T	F	F	F	T
F	T	T	T	T
F	F	T	T	T

Contingency; the last column has both T and F entries.

47.

p	q	$\neg p$	$p \vee q$	$\neg p \to (p \vee q)$
T	T	F	T	T
T	F	F	T	T
F	T	T	T	T
F	F	T	F	F

Contingency; the last column has both T and F entries.

49.

p	q	$\neg p$	$\neg p \wedge q$	$q \to (\neg p \wedge q)$
T	T	F	F	F
T	F	F	F	T
F	T	T	T	T
F	F	T	F	T

Contingency; the last column has both T and F entries.

51.

p	q	$\neg p \wedge q$	$q \to p$	$(\neg p \wedge q) \wedge (q \to p)$
T	T	F	T	F
T	F	F	T	F
F	T	T	F	F
F	F	F	T	F

Contradiction; each entry in the last column is F.

53.

p	q	$p \vee q$
T	T	T
T	F	T
F	T	T
F	F	F

A logical implication; whenever p is true, $p \vee q$ is also true.

55.

p	q	$\neg p$	$\neg p \wedge q$	$p \vee q$
T	T	F	F	T
T	F	F	F	T
F	T	T	T	T
F	F	T	F	F

$\neg p \wedge q$ logically implies $p \vee q$; whenever $\neg p \wedge q$ is true, $p \vee q$ is also true.

57.

p	q	$\neg p$	$\neg q$	$q \wedge \neg q$	$\neg p \to (q \wedge \neg q)$
T	T	F	F	F	T
T	F	F	T	F	T
F	T	T	F	F	F
F	F	T	T	F	F

$\neg p \to (q \wedge \neg q) \Rightarrow p$ since p is true whenever $\neg p \to (q \wedge \neg q)$ is true.

59.

p	q	$\neg p$	$\neg p \to (p \vee q)$	$p \vee q$
T	T	F	T	T
T	F	F	T	T
F	T	T	T	T
F	F	T	F	F

$\neg p \to (p \vee q)$ and $p \vee q$ have identical truth tables.

61.

p	q	$p \vee q$	$p \wedge q$	$q \wedge (p \vee q)$	$q \vee (p \wedge q)$
T	T	T	T	T	T
T	F	T	F	F	F
F	T	T	F	T	T
F	F	F	F	F	F

$q \wedge (p \vee q)$ and $q \vee (p \wedge q)$ have identical truth tables.

63.

p	q	$p \to q$	$p \vee q$	$p \vee (p \to q)$	$p \to (p \vee q)$
T	T	T	T	T	T
T	F	F	T	T	T
F	T	T	T	T	T
F	F	T	F	T	T

$p \vee (p \to q)$ and $p \to (p \vee q)$ have identical truth tables.

65. $p \to \neg q \equiv \neg p \vee \neg q$ by (4)

$\neg p \vee \neg q \equiv \neg(p \wedge q)$ by (6)

67. $\neg(p \to q) \equiv \neg(\neg p \vee q)$ by (4)

$\neg(\neg p \vee q) \equiv \neg(\neg p) \wedge \neg q$ by (5)

$\neg(\neg p) \wedge \neg q \equiv p \wedge \neg q$ by (1)

69. Yes. If a proposition P is a contingency, then its truth table contains both true and false entries. Therefore, its negative $\neg P$ contains both true and false entries

71. No. A conditional proposition and its contrapositive are equivalent; their truth tables are identical.

EXERCISE 7-2

Things to remember:

1. $a \in A$ means "a is an element of set A."

2. $a \notin A$ means "a is not an element of set A."

3. $\emptyset$ represents "the empty set" or "null set."

4. $S = \{x \mid P(x)\}$ means "S is the set of all x such that $P(x)$ is true."

5. $A \subset B$ means "A is a subset of B."

6. $A = B$ means "A and B have exactly the same elements."

7. $A \not\subset B$ means "A is not a subset of B."

8. $A \neq B$ means "A and B do not have exactly the same elements."

9.   SET OPERATIONS
   a.   UNION
       $A \cup B = \{x \mid x \in A \text{ or } x \in B\}$.

   b.   INTERSECTION
       $A \cap B = \{x \mid x \in A \text{ and } x \in B\}$.

   c.   COMPLEMENT
       $A' = \{x \in U \mid x \notin A\}$, U is a universal set}

---

1.   No

3.   No

5.   No; the complement of the set of negative integers is the set of nonnegative integers.

7. $\{1, 2\} \subset \{2, 1\}$, T; (Note: $\{1, 2\} = \{2, 1\}$)

9. $\{5, 10\} = \{10, 5\}$, T

11. $\{0\} \in \{0, \{0\}\}$, T

13. $8 \in \{1, 2, 4\}$, F

15. $\{1, 2, 3\} \cap \{2, 3, 4\} = \{2, 3\}$

17. $\{1, 2, 3\} \cup \{2, 3, 4\} = \{1, 2, 3, 4\}$

19. $\{1, 4, 7\} \cup \{10, 13\} = \{1, 4, 7, 10, 13\}$

21. $\{1, 4, 7\} \cap \{10, 13\} = \emptyset$

23. $\{x \mid x^2 = 25\} = \{5, -5\}$

25. $\{x \mid x^3 = -27\} = \{-3\}$

27. $\{x \mid x$ is an odd number between 1 and 9 inclusive$\} = \{1, 3, 5, 7, 9\}$.

29. $U = \{1, 2, 3, 4, 5\}$; $A = \{2, 3, 4\}$. Then $A' = \{1, 5\}$.

31. $n(U) = 17 + 52 + 9 + 22 = 100$

33. $n(B) = 52 + 9 = 61$

35. $n(A \cup B) = 17 + 52 + 9 = 78$

37. $n(A') = 9 + 22 = 31$

39. $n(B \cap A') = 9$

41. $n[(A \cup B)'] = 22$

43. $n(A \cup A') = 100$

45.   (A) $\{x \mid x \in R \text{ or } x \in T\}$.
         $= R \cup T$ ("or" translated
            as $\cup$, union)

47. $Q \cap R = \{2, 4, 6\} \cap \{3, 4, 5, 6\}$
         $= \{4, 6\}$
       $P \cup (Q \cap R) = \{1, 2, 3, 4\} \cup \{4, 6\}$

$$= \{1, 2, 3, 4\} \cup \{2, 4, 6\}$$
$$= \{1, 2, 3, 4, 6\}$$

$$= \{1, 2, 3, 4, 6\}$$

(B) $R \cup T = \{1, 2, 3, 4, 6\}$

49. $H' = \{n \in N \mid n \leq 100\};$ finite.

51. $E \cup P$ is infinite; E and P are each infinite.

53. $E \cap P = \{2\};$ finite.

55. E' is infinite.

57. $H' \cup T = T;$ finite.

59. $H' = \{n \in N \mid n \leq 100\}, \quad T' = \{n \in N \mid n \geq 1000\};$ disjoint.

61. $P' = \{n \in N \mid n \text{ is not prime}\};$ not disjoint (e.g., $500 > 100$ and 500 is not prime).

63. True

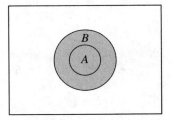

If $x \in A \cap B$, then $x \in A$ which implies
$A \cap B \subset A$
If $x \in A$, then $x \in B$ since $A \subset B$,
so $x \in A \cap B$. Thus $A \subset A \cap B$.
It now follows that $A \cap B = A$.

65. False

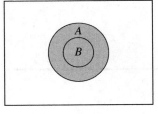

In this diagram, we have $B \subset A$.
Now $A \cup B = A$, but $A \not\subset B$.

Counterexample:
$A = \{1,2,3\}, \ B = \{1,2\}$

67. False

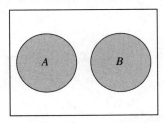

Here $A \cap B = \varnothing$, but $A \neq \varnothing$.

Counterexample:
$A = \{1,2,3\}, \ B = \{4,5\}$

69. False

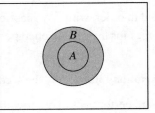

From the diagram it is clear that
$B' \subset A'$.

Counterexample:
$A = \{1,2\}, \ B = \{1,2,3\}, \ U = \{1,2,3,4,5\}$

71. False.   The empty set is **not** an "element".
The empty set **is a subset** of every set.

73. (A) 2, $\varnothing$, $\{a\}$; number of subsets $2 = 2^1$

(B) 4, $\varnothing$, $\{a\}$, $\{b\}$, $\{a,b\}$; number of subsets $4 = 2^2$

(C) 8, $\varnothing$, $\{a\}$, $\{b\}$, $\{c\}$, $\{a,b\}$, $\{a,c\}$, $\{b,c\}$, $\{a,b,c\}$; 2^3

(D) 16

75. $n(F) = 19 + 66 = 85$ **77.** $n(A) = 19 + 14 = 33$

79. $n(A \cap S) = 14$ **81.** $n(B \cap F) = 66$

83. $n(A \cup S) = 19 + 14 + 21 = 54$ **85.** $n(B \cup F) = 19 + 66 + 21 = 106$

87. $n(A \cap B) = 0$

89. The six two-person subsets that can be formed from the given set $\{P, V_1, V_2, V_3\}$ are:

$\{P, V_1\}$, $\{P, V_3\}$, $\{V_1, V_3\}$, $\{P, V_2\}$, $\{V_1, V_2\}$, $\{V_2, V_3\}$

91. From the given Venn diagram $A \cap Rh = \{A+, AB+\}$

93. Again, from the given Venn diagram: $A \cup Rh = \{A-, A+, B+, AB-, AB+, O+\}$

95. From the given Venn diagram: $(A \cup B)' = \{O+, O-\}$

97. $A' \cap B = \{B-, B+\}$

EXERCISE 7-3

Things to remember:

<u>1.</u>   Let A be a set with finitely many elements. Then $n(A)$ denotes the number of elements in A.

<u>2.</u>   ADDITION PRINCIPLE (for counting)

For any two sets A and B,
$$n(A \cup B) = n(A) + n(B) - n(A \cap B)$$

If A and B are disjoint, i.e., if $A \cap B = \varnothing$, then
$$n(A \cup B) = n(A) + n(B)$$

<u>3.</u>   MULTIPLICATION PRINCIPLE (for counting)

(a)   If two operations O_1 and O_2 are performed in order, with N_1 possible outcomes for the first operation and N_2 possible outcomes for the second operation, then there are
$$N_1 \cdot N_2$$
possible combined outcomes of the first operation followed by the second.

(b)   In general, if n operations $O_1, O_2, \ldots, O_n$ are performed in order with possible number of   outcomes $N_1, N_2, \ldots, N_n$,   respectively, then there are
$$N_1 \cdot N_2 \cdot \ldots \cdot N_n$$
possible combined outcomes of the operations performed in the given order.

---

1. $50 = 34 + 29 - x,\ x = 63 - 50 = 13$ **3.** $4x = 23 + 42 - x,\ 5x = 65,\ x = 13$

5. $3(x + 11) = 65 + x - 14,\ 3x + 33 = 51 + x,\ 2x = 18,\ x = 9$

7.   (A)   Tree Diagram

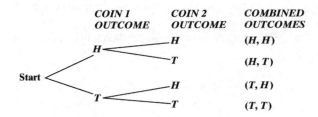

COIN 1 OUTCOME   COIN 2 OUTCOME   COMBINED OUTCOMES

Start

H — H (H, H)
H — T (H, T)
T — H (T, H)
T — T (T, T)

Thus, there are 4 ways.

(B)   Multiplication Principle
O_1: 1st coin
N_1: 2 ways
O_2: 2nd coin
N_2: 2 ways

Thus, there are
$N_1 \cdot N_2 = 2 \cdot 2 = 4$ ways.

9.   (A)   Tree Diagram

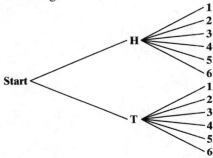

Start

H — 1, 2, 3, 4, 5, 6
T — 1, 2, 3, 4, 5, 6

Thus there are 12 combined outcomes

(B)   Multiplication Principle
O_1: Coin
N_1: 2 outcomes
O_2: Die
N_2: 6 outcomes

Thus, there are
$N_1 \cdot N_2 = 2 \cdot 6 = 12$ combined outcomes

11.   (A)   6 restaurants and 3 plays. If a couple goes to a dinner or a play, but not both, there are 9 possible selections.

(B)   If a couple goes to a restaurant and then to a play, there are $6 \cdot 3 = 18$ possible selections.

13.

No letter repeated:
O_1: select first letter; N_1: 5 ways
O_2: select second letter; N_2: 4 ways
O_3: select third letter; N_3: 3 ways
Multiplication principle: $N_1 N_2 N_3 = 60$

Letters repeated
O_1: select first letter; N_1: 5 ways
O_2: select second letter; N_2: 5 ways
O_3: select third letter; N_3: 5 ways
Multiplication principle: $N_1 N_2 N_3 = 5^3 = 125$

Adjacent letters different
O_1: select first letter; N_1: 5 ways
O_2: select second letter; N_2: 4 ways
O_3: select third letter; N_3: 4 ways
Multiplication principle: $N_1 N_2 N_3 = 80$

15. First, we find the number of bronze and silver courses.
    Let x = number of silver courses. Then $2x$ = number of bronze courses, and
    $$x + 2x = 18, \quad 3x = 18, \quad x = 6.$$

    Thus, there are 12 bronze courses, 6 silver courses and 2 gold courses.

    (A) If a golfer decides to play a round on a silver or gold course, but not both, there are $6 + 2 = 8$ possible selections.

    (B) If a golfer decides to play one round per week, first at a bronze course, then silver, then gold, there are $12 \cdot 6 \cdot 2 = 144$ possible selections.

17. $A = (A \cap B') \cup (A \cap B)$ and $B = (A' \cap B) \cup (A \cap B)$
    Thus,

    $n(A) = n(A \cap B') + n(A \cap B)$, so
    $n(A \cap B') = n(A) - n(A \cap B) = 80 - 20 = 60$
    $n(A \cap B) = 20$.

    $n(B) = n(A' \cap B) + n(A \cap B)$ so
    $n(A' \cap B) = n(B) - n(A \cap B) = 50 - 20 = 30$

    Also, $A \cup B = (A' \cap B) \cup (A \cap B) \cup (A \cap B')$ and
    $U = (A \cup B) \cup (A' \cap B')$ where $(A \cup B) \cap (A' \cap B') = \varnothing$

    Thus, $n(U) = n(A \cup B) + n(A' \cap B')$ so
    $n(A' \cap B') = n(U) - n(A \cup B) = 200 - (60 + 20 + 30) = 200 - 110 = 90$

19. $n(A \cup B) = n(A) + n(B) - n(A \cap B)$
    So $n(A \cap B) = n(A) + n(B) - n(A \cup B) = 25 + 55 - 60 = 20$
    $n(A \cap B') = n(A) - n(A \cap B) = 25 - 20 = 5$
    $n(A' \cap B) = n(B) - n(A \cap B) = 55 - 20 = 35$
    and $n(A' \cap B') = n(U) - n(A \cup B) = 100 - 60 = 40$

21. $A' = (A' \cap B) \cup (A' \cap B')$ so
    $n(A') = n(A' \cap B) + n(A' \cap B')$ and
    $n(A' \cap B) = n(A') - n(A' \cap B') = 65 - 25 = 40$

    $B' = (A \cap B') \cup (A' \cap B')$ so
    $n(B') = n(A \cap B') + n(A' \cap B')$ and
    $n(A \cap B') = n(B') - n(A' \cap B') = 40 - 25 = 15$

    $U = (A \cup B) \cup (A' \cap B')$ so
    $n(U) = n(A \cup B) \cup n(A' \cap B')$ and
    $n(A \cup B) = n(U) - n(A' \cap B') = 150 - 25 = 125$.

    $A \cup B = (A \cap B') \cup (A \cap B) \cup (A' \cap B)$ so
    $n(A \cup B) = n(A \cap B') + n(A \cap B) + n(A' \cap B)$ and
    $n(A \cap B) = n(A \cup B) - n(A \cap B') + n(A' \cap B) = 125 - 15 - 40 = 70$

23. $n(A \cap B') = n(A) - n(A \cap B) = 48 - 0 = 48$

$n(A \cap B) = 0$

$n(A' \cap B) = n(B) - n(A \cap B) = 62 - 0 = 62$

$n(A' \cap B') = n(U) - n(A \cup B) = 180 - 48 - 62 = 70$

25. $n(A \cap B) = 30$

$n(A \cap B') = n(A) - n(A \cap B) = 70 - 30 = 40$

$n(A' \cap B) = n(B) - n(A \cap B) = 90 - 30 = 60$

$n(A' \cap B') = n(U) - [n(A \cap B') + n(A \cap B) + n(A' \cap B)] = 200 - [40 + 30 + 60] = 200 - 130 = 70$

Therefore,

	A	A'	Totals
B	30	60	90
B'	40	70	110
Totals	70	130	200

27. $n(A \cap B) = n(A) + n(B) - n(A \cup B) = 45 + 55 - 80 = 20$

$n(A \cap B') = n(A) - n(A \cap B) = 45 - 20 = 25$

$n(A' \cap B) = n(B) - n(A \cap B) = 55 - 20 = 35$

$n(A' \cap B') = n(U) - n(A \cup B) = 100 - 80 = 20$

Therefore,

	A	A'	Totals
B	20	35	55
B'	25	20	45
Totals	45	55	100

29. $A \cup A' = U$ and $B \cup B' = U$. Therefore

$n(A) = n(U) - n(A') = 90 - 15 = 75;$

$n(B) = n(U) - n(B') = 90 - 24 = 66.$

Also, $(A' \cup B') \cup (A \cap B) = U,$ so

$n(A \cap B) = n(U) - n(A' \cup B') = 90 - 32 = 58$

Thus, we have

	A	A'	Totals
B	58		66
B'			24
Totals	75	15	90

From which is follows that

	A	A'	Totals
B	58	8	66
B'	17	7	24
Totals	75	15	90

31. $n(A \cap B) = n(A) + n(B) - n(A \cup B) = 110 + 145 - 255 = 0$

Thus,

	A	A'	Totals
B	0	145	145
B'	110	45	155
Totals	110	190	300

33.   (A)   True. Suppose $A = \varnothing$ is the empty set. Then $A \cap B = \varnothing \cap B = \varnothing$.
   Similarly if $B = \varnothing$ is the empty set.
   Thus, if either A or B is the empty set, then $A \cap B = \varnothing$, and A and B are disjoint.

   (B)   False. Let $A = \{1, 2, 3\}$ and $B = \{4, 5, 6\}$. Then $A \cap B = \varnothing$ and
   neither A nor B is the empty set.

35.   Using the Multiplication Principle:
O_1: Choose the color O_3: Choose the interior

N_1: 5 ways N_3: 4 ways

O_2: Choose the transmission O_4: Choose the engine

N_2: 3 ways N_4: 2 ways

Thus, there are

$N_1 \cdot N_2 \cdot N_3 \cdot N_4 = 5 \cdot 3 \cdot 4 \cdot 2 = 120$ different variations of this model car.

37.   Counting upper and lower case letters, there are 52 possible choices for each of the five characters.
Therefore there is a total of $52^5 = 380,204,032$ possible passwords

39.   (A)   Number of five-digit combinations, no digit repeated.
   O_1: Selecting the first digit O_4: Selecting the fourth digit
   N_1: 10 ways N_4: 7 ways

   O_2: Selecting the second digit O_5: Selecting the fifth digit
   N_2: 9 ways N_5: 6 ways

   O_3: Selecting the third digit
   N_3: 8 ways

   Thus, there are
   $N_1 \cdot N_2 \cdot N_3 \cdot N_4 \cdot N_5 = 10 \cdot 9 \cdot 8 \cdot 7 \cdot 6 = 30,240$
   possible combinations.

   (B)   Number of five-digit combinations, allowing repetition.
   O_1: Selecting the first digit O_4: Selecting the fourth digit
   N_1: 10 ways N_4: 10 ways

   O_2: Selecting the second digit O_5: Selecting the fifth digit
   N_2: 10 ways N_5: 10 ways

   O_3: Selecting the third digit
   N_3: 10 ways

Thus, there are
$$N_1 \cdot N_2 \cdot N_3 \cdot N_4 \cdot N_5 = 10 \cdot 10 \cdot 10 \cdot 10 \cdot 10 = 10^5 = 100{,}000$$
possible combinations.

(C) Number of five digit combinations, if successive digits must be different.

O_1: Selecting the first digit

N_1: 10 ways

O_2: Selecting the second digit

N_2: 9 ways

O_3: Selecting the third digit

N_3: 9 ways

O_4: Selecting the fourth digit

N_4: 9 ways

O_5: Selecting the fifth digit

N_5: 9 ways

Thus, there are

$N_1 \cdot N_2 \cdot N_3 \cdot N_4 \cdot N_5 = 10 \cdot 9 \cdot 9 \cdot 9 \cdot 9 = 10 \cdot 9^4 = 65{,}610$ possible combinations.

41. (A) Letters and/or digits may be repeated.

O_1: Selecting the first letter

N_1: 26 ways

O_2: Selecting the second letter

N_2: 26 ways

O_3: Selecting the third letter

N_3: 26 ways

O_4: Selecting the first digit

N_4: 10 ways

O_5: Selecting the second digit

N_5: 10 ways

O_6: Selecting the third digit

N_6: 10 ways

Thus, there are
$$N_1 \cdot N_2 \cdot N_3 \cdot N_4 \cdot N_5 \cdot N_6 = 26 \cdot 26 \cdot 26 \cdot 10 \cdot 10 \cdot 10 = 17{,}576{,}000$$
different license plates.

(B) No repeated letters and no repeated digits are allowed.

O_1: Select the three letters, no letter repeated

N_1: $26 \cdot 25 \cdot 24 = 15{,}600$ ways

O_2: Select the three numbers, no number repeated

N_2: $10 \cdot 9 \cdot 8 = 720$ ways

Thus, there are
$$N_1 \cdot N_2 = 15{,}600 \cdot 720 = 11{,}232{,}000$$
different license plates with no letter or digit repeated.

43.

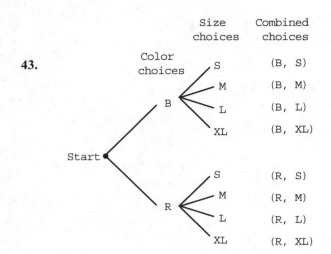

There are 8 combined choices; the color can be chosen in 2 ways and the size can be chosen in 4 ways. The number of possible combined choices is $2 \cdot 4 = 8$ just as in Example 3.

45. $z = 0$; if $B \subset A$, then there are no elements in B which are not in A.

47. $x = 0$ and $z = 0$; $A \varnothing B = A \cap B$ implies $A = B$.

49. Let T = the people who play tennis, and
G = the people who play golf.

Then $n(T) = 32$, $n(G) = 37$, $n(T \cap G) = 8$ and $n(U) = 75$.
Thus, $n(T \cup G) = n(T) + n(G) - n(T \cap G) = 32 + 37 - 8 = 61$

The set of people who play neither tennis nor golf is represented by
$T' \cap G'$. Since $U = (T \cup G) \cup (T' \cap G')$ and $(T \cup G) \cap (T' \cap G') = \varnothing$, it follows that

$n(T' \cap G') = n(U) - n(T \cup G) = 75 - 61 = 14$.

There are 14 people who play neither tennis nor golf.

51. Let F = the people who speak French, and
G = the people who speak German.

Then $n(F) = 42$, $n(G) = 55$, $n(F' \cap G') = 17$ and $n(U) = 100$. Since
$U = (F \cup G) \cup (F' \cap G')$ and $(F \cup G) \cap (F' \cap G') = \varnothing$, it follows that
$n(F \cup G) = n(U) - n(F' \cap G') = 100 - 17 = 83$

Now $n(F \cup G) = n(F) + n(G) - n(F \cap G)$, so
$n(F \cap G) = n(F) + n(G) - n(F \cup G) = 42 + 55 - 83 = 14$
There are 14 people who speak both French and German.

53. (A)

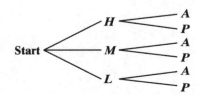

(B)   Operation 1: Test scores can be classified into three groups, high, middle, or low:

$$N_1 = 3$$

Operation 2: Interviews can be classified into two groups, aggressive or passive:

$$N_2 = 2$$

The total possible combined classifications is:

$$N_1 \cdot N_2 = 3 \cdot 2 = 6$$

55. O_1: Travel from home to airport and back O_3: Fly to second city

N_1: 2 ways N_3: 2 ways

O_2: Fly to first city O_4: Fly to third city

N_2: 3 ways N_4: 1 way

Thus, there are

$$N_1 \cdot N_2 \cdot N_3 \cdot N_4 = 2 \cdot 3 \cdot 2 \cdot 1 = 12$$

different travel plans.

57. Let U = the group of people surveyed

    H = people who own an HDTV, and

    D = people who own a DVD.

Then $n(U) = 1200$, $n(H) = 850$,
$n(D) = 740$ and $n(H \cap D) = 580$.

Now draw a Venn diagram.

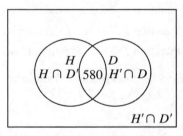

From this diagram, we see that

$$n(H \cap D') = n(H) - n(H \cap D) = 850 - 580 = 270$$
$$n(H' \cap D) = n(D) - n(H \cap D) = 740 - 580 = 160$$
$$n(H \cup D) = n(H \cap D') + n(H \cap D) + n(H' \cap D) = 580 + 270 + 160 = 1010$$

and

$$n(H' \cap D') = n(U) - n(H \cup D) = 1200 - 1010 = 190$$

Thus,
(A) $n(H \cup D) = 1010$ (B) $n(H' \cap D') = 190$ (C) $n(H \cap D') = 270$

59. Let U = group of people surveyed

    H = group of people who receive HBO

    S = group of people who receive Showtime.

Then, $n(U) = 8,000$, $n(H) = 2,450$,
$n(S) = 1,940$ and $n(H' \cap S') = 5,180$

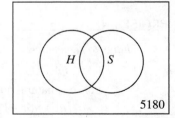

Now, $n(H \cup S) = n(U) - n(H' \cap S') = 8,000 - 5,180 = 2,820$

Since $n(H \cup S) = n(H) + n(S) - n(H \cap S)$, we have

$$n(H \cap S) = n(H) + n(S) - n(H \cup S) = 2,450 + 1,940 - 2,820 = 1,570$$

Thus, 1,570 subscribers receive both channels.

61. From the table:

(A)   The number of males aged 20-24 *and* below minimum wage is: 102 (the element in the (2, 2) position in the body of the table); 102,000.

(B)   The number of females aged 20 or older *and* at minimum wage is:
       186 + 503 = 689 (the sum of the elements in the (3, 2) and (3, 3) positions);   689,000.

(C)   The number of workers who are *either* aged 16-19 *or* are males at minimum wage is:
       343 + 118 + 367 + 251 + 154 + 237 = 1,470;   1,470,000.

(D)   The number of workers below minimum wage is: 379 + 993 = 1,372;   1,372,000.

63.   (A)

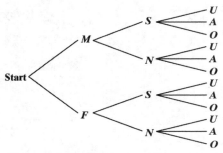

(B)   Operation 1: Two classifications, male and female; $N_1 = 2$.

       Operation 2: Two classifications, smoker and nonsmoker; $N_2 = 2$.

       Operation 3: Three classifications, underweight, average weight,   and overweight; $N_3 = 3$.

       Thus the total possible combined classifications
       $= N_1 \cdot N_2 \cdot N_3 = 2 \cdot 2 \cdot 3 = 12$

65. F = number of individuals who contributed to the first campaign
       S = number of individuals who contributed to the second campaign.

       Then $n(F) = 1{,}475$, $n(S) = 2{,}350$ and $n(F \cap S) = 920$.

       Now, $n(F \cup S) = n(F) + n(S) - n(F \cap S) = 1{,}475 + 2{,}350 - 920 = 2{,}905$

       Thus, 2,905 individuals contributed to either the first campaign or the second campaign.

<u>EXERCISE 7-4</u>

Things to remember:

<u>1.</u>   FACTORIAL

       For n a natural number,
              $n! = n(n-1)(n-2) \cdot \ \ ... \ \ \cdot 3 \cdot 2 \cdot 1$
              $0! = 1$
              $n! = n(n-1)!$

       [<u>NOTE</u>: Many calculators have an $\boxed{n!}$ key or its equivalent.]

2.   PERMUTATIONS

A PERMUTATION of a set of distinct objects is an arrangement of the objects in a specific order, without repetition. The NUMBER OF PERMUTATIONS of n distinct objects without repetition, denoted by $P_{n,n}$, is:

$$_nP_n = n(n-1)\cdot\ \ldots\ \cdot 3\cdot 2\cdot 1 = n!\ \ (n\text{ factors})$$

3.   PERMUTATIONS OF n OBJECTS TAKEN r AT A TIME

A permutation of a set of n distinct objects taken r at a time without repetition is an arrangement of the r objects in a specific order. The NUMBER OF PERMUTATIONS of n objects taken r at a time, denoted by $_nP_r$, is given by:

$$_nP_r = n(n-1)(n-2)\cdot\ \ldots\ \cdot (n-r+1)$$
$$(r\text{ factors})$$

or $_nP_r = \dfrac{n!}{(n-r)!}\quad 0 \le r \le n$

[Note: $_nP_n = \dfrac{n!}{(n-n)!} = \dfrac{n!}{0!} = n! = \dfrac{n!}{0!} = n!$, the number of permutations of n objects taken n at a time. Remember, by definition, $0! = 1$.]

4.   COMBINATIONS OF n OBJECTS TAKEN r AT A TIME

A combination of a set of n distinct objects taken r at a time without repetition is an r-element subset of the set of n objects. (The arrangement of the elements in the subset does not matter.) The NUMBER OF COMBINATIONS of n objects taken r at a time, denoted by $_nC_r$, or by $\dbinom{n}{r}$, is given by:

$$_nC_r = \binom{n}{r} = \frac{_nP_r}{r!} = \frac{n!}{r!(n-r)!}\quad 0 \le r \le n$$

5.   NOTE: In a permutation, the ORDER of the objects counts. In a combination, order does not count.

---

1. $\dfrac{12\cdot 11\cdot 10}{3\cdot 2\cdot 1} = 2\cdot 11\cdot 10 = 220$

3. $\dfrac{10\cdot 9\cdot 8\cdot 7\cdot 6}{5\cdot 4\cdot 3\cdot 2\cdot 1} = 3\cdot 2\cdot 7\cdot 6 = 252$

5. $\dfrac{100\cdot 99\cdot 98\cdot\ \cdots\ \cdot 3\cdot 2\cdot 1}{98\cdot 97\cdot 96\cdot\ \cdots\ \cdot 3\cdot 2\cdot 1} = 100\cdot 99 = 9,900$

7. $7! = 7\cdot 6\cdot 5\cdot 4\cdot 3\cdot 2\cdot 1 = 5,040$

9. $(5+6)! = 11! = 11\cdot 10\cdot 9\cdot 8\cdot 7\cdot 6\cdot 5\cdot 4\cdot 3\cdot 2\cdot 1 = 39,916,800$

11. $5! + 6! = 5\cdot 4\cdot 3\cdot 2\cdot 1 + 6\cdot 5\cdot 4\cdot 3\cdot 2\cdot 1 = 5\cdot 4\cdot 3\cdot 2\cdot 1(1+6) = 120\cdot 7 = 840$

13. $\dfrac{8!}{4!} = \dfrac{(8\cdot 7\cdot 6\cdot 5)4!}{4!} = 1,680$

15. $\dfrac{8!}{4!(8-4)!} = \dfrac{8!}{4!4!} = \dfrac{(8 \cdot 7 \cdot 6 \cdot 5)4!}{4!4!} = \dfrac{8 \cdot 7 \cdot 6 \cdot 5}{4 \cdot 3 \cdot 2 \cdot 1} = 70$

17. $\dfrac{500!}{498!} = \dfrac{(500 \cdot 499)498!}{498!} = 500 \cdot 499 = 249{,}500$

19. $_{13}C_8 = \dfrac{13!}{8!(13-8)!} = \dfrac{13!}{8!5!} = \dfrac{13 \cdot 12 \cdot 11 \cdot 10 \cdot 9}{5 \cdot 4 \cdot 3 \cdot 2 \cdot 1} = 13 \cdot 11 \cdot 9 = 1{,}287$

21. $_{18}P_6 = \dfrac{18!}{(18-6)!} = \dfrac{18!}{12!} = 18 \cdot 17 \cdot 16 \cdot 15 \cdot 14 \cdot 13 = 13{,}366{,}080$

23. $\dfrac{_{12}P_7}{12^7} = \dfrac{12!}{(12-7)!} \cdot \dfrac{1}{12^7} = \dfrac{12!}{5!} \cdot \dfrac{1}{12^7} = \dfrac{12 \cdot 11 \cdot 10 \cdot 9 \cdot 8 \cdot 7 \cdot 6}{12^7} = 0.1114$

25. $\dfrac{_{39}C_5}{_{52}C_5} = \dfrac{\dfrac{39!}{5!(39-5)!}}{\dfrac{52!}{5!(52-5)!}} = \dfrac{39!}{5! \cdot 34!} \cdot \dfrac{5! \cdot 47!}{52!} = \dfrac{39! \cdot 47!}{34! \cdot 52!} = \dfrac{39 \cdot 38 \cdot 37 \cdot 36 \cdot 35}{52 \cdot 51 \cdot 50 \cdot 49 \cdot 48} \approx 0.2215$

27. $\dfrac{n!}{(n-2)!} = \dfrac{n(n-1)(n-2)!}{(n-2)!} = n(n-1)$

29. $\dfrac{(n+1)!}{2!(n-1)!} = \dfrac{(n+1)(n)(n-1)!}{2 \cdot 1(n-1)!} = \dfrac{n(n+1)}{2}$

31. Permutation; order of selection counts.

33. Combination; order of selection does not count.

35. Neither a permutation nor a combination.

37. The number of different finishes (win, place, show) for the ten horses is the number of permutations of 10 objects 3 at a time. This is:

$$_{10}P_3 = \dfrac{10!}{(10-3)!} = \dfrac{10!}{7!} = \dfrac{10 \cdot 9 \cdot 8 \cdot 7!}{7!} = 720$$

39. (A) The number of ways that a three-person subcommittee can be selected   from a seven-member committee is the number of combinations (since order *is not* important in selecting a subcommittee) of 7 objects 3 at a time. This is:

$$_7C_3 = \dfrac{7!}{3!(7-3)!} = \dfrac{7!}{3!4!} = \dfrac{7 \cdot 6 \cdot 5 \cdot 4!}{3 \cdot 2 \cdot 1 \cdot 4!} = 35$$

(B) The number of ways a president, vice-president, and secretary can be chosen from a committee of 7 people is the number of permutations (since order *is* important in choosing 3 people for the positions) of 7 objects 3 at a time. This is:

$$_7P_3 = \dfrac{7!}{(7-3)!} = \dfrac{7!}{4!} = \dfrac{7 \cdot 6 \cdot 5 \cdot 4!}{4!} = 7 \cdot 6 \cdot 5 = 210$$

41. Calculate $x!$, 3^x, x^3 for $x = 1, 2, 3, \ldots$

For $x \geq 7$, $x! > 3^x$; for $x \geq 1$, $3^x > x^3$ (except for $x = 3$). In general, $x!$ grows much faster than 3^x, and 3^x grows much faster than x^3.

43. There are 26 red cards in a standard 52-card deck:

6-card hands, only red cards: $\quad {}_{26}C_6 = \dfrac{26!}{6!(26-6)!} = \dfrac{26!}{6!20!} = 230,230$

45. There are 12 face cards in a standard 52-card deck:

5-card hands, only face cards: $\quad {}_{12}C_5 = \dfrac{12!}{5!(12-5)!} = \dfrac{12!}{5!7!} = 792$

47. There are 4 kings in a standard 52-card deck. Subtracting the 4 kings, we choose 3 cards form the remaining 48. Seven-card hands containing 4 kings: $\quad {}_{48}C_3 = \dfrac{48!}{3!(48-3)!} = \dfrac{48!}{3!45!} = 17,296$

49. Select one card from each suit:

O_1: select a heart; N_1: 13 ways

O_2: select a spade; N_2: 13 ways

O_3: select a diamond; N_3: 13 ways

O_4: select a club; N_4: 13 ways

Multiplication principle: $N_1 N_2 N_3 N_4 = 13^4 = 28,561$

51. O_1: Selecting an appetizer; $\quad N_1 = {}_8C_3$

O_2: Selecting a main course: $\quad N_2 = {}_{10}C_4$

O_3: Selecting a dessert: $\quad N_3 = {}_7C_2$

Thus, there are $N_1 \cdot N_2 \cdot N_3 = {}_8C_3 \; {}_{10}C_4 \; {}_7C_2 = 246,960$

53. ${}_nC_{n-r} = \dfrac{n!}{(n-r)!(n-[n-r])!} = \dfrac{n!}{(n-r)!r!} = \dfrac{n!}{r!(n-r)!} = {}_nC_r$

The sequence of numbers is the same read top to bottom or bottom to top.

55. True; $(n+1)! = (n+1)n! > n!$

57. False; $\quad {}_nP_{n-1} = \dfrac{n!}{[n-(n-1)]!} = \dfrac{n!}{1!} = n!$

$\qquad\qquad {}_nP_n = \dfrac{n!}{(n-n)!} = \dfrac{n!}{0!} = \dfrac{n!}{1} = n!$

Note: ${}_nP_r < {}_nP_{r+1}$ for $0 \leq r < n-1$

59. True; $\quad {}_nC_{n-r} = \dfrac{n!}{(n-[n-r])!(n-r)!} = \dfrac{n!}{r!(n-r)!} = {}_nC_r$

61. (A)   A line segment joins two distinct points. Thus, the total number of line segments is given by:

$$_8C_2 = \frac{8!}{2!(8-2)!} = \frac{8!}{2!6!} = \frac{8 \cdot 7 \cdot 6!}{2 \cdot 1 \cdot 6!} = 28$$

(B)   Each triangle requires three distinct points. Thus, there are

$$_8C_3 = \frac{8!}{3!(8-3)!} = \frac{8!}{3!5!} = \frac{8 \cdot 7 \cdot 6 \cdot 5!}{3 \cdot 2 \cdot 1 \cdot 5!} = 56 \text{ triangles.}$$

(C)   Each quadrilateral requires four distinct points. Thus, there are

$$_8C_4 = \frac{8!}{4!(8-4)!} = \frac{8!}{4!4!} = \frac{8 \cdot 7 \cdot 6 \cdot 5 \cdot 4!}{4 \cdot 3 \cdot 2 \cdot 1 \cdot 4!} = 70 \text{ quadrilaterals.}$$

63. There will be $_6P_4$ ways to seat 4 people in a row of 6 chairs:

$$_6P_4 = \frac{6!}{(6-4)!} = \frac{6!}{2!} = 6 \cdot 5 \cdot 4 \cdot 3 = 360.$$

65. (A)   The distinct positions are taken into consideration.
The number of starting teams is given by:

$$_8P_5 = \frac{8!}{(8-5)!} = \frac{8!}{3!} = \frac{8 \cdot 7 \cdot 6 \cdot 5 \cdot 4 \cdot 3!}{3!} = 6,720$$

(B)   The distinct positions are not taken into consideration.
The number of starting teams is given by:

$$_8C_5 = \frac{8!}{5!(8-5)!} = \frac{8!}{5!3!} = \frac{8 \cdot 7 \cdot 6 \cdot 5!}{5! \cdot 3 \cdot 2 \cdot 1} = 56$$

(C)   Either Mike or Ken, but not both, must start; distinct positions are not taken into consideration.
O_1: Select either Mike or Ken
N_1: 2 ways

O_2: Select 4 players from the remaining 6
N_2: $_6C_4$

Thus, the number of starting teams is given by:

$$N_1 \cdot N_2 = 2 \cdot {_6C_4} = 2 \cdot \frac{6!}{4!(6-4)!} = 2 \cdot \frac{6 \cdot 5 \cdot 4!}{4! \cdot 2 \cdot 1} = 30$$

67. For many calculators, $k = 69$, but your calculator may be different. Note that $k!$ may also be calculated as $_kP_k$. On a TI-85, the largest integer k for which $k!$ can be calculated using $_kP_k$ is 449.

69. The largest value will be $_{24}C_{12} = \dfrac{24!}{12!12!} = 2,704,156.$

71. (A) Three printers are to be selected for the display. The *order* of selection does not count. Thus, the number of ways to select the 3 printers from 24 is:

$$_{24}C_3 = \frac{24!}{3!(24-3)!} = \frac{24 \cdot 23 \cdot 22(21!)}{3 \cdot 2 \cdot 1(21!)} = 2{,}024$$

(B) Nineteen of the 24 printers are not defective.

Thus, the number of ways to select 3 non-defective printers is:

$$_{19}C_3 = \frac{19!}{3!(19-3)!} = \frac{19 \cdot 18 \cdot 17(16!)}{3 \cdot 2 \cdot 1(16!)} = 969$$

73. (A) There are $8 + 12 + 10 = 30$ stores in all. The jewelry store chain   will select 10 of these stores to   close. Since order does not count here, the total number of ways to select the 10 stores to close is:

$$_{30}C_{10} = \frac{30!}{10!(30-10)!} = \frac{30 \cdot 29 \cdot 28 \cdot 27 \cdot 26 \cdot 25 \cdot 24 \cdot 23 \cdot 22 \cdot 21(20!)}{10 \cdot 9 \cdot 8 \cdot 7 \cdot 6 \cdot 5 \cdot 4 \cdot 3 \cdot 2 \cdot 1(20!)}$$
$$= 30{,}045{,}015$$

(B) The number of ways to close 2 stores in Georgia is: $_8C_2$

The number of ways to close 5 stores in Florida is: $_{12}C_5$

The number of ways to close 3 stores in Alabama is: $_{10}C_3$

By the multiplication principle, the total number of ways to select the 10 stores for closing is:

$$_8C_2 \cdot {}_{12}C_5 \cdot {}_{10}C_3 = \frac{8!}{2!(8-2)!} \cdot \frac{12!}{5!(12-5)!} \cdot \frac{10!}{3!(10-3)!}$$
$$= \frac{8 \cdot 7 \cdot 6!}{2 \cdot 1 \cdot 6!} \cdot \frac{12 \cdot 11 \cdot 10 \cdot 9 \cdot 8(7!)}{5 \cdot 4 \cdot 3 \cdot 2 \cdot 1(7!)} \cdot \frac{10 \cdot 9 \cdot 8(7!)}{3 \cdot 2 \cdot 1(7!)}$$
$$= 28 \cdot 792 \cdot 120 = 2{,}661{,}120$$

75. (A) Three females can be selected in $_6C_3$ ways. Two males can be selected in $_5C_2$ ways. Applying the Multiplication Principle, we have:

$$\text{Total number of ways} = {}_6C_3 \cdot {}_5C_2 = \frac{6!}{3!(6-3)!} \cdot \frac{5!}{2!(5-2)!} = 200$$

(B) Four females and one male can be selected in $_6C_4 \cdot {}_5C_1$ ways. Thus,

$$_6C_4 \cdot {}_5C_1 = \frac{6!}{4!(6-4)!} \cdot \frac{5!}{1!(5-1)!} = 75$$

(C) Number of ways in which 5 females can be selected is:

$$_6C_5 = \frac{6!}{5!(6-5)!} = 6$$

(D) Number of ways in which 5 people can be selected is:

$$_{11}C_5 = \frac{11!}{5!(11-5)!} = 462$$

(E) At least four females includes four females and five females. Four females and one male can be selected in 75 ways [see part (B)]. Five females can be selected in 6 ways [see part (C)]. Thus, total number of ways $= {}_6C_4 \cdot {}_5C_1 + {}_6C_5 = 75 + 6 = 81$

77. (A) Select 3 samples from 8 blood types, no two samples having the same type. This is a permutation problem. The number of different examinations is:

$$ {}_8P_3 = \frac{8!}{(8-3)!} = \frac{8!}{5!} = \frac{8 \cdot 7 \cdot 6 \cdot 5!}{5!} = 336 $$

(B) Select 3 samples from 8 blood types, repetition is allowed.

O_1: Select the first sample $\qquad\qquad O_3$: Select the third sample

N_1: 8 ways $\qquad\qquad\qquad\qquad\qquad\quad N_3$: 8 ways

O_2: Select the second sample

N_2: 8 ways

Thus, the number of different examinations in this case is:

$N_1 \cdot N_2 \cdot N_3 = 8 \cdot 8 \cdot 8 = 8^3 = 512$

79. This is a permutations problem. The number of buttons is given by:

$$ {}_4P_2 = \frac{4!}{(4-2)!} = \frac{4!}{2!} = \frac{4 \cdot 3 \cdot 2!}{2!} = 12 $$

CHAPTER 7 REVIEW

1. 3^4 is not less than 4^3; true.   (7-1)

2. 2^3 is less than 3^2 or 3^4 is less than 4^3; true.   (7-1)

3. 2^3 is less than 3^2 and 3^4 is less than 4^3; false.   (7-1)

4. If 2^3 is less than 3^2, then 3^4 is less than 4^3; false.   (7-1)

5. If 3^4 is less than 4^3, then 2^3 is less than 3^2; true.   (7-1)

6. If 3^4 is not less than 4^3, then 2^3 is not less than 3^2; false.   (7-1)

7. T; $\{5, 6, 7\}$ and $\{6, 7, 5\}$ have exactly the same elements.   (7-2)

8. F; $5 \notin \{55, 555\}$ (7-2)

9. T; $\{9, 27\}$ is a subset of $\{3, 9, 27, 81\}$.   (7-2)

10. F; $\{1, 2\} \in \{1, \{1, 2\}\}$.   (7-2)

11. *If* 9 is prime, *then* 10 is odd: conditional; true (9 is not prime so any conclusion is true).   (7-1)

12. 7 is even *or* 8 is odd: disjunction; false (7 is not even and 8 is not odd).   (7-1)

13. 53 is prime *and* 57 is prime: conjunction; false ($57 = 19 \cdot 3$ is not prime).   (7-1)

14. 51 is *not* prime: negation; true ($51 = 17 \cdot 3$ is not prime).   (7-1)

15. Converse: If the square matrix A does not have an inverse, then the square matrix A has a row of zeros. Contrapositive: If the square matrix A has an inverse, then the square matrix A does not have a row of zeros.   (7-1)

16. Converse: If the square matrix A has an inverse, then the square matrix A is an identity matrix.
Contrapositive: If the square matrix A does not have an inverse, then the square matrix A is not an identity matrix.   (7-1)

17. $\{1, 2, 3, 4\} \cup \{2, 3, 4, 5\} = \{1, 2, 3, 4, 5\}$ (7-2)

18. $\{1, 2, 3, 4\} \cap \{2, 3, 4, 5\} = \{2, 3, 4\}$ (7-2)

19. $\{1, 2, 3, 4\} \cap \{5, 6\} = \varnothing$ (7-2)

20. (A)   We construct the tree diagram
at the right for the experiment:

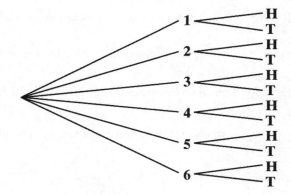

Total combined outcomes = 12.

(B)   Operation 1: Six possible outcomes, 1, 2, 3, 4, 5, or 6; $N_1 = 6$.

Operation 2: Two possible outcomes, heads (H) or tails (T); $N_2 = 2$.

Using the Multiplication Principle, the total combined outcomes = $N_1 \cdot N_2 = 6 \cdot 2 = 12$.   (7-3)

21. (A) $n(A) = 30 + 35 = 65$ (B) $n(B) = 35 + 40 = 75$

(C) $n(A \cap B) = 35$ (D) $n(A \cup B) = 65 + 75 - 35 = 105$

or $n(A \cup B) = 30 + 35 + 40 = 105$

(E) $n(U) = 30 + 35 + 40 + 45 = 150$

(F) $n(A') = n(U) - n(A) = 150 - 65 = 85$

(G) $n([A \cap B]') = n(U) - n(A \cap B)$ (H) $n([A \cup B]') = n(U) - n(A \cup B)$

$= 150 - 35 = 115$ $= 150 - 105 = 45$ (7-3)

22. $11! = 11 \cdot 10 \cdot 9 \cdot 8 \cdot 7 \cdot 6 \cdot 5 \cdot 4 \cdot 3 \cdot 2 \cdot 1 = 39,916,800$ (7-4)

23. $(10 - 6)! = 4! = 24$ (7-4)   **24.** $10! - 6! = 3,628,800 - 720 = 3,628,080$ (7-4)

25. $\dfrac{15!}{10!} = \dfrac{(15 \cdot 14 \cdot 13 \cdot 12 \cdot 11)10!}{10!} = 360,360$ (7-4)

26. $\dfrac{15!}{10!5!} = \dfrac{(15 \cdot 14 \cdot 13 \cdot 12 \cdot 11)10!}{10!(120)} = \dfrac{360,360}{120} = 3,003$ (7-4)

27. $_8C_5 = \dfrac{8!}{5!(8-5)!} = \dfrac{(8 \cdot 7 \cdot 6)5!}{5!3!} = \dfrac{8 \cdot 7 \cdot 6}{6} = 56$ (7-4)

28. $_8P_5 = \dfrac{8!}{(8-5)!} = \dfrac{8!}{3!} = \dfrac{(8 \cdot 7 \cdot 6 \cdot 5 \cdot 4)3!}{3!} = 6,720$ (7-4)

29.

Operation 1:	First person can choose the seat in 6 different ways; $N_1 = 6.$
Operation 2:	Second person can choose the seat in 5 different ways; $N_2 = 5.$
Operation 3:	Third person can choose the seat in 4 different ways; $N_3 = 4.$
Operation 4:	Fourth person can choose the seat in 3 different ways; $N_4 = 3.$
Operation 5:	Fifth person can choose the seat in 2 different ways; $N_5 = 2.$
Operation 6:	Sixth person can choose the seat in 1 way; $N_6 = 1.$

Using the Multiplication Principle, the total number of different arrangements that can be made is $6 \cdot 5 \cdot 4 \cdot 3 \cdot 2 \cdot 1 = 720.$ (7-3)

30. This is a permutations problem. The permutations of 6 objects taken 6 at a time is:

$$_6P_6 = \frac{6!}{(6-6)!} = 6! = 720 \qquad (7\text{-}4)$$

31.

p	q	$p \rightarrow q$	$q \rightarrow p$	$(p \rightarrow q) \wedge (q \rightarrow p)$
T	T	T	T	T
T	F	F	T	F
F	T	T	F	F
F	F	T	T	T

Contingency   (7-1)

32.

p	q	$q \rightarrow p$	$p \vee (q \rightarrow p)$
T	T	T	T
T	F	T	T
F	T	F	F
F	F	T	T

Contingency   (7-1)

33.

p	q	$\neg p$	$\neg q$	$p \vee \neg p$	$q \wedge \neg q$	$(p \vee \neg p) \rightarrow (q \wedge \neg q)$
T	T	F	F	T	F	F
T	F	F	T	T	F	F
F	T	T	F	T	F	F
F	F	T	T	T	F	F

Contradiction   (7-1)

34.

p	q	$\neg q$	$p \rightarrow q$	$\neg q \wedge (p \rightarrow q)$
T	T	F	T	F
T	F	T	F	F
F	T	F	T	F
F	F	T	T	T

Contingency   (7-1)

35.

p	q	$\neg p$	$p \rightarrow q$	$\neg p \rightarrow (p \rightarrow q)$
T	T	F	T	T
T	F	F	F	T
F	T	T	T	T
F	F	T	T	T

Tautology   (7-1)

36.

p	q	$\neg q$	$p \vee \neg q$	$\neg(p \vee \neg q)$
T	T	F	T	F
T	F	T	T	F
F	T	F	F	T
F	F	T	T	F

Contingency   (7-1)

37. $E \bigcup K$ is infinite; both E and K are infinite.   (7-2)

38. $M \bigcap K = \{n \in Z \mid 10^3 < n < 10^6\}$ is finite.   (7-2)

39. $K' = \{n \in Z \mid n \le 10^3\}$ is infinite.   (7-2)

40. $E \bigcap M = \{n \in Z \mid n$ is even and $n < 10^6\}$ is infinite.   (7-2)

41. $M' = \{n \in Z \mid n \ge 10^6\}$, $K' = \{n \in Z \mid n \le 10^3\}$; disjoint.   (7-2)

42. $M = \{n \in Z \mid n < 10^6\}$, $E' = \{n \in Z \mid n$ is odd$\}$
M and E' are not disjoint (e.g., $1 \in M \bigcap E'$)   (7-2)

43. K and K' are disjoint by definition.   (7-2)

44. Using the multiplication principle, the man has 5 children, $5 \cdot 3 = 15$ grandchildren, and $5 \cdot 3 \cdot 2 = 30$ great-grandchildren, for a total of $5 + 15 + 30 = 50$ descendents.   (7-3)

45.

Operation	Number of ways of completing operation under condition:		
	No letter repeated	Letters can be repeated	Adjacent letters not alike
O_1	8	8	8
O_2	7	8	7
O_3	6	8	7

Total outcomes, without repeating letters $= 8 \cdot 7 \cdot 6 = 336$.
Total outcomes, with repeating letters $= 8 \cdot 8 \cdot 8 = 512$.
Total outcomes, with adjacent letters not alike $= 8 \cdot 7 \cdot 7 = 392$.   (7-3)

46. (A)   This is a permutations problem.

$$_6P_3 = \frac{6!}{(6-3)!} = \frac{6 \cdot 5 \cdot 4 \cdot 3!}{3!} = 120$$

(B)   This is a combinations problem.

$$_5C_2 = \frac{5!}{2!(5-2)!} = \frac{5 \cdot 4 \cdot 3!}{2 \cdot 1 \cdot 3!} = 10 \qquad (7\text{-}4)$$

47. The largest value of $_{25}C_r$ is $_{25}C_{12} = {}_{25}C_{13} = 5,200,300$.   (7-4)

48. By the multiplication principle, there are
$$N_1 \cdot N_2 \cdot N_3$$
branches in the tree diagram.   (7-3)

49. $x^3 - x = 0$
$x(x^2 - 1) = 0$
$\qquad x = -1, 0, 1$
$\{x \mid x^3 - x = 0\} = \{-1, 0, 1\}$ (7-2)

50. $4! = 24, 5! = 120$
$\{x \mid x \text{ is a positive integer and } x! < 100\} = \{1, 2, 3, 4\}$.   (7-2, 7-4)

51. $\{x \mid x \text{ is a perfect square and } x < 50\} = \{1, 4, 9, 16, 25, 36, 49\}$.   (7-2)

52. (A)   This is a permutations problem.

$$_{10}P_3 = \frac{10!}{(10-3)!} = \frac{10!}{7!} = 10 \cdot 9 \cdot 8 = 720$$

(B)   The number of ways in which women are selected for all three positions is given by:

$$_6P_3 = \frac{6!}{(6-3)!} = \frac{6!}{3!} = 6 \cdot 5 \cdot 4 = 120$$

(C)   This is a combinations problem.

$$_{10}C_3 = \frac{10!}{3!(10-3)!} = \frac{10 \cdot 9 \cdot 8 \cdot 7!}{3 \cdot 2 \cdot 1 \cdot 7!} = 120 \qquad (7\text{-}3, 7\text{-}4)$$

53.   Draw a Venn diagram with: A = Chess players, B = Checker players.

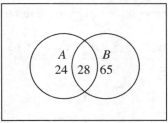

Now, $n(A \cap B) = 28$, $n(A \cap B') = n(A) - n(A \cap B) = 52 - 28 = 24$

$n(B \cap A') = n(B) - n(A \cap B) = 93 - 28 = 65$

Since there are 150 people in all,

$n(A' \cap B') = n(U) - [n(A \cap B') + n(A \cap B) + n(B \cap A')]$

$= 150 - (24 + 28 + 65) = 150 - 117 = 33 \qquad (7\text{-}3)$

54. $x = 0$; if $A \subset B$, then there are no elements in A which are not in B.   (7-3)

55. $x = z = w = 0$; if $A \cap B = U$, then $A = B = U$ and there are no elements which are not in A, or not in B, or not in $A \cup B$.   (7-3)

56.   True; for $1 < r < n$, $_nC_r = \frac{n!}{r!(n-r)!} < \frac{n!}{(n-r)!} = {_nP_r}$ (7-4)

57.   False; $_nP_{n-1} = \frac{n!}{[n-(n-1)]!} = \frac{n!}{1!} = n!$

Note: $_nP_r < n!$ for $0 \le r < n-1$ (7-4)

58.   True; for $1 < r < n$, $_nC_r = \frac{n!}{r!(n-r)!} < n!$ (7-4)

59.

p	q	$p \wedge q$
T	T	T
T	F	F
F	T	F
F	F	F

$p \wedge q \Rightarrow p$ since p is true whenever $p \wedge q$ is true.
(7-1)

60.

p	q	$p \rightarrow q$
T	T	T
T	F	F
F	T	T
F	F	T

$q \Rightarrow p \rightarrow q$ since $p \rightarrow q$ is true whenever q is true.
(7-1)

61.

p	q	$\neg p$	$q \wedge \neg q$	$\neg p \rightarrow (q \wedge \neg q)$
T	T	F	F	T
T	F	F	F	T
F	T	T	F	F
F	F	T	F	F

$p \equiv \neg p \rightarrow (q \wedge \neg q)$ since p and $\neg p \rightarrow (q \wedge \neg q)$ have the same truth tables.   (7-1)

62.

p	q	$p \vee q$	$\neg p$	$\neg p \rightarrow q$
T	T	T	F	T
T	F	T	F	T
F	T	T	T	T
F	F	F	T	F

$p \vee q \equiv \neg p \rightarrow q$ since $p \vee q$ and $\neg p \rightarrow q$ have the same truth tables.   (7-1)

63.

p	q	$p \rightarrow q$	$p \wedge (p \rightarrow q)$
T	T	T	T
T	F	F	F
F	T	T	F
F	F	T	F

$p \wedge (p \rightarrow q) \Rightarrow q$ since q is true whenever $p \wedge (p \rightarrow q)$ is true.   (7-1)

64.

p	q	$\neg q$	$p \wedge \neg q$	$\neg(p \wedge \neg q)$	$p \rightarrow q$
T	T	F	F	T	T
T	F	T	T	F	F
F	T	F	F	T	T
F	F	T	F	T	T

$\neg(p \wedge \neg q) \equiv p \rightarrow q$ since $\neg(p \wedge \neg q)$ and $p \rightarrow q$ have the same truth tables.   (7-1)

65. Operation 1: Two possible outcomes, boy or girl, $N_1 = 2$.

Operation 2: Two possible outcomes, boy or girl, $N_2 = 2$.

Operation 3: Two possible outcomes, boy or girl, $N_3 = 2$.

Operation 4: Two possible outcomes, boy or girl, $N_4 = 2$.

Operation 5: Two possible outcomes, boy or girl, $N_5 = 2$.

Using the Multiplication Principle, the total combined outcomes is:
$N_1 \cdot N_2 \cdot N_3 \cdot N_4 \cdot N_5 = 2 \cdot 2 \cdot 2 \cdot 2 \cdot 2 = 32$.

If order pattern is not taken into account, there would be only 6 possible outcomes: families with 0, 1, 2, 3, 4, or 5 boys.   (7-3)

66. $_nC_r = \dfrac{n!}{r!(n-r)!}$ and $_nP_r = \dfrac{n!}{(n-r)!}$;

$\dfrac{n!}{r!(n-r)!} = \dfrac{n!}{(n-r)!}$ when $r = 0$ or $r = 1$; otherwise $_nC_r \neq {}_nP_r$ (7-4)

67. The number of routes starting from A and visiting each of the 5 stores exactly once is the number of permutations of 5 objects taken 5 at a time, i.e.,

$_5P_5 = \dfrac{5!}{(5-5)!} = 120.$ (7-3)

68. Draw a Venn diagram with:

  S = people who have invested in stocks, and

  B = people who have invested in bonds.

Then $n(U) = 1000$, $n(S) = 340$,
$n(B) = 480$ and $n(S \cap B) = 210$

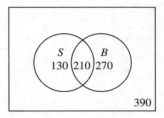

$n(S \cap B') = 340 - 210 = 130$ $n(B \cap S') = 480 - 210 = 270$

$n(S' \cap B') = 1000 - (130 + 210 + 270) = 1000 - 610 = 390$

(A) $n(S \cup B) = n(S) + n(B) - n(S \cap B) = 340 + 480 - 210 = 610.$

(B) $n(S' \cap B') = 390$

(C) $n(B \cap S') = 270$ (7-3)

69. Since the order of selection does not matter, the number of ways to select 6 from 40 is

$_{40}C_6 = \dfrac{40!}{6!34!} = \dfrac{40 \cdot 39 \cdot 38 \cdot 37 \cdot 36 \cdot 35 \cdot 34!}{6!34!} = \dfrac{40 \cdot 39 \cdot 38 \cdot 37 \cdot 36 \cdot 35}{6 \cdot 5 \cdot 4 \cdot 3 \cdot 2 \cdot 1} = 10 \cdot 39 \cdot 38 \cdot 37 \cdot 7 = 3{,}838{,}380$ (7-4)

70. (A) The total number of ways that the 67 names can be ordered is

  $67! \approx 3.647 \times 10^{94}.$

(B) It is not possible to print 67! ballots.   (7-4)

8   PROBABILITY

EXERCISE 8-1

Things to remember:

1.   SAMPLE SPACE AND EVENTS

If the formulation of a set S of outcomes (events) of an experiment is such that in each trial of the experiment one and only one of the outcomes in the set S will occur, then S is called a SAMPLE SPACE for the experiment. Each element in S is called a SIMPLE OUTCOME, or SIMPLE EVENT.

An EVENT E is any subset of S (including the empty set $\varnothing$ and the sample space S). Event E is a SIMPLE EVENT if it contains only one element and a COMPOUND EVENT if it contains more than one element. Event E OCCURS if the result of performing the experiment is one of the simple events in E.

2.   CHOOSING SAMPLE SPACES

There is no one correct sample space for a given experiment. When specifying a sample space for an experiment, include as much detail as necessary to answer all questions of interest regarding the outcomes of the experiment. When in doubt, choose a sample space with more elements rather than fewer.

3.   PROBABILITIES FOR SIMPLE EVENTS

Given a sample space
$$S = \{e_1, e_2, ..., e_n\}.$$
To each simple event e_i assign a real number denoted by $P(e_i)$, called the PROBABILITY OF THE EVENT e_i. These numbers can be assigned in an arbitrary manner provided the following two conditions are satisfied:

(a)   The probability of a simple event is a number between 0 and 1, inclusive. That is,
$$0 \le P(e_i) \le 1$$

(b)   The sum of the probabilities of all simple events in the sample space is 1. That is,
$$P(e_1) + P(e_2) + ... + P(e_n) = 1$$

Any probability assignment that satisfies these two conditions is called an ACCEPTABLE PROBABILITY ASSIGNMENT.

4.   PROBABILITY OF AN EVENT E

Given an acceptable probability assignment for the simple events in a sample space S, the probability of an arbitrary event E, denoted $P(E)$, is defined as follows:

(a) $P(E) = 0$ if E is the empty set.

(b)   If E is a simple event, then $P(E)$ has already been assigned.

(c)   If E is a compound event, then $P(E)$ is the sum of the probabilities of all the simple events in E.

(d)   If $E = S$, then $P(E) = P(S) = 1$ [this follows from 3(b)].

5. STEPS FOR FINDING THE PROBABILITY OF AN EVENT E

   (a) Set up an appropriate sample space S for the experiment.

   (b) Assign acceptable probabilities to the simple events in S.

   (c) To obtain the probability of an arbitrary event E, add the probabilities of the simple events in E.

6. EMPIRICAL PROBABILITY

   If an experiment is conducted n times and event E occurs with FREQUENCY $f(E)$, then the ratio $f(E)/n$ is called the RELATIVE FREQUENCY of the occurrence of event E in n trials. The EMPIRICAL PROBABILITY of E, denoted by $P(E)$, is given by the number (if it exists) that the relative frequency $f(E)/n$ approaches as n gets larger and larger. For any particular n, the relative frequency $f(E)/n$ is also called the APPROXIMATE EMPIRICAL PROBABILITY of event E:

$$P(E) \approx \frac{\text{Frequency of occurrence of } E}{\text{Total number of trials}} = \frac{f(E)}{n}$$

   (The larger n is, the better the approximation.)

7. PROBABILITIES UNDER AN EQUALLY LIKELY ASSUMPTION
   If, in a sample space

$$S = \{e_1, e_2, ..., e_n\},$$

   each simple event e_i is as likely to occur as any other, then $P(e_i) = \dfrac{1}{n}$, for $i = 1, 2, ..., n$, i.e. assign the same probability, $1/n$, to each simple event. The probability of an arbitrary event E in this case is:

$$P(E) = \frac{\text{Number of elements in } E}{\text{Number of elements in } S} = \frac{n(E)}{n(S)}$$

---

1. E is more likely; $\dfrac{5}{6} > \dfrac{4}{5}$

3. F is more likely; $0.4 > \dfrac{3}{8}$

5. F is more likely; $\dfrac{1}{6} > 0.15$

In Problems 7−14 the spinner will land on any one of 12 equally likely sectors.

7. $n(\text{blue}) = 4$; $P(\text{blue}) = \dfrac{4}{12} = \dfrac{1}{3}$.

9. $n(\text{yellow or green}) = 3$; $P(\text{yellow or green}) = \dfrac{3}{12} = \dfrac{1}{4}$.

11. $n(\text{orange}) = 0$; $P(\text{orange}) = \dfrac{0}{12} = 0$.

13. $n(\text{blue, red, yellow or green}) = 12$; $P(\text{blue, red, yellow or green}) = \dfrac{12}{12} = 1$.

In problems 15 − 24 there are 52 equally likely outcomes.

15. $n(\text{club}) = 13;\ P(\text{club}) = \dfrac{13}{52} = \dfrac{1}{4}.$

17. $n(\text{heart or diamond}) = 26;\ P(\text{heart or diamond}) = \dfrac{26}{52} = \dfrac{1}{2}.$

19. $P(\text{jack of clubs}) = \dfrac{1}{52}.$

21. $n(\text{king or spade}) = n(\text{kings}) + n(\text{spades}) - n(\text{king of spades}) = 4 + 13 - 1 = 16;$

$P(\text{king or spade}) = \dfrac{16}{52} = \dfrac{4}{13}.$

23. $n(\text{black diamonds}) = 0;\quad P(\text{black diamond}) = 0.$

25.   Let B = boy and G = girl. Then

$S = \{(B, B), (B, G), (G, B), (G, G)\}$

where (B, B) means both children are boys, (B, G) means the first child is a boy, the second is a girl, and so on. The event E corresponding to having two children of opposite sex is $E = \{(B, G), (G, B)\}$. Since the simple events are equally likely,

$P(E) = \dfrac{n(E)}{n(S)} = \dfrac{2}{4} = \dfrac{1}{2}.$

27.   (A)   Reject; $P(G) = -0.35$ — no probability can be negative.

(B)   Reject; $P(J) + P(G) + P(P) + P(S) = 0.32 + 0.28 + 0.24 + 0.30$
$$= 1.14 \neq 1$$

(C)   Acceptable; each probability is between 0 and 1 (inclusive), and the sum of the probabilities is 1.

29. $P(\{J, P\}) = P(J) + P(P) = 0.26 + 0.30 = 0.56.$

31. $S = \{(B, B, B), (B, B, G), (B, G, B), (B, G, G), (G, B, B), (G, B, G),$
$\qquad (G, G, B), (G, G, G)\}$
$E = \{(B, B, G)\}$

Since the events are equally likely and $n(S) = 8$, $P(E) = \dfrac{1}{8}.$

33.   The number of three-digit sequences with no digit repeated is $_{10}P_3$. Since the possible opening combinations are equally likely, the probability of guessing the right combination is:

$\dfrac{1}{_{10}P_3} = \dfrac{1}{10 \cdot 9 \cdot 8} = \dfrac{1}{720} \approx 0.0014$

35.   Let S = the set of five-card hands. Then $n(S) = {}_{52}C_5$

Let A = "five black cards." Then $n(A) = {}_{26}C_5.$

Since individual hands are equally likely to occur:

$P(A) = \dfrac{n(A)}{n(S)} = \dfrac{_{26}C_5}{_{52}C_5} = \dfrac{\frac{26!}{5!21!}}{\frac{52!}{5!47!}} = \dfrac{26 \cdot 25 \cdot 24 \cdot 23 \cdot 22}{52 \cdot 51 \cdot 50 \cdot 49 \cdot 48} \approx 0.0253$

37. S = set of five-card hands; $n(S) = {}_{52}C_5$.

F = "five face cards"; $n(F) = {}_{26}C_5$.

Since individual hands are equally likely to occur:

$$P(F) = \frac{n(F)}{n(S)} = \frac{{}_{26}C_5}{{}_{52}C_5} = \frac{\dfrac{12!}{5!7!}}{\dfrac{52!}{5!47!}} = \frac{12 \cdot 11 \cdot 10 \cdot 9 \cdot 8}{52 \cdot 51 \cdot 50 \cdot 49 \cdot 48} \approx 0.000305$$

39. S = {all the days in a year} (assume 365 days; exclude leap year); Equivalently, number the days of the year, beginning with January 1. Then $S = \{1, 2, 3, \ldots, 365\}$.

Assume that each day is as likely as any other day for a person to be born. Then the probability of each simple event is: $\dfrac{1}{365}$.

41. $n(S) = {}_5P_5 = 5! = 120$

Let A = all notes inserted into the correct envelopes. Then $n(A) = 1$ and

$$P(A) = \frac{n(A)}{n(S)} = \frac{1}{120} \approx 0.00833$$

43. Let E = "Sum being 6." Then $n(E) = 5$. Thus, $P(E) = \dfrac{n(E)}{n(S)} = \dfrac{5}{36}$.

45. Let E = "Sum being less than 5." Then $n(E) = 6$. Thus,

$$P(E) = \frac{n(E)}{n(S)} = \frac{6}{36} = \frac{1}{6}.$$

47. Let E = "Sum not 7 or 11." Then $n(E) = 28$ and $P(E) = \dfrac{n(E)}{n(S)} = \dfrac{28}{36} = \dfrac{7}{9}$.

49. E = "Sum being 1" is not possible. Thus, $P(E) = 0$.

51. Let E = "Sum is divisible by 3" = "Sum is 3, 6, 9, or 12." Then

$n(E) = 12$ and $P(E) = \dfrac{n(E)}{n(S)} = \dfrac{12}{36} = \dfrac{1}{3}$.

53. Let E = "Sum is 7 or 11." Then $n(E) = 8$. Thus, $P(E) = \dfrac{n(E)}{n(S)} = \dfrac{8}{36} = \dfrac{2}{9}$.

55. Let E = "Sum is divisible by 2 or 3" = "Sum is 2, 3, 4, 6, 8, 9, 10, 12." Then $n(E) = 24$, and

$$P(E) = \frac{n(E)}{n(S)} = \frac{24}{36} = \frac{2}{3}.$$

For Problems 57—61, the sample space S is given by:

$S = \{(H, H, H), (H, H, T), (H, T, H), (H, T, T)\}$

The outcomes are equally likely and n(S) = 4.

57. Let E = "1 head." Then $n(E) = 1$ and $P(E) = \dfrac{n(E)}{n(S)} = \dfrac{1}{4}$.

59. Let E = "3 heads." Then $n(E) = 1$ and $P(E) = \dfrac{n(E)}{n(S)} = \dfrac{1}{4}$.

61. Let E = "More than 1 head." Then $n(E) = 3$ and $P(E) = \dfrac{n(E)}{n(S)} = \dfrac{3}{4}$.

63. No. The events "heart," "not a heart" are not equally likely; $P(H) = \dfrac{1}{4}$, $P(N) = \dfrac{3}{4}$.

65. Yes. The events "even," "odd" are equally likely; $P(E) = P(O) = \dfrac{1}{2}$.

67. Yes. Since the seven sectors have equal area, the events are equally likely; each event has probability $\dfrac{1}{7}$.

69. (A) Yes. If we flip a fair coin 20 times, a representation of the sample space is $\{0, 1, 2, \dots , 19, 20\}$ where each element denotes the number of heads. Each of these outcomes is possible; 10 is the outcome with the highest probability.

(B) Yes. On average we would expect 20 heads in 40 flips of a fair coin. Based on the evidence, it would appear that

$$P(H) = \frac{37}{40} \text{ and } P(T) = \frac{3}{40}.$$

For Problems 71 – 77, the sample space S is given by:

$$S = \left\{ \begin{matrix} (1,1),(1,2),(1,3) \\ (2,1),(2,2),(2,3) \\ (3,1),(3,2),(3,3) \end{matrix} \right\}$$

The outcomes are equally likely and n(S) = 9.

71. Let E = "Sum is 2." Then $n(E) = 1$ and $P(E) = \dfrac{n(E)}{n(S)} = \dfrac{1}{9}$.

73. Let E = "Sum is 4." Then $n(E) = 3$ and $P(E) = \dfrac{n(E)}{n(S)} = \dfrac{3}{9} = \dfrac{1}{3}$.

75. Let E = "Sum is 6." Then $n(E) = 1$ and $P(E) = \dfrac{n(E)}{n(S)} = \dfrac{1}{9}$.

77. Let E = "Sum is odd" = "Sum is 3 or 5." Then $n(E) = 4$ and

$$P(E) = \frac{n(E)}{n(S)} = \frac{4}{9}.$$

For Problems 79—85, the sample space is the set of all 5-card hands.
The outcomes are equally likely and $n(S) = {}_{52}C_5$

79. Let E = "red cards." Then $n(E) = {}_{26}C_5$ and $P(E) = \dfrac{{}_{26}C_5}{{}_{52}C_5} = \dfrac{\dfrac{26}{5!21!}}{\dfrac{52!}{5!47!}} = \dfrac{26 \cdot 25 \cdot 24 \cdot 23 \cdot 22}{52 \cdot 51 \cdot 50 \cdot 49 \cdot 48} \approx 0.0253$

81. Let E = "6 cards with exactly 2 face cards." Then $n(E) = {}_{12}C_2 \cdot {}_{40}C_4$ and

$$P(E) = \frac{{}_{12}C_2 \cdot {}_{40}C_4}{{}_{52}C_6} = \frac{\dfrac{12!}{2!10!} \cdot \dfrac{40!}{4!36!}}{\dfrac{52!}{6!50!}} \approx 0.2963$$

83. Let E = "4 cards, no aces." Then $n(E) = 48$ and $P(E) = \dfrac{{}_{48}C_4}{{}_{52}C_4} = \dfrac{\dfrac{48!}{4!44!}}{\dfrac{52!}{4!48!}} \approx 0.7187$

85. Let E = "7 cards: exactly 2 diamonds, exactly 2 spades and 3 other cards."   Then

$n(E) = {}_{13}C_2 \cdot {}_{13}C_2 \cdot {}_{26}C_3$ and

$$P(E) = \dfrac{{}_{13}C_2 \cdot {}_{13}C_2 \cdot {}_{26}C_3}{{}_{52}C_7} \approx 0.1182$$

87. (A) From the plot, $P(6) = \dfrac{7}{50} = 0.14.$

   (B) If the outcomes are equally likely, $P(6) = \dfrac{1}{6} = 0.167.$

   (C) The answer here depends on the results of your simulation.

89. (A) Represent the outcomes H and T by the numbers 1 and 2, respectively, and select 500 random integers from the set $\{1, 2\}$.

   (B) The answer depends on the results of your simulation.

   (C) If the outcomes are equally likely, then $P(H) = P(T) = \dfrac{1}{2}$.

91. (A) The sample space S is the set of all possible permutations of the 12 brands taken 4 at a time, and $n(S) = P_{12,4}.$ Thus, the probability of selecting 4 brands and identifying them correctly, with no answer repeated, is:

   $$P(E) = \dfrac{1}{{}_{12}P_4} = \dfrac{1}{\dfrac{12!}{(12-4)!}} = \dfrac{1}{12 \cdot 11 \cdot 10 \cdot 9} \approx 0.000084$$

   (B) Allowing repetition, $n(S) = 12^4$ and the probability of identifying them correctly is:

   $$P(F) = \dfrac{1}{12^4} \approx 0.000048$$

93. (A) Total number of applicants = 6 + 5 = 11.

   $$n(S) = {}_{11}C_5 = \dfrac{11!}{5!(11-5)!} = 462$$

   The number of ways that three females and two males can be selected is:

   $${}_6C_3 \cdot {}_5C_2 = \dfrac{6!}{3!(6-3)!} \cdot \dfrac{5!}{2!(5-2)!} = 20 \cdot 10 = 200$$

   Thus, $P(A) = \dfrac{{}_6C_3 \cdot {}_5C_2}{{}_{11}C_5} = \dfrac{200}{462} = 0.433$

   (B) $P(\text{4 females and 1 male}) = \dfrac{{}_6C_4 \cdot {}_5C_1}{{}_{11}C_5} = 0.162$

   (C) $P(\text{5 females}) = \dfrac{{}_6C_5}{{}_{11}C_5} = 0.013$

   (D) $P(\text{at least four females}) = P(\text{4 females and 1 male}) + P(\text{5 females})$

   $$= \dfrac{{}_6C_4 \cdot {}_5C_1}{{}_{11}C_5} + \dfrac{{}_6C_5}{{}_{11}C_5}$$

   $$= 0.162 + 0.013 \text{ [refer to parts (B) and (C)]} = 0.175$$

95. (A) The sample space S consists of the number of permutations of the 8 blood types chosen 3 at a time. Thus, $n(S) = P_{8,3}$ and the probability of guessing the three types in a sample correctly is:

$$P(E) = \frac{1}{{_8P_3}} = \frac{1}{\dfrac{8!}{(8-3)!}} = \frac{1}{8 \cdot 7 \cdot 6} \approx 0.0030$$

(B) Allowing repetition, $n(S) = 8^3$ and the probability of guessing the three types in a sample correctly is:

$$P(E) = \frac{1}{8^3} \approx 0.0020$$

97. (A) The total number of ways of selecting a three-person committee from the 9 members of the council is:

$$_9C_3, \text{ i.e., } n(S) = {_9C_3}.$$

The total number of ways of selecting a three-person committee from the 5 Democrats is: $_5C_3$

Thus, if E is the event "three-person committee is composed solely of Democrats," then

$$P(E) = \frac{{_5C_3}}{{_9C_3}} = \frac{\dfrac{5!}{3!(5-3)!}}{\dfrac{9!}{3!(9-3)!}} = \frac{10}{84} = 0.1190$$

(B) The total number of ways of having a majority of the committee be Democrats is

$$_5C_3 + {_5C_2} \cdot {_4C_1} = \frac{5!}{3!(5-3)!} + \frac{5!}{2!(5-2)!} \cdot \frac{4!}{1!(4-1)!} = 10 + (10)4 = 50.$$

If we let F be the event "The majority of the members are Democrats," then

$$P(F) = \frac{n(F)}{n(S)} = \frac{50}{84} \approx 0.5952.$$

EXERCISE 8-2

Things to remember:

1. UNION AND INTERSECTION OF EVENTS

If A and B are two events in a sample space S, then the UNION of A and B, denoted by $A \cup B$, and the INTERSECTION of A and B, denoted by $A \cap B$, are defined as follows:

$A \cup B = \{e \in S \mid e \in A \text{ OR } e \in B\}$ $A \cap B = \{e \in S \mid e \in A \text{ AND } e \in B\}$

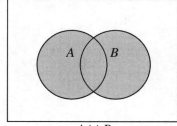

$A \cup B$

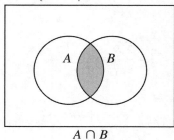

$A \cap B$

Furthermore, we define:

The **event A or B** to be $A \cup B$.

The **event A and B** to be $A \cap B$.

2.   PROBABILITY OF A UNION OF TWO EVENTS

For any events A and B,

(a) $P(A \cup B) = P(A) + P(B) - P(A \cap B)$.

If A and B are MUTUALLY EXCLUSIVE $(A \cap B = \varnothing)$, then

(b) $P(A \cup B) = P(A) + P(B)$.

3.   PROBABILITY OF COMPLEMENTS

For any event E, $E \cup E' = S$ and $E \cap E' = \varnothing$. Thus,

$$P(E) = 1 - P(E')$$

$$P(E') = 1 - P(E)$$

4.   PROBABILITY TO ODDS

If $P(E)$ is the probability of the event E, then:

(a)   Odds for $E = \dfrac{P(E)}{1 - P(E)} = \dfrac{P(E)}{P(E')}$ $[P(E) \neq 1]$

(b)   Odds against $E = \dfrac{P(E')}{P(E)}$ $[P(E) \neq 0]$

The ratio $\dfrac{P(E)}{P(E')}$, giving odds for E, is usually expressed as an equivalent ratio $\dfrac{a}{b}$ of whole numbers (by multiplying numerator and denominator by the same number), and written "a to b" or "a:b". In this case, the odds against E are written "b to a" or "b:a".

ODDS TO PROBABILTY

If the odds for an event E are "a to b", then the probability of E is:

$$P(E) = \frac{a}{a+b}$$

---

1. $\dfrac{\dfrac{3}{10}}{\dfrac{9}{10}} = \dfrac{3}{10} \cdot \dfrac{10}{9} = \dfrac{1}{3}$

3. $\dfrac{\dfrac{1}{8}}{\dfrac{3}{7}} = \dfrac{1}{8} \cdot \dfrac{7}{3} = \dfrac{7}{24}$

5. $\dfrac{\dfrac{2}{9}}{1 - \dfrac{2}{9}} = \dfrac{\dfrac{2}{9}}{\dfrac{7}{9}} = \dfrac{2}{9} \cdot \dfrac{9}{7} = \dfrac{2}{7}$

7. $P(A \cap B) = \dfrac{39}{100} = 0.39.$

9. $P(A' \cup B) = \dfrac{39 + 21 + 26}{100} = \dfrac{86}{100} = 0.86.$

11. $P(A \cup B) = \dfrac{14 + 39 + 21}{100} = \dfrac{74}{100} = 0.74;$ $P[(A \cup B)'] = 1 - P(A \cup B) = 0.26.$

13. $n(D) = 13$, $P(D) = \dfrac{n(D)}{n(S)} = \dfrac{13}{52} = \dfrac{1}{4}$

15. $n(F) = 12$, $n(F') = 40$; $P(F') = \dfrac{40}{52} = \dfrac{10}{13}$

17. $D \cap F = $ jack, queen, king of diamonds; $n(D \cap F) = 3$

$P(D \cap F) = \dfrac{3}{52}$

19. $n(D \cup F) = 13 + 12 - 3 = 22$ (number of diamonds + number of face cards) – (jack, queen , king of diamonds)

$P(D \cup F) = \dfrac{22}{52} = \dfrac{11}{26}$

21. $D \cap F' = 2 - 10$ of diamonds and the ace of diamonds.

$n(D \cap F') = 10$; $P(D \cap F') = \dfrac{10}{52} = \dfrac{5}{26}$

23. $n(D \cup F') = n(D) + n(F') - n(D \cap F') = 13 + 40 - 10 = 43$

$P(D \cup F') = \dfrac{n(D \cup F')}{n(S)} = \dfrac{43}{52}$

25. Let $E = $ number is odd; $F = $ number is multiple of 4.
Since $E \cap F = \varnothing$, we use equation (1):
$n(E) = 13$, $n(F) = 6$

$P(E \cup F) = P(E) + P(F) = \dfrac{13}{25} + \dfrac{6}{25} = \dfrac{19}{25} = 0.76$.

27. Let $E = $ number is prime $= \{2, 3, 5, 7, 11, 13, 17, 19, 23\}$
   $F = $ number is greater than 20 $= \{ 21, 22, 23, 24, 25\}$
Since $E \cap F \neq 0$, we use equation (1): $n(E) = 9$, $n(F) = 5$, $n(E \cap F) = 1$,

$P(E \cup F) = P(E) + P(F) - P(E \cap F) = \dfrac{9}{25} + \dfrac{5}{25} - \dfrac{1}{25} = \dfrac{13}{25} = 0.52$.

29. Let $E = $ number is a multiple of 2. $F = $ number is a multiple of 5.
Since $E \cap F \neq \varnothing$, we use equation (1):
$n(E) = 12$, $n(F) = 5$, $n(E \cap F) = 2$

$P(E \cup F) = P(E) + P(F) - P(E \cap F) = \dfrac{12}{25} + \dfrac{5}{25} - \dfrac{2}{25} = \dfrac{15}{25} = 0.60$.

31. Let $E = $ number is less than 5; $F = $ number is greater than 20.
Since $E \cap F = \varnothing$, we use equation (1):
$n(E) = 4$, $n(F) = 5$

$P(E \cup F) = P(E) + P(F) = \dfrac{4}{25} + \dfrac{5}{25} = \dfrac{9}{25} = 0.36$

33. $P(\text{loss}) = 1 - P(\text{win}) = 1 - 0.51 = 0.49$

35. $A = \text{sum} \leq 5,\ n(A) = 10,\ P(A) = \dfrac{10}{36} = \dfrac{5}{18}.$

37. $A = $ number on the first die $= 6,\ n(A) = 6;$

$B = $ number on the second die $= 3;\ n(B) = 6;\ n(A \cap B) = 1;$

$$P(A \cup B) = P(A) + P(B) - P(A \cap B) = \frac{6}{36} + \frac{6}{36} - \frac{1}{36} = \frac{11}{36}.$$

39. Use $\underline{4}$ to find the odds for Event E.

(A) $P(E) = \dfrac{3}{8},\ P(E') = 1 - P(E) = \dfrac{5}{8}$

Odds for $E = \dfrac{P(E)}{P(E')}$

$\qquad = \dfrac{3/8}{5/8} = \dfrac{3}{5}$ (3 to 5)

Odds against $E = \dfrac{P(E')}{P(E)}$

$\qquad = \dfrac{5/8}{3/8} = \dfrac{5}{3}$ (5 to 3)

(B) $P(E) = \dfrac{1}{4},\ P(E') = 1 - P(E) = \dfrac{3}{4}$

Odds for $E = \dfrac{P(E)}{P(E')}$

$\qquad = \dfrac{1/4}{3/4} = \dfrac{1}{3}$ (1 to 3)

Odds against $E = \dfrac{P(E')}{P(E)}$

$\qquad = \dfrac{3/4}{1/4} = \dfrac{3}{1}$ (3 to 1)

(C) $P(E) = .4,\ P(E') = 1 - P(E) = .6$

Odds for $E = \dfrac{P(E)}{P(E')}$

$\qquad = \dfrac{.4}{.6} = \dfrac{2}{3}$ (2 to 3)

Odds against $E = \dfrac{P(E')}{P(E)}$

$\qquad = \dfrac{.6}{.4} = \dfrac{3}{2}$ (3 to 2)

(D) $P(E) = .55,\ P(E') = 1 - P(E) = .45$

Odds for $E = \dfrac{P(E)}{P(E')}$

$\qquad = \dfrac{.55}{.45} = \dfrac{11}{9}$ (11 to 9)

Odds against $E = \dfrac{P(E')}{P(E)}$

$\qquad = \dfrac{.45}{.55} = \dfrac{9}{11}$

(9 to 11)

41. Use $\underline{4}$ to find the probability of event E.

(A)   Odds for $E = \dfrac{3}{8}$

$P(E) = \dfrac{3}{3+8} = \dfrac{3}{11}$

(C)   Odds for $E = \dfrac{4}{1}$

$P(E) = \dfrac{4}{4+1} = \dfrac{4}{5} = .8$

(B)   Odds for $E = \dfrac{11}{7}$

$P(E) = \dfrac{11}{11+7} = \dfrac{11}{18}$

(D)   Odds for $E = \dfrac{49}{51}$

$P(E) = \dfrac{49}{49+51} = \dfrac{49}{100} = .49$

43. False. Odds for $E = \dfrac{P(E)}{P(E')}$;

odds against $E' = \dfrac{P([E']')}{P(E')} = \dfrac{P(E)}{P(E')}$

So, odds for E = odds against E' independent of $P(E)$.

45. False. Since $P(E \cup F) = P(E) + P(F) - P(E \cap F)$ for *any* events E, F,
$P(E) + P(F) = P(E \cup F) + P(E \cap F)$
whether or not E and F are disjoint.

47. True. E and F are complementary, then $F = E'$ and $E \cap F = \varnothing$. Therefore, E and F are mutually exclusive.

49. Odds for $E = \dfrac{P(E)}{P(E')} = \dfrac{1/2}{1/2} = 1$.

The odds in favor of getting a head in a single toss of a coin are 1 to 1.

51. The sample space for this problem is:
S = {HHH, HHT, THH, HTH, TTH, HTT, THT, TTT}

Let Event E = "getting at least 1 head."
Let Event E' = "getting no heads."

Thus, $\dfrac{P(E)}{P(E')} = \dfrac{7/8}{1/8} = \dfrac{7}{1}$

The odds in favor of getting at least 1 head are 7 to 1.

53. Let Event E = "getting a number greater than 4."
Let Event E' = "not getting a number greater than 4."

Thus, $\dfrac{P(E')}{P(E)} = \dfrac{4/6}{2/6} = \dfrac{2}{1}$

The odds against getting a number greater than 4 in a single roll of a die are 2 to 1.

55. Let Event E = "getting 3 or an even number" = {2, 3, 4, 6}.
Let Event E' = "not getting 3 or an even number" = {1, 5}.

Thus, $\dfrac{P(E')}{P(E)} = \dfrac{2/6}{4/6} = \dfrac{1}{2}$

The odds against getting 3 or an even number are 1 to 2.

57. Let E = "rolling a sum of five." Then $P(E) = \dfrac{n(E)}{n(S)} = \dfrac{4}{36} = \dfrac{1}{9}$ and $P(E') = \dfrac{8}{9}$.

(A)   Odds for $E = \dfrac{1/9}{8/9} = \dfrac{1}{8}$ (1 to 8)

(B)   The house should pay \$8 for the game to be fair (see Example 6).

59. (A) Let E = "sum is less than 4 or greater than 9." Then

$$P(E) = \frac{10+30+120+80+70}{1000} = \frac{310}{1000} = \frac{31}{100} = 0.31 \text{ and } P(E') = \frac{69}{100}.$$

Thus,

Odds for $E = \dfrac{31/100}{69/100} = \dfrac{31}{69}$

(B) Let F = "sum is even or divisible by 5." Then

$$P(F) = \frac{10+50+110+170+120+70+70}{1000} = \frac{600}{1000} = \frac{6}{10} = .6$$

and $P(F') = \dfrac{4}{10}$. Thus,

Odds for $F = \dfrac{6/10}{4/10} = \dfrac{6}{4} = \dfrac{3}{2}$

61. Let A = "drawing a face card" (Jack, Queen, King)
and B = "drawing a club."

Then $P(A \cup B) = P(A) + P(B) - P(A \cap B) = \dfrac{12}{52} + \dfrac{13}{52} - \dfrac{3}{52} = \dfrac{22}{52} = \dfrac{11}{26}$

$P[(A \cup B)'] = \dfrac{15}{26}$

Odds for $A \cup B = \dfrac{11/26}{15/26} = \dfrac{11}{15}$

63. Let A = "drawing a black card"
and B = "drawing an ace."

$P(A \cup B) = P(A) + P(B) - P(A \cap B) = \dfrac{26}{52} + \dfrac{4}{52} - \dfrac{2}{52} = \dfrac{28}{52} = \dfrac{7}{13}$

$P[(A \cup B)'] = \dfrac{6}{13}$

Odds for $A \cup B = \dfrac{7/13}{6/13} = \dfrac{7}{6}$

65. The sample space S is the set of all 5-card hands and $n(S) = {}_{52}C_5$.
Let E = "getting at least one diamond."
Then E' = "no diamonds" and $n(E) = {}_{39}C_5$.

Thus, $P(E') = \dfrac{{}_{39}C_5}{{}_{52}C_5}$, and

$$P(E) = 1 - \frac{{}_{39}C_5}{{}_{52}C_5} = 1 - \frac{\dfrac{39!}{5!34!}}{\dfrac{52!}{5!47!}} = 1 - \frac{39 \cdot 38 \cdot 37 \cdot 36 \cdot 35}{52 \cdot 51 \cdot 50 \cdot 49 \cdot 48} \approx 1 - .22 = .78.$$

67. The number of numbers less than or equal to 1000 which are divisible by 6 is the largest integer in $\dfrac{1000}{6}$

or 166.

The number of numbers less than or equal to 1000 which are divisible by 8 is the largest integer in $\dfrac{1000}{8}$

or 125.

The number of numbers less than or equal to 1000 which are divisible by both 6 and 8 is the same as the

number of numbers which are divisible by 24. This is the largest integer in $\dfrac{1000}{24}$ or 41.

Thus, if A is the event "selecting a number which is divisible by either 6 or 8," then

$n(A) = 166 + 125 - 41 = 250$ and $P(A) = \dfrac{250}{1000} = .25$.

69. In general, for three events A, B, and C,

$$P(A \cup B \cup C) = P(A) + P(B) + P(C) - P(A \cap B) - P(A \cap C) - P(B \cap C) + P(A \cap B \cap C)$$

Therefore,

(*) $P(A \cup B \cup C) = P(A) + P(B) + P(C) - P(A \cap B)$

will hold if A and C, and B and C are mutually exclusive. Note that, if either $A \cap C = \varnothing$, or $B \cap C = \varnothing$,
then $A \cap B \cap C = \varnothing$.

Equation (*) will also hold if A, B, and C are mutually exclusive, in which case

$$P(A \cup B \cup C) = P(A) + P(B) + P(C)$$

71. From Example 5,

$$P(E) = 1 - \frac{365!}{365^n(365-n)!} = 1 - \frac{1}{365^n} \cdot \frac{365!}{(365-n)!}$$

$$= 1 - \frac{1}{(365)^n} \cdot P_{365,n}$$

$$= 1 - \frac{P_{365,n}}{(365)^n}$$

For calculators with a $P_{n,r}$ key, this form involves fewer calculator steps. Also, 365! produces an overflow
error on many calculators, while $P_{365,n}$ does not produce an overflow error for many values of n.

73. S = set of all lists of n birth months, $n \leq 12$. Then

$n(S) = 12 \cdot 12 \cdot \ ... \ \cdot 12$ (n times) $= 12^n$.

Let E = "at least two people have the same birth month."

Then E' = "no two people have the same birth month."

$n(E') = 12 \cdot 11 \cdot 10 \cdot \ ... \ \cdot [12 - (n-1)]$

$$= \frac{12 \cdot 11 \cdot 10 \cdot ... \cdot [12-(n-1)](12-n)[12-(n+1)] \cdot ... \cdot 3 \cdot 2 \cdot 1}{(12-n)[12-(n+1)] \cdot ... \cdot 3 \cdot 2 \cdot 1} = \frac{12!}{(12-n)!}$$

Thus, $P(E') = \dfrac{\dfrac{12!}{(12-n)!}}{12^n} = \dfrac{12!}{12^n(12-n)!}$ and $P(E) = 1 - \dfrac{12!}{12^n(12-n)!}$.

75.   Odds for $E = \dfrac{P(E)}{P(E')} = \dfrac{P(E)}{1 - P(E)} = \dfrac{a}{b}$.   Therefore,

$bP(E) = a[1 - P(E)] = a - aP(E)$.

Thus, $aP(E) + bP(E) = a$

$\qquad (a + b)P(E) = a$

$$P(E) = \frac{a}{a+b}$$

77.   (A)   From the plot, 7 and 8 each came up 10 times.

Therefore, $P(7 \text{ or } 8) = \dfrac{10}{50} + \dfrac{10}{50} = \dfrac{20}{50} = 0.4$.

(B)   Theoretical probability: $P(7) = \dfrac{1}{6}$, $P(8) = \dfrac{5}{36}$

so $P(7 \text{ or } 8) = \dfrac{1}{6} + \dfrac{5}{36} = \dfrac{6}{36} + \dfrac{5}{36} = \dfrac{11}{36} \approx 0.306$.

(C)   The answer depends on the results of your simulation.

79.   Venn diagram:

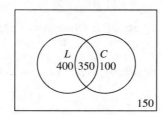

Let L be the event that the student owns a laptop and C be the event that the student owns a car.

The table corresponding to the given data is as follows:

	C	C'	Total
L	350	400	750
L'	100	150	250
Total	450	550	1000

The corresponding probabilities are:

	C	C'	Total
L	.35	.40	.75
L'	.10	.15	.25
Total	.45	.55	1.00

From the above table:

(A) $P(C \text{ or } L) = P(C \cup L) = P(C) + P(L) - P(C \cap L) = .45 + .75 - .35 = .85$

(B) $P(C' \cap L') = .15$

81.   (A)   Using the table, we have:

$P(M_1 \text{ or } A) = P(M_1 \cup A) = P(M_1) + P(A) - P(M_1 \cap A) = .2 + .3 - .05 = .45$

(B) $P[(M_2 \cap A') \cup (M_3 \cap A')] = P(M_2 \cap A') + P(M_3 \cap A')$

$\qquad\qquad\qquad\qquad\qquad = .2 + .35 \text{ (from the table)}$

$\qquad\qquad\qquad\qquad\qquad = .55$

83. The sample space S is the set of all possible 10-element samples from the 60 game players, and $n(S) = C_{60,10}$. Let E be the event that a sample contains at least one defective game player. Then E' is the event that a sample contains no game players. Now, $n(E') = {}_{51}C_{10}$.

Thus, $P(E') = \dfrac{{}_{51}C_{10}}{{}_{60}C_{10}} = \dfrac{\dfrac{51!}{10!41!}}{\dfrac{60!}{10!50!}} \approx .17$ and $P(E) \approx 1 - .17 = .83$.

Therefore, the probability that a sample will be returned is .83.

85. The given information is displayed in the Venn diagram:

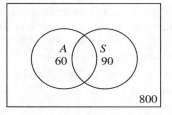

A = suffers from loss of appetite
S = suffers from loss of sleep

Thus, we can conclude that $n(A \cap S) = 1000 - (60 + 90 + 800) = 50$.

$P(A \cap S) = \dfrac{50}{1000} = .05$

87. (A)   "Unaffiliated or no preference" = $U \cup N$.

$P(U \cup N) = P(U) + P(N) - P(U \cap N)$

$= \dfrac{150}{1000} + \dfrac{85}{1000} - \dfrac{15}{1000} = \dfrac{220}{1000} = \dfrac{11}{50} = .22$

Therefore, $P[(U \cup N)'] = 1 - \dfrac{11}{50} = \dfrac{39}{50}$ and

Odds for $U \cup N = \dfrac{11/50}{39/50} = \dfrac{11}{39}$

(B)   "Affiliated with a party and prefers candidate A' = $(D \cup R) \cap A$.

$P[(D \cup R) \cap A] = \dfrac{300}{1000} = \dfrac{3}{10} = .3$

The odds against this event are:

$\dfrac{1 - 3/10}{3/10} = \dfrac{7/10}{3/10} = \dfrac{7}{3}$

EXERCISE 8-3

Things to remember:

1.   CONDITIONAL PROBABILITY

For events A and B in a sample space S, the CONDITIONAL PROBABILITY of A given B, denoted $P(A \mid B)$, is defined by

$$P(A \mid B) = \dfrac{P(A \cap B)}{P(B)}, \quad P(B) \neq 0$$

2.   PRODUCT RULE

For events A and B, $P(A) \neq 0$, $P(B) \neq 0$, in a sample space S,

$$P(A \cap B) = P(A) \cdot P(B \mid A) = P(B) \cdot P(A \mid B).$$

[Note: We can use either $P(A) \cdot P(B \mid A)$ or $P(B) \cdot P(A \mid B)$ to compute $P(A \cap B)$.]

3.   PROBABILITY TREES
Given a sequence of probability experiments. To compute the probabilities of combined outcomes:

Step 1. Draw a tree diagram corresponding to all combined outcomes of the sequence of experiments.

Step 2. Assign a probability to each tree branch. (This is the probability of the occurrence of the event on the right end of the branch subject to the occurrence of all events on the path leading to the event on the right end of the branch. The probability of the occurrence of a combined outcome that corresponds to a path through the tree is the product of all branch probabilities on the path.)

Step 3. Use the results in Steps 1 and 2 to answer various questions related to the sequence of experiments as a whole.

4.   INDEPENDENCE

Let A and B be any events in a sample space S. Then A and B are INDEPENDENT if and only if

$$P(A \cap B) = P(A) \cdot P(B).$$

Otherwise, A and B are DEPENDENT.

5.   INDEPENDENT SET OF EVENTS

A set of events is said to be INDEPENDENT if for each finite subset $\{E_1, E_2, \ldots, E_k\}$

$$P(E_1 \cap E_2 \cap \ldots \cap E_k) = P(E_1)P(E_2) \cdot \ldots \cdot P(E_k)$$

1.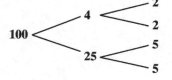

3.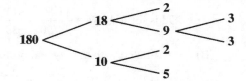

5.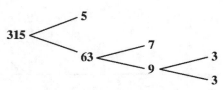

7. A = card is an ace, H = card is a heart. $P(A|H) = \dfrac{P(A \cap H)}{P(H)} = \dfrac{\frac{1}{52}}{\frac{13}{52}} = \dfrac{1}{13}$

9. H = card is a heart, A = card is an ace. $P(H|A) = \dfrac{P(H \cap A)}{P(A)} = \dfrac{\frac{1}{52}}{\frac{4}{52}} = \dfrac{1}{4}$

11. B = card is black, C = card is a club. $P(B|C) = \dfrac{P(B \cap C)}{P(C)} = \dfrac{\frac{13}{52}}{\frac{13}{52}} = 1$

13. C = card is a club, B = card is black. $P(C|B) = \dfrac{P(C \cap B)}{P(B)} = \dfrac{\frac{13}{52}}{\frac{26}{52}} = \dfrac{13}{52} \cdot \dfrac{52}{26} = \dfrac{1}{2}$

To find the conditional probabilities in Problems 15 – 21, construct a table similar to those in Section 8.2.

15. F = sum is less than 6, E = sum is even. $n(F) = 10$, $n(E) = 18$, $n(F \cap E) = 4$.

$$P(F|E) = \frac{P(F \cap E)}{P(E)} = \frac{\frac{4}{36}}{\frac{18}{36}} = \frac{4}{36} \cdot \frac{36}{18} = \frac{2}{9}$$

17. E = sum is even, F = sum is less than 6. $n(F) = 10$, $n(E) = 18$, $n(F \cap E) = 4$.

$$P(E|F) = \frac{P(E \cap F)}{P(F)} = \frac{\frac{4}{36}}{\frac{10}{36}} = \frac{4}{36} \cdot \frac{36}{10} = \frac{2}{5}$$

19. E = sum is greater than 7, F = neither die is a 6. $n(E) = 15$, $n(F) = 25$, $n(E \cap F) = 6$.

$$P(E|F) = \frac{P(E \cap F)}{P(F)} = \frac{\frac{6}{36}}{\frac{25}{36}} = \frac{6}{36} \cdot \frac{36}{25} = \frac{6}{25}$$

21. F = neither die is a 6, E = sum is greater than 7. $n(F) = 25$, $n(E) = 15$, $n(F \cap E) = 6$.

$$P(F|E) = \frac{P(F \cap E)}{P(E)} = \frac{\frac{6}{36}}{\frac{15}{36}} = \frac{6}{36} \cdot \frac{36}{15} = \frac{2}{5}$$

23. $P(B) = 0.03 + 0.05 + 0.02 = 0.10$ **25.** $P(B \cap D) = 0.03$

27. $P(D|B) = \dfrac{P(D \cap B)}{P(B)} = \dfrac{0.03}{0.10} = 0.30$ **29.** $P(B|D) = \dfrac{P(B \cap D)}{P(D)} = \dfrac{0.03}{0.30} = 0.10$

31. $P(D|C) = \dfrac{P(D \cap C)}{P(C)} = \dfrac{0.07}{0.20} = 0.35$

33. $A \cap C = \varnothing$. Therefore, $P(A \mid C) = \dfrac{P(A \cap C)}{P(C)} = \dfrac{0}{0.20} = 0.$

35. Events A and D are independent if $P(A \cap D) = P(A) \cdot P(D)$:
$P(A) = 0.70$, $P(D) = 0.30$, $P(A \cap D) = 0.20$; $P(A) \cdot P(D) = 0.21 \neq P(A \cap D)$; A and D are dependent.

37. $P(B) = 0.10$, $P(D) = 0.30$, $P(B \cap D) = 0.03$; $P(B) \cdot P(D) = 0.03 = P(B \cap D)$; B and D are independent.

39. $P(B) = 0.10$, $P(F) = 0.30$, $P(B \cap F) = 0.02$; $P(B) \cdot P(F) = 0.03 \neq P(B \cap F)$; B and F are dependent.

41. $P(A) = 0.70$, $P(B) = 0.10$, $A \cap B = \varnothing$, $P(A \cap B) = 0$; $P(A) \cdot P(B) = 0.07 \neq P(A \cap B)$; A and B are dependent.

43. (A) Let $H_8 =$ "a head on the eighth toss." Since each toss is independent of the other tosses, $P(H_8) = \dfrac{1}{2}$.

(B) Let $H_i =$ "a head on the ith toss." Since the tosses are independent,

$$P(H_1 \cap H_2 \cap \cdots \cap H_8) = P(H_1)P(H_2) \cdots P(H_8) = \left(\frac{1}{2}\right)^8 = \frac{1}{2^8} = \frac{1}{256}.$$

Similarly, if $T_i =$ "a tail on the ith toss," then

$$P(T_1 \cap T_2 \cap \cdots \cap T_8) = P(T_1)P(T_2) \cdots P(T_8) = \frac{1}{2^8} = \frac{1}{256}. \text{ Finally, if}$$

$H =$ "all heads" and $T =$ "all tails," then $H \cap T = \varnothing$ and

$$P(H \cup T) = P(H) + P(T) = \frac{1}{256} + \frac{1}{256} = \frac{2}{256} = \frac{1}{128} \approx .00781.$$

45. Given the table:

e_i	1	2	3	4	5
P_i	.3	.1	.2	.3	.1

$E =$ "pointer lands on an even number" $= \{2, 4\}$.
$F =$ "pointer lands on a number less than 4" $= \{1, 2, 3\}$.

(A) $P(F \mid E) = \dfrac{P(F \cap E)}{P(E)} = \dfrac{P(2)}{P(2) + P(4)} = \dfrac{.1}{.1 + .3} = \dfrac{.1}{.4} = \dfrac{1}{4}$

(B) $P(E \cap F) = P(2) = .1$,
$P(E) = .4$, $P(F) = P(1) + P(2) + P(3) = .3 + .1 + .2 = .6$,
and
$P(E)P(F) = (.4)(.6) = .24 \neq P(E \cap F)$.
Thus, E and F are dependent.

47. From the probability tree,
(A) $P(M \cap S) = (.3)(.6) = .18$

(B) $P(R) = P(N \cap R) + P(M \cap R) = (.7)(.2) + (.3)(.4) = .14 + .12 = .26$

49. $E_1 = \{HH, HT\}$ and $P(E_1) = \dfrac{1}{2}$

$E_2 = \{TH, TT\}$ and $P(E_2) = \dfrac{1}{2}$

$E_4 = \{HH, TH\}$ and $P(E_4) = \dfrac{1}{2}$

(A)   Since $E_1 \cap E_4 = \{HH\} \neq \emptyset$, E_1 and E_4 **are not** mutually exclusive.

Since $P(E_1 \cap E_4) = P(HH) = \dfrac{1}{4} = P(E_1) \cdot P(E_4)$, E_1 and E_4 are independent.

(B)   Since $E_1 \cap E_2 = \emptyset$, E_1 and E_2 **are** mutually exclusive.

Since $P(E_1 \cap E_2) = 0$ and $P(E_1) \cdot P(E_2) = \dfrac{1}{4}$, $P(E_1 \cap E_2) \neq P(E_1) \cdot P(E_2)$.

Therefore, E_1 and E_2 are dependent.

51.   Let E_i = "even number on the ith throw," $i = 1, 2$, and O_i = "odd number on the ith throw," $i = 1, 2$.

Then $P(E_i) = \dfrac{1}{2}$ and $P(O_i) = \dfrac{1}{2}$, $i = 1, 2$.

The probability tree for this experiment is shown at the right.

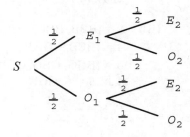

$$P(E_1 \cap E_2) = \left(\frac{1}{2}\right)\left(\frac{1}{2}\right) = \frac{1}{4}$$

$$P(E_1 \cup E_2) = P(E_1) + P(E_2) - P(E_1 \cap E_2) = \frac{1}{2} + \frac{1}{2} - \frac{1}{4} = \frac{3}{4}.$$

53.   Let C = "first card is a club," and H = "second card is a heart."

(A)   Without replacement, the probability tree is as shown at the right.

Thus, $P(C \cap H) = \left(\dfrac{1}{4}\right)\left(\dfrac{13}{51}\right) \approx .0637$.

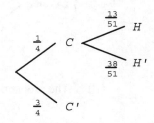

(B)   With replacement, the draws are independent and

$$P(C \cap H) = \left(\frac{1}{4}\right)\left(\frac{1}{4}\right) = \frac{1}{16} = 0.0625.$$

55. G = "the card is black" = {spade or club} and $P(G) = \dfrac{1}{2}$.

H = "the card is divisible by 3" = {3, 6, or 9}. $P(H) = \dfrac{12}{52} = \dfrac{3}{13}$

$P(H \cap G) = \{3, 6,$ or 9 of clubs or spades$\} = \dfrac{6}{52} = \dfrac{3}{26}$

(A) $P(H \mid G) = \dfrac{P(H \cap G)}{P(G)} = \dfrac{3/26}{1/2} = \dfrac{6}{26} = \dfrac{3}{13}$

(B) $P(H \cap G) = \dfrac{3}{26} = P(H) \cdot P(G)$

Thus, H and G *are* independent.

57. (A) $S = \{BB, BG, GB, GG\}$

$A = \{BB, GG\}$ and $P(A) = \dfrac{2}{4} = \dfrac{1}{2}$

$B = \{BG, GB, GG\}$ and $P(B) = \dfrac{3}{4}$

$A \cap B = \{GG\}$.

$P(A \cap B) = \dfrac{1}{4}$ and $P(A) \cdot P(B) = \dfrac{1}{2} \cdot \dfrac{3}{4} = \dfrac{3}{8}$

Thus, $P(A \cap B) \neq P(A) \cdot P(B)$ and the events are dependent.

(B) $S = \{BBB, BBG, BGB, BGG, GBB, GBG, GGB, GGG\}$
$A = \{BBB, GGG\}$
$B = \{BGG, GBG, GGB, GGG\}$
$A \cap B = \{GGG\}$

$P(A) = \dfrac{2}{8} = \dfrac{1}{4}$, $P(B) = \dfrac{4}{8} = \dfrac{1}{2}$, and $P(A \cap B) = \dfrac{1}{8}$

Since $P(A \cap B) = \dfrac{1}{8} = P(A) \cdot P(B)$, A and B are independent.

59. (A)   The probability tree with replacement is as follows:

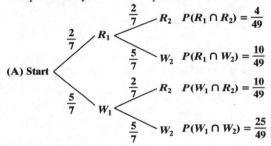

(B)   The probability tree without replacement is as follows:

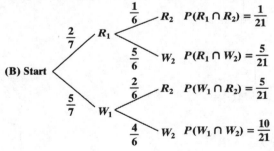

61.   Let E = At least one ball was red = $\{R_1 \cap R_2, R_1 \cap W_2, W_1 \cap R_2\}$.

(A)   With replacement [see the probability tree in Problem 59(A)]:
$P(E) = P(R_1 \cap R_2) + P(R_1 \cap W_2) + P(W_1 \cap R_2)$

$= \dfrac{4}{49} + \dfrac{10}{49} + \dfrac{10}{49} = \dfrac{24}{49}$

(B)   Without replacement [see the probability tree in Problem 59(B)]:
$P(E) = P(R_1 \cap R_2) + P(R_1 \cap W_2) + P(W_1 \cap R_2) = \dfrac{1}{21} + \dfrac{5}{21} + \dfrac{5}{21} = \dfrac{11}{21}$

63. False. Flip a fair coin twice. Let A = two heads and B = a head on the first toss. Then

$$P(A \mid B) = \frac{P(A \cap B)}{P(B)} = \frac{1/4}{1/2} = \frac{1}{2} = P(B)$$

But $P(A \cap B) = \frac{1}{4}$ and $P(A) \cdot P(B) = \frac{1}{4} \cdot \frac{1}{2} = \frac{1}{8}$ so A and B are not independent.

65. True. $P(A \mid B) = \dfrac{P(A \cap B)}{P(B)} = \dfrac{P(A)}{P(B)}$ since $A \subset B$ implies $A \cap B = A$

$\dfrac{P(A)}{P(B)} \geq \dfrac{P(A)}{1} = P(A)$ since $P(B) \leq 1$

Therefore, if $A \subset B$, then $P(A \mid B) \geq P(A)$.

67. False.   Suppose $0 < P(A) < 1$. Let $B = A'$. Then $0 < P(B) < 1$. Now
$P(A \cap B) = P(A \cap A') = 0$ and $P(A) \cdot P(B) = P(A) \cdot P(A') \neq 0$

69. True.

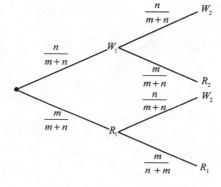

$$P(W_1 \cap R_2) = \frac{mn}{(m+n)^2} = P(R_1 \cap W_2)$$

71. Total number of balls $2 + 3 + 4 = 9$; $n(S) = C_{9,2} = \dfrac{9!}{2!(9-2)!} = \dfrac{9 \cdot 8 \cdot 7!}{2 \cdot 1 \cdot 7!} = 36$

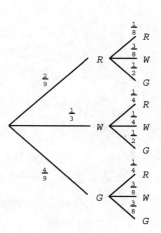

Let A = Both balls are the same color.

$n(A)$ = (No. of ways 2 red balls are selected)
   + (No. of ways 2 white balls are selected)
   + (No. of ways 2 green balls are selected)

$= C_{2,2} + C_{3,2} + C_{4,2}$

$= \dfrac{2!}{2!(2-2)!} + \dfrac{3!}{2!(3-2)!} + \dfrac{4!}{2!(4-2)!}$

$= 1 + 3 + 6 = 10$

$P(A) = \dfrac{n(A)}{n(S)} = \dfrac{10}{36} = \dfrac{5}{18}$

Alternatively, the probability tree for this experiment is shown above,
and $P(RR, WW, \text{or } GG) = P(RR) + P(WW) + P(GG)$

$$= \left(\frac{2}{9}\right)\left(\frac{1}{8}\right) + \left(\frac{1}{3}\right)\left(\frac{1}{4}\right) + \left(\frac{4}{9}\right)\left(\frac{3}{8}\right) = \frac{2}{72} + \frac{1}{12} + \frac{12}{72} = \frac{20}{72} = \frac{5}{18}$$

73. The probability tree for
this experiment is:

(A) $P(\$16) = \left(\frac{1}{4}\right)\left(\frac{2}{3}\right)\left(\frac{1}{2}\right) + \left(\frac{1}{2}\right)\left(\frac{1}{3}\right)\left(\frac{1}{2}\right)$

$= \frac{1}{12} + \frac{1}{12} = \frac{1}{6} \approx .167$

(B) $P(\$17) = \left(\frac{1}{2}\right)\left(\frac{1}{3}\right)\left(\frac{1}{2}\right) + \left(\frac{1}{2}\right)\left(\frac{1}{3}\right)\left(\frac{1}{2}\right) + \left(\frac{1}{4}\right)\left(\frac{2}{3}\right)\left(\frac{1}{2}\right)$

$= \frac{1}{12} + \frac{1}{12} + \frac{1}{12} = \frac{1}{4} = .25$

(C) Let A = "$10 on second draw." Then

$P(A) = \left(\frac{1}{4}\right)\left(\frac{1}{3}\right) + \left(\frac{1}{2}\right)\left(\frac{1}{3}\right) = \frac{1}{12} + \frac{1}{6} = \frac{1}{4} = .25$

75. Assume that A and B are independent events with $P(A) \neq 0$, $P(B) \neq 0$. Then, by definition
(*) $P(A \cap B) = P(A) \cdot P(B)$.
Now,

$P(A \mid B) = \dfrac{P(A \cap B)}{P(B)}$ (definition of conditional probability)

$= \dfrac{P(A) \cdot P(B)}{P(B)}$ (by *)

$= P(A)$

Also,

$P(B \mid A) = \dfrac{P(B \cap A)}{P(A)} = \dfrac{P(A \cap B)}{P(A)} = \dfrac{P(A) \cdot P(B)}{P(A)} = P(B)$

77. Assume $P(A) \neq 0$. Then $P(A \mid A) = \dfrac{P(A \cap A)}{P(A)} = \dfrac{P(A)}{P(A)} = 1$.

79. If A and B are mutually exclusive, then $A \cap B = \varnothing$ and $P(A \cap B) = P(\varnothing) = 0$. Also, if $P(A) \neq 0$ and $P(B) \neq 0$, then $P(A) \cdot P(B) \neq 0$. Therefore, $P(A \cap B) = 0 \neq P(A) \cdot P(B)$, and events A and B are dependent.

81. (A)

To strike	Hourly H	Salary S	Salary + bonus B	Total
Yes (Y)	.400	.180	.020	.600
No (N)	.150	.120	.130	.400
Totals	.550	.300	.150	1.000

[Note: The probability table above was derived from the table given in the problem by dividing each entry by 1000.]

Referring to the table in part (A):

(B) $P(Y \mid H) = \dfrac{P(Y \cap H)}{P(H)} = \dfrac{.400}{.55} \approx .727$

(C) $P(Y \mid B) = \dfrac{P(Y \cap B)}{P(B)} = \dfrac{.02}{.15} \approx .133$

(D) $P(S) = .300$

$$P(S \mid Y) = \frac{P(S \cap Y)}{P(Y)} = \frac{.180}{.60} = .300$$

(E) $P(H) = .550$

$$P(H \mid Y) = \frac{P(H \cap Y)}{P(Y)} = \frac{.400}{.600} \approx .667$$

(F) $P(B \cap N) = .130$

(G)   Yes, S and Y are **independent** since
$$P(S \mid Y) = P(S) = .300$$

(H)   No, H and Y are **dependent** since
$P(H \mid Y) \approx .667$ is not equal to
$P(H) = .550$.

(I) $P(B \mid N) = \dfrac{P(B \cap N)}{P(N)}$

$\qquad = \dfrac{.130}{.400}$ (from table)

$\qquad = .325$

and $P(B) = .150$. Since
$P(B \mid N) \neq P(B)$, B and N are
dependent.

83.   The probability tree for this experiment is:

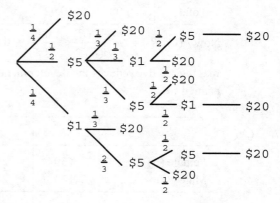

(A) $P(\$26,000) = \left(\dfrac{1}{2}\right)\left(\dfrac{1}{3}\right)\left(\dfrac{1}{2}\right) + \left(\dfrac{1}{4}\right)\left(\dfrac{2}{3}\right)\left(\dfrac{1}{2}\right)$

$\qquad\qquad\qquad = \dfrac{1}{12} + \dfrac{1}{12} = \dfrac{1}{6} \approx 0.167$

(B) $P(\$31,000) = \left(\dfrac{1}{2}\right)\left(\dfrac{1}{3}\right)\left(\dfrac{1}{2}\right) + \left(\dfrac{1}{2}\right)\left(\dfrac{1}{3}\right)\left(\dfrac{1}{2}\right) + \left(\dfrac{1}{4}\right)\left(\dfrac{2}{3}\right)\left(\dfrac{1}{2}\right) = \dfrac{3}{12} = \dfrac{1}{4} = 0.25$

(C)   Let $A = $ "$\$20$ on third draw." Then

$P(A) = \left(\dfrac{1}{4}\right)\left(\dfrac{2}{3}\right)\left(\dfrac{1}{2}\right) + \left(\dfrac{1}{2}\right)\left(\dfrac{1}{3}\right)\left(\dfrac{1}{2}\right) + \left(\dfrac{1}{2}\right)\left(\dfrac{1}{3}\right)\left(\dfrac{1}{2}\right) = \dfrac{3}{12} = \dfrac{1}{4} = 0.25$

85.   (A)

	C	C'	Totals

R	0.06	0.44	0.50
R'	0.02	0.48	0.50
Total	0.08	0.92	1.00

(B) $P(C) = 0.08$, $P(R) = 0.50$, $P(R \cap C) = 0.06$

Since $0.06 = P(R \cap C) \neq P(R) \cdot P(C) = 0.04$, R and C are **dependent.**

(C) $P(C \mid R) = \dfrac{P(C \cap R)}{P(R)} = \dfrac{0.06}{0.50} = 0.12$ and $P(C) = 0.08$

Since $P(C \mid R) > P(C)$, cancer is more likely to be developed if the red dye is used. The FDA should ban the use of the red dye.

(D) The new probability table is

	C	C'	Totals
R	0.02	0.48	0.50
R'	0.06	0.44	0.50
Total	0.08	0.92	1.00

Now $P(C \mid R) = \dfrac{P(C \cap R)}{P(R)} = \dfrac{0.02}{0.5} = 0.04$ and $P(C) = 0.08$.   Since $P(C \mid R) < P(C)$ it appears

that the red dye reduces the development of cancer. Therefore, the use of the dye should not be banned.

87.   (A)

	A	B	C	Total
Female (F)	.130	.286	.104	.520
Male (F')	.120	.264	.096	.480
Total	.250	.550	.200	1.000

[Note: The probability table above was derived from the table given in the problem by dividing each entry by 1000.]

Referring to the table in part (A):

(B) $P(A \mid F) = \dfrac{P(A \cap F)}{P(F)} = \dfrac{.130}{.520} \approx .250$, $P(A \mid F') = \dfrac{P(A \cap F')}{P(F')} = \dfrac{.120}{.480} = .250$

(C) $P(C \mid F) = \dfrac{P(C \cap F)}{P(F)} = \dfrac{.104}{.520} \approx .200$, $P(C \mid F') = \dfrac{P(C \cap F')}{P(F')} = \dfrac{.096}{.480} = .200$

(D) $P(A) = .250$

(E) $P(B) = .550$, $P(B \mid F') = \dfrac{P(B \cap F')}{P(F')} = \dfrac{.264}{.480} = .550$

(F) $P(F \cap C) = .104$

(G)   No, the results in parts (B), (C), (D), and (E) imply that A, B, and C are independent of F and F'.

Things to remember:

1. BAYES' FORMULA

Let U_1, U_2, ..., U_n be n mutually exclusive events whose union is the sample space S. Let

E be an arbitrary event in S such that $P(E) \neq 0$. Then

$$P(U_1|E) = \frac{P(U_1 \cap E)}{P(E)}$$

$$= \frac{P(U_1 \cap E)}{P(U_1 \cap E) + P(U_2 \cap E) + \cdots + P(U_n \cap E)}$$

$$= \frac{P(E \mid U_1)P(U_1)}{P(E \mid U_1)P(U_1) + \cdots + P(E \mid U_n)P(U_n)}$$

Similar results hold for U_2, U_3, ..., U_n.

2. BAYES' FORMULA AND PROBABILITY TREES

$$P(E \mid U_1) = \frac{\text{product of branch probabilities leading to } E \text{ through } U_1}{\text{sum of all branch probabilities leading to } E}$$

Similar results hold for U_2, U_3, ..., U_n.

---

1. $\dfrac{\frac{1}{3}}{\frac{1}{3} + \frac{1}{2}} = \dfrac{\frac{1}{3}}{\frac{5}{6}} = \dfrac{1}{3} \cdot \dfrac{6}{5} = \dfrac{2}{5}$

3. $\dfrac{\frac{1}{3}}{\frac{1}{3}} + \dfrac{1}{2} = 1 + \dfrac{1}{2} = \dfrac{3}{2}$

5. $\dfrac{\frac{4}{5} \cdot \frac{3}{4}}{\frac{1}{5} \cdot \frac{1}{3} + \frac{4}{5} \cdot \frac{3}{4}} = \dfrac{\frac{3}{5}}{\frac{1}{15} + \frac{3}{5}} = \dfrac{\frac{3}{5}}{\frac{10}{15}} = \dfrac{3}{5} \cdot \dfrac{15}{10} = \dfrac{9}{10}$

7. $P(M \cap A) = P(M) \cdot P(A \mid M) = (.6)(.8) = .48$

9. $P(A) = P(M \cap A) + P(N \cap A) = P(M)P(A \mid M) + P(N)P(A \mid N) = (.6)(.8) + (.4)(.3) = .60$

11. $P(M \mid A) = \dfrac{P(M \cap A)}{P(M \cap A) + P(N \cap A)} = \dfrac{.48}{.60} = \dfrac{4}{5} = .80$

13. Referring to the Venn diagram:

$$P(U_1 \mid R) = \frac{P(U_1 \cap R)}{P(R)} = \frac{\frac{25}{100}}{\frac{60}{100}} = \frac{25}{60} = \frac{5}{12} \approx .417$$

Using Bayes' formula:

$$P(U_1 \mid R) = \frac{P(U_1 \cap R)}{P(U_1 \cap R) + P(U_2 \cap R)} = \frac{P(U_1)P(R \mid U_1)}{P(U_1)P(R \mid U_1) + P(U_2)P(R \mid U_2)}$$

$$= \frac{\left(\frac{40}{100}\right)\left(\frac{25}{40}\right)}{\left(\frac{40}{100}\right)\left(\frac{25}{40}\right) + \left(\frac{60}{100}\right)\left(\frac{35}{60}\right)} = \frac{.25}{.25 + .35} = \frac{.25}{.60} = \frac{5}{12} \approx .417$$

15. $P(U_1 \mid R') = \dfrac{P(U_1 \cap R')}{P(R')} = \dfrac{\dfrac{15}{100}}{1 - P(R)}$ (from the Venn diagram)

$$= \frac{\dfrac{15}{100}}{1 - \dfrac{60}{100}} = \frac{\dfrac{15}{100}}{\dfrac{40}{100}} = \frac{3}{8} = .375$$

Using Bayes' formula:

$$P(U_1 \mid R') = \frac{P(U_1 \cap R')}{P(R')} = \frac{P(U_1)P(R' \mid U_1)}{P(U_1 \cap R') + P(U_2 \cap R')}$$

$$= \frac{P(U_1)P(R' \mid U_1)}{P(U_1)P(R' \mid U_1) + P(U_2)P(R' \mid U_2)} = \frac{\left(\dfrac{40}{100}\right)\left(\dfrac{15}{40}\right)}{\left(\dfrac{40}{100}\right)\left(\dfrac{15}{40}\right) + \left(\dfrac{60}{100}\right)\left(\dfrac{25}{60}\right)}$$

$$= \frac{.15}{.15 + .25} = \frac{15}{40} = \frac{3}{8} = .375$$

17. $P(U \mid C) = \dfrac{P(U \cap C)}{P(C)} = \dfrac{P(U \cap C)}{P(U \cap C) + P(V \cap C) + P(W \cap C)}$

$$= \frac{(.1)(.4)}{(.1)(.4) + (.6)(.2) + (.3)(.7)} = \frac{.04}{.37} \approx .108$$

[Note: Recall $P(A \cap B)$ $= P(A) \cdot P(B \mid A)$.]

19. $P(W \mid C) = \dfrac{P(W \cap C)}{P(C)} = \dfrac{P(W \cap C)}{P(U \cap C) + P(V \cap C) + P(W \cap C)}$

$$= \frac{(.3)(.7)}{(.1)(.4) + (.6)(.2) + (.3)(.7)} = \frac{.21}{.37} \approx .568$$

21. $P(V \mid C) = \dfrac{P(V \cap C)}{P(C)} = \dfrac{P(V \cap C)}{P(U \cap C) + P(V \cap C) + P(W \cap C)}$

$$= \frac{(.6)(.2)}{(.1)(.4) + (.6)(.2) + (.3)(.7)} = \frac{.12}{.37} \approx .324$$

23. From the Venn diagram,

$$P(U_1 \mid R) = \frac{5}{5 + 15 + 20} = \frac{5}{40} = \frac{1}{8} = .125$$

or

$$= \frac{P(U_1 \cap R)}{P(R)} = \frac{\dfrac{5}{100}}{\dfrac{40}{100}} = .125$$

Using Bayes' formula:

$$P(U_1 \mid R) = \frac{P(U_1 \cap R)}{P(U_1 \cap R) + P(U_2 \cap R) + P(U_3 \cap R)} = \frac{\dfrac{5}{100}}{\dfrac{5}{100} + \dfrac{15}{100} + \dfrac{20}{100}}$$

$$= \frac{.05}{.05 + .15 + .2} = \frac{.05}{.40} = .125$$

25. From the Venn diagram,

$$P(U_3 \mid R) = \frac{20}{5 + 15 + 20} = \frac{20}{40} = .5$$

Using Bayes' formula:

$$P(U_3 \mid R) = \frac{P(U_3 \cap R)}{P(U_1 \cap R) + P(U_2 \cap R) + P(U_3 \cap R)} = \frac{\dfrac{20}{100}}{\dfrac{5}{100} + \dfrac{15}{100} + \dfrac{20}{100}}$$

$$= \frac{.2}{.05 + .15 + .2} = \frac{.2}{.4} = .5$$

27. From the Venn diagram,

$$P(U_2 \mid R) = \frac{15}{5 + 15 + 20} = \frac{15}{40} = .375$$

Using Bayes' formula:

$$P(U_2 \mid R) = \frac{P(U_2 \cap R)}{P(U_1 \cap R) + P(U_2 \cap R) + P(U_3 \cap R)} = \frac{\dfrac{15}{100}}{\dfrac{5}{100} + \dfrac{15}{100} + \dfrac{20}{100}} = \frac{.15}{.05 + .15 + .2} = \frac{.15}{.40} = .375$$

29. From the given tree diagram, we have:

$$P(A) = \frac{1}{4} \qquad\qquad P(A') = \frac{3}{4}$$

$$P(B \mid A) = \frac{1}{5} \qquad\qquad P(B \mid A') = \frac{3}{5}$$

$$P(B' \mid A) = \frac{4}{5} \qquad\qquad P(B' \mid A') = \frac{2}{5}$$

We want to find the following:

$$P(B) = P(B \cap A) + P(B \cap A') = P(A)P(B \mid A) + P(A')P(B \mid A')$$

$$= \left(\frac{1}{4}\right)\left(\frac{1}{5}\right) + \left(\frac{3}{4}\right)\left(\frac{3}{5}\right) = \frac{1}{20} + \frac{9}{20} = \frac{10}{20} = \frac{1}{2}$$

$$P(B') = 1 - P(B) = 1 - \frac{1}{2} = \frac{1}{2}$$

$$P(A \mid B) = \frac{P(A \cap B)}{P(B)} = \frac{P(A)P(B \mid A)}{P(B)} = \frac{\left(\frac{1}{4}\right)\left(\frac{1}{5}\right)}{\frac{1}{2}} = \frac{\frac{1}{20}}{\frac{1}{2}} = \frac{1}{10}$$

Thus, $P(A' \mid B) = 1 - P(A \mid B) = 1 - \frac{1}{10} = \frac{9}{10}$.

$$P(A \mid B') = \frac{P(A \cap B')}{P(B')} = \frac{P(A)P(B' \mid A)}{P(B')}$$

$$= \frac{\left(\frac{1}{4}\right)\left(\frac{4}{5}\right)}{\frac{1}{2}} = \frac{\frac{4}{20}}{\frac{1}{2}} = \frac{2}{5}$$

Thus, $P(A' \mid B') = 1 - P(A \mid B') = 1 - \frac{2}{5} = \frac{3}{5}$.

Therefore, the tree diagram for this problem
is as shown at the right.

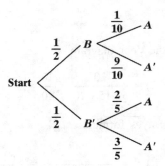

The following tree diagram is to be used for Problems 31 and 33.

```
                            1
                            5 = .2   W (white)
                    U₁ (urn 1) <
            .5                  4
                                5 = .8   R (red)
    Start <
                                3
                                5 = .6   W
            .5      U₂ (urn 2) <
                                2
                                5 = .4   R
```

31. $P(U_1 \mid W) = \dfrac{P(U_1 \cap W)}{P(W)} = \dfrac{P(U_1 \cap W)}{P(U_1 \cap W) + P(U_2 \cap W)}$

$$= \frac{P(U_1)P(W \mid U_1)}{P(U_1)P(W \mid U_1) + P(U_2)P(W \mid U_2)} = \frac{(.5)(.2)}{(.5)(.2) + (.5)(.6)} = \frac{.1}{.4} = .25$$

33. $P(U_2 \mid R) = \dfrac{P(U_2 \cap R)}{P(R)} = \dfrac{P(U_2 \cap R)}{P(U_2 \cap R) + P(U_1 \cap R)}$

$$= \frac{P(U_2)P(R \mid U_2)}{P(U_2)P(R \mid U_2) + P(U_1)P(R \mid U_1)} = \frac{(.5)(.4)}{(.5)(.4) + (.5)(.8)} = \frac{.4}{1.2} = \frac{1}{3} \approx .333$$

35. $P(W_1 \mid W_2) = \dfrac{P(W_1 \cap W_2)}{P(W_2)} = \dfrac{P(W_1)P(W_2 \mid W_1)}{P(R_1 \cap W_2) + P(W_1 \cap W_2)}$

$$= \frac{P(W_1)P(W_2 \mid W_1)}{P(R_1)P(W_2 \mid R_1) + P(W_1)P(W_2 \mid W_1)} = \frac{\left(\frac{5}{9}\right)\left(\frac{4}{8}\right)}{\left(\frac{4}{9}\right)\left(\frac{5}{8}\right) + \left(\frac{5}{9}\right)\left(\frac{4}{8}\right)} = \frac{\frac{20}{72}}{\frac{20}{72} + \frac{20}{72}}$$

$$= \frac{20}{40} = \frac{1}{2} \text{ or } .5$$

37. $P(U_{R_1} | U_{R_2}) = \dfrac{P(U_{R_1} \cap U_{R_2})}{P(U_{R_2})} = \dfrac{P(U_{R_1}) P(U_{R_2} | U_{R_1})}{P(U_{W_1} \cap U_{R_2}) + P(U_{R_1} \cap U_{R_2})}$

$\qquad\qquad\qquad = \dfrac{P(U_{R_1}) P(U_{R_2} | U_{R_1})}{P(U_{W_1}) P(U_{R_2} | U_{W_1}) + P(U_{R_1}) P(U_{R_2} | U_{R_1})}$

$\qquad\qquad\qquad = \dfrac{\left(\dfrac{7}{10}\right)\left(\dfrac{5}{10}\right)}{\left(\dfrac{3}{10}\right)\left(\dfrac{4}{10}\right) + \left(\dfrac{7}{10}\right)\left(\dfrac{5}{10}\right)} = \dfrac{.35}{.12 + .35} = \dfrac{.35}{.47} = \dfrac{35}{47} \approx .745$

The tree diagram follows:

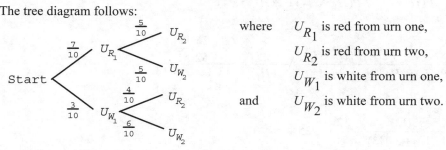

where U_{R_1} is red from urn one,

$\qquad\quad U_{R_2}$ is red from urn two,

$\qquad\quad U_{W_1}$ is white from urn one,

and U_{W_2} is white from urn two.

39. Suppose $c = e$. Then
$\quad P(M) = ac + be = ac + bc = c(a + b) = c \quad (a + b = 1)$
and

$\quad P(M | U) = \dfrac{P(M \cap U)}{P(U)} = \dfrac{ac}{a} = c$

Therefore, M and U are independent.
Alternatively, note that
$\quad P(M) = c, P(U) = a \quad$ and $\quad P(M \cap U) = ac = P(M) \cdot P(U)$,
which implies that M and U are independent.

41. Draw a tree diagram to verify the probabilities given below.

(A)   With replacement--True.

$\quad P(B_2 | B_1) = \dfrac{m}{m + n}$

$\quad P(B_1 | B_2) = \dfrac{P(B_1 \cap B_2)}{P(B_1 \cap B_2) + P(W_1 \cap B_2)}$

$\qquad\qquad\quad = \dfrac{\dfrac{m^2}{(m+n)^2}}{\dfrac{m^2}{(m+n)^2} + \dfrac{mn}{(m+n)^2}} = \dfrac{m^2}{m^2 + mn} = \dfrac{m}{m+n} = P(B_2 | B_1)$

(B)   Without replacement--True.

$\quad P(B_2 | B_1) = \dfrac{m - 1}{m + n - 1}$

$\quad P(B_1 | B_2) = \dfrac{P(B_1 \cap B_2)}{P(B_1 \cap B_2) + P(W_1 \cap B_2)} = \dfrac{\dfrac{m(m-1)}{(m+n)(m+n-1)}}{\dfrac{m(m-1)}{(m+n)(m+n-1)} + \dfrac{nm}{(m+n)(m+n-1)}}$

$\qquad\qquad\quad = \dfrac{m(m-1)}{m(m-1) + mn} = \dfrac{m-1}{m+n-1} = P(B_2 | B_1)$

43.

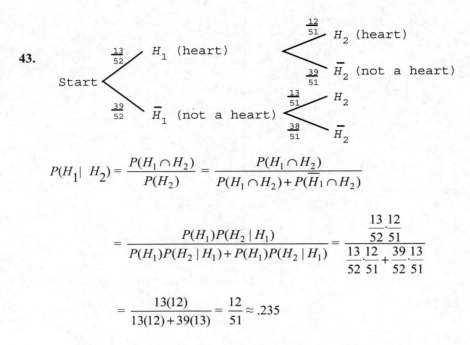

$$P(H_1 \mid H_2) = \frac{P(H_1 \cap H_2)}{P(H_2)} = \frac{P(H_1 \cap H_2)}{P(H_1 \cap H_2) + P(\overline{H_1} \cap H_2)}$$

$$= \frac{P(H_1)P(H_2 \mid H_1)}{P(H_1)P(H_2 \mid H_1) + P(H_1)P(H_2 \mid H_1)} = \frac{\dfrac{13}{52} \cdot \dfrac{12}{51}}{\dfrac{13}{52} \cdot \dfrac{12}{51} + \dfrac{39}{52} \cdot \dfrac{13}{51}}$$

$$= \frac{13(12)}{13(12) + 39(13)} = \frac{12}{51} \approx .235$$

For the 3-card hand in Problems 45 – 50 , let

E_3 = all three cards are clubs

E_2 = only two of the cards are clubs

E_1 = only one of the cards is a club

E_0 = none of the cards is a club

Let E = card chosen is a club.

The probabilities are:

$$P(E_3) = \frac{C_{13,3}}{C_{52,3}} = \frac{286}{22,100} \approx 0.0129$$

$$P(E_2) = \frac{C_{13,2} \cdot C_{39,1}}{C_{52,3}} = \frac{78 \cdot 39}{22,100} \approx 0.1377$$

$$P(E_1) = \frac{13 \cdot C_{39,2}}{C_{52,3}} = \frac{13 \cdot 741}{22,100} \approx 0.4359$$

$$P(E_0) = \frac{C_{39,3}}{C_{52,3}} = \frac{9,139}{22,100} \approx 0.4135$$

$$P(E \mid E_3) = 1, \quad P(E \mid E_2) = \frac{2}{3}, \quad P(E \mid E_1) = \frac{1}{3}, \quad P(E \mid E_0) = 0$$

$$P(E) = P(E_3) \cdot P(E \mid E_3) + P(E_2) \cdot P(E \mid E_2) + P(E_1) \cdot P(E \mid E_1)$$

$$= 0.0129(1) + 0.1377\left(\frac{2}{3}\right) + 0.4359\left(\frac{1}{3}\right) \approx 0.2500$$

45. $P(E \mid E_1) = \dfrac{1}{3}$

47. $P(E_1 \mid E) = \dfrac{P(E_1 \cap E)}{P(E)} = \dfrac{P(E_1) \cdot P(E \mid E_1)}{P(E)} = \dfrac{0.4359(1/3)}{0.2500} \approx 0.581$

49. $P(E_3 \mid E) = \dfrac{P(E_3 \cap E)}{P(E)} = \dfrac{P(E_3) \cdot P(E \mid E_3)}{P(E)} = \dfrac{0.0129(1)}{0.2500} \approx 0.052$

51. Consider the following Venn diagram:

$P(U_1 \mid R) = \dfrac{P(U_1 \cap R)}{P(U_1 \cap R) + P(U_1' \cap R)}$

and

$P(U_1' \mid R) = \dfrac{P(U_1' \cap R)}{P(U_1 \cap R) + P(U_1' \cap R)}$

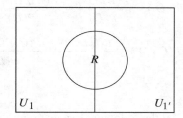

Adding these two equations, we obtain:

$P(U_1 \mid R) + P(U_1' \mid R) = \dfrac{P(U_1 \cap R)}{P(U_1 \cap R) + P(U_1' \cap R)} + \dfrac{P(U_1' \cap R)}{P(U_1 \cap R) + P(U_1' \cap R)}$

$= \dfrac{P(U_1 \cap R) + P(U_1' \cap R)}{P(U_1 \cap R) + P(U_1' \cap R)} = 1$

53. Consider the following tree diagram:

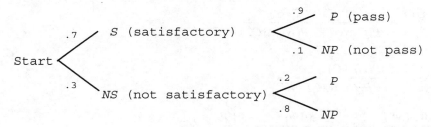

$P(S \mid P) = \dfrac{P(S \cap P)}{P(S \cap P) + P(NS \cap P)} = \dfrac{P(S)P(P \mid S)}{P(S)P(P \mid S) + P(NS)P(P \mid NS)}$

$= \dfrac{(.7)(.9)}{(.7)(.9) + (.3)(.2)} = \dfrac{.63}{.69} \approx .913$

$P(S \mid NP) = \dfrac{P(S \cap NP)}{P(NP)} = \dfrac{P(S \cap NP)}{P(S \cap NP) + P(NS \cap NP)} = \dfrac{(.7)(.1)}{(.7)(.1) + (.3)(.8)} = \dfrac{.07}{.31} \approx .226$

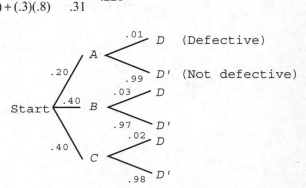

55. Consider the following tree diagram:

$P(A \mid D) = \dfrac{P(A \cap D)}{P(D)}$, where

$$P(D) = P(A \cap D) + P(B \cap D) + P(C \cap D)$$
$$= P(A)P(D \mid A) + P(B)P(D \mid B) + P(C)P(D \mid C)$$
$$= (.2)(.01) + (.40)(.03) + (.40)(.02) = .002 + .012 + .008 = .022$$

Thus, $P(A \mid D) = \dfrac{P(A \cap D)}{P(D)} = \dfrac{P(A)P(D \mid A)}{P(D)} = \dfrac{(.20)(.01)}{.022} = \dfrac{.002}{.022} = \dfrac{2}{22}$ or .091

Similarly,

$$P(B \mid D) = \dfrac{P(B \cap D)}{P(D)} = \dfrac{P(B)P(D \mid B)}{P(D)} = \dfrac{(.40)(.03)}{.022} = \dfrac{.012}{.022} = \dfrac{6}{11} \text{ or } .545,$$

and $P(C \mid D) = \dfrac{P(C \cap D)}{P(D)} = \dfrac{P(C)P(D \mid C)}{P(D)} = \dfrac{(.40)(.02)}{.022} = \dfrac{.008}{.022} = \dfrac{4}{11}$ or .364.

57. Consider the following tree diagram:

```
                                      .98   CT   (Cancer by new test)
                  .02    C (Cancer)  <
                  /                   .02   NCT  (No cancer by new test)
        Start  <
                  \      .01
                  .98   <      CT
                NC (No cancer) 
                      .99   NCT
```

$$P(C \mid CT) = \frac{P(C \cap CT)}{P(CT)} = \frac{P(C)P(CT \mid C)}{P(C \cap CT) + P(NC \cap CT)}$$

$$= \frac{P(C)P(CT \mid C)}{P(C)P(CT \mid C) + P(NC)P(CT \mid NC)}$$

$$= \frac{(.02)(.98)}{(.02)(.98) + (.98)(.01)} = \frac{.0196}{.0196 + .0098} = \frac{.0196}{.0294} = .667$$

$$P(C \mid NCT) = \frac{P(C \cap NCT)}{P(NCT)} = \frac{P(C)P(NCT \mid C)}{P(C)P(NCT \mid C) + P(NC)P(NCT \mid NC)}$$

$$= \frac{(.02)(.02)}{(.02)(.02) + (.98)(.99)} \approx .000412$$

59. Consider the following tree diagram.

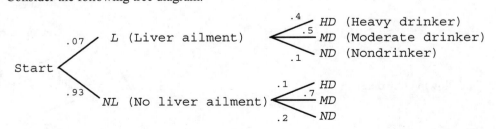

$$P(L \mid HD) = \frac{P(L \cap HD)}{P(HD)} = \frac{P(L)P(HD \mid L)}{P(L \cap HD) + P(NL \cap HD)}$$

$$= \frac{P(L)P(HD \mid L)}{P(L)P(HD \mid L) + P(NL)P(HD \mid NL)}$$

$$= \frac{(.07)(.4)}{(.07)(.4) + (.93)(.1)} \quad \text{(from the tree diagram)}$$

$$= \frac{.028}{.028 + .093} = \frac{.028}{.121} = \frac{28}{121} \approx .231$$

$$P(L \mid ND) = \frac{P(L \cap ND)}{P(ND)} = \frac{P(L)P(ND \mid L)}{P(L \cap ND) + P(NL \cap ND)} = \frac{P(L)P(ND \mid L)}{P(L)P(ND \mid L) + P(NL)P(ND \mid NL)}$$

$$= \frac{(.07)(.1)}{(.07)(.1) + (.93)(.2)} \quad \text{(from the tree diagram)}$$

$$= \frac{.007}{.007 + .186} = \frac{.007}{.193} = \frac{7}{193} \approx .036$$

61.   Consider the following tree diagram.

$$P(L \mid LT) = \frac{P(L \cap LT)}{P(LT)} = \frac{P(L \cap LT)}{P(L \cap LT) + P(\bar{L} \cap LT)} = \frac{(.5)(.8)}{(.5)(.8) + (.5)(.05)}$$

$$= \frac{.4}{.425} \approx .941 \qquad \text{If the test indicates that the subject was lying, then he was lying with a probability of 0.941.}$$

$$P(\bar{L} \mid LT) = \frac{P(\bar{L} \cap LT)}{P(LT)} = \frac{(.5)(.05)}{(.5)(.8) + (.5)(.05)}$$

$$= \frac{.05}{.85} \approx .0588 \qquad \text{If the test indicates that the subject was lying, there is still a probability of 0.0588 that he was not lying.}$$

EXERCISE 8-5

Things to remember:

<u>1.</u>   RANDOM VARIABLE

A RANDOM VARIABLE is a function that assigns a numerical value to each simple event in a sample space *S*.

2. PROBABILITY DISTRIBUTION OF A RANDOM VARIABLE X

   The PROBABILITY DISTRIBUTION OF A RANDOM VARIABLE X, denoted

   $$P(X = x) = p(x),$$

   satisfies

   (a) $0 \leq p(x) \leq 1, x \in \{x_1, x_2, ..., x_n\}$,

   (b) $p(x_1) + p(x_2) + \cdots + p(x_n) = 1$,

   where $\{x_1, x_2, ..., x_n\}$ are the (range) values of X.

3. EXPECTED VALUE OF A RANDOM VARIABLE X

   Given the probability distribution for the random variable X:

X_i	x_1	x_2	...	x_n
P_i	p_1	p_2	...	p_n

   where $p_i = p(x_i)$.

   The expected value of X, denoted by $E(X)$, is given by the formula:

   $$E(X) = x_1 p_1 + x_2 p_2 + \cdots + x_m \ p_m$$

4. STEPS FOR COMPUTING THE EXPECTED VALUE OF A RANDOM VARIABLE X.

   Step 1. Form the probability distribution for the random variable X.

   Step 2. Multiply each image value of X, x_i, by its corresponding probability of occurrence, p_i, then add the results.

---

1. Average $= \dfrac{73 + 89 + 45 + 82 + 66}{5} = \dfrac{355}{5} = 71$

3. Average $= \dfrac{77 + 93 + 49 + 86 + 70}{5} = \dfrac{375}{5} = 75$

5. Average $= \dfrac{146 + 178 + 90 + 164 + 132}{5} = \dfrac{710}{5} = 142$

7. Expected value of X: $E(X) = -3(.3) + 0(.5) + 4(.2) = -0.1$

9. Expected value of X: $E(X) = 5(.25) + 20(.25) + 50(.25) + 100(.25) = 43.75$; $\$43.75$

11. There are 50 coins in the bowl. Expected value of X:

    $$E(X) = (.01)\left(\frac{15}{50}\right) + (.1)\left(\frac{10}{50}\right) + (.25)\left(\frac{25}{50}\right) = \frac{7.4}{50} = .148$$

13. R = red card. $P(R) = \dfrac{25}{50} = .5$. $E(X) = (.5)(50) + .5(0) = .25$; expected value: $\$25$

15. Assign the number 0 to the event of observing zero heads, the number 1 to the event of observing one head, and the number 2 to the event of observing two heads. The probability distribution for X, then, is:

x_i	0	1	2
p_i	$\frac{1}{4}$	$\frac{1}{2}$	$\frac{1}{4}$

[Note: One head can occur two ways out of a total of four different ways (HT, TH).]

Hence, $E(X) = 0 \cdot \dfrac{1}{4} + 1 \cdot \dfrac{1}{2} + 2 \cdot \dfrac{1}{4} = 1.$

17. Assign a payoff of $1 to the event of observing a head and –$1 to the event of observing a tail. Thus, the probability distribution for X is:

x_i	1	−1
p_i	$\frac{1}{2}$	$\frac{1}{2}$

Hence, $E(X) = 1 \cdot \dfrac{1}{2} + (-1) \cdot \dfrac{1}{2} = 0.$ The game is fair.

19. The table shows a payoff or probability distribution for the game.

Net gain	x_i	− 3	− 2	− 1	0	1	2
	p_i	$\frac{1}{6}$	$\frac{1}{6}$	$\frac{1}{6}$	$\frac{1}{6}$	$\frac{1}{6}$	$\frac{1}{6}$

[Note: A payoff valued at –$3 is assigned to the event of observing a "1" on the die, resulting in a net gain of –$3, and so on.]

Hence, $E(X) = -3 \cdot \dfrac{1}{6} - 2 \cdot \dfrac{1}{6} - 1 \cdot \dfrac{1}{6} + 0 \cdot \dfrac{1}{6} + 1 \cdot \dfrac{1}{6} + 2 \cdot \dfrac{1}{6} = -\dfrac{1}{2}$ or –$0.50.
The game is not fair.

21. The probability distribution is:

Number of heads	Gain, x_i	Probability, p_i
0	2	$\frac{1}{4}$
1	−3	$\frac{1}{2}$
2	2	$\frac{1}{4}$

The expected value is:

$E(X) = 2 \cdot \dfrac{1}{4} + (-3) \cdot \dfrac{1}{2} + 2 \cdot \dfrac{1}{4} = 1 - \dfrac{3}{2} = -\dfrac{1}{2}$ or –$0.50.

23. In 4 rolls of a die, the total number of possible outcomes is $6 \cdot 6 \cdot 6 \cdot 6 = 6^4$. Thus, $n(S) = 6^4 = 1296$. The total number of outcomes that contain *no* 6's is $5 \cdot 5 \cdot 5 \cdot 5 = 5^4$. Thus, if E is the event "At least one 6," then $n(E) = 6^4 - 5^4 = 671$ and

$P(E) = \dfrac{n(E)}{n(S)} = \dfrac{671}{1296} \approx 0.5177.$

First, we compute the expected value to you.

The payoff table is:

x_i	−$1	$1
P_i	0.5177	0.4823

The expected value to you is:

$E(X) = (-1)(0.5177) + 1(0.4823) = -0.0354$ or −$0.035

The expected value to her is:

$E(X) = 1(0.5177) + (-1)(0.4823) = 0.035$ or $0.035

25. Let x = amount you should lose if a 6 turns up.

The payoff table is:

	1	2	3	4	5	6
x_i	$5	$5	$10	$10	$10	$x
P_i	$\frac{1}{6}$	$\frac{1}{6}$	$\frac{1}{6}$	$\frac{1}{6}$	$\frac{1}{6}$	$\frac{1}{6}$

Now $E(X) = 5\left(\dfrac{1}{6}\right) + 5\left(\dfrac{1}{6}\right) + 10\left(\dfrac{1}{6}\right) + 10\left(\dfrac{1}{6}\right) + 10\left(\dfrac{1}{6}\right) + x\left(\dfrac{1}{6}\right) = \dfrac{40}{6} + \dfrac{x}{6}$

The game is fair if and only if $E(X) = 0$:

solving $\dfrac{40}{6} + \dfrac{x}{6} = 0$

gives $x = -40$

Thus, you should **lose** $40 for the game to be fair.

27. $P(\text{sum} = 7) = \dfrac{6}{36} = \dfrac{1}{6}$

$P(\text{sum} = 11 \text{ or } 12) = P(\text{sum} = 11) + P(\text{sum} = 12) = \dfrac{2}{36} + \dfrac{1}{36} = \dfrac{3}{36} = \dfrac{1}{12}$

$P(\text{sum other than 7, 11, or 12}) = 1 - P(\text{sum} = 7, 11, \text{ or } 12) = 1 - \dfrac{9}{36} = \dfrac{27}{36} = \dfrac{3}{4}$

Let x_1 = sum is 7, x_2 = sum is 11 or 12, x_3 = sum is not 7, 11, or 12, and let t denote the amount you "win" if x_3 occurs. Then the payoff table is:

x_i	−$10	$11	t
p_i	$\frac{1}{6}$	$\frac{1}{12}$	$\frac{3}{4}$

The expected value is:

$E(X) = -10\left(\dfrac{1}{6}\right) + 11\left(\dfrac{1}{12}\right) + t\left(\dfrac{3}{4}\right) = \dfrac{-10}{6} + \dfrac{11}{12} + \dfrac{3t}{4}$

The game is fair if $E(X) = 0$, i.e., if

$\dfrac{-10}{6} + \dfrac{11}{12} + \dfrac{3}{4}t = 0$ or $\dfrac{3}{4}t = \dfrac{10}{6} - \dfrac{11}{12} = \dfrac{20}{12} - \dfrac{11}{12} = \dfrac{9}{12} = \dfrac{3}{4}$

Therefore, $t = 1.

29. Let K = card is a King. Then $P(K) = \dfrac{4}{52} = \dfrac{1}{13}$ and $P(K') = \dfrac{12}{13}$.

The payoff table is:

	K	K'
x_i	\$10	$-\$1$
p_i	$\frac{1}{13}$	$\frac{12}{13}$

Now, $E(X) = 10\left(\dfrac{1}{13}\right) - 1\left(\dfrac{12}{13}\right) = -\dfrac{2}{13} = -\$0.154$

31. Let K = hand contains at least one King. Then
K' = hand contains no Kings. The probabilities are:

$P(K') = \dfrac{C_{48,5}}{C_{52,5}} \approx 0.65884$

$P(K) = 1 - P(K') \approx 0.34116$
The payoff table is:

	K	K'
x_i	\$10	$-\$1$
p_i	0.34116	0.65884

Now, $E(X) = 10(0.34116) - 1(0.65884) \approx \$2.75$

33. Course A_1: $E(X) = (-200)(.1) + 100(.2) + 400(.4) + 100(.3)$
$$= -20 + 20 + 160 + 30$$
$$= \$190$$

Course A_2: $E(X) = (-100)(.1) + 200(.2) + 300(.4) + 200(.3)$
$$= -10 + 40 + 120 + 60$$
$$= \$210$$

A_2 will produce the largest expected value, and that value is \$210.

35. The probability of winning \$35 is $\dfrac{1}{38}$ and the probability of losing \$1 is $\dfrac{37}{38}$. Thus, the payoff table is:

x_i	\$35	$-\$1$
p_i	$\frac{1}{38}$	$\frac{37}{38}$

The expected value of the game is:

$$E(X) = 35\left(\frac{1}{38}\right) + (-1)\left(\frac{37}{38}\right) = \frac{35-37}{38} = \frac{-1}{19} \approx -0.0526 \quad \text{or} \quad -5.26¢$$

37. Let p = probability of winning. Then $1 - p$ is the probability of losing and the payoff table is:

	W	L
x_i	99,900	-100
p_i	p	$1-p$

Since $E(X) = 100$, we have
$$99,900(p) - 100(1 - p) = 100$$
$$99,900p - 100 + 100p = 100$$
$$100,000p = 200$$
$$p = 0.002$$

The probability of winning is 0.002. Since the expected value is positive, you should play the game. In the long run, you will win \$100 per game.

39.

p_i		x_i
$\dfrac{1}{5000}$	chance of winning	$499
$\dfrac{3}{5000}$	chance of winning	$99
$\dfrac{5}{5000}$	chance of winning	$19
$\dfrac{20}{5000}$	chance of winning	$4
$\dfrac{4971}{5000}$	chance of losing	$1 [<u>Note</u>: $5000 - (1 + 3 + 5 + 20) = 4971$.]

The payoff table is:

x_i	$499	$99	$19	$4	−$1
P_i	0.0002	0.0006	0.001	0.004	0.9942

Thus,

$E(X) = 499(0.0002) + 99(0.0006) + 19(0.001) + 4(0.004) - 1(0.9942) = -0.80$

or $E(X) = -\$0.80$ or $-80¢$

41.   (A)   Total number of simple events $= n(S) = C_{10,2} = \dfrac{10!}{2!(10-2)!} = \dfrac{10!}{2!8!} = \dfrac{10\cdot 9}{2} = 45$

$P(\text{zero defective}) = P(0) = \dfrac{C_{7,2}}{45}$ [<u>Note</u>: None defective means 2 selected from 7 nondefective.]

$= \dfrac{\dfrac{7!}{2!5!}}{45} = \dfrac{21}{45} = \dfrac{7}{15}$

$P(\text{one defective}) = P(1) = \dfrac{C_{3,1}\cdot C_{7,1}}{45} = \dfrac{21}{45} = \dfrac{7}{15}$

$P(\text{two defective}) = P(2) = \dfrac{C_{3,2}}{45}$ [<u>Note</u>: Two defectives selected from 3 defectives.]

$= \dfrac{3}{45} = \dfrac{1}{15}$

The probability distribution is as follows:

x_i	0	1	2
P_i	$\frac{7}{15}$	$\frac{7}{15}$	$\frac{1}{15}$

(B) $E(X) = 0\left(\dfrac{7}{15}\right) + 1\left(\dfrac{7}{15}\right) + 2\left(\dfrac{1}{15}\right) = \dfrac{9}{15} = \dfrac{3}{5} = 0.6$

43.   (A)   The total number of simple events $= n(S) = C_{1000,5}$.

$P(\text{0 winning tickets}) = P(0) = \dfrac{C_{997,5}}{C_{1000,5}}$

$= \dfrac{997\cdot 996\cdot 995\cdot 994\cdot 993}{1000\cdot 999\cdot 998\cdot 997\cdot 996} \approx 0.985$

$$P(\text{1 winning ticket}) = P(1) = \frac{C_{3,1} \cdot C_{997,4}}{C_{1000,5}} = \frac{3 \cdot \dfrac{997!}{4!(993)!}}{\dfrac{1000!}{5!(995)!}} \approx 0.0149$$

$$P(\text{2 winning tickets}) = P(2) = \frac{C_{3,2} \cdot C_{997,3}}{C_{1000,5}} = \frac{3 \cdot \dfrac{997!}{3!(994)!}}{\dfrac{1000!}{5!(995)!}} \approx 0.0000599$$

$$P(\text{3 winning tickets}) = P(3) = \frac{C_{3,3} \cdot C_{997,2}}{C_{1000,5}} = \frac{1 \cdot \dfrac{997!}{2!(995)!}}{\dfrac{1000!}{5!(995)!}} \approx 0.00000006$$

The payoff table is as follows:

x_i	−$5	$195	$395	$595
P_i	0.985	0.0149	0.0000599	0.00000006

(B)   The expected value to you is:

$$E(X) = (-5)(0.985) + 195(0.0149) + 395(0.0000599) + 595(0.00000006) \approx -\$2.00$$

45.   (A)   From the statistical plot, the number 13 came up 3 times in 200 games. If $1 was bet on 13 in each of the 200 games, then the result is:

$$3(35) - 1(197) = 105 - 197 = -\$92$$

(B)   Based on the simulation, the value per game is: $-\dfrac{92}{200} = -\$0.46$;   you lose 46 cents per game. The (theoretical) expected value of the game is:

$$E(X) = \frac{1}{38}(35) + \frac{37}{38}(-1) = \frac{35}{38} - \frac{37}{38} = -\frac{1}{19} \approx -\$0.0526;$$

You will lose 5 cents per game.

(C)   The simulated gain or loss depends on the results of your simulation. From (B), the expected loss is: $\dfrac{1}{19}(500) \approx \$26.32$.

47.   Let D_i = exactly i diamonds in the 3-card hand, $i = 0, 1, 2, 3$. The probabilities are

$$P(D_0) = \frac{{}_{39}C_3}{{}_{52}C_3} \approx 0.4135, \qquad P(D_1) = \frac{13 \cdot {}_{39}C_2}{{}_{52}C_3} \approx 0.4359$$

$$P(D_2) = \frac{{}_{13}C_2 \cdot 39}{{}_{52}C_3} \approx 0.1377, \qquad P(D_3) = \frac{{}_{13}C_3}{{}_{52}C_3} \approx 0.0129$$

Let x be the amount you should lose. The payoff table is:

x_i	$-x$	20	40	60
P_i	0.4135	0.4359	0.1377	0.0129

The expected value is
$$E(X) = -x(0.4135) + 20(0.4359) + 40(0.1377) + 60(0.0129)$$
$$= -x(0.4135) + 15$$

If the game is fair, $-x(0.4135) + 15 = 0$; $x = \dfrac{15}{0.4135} = \$36.27$.

49. The payoff table is as follows:

Gain

x_i	\$4850	$-\$150$
p_i	0.01	0.99

[Note: $5000 - 150 = 4850$, the gain with probability of 0.01 if stolen.]

Hence, $E(X) = 4850(0.01) - 150(0.99) = -\$100$

51. The payoff table for site A is as follows:

x_i	30 million	-3 million
p_i	0.2	0.8

Hence $E(X) = 30(0.2) - 3(0.8)$
$= 6 - 2.4$
$= \$3.6$ million

The payoff table for site B is as follows:

x_i	70 million	-4 million
p_i	0.1	0.9

Hence, $E(X) = 70(0.1) + (-4)(0.9)$
$= 7 - 3.6$
$= \$3.4$ million

The company should choose site A with $E(X) = \$3.6$ million.

53. Using **4**, $E(X) = 0(0.12) + 1(0.36) + 2(0.38) + 3(0.14) = 1.54$

55. Action A_1: $E(X) = 10(0.3) + 5(0.2) + 0(0.5) = \$4.00$

Action A_2: $E(X) = 15(0.3) + 3(0.1) + 0(0.6) = \$4.80$

Action A_2 is the better choice.

CHAPTER 8 REVIEW

1. First, we calculate the number of 5-card combinations that can be dealt from 52 cards:

$$n(S) = {}_{52}C_5 = \frac{52!}{5! \cdot 47!} = 2{,}598{,}960$$

We then calculate the number of 5-club combinations that can be obtained from 13 clubs:

$$n(E) = {}_{13}C_5 = \frac{13!}{5! \cdot 8!} = 1287$$

Thus, $P(5 \text{ clubs}) = P(E) = \dfrac{n(E)}{n(S)} = \dfrac{1287}{2{,}598{,}960} \approx 0.0005.$ (8-1)

2. $n(S)$ is computed by using the permutation formula:

$$n(S) = {}_{15}P_2 = \frac{15!}{(15-2)!} = 15 \cdot 14 = 210.$$

Thus, the probability that Brittani will be president and Ramon will be treasurer is:

$$\frac{n(E)}{n(S)} = \frac{1}{210} \approx 0.0048.$$ (8-1)

3. (A)   The total number of ways of drawing 3 cards from 10 with order taken into account is given by:

$${}_{10}P_3 = \frac{10!}{(10-3)!} = \frac{10 \cdot 9 \cdot 8 \cdot 7!}{7!} = 720$$

Thus, the probability of drawing the code word "dig" is:

$$P(\text{"dig"}) = \frac{1}{720} \approx 0.0014$$

(B)   The total number of ways of drawing 3 cards from 10 without regard to order is given by:

$$C_{10,3} = \frac{10!}{3!(10-3)!} = \frac{10 \cdot 9 \cdot 8 \cdot 7!}{3!7!} = 120$$

Thus, the probability of drawing the 3 cards "d," "i," and "g" (in some order) is:

$$P(\text{"d," "i," "g"}) = \frac{1}{120} \approx 0.0083. \qquad (8\text{-}1)$$

4. $P(\text{person having side effects}) = \dfrac{f(E)}{n} = \dfrac{50}{1000} = 0.05. \qquad (8\text{-}1)$

5.   The payoff table is as follows:

x_i	−\$2	−\$1	\$0	\$1	\$2
p_i	$\frac{1}{5}$	$\frac{1}{5}$	$\frac{1}{5}$	$\frac{1}{5}$	$\frac{1}{5}$

Hence, $E(X) = (-2) \cdot \dfrac{1}{5} + (-1) \cdot \dfrac{1}{5} + 0 \cdot \dfrac{1}{5} + 1 \cdot \dfrac{1}{5} + 2 \cdot \dfrac{1}{5} = 0$

The game is fair.   (8-5)

6. $P(A) = .3$, $P(B) = .4$, $P(A \cap B) = .1$

(A) $P(A') = 1 - P(A) = 1 - .3 = .7$

(B) $P(A \cup B) = P(A) + P(B) - P(A \cap B) = .3 + .4 - .1 = .6. \qquad (8\text{-}2)$

7.   Since the spinner cannot land on R and G simultaneously, $R \cap G = \varnothing$. Thus,
$P(R \cup G) = P(R) + P(G) = .3 + .5 = .8$

The odds for an event E are: $\dfrac{P(E)}{P(E')}$

Thus, the odds for landing on either R or G are: $\dfrac{P(R \cup G)}{P[(R \cup G)']} = \dfrac{.8}{.2} = \dfrac{8}{2}$ or the odds are 8 to 2.   (8-2)

8.   If the odds for an event E are a to b, then $P(E) = \dfrac{a}{a+b}$. Thus, the probability of rolling an 8 before rolling

a 7 is: $\dfrac{5}{11} \approx .455. \qquad (8\text{-}2)$

9. $P(T) = .27$ (8-3)   10. $P(Z) = .20$ (8-3)

11. $P(T \cap Z) = .02$ (8-3)   12. $P(R \cap Z) = .03$ (8-3)

13. $P(R \mid Z) = \dfrac{P(R \cap Z)}{P(Z)} = \dfrac{.03}{.20} = .15$ (8-3)   14. $P(Z \mid R) = \dfrac{P(Z \cap R)}{P(R)} = \dfrac{.03}{.23} \approx .1304$ (8-3)

15. $P(T \mid Z) = \dfrac{P(T \cap Z)}{P(Z)} = \dfrac{.02}{.20} = .10$ (8-3)

16.   No, because $P(T \cap Z) = .02 \neq P(T) \cdot P(Z) = (.27)(.20) = .054. \qquad (8\text{-}3)$

17.   Yes, because $P(S \cap X) = .10 = P(S) \cdot P(X) = (.5)(.2). \qquad (8\text{-}3)$

18. $P(A) = .4$ from the tree diagram.   (8-3)

19. $P(B \mid A) = .2$ from the tree diagram.   (8-3)

20. $P(B \mid A') = .3$ from the tree diagram.   (8-3)

21. $P(A \cap B) = P(A)P(B \mid A) = (.4)(.2) = .08$ (8-3)

22. $P(A' \cap B) = P(A')P(B \mid A) = (.6)(.3) = .18$ (8-3)

23. $P(B) = P(A \cap B) + P(A' \cap B)$

   $= P(A)P(B \mid A) + P(A')P(B \mid A')$

   $= (.4)(.2) + (.6)(.3) = .08 + .18 = .26$ (8-3)

24. $P(A \mid B) = \dfrac{P(A \cap B)}{P(B)} = \dfrac{P(A)P(B \mid A)}{P(A \cap B) + P(A' \cap B)} = \dfrac{P(A)P(B \mid A)}{P(A)P(B \mid A) + P(A')P(B \mid A')}$

$$= \frac{(.4)(.2)}{(.4)(.2) + (.6)(.3)} \quad \text{(from the tree diagram)}$$

$$= \frac{.08}{.26} = \frac{8}{26} \quad \text{or } .307 \approx .31 \qquad (8\text{-}4)$$

25. $P(A \mid B') = \dfrac{P(A \cap B')}{P(B')} = \dfrac{P(A)P(B' \mid A)}{1 - P(B)} = \dfrac{(.4)(.8)}{1 - .26}$ $[P(B) = .26,$ see Problem 23.$]$

$$= \frac{.32}{.74} = \frac{16}{37} \quad \text{or } .432 \qquad (8\text{-}4)$$

26. Let $E =$ "born in June, July or August."

(A)   Empirical Probability:

   $P(E) = \dfrac{f(E)}{n} = \dfrac{10}{32} = \dfrac{5}{16}$

(B)   Theoretical Probability:

   $P(E) = \dfrac{n(E)}{n(S)} = \dfrac{3}{12} = \dfrac{1}{4}$

(C)   As the sample size in part (A) increases, the approximate empirical probability of event E approaches the theoretical probability of event E.   (8-1)

27. No. The total number of 3-card hands is $C_{52,3}$. The number of hands containing 3 red cards is $C_{26,3} = 2600$; the number of hands containing 2 red cards and one black card is $C_{26,2} \cdot C_{26,1} = 8,450$. These events are not equally likely.   (8-1)

28. Yes. The number of hands containing either 2 or 3 red cards equals the number of hands containing 2 or 3 black cards.   (8-1)

29. $S = \{HH, HT, TH, TT\}$.

The probabilities for 2 "heads," 1 "head," and 0 "heads" are, respectively, $\frac{1}{4}$, $\frac{1}{2}$, and $\frac{1}{4}$. Thus, the payoff table is:

x_i	$5	−$4	$2
P_i	0.25	0.5	0.25

$E(X) = 0.25(5) + 0.5(-4) + 0.25(2) = -0.25$ or $-\$0.25$
The game is not fair.   (8-5)

30. $S = \{(1,1), (2,2), (3,3), (1,2), (2,1), (1,3), (3,1), (2,3), (3,2)\}$; $n(S) = 3\cdot3 = 9$

   (A) $P(A) = \dfrac{n(A)}{n(S)} = \dfrac{3}{9} = \dfrac{1}{3}$ $[A = \{(1,1), (2,2), (3,3)\}]$

   (B) $P(B) = \dfrac{n(B)}{n(S)} = \dfrac{2}{9}$ $[B = \{(2,3), (3,2)\}]$ (8-3)

31.   (A) $P(\text{jack or queen}) = P(\text{jack}) + P(\text{queen}) = \dfrac{4}{52} + \dfrac{4}{52} = \dfrac{8}{52} = \dfrac{2}{13}$

   [<u>Note</u>: jack $\cap$ queen $= \varnothing$.]

   The odds for drawing a jack or queen are 2 to 11.

   (B) $P(\text{jack or spade}) = P(\text{jack}) + P(\text{spade}) - P(\text{jack and spade})$

   $= \dfrac{4}{52} + \dfrac{13}{52} - \dfrac{1}{52} = \dfrac{16}{52} = \dfrac{4}{13}$

   The odds for drawing a jack or a spade are 4 to 9.

   (C) $P(\text{ace}) = \dfrac{4}{52} = \dfrac{1}{13}$. Thus,

   $P(\text{card other than an ace}) = 1 - P(\text{ace}) = 1 - \dfrac{1}{13} = \dfrac{12}{13}$

   The odds for drawing a card other than a ace are 12 to 1.   (8-2)

32.   (A)   The probability of rolling a 5 is $\dfrac{4}{36} = \dfrac{1}{9}$.
   Thus, the odds for rolling a five are 1 to 8.

   (B)   Let x = amount house should pay (and return the $1 bet).
   Then, for the game to be fair,

   $E(X) = x\left(\dfrac{1}{9}\right) + (-1)\left(\dfrac{8}{9}\right) = \dfrac{x}{9} - \dfrac{8}{9} = 0$

   $x = 8$

   Thus, the house should pay $8.   (8-2)

33.   Event $E_1 = 2$ heads; $f(E_1) = 210$.
   Event $E_2 = 1$ head; $f(E_2) = 480$.
   Event $E_3 = 0$ heads; $f(E_3) = 310$.

   Total number of trials = 1000.

(A)   The empirical probabilities for the events above are as follows:

$$P(E_1) = \frac{210}{1000} = 0.21$$

$$P(E_2) = \frac{480}{1000} = 0.48$$

$$P(E_3) = \frac{310}{1000} = 0.31$$

(B)   Sample space $S = \{HH, HT, TH, TT\}$.

$$P(2\text{ heads}) = \frac{1}{4} = 0.25$$

$$P(1\text{ head}) = \frac{2}{4} = 0.5$$

$$P(0\text{ heads}) = \frac{1}{4} = 0.25$$

(C)   Using part (B), the expected frequencies for each outcome are as follows:

$$2\text{ heads} = 1000 \cdot \frac{1}{4} = 250$$

$$1\text{ head} = 1000 \cdot \frac{2}{4} = 500$$

$$0\text{ heads} = 1000 \cdot \frac{1}{4} = 250 \qquad (8\text{-}1, 8\text{-}5)$$

34. The individual tosses of a coin are independent events (the coin has no memory). Therefore, $P(H) = \frac{1}{2}$.

(8-3)

35. (A)   The sample space S is given by:

$$
\begin{aligned}
S = \{ &(1,1), (1,2), (1,3), (1,4), (1,5), (1,6),\\
&(2,1), (2,2), (2,3), (2,4), (2,5), (2,6),\\
&(3,1), (3,2), (3,3), (3,4), (3,5), (3,6),\\
&(4,1), (4,2), (4,3), (4,4), (4,5), (4,6),\\
&(5,1), (5,2), (5,3), (5,4), (5,5), (5,6),\\
&(6,1), (6,2), (6,3), (6,4), (6,5), (6,6)\}
\end{aligned}
$$

Sum 2, Sum 3, Sum 4, Sum 5

[<u>Note</u>: Event (2,3) means 2 on the first die and 3 on the second die.]

The probability distribution corresponding to this sample space is:

Sum	x_i	2	3	3	5	6	7	8	9	10	11	12
Probability	p_i	$\frac{1}{36}$	$\frac{2}{36}$	$\frac{3}{36}$	$\frac{4}{36}$	$\frac{5}{36}$	$\frac{6}{36}$	$\frac{5}{36}$	$\frac{4}{36}$	$\frac{3}{36}$	$\frac{2}{36}$	$\frac{1}{36}$

(B) $E(X) = 2\left(\frac{1}{36}\right) + 3\left(\frac{2}{36}\right) + 4\left(\frac{3}{36}\right) + 5\left(\frac{4}{36}\right) + 6\left(\frac{5}{36}\right) + 7\left(\frac{6}{36}\right) + 8\left(\frac{5}{36}\right)$

$\qquad + 9\left(\frac{4}{36}\right) + 10\left(\frac{3}{36}\right) + 11\left(\frac{2}{36}\right) + 12\left(\frac{1}{36}\right) = 7 \qquad (8\text{-}5)$

36. The event A that corresponds to the sum being divisible by 4 includes sums 4, 8, and 12. This set is:

$A = \{(1, 3), (2, 2), (3, 1), (2, 6), (3, 5), (4, 4), (5, 3), (6, 2), (6, 6)\}$

The event B that corresponds to the sum being divisible by 6 includes sums 6 and 12. This set is:

$B = \{(1, 5), (2, 4), (3, 3), (4, 2), (5, 1), (6, 6)\}$

$$P(A) = \frac{n(A)}{n(S)} = \frac{9}{36} = \frac{1}{4}$$

$$P(B) = \frac{n(B)}{n(S)} = \frac{6}{36} = \frac{1}{6}$$

$$P(A \cap B) = \frac{1}{36} \text{ [Note: } A \cap B = \{(6, 6)\}\text{]}$$

$$P(A \cup B) = \frac{14}{36} \text{ or } \frac{7}{18} \qquad \begin{array}{l} \text{[Note: } A \cup B = \{(1, 3), (2, 2), (3, 1), (2, 6), \\ (3, 5), (4, 4), (5, 3), (6, 2), (6, 6), \\ (1, 5), (2, 4), (3, 3), (4, 2), (5, 1)\}\text{]} \qquad (8\text{-}2) \end{array}$$

37. The function P cannot be a probability function because:

(a) P cannot be negative. [Note: $P(e_2) = -0.2$.]

(b) P cannot have a value greater than 1. [Note: $P(e_4) = 2$.]

(c)   The sum of the values of P must equal 1.
      [Note: $P(e_1) + P(e_2) + P(e_3) + P(e_4) = 0.1 + (-0.2) + 0.6 + 2 = 2.5 \neq 1$.]   (8-1)

38. Since $n(A \cup B) = n(A) + n(B) - n(A \cap B)$, we have
$80 = 50 + 45 - n(A \cap B)$ which implies $n(A \cap B) = 15$
Now, $n(B') = n(U) - n(B) = 100 - 45 = 55$
   $n(A') = n(U) - n(A) = 100 - 50 = 50$, $n(A \cap B') = n(A) - n(A \cap B) = 50 - 15 = 35$
   $n(B \cap A') = 45 - 15 = 30$, $n(A' \cap B') = 55 - 35 = 20$
Thus,

	A	A'	Totals
B	15	30	45
B'	35	20	55
Totals	50	50	100

(8-2)

39. (A) $P(\text{odd number}) = P(1) + P(3) + P(5) = .2 + .3 + .1 = .6$

(B)   Let $E = $ "number less than 4,"
      and $F = $ "odd number."
      Now, $E \cap F = \{1, 3\}$, $F = \{1, 3, 5\}$.
      $$P(E \mid F) = \frac{P(E \cap F)}{P(F)} = \frac{.2 + .3}{.6} = \frac{5}{6} \qquad (8\text{-}3)$$

40. Let E = "card is red" and F = "card is an ace." Then $F \cap E$ = "card is a red ace."

(A) $\quad P(F \mid E) = \dfrac{P(F \cap E)}{P(E)} = \dfrac{2/52}{26/52} = \dfrac{1}{13}$

(B) $\quad P(F \cap E) = \dfrac{1}{26}$, and $P(E) = \dfrac{1}{2}$, $P(F) = \dfrac{1}{13}$. Thus,

$P(F \cap E) = P(E) \cdot P(F)$, and E and F are independent.   (8-3)

41. (A)   The tree diagram with replacement is: (B)   The tree diagram without replacement is:

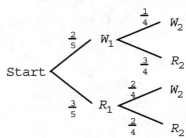

$$P(W_1 \cap R_2) = P(W_1)\,P(R_2)$$

$$= \frac{2}{5} \cdot \frac{3}{5} = \frac{6}{25} \approx .24$$

$$P(W_1 \cap R_2) = P(W_1)P(R_2|W_1)$$

$$= \frac{2}{5} \cdot \frac{3}{4} = \frac{6}{20} = \frac{3}{10}$$

(8-3)

42. Part (B) involves dependent events because

$$P(R_2 \mid W_1) = \frac{3}{4}$$

$$P(R_2) = P(W_1 \cap R_2) + P(R_1 \cap R_2) = \frac{6}{20} + \frac{6}{20} = \frac{12}{20} = \frac{3}{5}$$

and $\;P(R_2 \mid W_1) \neq P(R_2)$. The events in part (A) are independent.   (8-3)

43. (A)   Using the tree diagram in Problem 41(A), we have:

$P(\text{zero red balls}) = P(W_1 \cap W_2) = P(W_1)P(W_2) = \dfrac{2}{5} \cdot \dfrac{2}{5} = \dfrac{4}{25} = .16$

$P(\text{one red ball}) = P(W_1 \cap R_2) + P(R_1 \cap W_2) = P(W_1)P(R_2) + P(R_1)P(W_2)$

$$= \frac{2}{5} \cdot \frac{3}{5} + \frac{3}{5} \cdot \frac{2}{5} = \frac{12}{25} = .48$$

$P(\text{two red balls}) = P(R_1 \cap R_2) = P(R_1)P(R_2) = \dfrac{3}{5} \cdot \dfrac{3}{5} = \dfrac{9}{25} = .36$

Thus, the probability distribution is:

Number of red balls	Probability
x_i	p_i
0	.16
1	.48
2	.36

The expected number of red balls is:

$$E(X) = 0(.16) + 1(.48) + 2(.36) = .48 + .72 = 1.2$$

(B)   Using the tree diagram in Problem 41(B), we have:

$$P(\text{zero red balls}) = P(W_1 \cap W_2) = P(W_1)P(W_2|\ W_1) = \frac{2}{5} \cdot \frac{1}{4} = \frac{1}{10} = .1$$

$$\begin{aligned}
P(\text{one red ball}) &= P(W_1 \cap R_2) + P(R_1 \cap W_2) \\
&= P(W_1)P(R_2|\ W_1) + P(R_1)P(W_2|\ R_1) \\
&= \frac{2}{5} \cdot \frac{3}{4} + \frac{3}{5} \cdot \frac{2}{4} = \frac{12}{20} = \frac{3}{5} = .6
\end{aligned}$$

$$P(\text{two red balls}) = P(R_1 \cap R_2) = P(R_1)P(R_2|\ R_1) = \frac{3}{5} \cdot \frac{2}{4} = \frac{6}{20} = .3$$

Thus, the probability distribution is:

Number of red balls x_i	Probability p_i
0	.1
1	.6
2	.3

The expected number of red balls is:

$$E(X) = 0(.1) + 1(.6) + 2(.3) = 1.2. \qquad (8\text{-}5)$$

44.   The tree diagram for this problem is as follows:

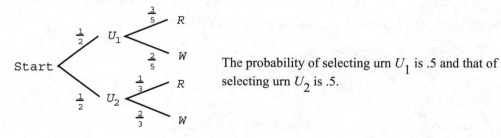

The probability of selecting urn U_1 is .5 and that of selecting urn U_2 is .5.

(A) $P(R \mid U_1) = \dfrac{3}{5}$ (B) $P(R \mid U_2) = \dfrac{1}{3}$

(C) $\begin{aligned}
P(R) &= P(R \cap U_1) + P(R \cap U_2) \\
&= P(U_1)P(R \mid U_1) + P(U_2)P(R \mid U_2) \\
&= \frac{1}{2} \cdot \frac{3}{5} + \frac{1}{2} \cdot \frac{1}{3} = \frac{28}{60} = \frac{7}{15} \approx .4667
\end{aligned}$

(D) $\begin{aligned}
P(U_1|\ R) &= \frac{P(U_1 \cap R)}{P(R)} = \frac{P(U_1)P(R \mid U_1)}{P(U_1)P(R \mid U_1) + P(U_2)P(R \mid U_2)} \\[2mm]
&= \frac{\dfrac{1}{2} \cdot \dfrac{3}{5}}{\dfrac{1}{2} \cdot \dfrac{3}{5} + \dfrac{1}{2} \cdot \dfrac{1}{3}} = \frac{\dfrac{3}{10}}{\dfrac{7}{15}} = \frac{9}{14} \approx .6429
\end{aligned}$

(E) $P(U_2 \mid W) = \dfrac{P(U_2 \cap W)}{P(W)} = \dfrac{P(U_2)P(W \mid U_2)}{P(U_2)P(W \mid U_2) + P(U_1)P(W \mid U_1)}$

$$= \dfrac{\dfrac{1}{2} \cdot \dfrac{2}{3}}{\dfrac{1}{2} \cdot \dfrac{2}{3} + \dfrac{1}{2} \cdot \dfrac{2}{5}} = \dfrac{\dfrac{2}{3}}{\dfrac{16}{15}} = \dfrac{5}{8} = .625$$

(F) $P(U_1 \cap R) = P(U_1)P(R \mid U_1) = \dfrac{1}{2} \cdot \dfrac{3}{5} = \dfrac{3}{10} = .3$

[Note: In parts (A)—(F), we derived the values of the probabilities from the tree diagram.]   (8-3, 8-4)

45. No, because $P(R \mid U_1) \neq P(R)$. (See Problem 44.)   (8-3)

46. $n(S) = C_{52,5}$

(A) Let A be the event "all diamonds." Then $n(A) = C_{13,5}$. Thus,

$$P(A) = \dfrac{n(A)}{n(S)} = \dfrac{C_{13,5}}{C_{52,5}} .$$

(B) Let B be the event "3 diamonds and 2 spades." Then $n(B) = C_{13,3} \cdot C_{13,2}$. Thus,

$$P(B) = \dfrac{n(B)}{n(S)} = \dfrac{C_{13,3} \cdot C_{13,2}}{C_{52,5}} .   (8-1)$$

47. $n(S) = C_{10,4} = \dfrac{10!}{4!(10-4)!} = \dfrac{10 \cdot 9 \cdot 8 \cdot 7 \cdot 6!}{4 \cdot 3 \cdot 2 \cdot 1 \cdot 6!} = 210$

Let A be the event "The married couple is in the group of 4 people." Then

$n(A) = C_{2,2} \cdot C_{8,2} = 1 \cdot \dfrac{8!}{2!(8-2)!} = \dfrac{8 \cdot 7 \cdot 6!}{2 \cdot 1 \cdot 6!} = 28.$

Thus, $P(A) = \dfrac{n(A)}{n(S)} = \dfrac{28}{210} = \dfrac{2}{15} \approx 0.1333.$ (8-1)

48. Events S and H are mutually exclusive. Hence, $P(S \cap H) = 0$, while

$P(S) \neq 0$ and $P(H) \neq 0$. Therefore,
$$P(S \cap H) \neq P(S) \cdot P(H)$$

which implies that S and F are dependent.   (8-3)

49. (A) From the plot, $P(2) = \dfrac{9}{50} = 0.18.$

(B) The event $A = $ "the minimum of the two numbers is 2" contains the simple events (2, 2), (2, 3), (3, 2), (2, 4), (4, 2), (2, 5),

(5, 2), (2, 6), (6, 2). Thus $n(A) = 9$ and $P(A) = \dfrac{9}{36} = \dfrac{1}{4} = 0.25.$

(C) The empirical probability depends on the results of your simulation. For the theoretical probability, let $A = $ "minimum of the two numbers is 4". Then $A = \{(4, 4), (4, 5), (5, 4), (4, 6), (6, 4)\}$,

$n(A) = 5$ and $P(A) = \dfrac{5}{36} \approx 0.139.$ (8-1)

50. The empirical probability depends on the results of your simulation.

Since there are 2 black jacks in a standard 52-card deck, the theoretical probability of drawing a black jack

is: $\dfrac{2}{52} = \dfrac{1}{26} \approx 0.038.$ (8-3)

51. False. If $P(E) = 1$, then $P(E') = 0$ and the odds for $E = \dfrac{P(E)}{P(E')} = \dfrac{1}{0}$; $\dfrac{1}{0}$ is undefined.   (8-2)

52. True. In general, $P(E \cup F) = P(E) + P(F) - P(E \cap F)$.
If $E = F'$, then $E \cap F = F' \cap F = \varnothing$ and $P(E \cap F) = 0$.   (8-2)

53. False. Let E and F be complementary events with $0 < P(E) < 1$, $0 < P(F) < 1$. Then $E \cap F = E \cap E' = \varnothing$
and $P(E \cap F) = 0$ while $P(E) \cdot P(F) = P(E)[1 - P(E)] \neq 0$.   (8-3)

54. False. Counterexample: Roll a fair die; $S = \{1, 2, 3, 4, 5, 6\}$.
Let E = the number that turns up is ≥ 2;
$\quad F$ = the number that turns up is ≤ 4.
Then $E \cup F = \{1, 2, 3, 4, 5, 6\}$ and $P(E \cup F) = 1$ but $F \neq E'$.   (8-2)

55. True. This is the definition of independent events.   (8-3)

56. False. If E and F are mutually exclusive, then $E \cap F = \varnothing$ and the example in Problem 53 is a
counterexample here.   (8-2)

57. Let E_2 be the event "2 heads."

(A)   From the table, $f(E_2) = 350$. Thus, the approximate empirical probability of obtaining 2 heads is:
$$P(E_2) \approx \frac{f(E_2)}{n} = \frac{350}{1000} = 0.350$$

(B) $S = \{$HHH, HHT, HTH, HTT, THH, THT, TTH, TTT$\}$
The theoretical probability of obtaining 2 heads is:
$$P(E_2) = \frac{n(E_2)}{n(S)} = \frac{3}{8} = 0.375$$

(C)   The expected frequency of obtaining 2 heads in 1000 tosses of 3 fair coins is:
$f(E_2) = 1000(0.375) = 375.$ (8-1)

58. On one roll of the dice, the probability of getting a double six is $\dfrac{1}{36}$ and the probability of not getting a

double six is $\dfrac{35}{36}$. On n independent rolls, the probability of no double sixes is $\left(\dfrac{35}{36}\right)^n$. In particular, we

conclude that, in 24 rolls of the die,

$P(E') = \left(\dfrac{35}{36}\right)^{24} \approx 0.5086$

Therefore, $P(E) = 1 - 0.5086 = 0.4914$.

The payoff table is:

x_i	1	−1
P_i	0.4914	0.5086

and $E(X) = 1(0.4914) + (-1)(0.5086) = 0.4914 - 0.5086 = -0.0172$

Thus, your expectation is −$0.0172. Your friend's expectation is $0.0172. The game is not fair.   (8-5)

59. The total number of ways that 3 people can be selected from a group of 10 is:

$$C_{10,3} = \frac{10!}{3!(10-3)!} = \frac{10 \cdot 9 \cdot 8 \cdot 7!}{3 \cdot 2 \cdot 1 \cdot 7!} = 120$$

The number of ways of selecting *no* women is:

$$C_{7,3} = \frac{7!}{3!(7-3)!} = \frac{7 \cdot 6 \cdot 5 \cdot 4!}{3 \cdot 2 \cdot 1 \cdot 4!} = 35$$

Thus, the number of samples of 3 people that contain at least one woman is $120 - 35 = 85$.

Therefore, if event *A* is "At least one woman is selected," then

$$P(A) = \frac{n(A)}{n(S)} = \frac{85}{120} = \frac{17}{24} \approx 0.708. \qquad \text{(8-1)}$$

60. $P(\text{second heart}|\text{first heart}) = P(H_2|H_1) = \frac{12}{51} \approx .235$

[Note: One can see that $P(H_2|H_1) = \frac{12}{51}$ directly.] $\qquad$ (8-3)

61. $P(\text{first heart}|\text{second heart}) = P(H_1|H_2) = \frac{P(H_1 \cap H_2)}{P(H_2)} = \frac{P(H_1)P(H_2 \mid H_1)}{P(H_2)}$

$$= \frac{P(H_1)P(H_2 \mid H_1)}{P(H_1 \cap H_2) + P(H_1{}' \cap H_2)} = \frac{P(H_1)P(H_2 \mid H_1)}{P(H_1)P(H_2 \mid H_1) + P(H_1{}')P(H_2 \mid H_1{}')}$$

$$= \frac{\dfrac{13}{52} \cdot \dfrac{12}{51}}{\dfrac{13}{52} \cdot \dfrac{12}{51} + \dfrac{39}{52} \cdot \dfrac{13}{51}} = \frac{12}{51} \approx .235 \qquad \text{(8-3)}$$

62. Since each die has 6 faces, there are $6 \cdot 6 = 36$ possible pairs for the two up faces.

A sum of 2 corresponds to having (1, 1) as the up faces. This sum can be obtained in $3 \cdot 3 = 9$ ways (3 faces on the first die, 3 faces on the second). Thus,

$$P(2) = \frac{9}{36} = \frac{1}{4}.$$

A sum of 3 corresponds to the two pairs (2, 1) and (1, 2). The number of such pairs is $2 \cdot 3 + 3 \cdot 2 = 12$. Thus,

$$P(3) = \frac{12}{36} = \frac{1}{3}.$$

A sum of 4 corresponds to the pairs (3, 1), (2, 2), (1, 3).

There are $1 \cdot 3 + 2 \cdot 2 + 3 \cdot 1 = 10$ such pairs. Thus,

$$P(4) = \frac{10}{36}.$$

A sum of 5 corresponds to the pairs (2, 3) and (3, 2).

There are $2 \cdot 1 + 1 \cdot 2 = 4$ such pairs. Thus,

$$P(5) = \frac{4}{36} = \frac{1}{9}.$$

A sum of 6 corresponds to the pair (3, 3) and there is one such pair. Thus,

$P(6) = \dfrac{1}{36}$.

(A)   The probability distribution for X is:

x_i	2	3	4	5	6
P_i	$\frac{9}{36}$	$\frac{12}{36}$	$\frac{10}{36}$	$\frac{4}{36}$	$\frac{1}{36}$

(B)   The expected value is:

$$E(X) = 2\left(\frac{9}{36}\right) + 3\left(\frac{12}{36}\right) + 4\left(\frac{10}{36}\right) + 5\left(\frac{4}{36}\right) + 6\left(\frac{1}{36}\right) = \frac{120}{36} = \frac{10}{3} \qquad (8\text{-}5)$$

63.   The payoff table is:

x_i	−$1.50	−$0.50	$0.50	$1.50	$2.50
P_i	$\frac{9}{36}$	$\frac{12}{36}$	$\frac{10}{36}$	$\frac{4}{36}$	$\frac{1}{36}$

and $E(X) = \dfrac{9}{36}\,(-1.50) + \dfrac{12}{36}\,(-0.50) + \dfrac{10}{36}\,(0.50) + \dfrac{4}{36}\,(1.50) + \dfrac{1}{36}\,(2.50)$

$\qquad = -0.375 - 0.167 + 0.139 + 0.167 + 0.069$

$\qquad = -0.167$ or $-\$0.167 \approx -\$0.17$

The game is not fair. The game would be fair if you paid $3.50 − $0.17 = $3.33 to play.   (8-5)

64.   The tree diagram for this experiment is:

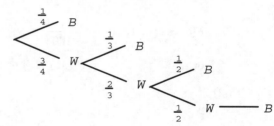

(A) P(black on the fourth draw)

$\qquad = \dfrac{3}{4} \cdot \dfrac{2}{3} \cdot \dfrac{1}{2} = \dfrac{1}{4}$

The odds for black on the fourth draw are 1 to 3.

(B)   Let x = amount house should pay (and return the $1 bet).
Then, for the game to be fair:

$$E(X) = x\left(\frac{1}{4}\right) + (-1)\left(\frac{3}{4}\right) = \frac{x}{4} - \frac{3}{4} = 0; \quad x = 3$$

Thus, the house should pay $3.   (8-2, 8-4)

65. $n(S) = 10 \cdot 10 \cdot 10 \cdot 10 \cdot 10 = 10^5$

Let event A = "at least two people identify the same book." Then A' = "each person identifies a different book," and

$$n(A') = 10 \cdot 9 \cdot 8 \cdot 7 \cdot 6 = \frac{10!}{5!}$$

Thus, $P(A') = \dfrac{\dfrac{10!}{5!}}{10^5} = \dfrac{10!}{5!10^5}$ and $P(A) = 1 - \dfrac{10!}{5!10^5} \approx 1 - .3 = .7.$ (8-2)

66. $P(A \mid B) = \dfrac{P(A \cap B)}{P(B)}$, $P(B \mid A) = \dfrac{P(A \cap B)}{P(A)}$.

Now, $P(A \mid B) = P(B \mid A)$ if and only if $\dfrac{P(A \cap B)}{P(B)} = \dfrac{P(A \cap B)}{P(A)}$

which implies $P(A) = P(B)$ or $P(A \cap B) = 0$.   (8-3)

67.

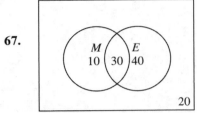

Event M = Reads the morning paper.

Event E = Watches evening news.

(A) $P(\text{reads the paper or watches the news}) = P(M \text{ or } E) = P(M \cup E)$
$$= P(M) + P(E) - P(M \cap E)$$
$$= \frac{40}{100} + \frac{70}{100} - \frac{30}{100} = .8$$

(B) $P(\text{does neither}) = \dfrac{20}{100} = .20$ (from the Venn diagram)

or
$$= 1 - P(M \cup E) \quad \{\text{i.e., } P[(M \cup E)'\} $$
$$= 1 - .8 = .20$$

(C) $P(\text{does exactly one}) = \dfrac{10 + 40}{100} = .50$ (from the Venn diagram)

or
$$= P[(M \cap E') \text{ or } (M' \cap E)] = P(M \cap E') + P(M' \cap E) = \frac{10}{100} + \frac{40}{100} = .50$$

(8-2)

68. Let A be the event that a person has seen the advertising and P be the event that the person purchased the product. Given:

$P(A) = .4$ and $P(P \mid A) = .85$

We want to find:

$P(A \cap P) = P(A)P(P \mid A) = (.4)(.85) = .34.$ (8-3)

69. (A) $P(A) = \dfrac{290}{1000} = 0.290$

$P(B) = \dfrac{290}{1000} = 0.290$

$P(A \cap B) = \dfrac{100}{1000} = 0.100$

$P(A \mid B) = \dfrac{100}{290} = 0.345$

$P(B \mid A) = \dfrac{100}{290} = 0.345$

(B) A and B are **not** independent because
$$0.100 = P(A \cap B) \neq P(A) \cdot P(B) = (0.290)(0.290) = 0.084$$

(C) $P(C) = \dfrac{880}{1000} = 0.880$

   $P(D) = \dfrac{120}{1000} = 0.120$

   $P(C \cap D) = 0; \quad P(C \mid D) = P(D \mid C) = 0$

(D) C and D are mutually exclusive since $C \cap D = \varnothing$. C and D are dependent since
   $0 = P(C \cap D) \neq P(C) \cdot P(D) = (0.120)(0.880) = 0.106.$ (8-3)

70. The payoff table for plan A is:

x_i	10 million	−2 million
p_1	0.8	0.2

Hence, $E(X) = 10(0.8) - 2(0.2) = 8 - 0.4 = \$7.6$ million.

The payoff table for plan B is:

x_i	12 million	−2 million
p_1	0.7	0.3

Hence, $E(X) = 12(0.7) - 2(0.3) = 8.4 - 0.6 = \$7.8$ million.
Plan B should be chosen.   (8-5)

71. The payoff table is:

x_i	1830	−170
p_1	0.8	0.92

[Note: $2000 - 170 = 1830$ is the "gain" if the bicycle is stolen]

Hence, $E(X) = 1830(0.08) - 170(0.92) = 146.4 - 156.4 = -\$10.$ (8-5)

72. $n(S) = C_{12,4} = \dfrac{12!}{4!(12-4)!} = \dfrac{12\cdot11\cdot10\cdot9\cdot8!}{4\cdot3\cdot2\cdot1\cdot8!} = 495$

The number of samples that contain *no* substandard parts is:

$C_{10,4} = \dfrac{10!}{4!(10-4)!} = \dfrac{10\cdot9\cdot8\cdot7\cdot6!}{4\cdot3\cdot2\cdot1\cdot6!} = 210$

Thus, the number of samples that have at least one defective part is
$495 - 210 = 285$. If E is the event "The shipment is returned," then

$P(E) = \dfrac{n(E)}{n(S)} = \dfrac{285}{495} \approx 0.576.$ (8-2)

73. $n(S) = C_{12,3} = \dfrac{12!}{3!(12-3)!} = \dfrac{12\cdot11\cdot10\cdot9!}{3\cdot2\cdot1\cdot9!} = 220$

A sample will either have 0, 1, or 2 tablet computers.

$P(0) = \dfrac{C_{10,3}}{C_{12,3}} = \dfrac{\frac{10!}{3!(10-3)!}}{220} = \dfrac{\frac{10\cdot9\cdot8\cdot7!}{3\cdot2\cdot1\cdot7!}}{220} = \dfrac{120}{220} = \dfrac{12}{22}$

$$P(1) = \frac{C_{2,1} \cdot C_{10,1}}{C_{12,3}} = \frac{2 \cdot \dfrac{10!}{2!(10-2)!}}{220} = \frac{90}{220} = \frac{9}{22}$$

$$P(2) = \frac{C_{2,2} \cdot C_{10,1}}{C_{12,3}} = \frac{10}{220} = \frac{1}{22}$$

(A)   The probability
distribution of X is:

x_i	0	1	2
p_i	$\frac{12}{22}$	$\frac{9}{22}$	$\frac{1}{22}$

(B) $E(X) = 0\left(\dfrac{12}{22}\right) + 1\left(\dfrac{9}{22}\right) + 2\left(\dfrac{1}{22}\right)$

$$= \frac{11}{22} = \frac{1}{2} \qquad (8\text{-}5)$$

74.   Let Event NH = individual with normal heart,

Event MH = individual with minor heart problem,

Event SH = individual with severe heart problem,

and Event P = individual passes the cardiogram test.

Then, using the notation given above, we have:

$P(NH) = .82$

$P(MH) = .11$

$P(SH) = .07$

$P(P \mid NH) = .95$

$P(P \mid MH) = .30$

$P(P \mid SH) = .05$

We want to find $P(NH \mid P) = \dfrac{P(NH \cap P)}{P(P)} = \dfrac{P(NH)P(P \mid NH)}{P(NH \cap P) + P(MH \cap P) + P(SH \cap P)}$

$$= \frac{P(NH)P(P \mid NH)}{P(NH)P(P \mid NH) + P(MN)P(P \mid MH) + P(SH)P(P \mid SH)}$$

$$= \frac{(.82)(.95)}{(.82)(.95) + (.11)(.30) + (.07)(.05)} = .955 \qquad (8\text{-}4)$$

75.   The tree diagram for this problem is as follows:

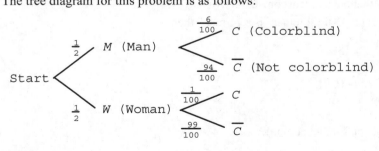

We now compute

$$P(M \mid C) = \frac{P(M \cap C)}{P(C)} = \frac{P(M \cap C)}{P(M \cap C) + P(W \cap C)} = \frac{P(M)P(C \mid M)}{P(M)P(C \mid M) + P(W)P(C \mid W)}$$

$$= \frac{\frac{1}{2} \cdot \frac{6}{100}}{\frac{1}{2} \cdot \frac{6}{100} + \frac{1}{2} \cdot \frac{1}{100}} = \frac{6}{7} \approx .857. \qquad (8\text{-}4)$$

76. According to the empirical probabilities, candidate *A* should have won the election. Since candidate *B* won the election one week later, either some of the students changed their minds during the week, or the 30 students in the math class were not representative of the student body.

9   MARKOV CHAINS

EXERCISE 9-1

Things to remember:

<u>1.</u>   MARKOV CHAINS

A MARKOV CHAIN, or PROCESS, is a sequence of experiments, trials, or observations such that the transition probability matrix from one state to the next is constant.

Given a Markov chain with n states, a kth STATE MATRIX is a matrix of the form
$$S_k = [s_{k1} \ s_{k2} \ \cdots \ s_{kn}]$$

Each entry s_{ki} is the proportion of the population that are in state i after the kth trial, or, equivalently, the probability of a randomly selected element of the population being in state i after the kth trial. The sum of all the entries in the kth state matrix S_k must be 1.

A TRANSITION MATRIX is a constant square matrix P of order n such that the entry in the ith row and jth column indicates the probability of the system moving from the ith state to the jth state on the next observation or trial. The sum of the entries in each row must be 1.

<u>2.</u>   COMPUTING STATE MATRICES FOR A MARKOV CHAIN

If S_0 is the initial state matrix and P is the transition matrix for a Markov chain, then the subsequent state matrices are given by:

$$S_1 = S_0 P \quad \text{First-state matrix}$$
$$S_2 = S_1 P \quad \text{Second-state matrix}$$
$$S_3 = S_2 P \quad \text{Third-state matrix}$$
$$\vdots$$
$$S_k = S_{k-1} P \quad k\text{th-state matrix}$$

<u>3.</u>   POWERS OF A TRANSITION MATRIX

If P is the transition matrix and S_0 is an initial state matrix for a Markov chain, then the kth state matrix is given by

$$S_k = S_0 P^k$$

The entry in the ith row and jth column of P^k indicates the probability of the system moving from the ith state to the jth state in k observations or trials. The sum of the entries in each row of P^k is 1.

---

1. $\begin{bmatrix} 2 & 5 \\ 4 & 1 \end{bmatrix} \begin{bmatrix} 3 \\ 2 \end{bmatrix} = \begin{bmatrix} 2 \cdot 3 + 5 \cdot 2 \\ 4 \cdot 3 + 1 \cdot 2 \end{bmatrix} = \begin{bmatrix} 16 \\ 14 \end{bmatrix}$ **3.** $\begin{bmatrix} 3 \\ 2 \end{bmatrix} \begin{bmatrix} 2 & 5 \\ 4 & 1 \end{bmatrix}$ not defined.

5. $\begin{bmatrix} 3 & 2 \end{bmatrix} \begin{bmatrix} 2 & 5 \\ 4 & 1 \end{bmatrix} = \begin{bmatrix} 3 \cdot 2 + 2 \cdot 4 \\ 3 \cdot 5 + 2 \cdot 1 \end{bmatrix} = \begin{bmatrix} 14 & 17 \end{bmatrix}$ **7.** $\begin{bmatrix} 2 & 5 \\ 4 & 1 \end{bmatrix} \begin{bmatrix} 3 & 2 \end{bmatrix}$ not defined.

9. $S_1 = S_0P = \begin{bmatrix} 1 & 0 \end{bmatrix} \begin{bmatrix} .8 & .2 \\ .4 & .6 \end{bmatrix} = \overset{A\ B}{\begin{bmatrix} .8 & .2 \end{bmatrix}}$

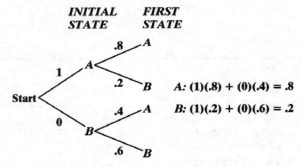

A: (1)(.8) + (0)(.4) = .8

B: (1)(.2) + (0)(.6) = .2

11. $S_1 = S_0P = \begin{bmatrix} .5 & .5 \end{bmatrix} \begin{bmatrix} .8 & .2 \\ .4 & .6 \end{bmatrix} = \overset{A\ B}{\begin{bmatrix} .6 & .4 \end{bmatrix}}$

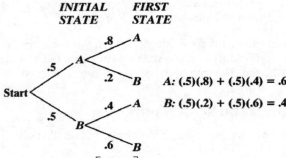

A: (.5)(.8) + (.5)(.4) = .6

B: (.5)(.2) + (.5)(.6) = .4

13. $S_2 = S_1P = \begin{bmatrix} .8 & .2 \end{bmatrix} \begin{bmatrix} .8 & .2 \\ .4 & .6 \end{bmatrix}$ (from Problem 9)

$= \overset{A\quad B}{\begin{bmatrix} .72 & .28 \end{bmatrix}}$

The probability of being in state *A* after two trials is .72; the probability of being in state *B* after two trials is .28.

15. $S_2 = S_1P = \begin{bmatrix} .6 & .4 \end{bmatrix} \begin{bmatrix} .8 & .2 \\ .4 & .6 \end{bmatrix}$ (from Problem 11)

$= \overset{A\quad B}{\begin{bmatrix} .64 & .36 \end{bmatrix}}$

The probability of being in state *A* after two trials is .64; the probability of being in state *B* after two trials is .36.

The transition matrix corresponding to the transition diagram for Problems 17 – 23 is:

$$P = \begin{bmatrix} .7 & .3 \\ .9 & .1 \end{bmatrix}$$

17. $S_1 = S_0 P = \begin{bmatrix} .2 & .8 \end{bmatrix} \begin{bmatrix} .7 & .3 \\ .9 & .1 \end{bmatrix} = \begin{bmatrix} .86 & .14 \end{bmatrix}.$ **19.** $S_1 = S_0 P = \begin{bmatrix} .9 & .1 \end{bmatrix} \begin{bmatrix} .7 & .3 \\ .9 & .1 \end{bmatrix} = \begin{bmatrix} .72 & .28 \end{bmatrix}.$

21. From Problem 17, $S_1 = \begin{bmatrix} .86 & .14 \end{bmatrix}$. Thus, $S_2 = \begin{bmatrix} .86 & .14 \end{bmatrix} \begin{bmatrix} .7 & .3 \\ .9 & .1 \end{bmatrix} = \begin{bmatrix} .728 & .272 \end{bmatrix}.$

23. From Problem 19, $S_1 = \begin{bmatrix} .72 & .28 \end{bmatrix}$. Thus, $S_2 = \begin{bmatrix} .72 & .28 \end{bmatrix} \begin{bmatrix} .7 & .3 \\ .9 & .1 \end{bmatrix} = \begin{bmatrix} .756 & .244 \end{bmatrix}.$

25. $\begin{bmatrix} .3 & .7 \\ 1 & 0 \end{bmatrix}$ Yes; each entry is nonnegative, the sum of the entries in each row is 1.

27. $\begin{bmatrix} .5 & .5 \\ .7 & -.3 \end{bmatrix}$ No; the entry in the (2, 2) position is not nonnegative, the sum of the entries in the second row is not 1.

29. No. A transition matrix must be a square matrix.

31. $\begin{bmatrix} .5 & .1 & .4 \\ 0 & .5 & .5 \\ .2 & .1 & .7 \end{bmatrix}$ Yes; each entry is nonnegative, the sum of the entries in each row is 1.

33. 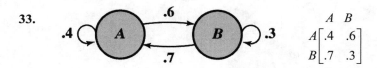 $\begin{array}{c} \quad A \quad\ B \\ \begin{array}{c} A \\ B \end{array}\begin{bmatrix} .4 & .6 \\ .7 & .3 \end{bmatrix} \end{array}$

35. No. Choose any x, $0 \le x \le 1$, then

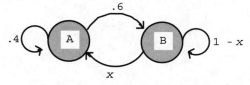

is an acceptable transition diagram, and

$\begin{array}{c} \ A \quad\ B \\ \begin{array}{c} A \\ B \end{array}\begin{bmatrix} .4 & .6 \\ x & 1-x \end{bmatrix} \end{array}$ is the corresponding transition matrix

37.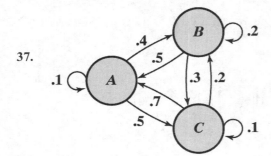

$$\begin{array}{c} \ \ A\ \ B\ \ C \\ \begin{array}{c} A \\ B \\ C \end{array}\!\! \begin{bmatrix} .1 & .4 & .5 \\ .5 & .2 & .3 \\ .7 & .2 & .1 \end{bmatrix} \end{array}$$

39. $0 + .5 + a = 1$ implies $a = .5$
$b + 0 + .4 = 1$ implies $b = .6$
$.2 + c + .1 = 1$ implies $c = .7$

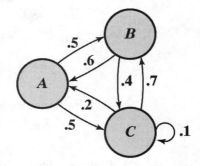

41. $0 + a + .3 = 1$ implies $a = .7$
$0 + b + 0 = 1$ implies $b = 1$
$c + .8 + 0 = 1$ implies $c = .2$

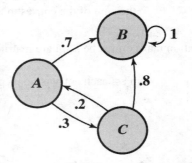

43.   No. Choose any x, $0 \le x \le .4$ and let $a = x$, $b = .1$,
$c = 1 - (x + .4) = .6 - x$

$$\begin{array}{c} \ \ A\ \ \ B\ \ \ \ C \\ \begin{array}{c} A \\ B \\ C \end{array}\!\! \begin{bmatrix} .2 & .1 & .7 \\ x & .4 & .6 - x \\ .5 & .1 & .4 \end{bmatrix} \end{array}$$ is a transition matrix.

45. The probability of staying in state A is .3.

The probability of staying in state B is .1.

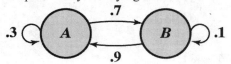

$$\begin{array}{c} \\ A \\ B \end{array}\begin{array}{cc} A & B \\ \begin{bmatrix} .3 & .7 \\ .9 & .1 \end{bmatrix} \end{array}$$

47. The probability of staying in state A is .6.

The probability of staying in state B is .3.

Since the probability of staying in state C is 1, the probability of going from state C to state A is 0 and the probability of going from state C to state B is 0.

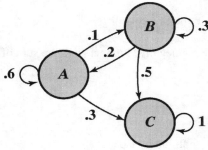

$$\begin{array}{c} \\ A \\ B \\ C \end{array}\begin{array}{ccc} A & B & C \\ \begin{bmatrix} .6 & .1 & .3 \\ .2 & .3 & .5 \\ 0 & 0 & 1 \end{bmatrix} \end{array}$$

49. Using P^2, the probability of going from state A to state B in two trials is the element in the (1,2) position is .35.

51. Using P^3, the probability of going from state C to state A in three trials is the number in the (3,1) position is .212.

53. $S_2 = S_0 P^2 = \begin{bmatrix} 1 & 0 & 0 \end{bmatrix}\begin{bmatrix} .43 & .35 & .22 \\ .25 & .37 & .38 \\ .17 & .27 & .56 \end{bmatrix} \begin{array}{ccc} A & B & C \end{array} = \begin{bmatrix} .43 & .35 & .22 \end{bmatrix}$

These are the probabilities of going from state A to states A, B, and C, respectively, in two trials.

55. $S_3 = S_0 P^3 = \begin{bmatrix} 0 & 0 & 1 \end{bmatrix}\begin{bmatrix} .35 & .348 & .302 \\ .262 & .336 & .402 \\ .212 & .298 & .49 \end{bmatrix} \begin{array}{ccc} A & B & C \end{array} = \begin{bmatrix} .212 & .298 & .49 \end{bmatrix}$

These are the probabilities of going from state C to states A, B, and C, respectively, in three trials.

57. $n = 9$

59. $P = \begin{bmatrix} .1 & .9 \\ .6 & .4 \end{bmatrix}$, $P^2 = \begin{bmatrix} .1 & .9 \\ .6 & .4 \end{bmatrix}\begin{bmatrix} .1 & .9 \\ .6 & .4 \end{bmatrix} = \begin{bmatrix} .55 & .45 \\ .3 & .7 \end{bmatrix}$;

$$P^4 = P^2 \cdot P^2 = \begin{bmatrix} .55 & .45 \\ .3 & .7 \end{bmatrix}\begin{bmatrix} .55 & .45 \\ .3 & .7 \end{bmatrix} = \begin{array}{c} \\ A \\ B \end{array}\begin{array}{cc} A & B \\ \begin{bmatrix} .4375 & .5625 \\ .375 & .625 \end{bmatrix} \end{array}$$

$$S_4 = S_0 P^4 = \begin{bmatrix} .8 & .2 \end{bmatrix}\begin{array}{cc} A & B \\ \begin{bmatrix} .4375 & .5625 \\ .375 & .625 \end{bmatrix} \end{array} = \begin{bmatrix} .425 & .575 \end{bmatrix}$$

61. $P = \begin{bmatrix} 0 & .4 & .6 \\ 0 & 0 & 1 \\ 1 & 0 & 0 \end{bmatrix}, P^2 = \begin{bmatrix} 0 & .4 & .6 \\ 0 & 0 & 1 \\ 1 & 0 & 0 \end{bmatrix} \begin{bmatrix} 0 & .4 & .6 \\ 0 & 0 & 1 \\ 1 & 0 & 0 \end{bmatrix} = \begin{bmatrix} .6 & 0 & .4 \\ 1 & 0 & 0 \\ 0 & .4 & .6 \end{bmatrix};$

$$P^4 = P^2 \cdot P^2 = \begin{bmatrix} .6 & 0 & .4 \\ 1 & 0 & 0 \\ 0 & .4 & .6 \end{bmatrix} \begin{bmatrix} .6 & 0 & .4 \\ 1 & 0 & 0 \\ 0 & .4 & .6 \end{bmatrix} = \begin{matrix} A \\ B \\ C \end{matrix} \begin{matrix} A & B & C \\ \begin{bmatrix} .36 & .16 & .48 \\ .6 & 0 & .4 \\ .4 & .24 & .36 \end{bmatrix} \end{matrix}$$

$$S_4 = S_0 P^4 = \begin{bmatrix} .2 & .3 & .5 \end{bmatrix} \begin{bmatrix} .36 & .16 & .48 \\ .6 & 0 & .4 \\ .4 & .24 & .36 \end{bmatrix} \begin{matrix} A & B & C \\ \end{matrix} = \begin{bmatrix} .452 & .152 & .396 \end{bmatrix}$$

63. $S_k = S_0 P^k = \begin{bmatrix} 1 & 0 \end{bmatrix} P^k$; The entries in S_k are the entries in the first row of P^k.

65. False. For P to be a transition matrix its entries must be nonnegative numbers and the sum of the entries in each row must be 1.

$$P = \begin{bmatrix} .7 & .3 \\ .9 & .1 \end{bmatrix}$$

is a transition matrix and the sum of the entries in the first column is 1.6.

67. True. The entries in a transition matrix are nonnegative numbers. Therefore the product of the entries in each column is nonnegative.

69. False. See the transition diagram for Problems 17 – 24. The sum of the probabilities on the arrows going into A is $.7 + .9 = 1.6 \neq 1$.

71. (A) $P^2 = \begin{bmatrix} .2 & .2 & .3 & .3 \\ 0 & 1 & 0 & 0 \\ .2 & .2 & .1 & .5 \\ 0 & 0 & 0 & 1 \end{bmatrix} \begin{bmatrix} .2 & .2 & .3 & .3 \\ 0 & 1 & 0 & 0 \\ .2 & .2 & .1 & .5 \\ 0 & 0 & 0 & 1 \end{bmatrix} = \begin{bmatrix} .1 & .3 & .09 & .51 \\ 0 & 1 & 0 & 0 \\ .06 & .26 & .07 & .61 \\ 0 & 0 & 0 & 1 \end{bmatrix}$

$P^4 = \begin{bmatrix} .1 & .3 & .09 & .51 \\ 0 & 1 & 0 & 0 \\ .06 & .26 & .07 & .61 \\ 0 & 0 & 0 & 1 \end{bmatrix} \begin{bmatrix} .1 & .3 & .09 & .51 \\ 0 & 1 & 0 & 0 \\ .06 & .26 & .07 & .61 \\ 0 & 0 & 0 & 1 \end{bmatrix}$

$$= \begin{matrix} A \\ B \\ C \\ D \end{matrix} \begin{matrix} A & B & C & D \\ \begin{bmatrix} .0154 & .3534 & .0153 & .6159 \\ 0 & 1 & 0 & 0 \\ .0102 & .2962 & .0103 & .6833 \\ 0 & 0 & 0 & 1 \end{bmatrix} \end{matrix}$$

(B)   The probability of going from state A to state D in 4 trials is the element in the (1,4) position: .6159.

(C)   The element in the (3,2) position: .2962.

(D)   The element in the (2,1) position: 0.

73. If $P = \begin{bmatrix} a & 1-a \\ 1-b & b \end{bmatrix}$ is a probability matrix, then $0 \le a \le 1, 0 \le b \le 1$

$$P^2 = \begin{bmatrix} a & 1-a \\ 1-b & b \end{bmatrix}\begin{bmatrix} a & 1-a \\ 1-b & b \end{bmatrix} = \begin{bmatrix} a^2 + (1-a)(1-b) & a(1-a)+(1-a)b \\ (1-b)a+b(1-b) & (1-b)(1-a)+b^2 \end{bmatrix}$$

Now, $a^2 + (1 - a)(1 - b) \ge 0$ and $a(1 - a) + (1 - a)b = (1 - a)(a + b) \ge 0$ since $0 \le a \le 1$ and $0 \le b \le 1$.

Also, $a^2 + (1 - a)(1 - b) + (1 - a)(a + b) = a^2 + (1 - a)[1 - b + a + b]$
$$= a^2 + (1 - a)(1 + a) = a^2 + 1 - a^2 = 1$$

Therefore, the elements in the first row of P^2 are nonnegative and their sum is 1. The same arguments apply to the elements in the second row of P^2. Thus, P^2 is a probability matrix.

75. $P = \begin{bmatrix} .4 & .6 \\ .2 & .8 \end{bmatrix}$

(A)   Let $S_0 = [0 \;\; 1]$. Then

$S_2 = S_0 P^2 = [.24 \;\; .76]$

$S_4 = S_0 P^4 = [.2496 \;\; .7504]$

$S_8 = S_0 P^8 = [.24999936 \;\; .7500006]$

S_k is approaching $[.25 \;\; .75]$

(B)   Let $S_0 = [1 \;\; 0]$. Then

$S_2 = S_0 P^2 = [.28 \;\; .72]$

$S_4 = S_0 P^4 = [.2512 \;\; .7488]$

$S_8 = S_0 P^8 = [.25000192 \;\; .7499980]$

S_k is approaching $[.25 \;\; .75]$

(C)   Let $S_0 = [.5 \;\; .5]$. Then

$S_2 = S_0 P^2 = [.26 \;\; .74]$

$S_4 = S_0 P^4 = [.2504 \;\; .7496]$

$S_8 = S_0 P^8 = [.25000064 \;\; .74999936]$

S_k is approaching $[.25 \;\; .75]$

(D) $[.25 \;\; .75]\begin{bmatrix} .4 & .6 \\ .2 & .8 \end{bmatrix} = [.25 \;\; .75]$

(E)   The state matrices S_k appear to approach the same matrix $[.25 \;\; .75]$, regardless of the values in the initial state matrix S_0.

77. $P^2 = \begin{bmatrix} .28 & .72 \\ .24 & .76 \end{bmatrix}$, $P^4 = \begin{bmatrix} .2512 & .7488 \\ .2496 & .7504 \end{bmatrix}$

$P^8 = \begin{bmatrix} .25000192 & .7499980 \\ .24999936 & .7500006 \end{bmatrix} \cdots$

The matrices P^k are approaching $Q = \begin{bmatrix} .25 & .75 \\ .25 & .75 \end{bmatrix}$; the rows of Q are the same as the matrix $S = \begin{bmatrix} .25 & .75 \end{bmatrix}$

in Problem 75.

79. Let R denote "rain" and R' "not rain".

(A)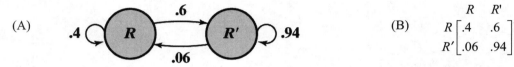

(B) $\begin{array}{c} \\ R \\ R' \end{array} \begin{array}{cc} R & R' \\ \begin{bmatrix} .4 & .6 \\ .06 & .94 \end{bmatrix} \end{array}$

(C) Rain on Saturday: $P^2 = \begin{array}{c} \\ R \\ R' \end{array} \begin{array}{cc} R & R' \\ \begin{bmatrix} .196 & .804 \\ .0804 & .9196 \end{bmatrix} \end{array}$

The probability that it will rain on Saturday is .196.

Rain on Sunday: $P^3 = \begin{array}{c} \\ R \\ R' \end{array} \begin{array}{cc} R & R' \\ \begin{bmatrix} .12664 & .87336 \\ .087336 & .912664 \end{bmatrix} \end{array}$

The probability that it will rain on Sunday is .12664.

81. (A)

(B) $\begin{array}{c} \\ X \\ X' \end{array} \begin{array}{cc} X & X' \\ \begin{bmatrix} .8 & .2 \\ .2 & .8 \end{bmatrix} \end{array}$

(C) $S = \begin{bmatrix} .2 & .8 \end{bmatrix}$

$S_1 = SP = \begin{bmatrix} .2 & .8 \end{bmatrix} \begin{bmatrix} .8 & .2 \\ .2 & .8 \end{bmatrix} = \begin{bmatrix} .32 & .68 \end{bmatrix}$

32% will be using brand X one week later.

$S_2 = SP^2 = \begin{bmatrix} .2 & .8 \end{bmatrix} \begin{bmatrix} .68 & .32 \\ .32 & .68 \end{bmatrix} = \begin{bmatrix} .392 & .608 \end{bmatrix}$

39.2% will be using brand X two weeks later.

83. (A)

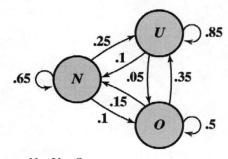

(B)
$$P = \begin{array}{c} \\ N \\ U \\ O \end{array}\begin{array}{ccc} N & U & O \\ \left[\begin{array}{ccc} .65 & .25 & .10 \\ .10 & .85 & .05 \\ .15 & .35 & .50 \end{array}\right] \end{array}$$

(C) $S = \begin{array}{ccc} N & U & O \\ [.50 & .30 & .20] \end{array}$

$\qquad\qquad N \quad\; U \quad\; O$

After one year: $SP = [.385\;\; .45\;\; .165]$

38.5% will be insured by National Property after one year.

$\qquad\qquad N \quad\quad U \quad\quad O$

After two years: $SP^2 = [.32\;\; .5365\;\; .1435]$

32% will be insured by National Property after two years.

(D)   45% of the homes will be insured by United Family after one year;

53.65% will be insured by United Family after two years.

85. (A)

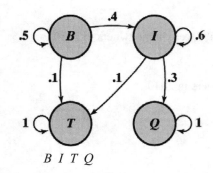

$B\; I\; T\; Q$

(B)
$$P = \begin{array}{c} \\ B \\ I \\ T \\ Q \end{array}\begin{array}{cccc} B & I & T & Q \\ \left[\begin{array}{cccc} .5 & .4 & .1 & 0 \\ 0 & .6 & .1 & .3 \\ 0 & 0 & 1 & 0 \\ 0 & 0 & 0 & 1 \end{array}\right] \end{array}$$

(C) $S = [1\; 0\; 0\; 0]$

$\qquad\qquad\quad B \quad\; I \quad\; T \quad\; Q$

After one year: $SP^2 = [.25\;\; .44\;\; .19\;\; .12]$

The probability that a beginning agent will be promoted to qualified agent within one year (i.e., after 2 reviews) is: .12.

$\qquad\qquad\quad B \quad\quad I \quad\quad T \quad\quad Q$

After two years: $SP^4 = [.0625\;\; .2684\;\; .3079\;\; .3612]$

The probability that a beginning agent will be promoted to qualified agent within two years (i.e., after 4 reviews) is: .3612.

87.   (A) $P = \begin{array}{c} \\ HMO \\ PPO \\ FFS \end{array} \begin{array}{ccc} HMO & PPO & FFS \\ \left[\begin{array}{ccc} .80 & .15 & .05 \\ .20 & .70 & .10 \\ .25 & .30 & .45 \end{array} \right] \end{array}$

(B) $\begin{array}{ccc} HMO & PPO & FFS \end{array}$
$S = [.20 \quad .25 \quad .55]$
$\begin{array}{ccc} HMO & PPO & FFS \end{array}$
$SP = [.3475 \quad .37 \quad .2825]$

34.75% were enrolled in the HMO; 37% were enrolled in the PPO; 28.25% were enrolled in the FFS.

(C) $\begin{array}{ccc} HMO & PPO & FFS \end{array}$
$SP^2 = [.422625 \quad .395875 \quad .1815]$

42.2625% will be enrolled in the HMO; 39.5875% will be enrolled in the PPO; 18.15% will be enrolled in the FFS.

89.   (A) $P = \begin{array}{c} \\ H \\ R \end{array} \begin{array}{cc} H & R \\ \left[\begin{array}{cc} .847 & .153 \\ .174 & .826 \end{array} \right] \end{array}$

(B) $\begin{array}{cc} H & R \end{array}$
$S = [0.419 \quad 0.581]; \quad SP = [0.456 \quad 0.544], \quad$ 45.6% were homeowners in 2010.

(C) $\begin{array}{cc} H & R \end{array}$
$SP^3 = [0.498 \quad 0.502], \quad$ 49.8% will be homeowners in 2030.

EXERCISE 9-2

Things to remember:

1.   STATIONARY MATRIX FOR A MARKOV CHAIN
     The state matrix $S = [s_1 \ s_2 \ \dots \ s_n]$ is a STATIONARY MATRIX for a Markov chain with transition matrix P if

     $SP = S$

     where $s_i \geq 0$, $i = 1, \dots , n$ and $s_1 + s_2 + \dots + s_n = 1$.

2.   REGULAR MARKOV CHAINS
     A transition matrix P is REGULAR if some power of P has only positive entries. A Markov chain is a REGULAR MARKOV CHAIN if its transition matrix is regular.

3.   PROPERTIES OF REGULAR MARKOV CHAINS
     Let P be the transition matrix for a regular Markov chain.

(A) There is a unique stationary matrix S which can be found by solving the equation $SP = S$

(B) Given any initial state matrix S_0, the state matrices S_k approach the stationary matrix S.

(C) The matrices P^k approach a limiting matrix $\overline{P}$ where each row of $\overline{P}$ is equal to the stationary matrix S.

---

1. $\begin{bmatrix} 1 & 0 \\ 0 & 0 \end{bmatrix}^2 = \begin{bmatrix} 1 & 0 \\ 0 & 0 \end{bmatrix}\begin{bmatrix} 1 & 0 \\ 0 & 0 \end{bmatrix} = \begin{bmatrix} 1 & 0 \\ 0 & 0 \end{bmatrix}$; therefore $\begin{bmatrix} 1 & 0 \\ 0 & 0 \end{bmatrix}^{100} = \begin{bmatrix} 1 & 0 \\ 0 & 0 \end{bmatrix}$

3. $\begin{bmatrix} 1 & 0 \\ 0 & 1 \end{bmatrix}^2 = \begin{bmatrix} 1 & 0 \\ 0 & 1 \end{bmatrix}\begin{bmatrix} 1 & 0 \\ 0 & 1 \end{bmatrix} = \begin{bmatrix} 1 & 0 \\ 0 & 1 \end{bmatrix}$; therefore $\begin{bmatrix} 1 & 0 \\ 0 & 1 \end{bmatrix}^{100} = \begin{bmatrix} 1 & 0 \\ 0 & 1 \end{bmatrix}$

5. $\begin{bmatrix} 1 & 0 & 0 \\ 0 & 0 & 0 \\ 0 & 0 & 1 \end{bmatrix}\begin{bmatrix} 1 & 0 & 0 \\ 0 & 0 & 0 \\ 0 & 0 & 1 \end{bmatrix} = \begin{bmatrix} 1 & 0 & 0 \\ 0 & 0 & 0 \\ 0 & 0 & 1 \end{bmatrix}$; therefore $\begin{bmatrix} 1 & 0 & 0 \\ 0 & 0 & 0 \\ 0 & 0 & 1 \end{bmatrix}^{100} = \begin{bmatrix} 1 & 0 & 0 \\ 0 & 0 & 0 \\ 0 & 0 & 1 \end{bmatrix}$

7. $\begin{bmatrix} 0 & 0 & 1 \\ 0 & 0 & 0 \\ 0 & 0 & 0 \end{bmatrix}\begin{bmatrix} 0 & 0 & 1 \\ 0 & 0 & 0 \\ 0 & 0 & 0 \end{bmatrix} = \begin{bmatrix} 0 & 0 & 0 \\ 0 & 0 & 0 \\ 0 & 0 & 0 \end{bmatrix}$; therefore $\begin{bmatrix} 0 & 0 & 1 \\ 0 & 0 & 0 \\ 0 & 0 & 0 \end{bmatrix}^{100} = \begin{bmatrix} 0 & 0 & 0 \\ 0 & 0 & 0 \\ 0 & 0 & 0 \end{bmatrix}$

9. $P = \begin{bmatrix} .6 & .4 \\ .4 & .6 \end{bmatrix}$ Yes; P is a transition matrix and all entries are positive.

11. $P = \begin{bmatrix} .1 & .9 \\ .5 & .4 \end{bmatrix}$ No; P is not a transition matrix, the sum of the entries in the second row is not 1.

13. $P = \begin{bmatrix} .4 & .6 \\ 0 & 1 \end{bmatrix}$ No; P is a transition matrix but the second row of every power of P is [0 1].

15. $P = \begin{bmatrix} 0 & 1 \\ .8 & .2 \end{bmatrix}$ Yes; P is a transition matrix and $P^2 = \begin{bmatrix} .8 & .2 \\ .16 & .84 \end{bmatrix}$ has only positive entries.

17. $P = \begin{bmatrix} .6 & .4 \\ .1 & .9 \\ .3 & .7 \end{bmatrix}$ No; P is not a square matrix.

19. $P = \begin{bmatrix} 0 & 1 & 0 \\ 0 & 0 & 1 \\ .5 & .5 & 0 \end{bmatrix}$ Yes; P is a transition matrix and

$$P^5 = \begin{bmatrix} .25 & .25 & .5 \\ .25 & .5 & .25 \\ .125 & .375 & .5 \end{bmatrix} \text{ has only positive entries.}$$

21. $P = \begin{bmatrix} .1 & .3 & .6 \\ .8 & .1 & .1 \\ 0 & 0 & 1 \end{bmatrix}$ No; P is a transition matrix but the third row of every power of P is $[0 \ \ 0 \ \ 1]$.

23. Let $S = [s_1, s_2]$, and solve the system:

$$[s_1 \ \ s_2] \begin{bmatrix} .1 & .9 \\ .6 & .4 \end{bmatrix} = [s_1 \ \ s_2], s_1 + s_2 = 1$$

which is equivalent to

$$\begin{array}{ll} .1s_1 + .6s_2 = s_1 & \qquad -.9s_1 + .6s_2 = 0 \\ .9s_1 + .4s_2 = s_2 \qquad \text{or} & \qquad .9s_1 - .6s_2 = 0 \\ s_1 + \ s_2 = 1 & \qquad s_1 + \ s_2 = 1 \end{array}$$

The solution is: $s_1 = .4, s_2 = .6$

The stationary matrix $S = [.4 \ \ .6]$; the limiting matrix $\overline{P} = \begin{bmatrix} .4 & .6 \\ .4 & .6 \end{bmatrix}$.

25. Let $S = [s_1, s_2]$, and solve the system:

$$[s_1 \ \ s_2] \begin{bmatrix} .5 & .5 \\ .3 & .7 \end{bmatrix} = [s_1 \ \ s_2], s_1 + s_2 = 1$$

which is equivalent to

$$\begin{array}{ll} .5s_1 + .3s_2 = s_1 & \qquad -.5s_1 + .3s_2 = 0 \\ .5s_1 + .7s_2 = s_2 \qquad \text{or} & \qquad .5s_1 - .3s_2 = 0 \\ s_1 + \ s_2 = 1 & \qquad s_1 + \ s_2 = 1 \end{array}$$

The solution is: $s_1 = \dfrac{3}{8} = .375, s_2 = \dfrac{5}{8} = .625$

The stationary matrix $S = [.375 \ \ .625]$; the limiting matrix

$$\overline{P} = \begin{bmatrix} .375 & .625 \\ .375 & .625 \end{bmatrix}.$$

27. Let $S = [s_1 \ \ s_2 \ \ s_3]$, and solve the system:

$$[s_1 \ \ s_2 \ \ s_3] \begin{bmatrix} .5 & .1 & .4 \\ .3 & .7 & 0 \\ 0 & .6 & .4 \end{bmatrix} = [s_1 \ \ s_2 \ \ s_3], \quad s_1 + s_2 + s_3 = 1$$

which is equivalent to

$$\begin{array}{ll} .5s_1 + .3s_2 = s_1 & \qquad -.5s_1 + .3s_2 = 0 \\ .1s_1 + .7s_2 + .6s_3 = s_2 \qquad \text{or} & \qquad .1s_1 - .3s_2 + .6s_3 = 0 \\ .4s_1 + .4s_3 = s_3 & \qquad .4s_1 - .6s_3 = 0 \\ s_1 + \ s_2 + \ s_3 = 1 & \qquad s_1 + \ s_2 + \ s_3 = 1 \end{array}$$

From the first and third equations, we have $s_2 = \frac{5}{3}s_1$, and $s_3 = \frac{2}{3}s_1$.

Substituting these values into the fourth equation, we get:

$$s_1 + \frac{5}{3}s_1 + \frac{2}{3}s_1 = 1 \text{ or } \frac{10}{3}s_1 = 1$$

Therefore, $s_1 = .3, s_2 = .5, s_3 = .2$.

The stationary matrix $S = [.3 \ .5 \ .2]$;

the limiting matrix $\overline{P} = \begin{bmatrix} .3 & .5 & .2 \\ .3 & .5 & .2 \\ .3 & .5 & .2 \end{bmatrix}$.

29. Let $S = [s_1 \ s_2 \ s_3]$, and solve the system:

$$[s_1 \ s_2 \ s_3] \begin{bmatrix} .8 & .2 & 0 \\ .5 & .1 & .4 \\ 0 & .6 & .4 \end{bmatrix} = [s_1 \ s_2 \ s_3], s_1 + s_2 + s_3 = 1$$

which is equivalent to

$$\begin{aligned}
.8s_1 + .5s_2 &= s_1 \\
.2s_1 + .1s_2 + .6s_3 &= s_2 \\
.4s_2 + .4s_3 &= s_3 \\
s_1 + s_2 + s_3 &= 1
\end{aligned} \quad \text{or} \quad \begin{aligned}
-.2s_1 + .5s_2 &= 0 \\
.2s_1 - .9s_2 + .6s_3 &= 0 \\
.4s_2 - .6s_3 &= 0 \\
s_1 + s_2 + s_3 &= 1
\end{aligned}$$

From the first and third equations, we have $s_1 = \frac{5}{2}s_2$, and $s_3 = \frac{2}{3}s_2$. Substituting these values into the fourth equation, we get:

$$\frac{5}{2}s_2 + s_2 + \frac{2}{3}s_2 = 1 \text{ or } \frac{25}{6}s_2 = 1$$

Therefore, $s_2 = \frac{6}{25} = .24, s_1 = .6, s_3 = .16$.

The stationary matrix $S = [.6 \ .24 \ .16]$;   the limiting matrix $\overline{P} = \begin{bmatrix} .6 & .24 & .16 \\ .6 & .24 & .16 \\ .6 & .24 & .16 \end{bmatrix}$.

31. False. For the identity matrix I, $I^k = I$ for all positive integers k; no power of I has all positive entries.

33. True. Suppose $P = \begin{bmatrix} a & b \\ c & d \end{bmatrix}$ has two entries equal to 0. Then either P has a row of zeros, a column of zeros, or P has one of the two forms $P = \begin{bmatrix} a & 0 \\ 0 & d \end{bmatrix}$, or $P = \begin{bmatrix} 0 & b \\ c & 0 \end{bmatrix}$. In each case, P^k has the same form as P.

35. False. Every state matrix $S = [s_1, s_2]$ is a stationary matrix for the transition matrix $I = \begin{bmatrix} 1 & 0 \\ 0 & 1 \end{bmatrix}$, and I is not regular.

37. $P = \begin{bmatrix} .51 & .49 \\ .27 & .73 \end{bmatrix}, P^2 = \begin{bmatrix} .3924 & .6076 \\ .3348 & .6652 \end{bmatrix}, P^4 \approx \begin{bmatrix} .3574 & .6426 \\ .3541 & .6459 \end{bmatrix}$

$P^8 \approx \begin{bmatrix} .3553 & .6447 \\ .3553 & .6447 \end{bmatrix}, P^{16} \approx \begin{bmatrix} .3553 & .6447 \\ .3553 & .6447 \end{bmatrix}$

Therefore, $S \approx [.3553 \ .6447]$.

39. $P = \begin{bmatrix} .5 & .5 & 0 \\ 0 & .5 & .5 \\ .8 & .1 & .1 \end{bmatrix}, P^2 = \begin{bmatrix} .25 & .5 & .25 \\ .4 & .3 & .3 \\ .48 & .46 & .06 \end{bmatrix}, P^4 = \begin{bmatrix} .3825 & .39 & .2275 \\ .364 & .428 & .208 \\ .3328 & .4056 & .2616 \end{bmatrix},$

$P^8 \approx \begin{bmatrix} .3640 & .4084 & .2277 \\ .3642 & .4095 & .2262 \\ .3620 & .4095 & .2285 \end{bmatrix}, P^{16} \approx \begin{bmatrix} .3636 & .4091 & .2273 \\ .3636 & .4091 & .2273 \\ .3636 & .4091 & .2273 \end{bmatrix}$

Therefore, $S \approx [.3636 \ .4091 \ .2273]$.

41. (A)

(B) $\begin{array}{cc} & \text{Red} \quad \text{Blue} \end{array}$
$P = \begin{array}{c} \text{Red} \\ \text{Blue} \end{array}\begin{bmatrix} .4 & .6 \\ .2 & .8 \end{bmatrix}$

(C) Let $S = [s_1 \ s_2]$ and solve the system:

$$[s_1 \ s_2]\begin{bmatrix} .4 & .6 \\ .2 & .8 \end{bmatrix} = [s_1 \ s_2], s_1 + s_2 = 1$$

which is equivalent to

$.4s_1 + .2s_2 = s_1 \qquad\qquad -.6s_1 + .2s_2 = 0$

$.6s_1 + .8s_2 = s_2 \qquad \text{or} \qquad .6s_1 - .2s_2 = 0$

$s_1 + \ s_2 = 1 \qquad\qquad\qquad s_1 + \ s_2 = 1$

The solution is: $s_1 = .25, \quad s_2 = .75$.

Thus, the stationary matrix $S = [.25 \ .75]$. In the long run, the red urn will be selected 25% of the time and the blue urn 75% of the time.

43. (A) $S_1 = [.2 \ .8]\begin{bmatrix} 0 & 1 \\ 1 & 0 \end{bmatrix} = [.8 \ .2]; \quad S_2 = S_1P = [.8 \ .2]\begin{bmatrix} 0 & 1 \\ 1 & 0 \end{bmatrix} = [.2 \ .8]; \quad S_3 = S_2P = [.8 \ .2]$

and so on.

The state matrices alternate between $[.2 \ .8]$ and $[.8 \ .2]$; they do not approach a "limiting" matrix.

(B) $S_1 = [.5 \ .5]\begin{bmatrix} 0 & 1 \\ 1 & 0 \end{bmatrix} = [.5 \ .5]; \quad S_2 = S_1P = [.5 \ .5]\begin{bmatrix} 0 & 1 \\ 1 & 0 \end{bmatrix} = [.5 \ .5]$

and so on.

Thus, $S_1 = S_2 = S_3 = \ldots = [.5 \ .5] = S_0$; S_0 is a stationary matrix.

(C) $P = \begin{bmatrix} 0 & 1 \\ 1 & 0 \end{bmatrix}, P^2 = \begin{bmatrix} 1 & 0 \\ 0 & 1 \end{bmatrix}, P^3 = \begin{bmatrix} 0 & 1 \\ 1 & 0 \end{bmatrix}, \ldots$

The powers of P alternate between P and the identity, I; they do not approach a limiting matrix.

(D)   Parts (B) and (C) are not valid for this matrix. Since P is not regular, this does not contradict Theorem 1.

45.   (A) $RP = [1 \ 0 \ 0] \begin{bmatrix} 1 & 0 & 0 \\ .2 & .2 & .6 \\ 0 & 0 & 1 \end{bmatrix} = [1 \ 0 \ 0]$

Therefore R is a stationary matrix for P.

$SP = [0 \ 0 \ 1] \begin{bmatrix} 1 & 0 & 0 \\ .2 & .2 & .6 \\ 0 & 0 & 1 \end{bmatrix} = [0 \ 0 \ 1]$

Therefore S is a stationary matrix for P.
The powers of P have the form

$$\begin{bmatrix} 1 & 0 & 0 \\ a & b & c \\ 0 & 0 & 1 \end{bmatrix}$$

Therefore P is not regular. As a result, P may have more than one stationary matrix.

(B)   Following the hint, let
$T = a[1 \ 0 \ 0] + (1 - a)[0 \ 0 \ 1], \quad 0 < a < 1.$
$= [a \ \ 0 \ \ 1 - a]$

Now, $TP = [a \ \ 0 \ \ 1 - a] \begin{bmatrix} 1 & 0 & 0 \\ .2 & .2 & .6 \\ 0 & 0 & 1 \end{bmatrix} = [a \ \ 0 \ \ 1 - a] = T$

Thus, $[a \ \ 0 \ \ 1 - a]$ is a stationary matrix for P for every a with $0 < a < 1$. Note that if $a = 1$, then $T = R$, and if $a = 0$, $T = S$. If we let $a = .5$, then $T = [.5 \ 0 \ .5]$ is a stationary matrix.

(C) P has infinitely many stationary matrices.

47. $\overline{P} = \begin{bmatrix} 1 & 0 & 0 \\ .25 & 0 & .75 \\ 0 & 0 & 1 \end{bmatrix}$

Each row of $\overline{P}$ is a stationary matrix for P. As we saw in Problem 45, part (B),
$T = [a \ \ 0 \ \ 1 - a]$
is a stationary matrix for P for each a where $0 \le a \le 1$; $a = 1$ gives the first row of $\overline{P}$, $a = .25$ gives the second row; $a = 0$ gives the third row.

49.   (A)   For $P^2, M_2 = .39$;   for $P^3, M_3 = .3$;   for $P^4, M_4 = .284$;   for $P^5, M_5 = .277$.

(B)   Each entry of the second column of P^{k+1} is a product of the form

$ap_{12}^k + bp_{22}^k + cp_{32}^k$

where $p_{12}^k, \ p_{22}^k, \ p_{32}^k$ are the entries in the second column of P^k;

$a, b, c \ge 0$ and $a + b + c = 1$. Thus,
$M_{k+1} = ap_{12}^k + bp_{22}^k + cp_{32}^k \le aM_k + bM_k + cM_k = (a + b + c)M_k = M_k$
Therefore, $M_k \ge M_{k+1}$ for all positive integers k.

51. The transition matrix is

$$P = \begin{matrix} & H & N \\ H & \\ N & \end{matrix} \begin{bmatrix} .89 & .11 \\ .29 & .71 \end{bmatrix} \qquad \begin{matrix} H = \text{home trackage} \\ N = \text{national pool} \end{matrix}$$

Calculating powers of P, we have

$$P^2 = \begin{bmatrix} .824 & .176 \\ .464 & .536 \end{bmatrix}, \quad P^4 \approx \begin{bmatrix} .7606 & .2394 \\ .6310 & .3690 \end{bmatrix}, \quad P^8 \approx \begin{bmatrix} .7296 & .2704 \\ .7128 & .2872 \end{bmatrix}, \quad P^{16} \approx \begin{bmatrix} .7251 & .2749 \\ .7248 & .2752 \end{bmatrix}$$

In the long run, 72.5% of the company's box cars will be on its home tracks.

53. (A) $S_1 = S_0 P = [.433 \quad .567] \begin{bmatrix} .92 & .08 \\ .2 & .8 \end{bmatrix} = [.51176 \quad .48824] \approx [.512 \quad .488]$

$S_2 = S_0 P^2 = [.433 \quad .567] \begin{bmatrix} .8624 & .1376 \\ .344 & .656 \end{bmatrix} = [.5684672 \quad .4315328] \approx [.568 \quad .432]$

$S_3 = S_0 P^3 = [.433 \quad .567] \begin{bmatrix} .820928 & .179072 \\ .44768 & .55232 \end{bmatrix} = [.609296384 \quad .390703616] \approx [.609 \quad .391]$

$S_4 = S_0 P^4 = [.433 \quad .567] \begin{bmatrix} .791068 & .208932 \\ .52233 & .47767 \end{bmatrix} = [.638693 \quad .361307] \approx [.639 \quad .361]$

(B)

Year	Data%	Model%
1970	43.3	43.3
1980	51.5	51.2
1990	57.5	56.8
2000	59.8	60.9
2010	58.5	63.9

(C) $P^4 \approx \begin{bmatrix} .7911 & .2089 \\ .5223 & .4777 \end{bmatrix}, \quad P^8 \approx \begin{bmatrix} .7349 & .2651 \\ .6627 & .3373 \end{bmatrix},$

$P^{16} \approx \begin{bmatrix} .7158 & .2842 \\ .7106 & .2894 \end{bmatrix}, \quad P^{32} \approx \begin{bmatrix} .7143 & .2857 \\ .7143 & .2857 \end{bmatrix}$

In the long run, 71.4% of the female population will be in the labor force.

55. The transition matrix for this problem is:

$$\begin{matrix} & GTT & NCJ & Dash \\ GTT \\ NCJ \\ Dash \end{matrix} \begin{bmatrix} .75 & .05 & .20 \\ .15 & .75 & .1 \\ .05 & .10 & .85 \end{bmatrix}$$

To find the steady-state matrix, we solve the system

$$[s_1 \quad s_2 \quad s_3] \begin{bmatrix} .75 & .05 & .20 \\ .15 & .75 & .10 \\ .05 & .10 & .85 \end{bmatrix} = [s_1 \quad s_2 \quad s_3], \quad s_1 + s_2 + s_3 = 1$$

which is equivalent to the system of equations

$$.75s_1 + .15s_2 + .05s_3 = s_1$$
$$.05s_1 + .75s_2 + .1s_3 = s_2$$
$$.20s_1 + .10s_2 + .85s_3 = s_3$$
$$s_1 + s_2 + s_3 = 1$$

The solution of this system is $s_1 = .25$, $s_2 = .25$, $s_3 = .5$.

Thus, the expected market share of each company is: GTT - 25%; NCJ - 25%; and Dash - 50%.

57. The transition matrix for this problem is:

$$
\begin{array}{c}
\\
\text{Poor} \\
\text{Satisfactory} \\
\text{Preferred}
\end{array}
\begin{array}{ccc}
\text{Poor} & \text{Satisfactory} & \text{Preferred} \\
\left[\begin{array}{ccc}
.60 & .40 & 0 \\
.20 & .60 & .20 \\
0 & .20 & .80
\end{array}\right]
\end{array}
$$

To find the steady-state matrix, we solve the system

$$[s_1 \ s_2 \ s_3]\begin{bmatrix} .60 & .40 & 0 \\ .20 & .60 & .20 \\ 0 & .20 & .80 \end{bmatrix} = [s_1 \ s_2 \ s_3], \quad s_1 + s_2 + s_3 = 1$$

which is equivalent to the system of equations:

$$.6s_1 + .2s_2 = s_1$$
$$.4s_1 + .6s_2 + .2s_3 = s_2$$
$$.2s_2 + .8s_3 = s_3$$
$$s_1 + s_2 + s_3 = 1$$

The solution of this system is $s_1 = .20$, $s_2 = .40$, and $s_3 = .40$.

Thus, the expected percentage in each category is: poor - 20%;   satisfactory - 40%;   preferred - 40%.

59. The transition matrix is:

$$P = \begin{bmatrix} .4 & .1 & .3 & .2 \\ .3 & .2 & .2 & .3 \\ .1 & .2 & .2 & .5 \\ .3 & .3 & .1 & .3 \end{bmatrix}$$

$$S_0 P = [.3 \ .3 \ .4 \ 0]P = [.25 \ .17 \ .23 \ .35] = S_1$$
$$S_1 P = [.25 \ .17 \ .23 \ .35]P = [.28 \ .21 \ .19 \ .32] = S_2$$
$$S_2 P = [.28 \ .21 \ .19 \ .32]P = [.29 \ .20 \ .20 \ .31] = S_3$$
$$S_3 P = [.29 \ .20 \ .20 \ .31]P = [.29 \ .20 \ .20 \ .31] = S_4$$

Thus, $S = [.29 \ .20 \ .20 \ .31]$ is the steady-state matrix. The expected market share for the two Acme soaps is $.20 + .31 = .51$ or 51%.

61. To find the stationary solution, we solve the system

$$[s_1 \ s_2 \ s_3] \begin{bmatrix} .5 & .5 & 0 \\ .25 & .5 & .25 \\ 0 & .5 & .5 \end{bmatrix} = [s_1 \ s_2 \ s_3], \quad s_1 + s_2 + s_3 = 1,$$

which is equivalent to:

$$\begin{array}{lll} .5s_1 + .25s_2 \quad\quad = s_1 & & -.5s_1 + .25s_2 \quad\quad = 0 \\ .5s_1 + .5s_2 + .5s_3 = s_2 & \text{or} & .5s_1 - .5s_2 + .5s_3 = 0 \\ \quad\quad .25s_2 + .5s_3 = s_3 & & \quad\quad .25s_2 - .5s_3 = 0 \\ s_1 + \quad s_2 + \quad s_3 = 1 & & s_1 + \quad s_2 + \quad s_3 = 1 \end{array}$$

The solution of this system is $s_1 = .25$, $s_2 = .5$, $s_3 = .25$.

Thus, the stationary matrix is $S = [.25 \ .5 \ .25]$.

63.

<table>
<tr><td></td><td></td><td></td><td>Rapid
transit</td><td>Auto</td></tr>
</table>

(A) Initial-state matrix $= [.25 \quad .75]$

(B) Second-state matrix $= [.25 \ .75]\begin{bmatrix} .8 & .2 \\ .3 & .7 \end{bmatrix} = [.425 \ .575]$

Thus, 42.5% will be using the new system after one month.

Third-state matrix $= [.425 \ .575]\begin{bmatrix} .8 & .2 \\ .3 & .7 \end{bmatrix} = [.5125 \ .4875]$

Thus, 51.25% will be using the new system after two months.

(C) To find the stationary solution, we solve the system

$$[s_1 \ s_2]\begin{bmatrix} .8 & .2 \\ .3 & .7 \end{bmatrix} = [s_1 \ s_2], \quad s_1 + s_2 = 1,$$

which is equivalent to:

$$\begin{array}{lll} .8s_1 + .3s_2 = s_1 & & -.2s_1 + .3s_2 = 0 \\ .2s_1 + .7s_2 = s_2 & \text{or} & .2s_1 - .3s_2 = 0 \\ s_1 + \quad s_2 = 1 & & s_1 + \quad s_2 = 1 \end{array}$$

The solution of this system of linear equations is $s_1 = .6$ and $s_2 = .4$. Thus, the stationary solution is

$S = [.6 \ .4]$, which means that 60% of the commuters will use rapid transit and 40% will travel by automobile after the system has been in service for a long time.

65. (A) $S_1 = S_0 P = [.309 \ .691]\begin{bmatrix} .61 & .39 \\ .21 & .79 \end{bmatrix} = [.3336 \ .6664] \approx [.334 \ .666]$

$S_2 = S_0 P^2 = [.309 \ .691]\begin{bmatrix} .454 & .546 \\ .294 & .706 \end{bmatrix} = [.34344 \ .65656] \approx [.343 \ .657]$

$S_3 = S_0 P^3 = [.309 \ .691]\begin{bmatrix} .3916 & .6084 \\ .3276 & .6724 \end{bmatrix} = [.347376 \ .652624] \approx [.347 \ .653]$

$S_4 = S_0 P^4 = [.309 \ .691]\begin{bmatrix} .36664 & .63336 \\ .34104 & .65896 \end{bmatrix} = [.34895 \ .65105] \approx [.349 \ .651]$

(B)

Year	Data%	Model%
1970	30.9	30.9
1980	33.3	33.4
1990	34.4	34.3
2000	35.6	34.7
2010	37.9	34.9

(C) $P^4 \approx \begin{bmatrix} .3666 & .6334 \\ .3410 & .6590 \end{bmatrix}$, $P^8 \approx \begin{bmatrix} .3504 & .6496 \\ .3498 & .6502 \end{bmatrix}$,

$P^{16} \approx \begin{bmatrix} .3500 & .6500 \\ .3500 & .6500 \end{bmatrix}$

In the long run, 35% of the population will live in the South region.

EXERCISE 9-3

Things to remember:

1. ABSORBING STATES

    A state in a Markov chain is an ABSORBING STATE if once the state is entered, it is impossible to leave.

2. ABSORBING STATES AND TRANSITION MATRICES

    A state in a Markov chain is ABSORBING if and only if the row of the transition matrix corresponding to the state has a 1 on the main diagonal and zeros elsewhere.

3. ABSORBING MARKOV CHAINS

    A Markov chain is an ABSORBING CHAIN if

    (A)   There is at least one absorbing state.

    (B)   It is possible to go from each nonabsorbing state to at least one absorbing state in a finite number of steps.

4. STANDARD FORMS FOR ABSORBING MARKOV CHAINS

    A transition matrix for an absorbing Markov chain is a STANDARD FORM if the rows and columns are labeled so that all the absorbing states precede all the nonabsorbing states. (There may be more than one standard form.) Any standard form can always be partitioned into four submatrices:

    $$\begin{matrix} & A & N \\ \begin{matrix} A \\ N \end{matrix} & \left[\begin{array}{c|c} I & 0 \\ \hline R & Q \end{array}\right] \end{matrix} \quad \left[\begin{array}{c} A = \text{all absorbing states} \\ \hline N = \text{all nonabsorbing states} \end{array}\right]$$

    where I is an identity matrix and 0 is a zero matrix.

5.   LIMITING MATRICES FOR ABSORBING MARKOV CHAINS

If a standard form P for an absorbing Markov chain is partitioned as

$$P = \left[\begin{array}{c|c} I & 0 \\ \hline R & Q \end{array}\right]$$

then P^k approaches a matrix $\overline{P}$ as k increases, where

$$\overline{P} = \left[\begin{array}{c|c} I & 0 \\ \hline FR & 0 \end{array}\right]$$

The matrix F is given by $F = (I - Q)^{-1}$ and is called the FUNDAMENTAL MATRIX for P.

The identity matrix used to form the fundamental matrix F must be the same size as the matrix Q.

6.   PROPERTIES OF THE LIMITING MATRIX $\overline{P}$

If P is a standard form transition matrix for an absorbing Markov chain, F is the fundamental matrix, and $\overline{P}$ is the limiting matrix, then

(A)   The entry in row i and column j of $\overline{P}$ is the long run probability of going from state i to state j. For the nonabsorbing states, these probabilities are also the entries in the matrix FR used to form $\overline{P}$.

(B)   The sum of the entries in each row of the fundamental matrix F is the average number of trials it will take to go from each nonabsorbing state to some absorbing state.

[Note that the rows of both F and FR correspond to the nonabsorbing states in the order given in the standard form P.]

---

1.   By 2, states B and C are absorbing states.

3.   By 2, there are no absorbing states.

5.   By 2, states A and D are absorbing states.

7. B is an absorbing state; the diagram represents an absorbing Markov chain since it is possible to go from states A and C to state B in a finite number of steps.

9. C is an absorbing state; the diagram does not represent an absorbing Markov chain since it is not possible to go from either states A or D to state C.

11. $P = \begin{bmatrix} 0 & 1 \\ 1 & 0 \end{bmatrix}$ No; P has no absorbing states.

13. $P = \begin{array}{c} \\ A \\ B \end{array}\begin{array}{c} A \quad B \\ \begin{bmatrix} .3 & .7 \\ 0 & 1 \end{bmatrix} \end{array}$ Yes; B is an absorbing state and it is possible for state A to go to state B in one step.

15. $P = \begin{array}{c} \\ A \\ B \\ C \end{array} \begin{array}{ccc} A & B & C \\ \left[\begin{array}{ccc} 1 & 0 & 0 \\ 0 & 1 & 0 \\ 0 & 0 & 1 \end{array}\right] \end{array}$ Yes; A, B and C are each absorbing states, there are no nonabsorbing states.

17. $P = \begin{array}{c} \\ A \\ B \\ C \end{array} \begin{array}{ccc} A & B & C \\ \left[\begin{array}{ccc} .9 & .1 & 0 \\ .9 & .1 & 0 \\ 0 & 0 & 1 \end{array}\right] \end{array}$ No; C is an absorbing state but it is impossible to go from either state A or state B to the absorbing state C.

19. $P = \begin{array}{c} \\ A \\ B \\ C \end{array} \begin{array}{ccc} A & B & C \\ \left[\begin{array}{ccc} .9 & 0 & .1 \\ 0 & 1 & 0 \\ 0 & .2 & .8 \end{array}\right] \end{array}$ Yes; B is an absorbing state and state C can go to state B in one step, state A can go to state B in two steps.

21. The transition diagram is represented by the matrix:

$$\begin{array}{c} \\ A \\ B \\ C \end{array} \begin{array}{ccc} A & B & C \\ \left[\begin{array}{ccc} .2 & .5 & .3 \\ 0 & 1 & 0 \\ .5 & .1 & .4 \end{array}\right] \end{array}$$

A standard form for this matrix is:

$$\begin{array}{c} \\ B \\ A \\ C \end{array} \begin{array}{ccc} B & A & C \\ \left[\begin{array}{ccc} 1 & 0 & 0 \\ .5 & .2 & .3 \\ .1 & .5 & .4 \end{array}\right] \end{array}$$

23. The transition diagram is represented by the matrix

$$\begin{array}{c} \\ A \\ B \\ C \\ D \end{array} \begin{array}{cccc} A & B & C & D \\ \left[\begin{array}{cccc} .3 & .4 & .2 & .1 \\ 0 & 1 & 0 & 0 \\ 0 & .4 & .3 & .3 \\ 0 & 0 & 0 & 1 \end{array}\right] \end{array}$$

A standard form for this matrix

$$\begin{array}{c} \\ B \\ D \\ A \\ C \end{array} \begin{array}{cccc} B & D & A & C \\ \left[\begin{array}{cccc} 1 & 0 & 0 & 0 \\ 0 & 1 & 0 & 0 \\ .4 & .1 & .3 & .2 \\ .4 & .3 & 0 & .3 \end{array}\right] \end{array}$$

25. A standard form for

$$P = \begin{array}{c} \\ A \\ B \\ C \end{array} \begin{array}{ccc} A & B & C \\ \left[\begin{array}{ccc} .2 & .3 & .5 \\ 1 & 0 & 0 \\ 0 & 0 & 1 \end{array}\right] \end{array}$$

is:

$$\begin{array}{c} \\ C \\ A \\ B \end{array} \begin{array}{ccc} C & A & B \\ \left[\begin{array}{ccc} 1 & 0 & 0 \\ .5 & .2 & .3 \\ 0 & 1 & 0 \end{array}\right] \end{array}$$

27. A standard form for

$$P = \begin{array}{c} \\ A \\ B \\ C \\ D \end{array} \begin{array}{cccc} A & B & C & D \\ \left[\begin{array}{cccc} .1 & .2 & .3 & .4 \\ 0 & 1 & 0 & 0 \\ .5 & .2 & .2 & .1 \\ 0 & 0 & 0 & 1 \end{array}\right] \end{array}$$

is:

$$\begin{array}{c} \\ B \\ D \\ A \\ C \end{array} \begin{array}{cccc} B & D & A & C \\ \left[\begin{array}{cccc} 1 & 0 & 0 & 0 \\ 0 & 1 & 0 & 0 \\ .2 & .4 & .1 & .3 \\ .2 & .1 & .5 & .2 \end{array}\right] \end{array}$$

29. For

$$P = \begin{array}{c} \\ A \\ B \\ C \end{array}\begin{array}{ccc} A & B & C \\ \left[\begin{array}{ccc} 1 & 0 & 0 \\ 0 & 1 & 0 \\ .1 & .4 & .5 \end{array}\right] \end{array}$$

we have $R = [.1 \ .4]$ and $Q = [.5]$.

The limiting matrix $\overline{P}$ has the form

$$\overline{P} = \left[\begin{array}{cc|c} 1 & 0 & 0 \\ 0 & 1 & 0 \\ \hline FR & & 0 \end{array}\right]$$

where $F = (I - Q)^{-1} = ([1] - [.5])^{-1} = [.5]^{-1} = [2]$ and $FR = [2][.1 \ .4] = [.2 \ .8]$.

Thus,

$$\overline{P} = \begin{array}{c} \\ A \\ B \\ C \end{array}\begin{array}{ccc} A & B & C \\ \left[\begin{array}{ccc} 1 & 0 & 0 \\ 0 & 1 & 0 \\ .2 & .8 & 0 \end{array}\right] \end{array}$$

Let $P(i \text{ to } j)$ denote the probability of going from state i to state j. Then $P(C \text{ to } A) = .2$, $P(C \text{ to } B) = .8$
Since $F = [2]$, it will take an average of 2 trials to go from C to either A or B.

31. For

$$P = \begin{array}{c} \\ A \\ B \\ C \end{array}\begin{array}{ccc} A & B & C \\ \left[\begin{array}{ccc} 1 & 0 & 0 \\ .2 & .6 & .2 \\ .4 & .2 & .4 \end{array}\right] \end{array}$$

we have $R = \left[\begin{array}{c} .2 \\ .4 \end{array}\right]$ and $Q = \left[\begin{array}{cc} .6 & .2 \\ .2 & .4 \end{array}\right]$.

The limiting matrix $\overline{P}$ has the form

$$\overline{P} = \left[\begin{array}{c|c} 1 & 0 & 0 \\ \hline FR & 0 \end{array}\right] \text{ where } F = (I-Q)^{-1} = \left(\left[\begin{array}{cc} 1 & 0 \\ 0 & 1 \end{array}\right] - \left[\begin{array}{cc} .6 & .2 \\ .2 & .4 \end{array}\right]\right)^{-1} = \left[\begin{array}{cc} .4 & -.2 \\ -.2 & .6 \end{array}\right]^{-1} = \left[\begin{array}{cc} \frac{2}{5} & -\frac{1}{5} \\ -\frac{1}{5} & \frac{3}{5} \end{array}\right]^{-1}$$

We use row operations to find the inverse:

$$\left[\begin{array}{cc|cc} \frac{2}{5} & -\frac{1}{5} & 1 & 0 \\ -\frac{1}{5} & \frac{3}{5} & 0 & 1 \end{array}\right] \sim \left[\begin{array}{cc|cc} 1 & -\frac{1}{2} & \frac{5}{2} & 0 \\ -\frac{1}{5} & \frac{3}{5} & 0 & 1 \end{array}\right] \sim \left[\begin{array}{cc|cc} 1 & -\frac{1}{2} & \frac{5}{2} & 0 \\ 0 & \frac{1}{2} & \frac{1}{2} & 1 \end{array}\right] \sim \left[\begin{array}{cc|cc} 1 & -\frac{1}{2} & \frac{5}{2} & 0 \\ 0 & 1 & 1 & 2 \end{array}\right]$$

$$\left(\frac{5}{2}\right)R_1 \rightarrow R_1 \qquad \left(\frac{1}{5}\right)R_1 + R_2 \rightarrow R_2 \qquad 2R_2 \rightarrow R_2 \qquad \left(\frac{1}{2}\right)R_2 + R_1 \rightarrow R_1$$

$$\sim \left[\begin{array}{cc|cc} 1 & 0 & 3 & 1 \\ 0 & 1 & 1 & 2 \end{array}\right]$$

Thus, $F = \left[\begin{array}{cc} 3 & 1 \\ 1 & 2 \end{array}\right]$ and $FR = \left[\begin{array}{cc} 3 & 1 \\ 1 & 2 \end{array}\right]\left[\begin{array}{c} .2 \\ .4 \end{array}\right] = \left[\begin{array}{c} 1 \\ 1 \end{array}\right]$

Now

$$\overline{P} = \begin{array}{c} \\ A \\ B \\ C \end{array}\begin{array}{ccc} A & B & C \\ \left[\begin{array}{ccc} 1 & 0 & 0 \\ 1 & 0 & 0 \\ 1 & 0 & 0 \end{array}\right] \end{array}$$

$P(B \text{ to } A) = 1$, $P(C \text{ to } A) = 1$
It will take an average of 4 trials to go from B to A; it will take an average of 3 trials to go from C to A.

33. For

$$P = \begin{array}{c} \\ A \\ B \\ C \\ D \end{array} \begin{array}{cccc} A & B & C & D \\ \begin{bmatrix} 1 & 0 & 0 & 0 \\ 0 & 1 & 0 & 0 \\ .1 & .2 & .6 & .1 \\ .2 & .2 & .3 & .3 \end{bmatrix} \end{array}$$

we have $R = \begin{bmatrix} .1 & .2 \\ .2 & .2 \end{bmatrix}$ and $Q = \begin{bmatrix} .6 & .1 \\ .3 & .3 \end{bmatrix}$

The limiting matrix $\overline{P}$ has the form

$$\overline{P} = \left[\begin{array}{c|cc} I & 0 & 0 \\ & 0 & 0 \\ \hline FR & 0 \end{array} \right]$$

where $F = (I - Q)^{-1} = \left(\begin{bmatrix} 1 & 0 \\ 0 & 1 \end{bmatrix} - \begin{bmatrix} .6 & .1 \\ .3 & .3 \end{bmatrix} \right)^{-1} = \begin{bmatrix} .4 & -.1 \\ -.3 & .7 \end{bmatrix}^{-1} = \begin{bmatrix} \frac{2}{5} & -\frac{1}{10} \\ -\frac{3}{10} & \frac{7}{10} \end{bmatrix}^{-1}$

We use row operations to find the inverse:

$$\begin{bmatrix} \frac{2}{5} & -\frac{1}{10} & 1 & 0 \\ -\frac{3}{10} & \frac{7}{10} & 0 & 1 \end{bmatrix} \sim \begin{bmatrix} 1 & -\frac{1}{4} & \frac{5}{2} & 0 \\ -\frac{3}{10} & \frac{7}{10} & 0 & 1 \end{bmatrix} \sim \begin{bmatrix} 1 & -\frac{1}{4} & \frac{5}{2} & 0 \\ 0 & \frac{5}{8} & \frac{3}{4} & 1 \end{bmatrix} \sim \begin{bmatrix} 1 & -\frac{1}{4} & \frac{5}{2} & 0 \\ 0 & 1 & \frac{6}{5} & \frac{8}{5} \end{bmatrix}$$

$$\left(\frac{5}{2} \right) R_1 \to R_1 \qquad \left(\frac{3}{10} \right) R_1 + R_2 \to R_2 \qquad \left(\frac{8}{5} \right) R_2 \to R_2 \qquad \left(\frac{1}{4} \right) R_2 + R_1 \to R_1$$

$$\sim \begin{bmatrix} 1 & 0 & \frac{14}{5} & \frac{2}{5} \\ 0 & 1 & \frac{6}{5} & \frac{8}{5} \end{bmatrix}$$

Thus, $F = \begin{bmatrix} \frac{14}{5} & \frac{2}{5} \\ \frac{6}{5} & \frac{8}{5} \end{bmatrix} = \begin{bmatrix} 2.8 & .4 \\ 1.2 & 1.6 \end{bmatrix}$,

$$FR = \begin{bmatrix} 2.8 & .4 \\ 1.2 & 1.6 \end{bmatrix} \begin{bmatrix} .1 & .2 \\ .2 & .2 \end{bmatrix} = \begin{bmatrix} .36 & .64 \\ .44 & .56 \end{bmatrix}$$

and

$$\overline{P} = \begin{array}{c} \\ A \\ B \\ C \\ D \end{array} \begin{array}{cccc} A & B & C & D \\ \begin{bmatrix} 1 & 0 & 0 & 0 \\ 0 & 1 & 0 & 0 \\ .36 & .64 & 0 & 0 \\ .44 & .56 & 0 & 0 \end{bmatrix} \end{array}$$

$P(C \text{ to } A) = .36, P(C \text{ to } B) = .64,$

$P(D \text{ to } A) = .44, P(D \text{ to } B) = .56$

It will take an average of 3.2 trials to go from C to either A or B; it will take an average of 2.8 trials to go from D to either A or B.

35. (A) $S_0 \overline{P} = [0\ 0\ 1] \begin{bmatrix} 1 & 0 & 0 \\ 0 & 1 & 0 \\ .2 & .8 & 0 \end{bmatrix} = [.2\ .8\ 0]$

  (B) $S_0 \overline{P} = [.2\ .5\ .3] \begin{bmatrix} 1 & 0 & 0 \\ 0 & 1 & 0 \\ .2 & .8 & 0 \end{bmatrix} = [.26\ .74\ 0]$

37. (A) $S_0 \overline{P} = [0\ 0\ 1] \begin{bmatrix} 1 & 0 & 0 \\ 1 & 0 & 0 \\ 1 & 0 & 0 \end{bmatrix} = [1\ 0\ 0]$

  (B) $S_0 \overline{P} = [.2\ .5\ .3] \begin{bmatrix} 1 & 0 & 0 \\ 1 & 0 & 0 \\ 1 & 0 & 0 \end{bmatrix} = [1\ 0\ 0]$

39. (A) $S_0 \overline{P} = [0\ 0\ 0\ 1] \begin{bmatrix} 1 & 0 & 0 & 0 \\ 0 & 1 & 0 & 0 \\ .36 & .64 & 0 & 0 \\ .44 & .56 & 0 & 0 \end{bmatrix} = [.44\ .56\ 0\ 0]$

  (B) $S_0 \overline{P} = [0\ 0\ 1\ 0] \begin{bmatrix} 1 & 0 & 0 & 0 \\ 0 & 1 & 0 & 0 \\ .36 & .64 & 0 & 0 \\ .44 & .56 & 0 & 0 \end{bmatrix} = [.36\ .64\ 0\ 0]$

  (C) $S_0 \overline{P} = [0\ 0\ .4\ .6] \begin{bmatrix} 1 & 0 & 0 & 0 \\ 0 & 1 & 0 & 0 \\ .36 & .64 & 0 & 0 \\ .44 & .56 & 0 & 0 \end{bmatrix} = [.408\ .592\ 0\ 0]$

  (D) $S_0 \overline{P} = [.1\ .2\ .3\ .4] \begin{bmatrix} 1 & 0 & 0 & 0 \\ 0 & 1 & 0 & 0 \\ .36 & .64 & 0 & 0 \\ .44 & .56 & 0 & 0 \end{bmatrix} = [.384\ .616\ 0\ 0]$

41. False. For the transition matrix:

$$\begin{array}{c} \\ A \\ B \\ C \end{array} \begin{array}{ccc} A & B & C \\ \begin{bmatrix} 1 & 0 & 0 \\ 0 & 0 & 1 \\ 0 & 1 & 0 \end{bmatrix} \end{array}$$

A is an absorbing state; B and C are nonabsorbing states and it is impossible to go from either state B or state C to state A.

43. False. The transition matrix in Problem 41 is a counterexample.

45. True. If every state is absorbing, then there are no nonabsorbing states so 3(<u>B</u>) is vacuously true.

47. False. A Markov chain whose transition matrix is I, the identity matrix, is absorbing but not regular.

49. By Theorem 2, P has a limiting matrix:

$$P^4 \approx \begin{bmatrix} 1 & 0 & 0 & 0 \\ 0 & 1 & 0 & 0 \\ .6364 & .362 & 0 & 0 \\ .7364 & .262 & 0 & 0 \end{bmatrix}, \quad P^8 \approx \begin{bmatrix} 1 & 0 & 0 & 0 \\ 0 & 1 & 0 & 0 \\ .6375 & .3625 & 0 & 0 \\ .7375 & .2625 & 0 & 0 \end{bmatrix}$$

$$P^{16} \approx \begin{bmatrix} 1 & 0 & 0 & 0 \\ 0 & 1 & 0 & 0 \\ .6375 & .3625 & 0 & 0 \\ .7375 & .2625 & 0 & 0 \end{bmatrix}; \quad \overline{P} = \begin{bmatrix} 1 & 0 & 0 & 0 \\ 0 & 1 & 0 & 0 \\ .6375 & .3625 & 0 & 0 \\ .7375 & .2625 & 0 & 0 \end{bmatrix}$$

51. By Theorem 2, P has a limiting matrix:

$$P^4 \approx \begin{bmatrix} 1 & 0 & 0 & 0 & 0 \\ 0 & 1 & 0 & 0 & 0 \\ .0724 & .8368 & .0625 & .011 & .0173 \\ .174 & .7792 & 0 & .0279 & .0189 \\ .4312 & .5472 & 0 & .0126 & .009 \end{bmatrix},$$

$$P^{16} \approx \begin{bmatrix} 1 & 0 & 0 & 0 & 0 \\ 0 & 1 & 0 & 0 & 0 \\ .0875 & .9125 & 0 & 0 & 0 \\ .1875 & .8125 & 0 & 0 & 0 \\ .4375 & .5625 & 0 & 0 & 0 \end{bmatrix}, \quad P^{32} \approx \begin{bmatrix} 1 & 0 & 0 & 0 & 0 \\ 0 & 1 & 0 & 0 & 0 \\ .0875 & .9125 & 0 & 0 & 0 \\ .1875 & .8125 & 0 & 0 & 0 \\ .4375 & .5625 & 0 & 0 & 0 \end{bmatrix};$$

$$\overline{P} = \begin{bmatrix} 1 & 0 & 0 & 0 & 0 \\ 0 & 1 & 0 & 0 & 0 \\ .0875 & .9125 & 0 & 0 & 0 \\ .1875 & .8125 & 0 & 0 & 0 \\ .4375 & .5625 & 0 & 0 & 0 \end{bmatrix}$$

53. *Step 1.*   Transition diagram:

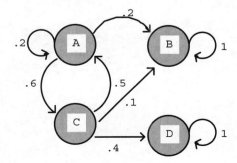

Standard form:

$$M = \begin{array}{c} \\ B \\ D \\ A \\ C \end{array} \begin{array}{cccc} B & D & A & C \\ \begin{bmatrix} 1 & 0 & 0 & 0 \\ 0 & 1 & 0 & 0 \\ .2 & 0 & .2 & .6 \\ .1 & .4 & .5 & 0 \end{bmatrix} \end{array}$$

Step 2.   Limiting matrix:

For M, we have $R = \begin{bmatrix} .2 & 0 \\ .1 & .4 \end{bmatrix}$ and $Q = \begin{bmatrix} .2 & .6 \\ .5 & 0 \end{bmatrix}$

The limiting matrix $\overline{M}$ has the form:

$$\overline{M} = \left[\begin{array}{c|c} I & 0 \\ \hline FR & 0 \end{array} \right]$$

where $F = (I - Q)^{-1} = \left(\begin{bmatrix} 1 & 0 \\ 0 & 1 \end{bmatrix} - \begin{bmatrix} .2 & .6 \\ .5 & 0 \end{bmatrix} \right)^{-1} = \begin{bmatrix} .8 & -.6 \\ -.5 & 1 \end{bmatrix}^{-1} = \begin{bmatrix} \frac{4}{5} & -\frac{3}{5} \\ -\frac{1}{2} & 1 \end{bmatrix}^{-1}$

We use row operations to find the inverse:

$$\left[\begin{array}{cc|cc} \frac{4}{5} & -\frac{3}{5} & 1 & 0 \\ -\frac{1}{2} & 1 & 0 & 1 \end{array} \right] \sim \left[\begin{array}{cc|cc} 1 & -\frac{3}{4} & \frac{5}{4} & 0 \\ -\frac{1}{2} & 1 & 0 & 1 \end{array} \right] \sim \left[\begin{array}{cc|cc} 1 & -\frac{3}{4} & \frac{5}{4} & 0 \\ 0 & \frac{5}{8} & \frac{5}{8} & 1 \end{array} \right]$$

$$\left(\frac{5}{4} \right) R_1 \to R_1 \qquad \left(\frac{1}{2} \right) R_1 + R_2 \to R_2 \qquad \left(\frac{8}{5} \right) R_2 \to R_2$$

$$\sim \left[\begin{array}{cc|cc} 1 & -\frac{3}{4} & \frac{5}{4} & 0 \\ 0 & 1 & 1 & \frac{8}{5} \end{array} \right] \sim \left[\begin{array}{cc|cc} 1 & 0 & 2 & \frac{6}{5} \\ 0 & 1 & 1 & \frac{8}{5} \end{array} \right]$$

$$\left(\frac{3}{4} \right) R_2 + R_1 \to R_1$$

Thus, $F = \begin{bmatrix} 2 & 1.2 \\ 1 & 1.6 \end{bmatrix}$ and $FR = \begin{bmatrix} 2 & 1.2 \\ 1 & 1.6 \end{bmatrix} \begin{bmatrix} .2 & 0 \\ .1 & .4 \end{bmatrix} = \begin{bmatrix} .52 & .48 \\ .36 & .64 \end{bmatrix}$

Therefore, $\overline{M} = \begin{array}{c} \\ B \\ D \\ A \\ C \end{array} \begin{array}{cccc} B & D & A & C \\ \begin{bmatrix} 1 & 0 & 0 & 0 \\ 0 & 1 & 0 & 0 \\ .52 & .48 & 0 & 0 \\ .36 & .64 & 0 & 0 \end{bmatrix} \end{array}$

Step 3.   Transition diagram for $\overline{M}$:

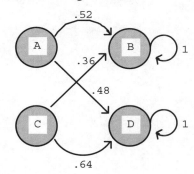

Limiting matrix for P:

$$\overline{P} = \begin{array}{c} \\ A \\ B \\ C \\ D \end{array} \begin{array}{c} \begin{array}{cccc} A & B & C & D \end{array} \\ \begin{bmatrix} 0 & .52 & 0 & .48 \\ 0 & 1 & 0 & 0 \\ 0 & .36 & 0 & .64 \\ 0 & 0 & 0 & 1 \end{bmatrix} \end{array}$$

55. $P^4 \approx \begin{bmatrix} .1276 & .426 & .0768 & .3696 \\ 0 & 1 & 0 & 0 \\ .064 & .29 & .102 & .544 \\ 0 & 0 & 0 & 1 \end{bmatrix}$, $P^8 \approx \begin{bmatrix} .0212 & .5026 & .0176 & .4585 \\ 0 & 1 & 0 & 0 \\ .0147 & .3468 & .0153 & .6231 \\ 0 & 0 & 0 & 0 \end{bmatrix}$

$P^{32} \approx \begin{bmatrix} 0 & .52 & 0 & .48 \\ 0 & 1 & 0 & 0 \\ 0 & .36 & 0 & .64 \\ 0 & 0 & 0 & 1 \end{bmatrix}$

57. Let $S = [x \quad 1-x \quad 0]$, $0 \le x \le 1$. Then

$$SP = [x \quad 1-x \quad 0] \begin{bmatrix} 1 & 0 & 0 \\ 0 & 1 & 0 \\ .1 & .5 & .4 \end{bmatrix} = [x \quad 1-x \quad 0]$$

Thus, S is a stationary matrix for P.

A stationary matrix for an absorbing Markov chain with two absorbing states and one nonabsorbing state will have one of the forms

   $[x \quad 1-x \quad 0]$, $[x \quad 0 \quad 1-x]$, $[0 \quad x \quad 1-x]$

59. (A) For P^2, $w_2 = .370$; for P^4, $w_4 = .297$; for P^8, $w_8 = .227$;

   for P^{16}, $w_{16} = .132$; for P^{32}, $w_{32} = .045$

(B) For large k, the entries of Q^k are close to 0.

61. A transition matrix for this problem is:

$$P = \begin{array}{c} \\ F \\ G \\ A \\ B \end{array} \begin{array}{cccc} F & G & A & B \\ \left[\begin{array}{cccc} 1 & 0 & 0 & 0 \\ .1 & .8 & .1 & 0 \\ .1 & .4 & .4 & .1 \\ 0 & 0 & 0 & 1 \end{array}\right] \end{array}$$

A standard form for this matrix is:

$$M = \begin{array}{c} \\ F \\ B \\ G \\ A \end{array} \begin{array}{cccc} F & B & G & A \\ \left[\begin{array}{cccc} 1 & 0 & 0 & 0 \\ 0 & 1 & 0 & 0 \\ .1 & 0 & .8 & .1 \\ .1 & .1 & .4 & .4 \end{array}\right] \end{array}$$

For this matrix, we have:

$$R = \begin{bmatrix} .1 & 0 \\ .1 & .1 \end{bmatrix} \text{ and } Q = \begin{bmatrix} .8 & .1 \\ .4 & .4 \end{bmatrix}$$

The limiting matrix for M has the form:

$$\overline{M} = \left[\begin{array}{c|c} I & 0 \\ \hline FR & 0 \end{array}\right]$$

where $F = (I - Q)^{-1} = \left(\begin{bmatrix} 1 & 0 \\ 0 & 1 \end{bmatrix} - \begin{bmatrix} .8 & .1 \\ .4 & .4 \end{bmatrix}\right)^{-1} = \begin{bmatrix} .2 & -.1 \\ -.4 & .6 \end{bmatrix}^{-1} = \begin{bmatrix} \frac{1}{5} & -\frac{1}{10} \\ -\frac{2}{5} & \frac{3}{5} \end{bmatrix}^{-1}$

We use row operations to find the inverse:

$$\begin{bmatrix} \frac{1}{5} & -\frac{1}{10} & 1 & 0 \\ -\frac{2}{5} & \frac{3}{5} & 0 & 1 \end{bmatrix} \sim \begin{bmatrix} 1 & -\frac{1}{2} & 5 & 0 \\ -\frac{2}{5} & \frac{3}{5} & 0 & 1 \end{bmatrix} \sim \begin{bmatrix} 1 & -\frac{1}{2} & 5 & 0 \\ 0 & \frac{2}{5} & 2 & 1 \end{bmatrix} \sim \begin{bmatrix} 1 & -\frac{1}{2} & 5 & 0 \\ 0 & 1 & 5 & \frac{5}{2} \end{bmatrix}$$

$$5R_1 \to R_1 \qquad \left(\frac{2}{5}\right)R_1 + R_2 \to R_2 \qquad \left(\frac{5}{2}\right)R_2 \to R_2 \qquad \left(\frac{1}{2}\right)R_2 + R_1 \to R_1$$

$$\sim \begin{bmatrix} 1 & 0 & \frac{15}{2} & \frac{5}{4} \\ 0 & 1 & 5 & \frac{5}{2} \end{bmatrix}$$

Thus, $F = \begin{bmatrix} 7.5 & 1.25 \\ 5 & 2.5 \end{bmatrix}$ and $FR = \begin{bmatrix} .875 & .125 \\ .75 & .25 \end{bmatrix}$

Therefore,

$$\overline{M} = \begin{array}{c} \\ F \\ B \\ G \\ A \end{array} \begin{array}{cccc} F & B & G & A \\ \left[\begin{array}{cccc} 1 & 0 & 0 & 0 \\ 0 & 1 & 0 & 0 \\ .875 & .125 & 0 & 0 \\ .75 & .25 & 0 & 0 \end{array}\right] \end{array}$$

(A) In the long run, 75% of the accounts in arrears will pay in full.

(B) In the long run, 12.5% of the accounts in good standing will become bad debts.

(C) The average number of months that an account in arrears will either be paid in full or classified as a bad debt is: $5 + 2.5 = 7.5$ months

63. A transition matrix in standard form for this problem is:

$$P = \begin{array}{c} \\ A \\ B \\ C \\ N \end{array} \begin{array}{cccc} A & B & C & N \\ \left[\begin{array}{cccc} 1 & 0 & 0 & 0 \\ 0 & 1 & 0 & 0 \\ 0 & 0 & 1 & 0 \\ .06 & .03 & .11 & .8 \end{array} \right] \end{array}$$

For this matrix, we have $R = [.06\ .03\ .11]$ and $Q = [.8]$.

The limiting matrix for P has the form:

$$\overline{P} = \left[\begin{array}{c|c} I & 0 \\ \hline FR & 0 \end{array} \right]$$

where $F = (I - Q)^{-1} = ([1] - [.8])^{-1} = [.2]^{-1} = 5$

Now, $FR = [5][.06\ .03\ .11] = [.3\ .15\ .55]$

and

$$\overline{P} = \begin{array}{c} \\ A \\ B \\ C \\ N \end{array} \begin{array}{cccc} A & B & C & N \\ \left[\begin{array}{cccc} 1 & 0 & 0 & 0 \\ 0 & 1 & 0 & 0 \\ 0 & 0 & 1 & 0 \\ .3 & .15 & .55 & 0 \end{array} \right] \end{array}$$

(A) In the long run, the market share of each company is:
Company A: 30%; Company B: 15%; and Company C: 55%.

(B) On the average, it will take 5 years for a department to decide to
use a calculator from one of these companies in their courses.

65. Let I denote ICU, C denote CCW, D denote "died", and R denote "released". A transition matrix in standard form for this problem is:

$$P = \begin{array}{c} \\ D \\ R \\ I \\ C \end{array} \begin{array}{cccc} D & R & I & C \\ \left[\begin{array}{cccc} 1 & 0 & 0 & 0 \\ 0 & 1 & 0 & 0 \\ .02 & 0 & .46 & .52 \\ .01 & .22 & .04 & .73 \end{array} \right] \end{array}$$

For this matrix, we have

$$R = \left[\begin{array}{cc} .02 & 0 \\ .01 & .22 \end{array} \right] \text{ and } Q = \left[\begin{array}{cc} .46 & .52 \\ .04 & .73 \end{array} \right]$$

The limiting matrix for P has the form:

$$\overline{P} = \left[\begin{array}{c|c} I & 0 \\ \hline FR & 0 \end{array} \right]$$

where $F = (I - Q)^{-1} = \left(\left[\begin{array}{cc} 1 & 0 \\ 0 & 1 \end{array} \right] - \left[\begin{array}{cc} .46 & .52 \\ .04 & .73 \end{array} \right] \right)^{-1} = \left[\begin{array}{cc} .54 & -.52 \\ -.04 & .27 \end{array} \right]^{-1} = \left[\begin{array}{cc} 2.16 & 4.16 \\ .32 & 4.32 \end{array} \right]$

Now, $FR = \begin{bmatrix} 2.16 & 4.16 \\ .32 & 4.32 \end{bmatrix} \begin{bmatrix} .02 & 0 \\ .01 & .22 \end{bmatrix} = \begin{bmatrix} .0848 & .9152 \\ .0496 & .9504 \end{bmatrix}$

and

$$\overline{P} = \begin{array}{c} \\ D \\ R \\ I \\ C \end{array} \begin{array}{cccc} D & R & I & C \\ \begin{bmatrix} 1 & 0 & 0 & 0 \\ 0 & 1 & 0 & 0 \\ .0848 & .9152 & 0 & 0 \\ .0496 & .9504 & 0 & 0 \end{bmatrix} \end{array}$$

(A)   In the long run, 91.52% of the patients are released from the hospital.

(B)   In the long run, 4.96% of the patients in the CCW die without being released from the hospital.

(C)   The average number of days a patient in the ICU will stay in the hospital is:
$$2.16 + 4.16 = 6.32 \text{ days}$$

67.   A transition matrix in standard form for this problem is:

$$P = \begin{array}{c} \\ L \\ R \\ F \\ B \end{array} \begin{array}{cccc} L & R & F & B \\ \begin{bmatrix} 1 & 0 & 0 & 0 \\ 0 & 1 & 0 & 0 \\ \frac{1}{4} & \frac{1}{4} & 0 & \frac{1}{2} \\ \frac{2}{5} & \frac{1}{5} & \frac{2}{5} & 0 \end{bmatrix} \end{array}$$

For this matrix we have:

$$R = \begin{bmatrix} \frac{1}{4} & \frac{1}{4} \\ \frac{2}{5} & \frac{1}{5} \end{bmatrix} \text{ and } Q = \begin{bmatrix} 0 & \frac{1}{2} \\ \frac{2}{5} & 0 \end{bmatrix}$$

The limiting matrix for P has the form:

$$\overline{P} = \begin{bmatrix} I & 0 \\ \hline FR & 0 \end{bmatrix}$$

where $F = (I - Q)^{-1} = \left(\begin{bmatrix} 1 & 0 \\ 0 & 1 \end{bmatrix} - \begin{bmatrix} 0 & \frac{1}{2} \\ \frac{2}{5} & 0 \end{bmatrix} \right)^{-1} = \begin{bmatrix} 1 & -\frac{1}{2} \\ -\frac{2}{5} & 1 \end{bmatrix}^{-1}$

We use row operations to find the inverse:

$$\begin{bmatrix} 1 & -\frac{1}{2} & 1 & 0 \\ -\frac{2}{5} & 1 & 0 & 1 \end{bmatrix} \sim \begin{bmatrix} 1 & -\frac{1}{2} & 1 & 0 \\ 0 & \frac{4}{5} & \frac{2}{5} & 1 \end{bmatrix} \sim \begin{bmatrix} 1 & -\frac{1}{2} & 1 & 0 \\ 0 & 1 & \frac{1}{2} & \frac{5}{4} \end{bmatrix} \sim \begin{bmatrix} 1 & 0 & \frac{5}{4} & \frac{5}{8} \\ 0 & 1 & \frac{1}{2} & \frac{5}{4} \end{bmatrix}$$

$$\left(\frac{2}{5} \right) R_1 + R_2 \rightarrow R_2 \quad \left(\frac{5}{4} \right) R_2 \rightarrow R_2 \quad \left(\frac{1}{2} \right) R_2 + R_1 \rightarrow R_1$$

Thus, $F = \begin{bmatrix} \frac{5}{4} & \frac{5}{8} \\ \frac{1}{2} & \frac{5}{4} \end{bmatrix}$ and $FR = \begin{bmatrix} \frac{5}{4} & \frac{5}{8} \\ \frac{1}{2} & \frac{5}{4} \end{bmatrix} \begin{bmatrix} \frac{1}{4} & \frac{1}{4} \\ \frac{2}{5} & \frac{1}{5} \end{bmatrix} = \begin{bmatrix} \frac{9}{16} & \frac{7}{16} \\ \frac{5}{8} & \frac{3}{8} \end{bmatrix}$.

Now,

$$\overline{P} = \begin{array}{c} \\ L \\ R \\ F \\ B \end{array} \begin{array}{cccc} \hspace{0.5em} L & R & F & B \\ \left[\begin{array}{cccc} 1 & 0 & 0 & 0 \\ 0 & 1 & 0 & 0 \\ \frac{9}{16} & \frac{7}{16} & 0 & 0 \\ \frac{5}{8} & \frac{3}{8} & 0 & 0 \end{array} \right] \end{array}$$

(A)   The long run probability that a rat placed in room B will end up in room R is $\dfrac{3}{8} = .375$.

(B)   The average number of exits that a rat placed in room B will choose until it finds food is:

$$\frac{1}{2} + \frac{5}{4} = \frac{7}{4} = 1.75$$

CHAPTER 9 REVIEW

1. $S_1 = S_0 P = [.3 \ \ .7] \begin{bmatrix} .6 & .4 \\ .2 & .8 \end{bmatrix} \overset{A \quad\ \ B}{= [.32 \ \ .68]}$

   $S_2 = S_1 P = [.32 \ \ .68] \begin{bmatrix} .6 & .4 \\ .2 & .8 \end{bmatrix} \overset{A \quad\quad\ B}{= [.328 \ \ .672]}$

   The probability of being in state A after one trial is .32; after two trials .328. The probability of being in state B after one trial is .68; after two trials .672.   (9-1)

2. A is an absorbing state; the chain is absorbing since it is possible to go from state B to state A.

3.   There are no absorbing states since there are no 1's on the main diagonal. P is regular since

   $P^2 = \begin{bmatrix} .7 & .3 \\ .21 & .79 \end{bmatrix}$ has only positive entries.   (9-2, 9-3)

4. $P = \begin{bmatrix} 0 & 1 \\ 1 & 0 \end{bmatrix}$ has no absorbing states. Since P^k, $k = 1, 2, 3, \ldots$, alternates between $\begin{bmatrix} 0 & 1 \\ 1 & 0 \end{bmatrix}$ and $\begin{bmatrix} 1 & 0 \\ 0 & 1 \end{bmatrix}$,
   P is not regular.   (9-2, 9-3)

5.   States B and C are absorbing. The chain is absorbing; it is possible to go from the nonabsorbing state A to the absorbing state C in one step.   (9-2, 9-3)

6.   States A and B are absorbing. The chain is neither absorbing (it is not possible to go from States C and D to either state A or state B) nor regular (all powers of P will have the same form).   (9-2, 9-3)

7. $P = \begin{matrix} A \\ B \\ C \end{matrix} \begin{bmatrix} 0 & 1 & 0 \\ .1 & 0 & .9 \\ 0 & 1 & 0 \end{bmatrix}$ There are no absorbing states.

P^k, $k = 1, 2, 3, \dots$, alternates between $\begin{bmatrix} 0 & 1 & 0 \\ .1 & 0 & .9 \\ 0 & 1 & 0 \end{bmatrix}$ and $\begin{bmatrix} .1 & 0 & .9 \\ 0 & 1 & 0 \\ .1 & 0 & .9 \end{bmatrix}$. Thus, P is not regular.

(9-1, 9-2, 9-3)

8. $P = \begin{matrix} & A & B & C \\ A \\ B \\ C \end{matrix} \begin{bmatrix} 0 & 1 & 0 \\ .1 & .2 & .7 \\ 0 & 0 & 1 \end{bmatrix}$

C is an absorbing state. The chain is absorbing since it is possible to go from state A to state C (via B) and from state B to state C.   (9-1, 9-2, 9-3)

9. $P = \begin{matrix} & A & B & C \\ A \\ B \\ C \end{matrix} \begin{bmatrix} 0 & 0 & 1 \\ .1 & .2 & .7 \\ 0 & 1 & 0 \end{bmatrix}$

There are no absorbing states since there are no 1's on the main diagonal.

P is regular since $P^3 = \begin{bmatrix} .1 & .2 & .7 \\ .074 & .388 & .538 \\ .02 & .74 & .24 \end{bmatrix}$ has only positive entries.   (9-1, 9-2, 9-3)

10. $P = \begin{matrix} & A & B & C & D \\ A \\ B \\ C \\ D \end{matrix} \begin{bmatrix} .3 & .2 & 0 & .5 \\ 0 & 1 & 0 & 0 \\ 0 & 0 & .2 & .8 \\ 0 & 0 & .3 & .7 \end{bmatrix}$

B is an absorbing state. The chain is not absorbing since it is not possible to go from state C to B, nor is it possible to go from state D to state B.   (9-1, 9-2, 9-3)

11.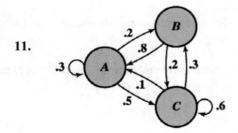

$P = \begin{matrix} & A & B & C \\ A \\ B \\ C \end{matrix} \begin{bmatrix} .3 & .2 & .5 \\ .8 & 0 & .2 \\ .1 & .3 & .6 \end{bmatrix}$

(9-1)

12. $P = \begin{array}{c} \\ A \\ B \end{array}\begin{array}{cc} A & B \\ \left[.4 \right. & \left. .6 \right] \\ \left[.9 \right. & \left. .1 \right] \end{array}$

(A) $P^2 = \begin{array}{c} \\ A \\ B \end{array}\begin{array}{cc} A & B \\ \left[.7 \right. & \left. .3 \right] \\ \left[.45 \right. & \left. .55 \right] \end{array}$

The probability of going from state A to state B in two trials is .3.

(B) $P^3 = \begin{array}{c} \\ A \\ B \end{array}\begin{array}{cc} A & B \\ \left[.55 \right. & \left. .45 \right] \\ \left[.675 \right. & \left. .325 \right] \end{array}$

The probability of going from state B to state A in three trials is .675.

(9-1)

13. Let $S = [s_1 \ \ s_2]$ and solve the system:

$$[s_1 \ \ s_2]\begin{bmatrix} .4 & .6 \\ .2 & .8 \end{bmatrix} = [s_1 \ \ s_2], \quad s_1 + s_2 = 1$$

which is equivalent to

$$\begin{array}{ccc} .4s_1 + .2s_2 = s_1 & & -.6s_1 + .2s_2 = 0 \\ .6s_1 + .8s_2 = s_2 & \text{or} & .6s_1 - .2s_2 = 0 \\ s_1 + s_2 = 1 & & s_1 + s_2 = 1 \end{array}$$

The solution is: $s_1 = .25, \quad s_2 = .75$.

The stationary matrix $S = [.25 \ \ .75]$

The limiting matrix $\overline{P} = \begin{array}{c} \\ A \\ B \end{array}\begin{array}{cc} A & B \\ \left[.25 \right. & \left. .75 \right] \\ \left[.25 \right. & \left. .75 \right] \end{array}$

14. Let $S = [s_1 \ \ s_2 \ \ s_3]$ and solve the system:

$$[s_1 \ \ s_2 \ \ s_3]\begin{bmatrix} .4 & .6 & 0 \\ .5 & .3 & .2 \\ 0 & .8 & .2 \end{bmatrix} = [s_1 \ \ s_2 \ \ s_3], \quad s_1 + s_2 + s_3 = 1$$

which is equivalent to

$$\begin{array}{ccc} .4s_1 + .5s_2 = s_1 & & -.6s_1 + .5s_2 = 0 \\ .6s_1 + .3s_2 + .8s_3 = s_2 & \text{or} & .6s_1 - .7s_2 + .8s_3 = 0 \\ .2s_2 + .2s_3 = s_3 & & .2s_2 - .8s_3 = 0 \\ s_1 + s_2 + s_3 = 1 & & s_1 + s_2 + s_3 = 1 \end{array}$$

From the first and third equations, we have $s_1 = \dfrac{5}{6}s_2$ and $s_3 = \dfrac{1}{4}s_2$.

Substituting these values into the fourth equation, we get

$$\frac{5}{6}s_2 + s_2 + \frac{1}{4}s_2 = 1 \ \text{ or } \ \frac{25}{12}s_2 = 1 \ \text{ and } \ s_2 = .48$$

Therefore, $s_1 = .4, s_2 = .48, s_3 = .12$.

The stationary matrix $S = [.4\ .48\ .12]$.

The limiting matrix $\overline{P} = \begin{array}{c} \\ A \\ B \\ C \end{array} \begin{array}{ccc} A & B & C \\ \left[\begin{array}{ccc} .4 & .48 & .12 \\ .4 & .48 & .12 \\ .4 & .48 & .12 \end{array}\right] \end{array}$ (9-2)

15. For $P = \begin{array}{c} \\ A \\ B \\ C \end{array} \begin{array}{ccc} A & B & C \\ \left[\begin{array}{ccc} 1 & 0 & 0 \\ 0 & 1 & 0 \\ .3 & .1 & .6 \end{array}\right] \end{array}$

we have $R = [.3\ .1]$ and $Q = [.6]$. The limiting matrix $\overline{P}$ has the form

$$\overline{P} = \left[\begin{array}{cc|c} 1 & 0 & 0 \\ 0 & 1 & 0 \\ \hline FR & & 0 \end{array}\right]$$

where $F = (I - Q)^{-1} = ([1] - [.6])^{-1} = [.4]^{-1} = \left[\dfrac{5}{2}\right]$

and $FR = \left[\dfrac{5}{2}\right][.3\ .1] = [.75\ .25]$.

Thus, $\overline{P} = \begin{array}{c} \\ A \\ B \\ C \end{array} \begin{array}{ccc} A & B & C \\ \left[\begin{array}{ccc} 1 & 0 & 0 \\ 0 & 1 & 0 \\ .75 & .25 & 0 \end{array}\right] \end{array}$

$P(C \text{ to } A) = .75$, $P(C \text{ to } B) = .25$. Since $F = \left[\dfrac{5}{2}\right]$, it will take an average of 2.5 trials to go from C to

either A or B.   (9-3)

16. For $P = \begin{array}{c} \\ A \\ B \\ C \\ D \end{array} \begin{array}{cccc} A & B & C & D \\ \left[\begin{array}{cccc} 1 & 0 & 0 & 0 \\ 0 & 1 & 0 & 0 \\ .1 & .5 & .2 & .2 \\ .1 & .1 & .4 & .4 \end{array}\right] \end{array}$

we have $R = \left[\begin{array}{cc} .1 & .5 \\ .1 & .1 \end{array}\right]$ and $Q = \left[\begin{array}{cc} .2 & .2 \\ .4 & .4 \end{array}\right]$.

The limiting matrix $\overline{P}$ has the form

$$\overline{P} = \left[\begin{array}{c|c} I & 0 \\ \hline FR & 0 \end{array}\right]$$

where $F = (I - Q)^{-1} = \left(\left[\begin{array}{cc} 1 & 0 \\ 0 & 1 \end{array}\right] - \left[\begin{array}{cc} .2 & .2 \\ .4 & .4 \end{array}\right]\right)^{-1} = \left[\begin{array}{cc} .8 & -.2 \\ -.4 & .6 \end{array}\right]^{-1} = \left[\begin{array}{cc} \frac{4}{5} & -\frac{1}{5} \\ -\frac{2}{5} & \frac{3}{5} \end{array}\right]^{-1}$

$$\begin{bmatrix} \frac{4}{5} & -\frac{1}{5} & 1 & 0 \\ -\frac{2}{5} & \frac{3}{5} & 0 & 1 \end{bmatrix} \sim \begin{bmatrix} 1 & -\frac{1}{4} & \frac{5}{4} & 0 \\ -\frac{2}{5} & \frac{3}{5} & 0 & 1 \end{bmatrix} \sim \begin{bmatrix} 1 & -\frac{1}{4} & \frac{5}{4} & 0 \\ 0 & \frac{1}{2} & \frac{1}{2} & 1 \end{bmatrix} \sim \begin{bmatrix} 1 & -\frac{1}{4} & \frac{5}{4} & 0 \\ 0 & 1 & 1 & 2 \end{bmatrix}$$

$$\left(\frac{5}{4}\right)R_1 \to R_1 \qquad \left(\frac{2}{5}\right)R_1 + R_2 \to R_2 \qquad 2R_2 \to R_2 \qquad \left(\frac{1}{4}\right)R_2 + R_1 \to R_1$$

$$\sim \begin{bmatrix} 1 & 0 & \frac{3}{2} & \frac{1}{2} \\ 0 & 1 & 1 & 2 \end{bmatrix}$$

Thus, $F = \begin{bmatrix} \frac{3}{2} & \frac{1}{2} \\ 1 & 2 \end{bmatrix} = \begin{bmatrix} 1.5 & .5 \\ 1 & 2 \end{bmatrix}$, $FR = \begin{bmatrix} 1.5 & .5 \\ 1 & 2 \end{bmatrix}\begin{bmatrix} .1 & .5 \\ .1 & .1 \end{bmatrix} = \begin{bmatrix} .2 & .8 \\ .3 & .7 \end{bmatrix}$.

and $\overline{P} = \begin{array}{c} \\ A \\ B \\ C \\ D \end{array}\begin{array}{cccc} A & B & C & D \\ \begin{bmatrix} 1 & 0 & 0 & 0 \\ 0 & 1 & 0 & 0 \\ .2 & .8 & 0 & 0 \\ .3 & .7 & 0 & 0 \end{bmatrix} \end{array}$

$P(C \text{ to } A) = .2$, $P(C \text{ to } B) = .8$, $P(D \text{ to } A) = .3$, $P(D \text{ to } B) = .7$.

It takes an average of 2 trials to go from C to either A or B; it takes an average of three trials to go from D to A or B. **(9-3)**

17. $P = \begin{array}{c} A \\ B \end{array}\begin{array}{cc} A & B \\ \begin{bmatrix} .4 & .6 \\ .2 & .8 \end{bmatrix} \end{array}$, $P^4 \approx \begin{bmatrix} .2512 & .7488 \\ .2496 & .7504 \end{bmatrix}$, $P^8 \approx \begin{bmatrix} .2500 & .7499 \\ .2499 & .7500 \end{bmatrix}$; $\overline{P} = \begin{array}{c} A \\ B \end{array}\begin{array}{cc} A & B \\ \begin{bmatrix} .25 & .75 \\ .25 & .75 \end{bmatrix} \end{array}$

(9-3)

18. $P = \begin{array}{c} A \\ B \\ C \end{array}\begin{array}{ccc} A & B & C \\ \begin{bmatrix} .4 & .6 & 0 \\ .5 & .3 & .2 \\ 0 & .8 & .2 \end{bmatrix} \end{array}$, $P^4 \approx \begin{bmatrix} .4066 & .4722 & .1212 \\ .3935 & .4895 & .117 \\ .404 & .468 & .128 \end{bmatrix}$,

$P^8 \approx \begin{bmatrix} .4001 & .4799 & .1200 \\ .3999 & .4802 & .1199 \\ .4001 & .4798 & .1201 \end{bmatrix}$; $\overline{P} = \begin{array}{c} A \\ B \\ C \end{array}\begin{array}{ccc} A & B & C \\ \begin{bmatrix} .4 & .48 & .12 \\ .4 & .48 & .12 \\ .4 & .48 & .12 \end{bmatrix} \end{array}$ **(9-3)**

19. $P = \begin{array}{c} A \\ B \\ C \end{array}\begin{array}{ccc} A & B & C \\ \begin{bmatrix} 1 & 0 & 0 \\ 0 & 1 & 0 \\ .3 & .1 & .6 \end{bmatrix} \end{array}$, $P^4 \approx \begin{bmatrix} 1 & 0 & 0 \\ 0 & 1 & 0 \\ .6528 & .2176 & .1296 \end{bmatrix}$, $P^8 \approx \begin{bmatrix} 1 & 0 & 0 \\ 0 & 1 & 0 \\ .7374 & .2458 & .01680 \end{bmatrix}$,

$P^{16} \approx \begin{bmatrix} 1 & 0 & 0 \\ 0 & 1 & 0 \\ .7498 & .2499 & 0 \end{bmatrix}$; $\overline{P} = \begin{array}{c} A \\ B \\ C \end{array}\begin{array}{ccc} A & B & C \\ \begin{bmatrix} 1 & 0 & 0 \\ 0 & 1 & 0 \\ .75 & .25 & 0 \end{bmatrix} \end{array}$ **(9-3)**

20.

$$P = \begin{array}{c} \\ A \\ B \\ C \\ D \end{array} \begin{array}{cccc} A & B & C & D \\ \end{array} \begin{bmatrix} 1 & 0 & 0 & 0 \\ 0 & 1 & 0 & 0 \\ .1 & .5 & .2 & .2 \\ .1 & .1 & .4 & .4 \end{bmatrix},$$

$$P^4 \approx \begin{bmatrix} 1 & 0 & 0 & 0 \\ 0 & 1 & 0 & 0 \\ .1784 & .7352 & .0432 & .0432 \\ .2568 & .5704 & .0864 & .0864 \end{bmatrix},$$

$$P^8 \approx \begin{bmatrix} 1 & 0 & 0 & 0 \\ 0 & 1 & 0 & 0 \\ .1972 & .7916 & .0056 & .0056 \\ .2944 & .6832 & .0112 & .0112 \end{bmatrix},$$

$$P^{16} \approx \begin{bmatrix} 1 & 0 & 0 & 0 \\ 0 & 1 & 0 & 0 \\ .2000 & .7999 & 0 & 0 \\ .2999 & .6997 & 0 & 0 \end{bmatrix},$$

$$\overline{P} = \begin{array}{c} \\ A \\ B \\ C \\ D \end{array} \begin{array}{cccc} A & B & C & D \\ \end{array} \begin{bmatrix} 1 & 0 & 0 & 0 \\ 0 & 1 & 0 & 0 \\ .2 & .8 & 0 & 0 \\ .3 & .7 & 0 & 0 \end{bmatrix} \quad (9\text{-}3)$$

21.   A standard form for the given matrix is:

$$P = \begin{array}{c} \\ B \\ D \\ A \\ C \end{array} \begin{array}{cccc} B & D & A & C \\ \end{array} \begin{bmatrix} 1 & 0 & 0 & 0 \\ 0 & 1 & 0 & 0 \\ .1 & .1 & .6 & .2 \\ .2 & .2 & .3 & .3 \end{bmatrix} \quad (9\text{-}3)$$

22.   We will determine the limiting matrix of:

$$P = \begin{array}{c} \\ A \\ B \\ C \end{array} \begin{array}{ccc} A & B & C \\ \end{array} \begin{bmatrix} 0 & 1 & 0 \\ 0 & 0 & 1 \\ .2 & .6 & .2 \end{bmatrix}$$

by solving

$$[s_1 \quad s_2 \quad s_3] \begin{bmatrix} 0 & 1 & 0 \\ 0 & 0 & 1 \\ .2 & .6 & .2 \end{bmatrix} = [s_1 \quad s_2 \quad s_3], \, s_1 + s_2 + s_3 = 1.$$

The corresponding system of equations is:

$$\begin{aligned} .2s_3 &= s_1 \\ s_1 + .6s_3 &= s_2 \\ s_2 + .2s_3 &= s_3 \\ s_1 + s_2 + s_3 &= 1 \end{aligned} \quad \text{or} \quad \begin{aligned} s_1 \qquad - .2s_3 &= 0 \\ s_1 - s_2 + .6s_3 &= 0 \\ s_2 - .8s_3 &= 0 \\ s_1 + s_2 + s_3 &= 0 \end{aligned}$$

From the first and third equations, we have $s_1 = .2s_3$ and $s_2 = .8s_3$. Substituting these values into the fourth equation gives

$.2s_3 + .8s_3 + s_3 = 1$ and $s_3 = .5$

It now follows that $s_1 = .1$ and $s_2 = .4$. Thus, $s = [.1 \ .4 \ .5]$ and

$$\overline{P} = \begin{array}{c} \\ A \\ B \\ C \end{array} \begin{array}{ccc} A & B & C \\ \begin{bmatrix} .1 & .4 & .5 \\ .1 & .4 & .5 \\ .1 & .4 & .5 \end{bmatrix} \end{array}$$

(A) $\begin{bmatrix} 0 & 0 & 1 \end{bmatrix} \begin{bmatrix} .1 & .4 & .5 \\ .1 & .4 & .5 \\ .1 & .4 & .5 \end{bmatrix} = \begin{array}{c} A \ B \ C \\ [.1 \ .4 \ .5] \end{array}$

(B) $\begin{bmatrix} .5 & .3 & .2 \end{bmatrix} \begin{bmatrix} .1 & .4 & .5 \\ .1 & .4 & .5 \\ .1 & .4 & .5 \end{bmatrix} = \begin{array}{c} A \ B \ C \\ [.1 \ .4 \ .5] \end{array}$ (9-3)

23. The transition matrix:

$$P = \begin{array}{c} \\ A \\ B \\ C \end{array} \begin{array}{ccc} A & B & C \\ \begin{bmatrix} 1 & 0 & 0 \\ 0 & 1 & 0 \\ .2 & .6 & .2 \end{bmatrix} \end{array}$$

is the standard form for an absorbing Markov chain with two absorbing and one nonabsorbing states. For this matrix, we have:

$R = [.2 \ .6]$ and $Q = [.2]$.

The limiting matrix has the form

$$\overline{P} = \left[\begin{array}{c|c} I & 0 \\ \hline FR & 0 \end{array} \right]$$

where $F = (I - Q)^{-1} = ([1] - [.2])^{-1} = [.8]^{-1} = [1.25]$

Thus, $FR = [1.25][.2 \ .6] = [.25 \ .75]$ and

$$\overline{P} = \begin{array}{c} \\ A \\ B \\ C \end{array} \begin{array}{ccc} A & B & C \\ \begin{bmatrix} 1 & 0 & 0 \\ 0 & 1 & 0 \\ .25 & .75 & 0 \end{bmatrix} \end{array}$$

(A) $\begin{bmatrix} 0 & 0 & 1 \end{bmatrix} \begin{bmatrix} 1 & 0 & 0 \\ 0 & 1 & 0 \\ .25 & .75 & 0 \end{bmatrix} = \begin{array}{c} A \ B \ C \\ [.25 \ .75 \ 0] \end{array}$

(B) $\begin{bmatrix} .5 & .3 & .2 \end{bmatrix} \begin{bmatrix} 1 & 0 & 0 \\ 0 & 1 & 0 \\ .25 & .75 & 0 \end{bmatrix} = \begin{array}{c} A \ B \ C \\ [.55 \ .45 \ 0] \end{array}$ (9-3)

24. No. If P is a transition matrix with 2 entries equal to 0, then P has one of the forms: $P_1 = \begin{bmatrix} 1 & 0 \\ 0 & 1 \end{bmatrix}$,

$P_2 = \begin{bmatrix} 0 & 1 \\ 1 & 0 \end{bmatrix}$, $\quad P_3 = \begin{bmatrix} 1 & 0 \\ 1 & 0 \end{bmatrix}$, $\quad P_4 = \begin{bmatrix} 0 & 1 \\ 0 & 1 \end{bmatrix}$

$P_1{}^k = P_1$, $P_2{}^k = P_2$, $P_3{}^k = P_3$ for all k, and $P_2{}^k = \begin{pmatrix} 1 & 0 \\ 0 & 1 \end{pmatrix}$ if k is even and $P_2{}^k = P_2$ if k is odd. No

power of P_i , $i = 1, 2, 3, 4$, has all positive entries.   (9-2)

25. Yes; $P = \begin{bmatrix} 0 & .5 & .5 \\ .5 & 0 & .5 \\ .5 & .5 & 0 \end{bmatrix}$ is regular since $P^2 = \begin{bmatrix} .5 & .25 & .25 \\ .25 & .5 & .25 \\ .25 & .25 & .5 \end{bmatrix}$

$P = \begin{bmatrix} 0 & 0 & 1 \\ 0 & 0 & 1 \\ .2 & .3 & .5 \end{bmatrix}$ is regular since $P^2 = \begin{bmatrix} .2 & .3 & .5 \\ .2 & .3 & .5 \\ .1 & .15 & .75 \end{bmatrix}$

26. (A)

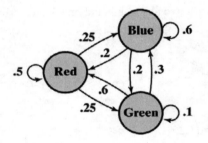

(B) $P = \begin{array}{c} \\ R \\ B \\ G \end{array} \begin{array}{ccc} R & B & G \\ \begin{bmatrix} .5 & .25 & .25 \\ .2 & .6 & .2 \\ .6 & .3 & .1 \end{bmatrix} \end{array}$

(C)   The chain is regular since it has only positive entries.
(D)   Let $S = [s_1 \quad s_2 \quad s_3]$ and solve the system:

$$[s_1 \quad s_2 \quad s_3] \begin{bmatrix} .5 & .25 & .25 \\ .2 & .6 & .2 \\ .6 & .3 & .1 \end{bmatrix} = [s_1 \quad s_2 \quad s_3], \quad s_1 + s_2 + s_3 = 1$$

which is equivalent to:

$$\begin{aligned} s_1 + s_2 + s_3 &= 1 \\ .5s_1 + .2s_2 + .6s_3 &= s_1 \\ .25s_1 + .6s_2 + .3s_3 &= s_2 \\ .25s_1 + .2s_2 + .1s_3 &= s_3 \end{aligned} \quad \text{or} \quad \begin{aligned} s_1 + s_2 + s_3 &= 1 \\ -.5s_1 + .2s_2 + .6s_3 &= 0 \\ .25s_1 - .4s_2 + .3s_3 &= 0 \\ .25s_1 + .2s_2 - .9s_3 &= 0 \end{aligned}$$

We use row operations to solve this system; but first multiply the second, third and fourth equations by 10 to simplify the calculations.

$$\begin{bmatrix} 1 & 1 & 1 & | & 1 \\ -5 & 2 & 6 & | & 0 \\ \frac{5}{2} & -4 & 3 & | & 0 \\ \frac{5}{2} & 2 & -9 & | & 0 \end{bmatrix} \sim \begin{bmatrix} 1 & 1 & 1 & | & 1 \\ 0 & 7 & 11 & | & 5 \\ 0 & -\frac{13}{2} & \frac{1}{2} & | & -\frac{5}{2} \\ 0 & -\frac{1}{2} & -\frac{23}{2} & | & -\frac{5}{2} \end{bmatrix} \sim \begin{bmatrix} 1 & 1 & 1 & | & 1 \\ 0 & 1 & 23 & | & 5 \\ 0 & -\frac{13}{2} & \frac{1}{2} & | & -\frac{5}{2} \\ 0 & 7 & 11 & | & 5 \end{bmatrix}$$

$$5R_1 + R_2 \rightarrow R_2 \qquad\qquad -2R_4 \rightarrow R_4 \qquad\qquad (-1)R_2 + R_1 \rightarrow R_1$$

$$\left(-\frac{5}{2}\right)R_1 + R_3 \rightarrow R_3 \qquad\qquad R_2 \leftrightarrow R_4 \qquad\qquad \left(\frac{13}{2}\right)R_2 + R_3 \rightarrow R_3$$

$$\left(-\frac{5}{2}\right)R_1 + R_4 \rightarrow R_4 \qquad\qquad\qquad\qquad (-7)R_2 + R_4 \rightarrow R_4$$

$$\sim \begin{bmatrix} 1 & 0 & -22 & | & -4 \\ 0 & 1 & 23 & | & 5 \\ 0 & 0 & 150 & | & 30 \\ 0 & 0 & -150 & | & -30 \end{bmatrix} \sim \begin{bmatrix} 1 & 0 & -22 & | & -4 \\ 0 & 1 & 23 & | & 5 \\ 0 & 0 & 1 & | & \frac{1}{5} \\ 0 & 0 & -150 & | & -30 \end{bmatrix} \sim \begin{bmatrix} 1 & 0 & 0 & | & \frac{2}{5} \\ 0 & 1 & 0 & | & \frac{2}{5} \\ 0 & 0 & 1 & | & \frac{1}{5} \\ 0 & 0 & 0 & | & 0 \end{bmatrix}$$

$$\frac{1}{150}R_3 \rightarrow R_3 \qquad\qquad 22R_3 + R_1 \rightarrow R_1$$

$$(-23)R_3 + R_2 \rightarrow R_2$$

$$150R_3 + R_4 \rightarrow R_4$$

The solution is $s_1 = 0.4$, $s_2 = 0.4$, $s_3 = 0.2$ and

$$\overline{P} = \begin{array}{c} \\ R \\ B \\ G \end{array} \begin{array}{c} R \quad B \quad G \\ \begin{bmatrix} .4 & .4 & .2 \\ .4 & .4 & .2 \\ .4 & .4 & .2 \end{bmatrix} \end{array}$$

In the long run, the red urn will be selected 40% of the time, the blue urn 40% of the time, and the green urn 20% of the time.   (9-2)

27. (A)

(B) $P = \begin{array}{c} \\ R \\ B \\ G \end{array} \begin{array}{c} R \quad B \quad G \\ \begin{bmatrix} 1 & 0 & 0 \\ .2 & .6 & .2 \\ .6 & .3 & .1 \end{bmatrix} \end{array}$

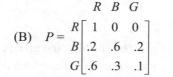

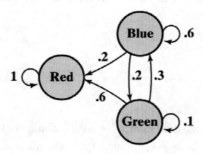

(C)   State R is an absorbing state. The chain is absorbing since it is possible to go from states B and G to state R in a finite number (namely 1) of steps.

(D)   For $P = \begin{bmatrix} 1 & 0 & 0 \\ .2 & .6 & .2 \\ .6 & .3 & .1 \end{bmatrix}$ we have $R = \begin{bmatrix} .2 \\ .6 \end{bmatrix}$ and $Q = \begin{bmatrix} .6 & .2 \\ .3 & .1 \end{bmatrix}$.

The limiting matrix $\overline{P}$ has the form:

$$\overline{P} = \left[\begin{array}{c|c} I & 0 \\ \hline FR & 0 \end{array} \right]$$

where $F = (I - Q)^{-1} = \left(\begin{bmatrix} 1 & 0 \\ 0 & 1 \end{bmatrix} - \begin{bmatrix} .6 & .2 \\ .3 & .1 \end{bmatrix} \right)^{-1} = \begin{bmatrix} .4 & -.2 \\ -.3 & .9 \end{bmatrix}^{-1} = \begin{bmatrix} \frac{2}{5} & -\frac{1}{5} \\ -\frac{3}{10} & \frac{9}{10} \end{bmatrix}^{-1}$

We use row operations to find the inverse:

$\left[\begin{array}{cc|cc} \frac{2}{5} & -\frac{1}{5} & 1 & 0 \\ -\frac{3}{10} & \frac{9}{10} & 0 & 1 \end{array} \right] \sim \left[\begin{array}{cc|cc} 1 & -\frac{1}{2} & \frac{5}{2} & 0 \\ -\frac{3}{10} & \frac{9}{10} & 0 & 1 \end{array} \right] \sim \left[\begin{array}{cc|cc} 1 & -\frac{1}{2} & \frac{5}{2} & 0 \\ 0 & \frac{3}{4} & \frac{3}{4} & 1 \end{array} \right]$

$\left(\frac{5}{2} \right) R_1 \rightarrow R_1 \qquad \left(\frac{3}{10} \right) R_1 + R_2 \rightarrow R_2 \qquad \frac{4}{3} R_2 \rightarrow R_2$

$\sim \left[\begin{array}{cc|cc} 1 & -\frac{1}{2} & \frac{5}{2} & 0 \\ 0 & 1 & 1 & \frac{4}{3} \end{array} \right] \sim \left[\begin{array}{cc|cc} 1 & 0 & 3 & \frac{2}{3} \\ 0 & 1 & 1 & \frac{4}{3} \end{array} \right]$

$\left(\frac{1}{2} \right) R_2 + R_1 \rightarrow R_1$

Thus, $F = \begin{bmatrix} 3 & \frac{2}{3} \\ 1 & \frac{4}{3} \end{bmatrix}$ and $FR = \begin{bmatrix} 3 & \frac{2}{3} \\ 1 & \frac{4}{3} \end{bmatrix} \begin{bmatrix} \frac{1}{5} \\ \frac{3}{5} \end{bmatrix} = \begin{bmatrix} 1 \\ 1 \end{bmatrix}$.

Now, $\overline{P} = \begin{array}{c} \\ R \\ B \\ G \end{array} \begin{array}{c} R \quad B \quad G \\ \begin{bmatrix} 1 & 0 & 0 \\ 1 & 0 & 0 \\ 1 & 0 & 0 \end{bmatrix} \end{array}$

Once the red urn is selected, the blue and green urns will never be selected again. It will take an average of 3.67 trials to reach the red urn from the blue urn and an average of 2.33 trials to reach the red urn from the green urn.   (9-3)

28. $[x \ y \ z \ 0] \begin{bmatrix} 1 & 0 & 0 & 0 \\ 0 & 1 & 0 & 0 \\ 0 & 0 & 1 & 0 \\ .1 & .3 & .4 & .2 \end{bmatrix} = [x \ y \ z \ 0]$

Thus, $[x \ y \ z \ 0]$ is a stationary matrix for P. If P is a transition matrix for an absorbing chain with three absorbing states and one nonabsorbing state, then P will have exactly three 1's on the main diagonal (and zeros elsewhere in the row containing the 1's) and one row with at least one nonzero entry off the main diagonal. One of the following matrices will be a stationary matrix for P:

$[x \ y \ z \ 0], \quad [x \ y \ 0 \ z], \quad [x \ 0 \ y \ z], \quad [0 \ x \ y \ z]$

where $x + y + z = 1$. The position of the zero corresponds to the one row of P which has a nonzero entry off the main diagonal.   (9-2, 9-3)

29. No such chain exists; if the chain has an absorbing state, then a corresponding transition matrix P will have a row containing a 1 on the main diagonal and zeros elsewhere. All powers of P will have the same row, and so the chain cannot be regular.   (9-2, 9-3)

30. No such chain exists. The reasoning in Problem 29 applies here as well.   (9-2, 9-3)

31. No such chain exists. By Theorem 1, Section 9-2 a regular Markov chain has a unique stationary matrix. (9-2)

32. $S = [1 \quad 0 \quad 0]$ and $S' = [0 \quad 1 \quad 0]$ are both stationary matrices for $P = \begin{matrix} & A & B & C \\ A & \\ B & \\ C & \end{matrix} \begin{bmatrix} 1 & 0 & 0 \\ 0 & 1 & 0 \\ .6 & .3 & .1 \end{bmatrix}$ (9-3)

33. $P = \begin{matrix} & A & B \\ A \\ B \end{matrix} \begin{bmatrix} 0 & 1 \\ 1 & 0 \end{bmatrix}$ has no limiting matrix; $P^{2k} = \begin{bmatrix} 1 & 0 \\ 0 & 1 \end{bmatrix}$ and $P^{2k+1} = \begin{bmatrix} 0 & 1 \\ 1 & 0 \end{bmatrix}$ for all positive integers k.

(9-2, 9-3)

34. No such chain exists. By Theorem 1, Section 9-2, a regular Markov chain has a unique limiting matrix. (9-2)

35. No such chain exists. By Theorem 2, Section 9-3, an absorbing Markov chain has a limiting matrix. (9-3)

36. $P = \begin{matrix} & A & B & C & D \\ A \\ B \\ C \\ D \end{matrix} \begin{bmatrix} .2 & .3 & .1 & .4 \\ 0 & 0 & 1 & 0 \\ 0 & .8 & 0 & .2 \\ 0 & 0 & 1 & 0 \end{bmatrix}$ No limiting matrix

For example

$$P^{200} = P^{202} = P^{204} = \dots = \begin{bmatrix} 0 & .2 & .75 & .05 \\ 0 & .8 & 0 & .2 \\ 0 & 0 & 1 & 0 \\ 0 & .8 & 0 & .2 \end{bmatrix}$$

$$P^{201} = P^{203} = P^{205} = \dots = \begin{bmatrix} 0 & .6 & .25 & .15 \\ 0 & 0 & 1 & 0 \\ 0 & .8 & 0 & .2 \\ 0 & 0 & 1 & 0 \end{bmatrix} \quad (9\text{-}2, 9\text{-}3)$$

37.

$$P = \begin{array}{c} \\ A \\ B \\ C \\ D \end{array} \begin{array}{cccc} A & B & C & D \\ \left[\begin{array}{cccc} .1 & 0 & .3 & .6 \\ .2 & .4 & .1 & .3 \\ .3 & .5 & 0 & .2 \\ .9 & .1 & 0 & 0 \end{array} \right] \end{array}$$

Limiting matrix

$$\begin{array}{c} \\ A \\ B \\ C \\ D \end{array} \begin{array}{cccc} A & B & C & D \\ \left[\begin{array}{cccc} .392 & .163 & .134 & .311 \\ .392 & .163 & .134 & .311 \\ .392 & .163 & .134 & .311 \\ .392 & .163 & .134 & .311 \end{array} \right] \end{array}$$

(9-2)

38. (A)

(B) $P = \begin{array}{c} \\ x \\ x' \end{array} \begin{array}{cc} x & x' \\ \left[\begin{array}{cc} .7 & .3 \\ .5 & .5 \end{array} \right] \end{array}$ (C) $S = [.2 \ \ .8]$

(D) $S_1 = SP = [.2 \ \ .8] \begin{bmatrix} .7 & .3 \\ .5 & .5 \end{bmatrix} = [.54 \ \ .46]$

54% of the consumers will use brand x on the next purchase.

(E) To find the stationary matrix $S = [s_1 \ \ s_2]$, we need to solve:

$$[s_1 \ \ s_2] \begin{bmatrix} .7 & .3 \\ .5 & .5 \end{bmatrix} = [s_1 \ \ s_2], s_1 + s_2 = 1$$

This yields the system of equations:

$$\begin{array}{lll} .7s_1 + .5s_2 = s_1 & & -.3s_1 + .5s_2 = 0 \\ .3s_1 + .5s_2 = s_2 & \text{or} & .3s_1 - .5s_2 = 0 \\ s_1 + s_2 = 1 & & s_1 + s_2 = 1 \end{array}$$

The solution is $s_1 = .625$, $s_2 = .375$. Thus, $S = [.625 \ \ .375]$.

(F) Brand X will have 62.5% of the market in the long run.   (9-2)

39. A transition matrix in standard form for this problem is:

$$P = \begin{array}{c} \\ A \\ B \\ C \\ M \end{array} \begin{array}{cccc} A & B & C & M \\ \left[\begin{array}{cccc} 1 & 0 & 0 & 0 \\ 0 & 1 & 0 & 0 \\ 0 & 0 & 1 & 0 \\ .06 & .08 & .11 & .75 \end{array} \right] \end{array}$$

For this matrix, $R = [.06 \ \ .08 \ \ .11]$ and $Q = [.75]$.

The limiting matrix for P has the form: $\overline{P} = \left[\begin{array}{c|c} I & 0 \\ \hline FR & 0 \end{array} \right]$

where $F = (I - Q)^{-1} = ([1] - [.75])^{-1} = [.25]^{-1} = [4]$.

Thus, $FR = [4][.06 \ \ .08 \ \ .11] = [.24 \ \ .32 \ \ .44]$

$$\text{and } \overline{P} = \begin{array}{c} \\ A \\ B \\ C \\ M \end{array} \begin{array}{cccc} A & B & C & M \\ \begin{bmatrix} 1 & 0 & 0 & 0 \\ 0 & 1 & 0 & 0 \\ 0 & 0 & 1 & 0 \\ .24 & .32 & .44 & 0 \end{bmatrix} \end{array}$$

(A)   In the long run, brand A will have 24% of the market, brand B will have 32% and brand C will have 44%.

(B)   A company will wait an average of 4 years before converting to one of the new milling machines. (9-3)

40.   (A) $P = \begin{bmatrix} .95 & .05 \\ .40 & .60 \end{bmatrix}, S_0 = [.14 \ .86]$

$S_1 = S_0 P = [.14 \ .86] \begin{bmatrix} .95 & .05 \\ .40 & .60 \end{bmatrix} = [.48 \ .52]$

$S_2 = S_1 P = [.48 \ .52] \begin{bmatrix} .95 & .05 \\ .40 & .60 \end{bmatrix} = [.66 \ .34]$

$S_3 = S_2 P = [.66 \ .34] \begin{bmatrix} .95 & .05 \\ .40 & .60 \end{bmatrix} = [.76 \ .24]$

Year	Data%	Model%
1995	14	14
2000	49	48
2005	68	66
2010	79	76

(B) is at the left of the table rows.

(C) $S_n = S_{n-1} P \approx [.89 \ .11]$ for large n;   89% will be online in the long run.   (9-2)

41.   A transition matrix in standard form for this problem is:

$$P = \begin{array}{c} \\ F \\ L \\ T \\ A \end{array} \begin{array}{cccc} F & L & T & A \\ \begin{bmatrix} 1 & 0 & 0 & 0 \\ 0 & 1 & 0 & 0 \\ 0 & .05 & .8 & .15 \\ .17 & .03 & 0 & .8 \end{bmatrix} \end{array}$$

Where F = "Fellow", A = "Associate", T = "Trainee", L = leaves

For this matrix, $R = \begin{bmatrix} 0 & .05 \\ .17 & .03 \end{bmatrix}$ and $Q = \begin{bmatrix} .8 & .15 \\ 0 & .8 \end{bmatrix}$.

The limiting matrix for P has the form

$$\overline{P} = \left[\begin{array}{c|c} I & 0 \\ \hline FR & 0 \end{array}\right]$$

where $F = (I - Q)^{-1} = \left(\begin{bmatrix} 1 & 0 \\ 0 & 1 \end{bmatrix} - \begin{bmatrix} .8 & .15 \\ 0 & .8 \end{bmatrix}\right)^{-1} = \begin{bmatrix} .2 & -.15 \\ 0 & .2 \end{bmatrix}^{-1}$

We use row operations to calculate the inverse:

$$\left[\begin{array}{cc|cc} .2 & -.15 & 1 & 0 \\ 0 & .2 & 0 & 1 \end{array}\right] \sim \left[\begin{array}{cc|cc} 1 & -.75 & 5 & 0 \\ 0 & 1 & 0 & 5 \end{array}\right] \sim \left[\begin{array}{cc|cc} 1 & 0 & 5 & 3.75 \\ 0 & 1 & 0 & 5 \end{array}\right]$$

$5R_1 \to R_1 \qquad (.75)R_2 + R_1 \to R_1$

$5R_2 \to R_2$

Thus, $F = \begin{bmatrix} 5 & 3.75 \\ 0 & 5 \end{bmatrix}$ and $FR = \begin{bmatrix} 5 & 3.75 \\ 0 & 5 \end{bmatrix} \begin{bmatrix} 0 & .05 \\ .17 & .03 \end{bmatrix} = \begin{bmatrix} .6375 & .3625 \\ .85 & .15 \end{bmatrix}$

The limiting matrix is:

$$\overline{P} = \begin{array}{c} \\ F \\ L \\ T \\ A \end{array} \begin{array}{c} \begin{array}{cccc} F & L & T & A \end{array} \\ \left[\begin{array}{cccc} 1 & 0 & 0 & 0 \\ 0 & 1 & 0 & 0 \\ .6375 & .3625 & 0 & 0 \\ .85 & .15 & 0 & 0 \end{array}\right] \end{array}$$

(A)   In the long run, 63.75% of the trainees will become Fellows.

(B)   In the long run, 15% of the Associates will leave the company.

(C)   A trainee remains in the program an average of $5 + 3.75 = 8.75$ yrs.   (9-3)

42.   We shall find the limiting matrix for:

$$P = \begin{array}{c} \\ R \\ P \\ W \end{array} \begin{array}{c} \begin{array}{ccc} R & P & W \end{array} \\ \left[\begin{array}{ccc} 1 & 0 & 0 \\ .5 & .5 & 0 \\ 0 & 1 & 0 \end{array}\right] \end{array} \text{ We have } R = \begin{bmatrix} .5 \\ 0 \end{bmatrix} \text{ and } Q = \begin{bmatrix} .5 & 0 \\ 1 & 0 \end{bmatrix}.$$

The limiting matrix for P will have the form:

$$\overline{P} = \left[\begin{array}{c|c} I & 0 \\ \hline FR & 0 \end{array}\right]$$

where $F = (I - Q)^{-1} = \left(\begin{bmatrix} 1 & 0 \\ 0 & 1 \end{bmatrix} - \begin{bmatrix} .5 & 0 \\ 1 & 0 \end{bmatrix}\right)^{-1} = \begin{bmatrix} .5 & 0 \\ -1 & 1 \end{bmatrix}^{-1} = \begin{bmatrix} \frac{1}{2} & 0 \\ -1 & 1 \end{bmatrix}^{-1}$

We use row operations to find the inverse:

$$\left[\begin{array}{cc|cc} \frac{1}{2} & 0 & 1 & 0 \\ -1 & 1 & 0 & 1 \end{array}\right] \sim \left[\begin{array}{cc|cc} 1 & 0 & 2 & 0 \\ -1 & 1 & 0 & 1 \end{array}\right] \sim \left[\begin{array}{cc|cc} 1 & 0 & 2 & 0 \\ 0 & 1 & 2 & 1 \end{array}\right]$$

$2R_1 \to R_1 \qquad R_1 + R_2 \to R_2$

Thus, $F = \begin{bmatrix} 2 & 0 \\ 2 & 1 \end{bmatrix}$ and $FR = \begin{bmatrix} 2 & 0 \\ 2 & 1 \end{bmatrix} \begin{bmatrix} .5 \\ 0 \end{bmatrix} = \begin{bmatrix} 1 \\ 1 \end{bmatrix}$

The limiting matrix $\overline{P}$ is:

$$\begin{array}{ccc} R & P & W \end{array}$$

$$\overline{P} = \begin{matrix} R \\ P \\ W \end{matrix} \begin{bmatrix} 1 & 0 & 0 \\ 1 & 0 & 0 \\ 1 & 0 & 0 \end{bmatrix}$$

From this matrix, we conclude that eventually all of the flowers will be red.   (9-3)

43. (A) $P = \begin{bmatrix} .74 & .26 \\ .03 & .97 \end{bmatrix}$, $S_0 = [.301 \ .699]$

$S_1 = S_0 P = [.301 \ .699] \begin{bmatrix} .74 & .26 \\ .03 & .97 \end{bmatrix} = [.244 \ .756]$

$S_2 = S_1 P = [.244 \ .756] \begin{bmatrix} .74 & .26 \\ .03 & .97 \end{bmatrix} = [.203 \ .797]$

$S_3 = S_2 P = [.203 \ .797] \begin{bmatrix} .74 & .26 \\ .03 & .97 \end{bmatrix} = [.174 \ .826]$

	Year	Data%	Model%
	1985	30.1	30.1
(B)	1995	24.7	24.4
	2005	20.9	20.3
	2010	19.3	17.4

(C) $S_n = S_{n-1} P \approx [.103 \ .897]$ for large n; 10.3% of the adult

U.S. population will be smokers in the long run.   (9-2)

10 LIMITS AND THE DERIVATIVE

EXERCISE 10-1
_____

Things to remember:

1. LIMIT

   We write

   $$\lim_{x \to c} f(x) = L \text{ or } f(x) \to L \text{ as } x \to c$$

   if the functional value $f(x)$ is close to the single real number L whenever x is close to but not equal to c (on either side of c).

   [Note: The existence of a limit at c has nothing to do with the value of the function at c. In fact, c may not even be in the domain of f. However, the function must be defined on both sides of c.]

2. ONE-SIDED LIMITS

   We write $\lim_{x \to c^-} f(x) = K$ [$x \to c^-$ is read "x approaches c from the left" and means $x \to c$
   and $x < c$] and call K the LIMIT FROM THE LEFT or LEFT-HAND LIMIT if $f(x)$ is close to K whenever x is close to c, but to the left of c on the real number line.

   We write $\lim_{x \to c^+} f(x) = L$ [$x \to c^+$ is read "x approaches c from the right" and means $x \to c$
   and $x > c$] and call L the LIMIT FROM THE RIGHT or RIGHT-HAND LIMIT if $f(x)$ is close to L whenever x is close to c, but to the right of c on the real number line.

3. EXISTENCE OF A LIMIT

   In order for a limit to exist, the limit from the left and the limit from the right must both exist, and must be equal. That is,

   $$\lim_{x \to c} f(x) = L \quad \text{if and only if} \quad \lim_{x \to c^-} f(x) = \lim_{x \to c^+} f(x) = L.$$

4. PROPERTIES OF LIMITS

   (a) $\lim_{x \to c} k = k$ for any constant k

   (b) $\lim_{x \to c} x = c$

   Let f and g be two functions and assume that

   $$\lim_{x \to c} f(x) = L \qquad \lim_{x \to c} g(x) = M$$

   where L and M are real numbers (both limits exist). Then:

   (c) $\lim_{x \to c} [f(x) + g(x)] = \lim_{x \to c} f(x) + \lim_{x \to c} g(x) = L + M.$

   (d) $\lim_{x \to c} [f(x) - g(x)] = \lim_{x \to c} f(x) - \lim_{x \to c} g(x) = L - M.$

   (e) $\lim_{x \to c} kf(x) = k \lim_{x \to c} f(x) = kL$ for any constant k.

   (f) $\lim_{x \to c} [f(x)g(x)] = \left(\lim_{x \to c} f(x)\right)\left(\lim_{x \to c} g(x)\right) = LM.$

(g) $\lim\limits_{x \to c} \dfrac{f(x)}{g(x)} = \dfrac{L}{M}$ if $M \neq 0$; $\lim\limits_{x \to c} \dfrac{f(x)}{g(x)}$ does not exist if $L \neq 0$

and $M = 0$; $\lim\limits_{x \to c} \dfrac{f(x)}{g(x)}$ is a 0/0 INDETERMINATE FORM if $L = M = 0$.

(h) $\lim\limits_{x \to c} \sqrt[n]{f(x)} = \sqrt[n]{\lim\limits_{x \to c} f(x)} = \sqrt[n]{L}$ ($L \geq 0$ for n even).

5. LIMITS OF POLYNOMIAL AND RATIONAL FUNCTIONS

(a) $\lim\limits_{x \to c} f(x) = f(c)$ for f any polynomial function

(b) $\lim\limits_{x \to c} r(x) = r(c)$ for r any rational function with nonzero denominator at $x = c$.

6. DIFFERENCE QUOTIENT

Let the function f be defined in an open interval containing the number a. The expression

$$\frac{f(a+h) - f(a)}{h}$$

is called the DIFFERENCE QUOTIENT. One of the most important limits in calculus is the limit of the difference quotient:

$$\lim_{h \to 0} \frac{f(a+h) - f(a)}{h}$$

---

1. $x^2 - 81 = (x - 9)(x + 9)$

3. $x^2 - 4x - 21 = (x - 7)(x + 3)$

5. $x^3 - 7x^2 + 12x = x(x^2 - 7x + 12) = x(x - 3)(x - 4)$

7. $6x^2 - x - 1 = (2x - 1)(3x + 1)$

9. $f(-0.5) = 2$

11. $f(1.75) = 1.25$

13. (A) $\lim\limits_{x \to 0^-} f(x) = 2$ (B) $\lim\limits_{x \to 0^+} f(x) = 2$ (C) $\lim\limits_{x \to 0} f(x) = 2$ (D) $f(0) = 2$

15. (A) $\lim\limits_{x \to 2^-} f(x) = 1$ (B) $\lim\limits_{x \to 2^+} f(x) = 2$ (C) $\lim\limits_{x \to 2} f(x)$ does not exist

(D) $f(2) = 2$ (E) No, because $\lim\limits_{x \to 2^-} f(x) = 1 \neq \lim\limits_{x \to 2^+} f(x) = 2$

17. $g(1.9) = 2$

19. $g(3.5) = 0.5$

21. (A) $\lim\limits_{x \to 1^-} g(x) = 1$ (B) $\lim\limits_{x \to 1^+} g(x) = 2$ (C) $\lim\limits_{x \to 1} g(x) =$ does not exist

(D) $g(1)$ does not exist (E) No, because $\lim\limits_{x \to 1^-} g(x) = 1 \neq \lim\limits_{x \to 1^+} g(x) = 2$

23. (A) $\lim\limits_{x \to 3^-} g(x) = 1$ (B) $\lim\limits_{x \to 3^+} g(x) = 1$ (C) $\lim\limits_{x \to 3} g(x) = 1$ (D) $g(3) = 3$

(E) Yes, define $g(3) = 1$.

25. (A) $\lim\limits_{x \to -3^+} f(x) = -2$ (B) $\lim\limits_{x \to -3^-} f(x) = -2$ (C) $\lim\limits_{x \to -3} f(x) = -2$

(D) $f(-3) = 1$ (E) Yes, set $f(-3) = -2$.

27. (A) $\lim\limits_{x\to 0^+} f(x) = 2$ (B) $\lim\limits_{x\to 0^-} f(x) = 2$ (C) $\lim\limits_{x\to 0} f(x) = 2$

(D) $f(0)$ does not exist. (E) Yes, define $f(0) = 2$.

29. $\lim\limits_{x\to 3} 4x = 4\cdot3 = 12$ (use $\underline{5}$)

31. $\lim\limits_{x\to -4} (x+5) = -4+5 = 1$ (use $\underline{5}$)

33. $\lim\limits_{x\to 2} x(x-4) = 2(2-4) = 2(-2) = -4$ (use $\underline{4}$f and $\underline{5}$)

35. $\lim\limits_{x\to -3} \dfrac{x}{x+5} = \dfrac{-3}{-3+5} = -\dfrac{3}{2} = -1.5$ (use $\underline{4}$g and $\underline{5}$)

37. $\lim\limits_{x\to 1} \sqrt{5x+4} = \sqrt{5+4} = \sqrt{9} = 3$ (use $\underline{4}$h and $\underline{5}$)

39. $\lim\limits_{x\to 1} -3f(x) = -3 \lim\limits_{x\to 1} f(x) = -3(-5) = 15$

41. $\lim\limits_{x\to 1} [2f(x) + g(x)] = 2\lim\limits_{x\to 1} f(x) + \lim\limits_{x\to 1} g(x) = 2(-5) + 4 = -6$

43. $\lim\limits_{x\to 1} \dfrac{2-f(x)}{x+g(x)} = \dfrac{\lim\limits_{x\to 1}[2-f(x)]}{\lim\limits_{x\to 1}[x+g(x)]} = \dfrac{2-\lim\limits_{x\to 1} f(x)}{1+\lim\limits_{x\to 1} g(x)} = \dfrac{2-(-5)}{1+4} = \dfrac{7}{5}$

45. $\lim\limits_{x\to 1} \sqrt{g(x)-f(x)} = \sqrt{\lim\limits_{x\to 1}[g(x)-f(x)]} = \sqrt{\lim\limits_{x\to 1} g(x) - \lim\limits_{x\to 1} f(x)} = \sqrt{4-(-5)} = \sqrt{9} = 3$

Note: Answers for Problems 47 and 49 may vary.

47.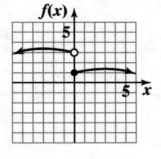

49.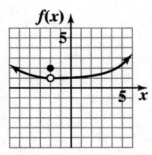

51. $f(x) = \begin{cases} 1-x^2 & \text{if } x \le 0 \\ 1+x^2 & \text{if } x > 0 \end{cases}$

(A) $\lim\limits_{x\to 0^+} f(x) = \lim\limits_{x\to 0^+} (1+x^2) = 1$

(B) $\lim\limits_{x\to 0^-} f(x) = \lim\limits_{x\to 0^-} (1-x^2) = 1$

(C) $\lim\limits_{x\to 0} f(x) = 1$

(D) $f(0) = 1$

53. $f(x) = \begin{cases} x^2 & \text{if } x < 1 \\ 2x & \text{if } x > 1 \end{cases}$

(A) $\lim\limits_{x\to 1^+} f(x) = \lim\limits_{x\to 1^+} 2x = 2$

(B) $\lim\limits_{x\to 1^-} f(x) = \lim\limits_{x\to 1^-} x^2 = 1$

(C) $\lim\limits_{x\to 1} f(x)$ does not exist
(D) $f(1)$ does not exist

55. $f(x) = \begin{cases} \dfrac{x^2-9}{x+3} & \text{if } x<0 \\ \dfrac{x^2-9}{x-3} & \text{if } x>0 \end{cases}$

(A) $\lim\limits_{x\to -3^-} f(x) = \lim\limits_{x\to -3} \dfrac{x^2-9}{x+3} = \lim\limits_{x\to -3} \dfrac{(x-3)(x+3)}{x+3} = \lim\limits_{x\to -3}(x-3) = -6$

(B) $\lim\limits_{x\to 0^-} f(x) = \lim\limits_{x\to 0^-} \dfrac{x^2-9}{x+3} = \dfrac{\lim\limits_{x\to 0^-}(x^2-9)}{\lim\limits_{x\to 0^-}(x+3)} = \dfrac{-9}{3} = -3$

$\lim\limits_{x\to 0^+} f(x) = \lim\limits_{x\to 0^+} \dfrac{x^2-9}{x-3} = \dfrac{\lim\limits_{x\to 0^+}(x^2-9)}{\lim\limits_{x\to 0^+}(x-3)} = \dfrac{-9}{-3} = 3$

$\lim\limits_{x\to 0} f(x)$ does not exist

(C) $\lim\limits_{x\to 3^-} f(x) = \lim\limits_{x\to 3} \dfrac{x^2-9}{x-3} = \lim\limits_{x\to 3} \dfrac{(x-3)(x+3)}{x-3} = \lim\limits_{x\to 3}(x+3) = 6$

57. $f(x) = \dfrac{|x-1|}{x-1}$

(A) For $x>1$, $|x-1| = x-1$.

Thus, $\lim\limits_{x\to 1^+} \dfrac{|x-1|}{x+1} = \lim\limits_{x\to 1^+} \dfrac{x-1}{x-1} = \lim\limits_{x\to 1^+} 1 = 1$.

(B) For $x<1$, $|x-1| = -(x-1)$.

Thus, $\lim\limits_{x\to 1^-} \dfrac{|x-1|}{x-1} = \lim\limits_{x\to 1^-} \dfrac{-(x-1)}{x-1} = \lim\limits_{x\to 1^-} -1 = -1$

(C) $\lim\limits_{x\to 1} f(x)$ does not exist

(D) $f(1)$ does not exist

59. $f(x) = \dfrac{x-2}{x^2-2x} = \dfrac{x-2}{x(x-2)} = \dfrac{1}{x}$, $x\neq 2$; $f(2)$ does not exist.

(A) $\lim\limits_{x\to 0} f(x) = \lim\limits_{x\to 0} \dfrac{1}{x}$ does not exist

(B) $\lim\limits_{x\to 2} f(x) = \lim\limits_{x\to 2} \dfrac{1}{x} = \dfrac{1}{2}$

(C) $\lim\limits_{x\to 4} f(x) = \lim\limits_{x\to 4} \dfrac{1}{x} = \dfrac{1}{4}$

61. $f(x) = \dfrac{x^2 - x - 6}{x + 2} = \dfrac{(x-3)(x+2)}{x+2} = x - 3, x \neq -2;$ $f(-2)$ does not exist

(A) $\lim\limits_{x \to -2} f(x) = \lim\limits_{x \to -2} (x - 3) = -5$

(B) $\lim\limits_{x \to 0} f(x) = \lim\limits_{x \to 0} (x - 3) = -3$

(C) $\lim\limits_{x \to 3} f(x) = \lim\limits_{x \to 3} (x - 3) = 0$

63. $f(x) = \dfrac{(x+2)^2}{x^2 - 4} = \dfrac{(x+2)^2}{(x-2)(x+2)} = \dfrac{x+2}{x-2}, x \neq -2;$ $f(-2)$ does not exist

(A) $\lim\limits_{x \to -2} f(x) = \lim\limits_{x \to -2} \dfrac{x+2}{x-2} = \dfrac{0}{-4} = 0$

(B) $\lim\limits_{x \to 0} f(x) = \lim\limits_{x \to 0} \dfrac{x+2}{x-2} = \dfrac{2}{-2} = -1$

(C) $\lim\limits_{x \to 2} f(x) = \lim\limits_{x \to 2} \dfrac{x+2}{x-2}$ does not exist

65. $f(x) = \dfrac{2x^2 - 3x - 2}{x^2 + x - 6} = \dfrac{(2x+1)(x-2)}{(x+3)(x-2)} = \dfrac{2x+1}{x+3}, x \neq 2;$ $f(2)$ does not exist

(A) $\lim\limits_{x \to 2} f(x) = \lim\limits_{x \to 2} \dfrac{2x+1}{x+3} = \dfrac{5}{5} = 1$

(B) $\lim\limits_{x \to 0} f(x) = \lim\limits_{x \to 0} \dfrac{2x+1}{x+3} = \dfrac{1}{3}$

(C) $\lim\limits_{x \to 1} f(x) = \lim\limits_{x \to 1} \dfrac{2x+1}{x+3} = \dfrac{3}{4}$

67. False. Set $f(x) = x^2 - 1,$ $g(x) = x - 1.$ Then $\lim\limits_{x \to 1} \dfrac{x^2 - 1}{x - 1} = \lim\limits_{x \to 1} x + 1 = 2.$

69. True. $\lim\limits_{x \to 0} f(x) = f(0)$ for any polynomial function f.

71. False. Set $f(x) = \dfrac{x^2}{x}$. Then $f(0)$ does not exist, but $\lim\limits_{x \to 0} \dfrac{x^2}{x} = \lim\limits_{x \to 0} x = 0.$

73. $\lim\limits_{x \to 7} \dfrac{(x-7)^2}{x^2 - 4x - 21}$ has the form $\dfrac{0}{0}$; $\dfrac{(x-7)^2}{x^2 - 4x - 21} = \dfrac{(x-7)^2}{(x-7)(x+3)} = \dfrac{x-7}{x+3}.$

Therefore, $\lim\limits_{x \to 7} \dfrac{(x-7)^2}{x^2 - 4x - 21} = \lim\limits_{x \to 7} \dfrac{x-7}{x+3} = 0.$

75. $\lim\limits_{x \to 4} \dfrac{x^2 + 4}{(x+4)^2}$ does not have the form $\dfrac{0}{0}$; $\lim\limits_{x \to 4} \dfrac{x^2 + 4}{(x+4)^2} = \dfrac{16+4}{8^2} = \dfrac{20}{64} = \dfrac{5}{16}.$

77. $\lim\limits_{x \to -6} \dfrac{x^2 + 36}{x + 6}$ does not have the form $\dfrac{0}{0}$; $\lim\limits_{x \to -6} \dfrac{x^2 + 36}{x + 6}$ has the form $\dfrac{72}{0}$, the limit does not exist.

79. $\lim\limits_{x \to 8} \dfrac{x-8}{x^2-64}$ has the form $\dfrac{x-8}{x^2-64} = \dfrac{x-8}{(x-8)(x+8)} = \dfrac{1}{x+8}$.

Therefore, $\lim\limits_{x \to 8} \dfrac{x-8}{x^2-64} = \lim\limits_{x \to 8} \dfrac{1}{x+8} = \dfrac{1}{16}$.

81. $f(x) = 3x + 1$

$\lim\limits_{h \to 0} \dfrac{f(2+h)-f(2)}{h} = \lim\limits_{h \to 0} \dfrac{3(2+h)+1-(3 \cdot 2+1)}{h} = \lim\limits_{h \to 0} \dfrac{6+3h+1-7}{h} = \lim\limits_{h \to 0} \dfrac{3h}{h} = \lim\limits_{h \to 0} 3 = 3$

83. $f(x) = x^2 + 1$

$\lim\limits_{h \to 0} \dfrac{f(2+h)-f(2)}{h} = \lim\limits_{h \to 0} \dfrac{(2+h)^2+1-(2^2+1)}{h} = \lim\limits_{h \to 0} \dfrac{4+4h+h^2+1-5}{h} = \lim\limits_{h \to 0} \dfrac{4h+h^2}{h} = \lim\limits_{h \to 0} (4+h) = 4$

85. $f(x) = -7x + 9$

$\lim\limits_{h \to 0} \dfrac{f(2+h)-f(2)}{h} = \lim\limits_{h \to 0} \dfrac{-7(2+h)+9-[-7(2)+9]}{h} = \dfrac{-7h}{h} = -7$.

87. $f(x+1) = |x+1|$. For $x \geq -1$, $f(x) = x+1$. Therefore,

$\lim\limits_{h \to 0} \dfrac{f(2+h)-f(2)}{h} = \lim\limits_{h \to 0} \dfrac{(2+h)+1-[2+1]}{h} = \dfrac{h}{h} = 1$.

89. (A) $\lim\limits_{x \to 1^-} f(x) = \lim\limits_{x \to 1^-} (1+x) = 2$

$\lim\limits_{x \to 1^+} f(x) = \lim\limits_{x \to 1^+} (4-x) = 3$

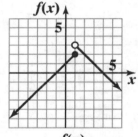

(B) $\lim\limits_{x \to 1^-} f(x) = \lim\limits_{x \to 1^-} (1+2x) = 3$

$\lim\limits_{x \to 1^+} f(x) = \lim\limits_{x \to 1^+} (4-2x) = 2$

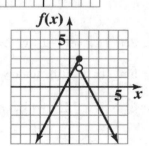

(C) $\lim\limits_{x \to 1^-} f(x) = \lim\limits_{x \to 1^-} (1+mx) = 1+m$

$\lim\limits_{x \to 1^+} f(x) = \lim\limits_{x \to 1^+} (4-mx) = 4-m$

$1+m = 4-m$

$2m = 3$

$m = \dfrac{3}{2}$

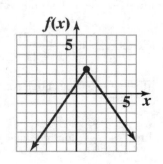

(D)   The graph in (A) is broken at $x = 1$; it jumps up from $(1, 2)$ to $(1, 3)$.

The graph in (B) is also broken at $x = 1$; it jumps down from $(1, 3)$ to $(1, 2)$.

The graph in (C) is not broken; the two pieces meet at $\left(1, \dfrac{5}{2}\right)$.

91.   (A)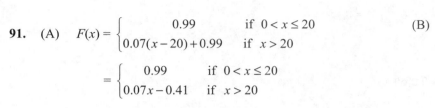

(B)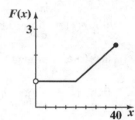

(C) $\lim\limits_{x \to 20^-} F(x) = 0.99 = \lim\limits_{x \to 20^+} F(x)$.   Therefore, $\lim\limits_{x \to 20} F(x) = 0.99$

93.   At $x = 20$ minutes, the first service charge is \$0.99 and the second service charge is \$2.70. Also, after 20 minutes, the first service charge is \$0.07 per minute versus \$0.09 per minute for the second service. The second service is much more expensive than the first (unless the call is 10 minutes or less).

95.   (A) $D(x) = \begin{cases} x & \text{if} & 0 \le x < 300 \\ 0.97x & \text{if} & 300 \le x < 1{,}000 \\ 0.95x & \text{if} & 1{,}000 \le x < 3{,}000 \\ 0.93x & \text{if} & 3{,}000 \le x < 5{,}000 \\ 0.90x & \text{if} & 5{,}000 \le x \end{cases}$

(B) $\lim\limits_{x \to 1000^-} D(x) = \lim\limits_{x \to 1000^-} 0.97x = 970,$

$\lim\limits_{x \to 1000^+} D(x) = \lim\limits_{x \to 1000^+} 0.95x = 950,$

$\lim\limits_{x \to 1000} D(x)$ does not exist;

$\lim\limits_{x \to 3000^-} D(x) = \lim\limits_{x \to 3000^-} 0.95x = 2850,$

$\lim\limits_{x \to 3000^+} D(x) = \lim\limits_{x \to 3000^+} 0.93x = 2790,$

$\lim\limits_{x \to 3000} D(x)$ does not exist.

97.   (A) $F(x) = \begin{cases} 20x & \text{if} & 0 \le x \le 4{,}000 \\ 80{,}000 & \text{if} & x > 4{,}000 \end{cases}$

(B) $\lim\limits_{x \to 4000^-} F(x) = \lim\limits_{x \to 4000^-} 20x = 80{,}000,$

$\lim\limits_{x \to 4000^+} F(x) = \lim\limits_{x \to 4000^+} 80{,}000 = 80{,}000.$

Therefore, $\lim\limits_{x \to 4000} F(x) = 80{,}000.$

$\lim\limits_{x \to 8000} F(x) = \lim\limits_{x \to 8000} 80{,}000 = 80{,}000.$

99. $\lim\limits_{x \to 5^-} f(x) = \lim\limits_{x \to 5^-} 0 = 0,$

$\lim\limits_{x \to 5^+} f(x) = \lim\limits_{x \to 5^+} (0.8 - 0.08x) = 0.4.$

Therefore, $\lim\limits_{x \to 5} f(x)$ does not exist.

$\lim\limits_{x \to 5^-} g(x) = \lim\limits_{x \to 5^-} 0 = 0,$

$\lim\limits_{x \to 5^+} g(x) = \lim\limits_{x \to 5^+} (0.8x - 0.04x^2 - 3) = 0.$

Therefore, $\lim\limits_{x \to 5} g(x) = 0.$

$\lim\limits_{x \to 10^-} f(x) = \lim\limits_{x \to 10^-} (0.8 - 0.08x) = 0,$

$\lim\limits_{x \to 10^+} f(x) = \lim\limits_{x \to 10^+} 0 = 0.$

Therefore, $\lim\limits_{x \to 10} f(x) = 0.$

$\lim\limits_{x \to 10^-} g(x) = \lim\limits_{x \to 10^-} (0.8x - 0.04x^2 - 3) = 1,$

$\lim\limits_{x \to 10^+} g(x) = \lim\limits_{x \to 0^+} 1 = 1.$

Therefore, $\lim\limits_{x \to 10} g(x) = 1.$

EXERCISE 10-2

Things to remember:

1. VERTICAL ASYMPTOTES

    The vertical line $x = a$ is a VERTICAL ASYMPTOTE for the graph of $y = f(x)$ if
    $$f(x) \to \infty \text{ or } f(x) \to -\infty \text{ as } x \to a^+ \text{ or } x \to a^-$$
    That is, if $f(x)$ either increases or decreases without bound as x approaches a from the right or from the left.

2. LOCATING VERTICAL ASYMPTOTES

    A polynomial function has no vertical asymptotes.

    If $f(x) = n(x)/d(x)$ is a rational function, $d(c) = 0$ and
    $n(c) \neq 0,$ then the line $x = c$ is a vertical asymptote of the graph of $f.$

3. HORIZONTAL ASYMPTOTES

    The horizontal line $y = b$ is a HORIZONTAL ASYMPTOTE for the graph of $y = f(x)$ if
    $$\lim\limits_{x \to -\infty} f(x) = b \text{ or } \lim\limits_{x \to \infty} f(x) = b.$$

4. LIMITS AT INFINITY FOR POWER FUNCTIONS

    If p is a positive real number and k is a nonzero constant, then

(a) $\lim\limits_{x \to -\infty} \dfrac{k}{x^p} = 0$

(b) $\lim\limits_{x \to \infty} \dfrac{k}{x^p} = 0$

(c) $\lim\limits_{x \to -\infty} kx^p = \pm\infty$

(d) $\lim\limits_{x \to \infty} kx^p = \pm\infty$

provided that x^p is defined for negative values of x. The limits in (c) and (d) will be either $-\infty$ or ∞, depending on k and p.

<u>5.</u>   LIMITS AT INFINITY FOR POLYNOMIAL FUNCTIONS

If

$$p(x) = a_n x^n + a_{n-1} x^{n-1} + \ldots + a_1 x + a_0, \, a_n \neq 0, n \geq 1,$$

then

$$\lim\limits_{x \to \infty} p(x) = \lim\limits_{x \to \infty} a_n x^n = \pm\infty$$

and

$$\lim\limits_{x \to -\infty} p(x) = \lim\limits_{x \to -\infty} a_n x^n = \pm\infty$$

Each limit will be either $-\infty$ or ∞, depending on a_n and n.

<u>6.</u>   LIMITS AT INFINITY AND HORIZONTAL ASYMPTOTES FOR RATIONAL FUNCTIONS

If $f(x) = \dfrac{a_m x^m + a_{m-1} x^{m-1} + \ldots + a_1 x + a_0}{b_n x^n + b_{n-1} x^{n-1} + \ldots + b_1 x + b_0}$, $a_m \neq 0, b_n \neq 0$

then $\lim\limits_{x \to \infty} f(x) = \lim\limits_{x \to \infty} \dfrac{a_m x^m}{b_n x^n}$ and $\lim\limits_{x \to -\infty} f(x) = \lim\limits_{x \to -\infty} \dfrac{a_m x^m}{b_n x^n}$

(a)   If $m < n$, then $\lim\limits_{x \to \infty} f(x) = \lim\limits_{x \to -\infty} f(x) = 0$ and the line

$y = 0$ (the x-axis) is a horizontal asymptote for $f(x)$.

(b)   If $m = n$, then $\lim\limits_{x \to \infty} f(x) = \lim\limits_{x \to -\infty} f(x) = \dfrac{a_m}{b_n}$ and the line $y = \dfrac{a_m}{b_n}$ is a horizontal

asymptote for $f(x)$.

(c)   If $m > n$, then each limit will be ∞ or $-\infty$, depending on m, n, a_m, and b_n, and $f(x)$

does not have a horizontal asymptote

---

1. $y = 4$

3. $x = -6$

5. $y - 9 = 2(x + 2)$ (point-slope form); $2x - y = -13$.

7.   **Slope:** $m = \dfrac{7 - 0}{0 - 9} = \dfrac{-7}{9}$; $y - 0 = \dfrac{-7}{9}(x - 9)$ (point-slope form); $7x + 9y = 63$.

9. $\lim\limits_{x \to \infty} f(x) = -2$

11. $\lim\limits_{x \to -2^+} f(x) = -\infty$

13. $\lim_{x \to -2} f(x)$ does not exist

15. $\lim_{x \to 2^-} f(x) = 0$

17. $f(x) = \dfrac{x}{x-5}$

    (A) $\lim_{x \to 5^-} f(x) = -\infty$ (B) $\lim_{x \to 5^+} f(x) = \infty$

    (C) $\lim_{x \to 5} f(x)$ does not exist.

19. $f(x) = \dfrac{2x-4}{(x-4)^2}$

    (A) $\lim_{x \to 4^-} f(x) = \infty$ (B) $\lim_{x \to 4^+} f(x) = \infty$ (C) $\lim_{x \to 4} f(x) = \infty$

21. $f(x) = \dfrac{x^2 + x - 2}{(x-1)} = \dfrac{(x-1)(x+2)}{x-1} = x+2$, provided $x \neq 1$

    (A) $\lim_{x \to 1^-} f(x) = \lim_{x \to 1^-} (x+2) = 3$

    (B) $\lim_{x \to 1^+} f(x) = \lim_{x \to 1^+} (x+2) = 3$

    (C) $\lim_{x \to 1} f(x) = \lim_{x \to 1} (x+2) = 3$

23. $f(x) = \dfrac{x^2 - 3x + 2}{x+2}$

    (A) $\lim_{x \to -2^-} f(x) = -\infty$ (B) $\lim_{x \to -2^+} f(x) = \infty$

    (C) $\lim_{x \to -2} f(x)$ does not exist.

25. $p(x) = 15 + 3x^2 - 5x^3 = -5x^3 + 3x^2 + 15$

    (A) Leading term: $-5x^3$ (B) $\lim_{x \to \infty} p(x) = \lim_{x \to \infty} (-5x^3) = -\infty$ (C) $\lim_{x \to -\infty} p(x) = \lim_{x \to -\infty} (-5x^3) = \infty$

27. $p(x) = 9x^2 - 6x^4 + 7x = -6x^4 + 9x^2 + 7x$

    (A) Leading term: $-6x^4$ (B) $\lim_{x \to \infty} p(x) = \lim_{x \to \infty} (-6x^4) = -\infty$ (C) $\lim_{x \to -\infty} p(x) = \lim_{x \to -\infty} (-6x^4) = -\infty$

29. $p(x) = x^2 + 7x + 12$

    (A) Leading term: x^2 (B) $\lim_{x \to \infty} p(x) = \lim_{x \to \infty} (x^2) = \infty$ (C) $\lim_{x \to -\infty} p(x) = \lim_{x \to -\infty} (x^2) = \infty$

31. $p(x) = x^4 + 2x^5 - 11x = 2x^5 + x^4 - 11x$

    (A) Leading term: $2x^5$ (B) $\lim_{x \to \infty} p(x) = \lim_{x \to \infty} (2x^5) = \infty$ (C) $\lim_{x \to -\infty} p(x) = \lim_{x \to -\infty} (2x^5) = -\infty$

33. $f(x) = \dfrac{1}{x+3}$; f is discontinuous at $x = -3$.

    $\lim_{x \to -3^-} f(x) = -\infty$, $\lim_{x \to -3^+} f(x) = \infty$; $x = -3$ is a vertical asymptote.

35. $h(x) = \dfrac{x^2 + 4}{x^2 - 4} = \dfrac{x^2 + 4}{(x-2)(x+2)}$; h is discontinuous at $x = -2$, $x = 2$.

At $x = -2$:
$\displaystyle\lim_{x \to -2^-} h(x) = \infty$, $\displaystyle\lim_{x \to -2^+} h(x) = -\infty$; $x = -2$ is a vertical asymptote.

At $x = 2$:
$\displaystyle\lim_{x \to 2^-} h(x) = -\infty$, $\displaystyle\lim_{x \to 2^+} h(x) = \infty$; $x = 2$ is a vertical asymptote.

37. $F(x) = \dfrac{x^2 - 4}{x^2 + 4}$. Since $x^2 + 4 \neq 0$ for all x, F is continuous for all x; there are no vertical asymptotes.

39. $H(x) = \dfrac{x^2 - 2x - 3}{x^2 - 4x + 3} = \dfrac{(x-3)(x+1)}{(x-3)(x-1)}$; H is discontinuous at $x = 1$, $x = 3$.

At $x = 1$:
$\displaystyle\lim_{x \to 1^-} H(x) = -\infty$, $\displaystyle\lim_{x \to 1^+} H(x) = \infty$; $x = 1$ is a vertical asymptote.

At $x = 3$:

Since $\dfrac{(x-3)(x+1)}{(x-3)(x-1)} = \dfrac{x+1}{x-1}$ provided $x \neq 3$,

$\displaystyle\lim_{x \to 3} H(x) = \lim_{x \to 3}\left(\dfrac{x+1}{x-1}\right) = \dfrac{4}{2} = 2$; H does not have a vertical asymptote at $x = 3$.

41. $T(x) = \dfrac{8x - 16}{x^4 - 8x^3 + 16x^2} = \dfrac{8(x-2)}{x^2(x^2 - 8x - 16)} = \dfrac{8(x-2)}{x^2(x-4)^2}$

T is discontinuous at $x = 0$, $x = 4$.

At $x = 0$:
$\displaystyle\lim_{x \to 0^-} T(x) = -\infty$, $\displaystyle\lim_{x \to 0^+} T(x) = -\infty$, $x = 0$ is a vertical asymptote.

At $x = 4$:
$\displaystyle\lim_{x \to 4^-} T(x) = \infty$, $\displaystyle\lim_{x \to 4^+} T(x) = \infty$; $x = 4$ is a vertical asymptote.

43. $f(x) = \dfrac{4x + 7}{5x - 9}$

(A) $f(10) = \dfrac{4(10) + 7}{5(10) - 9} = \dfrac{47}{41} \approx 1.146$ (B) $f(100) = \dfrac{4(100) + 7}{5(100) - 9} = \dfrac{407}{491} \approx 0.829$

(C) $\displaystyle\lim_{x \to \infty} \dfrac{4x + 7}{5x - 9} = \dfrac{4}{5} = 0.8$

45. $f(x) = \dfrac{5x^2 + 11}{7x - 2}$

(A) $f(20) = \dfrac{5(20)^2 + 11}{7(20) - 2} = \dfrac{2011}{138} \approx 14.572$ (B) $f(50) = \dfrac{5(50)^2 + 11}{7(50) - 2} = \dfrac{12{,}511}{348} \approx 35.951$

(C) $\lim\limits_{x \to \infty} \dfrac{5x^2 + 11}{7x - 2} = \lim\limits_{x \to \infty} \dfrac{5x^2}{7x} = \infty$ since $m = 2 > n = 1$

47. $f(x) = \dfrac{7x^4 - 14x^2}{6x^5 + 3}$

(A) $f(-6) = \dfrac{7(-6)^4 - 14(-6)^2}{6(-6)^5 + 3} = \dfrac{9,072 - 504}{-46,656 + 3} = -\dfrac{8,568}{46,653} \approx -0.184$

(B) $f(-12) = \dfrac{7(-12)^4 - 14(-12)^2}{6(-12)^5 + 3} = \dfrac{145,152 - 2,016}{-1,492,992 + 3} = -\dfrac{143,136}{1,492,989} \approx -0.096$

(C) $\lim\limits_{x \to \infty} \dfrac{7x^4 - 14x^2}{6x^5 + 3} = \lim\limits_{x \to \infty} \dfrac{7x^4}{6x^5} = 0$ since $m = 4 < n = 5$

49. $f(x) = \dfrac{10 - 7x^3}{4 + x^3}$

(A) $f(-10) = \dfrac{10 - 7(-10)^3}{4 + (-10)^3} = -\dfrac{7,010}{996} \approx -7.038$

(B) $f(-20) = \dfrac{10 - 7(-20)^3}{4 + (-20)^3} = -\dfrac{56,010}{7,996} \approx -7.005$

(C) $\lim\limits_{x \to -\infty} \dfrac{10 - 7x^3}{4 + x^3} = \lim\limits_{x \to -\infty} \dfrac{-7x^3}{x^3} = -7$

51. $f(x) = \dfrac{2x}{x + 2}$; f is discontinuous at $x = -2$

$\lim\limits_{x \to -2^-} f(x) = \infty$, $\lim\limits_{x \to -2^+} f(x) = -\infty$; $x = -2$ is a vertical asymptote.

$\lim\limits_{x \to \infty} \dfrac{2x}{x + 2} = \lim\limits_{x \to \infty} \dfrac{2x}{x} = 2$; $y = 2$ is a horizontal asymptote.

53. $f(x) = \dfrac{x^2 + 1}{x^2 - 1} = \dfrac{x^2 + 1}{(x - 1)(x + 1)}$; f is discontinuous at $x = -1$, $x = 1$.

At $x = -1$:

$\lim\limits_{x \to -1^-} f(x) = \infty$, $\lim\limits_{x \to -1^+} f(x) = -\infty$; $x = -1$ is a vertical asymptote.

At $x = 1$:

$\lim\limits_{x \to 1^-} f(x) = -\infty$, $\lim\limits_{x \to 1^+} f(x) = \infty$; $x = 1$ is a vertical asymptote.

$\lim\limits_{x \to \infty} \dfrac{x^2 + 1}{x^2 - 1} = \lim\limits_{x \to \infty} \dfrac{x^2}{x^2} = 1$; $y = 1$ is a horizontal asymptote.

55. $f(x) = \dfrac{x^3}{x^2+6}$. Since $x^2 + 6 \neq 0$ for all x, f is continuous for all x; there are no vertical asymptotes.

$\lim\limits_{x \to \infty} \dfrac{x^3}{x^2+6} = \lim\limits_{x \to \infty} \dfrac{x^3}{x^2} = \lim\limits_{x \to \infty} x = \infty$; there are no horizontal asymptotes.

57. $f(x) = \dfrac{x}{x^2+4}$. Since $x^2 + 4 \neq 0$ for all x, f is continuous for all x; there are no vertical asymptotes.

$\lim\limits_{x \to \infty} \dfrac{x}{x^2+4} = \lim\limits_{x \to \infty} \dfrac{x}{x^2} = \lim\limits_{x \to \infty} \dfrac{1}{x} = 0$; $y = 0$ is a horizontal asymptote.

59. $f(x) = \dfrac{x^2}{x-3}$; f is discontinuous at $x = 3$.

At $x = 3$:

$\lim\limits_{x \to 3^-} f(x) = -\infty$, $\lim\limits_{x \to 3^+} f(x) = \infty$; $x = 3$ is a vertical asymptote.

$\lim\limits_{x \to \infty} \dfrac{x^2}{x-3} = \lim\limits_{x \to \infty} \dfrac{x^2}{x} = \lim\limits_{x \to \infty} x = \infty$; there are no horizontal asymptotes.

61. $f(x) = \dfrac{2x^2+3x-2}{x^2-x-2} = \dfrac{(x+2)(2x-1)}{(x-2)(x+1)}$; f is discontinuous at $x = -1$, $x = 2$.

At $x = -1$:

$\lim\limits_{x \to -1^-} f(x) = -\infty$, $\lim\limits_{x \to -1^+} f(x) = \infty$; $x = -1$ is a vertical asymptote.

At $x = 2$:

$\lim\limits_{x \to 2^-} f(x) = -\infty$, $\lim\limits_{x \to 2^+} f(x) = \infty$; $x = 2$ is a vertical asymptote.

$\lim\limits_{x \to \infty} \dfrac{2x^2+3x-2}{x^2-x-2} = \lim\limits_{x \to \infty} \dfrac{2x^2}{x^2} = 2$; $y = 2$ is a horizontal asymptote.

63. $f(x) = \dfrac{2x^2-5x+2}{x^2-x-2} = \dfrac{(x-2)(2x-1)}{(x-2)(x+1)}$; f is discontinuous at $x = -1, x = 2$.

At $x = -1$:

$\lim\limits_{x \to -1^-} f(x) = \infty$, $\lim\limits_{x \to -1^+} f(x) = -\infty$; $x = -1$ is a vertical asymptote.

At $x = 2$:

$\lim\limits_{x \to 2} f(x) = \lim\limits_{x \to 2} \left(\dfrac{2x-1}{x+1} \right) = 1$

$\lim\limits_{x \to \infty} \dfrac{2x^2-5x+2}{x^2-x-2} = \lim\limits_{x \to \infty} \dfrac{2x^2}{x^2} = 2$; $y = 2$ is a horizontal asymptote.

65. $f(x) = \dfrac{x+3}{x^2-5}$; $\lim\limits_{x \to \infty} f(x) = \lim\limits_{x \to \infty} \dfrac{x+3}{x^2-5} = \lim\limits_{x \to \infty} \dfrac{x}{x^2} = \lim\limits_{x \to \infty} \dfrac{1}{x} = 0$

67. $f(x) = \dfrac{x^2-5}{x+3}$; $\lim\limits_{x \to \infty} f(x) = \lim\limits_{x \to \infty} \dfrac{x^2-5}{x+3} = \lim\limits_{x \to \infty} \dfrac{x^2}{x} = \lim\limits_{x \to 0} x = \infty$

69. $f(x) = \dfrac{5 - 2x^2}{1 + 8x^2}$; $\displaystyle\lim_{x\to-\infty} f(x) = \lim_{x\to-\infty} \dfrac{5 - 2x^2}{1 + 8x^2} = \lim_{x\to-\infty} \dfrac{-2x^2}{8x^2} = -\dfrac{1}{4}$

71. $f(x) = \dfrac{x^2 + 4x}{3x + 2}$; $\displaystyle\lim_{x\to-\infty} f(x) = \lim_{x\to-\infty} \dfrac{x^2 + 4x}{3x + 2} = \lim_{x\to-\infty} \dfrac{x^2}{3x} = \lim_{x\to-\infty} \dfrac{x}{3} = -\infty$

73. $f(x) = x^3 - 3x + 1$;

$\displaystyle\lim_{x\to\infty} f(x) = \lim_{x\to\infty} (x^3 - 3x + 1) = \lim_{x\to\infty} (x^3) = \infty$; $\displaystyle\lim_{x\to-\infty} f(x) = \lim_{x\to-\infty} (x^3 - 3x + 1) = \lim_{x\to-\infty} (x^3) = -\infty$

75. $f(x) = \dfrac{2 + 5x}{1 - x}$;

$\displaystyle\lim_{x\to\infty} f(x) = \lim_{x\to\infty} \dfrac{2 + 5x}{1 - x} = \lim_{x\to\infty} \dfrac{5x}{-x} = -5$; $\displaystyle\lim_{x\to-\infty} f(x) = \lim_{x\to-\infty} \dfrac{2 + 5x}{1 - x} = \lim_{x\to-\infty} \dfrac{5x}{-x} = -5$

77. False. $f(x) = \dfrac{1}{x^2 + 1}$ has no vertical asymptotes.

79. False. $f(x) = \dfrac{x^2 + 1}{x + 1}$ has no horizontal asymptote.

81. True. Since the domain of a polynomial function is all real numbers, a polynomial has no vertical asymptotes. Also, a polynomial function of degree $n \geq 1$ has no horizontal asymptotes.

83. If $n \geq 1$ and $a_n > 0$, then

$$\lim_{n\to\infty} (a_n x^n + a_{n-1} x^{n-1} + \ldots + a_0) = \infty$$

If $n \geq 1$ and $a_n < 0$, then

$$\lim_{n\to\infty} (a_n x^n + a_{n-1} x^{n-1} + \ldots + a_0) = -\infty$$

85. (A) Since $C(x)$ is a linear function of x, it can be written in the form

$$C(x) = mx + b$$

Since the fixed costs are \$200, $b = 200$.

Also, $C(20) = 3800$, so

$$3800 = m(20) + 200$$

$$20m = 3600$$

$$m = 180$$

Therefore, $C(x) = 180x + 200$

(B) $\overline{C}(x) = \dfrac{C(x)}{x} = \dfrac{180x + 200}{x}$

(C)

(D) $\overline{C}(x) = \dfrac{180x + 200}{x} = \dfrac{180 + \dfrac{200}{x}}{1}$

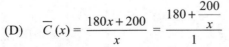

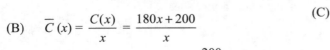

As x increases, the numerator tends to 180 and the denominator is 1. Therefore, $\overline{C}(x)$ tends to 180 or \$180 per board.

87. (A) $C_e(x) = 950 + 56x$; $\overline{C}_e(x) = \dfrac{C_e(x)}{x} = \dfrac{950 + 56x}{x} = \dfrac{950}{x} + 56$

(B) $C_c(x) = 900 + 66x$; $\overline{C}_c(x) = \dfrac{C_c(x)}{x} = \dfrac{900 + 66x}{x} = \dfrac{900}{x} + 66$

(C) Set $C_c(x) = C_e(x)$ and solve for x:

$900 + 66x = 950 + 56x$

$\qquad 10x = 50$

$\qquad\quad x = 5$

The total costs for the two models are equal at $x = 5$ years.

(D) Set $\overline{C}_c(x) = \overline{C}_e(x)$ and solve for x.

$\dfrac{900}{x} + 66 = \dfrac{950}{x} + 56$

$\dfrac{900 - 950}{x} = -10$

$-\dfrac{50}{x} = -10$

$-10x = -50$

$\qquad x = 5$

The average costs for the two models are equal at $x = 5$ years.

(E) $\displaystyle\lim_{x \to \infty} \overline{C}_e(x) = \lim_{x \to \infty} \left(\dfrac{950}{x} + 56 \right) = 56$

$\displaystyle\lim_{x \to \infty} \overline{C}_c(x) = \lim_{x \to \infty} \left(\dfrac{900}{x} + 66 \right) = 66$

For large x, the energy efficient model is approximately \$10
per year cheaper to operate than the conventional model.

89. $C(t) = \dfrac{5t^2(t + 50)}{t^3 + 100} = \dfrac{5t^3 + 250t^2}{t^3 + 100}$

$\displaystyle\lim_{t \to \infty} C(t) = \lim_{t \to \infty} \dfrac{5t^3}{t^3} = 5$; the long-term drug concentration is 5 mg/ml.

91. $P(x) = \dfrac{2x}{1 - x}$, $0 \le x < 1$

(A) $P(0.9) = \dfrac{2(0.9)}{1 - 0.9} = \dfrac{1.8}{0.1} = 18$; \$18 million

(B) $P(0.95) = \dfrac{2(0.95)}{1 - 0.95} = \dfrac{1.9}{0.05} = 38$; \$38 million

(C) $\displaystyle\lim_{x \to 1^-} P(x) = \lim_{x \to 1^-} \dfrac{2x}{1 - x} = \infty$; removal of 100% of the contamination would require an infinite amount
of money; impossible.

93. $V(s) = \dfrac{V_{\max} s}{K_M + s}$

(A) $\lim\limits_{s \to \infty} V(s) = \lim\limits_{s \to \infty} \dfrac{V_{\max} s}{K_M + s} = \lim\limits_{s \to \infty} \dfrac{V_{\max} s}{s} = V_{\max}$

(B) $V(K_M) = \dfrac{V_{\max} \cdot K_M}{K_M + K_M} = \dfrac{V_{\max} K_M}{2 K_M} = \dfrac{V_{\max}}{2}$

(C)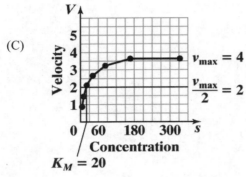

(D) $V(s) = \dfrac{4s}{20 + s}$

(E) $V(15) = \dfrac{4(15)}{20 + 15} = \dfrac{60}{35} = \dfrac{12}{7}$

Set $V = 3$ and solve for s:

$3 = \dfrac{4s}{20 + s}$

$60 + 3s = 4s$

$s = 60$

Thus, $s = 60$ when $V = 3$.

95. (A) $C_{\max} = 18$, $M = 150$

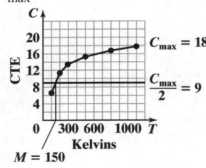

(B) $C(T) = \dfrac{18T}{150 + T}$

(C) $C(600) = \dfrac{18(600)}{150 + 600} = 14.4$

To find T when $C(T) = 12$, solve $\dfrac{18T}{150 + T} = 12$ for T:

$18T = 1800 + 12T$

$6T = 1800$

$T = 300$

Thus, $C(T) = 12$ at $T = 300$K

EXERCISE 10-3
_____

Things to remember:

1.   CONTINUITY
      A function f is CONTINUOUS AT THE POINT $x = c$ if:

      (a) $\lim\limits_{x \to c} f(x)$ exists;   (b) $f(c)$ exists;   (c) $\lim\limits_{x \to c} f(x) = f(c)$

      If one or more of the three conditions fails, then f is DISCONTINUOUS at $x = c$.

      A function is CONTINUOUS ON THE OPEN INTERVAL (a, b) if it is continuous at each point on the interval.

2.   ONE-SIDED CONTINUITY
      A function f is CONTINUOUS ON THE LEFT AT $x = c$ if $\lim\limits_{x \to c^-} f(x) = f(c)$; f is

      CONTINUOUS ON THE RIGHT AT $x = c$ if $\lim\limits_{x \to c^+} f(x) = f(c)$.

      The function f is continuous on the closed interval $[a, b]$ if it is continuous on the open interval (a, b), and is continuous on the right at a and continuous on the left at b.

3.   CONTINUITY PROPERTIES OF SOME SPECIFIC FUNCTIONS
      (a)   A constant function, $f(x) = k$, is continuous for all x.

      (b)   For n a positive integer, $f(x) = x^n$ is continuous for all x.

      (c)   A polynomial function
             $$P(x) = a_n x^n + a_{n-1} x^{n-1} + \ldots + a_1 x + a_0$$
             is continuous for all x.

      (d)   A rational function
             $$R(x) = \frac{P(x)}{Q(x)},$$

             P and Q polynomial functions, is continuous for all x except those numbers $x = c$ such that $Q(c) = 0$.

      (e)   For n an odd positive integer, $n > 1$, $\sqrt[n]{f(x)}$ is continuous wherever f is continuous.

      (f)   For n an even positive integer, $\sqrt[n]{f(x)}$ is continuous wherever f is continuous and non-negative.

4.   SIGN PROPERTIES ON AN INTERVAL (a, b)

      If f is continuous or (a, b) and $f(x) \neq 0$ for all x in (a, b), then either $f(x) > 0$ for all x in (a, b) or $f(x) < 0$ for all x in (a, b).

5.   CONSTRUCTING SIGN CHARTS

      Given a function f:

Step 1. Find all partition numbers. That is:

(A) Find all numbers where f is discontinuous. (Rational functions are discontinuous for values of x that make a denominator 0.)

(B) Find all numbers where $f(x) = 0$. (For a rational function, this occurs where the numerator is 0 and the denominator is not 0.)

Step 2. Plot the numbers found in step 1 on a real number line, dividing the number line into intervals.

Step 3. Select a test number in each open interval determined in step 2, and evaluate $f(x)$ at each test number to determine whether $f(x)$ is positive (+) or negative (–) in each interval.

Step 4. Construct a sign chart using the real number line in step 2. This will show the sign of $f(x)$ on each open interval.

[*Note*: From the sign chart, it is easy to find the solution for the inequality $f(x) < 0$ or $f(x) > 0$.]

---

1. $[-3,5]$ 3 $(-10,100)$ 5. $(-\infty,-5)\cup(5,\infty)$ 7. $(-\infty,-1]\cup(2,\infty)$

Note: Answers to problems 9, 11, 13 may vary.

9. f is continuous at $x = 1$, since $\lim\limits_{x\to1}f(x) = f(1) = 2$

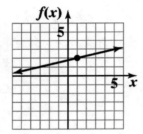

11. f is discontinuous at $x = 1$, since $\lim\limits_{x\to1}f(x) \neq f(1)$

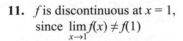
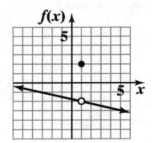

13. $\lim\limits_{x\to1^-}f(x) = 2, \ \lim\limits_{x\to1^+}f(x) = -2$

implies $\lim\limits_{x\to1}f(x)$ does not exist;

f is discontinuous at $x = 1$, since $\lim\limits_{x\to1}f(x)$ does not exist

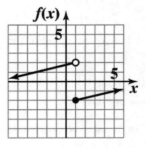

15. $f(0.9) \approx 1.9$

17. $f(-1.9) \approx 0.9$

19. (A) $\lim\limits_{x\to1^-}f(x) = 2$ (B) $\lim\limits_{x\to1^+}f(x) = 1$ (C) $\lim\limits_{x\to1}f(x)$ does not exist   (D) $f(1) = 1$

(E) No, because $\lim\limits_{x\to1}f(x)$ does not exist.

21. (A) $\lim\limits_{x\to-2^-} f(x) = 1$ (B) $\lim\limits_{x\to-2^+} f(x) = 1$ (C) $\lim\limits_{x\to-2} f(x) = 1$ (D) $f(-2) = 3$

(E) No, because $\lim\limits_{x\to-2} f(x) \neq f(-2)$.

23. $g(-3.1) \approx 0.9$ **25.** $g(1.9) \approx 2.05$

27. (A) $\lim\limits_{x\to-3^-} g(x) = 1$ (B) $\lim\limits_{x\to-3^+} g(x) = 1$ (C) $\lim\limits_{x\to-3} g(x) = 1$ (D) $g(-3) = 3$

(E) No, because $\lim\limits_{x\to-3} g(x) \neq g(-3)$.

29. (A) $\lim\limits_{x\to2^-} g(x) = 2$ (B) $\lim\limits_{x\to2^+} g(x) = -1$ (C) $\lim\limits_{x\to2} g(x)$ does not exist   (D) $g(2) = 2$

(E) No, because $\lim\limits_{x\to2} g(x)$ does not exist.

31. $f(x) = 3x - 4$ is a polynomial function. Therefore, f is continuous for all x [<u>3</u>(c)].

33. $g(x) = \dfrac{3x}{x+2}$ is a rational function and the denominator $x + 2$ is 0 at

$x = -2$. Thus, g is continuous for all x except $x = -2$ [<u>3</u>(d)].

35. $m(x) = \dfrac{x+1}{(x-1)(x+4)}$ is a rational function and the denominator

$(x - 1)(x + 4)$ is 0 at $x = 1$ or $x = -4$. Thus, m is continuous for all x except $x = 1$, $x = -4$ [<u>3</u>(d)].

37. $F(x) = \dfrac{2x}{x^2+9}$ is a rational function and the denominator $x^2 + 9 \neq 0$ for all x. Thus, F is continuous for all x.

39. $M(x) = \dfrac{x-1}{4x^2-9}$ is a rational function and the denominator $4x^2 - 9 = 0$ at $x = \dfrac{3}{2}, -\dfrac{3}{2}$. Thus, M is

continuous for all x except $x = \pm\dfrac{3}{2}$.

41. $f(x) = \dfrac{3x+8}{x-4}$; f is discontinuous at $x = 4$; $f(x) = 0$ at $x = \dfrac{-8}{3}$. Partition numbers $4, \dfrac{-8}{3}$.

43. $f(x) = \dfrac{1-x^2}{1+x^2}$; $f(x) = 0$ at $x = 1, -1$; $1 + x^2 \neq 0$ for all x. Partition numbers $1, -1$.

45. $f(x) = \dfrac{x^2+4x-45}{x^2+6x} = \dfrac{(x+9)(x-5)}{x(x+6)}$; f is discontinuous at $x = 0, -6$; $f(x) = 0$ at $x = -9, 5$.
Partition numbers $-9, -6, 0, 5$.

47. $x^2 - x - 12 < 0$

Let $f(x) = x^2 - x - 12 = (x - 4)(x + 3)$. Then f is continuous for all x and $f(-3) = f(4) = 0$. Thus, $x = -3$ and $x = 4$ are partition numbers.

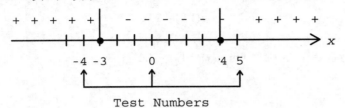

Test Numbers	
x	$f(x)$
-4	$8(+)$
0	$-12(-)$
5	$8(+)$

Thus, $x^2 - x - 12 < 0$ for:

$-3 < x < 4$ (inequality notation)

$(-3, 4)$ (interval notation)

49. $x^2 + 21 > 10x$ or $x^2 - 10x + 21 > 0$

Let $f(x) = x^2 - 10x + 21 = (x - 7)(x - 3)$. Then f is continuous for all x and $f(3) = f(7) = 0$. Thus, $x = 3$ and $x = 7$ are partition numbers.

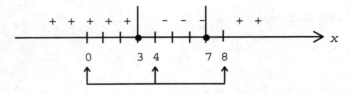

Test Numbers	
x	$f(x)$
0	$21(+)$
4	$-3(-)$
8	$5(+)$

Thus, $x^2 - 10x + 21 > 0$ for:

$x < 3$ or $x > 7$ (inequality notation)

$(-\infty, 3) \cup (7, \infty)$ (interval notation)

51. $x^3 < 4x$ or $x^3 - 4x < 0$

Let $f(x) = x^3 - 4x = x(x^2 - 4) = x(x - 2)(x + 2)$. Then f is continuous for all x and $f(-2) = f(0) = f(2) = 0$. Thus, $x = -2$, $x = 0$ and $x = 2$ are partition numbers.

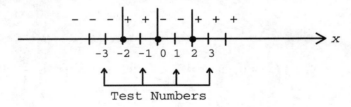

Test Numbers	
x	$f(x)$
-3	$-15(-)$
-1	$3(+)$
1	$-3(-)$
3	$15(+)$

Thus, $x^3 < 4x$ for:

$-\infty < x < -2$ or $0 < x < 2$ (inequality notation)

$(-\infty, -2) \cup (0, 2)$ (interval notation)

53. $\dfrac{x^2 + 5x}{x - 3} > 0$

Let $f(x) = \dfrac{x^2 + 5x}{x - 3} = \dfrac{x(x + 5)}{x - 3}$. Then f is discontinuous at $x = 3$ and $f(0) = f(-5) = 0$. Thus, $x = -5$, $x = 0$,

and $x = 3$ are partition numbers.

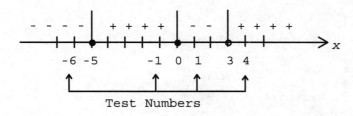

Thus, $\dfrac{x^2 + 5x}{x - 3} > 0$ for: $-5 < x < 0$ or $x > 3$ (inequality notation)

$(-5, 0) \cup (3, \infty)$ (interval notation)

55. (A) $f(x) > 0$ on $(-4, -2) \cup (0, 2) \cup (4, \infty)$

(B) $f(x) < 0$ on $(-\infty, -4) \cup (-2, 0) \cup (2, 4)$

57. $f(x) = x^4 - 6x^2 + 3x + 5$

Partition numbers: $x_1 \approx -2.5308$, $x_2 \approx -0.7198$

(A) $f(x) > 0$ on $(-\infty, -2.5308) \cup (-0.7198, \infty)$

(B) $f(x) < 0$ on $(-2.5308, -0.7198)$

59. $f(x) = \dfrac{3 + 6x - x^3}{x^2 - 1}$

Partition numbers: $x_1 \approx -2.1451, x_2 = -1, x_3 \approx -0.5240,$
$x_4 = 1, x_5 \approx 2.6691$

(A) $f(x) > 0$ on $(-\infty, -2.1451) \cup (-1, -0.5240) \cup (1, 2.6691)$

(B) $f(x) < 0$ on $(-2.1451, -1) \cup (-0.5240, 1) \cup (2.6691, \infty)$

61. $f(x) = x - 6$ is continuous for all x since it is a polynomial function. Therefore, $g(x) = \sqrt{x - 6}$ is continuous for all x such that $x - 6 \geq 0$, that is, for all x in $[6, \infty)$ [see $\underline{3}$(f)].

63. $f(x) = 5 - x$ is continuous for all x since it is a polynomial function. Therefore, $F(x) = \sqrt[3]{5 - x}$ is continuous for all x, that is, for all x in $(-\infty, \infty)$.

65. $f(x) = x^2 - 9$ is continuous for all x since it is a polynomial function. Therefore, $g(x) = \sqrt{x^2 - 9}$ is continuous for all x such that
$x^2 - 9 = (x - 3)(x + 3) \geq 0$.

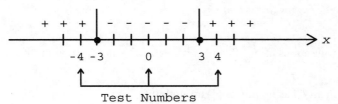

Test Numbers	
x	$f(x)$
-4	7
0	-9
4	7

$\sqrt{x^2 - 9}$ is continuous on $(-\infty, -3] \cup [3, \infty)$.

67. $f(x) = x^2 + 1$ is continuous for all x since it is a polynomial function. Also $x^2 + 1 \geq 1 > 0$ for all x.

Therefore, $\sqrt{x^2 + 1}$ is continuous for all x, that is, for all x in $(-\infty, \infty)$.

69. The graph of f is shown at the right. This function is discontinuous at $x = 1$. [$\lim\limits_{x \to 1} f(x)$ does not exist.]

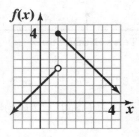

71. The graph of f is:

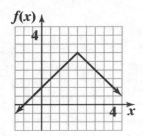

This function is continuous for all x.

$$\left[\lim_{x \to 2} f(x) = f(2) = 3.\right]$$

73. The graph of f is:

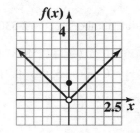

This function is discontinuous at $x = 0$.

$$\left[\lim_{x \to 0} f(x) = 0 \neq f(0) = 1.\right]$$

75. (A)   Since $\lim\limits_{x \to 0^+} f(x) = f(0) = 0$, f is continuous from the right

at $x = 0$.

(B)   Since $\lim\limits_{x \to 0^-} f(x) = -1 \neq f(0) = 0$, f is not continuous from the left

at $x = 0$.

(C) f is continuous on the open interval $(0, 1)$.

(D) f is *not* continuous on the closed interval $[0, 1]$ since

$\lim\limits_{x \to 1^-} f(x) = 0 \neq f(1) = 1$, i.e., f is not continuous from the left at $x = 1$.

(E) f is continuous on the half-closed interval $[0, 1)$.

77. True. Theorem 1(c) in the text and <u>3</u>c in *Things to Remember*.

79. False. Set $f(x) = \dfrac{1}{x - 1}$. Then f is continuous at $x = 0$ and $x = 2$, but f is not continuous at $x = 1$.

81. True. If f has no partition numbers, then f has no points of discontinuity.

83. *x* intercepts:
$x = -5, 2$

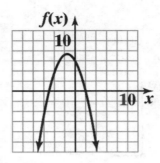

85. *x* intercepts:
$x = -6, -1, 4$

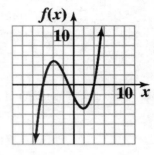

87. $f(x) = \dfrac{2}{1-x} \neq 0$ for all *x*. This does not contradict Theorem 2 because *f* is not continuous on $(-1, 3)$; *f* is discontinuous at $x = 1$.

89. (A)

$$P(x) = \begin{cases} 0.44 & \text{if } 0 < x \le 1 \\ 0.61 & \text{if } 1 < x \le 2 \\ 0.78 & \text{if } 2 < x \le 3 \\ 0.95 & \text{if } 3 < x < 3.5 \end{cases}$$

(B)

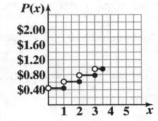

(C) *P* is continuous at $x = 2.5$ since *P* is a continuous function
on $(2, 3]$; $P(x) = 0.88$ for $2 < x \le 3$.
P is not continuous at $x = 3$ since $\lim\limits_{x \to 3^-} P(x) = 0.78 = P(3) \neq \lim\limits_{x \to 3^+} P(x) = 0.95$.

91. *Q* is defined for all real numbers whereas *P* is defined only for $x \in (0, 3.5]$.

93. (A) $S(x) = 5.00 + 0.63x$ if $0 \le x \le 50$;
$S(50) = 36.50$;
$S(x) = 36.50 + 0.45(x - 50)$
$= 14 + 0.45x$ if $x > 50$

Therefore, $S(x) = \begin{cases} 5.00 + 0.63x & \text{if } 0 \le x \le 50 \\ 14.00 + 0.45x & \text{if } x > 50 \end{cases}$

(B)

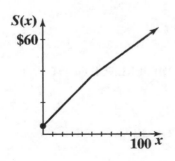

(C) $S(x)$ is continuous at $x = 50$;

$$\lim_{x \to 50^-} S(x) = \lim_{x \to 50^+} S(x)$$
$$= \lim_{x \to 50} S(x) = S(50) = 36.5.$$

95.   (A) $E(s) = \begin{cases} 1000, 0 \le s \le 10{,}000 \\ 1000 + 0.05(s - 10{,}000), 10{,}000 < s < 20{,}000 \\ 1500 + 0.05(s - 10{,}000), s \ge 20{,}000 \end{cases}$

The graph of E is:

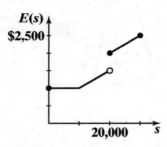

(B)   From the graph, $\lim\limits_{s \to 10{,}000} E(s) = \$1000$ and $E(10{,}000) = \$1000$.

(C)   From the graph, $\lim\limits_{s \to 20{,}000} E(s)$ does not exist. $E(20{,}000) = \$2000$.

(D) E is continuous at 10,000; E is not continuous at 20,000.

97.   (A)   From the graph, N is discontinuous at $t = t_2$, $t = t_3$, $t = t_4$, $t = t_6$, and $t = t_7$.

(B)   From the graph, $\lim\limits_{t \to t_5} N(t) = 7$ and $N(t_5) = 7$.

(C)   From the graph, $\lim\limits_{t \to t_3} N(t)$ does not exist; $N(t_3) = 4$.

EXERCISE 10-4

Things to remember:

1.   AVERAGE RATE OF CHANGE

For $y = f(x)$, the AVERAGE RATE OF CHANGE FROM $x = a$ TO $x = a + h$ is

$$\frac{f(a + h) - f(a)}{(a + h) - a} = \frac{f(a + h) - f(a)}{h} \quad h \neq 0$$

The expression $\dfrac{f(a + h) - f(a)}{h}$ is called the DIFFERENCE QUOTIENT.

2.   INSTANTANEOUS RATE OF CHANGE

For $y = f(x)$, the INSTANTANEOUS RATE OF CHANGE AT $x = a$ is

$$\lim\limits_{h \to 0} \frac{f(a + h) - f(a)}{h}$$

if the limit exists.

3.   SECANT LINE

A line through two points on the graph of a function is called a SECANT LINE. If $(a, f(a))$ and $((a + h), f(a + h))$ are two points on the graph of $y = f(x)$, then

$$\text{Slope of secant line} = \frac{f(a + h) - f(a)}{h} \quad \text{[Difference quotient]}$$

4.   SLOPE OF A GRAPH

For $y = f(x)$, the SLOPE OF THE GRAPH at the point $(a, f(a))$ is given by

$$\lim_{h \to 0} \frac{f(a + h) - f(a)}{h}$$

provided the limit exists. The slope of the graph is also the SLOPE OF THE TANGENT LINE at the point $(a, f(a))$.

5.   THE DERIVATIVE

For $y = f(x)$, we define THE DERIVATIVE OF f AT x, denoted by $f'(x)$, to be

$$f'(x) = \lim_{h \to 0} \frac{f(x + h) - f(x)}{h} \quad \text{if the limit exists.}$$

If $f'(x)$, exists for each x in the open interval (a, b), then f is said to be DIFFERENTIABLE OVER (a, b).

6.   INTERPRETATIONS OF THE DERIVATIVE

The derivative of a function f is a new function f'. The domain of f' is a subset of the domain of f. Interpretations of the derivative are:

a.   Slope of the tangent line. For each x in the domain of f', $f'(x)$ is the slope of the line tangent to the graph of f at the point $(x, f(x))$.

b.   Instantaneous rate of change. For each x in the domain of f', $f'(x)$ is the instantaneous rate of change of $y = f(x)$ with respect to x.

c.   Velocity. If $f(x)$ is the position of a moving object at time x, then $v = f'(x)$, is the velocity of the object at that time.

7.   THE FOUR STEP PROCESS FOR FINDING THE DERIVATIVE OF A FUNCTION f.

Step 1.   Find $f(x + h)$.

Step 2.   Find $f(x + h) - f(x)$.

Step 3.   Find $\dfrac{f(x + h) - f(x)}{h}$.

Step 4.   Find $\lim\limits_{h \to 0} \dfrac{f(x + h) - f(x)}{h}$.

---

1.   Slope $m = \dfrac{16 - 7}{6 - 2} = \dfrac{9}{4}$, 2.25

3.   Slope $m = \dfrac{68 - 14}{0 - 10} = \dfrac{-54}{10} = \dfrac{-27}{5}$; -5.4

5. $\dfrac{1}{\sqrt{3}} = \dfrac{1}{\sqrt{3}} \cdot \dfrac{\sqrt{3}}{\sqrt{3}} = \dfrac{\sqrt{3}}{3}$

7. $\dfrac{5}{3+\sqrt{7}} = \dfrac{5}{3+\sqrt{7}} \cdot \dfrac{3-\sqrt{7}}{3-\sqrt{7}} = \dfrac{15-5\sqrt{7}}{2} = \dfrac{15}{2} - \dfrac{5}{2}\sqrt{7}$

9. (A) $\dfrac{f(2)-f(1)}{2-1} = \dfrac{1-4}{1} = -3$ is the slope of the secant line through $(1, f(1))$ and $(2, f(2))$.

(B) $\dfrac{f(1+h)-f(1)}{h} = \dfrac{5-(1+h)^2 - 4}{h} = \dfrac{5-[1+2h+h^2]-4}{h} = \dfrac{-2h-h^2}{h} = -2-h;$

slope of the secant line through $(1, f(1))$ and $(1+h, f(1+h))$

(C) $\lim\limits_{h\to 0} \dfrac{f(1+h)-f(1)}{h} = \lim\limits_{h\to 0} (-2-h) = -2;$

slope of the tangent line at $(1, f(1))$

11. $f(x) = 3x^2$

(A) Slope of secant line through $(1, f(1))$ and $(4, f(4))$:

$\dfrac{f(4)-f(1)}{4-1} = \dfrac{3(4)^2 - 3(1)^2}{4-1} = \dfrac{48-3}{3} = \dfrac{45}{3} = 15.$

(B) Slope of secant line through $(1, f(1))$ and $(1+h, f(1+h))$:

$\dfrac{3(1+h)^2 - 3(1)^2}{1+h-1} = \dfrac{3(1+2h+h^2)-3}{h} = \dfrac{6h+3h^2}{h} = 6+3h$

(C) Slope of the graph at $(1, f(1))$: $\lim\limits_{h\to 0} \dfrac{f(1+h)-f(1)}{h} = \lim\limits_{h\to 0}(6+3h) = 6.$

13. (A) Distance traveled for $0 \le t \le 2$: 80 km; average velocity: $v = \dfrac{80}{2} = 40$ km/h..

(B) $\dfrac{f(2)-f(0)}{2-0} = \dfrac{80}{2} = 40.$

(C) Slope at $x=2$: m = 45. Equation of tangent line at $(2, f(2))$: $y-80 = 45(x-2).$

15. $f(x) = \dfrac{1}{1+x^2}$; $f(1) = \dfrac{1}{2}$. Equation of tangent line: $y - \dfrac{1}{2} = -\dfrac{1}{2}(x-1).$

17. $f(x) = x^4$; $f(-2) = 16$. Equation of tangent line: $y - 16 = -32(x+2)$ or $y = -32x - 48.$

19. $f(x) = -5$

Step 1. Find $f(x+h)$:

$f(x+h) = -5$

Step 2. Find $f(x+h) - f(x)$:

$f(x+h) - f(x) = -5 - (-5) = 0$

Step 3.   Find $\dfrac{f(x+h)-f(x)}{h}$:

$$\dfrac{f(x+h)-f(x)}{h} = \dfrac{0}{h} = 0$$

Step 4.   Find $\lim\limits_{h\to 0} \dfrac{f(x+h)-f(x)}{h}$:

$$\lim\limits_{h\to 0} \dfrac{f(x+h)-f(x)}{h} = \lim\limits_{h\to 0} 0 = 0$$

Thus, $f'(x) = 0$.

$f'(1) = 0$, $f'(2) = 0$, $f'(3) = 0$

21. $f(x) = 3x - 7$

Step 1.   Find $f(x + h)$:
$f(x + h) = 3(x + h) - 7 = 3x + 3h - 7$

Step 2.   Find $f(x + h) - f(x)$:
$f(x + h) - f(x) = 3x + 3h - 7 - (3x - 7) = 3h$

Step 3.   Find $\dfrac{f(x+h)-f(x)}{h}$:

$$\dfrac{f(x+h)-f(x)}{h} = \dfrac{3h}{h} = 3$$

Step 4.   Find $\lim\limits_{h\to 0} \dfrac{f(x+h)-f(x)}{h}$:

$$\lim\limits_{h\to 0} \dfrac{f(x+h)-f(x)}{h} = \lim\limits_{h\to 0} 3 = 3$$

Thus, $f'(x) = 3$.

$f'(1) = 3$, $f'(2) = 3$, $f'(3) = 3$

23. $f(x) = 2 - 3x^2$

Step 1.   Find $f(x + h)$:
$f(x + h) = 2 - 3(x + h)^2 = 2 - 3(x^2 + 2xh + h^2) = 2 - 3x^2 - 6xh - 3h^2$

Step 2.   Find $f(x + h) - f(x)$:
$f(x + h) - f(x) = 2 - 3x^2 - 6xh - 3h^2 - (2 - 3x^2) = -6xh - 3h^2$

Step 3.   Find $\dfrac{f(x+h)-f(x)}{h}$:

$$\dfrac{f(x+h)-f(x)}{h} = \dfrac{-6xh - 3h^2}{h} = -6x - 3h$$

Step 4.   Find $\lim\limits_{h\to 0} \dfrac{f(x+h)-f(x)}{h}$:

$$\lim\limits_{h\to 0} \dfrac{f(x+h)-f(x)}{h} = \lim\limits_{h\to 0} (-6x - 3h) = -6x$$

Thus, $f'(x) = -6x$.

$f'(1) = -6$, $f'(2) = -12$, $f'(3) = -18$

25. $f(x) = x^2 + 6x - 10$

Step 1. Find $f(x + h)$:
$$f(x + h) = (x + h)^2 + 6(x + h) - 10 = x^2 + 2xh + h^2 + 6x + 6h - 10$$

Step 2. Find $f(x + h) - f(x)$:
$$f(x + h) - f(x) = x^2 + 2xh + h^2 + 6x + 6h - 10 - (x^2 + 6x - 10) = 2xh + h^2 + 6h$$

Step 3. Find $\dfrac{f(x+h) - f(x)}{h}$:
$$\frac{f(x+h) - f(x)}{h} = \frac{2xh + h^2 + 6h}{h} = 2x + h + 6$$

Step 4. Find $\displaystyle\lim_{h \to 0} \dfrac{f(x+h) - f(x)}{h}$:
$$\lim_{h \to 0} \frac{f(x+h) - f(x)}{h} = \lim_{h \to 0} (2x + h + 6) = 2x + 6$$
Thus, $f'(x) = 2x + 6$.
$f'(1) = 8$, $f'(2) = 10$, $f'(3) = 12$

27. $f(x) = 2x^2 - 7x + 3$

Step 1. Find $f(x + h)$:
$$f(x + h) = 2(x + h)^2 - 7(x + h) + 3 = 2(x^2 + 2xh + h^2) - 7x - 7h + 3$$
$$= 2x^2 + 4xh + 2h^2 - 7x - 7h + 3$$

Step 2. Find $f(x + h) - f(x)$:
$$f(x + h) - f(x) = 2x^2 + 4xh + 2h^2 - 7x - 7h + 3 - (2x^2 - 7x + 3)$$
$$= 4xh + 2h^2 - 7h$$

Step 3. Find $\dfrac{f(x+h) - f(x)}{h}$:
$$\frac{f(x+h) - f(x)}{h} = \frac{4xh + 2h^2 - 7h}{h} = 4x + 2h - 7$$

Step 4. Find $\displaystyle\lim_{h \to 0} \dfrac{f(x+h) - f(x)}{h}$:
$$\lim_{h \to 0} \frac{f(x+h) - f(x)}{h} = \lim_{h \to 0} (4x + 2h - 7) = 4x - 7$$
Thus, $f'(x) = 4x - 7$.
$f'(1) = -3$, $f'(2) = 1$, $f'(3) = 5$

29. $f(x) = -x^2 + 4x - 9$

Step 1.   Find $f(x + h)$:

$$f(x + h) = -(x + h)^2 + 4(x + h) - 9 = -(x^2 + 2xh + h^2) + 4x + 4h - 9$$
$$= -x^2 - 2xh - h^2 + 4x + 4h - 9$$

Step 2.   Find $f(x + h) - f(x)$:

$$f(x + h) - f(x) = -x^2 - 2xh - h^2 + 4x + 4h - 9 - (-x^2 + 4x - 9)$$
$$= -2xh - h^2 + 4h$$

Step 3.   Find $\dfrac{f(x + h) - f(x)}{h}$:

$$\frac{f(x + h) - f(x)}{h} = \frac{-2xh - h^2 + 4h}{h} = -2x - h + 4$$

Step 4.   Find $\lim\limits_{h \to 0} \dfrac{f(x + h) - f(x)}{h}$:

$$\lim_{h \to 0} \frac{f(x + h) - f(x)}{h} = \lim_{h \to 0} (-2x - h + 4) = -2x + 4$$

Thus, $f'(x) = -2x + 4$.

$f'(1) = 2, \ f'(2) = 0, \ f'(3) = -2$

31. $f(x) = 2x^3 + 1$

Step 1.   Find $f(x + h)$:

$$f(x + h) = 2(x + h)^3 + 1 = 2(x^3 + 3x^2h + 3xh^2 + h^3) + 1$$
$$= 2x^3 + 6x^2h + 6xh^2 + 2h^3 + 1$$

Step 2.   Find $f(x + h) - f(x)$:

$$f(x + h) - f(x) = 2x^3 + 6x^2h + 6xh^2 + 2h^3 + 1 - (2x^3 + 1)$$
$$= 6x^2h + 6xh^2 + 2h^3$$

Step 3.   Find $\dfrac{f(x + h) - f(x)}{h}$:

$$\frac{f(x + h) - f(x)}{h} = \frac{6x^2h + 6xh^2 + 2h^3}{h} = 6x^2 + 6xh + 2h^2$$

Step 4.   Find $\lim\limits_{h \to 0} \dfrac{f(x + h) - f(x)}{h}$:

$$\lim_{h \to 0} \frac{f(x + h) - f(x)}{h} = \lim_{h \to 0} (6x^2 + 6xh + 2h^2) = 6x^2$$

Thus, $f'(x) = 6x^2$.

$f'(1) = 6, \ f'(2) = 24, \ f'(3) = 54$

33. $f(x) = 4 + \dfrac{4}{x}$

Step 1. Find $f(x + h)$:

$$f(x + h) = 4 + \dfrac{4}{x + h}$$

Step 2. Find $f(x + h) - f(x)$:

$$f(x + h) - f(x) = 4 + \dfrac{4}{x + h} - \left(4 + \dfrac{4}{x}\right) = \dfrac{4}{x + h} - \dfrac{4}{x}$$

$$= \dfrac{4x - 4(x + h)}{x(x + h)} = -\dfrac{4h}{x(x + h)}$$

Step 3. Find $\dfrac{f(x + h) - f(x)}{h}$:

$$\dfrac{f(x + h) - f(x)}{h} = \dfrac{-\dfrac{4h}{x(x + h)}}{h} = -\dfrac{4}{x(x + h)}$$

Step 4. Find $\lim\limits_{h \to 0} \dfrac{f(x + h) - f(x)}{h}$:

$$\lim_{h \to 0} \dfrac{f(x + h) - f(x)}{h} = \lim_{h \to 0} -\dfrac{4}{x(x + h)} = -\dfrac{4}{x^2}$$

Thus, $f'(x) = -\dfrac{4}{x^2}$.

$$f'(1) = -4, \quad f'(2) = -1, \quad f'(3) = -\dfrac{4}{9}$$

35. $f(x) = 5 + 3\sqrt{x}$

Step 1. Find $f(x + h)$:
$$f(x + h) = 5 + 3\sqrt{x + h}$$

Step 2. Find $f(x + h) - f(x)$:
$$f(x + h) - f(x) = 5 + 3\sqrt{x + h} - (5 + 3\sqrt{x}) = 3(\sqrt{x + h} - \sqrt{x})$$

Step 3. Find $\dfrac{f(x + h) - f(x)}{h}$:
$$\dfrac{f(x + h) - f(x)}{h} = \dfrac{3(\sqrt{x + h} - \sqrt{x})}{h} = \dfrac{3(\sqrt{x + h} - \sqrt{x})}{h} \cdot \dfrac{(\sqrt{x + h} + \sqrt{x})}{(\sqrt{x + h} + \sqrt{x})}$$

$$= \dfrac{3(x + h - x)}{h(\sqrt{x + h} + \sqrt{x})} = \dfrac{3h}{h(\sqrt{x + h} + \sqrt{x})} = \dfrac{3}{\sqrt{x + h} + \sqrt{x}}$$

Step 4. Find $\lim\limits_{h \to 0} \dfrac{f(x + h) - f(x)}{h}$:

$$\lim_{h \to 0} \dfrac{f(x + h) - f(x)}{h} = \lim_{h \to 0} \dfrac{3}{\sqrt{x + h} + \sqrt{x}} = \dfrac{3}{2\sqrt{x}}$$

Thus, $f'(x) = \dfrac{3}{2\sqrt{x}}$.

$f'(1) = \dfrac{3}{2}$, $f'(2) = \dfrac{3}{2\sqrt{2}} = \dfrac{3\sqrt{2}}{4}$, $f'(3) = \dfrac{3}{2\sqrt{3}} = \dfrac{\sqrt{3}}{2}$

37. $f(x) = 10\sqrt{x+5}$

Step 1.   Find $f(x+h)$:
$f(x+h) = 10\sqrt{x+h+5}$

Step 2.   Find $f(x+h) - f(x)$:
$f(x+h) - f(x) = 10\sqrt{x+h+5} - 10\sqrt{x+5} = 10\left(\sqrt{x+h+5} - \sqrt{x+5}\right)$

Step 3.   Find $\dfrac{f(x+h) - f(x)}{h}$:

$$\dfrac{f(x+h) - f(x)}{h} = \dfrac{10\left(\sqrt{x+h+5} - \sqrt{x+5}\right)}{h}$$

$$= \dfrac{10\left(\sqrt{x+h+5} - \sqrt{x+5}\right)}{h} \cdot \dfrac{\left(\sqrt{x+h+5} + \sqrt{x+5}\right)}{\left(\sqrt{x+h+5} + \sqrt{x+5}\right)}$$

$$= \dfrac{10[x+h+5-(x+5)]}{h\left(\sqrt{x+h+5} + \sqrt{x+5}\right)} = \dfrac{10h}{h\left(\sqrt{x+h+5} + \sqrt{x+5}\right)}$$

$$= \dfrac{10}{\sqrt{x+h+5} + \sqrt{x+5}}$$

Step 4.   Find $\lim\limits_{h \to 0} \dfrac{f(x+h) - f(x)}{h}$:

$$\lim\limits_{h \to 0} \dfrac{f(x+h) - f(x)}{h} = \lim\limits_{h \to 0} \dfrac{10}{\sqrt{x+h+5} + \sqrt{x+5}} = \dfrac{10}{2\sqrt{x+5}} = \dfrac{5}{\sqrt{x+5}}$$

Thus, $f'(x) = \dfrac{5}{\sqrt{x+5}}$.

$f'(1) = \dfrac{5}{\sqrt{6}} = \dfrac{5\sqrt{6}}{6}$, $f'(2) = \dfrac{5}{\sqrt{7}} = \dfrac{5\sqrt{7}}{7}$, $f'(3) = \dfrac{5}{\sqrt{8}} = \dfrac{5}{2\sqrt{2}} = \dfrac{5\sqrt{2}}{4}$

39. $f(x) = \dfrac{1}{x-4}$.

Step 1. $f(x+h) = \dfrac{1}{x-4+h}$

Step 2. $f(x+h) - f(x) = \dfrac{1}{x-4+h} - \dfrac{1}{x-4} = \dfrac{x-4-(x-4+h)}{(x-4+h)(x-4)} = \dfrac{-h}{(x-4+h)(x-4)}$

Step3. $\dfrac{f(x+h) - f(x)}{h} = \dfrac{-h}{h(x-4+h)(x-4)} = \dfrac{-1}{(x-4+h)(x-4)}$

Step 4. $f'(x) = \lim\limits_{h \to 0} \dfrac{f(x+h) - f(x)}{h} = \lim\limits_{h \to 0} \dfrac{-1}{(x-4+h)(x-4)} = \dfrac{-1}{(x-4)^2}$.

$$f'(1) = \dfrac{-1}{9}, \quad f'(2) = \dfrac{-1}{4}, \quad f'(3) = -1$$

41. $f(x) = \dfrac{x}{x+1}$

Step1. $f(x+h) = \dfrac{x+h}{x+1+h}$

Step 2. $f(x+h) - f(x) = \dfrac{x+h}{x+1+h} - \dfrac{x}{x+1} = \dfrac{(x+h)(x+1) - x(x+1+h)}{(x+1+h)(x+1)} = \dfrac{h}{(x+1+h)(x+1)}$

Step 3. $\dfrac{f(x+h) - f(x)}{h} = \dfrac{h}{h(x+1+h)(x+1)} = \dfrac{1}{(x+1+h)(x+1)}$

Step 4. $f'(x) = \lim\limits_{h \to 0} \dfrac{f(x+h) - f(x)}{h} = \lim\limits_{h \to 0} \dfrac{1}{(x+1+h)(x+1)} = \dfrac{1}{(x+1)^2}$.

$$f'(1) = \dfrac{1}{4}, \quad f'(2) = \dfrac{1}{9}, \quad f'(3) = \dfrac{1}{16}$$

43. $y = f(x) = x^2 + x$

(A) $f(1) = 1^2 + 1 = 2, f(3) = 3^2 + 3 = 12$

Slope of secant line: $\dfrac{f(3) - f(1)}{3 - 1} = \dfrac{12 - 2}{2} = 5$

(B) $f(1) = 2, f(1+h) = (1+h)^2 + (1+h) = 1 + 2h + h^2 + 1 + h = 2 + 3h + h^2$

Slope of secant line: $\dfrac{f(1+h) - f(1)}{h} = \dfrac{2 + 3h + h^2 - 2}{h} = 3 + h$

(C) Slope of tangent line at $(1, f(1))$:

$\lim\limits_{h \to 0} \dfrac{f(1+h) - f(1)}{h} = \lim\limits_{h \to 0} (3 + h) = 3$

(D) Equation of tangent line at $(1, f(1))$:

$y - f(1) = f'(1)(x - 1)$ or $y - 2 = 3(x - 1)$ and $y = 3x - 1$.

45. $f(x) = x^2 + x$

(A) Average velocity: $\dfrac{f(3) - f(1)}{3 - 1} = \dfrac{3^2 + 3 - (1^2 + 1)}{2} = \dfrac{12 - 2}{2} = 5$ meters/sec.

(B) Average velocity: $\dfrac{f(1+h) - f(1)}{h} = \dfrac{(1+h)^2 + (1+h) - (1^2 + 1)}{h} = \dfrac{1 + 2h + h^2 + 1 + h - 2}{h}$

$$= \dfrac{3h + h^2}{h} = 3 + h \text{ meters/sec.}$$

(C) Instantaneous velocity: $\lim\limits_{h \to 0} \dfrac{f(1+h) - f(1)}{h} = \lim\limits_{h \to 0} (3 + h) = 3$ m/sec.

47. $F'(x)$ does exist at $x = a$.

49. $F'(x)$ does not exist at $x = c$; the graph has a vertical tangent line at $(c, F(c))$.

51. $F'(x)$ does exist at $x = e$; $F'(e) = 0$.

53. $F'(x)$ does exist at $x = g$.

55. $f(x) = x^2 - 4x$

(A) Step 1.   Find $f(x + h)$:
$$f(x + h) = (x + h)^2 - 4(x + h) = x^2 + 2xh + h^2 - 4x - 4h$$

Step 2.   Find $f(x + h) - f(x)$:
$$f(x + h) - f(x) = x^2 + 2xh + h^2 - 4x - 4h - (x^2 - 4x) = 2xh + h^2 - 4h$$

Step 3.   Find $\dfrac{f(x+h) - f(x)}{h}$:
$$\frac{f(x+h) - f(x)}{h} = \frac{2xh + h^2 - 4h}{h} = 2x + h - 4$$

Step 4.   Find $\lim\limits_{h \to 0} \dfrac{f(x+h) - f(x)}{h}$:
$$\lim_{h \to 0} \frac{f(x+h) - f(x)}{h} = \lim_{h \to 0} (2x + h - 4) = 2x - 4$$
Thus, $f'(x) = 2x - 4$.

(B) $f'(0) = -4$, $f'(2) = 0$, $f'(4) = 4$

(C) Since f is a quadratic function, the graph of f is a parabola.

y intercept: $y = 0$
x intercepts: $x = 0$, $x = 4$
Vertex: $(2, -4)$

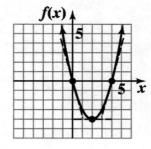

57. To find $v = f'(x)$, use the four-step process on the position function $f(x) = 4x^2 - 2x$.

Step 1.   Find $f(x + h)$:
$$f(x + h) = 4(x + h)^2 - 2(x + h) = 4(x^2 + 2xh + h^2) - 2x - 2h$$
$$= 4x^2 + 8xh + 4h^2 - 2x - 2h$$

Step 2.   Find $f(x + h) - f(x)$:
$$f(x + h) - f(x) = 4x^2 + 8xh + 4h^2 - 2x - 2h - (4x^2 - 2x) = 8xh + 4h^2 - 2h$$

Step 3.   Find $\dfrac{f(x+h) - f(x)}{h}$:
$$\frac{f(x+h) - f(x)}{h} = \frac{8xh + 4h^2 - 2h}{h} = 8x + 4h - 2$$

Step 4. Find $\lim\limits_{h \to 0} \dfrac{f(x+h) - f(x)}{h}$:

$$\lim\limits_{h \to 0} \dfrac{f(x+h) - f(x)}{h} = \lim\limits_{h \to 0} (8x + 4h - 2) = 8x - 2$$

Thus, the velocity, $v(x) = f'(x) = 8x - 2$

$f'(1) = 8 \cdot 1 - 2 = 6$ ft/sec, $f'(3) = 8 \cdot 3 - 2 = 22$ ft/sec, $f'(5) = 8 \cdot 5 - 2 = 38$ ft/sec

59. (A) The graphs of g and h are vertical translations of the graph of f. All three functions should have the same derivative.

(B) $m(x) = x^2 + C$

Step 1. Find $m(x + h)$:

$$m(x + h) = (x + h)^2 + C$$

Step 2. Find $m(x + h) - m(x)$:

$$m(x + h) - m(x) = (x + h)^2 + C - (x^2 + C) = x^2 + 2xh + h^2 + C - x^2 - C$$
$$= 2xh + h^2$$

Step 3. Find $\dfrac{m(x+h) - m(x)}{h}$:

$$\dfrac{m(x+h) - m(x)}{h} = \dfrac{2xh + h^2}{h} = 2x + h$$

Step 4. $\lim\limits_{h \to 0} \dfrac{m(x+h) - m(x)}{h}$:

$$\lim\limits_{h \to 0} \dfrac{m(x+h) - m(x)}{h} = \lim\limits_{h \to 0} (2x + h) = 2x$$

Thus, $m'(x) = 2x$.

61. True. The graph of a constant function is a horizontal line. The slope of a horizontal line is 0.

63. False. The function $f(x) = |x|$ is continuous on (-1,1) but it is not differentiable at $x = 0$.

65. False. Set $f(x) = x^3$ on [0,1]. The average rate of change of f:

$$\dfrac{f(1) - f(0)}{1} = 1;$$

$f'(x) = 3x^2$. The instantaneous rate of change of f at $x = 1/2$ is $f'(1/2) = 3/4 < 1$.

67. The graph of $f(x) = \begin{Bmatrix} 2x, x < 1 \\ 2, x \ge 1 \end{Bmatrix}$ is: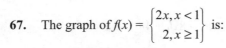

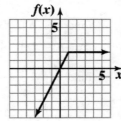

f is not differentiable at $x = 1$ because the graph of f has a sharp corner at this point.

69. $f(x) = \begin{cases} x^2 + 1 & \text{if } x < 0 \\ 1 & \text{if } x \geq 0 \end{cases}$

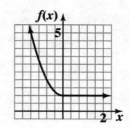

It is clear that $f'(x) = \begin{cases} 2x & \text{if } x < 0 \\ 0 & \text{if } x > 0 \end{cases}$

Thus, the only question is $f'(0)$. Since

$$\lim_{x \to 0^-} f'(x) = \lim_{x \to 0^-} 2x = 0 \text{ and } \lim_{x \to 0^+} f'(x) = \lim_{x \to 0^+} 0 = 0$$

f is differentiable at 0 as well; f is differentiable for all real numbers.

71. $f(x) = |x|$

$$\lim_{h \to 0} \frac{f(0+h) - f(0)}{h} = \lim_{h \to 0} \frac{|0+h| - |0|}{h} = \lim_{h \to 0} \frac{|h|}{h}$$

The limit does not exist. Thus, f is not differentiable at $x = 0$.

73. $f(x) = \sqrt[3]{x} = x^{1/3}$

$$\lim_{h \to 0} \frac{f(0+h) - f(0)}{h} = \lim_{h \to 0} \frac{(0+h)^{1/3} - 0^{1/3}}{h} = \lim_{h \to 0} \frac{h^{1/3}}{h} = \lim_{h \to 0} \frac{1}{h^{2/3}}$$

The limit does not exist. Thus, f is not differentiable at $x = 0$.

75. $f(x) = \sqrt{1 - x^2}$

$$\frac{f(0+h) - f(0)}{h} = \frac{\sqrt{1-h^2} - 1}{h} = \frac{\sqrt{1-h^2} - 1}{h} \cdot \frac{\sqrt{1-h^2} + 1}{\sqrt{1-h^2} + 1} = \frac{1 - h^2 - 1}{h\left(\sqrt{1-h^2} + 1\right)} = \frac{-h}{\sqrt{1-h^2} + 1}$$

$$\lim_{h \to 0} \frac{f(0+h) - f(0)}{h} = \lim_{h \to 0} \frac{-h}{\sqrt{1-h^2} + 1} = 0$$

f is differentiable at 0; $f'(0) = 0$.

77. The height of the ball at x seconds is $h(x) = 576 - 16x^2$. To find when the ball hits the ground, we solve:

$$576 - 16x^2 = 0$$
$$16x^2 = 576$$
$$x^2 = 36$$
$$x = 6 \text{ seconds}$$

The velocity of the ball is given by $h'(x) = -32x$. The velocity at impact is $h'(6) = -32(6) = -192$; the ball hits the ground at 192 ft/sec.

79. $R(x) = 60x - 0.025x^2$ $0 \le x \le 2,400.$

   (A)   Average rate of change:

$$\frac{R(1,050) - R(1,000)}{1,050 - 1,000}$$

$$= \frac{60(1,050) - 0.025(1,050)^2 - [60(1,000) - 0.025(1,000)^2]}{50}$$

$$= \frac{35,437.50 - 35,000}{50} = \$8.75 \text{ per car seat}$$

   (B)   <u>Step 1</u>.   Find $R(x + h)$:

$$R(x + h) = 60(x + h) - 0.025(x + h)^2$$
$$= 60x + 60h - 0.025(x^2 + 2xh + h^2)$$
$$= 60x + 60h - 0.025x^2 - 0.050xh - 0.025h^2$$

   <u>Step 2</u>.   Find $R(x + h) - R(x)$:

$$R(x + h) - R(x) = 60x + 60h - 0.025x^2 - 0.050xh - 0.025h^2 - (60x - 0.025x^2)$$
$$= 60h - 0.050xh - 0.025h^2$$

   <u>Step 3</u>.   Find $\dfrac{R(x + h) - R(x)}{h}$:

$$\frac{R(x + h) - R(x)}{h} = \frac{60h - 0.050xh - 0.025h^2}{h} = 60 - 0.050x - 0.025h$$

   <u>Step 4</u>.   Find $\displaystyle\lim_{h \to 0} \dfrac{R(x + h) - R(x)}{h}$:

$$\lim_{h \to 0} \frac{R(x + h) - R(x)}{h} = \lim_{h \to 0}(60 - 0.050x - 0.025h) = 60 - 0.050x$$

   Thus, $R'(x) = 60 - 0.050x.$

   (C) $R(1,000) = 60(1,000) - 0.025(1,000)^2 = \$35,000;$
   $R'(1,000) = 60 - 0.05(1,000) = \$10;$
   at a production level of 1,000 car seats, the revenue is \$35,000 and is increasing at the rate of \$10 per   car seat.

81. (A) $S(t) = 2\sqrt{t + 10}$

   <u>Step 1</u>.   Find $S(t + h)$:
   $S(t + h) = 2\sqrt{t + h + 10}$

   <u>Step 2</u>.   Find $S(t + h) - S(t)$:
   $S(t + h) - S(t) = 2\sqrt{t + h + 10} - 2\sqrt{t + 10} = 2(\sqrt{t + h + 10} - \sqrt{t + 10})$

   <u>Step 3</u>.   Find $\dfrac{S(t + h) - S(t)}{h}$:

$$\frac{S(t + h) - S(t)}{h} = \frac{2\left(\sqrt{t + h + 10} - \sqrt{t + 10}\right)}{h}$$

$$= \frac{2\left(\sqrt{t+h+10} - \sqrt{t+10}\right)}{h} \cdot \frac{\left(\sqrt{t+h+10} + \sqrt{t+10}\right)}{\left(\sqrt{t+h+10} + \sqrt{t+10}\right)}$$

$$= \frac{2[t+h+10 - (t+10)]}{h\left(\sqrt{t+h+10} + \sqrt{t+10}\right)} = \frac{2h}{h\left(\sqrt{t+h+10} + \sqrt{t+10}\right)}$$

$$= \frac{2}{\sqrt{t+h+10} + \sqrt{t+10}}$$

Step 4.   Find $\lim\limits_{h\to 0} \dfrac{S(t+h) - S(t)}{h}$:

$$\lim_{h\to 0} \frac{S(t+h) - S(t)}{h} = \lim_{h\to 0} \frac{2}{\sqrt{t+h+10} + \sqrt{t+10}} = \frac{1}{\sqrt{t+10}}$$

Thus, $S'(t) = \dfrac{1}{\sqrt{t+10}}$.

(B) $S(15) = 2\sqrt{15+10} = 2\sqrt{25} = 10;$

$S'(15) = \dfrac{1}{\sqrt{15+10}} = \dfrac{1}{\sqrt{25}} = \dfrac{1}{5} = 0.2$

After 15 months, the total sales are \$10 million and are INCREASING at the rate of \$0.2 million or \$200,000 per month.

(C)   The estimated total sales are \$10.2 million after 16 months and \$10.4 million after 17 months.

83. $p(t) = 138t^2 + 1,072t + 14,917$

(A)   Step 1.   Find $p(t+h)$:

$p(t+h) = 138(t+h)^2 + 1,072(t+h) + 14,917$

Step 2.   Find $p(t+h) - p(t)$:

$p(t+h) - p(t) = 138(t+h)^2 + 1,072(t+h) + 14,917 - (138t^2 + 1,072t + 14,917)$

$= 276th + 138h^2 + 1,072h$

Step 3.   Find $\dfrac{p(t+h) - p(t)}{h}$:

$\dfrac{p(t+h) - p(t)}{h} = \dfrac{276th + 138h^2 + 1,072h}{h} = 276t + 138h + 1,072$

Step 4.   Find $\lim\limits_{h\to 0} \dfrac{p(t+h) - p(t)}{h}$:

$\lim\limits_{h\to 0} \dfrac{p(t+h) - p(t)}{h} = \lim\limits_{h\to 0} (276t + 138h + 1,072) = 276t + 1,072$

Thus, $p'(t) = 276t + 1,072$

(B)   The year 2020 corresponds to $t = 10$.

$p(10) = 138(10)^2 + 1,072(10) + 14,917 = 39,437$ metric tons;

$p'(t) = 276t + 1,072,$

$p'(10) = 276(10) + 1,072 = 3,832.$

In 2020 the US will produce 39,437 metric tons of tungsten and this quantity is increasing at the rate of 3,832 metric tons/year.

85. (A) Quadratic regression
model

(B)

$R(x) \approx -0.1339205714\, x^2 + 25.225\, x + 1199.785714$;
$R(20) \approx 1{,}650.7$;
$R'(x) \approx -0.267857142\, x + 25.225$;
$R'(20) \approx 19.9$

In 2020, 1,650.7 billion residential kilowatts will be sold and
the amount sold is increasing at the rate of 19.9 billion kilowatts
per year.

87. (A) $P(t) = 80 + 12t - t^2$

Step 1. Find $P(t + h)$:
$$P(t + h) = 80 + 12(t + h) - (t + h)^2 = 80 + 12t + 12h - (t + h)^2$$

Step 2. Find $P(t + h) - P(t)$:
$$P(t + h) - P(t) = 80 + 12t + 12h - (t + h)^2 - (80 + 12t - t^2)$$
$$= 12h - 2th - h^2$$

Step 3. Find $\dfrac{P(t + h) - P(t)}{h}$:

$$\frac{P(t + h) - P(t)}{h} = \frac{12h - 2th - h^2}{h} = 12 - 2t - h$$

Step 4. Find $\displaystyle\lim_{h \to 0} \frac{P(t + h) - P(t)}{h}$:

$$\lim_{h \to 0} \frac{P(t + h) - P(t)}{h} = \lim_{h \to 0} (12 - 2t - h) = 12 - 2t$$

Thus, $P'(t) = 12 - 2t$.

(B) $P(3) = 80 + 12(3) - (3)^2 = 107; \ P'(3) = 12 - 2(3) = 6$

After 3 hours, the ozone level is 107 ppb and is INCREASING at the
rate of 6 ppb per hour.

EXERCISE 10-5

Things to remember:

<u>1</u>.   DERIVATIVE NOTATION

Given $y = f(x)$, then

$$f'(x), \quad y', \quad \frac{dy}{dx}$$

all represent the derivative of f at x.

<u>2</u>.   CONSTANT FUNCTION RULE

If $f(x) = C$, C a constant, then $f'(x) = 0$. Also

$$y' = 0 \text{ and } \frac{dy}{dx} = 0.$$

<u>3</u>.   POWER RULE

If $f(x) = x^n$, n any real number, then

$$f'(x) = nx^{n-1}.$$

Also, $y' = nx^{n-1}$ and $\dfrac{dy}{dx} = nx^{n-1}$

<u>4</u>.   CONSTANT MULTIPLE PROPERTY

If $y = f(x) = ku(x)$, where k is a constant, then

$$f'(x) = ku'(x).$$

Also,

$$y' = ku' \text{ and } \frac{dy}{dx} = k\frac{du}{dx}.$$

<u>5</u>.   SUM AND DIFFERENCE PROPERTY

If $y = f(x) = u(x) \pm v(x)$, then

$$f'(x) = u'(x) \pm v'(x).$$

Also,

$$y' = u' \pm v' \text{ and } \frac{dy}{dx} = \frac{du}{dx} \pm \frac{dv}{dx}$$

[<u>Note</u>: This rule generalizes to the sum and difference of any given number of functions.]

---

1. $\sqrt{x} = x^{1/2}$ **3.** $\dfrac{1}{x^5} = x^{-5}$ **5.** $(x^4)^3 = x^{12}$ **7.** $\dfrac{1}{\sqrt[4]{x}} = \dfrac{1}{x^{1/4}} = x^{-1/4}$

9. $f(x) = 7; f'(x) = 0$ (using <u>2</u>) **11.** $y = x^9; \dfrac{dy}{dx} = 9x^8$ (using <u>3</u>)

13. $\dfrac{d}{dx}x^3 = 3x^2$ (using $\underline{3}$)

15. $y = x^{-4}; y' = -4x^{-5}$ (using $\underline{3}$)

17. $g(x) = x^{8/3}; g'(x) = \dfrac{8}{3}x^{5/3}$ (using $\underline{3}$)

19. $y = \dfrac{1}{x^{10}}; \dfrac{dy}{dx} = -10x^{-11} = \dfrac{-10}{x^{11}}$

21. $f(x) = 5x^2; f'(x) = 5(2x) = 10x$ (using $\underline{4}$)

23. $y = 0.4x^7; y' = 0.4(7x^6) = 2.8x^6$

25. $\dfrac{d}{dx}\left(\dfrac{x^3}{18}\right) = \dfrac{1}{18}(3x^2) = \dfrac{1}{6}x^2$

27. $h(x) = 4f(x); \quad h'(2) = 4 \cdot f'(2) = 4(3) = 12$

29. $h(x) = f(x) + g(x); h'(2) = f'(2) + g'(2) = 3 + (-1) = 2$

31. $h(x) = 2f(x) - 3g(x) + 7; h'(2) = 2f'(2) - 3g'(2) = 2(3) - 3(-1) = 9$

33. $\dfrac{d}{dx}(2x - 5) = \dfrac{d}{dx}(2x) - \dfrac{d}{dx}(5) = 2$

35. $f(t) = 2t^2 - 3t + 1; f'(t) = (2t^2)' - (3t)' + (1)' = 4t - 3$

37. $y = 5x^{-2} + 9x^{-1}; y' = -10x^{-3} - 9x^{-2}$

39. $\dfrac{d}{du}(5u^{0.3} - 4u^{2.2}) = \dfrac{d}{du}(5u^{0.3}) - \dfrac{d}{du}(4u^{2.2}) = 1.5u^{-0.7} - 8.8u^{1.2}$

41. $h(t) = 2.1 + 0.5t - 1.1t^3; h'(t) = 0.5 - (1.1)3t^2 = 0.5 - 3.3t^2$

43. $y = \dfrac{2}{5x^4} = \dfrac{2}{5}x^{-4}; y' = \dfrac{2}{5}(-4x^{-5}) = -\dfrac{8}{5}x^{-5} = \dfrac{-8}{5x^5}$

45. $\dfrac{d}{dx}\left(\dfrac{3x^2}{2} - \dfrac{7}{5x^2}\right) = \dfrac{d}{dx}\left(\dfrac{3}{2}x^2\right) - \dfrac{d}{dx}\left(\dfrac{7}{5}x^{-2}\right) = 3x + \dfrac{14}{5}x^{-3} = 3x + \dfrac{14}{5x^3}$

47. $G(w) = \dfrac{5}{9w^4} + 5\sqrt[3]{w} = \dfrac{5}{9}w^{-4} + 5w^{1/3};$

$G'(w) = -\dfrac{20}{9}w^{-5} + \dfrac{5}{3}w^{-2/3} = \dfrac{-20}{9w^5} + \dfrac{5}{3w^{2/3}}$

49. $\dfrac{d}{du}(3u^{2/3} - 5u^{1/3}) = \dfrac{d}{du}(3u^{2/3}) - \dfrac{d}{du}(5u^{1/3}) = 2u^{-1/3} - \dfrac{5}{3}u^{-2/3} = \dfrac{2}{u^{1/3}} - \dfrac{5}{3u^{2/3}}$

51. $h(t) = \dfrac{3}{t^{3/5}} - \dfrac{6}{t^{1/2}} = 3t^{-3/5} - 6t^{-1/2};$

$h'(t) = 3\left(-\dfrac{3}{5}t^{-8/5}\right) - 6\left(-\dfrac{1}{2}t^{-3/2}\right) = -\dfrac{9}{5}t^{-8/5} + 3t^{-3/2} = \dfrac{-9}{5t^{8/5}} + \dfrac{3}{t^{3/2}}$

53. $y = \dfrac{1}{\sqrt[3]{x}} = \dfrac{1}{x^{1/3}} = x^{-1/3}; y' = -\dfrac{1}{3}x^{-4/3} = \dfrac{-1}{3x^{4/3}}$

55. $\dfrac{d}{dx}\left(\dfrac{1.2}{\sqrt{x}} - 3.2x^{-2} + x\right) = \dfrac{d}{dx}(1.2x^{-1/2} - 3.2x^{-2} + x) = \dfrac{d}{dx}(1.2x^{-1/2}) - \dfrac{d}{dx}(3.2x^{-2}) + \dfrac{d}{dx}(x)$

$= -0.6x^{-3/2} + 6.4x^{-3} + 1 = \dfrac{-0.6}{x^{3/2}} + \dfrac{6.4}{x^3} + 1$

57. $f(x) = 6x - x^2$

(A) $f'(x) = 6 - 2x$

(B) Slope of the graph of f at $x = 2$: $f'(2) = 6 - 2(2) = 2$
Slope of the graph of f at $x = 4$: $f'(4) = 6 - 2(4) = -2$

(C) Tangent line at $x = 2$: $y - y_1 = m(x - x_1)$

$x_1 = 2, \quad y_1 = f(2) = 6(2) - 2^2 = 8$

$m = f'(2) = 2$
Thus, $y - 8 = 2(x - 2)$ or $y = 2x + 4$.
Tangent line at $x = 4$: $y - y_1 = m(x - x_1)$

$x_1 = 4, \quad y_1 = f(4) = 6(4) - 4^2 = 8$

$m = f'(4) = -2$
Thus, $y - 8 = -2(x - 4)$ or $y = -2x + 16$

(D) The tangent line is horizontal at the values $x = c$ such that $f'(c) = 0$. Thus, we must solve the following:

$$f'(x) = 6 - 2x = 0$$
$$2x = 6$$
$$x = 3$$

59. $f(x) = 3x^4 - 6x^2 - 7$

(A) $f'(x) = 12x^3 - 12x$

(B) Slope of the graph at $x = 2$: $f'(2) = 12(2)^3 - 12(2) = 72$

Slope of the graph of $x = 4$: $f'(4) = 12(4)^3 - 12(4) = \dfrac{72}{0}, 720$

(C) Tangent line at $x = 2$: $y - y_1 = m(x - x_1)$, where $x_1 = 2$,

$y_1 = f(2) = 3(2)^4 - 6(2)^2 - 7 = 17, \quad m = 72.$
$y - 17 = 72(x - 2)$ or $y = 72x - 127$

Tangent line at $x = 4$: $y - y_1 = m(x - x_1)$, where $x_1 = 4$,

$y_1 = f(4) = 3(4)^4 - 6(4)^2 - 7 = 665$, $m = 720$.

$y - 665 = 720(x - 4)$ or $y = 720x - 2215$

(D)   Solve $f'(x) = 0$ for x:

$$12x^3 - 12x = 0$$
$$12x(x^2 - 1) = 0$$
$$12x(x - 1)(x + 1) = 0$$
$$x = -1, \ x = 0, \ x = 1$$

61. $f(x) = 176x - 16x^2$

(A) $v = f'(x) = 176 - 32x$

(B) $v\big|_{x=0} = f'(0) = 176$ ft/sec.

$v\big|_{x=3} = f'(3) = 176 - 32(3) = 80$ ft/sec.

(C)   Solve $v = f'(x) = 0$ for x:

$$176 - 32x = 0$$
$$32x = 176$$
$$x = 5.5 \text{ sec.}$$

63. $f(x) = x^3 - 9x^2 + 15x$

(A) $v = f'(x) = 3x^2 - 18x + 15$

(B) $v\big|_{x=0} = f'(0) = 15$ feet/sec.

$v\big|_{x=3} = f'(3) = 3(3)^2 - 18(3) + 15 = -12$ feet/sec.

(C)   Solve $v = f'(x) = 0$ for x:

$$3x^2 - 18x + 15 = 0$$
$$3(x^2 - 6x + 5) = 0$$
$$3(x - 5)(x - 1) = 0$$
$$x = 1, \ x = 5$$

65. $f(x) = x^2 - 3x - 4\sqrt{x} = x^2 - 3x - 4x^{1/2}$

$f'(x) = 2x - 3 - 2x^{-1/2}$

The graph of f has a horizontal tangent line at the value(s) of x where $f'(x) = 0$. Thus, we need to solve the equation

$$2x - 3 - 2x^{-1/2} = 0$$

By graphing the function $y = 2x - 3 - 2x^{-1/2}$, we see that there is one zero. To four decimal places, it is $x = 2.1777$.

67. $f(x) = 3\sqrt[3]{x^4} - 1.5x^2 - 3x = 3x^{4/3} - 1.5x^2 - 3x$

$f'(x) = 4x^{1/3} - 3x - 3$

The graph of f has a horizontal tangent line at the value(s) of x where $f'(x) = 0$. Thus, we need to solve the equation

$$4x^{1/3} - 3x - 3 = 0$$

Graphing the function $y = 4x^{1/3} - 3x - 3$, we see that there is one zero. To four decimal places, it is $x = -2.9018$.

69. $f(x) = 0.05x^4 - 0.1x^3 - 1.5x^2 - 1.6x + 3$

$f'(x) = 0.2x^3 + 0.3x^2 - 3x - 1.6$

The graph of f has a horizontal tangent line at the value(s) of x where $f'(x) = 0$. Thus, we need to solve the equation

$$0.2x^3 + 0.3x^2 - 3x - 1.6 = 0$$

By graphing the function $y = 0.2x^3 + 0.3x^2 - 3x - 1.6$, we see that there are three zeros. To four decimal places, they are

$$x_1 = -4.4607, \quad x_2 = -0.5159, \quad x_3 = 3.4765$$

71. $f(x) = 0.2x^4 - 3.12x^3 + 16.25x^2 - 28.25x + 7.5$

$f'(x) = 0.8x^3 - 9.36x^2 + 32.5x - 28.25$

The graph of f has a horizontal tangent line at the value(s) of x where $f'(x) = 0$. Thus, we need to solve the equation

$$0.8x^3 - 9.36x^2 + 32.5x - 28.25 = 0$$

Graphing the function $y = 0.8x^3 - 9.36x^2 + 32.5x - 28.25$, we see that there is one zero. To four decimal places, it is $x = 1.3050$.

73. $f(x) = ax^2 + bx + c; f'(x) = 2ax + b$.

The derivative is 0 at the vertex of the parabola:

$2ax + b = 0$

$$x = -\frac{b}{2a}$$

75. (A) $f(x) = x^3 + x$ (B) $f(x) = x^3$ (C) $f(x) = x^3 - x$

77. $f(x) = (2x - 1)^2 = 4x^2 - 4x + 1; \quad f'(x) = 8x - 4$

79. $\dfrac{d}{dx}\left(\dfrac{10x + 20}{x}\right) = \dfrac{d}{dx}\left(10 + \dfrac{20}{x}\right) = \dfrac{d}{dx}(10) + \dfrac{d}{dx}(20x^{-1}) = -20x^{-2} = -\dfrac{20}{x^2}$

81. $y = \dfrac{3x - 4}{12x^2} = \dfrac{3x}{12x^2} - \dfrac{4}{12x^2} = \dfrac{1}{4}x^{-1} - \dfrac{1}{3}x^{-2}$

$\dfrac{dy}{dx} = -\dfrac{1}{4}x^{-2} + \dfrac{2}{3}x^{-3} = -\dfrac{1}{4x^2} + \dfrac{2}{3x^3}$

83. False. Set $f(x) = x^2$, $g(x) = x^3$. Then $f(x) \cdot g(x) = x^5$ and $[f(x) \cdot g(x)]' = 5x^4$;

$f'(x) = 2x, g'(x) = 3x^2$ and $f'(x) \cdot g'(x) = 6x^3 \neq [f(x) \cdot g(x)]'$.

85. True. Theorem 1, which is also $\underline{2}$ in *Things to Remember*.

87. $f(x) = u(x) + v(x)$

Step 1. $f(x + h) = u(x + h) + v(x + h)$

Step 2. $f(x + h) - f(x) = u(x + h) + v(x + h) - [u(x) + v(x)] = u(x + h) - u(x) + [v(x + h) - v(x)]$

Step 3. $\dfrac{f(x+h)-f(x)}{h}=\dfrac{u(x+h)-u(x)+[v(x+h)-v(x)]}{h}=\dfrac{u(x+h)-u(x)}{h}+\dfrac{v(x+h)-v(x)}{h}$

Step 4. $\displaystyle\lim_{h\to0}\dfrac{f(x+h)-f(x)}{h}=\lim_{h\to0}\left[\dfrac{u(x+h)-u(x)}{h}+\dfrac{v(x+h)-v(x)}{h}\right]$

$=\displaystyle\lim_{h\to0}\dfrac{u(x+h)-u(x)}{h}+\lim_{h\to0}\dfrac{v(x+h)-v(x)}{h}=u'(x)+v'(x)$

89. (A) $S(t)=0.03t^3+0.5t^2+2t+3$

$S'(t)=0.09t^2+t+2$

(B) $S(5)=0.03(5)^3+0.5(5)^2+2(5)+3=29.25$

$S'(5)=0.09(5)^2+5+2=9.25$

After 5 months, sales are \$29.25 million and are increasing
at the rate of \$9.25 million per month.

(C) $S(10)=0.03(10)^3+0.5(10)^2+2(10)+3=103$

$S'(10)=0.09(10)^2+10+2=21$

After 10 months, sales are \$103 million and are increasing
at the rate of \$21 million per month.

91. (A) $N(x)=1,000-\dfrac{3,780}{x}=1,000-3,780x^{-1}$

$N'(x)=3,780x^{-2}=\dfrac{3,780}{x^2}$

(B) $N'(10)=\dfrac{3,780}{(10)^2}=37.8$

At the \$10,000 level of advertising, sales are INCREASING at the
rate of 37.8 boats per \$1000 spent on advertising.

$N'(20)=\dfrac{3,780}{(20)^2}=9.45$

At the \$20,000 level of advertising, sales are INCREASING at the
rate of 9.45 boats per \$1000 spent on advertising.

93. (A) Cubic regression model

```
CubicReg
y=ax3+bx2+cx+d
a=-8.083333E-4
b=.0624285714
c=-1.081309524
d=40.57571429
■
```

(B) $M(x)\approx-0.0008083x^3+0.0624x^2-1.081x+40.576$

$M'(x)\approx-0.0024249x^2+0.1248x-1.081$

In 2020, 41.5% of male high school graduates will enroll in college
and the percentage is decreasing at the rate of 0.9% per year.

95. $y=590x^{-1/2},\ 30\le x\le75$

First, find $\dfrac{dy}{dx}=\dfrac{d}{dx}590x^{-1/2}=-295x^{-3/2}=\dfrac{-295}{x^{3/2}}$, the instantaneous rate of change of pulse when a
person is x inches tall.

(A)   The instantaneous rate of change of pulse rate at $x = 36$ is:

$$\frac{-295}{(36)^{3/2}} = \frac{-295}{216} = -1.37 \text{ (1.37 decrease in pulse rate)}$$

(B)   The instantaneous rate of change of pulse rate at $x = 64$ is:

$$\frac{-295}{(64)^{3/2}} = \frac{-295}{512} = -0.58 \text{ (0.58 decrease in pulse rate)}$$

97. $y = 50\sqrt{x}$, $0 \le x \le 9$

First, find $y' = (50\sqrt{x})' = (50x^{1/2})' = 25x^{-1/2}$

$$= \frac{25}{\sqrt{x}}, \text{ the rate of learning at the end of } x \text{ hours.}$$

(A)   Rate of learning at the end of 1 hour: $\dfrac{25}{\sqrt{1}} = 25$ items/hr

(B)   Rate of learning at the end of 9 hours: $\dfrac{25}{\sqrt{9}} = \dfrac{25}{3} = 8.33$ items/hr

EXERCISE 10-6

Things to remember:

1.   INCREMENTS

For $y = f(x)$, $\Delta x = x_2 - x_1$, $\Delta y = y_2 - y_1$, so $x_2 = x_1 + \Delta x$, and

$$\Delta y = y_2 - y_1$$
$$= f(x_2) - f(x_1)$$
$$= f(x_1 + \Delta x) - f(x_1)$$

Δy represents the change in y corresponding to a Δx change in x. Δx can be either positive or negative.

[*Note*: Δy depends on the function f, the input x, and the increment Δx.]

2.   DIFFERENTIALS

If $y = f(x)$ defines a differentiable function, then the **differential dy, or df**, is defined as the product of $f'(x)$ and dx, where $dx = \Delta x$. Symbolically,

$$dy = f'(x)dx \quad \text{or} \quad df = f'(x)dx$$

where

$$dx = \Delta x$$

[*Note*: The differential dy (or df) is actually a function involving two independent variables, x and dx; a change in either one or both will affect dy (or df).]

1. $f(x) = 0.1x + 3$; $f(0) = 3$, $f(0.1) = 0.1(0.1) + 3 = 3.01$

3. $f(x) = 0.1x + 3$; $f(-2) = 0.1(-2) + 3 = 2.8$, $f(-2.1) = 0.1(-2.1) + 3 = 2.79$

5. $g(x) = x^2$; $g(0) = 0^2 = 0$, $g(0.1) = (0.1)^2 = 0.01$

7. $g(x) = x^2;$ $g(10) = (10)^2 = 100,$ $g(10.1) = (10.1)^2 = (10 + 0.1)^2 = 102.01$

9. $\Delta x = x_2 - x_1 = 4 - 1 = 3$

$\Delta y = f(x_2) - f(x_1) = 3 \cdot 4^2 - 3 \cdot 1^2 = 48 - 3 = 45$

$\dfrac{\Delta y}{\Delta x} = \dfrac{45}{3} = 15$

11. $\dfrac{f(x_1 + \Delta x) - f(x_1)}{\Delta x} = \dfrac{f(1 + 2) - f(1)}{2} = \dfrac{3 \cdot 3^2 - 3 \cdot 1^2}{2} = \dfrac{24}{2} = 12$

13. $\Delta y = f(x_2) - f(x_1) = f(3) - f(1) = 3 \cdot 3^2 - 3 \cdot 1^2 = 27 - 3 = 24$

$\Delta x = x_2 - x_1 = 3 - 1 = 2$

$\dfrac{\Delta y}{\Delta x} = \dfrac{24}{2} = 12$

15. $y = 30 + 12x^2 - x^3$

$dy = (30 + 12x^2 - x^3)' dx = (24x - 3x^2) dx$

17. $y = x^2 \left(1 - \dfrac{x}{9}\right) = x^2 - \dfrac{x^3}{9}$

$dy = \left[x^2 \left(1 - \dfrac{x}{9}\right)\right]' dx = \left[x^2 - \dfrac{x^3}{9}\right]' dx = \left(2x - \dfrac{1}{3}x^2\right) dx$

19. $y = \dfrac{590}{\sqrt{x}} = 590 x^{-1/2}$

$dy = 590 \left(-\dfrac{1}{2}\right) x^{-3/2} dx = \dfrac{-295}{x^{3/2}} dx$

21.   (A) $\dfrac{f(2 + \Delta x) - f(2)}{\Delta x} = \dfrac{3(2 + \Delta x)^2 - 3 \cdot 2^2}{\Delta x}$

$= \dfrac{3(2^2 + 4\Delta x + \Delta x^2) - 12}{\Delta x}$

$= \dfrac{\Delta x(12 + 3\Delta x)}{\Delta x}$

$= 12 + 3\Delta x, \; \Delta x \neq 0$

(B)   As Δx tends to zero, then, clearly, $12 + 3\Delta x$ tends to 12. Note the values in the following table:

Δx	$12 + 3\Delta x$
1	15
0.1	12.3
0.01	12.03
0.001	12.003

23. $y = (2x + 1)^2 = 4x^2 + 4x + 1$

$dy = (8x + 4) \, dx$

25. $y = \dfrac{x^2 + 9}{x} = x + \dfrac{9}{x} = x + 9x^{-1}$

$dy = \left(1 - 9x^{-2}\right)dx = \left(1 - \dfrac{9}{x^2}\right)dx$

27. $y = f(x) = x^2 - 3x + 2$

$\Delta y = f(5 + 0.2) - f(5)$ (using 1)

$= f(5.2) - f(5)$

$= (5.2)^2 - 3(5.2) + 2 - (5^2 - 3 \cdot 5 + 2) = 1.44$

$dy = (x^2 - 3x)'\big|_{x=5} \Delta x = (2x - 3)\big|_{x=5} \Delta x = 7(0.2) = 1.4$

29. $y = f(x) = 75\left(1 - \dfrac{2}{x}\right)$

$\Delta y = f[5 + (-0.5)] - f(5) = f(4.5) - f(5) = 75\left(1 - \dfrac{2}{4.5}\right) - 75\left(1 - \dfrac{2}{5}\right) = 41.67 - 45 = -3.33$

$dy = \left[75\left(1 - \dfrac{2}{x}\right)\right]'_{x=5}(-0.5) = \dfrac{150}{x^2}\bigg]_{x=5}(-0.5) = 6\left(-\dfrac{1}{2}\right) = -3$

31. A cube with sides of length x has volume $V(x) = x^3$. If we increase the length of each side by an amount $\Delta x = dx$, then the approximate change in volume is:

$dV = 3x^2\,dx$

Therefore, letting $x = 10$ and $dx = 0.4$ [$= 2(0.2)$], since each face has a 0.2 inch coating], we have

$dV = 3(10)^2(0.4) = 120$ cubic inches,

which is the approximate volume of the fiberglass shell.

33. $f(x) = x^2 + 2x + 3;\ f'(x) = 2x + 2;\ x = -0.5;\ \Delta x = dx$

(A) $\Delta y = f(-0.5 + \Delta x) - f(-0.5)$

$= (-0.5 + \Delta x)^2 + 2(-0.5 + \Delta x) + 3 - [(-0.5)^2 + 2(-0.5) + 3]$

$= (-0.5 + \Delta x)^2 + 2(-0.5 + \Delta x) + 0.75 = \Delta x + (\Delta x)^2$

$dy = f'(-0.5)dx = 1 \cdot dx = dx = \Delta x$

(B)
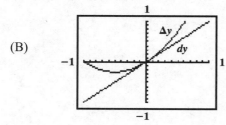

(C) $\Delta y(0.1) = (-0.5 + 0.1)^2 + 2(-0.5 + 0.1) + 0.75 = 0.11$

$dy(0.1) = 0.1$

$\Delta y(0.2) = (-0.5 + 0.2)^2 + 2(-0.5 + 0.2) + 0.75 = 0.24$

$dy(0.2) = 0.2$

$\Delta y(0.3) = (-0.5 + 0.3)^2 + 2(-0.5 + 0.3) + 0.75 = 0.39$

$dy(0.3) = 0.3$

Δx	Δy	dy
.1	.11	.1
.2	.24	.2
.3	.39	.3

35. $f(x) = x^3 - 2x^2; \; f'(x) = 3x^2 - 4x; \; x = 1; \; \Delta x = dx$

     (A) $\Delta y = f(1 + \Delta x) - f(1) = (1 + \Delta x)^3 - 2(1 + \Delta x)^2 - [1^3 - 2(1)^2]$

             $= (1 + \Delta x)^3 - 2(1 + \Delta x)^2 + 1$

             $= -\Delta x + (\Delta x)^2 + (\Delta x)^3$

          $dy = f'(1)dx = (-1)dx = -dx$

     (B)

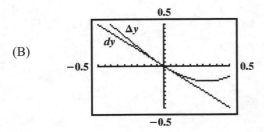

     (C) $\Delta y(0.05) = -0.05 + (0.05)^2 + (0.05)^3 = -0.047375$

          $dy(0.05) = -0.05$

          $\Delta y(0.10) = -0.10 + (0.10)^2 + (0.10)^3 = -0.089$

          $dy(0.10) = -0.10$

          $\Delta y(0.15) = -0.15 + (0.15)^2 + (0.15)^3 = -0.124125$

          $dy(0.15) = -0.15 \quad 0.3$

37.   True

     $f(x) = mx + b; \; f'(x) = m; \; x = 3; \; \Delta x = dx$

     $\Delta y = f(3 + \Delta x) - f(3) = m(3 + \Delta x) + b - (3m + b) = m \, \Delta x$

     $dy = (3)dx = m \, dx$

     Thus, $\Delta y = dy$

39.   False.

     At $x = 2$, $\; dy = f'(2)dx$. $\; Dy = 0$ for all dx implies only that $f'(2) = 0$.

     Example. Let $f(x) = x^2 - 4x$

41. $y = (1 - 2x)\sqrt[3]{x^2} = (1 - 2x)x^{2/3} = x^{2/3} - 2x^{5/3}$

     $Dy = \left(\dfrac{2}{3}x^{-1/3} - \dfrac{10}{3}x^{2/3} \right)dx$

43. $y = f(x) = 52\sqrt{x} = 52x^{1/2}; \; x = 4, \; \Delta x = dx = 0.3$

     $\Delta y = f(x + \Delta x) - f(x)$ $dy = f'(x)dx = \dfrac{26}{x^{1/2}}\, dx$

     Let $x = 4$, $\Delta x = 0.3$. Then: Let $x = 4$, $dx = 0.3$. Then:

     $\Delta y = 52(4 + 0.3)^{1/2} - 52(4)^{1/2}$ $dy = \dfrac{26}{4^{1/2}}(0.3)$

$$= 52(4.3)^{1/2} - 104 \qquad\qquad = \frac{26}{2}(0.3) = 3.9$$

$$\approx 107.83 - 104$$

$$\approx 3.83$$

45. Given $N(x) = 60x - x^2$, $5 \le x \le 30$. Then $N'(x) = 60 - 2x$. The approximate change in the sales dN corresponding to a change $\Delta x = dx$ in the amount x (in thousands of dollars) spent on advertising is:

$$dN = N'(x)dx$$

Thus, letting $x = 10$ and $dx = 1$, we get:

$$dN = N'(10) \cdot 1 = (60 - 2 \cdot 10) = 40$$

There will be a 40-unit increase in sales (approximately) when the advertising budget is increased from $10,000 to $11,000.

Similarly, letting $x = 20$ and $dx = 1$, we get:

$$dN = N'(20) \cdot 1 = (60 - 2 \cdot 20) = 20\text{-unit increase.}$$

47. $\overline{C}(x) = \dfrac{400}{x} + 5 + \dfrac{1}{2}x$, $x \ge 1$.

If we increase production per hour by an amount $\Delta x = dx$, then the approximate change in average cost is:

$$d\overline{C} = \overline{C}'(x)\,dx = \left(-\frac{400}{x^2} + \frac{1}{2}\right)dx$$

Thus, letting $x = 20$ and $dx = 5$, we have

$$d\overline{C} = \left(-\frac{400}{(20)^2} + \frac{1}{2}\right)5 = \left(-1 + \frac{1}{2}\right)5 = -2.50,$$

that is, the average cost per racket will *decrease* $2.50.

Letting $x = 40$ and $dx = 5$, we have

$$d\overline{C} = \left(-\frac{400}{(40)^2} + \frac{1}{2}\right)5 = \left(-\frac{1}{4} + \frac{1}{2}\right)5 = 1.25,$$

that is, the average cost per racket will *increase* $1.25.

49. $y = \dfrac{590}{\sqrt{x}}$, $30 \le x \le 75$. Thus, $y = 590x^{-1/2}$ and $y' = -295x^{-3/2} = \dfrac{-295}{x^{3/2}}$.

The approximate change in pulse rate for a height change from 36 to 37 inches is given by:

$$dy = \frac{-295}{x^{3/2}}\bigg]_{x=36} = -1.37 \quad\text{per minute.} \qquad [\underline{\text{Note}}: \Delta x = dx = 1.]$$

Similarly, the approximate change in pulse rate for a height change from 64 to 65 inches is given by:

$$dy = \frac{-295}{x^{3/2}}\bigg]_{x=64} = -0.58 \quad\text{per minute.}$$

51. Area: $A(r) = \pi r^2 \approx 3.14r^2$; $A'(r) = 6.28r$.

An approximate increase in cross-sectional area when the radius of an artery is increased from 2 mm to 2.1 mm is given by:

$dA = A'(r)\big|_{r=2} \times 0.1$ [<u>Note</u>: $\Delta r = 0.1$.]

$= \left(6.28r\big|_{r=2}\right) \times 0.1 = 12.56 \times 0.1 = 1.256 \text{ mm}^2 \approx 1.26 \text{ mm}^2$

53. $N(t) = 75\left(1 - \dfrac{2}{t}\right)$, $3 \le t \le 20$; $N'(t) = \dfrac{150}{t^2}$.

The approximate improvement from 5 to 5.5 weeks of practice is given by:

$dN = N'(t)\big|_{t=5} \times 0.5$ [<u>Note</u>: $\Delta t = 0.5$.]

$= \dfrac{150}{t^2}\Big|_{t=5} \times 0.5 = 6 \times 0.5 = 3$ words per minute

55. $N(t) = 30 + 12t^2 - t^3$, $0 \le t \le 8$; $N'(t) = 24t - 3t^2$.

(A)   The approximate change in votes when time changes from 1 to 1.1 years is

$= N'(t)\big|_{t=1} \times 0.1$ [<u>Note</u>: $\Delta t = 0.1$.]

$= \left(24t - 3t^2\big|_{t=1}\right) \times 0.1 = 21 \times 0.1 = 2.1$ thousand or 2100 increase.

(B)   The approximate change in votes when time changes from 4 to 4.1 years is

$= N'(t)\big|_{t=4} \times 0.1$ [<u>Note</u>: $\Delta t = 0.1$.]

$= \left(24t - 3t^2\big|_{t=4}\right) \times 0.1 = 48 \times 0.1 = 4.8$ thousand or 4800 increase.

(C)   The approximate change in votes when time changes from 7 to 7.1 years is

$= N'(t)\big|_{t=7} \times 0.1$ [<u>Note</u>: $\Delta t = 0.1$.]

$= \left(24t - 3t^2\big|_{t=7}\right) \times 0.1 = 2.1$ thousand or 2100 increase.

EXERCISE 10-7

Things to remember:

<u>1.</u>   MARGINAL COST, REVENUE, AND PROFIT

If x is the number of units of a product produced in some time interval, then:

Total Cost = $C(x)$
Marginal Cost = $C'(x)$
Total Revenue = $R(x)$
Marginal Revenue = $R'(x)$
Total Profit = $P(x) = R(x) - C(x)$
Marginal Profit = $P'(x) = R'(x) - C'(x)$
    = (Marginal Revenue) − (Marginal Cost)

Marginal cost (or revenue or profit) is the instantaneous rate of change of cost (or revenue or profit) relative to production at a given production level.

2.   MARGINAL COST AND EXACT COST

If $C(x)$ is the cost of producing x items, then the marginal cost function approximates the exact cost of producing the
$(x + 1)$st item:

Marginal Cost		Exact Cost
$C'(x)$	$\approx$	$C(x+1) - C(x)$

Similar interpretations can be made for total revenue and total profit functions.

3.   BREAK-EVEN POINTS

The BREAK-EVEN POINTS are the points where total revenue equals total cost.

4.   MARGINAL AVERAGE COST, REVENUE, AND PROFIT

If x is the number of units of a product produced in some time interval, then:

Average Cost = $\overline{C}(x) = \dfrac{C(x)}{x}$ Cost per unit

Marginal Average Cost = $\overline{C}'(x)$

Average Revenue = $\overline{R}(x) = \dfrac{R(x)}{x}$ Revenue per unit

Marginal Average Revenue = $\overline{R}'(x)$

Average Profit = $\overline{P}(x) = \dfrac{P(x)}{x}$ Profit per unit

Marginal Average Profit = $\overline{P}'(x)$

---

1. $C(99) = 10,000 + 150(99) - 0.2(99)^2 = 22,889.80$, $\$22,889.80$

3. $C(100) = 10,000 + 150(100) - 0.2(100)^2 = 23,000;$ $C(100) - C(99) = 23,000 - 22889.80 = 110.20,$
      $\$110.20$

5. $C(200) = 10,000 + 150(200) - 0.2(200)^2 = 32,000,$ $\$32,000$

7.   Average cost of producing 100 bicycles: $\dfrac{C(100)}{100} = \dfrac{23,000}{100} = 230,$ $\$230$

9. $C(x) = 175 + 0.8x;$ $C'(x) = 0.8$

11. $C(x) = 210 + 4.6x - 0.01x^2;$ $C'(x) = 4.6 - 0.02x$

13. $R(x) = 4x - 0.01x^2;$ $R'(x) = 4 - 0.02x$

15. $R(x) = x(12 - 0.04x) = 12x - 0.04x^2;$ $R'(x) = 12 - 0.08x$

17. $P(x) = R(x) - C(x) = 4x - 0.01x^2 - [175 + 0.8x] = 3.2x - 0.01x^2 - 175;$
      $P'(x) = 3.2 - 0.02x$

19. $P(x) = R(x) - C(x) = x(12 - 0.04x) - [210 + 4.6x - 0.01x^2] = 7.4x - 0.03x^2 - 210;$

$P'(x) = 7.4 - 0.06x$

21. $C(x) = 145 + 1.1x;$ $\overline{C}(x) = \dfrac{145 + 1.1x}{x} = 1.1 + \dfrac{145}{x}$

23. $\overline{C}(x) = 1.1 + \dfrac{145}{x};$ $\overline{C}'(x) = -\dfrac{145}{x^2}$

25. $P(x) = R(x) - C(x) = 5x - 0.02x^2 - [145 + 1.1x] = 3.9x - 0.02x^2 - 145$

27. $\overline{P}(x) = \dfrac{P(x)}{x} = 3.9 - 0.02x - \dfrac{145}{x}$

29.   True. $C(x) = ax + b,$ $C'(x) = a,$ constant.

31.   False. $P(x) = R(x) - C(x);$ $P'(x) = R'(x) - C'(x) =$ marginal revenue − marginal cost.

33. $C(x) = 2000 + 50x - 0.5x^2$

   (A)   The exact cost of producing the 21st food processor is:

$$C(21) - C(20) = 2000 + 50(21) - \dfrac{(21)^2}{2} - \left[2000 + 50(20) - \dfrac{(20)^2}{2} \right]$$
$$= 2829.50 - 2800$$
$$= 29.50 \text{ or } \$29.50$$

   (B) $C'(x) = 50 - x$
   $C'(20) = 50 - 20 = 30$ or \$30

35. $C(x) = 60,000 + 300x$

   (A) $\overline{C}(x) = \dfrac{60,000 + 300x}{x} = \dfrac{60,000}{x} + 300 = 60,000x^{-1} + 300$

   $\overline{C}(500) = \dfrac{60,000 + 300(500)}{500} = \dfrac{210,000}{500} = 420$ or \$420

   (B) $\overline{C}'(x) = -60,000x^{-2} = \dfrac{-60,000}{x^2}$

   $\overline{C}'(500) = \dfrac{-60,000}{(500)^2} = -0.24$ or −\$0.24

   Interpretation: At a production level of 500 frames, average cost is decreasing at the rate of 24¢ per frame.

   (C)   The average cost per frame if 501 frames are produced is approximately \$420 − \$0.24 = \$419.76.

37. $P(x) = 30x - 0.3x^2 - 250, 0 \leq x \leq 100$

   (A)   The exact profit from the sale of the 26th skateboard is:
   $P(26) - P(25) = 30(26) - 0.3(26)^2 - 250 - [30(25) - 0.3(25)^2 - 250]$
   $= 327.20 - 312.50 = \$14.70$

   (B)   Marginal profit: $P'(x) = 30 - 0.6x;$ $P'(25) = \$15$

39. $P(x) = 5x - \dfrac{x^2}{200} - 450,\ 0 \leq x \leq 1000;$ $P'(x) = 5 - \dfrac{x}{100}$

(A) $P'(450) = 5 - \dfrac{450}{100} = 0.5$ or $\$0.50$

Interpretation: At a production level of 450 DVD's, profit is increasing at the rate of 50¢ per DVD.

(B) $P'(750) = 5 - \dfrac{750}{100} = -2.5$ or $-\$2.50$

Interpretation: At a production level of 750 DVD's, profit is decreasing at the rate of $\$2.50$ per DVD.

41. $P(x) = 30x - 0.03x^2 - 750, \quad 0 \le x \le 1000$

Average profit: $\overline{P}(x) = \dfrac{P(x)}{x} = 30 - 0.03x - \dfrac{750}{x} = 30 - 0.03x - 750x^{-1}$

(A)   At $x = 50$, $\overline{P}(50) = 30 - (0.03)50 - \dfrac{750}{50} = 13.50$ or $\$13.50$.

(B) $\overline{P}'(x) = -0.03 + 750x^{-2} = -0.03 + \dfrac{750}{x^2}$

$\overline{P}'(50) = -0.03 + \dfrac{750}{(50)^2} = -0.03 + 0.3 = 0.27$ or $\$0.27$; at a

production level of 50 mowers, the average profit per mower is INCREASING at the rate of $\$0.27$ per mower.

(C)   The average profit per mower if 51 mowers are produced is approximately $\$13.50 + \$0.27 = \$13.77$.

43. $x = 4{,}000 - 40p$

(A)   Solving the given equation for p, we get $40p = 4{,}000 - x$

and $p = 100 - \dfrac{1}{40}x$ or $p = 100 - 0.025x$

Since $p \ge 0$, the domain is: $0 \le x \le 4{,}000$

(B) $R(x) = xp = 100x - 0.025x^2, \quad 0 \le x \le 4{,}000$

(C) $R'(x) = 100 - 0.05x; \quad R'(1{,}600) = 100 - 80 = 20$

At a production level of 1,600 pairs of running shoes, revenue is INCREASING at the rate of $\$20$ per pair.

(D) $R'(2{,}500) = 100 - 125 = -25$

At a production level of 2,500 pairs of running shoes, revenue is DECREASING at the rate of $\$25$ per pair.

45. Price-demand equation: $x = 6{,}000 - 30p$

Cost function: $C(x) = 72{,}000 + 60x$

(A)   Solving the price-demand equation for p, we get

$p = 200 - \dfrac{1}{30}x; \quad$ domain: $0 \le x \le 6{,}000$

(B)   Marginal cost: $C'(x) = 60$

(C)   Revenue function: $R(x) = 200x - \dfrac{1}{30}x^2$; domain: $0 \le x \le 6{,}000$

(D)   Marginal revenue: $R'(x) = 200 - \dfrac{1}{15}x$

(E) $R'(1{,}500) = 100$; at a production level of 1,500 saws, revenue is INCREASING at the rate of $100 per saw.

   $R'(4{,}500) = -100$; at a production level of 4,500 saws, revenue is DECREASING at the rate of $100 per saw.

(F)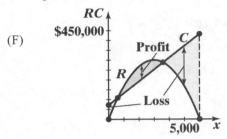

(G)   Profit function: $P(x) = R(x) - C(x) = 200x - \dfrac{1}{30}x^2 - [72{,}000 + 60x] = 140x - \dfrac{1}{30}x^2 - 72{,}000$

(H)   Marginal profit: $P'(x) = 140 - \dfrac{1}{15}x$

(I) $P'(1{,}500) = 140 - 100 = 40$; at a production level of 1,500 saws, profit is INCREASING at the rate of $40 per saw.

   $P'(3{,}000) = 140 - 200 = -60$; at a production level of 3,000 saws, profit is DECREASING at the rate of $60 per saw.

47. (A)   Assume $p = mx + b$. We are given
   $16 = m \cdot 200 + b$
   and $14 = m \cdot 300 + b$

Subtracting the second equation from the first, we get
   $-100m = 2$ so $m = -\dfrac{1}{50} = -0.02$

Substituting this value into either equation yields $b = 20$. Therefore,
   $P = 20 - 0.02x$; domain: $0 \le x \le 1{,}000$

(B)   Revenue function: $R(x) = xp = 20x - 0.02x^2$, domain: $0 \le x \le 1{,}000$.

(C) $C(x) = mx + b$. From the finance department's estimates, $m = 4$ and $b = 1{,}400$. Thus,
   $C(x) = 4x + 1{,}400$.

(D)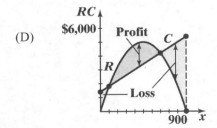

(E)   Profit function: $P(x) = R(x) - C(x) = 20x - 0.02x^2 - [4x + 1,400]$

$$= 16x - 0.02x^2 - 1,400$$

(F)   Marginal profit: $P'(x) = 16 - 0.04x$

$P'(250) = 16 - 10 = 6$; at a production level of 250 toasters, profit is INCREASING at the rate of $6 per toaster.

$P'(475) = 16 - 19 = -3$; at a production level of 475 toasters, profit is DECREASING at the rate of $3 per toaster.

49.   Total cost: $C(x) = 24x + 21,900$

Total revenue: $R(x) = 200x - 0.2x^2, 0 \leq x \leq 1,000$

(A) $R'(x) = 200 - 0.4x$

The graph of R has a horizontal tangent line at the value(s) of x where $R'(x) = 0$, i.e.,

$$200 - 0.4x = 0 \quad \text{or} \quad x = 500$$

(B) $P(x) = R(x) - C(x) = 200x - 0.2x^2 - (24x + 21,900) = 176x - 0.2x^2 - 21,900$

(C) $P'(x) = 176 - 0.4x$. Setting $P'(x) = 0$, we have

$$176 - 0.4x = 0 \quad \text{or} \quad x = 440$$

(D)   The graphs of C, R and P are shown below.

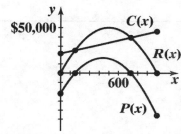

Break-even points: $R(x) = C(x)$

$$200x - 0.2x^2 = 24x + 21,900$$

$$0.2x^2 - 176x + 21,900 = 0$$

$$x = \frac{176 \pm \sqrt{(-176)^2 - (4)(0.2)(21,900)}}{2(0.2)} \quad \text{(quadratic formula)}$$

$$= \frac{176 \pm \sqrt{30,976 - 17,520}}{0.4} = \frac{176 \pm \sqrt{13,456}}{0.4} = \frac{176 \pm 116}{0.4} = 730, \ 150$$

Thus, the break-even points are: (730, 39,420) and (150, 25,500).

x-intercepts for P: $-0.2x^2 + 176x - 21,900 = 0$ or $0.2x^2 - 176x + 21,900 = 0$
which is the same as the equation above. Thus, $x = 150$ and $x = 730$.

51.   Demand equation: $p = 20 - \sqrt{x} = 20 - x^{1/2}$

Cost equation: $C(x) = 500 + 2x$

(A)   Revenue $R(x) = xp = x(20 - x^{1/2})$

or $R(x) = 20x - x^{3/2}$

(B)   The graphs for R and C for $0 \leq x \leq 400$
are shown at the right.

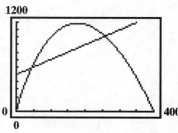

Break-even points (44, 588) and (258, 1,016).

53. **(A)**

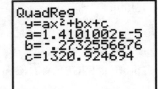

(B) Fixed costs ≈ \$721,680; variable costs ≈ \$121 per projector

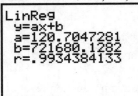

(C) Let $y = p(x)$ be the quadratic regression equation found in part (A) and let $y = C(x)$ be the linear regression equation found in part (B). Then revenue $R(x) = xp(x)$, and the break-even points are the points where $R(x) = C(x)$.

break-even points: (713, 807,703), (5,423, 1,376,227)

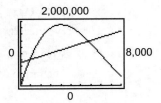

(D) The company will make a profit when $713 \leq x \leq 5,423$. From part (A), $p(713) \approx 1,133$ and $p(5,423) \approx 254$. Thus, the company will make a profit for the price range $\$254 \leq p \leq \$1,133$.

CHAPTER 10 REVIEW

1. $f(x) = 2x^2 + 5$

(A) $f(3) - f(1) = 2(3)^2 + 5 - [2(1)^2 + 5] = 16$

(B) Average rate of change: $\dfrac{f(3) - f(1)}{3 - 1} = \dfrac{16}{2} = 8$

(C) Slope of secant line: $\dfrac{f(3) - f(1)}{3 - 1} = \dfrac{16}{2} = 8$

(D) Instantaneous rate of change at $x = 1$:

Step 1. $\dfrac{f(1+h) - f(1)}{h} = \dfrac{2(1+h)^2 + 5 - [2(1)^2 + 5]}{h} = \dfrac{2(1 + 2h + h^2) + 5 - 7}{h} = \dfrac{4h + 2h^2}{h} = 4 + 2h$

Step 2. $\displaystyle\lim_{h \to 0} \dfrac{f(1+h) - f(1)}{h} = \lim_{h \to 0} (4 + 2h) = 4$

(E) Slope of the tangent line at $x = 1$: 4

(F) $f'(1) = 4$ (2-2)

2. $f(x) = -3x + 2$

Step 1.   Find $f(x + h)$

$$f(x + h) = -3(x + h) + 2 = -3x - 3h + 2$$

Step 2.   Find $f(x + h) - f(x)$

$$f(x + h) - f(x) = -3x - 3h + 2 - (-3x + 2) = -3x - 3h + 2 + 3x - 2 = -3h$$

Step 3.   Find $\dfrac{f(x + h) - f(x)}{h}$

$$\frac{f(x + h) - f(x)}{h} = \frac{-3h}{h} = -3$$

Step 4.   Find $\lim\limits_{h \to 0} \dfrac{f(x + h) - f(x)}{h}$.

$$\lim_{h \to 0} \frac{f(x + h) - f(x)}{h} = \lim_{h \to 0} (-3) = -3 \qquad (2\text{-}2)$$

3. (A) $\lim\limits_{x \to 1} (5f(x) + 3g(x)) = 5 \lim\limits_{x \to 1} f(x) + 3 \lim\limits_{x \to 1} g(x) = 5 \cdot 2 + 3 \cdot 4 = 22$

(B) $\lim\limits_{x \to 1} [f(x)g(x)] = [\lim\limits_{x \to 1} f(x)][\lim\limits_{x \to 1} g(x)] = 2 \cdot 4 = 8$

(C) $\lim\limits_{x \to 1} \dfrac{g(x)}{f(x)} = \dfrac{\lim\limits_{x \to 1} g(x)}{\lim\limits_{x \to 1} f(x)} = \dfrac{4}{2} = 2$

(D) $\lim\limits_{x \to 1} [5 + 2x - 3g(x)] = \lim\limits_{x \to 1} 5 + \lim\limits_{x \to 1} 2x - 3 \lim\limits_{x \to 1} g(x) = 5 + 2 - 3(4) = -5 \qquad (2\text{-}1)$

4. $f(1.5) \approx 1.5 \qquad (2\text{-}1)$ **5.** $f(2.5) \approx 3.5 \qquad (2\text{-}1)$

6. $f(2.75) \approx 3.75 \qquad (2\text{-}1)$ **7.** $f(3.25) \approx 3.75 \qquad (2\text{-}1)$

8. (A) $\lim\limits_{x \to 1^-} f(x) = 1$ (B) $\lim\limits_{x \to 1^+} f(x) = 1$ (C) $\lim\limits_{x \to 1} f(x) = 1$ (D) $f(1) = 1 \qquad (2\text{-}1)$

9. (A) $\lim\limits_{x \to 2^-} f(x) = 2$ (B) $\lim\limits_{x \to 2^+} f(x) = 3$ (C) $\lim\limits_{x \to 2} f(x)$ does not exist (D) $f(2) = 3 \qquad (2\text{-}1)$

10. (A) $\lim\limits_{x \to 3^-} f(x) = 4$ (B) $\lim\limits_{x \to 3^+} f(x) = 4$ (C) $\lim\limits_{x \to 3} f(x) = 4$ (D) $f(3)$ does not exist $\qquad (2\text{-}1)$

11. (A) From the graph, $\lim\limits_{x \to 1} f(x)$ does not exist since

$$\lim_{x \to 1^-} f(x) = 2 \neq \lim_{x \to 1^+} f(x) = 3.$$

(B) $f(1) = 3$

(C) f is NOT continuous at $x = 1$, since $\lim\limits_{x \to 1} f(x)$ does not exist. $\qquad (2\text{-}3)$

12. (A) $\lim\limits_{x \to 2^-} f(x) = 2$ (B) $f(2)$ is not defined

(C) f is NOT continuous at $x = 2$ since $f(2)$ is not defined. $\qquad (2\text{-}3)$

13. (A) $\lim_{x \to 3} f(x) = 1$ (B) $f(3) = 1$

 (C) f is continuous at $x = 3$ since $\lim_{x \to 3} f(x) = f(3)$.   (2-3)

14. $\lim_{x \to \infty} f(x) = 5$ (2-2)

15. $\lim_{x \to -\infty} f(x) = 5$ (2-2)

16. $\lim_{x \to 2^+} f(x) = \infty$ (2-2)

17. $\lim_{x \to 2^-} f(x) = -\infty$ (2-2)

18. $\lim_{x \to 0^-} f(x) = 0$ (2-1)

19. $\lim_{x \to 0^+} f(x) = 0$ (2-1)

20. $\lim_{x \to 0} f(x) = 0$ (2-1)

21. $x = 2$ is a vertical asymptote   (2-3)

22. $y = 5$ is a horizontal asymptote   (2-2)

23. f is discontinuous at $x = 2$ (2-3)

24. $f(x) = 5x^2$

 <u>Step 1.</u>   Find $f(x + h)$:

 $$f(x + h) = 5(x + h)^2 = 5(x^2 + 2xh + h^2) = 5x^2 + 10xh + 5h^2$$

 <u>Step 2.</u>   Find $f(x + h) - f(x)$:

 $$f(x + h) - f(x) = 5x^2 + 10xh + 5h^2 - 5x^2 = 10xh + 5h^2$$

 <u>Step 3.</u>   Find $\dfrac{f(x+h) - f(x)}{h}$:

 $$\frac{f(x+h) - f(x)}{h} = \frac{10xh + 5h^2}{h} = 10x + 5h$$

 <u>Step 4.</u>   Find $\lim_{h \to 0} \dfrac{f(x+h) - f(x)}{h}$

 $$\lim_{h \to 0} \frac{f(x+h) - f(x)}{h} = \lim_{h \to 0} (10x + 5h) = 10x$$

 Thus, $f'(x) = 10x$.   (2-4)

25. (A) $h'(x) = (3f(x))' = 3 f'(x);\ h'(5) = 3 f'(5) = 3(-1) = -3$

 (B) $h'(x) = (-2g(x))' = -2g'(x);\ h'(5) = -2g'(5) = -2(-3) = 6$

 (C) $h'(x) = 2 f'(x);\ h'(5) = 2(-1) = -2$

 (D) $h'(x) = -g'(x);\ h'(5) = -(-3) = 3$

 (E) $h'(x) = 2 f'(x) + 3g'(x);\ h'(5) = 2(-1) + 3(-3) = -11$ (2-5)

26. $f(x) = \dfrac{1}{3}x^3 - 5x^2 + 1;\ f'(x) = x^2 - 10x$ (2-5)

27. $f(x) = 2x^{1/2} - 3x;\ f'(x) = 2 \cdot \dfrac{1}{2} x^{-1/2} - 3 = \dfrac{1}{x^{1/2}} - 3$ (2-5)

28. $f(x) = 5$
$f'(x) = 0$
    (2-5)

29. $f(x) = \dfrac{3}{2x} + \dfrac{5x^3}{4} = \dfrac{3}{2}x^{-1} + \dfrac{5}{4}x^3;$

$f'(x) = -\dfrac{3}{2}x^{-2} + \dfrac{15}{4}x^2 = -\dfrac{3}{2x^2} + \dfrac{15}{4}x^2$ (2-5)

30. $f(x) = \dfrac{0.5}{x^4} + 0.25x^4 = 0.5x^{-4} + 0.25x^4$

$f'(x) = 0.5(-4)x^{-5} + 0.25(4x^3) = -2x^{-5} + x^3 = -\dfrac{2}{x^5} + x^3$ (2-5)

31. $f(x) = (3x^3 - 2)(x + 1) = 3x^4 + 3x^3 - 2x - 2$
$f'(x) = 12x^3 + 9x^2 - 2$ (2-5)

For Problems 32 – 35, $f(x) = x^2 + x$.

32. $\Delta x = x_2 - x_1 = 3 - 1 = 2, \Delta y = f(x_2) - f(x_1) = 12 - 2 = 10,$

$\dfrac{\Delta y}{\Delta x} = \dfrac{10}{2} = 5.$ (2-6)

33. $\dfrac{f(x_1 + \Delta x) - f(x_1)}{\Delta x} = \dfrac{f(1 + 2) - f(1)}{2} = \dfrac{f(3) - f(1)}{2} = \dfrac{12 - 2}{2} = 5$ (2-6)

34. $dy = f'(x)dx = (2x + 1)dx.$ For $x_1 = 1, x_2 = 3,$
$dx = \Delta x = 3 - 1 = 2, dy = (2 \cdot 1 + 1) \cdot 2 = 3 \cdot 2 = 6$ (2-6)

35. $\Delta y = f(x + \Delta x) - f(x);$ at $x = 1, \Delta x = 0.2,$
$\Delta y = f(1.2) - f(1) = 0.64$
$dy = f'(x)dx$ where $f'(x) = 2x + 1;$ at $x = 1$
$dy = 3(0.2) = 0.6$ (2-6)

36. From the graph:
(A) $\displaystyle\lim_{x \to 2^-} f(x) = 4$

(B) $\displaystyle\lim_{x \to 2^+} f(x) = 6$

(C) $\displaystyle\lim_{x \to 2} f(x)$ does not exist since $\displaystyle\lim_{x \to 2^-} f(x) \neq \displaystyle\lim_{x \to 2^+} f(x)$

(D) $f(2) = 6$

(E) No, since $\displaystyle\lim_{x \to 2} f(x)$ does not exist.   (2-3)

37. From the graph:
(A) $\displaystyle\lim_{x \to 5^-} f(x) = 3$ (B) $\displaystyle\lim_{x \to 5^+} f(x) = 3$ (C) $\displaystyle\lim_{x \to 5} f(x) = 3$ (D) $f(5) = 3$

(E)   Yes, since $\displaystyle\lim_{x \to 5} f(x) = f(5) = 3.$ (2-3)

38. (A) $f(x) < 0$ on $(8, \infty)$ (B) $f(x) \geq 0$ on $[0, 8]$ (2-3)

39. $x^2 - x < 12$ or $x^2 - x - 12 < 0$

Let $f(x) = x^2 - x - 12 = (x + 3)(x - 4)$. Then f is continuous for all x and $f(-3) = f(4) = 0$. Thus, $x = -3$ and $x = 4$ are partition numbers.

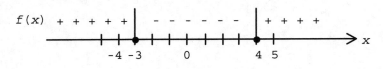

Test Numbers	
x	$f(x)$
-4	8 (+)
0	-12 (-)
5	8 (+)

Thus, $x^2 - x < 12$ for: $-3 < x < 4$ or $(-3, 4)$.   (2-3)

40. $\dfrac{x-5}{x^2 + 3x} > 0$ or $\dfrac{x-5}{x(x+3)} > 0$

Let $f(x) = \dfrac{x-5}{x(x+3)}$. Then f is discontinuous at $x = 0$ and $x = -3$, and $f(5) = 0$. Thus, $x = -3, x = 0,$ and $x = 5$ are partition numbers.

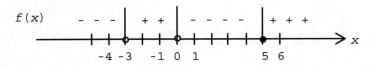

Test Numbers	
x	$f(x)$
-4	$-\frac{9}{4}(-)$
-1	$3(+)$
1	$-1(-)$
6	$\frac{1}{54}(+)$

Thus, $\dfrac{x-5}{x^2+3x} > 0$ for $-3 < x < 0$ or $x > 5$, or $(-3, 0) \cup (5, \infty)$.   (2-3)

41. $x^3 + x^2 - 4x - 2 > 0$

Let $f(x) = x^3 + x^2 - 4x - 2$. Then f is continuous for all x and $f(x) = 0$ at $x = -2.3429, -0.4707$ and 1.8136.

```
f(x)      - - - 0 + + + 0 - - - - - 0 + + +
        |—◆—|—|—◆—|—|—|—|—◆—|—|——▶ x
        -2.34  -0.47 0    1.81
```

Thus, $x^3 + x^2 - 4x - 2 > 0$ for $-2.3429 < x < -0.4707$ or $1.8136 < x < \infty$, or $(-2.3429, -0.4707) \cup (1.8136, \infty)$.   (2-3)

42. $f(x) = 0.5x^2 - 5$

(A) $\dfrac{f(4) - f(2)}{4 - 2} = \dfrac{0.5(4)^2 - 5 - [0.5(2)^2 - 5]}{2} = \dfrac{8 - 2}{2} = 3$

(B) $\dfrac{f(2+h)-f(2)}{h} = \dfrac{0.5(2+h)^2-5-[0.5(2)^2-5]}{h} = \dfrac{0.5(4+4h+h^2)-5+3}{h}$

$= \dfrac{2h+0.5h^2}{h} = \dfrac{h(2+0.5h)}{h} = 2+0.5h$

(C) $\displaystyle\lim_{h\to 0}\dfrac{f(2+h)-f(2)}{h} = \lim_{h\to 0}(2+0.5h) = 2$ (2-4)

43. $y = \dfrac{1}{3}x^{-3} - 5x^{-2} + 1;$

$\dfrac{dy}{dx} = \dfrac{1}{3}(-3)x^{-4} - 5(-2)x^{-3} = -x^{-4} + 10x^{-3}$ (2-5)

44. $y = \dfrac{3\sqrt{x}}{2} + \dfrac{5}{3\sqrt{x}} = \dfrac{3}{2}x^{1/2} + \dfrac{5}{3}x^{-1/2};$

$y' = \dfrac{3}{2}\left(\dfrac{1}{2}x^{-1/2}\right) + \dfrac{5}{3}\left(-\dfrac{1}{2}x^{-3/2}\right) = \dfrac{3}{4x^{1/2}} - \dfrac{5}{6x^{3/2}} = \dfrac{3}{4\sqrt{x}} - \dfrac{5}{6\sqrt{x^3}}$ (2-5)

45. $g(x) = 1.8\sqrt[3]{x} + \dfrac{0.9}{\sqrt[3]{x}} = 1.8x^{1/3} + 0.9x^{-1/3}$

$g'(x) = 1.8\left(\dfrac{1}{3}x^{-2/3}\right) + 0.9\left(-\dfrac{1}{3}x^{-4/3}\right) = 0.6x^{-2/3} - 0.3x^{-4/3} = \dfrac{0.6}{x^{2/3}} - \dfrac{0.3}{x^{4/3}}$ (2-5)

46. $y = \dfrac{2x^3-3}{5x^3} = \dfrac{2}{5} - \dfrac{3}{5}x^{-3}; \; y' = -\dfrac{3}{5}(-3x^{-4}) = \dfrac{9}{5x^4}$ (2-5)

47. $f(x) = x^2 + 4$
$f'(x) = 2x$

(A)   The slope of the graph at $x = 1$ is $m = f'(1) = 2$.

(B) $f(1) = 1^2 + 4 = 5$
The tangent line at (1, 5), where the slope $m = 2$, is:
$(y - 5) = 2(x - 1)$ [Note: $(y - y_1) = m(x - x_1)$.]

$y = 5 + 2x - 2$
$y = 2x + 3$ (2-4, 2-5)

48. $f(x) = 10x - x^2$
$f'(x) = 10 - 2x$
The tangent line is horizontal at the values of x such that
$f'(x) = 0$:
$10 - 2x = 0$
$x = 5$ (2-4)

49. $f(x) = x^3 + 3x^2 - 45x - 135$

$f'(x) = 3x^2 + 6x - 45$
Set $f'(x) = 0$:
$3x^2 + 6x - 45 = 0$
$x^2 + 2x - 15 = 0$
$(x - 3)(x + 5) = 0$
$\qquad\qquad x = 3, \quad x = -5 \qquad$ (2-5)

50. $f(x) = x^4 - 2x^3 - 5x^2 + 7x$

$f'(x) = 4x^3 - 6x^2 - 10x + 7$

Set $f'(x) = 4x^3 - 6x^2 - 10x + 7 = 0$ and solve for x using a root-approximation routine on a graphing utility:

$f'(x) = 0$ at $x = -1.34, \quad x = 0.58, \quad x = 2.26 \qquad$ (2-5)

51. $f(x) = x^5 - 10x^3 - 5x + 10$

$f'(x) = 5x^4 - 30x^2 - 5 = 5(x^4 - 6x^2 - 1)$

Let $f'(x) = 5(x^4 - 6x^2 - 1) = 0$ and solve for x using a root-approximation routine on a graphing utility; that is, solve $x^4 - 6x^2 - 1 = 0$ for x.

$f'(x) = 0$ at $x = \pm 2.4824 \qquad$ (2-5)

52. $y = f(x) = 8x^2 - 4x + 1$

(A) Instantaneous velocity function; $v(x) = f'(x) = 16x - 4$.
(B) $v(3) = 16(3) - 4 = 44$ ft/sec. $\qquad$ (2-5)

53. $y = f(x) = -5x^2 + 16x + 3$
(A) Instantaneous velocity function: $v(x) = f'(x) = -10x + 16$.
(B) $v(x) = 0$ when $-10x + 16 = 0$
$\qquad\qquad\qquad\quad 10x = 16$
$\qquad\qquad\qquad\qquad x = 1.6$ sec $\qquad$ (2-5)

54. (A) $f(x) = x^3$, $g(x) = (x-4)^3$, $h(x) = (x+3)^3$

The graph of g is the graph of f shifted 4 units to the right;
the graph of h is the graph of f shifted 3 units to the left.

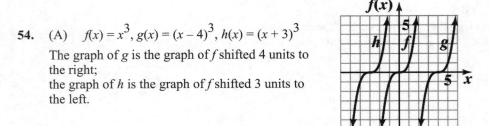

(B) $f'(x) = 3x^2$, $g'(x) = 3(x-4)^2$,
$\quad h'(x) = 3(x+3)^2$

The graph of g' is the graph of f' shifted 4 units to the right; the graph of h' is the graph of f' shifted 3 units to the left.

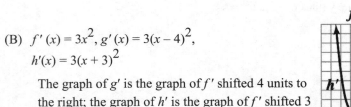
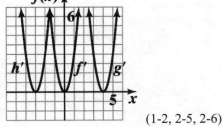

(1-2, 2-5, 2-6)

55. $f(x) = x^2 - 4$ is a polynomial function; f is continuous on $(-\infty, \infty)$. $\qquad$ (2-3)

56. $f(x) = \dfrac{x+1}{x-2}$ is a rational function and the denominator $x-2$ is 0 at $x=2$. Thus f is continuous for all x such that $x \neq 2$, i.e., on $(-\infty, 2) \cup (2, \infty)$.   (2-3)

57. $f(x) = \dfrac{x+4}{x^2+3x-4}$ is a rational function and the denominator

$x^2 + 3x - 4 = (x+4)(x-1)$ is 0 at $x = -4$ and $x = 1$. Thus, f is continuous for all x except $x = -4$ and $x = 1$, i.e., on $(-\infty, -4) \cup (-4, 1) \cup (1, \infty)$.   (2-2)

58. $f(x) = \sqrt[3]{4-x^2}$; $g(x) = 4 - x^2$ is continuous for all x since it is a polynomial function. Therefore, $f(x) = \sqrt[3]{g(x)}$ is continuous for all x, i.e., on $(-\infty, \infty)$.   (2-3)

59. $f(x) = \sqrt{4-x^2}$; $g(x) = 4 - x^2$ is continuous for all x and $g(x)$ is nonnegative for $-2 \le x \le 2$. Therefore, $f(x) = \sqrt{g(x)}$ is continuous for $-2 \le x \le 2$, i.e., on $[-2, 2]$.   (2-3)

60. $f(x) = \dfrac{2x}{x^2 - 3x} = \dfrac{2x}{x(x-3)} = \dfrac{2}{x-3}, x \neq 0$

(A) $\displaystyle\lim_{x\to 1} f(x) = \lim_{x\to 1} \dfrac{2}{x-3} = \dfrac{\displaystyle\lim_{x\to 1} 2}{\displaystyle\lim_{x\to 1}(x-3)} = \dfrac{2}{-2} = -1$

(B) $\displaystyle\lim_{x\to 3} f(x) = \lim_{x\to 3} \dfrac{2}{x-3}$ does not exist since $\displaystyle\lim_{x\to 3} 2 = 2$ and
$\displaystyle\lim_{x\to 3} (x-3) = 0$

(C) $\displaystyle\lim_{x\to 0} f(x) = \lim_{x\to 0} \dfrac{2}{x-3} = -\dfrac{2}{3}$ (2-1)

61. $f(x) = \dfrac{x+1}{(3-x)^2}$

(A) $\displaystyle\lim_{x\to 1} \dfrac{x+1}{(3-x)^2} = \dfrac{\displaystyle\lim_{x\to 1}(x+1)}{\displaystyle\lim_{x\to 1}(3-x)^2} = \dfrac{2}{2^2} = \dfrac{1}{2}$

(B) $\displaystyle\lim_{x\to -1} \dfrac{x+1}{(3-x)^2} = \dfrac{\displaystyle\lim_{x\to -1}(x+1)}{\displaystyle\lim_{x\to -1}(3-x)^2} = \dfrac{0}{4^2} = 0$

(C) $\displaystyle\lim_{x\to 3} \dfrac{x+1}{(3-x)^2}$ does not exist since $\displaystyle\lim_{x\to 3}(x+1) = 4$ and $\displaystyle\lim_{x\to 3}(3-x)^2 = 0$ (2-1)

62. $f(x) = \dfrac{|x-4|}{x-4} = \begin{cases} -1 & \text{if } x < 4 \\ 1 & \text{if } x > 4 \end{cases}$

(A) $\displaystyle\lim_{x\to 4^-} f(x) = -1$ (B) $\displaystyle\lim_{x\to 4^+} f(x) = 1$ (C) $\displaystyle\lim_{x\to 4} f(x)$ does not exist.   (2-1)

63. $f(x) = \dfrac{x-3}{9-x^2} = \dfrac{x-3}{(3+x)(3-x)} = \dfrac{-(3-x)}{(3+x)(3-x)} = \dfrac{-1}{3+x}, x \neq 3$

(A) $\lim\limits_{x\to3} f(x) = \lim\limits_{x\to3} \dfrac{-1}{3+x} = -\dfrac{1}{6}$

(B) $\lim\limits_{x\to-3} f(x) = \lim\limits_{x\to-3} \dfrac{-1}{3+x}$ does not exist

(C) $\lim\limits_{x\to0} f(x) = \lim\limits_{x\to0} \dfrac{-1}{3+x} = -\dfrac{1}{3}$ (2-1)

64. $f(x) = \dfrac{x^2-x-2}{x^2-7x+10} = \dfrac{(x-2)(x+1)}{(x-2)(x-5)} = \dfrac{x+1}{x-5}, x \neq 2$

(A) $\lim\limits_{x\to-1} f(x) = \lim\limits_{x\to-1} \dfrac{x+1}{x-5} = 0$

(B) $\lim\limits_{x\to2} f(x) = \lim\limits_{x\to2} \dfrac{x+1}{x-5} = \dfrac{3}{-3} = -1$

(C) $\lim\limits_{x\to5} f(x) = \lim\limits_{x\to5} \dfrac{x+1}{x-5}$ does not exist   (2-1)

65. $f(x) = \dfrac{2x}{3x-6} = \dfrac{2x}{3(x-2)}$

(A) $\lim\limits_{x\to\infty} \dfrac{2x}{3x-6} = \lim\limits_{x\to\infty} \dfrac{2x}{3x} = \dfrac{2}{3}$

(B) $\lim\limits_{x\to-\infty} \dfrac{2x}{3x-6} = \lim\limits_{x\to-\infty} \dfrac{2x}{3x} = \dfrac{2}{3}$

(C) $\lim\limits_{x\to2^-} \dfrac{2x}{3x-6} = \lim\limits_{x\to2^-} \dfrac{2x}{3(x-2)} = -\infty$

$\lim\limits_{x\to2^+} \dfrac{2x}{3(x-2)} = \infty$; $\lim\limits_{x\to2} \dfrac{2x}{3x-6}$ does not exist.   (2-2)

66. $f(x) = \dfrac{2x^3}{3(x-2)^2} = \dfrac{2x^3}{3x^2-12x+12}$

(A) $\lim\limits_{x\to\infty} \dfrac{2x^3}{3x^2-12x+12} = \lim\limits_{x\to\infty} \dfrac{2x^3}{3x^2} = \lim\limits_{x\to\infty} \dfrac{2x}{3} = \infty$

(B) $\lim\limits_{x\to-\infty} \dfrac{2x^3}{3x^2-12x+12} = \lim\limits_{x\to-\infty} \dfrac{2x^3}{3x^2} = \lim\limits_{x\to-\infty} \dfrac{2x}{3} = -\infty$

(C) $\lim\limits_{x\to2^-} \dfrac{2x^3}{3(x-2)^2} = \lim\limits_{x\to2^+} \dfrac{2x^3}{3(x-2)^2} = \infty$; $\lim\limits_{x\to2} \dfrac{2x^3}{3(x-2)^2} = \infty$ (2-2)

67. $f(x) = \dfrac{2x}{3(x-2)^3}$

(A) $\displaystyle\lim_{x\to\infty} \dfrac{2x}{3(x-2)^3} = \lim_{x\to\infty} \dfrac{2x}{3x^3} = \lim_{x\to\infty} \dfrac{2}{3x^2} = 0$

(B) $\displaystyle\lim_{x\to-\infty} \dfrac{2x}{3(x-2)^3} = \lim_{x\to-\infty} \dfrac{2x}{3x^3} = \lim_{x\to-\infty} \dfrac{2}{3x^2} = 0$

(C) $\displaystyle\lim_{x\to 2^-} \dfrac{2x}{3(x-2)^3} = -\infty, \quad \lim_{x\to 2^+} \dfrac{2x}{3(x-2)^3} = \infty; \quad \lim_{x\to 2} \dfrac{2x}{3(x-2)^3}$ does not exist.   (2-2)

68. $f(x) = x^2 + 4$

$\displaystyle\lim_{h\to 0} \dfrac{f(2+h) - f(2)}{h} = \lim_{h\to 0} \dfrac{[(2+h)^2 + 4] - [2^2 + 4]}{h} = \lim_{h\to 0} \dfrac{4 + 4h + h^2 + 4 - 8}{h} = \lim_{h\to 0} \dfrac{4h + h^2}{h}$

$\qquad\qquad = \displaystyle\lim_{h\to 0}(4+h) = 4$ (2-1)

69. Let $f(x) = \dfrac{1}{x+2}$

$\displaystyle\lim_{h\to 0} \dfrac{f(x+h) - f(x)}{h} = \lim_{h\to 0} \dfrac{\dfrac{1}{(x+h)+2} - \dfrac{1}{x+2}}{h} = \lim_{h\to 0} \dfrac{x+2-(x+h+2)}{h(x+h+2)(x+2)} = \lim_{h\to 0} \dfrac{-h}{h(x+h+2)(x+2)}$

$\qquad\qquad = \displaystyle\lim_{h\to 0} \dfrac{-1}{(x+h+2)(x+2)} = \dfrac{-1}{(x+2)^2}$ (2-1)

70. $f(x) = x^2 - x$

<u>Step 1.</u>   Find $f(x+h)$.

$\qquad f(x+h) = (x+h)^2 - (x+h) = x^2 + 2xh + h^2 - x - h$

<u>Step 2.</u>   Find $f(x+h) - f(x)$

$\qquad f(x+h) - f(x) = x^2 + 2xh + h^2 - x - h - (x^2 - x) = x^2 + 2xh + h^2 - x - h - (x^2 - x)$

$\qquad\qquad = x^2 + 2xh + h^2 - x - h - x^2 + x = 2xh + h^2 - h$

<u>Step 3.</u>   Find $\dfrac{f(x+h) - f(x)}{h}$.

$\qquad \dfrac{f(x+h) - f(x)}{h} = \dfrac{2xh + h^2 - h}{h} = 2x + h - 1$

<u>Step 4.</u>   Find $\displaystyle\lim_{h\to 0} \dfrac{f(x+h) - f(x)}{h}$.

$\qquad \displaystyle\lim_{h\to 0} \dfrac{f(x+h) - f(x)}{h} = \lim_{h\to 0}(2x + h - 1) = 2x - 1$

$\qquad$ Thus, $f'(x) = 2x - 1$.   (2-4)

71. $f(x) = \sqrt{x} - 3$

<u>Step 1.</u>   Find $f(x+h)$.

$\qquad f(x+h) = \sqrt{x+h} - 3$

Step 2.   Find $f(x+h) - f(x)$

$$f(x+h) - f(x) = \sqrt{x+h} - 3 - (\sqrt{x} - 3) = \sqrt{x+h} - 3 - \sqrt{x} + 3 = \sqrt{x+h} - \sqrt{x}$$

Step 3.   Find $\dfrac{f(x+h) - f(x)}{h}$.

$$\frac{f(x+h) - f(x)}{h} = \frac{\sqrt{x+h} - \sqrt{x}}{h} = \frac{\sqrt{x+h} - \sqrt{x}}{h} \cdot \frac{\sqrt{x+h} + \sqrt{x}}{\sqrt{x+h} + \sqrt{x}} = \frac{1}{\sqrt{x+h} + \sqrt{x}}$$

Step 4.   Find $\displaystyle\lim_{h\to 0} \dfrac{f(x+h) - f(x)}{h}$.

$$\lim_{h\to 0} \frac{f(x+h) - f(x)}{h} = \lim_{h\to 0} \frac{1}{\sqrt{x+h} + \sqrt{x}} = \frac{1}{2\sqrt{x}} \qquad (2\text{-}4)$$

72. Yes, $f'(-1) = 0$.   (2-4)

73. No. f is not differentiable at $x = 0$ since it is not continuous at $x = 0$.   (2-4)

74. No. f has a vertical tangent at $x = 1$.   (2-4)

75. No. f is not differentiable at $x = 2$; the curve has a "corner" at this point.   (2-4)

76. Yes. f is differentiable at $x = 3$. In fact, $f'(3) = 0$.   (2-4)

77. Yes. f is differentiable at $x = 4$.   (2-4)

78. $f(x) = \dfrac{5x}{x-7}$; f is discontinuous at $x = 7$

$\displaystyle\lim_{x\to 7^-} \frac{5x}{x-7} = -\infty, \ \lim_{x\to 7^+} \frac{5x}{x-7} = \infty; \ x = 7$ is a vertical asymptote

$\displaystyle\lim_{x\to\infty} f(x) = \lim_{x\to\infty} \frac{5x}{x-7} = \lim_{x\to\infty} \frac{5x}{x} = 5; \ y = 5$ is a horizontal asymptote.   (2-2)

$(2\text{-}3)$

79. $f(x) = \dfrac{-2x+5}{(x-4)^2}$; f is discontinuous at $x = 4$.

$\displaystyle\lim_{x\to 4^-} \frac{-2x+5}{(x-4)^2} = -\infty, \ \lim_{x\to 4^+} \frac{-2x+5}{(x-4)^2} = -\infty; \ x = 4$ is a vertical asymptote.

$\displaystyle\lim_{x\to\infty} \frac{-2x+5}{(x-4)^2} = \lim_{x\to\infty} \frac{-2x}{x^2} = \lim_{x\to\infty} \frac{-2}{x} = 0 ; \ y = 0$ is a horizontal asymptote.   (2-2)

80. $f(x) = \dfrac{x^2 + 9}{x - 3}$; f is discontinuous at $x = 3$.

$\displaystyle\lim_{x \to 3^-} \dfrac{x^2 + 9}{x - 3} = -\infty, \quad \lim_{x \to 3^+} \dfrac{x^2 + 9}{x - 3} = \infty; \quad x = 3$ is a vertical asymptote.

$\displaystyle\lim_{x \to \infty} \dfrac{x^2 + 9}{x - 3} = \lim_{x \to \infty} \dfrac{x^2}{x} = \lim_{x \to \infty} x = \infty; \quad$ no horizontal asymptotes.   (2-2)

81. $f(x) = \dfrac{x^2 - 9}{x^2 + x - 2} = \dfrac{x^2 - 9}{(x + 2)(x - 1)}$; f is discontinuous at $x = -2, \; x = 1$.

At $x = -2$:

$\displaystyle\lim_{x \to -2^-} \dfrac{x^2 - 9}{(x + 2)(x - 1)} = -\infty, \quad \lim_{x \to -2^+} \dfrac{x^2 - 9}{(x + 2)(x - 1)} = \infty; \quad x = -2$ is a vertical asymptote.

At $x = 1$

$\displaystyle\lim_{x \to 1^-} \dfrac{x^2 - 9}{(x + 2)(x - 1)} = \infty, \quad \lim_{x \to 1^+} \dfrac{x^2 - 9}{(x + 2)(x - 1)} = -\infty; \quad x = 1$ is a vertical asymptote.

$\displaystyle\lim_{x \to \infty} \dfrac{x^2 - 9}{x^2 + x - 2} = \lim_{x \to \infty} \dfrac{x^2}{x^2} = \lim_{x \to \infty} 1 = 1; \quad y = 1$ is a horizontal asymptote.   (2-2)

82. $f(x) = \dfrac{x^3 - 1}{x^3 - x^2 - x + 1} = \dfrac{(x - 1)(x^2 + x + 1)}{(x - 1)(x^2 - 1)} = \dfrac{(x - 1)(x^2 + x + 1)}{(x - 1)^2(x + 1)} = \dfrac{x^2 + x + 1}{(x - 1)(x + 1)}, \; x \neq 1$.

f is discontinuous at $x = 1, x = -1$.

At $x = 1$:

$\displaystyle\lim_{x \to 1^-} f(x) = \lim_{x \to 1^-} \dfrac{x^2 + x + 1}{(x - 1)(x + 1)} = -\infty, \quad \lim_{x \to 1^+} f(x) = \infty; \quad x = 1$ is a vertical asymptote.

At $x = -1$:

$\displaystyle\lim_{x \to -1^-} \dfrac{x^2 + x + 1}{(x - 1)(x + 1)} = \infty, \quad \lim_{x \to -1^+} \dfrac{x^2 + x + 1}{(x - 1)(x + 1)} = -\infty; \quad x = -1$ is a vertical asymptote.

$\displaystyle\lim_{x \to \infty} \dfrac{x^3 - 1}{x^3 - x^2 - x + 1} = \lim_{x \to \infty} \dfrac{x^3}{x^3} = \lim_{x \to \infty} 1 = 1; \quad y = 1$ is a horizontal asymptote.   (2-2)

83. $f(x) = x^{1/5}; \; f'(x) = \dfrac{1}{5} x^{-4/5} = \dfrac{1}{5x^{4/5}}$

The domain of f' is all real numbers except $x = 0$. At $x = 0$, the graph of f is smooth, but the tangent line to the graph at $(0, 0)$ is vertical.   (2-4)

84. $f(x) = \begin{cases} x^2 - m & \text{if } x \le 1 \\ -x^2 + m & \text{if } x > 1 \end{cases}$

(A)

(B)

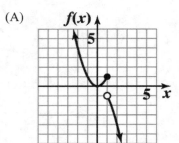

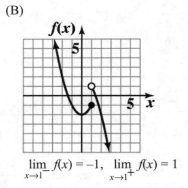

$\displaystyle\lim_{x \to 1^-} f(x) = 1, \quad \lim_{x \to 1^+} f(x) = -1$

$\displaystyle\lim_{x \to 1^-} f(x) = -1, \quad \lim_{x \to 1^+} f(x) = 1$

(C) $\displaystyle\lim_{x \to 1^-} f(x) = 1 - m, \quad \lim_{x \to 1^+} f(x) = -1 + m$

We want $1 - m = -1 + m$ which implies $m = 1$.

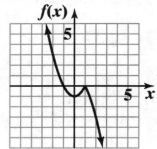

(D)   The graphs in (A) and (B) have jumps at $x = 1$; the graph in (C) does not.   (2-2)

85. $f(x) = 1 - |x - 1|, \, 0 \le x \le 2$

(A) $\displaystyle\lim_{h \to 0^-} \frac{f(1+h) - f(1)}{h} = \lim_{h \to 0^-} \frac{1 - |1 + h - 1| - 1}{h} = \lim_{h \to 0^-} \frac{-|h|}{h} = \lim_{h \to 0^-} \frac{h}{h} = 1$ $(|h| = -h \text{ if } h < 0)$

(B) $\displaystyle\lim_{h \to 0^+} \frac{f(1+h) - f(1)}{h} = \lim_{h \to 0^+} \frac{1 - |1 + h - 1| - 1}{h} = \lim_{h \to 0^+} \frac{-|h|}{h} = \lim_{h \to 0^+} \frac{-h}{h} = -1$ $(|h| = h \text{ if } h > 0)$

(C) $\displaystyle\lim_{h \to 0} \frac{f(1+h) - f(1)}{h}$ does not exist, since the left limit and the right limit are not equal.

(D) $f'(1)$ does not exist.   (2-4)

86. (A) $S(x) = 7.47 + 0.4000x$ for $0 \le x \le 90$; $S(90) = 43.47$;
       $S(x) = 43.47 + 0.2076 (x - 90) = 24.786 + 0.2076x, \, x > 90$
       Therefore,
       $S(x) = \begin{cases} 7.47 + 0.4000x & \text{if } 0 \le x \le 90 \\ 24.786 + 0.2076x & \text{if } x > 90 \end{cases}$

(B)

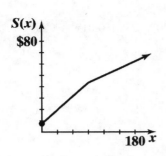

(C) $\lim\limits_{x \to 90^-} S(x) = \lim\limits_{x \to 90^+} S(x) = 43.47 = S(90);$

$S(x)$ is continuous at $x = 90$.

(2-2)

87. $C(x) = 10,000 + 200x - 0.1x^2$

(A) $C(101) - C(100) = 10,000 + 200(101) - 0.1(101)^2 - [10,000 + 200(100) - 0.1(100)^2]$
$= 29,179.90 - 29,000 = \$179.90$

(B) $C'(x) = 200 - 0.2x$
$C'(100) = 200 - 0.2(100) = 200 - 20 = \$180$ (2-7)

88. $C(x) = 5,000 + 40x + 0.05x^2$

(A) Cost of producing 100 bicycles:

$C(100) = 5,000 + 40(100) + 0.05(100)^2 = 9000 + 500 = 9500$
Marginal cost:
$C'(x) = 40 + 0.1x$
$C'(100) = 40 + 0.1(100) = 40 + 10 = 50$
Interpretation: At a production level of 100 bicycles, the total cost is \$9,500 and is increasing at the rate of \$50 per additional bicycle.

(B) Average cost: $\overline{C}(x) = \dfrac{C(x)}{x} = \dfrac{5000}{x} + 40 + 0.05x$

$\overline{C}(100) = \dfrac{5000}{100} + 40 + 0.05(100) = 50 + 40 + 5 = 95$

Marginal average cost: $\overline{C}'(x) = -\dfrac{5000}{x^2} + 0.05$ and

$\overline{C}'(100) = -\dfrac{5000}{(100)^2} + 0.05 = -0.5 + 0.05 = -0.45$

Interpretation: At a production level of 100 bicycles, the average cost is \$95 and the average cost is decreasing at a rate of \$0.45 per additional bicycle.   (2-7)

89. The approximate cost of producing the 201st printer is greater than that of producing the 601st printer (the slope of the tangent line at $x = 200$ is greater than the slope of the tangent line at $x = 600$). Since the marginal costs are decreasing, the manufacturing process is becoming more efficient.   (2-7)

90. $p = 25 - 0.01x$, $C(x) = 2x + 9,000$

(A) Marginal cost: $C'(x) = 2$

Average cost: $\overline{C}(x) = \dfrac{C(x)}{x} = 2 + \dfrac{9,000}{x}$

Marginal average cost: $\overline{C}' = -\dfrac{9,000}{x^2}$

(B) Revenue: $R(x) = xp = 25x - 0.01x^2$
Marginal revenue: $R'(x) = 25 - 0.02x$

Average revenue: $\overline{R}(x) = \dfrac{R(x)}{x} = 25 - 0.01x$

Marginal average revenue: $\overline{R}'(x) = -0.01$

(C) Profit: $P(x) = R(x) - C(x) = 25x - 0.01x^2 - (2x + 9,000) = 23x - 0.01x^2 - 9,000$

Marginal profit: $P'(x) = 23 - 0.02x$

Average profit: $\overline{P}(x) = \dfrac{P(x)}{x} = 23 - 0.01x - \dfrac{9,000}{x}$

Marginal average profit: $\overline{P}'(x) = -0.01 + \dfrac{9,000}{x^2}$

(D) Break-even points: $R(x) = C(x)$

$$25x - 0.01x^2 = 2x + 9,000$$

$$0.01x^2 - 23x + 9,000 = 0$$

$$x^2 - 2,300x + 900,000 = 0$$

$$(x - 500)(x - 1,800) = 0$$

Thus, the break-even points are at $x = 500$, $x = 1,800$;

break-even points: (500, 10,000), (1,800, 12,600).

(E) $P'(1,000) = 23 - 0.02(1000) = 3$; profit is increasing at the rate of \$3 per umbrella.

$P'(1,150) = 23 - 0.02(1,150) = 0$; profit is flat.

$P'(1,400) = 23 - 0.02(1,400) = -5$; profit is decreasing at the rate of \$5 per umbrella.

(F)

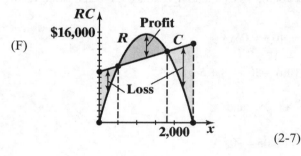

(2-7)

91. $N(t) = \dfrac{40t - 80}{t} = 40 - \dfrac{80}{t}, \; t \geq 2$

(A) Average rate of change from $t = 2$ to $t = 5$:

$$\dfrac{N(5) - N(2)}{5 - 2} = \dfrac{\dfrac{40(5) - 80}{5} - \dfrac{40(2) - 80}{2}}{3} = \dfrac{120}{15} = 8 \text{ components per day.}$$

(B) $N(t) = 40 - \dfrac{80}{t} = 40 - 80t^{-1}; \quad N'(t) = 80t^{-2} = \dfrac{80}{t^2}.$

$N'(2) = \dfrac{80}{4} = 20$ components per day.   (2-5)

92. $N(t) = 2t + \dfrac{1}{3}t^{3/2}, \quad N'(t) = 2 + \dfrac{1}{2}t^{1/2} = \dfrac{4 + \sqrt{t}}{2}$

$N(9) = 18 + \dfrac{1}{3}(9)^{3/2} = 27, \quad N'(9) = \dfrac{4 + \sqrt{9}}{2} = \dfrac{7}{2} = 3.5$

After 9 months, 27,000 pools have been sold and the total sales are increasing at the rate of 3,500 pools per month.   (2-5)

93. (A)

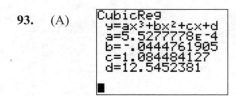

(B) $N(x) \approx 0.0005528x^3 - 0.044x^2 + 1.084x + 12.545$

$N'(x) \approx 0.0016584x^2 - 0.088x + 1.084$

$N(60) \approx 36.9, \quad N'(60) \approx 1.7.$ In 2020, natural gas consumption

will be 36.9 trillion cubic feet and will be INCREASING at the

rate of 1.7 trillion cubic feet per year.   (2-4)

94. (A)

```
LinReg
 y=ax+b
 a=-.0384180791
 b=13.59887006
 r=-.9897782666
```

(B)   Fixed costs: \$484.21; variable cost per kringle: \$2.11.

```
LinReg
 y=ax+b
 a=2.107344633
 b=484.2090395
 r=.9939318704
```

(C)   Let $p(x)$ be the linear regression equation found in
part (A) and let $C(x)$ be the linear regression equation found in part (B).
Then revenue $R(x) = xp(x)$ and the break-even points are the points
where $R(x) = C(x)$.
Using an intersection routine on a graphing utility,
the break-even points are: (51, 591.15) and (248, 1,007.62).

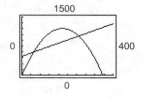

(D)   The bakery will make a profit when $51 < x < 248$. From the regression equation in part (A),
$p(51) = 11.64$ and $p(248) = 4.07$. Thus, the bakery will make a profit for the price range
\$4.07 $< p <$ \$11.64.   (2-7)

95. $C(x) = \dfrac{500}{x^2} = 500x^{-2}, x \geq 1.$

The instantaneous rate of change of concentration at x meters is:

$C'(x) = 500(-2)x^{-3} = \dfrac{-1000}{x^3}$

The rate of change of concentration at 10 meters is:

$C'(10) = \dfrac{-1000}{10^3} = -1$ parts per million per meter

The rate of change of concentration at 100 meters is:

$C'(100) = \dfrac{-1000}{(100)^3} = \dfrac{-1000}{100,000,000} = -\dfrac{1}{1000} = -0.001$ parts per million per meter.   (2-5)

96. $F(t) = 0.16t^2 - 1.6t + 102, F'(t) = 0.32t - 1.6$
$F(4) = 98.16, F'(4) = -0.32.$
After 4 hours the patient's temperature is 98.16°F and is decreasing at the rate of 0.32°F per hour.
(2-5)

97. $N(t) = 20\sqrt{t} = 20t^{1/2}$

The rate of learning is $N'(t) = 20\left(\dfrac{1}{2}\right)t^{-1/2} = 10t^{-1/2} = \dfrac{10}{\sqrt{t}}.$

(A)   The rate of learning after one hour is $N'(1) = \dfrac{10}{\sqrt{1}} = 10$ items per hour.

(B)   The rate of learning after four hours is $N'(4) = \dfrac{10}{\sqrt{4}} = \dfrac{10}{2} = 5$ items per hour.   (2-5)

98.   (A)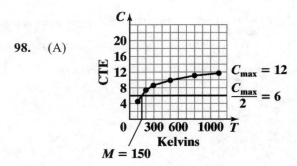

(B) $C(T) = \dfrac{12T}{150+T}$

(C) $C(600) = \dfrac{12(600)}{150+600} = 9.6$

To find T when $C = 10$, solve $\dfrac{12T}{150+T} = 10$ for T.

$$\frac{12T}{150+T} = 10$$
$$12T = 1500 + 10T$$
$$2T = 1500$$
$$T = 750$$

$T = 750$ when $C = 10$.   (2-3)

11 ADDITIONAL DERIVATIVE TOPICS

EXERCISE 11-1

Things to remember:

1. THE NUMBER e

    The irrational number e is defined by

    $$e = \lim_{n \to \infty} \left(1 + \frac{1}{n}\right)^n$$

    or alternatively,

    $$e = \lim_{s \to 0} (1 + s)^{1/s}$$

    Both limits are equal to $e = 2.718\ 281\ 828\ 459\ \ldots$

2. CONTINUOUS COMPOUND INTEREST FORMULA

    $A = Pe^{rt}$

    where P = Principal
       r = Annual nominal interest rate compounded continuously
       t = Time in years
       A = Amount at time t

---

1. $A = 1,200\,e^{0.04(5)} = 1,200\,e^{0.2} \approx 1,465.68$

3. $9,827.30 = Pe^{0.025(3)}; \quad P = \dfrac{9,827.30}{e^{0.075}} \approx 9,117.21$

5. $6,000 = 5,000e^{0.0325t}, \quad e^{0.0325t} = \dfrac{6,000}{5,000}, \quad 0.0325t = \ln\left(\dfrac{6,000}{5,000}\right), \quad t = \dfrac{\ln 1.2}{0.0325} \approx 5.61$

7. $956 = 900e^{1.5r}, \quad e^{1.5r} = \dfrac{956}{900}, \quad 1.5r = \ln\left(\dfrac{956}{900}\right), \quad r \approx 0.04$

9. $A = \$1000e^{0.1t}$
    When $t = 2$, $A = \$1000e^{(0.1)2} = \$1000e^{0.2} = \$1221.40$.
    When $t = 5$, $A = \$1000e^{(0.1)5} = \$1000e^{0.5} = \$1648.72$.
    When $t = 8$, $A = \$1000e^{(0.1)8} = \$1000e^{0.8} = \$2225.54$

11.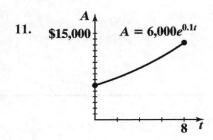

13. $2 = e^{0.06t}$

Take the natural log of both sides of this equation

$\ln(e^{0.06t}) = \ln 2$

$0.06t \ln e = \ln 2$

$0.06t = \ln 2 \quad (\ln e = 1)$

$t = \dfrac{\ln 2}{0.06} \approx 11.55$

15. $3 = e^{0.1t}$

$\ln(e^{0.1t}) = \ln 3$

$0.1t = \ln 3$

$t = \dfrac{\ln 3}{0.1} \approx 10.99$

17. $2 = e^{5r}$

$\ln(e^{5r}) = \ln 2$

$5r = \ln 2$

$r = \dfrac{\ln 2}{5} \approx 0.14$

19.

n	$\left(1+\dfrac{1}{n}\right)^n$
10	2.59374
100	2.70481
1000	2.71692
10,000	2.71815
100,000	2.71827
1,000,000	2.71828
10,000,000	2.71828
$\downarrow$	$\downarrow$
∞	$e = 2.7182818\ldots$

21.

n	4	16	64	256	1024	4096
$(1+n)^{1/n}$	1.495349	1.193722	1.067399	1.021913	1.006793	1.002033

$\lim\limits_{n\to\infty} (1+n)^{1/n} = 1$

23. The graphs of $y_1 = \left(1+\dfrac{1}{n}\right)^n$, $y_2 = 2.718281828 \approx e$,

and

$y_3 = \left(1+\dfrac{1}{n}\right)^{n+1}$ for $0 \le n \le 20$ are given at the right.

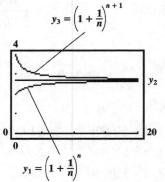

25. (A) $A = Pe^{rt}$; $P = \$10{,}000$, $r = 2.15\% = 0.0215$, $t = 10$:

$A = 10{,}000e^{(0.0215)10} = 10{,}000e^{0.215} = \$12{,}398.62$

(B) $A = \$18,000,\ P = \$10,000,\ r = 0.0215$:

$$18,000 = 10,000e^{0.0215t}$$

$$e^{0.0215t} = 1.8$$

$$0.0215t = \ln(1.8)$$

$$t = \frac{\ln(1.8)}{0.0215} \approx 27.34 \text{ years}$$

27. $A = Pe^{rt};\ A = \$20,000,\ r = 0.052,\ t = 10$:

$$20,000 = Pe^{(0.052)10} = Pe^{0.52}$$

$$P = \frac{20,000}{e^{0.52}} = 20,000e^{-0.52} \approx \$11,890.41$$

29. $30,000 = 20,000e^{5r}$

$$e^{5r} = 1.5$$

$$5r = \ln(1.5)$$

$$r = \frac{\ln 1.5}{5} \approx 0.0811 \text{ or } 8.11\%$$

31. $P = 10,000e^{-0.08t},\ 0 \le t \le 50$

(A)
t	0	10	20	30	40	50
P	10,000	4493.30	2019	907.18	407.62	183.16

The graph of P is shown at   the right.

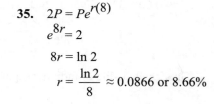

(B) $\displaystyle \lim_{t \to \infty} 10,000e^{-0.08t} = 0$

33.
$$2P = Pe^{0.04t}$$
$$e^{0.04t} = 2$$
$$0.04t = \ln 2$$
$$t = \frac{\ln 2}{0.04} \approx 17.33 \text{ years}$$

35. $2P = Pe^{r(8)}$
$$e^{8r} = 2$$
$$8r = \ln 2$$
$$r = \frac{\ln 2}{8} \approx 0.0866 \text{ or } 8.66\%$$

37. The total investment in the two accounts is given by
$$A = 10,000e^{0.072t} + 10,000(1 + 0.084)^t$$
$$= 10,000[e^{0.072t} + (1.084)^t]$$

On a graphing utility, locate the intersection point of
$y_1 = 10,000[e^{0.072x} + (1.084)^x]$ and $y_2 = 35,000$.

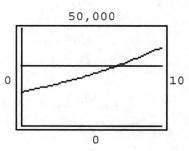

The result is: $x = t \approx 7.3$ years.

39. (A) $A = Pe^{rt}$; set $A = 2P$

(B) $2P = Pe^{rt}$

$e^{rt} = 2$

$rt = \ln 2$

$t = \dfrac{\ln 2}{r}$

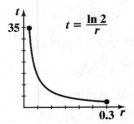

In theory, r could be any positive number. However, the restrictions on r are reasonable in the sense that most investments would be expected to earn between 2% and 30%.

(C) $r = 5\%$; $t = \dfrac{\ln 2}{0.05} \approx 13.86$ years

$r = 10\%$; $t = \dfrac{\ln 2}{0.10} \approx 6.93$ years

$r = 15\%$; $t = \dfrac{\ln 2}{0.15} \approx 4.62$ years

$r = 20\%$; $t = \dfrac{\ln 2}{0.20} \approx 3.47$ years

$r = 25\%$; $t = \dfrac{\ln 2}{0.25} \approx 2.77$ years

$r = 30\%$; $t = \dfrac{\ln 2}{0.30} \approx 2.31$ years

41.
$$Q = Q_0 e^{-0.0004332t}$$

$$\frac{1}{2}Q_0 = Q_0 e^{-0.0004332t}$$

$$e^{-0.0004332t} = \frac{1}{2}$$

$$\ln(e^{-0.0004332t}) = \ln\left(\frac{1}{2}\right) = \ln 1 - \ln 2$$

$$-0.0004332t = -\ln 2 \ \ (\ln 1 = 0)$$

$$t = \frac{\ln 2}{0.0004332} \approx \frac{0.6931}{0.0004332} \approx 1599.95$$

Thus, the half-life of radium is approximately 1600 years.

43.
$$Q = Q_0 e^{rt} \ \ (r < 0)$$

$$\frac{1}{2}Q_0 = Q_0 e^{r(30)}$$

$$e^{30r} = \frac{1}{2}$$

$$\ln(e^{30r}) = \ln\frac{1}{2} = \ln 1 - \ln 2$$

$30r = -\ln 2$ $(\ln 1 = 0)$

$$r = \frac{-\ln 2}{30} \approx \frac{-0.6931}{30} \approx -0.0231$$

Thus, the continuous compound rate of decay of the cesium isotope is approximately -0.0231.

45. $2P_0 = P_0 e^{0.013t}$

$e^{0.013t} = 2$

$0.013t = \ln 2$

$$t = \frac{\ln 2}{0.013} \approx 53.3$$

It will take approximately 53.3 years.

47. $2P_0 = P_0 e^{r(50)}$

$e^{50r} = 2$

$50r = \ln 2$

$$r = \frac{\ln 2}{50} \approx 0.0139 \quad \text{or} \quad 1.39\%$$

EXERCISE 11-2

Things to remember:

<u>1.</u>   DERIVATIVES OF EXPONENTIAL FUNCTIONS

(a) $\dfrac{d}{dx} e^x = e^x$ (b) $\dfrac{d}{dx} b^x = b^x \ln b$

<u>2.</u>   LOGARITHMIC FUNCTIONS

The inverse of an exponential function is called a LOGARITHMIC FUNCTION. For $b > 0$, $b \neq 1$,

Logarithmic form		Exponential form
$y = \log_b x$	is equivalent to	$x = b^y$

Domain: $(0, \infty)$ \qquad\qquad\qquad\qquad Domain: $(-\infty, \infty)$

Range: $(-\infty, \infty)$ \qquad\qquad\qquad\qquad Range: $(0, \infty)$

The graphs of $y = \log_b x$ and $y = b^x$ are symmetric with respect to the line $y = x$.

The two most commonly used logarithmic functions are:

$\log x = \log_{10} x$ Common logarithm (base 10)

$\ln x = \log_e x$ Natural logarithm (base e)

<u>3.</u>   DERIVATIVES OF LOGARITHMIC FUNCTIONS

(a) $\dfrac{d}{dx} \ln x = \dfrac{1}{x}$ (b) $\dfrac{d}{dx} \log_b x = \dfrac{1}{\ln b} \cdot \dfrac{1}{x}$

1. $y = \log_2 128 = \log_2 2^7 = 7$

3. $\log_3 x = 4, \quad x = 3^4 = 81$

5. $\log_b 64 = 2, \quad b^2 = 64, \quad b = 8$

7. $y = \ln \sqrt{e} = \ln e^{1/2} = 1/2$

9. $f(x) = 5e^x + 3x + 1$

$f'(x) = 5e^x + 3$

11. $f(x) = -2 \ln x + x^2 - 4$

$f'(x) = -2\left(\dfrac{1}{x}\right) + 2x = \dfrac{-2}{x} + 2x$

13. $f(x) = x^3 - 6e^x$

$f'(x) = 3x^2 - 6e^x$

15. $f(x) = e^x + x - \ln x$

$f'(x) = e^x + 1 - \dfrac{1}{x}$

17. $f(x) = \ln x^3 = 3 \ln x$

$f'(x) = 3\left(\dfrac{1}{x}\right) = \dfrac{3}{x}$

19. $f(x) = 5x - \ln x^5 = 5x - 5 \ln x$

$f'(x) = 5 - 5\left(\dfrac{1}{x}\right) = 5 - \dfrac{5}{x}$

21. $f(x) = \ln x^2 + 4e^x = 2 \ln x + 4e^x$

$f'(x) = \dfrac{2}{x} + 4e^x$

23. $f(x) = e^x + x^e$

$f'(x) = e^x + ex^{e-1}$

25. $f(x) = x\,x^e = x^{e+1}$, $f'(x) = (e+1)x^e$

27. $f(x) = 3 + \ln x;\ f(1) = 3$

$f'(x) = \dfrac{1}{x};\ f'(1) = 1$

Equation of the tangent line: $y - 3 = 1(x - 1)$ or $y = x + 2$

29. $f(x) = 3e^x;\ f(0) = 3$

$f'(x) = 3e^x;\ f'(0) = 3$

Tangent line: $y - 3 = 3(x - 0)$ or $y = 3x + 3$

31. $f(x) = \ln x^3 = 3 \ln x;\ f(e) = 3 \ln e = 3$

$f'(x) = \dfrac{3}{x};\ f'(e) = \dfrac{3}{e}$

Tangent line: $y - 3 = \dfrac{3}{e}(x - e)$ or $y = \dfrac{3}{e}x$

33. $f(x) = 2 + e^x;\ f(1) = 2 + e$

$f'(x) = e^x;\ f'(1) = e$

Tangent line: $y - (2 + e) = e(x - 1)$ or $y = ex + 2$

35. An equation for the tangent line to the graph of $f(x) = e^x$ at the point $(3, f(3)) = (3, e^3)$ is:

$$y - e^3 = e^3(x - 3)$$

or $\quad y = xe^3 - 2e^3 = e^3(x - 2)$

Clearly, $y = 0$ when $x = 2$, that is the tangent line passes through the point $(2, 0)$. In general, an equation for the tangent line to the graph of $f(x) = e^x$ at the point $(c, f(c)) = (c, e^c)$ is:

$$y - e^c = e^c(x - c)$$

or $\quad y = e^c(x - [c - 1])$

Thus, the tangent line at the point (c, e^c) passes through $(c - 1, 0)$; the tangent line at the point $(4, e^4)$ passes through $(3, 0)$.

37. An equation for the tangent line to the graph of $g(x) = \ln x$ at the point $(3, g(3)) = (3, \ln 3)$ is:

$$y - \ln 3 = m(x - 3) \text{ where } m = g'(3)$$

$g'(x) = \dfrac{d}{dx} \ln x = \dfrac{1}{x}$; $g'(3) = \dfrac{1}{3}$. Thus,

$y - \ln 3 = \dfrac{1}{3} (x - 3)$

For $x = 0$, $y = \ln 3 - 1$, so this tangent line does not pass through the origin. In fact, for any real number c, the tangent line to $g(x) = \ln x$ at the point $(c, \ln c)$ has equation $y - \ln c = \dfrac{1}{c}(x - c)$, and thus the only tangent line which passes through the origin is the tangent line at $(e, 1)$.

39. $f(x) = 10x + \ln 10x = 10x + \ln 10 + \ln x$

$f'(x) = 10 + \dfrac{1}{x}$

41. $f(x) = \ln\left(\dfrac{4}{x^3}\right) = \ln 4 - \ln x^3 = \ln 4 - 3 \ln x$; $f'(x) = -\dfrac{3}{x}$

43. $y = \log_2 x$; $\dfrac{dy}{dx} = \dfrac{1}{\ln 2} \cdot \dfrac{1}{x} = \dfrac{1}{x \ln 2}$

45. $y = 3^x$; $\dfrac{dy}{dx} = 3^x \ln 3$

47. $y = 2x - \log x = 2x - \log_{10} x$; $\dfrac{dy}{dx} = 2 - \dfrac{1}{x \ln 10}$

49. $y = 10 + x + 10^x$; $y' = 1 + 10^x \ln 10$

51. $y = 3 \ln x + 2 \log_3 x$; $y' = \dfrac{3}{x} + \dfrac{2}{x \ln 3}$

53. $y = 2^x + e^2$; $y' = 2^x \ln 2$ (e^2 is a constant; $\dfrac{d}{dx}(e^2) = 0$)

55.   On a graphing utility, graph $y_1 = e^x$ and $y_2 = x^4$. Rounded off to two decimal places, the points of intersection are: $(-0.82, 0.44)$, $(1.43, 4.18)$, $(8.61, 5503.66)$.

57.   On a graphing utility, graph $y_1 = (\ln x)^2$ and $y_2(x) = x$. The curves intersect at $(0.49, 0.49)$ (two decimal places).

59.   On a graphing utility, graph $y_1 = \ln x$ and $y_2 = x^{1/5}$. There is a point of intersection at $(3.65, 1.30)$ (two decimal places). Using the hint that $\ln x < x^{1/5}$ for large x, we find a second point of intersection at $(332{,}105.11, 12.71)$ (two decimal places).

61.   Assume $c \neq 0$. Then $\dfrac{e^{ch} - 1}{h} = c \dfrac{e^{ch} - 1}{ch}$.

Let $t = ch$. Then $t \to 0$ if and only if $h \to 0$. Therefore,

$$\lim_{h \to 0} \dfrac{e^{ec} - 1}{h} = \lim_{h \to 0} c \dfrac{e^{ec} - 1}{ch} = c \lim_{h \to 0} \dfrac{e^{ec} - 1}{ch} = c \lim_{t \to 0} \dfrac{e^t - 1}{t} = c \cdot 1 = c$$

63. $S(t) = 300,000(0.9)^t, \quad t \geq 0$

$S'(t) = 300,000(0.9)^t \ln(0.9) = -31,608.15(0.9)^t$

The rate of depreciation after 1 year is:

$S'(1) = -31,608.15(0.9) \approx -28,447.34;$ rate of depreciation $28,447.34$.

The rate of depreciation after 5 years is:

$S'(5) = -31,608.15(0.9)^5 \approx -18,664.30;$ rate of depreciation $18,664.30$.

The rate of depreciation after 10 years is:

$S'(10) = -31,608.15(0.9)^{10} \approx -11,021.08;$ rate of depreciation $11,021.08$

65. $A(t) = 5,000 \cdot 4^t; \quad A'(t) = 5,000 \cdot 4^t (\ln 4)$

$A'(1) = 5,000 \cdot 4(\ln 4) \approx 27,726$ – the rate of change of the bacteria population at the end of the first hour.

$A'(5) = 5,000 \cdot 4^5(\ln 4) \approx 7,097,827$ – the rate of change of the bacteria population at the end of the fifth hour.

67. $P(x) = 17.5(1 + \ln x), \quad 10 \leq x \leq 100, \quad P'(x) = \dfrac{17.5}{x}$

$P'(40) = \dfrac{17.5}{40} \approx 0.44$

$P'(90) = \dfrac{17.5}{90} \approx 0.19$

Thus, at the 40 pound weight level, blood pressure would increase at the rate of 0.44 mm of mercury per pound of weight gain; at the 90 pound weight level, blood pressure would increase at the rate of 0.19 mm of mercury per pound of weight gain.

69. $R = k \ln(S/S_0) = k[\ln S - \ln S_0]; \quad \dfrac{dR}{dS} = \dfrac{k}{S}$

71. $P(t) = 10,000\,e^{0.075t}, \quad P'(t) = 750\,e^{0.075t}$

(A) $P'(1) = 750\,e^{0.075(1)} \approx 808.41;$ 808.41 per year.

(B)   First solve $10,000\,e^{0.075t} = 12,500$ for t:

$e^{0.075t} = \dfrac{12,500}{10,000} = 1.25$

$0.075t = \ln 1.25$

$t = \dfrac{\ln 1.25}{0.075} \approx 2.975247351$

$P'(2.98) = 750\,e^{0.075(2.975247351)} \approx 937.50;$ 937.50 per year.

EXERCISE 11-3

Things to remember:

1.   PRODUCT RULE

If

$$y = f(x) = F(x)S(x)$$

and if $F'(x)$ and $S'(x)$ exist, then

$$f'(x) = F(x)\ S'(x) + S(x)\ F'(x).$$

Also,

$$y' = F\,S'\, + S\,F';$$

$$\frac{dy}{dx} = F\frac{dS}{dx} + S\frac{dF}{dx}$$

2.   QUOTIENT RULE

If

$$y = f(x) = \frac{T(x)}{B(x)}$$

and if $T'(x)$ and $B'(x)$ exist, then

$$f'(x) = \frac{B(x)T'(x) - T(x)B'(x)}{[B(x)]^2}.$$

Also,

$$y' = \frac{BT' - TB'}{B^2};$$

$$\frac{dy}{dx} = \frac{B\frac{dT}{dx} - T\frac{dB}{dx}}{B^2}.$$

---

1. $f(x) = 5x^3 - 4x^3\ln x = x^3(5 - 4\ln x)$, $F(x) = x^3$, $S(x) = 5 - 4\ln x$

3. $f(x) = x^3 e^x + 2x^3 + 3e^x + 6 = x^3(e^x + 2) + 3(e^x + 2) = (x^3 + 3)(e^x + 2)$

    $F(x) = x^3 + 3$, $S(x) = e^x + 2$

5. $f(x) = 9x^2 e^{-5x} = \dfrac{9x^2}{e^{5x}}$, $T(x) = 9x^2$, $B(x) = e^{5x}$

7. $f(x) = \dfrac{3}{x^2} + \dfrac{e^x}{x^4} = \dfrac{3x^2 + e^x}{x^4}$; $T(x) = 3x^2 + e^x$, $B(x) = x^4$

9. $f(x) = 2x^3(x^2 - 2)$

    $f'(x) = 2x^3(x^2 - 2)' + (x^2 - 2)(2x^3)'$ [using 1 with $F(x) = 2x^3$, $S(x) = x^2 - 2$]

       $= 2x^3(2x) + (x^2 - 2)(6x^2) = 4x^4 + 6x^4 - 12x^2 = 10x^4 - 12x^2$

11. $f(x) = (x - 3)(2x - 1)$

$f'(x) = (x - 3)(2x - 1)' + (2x - 1)(x - 3)'$ (using $\underline{1}$)

$= (x - 3)(2) + (2x - 1)(1) = 2x - 6 + 2x - 1 = 4x - 7$

13. $f(x) = \dfrac{x}{x - 3}$

$f'(x) = \dfrac{(x - 3)(x)' - x(x - 3)'}{(x - 3)^2}$ [using $\underline{2}$ with $T(x) = x$, $B(x) = x - 3$]

$= \dfrac{(x - 3)(1) - x(1)}{(x - 3)^2} = \dfrac{-3}{(x - 3)^2}$

15. $f(x) = \dfrac{2x + 3}{x - 2}$

$f'(x) = \dfrac{(x - 2)(2x + 3)' - (2x + 3)(x - 2)'}{(x - 2)^2}$ (using 2)

$= \dfrac{(x - 2)(2) - (2x + 3)(1)}{(x - 2)^2} = \dfrac{2x - 4 - 2x - 3}{(x - 2)^2} = \dfrac{-7}{(x - 2)^2}$

17. $f(x) = 3xe^x$

$f'(x) = 3x(e^x)' + e^x(3x)'$ (using $\underline{1}$)

$= 3xe^x + e^x(3) = 3(x + 1)e^x$

19. $f(x) = x^3 \ln x$

$f'(x) = x^3(\ln x)' + \ln x(x^3)'$ (using $\underline{1}$)

$= x^3 \cdot \dfrac{1}{x} + \ln x(3x^2) = x^2(1 + 3 \ln x)$

21. $f(x) = (x^2 + 1)(2x - 3)$

$f'(x) = (x^2 + 1)(2x - 3)' + (2x - 3)(x^2 + 1)'$ (using $\underline{1}$)

$= (x^2 + 1)(2) + (2x - 3)(2x) = 2x^2 + 2 + 4x^2 - 6x = 6x^2 - 6x + 2$

23. $f(x) = (0.4x + 2)(0.5x - 5)$

$f'(x) = (0.4x + 2)(0.5x - 5)' + (0.5x - 5)(0.4x + 2)'$

$= (0.4x + 2)(0.5) + (0.5x - 5)(0.4) = 0.2x + 1 + 0.2x - 2 = 0.4x - 1$

25. $f(x) = \dfrac{x^2 + 1}{2x - 3}$

$f'(x) = \dfrac{(2x - 3)(x^2 + 1)' - (x^2 + 1)(2x - 3)'}{(2x - 3)^2}$ (using 2)

$= \dfrac{(2x - 3)(2x) - (x^2 + 1)(2)}{(2x - 3)^2} = \dfrac{4x^2 - 6x - 2x^2 - 2}{(2x - 3)^2} = \dfrac{2x^2 - 6x - 2}{(2x - 3)^2}$

27. $f(x) = (x^2 + 2)(x^2 - 3)$

$f'(x) = (x^2 + 2)(x^2 - 3)' + (x^2 - 3)(x^2 + 2)'$

$= (x^2 + 2)(2x) + (x^2 - 3)(2x) = 2x^3 + 4x + 2x^3 - 6x = 4x^3 - 2x$

29. $f(x) = \dfrac{x^2 + 2}{x^2 - 3}$

$f'(x) = \dfrac{(x^2 - 3)(x^2 + 2)' - (x^2 + 2)(x^2 - 3)'}{(x^2 - 3)^2}$

$\qquad = \dfrac{(x^2 - 3)(2x) - (x^2 + 2)(2x)}{(x^2 - 3)^2} = \dfrac{2x^3 - 6x - 2x^3 - 4x}{(x^2 - 3)^2} = \dfrac{-10x}{(x^2 - 3)^2}$

31. $f(x) = \dfrac{e^x}{x^2 + 1}$

$f'(x) = \dfrac{(x^2 + 1)(e^x)' - e^x(x^2 + 1)'}{(x^2 + 1)^2} = \dfrac{(x^2 + 1)e^x - e^x(2x)}{(x^2 + 1)^2} = \dfrac{e^x(x^2 - 2x + 1)}{(x^2 + 1)^2} = \dfrac{e^x(x - 1)^2}{(x^2 + 1)^2}$

33. $f(x) = \dfrac{\ln x}{x + 1}$

$f'(x) = \dfrac{(x + 1)(\ln x)' - \ln x(x + 1)'}{(x + 1)^2} = \dfrac{(x + 1)\dfrac{1}{x} - \ln x(1)}{(x + 1)^2} = \dfrac{x + 1 - x \ln x}{x(x + 1)^2}$

35. $h(x) = xf(x);\quad h'(x) = xf'(x) + f(x)$

37. $h(x) = x^3 f(x);\quad h'(x) = x^3 f'(x) + f(x)(3x^2) = x^3 f'(x) + 3x^2 f(x)$

39. $h(x) = \dfrac{f(x)}{x^2};\quad h'(x) = \dfrac{x^2 f'(x) - f(x)(2x)}{(x^2)^2} = \dfrac{x^2 f'(x) - 2xf(x)}{x^4} = \dfrac{xf'(x) - 2f(x)}{x^3}$

$\quad$ or $\quad h(x) = x^{-2} f(x);\quad h'(x) = x^{-2} f'(x) + f(x)(-2x^{-3}) = \dfrac{xf'(x) - 2f(x)}{x^3}$

41. $h(x) = \dfrac{x}{f(x)};\quad h'(x) = \dfrac{f(x) - xf'(x)}{[f(x)]^2}$

43. $h(x) = e^x f(x)$

$\quad h'(x) = e^x f'(x) + f(x)e^x = e^x[f'(x) + f(x)]$

45. $h(x) = \dfrac{\ln x}{f(x)}$

$\quad h'(x) = \dfrac{f(x)\dfrac{1}{x} - \ln x(f'(x))}{[f(x)]^2} = \dfrac{f(x) - (x \ln x)f'(x)}{x[f(x)]^2}$

47. $f(x) = (2x + 1)(x^2 - 3x)$

$f'(x) = (2x + 1)(x^2 - 3x)' + (x^2 - 3x)(2x + 1)'$

$\qquad = (2x + 1)(2x - 3) + (x^2 - 3x)(2) = 6x^2 - 10x - 3$

49. $y = (2.5t - t^2)(4t + 1.4)$

$$\frac{dy}{dt} = (2.5t - t^2) \frac{d}{dt}(4t + 1.4) + (4t + 1.4) \frac{d}{dt}(2.5t - t^2)$$

$$= (2.5t - t^2)(4) + (4t + 1.4)(2.5 - 2t) = 10t - 4t^2 + 10t - 2.8t + 3.5 - 8t^2 = -12t^2 + 17.2t + 3.5$$

51. $y = \dfrac{5x - 3}{x^2 + 2x}$

$$y' = \frac{(x^2 + 2x)(5x - 3)' - (5x - 3)(x^2 + 2x)'}{(x^2 + 2x)^2}$$

$$= \frac{(x^2 + 2x)(5) - (5x - 3)(2x + 2)}{(x^2 + 2x)^2} = \frac{-5x^2 + 6x + 6}{(x^2 + 2x)^2}$$

53. $\dfrac{d}{dw}\left[\dfrac{w^2 - 3w + 1}{w^2 - 1}\right] = \dfrac{(w^2 - 1)\dfrac{d}{dw}(w^2 - 3w + 1) - (w^2 - 3w + 1)\dfrac{d}{dw}(w^2 - 1)}{(w^2 - 1)^2}$

$$= \frac{(w^2 - 1)(2w - 3) - (w^2 - 3w + 1)(2w)}{(w^2 - 1)^2} = \frac{3w^2 - 4w + 3}{(w^2 - 1)^2}$$

55. $y = (1 + x - x^2)\, e^x$
$y' = (1 + x - x^2)e^x + e^x(1 - 2x)$
$\quad = e^x(1 + x - x^2 + 1 - 2x) = (2 - x - x^2)e^x$

57. $f(x) = \dfrac{1}{x}$

(A) $f'(x) = \dfrac{x\dfrac{d}{dx}(1) - 1\dfrac{d}{dx}(x)}{x^2} = \dfrac{-1}{x^2}$

(B) $f(x) = \dfrac{1}{x} = x^{-1}, \quad f'(x) = -x^{-2} = \dfrac{-1}{x^2}$ (power rule)

59. $f(x) = \dfrac{-3}{x^4}$

(A) $f'(x) = \dfrac{x^4 \dfrac{d}{dx}(-3) - (-3)\dfrac{d}{dx}(x^4)}{(x^4)^2} = \dfrac{12x^3}{x^8} = \dfrac{12}{x^5}$

(B) $f(x) = \dfrac{-3}{x^4} = -3x^{-4}, \quad f'(x) = 12x^{-5} = \dfrac{12}{x^5}$ (power rule)

61. $f(x) = (1 + 3x)(5 - 2x)$

First find $f'(x)$:

$f'(x) = (1 + 3x)(5 - 2x)' + (5 - 2x)(1 + 3x)'$
$\quad = (1 + 3x)(-2) + (5 - 2x)(3) = -2 - 6x + 15 - 6x = 13 - 12x$

An equation for the tangent line at $x = 2$ is:

$y - y_1 = m(x - x_1)$

where $x_1 = 2$, $y_1 = f(x_1) = f(2) = 7$, and $m = f'(x_1) = f'(2) = -11$.

Thus, we have:

$y - 7 = -11(x - 2)$ or $y = -11x + 29$

63. $f(x) = \dfrac{x - 8}{3x - 4}$

First find $f'(x)$:

$$f'(x) = \frac{(3x - 4)(x - 8)' - (x - 8)(3x - 4)'}{(3x - 4)^2}$$

$$= \frac{(3x - 4)(1) - (x - 8)(3)}{(3x - 4)^2} = \frac{20}{(3x - 4)^2}$$

An equation for the tangent line at $x = 2$ is: $y - y_1 = m(x - x_1)$

where $x_1 = 2$, $y_1 = f(x_1) = f(2) = -3$, and $m = f'(x_1) = f'(2) = 5$.

Thus, we have: $y - (-3) = 5(x - 2)$ or $y = 5x - 13$

65. $f(x) = \dfrac{x}{2^x}$; $f'(x) = \dfrac{2^x(1) - x \cdot 2^x \cdot \ln 2}{[2^x]^2} = \dfrac{2^x(1 - x \ln 2)}{[2^x]^2} = \dfrac{1 - x \ln 2}{2^x}$

$f(2) = \dfrac{2}{4} = \dfrac{1}{2}$; $f'(2) = \dfrac{1 - 2 \ln 2}{4}$

Tangent line: $y - \dfrac{1}{2} = \dfrac{1 - 2 \ln 2}{4}(x - 2)$ or $y = \dfrac{1 - 2 \ln 2}{4}x + \ln 2$.

67. $f(x) = (2x - 15)(x^2 + 18)$

$f'(x) = (2x - 15)(x^2 + 18)' + (x^2 + 18)(2x - 15)'$

$= (2x - 15)(2x) + (x^2 + 18)(2) = 6x^2 - 30x + 36$

To find the values of x where $f'(x) = 0$, set: $f'(x) = 6x^2 - 30x + 36 = 0$

or $\qquad x^2 - 5x + 6 = 0$

$\qquad (x - 2)(x - 3) = 0$

Thus, $x = 2$, $x = 3$.

69. $f(x) = \dfrac{x}{x^2 + 1}$

$$f'(x) = \frac{(x^2 + 1)(x)' - x(x^2 + 1)'}{(x^2 + 1)^2} = \frac{(x^2 + 1)(1) - x(2x)}{(x^2 + 1)^2} = \frac{1 - x^2}{(x^2 + 1)^2}$$

Now, set $f'(x) = \dfrac{1 - x^2}{(x^2 + 1)^2} = 0$

or $\qquad\qquad 1 - x^2 = 0$

$\qquad (1 - x)(1 + x) = 0$

Thus, $x = 1$, $x = -1$.

71. $f(x) = x^3(x^4 - 1)$

First, we use the product rule:

$$f'(x) = x^3(x^4 - 1)' + (x^4 - 1)(x^3)'$$
$$= x^3(4x^3) + (x^4 - 1)(3x^2) = 7x^6 - 3x^2$$

Next, simplifying $f(x)$, we have $f(x) = x^7 - x^3$.

Thus, $f'(x) = 7x^6 - 3x^2$.

73. $f(x) = \dfrac{x^3 + 9}{x^3}$

First, we use the quotient rule:

$$f'(x) = \frac{x^3(x^3 + 9)' - (x^3 + 9)(x^3)'}{(x^3)^2} = \frac{x^3(3x^2) - (x^3 + 9)(3x^2)}{x^6} = \frac{-27x^2}{x^6} = \frac{-27}{x^4}$$

Next, simplifying $f(x)$, we have $f(x) = \dfrac{x^3 + 9}{x^3} = 1 + \dfrac{9}{x^3} = 1 + 9x^{-3}$

Thus, $f'(x) = -27x^{-4} = -\dfrac{27}{x^4}$.

75. $f(w) = (w + 1)2^w$

$$f'(w) = (w + 1)2^w(\ln 2) + 2^w(1) = [(w + 1)\ln 2 + 1]2^w = 2^w(w\ln 2 + \ln 2 + 1)$$

77. $y = 9x^{1/3}(x^3 + 5)$

$$\frac{dy}{dx} = 9x^{1/3}\frac{d}{dx}(x^3 + 5) + (x^3 + 5)\frac{d}{dx}(9x^{1/3})$$

$$= 9x^{1/3}(3x^2) + (x^3 + 5)\left(9 \cdot \frac{1}{3}x^{-2/3}\right) = 27x^{7/3} + (x^3 + 5)(3x^{-2/3}) = 27x^{7/3} + \frac{3x^3 + 15}{x^{2/3}} = \frac{30x^3 + 15}{x^{2/3}}$$

79. $y = \dfrac{\log_2 x}{1 + x^2}$

$$y' = \frac{(1 + x^2) \cdot \dfrac{1}{x\ln 2} - \log_2 x(2x)}{(1 + x^2)^2}$$

$$= \frac{1 + x^2 - 2x^2\ln 2\log_2 x}{x(1 + x^2)^2\ln 2} = \frac{1 + x^2 - 2x^2\ln x}{x(1 + x^2)^2\ln 2}$$

81. $f(x) = \dfrac{6\sqrt[3]{x}}{x^2 - 3} = \dfrac{6x^{1/3}}{x^2 - 3}$

$$f'(x) = \frac{(x^2 - 3)(6x^{1/3})' - 6x^{1/3}(x^2 - 3)'}{(x^2 - 3)^2}$$

$$= \frac{(x^2 - 3)\left(6 \cdot \dfrac{1}{3}x^{-2/3}\right) - 6x^{1/3}(2x)}{(x^2 - 3)^2} = \frac{(x^2 - 3)(2x^{-2/3}) - 12x^{4/3}}{(x^2 - 3)^2}$$

$$= \frac{\dfrac{2(x^2 - 3)}{x^{2/3}} - 12x^{4/3}}{(x^2 - 3)^2} = \frac{2x^2 - 6 - 12x^2}{(x^2 - 3)^2 x^{2/3}} = \frac{-10x^2 - 6}{(x^2 - 3)^2 x^{2/3}}$$

83. $g(t) = \dfrac{0.2t}{3t^2 - 1}$; $g'(t) = \dfrac{(3t^2 - 1)(0.2) - (0.2t)(6t)}{(3t^2 - 1)^2} = \dfrac{-0.6t^2 - 0.2}{(3t^2 - 1)^2}$

85. $\dfrac{d}{dx}[4x \log x^5] = 4x \dfrac{d}{dx}[\log x^5] + \log x^5 \dfrac{d}{dx}[4x] = 4x \dfrac{d}{dx}[5 \log x] + 4 \log x^5$

$$= 4x \cdot \left(\dfrac{5}{x \ln 10}\right) + 4 \log x^5 = \dfrac{20}{\ln 10} + 20 \log x = \dfrac{20(1 + \ln x)}{\ln 10}$$

87. $\dfrac{d}{dx} \dfrac{x^3 - 2x^2}{\sqrt[3]{x^2}} = \dfrac{d}{dx} \dfrac{x^3 - 2x^2}{x^{2/3}} = \dfrac{x^{2/3} \dfrac{d}{dx}(x^3 - 2x^2) - (x^3 - 2x^2)\dfrac{d}{dx}(x^{2/3})}{(x^{2/3})^2}$

$$= \dfrac{x^{2/3}(3x^2 - 4x) - (x^3 - 2x^2)\left(\dfrac{2}{3}x^{-1/3}\right)}{x^{4/3}} = x^{-2/3}(3x^2 - 4x) - \dfrac{2}{3}x^{-5/3}(x^3 - 2x^2)$$

$$= 3x^{4/3} - 4x^{1/3} - \dfrac{2}{3}x^{4/3} + \dfrac{4}{3}x^{1/3} = -\dfrac{8}{3}x^{1/3} + \dfrac{7}{3}x^{4/3}$$

89. $f(x) = \dfrac{(2x^2 - 1)(x^2 + 3)}{x^2 + 1}$

$f'(x) = \dfrac{(x^2 + 1)[(2x^2 - 1)(x^2 + 3)]' - (2x^2 - 1)(x^2 + 3)(x^2 + 1)'}{(x^2 + 1)^2}$

$$= \dfrac{(x^2 + 1)[(2x^2 - 1)(x^2 + 3)' + (x^2 + 3)(2x^2 - 1)'] - (2x^2 - 1)(x^2 + 3)(2x)}{(x^2 + 1)^2}$$

$$= \dfrac{(x^2 + 1)[(2x^2 - 1)(2x) + (x^2 + 3)(4x)] - (2x^2 - 1)(x^2 + 3)(2x)}{(x^2 + 1)^2}$$

$$= \dfrac{(x^2 + 1)[4x^3 - 2x + 4x^3 + 12x] - [2x^4 + 5x^2 - 3](2x)}{(x^2 + 1)^2} = \dfrac{(x^2 + 1)(8x^3 + 10x) - 4x^5 - 10x^3 + 6x}{(x^2 + 1)^2}$$

$$= \dfrac{8x^5 + 10x^3 + 8x^3 + 10x - 4x^5 - 10x^3 + 6x}{(x^2 + 1)^2} = \dfrac{4x^5 + 8x^3 + 16x}{(x^2 + 1)^2}$$

91. $y = \dfrac{t \ln t}{e^t}$

$$y' = \dfrac{e^t\left[t\left(\dfrac{1}{t}\right) + \ln t\right] - t \ln t(e^t)}{[e^t]^2} = \dfrac{e^t(1 + \ln t - t \ln t)}{[e^t]^2} = \dfrac{1 + \ln t - t \ln t}{e^t}$$

93. $S(t) = \dfrac{90t^2}{t^2 + 50}$

(A) $S'(t) = \dfrac{(t^2 + 50)(180t) - 90t^2(2t)}{(t^2 + 50)^2} = \dfrac{9000t}{(t^2 + 50)^2}$

(B) $S(10) = \dfrac{90(10)^2}{(10)^2 + 50} = \dfrac{9000}{150} = 60$;

$$S'(10) = \dfrac{9000(10)}{[(10)^2 + 50]^2} = \dfrac{90{,}000}{22{,}500} = 4$$

After 10 months, the total sales are 60,000 DVD's and the sales are INCREASING at the rate of 4,000 DVD's per month.

(C) The total sales after 11 months will be approximately 64,000 DVD's.

95. $x = \dfrac{4,000}{0.1p+1}, \ 10 \le p \le 70$

(A) $\dfrac{dx}{dp} = \dfrac{(0.1p+1)(0) - 4,000(0.1)}{(0.1p+1)^2} = \dfrac{-400}{(0.1p+1)^2}$

(B) $x(40) = \dfrac{4,000}{0.1(40)+1} = \dfrac{4,000}{5} = 800;$

$\dfrac{dx}{dp} = \dfrac{-400}{[0.1(40)+1]^2} = \dfrac{-400}{25} = -16$

At a price level of \$40, the demand is 800 DVD players and the demand is DECREASING at the rate of 16 CD players per dollar.

(C)   At a price of \$41, the demand will be approximately 784 CD players.

97. $C(t) = \dfrac{0.14t}{t^2+1}$

(A) $C'(t) = \dfrac{(t^2+1)(0.14t)' - (0.14t)(t^2+1)'}{(t^2+1)^2}$

$= \dfrac{(t^2+1)(0.14) - (0.14t)(2t)}{(t^2+1)^2} = \dfrac{0.14 - 0.14t^2}{(t^2+1)^2} = \dfrac{0.14(1-t^2)}{(t^2+1)^2}$

(B) $C'(0.5) = \dfrac{0.14(1-[0.5]^2)}{([0.5]^2+1)^2} = \dfrac{0.14(1-0.25)}{(1.25)^2} = 0.0672$

Interpretation: At $t = 0.5$ hours, the concentration is increasing at the rate of 0.0672 mg/cm^3 per hour.

$C'(3) = \dfrac{0.14(1-3^2)}{(3^2+1)^2} = \dfrac{0.14(-8)}{100} = -0.0112$

Interpretation: At $t = 3$ hours, the concentration is decreasing at the rate of 0.0112 mg/cm^3 per hour.

EXERCISE 11-4

Things to remember:

<u>1.</u>   COMPOSITE FUNCTIONS

A function m is a COMPOSITE of functions f and g if
$$m(x) = f[g(x)]$$
The domain of m is the set of all numbers x such that x is in the domain of g and $g(x)$ is in the domain of f.

2.   GENERAL POWER RULE

If $u(x)$ is a differentiable function, n is any real number, and

$$y = f(x) = [u(x)]^n$$

then

$$f'(x) = n[u(x)]^{n-1} u'(x)$$

This rule is often written more compactly as

$$y' = nu^{n-1}u' \quad \text{or} \quad \frac{d}{dx}(u^n) = nu^{n-1}\frac{du}{dx}, \quad u = u(x)$$

3.   THE CHAIN RULE: GENERAL FORM

If $y = f(u)$ and $u = g(x)$, define the composite function
$$y = m(x) = f[g(x)],$$

then

$$\frac{dy}{dx} = \frac{dy}{du}\frac{du}{dx} \quad \text{provided that} \quad \frac{dy}{du} \text{ and } \frac{du}{dx} \text{ exist.}$$

Or, equivalently,

$$m'(x) = f'[g(x)]g'(x) \text{ provided that } f'[g(x)] \text{ and } g'(x) \text{ exist.}$$

4.   GENERAL DERIVATIVE RULES

(a) $\dfrac{d}{dx}[f(x)]^n = n[f(x)]^{n-1}f'(x)$

(b) $\dfrac{d}{dx}\ln[f(x)] = \dfrac{1}{f(x)}f'(x)$

(c) $\dfrac{d}{dx}e^{f(x)} = e^{f(x)}f'(x)$

---

1. $y = f(u) = 3u + 5, \quad u = g(x) = x^3; \quad y = f(u) = f[g(x)] = 3x^3 + 5$

3. $y = f(u) = 2u + \ln u, \quad u = g(x) = x^2 e^x; \quad y = f(u) = f[g(x)] = 2x^2 e^x + \ln(x^2 e^x)$

5. $y = \ln(x^3 - 6x + 10)$. Let $y = E(u) = \ln u, \quad u = I(x) = x^3 - 6x + 10$.
      Then $y = E(u) = E[I(x)] = \ln(x^3 - 6x + 10)$

7. $y = \sqrt{x^2 + 4}$. Let $y = E(u) = \sqrt{u}, \quad u = I(x) = x^2 + 4$. Then $y = E(u) = E[I(x)] = \sqrt{x^2 + 4}$

9. $3; \dfrac{d}{dx}(3x + 4)^4 = 4(3x + 4)^3(3) = 12(3x + 4)^3$

11. $-4x; \dfrac{d}{dx}(4 - 2x^2)^3 = 3(4 - 2x^2)^2(-4x) = -12x(4 - 2x^2)^2$

13. $2x; \dfrac{d}{dx}(e^{x^2+1}) = e^{x^2+1}\dfrac{d}{dx}(x^2 + 1) = e^{x^2+1}(2x) = 2xe^{x^2+1}$

15. $4x^3$; $\dfrac{d}{dx}[\ln(x^4+1)] = \dfrac{1}{x^4+1}\dfrac{d}{dx}(x^4+1) = \dfrac{1}{x^4+1}(4x^3) = \dfrac{4x^3}{x^4+1}$

17. $f(x) = (5-2x)^4$

$f'(x) = 4(5-2x)^3(5-2x)' = 4(5-2x)^3(-2) = -8(5-2x)^3$

19. $f(x) = (4+0.2x)^5$

$f'(x) = 5(4+0.2x)^4(4+0.2x)' = 5(4+0.2x)^4(0.2) = (4+0.2x)^4$

21. $f(x) = (3x^2+5)^5$

$f'(x) = 5(3x^2+5)^4(3x^2+5)' = 5(3x^2+5)^4(6x) = 30x(3x^2+5)^4$

23. $f(x) = 5e^x$

$f'(x) = 5e^x + e^x(0) = 5e^x$

25. $f(x) = e^{5x}$

$f'(x) = e^{5x}(5x)' = e^{5x}(5) = 5e^{5x}$

27. $f(x) = 3e^{-6x}$

$f'(x) = 3e^{-6x}(-6x)' = 3e^{-6x}(-6) = -18e^{-6x}$

29. $f(x) = (2x-5)^{1/2}$

$f'(x) = \dfrac{1}{2}(2x-5)^{-1/2}(2x-5)' = \dfrac{1}{2}(2x-5)^{-1/2}(2) = \dfrac{1}{(2x-5)^{1/2}}$

31. $f(x) = (x^4+1)^{-2}$

$f'(x) = -2(x^4+1)^{-3}(x^4+1)' = -2(x^4+1)^{-3}(4x^3) = -8x^3(x^4+1)^{-3} = \dfrac{-8x^3}{(x^4+1)^3}$

33. $f(x) = 4 - 2\ln x$

$f'(x) = -\dfrac{2}{x}$

35. $f(x) = 3\ln(1+x^2)$

$f'(x) = 3 \cdot \dfrac{1}{1+x^2} \cdot (1+x^2)' = \dfrac{3}{1+x^2}(2x) = \dfrac{6x}{1+x^2}$

37. $f(x) = (1+\ln x)^3$

$f'(x) = 3(1+\ln x)^2(1+\ln x)' = 3(1+\ln x)^2 \cdot \dfrac{1}{x} = \dfrac{3}{x}(1+\ln x)^2$

39. $f(x) = (2x-1)^3$

$f'(x) = 3(2x-1)^2(2) = 6(2x-1)^2$

Tangent line at $x=1$: $y - y_1 = m(x-x_1)$ where $x_1 = 1$, $y_1 = f(1) = (2(1)-1)^3 = 1$, $m = f'(1)$

$= 6[2(1)-1]^2 = 6$. Thus, $y - 1 = 6(x-1)$ or $y = 6x - 5$.

The tangent line is horizontal at the value(s) of x such that $f'(x) = 0$:

$$6(2x - 1)^2 = 0$$
$$2x - 1 = 0$$
$$x = \frac{1}{2}$$

41. $f(x) = (4x - 3)^{1/2}$

$$f'(x) = \frac{1}{2}(4x - 3)^{-1/2}(4) = \frac{2}{(4x - 3)^{1/2}}$$

Tangent line at $x = 3$: $y - y_1 = m(x - x_1)$ where $x_1 = 3, y_1 = f(3) = (4 \cdot 3 - 3)^{1/2} = 3$,

$f'(3) = \dfrac{2}{(4 \cdot 3 - 3)^{1/2}} = \dfrac{2}{3}$. Thus, $y - 3 = \dfrac{2}{3}(x - 3)$ or $y = \dfrac{2}{3}x + 1$.

The tangent line is horizontal at the value(s) of x such that

$f'(x) = 0$. Since $\dfrac{2}{(4x - 3)^{1/2}} \neq 0$ for all $x \left(x \neq \dfrac{3}{4} \right)$, there are no values of x where the tangent line is horizontal.

43. $f(x) = 5e^{x^2 - 4x + 1}$

$$f'(x) = 5e^{x^2 - 4x + 1}(2x - 4) = 10(x - 2)e^{x^2 - 4x + 1}$$
Tangent line at $x = 0$: $y - y_1 = m(x - x_1)$ where $x_1 = 0$, $y_1 = f(0) = 5e$, $f'(0) = -20e$.

Thus, $y - 5e = -20ex$ or $y = -20ex + 5e$.

The tangent line is horizontal at the value(s) of x such that $f'(x) = 0$:

$$10(x - 2)e^{x^2 - 4x + 1} = 0$$
$$x - 2 = 0$$
$$x = 2$$

45. $y = 3(x^2 - 2)^4$

$$\frac{dy}{dx} = 3 \cdot 4(x^2 - 2)^3(2x) = 24x(x^2 - 2)^3$$

47. $\dfrac{d}{dt}[2(t^2 + 3t)^{-3}] = 2(-3)(t^2 + 3t)^{-4}(2t + 3) = \dfrac{-6(2t + 3)}{(t^2 + 3t)^4}$

49. $h(w) = \sqrt{w^2 + 8} = (w^2 + 8)^{1/2}$;

$$h'(w) = \frac{1}{2}(w^2 + 8)^{-1/2}(2w) = \frac{w}{(w^2 + 8)^{1/2}} = \frac{w}{\sqrt{w^2 + 8}}.$$

51. $g(x) = 4xe^{3x}$

$$g'(x) = 4x \cdot e^{3x}(3) + e^{3x} \cdot 4 = 12xe^{3x} + 4e^{3x} = 4(3x + 1)e^x$$

53. $\dfrac{d}{dx}\left[\dfrac{\ln(1 + x)}{x^3} \right] = \dfrac{x^3 \cdot \dfrac{1}{1 + x}(1) - \ln(1 + x)3x^2}{(x^3)^2} = \dfrac{x^2\left[\dfrac{x}{1 + x} - 3\ln(1 + x) \right]}{x^6} = \dfrac{x - 3(1 + x)\ln(1 + x)}{x^4(1 + x)}$

55. $F(t) = (e^{t^2+1})^3 = e^{3t^2+3}$

$F'(t) = e^{3t^2+3}(6t) = 6te^{3(t^2+1)}$

57. $y = \ln(x^2+3)^{3/2} = \dfrac{3}{2}\ln(x^2+3)$

$y' = \dfrac{3}{2} \cdot \dfrac{1}{x^2+3}(2x) = \dfrac{3x}{x^2+3}$

59. $\dfrac{d}{dw}\left[\dfrac{1}{(w^3+4)^5}\right] = \dfrac{d}{dw}[(w^3+4)^{-5}] = -5(w^3+4)^{-6}(3w^2) = \dfrac{-15w^2}{(w^3+4)^6}.$

61. $f(x) = x(4-x)^3$

$f'(x) = x[(4-x)^3]' + (4-x)^3(x)'$

$\quad = x(3)(4-x)^2(-1) + (4-x)^3(1) = (4-x)^3 - 3x(4-x)^2 = (4-x)^2[4-x-3x] = 4(4-x)^2(1-x)$

An equation for the tangent line to the graph of f at $x = 2$ is:

$y - y_1 = m(x - x_1)$ where $x_1 = 2$, $y_1 = f(x_1) = f(2) = 16$, and $m = f'(x_1) = f'(2) = -16$.

Thus, $y - 16 = -16(x - 2)$ or $y = -16x + 48$.

63. $f(x) = \dfrac{x}{(2x-5)^3}$

$f'(x) = \dfrac{(2x-5)^3(1) - x(3)(2x-5)^2(2)}{[(2x-5)^3]^2}$

$\quad = \dfrac{(2x-5)^3 - 6x(2x-5)^2}{(2x-5)^6} = \dfrac{(2x-5)-6x}{(2x-5)^4} = \dfrac{-4x-5}{(2x-5)^4}$

An equation for the tangent line to the graph of f at $x = 3$ is:

$y - y_1 = m(x - x_1)$ where $x_1 = 3$, $y_1 = f(x_1) = f(3) = 3$, and $m = f'(x_1) = f'(3) = -17$.

Thus, $y - 3 = -17(x - 3)$ or $y = -17x + 54$.

65. $f(x) = \sqrt{\ln x} = (\ln x)^{1/2}$

$f'(x) = \dfrac{1}{2}(\ln x)^{-1/2} \cdot \dfrac{1}{x} = \dfrac{1}{2x\sqrt{\ln x}}$

Tangent line at $x = e$:

$f(e) = \sqrt{\ln e} = \sqrt{1} = 1$, $f'(e) = \dfrac{1}{2e\sqrt{\ln e}} = \dfrac{1}{2e}$

$y - 1 = \dfrac{1}{2e}(x - e)$ or $y = \dfrac{1}{2e}x + \dfrac{1}{2}$

67. $f(x) = x^2(x-5)^3$

$f'(x) = x^2[(x-5)^3]' + (x-5)^3(x^2)' = x^2(3)(x-5)^2(1) + (x-5)^3(2x)$

$\quad = 3x^2(x-5)^2 + 2x(x-5)^3 = x(x-5)^2[3x + 2(x-5)] = x(x-5)^2[5x-10] = 5x(x-5)^2(x-2)$

The tangent line to the graph of f is horizontal at the values of x such that $f'(x) = 0$. Thus, we set $5x(x-5)^2(x-2) = 0$ which implies $x = 0$, $x = 2$, $x = 5$.

69. $f(x) = \dfrac{x}{(2x+5)^2}$

$f'(x) = \dfrac{(2x+5)^2(x)' - x[(2x+5)^2]'}{[(2x+5)^2]^2} = \dfrac{(2x+5)^2(1) - x(2)(2x+5)(2)}{(2x+5)^4} = \dfrac{2x+5-4x}{(2x+5)^3} = \dfrac{5-2x}{(2x+5)^3}$

The tangent line to the graph of f is horizontal at the values of x such that $f'(x) = 0$. Thus, we set

$\dfrac{5-2x}{(2x+5)^3} = 0$ which implies $5 - 2x = 0$ and $x = \dfrac{5}{2}$.

71. $f(x) = \sqrt{x^2 - 8x + 20} = (x^2 - 8x + 20)^{1/2}$

$f'(x) = \dfrac{1}{2}(x^2 - 8x + 20)^{-1/2}(2x - 8) = \dfrac{x-4}{(x^2 - 8x + 20)^{1/2}}$

The tangent line to the graph of f is horizontal at the values of x such that $f'(x) = 0$. Thus, we set

$\dfrac{x-4}{(x^2 - 8x + 20)^{1/2}} = 0$ which implies $x - 4 = 0$ and $x = 4$.

73. $f'(x) = \dfrac{1}{5(x^2+3)^4}[20(x^2+3)^3](2x) = \dfrac{8x}{x^2+3}$;

$g'(x) = 4 \cdot \dfrac{1}{x^2+3}(2x) = \dfrac{8x}{x^2+3}$

For another way to see this, recall the properties of logarithms discussed in Section 2-3:

$f(x) = \ln[5(x^2+3)^4] = \ln 5 + \ln(x^2+3)^4 = \ln 5 + 4\ln(x^2+3) = \ln 5 + g(x)$

Now $\dfrac{d}{dx}f(x) = \dfrac{d}{dx}\ln 5 + \dfrac{d}{dx}g(x) = 0 + \dfrac{d}{dx}g(x) = \dfrac{d}{dx}g(x)$

Conclusion: f and g differ by a constant; $f'(x)$ and g'(x) ARE the same function.

75. $f(u) = \ln u$, domain of $f : (0,\infty)$; $g(x) = 4 - x^2$, domain of $g : (-\infty,\infty)$

$m(x) = f[g(x)] = \ln(4 - x^2)$, domain of $m : (-2,2)$.

77. $f(u) = \dfrac{1}{u^2 - 1}$, domain of $f :$ all real numbers except $x = \pm 1$; $g(x) = \ln x$, domain of $g : (0,\infty)$

$m(x) = f[g(x)] = \dfrac{1}{(\ln x)^2 - 1}$, domain of $m :$ all real numbers except $x = e,\ e^{-1}$.

79. $\dfrac{d}{dx}[3x(x^2+1)^3] = 3x\dfrac{d}{dx}(x^2+1)^3 + (x^2+1)^3\dfrac{d}{dx}3x = 3x\cdot 3(x^2+1)^2(2x) + (x^2+1)^3(3)$

$= 18x^2(x^2+1)^2 + 3(x^2+1)^3 = (x^2+1)^2[18x^2 + 3(x^2+1)]$

$= (x^2+1)^2(21x^2+3) = 3(x^2+1)^2(7x^2+1)$

81. $\dfrac{d}{dx}\dfrac{(x^3-7)^4}{2x^3} = \dfrac{2x^3 \dfrac{d}{dx}(x^3-7)^4 - (x^3-7)^4 \dfrac{d}{dx}2x^3}{(2x^3)^2} = \dfrac{2x^3 \cdot 4(x^3-7)^3(3x^2) - (x^3-7)^4 6x^2}{4x^6}$

$\qquad = \dfrac{3(x^3-7)^3 x^2[8x^3-2(x^3-7)]}{4x^6} = \dfrac{3(x^3-7)^3(6x^3+14)}{4x^4} = \dfrac{3(x^3-7)^3(3x^3+7)}{2x^4}$

83. $\dfrac{d}{dx}\log_2(3x^2-1) = \dfrac{1}{\ln 2} \cdot \dfrac{1}{3x^2-1} \cdot 6x = \dfrac{1}{\ln 2} \cdot \dfrac{6x}{3x^2-1}$

85. $\dfrac{d}{dx}10^{x^2+x} = 10^{x^2+x}(\ln 10)(2x+1) = (2x+1)10^{x^2+x}\ln 10$

87. $\dfrac{d}{dx}\log_3(4x^3+5x+7) = \dfrac{1}{\ln 3} \cdot \dfrac{1}{4x^3+5x+7}(12x^2+5) = \dfrac{12x^2+5}{\ln 3(4x^3+5x+7)}$

89. $\dfrac{d}{dx}2^{x^3-x^2+4x+1} = 2^{x^3-x^2+4x+1}\ln 2(3x^2-2x+4) = \ln 2(3x^2-2x+4)2^{x^3-x^2+4x+1}$

91. $C(x) = 10 + \sqrt{2x+16} = 10 + (2x+16)^{1/2}, \ 0 \le x \le 50$

(A) $\quad C'(x) = \dfrac{1}{2}(2x+16)^{-1/2}(2) = \dfrac{1}{(2x+16)^{1/2}}$

(B) $\quad C'(24) = \dfrac{1}{[2(24)+16]^{1/2}} = \dfrac{1}{(64)^{1/2}} = \dfrac{1}{8}$ or \$12.50; at a production level of 24 cell phones, total costs

are INCREASING at the rate of \$12.50 per phone; also, the cost of producing the 25th phone is approximately \$12.50.

$\quad C'(42) = \dfrac{1}{[2(42)+16]^{1/2}} = \dfrac{1}{(100)^{1/2}} = \dfrac{1}{10}$ or \$10.00; at a production level of 42 cell phones, total

costs are INCREASING at the rate of \$10.00 per phone; also the cost of producing the 43rd phone is approximately \$10.00.

93. $x = 80\sqrt{p+25} - 400 = 80(p+25)^{1/2} - 400, \ 20 \le p \le 100$

(A) $\quad \dfrac{dx}{dp} = 80\left(\dfrac{1}{2}\right)(p+25)^{-1/2}(1) = \dfrac{40}{(p+25)^{1/2}}$

(B) At $p = 75$, $x = 80\sqrt{75+25} - 400 = 400$ and

$\quad \dfrac{dx}{dp} = \dfrac{40}{(75+25)^{1/2}} = \dfrac{40}{(100)^{1/2}} = 4.$

At a price of \$75, the supply is 400 bicycle helmets, and the supply is INCREASING at a rate of 4 helmet per dollar.

95. $C(t) = 4.35e^{-t}, \quad 0 \le t \le 5$

(A) $C'(t) = -4.35e^{-t}$

$C'(1) = -4.35e^{-1} \approx -1.60$

$C'(4) = -4.35e^{-4} \approx -0.08$

Thus, after one hour, the concentration is decreasing at the rate of 1.60 mg/ml per hour; after four hours, the concentration is decreasing at the rate of 0.08 mg/ml per hour.

(B) $C'(t) = -4.35e^{-t} < 0$ on (0, 5)

Thus, C is decreasing on (0, 5); there are no local extrema.

The graph of C is shown at the right.

t	$C(t)$
0	4.35
1	1.60
4	0.08
5	0.03

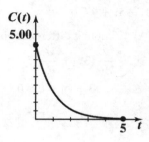

97. $P(x) = 40 + 25 \ln(x + 1)$ $0 \le x \le 65$

$P'(x) = 25\left(\dfrac{1}{x+1}\right)(1) = \dfrac{25}{x+1}$

$P'(10) = \dfrac{25}{11} \approx 2.27$

$P'(30) = \dfrac{25}{31} \approx 0.81$

$P'(60) = \dfrac{25}{61} \approx 0.41$

Thus, the rate of change of pressure at the end of 10 years is 2.27 millimeters of mercury per year; at the end of 30 years the rate of change is 0.81 millimeters of mercury per year; at the end of 60 years the rate of change is 0.41 millimeters of mercury per year.

EXERCISE 11-5

Things to remember:

1.   Let $y = y(x)$. Then

(a) $\dfrac{d}{dx} y^n = ny^{n-1}y'$ (General Power Rule)

(b) $\dfrac{d}{dx} \ln y = \dfrac{1}{y} \cdot y' = \dfrac{y'}{y}$

(c) $\dfrac{d}{dx} e^y = e^y \cdot y' = y'e^y$

1. $3x + 2y - 20 = 0;$ $2y = 20 - 3x,$ $y = 10 - \dfrac{3}{2}x$

3. $\dfrac{x^2}{9} + \dfrac{y^2}{16} = 1;$ $\dfrac{y^2}{16} = 1 - \dfrac{x^2}{9} = \dfrac{1}{9}(9 - x^2),$ $y^2 = \dfrac{16}{9}(9 - x^2),$ $y = \pm\dfrac{4}{3}\sqrt{9 - x^2}$

5. $x^2 + xy + y^2 = 1;$ $y^2 + xy + x^2 - 1 = 0,$

$$y = \frac{-x \pm \sqrt{x^2 - 4(x^2 - 1)}}{2} = \frac{-x \pm \sqrt{4 - 3x^2}}{2} \quad \text{(quadratic formula)}$$

7. $5x + 3y = e^y,$ impossible, cannot be solved for $y.$

9. $3x + 5y + 9 = 0$

(A) Implicit differentiation:

$$\frac{d}{dx}(3x) + \frac{d}{dx}(5y) + \frac{d}{dx}(9) = \frac{d}{dx}(0)$$

$$3 + 5y' + 0 = 0$$

$$y' = -\frac{3}{5}$$

(B) Solve for y:

$$5y = -9 - 3x$$

$$y = -\frac{9}{5} - \frac{3}{5}x$$

$$y' = -\frac{3}{5}$$

11. $3x^2 - 4y - 18 = 0$

(A) Implicit differentiation:

$$\frac{d}{dx}(3x^2) - \frac{d}{dx}(4y) - \frac{d}{dx}(18) = \frac{d}{dx}(0)$$

$$6x - 4y' - 0 = 0$$

$$y' = \frac{6}{4}x = \frac{3}{2}x$$

(B) Solve for y:

$$-4y = 18 - 3x^2$$

$$y = \frac{3}{4}x^2 - \frac{9}{2}$$

$$y' = \frac{6}{4}x = \frac{3}{2}x$$

13. $y - 5x^2 + 3 = 0; \ (1, 2)$

Using implicit differentiation:

$$\frac{d}{dx}(y) - \frac{d}{dx}(5x^2) + \frac{d}{dx}(3) = \frac{d}{dx}(0)$$

$$y' - 10x = 0$$

$$y' = 10x$$

$$y'\big|_{(1,2)} = 10(1) = 10$$

15. $x^2 - y^3 - 3 = 0; \ (2, 1)$

$$\frac{d}{dx}(x^2) - \frac{d}{dx}(y^3) - \frac{d}{dx}(3) = \frac{d}{dx}(0)$$

$$2x - 3y^2y' = 0$$

$$3y^2y' = 2x$$

$$y' = \frac{2x}{3y^2}$$

$$y'\big|_{(2,1)} = \frac{4}{3}$$

17. $y^2 + 2y + 3x = 0$; $(-1, 1)$

$$\frac{d}{dx}(y^2) + \frac{d}{dx}(2y) + \frac{d}{dx}(3x) = \frac{d}{dx}(0)$$

$$2yy' + 2y' + 3 = 0$$

$$2y'(y + 1) = -3$$

$$y' = -\frac{3}{2(y+1)}; \quad y'\Big|_{(-1,1)} = \frac{-3}{2(2)} = -\frac{3}{4}$$

19. $xy - 6 = 0$

$$\frac{d}{dx}xy - \frac{d}{dx}6 = \frac{d}{dx}(0)$$

$$xy' + y - 0 = 0$$

$$xy' = -y$$

$$y' = -\frac{y}{x}; \quad y' \text{ at } (2, 3) = -\frac{3}{2}$$

21. $2xy + y + 2 = 0$

$$2\frac{d}{dx}xy + \frac{d}{dx}y + \frac{d}{dx}2 = \frac{d}{dx}(0)$$

$$2xy' + 2y + y' + 0 = 0$$

$$y'(2x + 1) = -2y$$

$$y' = \frac{-2y}{2x+1}$$

$$y' \text{ at } (-1, 2) = \frac{-2(2)}{2(-1)+1} = 4$$

23. $x^2y - 3x^2 - 4 = 0$

$$\frac{d}{dx}x^2y - \frac{d}{dx}3x^2 - \frac{d}{dx}4 = \frac{d}{dx}(0)$$

$$x^2y' + y\frac{d}{dx}(x^2) - 6x - 0 = 0$$

$$x^2y' + y2x - 6x = 0$$

$$x^2y' = 6x - 2yx$$

$$y' = \frac{6x - 2yx}{x^2} \text{ or } \frac{6 - 2y}{x}$$

$$y'\Big|_{(2,4)} = \frac{6 - 2\cdot 4}{2^2} = \frac{6-8}{4} = -1$$

25. $e^y = x^2 + y^2$

$$\frac{d}{dx}e^y = \frac{d}{dx}x^2 + \frac{d}{dx}y^2$$

$$e^y y' = 2x + 2yy'$$

$$y'(e^y - 2y) = 2x$$

$$y' = \frac{2x}{e^y - 2y}$$

$$y'\bigg|_{(1,0)} = \frac{2\cdot 1}{e^0 - 2\cdot 0} = \frac{2}{1} = 2$$

27. $x^3 - y = \ln y$

$$\frac{d}{dx}x^3 - \frac{d}{dx}y = \frac{d}{dx}\ln y$$

$$3x^2 - y' = \frac{y'}{y}$$

$$3x^2 = \left(1 + \frac{1}{y}\right)y'$$

$$3x^2 = \frac{y+1}{y}y'$$

$$y' = \frac{3x^2 y}{y+1}$$

$$y'\bigg|_{(1,1)} = \frac{3\cdot 1^2 \cdot 1}{1+1} = \frac{3}{2}$$

29. $x \ln y + 2y = 2x^3$

$$\frac{d}{dx}[x \ln y] + \frac{d}{dx}2y = \frac{d}{dx}2x^3$$

$$\ln y \cdot \frac{d}{dx}x + x\frac{d}{dx}\ln y + 2y' = 6x^2$$

$$\ln y \cdot 1 + x \cdot \frac{y'}{y} + 2y' = 6x^2$$

$$y'\left(\frac{x}{y} + 2\right) = 6x^2 - \ln y$$

$$y' = \frac{6x^2 y - y \ln y}{x + 2y}$$

$$y'\bigg|_{(1,1)} = \frac{6\cdot 1^2 \cdot 1 - 1\cdot \ln 1}{1 + 2\cdot 1} = \frac{6}{3} = 2$$

31. $x^2 - t^2 x + t^3 + 11 = 0$

$$\frac{d}{dt}x^2 - \frac{d}{dt}(t^2 x) + \frac{d}{dt}t^3 + \frac{d}{dt}11 = \frac{d}{dt}0$$

$$2xx' - [t^2 x' + x(2t)] + 3t^2 + 0 = 0$$

$$2xx' - t^2 x' - 2tx + 3t^2 = 0$$

$$x'(2x - t^2) = 2tx - 3t^2$$

$$x' = \frac{2tx - 3t^2}{2x - t^2}$$

$$x'\bigg|_{(-2,1)} = \frac{2(-2)(1) - 3(-2)^2}{2(1) - (-2)^2} = \frac{-4 - 12}{2 - 4} = \frac{-16}{-2} = 8$$

33. $(x-1)^2 + (y-1)^2 = 1.$

Differentiating implicitly, we have:

$$\frac{d}{dx}(x-1)^2 + \frac{d}{dx}(y-1)^2 = \frac{d}{dx}(1)$$

$$2(x-1) + 2(y-1)y' = 0$$

$$y' = -\frac{(x-1)}{(y-1)}$$

To find the points on the graph where $x = 1.6$, we solve the given equation for y:

$$(y-1)^2 = 1 - (x-1)^2$$

$$y - 1 = \pm\sqrt{1-(x-1)^2}$$

$$y = 1 \pm \sqrt{1-(x-1)^2}$$

Now, when $x = 1.6$, $y = 1 + \sqrt{1-0.36} = 1 + \sqrt{0.64} = 1.8$ and $y = 1 - \sqrt{0.64} = 0.2$. Thus, the points are $(1.6, 1.8)$ and $(1.6, 0.2)$. These values can be verified on the graph.

$$y' \Big|_{(1.6,1.8)} = -\frac{(1.6-1)}{(1.8-1)} = -\frac{0.6}{0.8} = -\frac{3}{4}$$

$$y' \Big|_{(1.6,0.2)} = -\frac{(1.6-1)}{(0.2-1)} = -\frac{0.6}{(-0.8)} = \frac{3}{4}$$

35. $xy - x - 4 = 0$

When $x = 2$, $2y - 2 - 4 = 0$, so $y = 3$. Thus, we want to find the equation of the tangent line at $(2, 3)$.

First, find y'.

$$\frac{d}{dx}xy - \frac{d}{dx}x - \frac{d}{dx}4 = \frac{d}{dx}0$$

$$xy' + y - 1 - 0 = 0$$

$$xy' = 1 - y$$

$$y' = \frac{1-y}{x}$$

$$y' \Big|_{(2,3)} = \frac{1-3}{2} = -1$$

Thus, the slope of the tangent line at $(2, 3)$ is $m = -1$. The equation of the line through $(2, 3)$ with slope $m = -1$ is:

$(y-3) = -1(x-2)$ or $y = -x + 5$

37. $y^2 - xy - 6 = 0$

When $x = 1$,

$$y^2 - y - 6 = 0$$

$$(y - 3)(y + 2) = 0$$

$$y = 3 \text{ or } -2.$$

Thus, we want to find the equations of the tangent lines at $(1, 3)$ and $(1, -2)$. First, find y'.

$$\frac{d}{dx} y^2 - \frac{d}{dx} xy - \frac{d}{dx} 6 = \frac{d}{dx} 0$$

$$2yy' - xy' - y - 0 = 0$$

$$y'(2y - x) = y$$

$$y' = \frac{y}{2y - x}$$

$$y'\Big|_{(1,3)} = \frac{3}{2(3) - 1} = \frac{3}{5} \quad \text{[Slope at } (1, 3)\text{]}$$

The equation of the tangent line at $(1, 3)$ with $m = \dfrac{3}{5}$ is:

$$(y - 3) = \frac{3}{5}(x - 1)$$

$$y - 3 = \frac{3}{5}x - \frac{3}{5}$$

$$y = \frac{3}{5}x + \frac{12}{5}$$

$$y'\Big|_{(1,-2)} = \frac{-2}{2(-2) - 1} = \frac{2}{5} \quad \text{[Slope at } (1, -2)\text{]}$$

Thus, the equation of the tangent line at $(1, -2)$ with $m = \dfrac{2}{5}$ is:

$$(y + 2) = \frac{2}{5}(x - 1)$$

$$y + 2 = \frac{2}{5}x - \frac{2}{5}$$

$$y = \frac{2}{5}x - \frac{12}{5}$$

39. $xe^y = 1$

Implicit differentiation: $\quad x \cdot \dfrac{d}{dx} e^y + e^y \dfrac{d}{dx} x = \dfrac{d}{dx} 1$

$$xe^y y' + e^y = 0$$

$$y' = -\frac{e^y}{xe^y} = -\frac{1}{x}$$

Solve for y: $\quad e^y = \dfrac{1}{x}$

$$y = \ln\left(\frac{1}{x}\right) = -\ln x \quad \text{(see Section 2-3)}$$

$$y' = -\frac{1}{x}$$

In this case, solving for y first and then differentiating is a little easier than differentiating implicitly.

41. $(1 + y)^3 + y = x + 7$

$$\frac{d}{dx}(1 + y)^3 + \frac{d}{dx}y = \frac{d}{dx}x + \frac{d}{dx}7$$

$$3(1 + y)^2 y' + y' = 1$$

$$y'[3(1 + y)^2 + 1] = 1$$

$$y' = \frac{1}{3(1 + y)^2 + 1}$$

$$y'\Big|_{(2,1)} = \frac{1}{3(1 + 1)^2 + 1} = \frac{1}{13}$$

43. $(x - 2y)^3 = 2y^2 - 3$

$$\frac{d}{dx}(x - 2y)^3 = \frac{d}{dx}(2y^2) - \frac{d}{dx}(3)$$

$$3(x - 2y)^2(1 - 2y') = 4yy' - 0 \qquad \text{[Note:The chain rule is applied to the left-hand side.]}$$

$$3(x - 2y)^2 - 6(x - 2y)^2 y' = 4yy'$$

$$-6(x - 2y)^2 y' - 4yy' = -3(x - 2y)^2$$

$$-y'[6(x - 2y)^2 + 4y] = -3(x - 2y)^2$$

$$y' = \frac{3(x - 2y)^2}{6(x - 2y)^2 + 4y}$$

$$y'\Big|_{(1,1)} = \frac{3(1 - 2 \cdot 1)^2}{6(1 - 2)^2 + 4} = \frac{3}{10}$$

45. $\sqrt{7 + y^2} - x^3 + 4 = 0 \ \text{ or } \ (7 + y^2)^{1/2} - x^3 + 4 = 0$

$$\frac{d}{dx}(7 + y^2)^{1/2} - \frac{d}{dx}x^3 + \frac{d}{dx}4 = \frac{d}{dx}0$$

$$\frac{1}{2}(7 + y^2)^{-1/2}\frac{d}{dx}(7 + y^2) - 3x^2 + 0 = 0$$

$$\frac{1}{2}(7 + y^2)^{-1/2}2yy' - 3x^2 = 0$$

$$\frac{yy'}{(7 + y^2)^{1/2}} = 3x^2$$

$$y' = \frac{3x^2(7 + y^2)^{1/2}}{y}$$

$$y'\Big|_{(2,3)} = \frac{3 \cdot 2^2(7 + 3^2)^{1/2}}{3} = \frac{12(16)^{1/2}}{3} = 16$$

47. $\ln(xy) = y^2 - 1$

$$\frac{d}{dx}[\ln(xy)] = \frac{d}{dx}y^2 - \frac{d}{dx}1$$

$$\frac{1}{xy} \cdot \frac{d}{dx}(xy) = 2yy'$$

$$\frac{1}{xy}(xy' + y) = 2yy'$$

$$\frac{1}{y} \cdot y' - 2yy' + \frac{1}{x} = 0$$

$$xy' - 2xy^2y' + y = 0$$

$$y'(x - 2xy^2) = -y$$

$$y' = \frac{-y}{x - 2xy^2} = \frac{y}{2xy^2 - x}$$

$$y'\Big|_{(1,1)} = \frac{1}{2 \cdot 1 \cdot 1^2 - 1} = 1$$

49.   First find point(s) on the graph of the equation with abscissa $x = 1$:
Setting $x = 1$, we have
$$y^3 - y - 1 = 2 \text{ or } y^3 - y - 3 = 0$$
Graphing this equation on a graphing utility, we get $y \approx 1.67$.

Now, differentiate implicitly to find the slope of the tangent line at the point (1, 1.67):
$$3y^2y' - \qquad\qquad 3y^2y' - xy' - y - 3x^2 = 0$$

$$(3y^2 - x)y' = 3x^2 + y$$

$$y' = \frac{3x^2 + y}{3y^2 - x};$$

$$y'\Big|_{(1,1.67)} = \frac{3 + 1.67}{3(1.67)^2 - 1} = \frac{4.67}{7.37} \approx 0.63$$

Tangent line: $y - 1.67 = 0.63(x - 1)$ or $y = 0.63x + 1.04$

51. $x = p^2 - 2p + 1000$

$$\frac{d(x)}{dx} = \frac{d(p^2)}{dx} - \frac{d(2p)}{dx} + \frac{d(1000)}{dx}$$

$$1 = 2p\frac{dp}{dx} - 2\frac{dp}{dx} + 0$$

$$1 = (2p - 2)\frac{dp}{dx}$$

Thus, $\dfrac{dp}{dx} = p' = \dfrac{1}{2p - 2}$.

53. $x = \sqrt{10,000 - p^2} = (10,000 - p^2)^{1/2}$

$$\frac{d}{dx} x = \frac{d}{dx} (10,000 - p^2)^{1/2}$$

$$1 = \frac{1}{2} (10,000 - p^2)^{-1/2} \frac{d}{dx} [10,000 - p^2]$$

$$1 = \frac{1}{2(10,000 - p^2)^{1/2}} \cdot (-2pp')$$

$$1 = \frac{-pp'}{\sqrt{10,000 - p^2}}$$

$$p' = \frac{-\sqrt{10,000 - p^2}}{p}$$

55. $(L + m)(V + n) = k$

$$(L + m)(1) + (V + n) \frac{dL}{dV} = 0$$

$$\frac{dL}{dV} = -\frac{(L + m)}{V + n}$$

57. $v = \sqrt{kT} = \sqrt{k}\, T^{1/2}$

$$\frac{d}{dv}(v) = \frac{d}{dv} \left(\sqrt{k}\, T^{1/2} \right)$$

$$1 = \frac{1}{2} \sqrt{k}\, T^{-1/2} \cdot \frac{dT}{dv} = \frac{\sqrt{k}}{2\sqrt{T}} \cdot \frac{dT}{dv}$$

$$\frac{dT}{dv} = \frac{2\sqrt{T}}{\sqrt{k}}$$

59. $v = \sqrt{kT} = \sqrt{k}\, T^{1/2}$

$$\frac{dv}{dT} = \frac{1}{2} \sqrt{k}\, T^{-1/2} = \frac{\sqrt{k}}{2\sqrt{T}}$$

$\dfrac{dT}{dv}$ is the reciprocal of $\dfrac{dv}{dT}$;   that is $\dfrac{dT}{dv} = \dfrac{1}{dv / dT}$

EXERCISE 11-6

Things to remember:

1.   SUGGESTIONS FOR SOLVING RELATED RATE PROBLEMS

Step 1.   Sketch a figure.

Step 2.   Identify all relevant variables, including those whose rates are given and those whose rates are to be found.

Step 3.   Express all given rates and rates to be found as derivatives.

Step 4.   Find an equation connecting the variables in Step 2.

Step 5.   Implicitly differentiate the equation found in Step 4, using the chain rule

where appropriate,
and substitute in all given values.

Step 6.   Solve for the derivative that will give the unknown
rate.

---

1. $A = \pi r^2;\quad \pi r^2 = 300,\quad r^2 = \dfrac{300}{\pi},\quad r = \sqrt{\dfrac{300}{\pi}} \approx 9.77$ ft, diameter ≈ 19.5 ft.

3. $a^2 + b^2 = c^2,\ a = 20,\ c = 50;\quad b = \sqrt{(50)^2 - (20)^2} = \sqrt{2100} \approx 46$ m.

5.   Let h be the height of the streetlight. The distance from the base of the streetlight to the tip of the shadow is 40 ft $+ 96$ in $= 48$ ft. By similar triangles:

$\dfrac{h}{48} = \dfrac{69}{96};\quad h = 48 \cdot \dfrac{69}{96} = \dfrac{69}{2} = 34.5$ ft.

7.   Sphere: $V = \dfrac{4}{3}\pi r^3 = \dfrac{4}{3}\pi(12)^3 = 2304\pi.$ Cylinder: $V = 2304\pi(2) = 4608\pi$

$V = \pi r^2 h = \pi(12)^2 h = 4608\pi;\quad h = \dfrac{4608\pi}{144\pi} = 32$ ft.

9. $y = x^2 + 2$

Differentiating with respect to t:

$\dfrac{dy}{dt} = 2x\dfrac{dx}{dt};\quad \dfrac{dy}{dt} = 2(5)(3) = 30$ when $x = 5,\ \dfrac{dx}{dt} = 3$

11. $x^2 + y^2 = 1$

Differentiating with respect to t:

$2x\dfrac{dx}{dt} + 2y\dfrac{dy}{dt} = 0$

$2x\dfrac{dx}{dt} = -2y\dfrac{dy}{dt}$

$\dfrac{dx}{dt} = -\dfrac{y}{x}\dfrac{dy}{dt};\quad \dfrac{dx}{dt} = -\dfrac{0.8}{(-0.6)}(-4) = -\dfrac{16}{3},$

when $x = -0.6,\ y = 0.8,\ \dfrac{dy}{dt} = -4$

13. $x^2 + 3xy + y^2 = 11$

Differentiating with respect to t:

$2x\dfrac{dx}{dt} + 3x\dfrac{dy}{dt} + 3y\dfrac{dx}{dt} + 2y\dfrac{dy}{dt} = 0$

$(3x + 2y)\dfrac{dy}{dt} = -(2x + 3y)\dfrac{dx}{dt}$

$$\frac{dy}{dt} = -\frac{(2x+3y)}{3x+2y}\frac{dx}{dt}; \quad \frac{dy}{dt} = -\frac{(2\cdot1+3\cdot2)}{(3\cdot1+2\cdot2)}2 = -\frac{16}{7}$$

when $x = 1$, $y = 2$, $\frac{dx}{dt} = 2$

15. $xy = 36$

Differentiate with respect to t:

$$\frac{d(xy)}{dt} = \frac{d(36)}{dt}$$

$$x\frac{dy}{dt} + y\frac{dx}{dt} = 0$$

Given: $\frac{dx}{dt} = 4$ when $x = 4$ and $y = 9$. Therefore,

$$4\frac{dy}{dt} + 9(4) = 0$$

$$4\frac{dy}{dt} = -36 \quad \text{and} \quad \frac{dy}{dt} = -9.$$

The y coordinate is decreasing at 9 units per second.

17.

From the triangle,

$$x^2 + y^2 = z^2$$

or $x^2 + 16 = z^2$, since $y = 4$.

Differentiate with respect to t:

$$2x\frac{dx}{dt} = 2z\frac{dz}{dt}$$

or $x\frac{dx}{dt} = z\frac{dz}{dt}$

Given: $\frac{dz}{dt} = -3$. Also, when $x = 30$, $900 + 16 = z^2$ or $z = \sqrt{916}$.

Therefore,

$$30\frac{dx}{dt} = \sqrt{916}\,(-3) \quad \text{and} \quad \frac{dx}{dt} = \frac{-3\sqrt{916}}{30} = \frac{-\sqrt{916}}{10} \approx \frac{-30.27}{10} \approx -3.03 \text{ feet/second.}$$

[Note: The negative sign indicates that the distance between the boat and the dock is decreasing.]

19. Area: $A = \pi R^2$

$$\frac{dA}{dt} = \frac{d\pi R^2}{dt} = \pi\cdot2R\frac{dR}{dt}$$

Given: $\frac{dR}{dt} = 2$ ft/sec

$$\frac{dA}{dt} = 2\pi R\cdot2 = 4\pi R$$

21. $V = \frac{4}{3}\pi R^3$

$$\frac{dV}{dt} = \frac{4}{3}\pi3R^2\frac{dR}{dt} = 4\pi R^2\frac{dR}{dt}$$

Given: $\frac{dR}{dt} = 3$ cm/min

$$\frac{dV}{dt} = 4\pi R^2 3 = 12\pi R^2$$

$$\frac{dA}{dt}\bigg|_{R=10\,\text{ft}} = 4\pi(10) = 40\pi \ \text{ft}^2/\text{sec}$$
$$\approx 126 \ \text{ft}^2/\text{sec}$$

$$\frac{dV}{dt}\bigg|_{R=10\,\text{cm}} = 12\pi(10)^2 = 1200\pi$$
$$\approx 3770 \ \text{cm}^3/\text{min}$$

23. $\dfrac{P}{T} = k$ $\qquad\qquad$ (1)

$P = kT$

Differentiate with respect to t:

$$\frac{dP}{dt} = k\frac{dT}{dt}$$

Given: $\dfrac{dT}{dt} = 3$ degrees per hour, $T = 250°$, $P = 500$ pounds per square inch.

From (1), for $T = 250$ and $P = 500$,

$$k = \frac{500}{250} = 2.$$

Thus, we have

$$\frac{dP}{dt} = 2\frac{dT}{dt}$$

$$\frac{dP}{dt} = 2(3) = 6$$

Pressure increases at 6 pounds per square inch per hour.

25.   By the Pythagorean theorem,

$$x^2 + y^2 = 10^2$$

or $\ x^2 + y^2 = 100$ $\qquad$ (1)

Differentiate with respect to t:

$$2x\frac{dx}{dt} + 2y\frac{dy}{dt} = 0$$

Therefore, $\dfrac{dy}{dt} = -\dfrac{x}{y}\dfrac{dx}{dt}$. Given: $\dfrac{dx}{dt} = 3$. Thus, $\dfrac{dy}{dt} = \dfrac{-3x}{y}$.

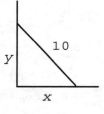

From (1), $y^2 = 100 - x^2$ and, when $x = 6$,

$y^2 = 100 - 6^2$
$\quad = 100 - 36 = 64.$

Thus, $y = 8$ when $x = 6$, and

$$\frac{dy}{dt}\bigg|_{(6,8)} = \frac{-3(6)}{8} = \frac{-18}{8} = \frac{-9}{4} \ \text{ft/sec.}$$

27. $y =$ length of shadow
$x =$ distance of man from light
$z =$ distance of tip of shadow from light

We want to compute $\dfrac{dz}{dt}$. Triangles ABE and CDE are similar triangles; thus,

the ratios of corresponding sides are equal.

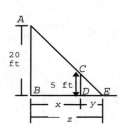

Therefore, $\dfrac{z}{20} = \dfrac{y}{5} = \dfrac{z-x}{5}$ [Note: $y = z - x$.]

or $\dfrac{z}{20} = \dfrac{z-x}{5}$

$\qquad z = 4(z - x)$

$\qquad z = 4z - 4x$

$\qquad 4x = 3z$

Differentiate with respect to t:

$4\dfrac{dx}{dt} = 3\dfrac{dz}{dt}$

$\dfrac{dz}{dt} = \dfrac{4}{3}\dfrac{dx}{dt}$

Given: $\dfrac{dx}{dt} = 5$. Thus, $\dfrac{dz}{dt} = \dfrac{4}{3}(5) = \dfrac{20}{3}$ ft/sec.

29. $V = \dfrac{4}{3}\pi r^3$ (1)

Differentiate with respect to t:

$\dfrac{dV}{dt} = 4\pi r^2 \dfrac{dr}{dt}$ and $\dfrac{dr}{dt} = \dfrac{1}{4\pi r^2}\cdot\dfrac{dV}{dt}$

Since $\dfrac{dV}{dt} = 4$ cu ft/sec,

$\dfrac{dr}{dt} = \dfrac{1}{4\pi r^2}(4) = \dfrac{1}{\pi r^2}$ ft/sec   (2)

At $t = 1$ minute = 60 seconds,

$\quad V = 4(60) = 240$ cu ft and, from (1),

$\quad r^3 = \dfrac{3V}{4\pi} = \dfrac{3(240)}{4\pi} = \dfrac{180}{\pi}; r = \left(\dfrac{180}{\pi}\right)^{1/3} \approx 3.855.$

From (2)

$\dfrac{dr}{dt} = \dfrac{1}{\pi(3.855)^2} \approx 0.0214$ ft/sec

At $t = 2$ minutes = 120 seconds,

$\quad V = 4(120) = 480$ cu ft and

$\quad r^3 = \dfrac{3V}{4\pi} = \dfrac{3(480)}{4\pi} = \dfrac{360}{\pi}; r = \left(\dfrac{360}{\pi}\right)^{1/3} \approx 4.857$

From (2),

$\dfrac{dr}{dt} = \dfrac{1}{\pi(4.857)^2} \approx 0.0135$ ft/sec

To find the time at which $\dfrac{dr}{dt} = 100$ ft/sec, solve

$$\frac{1}{\pi r^2} = 100$$

$$r^2 = \frac{1}{100\pi} \quad \text{and} \quad r = \frac{1}{\sqrt{100\pi}} = \frac{1}{10\sqrt{\pi}}$$

Now, when $r = \dfrac{1}{10\sqrt{\pi}}$,

$$V = \frac{4}{3}\pi\left(\frac{1}{10\sqrt{\pi}}\right)^3 = \frac{4}{3}\cdot\frac{1}{1000\sqrt{\pi}} = \frac{1}{750\sqrt{\pi}}$$

Since the volume at time t is $4t$, we have

$$4t = \frac{1}{750\sqrt{\pi}} \quad \text{and} \quad t = \frac{1}{3000\sqrt{\pi}} \approx 0.00019 \text{ secs.}$$

31. $y = e^x + x + 1;\ \dfrac{dx}{dt} = 3.$

Differentiate with respect to t:

$$\frac{dy}{dt} = e^x\frac{dx}{dt} + \frac{dx}{dt} = e^x(3) + 3 = 3(e^x + 1)$$

To find where the point crosses the x axis, use a graphing utility to solve

$e^x + x + 1 = 0$

The result is $x \approx -1.278$.

Now, at $x = -1.278$,

$$\frac{dy}{dt} = 3(e^{-1.278} + 1) \approx 3.835 \text{ units/sec.}$$

33. $C = 90{,}000 + 30x$ (1)

$R = 300x - \dfrac{x^2}{30}$ (2)

$P = R - C$ (3)

(A)   Differentiating (1) with respect to t:

$$\frac{dC}{dt} = \frac{d(90{,}000)}{dt} + \frac{d(30x)}{dt}$$

$$\frac{dC}{dt} = 30\frac{dx}{dt}$$

Thus, $\dfrac{dC}{dt} = 30(500) \quad \left(\dfrac{dx}{dt} = 500\right)$

$= \$15{,}000$ per week.

Costs are increasing at the rate of \$15,000 per week at this production level.

(B)   Differentiating (2) with respect to t:

$$\frac{dR}{dt} = \frac{d(300x)}{dt} - \frac{d\frac{x^2}{30}}{dt}$$

$$= 300\frac{dx}{dt} - \frac{2x}{30}\frac{dx}{dt}$$

$$= \left(300 - \frac{x}{15}\right)\frac{dx}{dt}$$

Thus, $\frac{dR}{dt} = \left(300 - \frac{6000}{15}\right)(500)$ $\left(x = 6000, \frac{dx}{dt} = 500\right)$

$$= (-100)500 = -50,000.$$

Revenue is decreasing at the rate of \$50,000 per week at this production level.

(C)   Differentiating (3) with respect to t:

$$\frac{dP}{dt} = \frac{dR}{dt} - \frac{dC}{dt}$$

Thus, from parts (A) and (B), we have:

$$\frac{dP}{dt} = -50,000 - 15,000 = -\$65,000$$

Profits are decreasing at the rate of \$65,000 per week at this production level.

35. $s = 60,000 - 40,000e^{-0.0005x}$

Differentiating implicitly with respect to t, we have

$$\frac{ds}{dt} = -40,000(-0.0005)e^{-0.0005x}\frac{dx}{dt} \text{ and } \frac{ds}{dt} = 20e^{-0.0005x}\frac{dx}{dt}$$

Now, for $x = 2000$ and $\frac{dx}{dt} = 300,$ we have

$$\frac{ds}{dt} = 20(300)e^{-0.0005(2000)}$$

$$= 6000e^{-1} = 2,207$$

Thus, sales are increasing at the rate of \$2,207 per week.

37.   Price p and demand x are related by the equation

$$2x^2 + 5xp + 50p^2 = 80,000 \tag{1}$$

Differentiating implicitly with respect to t, we have

$$4x\frac{dx}{dt} + 5x\frac{dp}{dt} + 5p\frac{dx}{dt} + 100p\frac{dp}{dt} = 0 \tag{2}$$

(A) From (2), $\dfrac{dx}{dt} = \dfrac{-(5x+100p)\frac{dp}{dt}}{4x+5p}$

Setting $p = 30$ in (1), we get

$$2x^2 + 150x + 45,000 = 80,000$$

or $x^2 + 75x - 17,500 = 0$

Thus, $x = \dfrac{-75 \pm \sqrt{(75)^2 + 70,000}}{2} = \dfrac{-75 \pm 275}{2} = 100,\ -175$

Since $x \geq 0,\ \ x = 100$.

Now, for $x = 100,\ p = 30$ and $\dfrac{dp}{dt} = 2$, we have

$\dfrac{dx}{dt} = \dfrac{-[5(100) + 100(30)] \cdot 2}{4(100) + 5(30)} = -\dfrac{7000}{550}$ and $\dfrac{dx}{dt} = -12.73$

The demand is decreasing at the rate of 12.73 units/month.

(B) From (2), $\dfrac{dp}{dt} = \dfrac{-(4x + 5p)\dfrac{dx}{dt}}{(5x + 100p)}$

Setting $x = 150$ in (1), we get

$\qquad 45,000 + 750p + 50p^2 = 80,000$

or $\qquad p^2 + 15p - 700 = 0$

and $p = \dfrac{-15 \pm \sqrt{225 + 2800}}{2} = \dfrac{-15 \pm 55}{2} = -35,\ 20$

Since $p \geq 0,\ p = 20$.

Now, for $x = 150,\ p = 20$ and $\dfrac{dx}{dt} = -6$, we have

$\dfrac{dp}{dt} = -\dfrac{[4(150) + 5(20)](-6)}{5(150) + 100(20)} = \dfrac{4200}{2750} \approx 1.53$

Thus, the price is increasing at the rate of $1.53 per month.

39. Volume $V = \pi R^2 h$, where h = thickness of the circular oil slick.

Since $h = 0.1 = \dfrac{1}{10}$, we have:

$V = \dfrac{\pi}{10} R^2$

Differentiating with respect to t:

$\dfrac{dV}{dt} = \dfrac{d\left(\dfrac{\pi}{10} R^2\right)}{dt} = \dfrac{\pi}{10} 2R \dfrac{dR}{dt} = \dfrac{\pi}{5} R \dfrac{dR}{dt}$

Given: $\dfrac{dR}{dt} = 0.32$ when $R = 500$. Therefore,

$\dfrac{dV}{dt} = \dfrac{\pi}{5} (500)(0.32) = 100\pi(0.32) \approx 100.53$ cubic feet per minute.

Things to remember:

<u>1</u>.   RELATIVE AND PERCENTAGE RATES OF CHANGE

The RELATIVE RATE OF CHANGE of a function $f(x)$ is $\dfrac{f'(x)}{f(x)}$.

The PERCENTAGE RATE OF CHANGE is $100 \times \dfrac{f'(x)}{f(x)}$.

<u>2</u>.   ELASTICITY OF DEMAND

If price and demand are related by $x = f(p)$, then the ELASTICITY OF DEMAND is given by

$$E(p) = -\frac{pf'(p)}{f(p)}$$

<u>3</u>.   INTERPRETATION OF ELASTICITY OF DEMAND

$E(p)$	Demand	Interpretation
$0 < E(p) < 1$	Inelastic	Demand is not sensitive to changes in price. A change in price produces a smaller change in demand.
$E(p) > 1$	Elastic	Demand is sensitive to changes in price. A change in price produces a larger change in demand.
$E(p) = 1$	Unit	A change in price produces the same change in demand.

<u>4</u>.   REVENUE AND ELASTICITY OF DEMAND

If $R(p) = pf(p)$ is the revenue function, then $R'(p)$ and $[1 - E(p)]$ always have the same sign.

Demand is inelastic $[E(p) < 1, R'(p) > 0]$:

   A price increase will increase revenue.

   A price decrease will decrease revenue.

Demand is elastic $[E(p) > 1, R'(p) < 0]$:

   A price increase will decrease revenue.

   A price decrease will increase revenue.

1. $p = 42 - 0.4x,\ \ 0 \le x \le 105;\ \ \ x = f(p) = \dfrac{42 - p}{0.4} = 105 - 2.5p,\ \ 0 \le p \le 42.$

3. $p = 50 - 0.5x^2,\ \ 0 \le x \le 10;\ \ \ x^2 = \dfrac{50 - p}{0.5} = 100 - 2p,\ \ \ x = f(p) = \sqrt{100 - 2p},\ \ 0 \le p \le 50.$

5. $p = 25e^{-x/20},\ \ 0 \le x \le 20;\ \ e^{-x/20} = \dfrac{p}{25},\ \ -\dfrac{x}{20} = \ln\left(\dfrac{p}{25}\right) = \ln p - \ln 25,$

   $x = f(p) = 20(\ln 25 - \ln p),\ \ \dfrac{25}{e} \approx 9.2 \le p \le 25.$

7. $p = 80 - 10\ln x,\ \ 1 \le x \le 30;\ \ \ln x = \dfrac{80 - p}{10} = 8 - 0.1p,\ \ x = e^{8-0.1p},$

   $x = f(p) = e^{8-0.1p},\ \ 80 - 10\ln 30 \approx 46 \le p \le 80.$

9. $f(x) = 35x - 0.4x^2$

$f'(x) = 35 - 0.8x$

Relative rate of change of f: $\dfrac{f'(x)}{f(x)} = \dfrac{35 - 0.8x}{35x - 0.4x^2}$

11. $f(x) = 7 + 4e^{-x}$

$f'(x) = -4e^{-x}$

Relative rate of change of f: $\dfrac{f'(x)}{f(x)} = \dfrac{-4e^{-x}}{7 + 4e^{-x}}$

13. $f(x) = 12 + 5\ln x$

$f'(x) = \dfrac{5}{x}$

Relative rate of change of f: $\dfrac{f'(x)}{f(x)} = \dfrac{5/x}{12 + 5\ln x} = \dfrac{5}{x(12 + 5\ln x)}$

15. $f(x) = 45, \quad f'(x) = 0$

$\dfrac{f'(x)}{f(x)} = 0$; relative rate of change of f at $x = 100$: 0

17. $f(x) = 420 - 5x, \quad f'(x) = -5$

$\dfrac{f'(x)}{f(x)} = \dfrac{-5}{420 - 5x}$; relative rate of change of f at $x = 25$: $\dfrac{-5}{420 - 5(25)} = \dfrac{-5}{295} \approx -0.017$

19. $f(x) = 420 - 5x, \quad f'(x) = -5$

$\dfrac{f'(x)}{f(x)} = \dfrac{-5}{420 - 5x}$; relative rate of change of f at $x = 55$: $\dfrac{-5}{420 - 5(55)} = \dfrac{-5}{145} \approx -0.034$

21. $f(x) = 4x^2 - \ln x, \quad f'(x) = 8x - \dfrac{1}{x}$

$\dfrac{f'(x)}{f(x)} = \dfrac{8x - \dfrac{1}{x}}{4x^2 - \ln x}$, relative rate of change of f at $x = 2$: $\dfrac{16 - \dfrac{1}{2}}{16 - \ln 2} \approx 1.013$

23. $f(x) = 4x^2 - \ln x, \quad f'(x) = 8x - \dfrac{1}{x}$

$\dfrac{f'(x)}{f(x)} = \dfrac{8x - \dfrac{1}{x}}{4x^2 - \ln x}$, relative rate of change of f at $x = 5$: $\dfrac{40 - \dfrac{1}{5}}{100 - \ln 5} \approx 0.405$

25. $f(x) = 225 + 65x, \quad f'(x) = 65$

$\dfrac{f'(x)}{f(x)} = \dfrac{65}{225 + 65x}$; percentage rate of change at $x = 5$: $100 \cdot \dfrac{65}{225 + 65(5)} = 100 \cdot \dfrac{65}{550} \approx 11.8$; 11.8%

27. $f(x) = 225 + 65x, \quad f'(x) = 65$

$\dfrac{f'(x)}{f(x)} = \dfrac{65}{225 + 65x}$; percentage rate of change at $x = 15$: $100 \cdot \dfrac{65}{225 + 65(15)} = 100 \cdot \dfrac{65}{1200} \approx 5.4$; 5.4%

29. $f(x) = 5,100 - 3x^2$, $\ f'(x) = -6x$

$$\frac{f'(x)}{f(x)} = \frac{-6x}{5,100 - 3x^2};\ \text{ percentage rate of change at } x = 35:$$

$$100 \cdot \frac{-6(35)}{5,100 - 3(35)^2} = 100 \cdot \frac{-210}{1425} \approx -14.7;\ \ -14.7\%$$

31. $f(x) = 5,100 - 3x^2$, $\ f'(x) = -6x$

$$\frac{f'(x)}{f(x)} = \frac{-6x}{5,100 - 3x^2};\ \text{ percentage rate of change at } x = 41:$$

$$100 \cdot \frac{-6(41)}{5,100 - 3(41)^2} = 100 \cdot \frac{-246}{57} \approx -431.6;\ \ -431.6\%$$

33. $x = f(p) = 25,000 - 450p$, $\ f'(p) = -450$

$$E(p) = \frac{-p\, f'(p)}{f(p)} = \frac{450p}{25,000 - 450p} = \frac{9p}{5,000 - 9p}$$

35. $x = f(p) = 4,800 - 4p^2$, $\ f'(p) = -8p$

$$E(p) = \frac{-p\, f'(p)}{f(p)} = \frac{8p^2}{4,800 - 4p^2} = \frac{2p^2}{1,200 - p^2}$$

37. $x = f(p) = 98 - 0.6e^p$, $\ f'(p) = -0.6e^p$

$$E(p) = \frac{-p\, f'(p)}{f(p)} = \frac{0.6p\, e^p}{98 - 0.6e^p}$$

39. $A(t) = 500\, e^{0.07t}$, $\ A'(t) = 35\, e^{0.07t}$, $\ \dfrac{A'(t)}{A(t)} = \dfrac{35\, e^{0.07t}}{500\, e^{0.07t}} = \dfrac{35}{500} = 0.07$

41. $A(t) = 3,500\, e^{0.15t}$, $\ A'(t) = 525\, e^{0.15t}$, $\ \dfrac{A'(t)}{A(t)} = \dfrac{525\, e^{0.15t}}{3,500\, e^{0.15t}} = \dfrac{525}{3,500} = 0.15$

43. $f(x) = xe^x$, $\ f'(x) = xe^x + e^x$; $\ \dfrac{f'(x)}{f(x)} = \dfrac{xe^x + e^x}{xe^x} = \dfrac{x+1}{x}$

45. $f(x) = \ln x$, $\ f'(x) = \dfrac{1}{x}$; $\ \dfrac{f'(x)}{f(x)} = \dfrac{1/x}{\ln x} = \dfrac{1}{x \ln x}$

47. $x = f(p) = 12,000 - 10p^2$
$f'(p) = -20p$

Elasticity of demand: $E(p) = \dfrac{-pf'(p)}{f(p)} = \dfrac{20p^2}{12,000 - 10p^2}$

(A) At $p = 10$: $E(10) = \dfrac{2000}{12,000 - 1000} = \dfrac{2000}{11,000} = \dfrac{2}{11}$

   Demand is inelastic.

(B) At $p = 20$: $E(20) = \dfrac{8000}{12,000 - 4000} = \dfrac{8000}{8000} = 1$; unit elasticity.

(C) At $p = 30$: $E(30) = \dfrac{18,000}{12,000 - 9,000} = \dfrac{18,000}{3,000} = 6$

Demand is elastic.

49. $x = f(p) = 950 - 2p - 0.1p^2$
$f'(p) = -2 - 0.2p$

Elasticity of demand: $E(p) = \dfrac{-pf'(p)}{f(p)} = \dfrac{2p + 0.2p^2}{950 - 2p - 0.1p^2}$

(A) At $p = 30$: $E(30) = \dfrac{60 + 180}{950 - 60 - 90} = \dfrac{240}{800} = \dfrac{3}{10}$

Demand is inelastic.

(B) At $p = 50$: $E(50) = \dfrac{100 + 500}{950 - 100 - 250} = \dfrac{600}{600} = 1$; unit elasticity.

(C) At $p = 70$: $E(70) = \dfrac{140 + 980}{950 - 140 - 490} = \dfrac{1120}{320} = 3.5$

Demand is elastic.

51. $p + 0.005x = 30$

(A) $x = \dfrac{30 - p}{0.005} = 6000 - 200p,\ 0 \le p \le 30$

(B) $f(p) = 6000 - 200p$
$f'(p) = -200$

Elasticity of demand: $E(p) = \dfrac{-pf'(p)}{f(p)} = \dfrac{200p}{6000 - 200p} = \dfrac{p}{30 - p}$

(C) At $p = 10$: $E(10) = \dfrac{10}{30 - 10} = \dfrac{1}{2} = 0.5$

If the price increases by 10%, the demand will decrease by approximately $0.5(10\%) = 5\%$.

(D) At $p = 25$: $E(25) = \dfrac{25}{30 - 25} = 5$

If the price increases by 10%, the demand will decrease by approximately $5(10\%) = 50\%$.

(E) At $p = 15$: $E(15) = \dfrac{15}{30 - 15} = 1$

If the price increases by 10%, the demand will decrease by approximately 10%.

53. $0.02x + p = 60$

(A) $x = \dfrac{60 - p}{0.02} = 3000 - 50p,\ 0 \le p \le 60$

(B) $R(p) = p(3000 - 50p) = 3000p - 50p^2$

(C) $f(p) = 3000 - 50p$
       $f'(p) = -50$

   Elasticity of demand: $E(p) = \dfrac{-pf'(p)}{f(p)} = \dfrac{50p}{3000 - 50p} = \dfrac{p}{60 - p}$

(D)   Elastic: $E(p) = \dfrac{p}{60 - p} > 1$

$$p > 60 - p$$
$$p > 30, \quad 30 < p < 60$$

   Inelastic: $E(p) = \dfrac{p}{60 - p} < 1$

$$p < 60 - p$$
$$p < 30, \quad 0 < p < 30$$

(E) $R'(p) = f(p)\,[1 - E(p)]$
       $R'(p) > 0$ if $E(p) < 1$; $R'(p) < 0$ if $E(p) > 1$
       Therefore, revenue is increasing for $0 < p < 30$ and decreasing for $30 < p < 60$.

(F)   If $p = \$10$ and the price is decreased, revenue will also decrease.

(G)   If $p = \$40$ and the price is decreased, revenue will increase.

55. $x = f(p) = 210 - 30p, \ 0 < p < 7; \ f'(p) = -30$

$$E(p) = \dfrac{-pf'(p)}{f(p)} = \dfrac{30p}{210 - 30p} = \dfrac{p}{7 - p}$$

Elastic:

$$E(p) = \dfrac{p}{7 - p} > 1$$
$$p > 7 - p$$
$$2p > 7$$
$$p > 3.5$$

Demand is elastic for $3.5 < p < 7$; demand is inelastic for $0 < p < 3.5$.

57. $x = f(p) = 3{,}125 - 5p^2, \ 0 < p < 25; \ f'(p) = -10p$

$$E(p) = \dfrac{-pf'(p)}{f(p)} = \dfrac{10p^2}{3{,}125 - 5p^2} = \dfrac{2p^2}{625 - p^2}$$

Elastic:

$$E(p) = \frac{2p^2}{625 - p^2} > 1$$

$$2p^2 > 625 - p^2$$

$$3p^2 > 625$$

$$p^2 > \frac{625}{3}$$

$$p > \frac{25}{\sqrt{3}} = \frac{25\sqrt{3}}{3}$$

Demand is elastic for $\dfrac{25\sqrt{3}}{3} < p < 25$; demand is inelastic for $0 < p < \dfrac{25\sqrt{3}}{3}$.

59. $x = f(p) = \sqrt{144 - 2p}$, $0 \le p \le 72$

$$f'(p) = \frac{1}{2}(144 - 2p)^{-1/2}(-2) = \frac{-1}{\sqrt{144 - 2p}}$$

Elasticity of demand: $E(p) = \dfrac{p}{144 - 2p}$

Elastic: $E(p) = \dfrac{p}{144 - 2p} > 1$

$$p > 144 - 2p$$

$$3p > 144$$

$$p > 48, \quad 48 < p < 72$$

Inelastic: $E(p) = \dfrac{p}{144 - 2p} < 1$

$$p < 144 - 2p$$

$$3p < 144$$

$$p < 48, \quad 0 < p < 48$$

61. $x = f(p) = \sqrt{2,500 - 2p^2}$ $0 \le p \le 25\sqrt{2}$

$$f'(p) = \frac{1}{2}(2,500 - 2p^2)^{-1/2}(-4p) = \frac{-2p}{(2,500 - 2p^2)^{1/2}}$$

Elasticity of demand: $E(p) = \dfrac{2p^2}{2,500 - 2p^2} = \dfrac{p^2}{1,250 - p^2}$

Elastic: $E(p) = \dfrac{p^2}{1,250 - p^2} > 1$

$$p^2 > 1,250 - p^2$$

$$2p^2 > 1,250$$

$$p^2 > 625$$

$$p > 25, \quad 25 < p < 25\sqrt{2}$$

Inelastic: $E(p) = \dfrac{p^2}{1,250 - p^2} < 1$

$$p^2 < 1,250 - p^2$$

$$2p^2 < 1,250$$

$$p^2 < 625$$

$$p < 25, \quad 0 < p < 25$$

63. $x = f(p) = 20(10 - p) \quad 0 \le p \le 10$

$R(p) = pf(p) = 20p(10 - p) = 200p - 20p^2$

$R'(p) = 200 - 40p$

Critical value: $R'(p) = 200 - 40p = 0;\ p = 5$

Sign chart for $R'(p)$:

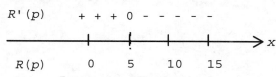

Test Numbers	
p	$R'(p)$
0	$200(+)$
10	$-200(-)$

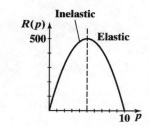

65. $x = f(p) = 40(p - 15)^2 \quad 0 \le p \le 15$

$R(p) = pf(p) = 40p(p - 15)^2$

$R'(p) = 40(p - 15)^2 + 40p(2)(p - 15)$

$\quad = 40(p - 15)[p - 15 + 2p]$

$\quad = 40(p - 15)(3p - 15)$

$\quad = 120(p - 15)(p - 5)$

Critical values [in (0, 15)]: $p = 5$

Sign chart for $R'(p)$:

Test Numbers	
p	$R'(p)$
0	$(+)$
10	$(-)$

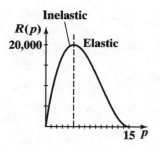

67. $x = f(p) = 30 - 10\sqrt{p} \quad 0 \le p \le 9$

$R(p) = pf(p) = 30p - 10p\sqrt{p}$

$R'(p) = 30 - 10\sqrt{p} - 10p \cdot \dfrac{1}{2}p^{-1/2}$

$\quad = 30 - 10\sqrt{p} - \dfrac{5p}{\sqrt{p}} = 30 - 15\sqrt{p}$

Critical values: $R'(p) = 30 - 15\sqrt{p} = 0$

$$\sqrt{p} = 2; \quad p = 4$$

Sign chart for $R'(p)$:

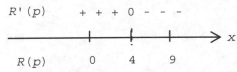

Test Numbers

p	$R'(p)$
0	30(+)
5	(−)

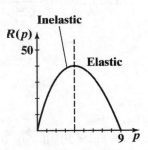

69. $p = g(x) = 50 - 0.1x$

$g'(x) = -0.1$

$E(x) = -\dfrac{g(x)}{xg'(x)} = -\dfrac{50 - 0.1x}{-0.1x} = \dfrac{500}{x} - 1$

$E(200) = \dfrac{500}{200} - 1 = \dfrac{3}{2}$

71. $p = g(x) = 50 - 2\sqrt{x}$

$g'(x) = -\dfrac{1}{\sqrt{x}}$

$E(x) = -\dfrac{g(x)}{xg'(x)} = -\dfrac{50 - 2\sqrt{x}}{x\left(-\frac{1}{\sqrt{x}}\right)} = \dfrac{50}{\sqrt{x}} - 2$

$E(400) = \dfrac{50}{20} - 2 = \dfrac{1}{2}$

73. $p = g(x) = 180 - 0.3x, \ \ 0 < x < 600; \ \ g'(x) = -0.3$

$E(x) = \dfrac{-g(x)}{xg'(x)} = \dfrac{-(180 - 0.3x)}{-0.3x} = \dfrac{180 - 0.3x}{0.3x} = \dfrac{600 - x}{x}$

Elastic:

$E(x) = \dfrac{600 - x}{x} > 1$

$600 - x > x$

$2x < 600$

$x < 300$

Demand is elastic for $0 < x < 300$; demand is inelastic for $300 < x < 600$.

75. $p = g(x) = 90 - 0.1x^2, \ \ 0 < x < 30; \ \ g'(x) = -0.2x$

$E(x) = \dfrac{-g(x)}{xg'(x)} = \dfrac{-(90 - 0.1x^2)}{-0.2x^2} = \dfrac{90 - 0.1x^2}{0.2x^2} = \dfrac{900 - x^2}{2x^2}$

Elastic:

$$E(x) = \frac{900 - x^2}{2x^2} > 1$$

$$900 - x^2 > 2x^2$$

$$3x^2 < 900$$

$$x^2 < 300$$

$$x < 10\sqrt{3}$$

Demand is elastic for $0 < x < 10\sqrt{3}$; demand is inelastic for $10\sqrt{3} < x < 30$.

77. $x = f(p) = Ap^{-k}$, A, k positive constants

$$f'(p) = -Akp^{-k-1}$$

$$E(p) = \frac{-pf'(p)}{f(p)} = \frac{Akp^{-k}}{Ap^{-k}} = k$$

79. The company's daily cost is increasing by $2.50(30) = \$75$ per day.

81. $x + 400p = 3,000$

$x = f(p) = 3,000 - 400p$

$f'(p) = -400$

Elasticity of demand: $E(p) = \dfrac{400p}{3,000 - 400p} = \dfrac{2p}{15 - 2p}$

$E(3) = \dfrac{6}{9} = \dfrac{2}{3} < 1$

The demand is inelastic; a price increase will increase revenue.

83. $x + 1,000p = 2,500$

$x = f(p) = 2,500 - 1,000p$

$f'(p) = -1,000$

Elasticity of demand: $E(p) = \dfrac{1,000p}{2,500 - 1,000p} = \dfrac{2p}{50 - 2p}$

$$E(0.99) = \frac{1.98}{5 - 1.98} \approx 0.66 < 1$$

The demand is inelastic; a price decrease will decrease revenue.

85. From Problem 83, $R(p) = pf(p) = 3,000p - 400p^2$

$R'(p) = 3,000 - 800p$

Critical values: $R'(p) = 3,000 - 800p = 0$

$$800p = 3000$$

$$p = 3.75$$

$R''(p) = -800$

Since $p = 3.75$ is the only critical value and $R''(3.75) = -800 < 0$, the maximum revenue occurs when the price $p = \$3.75$.

87. $f(t) = 0.31t + 18.5, \; 0 \le t \le 50$

$f'(t) = 0.31$

Percentage rate of change:

$$100 \frac{f'(t)}{f(t)} = \frac{31}{0.31t + 18.5}$$

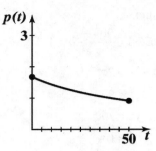

89. $r(t) = 3.3 - 0.7 \ln t, \quad r'(t) = -\dfrac{0.7}{t}$

Relative rate of change of $r(t)$: $\dfrac{r'(t)}{r(t)} = \dfrac{\frac{-0.7}{t}}{3.3 - 0.7 \ln t} = \dfrac{-0.7}{3.3t - 0.7t \ln t} = C(t)$.

Relative rate of change in 2020: $C(30) = \dfrac{-0.7}{3.3(30) - 0.7(30) \ln(30)} \approx -0.025$

The relative rate of change for robberies annually per 1,000 population is approximately -0.025.

CHAPTER 11 REVIEW

1. $A(t) = 2000e^{0.09t}$

$A(5) = 2000e^{0.09(5)} = 2000e^{0.45} \approx 3136.62 \;$ or $\; \$3136.62$

$A(10) = 2000e^{0.09(10)} = 2000e^{0.9} \approx 4919.21 \;$ or $\; \$4919.21$

$A(20) = 2000e^{0.09(20)} = 2000e^{1.8} \approx 12{,}099.29 \;$ or $\; \$12{,}099.29$ (3-1)

2. $f(x) = (6x + 5)^{3/2}$. Let $y = E(u) = u^{3/2}, \; u = I(x) = 6x + 5$.

Then $y = E(u) = E[I(x)] = (6x + 5)^{3/2}$. (3-4)

3. $f(x) = \ln(x^2 + 4)$. Let $y = E(u) = \ln u, \; u = I(x) = x^2 + 4$.

Then $y = E(u) = E[I(x)] = \ln(x^2 + 4)$. (3-4)

4. $f(x) = e^{0.02x}$. Let $y = E(u) = e^u, \; u = I(x) = 0.02x$. Then $y = E(u) = E[I(x)] = e^{0.02x}$. (3-4)

5. $\dfrac{d}{dx}(2 \ln x + 3e^x) = 2\dfrac{d}{dx}\ln x + 3\dfrac{d}{dx}e^x = \dfrac{2}{x} + 3e^x$ (3-2)

6. $\dfrac{d}{dx}e^{2x-3} = e^{2x-3}\dfrac{d}{dx}(2x - 3)$ (by the chain rule)

$\qquad = 2e^{2x-3}$ (3-4)

7. $y = \ln(2x + 7)$

$y' = \dfrac{1}{2x + 7}(2)$ (by the chain rule)

$\; = \dfrac{2}{2x + 7}$ (3-4)

8. $f(x) = \ln(3 + e^x)$

$f'(x) = \dfrac{1}{3 + e^x} \cdot e^x$ (by the chain rule)

$\qquad = \dfrac{e^x}{3 + e^x}$ (3-4)

9. $\dfrac{d}{dx} 2y^2 - \dfrac{d}{dx} 3x^3 - \dfrac{d}{dx} 5 = \dfrac{d}{dx}(0)$

$\qquad 4yy' - 9x^2 - 0 = 0$

$\qquad\qquad y' = \dfrac{9x^2}{4y}$

$\qquad\qquad \left.\dfrac{dy}{dx}\right|_{(1,2)} = \dfrac{9 \cdot 1^2}{4 \cdot 2} = \dfrac{9}{8}$ (3-5)

10. $y = 3x^2 - 5$

$\qquad \dfrac{dy}{dt} = \dfrac{d(3x^2)}{dt} - \dfrac{d(5)}{dt}$

$\qquad \dfrac{dy}{dt} = 6x\dfrac{dx}{dt}$

$\qquad x = 12; \dfrac{dx}{dt} = 3$

$\qquad \dfrac{dy}{dt} = 6 \cdot 12 \cdot 3 = 216$ (3-6)

11. $25p + x = 1{,}000$

(A) $x = 1{,}000 - 25p$

(B) $x = f(p) = 1{,}000 - 25p$

$\qquad f'(p) = -25$

$\qquad E(p) = -\dfrac{pf'(p)}{f(p)} = \dfrac{25p}{1{,}000 - 25p} = \dfrac{p}{40 - p}$

(C) $E(15) = \dfrac{15}{40 - 15} = \dfrac{15}{25} = \dfrac{3}{5} = 0.6$

Demand is inelastic and insensitive to small changes in price.

(D) Revenue: $R(p) = pf(p) = 1{,}000p - 25p^2$

(E) From (B), $E(25) = \dfrac{25}{40 - 25} = \dfrac{25}{15} = \dfrac{5}{3} = 1.6$

Demand is elastic; a price cut will increase revenue. (3-7)

12. $y = 100e^{-0.1x}$

$y' = 100(-0.1)e^{-0.1x}; \; y'(0) = 100(-0.1) = -10$ (3-2)

13.

n	1000	100,000	10,000,000	100,000,000
$\left(1+\dfrac{2}{n}\right)^n$	7.374312	7.388908	7.389055	7.389056

$\displaystyle\lim_{n\to\infty}\left(1+\frac{2}{n}\right)^n \approx 7.38906$ (5 decimal places);

$\displaystyle\lim_{n\to\infty}\left(1+\frac{2}{n}\right)^n = e^2$ (3-1)

14. $\dfrac{d}{dz}[(\ln z)^7 + \ln z^7] = \dfrac{d}{dz}[\ln z]^7 + \dfrac{d}{dz}\,7\ln z = 7[\ln z]^6\,\dfrac{d}{dz}\ln z + 7\dfrac{d}{dz}\ln z = 7[\ln z]^6\,\dfrac{1}{z} + \dfrac{7}{z}$

$\qquad\qquad\qquad\qquad\qquad = \dfrac{7(\ln z)^6 + 7}{z} = \dfrac{7[(\ln z)^6 + 1]}{z}$ (3-4)

15. $\dfrac{d}{dx}\,x^6 \ln x = x^6\,\dfrac{d}{dx}\ln x + (\ln x)\dfrac{d}{dx}\,x^6 = x^6\left(\dfrac{1}{x}\right) + (\ln x)6x^5 = x^5(1 + 6\ln x)$ (3-3)

16. $\dfrac{d}{dx}\left(\dfrac{e^x}{x^6}\right) = \dfrac{x^6\,\dfrac{d}{dx}\,e^x - e^x\,\dfrac{d}{dx}\,x^6}{(x^6)^2} = \dfrac{x^6 e^x - 6x^5 e^x}{x^{12}} = \dfrac{xe^x - 6e^x}{x^7} = \dfrac{e^x(x-6)}{x^7}$ (3-3)

17. $y = \ln(2x^3 - 3x)$

$y' = \dfrac{1}{2x^3 - 3x}(6x^2 - 3) = \dfrac{6x^2 - 3}{2x^3 - 3x}$ (3-4)

18. $f(x) = e^{x^3 - x^2}$

$f'(x) = e^{x^3 - x^2}(3x^2 - 2x)$

$\qquad = (3x^2 - 2x)e^{x^3 - x^2}$ (3-4)

19. $y = e^{-2x}\ln 5x$

$\dfrac{dy}{dx} = e^{-2x}\left(\dfrac{1}{5x}\right)(5) + (\ln 5x)(e^{-2x})(-2) = e^{-2x}\left(\dfrac{1}{x} - 2\ln 5x\right) = \dfrac{1 - 2x\ln 5x}{xe^{2x}}$ (3-4)

20. $f(x) = 1 + e^{-x}$

$f'(x) = e^{-x}(-1) = -e^{-x}$

An equation for the tangent line to the graph of f at $x = 0$ is:

$y - y_1 = m(x - x_1)$,

where $x_1 = 0$, $y_1 = f(0) = 1 + e^0 = 2$, and $m = f'(0) = -e^0 = -1$.

Thus, $y - 2 = -1(x - 0)$ or $y = -x + 2$.

An equation for the tangent line to the graph of f at $x = -1$ is:

$y - y_1 = m(x - x_1)$,

where $x_1 = -1$, $y_1 = f(-1) = 1 + e$, and $m = f'(-1) = -e$. Thus,

$y - (1 + e) = -e[x - (-1)]$ or $y - 1 - e = -ex - e$ and $y = -ex + 1$.   (3-4)

21. $x^2 - 3xy + 4y^2 = 23$

Differentiate implicitly:

$2x - 3(xy' + y \cdot 1) + 8yy' = 0$

$\quad 2x - 3xy' - 3y + 8yy' = 0$

$\qquad\qquad 8yy' - 3xy' = 3y - 2x$

$\qquad\qquad (8y - 3x)y' = 3y - 2x$

$\qquad\qquad\qquad\quad y' = \dfrac{3y - 2x}{8y - 3x}$

$\qquad y'\Big|_{(-1,2)} = \dfrac{3 \cdot 2 - 2(-1)}{8 \cdot 2 - 3(-1)} = \dfrac{8}{19}$ [Slope at $(-1, 2)$]   (3-5)

22. $x^3 - 2t^2x + 8 = 0$

Differentiate implicitly:

$3x^2x' - (2t^2x' + x \cdot 4t) + 0 = 0$

$\quad 3x^2x' - 2t^2x' - 4xt = 0$

$\qquad (3x^2 - 2t^2)x' = 4xt$

$\qquad\qquad\qquad x' = \dfrac{4xt}{3x^2 - 2t^2}$

$\qquad x'\Big|_{(-2,2)} = \dfrac{4 \cdot 2 \cdot (-2)}{3(2^2) - 2(-2)^2} = \dfrac{-16}{12 - 8} = \dfrac{-16}{4} = -4$ (3-5)

23. $x - y^2 = e^y$

Differentiate implicitly:

$1 - 2yy' = e^y y'$

$\quad 1 = e^y y' + 2yy'$

$\quad 1 = y'(e^y + 2y)$

$\quad y' = \dfrac{1}{e^y + 2y}$

$y'\Big|_{(1,0)} = \dfrac{1}{e^0 + 2 \cdot 0} = 1$ (3-5)

24. $\ln y = x^2 - y^2$

Differentiate implicitly:

$\dfrac{y'}{y} = 2x - 2yy'$

$y'\left(\dfrac{1}{y} + 2y\right) = 2x$

$y'\left(\dfrac{1 + 2y^2}{y}\right) = 2x$

$\qquad y' = \dfrac{2xy}{1 + 2y^2}$

$\qquad y'\Big|_{(1,1)} = \dfrac{2 \cdot 1 \cdot 1}{1 + 2(1)^2} = \dfrac{2}{3}$ (3-5)

25. $A(t) = 400e^{0.049t}$, $A'(t) = 19.6e^{0.049t}$;   logarithmic derivative: $\dfrac{A'(t)}{A(t)} = \dfrac{19.6e^{0.049t}}{400e^{0.049t}} = \dfrac{19.6}{400} = 0.049$

(3-7)

26. $f(p) = 100 - 3p$, $f'(p) = -3$;   logarithmic derivative: $\dfrac{f'(p)}{f(p)} = \dfrac{-3}{100 - 3p}$.   (3-7)

27. $f(x) = 1 + x^2$, $f'(x) = 2x$;   logarithmic derivative: $\dfrac{f'(x)}{f(x)} = \dfrac{2x}{1 + x^2}$.   (3-7)

28. $y^2 - 4x^2 = 12$

Differentiate with respect to t:

$$2y\frac{dy}{dt} - 8x\frac{dx}{dt} = 0$$

Given: $\frac{dx}{dt} = -2$ when $x = 1$ and $y = 4$. Therefore,

$$2\cdot4\frac{dy}{dt} - 8\cdot1\cdot(-2) = 0$$

$$8\frac{dy}{dt} + 16 = 0$$

$$\frac{dy}{dt} = -2.$$

The y coordinate is decreasing at 2 units per second.   (3-6)

29. From the figure, $x^2 + y^2 = 17^2$.

Differentiate with respect to t:

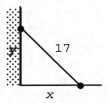

$$2x\frac{dx}{dt} + 2y\frac{dy}{dt} = 0 \quad \text{or} \quad x\frac{dx}{dt} + y\frac{dy}{dt} = 0$$

We are given $\frac{dx}{dt} = -0.5$ feet per second. Therefore,

$$x(-0.5) + y\frac{dy}{dt} = 0 \quad \text{or} \quad \frac{dy}{dt} = \frac{0.5x}{y} = \frac{x}{2y}$$

Now, when $x = 8$, we have: $8^2 + y^2 = 17^2$

$$y^2 = 289 - 64 = 225$$

$$y = 15$$

Therefore, $\frac{dy}{dt}\bigg|_{(8,15)} = \frac{8}{2(15)} = \frac{4}{15} \approx 0.27$ ft/sec.   (3-5)

30. $A = \pi R^2$. Given: $\frac{dA}{dt} = 24$ square inches per minute.

Differentiate with respect to t:

$$\frac{dA}{dt} = 2\pi R\frac{dR}{dt}$$

$$24 = 2\pi R\frac{dR}{dt}$$

Therefore, $\frac{dR}{dt} = \frac{24}{2\pi R} = \frac{12}{\pi R}$.

$$\frac{dR}{dt}\bigg|_{R=12} = \frac{12}{\pi\cdot12} = \frac{1}{\pi} \approx 0.318 \text{ inches per minute}$$ (3-6)

31. $x = f(p) = 20(p - 15)^2 \quad 0 \le p \le 15$

$f'(p) = 40(p - 15)$

$$E(p) = -\frac{pf'(p)}{f(p)} = \frac{-40p(p-15)}{20(p-15)^2} = \frac{-2p}{p-15}$$

Elastic: $E(p) = \dfrac{-2p}{p-15} > 1$

$$-2p < p - 15 \quad (p - 15 < 0 \text{ reverses inequality})$$
$$-3p < -15$$
$$p > 5; \quad 5 < p < 15$$

Inelastic: $E(p) = \dfrac{-2p}{p-15} < 1$

$$-2p > p - 15 \quad (p - 15 < 0 \text{ reverses inequality})$$
$$-3p > -15$$
$$p < 5; \quad 0 < p < 5 \qquad (3\text{-}7)$$

32. $x = f(p) = 5(20 - p)\ \ 0 \le p \le 20$

$R(p) = pf(p) = 5p(20 - p) = 100p - 5p^2$
$R'(p) = 100 - 10p = 10(10 - p)$

Critical values: $p = 10$

Sign chart for $R'(p)$:

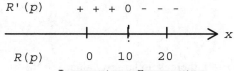

Test Numbers	
p	$R'(p)$
5	$50(+)$
15	$-50(-)$

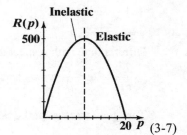
(3-7)

33. $y = w^3,\ w = \ln u,\ u = 4 - e^x$

(A) $y = [\ln(4 - e^x)]^3$

(B) $\dfrac{dy}{dx} = \dfrac{dy}{dw} \cdot \dfrac{dw}{du} \cdot \dfrac{du}{dx}$

$= 3w^2 \cdot \dfrac{1}{u} \cdot (-e^x) = 3[\ln(4 - e^x)]^2 \left(\dfrac{1}{4 - e^x} \right)(-e^x) = \dfrac{-3e^x[\ln(4 - e^x)]^2}{4 - e^x} \qquad (3\text{-}4)$

34. $y = 5^{x^2 - 1}$

$y' = 5^{x^2 - 1}(\ln 5)(2x) = 2x5^{x^2 - 1}(\ln 5) \qquad (3\text{-}4)$

35. $\dfrac{d}{dx} \log_5(x^2 - x) = \dfrac{1}{x^2 - x} \cdot \dfrac{1}{\ln 5} \cdot \dfrac{d}{dx}(x^2 - x) = \dfrac{1}{\ln 5} \cdot \dfrac{2x - 1}{x^2 - x} \qquad (3\text{-}4)$

36. $\dfrac{d}{dx} \sqrt{\ln(x^2 + x)} = \dfrac{d}{dx}[\ln(x^2 + x)]^{1/2} = \dfrac{1}{2}[\ln(x^2 + x)]^{-1/2} \dfrac{d}{dx} \ln(x^2 + x)$

$= \dfrac{1}{2}[\ln(x^2 + x)]^{-1/2} \dfrac{1}{x^2 + x} \dfrac{d}{dx}(x^2 + x)$

$= \dfrac{1}{2}[\ln(x^2 + x)]^{-1/2} \cdot \dfrac{2x + 1}{x^2 + x} = \dfrac{2x + 1}{2(x^2 + x)[\ln(x^2 + x)]^{1/2}} \qquad (3\text{-}4)$

37. $e^{xy} = x^2 + y + 1$

Differentiate implicitly:

$$\frac{d}{dx} e^{xy} = \frac{d}{dx} x^2 + \frac{d}{dx} y + \frac{d}{dx} 1$$

$$e^{xy}(xy' + y) = 2x + y'$$

$$xe^{xy}y' - y' = 2x - ye^{xy}$$

$$y' = \frac{2x - ye^{xy}}{xe^{xy} - 1}$$

$$y'\Big|_{(0,0)} = \frac{2 \cdot 0 - 0 \cdot e^0}{0 \cdot e^0 - 1} = 0 \qquad (3\text{-}5)$$

38. $A = \pi r^2, r \geq 0$

Differentiate with respect to t:

$$\frac{dA}{dt} = 2\pi r \frac{dr}{dt} = 6\pi r \ \text{ since } \ \frac{dr}{dt} = 3$$

The area increases at the rate $6\pi r$. This is smallest when $r = 0$; there is no largest value. $\qquad (3\text{-}6)$

39. $y = x^3$

Differentiate with respect to t:

$$\frac{dy}{dt} = 3x^2 \frac{dx}{dt}$$

Solving for $\dfrac{dx}{dt}$, we get

$$\frac{dx}{dt} = \frac{1}{3x^2} \cdot \frac{dy}{dt} = \frac{5}{3x^2} \ \text{ since } \ \frac{dy}{dt} = 5$$

To find where $\dfrac{dx}{dt} > \dfrac{dy}{dt}$, solve the inequality

$$\frac{5}{3x^2} > 5$$

$$\frac{1}{3x^2} > 1$$

$$3x^2 < 1$$

$$-\frac{1}{\sqrt{3}} < x < \frac{1}{\sqrt{3}} \quad \text{or} \quad \frac{-\sqrt{3}}{3} < x < \frac{\sqrt{3}}{3} \qquad (3\text{-}6)$$

40. (A) The compound interest formula is: $A = P(1 + r)^t$. Thus, the time for P to double when $r = 0.05$ and interest is compounded annually can be found by solving

$$2P = P(1 + 0.05)^t \ \text{ or } \ 2 = (1.05)^t \ \text{ for } t.$$

$$\ln(1.05)^t = \ln 2$$

$$t \ln(1.05) = \ln 2$$

$$t = \frac{\ln 2}{\ln(1.05)} \approx 14.2 \text{ or } 15 \text{ years}$$

(B)   The continuous compound interest formula is: $A = Pe^{rt}$. Proceeding as above, we have

$$2P = Pe^{0.05t} \text{ or } e^{0.05t} = 2.$$

Therefore, $0.05t = \ln 2$ and $t = \dfrac{\ln 2}{.05} \approx 13.9$ years   (3-1)

41. $A(t) = 100e^{0.1t}$

$A'(t) = 100(0.1)e^{0.1t} = 10e^{0.1t}$

$A'(1) = 11.05$ or \$11.05 per year

$A'(10) = 27.18$ or \$27.18 per year   (3-1)

42. $P(t) = 12,000\,e^{0.0395t}$.   First solve $12,000\,e^{0.0395t} = 25,000$ for t:

$$e^{0.0395t} = \frac{25,000}{12,000} = \frac{25}{12}$$

$$0.0395t = \ln(25/12)$$

$$t = \frac{\ln(25/12)}{0.0395} \approx 18.58$$

$$P'(18.58) = 474\,e^{0.0395(18.58)} \approx 987.50\,; \quad \$987.50 \text{ per year.} \quad (3-2)$$

43. $R(x) = xp(x) = 1000xe^{-0.02x}$

$R'(x) = 1000[xD_x e^{-0.02x} + e^{-0.02x}D_x x]$

$\qquad = 1000[x(-0.02)e^{-0.02x} + e^{-0.02x}] = (1000 - 20x)e^{-0.02x}$ (3-4)

44. $x = \sqrt{5000 - 2p^3} = (5000 - 2p^3)^{1/2}$

Differentiate implicitly with respect to x:

$$1 = \frac{1}{2}(5000 - 2p^3)^{-1/2}(-6p^2)\frac{dp}{dx}$$

$$1 = \frac{-3p^2}{(5000 - 2p^3)^{1/2}}\frac{dp}{dx}$$

$$\frac{dp}{dx} = \frac{-(5000 - 2p^3)^{1/2}}{3p^2} \qquad (3-5)$$

45.   Given: $R(x) = 750x - \dfrac{x^2}{30}$ and $\dfrac{dx}{dt} = 3$ when $x = 40$.

Differentiate with respect to t:

$$\frac{dR}{dt} = 750\frac{dx}{dt} - \frac{1}{30}(2x)\frac{dx}{dt} = 750\frac{dx}{dt} - \frac{x}{15}\cdot\frac{dx}{dt}$$

Thus, $\dfrac{dR}{dt}\bigg|_{x=40 \text{ and } \frac{dx}{dt}=3} = 750(3) - \dfrac{40}{15}\cdot 3 = \$2,242$ (3-6)

46. $p = 38.2 - 0.002x$

$$x = f(p) = \frac{38.2}{0.002} - \frac{1}{0.002}p = 19,100 - 500p$$

$f'(p) = -500$

Elasticity of demand: $E(p) = \dfrac{-pf'(p)}{f(p)} = \dfrac{500p}{19,100 - 500p} = \dfrac{5p}{191 - 5p}$

$$E(21) = \frac{105}{191 - 105} = \frac{105}{86} > 1$$

Demand is elastic, a (small) price decrease will increase revenue.   (3-7)

47. $f(t) = 1,700t + 20,500$

$f'(t) = 1,700$

Relative rate of change: $\dfrac{f'(t)}{f(t)} = \dfrac{1,700}{1,700t + 20,500}$

Relative rate of change at $t = 35$: $\dfrac{1,700}{1,700(35) + 20,500} \approx 0.02125$ (3-7)

48. $C(t) = 5e^{-0.3t}$

$C'(t) = 5e^{-0.3t}(-0.3) = -1.5e^{-0.3t}$

After one hour, the rate of change of concentration is

$C'(1) = -1.5e^{-0.3(1)} = -1.5e^{-0.3} \approx -1.111$ mg/ml per hour.

After five hours, the rate of change of concentration is

$C'(5) = -1.5e^{-0.3(5)} = -1.5e^{-1.5} \approx -0.335$ mg/ml per hour.   (3-4)

49.   Given: $A = \pi R^2$ and $\dfrac{dA}{dt} = -45$ mm^2 per day (negative because the area is decreasing).

Differentiate with respect to t:

$$\frac{dA}{dt} = \pi 2R \frac{dR}{dt}$$

$$-45 = 2\pi R \frac{dR}{dt}$$

$$\frac{dR}{dt} = -\frac{45}{2\pi R}$$

$$\left.\frac{dR}{dt}\right|_{R=15} = \frac{-45}{2\pi \cdot 15} = \frac{-3}{2\pi} \approx -0.477 \text{ mm per day} \qquad (3\text{-}6)$$

50. $N(t) = 10(1 - e^{-0.4t})$

(A) $N'(t) = -10e^{-0.4t}(-0.4) = 4e^{-0.4t}$

$N'(1) = 4e^{-0.4(1)} = 4e^{-0.4} \approx 2.68.$

Thus, learning is increasing at the rate of 2.68 units per day after 1 day.

$N'(5) = 4e^{-0.4(5)} = 4e^{-2} = 0.54$

Thus, learning is increasing at the rate of 0.54 units per day after 5 days.

(B)   We solve $N'(t) = 0.25 = 4e^{-0.4t}$ for t:

$$e^{-0.4t} = \frac{0.25}{4} = 0.0625$$

$$-0.4t = \ln(0.0625)$$

$$t = \frac{\ln(0.0625)}{-0.4} \approx 6.93$$

The rate of learning is less than 0.25 after 7 days.   (3-4)

51.   Given: $T = 2\left(1 + \dfrac{1}{x^{3/2}}\right) = 2 + 2x^{-3/2}$, and $\dfrac{dx}{dt} = 3$ when $x = 9$.

Differentiate with respect to t:

$$\frac{dT}{dt} = 0 + 2\left(-\frac{3}{2}x^{-5/2}\right)\frac{dx}{dt} = -3x^{-5/2}\frac{dx}{dt}$$

$$\left.\frac{dT}{dt}\right|_{x=9 \text{ and } \frac{dx}{dt}=3} = -3(9)^{-5/2}(3) = -3 \cdot 3^{-5} \cdot 3 = -3^{-3} = \frac{-1}{27} \approx -0.037 \text{ minutes per operation}$$

per hour.   (3-6)

12 GRAPHING AND OPTIMIZATION

EXERCISE 12-1

Things to remember:

1.   INCREASING AND DECREASING FUNCTIONS

   For the interval (a, b):

$f'(x)$	$f(x)$	Graph of f	Examples
+	Increases↗	Rises↗	
–	Decreases↘	Falls↘	

2.   CRITICAL VALUES

   The values of x in the domain of f where $f'(x) = 0$ or where $f'(x)$ does not exist are called the CRITICAL VALUES of f.

   The critical values of f are always in the domain of f and are also partition numbers for f', but f' may have partition numbers that are not critical values.

   If f is a polynomial, then both the partition numbers for f' and the critical values of f are the solutions of $f'(x) = 0$.

3.   LOCAL EXTREMA

   Given a function f. The value $f(c)$ is a LOCAL MAXIMUM of f if there is an interval (m, n) containing c such that $f(x) \leq f(c)$ for all x in (m, n). The value $f(e)$ is a LOCAL MINIMUM of f if there is an interval (p, q) containing e such that $f(x) \geq f(e)$ for all x in (p, q). Local maxima and local minima are called LOCAL EXTREMA.

   A point on the graph where a local extremum occurs is also called a TURNING POINT.

4.   EXISTENCE OF LOCAL EXTREMA

   If f is continuous on the interval (a, b), c is a number in (a, b) and $f(c)$ is a local extremum, then either $f'(c) = 0$ or $f'(c)$ does not exist (is not defined).

5.   FIRST DERIVATIVE TEST FOR LOCAL EXTREMA

   Let c be a critical value of f [$f(c)$ is defined and either $f'(c) = 0$ or $f'(c)$ is not defined.]

   Construct a sign chart for $f'(x)$ close to and on either side of c.

Sign Chart	$f(c)$
$f'(x)$ ⟵————(— — — ┆ + + +)————⟶ x m c n $f(x)$ Decreasing ┆ Increasing	$f(c)$ is a local minimum. If $f'(x)$ changes from negative to positive at c, then $f(c)$ is a local minimum.
$f'(x)$ ⟵————(+ + + ┆ — — —)————⟶ x m c n $f(x)$ Increasing ┆ Decreasing	$f(c)$ is a local maximum. If $f'(x)$ changes from positive to negative at c, then $f(c)$ is a local maximum.
$f'(x)$ ⟵————(— — — ┆ — — —)————⟶ x m c n $f(x)$ Decreasing ┆ Decreasing	$f(c)$ is not a local extremum. If $f'(x)$ does not change sign at c, then $f(c)$ is neither a local maximum nor a local minimum.
$f'(x)$ ⟵————(+ + + ┆ + + +)————⟶ x m c n $f(x)$ Increasing ┆ Increasing	$f(c)$ is not a local extremum. If $f'(x)$ does not change sign at c, then $f(c)$ is neither a local maximum nor a local minimum.

<u>6.</u>   INTERCEPTS AND LOCAL EXTREMA FOR POLYNOMIAL FUNCTIONS

If $f(x) = a_n x^n + a_{n-1} x^{n-1} + \ldots + a_1 x + a_0$, $a_n \neq 0$ is an nth degree polynomial then f has at most n x-intercepts and at most $n - 1$ local extrema.

---

1. $g(x) = |x|$ on $(-\infty, 0)$: decreasing

3. $f(x) = x$ on $(-\infty, \infty)$: increasing

5. $p(x) = \sqrt[3]{x}$ on $(-\infty, 0)$: increasing

7. $r(x) = 4 - \sqrt{x}$ on $(0, \infty)$: decreasing

9. $(a, b), (d, f), (g, h)$

11. $(b, c), (c, d), (f, g)$

13. $x = c, d, f$

15. $x = b, f$

17. f has a local maximum at $x = a$, and a local minimum at $x = c$; f does not have a local extremum at $x = b$ or at $x = d$.

19. $f(3) = 5$ is a local maximum; (e)

21.   No local extrema; (d)

23. $f(3) = 5$ is a local maximum; (f)

25.   No local extrema; (c)

27. $f(x) = x^3 - 12x + 8$

(A) $f'(x) = 3x^2 - 12$

(C)   Partition numbers: $x = -2, \ 2$

(B)   Critical values:
$$f'(x) = 0$$
$$3x^2 - 12 = 0$$
$$x^2 = 4$$
$$x = -2, \ 2$$

29. $f(x) = \dfrac{6}{x+2} = 6(x+2)^{-1}$

(A) $f'(x) = (-1)6(x+2)^{-2} = \dfrac{-6}{(x+2)^2}$ is

(B) Critical values: There are no critical values ($x = -2$ is not a critical value since -2 not in the domain of f.)

(C) Partition numbers: $x = -2$

31. $f(x) = |x| = \begin{Bmatrix} x, & x \ge 0 \\ -x, & x < 0 \end{Bmatrix}$

(A)
$$f'(x) = \begin{Bmatrix} 1, & x > 0 \\ -1, & x < 0 \end{Bmatrix}$$

(B) Critical values: $x = 0$ ($f'(0)$ does not exist)

(C) Partition numbers: $x = 0$.

33. $f(x) = 2x^2 - 4x$; domain of f: $(-\infty, \infty)$

$f'(x) = 4x - 4$; f' is continuous for all x.

$f'(x) = 4x - 4 = 0$

$\qquad\qquad x = 1$

Thus, $x = 1$ is a partition number for f', and since 1 is in the domain of f, $x = 1$ is a critical value of f.

Sign chart for f':

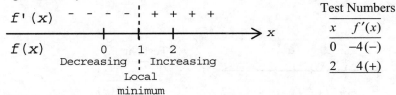

Test Numbers	
x	$f'(x)$
0	$-4(-)$
2	$4(+)$

Therefore, f is decreasing on $(-\infty, 1)$; f is increasing on $(1, \infty)$; $f(1) = -2$ is a local minimum.

35. $f(x) = -2x^2 - 16x - 25$; domain of f: $(-\infty, \infty)$

$f'(x) = -4x - 16$; f' is continuous for all x and

$\quad f'(x) = -4x - 16 = 0$

$\qquad\qquad x = -4$

Thus, $x = -4$ is a partition number for f', and since -4 is in the domain of f, $x = -4$ is a critical value for f.

Sign chart for f':

	$f'(x)$	+ + + + - - - - - -

Test Numbers	
x	$f'(x)$
-5	$4(+)$
0	$-16(-)$

Therefore, f is increasing on $(-\infty, -4)$; f is decreasing on $(-4, \infty)$; f has a local maximum at $x = -4$.

37. $f(x) = x^3 + 4x - 5$; domain of f: $(-\infty, \infty)$

$f'(x) = 3x^2 + 4 \ge 4 > 0$ for all x; f is increasing on $(-\infty, \infty)$; f has no local extrema.

39. $f(x) = 2x^3 - 3x^2 - 36x;$ domain of f: $(-\infty, \infty)$

$f'(x) = 6x^2 - 6x - 36;$ f is continuous for all x and

$\quad f'(x) = 6(x^2 - x - 6) = 0$

$\qquad 6(x - 3)(x + 2) = 0$

$\qquad\qquad x = -2, 3$

The partition numbers for f are $x = -2$ and $x = 3$. Since -2 and 3 are in the domain of f, $x = -2, x = 3$ are critical values for f.

Sign chart for f':

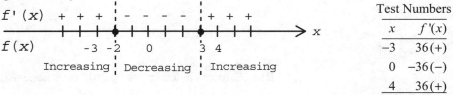

Test Numbers	
x	$f'(x)$
-3	$36(+)$
0	$-36(-)$
4	$36(+)$

Therefore, f is increasing on $(-\infty, -2)$ and $(3, \infty);$ f is decreasing on $(-2, 3);$ f has a local maximum at $x = -2$ and a local minimum at $x = 3$.

41. $f(x) = 3x^4 - 4x^3 + 5;$ domain of f: $(-\infty, \infty)$

$f'(x) = 12x^3 - 12x^2;$ f' is continuous for all x.

$f'(x) = 12x^3 - 12x^2 = 0$

$\qquad 12x^2(x - 1) = 0$

$\qquad\qquad x = 0, 1$

The partition numbers for f' are $x = 0$ and $x = 1$. Since 0 and 1 are in the domain of f, $x = 0, x = 1$ are critical values of f.

Sign chart for f:

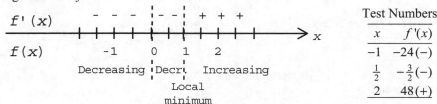

Test Numbers	
x	$f'(x)$
-1	$-24(-)$
$\frac{1}{2}$	$-\frac{3}{2}(-)$
2	$48(+)$

Therefore, f is decreasing on $(-\infty, 1);$ f is increasing on $(1, \infty);$ $f(1) = 4$ is a local minimum.

43. $f(x) = (x - 1)e^{-x};$ domain of f: $(-\infty, \infty)$

$f'(x) = (x - 1)e^{-x}(-1) + e^{-x} = 2e^{-x} - xe^{-x} = (2 - x)e^{-x}$

f' is continuous for all x and

$f'(x) = (2 - x)e^{-x} = 0$

$\qquad 2 - x = 0$

$\qquad\qquad x = 2$

$x = 2$ is a partition number for f'. Since 2 is in the domain of f, $x = 2$ is a critical value of f.

Sign chart for f':

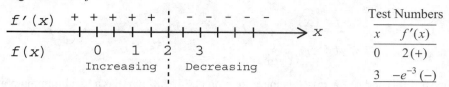

Test Numbers

x	$f'(x)$
0	$2(+)$
3	$-e^{-3}(-)$

Therefore, f is increasing on $(-\infty, 2)$; f is decreasing on $(2, \infty)$; f has a local maximum at $x = 2$.

45. $f(x) = 4x^{1/3} - x^{2/3}$; domain of f: $(-\infty, \infty)$

$$f'(x) = \frac{4}{3}x^{-2/3} - \frac{2}{3}x^{-1/3} = \frac{2}{3}\left[\frac{2}{x^{2/3}} - \frac{1}{x^{1/3}}\right] = \frac{2}{3}\left[\frac{2 - x^{1/3}}{x^{2/3}}\right]$$

f' is continuous for all x except $x = 0$.

$$f'(x) = \frac{2}{3}\left[\frac{2 - x^{1/3}}{x^{2/3}}\right] = 0$$
$$2 - x^{1/3} = 0$$
$$x = 8$$

$x = 0$ and $x = 8$ are partition numbers for f'. Since 0 and 8 are in the domain of f, $x = 0$, $x = 8$ are critical values of f ($f'(0)$ does not exist, $f'(8) = 0$).

Sign chart for f':

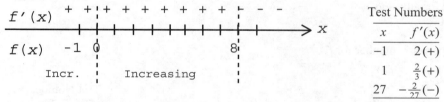

Test Numbers

x	$f'(x)$
-1	$2(+)$
1	$\frac{2}{3}(+)$
27	$-\frac{2}{27}(-)$

Therefore, f is increasing on $(-\infty, 8)$; f is decreasing on $(8, \infty)$; f has a local maximum at $x = 8$; $f(0)$ is not a local extremum.

47. $f(x) = x^4 - 4x^3 + 9x$; domain of f: $(-\infty, \infty)$

$f'(x) = 4x^3 - 12x^2 + 9$; f' is continuous for all x. Using a root-approximation routine, $f'(x) = 0$ at $x = -0.77$, $x = 1.08$, and $x = 2.69$; critical values.

Sign chart for f':

$$
\begin{array}{c}
f'(x) \quad - \ - \ | \ + \ + \ + \ | \ - \ - \ - \ | \ + \ + \ + \\
\text{———————————————→} \ x \\
f(x) \qquad\ \bullet \qquad\quad \bullet \qquad\quad \bullet \\
\quad -0.77 \ \ 0 \ \ \ 1.08 \ \ \ 2.69 \\
\text{Decr.} \ | \ \text{Incr.} \ | \ \text{Decr.} \ | \ \text{Incr.}
\end{array}
$$

f is decreasing on $(-\infty, -0.77)$ and $(1.08, 2.69)$; increasing on $(-0.77, 1.08)$ and $(2.69, \infty)$; f has a local minima at $x = -0.77$ and $x = 2.69$, f has a local maximum at $x = 1.08$.

49. $f(x) = x \ln x - (x - 2)^3$; domain of f: $(0, \infty)$

$f'(x) = 1 + \ln x - 3(x - 2)^2$; f' is continuous on $(0, \infty)$.

Using a root-approximation routine, $f'(x) = 0$ at $x = 1.34$ and $x = 2.82$; critical values.

Sign chart for f':

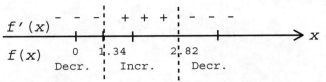

f is decreasing on (0, 1.34) and (2.82, ∞); f is increasing on (1.34, 2.82); f has a local minimum at $x = 1.34$; f has a local maximum at $x = 2.82$.

51. $f(x) = e^x - 2x^2$; domain of f: (−∞, ∞)

$f'(x) = e^x - 4x$; f' is continuous for all x.

Using a root-approximation routine, $f'(x) = 0$ at $x = 0.36$ and $x = 2.15$; critical values.

Sign chart for f' :

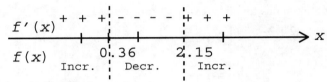

f is increasing on (−∞, 0.36) and (2.15, ∞); f is decreasing on (0.36, 2.15); f has a local maximum at $x = 0.36$; f has a local minimum at $x = 2.15$.

53. $f(x) = 4 + 8x - x^2$
$f'(x) = 8 - 2x$
f' is continuous for all x and
$f'(x) = 8 - 2x = 0$
$\qquad x = 4$
Thus, $x = 4$ is a partition number for f.

The sign chart for f' is:

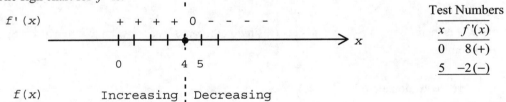

Test Numbers	
x	$f'(x)$
0	8(+)
5	−2(−)

Therefore, f is increasing on (−∞, 4) and decreasing on (4, ∞); f has a local maximum at $x = 4$.

x	$f'(x)$	f	GRAPH OF f
(−∞, 4)	+	Increasing	Rising
$x = 4$	0	Local maximum	Horizontal tangent
(4, ∞)	−	Decreasing	Falling

x	$f(x)$
0	4
4	20

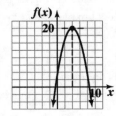

55. $f(x) = x^3 - 3x + 1$

$f'(x) = 3x^2 - 3$ is continuous for all x and

$f'(x) = 3x^2 - 3 = 0$

$\qquad 3(x^2 - 1) = 0$

$3(x + 1)(x - 1) = 0$

Thus, $x = -1$ and $x = 1$ are partition numbers for f'.

The sign chart for f' is:

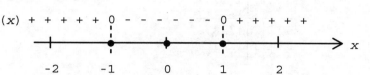

Test Numbers

x	$f(x)$
−2	9 (+)
0	−3 (−)
2	9 (+)

Therefore, f is increasing on $(-\infty, -1)$ and on $(1, \infty)$, f is decreasing on $(-1, 1)$; f has a local maximum at $x = -1$ and a local minimum at $x = 1$.

x	$f'(x)$	f	Graph of f
$(-\infty, -1)$	+	Increasing	Rising
$x = -1$	0	Local maximum	Horizontal tangent
$(-1, 1)$	−	Decreasing	Falling
$x = 1$	0	Local minimum	Horizontal tangent
$(1, \infty)$	+	Increasing	Rising

x	$f(x)$
−1	3
0	1
1	−1

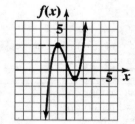

57. $f(x) = 10 - 12x + 6x^2 - x^3$

$f'(x) = -12 + 12x - 3x^2$

f' is continuous for all x and

$f'(x) = -12 + 12x - 3x^2 = 0$

$\qquad -3(x^2 - 4x + 4) = 0$

$\qquad\qquad -3(x - 2)^2 = 0$

Thus, $x = 2$ is a partition number for f'.

The sign chart for f' is:

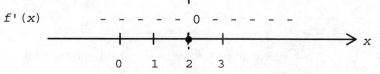

Test Numbers

x	$f'(x)$
0	−12(−)
3	−3(−)

Therefore, f is decreasing for all x, i.e., on $(-\infty, \infty)$, and there is a horizontal tangent line at $x = 2$.

x	$f'(x)$	f	GRAPH of f
$(-\infty, 2)$	-	Decreasing	Falling
$x = 2$	0		Horizontal tangent
$x > 2$	-	Decreasing	Falling

x	$f(x)$
0	10
2	2

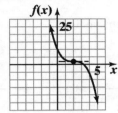

59. $f(x) = x^4 - 18x^2$

$f'(x) = 4x^3 - 36x$

f' is continuous for all x and

$f'(x) = 4x^3 - 36x = 0$

$\qquad 4x(x^2 - 9) = 0$

$\quad 4x(x - 3)(x + 3) = 0$

Thus, $x = -3$, $x = 0$, and $x = 3$ are partition numbers for f'.

Sign chart for f':

Test Numbers	
x	$f'(x)$
-4	$-112\,(-)$
-1	$32\,(+)$
1	$-32\,(-)$
4	$112\,(+)$

$$f'(x) \quad \text{---} \; 0 \;+++\; 0 \;\text{---}\; 0 \;+++ \longrightarrow x$$
$$f(x) \qquad\quad -3 \qquad 0 \qquad 3$$
$$\text{Decr.} \;\; \text{Incr.} \;\; \text{Decr.} \;\; \text{Incr.}$$

Therefore, f is increasing on $(-3, 0)$ and on $(3, \infty)$; f is decreasing on $(-\infty, -3)$ and on $(0, 3)$; f has a local maximum at $x = 0$ and local minima at $x = -3$ and $x = 3$.

x	$f'(x)$	f	GRAPH of f
$(-\infty, -3)$	-	Decreasing	Falling
$x = -3$	0	Local minimum	Horizontal tangent
$(-3, 0)$	+	Increasing	Rising
$x = 0$	0	Local maximum	Horizontal tangent
$(0, 3)$	-	Decreasing	Falling
$x = 3$	0	Local minimum	Horizontal tangent
$(3, \infty)$	+	Increasing	Rising

x	$f(x)$
0	0
-3	-81
3	-81

61.

x	$f'(x)$	$f(x)$	GRAPH of f
$(-\infty, -1)$	-	Increasing	Rising
$x = -1$	0	Neither local maximum nor local minimum	Horizontal tangent
$(-1, 1)$	+	Increasing	Rising
$x = 1$	0	Local maximum	Horizontal tangent
$(1, \infty)$	-	Decreasing	Falling

Using this information together with the points $(-2, -1)$, $(-1, 1)$, $(0, 2)$, $(1, 3)$, $(2, 1)$ on the graph, we have

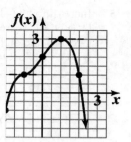

63.

x	$f'(x)$	$f(x)$	GRAPH of $f(x)$
$(-\infty, -1)$	-	Decreasing	Falling
$x = -1$	0	Local minimum	Horizontal tangent
$(-1, 0)$	+	Increasing	Rising
$x = 0$	Not defined	Local maximum	Vertical tangent line
$(0, 2)$	-	Decreasing	Falling
$x = 2$	0	Neither local maximum nor local minimum	Horizontal tangent
$(2, \infty)$	-	Decreasing	Falling

Using this information together with the points $(-2, 2)$, $(-1, 1)$, $(0, 2)$, $(2, 1)$, $(4, 0)$ on the graph, we have

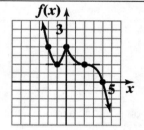

65.

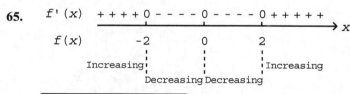

x	-2	0	2
$f(x)$	4	0	-4

67.

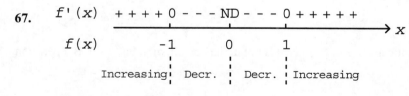

x	-1	0	1
$f(x)$	2	0	2

69. $f_1' = g_4$ **71.** $f_3' = g_6$ **73.** $f_5' = g_2$

75. Increasing on $(-1, 2)$
$[f'(x) > 0]$; decreasing on
$(-\infty, -1)$ and on $(2, \infty)$
$[f'(x) < 0]$; local minimum at
$x = -1$; local maximum at $x = 2$.

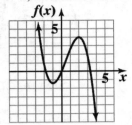

77. Increasing on $(-1, 2)$ and on
$(2, \infty)$ $[f'(x) > 0]$; decreasing on $(-\infty, -1)$
$[f'(x) < 0]$; local minimum at $x = -1$.

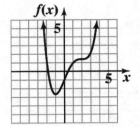

79. Increasing on $(-2, 0)$ and $(3, \infty)$ $[f'(x) > 0]$; decreasing on $(-\infty, -2)$
and $(0, 3)$ $[f'(x) < 0]$; local minima at $x = -2$ and $x = 3$, local
maximum at $x = 0$.

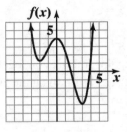

81. $f'(x) > 0$ on $(-\infty, -1)$ and on
$(3, \infty)$; $f'(x) < 0$ on $(-1, 3)$; $f'(x) = 0$ at $x = -1$ and
$x = 3$.

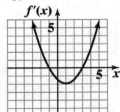

83. $f'(x) > 0$ on $(-2, 1)$ and on
$(3, \infty)$; $f'(x) < 0$ on $(-\infty, -2)$ and on $(1, 3)$:
$f'(x) = 0$ at $x = -2$, $x = 1$, and $x = 3$.

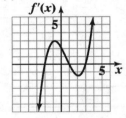

85. $f(x) = x + \dfrac{4}{x}$ [Note: f is not defined at $x = 0$.]

$$f'(x) = 1 - \frac{4}{x^2}$$

Critical values: $x = 0$ is *not* a critical value of f since 0 is not in the domain of f, but $x = 0$ is a partition number for f'.

$$f'(x) = 1 - \frac{4}{x^2} = 0$$
$$x^2 - 4 = 0$$
$$(x + 2)(x - 2) = 0$$

Thus, the critical values are $x = -2$ and $x = 2$; $x = -2$ and $x = 2$ are also partition numbers for f'.

The sign chart for f' is:

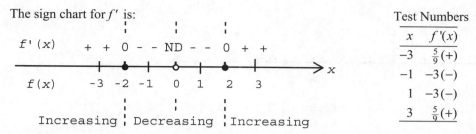

Test Numbers	
x	$f'(x)$
-3	$\frac{5}{9}(+)$
-1	$-3(-)$
1	$-3(-)$
3	$\frac{5}{9}(+)$

Therefore, f is increasing on $(-\infty, -2)$ and on $(2, \infty)$, f is decreasing on $(-2, 0)$ and on $(0, 2)$; f has a local maximum at $x = -2$ and a local minimum at $x = 2$.

87. $f(x) = 1 + \dfrac{1}{x} + \dfrac{1}{x^2}$ [Note: f is not defined at $x = 0$.]

$f'(x) = -\dfrac{1}{x^2} - \dfrac{2}{x^3}$

Critical values: $x = 0$ is not a critical value of f since 0 is not in the domain of f; $x = 0$ is a partition number for f'.

$$f'(x) = -\dfrac{1}{x^2} - \dfrac{2}{x^3} = 0$$
$$-x - 2 = 0$$
$$x = -2$$

Thus, the critical value is $x = -2$; -2 is also a partition number for f'.

The sign chart for f' is:

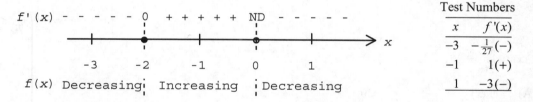

Test Numbers	
x	$f'(x)$
-3	$-\frac{1}{27}(-)$
-1	$1(+)$
1	$-3(-)$

Therefore, f is increasing on $(-2, 0)$ and f is decreasing on $(-\infty, -2)$ and on $(0, \infty)$; f has a local minimum at $x = -2$.

89. $f(x) = \dfrac{x^2}{x - 2}$ [Note: f is not defined at $x = 2$.]

$f'(x) = \dfrac{(x - 2)(2x) - x^2(1)}{(x - 2)^2} = \dfrac{x^2 - 4x}{(x - 2)^2}$

Critical values: $x = 2$ is *not* a critical value of f since 2 is not in the domain of f; $x = 2$ is a partition number for f'.

$$f'(x) = \dfrac{x^2 - 4x}{(x - 2)^2} = 0$$
$$x^2 - 4x = 0$$
$$x(x - 4) = 0$$

Thus, the critical values are $x = 0$ and $x = 4$; 0 and 4 are also partition numbers for f'.

The sign chart for f' is:

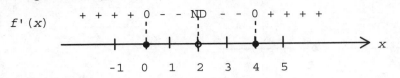

Test Numbers

x	$f'(x)$
−1	$\frac{5}{9}(+)$
1	$-3(-)$
3	$-3(-)$
5	$\frac{5}{9}(+)$

Therefore, f is increasing on $(-\infty, 0)$ and on $(4, \infty)$, f is decreasing on $(0, 2)$ and on $(2, 4)$; f has a local maximum at $x = 0$ and a local minimum at $x = 4$.

91. **(A)** The marginal profit function, P', is positive on $(0, 600)$, zero at $x = 600$, and negative on $(600, 1{,}000)$.

(B)

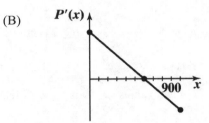

93. **(A)** The price function, $B(t)$, decreases for the first 15 months to a local minimum, increases for the next 40 months to a local maximum, and then decreases for the remaining 15 months.

(B)

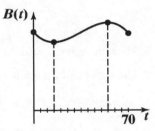

95. $C(x) = \dfrac{x^2}{20} + 20x + 320$

(A) $\overline{C}(x) = \dfrac{C(x)}{x} = \dfrac{x}{20} + 20 + \dfrac{320}{x}$

(B) Critical values:

$$\overline{C}'(x) = \frac{1}{20} - \frac{320}{x^2} = 0$$

$$x^2 - 320(20) = 0$$

$$x^2 - 6400 = 0$$

$$(x - 80)(x + 80) = 0$$

Thus, the critical value of $\overline{C}$ on the interval $(0, 150)$ is $x = 80$.

Next, construct the sign chart for $\overline{C}'$ ($x = 80$ is a partition number for $\overline{C}'$).

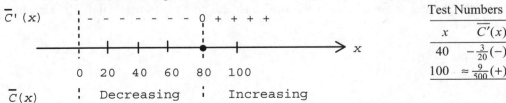

Test Numbers

x	$\overline{C}'(x)$
40	$-\frac{3}{20}(-)$
100	$\approx \frac{9}{500}(+)$

Therefore, $\overline{C}$ is increasing for $80 < x < 150$ and decreasing for $0 < x < 80$; $\overline{C}$ has a local minimum at $x = 80$.

97. $C(t) = \dfrac{0.28t}{t^2 + 4},\ 0 < t < 24$

$C'(t) = \dfrac{(t^2 + 4)(0.28) - 0.28t(2t)}{(t^2 + 4)^2} = \dfrac{0.28(4 - t^2)}{(t^2 + 4)^2}$

Critical values: C' is continuous for all t on the interval $(0, 24)$.

$C'(t) = \dfrac{0.28(4 - t^2)}{(t^2 + 4)^2} = 0$

$4 - t^2 = 0$

$(2 - t)(2 + t) = 0$

Thus, the critical value of C on the interval $(0, 24)$ is $t = 2$.

The sign chart for C' ($t = 2$ is a partition number) is:

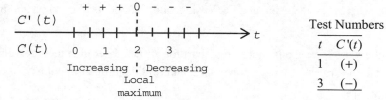

Test Numbers	
t	$C'(t)$
1	(+)
3	(−)

Therefore, C is increasing on $(0, 2)$ and decreasing on $(2, 24)$; $C(2) = 0.07$ is a local maximum.

EXERCISE 12-2

Things to remember:

1. **CONCAVITY**

   The graph of a function f is CONCAVE UPWARD on the interval (a, b) if $f'(x)$ is *increasing* on (a, b) and is CONCAVE DOWNWARD on the interval (a, b) if $f'(x)$ is *decreasing* on (a, b).

2. **SECOND DERIVATIVE**

   For $y = f(x)$, the SECOND DERIVATIVE of f, provided it exists, is:

   $$f''(x) = \frac{d}{dx}\, f'(x)$$

   Other notations for $f''(x)$ are:

   $$\frac{d^2 y}{dx^2} \quad \text{and} \quad y''.$$

3. **SUMMARY**

   For the interval (a, b):

$f''(x)$	$f'(x)$	Graph of $y = f(x)$	Example
+	Increasing	Concave upward	⌣
−	Decreasing	Concave downward	⌢

4.   INFLECTION POINT

An INFLECTION POINT is a point on the graph of a function where the concavity changes (from upward to downward, or from downward to upward). If f is continuous on (a, b) and has an inflection point at $x = c$, then either $f''(c) = 0$ or $f''(c)$ does not exist.

5.   GRAPHING STRATEGY

Step 1. Analyze $f(x)$.

Find the domain and the intercepts. The x intercepts are the solutions to $f(x) = 0$ and the y intercept is $f(0)$.

Step 2. Analyze $f'(x)$.

Find the partition numbers of $f'(x)$ and the critical values of $f(x)$. Construct a sign chart for $f'(x)$, determine the intervals where f is increasing and decreasing, and find the local maxima and minima.

Step 3. Analyze $f''(x)$.

Find the partition numbers of $f''(x)$. Construct a sign chart for $f''(x)$, determine the intervals where the graph of f is concave upward and concave downward, and find the inflection points.

Step 4. Sketch the graph of f.

Locate intercepts, local maxima and minima, and inflection points. Sketch in what you know from steps 1 – 3. Plot additional points as needed and complete the sketch.

---

1. $f(x) = x^2$ on $(-\infty, \infty)$; concave up

3. $m(x) = x^3$ on $(-\infty, 0)$; concave down

5. $p(x) = \sqrt{x}$ on $(0, \infty)$; concave down

7. $g(x) = |x|$ on $(-\infty, 0)$; neither

9.   (A) $(a,c), (c,d), (e,g)$ (B) $(d,e), (g,h)$ (C) $(d,e), (g,h)$

(D) $(a,c), (c,d), (e,g)$ (E) $(a,c), (c,d), (e,g)$ (F) $(d,e), (g,h)$

11.   (A) $f(-2) = 3$ is a local maximum; $f(2) = -1$ is a local minimum .

(B) $(0,1)$ is an inflection point.

(C) $f'(x)$ has local extremum at $x = 0$.

13. $f'(x) > 0$, $f''(x) > 0$; (C)

15. $f'(x) < 0, f''(x) > 0$; (D)

17. $f(x) = 2x^3 - 4x^2 + 5x - 6$
$f'(x) = 6x^2 - 8x + 5$
$f''(x) = 12x - 8$

19. $h(x) = 2x^{-1} - 3x^{-2}$
$h'(x) = -2x^{-2} + 6x^{-3}$
$h''(x) = 4x^{-3} - 18x^{-4}$

21. $y = x^2 - 18x^{1/2}$
$\dfrac{dy}{dx} = 2x - 9x^{-1/2}$
$\dfrac{d^2y}{dx^2} = 2 + \dfrac{9}{2}x^{-3/2}$

23. $y = (x^2 + 9)^4$
$y' = 4(x^2 + 9)^3(2x) = 8x(x^2 + 9)^3$
$y'' = 24x(x^2 + 9)^2(2x) + 8(x^2 + 9)^3$
$= 48x^2(x^2+9)^2 + 8(x^2+9)^3 = 8(x^2+9)^2(7x^2+9)$

25. $f(x) = x^3 + 30x^2$; $f'(x) = 3x^2 + 60x$; $f''(x) = 6x + 60$

$f''(x) = 0$: $6x + 60 = 0$

$x = -10$

$f(-10) = (-10)^3 + 30(-10)^2 = -1{,}000 + 3{,}000 = 2{,}000$. Inflection point: $(-10,\ 2{,}000)$

27. $f(x) = x^{5/3} + 2$; $f'(x) = \dfrac{5}{3}x^{2/3}$; $f''(x) = \dfrac{10}{9}x^{-1/3} = \dfrac{10}{9x^{1/3}}$.

$f''(x)$ does not exist at $x = 0$; $f(0) = 2$. Inflection point: $(0, 2)$.

29. $f(x) = 1 + x + x^{2/5}$; $f'(x) = 1 + \dfrac{2}{5}x^{-3/5}$; $f''(x) = -\dfrac{6}{25}x^{-8/5} = -\dfrac{6}{25x^{8/5}} < 0$ for all $x \neq 0$.

The graph is concave downward; no inflection points.

31. $f(x) = x^4 + 6x^2$; $f'(x) = 4x^3 + 12x$; $f''(x) = 12x^2 + 12 \geq 12 > 0$

The graph of f is concave upward for all x; there are no inflection points.

33. $f(x) = x^3 - 4x^2 + 5x - 2$; $f'(x) = 3x^2 - 8x + 5$; $f''(x) = 6x - 8$

$f''(x) = 0$: $6x - 8 = 0$

$x = \dfrac{4}{3}$

Sign chart for f'' $\left(\text{partition number is } \dfrac{4}{3}\right)$:

Therefore, the graph of f is concave downward on $\left(-\infty, \dfrac{4}{3}\right)$ and concave upward on $\left(\dfrac{4}{3}, \infty\right)$; there is an

inflection point at $x = \dfrac{4}{3}$.

35. $f(x) = -x^4 + 12x^3 - 12x + 24$; $f'(x) = -4x^3 + 36x^2 - 12$; $f''(x) = -12x^2 + 72x$

$f''(x) = 0$: $-12x^2 + 72x = 0$

$-12x(x - 6) = 0$

$x = 0, 6$

Sign chart for f'' (partition numbers 0, 6):

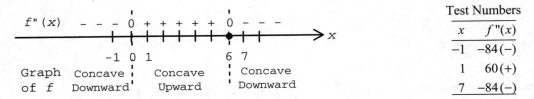

Test Numbers	
x	$f''(x)$
-1	$-84\,(-)$
1	$60\,(+)$
7	$-84\,(-)$

Therefore, the graph of f is concave downward on $(-\infty, 0)$ and $(6, \infty)$; concave upward on $(0, 6)$; there are inflection points at $x = 0$ and $x = 6$.

37. $f(x) = \ln(x^2 - 2x + 10)$ (Note: $x^2 - 2x + 10 = (x - 1)^2 + 9 > 0$ for all x)

$$f'(x) = \frac{1}{x^2 - 2x + 10}(2x - 2) = \frac{2x - 2}{x^2 - 2x + 10}$$

$$f''(x) = \frac{(x^2 - 2x + 10)(2) - (2x - 2)(2x - 2)}{(x^2 - 2x + 10)^2} = \frac{2x^2 - 4x + 20 - [4x^2 - 8x + 4]}{(x^2 - 2x + 10)^2} = \frac{-2x^2 + 4x + 16}{(x^2 - 2x + 10)^2}$$

$$= \frac{-2(x^2 - 2x - 8)}{(x^2 - 2x + 10)^2};$$

$$f''(x) = \frac{-2(x - 4)(x + 2)}{(x^2 - 2x + 10)^2}$$

$$f''(x) = 0: \quad -2(x - 4)(x + 2) = 0$$

$$x = 4, \, -2$$

Sign chart for f'' (partition numbers 4, –2):

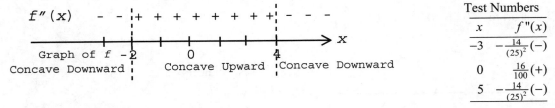

Test Numbers	
x	$f''(x)$
-3	$-\dfrac{14}{(25)^2}\,(-)$
0	$\dfrac{16}{100}\,(+)$
5	$-\dfrac{14}{(25)^2}\,(-)$

The graph of f is concave downward on $(-\infty, -2)$ and $(4, \infty)$; the graph of f is concave upward on $(-2, 4)$; there are inflection points at $x = -2$ and $x = 4$.

39. $f(x) = 8e^x - e^{2x}; \quad f'(x) = 8e^x - 2e^{2x}; \quad f''(x) = 8e^x - 4e^{2x}$

$$f''(x) = 0: \quad 8e^x - 4e^{2x} = 0$$

$$4e^x(2 - e^x) = 0$$

$$e^x = 2$$

$$x = \ln 2$$

Sign chart for f'' (partition number $\ln 2 \approx 0.69$):

$f''(x)$ + + + + | – – – – –

Graph of f ⎯⎯⎯⎯⎯⎯⎯ x
  0   0.69   1
Concave Upward   Concave Downward

Test Numbers	
x	$f''(x)$
0	$4\,(+)$
1	$8e - 4e^2\,(-)$

The graph of f is concave upward on $(-\infty, \ln 2)$ and concave downward on $(\ln 2, \infty)$; there is an inflection point at $x = \ln 2$.

41.

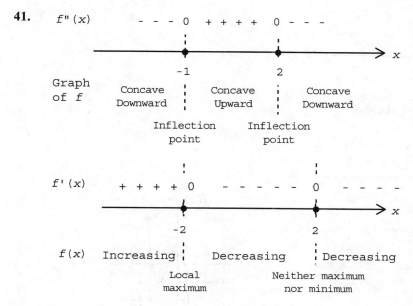

Using this information together with the points (−4, 0), (−2, 3), (−1, 1.5), (0, 0), (2, −1), (4, −3) on the graph, we have

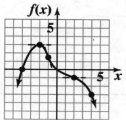

43.

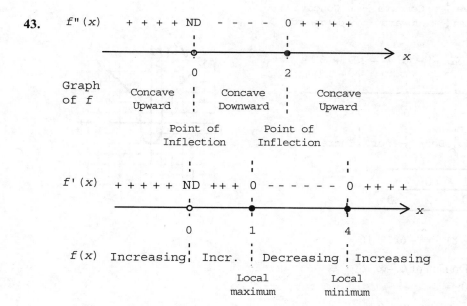

Using this information together with the points (−3, −4), (0, 0), (1, 2), (2, 1), (4, −1), (5, 0) on the graph, we have

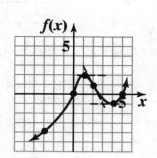

45.

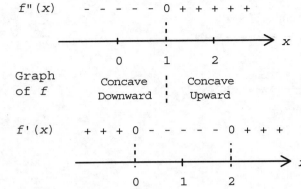

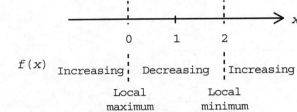

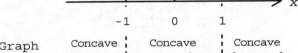

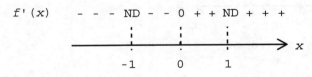

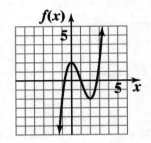

x	0	1	2
f(x)	2	0	-2

47.

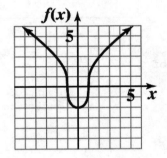

x	-1	0	1
f(x)	0	-2	0

49. $f(x) = (x-2)(x^2 - 4x - 8) = x^3 - 6x^2 + 16$

Step 1. Analyze $f(x)$. Domain of f: $(-\infty, \infty)$

x-intercept(s): $f(x) = 0$:

$$(x-2)(x^2 - 4x - 8) = 0$$
$$x = 2, 2 \pm 2\sqrt{3}$$

y-intercept: $f(0) = 16$

Step 2. Analyze $f'(x)$. $f'(x) = 3x^2 - 12x = 3x(x - 4)$

Critical values for f: $x = 0$, $x = 4$. Partition numbers for f' : 0, 4.

Sign chart for $f'(x)$:

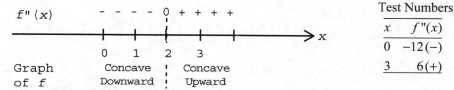

x	f'(x)
−1	15(+)
2	−12(−)
5	15(+)

Thus, f is increasing on $(-\infty, 0)$ and on $(4, \infty)$; $f(x)$ is decreasing on $(0, 4)$; f has a local maximum at $x = 0$ and a local minimum at

$x = 4$.

Step 3. Analyze $f''(x)$. $f''(x) = 6x - 12 = 6(x - 2)$

Partition number for $f''(x)$: $x = 2$

Sign chart for $f''(x)$:

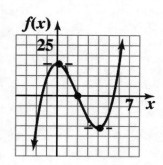

x	f''(x)
0	−12(−)
3	6(+)

The graph of f is concave upward on $(2, \infty)$, concave downward on $(-\infty, 2)$; and has an inflection point at $x = 2$.

Step 4. Sketch the graph of f.

x	f(x)
0	16
2	0
4	−16

51. $f(x) = (x + 1)(x^2 - x + 2) = x^3 + x + 2$

Step 1. Analyze $f(x)$. Domain of f: $(-\infty, \infty)$

x-intercept(s): $f(x) = 0$

$(x + 1)(x^2 - x + 2) = 0$

$x = -1$ (the quadratic factor does not have real roots)

y-intercept: $f(0) = 2$

Step 2. Analyze $f'(x)$. $f'(x) = 3x^2 + 1 > 0$ for all x.

Zeros of $f'(x)$: $f'(x)$ does not have any zeros.

Sign chart for $f'(x)$:

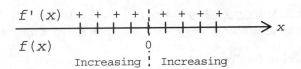

Thus, $f(x)$ is increasing on $(-\infty, \infty)$.

Step 3. Analyze $f''(x)$: $\quad f''(x) = 6x$
Partition numbers for $f''(x)$: $x = 0$
Sign chart for $f''(x)$:

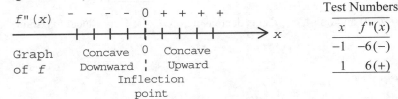

Test Numbers	
x	$f''(x)$
-1	$-6(-)$
1	$6(+)$

Thus, the graph of f is concave upward on $(0, \infty)$ and concave downward on $(-\infty, 0)$; the graph has an inflection point at $x = 0$.

Step 4. Sketch the graph of f.

x	$f(x)$
-1	0
0	2
1	4

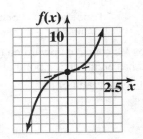

53. $f(x) = -0.25x^4 + x^3 = -\dfrac{1}{4}x^4 + x^3$

Step 1. Analyze $f(x)$. Domain of f: $(-\infty, \infty)$.
x-intercept(s): $\qquad f(x) = 0$:

$$-\frac{1}{4}x^4 + x^3 = 0$$

$$x^3\left(-\frac{1}{4}x + 1\right) = 0$$

$$x = 0, 4$$

y-intercept: $f(0) = 0$

Step 2. Analyze $f'(x)$. $f'(x) = -x^3 + 3x^2 = -x^2(x - 3)$
Critical values for f: $x = 0, 3$. Partition numbers for f': $0, 3$.
Sign chart for $f'(x)$:

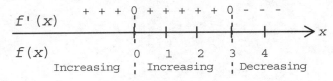

Test Numbers	
x	$f'(x)$
-1	$4(+)$
2	$4(+)$
4	$-16(-)$

Thus, f is increasing on $(-\infty, 3)$; f is decreasing on $(3, \infty)$; f has a local maximum at $x = 3$.

Step 3. Analyze $f''(x)$. $f''(x) = -3x^2 + 6x = -3x(x-2)$

Partition numbers for $f''(x)$: $x = 0, 2$
Sign chart for $f''(x)$:

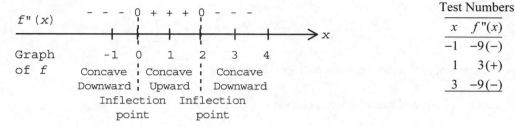

Test Numbers	
x	$f''(x)$
-1	$-9(-)$
1	$3(+)$
3	$-9(-)$

Thus, the graph of f is concave downward on $(-\infty, 0)$ and on $(2, \infty)$; concave upward on $(0, 2)$, and has inflection points at $x = 0, 2$.

Step 4. Sketch the graph of f.

x	$f(x)$
0	0
2	4
3	$\frac{27}{4}$
4	0

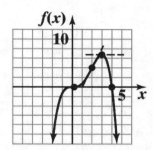

55. $f(x) = 16x(x-1)^3$
Step 1. Analyze $f(x)$. Domain of f: $(-\infty, \infty)$.
x-intercept(s):
$$f(x) = 0$$
$$16x(x-1)^3 = 0$$
$$x = 0, 1$$

y-intercept: $f(0) = 0$

Step 2. Analyze $f'(x)$. $f'(x) = 16x(3)(x-1)^2 + 16(x-1)^3$
$$= 16(x-1)^2(3x + x - 1)$$
$$= 16(x-1)^2(4x - 1)$$

Critical values for f: $x = 1$, $x = \frac{1}{4}$. Partition numbers for f': $1, \frac{1}{4}$.
Sign chart for $f'(x)$:

Test Numbers	
x	$f'(x)$
0	$-16(-)$
$\frac{1}{2}$	$4(+)$
$\frac{3}{2}$	$20(+)$

Thus, f is increasing on $\left(\frac{1}{4}, \infty\right)$; decreasing on $\left(-\infty, \frac{1}{4}\right)$; and has a local minimum at $x = \frac{1}{4}$.

Step 3. Analyze $f''(x)$. $f''(x) = 16(x-1)^2 4 + 32(x-1)(4x-1)$
$$= 32(x-1)[2(x-1) + 4x - 1]$$
$$= 32(x-1)(6x - 3)$$

Partition numbers for f'': $x = 1, \frac{1}{2}$

Sign chart for $f''(x)$:

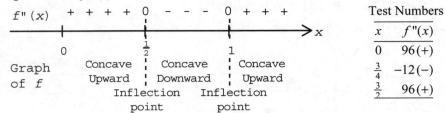

Test Numbers	
x	$f''(x)$
0	96(+)
$\frac{3}{4}$	$-12(-)$
$\frac{3}{2}$	96(+)

Thus, the graph of f is concave upward on $\left(-\infty, \frac{1}{2}\right)$ and on $(1, \infty)$, concave

downward on $\left(\frac{1}{2}, 1\right)$, and has inflection points at $x = \frac{1}{2}$, 1.

Step 4. Sketch the graph of f.

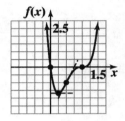

57. $f(x) = (x^2 + 3)(9 - x^2)$

Step 1. Analyze $f(x)$. Domain of f: $(-\infty, \infty)$.

Intercepts: y-intercept: $f(0) = 3(9) = 27$

x-intercepts: $(x^2 + 3)(9 - x^2) = 0$

$$(3 - x)(3 + x) = 0$$

$$x = 3, -3$$

Step 2. Analyze $f'(x)$. $f'(x) = (x^2 + 3)(-2x) + (9 - x^2)(2x)$

$$= 2x[9 - x^2 - (x^2 + 3)]$$

$$= 2x(6 - 2x^2)$$

$$= 4x(\sqrt{3} + x)(\sqrt{3} - x)$$

Critical values for f: $x = 0, x = -\sqrt{3}, x = \sqrt{3}$

Partition numbers for f': $0, -\sqrt{3}, \sqrt{3}$.

Sign chart for f':

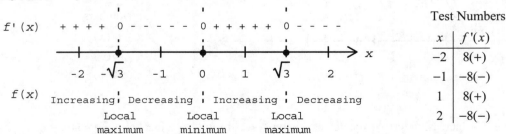

Test Numbers	
x	$f'(x)$
-2	8(+)
-1	$-8(-)$
1	8(+)
2	$-8(-)$

Thus, f is increasing on $(-\infty, -\sqrt{3})$ and on $(0, \sqrt{3})$; f is decreasing on $(-\sqrt{3}, 0)$ and on $(\sqrt{3}, \infty)$; f has local maxima at $x = -\sqrt{3}$ and $x = \sqrt{3}$ and a local minimum at $x = 0$.

Step 3. Analyze $f''(x)$. $f''(x) = 2x(-4x) + (6 - 2x^2)(2) = 12 - 12x^2 = -12(x - 1)(x + 1)$

Partition numbers for f'': $x = 1, x = -1$

Sign chart for f'':

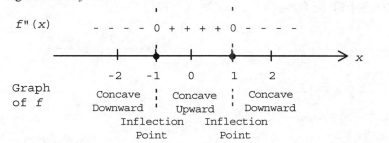

Test Numbers	
x	$f''(x)$
-2	$-36(-)$
0	$12(+)$
2	$-36(-)$

Thus, the graph of f is concave downward on $(-\infty, -1)$ and on $(1, \infty)$; the graph of f is concave upward on $(-1, 1)$; the graph has inflection points at $x = -1$ and $x = 1$.

Step 4. Sketch the graph of f:

x	$f(x)$
$-\sqrt{3}$	36
-1	32
0	27
1	32
$\sqrt{3}$	36

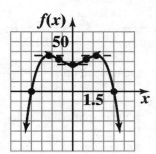

59. $f(x) = (x^2 - 4)^2$

Step 1. Analyze $f(x)$. Domain of f: $(-\infty, \infty)$.

Intercepts: y-intercept: $f(0) = (-4)^2 = 16$

              x-intercepts: $(x^2 - 4)^2 = 0$

                  $[(x - 2)(x + 2)]^2 = 0$

                  $(x - 2)^2(x + 2)^2 = 0$

                    $x = 2, -2$

Step 2. Analyze $f'(x)$. $f'(x) = 2(x^2 - 4)(2x) = 4x(x - 2)(x + 2)$

Critical values for f: $x = 0, x = 2, x = -2$

Partition numbers for f': $0, 2, -2$

Sign chart for f':

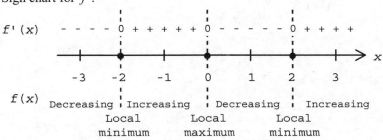

Test Numbers	
x	$f'(x)$
-3	$-60(-)$
-1	$12(+)$
1	$-12(-)$
3	$60(+)$

Thus, f is decreasing on $(-\infty, -2)$ and on $(0, 2)$; f is increasing on $(-2, 0)$ and on $(2, \infty)$; f has local minima at $x = -2$ and $x = 2$ and a local maximum at $x = 0$.

<u>Step 3. Analyze $f''(x)$.</u> $f''(x) = 4x(2x) + (x^2 - 4)(4) = 12x^2 - 16 = 12\left(x^2 - \dfrac{4}{3}\right)$

$$= 12\left(x - \dfrac{2\sqrt{3}}{3}\right)\left(x + \dfrac{2\sqrt{3}}{3}\right)$$

Partition numbers for f'': $x = \dfrac{2\sqrt{3}}{3}$, $x = \dfrac{-2\sqrt{3}}{3}$

Sign chart for f'':

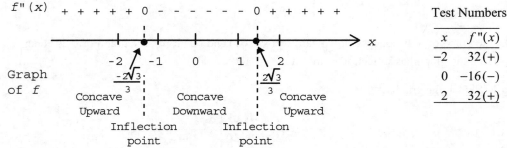

Test Numbers	
x	$f''(x)$
-2	$32\,(+)$
0	$-16\,(-)$
2	$32\,(+)$

Thus, the graph of f is concave upward on $\left(-\infty, \dfrac{-2\sqrt{3}}{3}\right)$ and on $\left(\dfrac{2\sqrt{3}}{3}, \infty\right)$; the graph of f is concave downward on $\left(\dfrac{-2\sqrt{3}}{3}, \dfrac{2\sqrt{3}}{3}\right)$; the graph has inflection points at $x = \dfrac{-2\sqrt{3}}{3}$ and $x = \dfrac{2\sqrt{3}}{3}$.

<u>Step 4. Sketch the graph of f.</u>

x	$f(x)$
-2	0
$-\dfrac{2\sqrt{3}}{3}$	$\dfrac{64}{9}$
0	16
$\dfrac{2\sqrt{3}}{3}$	$\dfrac{64}{9}$
2	0

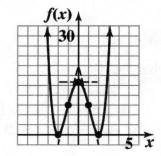

61. $f(x) = 2x^6 - 3x^5$

<u>Step 1. Analyze $f(x)$.</u> Domain of f: $(-\infty, \infty)$.

Intercepts: y-intercept: $f(0) = 2\cdot 0^6 - 3\cdot 0^5 = 0$

           x-intercepts: $2x^6 - 3x^5 = 0$

$$x^5(2x - 3) = 0$$

$$x = 0, \ \frac{3}{2}$$

<u>Step 2. Analyze $f'(x)$.</u> $f'(x) = 12x^5 - 15x^4 = 12x^4\left(x - \dfrac{5}{4}\right)$

Critical values for f: $x = 0$, $x = \dfrac{5}{4}$

Partition numbers for f': $0, \dfrac{5}{4}$

Sign chart for f':

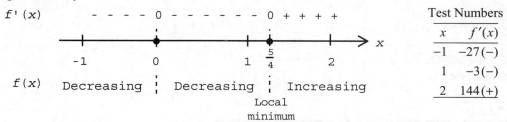

	Test Numbers	
x	$f'(x)$	
-1	$-27(-)$	
1	$-3(-)$	
2	$144(+)$	

Thus, f is decreasing on $(-\infty, 0)$ and $\left(0, \dfrac{5}{4}\right)$; f is increasing on $\left(\dfrac{5}{4}, \infty\right)$; f has a local minimum at $x = \dfrac{5}{4}$.

<u>Step 3. Analyze $f''(x)$.</u> $f''(x) = 60x^4 - 60x^3 = 60x^3(x-1)$

Partition numbers for f'': $x = 0, x = 1$

Sign chart for f'':

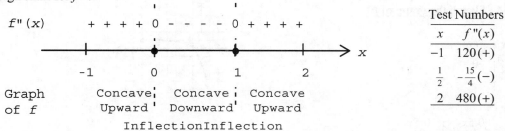

	Test Numbers	
x	$f''(x)$	
-1	$120(+)$	
$\frac{1}{2}$	$-\frac{15}{4}(-)$	
2	$480(+)$	

Thus, the graph of f is concave upward on $(-\infty, 0)$ and on $(1, \infty)$; the graph of f is concave downward on $(0, 1)$; the graph has inflection points at $x = 0$ and $x = 1$.

<u>Step 4. Sketch the graph of f.</u>

x	$f(x)$
0	0
1	-1
$\frac{5}{4}$	≈ -1.5

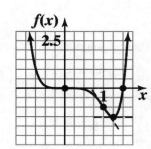

63. $f(x) = 1 - e^{-x}$

<u>Step 1. Analyze $f(x)$.</u> Domain of f: $(-\infty, \infty)$

Intercepts: x-intercepts: $f(x) = 0$

$$1 - e^{-x} = 0$$
$$e^{-x} = 1$$
$$-x = \ln 1 = 0$$
$$x = 0$$

y-intercept: $f(0) = 0$

<u>Step 2. Analyze $f'(x)$.</u> $f'(x) = e^{-x} > 0$ for all x.

Sign chart for $f'(x)$:

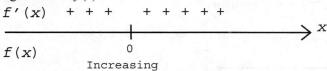

Thus, f is increasing on $(-\infty, \infty)$.

Step 3. Analyze $f''(x)$. $f''(x) = -e^{-x} < 0$ for all x.

Sign chart for $f''(x)$:

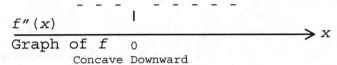

Thus, the graph of f is concave downward on $(-\infty, \infty)$.

Step 4. Sketch the graph of f:

x	$f(x)$
0	0

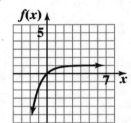

65. $f(x) = e^{0.5x} + 4e^{-0.5x}$

Step 1. Analyze $f(x)$. Domain of f: $(-\infty, \infty)$

Intercepts: x-intercepts: $f(x) = 0$

$$e^{0.5x} + 4e^{-0.5x} = 0$$

$$e^{-0.5x}[e^x + 4] = 0$$

This equation has no solutions; $e^{-0.5x} > 0$ and $e^x > 0$ for all x; there are no x-intercepts.

y-intercept: $f(0) = 5$

Step 2. Analyze $f'(x)$. $f'(x) = 0.5e^{0.5x} - 2e^{-0.5x}$; $f'(x) = 0$:

$$0.5e^{0.5x} - 2e^{-0.5x} = 0$$
$$e^{-0.5x}[0.5e^x - 2] = 0$$
$$e^x = 4$$
$$x = \ln 4$$

Critical values for f: $x = \ln 4$
Partition numbers for f': $\ln 4$
Sign chart for $f'(x)$:

$$
\begin{array}{c}
f'(x) \quad - \ - \ - \ - \ | \ + \ + \ + \ + \\
\xrightarrow{\hspace{6cm}} x \\
f(x) \qquad 0 \quad\ 1 \underset{\ln 4}{\bullet}\ 2 \\
\text{Decreasing} \ | \ \text{Increasing}
\end{array}
$$

Test Numbers

x	$f'(x)$
0	$-\frac{3}{2}(-)$
2	$\frac{e^2-4}{2e}(+)$

Thus, f is decreasing on $(-\infty, \ln 4)$; f is increasing on $(\ln 4, \infty)$; f has a local minimum at $x = \ln 4$.

Step 3. Analyze $f''(x)$. $f''(x) = 0.25e^{0.5x} + e^{-0.5x} > 0$ for all x.

Sign chart for $f''(x)$:

$$
\begin{array}{c}
f''(x) \\
\xrightarrow{\hspace{6cm}} x \\
\text{Graph of } f
\end{array}
$$

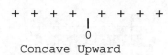

Concave Upward

Thus, the graph of f is concave upward on $(-\infty, \infty)$.

Step 4. Sketch the graph of f:

x	$f(x)$
0	5
$\ln 4$	4

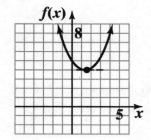

67. $f(x) = 2 \ln x - 4$

Step 1. Analyze $f(x)$. Domain of f: $(0, \infty)$

Intercepts: x-intercepts: $f(x) = 0$

$$2 \ln x - 4 = 0$$
$$\ln x = 2$$
$$x = e^2$$

y-intercept: no y-intercept; $f(0)$ is not defined.

Step 2. Analyze $f'(x)$. $f'(x) = \dfrac{2}{x} > 0$ on $(0, \infty)$

Sign chart for $f'(x)$:

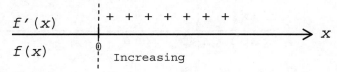

Thus, f is increasing on $(0, \infty)$.

Step 3. Analyze $f''(x)$. $f''(x) = -\dfrac{2}{x^2} < 0$ on $(0, \infty)$.

Sign chart for $f''(x)$:

f''(x)   - - - - - - -
Graph of 0 f   x
         Concave Downward

Step 4. Sketch the graph of f:

x	$f(x)$
e^2	0
e^3	2

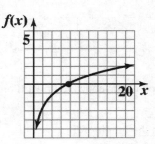

69. $f(x) = \ln(x + 4) - 2$

Step 1. Analyze $f(x)$. Domain of f: $(-4, \infty)$

Intercepts: x-intercepts: $f(x) = 0$

$$\ln(x + 4) - 2 = 0$$
$$\ln(x + 4) = 2$$
$$x + 4 = e^2$$
$$x = e^2 - 4 \approx 3.4$$

y-intercept: $f(0) = \ln 4 - 2 \approx -0.61$

Step 2. Analyze $f'(x)$. $f'(x) = \dfrac{1}{x+4} > 0$ on $(-4, \infty)$

Sign chart for $f'(x)$:

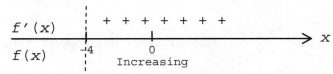

Thus, f is increasing on $(-4, \infty)$.

Step 3. Analyze $f''(x)$. $f''(x) = \dfrac{-1}{(x+4)^2} < 0$ on $(-4, \infty)$.

Sign chart for $f''(x)$:

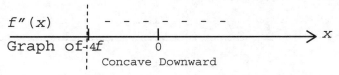

Step 4. Sketch the graph of f:

x	$f(x)$
0	-0.61
$e^2 - 4$	0

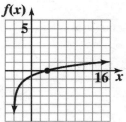

71.

x	$f'(x)$	$f(x)$
$-\infty < x < -1$	Positive and decreasing	Increasing and concave downward
$x = -1$	x-intercept	Local maximum
$-1 < x < 0$	Negative and decreasing	Decreasing and concave downward
$x = 0$	Local minimum	Inflection point
$0 < x < 2$	Negative and increasing	Decreasing and concave upward
$x = 2$	Local maximum	Inflection point
$2 < x < \infty$	Negative and decreasing	Decreasing and concave downward

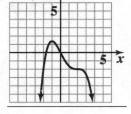

73.

x	$f'(x)$	$f(x)$

$f(x)$

$-\infty < x < -2$	Negative and increasing	Decreasing and concave upward
$x = -2$	Local maximum	Inflection point
$-2 < x < 0$	Negative and decreasing	Decreasing and concave downward
$x = 0$	Local minimum	Inflection point
$0 < x < 2$	Negative and increasing	Decreasing and concave upward
$x = 2$	Local maximum	Inflection point
$2 < x < \infty$	Negative and decreasing	Decreasing and concave downward

75. $f(x) = x^4 - 5x^3 + 3x^2 + 8x - 5$

Step 1. Analyze $f(x)$. Domain of f: $(-\infty, \infty)$.

Intercepts: y-intercept: $f(0) = -5$

x-intercepts: $x \approx -1.18, 0.61, 1.87, 3.71$

Step 2. Analyze $f'(x)$. $f'(x) = 4x^3 - 15x^2 + 6x + 8$
Critical values for f: $x \approx -0.53, 1.24, 3.04$
f is decreasing on $(-\infty, -0.53)$ and $(1.24, 3.04)$; f is increasing on
$(-0.53, 1.24)$ and $(3.04, \infty)$; f has local minima at $x = -0.53$ and 3.04; f has a local maximum at $x = 1.24$

Step 3. Analyze $f''(x)$. $f''(x) = 12x^2 - 30x + 6$
The graph of f is concave upward on $(-\infty, 0.22)$ and $(2.28, \infty)$; the graph of f is concave downward on
$(0.22, 2.28)$; the graph has inflection points at $x = 0.22$ and 2.28.

77. $f(x) = x^4 - 21x^3 + 100x^2 + 20x + 100$

Step 1. Analyze $f(x)$. Domain of f: $(-\infty, \infty)$.
Intercepts: y-intercept: $f(0) = 100$

x-intercept: $x \approx 8.01, 13.36$

Step 2. Analyze $f'(x)$. $f'(x) = 4x^3 - 63x^2 + 200x + 20$
Critical values of f:
  $x \approx -0.10, 4.57, 11.28$

f is increasing on $(-0.10, 4.57)$ and $(11.28, \infty)$;
f is decreasing on $(-\infty, -0.10)$ and $(4.57, 11.28)$;
f has a local maximum at $x = 4.57$; f has local minima at $x = -0.10$
and 11.28.

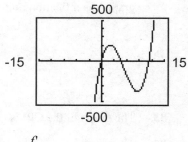

f

Step 3. Analyze $f''(x)$. $f''(x) = 12x^2 - 126x + 200$

The graph of f is concave upward on $(-\infty, 1.95)$ and $(8.55, \infty)$; the
graph of f is concave downward on $(1.95, 8.55)$; the graph has
inflection points at $x = 1.95$ and $x = 8.55$.

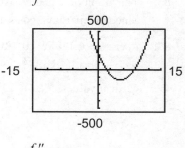

f''

79. $f(x) = -x^4 - x^3 + 2x^2 - 2x + 3$

Step 1. Analyze $f(x)$. Domain of f: $(-\infty, \infty)$.

Intercepts: y-intercept: $f(0) = 3$

x-intercepts: $x \approx -2.40, 1.16$

Step 2. Analyze $f'(x)$. $f'(x) = -4x^3 - 3x^2 + 4x - 2$

Critical value for f: $x \approx -1.58$

f is increasing on $(-\infty, -1.58)$; f is decreasing on $(-1.58, \infty)$; f has a local maximum at $x = -1.58$

Step 3. Analyze $f''(x)$. $f''(x) = -12x^2 - 6x + 4$

The graph of f is concave downward on $(-\infty, -0.88)$ and $(0.38, \infty)$; the graph of f is concave upward on $(-0.88, 0.38)$; the graph has inflection points at $x = -0.88$ and $x = 0.38$.

81. $f(x) = 0.1x^5 + 0.3x^4 - 4x^3 - 5x^2 + 40x + 30$

Step 1. Analyze $f(x)$. Domain of f: $(-\infty, \infty)$.

Intercepts: y-intercept: $f(0) = 3$

x-intercepts: $x \approx -6.68, -3.64, -0.72$

Step 2. Analyze $f'(x)$. $f'(x) = 0.5x^4 + 1.2x^3 - 12x^2 - 10x + 40$

Critical values for f: $x \approx -5.59, -2.27, 1.65, 3.82$

f is increasing on $(-\infty, -5.59)$, $(-2.27, 1.65)$, and $(3.82, \infty)$; f is decreasing on $(-5.59, -2.27)$ and $(1.65, 3.82)$; f has local minima at $x = -2.27$ and 3.82; f has local maxima at $x = -5.59$ and 1.65

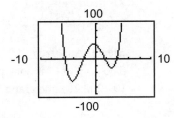

f'

Step 3. Analyze $f''(x)$. $f''(x) = 2x^3 + 3.6x^2 - 24x - 10$

The graph of f is concave downward on $(-\infty, -4.31)$ and $(-0.40, 2.91)$; the graph of f is concave upward on $(-4.31, -0.40)$ and $(2.91, \infty)$; the graph has inflection points at $x = -4.31, -0.40$ and 2.91.

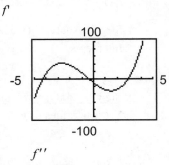

f''

83.   The graph of the CPI is concave up.

85.   The graph of C is increasing and concave down. Therefore, the graph of C' is positive and decreasing. Since the marginal costs are decreasing, the production process is becoming more efficient.

87. $R(x) = xp = 1296x - 0.12x^3, 0 < x < 80$

$R'(x) = 1296 - 0.36x^2$

Critical values: $R'(x) = 1296 - 0.36x^2 = 0$

$$x^2 = \frac{1296}{0.36} = 3600$$

$$x = \pm 60$$

Thus, $x = 60$ is the only critical value on the interval $(0, 80)$.

$R''(x) = -0.72x$

$R''(60) = -43.2 < 0$

(A) R has a local maximum at $x = 60$.

(B)   Since $R''(x) = -0.72x < 0$ for $0 < x < 80$, R is concave downward on this interval.

89.   Demand: $p = 10e^{-x}$, $0 \le x \le 5$

Revenue function: $R(x) = xp(x) = 10xe^{-x}$, $0 \le x \le 5$

(A) $R'(x) = -10xe^{-x} + 10e^{-x} = 10e^{-x}(1 - x)$

$R'(x) = 0$:

$10e^{-x}(1 - x) = 0$

$x = 1$

Sign chart for $R'(x)$:

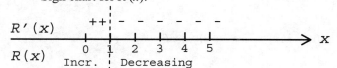

Test Numbers	
x	$R'(x)$
0	10 (+)
2	$-\frac{10}{e^2}$ (−)

Thus, R is increasing on [0, 1]; R is decreasing on (1, 5]; R has a local maximum at $x = 1$;

$R(1) = \dfrac{10}{e} \approx 3.68$.

(B) $R''(x) = 10xe^{-x} - 10e^{-x} - 10e^{-x} = 10e^{-x}(x - 2)$

$R''(x) = 0$:

$10e^{-x}(x - 2) = 0$

$x = 2$

Sign chart for R'':

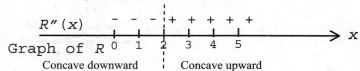

Test Numbers	
x	$R''(x)$
0	−20 (−)
3	$\frac{10}{e^3}$ (+)

Thus, the graph of R is concave downward on (0, 2) and concave upward on (2, 5); R has an inflection point at $x = 2$.

91. $T(x) = -0.25x^4 + 5x^3 = -\dfrac{1}{4}x^4 + 5x^3$, $0 \le x \le 15$

$T'(x) = -x^3 + 15x^2$

$T''(x) = -3x^2 + 30x = -3x(x - 10)$

Partition numbers for $T''(x)$: $x = 10$

Sign chart for $T''(x)$:

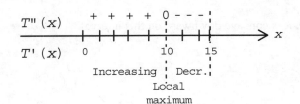

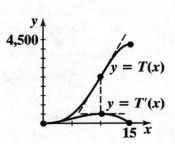

Test Numbers

x	$T''(x)$
9	27(+)
11	−33(−)

Thus, T is increasing on $(0, 10)$ and decreasing on $(10, 15)$; the point of diminishing returns is $x = 10$; the maximum rate of change is $T'(10) = 500$.

93. $N(x) = -0.25x^4 + 23x^3 - 540x^2 + 80,000,\ 24 \le x \le 45$

$N'(x) = -x^3 + 69x^2 - 1080x$

$N''(x) = -3x^2 + 138x - 1080 = -3(x^2 - 46x + 360) = -3(x - 10)(x - 36)$

Partition numbers for $N''(x)$: $x = 36$

Sign chart for $N''(x)$:

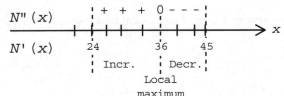

Test Numbers

x	$N''(x)$
35	75(+)
37	−81(−)

Thus, N' is increasing on $(24, 36)$ and decreasing on $(36, 45)$; the point of diminishing returns is $x = 36$; the maximum rate of change is $N'(36) = -(36)^3 + 69(36)^2 - 1080(36) = 3888$.

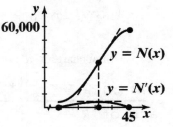

95. (A)

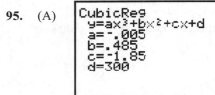

(B) From part (A),

$$y(x) = -0.005x^3 + 0.485x^2 - 1.85x + 300$$

so $y'(x) = -0.015x^2 + 0.970x - 1.85$

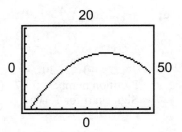

The graph of $y'(x)$ is shown at the right and the maximum value of y' occurs at $x \approx 32$; and $y(32) \approx 574$.

The manager should place 32 ads each month to maximize the rate of change of sales; the manager can expect to sell 574 cars.

97. $N(t) = 1000 + 30t^2 - t^3,\ 0 \le t \le 20$

$N'(t) = 60t - 3t^2$

$N''(t) = 60 - 6t$

(A) To determine when N' is increasing or decreasing, we must solve the inequalities $N''(t) > 0$ and $N''(t) < 0$, respectively. Now

$N''(t) = 60 - 6t = 0$

$t = 10$

The sign chart for N'' (partition number is 10) is:

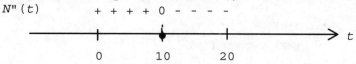

$N''(t)$ + + + + 0 – – – –

Test Numbers	
t	$N''(t)$
0	60(+)
20	–60(–)

$N'(t)$ Increasing ¦ Decreasing

Thus, N' is increasing on (0, 10) and decreasing on (10, 20).

(B)   From the results in (A), the graph of N has an inflection point at $t = 10$.

(C)

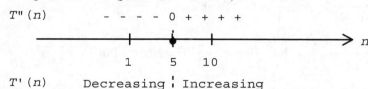

(D)   Using the results in (A),
         N' has a local maximum at
         $t = 10$:
         $N'(10) = 300$

99. $T(n) = 0.08n^3 - 1.2n^2 + 6n,\ n \geq 0$

   $T(n) = 0.24n^2 - 2.4n + 6,\ n \geq 0$

   $T''(n) = 0.48n - 2.4$

(A)   To determine when the rate of change of T, i.e., T', is increasing or decreasing, we must solve the
       inequalities $T''(n) > 0$ and $T''(n) < 0$, respectively. Now

       $T''(n) = 0.48n - 2.4 = 0$

                    $n = 5$

       The sign chart for T'' (partition number is 5) is:

       $T''(n)$ – – – – 0 + + + +

Test Numbers	
t	$T''(t)$
1	–1.92(–)
20	2.4(+)

       $T'(n)$ Decreasing ¦ Increasing

       Thus, T is increasing on (5, ∞) and decreasing on (0, 5).

(B)   Using the results in (A), the graph of T has an inflection point
       at $n = 5$. The graphs of T and T' areshown at the right.

(C)   Using the results in (A), T' has a local minimum at $n = 5$:
       $T'(5) = 0.24(5)^2 - 2.4(5) + 6 = 0$

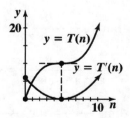

EXERCISE 12-3

Things to remember:

1. L'HÔPITAL'S RULE FOR 0/0 INDETERMINATE FORMS: VERSION 1

   For c a real number, if $\lim\limits_{x \to c} f(x) = 0$ and $\lim\limits_{x \to c} g(x) = 0$, then

   $$\lim_{x \to c} \frac{f(x)}{g(x)} = \lim_{x \to c} \frac{f'(x)}{g'(x)}.$$

   provided the second limit exists or is $+\infty$ or $-\infty$.

2. L'HÔPITAL'S RULE FOR 0/0 INDETERMINATE FORMS: VERSION 2

   (For One-Sided Limits and Limits at Infinity)

   The first version of L'Hôpital's rule remains valid if the symbol $x \to c$ is replaced everywhere it occurs with one of the following symbols:

   $$x \to c^+, \ x \to c^-, \ x \to \infty, \ x \to -\infty.$$

3. L'HÔPITAL'S RULE FOR THE INDETERMINATE FORM ∞/∞: VERSION 3

   Versions 1 and 2 of L'Hôpital's rule for the indeterminate form 0/0 are also valid if the limit of f and the limit of g are both infinite; that is both $+\infty$ and $-\infty$ are permissible for either limit.

Note: Throughout this section D_x denotes $\dfrac{d}{dx}$.

1. $\dfrac{5}{0.01} = 500; \ 500$

3. $\dfrac{3}{1,000} = 0.003; \ 0$

5. $\dfrac{1}{2(1.01-1)} = \dfrac{1}{0.02} = 50; \ 50$

7. $\dfrac{\ln 100}{100} < \dfrac{5}{100}; \ 0$

9. $\lim\limits_{x \to 3} \dfrac{x^2-9}{x-3} = \lim\limits_{x \to 3} \dfrac{D_x(x^2-9)}{D_x(x-3)} = \lim\limits_{x \to 3} \dfrac{2x}{1} = 6$. Therefore $\lim\limits_{x \to 3} \dfrac{x^2-9}{x-3} = 6$.

11. $\lim\limits_{x \to -5} \dfrac{x+5}{x^2-25} = \lim\limits_{x \to -5} \dfrac{D_x(x+5)}{D_x(x^2-25)} = \lim\limits_{x \to -5} \dfrac{1}{2x} = -\dfrac{1}{10}$. Therefore, $\lim\limits_{x \to -5} \dfrac{x+5}{x^2-25} = -\dfrac{1}{10}$.

13. $\lim\limits_{x \to 1} \dfrac{x^2+5x-6}{x-1} = \lim\limits_{x \to 1} \dfrac{D_x(x^2+5x-6)}{D_x(x-1)} = \lim\limits_{x \to 1} \dfrac{2x+5}{1} = 7$. Therefore $\lim\limits_{x \to 1} \dfrac{x^2+5x-6}{x-1} = 7$.

15. $\lim\limits_{x \to -9} \dfrac{x+9}{x^2+13x+36} = \lim\limits_{x \to -9} \dfrac{D_x(x+9)}{D_x(x^2+13x+36)} = \lim\limits_{x \to -9} \dfrac{1}{2x+13} = -\dfrac{1}{5}$. Therefore

   $\lim\limits_{x \to -9} \dfrac{x+9}{x^2+13x+36} = -\dfrac{1}{5}$.

17. $\lim\limits_{x \to \infty} \dfrac{2x+3}{5x-1} = \lim\limits_{x \to \infty} \dfrac{D_x(2x+3)}{D_x(5x-1)} = \lim\limits_{x \to \infty} \dfrac{2}{5} = \dfrac{2}{5}$. Therefore $\lim\limits_{x \to \infty} \dfrac{2x+3}{5x-1} = \dfrac{2}{5}$.

19. $\lim\limits_{x\to\infty}\dfrac{3x^2-1}{x^3+4}=\lim\limits_{x\to\infty}\dfrac{D_x(3x^2-1)}{D_x(x^3+4)}=\lim\limits_{x\to\infty}\dfrac{6x}{3x^2}=\lim\limits_{x\to\infty}\dfrac{2}{x}=0.$ Therefore $\lim\limits_{x\to\infty}\dfrac{3x^2-1}{x^3+4}=0.$

21. $\lim\limits_{x\to-\infty}\dfrac{x^2-9}{x-3}=\lim\limits_{x\to-\infty}\dfrac{D_x(x^2-9)}{D_x(x-3)}=\lim\limits_{x\to-\infty}\dfrac{2x}{1}=-\infty.$ Therefore $\lim\limits_{x\to-\infty}\dfrac{x^2-9}{x-3}=-\infty.$

23. $\lim\limits_{x\to\infty}\dfrac{2x^2+3x+1}{3x^2-2x+1}=\lim\limits_{x\to\infty}\dfrac{D_x(2x^2+3x+1)}{D_x(3x^2-2x+1)}=\lim\limits_{x\to\infty}\dfrac{4x+3}{6x-2}=\lim\limits_{x\to\infty}\dfrac{4x}{6x}=\dfrac{2}{3}.$ Therefore

$\lim\limits_{x\to\infty}\dfrac{2x^2+3x+1}{3x^2-2x+1}=\dfrac{2}{3}.$

25. $\lim\limits_{x\to0}\dfrac{e^x-1}{2x}$ (0/0 form)

$\lim\limits_{x\to0}\dfrac{e^x-1}{2x}=\lim\limits_{x\to0}\dfrac{D_x(e^x-1)}{D_x(2x)}=\lim\limits_{x\to0}\dfrac{e^x}{2}=\dfrac{1}{2}.$ Therefore, $\lim\limits_{x\to0}\dfrac{e^x-1}{2x}=\dfrac{1}{2}.$

27. $\lim\limits_{x\to1}\dfrac{x-1}{\ln x}.$ (0/0 form)

$\lim\limits_{x\to1}\dfrac{x-1}{\ln x}=\lim\limits_{x\to1}\dfrac{D_x(x-1)}{D_x(\ln x)}=\lim\limits_{x\to1}\dfrac{1}{(1/x)}=\lim\limits_{x\to1}\dfrac{x}{1}=1.$ Therefore $\lim\limits_{x\to1}\dfrac{x-1}{\ln x}=1.$

29. $\lim\limits_{x\to\infty}\dfrac{x^2}{e^x}$ (∞/∞ form)

$\lim\limits_{x\to\infty}\dfrac{x^2}{e^x}=\lim\limits_{x\to\infty}\dfrac{D_x(x^2)}{D_x(e^x)}=\lim\limits_{x\to\infty}\dfrac{2x}{e^x}=\lim\limits_{x\to\infty}\dfrac{2}{e^x}=0.$ Therefore $\lim\limits_{x\to\infty}\dfrac{x^2}{e^x}=0.$

31. $\lim\limits_{x\to0}\dfrac{e^{4x}-1}{x}$ (0/0 form)

$\lim\limits_{x\to0}\dfrac{e^{4x}-1}{x}=\lim\limits_{x\to0}\dfrac{D_x[e^{4x}-1]}{D_x(x)}=\lim\limits_{x\to0}\dfrac{4e^{4x}}{1}=4.$ Therefore $\lim\limits_{x\to0}\dfrac{e^{4x}-1}{x}=4.$

33. $\lim\limits_{x\to1}\dfrac{x^2+5x+4}{x^3+1}$; $\lim\limits_{x\to1}x^3+1=2.$ Therefore L'Hôpital's Rule does not apply. Use the rule for the limit of a

quotient:

$\lim\limits_{x\to1}\dfrac{x^2+5x+4}{x^3+1}=\dfrac{\lim\limits_{x\to1}(x^2+5x+4)}{\lim\limits_{x\to1}(x^3+1)}=\dfrac{10}{2}=5.$

35. $\lim\limits_{x\to2}\dfrac{x+2}{(x-2)^4}$; $\lim\limits_{x\to2}(x-2)^4=0$ and $\lim\limits_{x\to2}(x+2)=4$

Therefore the limit is not an indeterminate form.

$\lim\limits_{x\to2}\dfrac{x+2}{(x-2)^4}=\infty$

37. $\lim\limits_{x\to0}\dfrac{e^{4x}-1-4x}{x^2}$

Step 1:
$$\lim_{x \to 0} (e^{4x} - 1 - 4x) = e^0 - 1 = 0 \text{ and } \lim_{x \to 0} x^2 = 0.$$

Thus, L'Hôpital's rule 3a applies.

Step 2:
$$\lim_{x \to 0} \frac{D_x(e^{4x} - 1 - 4x)}{D_x x^2} = \lim_{x \to 0} \frac{4e^{4x} - 4}{2x}.$$

Since $\lim_{x \to 0} (4e^{4x} - 4) = 4e^0 - 4 = 0$ and $\lim_{x \to 0} 2x = 0$, $\lim_{x \to 0} \frac{4e^{4x} - 4}{2x}$ is a 0/0 indeterminate form and 3a

applies.

Step 3: Apply L'Hôpital's rule again.
$$\lim_{x \to 0} \frac{D_x(4e^{4x} - 4)}{D_x 2x} = \lim_{x \to 0} \frac{16e^{4x}}{2} = 8e^0 = 8.$$

Thus, $\lim_{x \to 0} \frac{e^{4x} - 1 - 4x}{x^2} = \lim_{x \to 0} \frac{4e^{4x} - 4}{2x} = \lim_{x \to 0} \frac{16e^{4x}}{2} = 8.$

39. $\lim_{x \to 2} \dfrac{\ln(x - 1)}{x - 1}$

Step 1:
$$\lim_{x \to 2} \ln(x - 1) = \ln(1) = 0 \text{ and } \lim_{x \to 2} (x - 1) = 1. \text{ Thus, L'Hôpital's rule does not apply.}$$

Step 2: Use the quotient property for limits.
$$\lim_{x \to 2} \frac{\ln(x - 1)}{x - 1} = \frac{\ln 1}{1} = \frac{0}{1} = 0.$$

41. $\lim_{x \to 0^+} \dfrac{\ln(1 + x^2)}{x^3}$

Step 1:
$$\lim_{x \to 0^+} \ln(1 + x^2) = \ln 1 = 0 \text{ and } \lim_{x \to 0^+} x^3 = 0.$$

Thus, L'Hôpital's rule 3b applies.

Step 2:
$$\lim_{x \to 0^+} \frac{D_x \ln(1 + x^2)}{D_x x^3} = \lim_{x \to 0^+} \frac{\frac{2x}{1 + x^2}}{3x^2} = \lim_{x \to 0^+} \frac{2}{3x(1 + x^2)} = \infty.$$

Thus, $\lim_{x \to 0^+} \dfrac{\ln(1 + x^2)}{x^3} = \infty.$

43. $\lim\limits_{x\to 0^+} \dfrac{\ln(1+\sqrt{x})}{x}$

Step 1:

$\lim\limits_{x\to 0^+} \ln(1+\sqrt{x}) = \ln 1 = 0$ and $\lim\limits_{x\to 0^+} x = 0$.

Thus, L'Hôpital's rule $\underline{3b}$ applies.

Step 2:

$$\lim_{x\to 0^+} \frac{D_x \ln(1+\sqrt{x})}{D_x x} = \lim_{x\to 0^+} \frac{\dfrac{1}{1+\sqrt{x}} \cdot \dfrac{1}{2} x^{-1/2}}{1} = \lim_{x\to 0^+} \frac{1}{2\sqrt{x}(1+\sqrt{x})} = \infty.$$

Thus, $\lim\limits_{x\to 0^+} \dfrac{\ln(1+\sqrt{x})}{x} = \infty$.

45. $\lim\limits_{x\to -2} \dfrac{x^2+2x+1}{x^2+x+1}$

Step 1:

Since $\lim\limits_{x\to -2}(x^2+x+1) = 4-2+1 = 3$, L'Hôpital's rule does not apply.

Step 2:

Using the limit properties, we have:

$$\lim_{x\to -2} \frac{x^2+2x+1}{x^2+x+1} = \frac{(-2)^2+2(-2)+1}{(-2)^2+(-2)+1} = \frac{4-4+1}{4-2+1} = \frac{1}{3}.$$

47. $\lim\limits_{x\to -1} \dfrac{x^3+x^2-x-1}{x^3+4x^2+5x+2}$

Step 1:

$\lim\limits_{x\to -1}(x^3+x^2-x-1) = -1+1+1-1 = 0$ and $\lim\limits_{x\to -1}(x^3+4x^2+5x+2) = -1+4-5+2 = 0$.

Thus, L'Hôpital's rule $\underline{3a}$ applies.

Step 2:

$$\lim_{x\to -1} \frac{D_x(x^3+x^2-x-1)}{D_x(x^3+4x^2+5x+2)} = \lim_{x\to -1} \frac{3x^2+2x-1}{3x^2+8x+5}.$$

Since, $\lim\limits_{x\to -1}(3x^2+2x-1) = 3-2-1 = 0$ and $\lim\limits_{x\to -1}(3x^2+8x+5) = 3-8+5 = 0$, $\lim\limits_{x\to -1} \dfrac{3x^2+2x-1}{3x^2+8x+5}$

is a 0/0 indeterminate form and $\underline{3a}$ applies again.

Step 3:

$$\lim_{x\to -1} \frac{D_x(3x^2+2x-1)}{D_x(3x^2+8x+5)} = \lim_{x\to -1} \frac{6x+2}{6x+8} = \frac{-4}{2} = -2.$$

Thus,

$$\lim_{x\to -1} \frac{x^3+x^2-x-1}{x^3+4x^2+5x+2} = \lim_{x\to -1} \frac{3x^2+2x-1}{3x^2+8x+5} = \lim_{x\to -1} \frac{6x+2}{6x+8} = -2.$$

49. $\displaystyle\lim_{x\to 2^-}\frac{x^3-12x+16}{x^3-6x^2+12x-8}$

Step 1:

$\displaystyle\lim_{x\to 2^-}(x^3-12x+16)=8-24+16=0$ and $\displaystyle\lim_{x\to 2^-}(x^3-6x^2+12x-8)=8-24+24-8=0$.

Thus L'Hôpital's 3b applies.

Step 2:

$\displaystyle\lim_{x\to 2^-}\frac{D_x(x^3-12x+16)}{D_x(x^3-6x^2+12x-8)}=\lim_{x\to 2^-}\frac{3x^2-12}{3x^2-12x+12}$.

Since $\displaystyle\lim_{x\to 2^-}(3x^2-12)=12-12=0$ and $\displaystyle\lim_{x\to 2^-}(3x^2-12x+12)=12-24+12=0$, $\displaystyle\lim_{x\to 2^-}\frac{3x^2-12}{3x^2-12x+12}$ is

a 0/0 indeterminate form and 3b applies again.

Step 3:

$\displaystyle\lim_{x\to 2^-}\frac{D_x(3x^2-12)}{D_x(3x^2-12x+12)}=\lim_{x\to 2^-}\frac{6x}{6x-12}=\lim_{x\to 2^-}\frac{x}{x-2}=-\infty$.

Thus,

$\displaystyle\lim_{x\to 2^-}\frac{x^3-12x+16}{x^3-6x^2+12x-8}=\lim_{x\to 2^-}\frac{3x^2-12}{3x^2-12x+12}=\lim_{x\to 2^-}\frac{x}{x-2}=-\infty$.

51. $\displaystyle\lim_{x\to\infty}\frac{3x^2+5x}{4x^3+7}$

Step 1:

$\displaystyle\lim_{x\to\infty}(3x^2+5x)=\infty$ and $\displaystyle\lim_{x\to\infty}(4x^3+7)=\infty$.

Thus, L'Hôpital's rule 4 applies.

Step 2:

$\displaystyle\lim_{x\to\infty}\frac{D_x(3x^2+5x)}{D_x(4x^3+7)}=\lim_{x\to\infty}\frac{6x+5}{12x^2}$.

Since $\displaystyle\lim_{x\to\infty}(6x+5)=\infty$ and $\displaystyle\lim_{x\to\infty}12x^2=\infty$, $\displaystyle\lim_{x\to\infty}\frac{6x+5}{12x^2}$ is an ∞/∞ indeterminate form and 4 applies again.

Step 3:

$\displaystyle\lim_{x\to\infty}\frac{D_x(6x+5)}{D_x(12x^2)}=\lim_{x\to\infty}\frac{6}{24x}=\frac{1}{4}\lim_{x\to\infty}\frac{1}{x}=0$.

Thus, $\displaystyle\lim_{x\to\infty}\frac{3x^2+5x}{4x^3+7}=\lim_{x\to\infty}\frac{6x+5}{12x^2}=\frac{1}{4}\lim_{x\to\infty}\frac{1}{x}=0$.

An alternative approach is:

$$\lim_{x\to\infty}\frac{3x^2+5x}{4x^3+7}=\lim_{x\to\infty}\frac{x^2\left(3+\dfrac{5}{x}\right)}{x^2\left(4x+\dfrac{7}{x^2}\right)}=\lim_{x\to\infty}\frac{3+\dfrac{5}{x}}{4x+\dfrac{7}{x^2}}=0.$$

53. $\displaystyle\lim_{x\to\infty}\frac{x^2}{e^{2x}}$

Step 1

$\displaystyle\lim_{x\to\infty}x^2=\infty$ and $\displaystyle\lim_{x\to\infty}e^{2x}=\infty$. Thus, L'Hôpital's rule $\underline{4}$ applies.

Step 2:

$$\lim_{x\to\infty}\frac{D_x x^2}{D_x e^{2x}}=\lim_{x\to\infty}\frac{2x}{2e^{2x}}=\lim_{x\to\infty}\frac{x}{e^{2x}}.$$

Since $\displaystyle\lim_{x\to\infty}x=\infty$ and $\displaystyle\lim_{x\to\infty}e^{2x}=\infty$, $\displaystyle\lim_{x\to\infty}\frac{x}{e^{2x}}$ is an ∞/∞ indeterminate form and $\underline{4}$ applies again.

Step 3:

$$\lim_{x\to\infty}\frac{D_x x}{D_x e^{2x}}=\lim_{x\to\infty}\frac{1}{2e^{2x}}=0. \text{ Thus, } \lim_{x\to\infty}\frac{x^2}{e^{2x}}=\lim_{x\to\infty}\frac{x}{e^{2x}}=\lim_{x\to\infty}\frac{1}{2e^{2x}}=0.$$

55. $\displaystyle\lim_{x\to\infty}\frac{1+e^{-x}}{1+x^2}$

Step 1:

$\displaystyle\lim_{x\to\infty}(1+e^{-x})=1$ and $\displaystyle\lim_{x\to\infty}(1+x^2)=\infty$. Thus, this limit is *not* an indeterminate form.

Step 2:

Since $\displaystyle\lim_{x\to\infty}(1+e^{-x})=1$ and $\displaystyle\lim_{x\to\infty}(1+x^2)=\infty$, $\displaystyle\lim_{x\to\infty}\frac{1+e^{-x}}{1+x^2}=0.$

57. $\displaystyle\lim_{x\to\infty}\frac{e^{-x}}{\ln(1+4e^{-x})}$

Step 1:

$\displaystyle\lim_{x\to\infty}e^{-x}=0$ and $\displaystyle\lim_{x\to\infty}\ln(1+4e^{-x})=\ln 1=0.$

Thus, L'Hôpital's rule $\underline{3b}$ applies.

Step 2:

$$\lim_{x\to\infty}\frac{D_x(e^{-x})}{D_x\ln(1+4e^{-x})}=\lim_{x\to\infty}\frac{-e^{-x}}{\dfrac{1}{1+4e^{-x}}\cdot(-4e^{-x})}=\lim_{x\to\infty}\frac{1+4e^{-x}}{4}=\frac{1}{4}.$$

Thus, $\displaystyle\lim_{x\to\infty}\frac{e^{-x}}{\ln(1+4e^{-x})}=\frac{1}{4}.$

59. $\displaystyle\lim_{x\to 0} \frac{e^x - e^{-x} - 2x}{x^3}$

<u>Step 1</u>:

$\displaystyle\lim_{x\to 0}(e^x - e^{-x} - 2x) = e^0 - e^0 = 0$ and $\displaystyle\lim_{x\to 0} x^3 = 0$.

Thus, L'Hôpital's rule <u>3</u>a applies.

<u>Step 2</u>:

$\displaystyle\lim_{x\to 0} \frac{D_x(e^x - e^{-x} - 2x)}{D_x x^3} = \lim_{x\to 0} \frac{(e^x + e^{-x} - 2)}{3x^2}$.

Since $\displaystyle\lim_{x\to 0}(e^x + e^{-x} - 2) = e^0 + e^0 - 2 = 0$ and $\displaystyle\lim_{x\to 0} 3x^2 = 0$, L'Hôpital's rule <u>3</u>a applies again.

<u>Step 3</u>:

$\displaystyle\lim_{x\to 0} \frac{D_x(e^x + e^{-x} - 2)}{D_x 3x^2} = \lim_{x\to 0} \frac{e^x - e^{-x}}{6x}$. Since $\displaystyle\lim_{x\to 0}(e^x - e^{-x}) = e^0 - e^0 = 0$

and $\displaystyle\lim_{x\to 0} 6x = 0$, we apply <u>3</u>a a third time.

<u>Step 4</u>:

$\displaystyle\lim_{x\to 0} \frac{D_x(e^x - e^{-x})}{D_x 6x} = \lim_{x\to 0} \frac{e^x + e^{-x}}{6} = \frac{e^0 + e^0}{6} = \frac{2}{6} = \frac{1}{3}$.

Therefore, $\displaystyle\lim_{x\to 0} \frac{e^x - e^{-x} - 2x}{x^3} = \frac{1}{3}$.

61. $\displaystyle\lim_{x\to 0^+} x \ln x = \lim_{x\to 0^+} \frac{\ln x}{\dfrac{1}{x}}$

<u>Step 1</u>:

$\displaystyle\lim_{x\to 0^+} \ln x = -\infty$ and $\displaystyle\lim_{x\to 0^+} \frac{1}{x} = \infty$.

Therefore, L'Hôpital's rule <u>4</u> applies.

<u>Step 2</u>:

$\displaystyle\lim_{x\to 0^+} \frac{D_x \ln x}{D_x \dfrac{1}{x}} = \lim_{x\to 0^+} \frac{\dfrac{1}{x}}{-\dfrac{1}{x^2}} = \lim_{x\to 0^+} (-x) = 0$. Thus, $\displaystyle\lim_{x\to 0^+} x \ln x = 0$.

63. $\displaystyle\lim_{x\to \infty} \frac{\ln x}{x^n}$, n a positive integer.

<u>Step 1</u>:

$\displaystyle\lim_{x\to \infty} \ln x = \infty$ and $\displaystyle\lim_{x\to \infty} x^n = \infty$. Therefore, L'Hôpital's rule <u>4</u> applies.

<u>Step 2</u>:

$\displaystyle\lim_{x\to \infty} \frac{D_x \ln x}{D_x x^n} = \lim_{x\to \infty} \frac{\dfrac{1}{x}}{nx^{n-1}} = \lim_{x\to \infty} \frac{1}{nx^n} = 0$. Thus, $\displaystyle\lim_{x\to \infty} \frac{\ln x}{x^n} = 0$.

65. $\lim\limits_{x\to\infty}\dfrac{e^x}{x^n}$, n a positive integer.

Step 1:

$\lim\limits_{x\to\infty} e^x = \infty$ and $\lim\limits_{x\to\infty} x^n = \infty$. Therefore, L'Hôpital's rule $\underline{4}$ applies.

Step 2:

$\lim\limits_{x\to\infty}\dfrac{D_x e^x}{D_x x^n} = \lim\limits_{x\to\infty}\dfrac{e^x}{nx^{n-1}}$

If $n = 1$, this limit is $\lim\limits_{x\to\infty}\dfrac{e^x}{1} = \infty$. If $n > 1$, then L'Hôpital's rule $\underline{4}$ applies again.

Step 3:

$\lim\limits_{x\to\infty}\dfrac{D_x e^x}{D_x nx^{n-1}} = \lim\limits_{x\to\infty}\dfrac{e^x}{n(n-1)x^{n-2}}$.

This limit is ∞ if $n = 2$ and has the indeterminate form ∞/∞ if $n > 2$. Applying L'Hôpital's rule $\underline{4}$ n-times,

we have $\lim\limits_{x\to\infty}\dfrac{e^x}{n!} = \infty$. Thus $\lim\limits_{x\to\infty}\dfrac{e^x}{x^n} = \infty$.

67. $\lim\limits_{x\to\infty}\dfrac{\sqrt{1+x^2}}{x}$

Step 1:

$\lim\limits_{x\to\infty}\sqrt{1+x^2} = \infty$ and $\lim\limits_{x\to\infty} x = \infty$.

Thus, L'Hôpital's rule applies.

Step 2:

$\lim\limits_{x\to\infty}\dfrac{D_x(\sqrt{1+x^2})}{D_x(x)} = \lim\limits_{x\to\infty}\dfrac{\frac{x}{\sqrt{1+x^2}}}{1} = \lim\limits_{x\to\infty}\dfrac{x}{\sqrt{1+x^2}}$

Since $\lim\limits_{x\to\infty} x = \lim\limits_{x\to\infty}\sqrt{1+x^2} = \infty$, L'Hôpital's rule applies again.

Step 3.

$\lim\limits_{x\to\infty}\dfrac{D_x(x)}{D_x\sqrt{1+x^2}} = \lim\limits_{x\to\infty}\dfrac{1}{\frac{x}{\sqrt{1+x^2}}} = \lim\limits_{x\to\infty}\dfrac{\sqrt{1+x^2}}{x}$, the original limit.

Algebraic manipulation: for $x > 0$

$$\dfrac{\sqrt{1+x^2}}{x} = \sqrt{\dfrac{1+x^2}{x^2}} = \sqrt{\dfrac{1}{x^2}+1}.$$

Therefore,

$$\lim\limits_{x\to\infty}\dfrac{\sqrt{1+x^2}}{x} = \lim\limits_{x\to\infty}\sqrt{\dfrac{1}{x^2}+1} = 1$$

69. $\lim\limits_{x \to -\infty} \dfrac{\sqrt[3]{x^3 + 1}}{x} = \lim\limits_{x \to -\infty} \dfrac{(x^3 + 1)^{1/3}}{x}$

Step 1:

$\lim\limits_{x \to -\infty} (x^3 + 1)^{1/3} = -\infty$ and $\lim\limits_{x \to -\infty} x = -\infty$

Thus, L'Hôpital's rule applies.

Step 2:

$\lim\limits_{x \to -\infty} \dfrac{D_x (x^3 + 1)^{1/3}}{D_x (x)} = \lim\limits_{x \to -\infty} \dfrac{\frac{1}{3}(x^3 + 1)^{-2/3} \cdot 3x^2}{1} = \lim\limits_{x \to -\infty} \dfrac{x^2}{(x^3 + 1)^{2/3}}$

Since $\lim\limits_{x \to -\infty} x^2 = \lim\limits_{x \to -\infty} (x^3 + 1)^{2/3} = \infty$, L'Hôpital's rule applies again.

Step 3:

$\lim\limits_{x \to -\infty} \dfrac{D_x (x^2)}{D_x (x^3 + 1)^{2/3}} = \lim\limits_{x \to -\infty} \dfrac{2x}{\frac{2}{3}(x^3 + 1)^{-1/3}(3x^2)} = \lim\limits_{x \to -\infty} \dfrac{(x^3 + 1)^{1/3}}{x}$, the original limit.

Algebraic manipulation: for $x < 0$,

$\dfrac{\sqrt[3]{x^3 + 1}}{x} = \sqrt[3]{\dfrac{x^3 + 1}{x^3}} = \sqrt[3]{1 + \dfrac{1}{x^3}}$.

Therefore,

$\lim\limits_{x \to -\infty} \dfrac{\sqrt[3]{x^3 + 1}}{x} = \lim\limits_{x \to -\infty} \sqrt[3]{1 + \dfrac{1}{x^3}} = 1$.

EXERCISE 12-4

Things to remember:

1.   **GRAPHING STRATEGY**

Step 1. Analyze $f(x)$. $f''(x)$
(A) Find the domain of f.
(B) Find the intercepts.
(C) Find asymptotes.

Step 2.   Analyze $f'(x)$. Find the partition numbers and critical values of $f'(x)$. Construct
a   sign chart for $f'(x)$, determine the intervals where f is increasing and
decreasing, and find local maxima and minima.

Step 3.   Analyze $f''(x)$. Find the partition numbers of $f''(x)$.
Construct a sign chart for $f''(x)$, determine the intervals where the graph of f is
concave upward and concave downward, and find inflection points.

Step 4.   Sketch the graph of f. Draw asymptotes and locate intercepts, local maxima and
minima, and inflection points. Sketch in what you know from steps 1—3. Plot
additional points as needed and complete the sketch.

---

1. $f(x) = 3x + 36$. Domain: All real numbers; x-intercept: $3x + 36 = 0,\ x = -12$; y-intercept: $f(0) = 36$.

3. $f(x) = \sqrt{25 - x}$. Domain: $(-\infty, 5]$; x-intercepts: $\sqrt{25 - x} = 0,\ x = 25$; y-intercept: $f(0) = 5$.

5. $f(x) = \dfrac{x + 1}{x - 2}$. Domain: All real numbers except $x = 2$; x-intercepts: $f(x) = \dfrac{x + 1}{x - 2} = 0,\ x = -1$; y-intercept: $f(0) = -\dfrac{1}{2}$.

7. $f(x) = \dfrac{3}{x^2 - 1}$. Domain: All real numbers except $x = -1, 1$; x-intercepts: none, y-intercept: $f(0) = -3$.

9. (A) $f'(x) < 0$ on $(-\infty, b), (0, e), (e, g)$

   (B) $f'(x) > 0$ on $(b, d), (d, 0), (g, \infty)$

   (C) $f(x)$ is increasing on $(b, d), (d, 0), (g, \infty)$

   (D) $f(x)$ is decreasing on $(-\infty, b), (0, e), (e, g)$

   (E) $f(x)$ has a local maximum at $x = 0$

   (F) $f(x)$ has local minima at $x = b$ and $x = g$

   (G) $f''(x) < 0$ on $(-\infty, a), (d, e), (h, \infty)$

   (H) $f''(x) > 0$ on $(a, d), (e, h)$

   (I) The graph of f is concave upward on (a, d) and (e, h).

   (J) The graph of f is concave downward on $(-\infty, a), (d, e)$, and (h, ∞).

   (K) Inflection points at $x = a, x = h$

   (L) Horizontal asymptote: $y = L$

   (M) Vertical asymptotes: $x = d, x = e$

11. Step 1. Analyze $f(x)$:

   (A) Domain: All real numbers

   (B) Intercepts: y-intercept: 0
            x-intercepts: $-4, 0, 4$

   (C) Asymptotes: Horizontal asymptote: $y = 2$

Step 2. Analyze $f'(x)$:

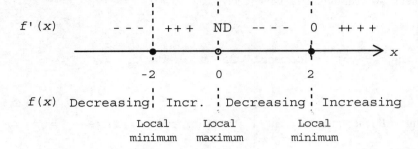

$f(x)$ Decreasing ┊ Incr. ┊ Decreasing ┊ Increasing

Local   Local   Local
minimum   maximum   minimum

Step 3. Analyze $f''(x)$:

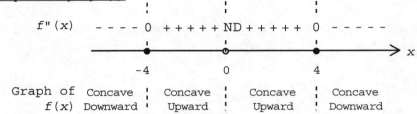

Graph of   Concave ┊ Concave ┊ Concave ┊ Concave
$f(x)$ Downward ┊ Upward ┊ Upward ┊ Downward

Step 4. Sketch the graph of f:

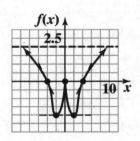

13. Step 1. Analyze $f(x)$:

(A) Domain: All real numbers except $x = -2$

(B) Intercepts: y-intercept: 0

                x-intercepts: $-4, 0$

(C) Asymptotes: Horizontal asymptote: $y = 1$

                Vertical asymptote: $x = -2$

Step 2. Analyze $f'(x)$:

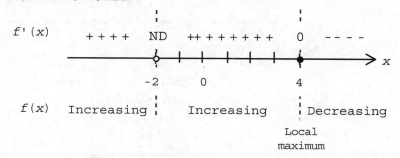

$f'(x)$ + + + + ND ++ + + + + + +   0   - - - -

-2   0   4

$f(x)$ Increasing   Increasing   Decreasing

Local
maximum

Step 3. Analyze $f''(x)$:

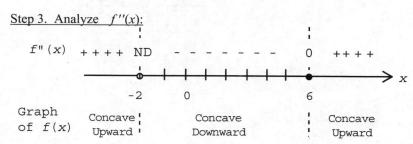

$f''(x)$ + + + + ND - - - - - - - -   0   ++ + +

-2   0   6

Graph
of $f(x)$ Concave   Concave   Concave
            Upward   Downward   Upward

Step 4. Sketch the graph of f:

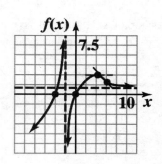

15. Step 1. Analyze $f(x)$:

(A) Domain: All real numbers except $x = -1$

(B) Intercepts: y-intercept: -1

x-intercept: 1

(C) Asymptotes: Horizontal asymptote: $y = 1$

Vertical asymptote: $x = -1$

Step 2. Analyze $f'(x)$:

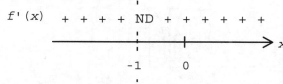

$f'(x)$ + + + + + ND + + + + + +

-1   0

$f(x)$ Increasing   Increasing

Step 3. Analyze $f''(x)$:

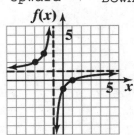

$f''(x)$ + + + ND − − −

Graph
of f

−1   0

Concave   Concave
Upward   Downward

Step 4. Sketch the graph of f:

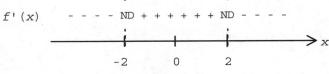

17.   Step 1. Analyze $f(x)$:

(A) Domain: All real numbers except $x = -2$, $x = 2$

(B) Intercepts: y-intercept: 0

            x-intercept: 0

(C) Asymptotes: Horizontal asymptote: $y = 0$

            Vertical asymptotes: $x = -2$, $x = 2$

Step 2. Analyze $f'(x)$:

$f'(x)$ − − − − ND + + + + + ND − − − −

−2   0   2

$f(x)$ Decreasing   Increasing   Decreasing

Step 3. Analyze $f''(x)$:

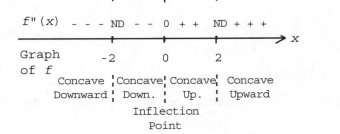

$f''(x)$ − − − ND − − 0 + + ND + + +

Graph   −2   0   2
of f

Concave   Concave   Concave   Concave
Downward   Down.   Up.   Upward

Inflection
Point

Step 4. Sketch the graph of f:

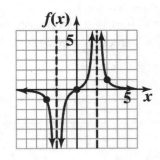

19. $f(x) = \dfrac{x+3}{x-3}$

<u>Step 1. Analyze $f(x)$</u>:

(A)   Domain: All real numbers except $x = 3$.

(B)   Intercepts: y-intercept: $f(0) = \dfrac{3}{-3} = -1$

x-intercepts: $\dfrac{x+3}{x-3} = 0$

$x + 3 = 0$

$x = -3$

(C)   Asymptotes:

<u>Horizontal asymptote</u>: $\displaystyle\lim_{x \to \infty} \frac{x+3}{x-3} = \lim_{x \to \infty} \frac{x\left(1 + \frac{3}{x}\right)}{x\left(1 - \frac{3}{x}\right)} = 1$.

Thus, $y = 1$ is a horizontal asymptote.

<u>Vertical asymptote</u>: The denominator is 0 at $x = 3$ and the numerator is not 0 at $x = 3$. Thus, $x = 3$ is a vertical asymptote.

<u>Step 2. Analyze $f'(x)$</u>:

$f'(x) = \dfrac{(x-3)(1) - (x+3)(1)}{(x-3)^2} = \dfrac{-6}{(x-3)^2} = -6(x-3)^{-2}$

Critical values: None

Partition number: $x = 3$

Sign chart for f':

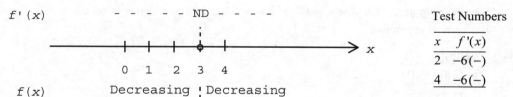

Test Numbers	
x	$f'(x)$
2	$-6(-)$
4	$-6(-)$

Thus, f is decreasing on $(-\infty, 3)$ and on $(3, \infty)$; there are no local extrema.

<u>Step 3. Analyze $f''(x)$</u>:

$f''(x) = 12(x-3)^{-3} = \dfrac{12}{(x-3)^3}$

Partition number for f'': $x = 3$

Sign chart for f'':

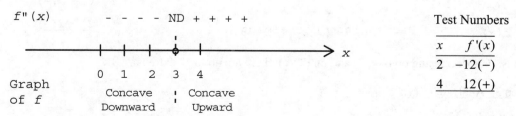

Test Numbers	
x	$f'(x)$
2	$-12(-)$
4	$12(+)$

Thus, the graph of f is concave downward on $(-\infty, 3)$ and concave upward on $(3, \infty)$.

<u>Step 4. Sketch the graph of f:</u>

x	$f(x)$
-3	0
0	-1
5	4

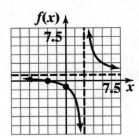

21. $f(x) = \dfrac{x}{x-2}$

<u>Step 1. Analyze $f(x)$:</u>

(A) Domain: All real numbers except $x = 2$.

(B) Intercepts: y-intercept: $f(0) = \dfrac{0}{-2} = 0$

x-intercepts: $\dfrac{x}{x-2} = 0$

$x = 0$

(C) Asymptotes:

<u>Horizontal asymptote</u>: $\displaystyle\lim_{x \to \infty} \dfrac{x}{x-2} = \lim_{x \to \infty} \dfrac{x}{x\left(1 - \frac{2}{x}\right)} = 1.$

Thus, $y = 1$ is a horizontal asymptote.

<u>Vertical asymptote</u>: The denominator is 0 at $x = 2$ and the numerator is not 0 at $x = 2$. Thus, $x = 2$ is a vertical asymptote.

<u>Step 2. Analyze $f'(x)$:</u>

$f'(x) = \dfrac{(x-2)(1) - x(1)}{(x-2)^2} = \dfrac{-2}{(x-2)^2} = -2(x-2)^{-2}$

Critical values: None

Partition number: $x = 2$

Sign chart for f' :

$f'(x)$ - - - - ND - - - -

Test Numbers

x	$f'(x)$
0	$-\frac{1}{2}\,(-)$
3	$-2\,(-)$

$f(x)$ Decreasing ┊ Decreasing

Thus, f is decreasing on $(-\infty, 2)$ and on $(2, \infty)$; there are no local extrema.

<u>Step 3. Analyze $f''(x)$:</u>

$f''(x) = 4(x-2)^{-3} = \dfrac{4}{(x-2)^3}$

Partition number for f'': $x = 2$

Sign chart for f'':

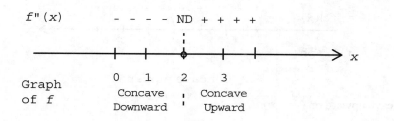

Thus, the graph of f is concave downward on $(-\infty, 2)$ and concave upward on $(2, \infty)$.

Step 4. Sketch the graph of f:

x	$f(x)$
0	0
4	2

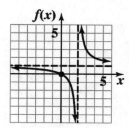

23. $f(x) = 5 + 5e^{-0.1x}$

Step 1. Analyze $f(x)$:

(A) Domain: All real numbers.

(B) Intercepts: y-intercept: $f(0) = 5 + 5e^0 = 10$

x-intercept: $5 + 5e^{-0.1x} = 0$

$e^{-0.1x} = -1$; no solutions

$e^{-0.1x} > 0$ for all x

(C) Asymptotes:

Vertical asymptotes: None

Horizontal asymptotes: $\lim\limits_{x\to\infty} (5 + 5e^{-0.1x}) = \lim\limits_{x\to\infty} \left(5 + \dfrac{5}{e^{0.1x}}\right) = 5$

$\lim\limits_{x\to-\infty} (5 + 5e^{-0.1x})$ does not exist

$y = 5$ is a horizontal asymptote.

Step 2. Analyze $f'(x)$:

$f'(x) = 5e^{-0.1x}(-0.1) = -0.5e^{-0.1x}$

Critical values: None

Partition numbers: None

Sign chart for f':

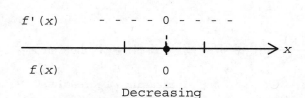

Thus, f decreases on $(-\infty, \infty)$.

Step 3. Analyze $f''(x)$:

$f''(x) = -0.5e^{-0.1x}(-0.1) = 0.05e^{-0.1x}$

Partition numbers for $f''(x)$: None

Sign chart for f'':

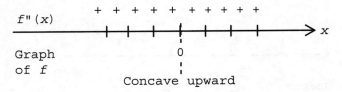

Thus, the graph of f is concave upward on $(-\infty, \infty)$.

Step 4. Sketch the graph of f:

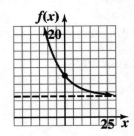

25. $f(x) = 5xe^{-0.2x}$

Step 1. Analyze $f(x)$:

(A)   Domain: All real numbers.

(B)   Intercepts: y-intercept: $f(0) = 5(0)e^0 = 0$

x-intercept: $5xe^{-0.2x} = 0$

$x = 0$

(C) Asymptotes:

Vertical asymptotes: None

Horizontal asymptotes:

x	10	20	30	$40 \to \infty$
$f(x)$	6.77	1.83	0.37	$0.067 \to 0$

x	-10	-20	$\to -\infty$
$f(x)$	-369.45	-5458.01	$\to -\infty$

$y = 0$ is a horizontal asymptote

Step 2. Analyze $f'(x)$:

$f'(x) = 5xe^{-0.2x}(-0.2) + e^{-0.2x} 5 = 5e^{-0.2x}[1 - 0.2x]$

Critical values: $x = 5$

Partition numbers: $x = 5$

Sign chart for f' :

$$
\begin{array}{c}
f'(x) \quad +\ +\ +\ +\ +\ +\ +\ 0\ -\ -\ -\ - \\
\longrightarrow x \\
f(x) \qquad\quad 0 \qquad\qquad 5 \\
\qquad\quad \text{Increasing} \mid \text{Decreasing} \\
\qquad\qquad\qquad \text{Local} \\
\qquad\qquad\quad \text{maximum}
\end{array}
$$

Test Numbers

x	$f'(x)$
0	$5(+)$
6	$-e^{-1.2} (-)$

Thus, $f(x)$ increases on $(-\infty, 5)$, has a local maximum at $x = 5$, and decreases on $(5, \infty)$.

Step 3. Analyze $f''(x)$:

$$f''(x) = 5e^{-0.2x}(-0.2) + [1 - 0.2x]5e^{-0.2x}(-0.2)$$
$$= -e^{-0.2x}[2 - 0.2x]$$

Partition numbers for f'': $x = 10$

Sign chart for f'':

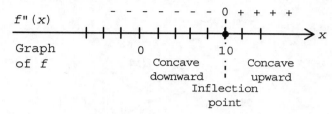

Test Numbers	
x	$f''(x)$
0	$-2\,(-)$
20	$2e^{-4}\,(+)$

Step 4. Sketch the graph of f:

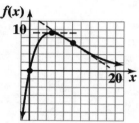

27. $f(x) = \ln(1 - x)$

Step 1. Analyze $f(x)$:
(A)   Domain: All real numbers x such that $1 - x > 0$, i.e., $x < 1$
          or $(-\infty, 1)$.

(B)   Intercepts: y-intercept: $f(0) = \ln(1 - 0) = \ln 1 = 0$
                       x-intercepts: $\ln(1 - x) = 0$
                                          $1 - x = 1$
                                              $x = 0$

(C)   Asymptotes:
       Horizontal asymptote: $\lim\limits_{x \to -\infty} f(x) = \lim\limits_{x \to -\infty} \ln(1 - x)$ does not exist. Thus, there are no horizontal
       asymptotes.
       Vertical asymptote: From the table,

x	0.9	0.99	0.99999	0.9999999	$\to 1$
$f(x)$	-2.30	-4.61	-11.51	-16.12	$\to -\infty$

       We conclude that $x = 1$ is a vertical asymptote.

Step 2. Analyze $f'(x)$:

$$f'(x) = \frac{1}{1 - x}(-1), \quad x < 1$$
$$= \frac{1}{x - 1}$$

Now, $f'(x) = \dfrac{1}{x - 1} < 0$ on $(-\infty, 1)$.

Thus, f is decreasing on $(-\infty, 1)$; there are no critical values and no local extrema.

Step 3. Analyze $f''(x)$:

$$f'(x) = (x-1)^{-1}$$

$$f''(x) = -1(x-1)^{-2} = \frac{-1}{(x-1)^2}$$

Since $f''(x) = \dfrac{-1}{(1-x)^2} < 0$ on $(-\infty, 1)$, the graph of f is concave downward on $(-\infty, 1)$; there are no inflection points.

Step 4. Sketch the graph of f:

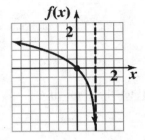

x	$f(x)$
0	0
-2	≈ 1.10
.9	≈ -2.30

29. $f(x) = x - \ln x$

Step 1. Analyze $f(x)$:

(A)   Domain: All positive real numbers, $(0, \infty)$.

   [Note: $\ln x$ is defined only for positive numbers.]

(B)   Intercepts: y-intercept: There is no y intercept; $f(0) = 0 - \ln(0)$ is not defined.

   x-intercept: $x - \ln x = 0$

   $\ln x = x$

   Since the graph of $y = \ln x$ is below the graph of $y = x$, there are no solutions to this equation; there are no x-intercepts.

(C)   Asymptotes:

   Horizontal asymptote: None

   Vertical asymptotes: Since $\lim\limits_{x \to 0^+} \ln x = -\infty$, $\lim\limits_{x \to 0^+} (x - \ln x) = \infty$. $x = 0$ is a vertical asymptote.

Step 2. Analyze $f'(x)$:

$$f'(x) = 1 - \frac{1}{x} = \frac{x-1}{x}, x > 0$$

Critical values: $\dfrac{x-1}{x} = 0, \quad x = 1$

Partition numbers: $x = 1$

Sign chart for $f'(x) = \dfrac{x-1}{x}$:

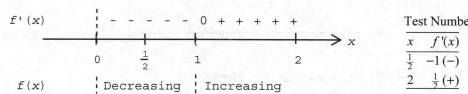

Test Numbers

x	$f'(x)$
$\frac{1}{2}$	$-1\,(-)$
2	$\frac{1}{2}\,(+)$

Thus, f is decreasing on $(0, 1)$ and increasing on $(1, \infty)$; f has a local minimum at $x = 1$.

Step 3. Analyze $f''(x)$:

$$f''(x) = \frac{1}{x^2},\ x > 0$$

Thus, $f''(x) > 0$ and the graph of f is concave upward on $(0, \infty)$.

Step 4. Sketch the graph of f:

x	$f(x)$
0.1	≈ 2.4
1	1
10	≈ 7.7

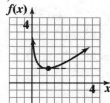

31. $f(x) = \dfrac{x}{x^2 - 4} = \dfrac{x}{(x-2)(x+2)}$

Step 1. Analyze $f(x)$:

(A) Domain: All real numbers except $x = 2$, $x = -2$.

(B) Intercepts: y-intercept: $f(0) = \dfrac{0}{-4} = 0$

x-intercept: $\dfrac{x}{x^2 - 4} = 0$

$x = 0$

(C) Asymptotes:

Horizontal asymptote:

$$\lim_{x\to\infty} \frac{x}{x^2 - 4} = \lim_{x\to\infty} \frac{x}{x^2\left(1 - \frac{4}{x^2}\right)} = \lim_{x\to\infty} \frac{1}{x}\left(\frac{1}{1 - \frac{4}{x^2}}\right) = 0$$

Thus, $y = 0$ (the x axis) is a horizontal asymptote.

Vertical asymptotes: The denominator is 0 at $x = 2$ and $x = -2$. The numerator is nonzero at each of these points. Thus, $x = 2$ and $x = -2$ are vertical asymptotes.

Step 2. Analyze $f'(x)$:

$$f'(x) = \frac{(x^2 - 4)(1) - x(2x)}{(x^2 - 4)^2} = \frac{-(x^2 + 4)}{(x^2 - 4)^2}$$

Critical values: None ($x^2 + 4 \neq 0$ for all x)

Partition numbers: $x = 2$, $x = -2$

Sign chart for f':

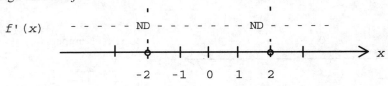

Thus, f is decreasing on $(-\infty, -2)$, on $(-2, 2)$, and on $(2, \infty)$; f has no local extrema.

Step 3. Analyze $f''(x)$:

$$f''(x) = \frac{(x^2-4)^2(-2x) - [-(x^2+4)](2)(x^2-4)(2x)}{(x^2-4)^4}$$

$$= \frac{(x^2-4)(-2x) + 4x(x^2+4)}{(x^2-4)^3} = \frac{2x^3 + 24x}{(x^2-4)^3} = \frac{2x(x^2+12)}{(x^2-4)^3}$$

Partition numbers for f'': $x = 0$, $x = 2$, $x = -2$

Sign chart for f'':

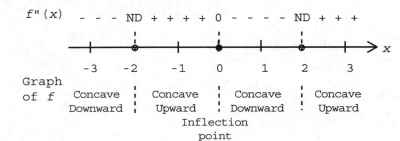

Test Numbers	
x	$f''(x)$
-3	$-\frac{126}{125}(-)$
-1	$\frac{26}{27}(+)$
1	$-\frac{26}{27}(-)$
3	$\frac{126}{125}(+)$

Thus, the graph of f is concave downward on $(-\infty, -2)$ and on $(0, 2)$; the graph of f is concave upward on $(-2, 0)$ and on $(2, \infty)$; the graph has an inflection point at $x = 0$.

Step 4. Sketch the graph of f:

x	$f(x)$
0	0
1	$-\frac{1}{3}$
-1	$\frac{1}{3}$
3	$\frac{3}{5}$
-3	$-\frac{3}{5}$

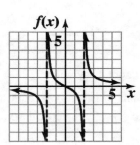

33. $f(x) = \dfrac{1}{1+x^2}$

Step 1. Analyze $f(x)$:

(A) Domain: All real numbers ($1 + x^2 \neq 0$ for all x).

(B) Intercepts: y-intercept: $f(0) = 1$

x-intercept: $\dfrac{1}{1+x^2} \neq 0$ for all x; no x intercepts

(C)   Asymptotes:

Horizontal asymptote: $\displaystyle\lim_{x\to\infty}\frac{1}{1+x^2}=0$.   Thus, $y=0$ (the x-axis) is a horizontal asymptote.

Vertical asymptotes: Since $1+x^2\neq 0$ for all x, there are no vertical asymptotes.

Step 2. Analyze $f'(x)$:

$$f'(x)=\frac{(1+x^2)(0)-1(2x)}{(1+x^2)^2}=\frac{-2x}{(1+x^2)^2}$$

Critical values: $x=0$
Partition numbers: $x=0$

Sign chart for f':

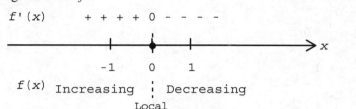

Test Numbers	
x	$f'(x)$
-1	$\frac{1}{2}(+)$
1	$-\frac{1}{2}(-)$

Thus, f is increasing on $(-\infty, 0)$; f is decreasing on $(0, \infty)$; f has a local maximum at $x=0$.

Step 3. Analyze $f''(x)$:

$$f''(x)=\frac{(1+x^2)^2(-2)-(-2x)(2)(1+x^2)2x}{(1+x^2)^4}=\frac{(-2)(1+x^2)+8x^2}{(1+x^2)^3}$$

$$=\frac{6x^2-2}{(1+x^2)^3}=\frac{6\left(x+\frac{\sqrt{3}}{3}\right)\left(x-\frac{\sqrt{3}}{3}\right)}{(1+x^2)^3}$$

Partition numbers for f'': $x=-\dfrac{\sqrt{3}}{3}$, $x=\dfrac{\sqrt{3}}{3}$

Sign chart for f'':

```
  f"(x)      + + + + 0 - - - 0 + + + +
         ┼──────┼───●───┼───●───┼──────┼──────→  x
        -2     -1  -√3  0  √3   1      2
                     3       3
  Graph
  of f         Concave   Concave   Concave
               Upward    Downward  Upward
                  Inflection Inflection
                    point      point
```

Test Numbers	
x	$f''(x)$
-1	$\frac{1}{2}(+)$
0	$-2(-)$
1	$\frac{1}{2}(+)$

Thus, the graph of f is concave upward on $\left(-\infty,\dfrac{-\sqrt{3}}{3}\right)$ and on $\left(\dfrac{\sqrt{3}}{3},\infty\right)$; the graph of f is concave downward on $\left(\dfrac{-\sqrt{3}}{3},\dfrac{\sqrt{3}}{3}\right)$; the graph has inflection points at $x=\dfrac{-\sqrt{3}}{3}$ and $x=\dfrac{\sqrt{3}}{3}$.

Step 4. Sketch the graph of f:

x	$f(x)$
$-\frac{\sqrt{3}}{3}$	$\frac{3}{4}$
0	1
$\frac{\sqrt{3}}{3}$	$\frac{3}{4}$

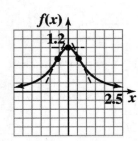

35. $f(x) = \dfrac{2x}{1-x^2}$

Step 1. Analyze $f(x)$:

(A) Domain: All real numbers except $x = -1$ and $x = 1$.

(B) Intercepts: y-intercept: $f(0) = \dfrac{0}{1} = 0$

$\qquad\qquad\qquad$ x-intercepts: $\dfrac{2x}{1-x^2} = 0$

$\qquad\qquad\qquad\qquad\qquad\qquad x = 0$

(C) Asymptotes:

$\qquad$ Horizontal asymptote: $\displaystyle\lim_{x\to\infty} \dfrac{2x}{1-x^2} = \lim_{x\to\infty} \dfrac{\frac{2}{x}}{\frac{1}{x^2}-1} = 0$. Thus, $y = 0$ (the x-axis) is a horizontal

$\qquad\qquad\qquad\qquad\qquad\qquad$ asymptote.

$\qquad$ Vertical asymptotes: $\qquad$ The denominator is 0 at $x = \pm1$ and the numerator is not 0 at $x = \pm1$.

$\qquad\qquad\qquad\qquad\qquad\qquad$ Thus, $x = -1$, $x = 1$ are vertical asymptotes.

Step 2. Analyze $f'(x)$:

$f'(x) = \dfrac{(1-x^2)2 - 2x(-2x)}{(1-x^2)^2} = \dfrac{2x^2+2}{(1-x^2)^2}$

Critical values: none

Partition numbers: $x = -1$, $x = 1$

Sign chart for f':

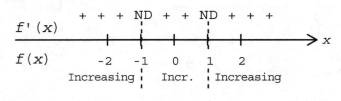

Test Numbers	
x	$f'(x)$
-2	$\frac{10}{9}(+)$
0	$2(+)$
2	$\frac{10}{9}(+)$

Thus, f is increasing on $(-\infty, -1)$, $(-1, 1)$, and $(1, \infty)$.

Step 3. Analyze $f''(x)$:

$f''(x) = \dfrac{(1-x^2)^2 4x - (2x^2+2)(2)(1-x^2)(-2x)}{(1-x^2)^4} = \dfrac{4x(x^2+3)}{(1-x^2)^3}$

Partition numbers for $f''(x)$: $x = -1$, $x = 0$, $x = 1$

Sign chart for f'':

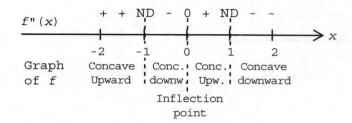

Thus, the graph of f is concave upward on $(-\infty, -1)$ and $(0, 1)$, concave downward on $(-1, 0)$ and $(1, \infty)$, and has an inflection point at $x = 0$.

Step 4. Sketch the graph of f:

x	$f(x)$
0	0

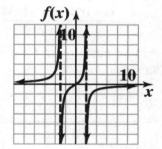

37. $f(x) = \dfrac{-5x}{(x-1)^2} = \dfrac{-5x}{x^2 - 2x + 1}$

Step 1. Analyze $f(x)$:

(A) Domain: All real numbers except $x = 1$.

(B) Intercepts: y-intercept: $f(0) = 0$

x-intercepts: $\dfrac{-5x}{(x-1)^2} = 0$

$x = 0$

(C) Asymptotes:

Horizontal asymptote: $\displaystyle\lim_{x \to \infty} \dfrac{-5x}{x^2 - 2x + 1} = \lim_{x \to \infty} \dfrac{-\frac{5}{x}}{1 - \frac{2}{x} + \frac{1}{x^2}} = 0$. Thus, $y = 0$ (the x-axis) is a

horizontal asymptote.

Vertical asymptotes: The denominator is 0 at $x = 1$ and the numerator is not 0 at $x = 1$. Thus, $x = 1$ is a vertical asymptote.

Step 2. Analyze $f'(x)$:

$f'(x) = \dfrac{(x-1)^2(-5) + 5x(2)(x-1)}{(x-1)^4} = \dfrac{5(x+1)}{(x-1)^3}$

Critical values: $x = -1$

Partition numbers: $x = -1, x = 1$

Sign chart for f':

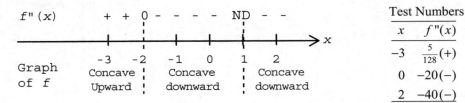

Thus, f is increasing on $(-\infty, -1)$, $(1, \infty)$, decreasing on $(-1, 1)$, and has a local maximum at $x = -1$.

Step 3. Analyze $f''(x)$:

$$f''(x) = \frac{(x-1)^3 5 - 5(x+1)(3)(x-1)^2}{(x-1)^6} = \frac{-10(x+2)}{(x-1)^4}$$

Partition numbers for f'': $x = -2, x = 1$

Sign chart for f'':

f" (x)	+ + 0 - - - - ND - -

Test Numbers

x	$f''(x)$
-3	$\frac{5}{128}(+)$
0	$-20(-)$
2	$-40(-)$

Thus, the graph of f is concave upward on $(-\infty, -2)$, concave downward on $(-2, 1)$ and $(1, \infty)$, and has an inflection point at $x = -2$.

Step 4. Sketch the graph of f:

x	$f(x)$
-2	$\frac{10}{9}$
-1	$\frac{5}{4}$
0	0

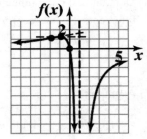

39. $f(x) = \dfrac{x^2 + x - 2}{x^2} = \dfrac{(x+2)(x-1)}{x^2}$

Step 1. Analyze $f(x)$:

(A) Domain: All real numbers except $x = 0$.

(B) Intercepts: y-intercept: $f(0)$ not defined; no y-intercept

x-intercepts: $\dfrac{(x+2)(x-1)}{x^2} = 0$

$$x = -2, 1$$

(C) Asymptotes:

Horizontal asymptote: $\displaystyle\lim_{x \to \infty} \frac{x^2 + x - 2}{x^2} = \lim_{x \to \infty} \frac{1 + \frac{1}{x} - \frac{2}{x^2}}{1} = 1$; $y = 1$ is a horizontal asymptote.

Vertical asymptotes: $x = 0$ (the x-axis)

Step 2. Analyze $f'(x)$:

$$f'(x) = \frac{x^2(2x+1)-(x^2+x-2)(2x)}{x^4} = \frac{4-x}{x^3}$$

Critical values: $x = 4$

Partition numbers: $x = 0$, $x = 4$

Sign chart for f':

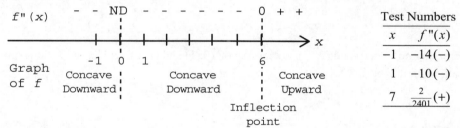

Test Numbers	
x	$f'(x)$
-1	$-5(-)$
1	$3(+)$
5	$\frac{-1}{125}(-)$

Thus, f is increasing on $(0, 4)$, decreasing on $(-\infty, 0)$ and $(4, \infty)$, and has a local maximum at $x = 4$.

Step 3. Analyze $f''(x)$:

$$f''(x) = \frac{x^3(-1)-(4-x)(3x^2)}{x^6} = \frac{2x-12}{x^4} = \frac{2(x-6)}{x^4}$$

Partition numbers for f'': $x = 0$, $x = 6$

Sign chart for f'':

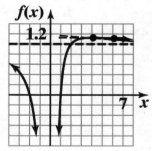

Test Numbers	
x	$f''(x)$
-1	$-14(-)$
1	$-10(-)$
7	$\frac{2}{2401}(+)$

Thus, the graph of f is concave upward on $(6, \infty)$, concave downward on $(-\infty, 0)$ and $(0, 6)$, and has an inflection point at $x = 6$.

Step 4. Sketch the graph of f:

x	$f(x)$
4	$\frac{9}{8}$
6	$\frac{10}{9}$

41. $f(x) = \dfrac{x^2}{x-1}$

Step 1. Analyze $f(x)$:

(A) Domain: All real numbers except $x = 1$.

(B) Intercepts: y-intercept: $f(0) = 0$

x-intercepts: $\dfrac{x^2}{x-1} = 0$

$x = 0$

(C) Asymptotes:

Horizontal asymptote: $\dfrac{x^2}{x} = x$; no horizontal asymptote

Vertical asymptote: $x = 1$

Oblique asymptote: It follows from above that $y = x$ is an oblique asymptote.

Step 2. Analyze $f'(x)$:

$$f'(x) = \frac{(x-1)(2x) - x^2}{(x-1)^2} = \frac{x^2 - 2x}{(x-1)^2} = \frac{x(x-2)}{(x-1)^2}$$

Critical values: $x = 0$, $x = 2$

Partition numbers: $x = 0$, $x = 1$, $x = 2$

Sign chart for f':

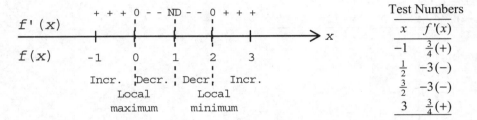

x	f'(x)
-1	$\frac{3}{4}(+)$
$\frac{1}{2}$	$-3(-)$
$\frac{3}{2}$	$-3(-)$
3	$\frac{3}{4}(+)$

Test Numbers

Thus, f is increasing on $(-\infty, 0)$ and $(2, \infty)$, decreasing on $(0, 1)$ and $(1, 2)$, and has·a local maximum at $x = 0$ and a local minimum at $x = 2$.

Step 3. Analyze $f''(x)$:

$$f''(x) = \frac{(x-1)^2(2x-2) - (x^2 - 2x)(2)(x-1)}{(x-1)^4} = \frac{2}{(x-1)^3}$$

Partition numbers for f'': $x = 1$

Sign chart for f'':

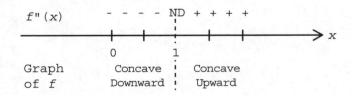

Thus, the graph of f is concave upward on $(1, \infty)$ and concave downward on $(-\infty, 1)$.

Step 4. Sketch the graph of f:

x	f(x)
0	0
2	4

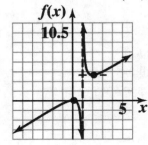

43. $f(x) = \dfrac{3x^2 + 2}{x^2 - 9}$

Step 1. Analyze $f(x)$:

(A) Domain: All real numbers except $x = -3$, $x = 3$.

(B) Intercepts: y-intercept: $f(0) = -\dfrac{2}{9}$

$\qquad\qquad$ x-intercepts: $3x^2 + 2 \neq 0$ for all x; no x-intercepts

(C)   Asymptotes:

$\qquad$ Horizontal asymptote: $\dfrac{3x^2}{x^2} = 3$; $y = 3$ is a horizontal asymptote

$\qquad$ Vertical asymptotes: $x = -3$, $x = 3$

Step 2. Analyze $f'(x)$:

$f'(x) = \dfrac{(x^2 - 9)(6x) - (3x^2 + 2)(2x)}{(x^2 - 9)^2} = \dfrac{-58x}{(x^2 - 9)^2}$

Critical values: $x = 0$

Partition numbers: $x = -3$, $x = 0$, $x = 3$

Sign chart for f':

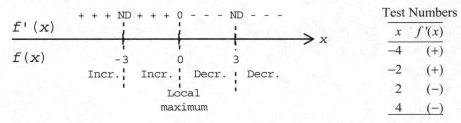

Test Numbers	
x	$f'(x)$
−4	(+)
−2	(+)
2	(−)
4	(−)

Thus, f is increasing on $(-\infty, -3)$ and $(-3, 0)$, decreasing on $(0, 3)$ and $(3, \infty)$, and has a local maximum at $x = 0$.

Step 3. Analyze $f''(x)$:

$f''(x) = \dfrac{(x^2 - 9)^2(-58) + 58x(2)(x^2 - 9)(2x)}{(x^2 - 9)^4} = \dfrac{174(x^2 + 3)}{(x^2 - 9)^3}$

Partition numbers for f'': $x = -3$, $x = 3$

Sign chart for f'':

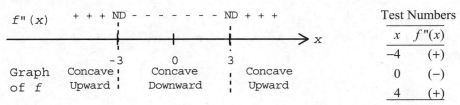

Test Numbers	
x	$f''(x)$
−4	(+)
0	(−)
4	(+)

Thus, the graph of f is concave upward on $(-\infty, -3)$ and $(3, \infty)$, and concave downward on $(-3, 3)$.

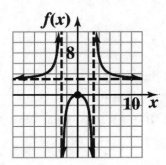

Step 4. Sketch the graph of f:

x	$f(x)$
0	$-\frac{2}{9}$

45. $f(x) = \dfrac{x^3}{x-2}$

Step 1. Analyze $f(x)$:

(A) Domain: All real numbers except $x = 2$.

(B) Intercepts: y-intercept: $f(0) = 0$

$\qquad\qquad$ x-intercepts: $\dfrac{x^3}{x-2} = 0$

$\qquad\qquad\qquad\qquad\qquad\quad x = 0$

(C)   Asymptotes:

$\qquad$ Horizontal asymptote: $\dfrac{x^3}{x} = x^2$; no horizontal asymptote

$\qquad$ Vertical asymptote: $x = 2$

Step 2. Analyze $f'(x)$:

$$f'(x) = \frac{(x-2)(3x^2) - x^3}{(x-2)^2} = \frac{2x^2(x-3)}{(x-2)^2}$$

Critical values: $x = 0$, $x = 3$

Partition numbers: $x = 0$, $x = 2$, $x = 3$

Sign chart for f':

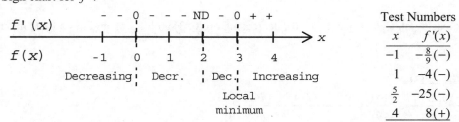

Test Numbers	
x	$f'(x)$
-1	$-\frac{8}{9}(-)$
1	$-4(-)$
$\frac{5}{2}$	$-25(-)$
4	$8(+)$

Thus, f is increasing on $(3, \infty)$, decreasing on $(-\infty, 2)$ and $(2, 3)$, and has a local minimum at $x = 3$.

Step 3. Analyze $f''(x)$:

$$f''(x) = \frac{(x-2)^2[2x^2 + 4x(x-3)] - 2x^2(x-3)(2)(x-2)}{(x-2)^4} = \frac{2x(x^2 - 6x + 12)}{(x-2)^3}$$

Partition numbers for f'': $x = 0$, $x = 2$ ($x^2 - 6x + 12$ has no real roots)

Sign chart for f'':

$f''(x)$ + + + 0 - - - - ND + + +

Test Numbers

x	$f''(x)$
−1	$\frac{38}{27}$ (+)
1	−14 (−)
3	18 (+)

Graph of f:

	0	2	3
	Concave Upward	Concave Downward	Concave Upward
		Inflection Point	

Thus, the graph of f is concave upward on $(-\infty, 0)$ and $(2, \infty)$, concave downward on $(0, 2)$, and has an inflection point at $x = 0$.

Step 4. Sketch the graph of f:

x	$f(x)$
0	0
3	27

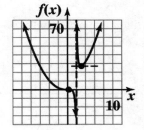

47. $f(x) = (3 - x)e^x$

Step 1. Analyze $f(x)$:

(A)   Domain: All real numbers, $(-\infty, \infty)$.

(B)   Intercepts: y-intercept: $f(0) = (3 - 0)e^0 = 3$

x-intercept: $(3 - x)e^x = 0$

$$3 - x = 0$$
$$x = 3$$

(C)   Asymptotes:
Horizontal asymptote: Consider the behavior of f as $x \to \infty$ and as $x \to -\infty$.
Using the following tables,

x	−1	−10	−20
$f(x)$	1.47	0.00059	0.000000047

x	5	10
$f(x)$	−296.83	−154,185.26

we conclude that $\lim\limits_{x \to -\infty} f(x) = 0$ and $\lim\limits_{x \to \infty} f(x)$ does not exist. Because of the first limit, $y = 0$ is a

horizontal asymptote.

Vertical asymptotes: There are no vertical asymptotes.

Step 2. Analyze $f'(x)$:
$f'(x) = (3 - x)e^x + e^x(-1) = (2 - x)e^x$

Critical values: $(2 - x)e^x = 0$

$$x = 2 \quad [\underline{\text{Note}}: e^x > 0]$$

Partition numbers: $x = 2$

Sign chart for f':

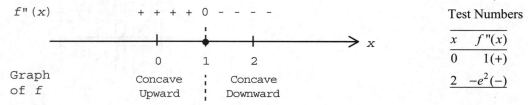

x	$f'(x)$
0	2(+)
3	$-e^3(-)$

Thus, f is increasing on $(-\infty, 2)$ and decreasing on $(2, \infty)$; f has a local maximum at $x = 2$.

Step 3. Analyze $f''(x)$:

$f''(x) = (2-x)e^x + e^x(-1) = (1-x)e^x$
Partition number for f'': $x = 1$
Sign chart for f'':

$f''(x)$ + + + + 0 - - - - → x

         0   1   2

Graph   Concave ┆ Concave
of f Upward ┆ Downward

x	$f''(x)$
0	1(+)
2	$-e^2(-)$

Thus, the graph of f is concave upward on $(-\infty, 1)$ and concave downward on $(1, \infty)$; the graph has an inflection point at $x = 1$.

Step 4. Sketch the graph of f:

x	$f(x)$
0	3
2	$e^2 \approx 7.4$
3	0

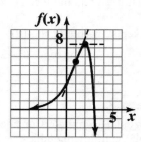

49. $f(x) = e^{-(1/2)x^2}$

Step 1. Analyze $f(x)$

(A) Domain: All real numbers, $(-\infty, \infty)$.

(B) Intercepts: y-intercept: $f(0) = e^{-(1/2)0} = e^0 = 1$

x-intercepts: Since $e^{-(1/2)x^2} \neq 0$ for all x, there are no x-intercepts.

(C) Asymptotes: $\displaystyle\lim_{x\to\infty} f(x) = \lim_{x\to\infty} e^{-(1/2)x^2} = \lim_{x\to\infty} \frac{1}{e^{(1/2)x^2}} = 0$

$\displaystyle\lim_{x\to-\infty} f(x) = \lim_{x\to-\infty} e^{-(1/2)x^2} = \lim_{x\to-\infty} \frac{1}{e^{(1/2)x^2}} = 0$

Thus, $y = 0$ is a horizontal asymptote.

Since $f(x) = e^{-(1/2)x^2} = \dfrac{1}{e^{(1/2)x^2}}$ and $e^{(1/2)x^2} \neq 0$ for all x, there are no vertical asymptotes.

Step 2. Analyze $f'(x)$:

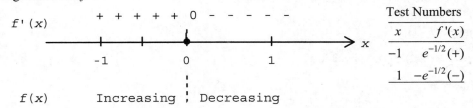

$f'(x) = e^{-(1/2)x^2}(-x) = -xe^{-(1/2)x^2}$

Critical values: $-xe^{-(1/2)x^2} = 0$

$x = 0$

Partition numbers: $x = 0$
Sign chart for f':

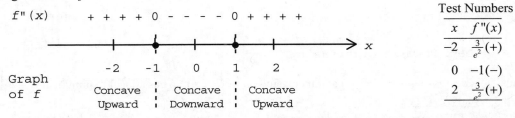

$f'(x)$ + + + + + 0 - - - - -

-1   0   1

$f(x)$ Increasing ⫶ Decreasing

Test Numbers	
x	$f'(x)$
-1	$e^{-1/2}(+)$
1	$-e^{-1/2}(-)$

Thus, f is increasing on $(-\infty, 0)$ and decreasing on $(0, \infty)$; f has a local maximum at $x = 0$.

Step 3. Analyze $f''(x)$:

$f''(x) = -xe^{-(1/2)x^2}(-x) - e^{-(1/2)x^2} = e^{-(1/2)x^2}(x^2 - 1) = e^{-(1/2)x^2}(x - 1)(x + 1)$

Partition numbers for f'': $e^{-(1/2)x^2}(x - 1)(x + 1) = 0$

$(x - 1)(x + 1) = 0$

$x = -1, 1$

Sign chart for f'':

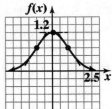

$f''(x)$ + + + + 0 - - - - 0 + + + +

-2   -1   0   1   2

Graph
of f Concave ⫶ Concave ⫶ Concave
          Upward ⫶ Downward ⫶ Upward

Test Numbers	
x	$f''(x)$
-2	$\frac{3}{e^2}(+)$
0	$-1(-)$
2	$\frac{3}{e^2}(+)$

Thus, the graph of f is concave upward on $(-\infty, -1)$ and on $(1, \infty)$; the graph of f is concave downward on $(-1, 1)$; the graph has inflection points at $x = -1$ and at $x = 1$.

Step 4. Sketch the graph of f:

x	$f'(x)$
0	1
-1	≈ 0.61
1	≈ 0.61

51. $f(x) = x^2 \ln x$.

Step 1. Analyze $f(x)$:

(A)   Domain: All positive numbers, $(0, \infty)$.

(B)   Intercepts: y-intercept: There is no y intercept.

x-intercept: $x^2 \ln x = 0$

$\ln x = 0$

$x = 1$

(C)   Asymptotes: Consider the behavior of f as $x \to \infty$ and as $x \to 0$. It is clear that $\lim\limits_{x \to \infty} f(x)$ does

not exist; f is unbounded as x approaches ∞.

The following table indicates that f approaches 0 as x approaches 0.

x	1	0.1	0.01	0.001
$f(x)$	0	−0.023	−0.00046	−0.000007

Thus, there are no vertical or horizontal asymptotes.

<u>Step 2. Analyze $f'(x)$:</u>

$$f'(x) = x^2 \left(\frac{1}{x} \right) + (\ln x)(2x) = x(1 + 2 \ln x)$$

Critical values: $x(1 + 2 \ln x) = 0$

$$1 + 2 \ln x = 0 \qquad [\text{Note: } x > 0]$$

$$\ln x = -\frac{1}{2}$$

$$x = e^{-1/2} = \frac{1}{\sqrt{e}} \approx 0.6065$$

Partition number: $x = \dfrac{1}{\sqrt{e}} \approx 0.6065$

Sign chart for f' :

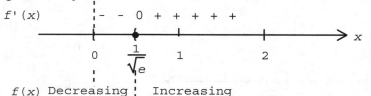

Test Numbers	
x	$f'(x)$
$\frac{1}{2}$	$\approx -.19\,(-)$
1	$1\,(+)$

Thus, f is decreasing on $(0, e^{-1/2})$ and increasing on $(e^{-1/2}, \infty)$; f has a local minimum at $x = e^{-1/2}$.

<u>Step 3. Analyze $f''(x)$:</u>

$$f''(x) = x \left(\frac{2}{x} \right) + (1 + 2 \ln x) = 3 + 2 \ln x$$

Partition number for f'': $3 + 2 \ln x = 0$

$$\ln x = -\frac{3}{2}$$

$$x = e^{-3/2} \approx 0.2231$$

Sign chart for f'':

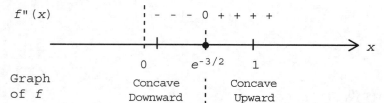

Test Numbers	
x	$f''(x)$
$\frac{1}{10}$	$\approx -1.61\,(-)$
1	$3\,(+)$

Thus, the graph of f is concave downward on $(0, e^{-3/2})$ and concave upward on $(e^{-3/2}, \infty)$; the graph has an inflection point at $x = e^{-3/2}$.

Step 4. Sketch the graph of f:

x	$f(x)$
$e^{-3/2}$	≈ -0.075
$e^{-1/2}$	≈ -0.18
1	0

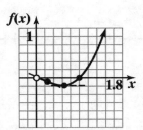

53. $f(x) = (\ln x)^2$

Step 1. Analyze $f(x)$:
(A)   Domain: All positive numbers, $(0, \infty)$.
(B)   Intercepts: y-intercept: There is no y-intercept.

$$x\text{-intercept:}\quad (\ln x)^2 = 0$$
$$\ln x = 0$$
$$x = 1$$

(C)   Asymptotes:
Consider the behavior of f as $x \to \infty$ and as $x \to 0$. It is clear that $\lim\limits_{x\to\infty} f(x)$ does not exist;

$f(x) \to \infty$ as $x \to \infty$. Thus, there is no horizontal asymptote.
The following table indicates that $f(x) \to \infty$ as $x \to 0$;
$x = 0$ (the y-axis) is a vertical asymptote.

x	1	0.01	0.0001	0.000001
$f(x)$	0	21.21	84.83	190.87

Step 2. Analyze $f'(x)$:

$$f'(x) = 2(\ln x)\frac{d}{dx}\ln x = \frac{2\ln x}{x}$$

Critical values: $\dfrac{2\ln x}{x} = 0$
$$\ln x = 0$$
$$x = 1$$

Partition numbers: $x = 1$
Sign chart for f':

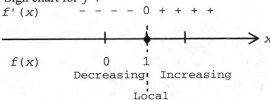

Test Numbers	
x	$f'(x)$
0.5	$-2.77(-)$
2	$0.69(+)$

Thus, f is decreasing on $(0, 1)$ and increasing on $(1, \infty)$; f has a local minimum at $x = 1$.

Step 3. Analyze $f''(x)$:

$$f''(x) = \frac{x\left(\frac{2}{x}\right) - 2\ln x}{x^2} = \frac{2(1 - \ln x)}{x^2}$$

Partition numbers for f'': $\dfrac{2(1 - \ln x)}{x^2} = 0$
$$\ln x = 1$$
$$x = e$$

Sign chart for f'':

$f''(x)$ + + + + + + + 0 − − − − →x

Graph of f 0   1   2   e 3

Concave upward   Concave downward

Inflection point

Test Numbers

x	$f''(x)$
1	2(+)
4	−0.048(−)

Thus, the graph of f is concave upward on $(0, e)$ and concave downward on (e, ∞); the graph has an inflection point at $x = e$.

Step 4. Sketch the graph of f:

x	$f(x)$
1	0
e	1

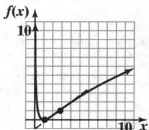

55. $f(x) = \dfrac{1}{x^2 + 2x - 8} = \dfrac{1}{(x+4)(x-2)}$

Step 1. Analyze $f(x)$:

(A) Domain: All real numbers except $x = -4$, $x = 2$.

(B) Intercepts: y-intercept: $f(0) = -\dfrac{1}{8}$

   x-intercepts: no x-intercept

(C) Asymptotes:

   Horizontal asymptote: $\dfrac{1}{x^2} \to 0$ as $x \to \infty$; $y = 0$ (the x-axis) is a horizontal asymptote.

   Vertical asymptote: $x = -4$, $x = 2$

Step 2. Analyze $f'(x)$:

$$f'(x) = \frac{-(2x+2)}{(x^2+2x-8)^2} = \frac{-2(x+1)}{(x^2+2x-8)^2} = \frac{-2(x+1)}{(x-2)^2(x+4)^2}$$

Critical values: $x = -1$

Partition numbers: $x = -4$, $x = -1$, $x = 2$

Sign chart for f' :

$f'(x)$ + + + ND + + + 0 − − − ND − − − →x

$f(x)$ −4   −1 0 1 2 3 4

   Incr.   Incr.   Decr.   Decr.

Local maximum

Test Numbers

x	$f'(x)$
−5	(+)
−2	(+)
0	$-\frac{1}{32}$(−)
3	(−)

Thus, f is increasing on $(-\infty, -4)$ and $(-4, -1)$, decreasing on $(-1, 2)$ and $(2, \infty)$, and has a local maximum at $x = -1$.

Step 3. Analyze $f''(x)$:

$$f''(x) = \frac{(x^2+2x-8)^2(-2)+(2x+2)(2)(x^2+2x-8)(2x+2)}{(x^2+2x-8)^4} = \frac{6(x^2+2x+4)}{(x^2+2x-8)^3}$$

Partition numbers for f'': $x = -4$, $x = 2$ ($x^2 + 2x + 4$ has no real roots)

Sign chart for f'':

```
f"(x)        + + ND - - - - - - - ND + +
                 |                 |
─────────────────┼────────┼────────┼──────────▶ x
                -4        0        2
Graph     Concave      Concave      Concave
of f      Upward       Downward     Upward
```

Test Numbers

x	$f''(x)$
-5	$(+)$
0	$(-)$
3	$(+)$

Thus, the graph of f is concave upward on $(-\infty, -4)$ and $(2, \infty)$, and concave downward on $(-4, 2)$.

Step 4. Sketch the graph of f:

x	$f(x)$
-1	$-\frac{1}{9}$
0	$-\frac{1}{8}$

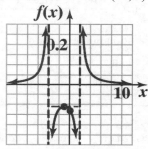

57. $f(x) = \dfrac{x^3}{3-x^2}$

Step 1. Analyze $f(x)$:

(A) Domain: All real numbers except $x = -\sqrt{3}$, $x = \sqrt{3}$.

(B) Intercepts: y-intercept: $f(0) = 0$

$\qquad\qquad$ x-intercepts: $\dfrac{x^3}{3-x^2} = 0$, $x = 0$

(C) Asymptotes:

$\qquad$ Horizontal asymptote: $\dfrac{x^3}{-x^2} = -x$; no horizontal asymptote

$\qquad$ Vertical asymptote: $x = -\sqrt{3}$, $x = \sqrt{3}$

$\qquad$ Oblique asymptote: It follows from above that $y = -x$ is an oblique asymptote.

Step 2. Analyze $f''(x)$:

$$f'(x) = \frac{(3-x^2)(3x^2)-x^3(-2x)}{(3-x^2)^2} = \frac{x^2(9-x^2)}{(3-x^2)^2}$$

Critical values: $x = -3, x = 0, x = 3$

Partition numbers: $x = -3, x = -\sqrt{3}, x = 0, x = \sqrt{3}, x = 3$

Sign chart for f':

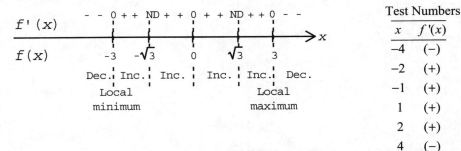

x	f'(x)
−4	(−)
−2	(+)
−1	(+)
1	(+)
2	(+)
4	(−)

Thus, f is increasing on $(-3, -\sqrt{3}), (-\sqrt{3}, \sqrt{3}), (\sqrt{3}, 3)$, decreasing on $(-\infty, -3)$ and $(3, \infty)$, and has a local minimum at $x = -3$ and a local maximum at $x = 3$.

Step 3. Analyze $f''(x)$:

$$f''(x) = \frac{(3-x^2)^2(18x-4x^3) - x^2(9-x^2)2(3-x^2)(-2x)}{(3-x^2)^4} = \frac{6x(9+x^2)}{(3-x^2)^3}$$

Partition numbers for f'': $x = -\sqrt{3}, \; x = 0, \; x = \sqrt{3}$

Sign chart for f'':

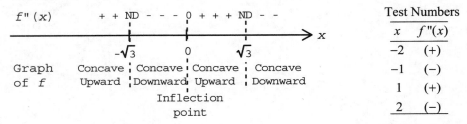

x	f''(x)
−2	(+)
−1	(−)
1	(+)
2	(−)

Thus, the graph of f is concave upward on $(-\infty, -\sqrt{3})$ and $(0, \sqrt{3})$, concave downward on $(-\sqrt{3}, 0)$ and $(\sqrt{3}, \infty)$, and has an inflection point at $x = 0$.

Step 4. Sketch the graph of f:

x	f(x)
−3	$\frac{9}{2}$
0	0
3	$-\frac{9}{2}$

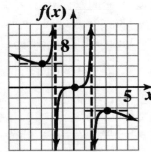

59. $f(x) = x + \dfrac{4}{x} = \dfrac{x^2 + 4}{x}$

Step 1. Analyze $f(x)$:

(A) Domain: All real numbers except $x = 0$.

(B) Intercepts: y-intercept: no y-intercept

                      x-intercepts: no x-intercepts

(C)   Asymptotes:

Horizontal asymptote: $\dfrac{x^2}{x} = x$; no horizontal asymptote

Vertical asymptote: $x = 0$ (the y-axis) is a vertical asymptote

Oblique asymptote: $\lim\limits_{x \to \infty} \left(x + \dfrac{4}{x} \right) = x$; $y = x$ is an oblique asymptote.

Step 2. Analyze $f'(x)$:

$f'(x) = 1 - \dfrac{4}{x^2} = \dfrac{x^2 - 4}{x^2}$

Critical values: $x = -2, x = 2$
Partition numbers: $x = -2, x = 0, x = 2$

Sign chart for f':

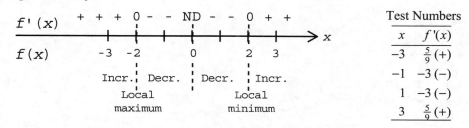

Test Numbers	
x	$f'(x)$
-3	$\frac{5}{9}\,(+)$
-1	$-3\,(-)$
1	$-3\,(-)$
3	$\frac{5}{9}\,(+)$

Thus, f is increasing on $(-\infty, -2)$ and $(2, \infty)$, decreasing on $(-2, 0)$ and $(0, 2)$, and has a local maximum at $x = -2$ and a local minimum at $x = 2$.

Step 3. Analyze $f''(x)$:

$f''(x) = \dfrac{8}{x^3}$

Partition numbers for f'': $x = 0$

Sign chart for f'':

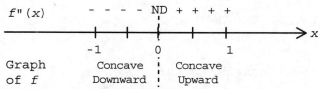

Thus, the graph of f is concave upward on $(0, \infty)$ and concave downward on $(-\infty, 0)$.

Step 4. Sketch the graph of f:

x	$f(x)$
-2	-4
2	4

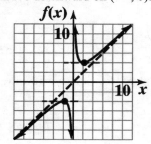

61. $f(x) = x - \dfrac{4}{x^2} = \dfrac{x^3 - 4}{x^2}$

Step 1. Analyze $f(x)$:

(A) Domain: All real numbers except $x = 0$.

(B) Intercepts: y-intercept: no y-intercept

x-intercepts: $\dfrac{x^3 - 4}{x} = 0$, $x = \sqrt[3]{4}$

(C) Asymptotes:

Horizontal asymptote: $\dfrac{x^3}{x^2} = x$; no horizontal asymptote

Vertical asymptote: $x = 0$ (the y-axis) is a vertical asymptote

Oblique asymptote: $\lim\limits_{x \to \infty} \left(x - \dfrac{4}{x^2} \right) = x$; $y = x$ is an oblique asymptote

Step 2. Analyze $f'(x)$:

$f'(x) = 1 + \dfrac{8}{x^3} = \dfrac{x^3 + 8}{x^3}$

Critical values: $x = -2$

Partition numbers: $x = -2$, $x = 0$

Sign chart for f':

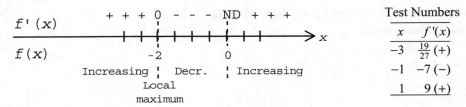

Test Numbers	
x	$f'(x)$
-3	$\frac{19}{27}$ (+)
-1	-7 (−)
1	9 (+)

Thus, f is increasing on $(-\infty, -2)$ and $(0, \infty)$, decreasing on $(-2, 0)$; f has a local maximum at $x = -2$.

Step 3. Analyze $f''(x)$:

$f''(x) = -\dfrac{24}{x^4}$

Partition numbers for $f''(x)$: $x = 0$

Sign chart for f'':

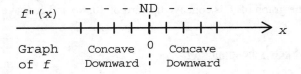

Thus, the graph of f is concave downward on $(-\infty, 0)$ and $(0, \infty)$.

Step 4. Sketch the graph of f:

x	$f(x)$
-2	-3

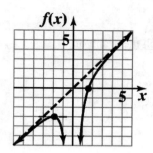

63. $f(x) = x - \dfrac{9}{x^3} = \dfrac{x^4 - 9}{x^3} = \dfrac{(x^2 - 3)(x^2 + 3)}{x^3}$

Step 1. Analyze $f(x)$:

(A) Domain: All real numbers except $x = 0$.

(B) Intercepts: y-intercept: no y-intercept

$\qquad\qquad\quad$ x-intercepts: $x = -\sqrt{3}$, $x = \sqrt{3}$

(C)   Asymptotes:

$\qquad$ Horizontal asymptote: $\dfrac{x^4}{x^3} = x$; no horizontal asymptote

$\qquad$ Vertical asymptote: $x = 0$ (the y-axis) is a vertical asymptote

$\qquad$ Oblique asymptote: $\displaystyle\lim_{x \to \infty} \left(x + \dfrac{9}{x^3} \right) = x$; $y = x$ is an oblique asymptote

Step 2. Analyze $f'(x)$:

$f'(x) = 1 + \dfrac{27}{x^4} = \dfrac{x^4 + 27}{x^4}$

Critical values: none

Partition numbers: $x = 0$

Sign chart for f':

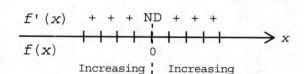

Thus, f is increasing on $(-\infty, 0)$ and $(0, \infty)$.

Step 3. Analyze $f''(x)$:

$f''(x) = -\dfrac{108}{x^5}$

Partition numbers for f'': $x = 0$

Sign chart for f'':

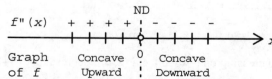

Thus, the graph of f is concave upward on $(-\infty, 0)$ and concave downward on $(0, \infty)$.

Step 4. Sketch the graph of f:

x	$f(x)$
$-\sqrt{3}$	0
$\sqrt{3}$	0

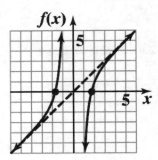

65. $f(x) = x + \dfrac{1}{x} + \dfrac{4}{x^3} = \dfrac{x^4 + x^2 + 4}{x^3}$

Step 1. Analyze $f(x)$:

(A) Domain: All real numbers except $x = 0$.

(B) Intercepts: y-intercept: no y-intercept

   x-intercepts: no x-intercepts

(C)   Asymptotes:

   Horizontal asymptote: $\dfrac{x^4}{x^3} = x$; no horizontal asymptote

   Vertical asymptote: $x = 0$ (the y-axis) is a vertical asymptote

   Oblique asymptote: $\displaystyle\lim_{x \to \infty}\left(x + \dfrac{1}{x} + \dfrac{4}{x^3}\right) = x$; $y = x$ is an oblique asymptote

Step 2. Analyze $f'(x)$:

$f'(x) = 1 - \dfrac{1}{x^2} - \dfrac{12}{x^4} = \dfrac{x^4 - x^2 - 12}{x^4} = \dfrac{(x^2 - 4)(x^2 + 3)}{x^4}$

Critical values: $x = -2, x = 2$

Partition numbers: $x = -2,\ x = 0,\ x = 2$

Sign chart for f':

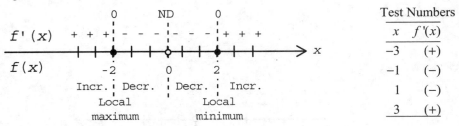

Test Numbers	
x	$f'(x)$
-3	$(+)$
-1	$(-)$
1	$(-)$
3	$(+)$

Thus, f is increasing on $(-\infty, -2)$ and $(2, \infty)$, decreasing on $(-2, 0)$ and $(0, 2)$; f has a local maximum at $x = -2$ and a local minimum at $x = 2$.

Step 3. Analyze $f''(x)$:

$$f''(x) = \frac{2}{x^3} + \frac{48}{x^5} = \frac{2x^2 + 48}{x^5}$$

Partition numbers for $f''(x)$: $x = 0$

Sign chart for f'':

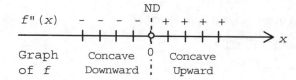

Thus, the graph of f is concave upward on $(0, \infty)$ and concave downward on $(-\infty, 0)$.

Step 4. Sketch the graph of f:

x	$f(x)$
-2	-3
2	3

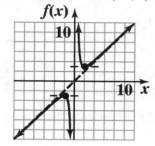

67. $C(x) = 10,000 + 90x + 0.02x^2$.

Average cost function: $\overline{C}(x) = \frac{C(x)}{x} = \frac{10,000}{x} + 90 + 0.02x \approx 90 + 0.02x$ for very large x.

69. $C(x) = 95,000 + 210x + 0.1x^2$.

Average cost function: $\overline{C}(x) = \frac{C(x)}{x} = \frac{95,000}{x} + 210 + 0.1x \approx 210 + 0.1x$ for very large x.

71. $f(x) = \frac{x^2 + x - 6}{x^2 - 6x + 8} = \frac{(x+3)(x-2)}{(x-4)(x-2)} = \frac{x+3}{x-4}, x \neq 2$

Step 1. Analyze $f(x)$:

(A) Domain: All real numbers except $x = 2$, $x = 4$.

(B) Intercepts: y-intercept: $f(0) = -\frac{3}{4}$

   x-intercepts: $x = -3$

(C)   Asymptotes:

   Horizontal asymptote: $\frac{x^2}{x^2} = 1$, $y = 1$ is a horizontal asymptote

   Vertical asymptote: $x = 4$ is a vertical asymptote

Step 2. Analyze $f'(x)$:

$$f'(x) = \frac{x - 4 - (x + 3)}{(x-4)^2} = -\frac{7}{(x-4)^2}$$

Critical values: None

Partition numbers: $x = 4$

Sign chart for f':

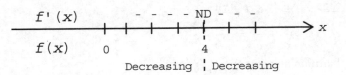

Thus, f is decreasing on $(-\infty, 4)$ and $(4, \infty)$.

Step 3. Analyze $f''(x)$:

$$f''(x) = \frac{14}{(x-4)^3}$$

Partition numbers for f'': $x = 4$

Sign chart for f'':

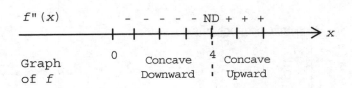

The graph of f is concave upward on $(4, \infty)$ and concave downward on $(-\infty, 4)$.

Step 4. Sketch the graph of f:

x	$f(x)$
-3	0
0	$-\frac{3}{4}$

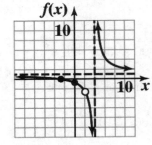

73. $f(x) = \dfrac{2x^2 + x - 15}{x^2 - 9} = \dfrac{(2x-5)(x+3)}{(x-3)(x+3)} = \dfrac{2x-5}{x-3}, x \neq -3$

Step 1. Analyze $f(x)$:
(A) Domain: All real numbers except $x = -3$, $x = 3$.

(B) Intercepts: y-intercept: $f(0) = \dfrac{5}{3}$

  x-intercepts: $x = \dfrac{5}{2}$

(C)   Asymptotes:

  Horizontal asymptote: $\dfrac{2x^2}{x^2} = 2$, $y = 2$ is a horizontal asymptote

  Vertical asymptote: $x = 3$ is a vertical asymptote

Step 2. Analyze $f'(x)$:

$$f'(x) = \frac{(x-3)2 - (2x-5)}{(x-3)^2} = \frac{-1}{(x-3)^2}$$

Critical values: None
Partition numbers: $x = 3$

Sign chart for f' :

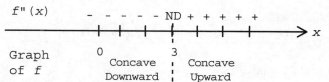

Thus, f is decreasing on $(-\infty, 3)$ and $(3, \infty)$.

Step 3. Analyze $f''(x)$:

$$f''(x) = \frac{2}{(x-3)^3}$$

Partition numbers for f'': $x = 3$

Sign chart for f'':

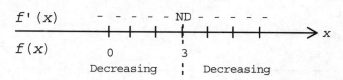

The graph of f is concave upward on $(3, \infty)$ and concave downward on $(-\infty, 3)$.

Step 4. Sketch the graph of f:

x	$f(x)$
$\frac{5}{2}$	0
0	$\frac{5}{3}$

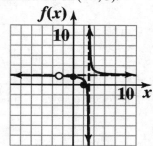

75. $f(x) = \dfrac{x^3 - 5x^2 + 6x}{x^2 - x - 2} = \dfrac{x(x-3)(x-2)}{(x-2)(x+1)} = \dfrac{x(x-3)}{x+1}$, $x \neq 2$

Step 1. Analyze $f(x)$:
(A) Domain: All real numbers except $x = -1$, $x = 2$.

(B) Intercepts: y-intercept: $f(0) = 0$
                  x-intercepts: $x = 0$, $x = 3$

(C)   Asymptotes:

Horizontal asymptote: $\dfrac{x^3}{x^2} = x$; no horizontal asymptote

Vertical asymptote: $x = -1$ is a vertical asymptote

Oblique asymptote: $\dfrac{x^2 - 3x}{x+1} = x - 4 + \dfrac{4}{x+1}$; $y = x - 4$ is an oblique asymptote

Step 2. Analyze $f'(x)$:

$$f'(x) = \frac{(x+1)(2x-3) - (x^2-3x)}{(x+1)^2} = \frac{x^2 + 2x - 3}{(x+1)^2} = \frac{(x+3)(x-1)}{(x+1)^2}$$

Critical values: $x = -3$, $x = 1$

Partition numbers: $x = -3$, $x = -1$, $x = 1$

Sign chart for f':

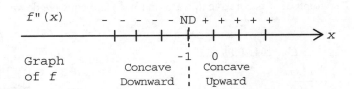

Test Numbers	
x	$f'(x)$
-4	$\frac{5}{9}(+)$
-2	$-3(-)$
0	$-3(-)$
2	$\frac{5}{9}(+)$

Thus, f is increasing on $(-\infty, -3)$ and $(1, \infty)$, f is decreasing on $(-3, -1)$ and $(-1, 1)$; f has a local maximum at $x = -3$ and a local minimum at $x = 1$.

Step 3. Analyze $f''(x)$:

$$f''(x) = \frac{(x+1)^2(2x+2) - (x^2 + 2x - 3)(2)(x+1)}{(x+1)^4} = \frac{8}{(x+1)^3}$$

Partition numbers for f'': $x = -1$

Sign chart for f'':

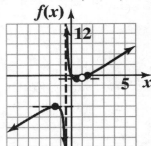

The graph of f is concave upward on $(-1, \infty)$ and concave downward on $(-\infty, -1)$.

Step 4. Sketch the graph of f:

x	$f(x)$
-3	-9
0	0
1	-1
3	0

77. $f(x) = \dfrac{x^2 + x - 2}{x^2 - 2x + 1} = \dfrac{(x+2)(x-1)}{(x-1)^2} = \dfrac{x+2}{x-1},\ x \neq 1$

Step 1. Analyze $f(x)$:

(A) Domain: All real numbers except $x = 1$.

(B) Intercepts: y-intercept: $f(0) = -2$

  x-intercepts: $x = -2$

(C)   Asymptotes:

  Horizontal asymptote: $\dfrac{x^2}{x^2} = 1$; $y = 1$ is a horizontal asymptote

  Vertical asymptote: $x = 1$ is a vertical asymptote

Step 2. Analyze $f'(x)$:

$$f'(x) = \frac{(x-1) - (x+2)}{(x-1)^2} = \frac{-3}{(x-1)^2}$$

Critical values: None

Partition numbers: $x = 1$

Sign chart for f':

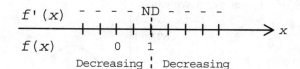

Thus, f is decreasing on $(-\infty, 1)$ and $(1, \infty)$.

<u>Step 3. Analyze $f''(x)$:</u>

$$f''(x) = \frac{6}{(x-1)^3}$$

Partition numbers for f'': $x = 1$

Sign chart for f'':

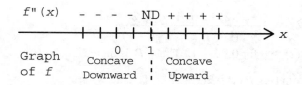

The graph of f is concave upward on $(1, \infty)$ and concave downward on $(-\infty, 1)$.

<u>Step 4. Sketch the graph of f:</u>

x	$f(x)$
-2	0
0	-2

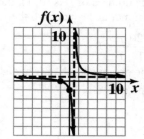

79. $R(x) = 1{,}296x - 0.12x^3,\ 0 \le x \le 80$

$R'(x) = 1{,}296 - 0.36x^2$

$R'(x) = 0$: $0.36x^2 = 1{,}296$

$\qquad\qquad x^2 = 3{,}600$

$\qquad\qquad\ x = 60$

The critical value or R is $x = 60$.

Sign chart for $R'(x)$:

Test Numbers	
x	$R'(x)$
30	972 (+)
80	$-1008\,(-)$

R is increasing on (0, 60) and decreasing on (60, 80).

$R''(x) = -0.72x < 0$ on (0, 80).

The graph of R is concave downward on (0, 80).

x	$f(x)$
0	0
60	51,840
80	42,240

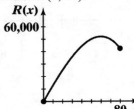

81. $P(x) = \dfrac{2x}{1-x}, 0 \le x < 1$

(A) $P'(x) = \dfrac{(1-x)(2) - 2x(-1)}{(1-x)^2} = \dfrac{2}{(1-x)^2}$

   $P'(x) > 0$ for $0 \le x < 1$. Thus, P is increasing on (0, 1).

(B) From (A), $P'(x) = 2(1-x)^{-2}$. Thus,

   $P''(x) = -4(1-x)^{-3}(-1) = \dfrac{4}{(1-x)^3}$.

   $P''(x) > 0$ for $0 \le x < 1$, and the graph of P is concave upward on (0, 1).

(C) Since the domain of P is [0, 1), there are no horizontal   asymptotes. The denominator is 0 at $x = 1$ and the numerator is nonzero there. Thus, $x = 1$ is a vertical asymptote.

(D) $P(0) = \dfrac{2(0)}{1-0} = 0$.

   Thus, the origin is both an x and a y intercept of the graph.

(E) The graph of P is:

x	$P(x)$
0	0
$\frac{1}{2}$	2
$\frac{3}{4}$	6

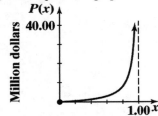

83. $C(n) = 3200 + 250n + 50n^2, 0 < n < \infty$

(A) Average cost per year:

   $\overline{C}(n) = \dfrac{C(n)}{n} = \dfrac{3200}{n} + 250 + 50n,\ 0 < n < \infty$

(B) Graph $\overline{C}(n)$:

   Step 1. Analyze $\overline{C}(n)$:

   Domain: $0 < n < \infty$

   Intercepts: C intercept: None ($n > 0$)

   n intercepts: $\dfrac{3200}{n} + 250 + 50n > 0$ on (0, ∞); there are no n intercepts.

Asymptotes: For large n, $C(n) = \dfrac{3200}{n} + 250 + 50n \approx 250 + 50n$. Thus, $y = 250 + 50n$ is an

oblique asymptote. As $n \to 0$, $\overline{C} \to \infty$. Thus, $n = 0$ is a vertical asymptote.

Step 2. Analyze $\overline{C}\,'(n)$:

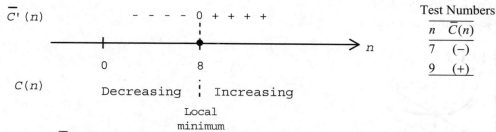

$\overline{C}\,'(n) = -\dfrac{3200}{n^2} + 50 = \dfrac{50n^2 - 3200}{n^2} = \dfrac{50(n^2 - 64)}{n^2} = \dfrac{50(n-8)(n+8)}{n^2},\ 0 < n < \infty$

Critical value: $n = 8$

Sign chart for $\overline{C}\,'$:

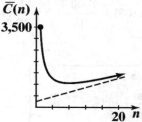

	Test Numbers
	$n \quad \overline{C}(n)$
	$7 \quad (-)$
	$9 \quad (+)$

$\overline{C}\,'(n)$ $- - - - \ 0 \ + + + +$

$0 \qquad\qquad 8$

$C(n)$ Decreasing ┊ Increasing

Local
minimum

Thus, $\overline{C}$ is decreasing on $(0, 8)$ and increasing on $(8, \infty)$; $n = 8$ is a local minimum.

Step 3: Analyze $C''(n)$:

$\overline{C}\,''(n) = \dfrac{6400}{n^3},\ 0 < n < \infty$

$\overline{C}\,''(n) > 0$ on $(0, \infty)$. Thus, the graph of $\overline{C}$ is concave upward on $(0, \infty)$.

Step 4. Sketch the graph of $\overline{C}$:

$\overline{C}(n)$

$3,500$

$20 \quad n$

(C)   The average cost per year is a minimum when $n = 8$ years.

85. $C(x) = 1000 + 5x + 0.1x^2,\ 0 < x < \infty.$

(A)   The average cost function is: $\overline{C}(x) = \dfrac{1000}{x} + 5 + 0.1x.$

Now, $\overline{C}\,'(x) = -\dfrac{1000}{x^2} + \dfrac{1}{10} = \dfrac{x^2 - 10,000}{10x^2} = \dfrac{(x+100)(x-100)}{10x^2}$

Sign chart for $\overline{C}\,'$:

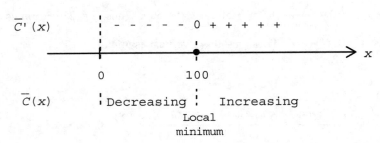

Test Numbers	
x	$\overline{C}'(x)$
1	$\approx\ -1000(-)$
101	$\approx\ \frac{1}{500}(+)$

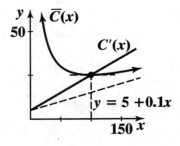

Thus, $\overline{C}$ is decreasing on $(0, 100)$ and increasing on $(100, \infty)$; $\overline{C}$ has a minimum at $x = 100$.

Since $\overline{C}''(x) = \dfrac{2000}{x^3} > 0$ for $0 < x < \infty$, the graph of $\overline{C}$ is concave upward on $(0, \infty)$. The line $x = 0$ is a

vertical asymptote and the line $y = 5 + 0.1x$ is an oblique asymptote for the graph of $\overline{C}$. The marginal cost function is $C'x) = 5 + 0.2x$.

The graphs of $\overline{C}$ and C' are:

(B)   The minimum average cost is:

$$\overline{C}(100) = \frac{1000}{100} + 5 + \frac{1}{10}(100) = 25$$

87.   (A)

```
QuadReg
y=ax²+bx+c
a=.0100714286
b=.7835714286
c=316
```

(B)   The average cost function $\overline{y} = \dfrac{y(x)}{x}$ where $y(x)$ is the

regression equation found in part (A).

The minimum average cost is $4.35 when 177 pizzas are produced.

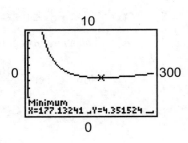

89. $C(t) = \dfrac{0.14t}{t^2 + 1}$

Step 1. Analyze $C(t)$:

Domain: $t \geq 0$, i.e., $[0, \infty)$

Intercepts: y intercept: $C(0) = 0$

t intercepts: $\dfrac{0.14t}{t^2 + 1} = 0$

$t = 0$

Asymptotes:

Horizontal asymptote: $\displaystyle\lim_{t \to \infty} \frac{0.14t}{t^2 + 1} = \lim_{t \to \infty} \frac{0.14t}{t^2\left(1 + \frac{1}{t^2}\right)} = \lim_{t \to \infty} \frac{0.14}{t\left(1 + \frac{1}{t^2}\right)} = 0$

Thus, $y = 0$ (the t axis) is a horizontal asymptote.

Vertical asymptotes: Since $t^2 + 1 > 0$ for all t, there are no vertical asymptotes.

Step 2. Analyze $C'(t)$:

$C'(t) = \dfrac{(t^2 + 1)(0.14) - 0.14t(2t)}{(t^2 + 1)^2} = \dfrac{0.14(1 - t^2)}{(t^2 + 1)^2} = \dfrac{0.14(1 - t)(1 + t)}{(t^2 + 1)^2}$

Critical values on $[0, \infty)$: $t = 1$

Sign chart for C':

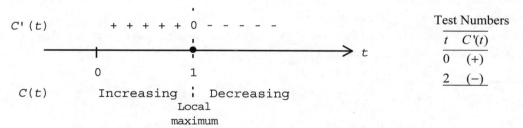

t	$C'(t)$
0	(+)
2	(−)

Test Numbers

Thus, C is increasing on $(0, 1)$ and decreasing on $(1, \infty)$; C has a maximum value at $t = 1$.

Step 3. Analyze $C''(t)$:

$C''(t) = \dfrac{(t^2 + 1)^2(-0.28t) - 0.14(1 - t^2)(2)(t^2 + 1)(2t)}{(t^2 + 1)^4} = \dfrac{(t^2 + 1)(-0.28t) - 0.56t(1 - t^2)}{(t^2 + 1)^3} = \dfrac{0.28t^3 - 0.84t}{(t^2 + 1)^3}$

$= \dfrac{0.28t(t^2 - 3)}{(t^2 + 1)^3} = \dfrac{0.28t(t - \sqrt{3})(t + \sqrt{3})}{(t^2 + 1)^3}$, $0 \leq t < \infty$

Partition numbers for C'' on $[0, \infty)$: $t = \sqrt{3}$

Sign chart for C'':

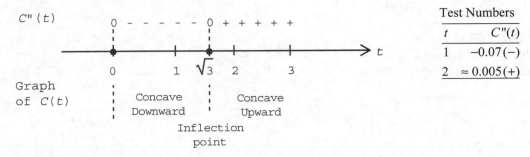

	Test Numbers	
t		$C''(t)$
1		$-0.07 (-)$
2		$\approx 0.005 (+)$

Thus, the graph of C is concave downward on $(0, \sqrt{3})$ and concave upward on $(\sqrt{3}, \infty)$; the graph has an inflection point at $t = \sqrt{3}$.

Step 4. Sketch the graph of $C(t)$:

t	$C(t)$
0	0
1	0.07
$\sqrt{3}$	≈ 0.06

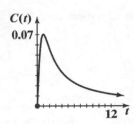

91. $N(t) = \dfrac{5t + 20}{t} = 5 + 20t^{-1}$, $1 \le t \le 30$

Step 1. Analyze $N(t)$:

Domain: $1 \le t \le 30$, or [1, 30].

Intercepts: There are no t or N intercepts.

Asymptotes: Since N is defined only for $1 \le t \le 30$, there are no horizontal asymptotes. Also, since $t \ne 0$ on [1, 30], there are no vertical asymptotes.

Step 2. Analyze $N'(t)$:

$N'(t) = -20t^{-2} = \dfrac{-20}{t^2}$, $1 \le t \le 30$

Since $N'(t) < 0$ for $1 \le t \le 30$, N is decreasing on (1, 30); N has no local extrema.

Step 3. Analyze $N''(t)$:

$N''(t) = \dfrac{40}{t^3}$, $1 \le t \le 30$

Since $N''(t) > 0$ for $1 \le t \le 30$, the graph of N is concave upward on (1, 30).

Step 4. Sketch the graph of N:

t	$N(t)$
1	25
5	9
10	7
30	5.67

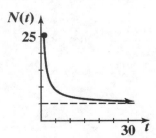

EXERCISE 12-5

Things to remember:

1.   ABSOLUTE MAXIMA AND MINIMA

If $f(c) \geq f(x)$ for all x in the domain of f, then $f(c)$ is called the ABSOLUTE MAXIMUM VALUE of f.

If $f(c) \leq f(x)$ for all x in the domain of f, then $f(c)$ is called the ABSOLUTE MINIMUM VALUE of f.

2.   A function f continuous on a closed interval $[a, b]$ has both an absolute maximum and an absolute minimum on that interval. Absolute extrema (if they exist) must always occur at critical values or at endpoints.

3.   PROCEDURE FOR FINDING ABSOLUTE EXTREMA ON A CLOSED INTERVAL

Step 1.   Check to make certain that f is continuous over $[a, b]$.

Step 2.   Find the critical values in the interval (a, b).

Step 3.   Evaluate f at the endpoints a and b and at the critical values found in Step 2.

Step 4.   The absolute maximum $f(x)$ on $[a, b]$ is the largest of the values found in Step 3.

Step 5.   The absolute minimum $f(x)$ on $[a, b]$ is the smallest of the values found in Step 3.

4.   SECOND DERIVATIVE TEST
Let c be a critical value for $f(x)$.

$f'(c)$	$f''(c)$	GRAPH OF f IS:	$f(c)$	EXAMPLE
0	+	Concave upward	Local minimum	$\smile$
0	−	Concave downward	Local maximum	$\frown$
0	0	?	Test does not apply	

5.   SECOND DERIVATIVE TEST FOR ABSOLUTE EXTREMUM
Let f be continuous on an interval I with only one critical value c on I:

If $f'(c) = 0$ and $f''(c) > 0$, then $f(c)$ is the absolute minimum of f on I.

If $f'(c) = 0$ and $f''(c) < 0$, then $f(c)$ is the absolute maximum of f on I.

1. Max $f(x) = f(3) = 3$; min $f(x) = f(-2) = -2$.

3. Max $h(x) = h(-5) = 25$; min $h(x) = h(0) = 0$.

5. Max $n(x) = n(4) = 2$; min $n(x) = n(3) = \sqrt{3}$.

7. Max $q(x) = q(27) = -3$; min $q(x) = q(64) = -4$.

9. Interval $[0, 10]$; absolute minimum: $f(0) = 0$;
   absolute maximum: $f(10) = 14$

11. Interval $[0, 8]$; absolute minimum: $f(0) = 0$;
    absolute maximum: $f(3) = 9$

13. Interval $[1, 10]$; absolute minimum: $f(1) = f(7) = 5$;
    absolute maximum: $f(10) = 14$

15. Interval $[1, 9]$; absolute minimum: $f(1) = f(7) = 5$;
    absolute maximum: $f(3) = f(9) = 9$

17. Interval $[2, 5]$; absolute minimum: $f(5) = 7$;
    absolute maximum: $f(3) = 9$

19. $f(x) = 2x - 5$.

    (A) On $[0,4]$: Max $f(x) = f(4) = 3$; min $f(x) = f(0) = -5$.

    (B) On $[0,10]$, Max $f(x) = f(10) = 15$; min $f(x) = f(0) = -5$.

    (C) On $[-5,10]$, Max $f(x) = f(10) = 15$; min $f(x) = f(-5) = -15$.

21. $f(x) = x^2$.

    (A) On $[-1,1]$, Max $f(x) = f(-1) = f(1) = 1$; min $f(x) = f(0) = 0$.

    (B) On $[1,5]$, Max $f(x) = f(5) = 25$; min $f(x) = f(1) = 1$.

    (C) On $[-5,5]$, Max $f(x) = f(-5) = f(5) = 25$; min $f(x) = f(0) = 0$.

23. $f(x) = e^{-x}$; $f'(x) = -e^{-x} < 0$; f is decreasing on $[-1,1]$.
    Absolute maximum: $f(-1) = e \approx 2.718$; absolute minimum: $f(1) = e^{-1} \approx 0.368$.

25. $f(x) = 9 - x^2$ on $[-4,4]$; $f'(x) = -2x$;   critical value: $-2x = 0$, $x = 0$.
    $f(-4) = -7$, $f(0) = 9$, $f(4) = -7$;
    absolute maximum: $f(0) = 9$; absolute minimum: $f(-4) = f(4) = -7$.

27. $f(x) = x^2 - 2x + 3$, $I = (-\infty, \infty)$
    $f'(x) = 2x - 2 = 2(x - 1)$
    $f'(x) = 0$: $2(x - 1) = 0$
    $\qquad\qquad\qquad x = 1$

    $x = 1$ is the ONLY critical value on I, and $f(1) = 1^2 - 2(1) + 3 = 2$.
    $f''(x) = 2$ and $f''(1) = 2 > 0$. Therefore, $f(1) = 2$ is the absolute minimum. The function does not have an
    absolute maximum since $\lim\limits_{x \to \pm\infty} f(x) = \infty$.

29. $f(x) = -x^2 - 6x + 9, I = (-\infty, \infty)$
$f'(x) = -2x - 6 = -2(x + 3)$
$f'(x) = 0: \quad -2(x + 3) = 0$
$\qquad\qquad\qquad x = -3$

$x = -3$ is the ONLY critical value on I, and $f(-3) = -(-3)^2 - 6(-3) + 9 = 18.$
$f''(x) = -2$ and $f''(-3) = -2 < 0$. Therefore, $f(-3) = 18$ is the absolute maximum. The function does not
have an absolute minimum since $\lim\limits_{x \to \pm\infty} f(x) = -\infty.$

31. $f(x) = x^3 + x, I = (-\infty, \infty)$
$f'(x) = 3x^2 + 1 \geq 1$ on I; f is increasing on I and $\lim\limits_{x \to -\infty} f(x) = -\infty, \ \lim\limits_{x \to \infty} f(x) = \infty$. Therefore, f does not have

any absolute extrema.

33. $f(x) = 8x^3 - 2x^4$; domain: all real numbers
$f'(x) = 24x^2 - 8x^3 = 8x^2(3 - x)$
$f''(x) = 48x - 24x^2 = 24x(2 - x)$
Critical values: $x = 0, \ x = 3$
$f''(0) = 0$ (second derivative test fails)
$f''(3) = -72$ f has a local maximum at $x = 3$.

Sign chart for $f'(x) = 8x^2(3 - x)$
(0 and 3 are partition numbers)

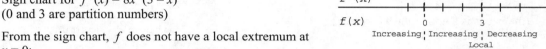

From the sign chart, f does not have a local extremum at
$x = 0;$
f has a local maximum at $x = 3$ which must be an absolute
maximum since f is increasing on $(-\infty, 3)$ and decreasing

on $(3, \infty)$; $f(3) = 54$ is the absolute maximum of f. f does not have an absolute minimum since
$\lim\limits_{x \to \infty} f(x) = \lim\limits_{x \to -\infty} f(x) = -\infty.$

35. $f(x) = x + \dfrac{16}{x}$; domain: all real numbers except $x = 0$.

$f'(x) = 1 - \dfrac{16}{x^2} = \dfrac{x^2 - 16}{x^2} = \dfrac{(x - 4)(x + 4)}{x^2}$

$f''(x) = \dfrac{32}{x^3}$

Critical values: $x = -4, \ x = 4$

$f''(-4) = -\dfrac{1}{2} < 0$; f has a local maximum at $x = -4$

$f''(4) = \dfrac{1}{2} > 0$; f has a local minimum at $x = 4$

$\lim\limits_{x \to \infty} f(x) = \lim\limits_{x \to \infty} \left(x + \dfrac{16}{x}\right) = \infty; \ \lim\limits_{x \to -\infty} f(x) = \lim\limits_{x \to -\infty} \left(x + \dfrac{16}{x}\right) = -\infty; \ f$ has no absolute extrema.

37. $f(x) = \dfrac{x^2}{x^2 + 1}$; domain: all real numbers

$f'(x) = \dfrac{(x^2 + 1)2x - x^2(2x)}{(x^2 + 1)^2} = \dfrac{2x}{(x^2 + 1)^2}$

$f''(x) = \dfrac{(x^2 + 1)^2(2) - 2x(2)(x^2 + 1)(2x)}{(x^2 + 1)^4} = \dfrac{2 - 6x^2}{(x^2 + 1)^3}$

Critical value: $x = 0$

Since f has only one critical value and $f''(0) = 2 > 0$, $f(0) = 0$ is the absolute minimum of f. Since

$$\lim_{x \to \infty} f(x) = \lim_{x \to \infty} \frac{x^2}{x^2 + 1} = 1, \ f \text{ has no absolute maximum; } y = 1 \text{ is a horizontal asymptote for the graph of}$$

f.

39. $f(x) = \dfrac{2x}{x^2 + 1}$; domain: all real numbers

$$f'(x) = \frac{(x^2 + 1)2 - 2x(2x)}{(x^2 + 1)^2} = \frac{2 - 2x^2}{(x^2 + 1)^2} = \frac{2(1 - x^2)}{(x^2 + 1)^2}$$

$$f''(x) = \frac{(x^2 + 1)^2(-4x) - 2(1 - x^2)(2)(x^2 + 1)(2x)}{(x^2 + 1)^4} = \frac{4x(x^2 - 3)}{(x^2 + 1)^3}$$

Critical values: $x = -1$, $x = 1$

$f''(-1) = 1 > 0$; f has a local minimum at $x = -1$

$f''(1) = -1 < 0$; f has a local maximum at $x = 1$

Sign chart for $f'(x)$

(partition numbers are -1 and 1)

$$\lim_{x \to \pm\infty} f(x) = \lim_{x \to \pm\infty} \frac{2x}{x^2 + 1} = 0$$

(the x-axis is a horizontal asymptote)

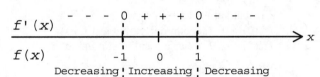

We can now conclude that $f(1) = 1$ is the absolute maximum of f and $f(-1) = -1$ is the absolute minimum of f.

41. $f(x) = \dfrac{x^2 - 1}{x^2 + 1}$; domain: all real numbers

$$f'(x) = \frac{(x^2 + 1)(2x) - (x^2 - 1)(2x)}{(x^2 + 1)^2} = \frac{4x}{(x^2 + 1)^2}$$

$$f''(x) = \frac{(x^2 + 1)^2(4) - 4x(2)(x^2 + 1)2x}{(x^2 + 1)^4} = \frac{4(1 - 3x^2)}{(x^2 + 1)^3}$$

Critical value: $x = 0$

$f''(0) = 4 > 0$; f has a local minimum at $x = 0$

Sign chart for $f'(x) = \dfrac{4x}{(x^2 + 1)^2}$ (0 is the partition number)

$\lim_{x \to \pm\infty} f(x) = \lim_{x \to \pm\infty} \dfrac{x^2 - 1}{x^2 + 1} = 1$; ($y = 1$ is a horizontal

asymptote)

We can now conclude that $f(0) = -1$ is the absolute minimum and f does not have an absolute maximum.

43. $f(x) = 2x^2 - 8x + 6$ on $I = [0, \infty)$

$f'(x) = 4x - 8 = 4(x - 2)$

$f''(x) = 4$

Critical value: $x = 2$

$f''(2) = 4 > 0$; f has a local minimum at $x = 2$

Since $x = 2$ is the only critical value of f on I, $f(2) = -2$ is the absolute minimum of f on I.

45. $f(x) = 3x^2 - x^3$ on $I = [0, \infty)$

$f'(x) = 6x - 3x^2 = 3x(2 - x)$

$f''(x) = 6 - 6x$

Critical value (in $(0, \infty)$): $x = 2$

$f''(2) = -6 < 0$; f has a local maximum at $x = 2$

Since $f(0) = 0$ and $x = 2$ is the only critical value of f in $(0, \infty)$, $f(2) = 4$ is the absolute maximum value of f on I.

47. $f(x) = (x + 4)(x - 2)^2$ on $I = [0, \infty)$

$f'(x) = (x + 4)(2)(x - 2) + (x - 2)^2 = (x - 2)[2x + 8 + x - 2] = (x - 2)(3x + 6) = 3x^2 - 12$

$f''(x) = 6x$

Critical value in I: $x = 2$

$f''(2) = 12 > 0$; f has a local minimum at $x = 2$

Since $f(0) = 16$ and $x = 2$ is the only critical value of f in $(0, \infty)$, $f(2) = 0$ is the absolute minimum of f on I.

49. $f(x) = 2x^4 - 8x^3$ on $I = (0, \infty)$

Since $\lim\limits_{x \to \infty} f(x) = \lim\limits_{x \to \infty} (2x^4 - 8x^3) = \infty$, f does not have an absolute maximum on I.

51. $f(x) = 20 - 3x - \dfrac{12}{x}$, $x > 0$; $I = (0, \infty)$

$f'(x) = -3 + \dfrac{12}{x^2}$

$f'(x) = 0$: $-3 + \dfrac{12}{x^2} = 0$

$\qquad\qquad 3x^2 = 12$

$\qquad\qquad x^2 = 4$

$\qquad\qquad x = 2 \quad (-2 \text{ is not in } I)$

$x = 2$ is the only critical value of f on I, and $f(2) = 20 - 3(2) - \dfrac{12}{2} = 8$.

$f''(x) = -\dfrac{24}{x^3}$; $f''(2) = -\dfrac{24}{8} = -3 < 0$. Therefore, $f(2) = 8$ is the absolute maximum of f. The function

does not have an absolute minimum since $\lim\limits_{x \to \infty} f(x) = -\infty$. (Also, $\lim\limits_{x \to 0^+} f(x) = -\infty$.)

53. $f(x) = 10 + 2x + \dfrac{64}{x^2}$, $x > 0$; $I = (0, \infty)$

$f'(x) = 2 - \dfrac{128}{x^3}$

$f'(x) = 0$: $2 - \dfrac{128}{x^3} = 0$

$\qquad\qquad 2x^3 = 128$

$$x^3 = 64$$
$$x = 4$$

$x = 4$ is the only critical value of f on I and $f(4) = 10 + 2(4) + \dfrac{64}{4^2} = 22$.

$f''(x) = \dfrac{384}{x^4}$; $f''(4) = \dfrac{384}{4^4} = \dfrac{3}{2} > 0$. Therefore, $f(4) = 22$ is the absolute minimum of f. The function

does not have an absolute maximum since $\lim\limits_{x \to \infty} f(x) = \infty$. (Also, $\lim\limits_{x \to 0^+} f(x) = \infty$.)

55. $f(x) = x + \dfrac{1}{x} + \dfrac{30}{x^3}$ on $I = (0, \infty)$

$f'(x) = 1 - \dfrac{1}{x^2} - \dfrac{90}{x^4} = \dfrac{x^4 - x^2 - 90}{x^4} = \dfrac{(x^2 - 10)(x^2 + 9)}{x^4}$

$f''(x) = \dfrac{2}{x^3} + \dfrac{360}{x^5}$

Critical value (in $(0, \infty)$): $x = \sqrt{10}$

$f''(\sqrt{10}) = \dfrac{2}{(10)^{3/2}} + \dfrac{360}{(10)^{5/2}} > 0$; f has a local minimum at $x = \sqrt{10}$.

Since $\sqrt{10}$ is the only critical value of f on I, $f(\sqrt{10}) = \dfrac{14}{\sqrt{10}}$ is the absolute minimum of f on I.

57. $f(x) = \dfrac{e^x}{x^2}$, $x > 0$

$f'(x) = \dfrac{x^2 \dfrac{d}{dx} e^x - e^x \dfrac{d}{dx} x^2}{x^4} = \dfrac{x^2 e^x - 2xe^x}{x^4} = \dfrac{xe^x(x-2)}{x^4} = \dfrac{e^x(x-2)}{x^3}$

Critical values: $f'(x) = \dfrac{e^x(x-2)}{x^3} = 0$

$$e^x(x - 2) = 0$$
$$x = 2 \quad \text{[Note: } e^x \neq 0 \text{ for all } x.]$$

Thus, $x = 2$ is the only critical value of f on $(0, \infty)$.

Sign chart for f':   [Note: This approach is a little easier than calculating $f''(x)$]

$f'(x)$	$-\ -\ -\ 0\ +\ +\ +$	Test Numbers
		x $f'(x)$
$f(x)$	$0 \quad 1 \quad 2 \quad 3$	$1 \quad -e(-)$
	Decr. Incr.	$3 \quad \frac{e^3}{27}(+)$

By the first derivative test, f has a minimum value at $x = 2$;

$f(2) = \dfrac{e^2}{2^2} = \dfrac{e^2}{4} \approx 1.847$ is the absolute minimum value of f.

59. $f(x) = \dfrac{x^3}{e^x}$

$f'(x) = \dfrac{\left(\dfrac{d}{dx} x^3\right) e^x - \left(\dfrac{d}{dx} e^x\right) x^3}{(e^x)^2} = \dfrac{3x^2 e^x - x^3 e^x}{e^{2x}} = \dfrac{x^2(3-x)e^x}{e^{2x}} = \dfrac{x^2(3-x)}{e^x}$

Critical values: $f'(x) = \dfrac{x^2(3-x)}{e^x} = 0$

$$x^2(3-x) = 0$$
$$x = 0 \text{ and } x = 3$$

Sign chart for f': [Note: This approach is a little easier than calculating $f''(x)$]:

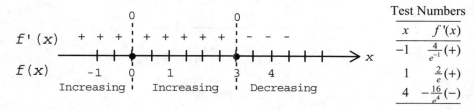

	Test Numbers	
x	$f'(x)$	
-1	$\dfrac{4}{e^{-1}}$	$(+)$
1	$\dfrac{2}{e}$	$(+)$
4	$-\dfrac{16}{e^4}$	$(-)$

By the first derivative test, f has a maximum value at $x = 3$; $f(3) = \dfrac{27}{e^3} \approx 1.344$ is the absolute maximum value of f.

61. $f(x) = 5x - 2x \ln x,\ x > 0$

$f'(x) = 5 - 2x\dfrac{d}{dx}(\ln x) - \ln x\,\dfrac{d}{dx}(2x) = 5 - 2x\left(\dfrac{1}{x}\right) - 2\ln x = 3 - 2\ln x,\ x > 0$

Critical values: $f'(x) = 3 - 2\ln x = 0$

$$\ln x = \dfrac{3}{2} = 1.5;\quad x = e^{1.5}$$

Thus, $x = e^{1.5}$ is the only critical value of f on $(0, \infty)$.

Now, $f''(x) = \dfrac{d}{dx}(3 - 2\ln x) = -\dfrac{2}{x}$, and $f''(e^{1.5}) = -\dfrac{2}{e^{1.5}} < 0$.

Therefore, f has a maximum value at $x = e^{1.5}$, and
$f(e^{1.5}) = 5e^{1.5} - 2e^{1.5}\ln(e^{1.5}) = 5e^{1.5} - 2(1.5)e^{1.5} = 2e^{1.5} \approx 8.963$
is the absolute maximum of f.

63. $f(x) = x^2(3 - \ln x),\ x > 0$

$f'(x) = x^2\dfrac{d}{dx}(3 - \ln x) + (3 - \ln x)\dfrac{d}{dx}x^2 = x^2\left(-\dfrac{1}{x}\right) + (3 - \ln x)2x = -x + 6x - 2x\ln x = 5x - 2x\ln x$

Critical values: $f'(x) = 5x - 2x\ln x = 0$
$$x(5 - 2\ln x) = 0$$
$$5 - 2\ln x = 0$$

$$\ln x = \dfrac{5}{2} = 2.5;\quad x = e^{2.5} \quad [\text{Note: } x \neq 0 \text{ on } (0, \infty)]$$

Now $f''(x) = 5 - 2x\left(\dfrac{1}{x}\right) - 2\ln x = 3 - 2\ln x$

and $f''(e^{2.5}) = 3 - 2\cdot\ln(e^{2.5}) = 3 - 2(2.5) = 3 - 5 = -2 < 0$

Therefore, f has a maximum value at $x = e^{2.5}$ and

$f(e^{2.5}) = (e^{2.5})^2(3 - \ln e^{2.5}) = e^5(3 - 2.5) = \dfrac{e^5}{2} \approx 74.207$

is the absolute maximum value of f.

65. $f(x) = \ln(xe^{-x})$, $x > 0$

$$f'(x) = \frac{1}{xe^{-x}} \frac{d}{dx}(xe^{-x}) = \frac{1}{xe^{-x}}[e^{-x} - xe^{-x}] = \frac{1-x}{x}$$

Critical values: $f'(x) = \dfrac{1-x}{x} = 0$; $x = 1$

Sign chart for $f'(x)$:

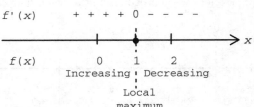

Test Numbers	
x	$f(x)$
$\frac{1}{2}$	$1(+)$
2	$-\frac{1}{2}(-)$

By the first derivative test, f has a maximum value at $x = 1$; $f(1) = \ln(e^{-1}) = -1$ is the absolute maximum value of f.

67. $f(x) = x^3 - 6x^2 + 9x - 6$

$f'(x) = 3x^2 - 12x + 9 = 3(x^2 - 4x + 3) = 3(x - 3)(x - 1)$

Critical values: $x = 1, 3$

(A)   On the interval $[-1, 5]$: $f(-1) = -1 - 6 - 9 - 6 = -22$

$f(1) = 1 - 6 + 9 - 6 = -2$

$f(3) = 27 - 54 + 27 - 6 = -6$

$f(5) = 125 - 150 + 45 - 6 = 14$

Thus, the absolute maximum of f is $f(5) = 14$, and the absolute minimum of f is $f(-1) = -22$.

(B)   On the interval $[-1, 3]$: $f(-1) = -22$

$f(1) = -2$

$f(3) = -6$

Absolute maximum of f: $f(1) = -2$; absolute minimum of f: $f(-1) = -22$

(C)   On the interval $[2, 5]$: $f(2) = 8 - 24 + 18 - 6 = -4$

$f(3) = -6$

$f(5) = 14$

Absolute maximum of f: $f(5) = 14$; absolute minimum of f: $f(3) = -6$

69. $f(x) = (x - 1)(x - 5)^3 + 1$

$f'(x) = (x - 1)3(x - 5)^2 + (x - 5)^3 = (x - 5)^2(3x - 3 + x - 5) = (x - 5)^2(4x - 8)$

Critical values: $x = 2, 5$

(A)   Interval $[0, 3]$: $f(0) = (-1)(-5)^3 + 1 = 126$

$f(2) = (2 - 1)(2 - 5)^3 + 1 = -26$

$f(3) = (3 - 1)(3 - 5)^3 + 1 = -15$

Absolute maximum of f: $f(0) = 126$; absolute minimum of f: $f(2) = -26$

(B)   Interval $[1, 7]$: $f(1) = 1$

$f(2) = -26$

$f(5) = 1$

$f(7) = (7 - 1)(7 - 5)^3 + 1 = 6 \cdot 8 + 1 = 49$

Absolute maximum of f: $f(7) = 49$; absolute minimum of f: $f(2) = -26$

(C)   Interval $[3, 6]$: $f(3) = (3 - 1)(3 - 5)^3 + 1 = -15$

$f(5) = 1$

$f(6) = (6 - 1)(6 - 5)^3 + 1 = 6$

Absolute maximum of f: $f(6) = 6$; absolute minimum of f: $f(3) = -15$

71. $f(x) = x^4 - 4x^3 + 5$

$f'(x) = 4x^3 - 12x^2 = 4x^2(x-3)$

Critical values: $x = 0$, $x = 3$

(A)   Interval $[-1, 2]$: $f(-1) = 10$

$f(0) = 5$

$f(2) = -11$

Absolute maximum of f: $f(-1) = 10$;   absolute minimum of f: $f(2) = -11$.

(B)   Interval $[0, 4]$: $f(0) = 5$

$f(3) = -22$

$f(4) = 5$

Absolute maximum of f: $f(0) = f(4) = 5$;   absolute minimum of f: $f(3) = -22$.

(C)   Interval $[-1, 1]$: $f(-1) = 10$

$f(0) = 5$

$f(1) = 2$

Absolute maximum of f: $f(-1) = 10$;   absolute minimum of f: $f(1) = 2$

73. f has a local minimum at $x = 2$.

75. Unable to determine from the given information $(f'(-3) = f''(-3) = 0)$.

77. Neither a local maximum nor a local minimum at $x = 6$; $x = 6$ is not a critical value of f.

79. f has a local maximum at $x = 2$.

EXERCISE 12-6

Things to remember:

STRATEGY FOR SOLVING OPTIMIZATION PROBLEMS

Step 1.   Introduce variables, look for relationships among these variables, and construct a mathematical model of the form: Maximize (or minimize) $f(x)$ on the interval I

Step 2.   Find the critical values of $f(x)$.

Step 3.   Use the procedures developed in Section 5-5 to find the absolute maximum (or minimum) value of $f(x)$ on the interval I and the value(s) of x where this occurs.

Step 4.   Use the solution to the mathematical model to answer all the questions asked in the problem.

1. $f = xy$ where $x + y = 28$. Since $y = 28 - x$, $f(x) = x(28 - x)$.

3.   Diameter x implies radius $r = \dfrac{x}{2}$. Therefore, $f(x) = \pi \left(\dfrac{x}{2} \right)^2 = \dfrac{\pi x^2}{4}$.

5.   Volume of a right circular cylinder of radius r and height h: $V = \pi r^2 h$. We have $h = x$. Therefore,

$f(x) = \pi (x)^2 (x) = \pi x^3$.

7. $f = xy$ where $2x + 2y = 120$, or $y = 60 - x$. Therefore, $f(x) = x(60 - x)$.

9.   Let x be one of the numbers and let y be the other.
Maximize $P = xy$ subject to $x + y = 15$.
$x + y = 15$ implies $y = 15 - x$ and $P(x) = x(15 - x) = 15x - x^2$.

Critical values:

$P'(x) = 15 - 2x$

$15 - 2x = 0;\ x = \dfrac{15}{2} = 7.5$

$P''(x) = -2 < 0$

P has a local maximum at $x = 7.5$. Since 7.5 is the only critical value, P has an absolute maximum at $x = 7.5$. The numbers are $x = 7.5$, $y = -7.5$.

11.   Let x be one of the numbers and let y be the other.
Minimize $P = xy$ subject to $x - y = 15$.
$x - y = 15$ implies $y = x - 15$ and $P(x) = x(x - 15) = x^2 - 15x$.

Critical values:

$P'(x) = 2x - 15$

$2x - 15 = 0;\ x = \dfrac{15}{2} = 7.5$

$P''(x) = 2 > 0$

P has a local minimum at $x = 7.5$. Since 7.5 is the only critical value, P has an absolute minimum at $x = 7.5$. The numbers are $x = y = 7.5$

13.   Let x be one of the numbers and let y be the other.
Minimize $P = x + y$ subject to $xy = 15$, $x \geq 0$, $y \geq 0$.

$xy = 15$ implies $y = \dfrac{15}{x}$ and $P(x) = x + \dfrac{15}{x}$

Critical values:

$P'(x) = 1 - \dfrac{15}{x^2}$

$1 - \dfrac{15}{x^2} = 0$

$x^2 = 15$

$x = \sqrt{15}$

$P''(x) = \dfrac{30}{x^3};\ P''(\sqrt{15}) = \dfrac{30}{15\sqrt{15}} = \dfrac{2}{\sqrt{15}} > 0$

P has a local minimum at $x = \sqrt{15}$. Since $\sqrt{15}$ is the only critical value, P has an absolute minimum at $x = \sqrt{15}$. The numbers are $x = y = \sqrt{15}$.

15.   Let x be the length and y the width of the rectangle.
Minimize $P = 2x + 2y$ subject to $xy = 200$, $x \geq 0$, $y \geq 0$.

$xy = 200$ implies $y = \dfrac{200}{x}$ and $P(x) = 2x + 2\left(\dfrac{200}{x}\right) = 2x + \dfrac{400}{x}$

Critical values:

$$P'(x) = 2 - \frac{400}{x^2}$$

$$2 - \frac{400}{x^2} = 0$$

$$2x^2 = 400$$

$$x^2 = 200$$

$$x = \sqrt{200} = 10\sqrt{2}$$

$$P''(x) = \frac{800}{x^3}; \quad P''(10\sqrt{2}) = \frac{800}{\left(10\sqrt{2}\right)^3} = \frac{2}{5\sqrt{2}} > 0$$

P has a local minimum at $x = 10\sqrt{2}$ and since this is the only critical value, P has an absolute minimum at $x = 10\sqrt{2}$.

At $x = 10\sqrt{2}$, $y = \frac{200}{10\sqrt{2}} = \frac{20}{\sqrt{2}} = 10\sqrt{2}$. The dimensions of the rectangle are: length $= 10\sqrt{2}$, width $= 10\sqrt{2}$.

17. Let x be the length and y the width of the rectangle.
Maximize $A = xy$ subject to $2x + 2y = 148$, $x \geq 0$, $y \geq 0$.
$2x + 2y = 148$ implies $y = 74 - x$ and $A(x) = x(74 - x) = 74x - x^2$.

Critical values:

$$A'(x) = 74 - 2x$$

$$74 - 2x = 0; \quad x = \frac{74}{2} = 37$$

$$A''(x) = -2 < 0$$

A has a local maximum at $x = 37$. Since 37 is the only critical value, A has an absolute maximum at $x = 37$.

At $x = 37$, $y = 74 - 37 = 37$. The dimensions of the rectangle are: length = 37, width = 37.

19. Price-demand: $p(x) = 500 - 0.5x$; cost: $C(x) = 20{,}000 + 135x$

(A) Revenue: $R(x) = x \cdot p(x) = 500x - 0.5x^2$, $0 \leq x < \infty$
$R'(x) = 500 - x$
$R'(x) = 500 - x = 0$ implies $x = 500$
$R''(x) = -1$; $R''(500) = -1 < 0$
R has an absolute maximum at $x = 500$.
$p(500) = 500 - 0.5(500) = 250$; $R(500) = (500)^2 - 0.5(500)^2 = 125{,}000$
The company should produce 500 phones each week at a price of $250 per phone to maximize their revenue. The maximum revenue is $125,000

(B) Profit: $P(x) = R(x) - C(x) = 500x - 0.5x^2 - (20{,}000 + 135x) = 365x - 0.5x^2 - 20{,}000$
$P'(x) = 365 - x$
$P'(x) = 365 - x = 0$ implies $x = 365$
$P''(x) = -1$; $P''(365) = -1 < 0$
P has an absolute maximum at $x = 365$

$p(365) = 500 - 0.5(365) = 317.50;$

$P(365) = (365)^2 - 0.5(365)^2 - 20,000 = 46,612.50$

To maximize profit, the company should produce 365 phones each week at a price of $317.50 per phone. The maximum profit is $46,612.50.

21.   (A)   Revenue $R(x) = x \cdot p(x) = x\left(200 - \dfrac{x}{30}\right) = 200x - \dfrac{x^2}{30}, 0 \le x \le 6,000$

$R'(x) = 200 - \dfrac{2x}{30} = 200 - \dfrac{x}{15}$

Now $R'(x) = 200 - \dfrac{x}{15} = 0$ implies $x = 3000.$

$R''(x) = -\dfrac{1}{15} < 0.$

Thus, $R''(3000) = -\dfrac{1}{15} < 0$ and we conclude that R has an absolute maximum at $x = 3000$. The

maximum revenue is $R(3000) = 200(3000) - \dfrac{(3000)^2}{30} = \$300,000$

(B)   Profit $P(x) = R(x) - C(x) = 200x - \dfrac{x^2}{30} - (72,000 + 60x)$

$= 140x - \dfrac{x^2}{30} - 72,000$

$P'(x) = 140 - \dfrac{x}{15}$

Now $140 - \dfrac{x}{15} = 0$ implies $x = 2,100.$ $P''(x) = -\dfrac{1}{15}$ and $P''(2,100) = -\dfrac{1}{15} < 0.$ Thus, the

maximum profit occurs when 2,100 television sets are produced. The maximum profit is

$P(2,100) = 140(2,100) - \dfrac{(2,100)^2}{30} - 72,000 = \$75,000$

the price that the company should charge is $p(2,100) = 200 - \dfrac{2,100}{30} = \$130$ for each set.

(C)   If the government taxes the company $5 for each set, then the profit $P(x)$ is given by

$P(x) = 200x - \dfrac{x^2}{30} - (72,000 + 60x) - 5x = 135x - \dfrac{x^2}{30} - 72,000.$

$P'(x) = 135 - \dfrac{x}{15}.$

Now $135 - \dfrac{x}{15} = 0$ implies $x = 2,025.$

$P''(x) = -\dfrac{1}{15}$ and $P''(2,025) = -\dfrac{1}{15} < 0.$ Thus, the maximum profit in   this case occurs when

2,025 television sets are produced. The maximum profit is

$P(2,025) = 135(2,025) - \dfrac{(2,025)^2}{30} - 72,000 = \$64,687.50$

and the company should charge $p(2,025) = 200 - \dfrac{2,025}{30} = \$132.50/\text{set}.$

23. (A) 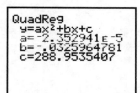   (B)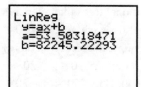

(C) The revenue at the demand level x is:
$$R(x) = xp(x)$$
where $p(x)$ is the quadratic regression equation in (A).

The cost at the demand level x is $C(x)$ given by the linear regression equation in (B). The profit $P(x) = R(x) - C(x)$.

The maximum profit is \$118,996 at the demand level $x = 1422$.

The price per sleeping bag at the demand level $x = 1422$ is \$195.

25. (A) Let x = number of 10¢ reductions in price. Then
$640 + 40x$ = number of sandwiches sold at x 10¢ reductions
$8 - 0.1x$ = price per sandwich, $0 \le x \le 80$
Revenue: $R(x) = (640 + 40x)(8 - 0.1x) = 5120 + 256x - 4x^2, 0 \le x \le 80$
$R'(x) = 256 - 8x$
$R'(x) = 256 - 8x = 0$ implies $x = 32$
$R(0) = 5120, R(32) = 9216, R(80) = 0$

Thus, the deli should charge $8 - 3.20 = \$4.80$ per sandwich to realize a maximum revenue of \$9216.

(B) Let x = number of 20¢ reductions in price. Then
$640 + 15x$ = number of sandwiches sold
$8 - 0.2x$ = price per sandwich $0 \le x \le 40$
Revenue: $R(x) = (640 + 15x)(8 - 0.2x) = 5120 - 8x - 3x^2, 0 \le x \le 40$
$R'(x) = -8 - 6x$
$R'(x) = -8 - 6x = 0$ has no solutions in $(0, 40)$
Now, $R(0) = 640 \cdot 8 = \$5120$, $R(40) = 0$

Thus, the deli should charge \$8 per sandwich to maximize their revenue under these conditions.

27. Let x = number of dollar increases in the rate per day. Then
$200 - 5x$ = total number of cars rented and $30 + x$ = rate per day.
Total income = (total number of cars rented)(rate)
$y(x) = (200 - 5x)(30 + x), 0 \le x \le 40$
$y'(x) = (200 - 5x)(1) + (30 + x)(-5) = 200 - 5x - 150 - 5x = 50 - 10x = 10(5 - x)$
Thus, $x = 5$ is the only critical value and
$y(5) = (200 - 25)(30 + 5) = 6125$.
$y''(x) = -10$
$y''(5) = -10 < 0$
Therefore, the absolute maximum income is $y(5) = \$6125$ when the rate is \$35 per day.

29.   Let x = number of additional trees planted per acre. Then
$30 + x$ = total number of trees per acre and $50 - x$ = yield per tree.

Yield per acre = (total number of trees per acre)(yield per tree)

$$y(x) = (30 + x)(50 - x), \ 0 \le x \le 20$$
$$y'(x) = (30 + x)(-1) + (50 - x) = 20 - 2x = 2(10 - x)$$

The only critical value is $x = 10$.

$y(10) = 40(40) = 1600$ pounds per acre.
$y''(x) = -2$
$y''(10) = -2 < 0$

Therefore, the absolute maximum yield is $y(10) = 1600$ pounds per acre when the number of trees per acre is 40.

31.   Volume = $V(x) = (12 - 2x)(8 - 2x)x, \ 0 \le x \le 4$

$$= 96x - 40x^2 + 4x^3$$

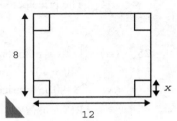

$V'(x) = 96 - 80x + 12x^2 = 4(24 - 20x + 3x^2)$

We solve $24 - 20x + 3x^2 = 0$ by using the quadratic formula:

$$x = \frac{20 \pm \sqrt{400 - 4 \cdot 24 \cdot 3}}{6} = \frac{10 \pm 2\sqrt{7}}{3}$$

Thus, $x = \dfrac{10 - 2\sqrt{7}}{3} \approx 1.57$ is the only critical value on the interval $[0, 4]$.

$V''(x) = -80 + 24x$
$V''(1.57) = -80 + 24(1.57) < 0$

Therefore, a square with a side of length $x = 1.57$ inches should be cut from each corner to obtain the maximum volume.

33.   Area = 800 square feet = xy (1)
Cost = $18x + 6(2y + x)$

From (1), we have $y = \dfrac{800}{x}$.

Hence, cost $C(x) = 18x + 6\left(\dfrac{1600}{x} + x\right)$, or

$$C(x) = 24x + \frac{9600}{x}, \ x > 0,$$

$$C'(x) = 24 - \frac{9600}{x^2} = \frac{24(x^2 - 400)}{x^2} = \frac{24(x - 20)(x + 20)}{x^2}.$$

Therefore, $x = 20$ is the only critical value.

$$C''(x) = \frac{19,200}{x^3}$$

$C''(20) = \dfrac{19,200}{8000} > 0$. Therefore, $x = 20$ for the minimum cost.

The dimensions of the fence are shown in the diagram at the right.

35.   (A)   Let x and y be the width and the length of the rectangle respectively. Then we have
$2x + y + (y - 100) = 240$ or $2x + 2y = 340$ and $x = 170 - y$ where $100 \le y \le 170$.

The Area $= xy = (170 - y)y$.

Let $f(y) = y(170 - y)$, $100 \le y \le 170$.

$f'(y) = 170 - 2y$ and $f''(y) = -2 < 0$. .

$f'(y) = 0$ implies $y = 85$ which is not in the domain of f. We note that $f(100) = 7,000$, $f(170) = 0$.

Thus, the maximum of f occurs when $y = 100$ and $x = 170 - y = 70$.

(B)   In this case, $2x + 2y - 100 = 400$ or $x + y = 250$ and $x = 250 - y$.

$f(y) = y(250 - y) = 250y - y^2$, $100 \le y \le 250$; $f'(y) = 250 - 2y$; $f''(y) = -2 < 0$.

$f'(y) = 0$ implies $y = 125$

Thus, f has an absolute maximum at $y = 125$, $x = 250 - y = 125$.

37.   Let $x =$ number of cans of paint produced in each production run. Then, number of production runs:

$$\frac{16,000}{x}, \ 1 \le x \le 16,000$$

Cost: $C(x) =$ cost of storage + cost of set up

$$= \frac{x}{2}(4) + \frac{16,000}{x}(500)$$

[Note: $\frac{x}{2}$ is the average number of cans of paint in storage per day.]

Thus,

$$C(x) = 2x + \frac{8,000,000}{x}, \ 1 \le x \le 16,000$$

$$C'(x) = 2 - \frac{8,000,000}{x^2} = \frac{2x^2 - 8,000,000}{x^2} = \frac{2(x^2 - 4,000,000)}{x^2}$$

Critical value: $x = 2000$

$$C''(x) = \frac{16,000,000}{x^3}; \ \ C''(2000) > 0.$$

Thus, the minimum cost occurs when $x = 2000$ and the number of production runs is $\dfrac{16,000}{2,000} = 8$.

39.   Let $x =$ number of books produced each printing. Then, the number of printings $= \dfrac{50,000}{x}$.

Cost $= C(x) =$ cost of storage + cost of printing

$$= \frac{x}{2} + \frac{50,000}{x}(1000), \ \ x > 0$$

[Note: $\frac{x}{2}$ is the average number in storage each day.]

$$C'(x) = \frac{1}{2} - \frac{50,000,000}{x^2} = \frac{x^2 - 100,000,000}{2x^2} = \frac{(x+10,000)(x-10,000)}{2x^2}$$

Critical value: $x = 10,000$

$$C''(x) = \frac{100,000,000}{x^3}$$

$$C''(10,000) = \frac{100,000,000}{(10,000)^3} > 0$$

Thus, the minimum cost occurs when $x = 10,000$ and the number of printings is $\dfrac{50,000}{10,000} = 5$.

41. Let x = number of hours it takes the train to travel 360 miles.

Then $360 = xv$ or $x = \dfrac{360}{v}$.

$$\text{Cost} = \left(300 + \frac{v^2}{4}\right)x = \left(300 + \frac{v^2}{4}\right)\left(\frac{360}{v}\right) = \frac{108,000}{v} + 90v$$

Let $C(v) = \dfrac{108,000}{v} + 90v$, $v > 0$. We want to minimize $C(v)$.

$$C'(v) = \frac{-108,000}{v^2} + 90 = \frac{-108,000 + 90v^2}{v^2}.$$

$C'(v) = 0$ implies $90v^2 = 108,000$ or $v = 34.64$

$$C''(v) = \frac{216,000}{v^3} > 0 \text{ for } v > 0$$

So, $C(v)$ has an absolute minimum at $v = 34.64$ miles per hour.

43. (A) Let the cost to lay the pipe on the land be 1 unit; then the cost to lay the pipe in the lake is 1.4 units.

$$C(x) = \text{total cost} = (1.4)\sqrt{x^2 + 25} + (1)(10 - x), 0 \le x \le 10 = (1.4)(x^2 + 25)^{1/2} + 10 - x$$

$$C'(x) = (1.4)\frac{1}{2}(x^2 + 25)^{-1/2}(2x) - 1 = (1.4)x(x^2 + 25)^{-1/2} - 1 = \frac{1.4x - \sqrt{x^2 + 25}}{\sqrt{x^2 + 25}}$$

$C'(x) = 0$ when $1.4x - \sqrt{x^2 + 25} = 0$ or $1.96x^2 = x^2 + 25$

$$.96x^2 = 25$$
$$x^2 = \frac{25}{.96} = 26.04$$
$$x = \pm 5.1$$

Thus, the critical value is $x = 5.1$.

$$C''(x) = (1.4)(x^2 + 25)^{-1/2} + (1.4)x\left(-\frac{1}{2}\right)(x^2 + 25)^{-3/2}2x = \frac{1.4}{(x^2 + 25)^{1/2}} - \frac{(1.4)x^2}{(x^2 + 25)^{3/2}}$$

$$= \frac{35}{(x^2 + 25)^{3/2}}$$

$$C''(5.1) = \frac{35}{[(5.1)^2 + 25]^{3/2}} > 0$$

Thus, the cost will be a minimum when $x = 5.1$.

Note that: $C(0) = (1.4)\sqrt{25} + 10 = 17$
$$C(5.1) = (1.4)\sqrt{51.01} + (10 - 5.1) = 14.9$$
$$C(10) = (1.4)\sqrt{125} = 15.65$$

Thus, the absolute minimum occurs when $x = 5.1$ miles.

(B) $C(x) = (1.1)\sqrt{x^2 + 25} + (1)(10 - x), 0 \le x \le 10$

$$C'(x) = \frac{(1.1)x - \sqrt{x^2 + 25}}{\sqrt{x^2 + 25}}$$

$C'(x) = 0$ when $1.1x - \sqrt{x^2 + 25} = 0$ or $(1.21)x^2 = x^2 + 25$

$$.21x^2 = 25$$

$$x^2 = \frac{25}{.21} = 119.05$$

$$x = \pm 10.91$$

Critical value: $x = 10.91 > 10$, i.e., there are no critical values on the interval [0, 10]. Now,

$C(0) = (1.1)\sqrt{25} + 10 = 15.5,$

$C(10) = (1.1)\sqrt{125} \approx 12.30.$

Therefore, the absolute minimum occurs when $x = 10$ miles.

45. $C(t) = 30t^2 - 240t + 500, \ 0 \le t \le 8$

$C'(t) = 60t - 240; \quad t = 4$ is the only critical value.

$C''(t) = 60$

$C''(4) = 60 > 0$

Now, $C(0) = 500$

$C(4) = 30(4)^2 - 240(4) + 500 = 20,$

$C(8) = 30(8)^2 - 240(8) + 500 = 500.$

Thus, 4 days after a treatment, the concentration will be minimum; the minimum concentration is 20 bacteria per cm^3.

47. $H(t) = 4t^{1/2} - 2t, \quad 0 \le t \le 2$

$H'(t) = 2t^{-1/2} - 2$

Thus, $t = 1$ is the only critical value.

Now, $H(0) = 4 \cdot 0^{1/2} - 2(0) = 0,$

$H(1) = 4 \cdot 1^{1/2} - 2(1) = 2,$

$H(2) = 4 \cdot 2^{1/2} - 4 \approx 1.66.$

Therefore, $H(1)$ is the absolute maximum, and after one month the maximum height will be 2 feet.

49. $N(t) = 30 + 12t^2 - t^3, \ 0 \le t \le 8$

The rate of increase $= R(t) = N'(t) = 24t - 3t^2,$ and

$R'(t) = N''(t) = 24 - 6t.$

Thus, $t = 4$ is the only critical value of $R(t)$.

Now, $R(0) = 0,$

$R(4) = 24 \cdot 4 - 3 \cdot 4^2 = 48,$

$R(8) = 24 \cdot 8 - 3 \cdot 8^2 = 0.$

Therefore, the absolute maximum value of R occurs when $t = 4$; the maximum rate of increase will occur four years from now.

CHAPTER 12 REVIEW

1. The function f is increasing on (a, c_1), (c_3, c_6).   (4-1, 4-2)

2. $f'(x) < 0$ on (c_1, c_3), (c_6, b).   (4-1, 4-2)

3. The graph of f is concave downward on (a, c_2), (c_4, c_5), (c_7, b).   (4-1, 4-2)

4. A local minimum occurs at $x = c_3$.   (4-1)

5. The absolute maximum occurs at $x = c_1, c_6$.   (4-1, 4-5)

6. $f'(x)$ appears to be zero at $x = c_1, c_3, c_5$.   (4-1)

7. $f'(x)$ does not exist at $x = c_4, c_6$.   (4-1)

8. $x = c_2, c_4, c_5, c_7$ are inflection points.   (4-2)

9.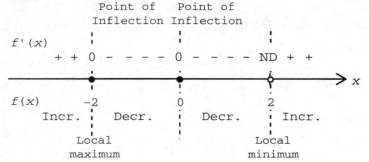

    Using this information together with the points $(-3, 0)$, $(-2, 3)$, $(-1, 2)$, $(0, 0)$, $(2, -3)$, $(3, 0)$ on the graph, we have

    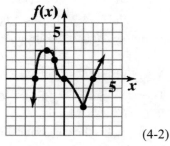

    (4-2)

10. Domain: all real numbers
    Intercepts: y-intercept: $f(0) = 0$
                  x-intercepts: $x = 0$
    Asymptotes: Horizontal asymptote: $y = 2$
                  no vertical asymptotes
    Critical values: $x = 0$

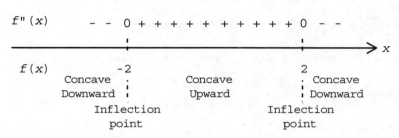

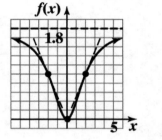

(4-1, 4-2)

11. $f(x) = x^4 + 5x^3$

$f'(x) = 4x^3 + 15x^2$

$f''(x) = 12x^2 + 30x$ (4-2)

12. $y = 3x + \dfrac{4}{x}$

$y' = 3 - \dfrac{4}{x^2}$

$y'' = \dfrac{8}{x^3}$ (4-2)

13. $f(x) = \dfrac{5+x}{4-x}$

Domain: All real numbers except $x = 4$.

y-intercept: $f(0) = \dfrac{5}{4}$

x-intercept: $f(x) = 0;\ \dfrac{5+x}{4-x} = 0,\ x = -5$ (4-1)

14. $f(x) = \ln(x+2)$

Domain: $x > -2,\ (-2, \infty)$

y-intercept: $f(0) = \ln 2$

x-intercept: $f(x) = 0;\ \ln(x+2) = 0,\ x + 2 = 1,\ x = -1.$ (4-1)

15. $f(x) = \dfrac{x+3}{x^2-4}$

Horizontal asymptote: $\displaystyle\lim_{x\to\infty}\dfrac{x+3}{x^2-4} = 0$, $y = 0$ is a horizontal asymptote.

Vertical asymptotes: $x^2 - 4 = 0$; $x = 2$, $x = -2$ are vertical asymptotes.   (4-4)

16. $f(x) = \dfrac{2x-7}{3x+10}$

Horizontal asymptote: $\displaystyle\lim_{x\to\infty}\dfrac{2x-7}{3x+10} = \dfrac{2}{3}$, $y = \dfrac{2}{3}$ is a horizontal asymptote.

Vertical asymptotes: $3x + 10 = 0$; $x = -\dfrac{10}{3}$ is a vertical asymptote.   (4-4)

17. $f(x) = x^4 - 12x^2$, $f'(x) = 4x^3 - 24x$, $f''(x) = 12x^2 - 24$

$\quad f''(x) = 0$

$12x^2 - 24 = 0$

$\quad\quad x^2 = 2$

$\quad\quad\quad x = \pm\sqrt{2}$

Inflection points: $(-\sqrt{2}, -20)$, $(\sqrt{2}, -20)$ (4-2)

18. $f(x) = (2x+1)^{1/3} - 6$, $f'(x) = \dfrac{1}{3}(2x+1)^{-2/3}(2) = \dfrac{2}{3}(2x+1)^{-2/3}$

$f''(x) = -\dfrac{4}{9}(2x+1)^{-5/3}(2) = \dfrac{-8}{9(2x+1)^{5/3}}$

$f''(x)$ does not exist at $x = -\dfrac{1}{2}$; inflection point $\left(-\frac{1}{2}, -6\right)$ (4-2)

19. $f(x) = x^{1/5}$

(A) $f'(x) = \dfrac{1}{5}x^{-4/5} = \dfrac{1}{5x^{4/5}}$

(B) Critical value: $x = 0$ ($f'(0)$ does not exist)

(C) Partition number: $x = 0$ (4-1)

20. $f(x) = x^{-1/5}$

(A) $f'(x) = -\dfrac{1}{5}x^{-6/5} = \dfrac{-1}{5x^{6/5}}$

(B) Critical value: none (0 is not in the domain of f).

(C) Partition number: $x = 0$ (4-1)

21. $f(x) = x^3 - 18x^2 + 81x$

Step 1. Analyze $f(x)$:

(A) Domain: All real numbers, $(-\infty, \infty)$

(B) Intercepts: y-intercept: $f(0) = 0^3 - 18(0)^2 + 81(0) = 0$

$\quad\quad\quad\quad\quad x$-intercepts: $x^3 - 18x^2 + 81x = 0$

$\quad\quad\quad\quad\quad\quad\quad\quad\quad x(x^2 - 18x + 81) = 0$

$$x(x-9)^2 = 0$$
$$x = 0, 9$$

(C) Asymptotes: No horizontal or vertical asymptotes.

Step 2. Analyze $f'(x)$:

$f'(x) = 3x^2 - 36x + 81 = 3(x^2 - 12x + 27) = 3(x-3)(x-9)$

Critical values: $x = 3$, $x = 9$
Partition numbers: $x = 3$, $x = 9$

Sign chart for f':

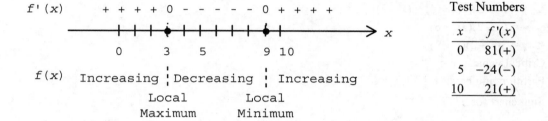

x	f'(x)
0	81(+)
5	−24(−)
10	21(+)

Test Numbers

Thus, f is increasing on $(-\infty, 3)$ and on $(9, \infty)$; f is decreasing on $(3, 9)$. There is a local maximum at $x = 3$ and a local minimum at $x = 9$.

Step 3. Analyze $f''(x)$:

$f''(x) = 6x - 36 = 6(x-6)$

Thus, $x = 6$ is a partition number for $f''(x)$.

Sign chart for f'':

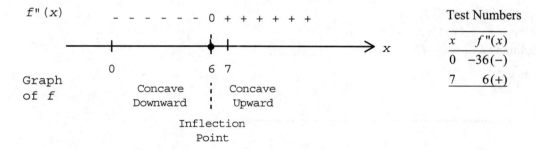

x	f''(x)
0	−36(−)
7	6(+)

Test Numbers

Thus, the graph of f is concave downward on $(-\infty, 6)$ and concave upward on $(6, \infty)$. The point $x = 6$ is an inflection point.

Step 4. Sketch the graph of f:

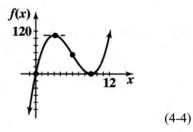

(4-4)

22. $f(x) = (x + 4)(x - 2)^2$

Step 1. Analyze $f(x)$:

(A)   Domain: All real numbers, $(-\infty, \infty)$.

(B)   Intercepts: y-intercept: $f(0) = 4(-2)^2 = 16$

   x-intercepts: $(x + 4)(x - 2)^2 = 0$

   $x = -4, 2$

(C)   Asymptotes: Since f is a polynomial, there are no horizontal or vertical asymptotes.

Step 2. Analyze $f'(x)$:

$f'(x) = (x + 4)2(x - 2)(1) + (x - 2)^2(1)$

   $= (x - 2)[2(x + 4) + (x - 2)]$

   $= (x - 2)(3x + 6)$

   $= 3(x - 2)(x + 2)$

Critical values: $x = -2, x = 2$

Partition numbers: $x = -2, x = 2$

Sign chart for f':

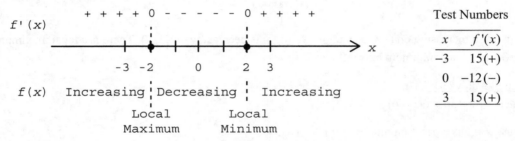

x	$f'(x)$
Test Numbers	
-3	$15(+)$
0	$-12(-)$
3	$15(+)$

Thus, f is increasing on $(-\infty, -2)$ and on $(2, \infty)$; f is decreasing on $(-2, 2)$; f has a local maximum at $x = -2$ and a local minimum at $x = 2$.

Step 3. Analyze $f''(x)$:

$f''(x) = 3(x + 2)(1) + 3(x - 2)(1) = 6x$

Partition number for f'': $x = 0$

Sign chart for f'':

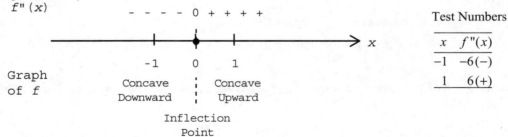

x	$f''(x)$
Test Numbers	
-1	$-6(-)$
1	$6(+)$

Thus, the graph of f is concave downward on $(-\infty, 0)$ and concave upward on $(0, \infty)$; there is an inflection point at $x = 0$.

Step 4. Sketch the graph of f:

x	$f(x)$
-2	32
0	16
2	0

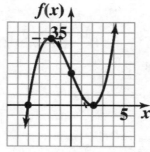

(4-4)

23. $f(x) = 8x^3 - 2x^4$

Step 1. Analyze $f(x)$:

(A)   Domain: All real numbers, $(-\infty, \infty)$.

(B)   Intercepts: y-intercept: $f(0) = 0$

x-intercepts: $8x^3 - 2x^4 = 0$

$$2x^3(4 - x) = 0$$

$$x = 0, 4$$

(C)   Asymptotes: No horizontal or vertical asymptotes.

Step 2. Analyze $f'(x)$:

$f'(x) = 24x^2 - 8x^3 = 8x^2(3 - x)$

Critical values: $x = 0,\ x = 3$

Partition numbers: $x = 0,\ x = 3$

Sign chart for f':

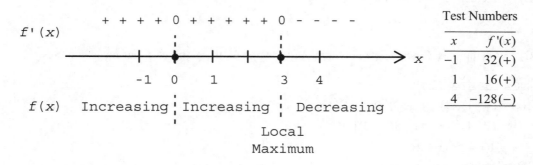

Thus, f is increasing on $(-\infty, 3)$ and decreasing on $(3, \infty)$; f has a local maximum at $x = 3$.

<u>Step 3. Analyze $f''(x)$</u>:

$f''(x) = 48x - 24x^2 = 24x(2 - x)$

Partition numbers for $f'' = 0$, $x = 2$

Sign chart for f'':

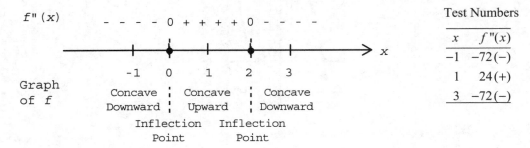

Test Numbers	
x	$f''(x)$
-1	$-72(-)$
1	$24(+)$
3	$-72(-)$

Thus, the graph of f is concave downward on $(-\infty, 0)$ and on $(2, \infty)$; the graph is concave upward on $(0, 2)$; there are inflection points at $x = 0$ and $x = 2$.

<u>Step 4. Sketch the graph of f</u>:

x	$f(x)$
0	0
2	32
3	54

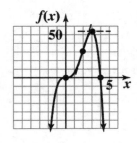

(4-4)

24. $f(x) = (x - 1)^3(x + 3)$

<u>Step 1. Analyze $f(x)$</u>:

(A)   Domain: All real numbers.

(B)   Intercepts: y-intercept: $f(0) = (-1)^3(3) = -3$

x-intercepts: $(x - 1)^3(x + 3) = 0$

$x = 1, -3$

(C)   Asymptotes: Since f is a polynomial (of degree 4), the graph of f has no asymptotes.

<u>Step 2. Analyze $f'(x)$</u>:

$f'(x) = (x - 1)^3(1) + (x + 3)(3)(x - 1)^2(1)$

$= (x - 1)^2[(x - 1) + 3(x + 3)]$

$= 4(x - 1)^2(x + 2)$

Critical values: $x = -2$, $x = 1$

Partition numbers: $x = -2$, $x = 1$

Sign chart for f':

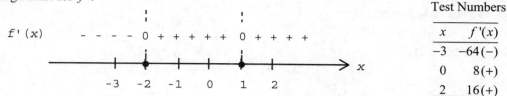

Test Numbers	
x	$f'(x)$
-3	$-64\,(-)$
0	$8\,(+)$
2	$16\,(+)$

Thus, f is decreasing on $(-\infty, -2)$; f is increasing on $(-2, 1)$ and $(1, \infty)$; f has a local minimum at $x = -2$.

Step 3. Analyze $f''(x)$:

$$f''(x) = 4(x-1)^2(1) + 4(x+2)(2)(x-1)(1)$$
$$= 4(x-1)[(x-1) + 2(x+2)]$$
$$= 12(x-1)(x+1)$$

Partition numbers for f'': $x = -1$, $x = 1$.
Sign chart for f'':

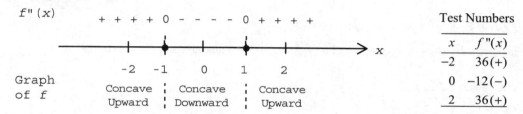

Test Numbers	
x	$f''(x)$
-2	$36\,(+)$
0	$-12\,(-)$
2	$36\,(+)$

Thus, the graph of f is concave upward on $(-\infty, -1)$ and on $(1, \infty)$; the graph of f is concave downward on $(-1, 1)$; the graph has inflection points at $x = -1$ and at $x = 1$.

Step 4. Sketch the graph of f:

x	$f(x)$
-2	-27
0	-3
1	0

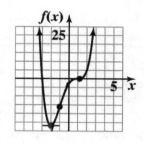

(4-4)

25. $f(x) = \dfrac{3x}{x+2}$

Step 1. Analyze $f(x)$:
The domain of f is all real numbers except $x = -2$.

Intercepts: y-intercept: $f(0) = \dfrac{3(0)}{0+2} = 0$

x-intercepts: $\dfrac{3x}{x+2} = 0$

$$3x = 0$$
$$x = 0$$

Asymptotes:
Horizontal asymptotes: $\dfrac{3x}{x} = 3$. Thus, the line $y = 3$ is a horizontal asymptote.

<u>Vertical asymptote(s):</u>   The denominator is 0 at $x = -2$ and the numerator is nonzero at $x = -2$. Thus, the line $x = -2$ is a vertical asymptote.

<u>Step 2. Analyze $f'(x)$:</u>

$$f'(x) = \frac{(x+2)(3) - 3x(1)}{(x+2)^2} = \frac{6}{(x+2)^2}$$

Critical values: $f'(x) = \frac{6}{(x+2)^2} \neq 0$ for all x ($x \neq -2$).

Thus, f does not have any critical values.

Partition numbers: $x = -2$ is a partition number for f'.

Sign chart for f':

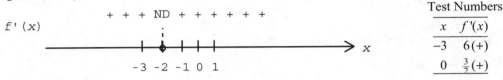

Test Numbers	
x	$f'(x)$
-3	$6(+)$
0	$\frac{3}{2}(+)$

Thus, f is increasing on $(-\infty, -2)$ and on $(-2, \infty)$; f does not have any local extrema.

<u>Step 3. Analyze $f''(x)$:</u>

$$f''(x) = -12(x+2)^{-3} = \frac{-12}{(x+2)^3}$$

Partition numbers for f'': $x = -2$

Sign chart for f'':

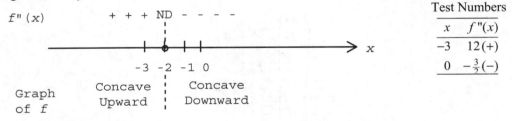

Test Numbers	
x	$f''(x)$
-3	$12(+)$
0	$-\frac{3}{2}(-)$

The graph of f is concave upward on $(-\infty, -2)$ and concave downward on $(-2, \infty)$. The graph of f does not have any inflection points.

<u>Step 4. Sketch the graph of f:</u>

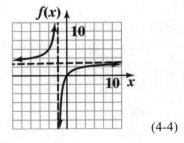

(4-4)

26. $f(x) = \dfrac{x^2}{x^2 + 27}$

Step 1. Analyze $f(x)$:

(A)   Domain: All real numbers.

(B)   Intercepts: y-intercepts: $f(0) = 0$

x-intercepts: $\dfrac{x^2}{x^2 + 27} = 0, \ \ x = 0$

(C)   Asymptotes:

Horizontal asymptote: $\dfrac{x^2}{x^2} = 1; \ \ y = 1$ is a horizontal asymptote

Vertical asymptote: no vertical asymptotes

Step 2. Analyze $f'(x)$:

$$f'(x) = \frac{(x^2 + 27)(2x) - x^2(2x)}{(x^2 + 27)^2} = \frac{54x}{(x^2 + 27)^2}$$

Critical values: $x = 0$
Partition numbers: $x = 0$

Sign chart for f':

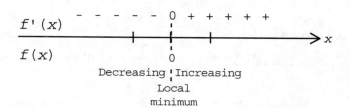

Thus, f is decreasing on $(-\infty, 0)$ and increasing on $(0, \infty)$; f has a local minimum at $x = 0$.

Step 3. Analyze $f''(x)$:

$$f''(x) = \frac{(x^2 + 27)^2(54) - 54x(2)(x^2 + 27)2x}{(x^2 + 27)^4} = \frac{162(9 - x^2)}{(x^2 + 27)^3}$$

Partition numbers for f'': $x = -3, \ x = 3$

Sign chart for f'':

Test Numbers	
x	$f''(x)$
-4	$(-)$
0	$(+)$
4	$(-)$

The graph of f is concave upward on $(-3, 3)$ and concave downward on $(-\infty, -3)$ and $(3, \infty)$; the graph has inflection points at $x = -3$ and $x = 3$.

Step 4. Sketch the graph of f:

x	$f(x)$
-3	$\frac{1}{4}$
0	0
3	$\frac{1}{4}$

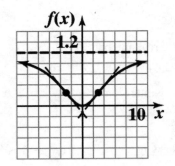

(4-4)

27. $f(x) = \dfrac{x}{(x+2)^2}$

Step 1. Analyze $f(x)$:

(A)   Domain: All real numbers except $x = -2$.

(B)   Intercepts: y-intercepts: $f(0) = 0$

x-intercepts: $\dfrac{x}{(x+2)^2} = 0$, $x = 0$

(C)   Asymptotes:

Horizontal asymptote: $\dfrac{x}{x^2} = \dfrac{1}{x}$; $y = 0$ (the x-axis) is a horizontal asymptote.

Vertical asymptote: $x = -2$ is a vertical asymptote

Step 2. Analyze $f'(x)$:

$f'(x) = \dfrac{(x+2)^2 - x(2)(x+2)}{(x+2)^4} = \dfrac{2-x}{(x+2)^3}$

Critical values: $x = 2$

Partition numbers: $x = -2$, $x = 2$

Sign chart for f':

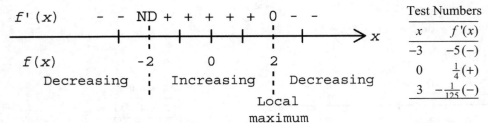

Thus, f is increasing on $(-2, 2)$ and decreasing on $(-\infty, -2)$ and $(2, \infty)$; f has a local maximum at $x = 2$.

Step 3. Analyze $f''(x)$:

$f''(x) = \dfrac{(x+2)^3(-1) - (2-x)(3)(x+2)^2}{(x+2)^6} = \dfrac{2(x-4)}{(x+2)^4}$

Partition numbers for f'': $x = -2$, $x = 4$

Sign chart for f'':

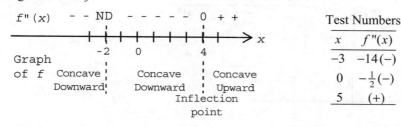

Test Numbers	
x	$f''(x)$
-3	$-14(-)$
0	$-\frac{1}{2}(-)$
5	$(+)$

The graph of f is concave upward on $(4, \infty)$ and concave downward on $(-\infty, -2)$ and $(-2, 4)$; the graph has an inflection point at $x = 4$.

Step 4. Sketch the graph of f:

x	$f(x)$
0	0
2	$\frac{1}{8}$
4	$\frac{1}{9}$

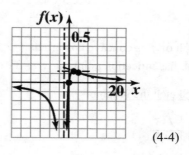

(4-4)

28. $f(x) = \dfrac{x^3}{x^2 + 3}$

Step 1. Analyze $f(x)$:

(A)   Domain: All real numbers.

(B)   Intercepts: y-intercepts: $f(0) = 0$

x-intercepts: $\dfrac{x^3}{x^2+3} = 0$, $x = 0$

(C)   Asymptotes:

Horizontal asymptote: $\dfrac{x^3}{x^2} = x$; no horizontal asymptote.

Vertical asymptote: no vertical asymptotes.

Oblique asymptote: It follows from above that $y = x$ is an oblique asymptote.

Step 2. Analyze $f'(x)$:

$$f'(x) = \frac{(x^2+3)(3x^2) - x^3(2x)}{(x^2+3)^2} = \frac{x^2(x^2+9)}{(x^2+3)^2}$$

Critical values: $x = 0$

Partition numbers: $x = 0$

Sign chart for f':

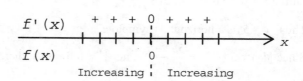

f is increasing on $(-\infty, \infty)$.

Step 3. Analyze $f''(x)$:

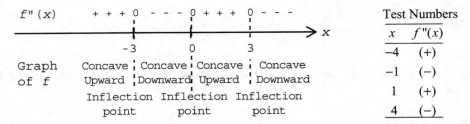

$$f''(x) = \frac{(x^2+3)^2(4x^3+18x) - x^2(x^2+9)(2)(x^2+3)2x}{(x^2+3)^4} = \frac{6x(9-x^2)}{(x^2+3)^3}$$

Partition numbers for f'': $x = -3$, $x = 0$, $x = 3$

Sign chart for f'':

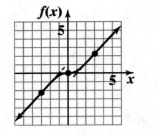

Test Numbers	
x	$f''(x)$
-4	$(+)$
-1	$(-)$
1	$(+)$
4	$(-)$

The graph of f is concave upward on $(-\infty, -3)$ and $(0, 3)$, and concave downward on $(-3, 0)$ and $(3, \infty)$; the graph has inflection points at $x = -3$, $x = 0$, $x = 3$.

Step 4. Sketch the graph of f:

x	$f(x)$
-3	$-\frac{9}{4}$
0	0
3	$\frac{9}{4}$

(4-4)

29. $f(x) = 5 - 5e^{-x}$

Step 1. Analyze $f(x)$:

(A)   Domain: All real numbers, $(-\infty, \infty)$.

(B)   Intercepts: y-intercept: $f(0) = 5 - 5e^{-0} = 0$

x-intercepts: $5 - 5e^{-x} = 0$

$$e^{-x} = 1$$

$$x = 0$$

(C)   Asymptotes:

$$\lim_{x \to \infty} (5 - 5e^{-x}) = \lim_{x \to \infty} \left(5 - \frac{5}{e^x}\right) = 5$$

$$\lim_{x \to -\infty} (5 - 5e^{-x}) \text{ does not exist.}$$

Thus, $y = 5$ is a horizontal asymptote.

Since $f(x) = 5 - \dfrac{5}{e^x} = \dfrac{5e^x - 5}{e^x}$ and $e^x \neq 0$ for all x, there are no vertical asymptotes.

Step 2. Analyze $f'(x)$:

$f'(x) = -5e^{-x}(-1) = 5e^{-x} > 0$ on $(-\infty, \infty)$

Thus, f is increasing on $(-\infty, \infty)$; there are no local extrema.

Step 3. Analyze $f''(x)$:

$f''(x) = -5e^{-x} < 0$ on $(-\infty, \infty)$.

Thus, the graph of f is concave downward on $(-\infty, \infty)$; there are no inflection points.

Step 4. Sketch the graph of f:

x	$f(x)$
0	0
-1	-8.59
2	4.32

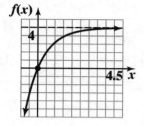

(4-4)

30. $f(x) = x^3 \ln x$

Step 1. Analyze $f(x)$:

(A)   Domain: all positive real numbers, $(0, \infty)$.

(B)   Intercepts: y-intercept: Since $x = 0$ is not in the domain, there   is no y-intercept.

x-intercepts: $x^3 \ln x = 0$

$\ln x = 0$

$x = 1$

(C)   Asymptotes:

$\lim\limits_{x\to\infty} (x^3 \ln x)$ does not exist.

It can be shown that $\lim\limits_{x\to 0^+} (x^3 \ln x) = 0$. Thus, there are no horizontal or vertical asymptotes.

Step 2. Analyze $f'(x)$:

$f'(x) = x^3 \left(\dfrac{1}{x}\right) + (\ln x)3x^2 = x^2[1 + 3\ln x], \;\; x > 0$

Critical values: $x^2[1 + 3\ln x] = 0$

$1 + 3\ln x = 0$ (since $x > 0$)

$\ln x = -\dfrac{1}{3}$

$x = e^{-1/3} \approx 0.72$

Partition numbers: $x = e^{-1/3}$

Sign chart for f':

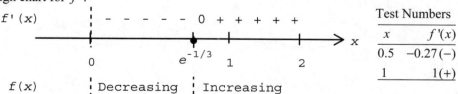

Test Numbers	
x	$f'(x)$
0.5	$-0.27\,(-)$
1	$1\,(+)$

Thus, f is decreasing on $(0, e^{-1/3})$ and increasing on $(e^{-1/3}, \infty)$; f has a local minimum at $x = e^{-1/3}$.

Step 3. Analyze $f''(x)$:

$f''(x) = x^2 \left(\dfrac{3}{x}\right) + (1 + 3\ln x)2x = x(5 + 6\ln x), \;\; x > 0$

Partition numbers: $x(5 + 6 \ln x) = 0$

$$5 + 6 \ln x = 0$$

$$\ln x = -\frac{5}{6}$$

$$x = e^{-5/6} \approx 0.43$$

Sign chart for f'':

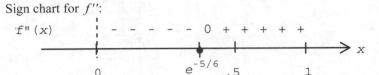

Test Numbers	
x	$f''(x)$
0.2	$-0.93(-)$
1	$5(+)$

Thus, the graph of f is concave downward on $(0, e^{-5/6})$ and concave upward on $(e^{-5/6}, \infty)$; the graph has an inflection point at $x = e^{-5/6}$.

Step 4. Sketch the graph of f:

x	$f(x)$
$e^{-5/6}$	-0.07
$e^{-1/3}$	-0.12
1	0

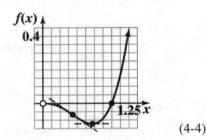

(4-4)

In Problems 31 – 40, D_x denotes $\dfrac{d}{dx}$.

31. $\displaystyle\lim_{x \to 0} \frac{e^{3x} - 1}{x}$

Step 1:

$\displaystyle\lim_{x \to 0} (e^{3x} - 1) = e^0 - 1 = 0$ and $\displaystyle\lim_{x \to 0} x = 0.$

Therefore, L'Hôpital's rule applies.

Step 2:

$\displaystyle\lim_{x \to 0} \frac{D_x(e^{3x} - 1)}{D_x x} = \lim_{x \to 0} \frac{3e^{3x}}{1} = 3.$ Thus, $\displaystyle\lim_{x \to 0} \frac{e^{3x} - 1}{x} = 3.$ (4-3)

32. $\displaystyle\lim_{x \to 2} \frac{x^2 - 5x + 6}{x^2 + x - 6}$

Step 1:

$\displaystyle\lim_{x \to 2} (x^2 - 5x + 6) = 2^2 - 10 + 6 = 0$ and $\displaystyle\lim_{x \to 2} (x^2 + x - 6) = 4 + 2 - 6 = 0.$

Therefore, L'Hôpital's rule applies.

Step 2:

$$\lim_{x \to 2} \frac{D_x(x^2 - 5x + 6)}{D_x(x^2 + x - 6)} = \lim_{x \to 2} \frac{(2x - 5)}{(2x + 1)} = \frac{-1}{5} \ . \ \text{Thus,} \ \lim_{x \to 2} \frac{x^2 - 5x + 6}{x^2 + x - 6} = \frac{-1}{5} \ . \qquad (4\text{-}3)$$

33. $\displaystyle\lim_{x \to 0^-} \frac{\ln(1 + x)}{x^2}$.

Step 1:

$$\lim_{x \to 0^-} \ln(1 + x) = \ln(1) = 0 \ \text{ and } \ \lim_{x \to 0^-} x^2 = 0.$$

Therefore, L'Hôpital's rule applies.

Step 2:

$$\lim_{x \to 0^-} \frac{D_x \ln(1 + x)}{D_x x^2} = \lim_{x \to 0^-} \frac{\dfrac{1}{1 + x}}{2x} = \lim_{x \to 0^-} \frac{1}{2x(1 + x)} = -\infty.$$

Thus, $\displaystyle\lim_{x \to 0^-} \frac{\ln(1 + x)}{x^2} = -\infty.$ (4-3)

34. $\displaystyle\lim_{x \to 0} \frac{\ln(1 + x)}{1 + x}$

Step 1:

$$\lim_{x \to 0} \ln(1 + x) = \ln(1) = 0 \ \text{ and } \ \lim_{x \to 0} (1 + x) = 1.$$

Therefore, L'Hôpital's rule does not apply.

Step 2:

Using the quotient property for limits

$$\lim_{x \to 0} \frac{\ln(1 + x)}{1 + x} = \frac{0}{1} = 0 \qquad (4\text{-}3)$$

35. $\displaystyle\lim_{x \to \infty} \frac{e^{4x}}{x^2}$

Step 1:

$$\lim_{x \to \infty} e^{4x} = \infty \ \text{ and } \ \lim_{x \to \infty} x^2 = \infty. \ \text{Therefore, L'Hôpital's rule applies.}$$

Step 2:

$$\lim_{x \to \infty} \frac{D_x e^{4x}}{D_x x^2} = \lim_{x \to \infty} \frac{4e^{4x}}{2x} = \lim_{x \to \infty} \frac{2e^{4x}}{x} \ .$$

Since $\displaystyle\lim_{x \to \infty} 2e^x = \infty$ and $\displaystyle\lim_{x \to \infty} x = \infty$, we apply L'Hôpital's rule again.

Step 3:

$$\lim_{x \to \infty} \frac{D_x 2e^{4x}}{D_x x} = \lim_{x \to \infty} \frac{8e^{4x}}{1} = \infty. \text{ Thus, } \lim_{x \to \infty} \frac{e^{4x}}{x^2} = \infty \qquad (4\text{-}3)$$

36. $\displaystyle\lim_{x \to 0} \frac{e^x + e^{-x} - 2}{x^2}$

Step 1:

$$\lim_{x \to 0} (e^x + e^{-x} - 2) = e^0 + e^0 - 2 = 0 \text{ and } \lim_{x \to 0} x^2 = 0.$$

Therefore, L'Hôpital's rule applies.

Step 2:

$$\lim_{x \to 0} \frac{D_x(e^x + e^{-x} - 2)}{D_x x^2} = \lim_{x \to 0} \frac{e^x - e^{-x}}{2x}. \text{ Since } \lim_{x \to 0} (e^x - e^{-x}) = e^0 - e^0 = 0 \text{ and } \lim_{x \to 0} 2x = 0, \text{ we apply}$$

L'Hôpital's again.

Step 3:

$$\lim_{x \to 0} \frac{D_x(e^x + e^{-x})}{D_x 2x} = \lim_{x \to 0} \frac{e^x + e^{-x}}{2} = \frac{e^0 + e^0}{2} = 1.$$

Thus, $\displaystyle\lim_{x \to 0} \frac{e^x + e^{-x} - 2}{x^2} = 1.$ \qquad (4-3)

37. $\displaystyle\lim_{x \to 0^+} \frac{\sqrt{1+x} - 1}{\sqrt{x}}$

Step 1:

$$\lim_{x \to 0^+} \left(\sqrt{1+x} - 1 \right) = 0 \text{ and } \lim_{x \to 0^+} \sqrt{x} = 0.$$

Therefore, L'Hôpital's rule applies.

Step 2:

$$\lim_{x \to 0^+} \frac{D_x \left[\sqrt{1+x} - 1 \right]}{D_x \sqrt{x}} = \lim_{x \to 0^+} \frac{\frac{1}{2}(1+x)^{-1/2}}{\frac{1}{2} x^{-1/2}} = \lim_{x \to 0^+} \frac{\sqrt{x}}{\sqrt{1+x}} = 0.$$

Thus, $\displaystyle\lim_{x \to 0^+} \frac{\sqrt{1+x} - 1}{\sqrt{x}} = 0.$ \qquad (4-3)

38. $\displaystyle\lim_{x \to \infty} \frac{\ln x}{x^5}$

Step 1:

$$\lim_{x \to \infty} \ln x = \infty \text{ and } \lim_{x \to \infty} x^5 = \infty. \text{ Therefore, L'Hôpital's rule applies.}$$

Step 2:

$$\lim_{x \to \infty} \frac{D_x \ln x}{D_x x^5} = \lim_{x \to \infty} \frac{\frac{1}{x}}{5x^4} = \lim_{x \to \infty} \frac{1}{5x^5} = 0. \text{ Thus, } \lim_{x \to \infty} \frac{\ln x}{x^5} = 0. \qquad (4\text{-}3)$$

39. $\displaystyle\lim_{x\to\infty} \frac{\ln(1+6x)}{\ln(1+3x)}$

Step 1:

$\displaystyle\lim_{x\to\infty} \ln(1+6x) = \infty$ and $\displaystyle\lim_{x\to\infty} \ln(1+3x) = \infty$.

Therefore, L'Hôpital's rule applies.

Step 2:

$$\lim_{x\to\infty} \frac{D_x \ln(1+6x)}{D_x \ln(1+3x)} = \lim_{x\to\infty} \frac{\dfrac{6}{1+6x}}{\dfrac{3}{1+3x}} = \lim_{x\to\infty} \frac{2(1+3x)}{1+6x} \ .$$

Since $\displaystyle\lim_{x\to\infty} 2(1+3x) = \infty$ and $\displaystyle\lim_{x\to\infty} (1+6x) = \infty$, we apply L'Hôpital's rule again.

Step 3:

$$\lim_{x\to\infty} \frac{D_x(2[1+3x])}{D_x(1+6x)} = \lim_{x\to\infty} \frac{D_x(2+6x)}{D_x(1+6x)} = \lim_{x\to\infty} \frac{6}{6} = 1.$$

Thus, $\displaystyle\lim_{x\to\infty} \frac{\ln(1+6x)}{\ln(1+3x)} = 1.$ (4-3)

40. $\displaystyle\lim_{x\to 0} \frac{\ln(1+6x)}{\ln(1+3x)}$

Step 1:

$\displaystyle\lim_{x\to 0} \ln(1+6x) = \ln 1 = 0$ and $\displaystyle\lim_{x\to 0} \ln(1+3x) = \ln 1 = 0$.

Therefore, L'Hôpital's rule applies.

Step 2:

$$\lim_{x\to 0} \frac{D_x \ln(1+6x)}{D_x \ln(1+3x)} = \lim_{x\to 0} \frac{\dfrac{6}{1+6x}}{\dfrac{3}{1+3x}} = \lim_{x\to 0} \frac{2(1+3x)}{1+6x} = 2.$$

Thus, $\displaystyle\lim_{x\to 0} \frac{\ln(1+6x)}{\ln(1+3x)} = 2.$ (4-3)

41.

x	$f'(x)$	$f(x)$
$-\infty < x < -2$	Negative and increasing	Decreasing and concave upward
$x = -2$	x-intercept	Local minimum
$-2 < x < -1$	Positive and increasing	Increasing and concave upward
$x = -1$	Local maximum	Inflection point
$-1 < x < 1$	Positive and decreasing	Increasing and concave downward
$x = 1$	Local minimum	Inflection point
$1 < x < \infty$	Positive and increasing	Increasing and concave upward

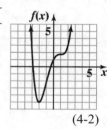

(4-2)

42. The graph in (C) could be the graph of $y = f'(x)$.   (4-2)

43. $f(x) = x^3 - 6x^2 - 15x + 12$

$f'(x) = 3x^2 - 12x - 15$

Critical values: f' is defined for all x:

$3x^2 - 12x - 15 = 0$

$3(x^2 - 4x - 5) = 0$

$3(x-5)(x+1) = 0$
Thus, $x = -1$ and $x = 5$ are critical values of f.
$f''(x) = 6x - 12$
Now, $f''(-1) = 6(-1) - 12 = -18 < 0$.
Thus, f has a local maximum at $x = -1$.
Also, $f''(5) = 6(5) - 12 = 18 > 0$ and f has a local minimum at $x = 5$.   (4-1, 4-2)

44. $y = f(x) = x^3 - 12x + 12$, $-3 \le x \le 5$
$f'(x) = 3x^2 - 12$
Critical values: f' is defined for all x:
$$3x^2 - 12 = 0$$
$$3(x^2 - 4) = 0$$
$$3(x-2)(x+2) = 0$$
Thus, the critical values of f are: $x = -2$, $x = 2$.
$f(-3) = (-3)^3 - 12(-3) + 12 = 21$
$f(-2) = (-2)^3 - 12(-2) + 12 = 28$
$f(2) = 2^3 - 12(2) + 12 = -4$ Absolute minimum
$f(5) = 5^3 - 12(5) + 12 = 77$ Absolute maximum   (4-5)

45. $y = f(x) = x^2 + \dfrac{16}{x^2}$, $x > 0$

$f'(x) = 2x - \dfrac{32}{x^3} = \dfrac{2x^4 - 32}{x^3} = \dfrac{2(x^4 - 16)}{x^3} = \dfrac{2(x-2)(x+2)(x^2+4)}{x^3}$

$f''(x) = 2 + \dfrac{96}{x^4}$

The only critical value of f in the interval $(0, \infty)$ is $x = 2$.
Since

$f''(2) = 2 + \dfrac{96}{2^4} = 8 > 0$,

$f(2) = 8$ is the absolute minimum of f on $(0, \infty)$.   (4-5)

46. $f(x) = 11x - 2x \ln x$, $x > 0$

$f'(x) = 11 - 2x\left(\dfrac{1}{x}\right) - (\ln x)(2) = 11 - 2 - 2\ln x = 9 - 2\ln x$, $x > 0$

Critical value(s):
$9 - 2\ln x = 0$
$2\ln x = 9$
$\ln x = \dfrac{9}{2}$
$x = e^{9/2}$

$f''(x) = -\dfrac{2}{x}$ and $f''(e^{9/2}) = -\dfrac{2}{e^{9/2}} < 0$

Since $x = e^{9/2}$ is the only critical value, and $f''(e^{9/2}) < 0$, f has an absolute maximum at $x = e^{9/2}$. The absolute maximum is:

$f(e^{9/2}) = 11e^{9/2} - 2e^{9/2}\ln(e^{9/2}) = 11e^{9/2} - 9e^{9/2} = 2e^{9/2} \approx 180.03$ (4-5)

47. $f(x) = 10xe^{-2x}, \quad x > 0$

$f'(x) = 10xe^{-2x}(-2) + 10e^{-2x}(1) = 10e^{-2x}(1 - 2x), \quad x > 0$

Critical value(s):

$10e^{-2x}(1 - 2x) = 0$

$1 - 2x = 0$

$x = \dfrac{1}{2}$

$f''(x) = 10e^{-2x}(-2) + 10(1 - 2x)e^{-2x}(-2) = -20e^{-2x}(1 + 1 - 2x) = -40e^{-2x}(1 - x)$

$f''\left(\dfrac{1}{2}\right) = -20e^{-1} < 0$

Since $x = \dfrac{1}{2}$ is the only critical value, and $f''\left(\dfrac{1}{2}\right) = -20e^{-1} < 0$, f has an absolute maximum at $x = \dfrac{1}{2}$.

The absolute maximum of f is: $f\left(\dfrac{1}{2}\right) = 10\left(\dfrac{1}{2}\right)e^{-2(1/2)} = 5e^{-1} \approx 1.84$ (4-5)

48. Yes. Consider f on the interval $[a, b]$. Since f is a polynomial, f is continuous on $[a, b]$. Therefore, f has an absolute maximum on $[a, b]$. Since f has a local minimum at $x = a$ and $x = b$, the absolute maximum of f on $[a, b]$ must occur at some point c in (a, b); f has a local maximum at $x = c$.   (4-5)

49. No, increasing/decreasing properties are stated in terms of intervals in the domain of f. A correct statement is: $f(x)$ is decreasing on $(-\infty, 0)$ and $(0, \infty)$.   (4-1)

50. A critical value for f is a partition number for f' that is also in the domain of f. However, f' may have partition numbers that are not in the domain of f and hence are not critical values for f. For example, let $f(x) = \dfrac{1}{x}$. Then $f'(x) = -\dfrac{1}{x^2}$ and 0 is a partition number for f', but 0 is NOT a critical value for f since it is not in the domain of f.   (4-1)

51. $f(x) = 6x^2 - x^3 + 8, \ 0 \le x \le 4$

$f'(x) = 12x - 3x^2$

$f''(x) = 12 - 6x$

Now, $f''(x)$ is defined for all x and $f''(x) = 12 - 6x = 0$ implies $x = 2$. Thus, f' has a critical value at $x = 2$. Since this is the only critical value of f' and $[f'(x)]'' = f'''(x) = -6$ so that $f'''(2) = -6 < 0$, it follows that $f'(2) = 12$ is the absolute maximum of f'. The graph is shown at the right.   (4-2, 4-5)

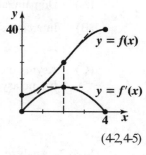

52. Let $x > 0$ be one of the numbers. Then $y = \dfrac{400}{x}$ is the other number. Now, we have:

$S(x) = x + \dfrac{400}{x}, \ x > 0,$

$S'(x) = 1 - \dfrac{400}{x^2} = \dfrac{x^2 - 400}{x^2} = \dfrac{(x - 20)(x + 20)}{x^2}$

Thus, $x = 20$ is the only critical value of S on $(0, \infty)$.

$$S''(x) = \frac{800}{x^3} \quad \text{and} \quad S''(20) = \frac{800}{8000} = \frac{1}{10} > 0$$

Therefore, $S(20) = 20 + \dfrac{400}{20} = 40$ is the absolute minimum sum, and this occurs when each number is

20.   (4-6)

53. $f(x) = x^4 + x^3 - 4x^2 - 3x + 4.$

Step 1. Analyze $f(x)$:

(A) Domain: All real numbers (f is a polynomial function)

(B) Intercepts: y-intercept: $f(0) = 4$

                     x-intercepts: $x \approx 0.79,\ 1.64$

(C) Asymptotes: Since f is a polynomial function (of degree 4), the graph of f has no asymptotes.

Step 2. Analyze $f'(x)$:

$= 4x^3 + 3x^2 - 8x - 3$

Critical values: $x \approx -1.68,\ -0.35,\ 1.28$;

f is increasing on $(-1.68, -0.35)$ and $(1.28, \infty)$; f is decreasing on $(-\infty, -1.68)$ and $(-0.35, 1.28)$. f has local minima at $x = -1.68$ and $x = 1.28$. f has a local maximum at $x = -0.35$.

Step 3. Analyze $f''(x)$:

$f''(x) = 12x^2 + 6x - 8$

The graph of f is concave downward on $(-1.10, 0.60)$; the graph of f is concave upward on $(-\infty, -1.10)$ and $(0.60, \infty)$; the graph has inflection points at $x \approx -1.10$ and 0.60.   (4-4)

54. $f(x) = 0.25x^4 - 5x^3 + 31x^2 - 70x$

Step 1. Analyze $f(x)$:

(A) Domain: all real numbers

(B) Intercepts: y-intercept: $f(0) = 0$

                     x-intercepts: $x = 0,\ 11.10$

(C) Asymptotes: since f is a polynomial function, the graph of f has no asymptotes; $\lim\limits_{x \to \pm\infty} f(x) = \infty$

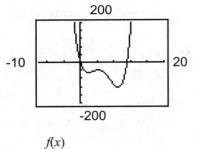

$f(x)$

Step 2. Analyze $f'(x)$:

$f'(x) = x^3 - 15x^2 + 62x - 70$

Critical values: $x \approx 1.87,\ 4.19,\ 8.94$

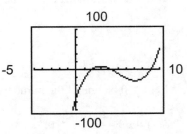

Sign chart for f':

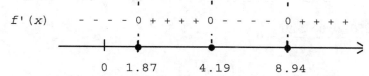

f is increasing on $(1.87, 4.19)$ and $(8.94, \infty)$; f is decreasing on $(-\infty, 1.87)$ and $(4.19, 8.94)$; f has local minima at $x = 1.87$ and $x = 8.94$; f has a local maximum at $x = 4.19$

Step 3. Analyze $f''(x)$:

$f''(x) = 3x^2 - 30x + 62$

Partition numbers for f'': $x \approx 2.92, 7.08$

Sign chart for f'':

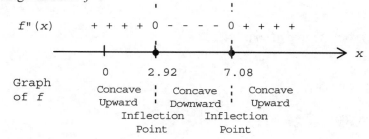

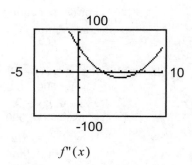

$f''(x)$

The graph of f is concave downward on $(2.92, 7.08)$ and concave upward on $(-\infty, 2.92)$ and $(7.08, \infty)$; the graph has inflection points at $x = 2.92$ and $x = 7.08$.   (4-4)

55. $f(x) = 3x - x^2 + e^{-x}$, $x > 0$

$f'(x) = 3 - 2x - e^{-x}$, $x > 0$

Critical value(s): $f'(x) = 3 - 2x - e^{-x} = 0$

$$x \approx 1.373$$

$f''(x) = -2 + e^{-x}$ and $f''(1.373) = -2 + e^{-1.373} < 0$

Since $x \approx 1.373$ is the only critical value, and $f''(1.373) < 0$, f has an absolute maximum at $x = 1.373$.

The absolute maximum of f is: $f(1.373) = 3(1.373) - (1.373)^2 + e^{-1.373} \approx 2.487$.   (4-5)

56. $f(x) = \dfrac{\ln x}{e^x}$, $x > 0$

$$f'(x) = \frac{e^x \left(\dfrac{1}{x}\right) - (\ln x)e^x}{(e^x)^2} = \frac{e^x \left(\dfrac{1}{x} - \ln x\right)}{e^{2x}} = \frac{1 - x\ln x}{xe^x}, \quad x > 0$$

Critical value(s): $f'(x) = \dfrac{1 - x\ln x}{xe^x} = 0$

$$1 - x\ln x = 0$$
$$x\ln x = 1$$
$$x \approx 1.763$$

$$f''(x) = \frac{xe^x[-1-\ln x]-(1-x\ln x)(xe^x+e^x)}{x^2e^{2x}} = \frac{-x(1+\ln x)-(x+1)(1-x\ln x)}{x^2e^x};$$

$$f''(1.763) \approx \frac{-1.763(1.567)-(2.763)(0.000349)}{(1.763)^2 e^{1.763}} < 0$$

Since $x = 1.763$ is the only critical value, and $f''(1.763) < 0$, f has an absolute maximum at $x = 1.763$.

The absolute maximum of f is: $f(1.763) = \dfrac{\ln(1.763)}{e^{1.763}} \approx 0.097.$ (4-5)

57. (A)   For the first 15 months, the price is increasing and concave down, with a local maximum at $t = 15$. For the next 15 months, the price is decreasing and concave down, with an inflection point at $t = 30$. For the next 15 months, the price is decreasing and concave up, with a local minimum at $t = 45$. For the remaining 15 months, the price is increasing and concave up.

(B)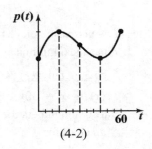

(4-2)

58. (A) $R(x) = xp(x) = 500x - 0.025x^2,\ 0 \le x \le 20{,}000$
$R'(x) = 500 - 0.05x;$
$500 - 0.05x = 0,\ \ x = 10{,}000$
Thus, $x = 10{,}000$ is a critical value.

Now, $R(0) = 0$
     $R(10{,}000) = 2{,}500{,}000$
     $R(20{,}000) = 0$

Thus, $R(10{,}000) = \$2{,}500{,}000$ is the absolute maximum of R.

(B) $P(x) = R(x) - C(x) = 500x - 0.025x^2 - (350x + 50{,}000)$
$\qquad\qquad\quad = 150x - 0.025x^2 - 50{,}000,\ \ \ 0 \le x \le 20{,}000$
$P'(x) = 150 - 0.05x;$
$150 - 0.05x = 0,\ \ x = 3{,}000$

Now, $P(0) = -50{,}000$
     $P(3{,}000) = 175{,}000$
     $P(20{,}000) = -7{,}050{,}000$

Thus, the maximum profit is $\$175{,}000$ when 3000 readers are manufactured and sold at $p(3{,}000) = \$425$ each.

(C)   If the government taxes the company $20 per reader, then the cost equation is:
$C(x) = 370x + 50{,}000$ and
$P(x) = 500x - 0.025x^2 - (370x + 50{,}000) = 130x - 0.025x^2 - 50{,}000,\ \ \ 0 \le x \le 20{,}000$
$P'(x) = 130 - 0.05x;$
$130 - 0.05x = 0,\ \ x = 2{,}600$

The maximum profit is $P(2{,}600) = \$119{,}000$ when 2,600 readers are produced and sold for $p(2{,}600) = \$435$ each.   (4-6)

59.

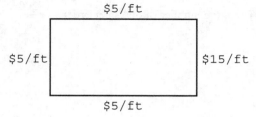

$5/ft

$5/ft

$15/ft

$5/ft

Let x be the length and y the width of the rectangle.

(A) $C(x, y) = 5x + 5x + 5y + 15y = 10x + 20y$

Also, Area $A = xy = 5000$, so $y = \dfrac{5000}{x}$.

Therefore, $C(x) = 10x + \dfrac{100,000}{x}$, $x > 0$.

Now, $C'(x) = 10 - \dfrac{100,000}{x^2}$ and

$10 - \dfrac{100,000}{x^2} = 0$ implies $10x^2 = 100,000$

$$x^2 = 10,000$$
$$x = \pm 100$$

Thus, $x = 100$ is the critical value.   [Note: $x > 0$, so $x = -100$ is not a critical value.]

Now, $C''(x) = \dfrac{200,000}{x^3}$ and $C''(100) = \dfrac{200,000}{1,000,000} = 0.2 > 0$. Thus, C has an absolute minimum when $x = 100$.

The most economical (i.e. least cost) fence will have dimensions: length $x = 100$ feet and width $y = \dfrac{5000}{100} = 50$ feet.

(B)   We want to maximize $A = xy$ subject to
$C(x, y) = 10x + 20y = 3000$ or $x = 300 - 2y$

Thus, $A = y(300 - 2y) = 300y - 2y^2$, $0 \le y \le 150$.

Now, $A'(y) = 300 - 4y$ and
$300 - 4y = 0$ implies $y = 75$.

Therefore, $y = 75$ is the critical value.

Now, $A''(y) = -4$ and $A''(75) = -4 < 0$. Thus, A has an absolute maximum when $y = 75$.
The dimensions of the rectangle that will enclose maximum area are: length $x = 300 - 2(75) = 150$ feet and width $y = 75$ feet.   (4-6)

60.   Let $x = $ the number of dollars increase in the nightly rate, $x \ge 0$. Then $200 - 4x$ rooms will be rented at $(40 + x)$ dollars per room. [Note: Since $200 - 4x \ge 0$, $x \le 50$.] The cost of service for $200 - 4x$ rooms at $8 per room is $8(200 - 4x)$. Thus:

Gross profit: $P(x) = (200 - 4x)(40 + x) - 8(200 - 4x) = (200 - 4x)(32 + x) = 6400 + 72x - 4x^2$, $0 \le x \le 50$
$P'(x) = 72 - 8x$
Critical value:
$72 - 8x = 0$, $x = 9$
Now, $P(0) = 6400$
    $P(9) = 6724$ (absolute maximum)
    $P(50) = 0$
Thus, the maximum gross profit is $6724 and this occurs at $x = 9$, i.e., the rooms should be rented at $49 per night.   (4-6)

61. Let x = number of times the company should order. Then, the number of disks per order = $\dfrac{7200}{x}$. The

average number of unsold disks is given by:

$$\frac{7200}{2x} = \frac{3600}{x}$$

Total cost: $C(x) = 5x + 0.2\left(\dfrac{3600}{x}\right) = 5x + \dfrac{720}{x}$, $x > 0$.

$$C'(x) = 5 - \frac{720}{x^2} = \frac{5x^2 - 720}{x^2} = \frac{5(x^2 - 144)}{x^2} = \frac{5(x+12)(x-12)}{x^2}$$

Critical value: $x = 12$ [Note: $x > 0$, so $x = -12$ is not a critical value.]

$C''(x) = \dfrac{1440}{x^3}$ and $C''(12) = \dfrac{1440}{12^3} > 0$

Therefore, $C(x)$ is a minimum when $x = 12$.   (4-6)

62. $C(x) = 4000 + 10x + 0.1x^2$, $x > 0$

Average cost = $\overline{C}(x) = \dfrac{4000}{x} + 10 + 0.1x$

Marginal cost = $C'(x) = 10 + \dfrac{2}{10}x = 10 + 0.2x$

The graph of C' is a straight line with slope $\dfrac{1}{5}$ and y intercept 10.

$$\overline{C}'(x) = \frac{-4000}{x^2} + \frac{1}{10} = \frac{-40,000 + x^2}{10x^2} = \frac{(x+200)(x-200)}{10x^2}$$

Thus, $\overline{C}'(x) < 0$ on $(0, 200)$ and $\overline{C}'(x) > 0$ on $(200, \infty)$. Therefore, $\overline{C}(x)$ is decreasing on $(0, 200)$, increasing on $(200, \infty)$, and a minimum occurs at $x = 200$.

Min = $\overline{C}(200) = \dfrac{4000}{200} + 10 + \dfrac{1}{10}(200) = 50$

$\overline{C}''(x) = \dfrac{8000}{x^3} > 0$ on $(0, \infty)$.

Therefore, the graph of $\overline{C}(x)$ is concave upward on $(0, \infty)$.

Using this information and point-by-point plotting (use a calculator), the graphs of $C(x)$ and $\overline{C}(x)$ are as shown in the diagram at the right. The line $y = 0.1x + 10$ is an oblique asymptote for $y = \overline{C}(x)$.

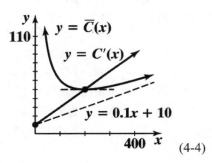

(4-4)

63. Cost: $C(x) = 200 + 50x - 50 \ln x$, $x \ge 1$

Average cost: $\overline{C} = \dfrac{C(x)}{x} = \dfrac{200}{x} + 50 - \dfrac{50}{x}\ln x$, $x \ge 1$

$\overline{C}'(x) = \dfrac{-200}{x^2} - \dfrac{50}{x}\left(\dfrac{1}{x}\right) + (\ln x)\dfrac{50}{x^2} = \dfrac{50(\ln x - 5)}{x^2}$, $x \ge 1$

Critical value(s): $\overline{C}'(x) = \dfrac{50(\ln x - 5)}{x^2} = 0$

$$\ln x = 5$$
$$x = e^5$$

Sign chart for $\overline{C}'$:

$$\overline{C}'(x) \qquad \underline{\quad - \; - \; - \; - \; - \; 0 \; + \; + \; + \; + \;} \; x$$

$$1$$

$$\overline{C}(x) \qquad \vdots \; \text{Decreasing} \; \vdots \; \text{Increasing}$$

Test Numbers

x	$\overline{C}'(x)$
1	$-250(-)$
e^6	$\frac{50}{e^{12}}(+)$

By the first derivative test, $\overline{C}$ has a local minimum at $x = e^5$. Since this is the only critical value of $\overline{C}$, $\overline{C}$ has as absolute minimum at $x = e^5$. Thus, the minimal average cost is:

$$\overline{C}(e^5) = \frac{200}{e^5} + 50 - \frac{50}{e^5}\ln(e^5) = 50 - \frac{50}{e^5} \approx 49.66 \quad \text{or} \quad \$49.66 \qquad (4\text{-}4)$$

64. $R(x) = xp(x) = 1000xe^{-0.02x}$

$R'(x) = 1000x \cdot e^{-0.02x}(-0.02) + 1000e^{-0.02x} = (1000 - 20x)e^{-0.02x}$

$R'(x) = 0: \quad (1000 - 20x)e^{-0.02x} = 0$

$$1000 - 20x = 0$$
$$x = 50$$

Critical value of R: $x = 50$

$R''(x) = (1000 - 20x)e^{-0.02x}(-0.02) + e^{-0.02x}(-20) = e^{-0.02x}[0.4x - 20 - 20] = e^{-0.02x}[0.4x - 40];$

$R''(50) = -20e^{-1} < 0.$

Since $x = 50$ is the only critical value and $R''(50) < 0$, R has an absolute maximum at a production level of 50 units. The maximum revenue is $R(50) = 1000(50)e^{-0.02(50)} = 50{,}000e^{-1} \approx \$18{,}394.$

The price per unit at the production level of 50 units is:

$$p(50) = 1000e^{-0.02(50)} = 1000e^{-1} \approx \$367.88. \qquad (4\text{-}6)$$

65. $R(x) = 1000xe^{-0.02x}, \; 0 \le x \le 100$

<u>Step 1. Analyze $R(x)$:</u>

(A)   Domain: $0 \le x \le 100$ or [0, 100]

(B)   Intercepts: y-intercept: $R(0) = 0$

$\qquad\qquad\qquad x$-intercepts: $100xe^{-0.02x} = 0$
$$x = 0$$

(C)   Asymptotes: There are no horizontal or vertical asymptotes.

<u>Step 2. Analyze $R'(x)$:</u>

$R'(x) = (1000 - 20x)e^{-0.02x}$ and $x = 50$ is a critical value.

Sign chart for R':

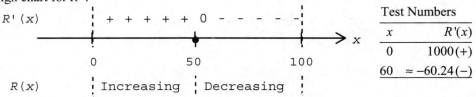

Test Numbers

x	$R'(x)$
0	$1000(+)$
60	$\approx -60.24(-)$

Thus, R is increasing on (0, 50) and decreasing on (50, 100); R has a maximum at $x = 50$.

<u>Step 3. Analyze $R''(x)$:</u>

$R''(x) = (0.4x - 40)e^{-0.02x} < 0$ on (0, 100).

Thus, the graph of R is concave downward on (0, 100).

Step 4. Sketch the graph of R:

x	$R(x)$
0	0
50	18,394
100	13,534

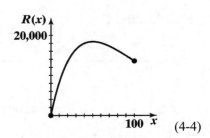

(4-4)

66.   Cost: $C(x) = 220x$

Price-demand equation: $p(x) = 1{,}000e^{-0.02x}$

Revenue: $R(x) = xp(x) = 1{,}000xe^{-0.02x}$

Profit: $P(x) = R(x) - C(x) = 1{,}000xe^{-0.02x} - 220x$

On a graphing utility, graph $P(x)$ and calculate its maximum value. The maximum value is \$9,864 at a demand level of 29.969082 (≈ 30). The price at this demand level is: $p = \$549.15$.   (4-6)

67.   Let x = the number of cream puffs.

Daily cost: $C(x) = x$ (dollars)

Daily revenue: $R(x) = xp(x)$, where

$p(x) = a + b \ln x$ is the logarithmic regression model for the given data.

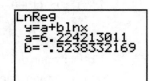

Profit: $P(x) = R(x) - C(x)$

Using a graphing utility, we find that the maximum profit is achieved at the demand level $x \approx 7888$. The price at this demand level is: $p(7888) = \$1.52$ (to the nearest cent).   (4-6)

68.   Let x be the length of the vertical portion of the chain. Then the length of each of the "arms" of the "Y" is $\sqrt{(10-x)^2 + 36} = \sqrt{x^2 - 20x + 136}$. Thus, the total length is given by:

$$L(x) = x + 2\sqrt{x^2 - 20x + 136}, \quad 0 \le x \le 10$$

Now, $L'(x) = 1 + 2\left(\dfrac{1}{2}\right)(x^2 - 20x + 136)^{-1/2}(2x - 20) = 1 + \dfrac{2x - 20}{(x^2 - 20x + 136)^{1/2}}$

$L'(x) = 0$: $1 + \dfrac{2x - 20}{(x^2 - 20x + 136)^{1/2}} = 0$

$$(x^2 - 20x + 136)^{1/2} + 2x - 20 = 0$$
$$(x^2 - 20x + 136)^{1/2} = 2(10 - x)$$
$$x^2 - 20x + 136 = 4(100 - 20x + x^2)$$
$$-3x^2 + 60x - 264 = 0$$

$$x = \frac{20 \pm \sqrt{48}}{2} = 10 \pm 2\sqrt{3}$$

Critical value (in (0, 10)): $x = 10 - 2\sqrt{3} \approx 6.54$

Sign chart for $L'(x)$:

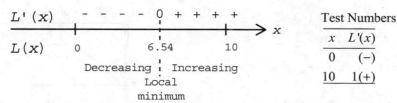

Thus, to minimize the length of the chain, the vertical portion should be 6.54 feet long. The total length of the chain will be $L(6.54) = 20.39$ feet.   (4-6)

69. (A)

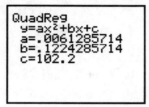

(B)   Let $C(x)$ be the regression equation from part (A). The average cost function $\bar{C}(x) = \dfrac{C(x)}{x}$.

Using the "find the minimum" routine on the graphing utility, we   find that
$\min \bar{C}(x) = \bar{C}(129) = 1.71$

The minimum average cost is \$1.71 at a production level of 129 dozen cookies.   (4-4)

70. $N(x) = -0.25x^4 + 11x^3 - 108x^2 + 3{,}000,\ 9 \le x \le 24$

$N'(x) = -x^3 + 33x^2 - 216x$

$N''(x) = -3x^2 + 66x - 216 = -3(x^2 - 22x + 72) = -3(x - 4)(x - 18)$

Partition numbers for N'': $x = 18$

Sign chart for $N''(x)$:

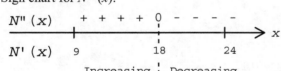

Test Numbers

x	$N''(x)$
17	39(+)
19	−45(−)

Thus, N' is increasing on (9, 18) and decreasing on (18, 24); the point of diminishing returns is $x = 18$; the maximum rate of change is $N'(18) = 972$.

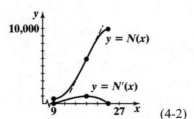

(4-2)

71. (A)

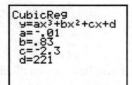

(B)   The regression equation found in (A) is:

$$y(x) = -0.01x^3 + 0.83x^2 - 2.3x + 221$$

The rate of change of sales with respect to the number of ads is:

$$y'(x) = -0.03x^2 + 1.66x - 2.3$$
$$y''(x) = -0.06x + 1.66$$

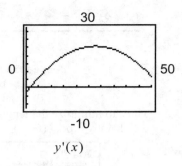

$y'(x)$

Critical value: $-0.06x + 1.66 = 0$
$$x \approx 27.667$$

From the graph, the absolute maximum of y' occurs at $x \approx 27.667$. Thus, 28 ads should be placed

each month. The expected number of sales is: $y(28) \approx 588$.   (4-6)

72. $C(t) = 20t^2 - 120t + 800, 0 \le t \le 9$

$C'(t) = 40t - 120 = 40(t - 3)$

Critical value: $t = 3$

$C''(t) = 40$ and $C''(3) = 40 > 0$

Therefore, a local minimum occurs at $t = 3$.

$C(0) = 800$

$C(3) = 20(3^2) - 120(3) + 800 = 620$ Absolute minimum

$C(9) = 20(81) - 120(9) + 800 = 1340$

Therefore, the bacteria count will be at a minimum three days after a treatment.   (4-6)

73. $N = 10 + 6t^2 - t^3, 0 \le t \le 5$

$$\frac{dN}{dt} = 12t - 3t^2$$

Now, find the critical values of the rate function $R(t)$:

$$R(t) = \frac{dN}{dt} = 12t - 3t^2$$

$$R'(t) = \frac{dR}{dt} = \frac{d^2N}{dt^2} = 12 - 6t$$

Critical value: $t = 2$

$R''(t) = -6$ and $R''(2) = -6 < 0$

$R(0) = 0$

$R(2) = 12$ Absolute maximum

$R(5) = -15$

Therefore, $R(t)$ has an absolute maximum at $t = 2$. The rate of increase will be a maximum after two years. (4-6)

13 INTEGRATION

EXERCISE 13-1

Things to remember:

<u>1</u>.   A function F is an ANTIDERIVATIVE of f if $F'(x) = f(x)$.

<u>2</u>.   THEOREM ON ANTIDERIVATIVES

If the derivatives of two functions are equal on an open interval (a, b), then the functions can differ by at most a constant. Symbolically: If F and G are differentiable functions on the interval (a, b) and $F'(x) = G'(x)$ for all x in (a, b), then $F(x) = G(x) + k$ for some constant k .

<u>3</u>.   The INDEFINITE INTEGRAL of $f(x)$, denoted

$$\int f(x)dx \, ,$$

represents all antiderivatives of $f(x)$ and is given by

$$\int f(x)dx = F(x) + C$$

where $F(x)$ is any antiderivative of $f(x)$ and C is an arbitrary constant. The symbol $\int$ is called an INTEGRAL SIGN, the function $f(x)$ is called the INTEGRAND, and C is called the CONSTANT OF INTEGRATION.

<u>4</u>.   Indefinite integration and differentiation are reverse operations (except for the addition of the constant of integration). This is expressed symbolically by:

(a) $\dfrac{d}{dx}\left(\int f(x)dx\right) = f(x)$

(b) $\int F'(x)dx = F(x) + C$

<u>5</u>.   INDEFINITE INTEGRAL FORMULAS:

(a) $\int x^n \, dx = \dfrac{x^{n+1}}{n+1} + C, \, n \neq -1$

(b) $\int e^x \, dx = e^x + C$

(c) $\int \dfrac{dx}{x} = \ln|x| + C, \, x \neq 0$

<u>6</u>.   PROPERTIES OF INDEFINITE INTEGRALS:

(a) $\int k \, f(x)dx = k \int f(x)dx \, , \quad k \text{ constant}$

(b) $\int [f(x) \pm g(x)] \, dx = \int f(x)dx \pm \int g(x)dx$

1. $\dfrac{5}{x^4} = 5x^{-4}$.

3. $\dfrac{3x-2}{x^5} = 3x^{-4} - 2x^{-5}$.

5. $\sqrt{x} + \dfrac{5}{\sqrt{x}} = x^{1/2} + \dfrac{5}{x^{1/2}} = x^{1/2} + 5x^{-1/2}$.

7. $\sqrt[3]{x}\left(4 + x - 3x^2\right) = 4x^{1/3} + x^{4/3} - 3x^{7/3}$.

9. $\displaystyle\int 7\,dx = 7x + C$

   Check: $\dfrac{d}{dx}\left(7x + C\right) = 7$

11. $\displaystyle\int 8x\,dx = 8\int x\,dx = 8\dfrac{x^2}{2} + C = 4x^2 + C$

    Check: $\dfrac{d}{dx}\left(4x^2 + C\right) = 8x$

13. $\displaystyle\int 9x^2\,dx = 9\int x^2\,dx = 9\dfrac{x^3}{3} + C = 3x^3 + C$

    Check: $\dfrac{d}{dx}(3x^3 + C) = 9x^2$

15. $\displaystyle\int x^5\,dx = \dfrac{1}{6}x^6 + C$

    Check: $\dfrac{d}{dx}\left(\dfrac{1}{6}x^6 + C\right) = x^5$

17. $\displaystyle\int x^{-3}\,dx = \dfrac{x^{-2}}{-2} + C$

    Check: $\dfrac{d}{dx}\left(\dfrac{x^{-2}}{-2} + C\right) = x^{-3}$

19. $\displaystyle\int 10x^{3/2}\,dx = 10\int x^{3/2}\,dx = 10\dfrac{x^{5/2}}{5/2} + C = 4x^{5/2} + C$

    Check: $\dfrac{d}{dx}\left(4x^{5/2} + C\right) = 10x^{3/2}$

21. $\displaystyle\int \dfrac{3}{z}\,dz = 3\int \dfrac{1}{z}\,dz = 3\ln|z| + C$

    Check: $\dfrac{d}{dz}\left(3\ln|z| + C\right) = \dfrac{3}{z}$

23. $\displaystyle\int 16e^u\,du = 16\int e^u\,du = 16e^u + C$

    Check: $\dfrac{d}{du}\left(16e^u + C\right) = 16e^u$

25. $F(x) = (x+1)(x+2) = x^2 + 3x + 2,\ F'(x) = 2x + 3 = f(x);$ yes.

27. $F(x) = 1 + x\ln x,\ F'(x) = 1 + \ln x = f(x);$ yes.

29. $F(x) = \dfrac{(2x+1)^3}{3},\ F'(x) = (2x+1)^2(2) \neq (2x+1)^2 = f(x);$ no.

31. $F(x) = e^{x^3/3},\ F'(x) = e^{x^3/3}(x^2) = x^2 e^{x^3/3} \neq e^{x^2} = f(x);$ no.

33. True: $f(x) = \pi,\ f'(x) = 0 = k(x)$

35. False: $f(x) = x^{-1}$ is a counter-example

37. True: $h(x) = 5e^x,\ h''(x) = 5e^x = h(x)$

39. The graphs in this set ARE NOT graphs from a family of antiderivative functions since the graphs are not vertical translations of each other.

41. The graphs in this set could be graphs from a family of antiderivative functions since they appear to be vertical translations of each other.

43. $\int 5x(1-x)\,dx = 5\int x(1-x)\,dx = 5\int (x-x^2)\,dx = 5\int x\,dx - 5\int x^2\,dx = \dfrac{5x^2}{2} - \dfrac{5x^3}{3} + C$

Check: $\dfrac{d}{dx}\left(\dfrac{5x^2}{2} - \dfrac{5x^3}{3} + C\right) = 5x - 5x^2 = 5x(1-x)$

45. $\int \dfrac{du}{\sqrt{u}} = \int \dfrac{du}{u^{1/2}} = \int u^{-1/2}\,du = \dfrac{u^{(-1/2)+1}}{-1/2+1} + C = \dfrac{u^{1/2}}{1/2} + C = 2u^{1/2} + C \ \ \text{or} \ \ 2\sqrt{u} + C$

Check: $\dfrac{d}{du}\left(2u^{1/2} + C\right) = 2\left(\dfrac{1}{2}\right)u^{-1/2} = \dfrac{1}{u^{1/2}} = \dfrac{1}{\sqrt{u}}$

47. $\int \dfrac{dx}{4x^3} = \dfrac{1}{4}\int x^{-3}\,dx = \dfrac{1}{4}\cdot\dfrac{x^{-2}}{-2} + C = \dfrac{-x^{-2}}{8} + C$

Check: $\dfrac{d}{dx}\left(\dfrac{-x^{-2}}{8} + C\right) = \dfrac{1}{8}(-2)(-x^{-3}) = \dfrac{1}{4}x^{-3} = \dfrac{1}{4x^3}$

49. $\int \dfrac{4+u}{u}\,du = \int\left(\dfrac{4}{u}+1\right)du = 4\int\dfrac{1}{u}\,du + \int 1\,du = 4\ln|u| + u + C$

Check: $\dfrac{d}{du}(4\ln|u| + u + C) = \dfrac{4}{u} + 1 = \dfrac{4+u}{u}$

51. $\int (5e^z + 4)\,dz = 5\int e^z\,dz + 4\int dz = 5e^z + 4z + C$

Check: $\dfrac{d}{dz}(5e^z + 4z + C) = 5e^z + 4$

53. $\int\left(3x^2 - \dfrac{2}{x^2}\right)dx = \int 3x^2\,dx - \int\dfrac{2}{x^2}\,dx = 3\int x^2\,dx - 2\int x^{-2}\,dx = 3\cdot\dfrac{x^3}{3} - \dfrac{2x^{-1}}{-1} + C = x^3 + 2x^{-1} + C$

Check: $\dfrac{d}{dx}(x^3 + 2x^{-1} + C) = 3x^2 - 2x^{-2} = 3x^2 - \dfrac{2}{x^2}$

55. $C'(x) = 6x^2 - 4x$

$C(x) = \int(6x^2 - 4x)\,dx = 6\int x^2\,dx - 4\int x\,dx = \dfrac{6x^3}{3} - \dfrac{4x^2}{2} + C = 2x^3 - 2x^2 + C$

Given $C(0) = 3000$: $3000 = 2(0^3) - 2(0^2) + C$. Hence, $C = 3000$ and $C(x) = 2x^3 - 2x^2 + 3000$.

57. $\dfrac{dx}{dt} = \dfrac{20}{\sqrt{t}}$

$x = \int\dfrac{20}{\sqrt{t}}\,dt = 20\int t^{-1/2}\,dt = 20\dfrac{t^{1/2}}{1/2} + C = 40\sqrt{t} + C$

Given $x(1) = 40$: $40 = 40\sqrt{1} + C$ or $40 = 40 + C$. Hence, $C = 0$ and $x = 40\sqrt{t}$.

59. $\dfrac{dy}{dx} = 2x^{-2} + 3x^{-1} - 1$

$$y = \int (2x^{-2} + 3x^{-1} - 1)\,dx = 2\int x^{-2}\,dx + 3\int x^{-1}\,dx - \int dx = \dfrac{2x^{-1}}{-1} + 3\,\ln|x| - x + C$$

$$= \dfrac{-2}{x} + 3\,\ln|x| - x + C$$

Given $y(1) = 0$: $0 = -\dfrac{2}{1} + 3\,\ln|1| - 1 + C$. Hence, $C = 3$ and $y = -\dfrac{2}{x} + 3\,\ln|x| - x + 3$.

61. $\dfrac{dx}{dt} = 4e^t - 2$

$$x = \int (4e^t - 2)\,dt = 4\int e^t\,dt - 2\int dt = 4e^t - 2t + C$$

Given $x(0) = 1$: $1 = 4e^0 - 2(0) + C = 4 + C$. Hence, $C = -3$ and $x = 4e^t - 2t - 3$.

63. $\dfrac{dy}{dx} = 4x - 3$

$$y = \int (4x - 3)\,dx = 4\int x\,dx - 3\int dx = \dfrac{4x^2}{2} - 3x + C = 2x^2 - 3x + C$$

Given $y(2) = 3$: $3 = 2 \cdot 2^2 - 3 \cdot 2 + C$. Hence, $C = 1$ and $y = 2x^2 - 3x + 1$.

65. $\displaystyle\int \dfrac{2x^4 - x}{x^3}\,dx = \int \left(\dfrac{2x^4}{x^3} - \dfrac{x}{x^3} \right) dx = 2\int x\,dx - \int x^{-2}\,dx = \dfrac{2x^2}{2} - \dfrac{x^{-1}}{-1} + C = x^2 + x^{-1} + C$

67. $\displaystyle\int \dfrac{x^5 - 2x}{x^4}\,dx = \int \left(\dfrac{x^5}{x^4} - \dfrac{2x}{x^4} \right) dx = \int x\,dx - 2\int x^{-3}\,dx = \dfrac{x^2}{2} - \dfrac{2x^{-2}}{-2} + C = \dfrac{x^2}{2} + x^{-2} + C$

69. $\displaystyle\int \dfrac{x^2 e^x - 2x}{x^2}\,dx = \int \left(\dfrac{x^2 e^x}{x^2} - \dfrac{2x}{x^2} \right) dx = \int e^x\,dx - 2\int x^{-1}\,dx = e^x - 2\,\ln|x| + C$

71. $\dfrac{d}{dx}\left[\displaystyle\int x^3\,dx \right] = x^3$ [by $\underline{4}$(a)]

73. $\displaystyle\int \dfrac{d}{dx}(x^4 + 3x^2 + 1)\,dx = x^4 + 3x^2 + 1 + C = x^4 + 3x^2 + C_1$ [by $\underline{4}$(b)]

($C_1 = 1 + C$ is an arbitrary constant since C is arbitrary)

75. $\dfrac{d}{dx}\left(\dfrac{x^{n+1}}{n+1} + C \right) = x^n$

77. Assume $x > 0$. Then $|x| = x$ and $\ln|x| = \ln x$.

Therefore, $\dfrac{d}{dx}(\ln|x| + C) = \dfrac{d}{dx}(\ln x + C) = \dfrac{1}{x}$.

79. Assume $\displaystyle\int f(x)\,dx = F(x) + C_1$ and $\displaystyle\int g(x)\,dx = G(x) + C_2$.

Then, $\dfrac{d}{dx}(F(x) + C_1) = f(x)$, $\dfrac{d}{dx}(G(x) + C_2) = g(x)$, and

$$\dfrac{d}{dx}(F(x) + C_1 + G(x) + C_2) = \dfrac{d}{dx}(F(x) + C_1) + \dfrac{d}{dx}(G(x) + C_2) = f(x) + g(x).$$

81. $\overline{C}'(x) = -\dfrac{1,000}{x^2}$

$\overline{C}(x) = \displaystyle\int \overline{C}'(x)\,dx = \int -\dfrac{1,000}{x^2}\,dx = -1,000\int x^{-2}\,dx = -1,000\,\dfrac{x^{-1}}{-1} + C = \dfrac{1,000}{x} + C$

Given $\overline{C}(100) = 25$: $\dfrac{1,000}{100} + C = 25$

$C = 15$

Thus, $\overline{C}(x) = \dfrac{1,000}{x} + 15.$

Cost function: $C(x) = x\,\overline{C}(x) = 15x + 1,000$

Fixed costs: $C(0) = \$1,000$

83. (A) The cost function increases from 0 to 8. The graph is concave downward from 0 to 4 and concave upward from 4 to 8. There is an inflection point at $x = 4$.

(B) $C(x) = \displaystyle\int C'(x)\,dx = \int (3x^2 - 24x + 53)\,dx = 3\int x^2\,dx - 24\int x\,dx + 53\int dx = x^3 - 12x^2 + 53x + K$

Since $C(0) = 30$, we have $K = 30$ and $C(x) = x^3 - 12x^2 + 53x + 30.$

$C(4) = 4^3 - 12(4)^2 + 53(4) + 30 = \$114$ thousand

$C(8) = 8^3 - 12(8)^2 + 53(8) + 30 = \$198$ thousand

(C)

(D) Manufacturing plants are often inefficient at low and high levels of production.

85. $S'(t) = -24t^{1/3}$

$S(t) = \displaystyle\int -24t^{1/3}\,dt = -24\int t^{1/3}\,dt = -24\,\dfrac{t^{4/3}}{4/3} + C = -18t^{4/3} + C$

Given $S(0) = 1200 = -18(0) + C$. Hence, $C = 1200$ and $S(t) = 1,200 - 18t^{4/3}.$

Now, we want to find t such that $S(t) = 300$, that is:

$1,200 - 18t^{4/3} = 300$

$-18t^{4/3} = -900$

$t^{4/3} = 50$

$t = 50^{3/4} \approx 18.803,$

Thus, the company should manufacture SUV's for 19 months.

87. $S'(t) = -24t^{1/3} - 70$

$S(t) = \int S'(t)\,dt = \int (-24t^{1/3} - 70)\,dt = -24\dfrac{t^{4/3}}{4/3} - 70t + C = -18t^{4/3} - 70t + C$

Given $S(0) = 1{,}200$ implies $C = 1{,}200$ and $S(t) = 1{,}200 - 18t^{4/3} - 70t$

Graphing $y_1 = 1{,}200 - 18t^{4/3} - 70t$, $y_2 = 800$ on $0 \le x \le 10$, $0 \le y \le 1000$, we see that the point of intersection is $t \approx 4.0527553$, $y = 800$. So we get $t \approx 4.05$ months.

89. $L'(x) = g(x) = 2400x^{-1/2}$

$L(x) = \int g(x)\,dx = \int 2{,}400x^{-1/2}\,dx = 2{,}400\int x^{-1/2}\,dx = 2{,}400\,\dfrac{x^{1/2}}{1/2} + C = 48{,}000\,x^{1/2} + C$

Given $L(16) = 19{,}200$: $19{,}200 = 4800(16)^{1/2} + C = 19{,}200 + C$. Hence, $C = 0$ and $L(x) = 4800x^{1/2}$.
$L(25) = 4800(25)^{1/2} = 4800(5) = 24{,}000$ labor hours.

91. $\dfrac{dW}{dh} = 0.0015h^2$

$W = \int 0.0015h^2\,dh = 0.0015\int h^2\,dh = 0.0015\,\dfrac{h^3}{3} + C = 0.0005h^3 + C$

Given $W(60) = 108$: $108 = 0.0005(60)^3 + C$ or $108 = 108 + C$.
Hence, $C = 0$ and $W(h) = 0.0005h^3$. Now $5'10'' = 70''$ and $W(70) = 0.0005(70)^3 = 171.5$ lb.

93. $\dfrac{dN}{dt} = 400 + 600\sqrt{t}$, $0 \le t \le 9$

$N = \int \left(400 + 600\sqrt{t}\right)dt = 400\int dt + 600\int t^{1/2}\,dt = 400t + 600\,\dfrac{t^{3/2}}{3/2} + C = 400t + 400t^{3/2} + C$

Given $N(0) = 5000$: $5000 = 400(0) + 400(0)^{3/2} + C$. Hence, $C = 5000$ and
$N(t) = 400t + 400t^{3/2} + 5000$.
$N(9) = 400(9) + 400(9)^{3/2} + 5000 = 3600 + 10{,}800 + 5000 = 19{,}400$

EXERCISE 13-2

Things to remember:

1.   REVERSING THE CHAIN RULE

   The chain rule formula for differentiating a composite function:

   $$\dfrac{d}{dx}f[g(x)] = f'[g(x)]g'(x),$$

   yields the integral formula

   $$\int f'[g(x)]g'(x)\,dx = f[g(x)] + C$$

2.   GENERAL INDEFINITE INTEGRAL FORMULAS (Version 1)

(a) $\int [f(x)]^n f'(x)\,dx = \dfrac{[f(x)]^{n+1}}{n+1} + C, n \neq -1$

(b) $\int e^{f(x)} f'(x)\,dx = e^{f(x)} + C$

(c) $\int \dfrac{1}{f(x)} f'(x)\,dx = \ln|f(x)| + C$

3.   DIFFERENTIALS

If $y = f(x)$ defines a differentiable function, then:

(a)   The DIFFERENTIAL dx of the independent variable x is an arbitrary real number.

(b)   The DIFFERENTIAL dy of the dependent variable y is defined as the product of $f'(x)$ and dx; that is: $dy = f'(x)dx$.

4.   GENERAL INDEFINITE INTEGRAL FORMULAS (Version 2)

(a) $\int u^n\,du = \dfrac{u^{n+1}}{n+1} + C, n \neq -1$

(b) $\int e^u\,du = e^u + C$

(c) $\int \dfrac{1}{u}\,du = \ln|u| + C$

These formulas are valid if u is an independent variable or if u is a function of another variable and du is its differential with respect to that variable.

5.   INTEGRATION BY SUBSTITUTION

Step 1.   Select a substitution that appears to simplify the integrand. In particular, try to select u so that du is a factor in the integrand.

Step 2.   Express the integrand entirely in terms of u and du, completely eliminating the original variable and its differential.

Step 3.   Evaluate the new integral, if possible.

Step 4.   Express the antiderivative found in Step 3 in terms of the original variable.

---

1. $f(x) = (5x+1)^{10}, \quad f'(x) = 10(5x+1)^9(5) = 50(5x+1)^9.$

3. $f(x) = (x^2+1)^7, \quad f'(x) = 7(x^2+1)^6(2x) = 14x(x^2+1)^6.$

5. $f(x) = e^{x^2}, \quad f'(x) = e^{x^2}(2x) = 2xe^{x^2}.$

7. $f(x) = \ln(x^4-10), \quad f'(x) = \dfrac{1}{x^4-10}(4x^3) = \dfrac{4x^3}{x^4-10}.$

9. $\int (3x+5)^2 (3)\,dx = \int u^2\,du = \dfrac{1}{3}u^3 + C = \dfrac{1}{3}(3x+5)^3 + C$ [Formula $\underline{4}$a]

Let $u = 3x + 5$

Then $du = 3\,dx$

Check: $\dfrac{d}{dx}\left[\dfrac{1}{3}(3x+5)^3 + C\right] = \dfrac{1}{3}\cdot 3(3x+5)^2 \dfrac{d}{dx}(3x+5) = (3x+5)^2(3)$

11. $\int\left(x^2 - 1\right)^5 (2x)\,dx = \int u^5\,du = \dfrac{1}{6}u^6 + C = \dfrac{1}{6}(x^2 - 1)^6 + C$ [Formula $\underline{4}$a]

Let $u = x^2 - 1$

Then $du = 2x\,dx$

Check: $\dfrac{d}{dx}\left[\dfrac{1}{6}(x^2 - 1)^6 + C\right] = \dfrac{1}{6}\cdot 6(x^2 - 1)^5 \dfrac{d}{dx}(x^2 - 1) = (x^2 - 1)^5(2x)$

13. $\int\left(5x^3 + 1\right)^{-3}\left(15x^2\right)\,dx = \int u^{-3}\,du = \dfrac{u^{-2}}{-2} + C = -\dfrac{1}{2}(5x^3 + 1)^{-2} + C$ [Formula $\underline{4}$a]

Let $u = 5x^3 + 1$

Then $du = 15x^2\,dx$

Check: $\dfrac{d}{dx}\left[-\dfrac{1}{2}(5x^3 + 1)^{-2} + C\right] = -\dfrac{1}{2}(-2)(5x^3 + 1)^{-3}\dfrac{d}{dx}(5x^3 + 1) = (5x^3 + 1)^{-3}(15x^2)$

15. $\int e^{5x}(5)\,dx = \int e^u\,du = e^u + C = e^{5x} + C$ [Formula $\underline{4}$b]

Let $u = 5x$

Then $du = 5\,dx$

Check: $\dfrac{d}{dx}(e^{5x} + C) = e^{5x}\dfrac{d}{dx}(5x) = e^{5x}(5)$

17. $\int\dfrac{1}{1+x^2}(2x)\,dx = \int\dfrac{1}{u}\,du = \ln|u| + C = \ln|1 + x^2| + C = \ln(1 + x^2) + C$ $(1 + x^2 > 0)$ [Formula $\underline{4}$c]

Let $u = 1 + x^2$.

Then $du = 2x\,dx$

Check: $\dfrac{d}{dx}(\ln(1 + x^2) + C) = \dfrac{1}{1+x^2}\dfrac{d}{dx}(1 + x^2) = \dfrac{1}{1+x^2}(2x)$

19. $\int\sqrt{1 + x^4}\,(4x^3)\,dx = \int\sqrt{u}\,du = \int u^{1/2}\,du = \dfrac{u^{3/2}}{3/2} + C = \dfrac{2}{3}u^{3/2} + C = \dfrac{2}{3}(1 + x^4)^{3/2} + C$

Let $u = 1 + x^4$.

Then $du = 4x^3\,dx$

Check: $\dfrac{d}{dx}\left[\dfrac{2}{3}(1 + x^4)^{3/2} + C\right] = \dfrac{3}{2}\cdot\dfrac{2}{3}(1 + x^4)^{1/2}\dfrac{d}{dx}(1 + x^4) = (1 + x^4)^{1/2}(4x^3) = \sqrt{1 + x^4}\,(4x^3)$

21. $\int (x + 3)^{10}\,dx = \int u^{10}\,du = \dfrac{1}{11}u^{11} + C = \dfrac{1}{11}(x + 3)^{11} + C$

Let $u = x + 3$

Then $du = dx$

Check: $\dfrac{d}{dx}\left[\dfrac{1}{11}(x + 3)^{11} + C\right] = \dfrac{1}{11}\cdot 11(x + 3)^{10}\dfrac{d}{dx}(x + 3) = (x + 3)^{10}$

23. $\displaystyle\int (6t-7)^{-2}\,dt = \int (6t-7)^{-2}\frac{6}{6}\,dt = \frac{1}{6}\int (6t-7)^{-2}\,6\,dt = \frac{1}{6}\int u^{-2}\,du = \frac{1}{6}\cdot\frac{u^{-1}}{-1} + C = -\frac{1}{6}(6t-7)^{-1} + C$

Let $u = 6t-7$
Then $du = 6\,dt$

Check: $\displaystyle\frac{d}{dt}\left[-\frac{1}{6}(6t-7)^{-1} + C\right] = -\frac{1}{6}(-1)(6t-7)^{-2}\frac{d}{dt}(6t-7) = \frac{1}{6}(6t-7)^{-2}(6) = (6t-7)^{-2}$

25. $\displaystyle\int \left(t^2+1\right)^5 t\,dt = \int \left(t^2+1\right)^5\frac{2}{2}\,t\,dt = \frac{1}{2}\int \left(t^2+1\right)^5\,2t\,dt = \frac{1}{2}\int u^5\,du = \frac{1}{2}\cdot\frac{1}{6}u^6 + C = \frac{1}{12}(t^2+1)^6 + C$

Let $u = t^2 + 1$
Then $du = 2t\,dt$

Check: $\displaystyle\frac{d}{dt}\left[\frac{1}{12}(t^2+1)^6 + C\right] = \frac{1}{12}\cdot 6(t^2+1)^5\frac{d}{dt}(t^2+1) = \frac{1}{2}(t^2+1)^5(2t) = (t^2+1)^5 t$

27. $\displaystyle\int x\,e^{x^2}\,dx = \int e^{x^2}\frac{2}{2}\,x\,dx = \frac{1}{2}\int e^{x^2}(2x)\,dx = \frac{1}{2}\int e^u\,du = \frac{1}{2}e^u + C = \frac{1}{2}e^{x^2} + C$

Let $u = x^2$
Then $du = 2x\,dx$

Check: $\displaystyle\frac{d}{dx}\left(\frac{1}{2}e^{x^2} + C\right) = \frac{1}{2}e^{x^2}\frac{d}{dx}(x^2) = \frac{1}{2}e^{x^2}(2x) = xe^{x^2}$

29. $\displaystyle\int\frac{1}{5x+4}\,dx = \int\frac{1}{5x+4}\cdot\frac{5}{5}\,dx = \frac{1}{5}\int\frac{1}{5x+4}\,5\,dx = \frac{1}{5}\int\frac{1}{u}\,du = \frac{1}{5}\ln|u| + C = \frac{1}{5}\ln|5x+4| + C$

Let $u = 5x + 4$
Then $du = 5\,dx$

Check: $\displaystyle\frac{d}{dx}\left[\frac{1}{5}\ln|5x+4| + C\right] = \frac{1}{5}\cdot\frac{1}{5x+4}\frac{d}{dx}(5x+4) = \frac{1}{5}\cdot\frac{1}{5x+4}\cdot 5 = \frac{1}{5x+4}$

31. $\displaystyle\int e^{1-t}\,dt = \int e^{1-t}\left(\frac{-1}{-1}\right)\,dt = \frac{1}{-1}\int e^{1-t}(-1)\,dt = -\int e^u\,du = -e^u + C = -e^{1-t} + C$

Let $u = 1 - t$
Then $du = -dt$

Check: $\displaystyle\frac{d}{dt}[-e^{1-t} + C] = -e^{1-t}\frac{d}{dt}(1-t) = -e^{1-t}(-1) = e^{1-t}$

33. $\displaystyle\int\frac{t}{(3t^2+1)^4}\,dt = \int (3t^2+1)^{-4}t\,dt = \int (3t^2+1)^{-4}\frac{6}{6}\,t\,dt = \frac{1}{6}\int (3t^2+1)^{-4}\,6t\,dt = \frac{1}{6}\int u^{-4}\,du$

Let $u = 3t^2 + 1$. Then $du = 6t\,dt$.
$\displaystyle = \frac{1}{6}\cdot\frac{u^{-3}}{-3} + C = \frac{-1}{18}(3t^2+1)^{-3} + C$

Check: $\displaystyle\frac{d}{dt}\left[\frac{-1}{18}(3t^2+1)^{-3} + C\right] = \left(\frac{-1}{18}\right)(-3)(3t^2+1)^{-4}(6t) = \frac{t}{(3t^2+1)^4}$

35. $\displaystyle\int x\sqrt{x+4}\,dx$

Let $u = x + 4$. Then $du = dx$ and $x = u - 4$.

$$\int x\sqrt{x+4}\, dx = \int (u-4)u^{1/2}\, du = \int \left(u^{3/2} - 4u^{1/2}\right) du = \frac{u^{5/2}}{5/2} - \frac{4u^{3/2}}{3/2} + C = \frac{2}{5}u^{5/2} - \frac{8}{3}u^{3/2} + C$$

$$= \frac{2}{5}(x+4)^{5/2} - \frac{8}{3}(x+4)^{3/2} + C \ \ (\text{since } u = x+4)$$

Check: $\dfrac{d}{dx}\left[\dfrac{2}{5}(x+4)^{5/2} - \dfrac{8}{3}(x+4)^{3/2} + C\right] = \dfrac{2}{5}\left(\dfrac{5}{2}\right)(x+4)^{3/2}(1) - \dfrac{8}{3}\left(\dfrac{3}{2}\right)(x+4)^{1/2}(1)$

$$= (x+4)^{3/2} - 4(x+4)^{1/2} = (x+4)^{1/2}[(x+4) - 4]$$
$$= x\sqrt{x+4}$$

37. $\displaystyle\int \frac{x}{\sqrt{x-3}}\, dx$

Let $u = x - 3$. Then $du = dx$ and $x = u + 3$.

$$\int \frac{x}{\sqrt{x-3}}\, dx = \int \frac{u+3}{u^{1/2}}\, du = \int \left(u^{1/2} + 3u^{-1/2}\right) du = \frac{u^{3/2}}{3/2} + \frac{3u^{1/2}}{1/2} + C = \frac{2}{3}u^{3/2} + 6u^{1/2} + C$$

$$= \frac{2}{3}(x-3)^{3/2} + 6(x-3)^{1/2} + C \ \ (\text{since } u = x-3)$$

Check: $\dfrac{d}{dx}\left[\dfrac{2}{3}(x-3)^{3/2} + 6(x-3)^{1/2} + C\right] = \dfrac{2}{3}\left(\dfrac{3}{2}\right)(x-3)^{1/2}(1) + 6\left(\dfrac{1}{2}\right)(x-3)^{-1/2}(1)$

$$= (x-3)^{1/2} + \frac{3}{(x-3)^{1/2}} = \frac{x-3+3}{(x-3)^{1/2}} = \frac{x}{\sqrt{x-3}}$$

39. $\displaystyle\int x(x-4)^9\, dx$

Let $u = x - 4$. Then $du = dx$ and $x = u + 4$.

$$\int x(x-4)^9\, dx = \int (u+4)u^9\, du = \int (u^{10} + 4u^9)\, du = \frac{u^{11}}{11} + \frac{4u^{10}}{10} + C = \frac{(x-4)^{11}}{11} + \frac{2}{5}(x-4)^{10} + C$$

Check: $\dfrac{d}{dx}\left[\dfrac{(x-4)^{11}}{11} + \dfrac{2}{5}(x-4)^{10} + C\right] = \dfrac{1}{11}(11)(x-4)^{10}(1) + \dfrac{2}{5}(10)(x-4)^9(1)$

$$= (x-4)^9[(x-4) + 4] = x(x-4)^9$$

41. $\displaystyle\int e^{2x}(1+e^{2x})^3\, dx$

Let $u = 1 + e^{2x}$. Then $du = 2e^{2x}dx$.

$$\int e^{2x}(1+e^{2x})^3\, dx = \int \left(1+e^{2x}\right)^3 \frac{2}{2}e^{2x}\, dx = \frac{1}{2}\int \left(1+e^{2x}\right)^3 2e^{2x}\, dx = \frac{1}{2}\int u^3\, du = \frac{1}{2} \cdot \frac{u^4}{4} + C$$

$$= \frac{1}{8}(1+e^{2x})^4 + C$$

Check: $\dfrac{d}{dx}\left[\dfrac{1}{8}(1+e^{2x})^4 + C\right] = \left(\dfrac{1}{8}\right)(4)(1+e^{2x})^3 e^{2x}(2) = e^{2x}(1+e^{2x})^3$

43. $\displaystyle\int \frac{1+x}{4+2x+x^2}\,dx$ Let $u = 4 + 2x + x^2$. Then $du = (2 + 2x)\,dx = 2(1 + x)\,dx$.

$$\int \frac{1+x}{4+2x+x^2}\,dx = \int \frac{1}{4+2x+x^2}\cdot\frac{2(1+x)}{2}\,dx = \frac{1}{2}\int \frac{1}{4+2x+x^2}2(1+x)\,dx = \frac{1}{2}\int\frac{1}{u}\,du = \frac{1}{2}\ \ln|u| + C$$

$$= \frac{1}{2}\ \ln|4 + 2x + x^2| + C$$

Check: $\displaystyle\frac{d}{dx}\left[\frac{1}{2}\ln\left|4 + 2x + x^2\right| + C\right] = \left(\frac{1}{2}\right)\frac{1}{4+2x+x^2}(2+2x) = \frac{1+x}{4+2x+x^2}$

45. $\displaystyle\int 5(5x+3)\,dx$

(1) By substitution: Let $u = 5x + 3$. Then $du = 5\,dx$.

$$\int 5(5x+3)\,dx = \int u\,du = \frac{u^2}{2} + C = \frac{1}{2}(5x+3)^2 + C = \frac{1}{2}(25x^2 + 30x + 9) + C$$

$$= \frac{25}{2}x^2 + 15x + K \quad (K = C + 9/2)$$

(2) Expanding the integrand: $\displaystyle\int 5(5x+3)\,dx = \int (25x+15)\,dx = \frac{25}{2}x^2 + 15x + C$.

47. $\displaystyle\int 2x(x^2-1)\,dx$

(1) By substitution: Let $u = x^2 - 1$. Then $du = 2x\,dx$.

$$\int 2x(x^2-1)\,dx = \int u\,du = \frac{u^2}{2} + C = \frac{1}{2}(x^2-1)^2 + C = \frac{1}{2}(x^4 - 2x^2 + 1) + C$$

$$= \frac{1}{2}x^4 - x^2 + K \quad (K = C + 1/2).$$

(2) Expanding the integrand: $\displaystyle\int 2x(x^2-1)\,dx = \int (2x^3 - 2x)\,dx = 2\frac{x^4}{4} - x^2 + C = \frac{1}{2}x^4 - x^2 + C$.

49. $\displaystyle\int 5x^4(x^5)^4\,dx$

(1) By substitution: Let $u = x^5$. Then $du = 5x^4\,dx$.

$$\int 5x^4(x^5)^4\,dx = \int u^4\,du = \frac{u^5}{5} + C = \frac{1}{5}(x^5)^5 + C = \frac{1}{5}x^{25} + C.$$

(2) Expanding the integrand: $\displaystyle\int 5x^4(x^5)^4\,dx = \int 5x^{24}\,dx = 5\frac{x^{25}}{25} + C = \frac{1}{5}x^{25} + C.$

51. $F(x) = x^2 e^x$, $F'(x) = x^2 e^x + e^x(2x) = x^2 e^x + 2xe^x \neq 2xe^x = f(x)$;   no.

53. $F(x) = (x^2 + 4)^6$, $F'(x) = 6(x^2+4)^5(2x) = 12x(x^2+4) = f(x)$;   yes.

55. $F(x) = e^{2x} + 4$, $F'(x) = e^{2x}(2) = 2e^{2x} \neq e^{2x} = f(x)$;   no.

57. $F(x) = 0.5(\ln x)^2 + 10$, $F'(x) = 2(0.5)(\ln x)^1\left(\frac{1}{x}\right) = \frac{\ln x}{x} = f(x)$;   yes.

59. $\int x\sqrt{3x^2+7}\,dx$

Let $u = 3x^2 + 7$. Then $du = 6x\,dx$.

$\int x\sqrt{3x^2+7}\,dx = \int (3x^2+7)^{1/2}x\,dx = \int (3x^2+7)^{1/2}\frac{6}{6}x\,dx = \frac{1}{6}\int u^{1/2}du = \frac{1}{6}\cdot\frac{u^{3/2}}{3/2} + C$

$$= \frac{1}{9}(3x^2+7)^{3/2} + C$$

Check: $\frac{d}{dx}\left[\frac{1}{9}(3x^2+7)^{3/2} + C\right] = \frac{1}{9}\left(\frac{3}{2}\right)(3x^2+7)^{1/2}(6x) = x(3x^2+7)^{1/2}$

61. $\int x(x^3+2)^2\,dx = \int x(x^6+4x^3+4)\,dx = \int(x^7+4x^4+4x)\,dx = \frac{x^8}{8} + \frac{4}{5}x^5 + 2x^2 + C$

Check: $\frac{d}{dx}\left[\frac{x^8}{8} + \frac{4}{5}x^5 + 2x^2 + C\right] = x^7 + 4x^4 + 4x = x(x^6+4x^3+4) = x(x^3+2)^2$

63. $\int x^2(x^3+2)^2\,dx$

Let $u = x^3 + 2$. Then $du = 3x^2\,dx$.

$\int x^2(x^3+2)^2\,dx = \int(x^3+2)^2\frac{3x^2}{3}\,dx = \frac{1}{3}\int(x^3+2)^2\,3x^2\,dx = \frac{1}{3}\int u^2\,du = \frac{1}{3}\cdot\frac{u^3}{3} + C$

$$= \frac{1}{9}u^3 + C = \frac{1}{9}(x^3+2)^3 + C$$

Check: $\frac{d}{dx}\left[\frac{1}{9}(x^3+2)^3 + C\right] = \frac{1}{9}(3)(x^3+2)^2(3x^2) = x^2(x^3+2)^2$

65. $\int\frac{x^3}{\sqrt{2x^4+3}}\,dx$

Let $u = 2x^4 + 3$. Then $du = 8x^3\,dx$.

$\int\frac{x^3}{\sqrt{2x^4+3}}\,dx = \int(2x^4+3)^{-1/2}x^3\,dx = \int(2x^4+3)^{-1/2}\frac{8}{8}x^3\,dx = \frac{1}{8}\int u^{-1/2}du = \frac{1}{8}\cdot\frac{u^{1/2}}{1/2} + C$

$$= \frac{1}{4}(2x^4+3)^{1/2} + C$$

Check: $\frac{d}{dx}\left[\frac{1}{4}(2x^4+3)^{1/2} + C\right] = \frac{1}{4}\left(\frac{1}{2}\right)(2x^4+3)^{-1/2}(8x^3) = \frac{x^3}{(2x^4+3)^{1/2}}$

67. $\int\frac{(\ln x)^3}{x}\,dx$

Let $u = \ln x$. Then $du = \frac{1}{x}\,dx$.

$\int\frac{(\ln x)^3}{x}\,dx = \int u^3\,du = \frac{u^4}{4} + C = \frac{(\ln x)^4}{4} + C$

Check: $\frac{d}{dx}\left[\frac{(\ln x)^4}{4} + C\right] = \frac{1}{4}(4)(\ln x)^3\cdot\frac{1}{x} = \frac{(\ln x)^3}{x}$

69. $\int \dfrac{1}{x^2} e^{-1/x}\, dx$

Let $u = \dfrac{-1}{x} = -x^{-1}$. Then $du = \dfrac{1}{x^2}\, dx$.

$\int \dfrac{1}{x^2} e^{-1/x}\, dx = \int e^u\, du = e^u + C = e^{-1/x} + C$

Check: $\dfrac{d}{dx}[e^{-1/x} + C] = e^{-1/x}\left(\dfrac{1}{x^2}\right) = \dfrac{1}{x^2} e^{-1/x}$

71. $\dfrac{dx}{dt} = 7t^2(t^3 + 5)^6$

Let $u = t^3 + 5$. Then $du = 3t^2\, dt$.

$x = \int 7t^2(t^3 + 5)^6\, dt = \dfrac{7}{3}\int (t^3 + 5)^6\, 3t^2\, dt = \dfrac{7}{3}\int u^6\, du = \dfrac{1}{3}u^7 + C = \dfrac{1}{3}(t^3 + 5)^7 + C$

73. $\dfrac{dy}{dt} = \dfrac{3t}{\sqrt{t^2 - 4}}$

Let $u = t^2 - 4$. Then $du = 2t\, dt$.

$y = \int \dfrac{3t}{(t^2 - 4)^{1/2}}\, dt = 3\int (t^2 - 4)^{-1/2}\, t\, dt = 3\int (t^2 - 4)^{-1/2}\dfrac{2}{2}t\, dt = \dfrac{3}{2}\int u^{-1/2}\, du = \dfrac{3}{2}\cdot\dfrac{u^{1/2}}{1/2} + C$

$$= 3(t^2 - 4)^{1/2} + C$$

75. $\dfrac{dp}{dx} = \dfrac{e^x + e^{-x}}{(e^x - e^{-x})^2}$

Let $u = e^x - e^{-x}$. Then $du = (e^x + e^{-x})\, dx$.

$p = \int \dfrac{e^x + e^{-x}}{(e^x - e^{-x})^2}\, dx = \int (e^x - e^{-x})^{-2}(e^x + e^{-x})\, dx = \int u^{-2}\, du = \dfrac{u^{-1}}{-1} + C = -(e^x - e^{-x})^{-1} + C$

77. $p'(x) = \dfrac{-6000}{(3x + 50)^2}$

Let $u = 3x + 50$. Then $du = 3\, dx$.

$p(x) = \int \dfrac{-6000}{(3x + 50)^2}\, dx = -6000\int (3x + 50)^{-2}\, dx = -6000\int (3x + 50)^{-2}\dfrac{3}{3}\, dx = -2000\int u^{-2}\, du$

$$= -2000\cdot\dfrac{u^{-1}}{-1} + C = \dfrac{2000}{3x + 50} + C$$

Given $p(150) = 8$:

$8 = \dfrac{2000}{(3\cdot 150 + 50)} + C = \dfrac{2000}{500} + C = 4 + C$.   Therefore, $C = 4$.

Thus, $p(x) = \dfrac{2000}{3x + 50} + 4$.

Now

$$6.50 = \frac{2000}{3x+50} + 4$$

$$2.50 = \frac{2000}{3x+50}$$

$$2.50(3x+50) = 2000$$

$$7.50x + 125 = 2000$$

$$7.50x = 1875$$

$$x = 250$$

Thus, the demand is 250 bottles when the price is $6.50.

79. $C'(x) = 12 + \dfrac{500}{x+1}, x > 0$

$$C(x) = \int\left(12 + \frac{500}{x+1}\right)dx = \int 12\,dx + 500\int\frac{1}{x+1}\,dx \quad (u = x+1,\ du = dx)$$

$$= 12x + 500\ln(x+1) + C$$

Now, $C(0) = 2000$. Thus, $C(x) = 12x + 500\ln(x+1) + 2000$. The average cost is:

$$\overline{C}(x) = 12 + \frac{500}{x}\ln(x+1) + \frac{2000}{x}$$

and

$$\overline{C}(1000) = 12 + \frac{500}{1000}\ln(1001) + \frac{2000}{1000} = 12 + \frac{1}{2}\ln(1001) + 2 \approx 17.45 \text{ or } \$17.45 \text{ per pair of shoes}$$

81. $S'(t) = 10 - 10e^{-0.1t}, 0 \le t \le 24$

(A) $S(t) = \int\left(10 - 10e^{-0.1t}\right)dt = \int 10\,dt - 10\int e^{-0.1t}\,dt = 10t - \dfrac{10}{-0.1}e^{-0.1t} + C = 10t + 100e^{-0.1t} + C$

Given $S(0) = 0$: $\quad 0 + 100e^0 + C = 0$

$$100 + C = 0$$

$$C = -100$$

Total sales at time t:

$$S(t) = 10t + 100e^{-0.1t} - 100, \quad 0 \le t \le 24.$$

(B) $S(12) = 10(12) + 100e^{-0.1(12)} - 100$

$$= 20 + 100e^{-1.2} \approx 50$$

Total estimated sales for the first twelve months: $50 million.

(C) On a graphing utility, solve

$$10t + 100e^{-0.1t} - 100 = 100$$

or $\qquad\qquad\qquad\qquad 10t + 100e^{-0.1t} = 200$

The result is: $t \approx 18.41$ months.

83. $Q(t) = \int R(t)\,dt = \int\left(\dfrac{100}{t+1} + 5\right)dt = 100\int\dfrac{1}{t+1}\,dt + \int 5\,dt = 100\ln(t+1) + 5t + C$

Given $Q(0) = 0$:

$0 = 100\ln(1) + 0 + C$

Thus, $C = 0$ and $Q(t) = 100\ln(t+1) + 5t, \ 0 \le t \le 20$.

$Q(9) = 100\ln(9+1) + 5(9) = 100\ln 10 + 45 \approx 275$ thousand barrels.

85. $W(t) = \int w(t)\,dt = \int 0.2\,e^{0.1t}\,dt = \dfrac{0.2}{0.1}\int e^{0.1t}\,(0.1)\,dt = 2e^{0.1t} + C$

Given $W(0) = 2$:

$2 = 2e^0 + C.$

Thus, $C = 0$ and $W(t) = 2e^{0.1t}.$

The weight of the culture after 8 hours is given by:

$W(8) = 2e^{0.1(8)} = 2e^{0.8} \approx 4.45$ grams.

87. $\dfrac{dN}{dt} = -\dfrac{2000t}{1+t^2}, \; 0 \le t \le 10$

(A) To find the minimum value of $\dfrac{dN}{dt}$, calculate

$\dfrac{d}{dt}\left(\dfrac{dN}{dt}\right) = \dfrac{d^2 N}{dt^2} = -\dfrac{(1+t^2)(2000) - 2000t\,(2t)}{(1+t^2)^2} = -\dfrac{2000[1-t^2]}{(1+t^2)^2} = \dfrac{-2000(1-t)(1+t)}{(1+t^2)^2}$

critical value: $t = 1$

Now $\dfrac{dN}{dt}\Big|_{t=0} = 0$

$\dfrac{dN}{dt}\Big|_{t=1} = -1{,}000$

$\dfrac{dN}{dt}\Big|_{t=10} = \dfrac{-20{,}000}{101} \approx -198.02$

Thus, the minimum value of $\dfrac{dN}{dt}$ is $-1{,}000$ bacteria/ml per day.

(B) $N(t) = \int \dfrac{-2{,}000t}{1+t^2}\,dt$

Let $u = 1 + t^2$. Then $du = 2t\,dt$

$N(t) = \int \dfrac{-2{,}000t}{1+t^2}\,dt = -1{,}000 \int \dfrac{2t}{1+t^2}\,dt = -1{,}000 \int \dfrac{1}{u}\,du = -1{,}000\,\ln|u| + C$

$= -1{,}000\,\ln(1+t^2) + C$

Given $N(0) = 5{,}000$:

$5{,}000 = -1{,}000\,\ln(1) + C = C \;\; (\ln 1 = 0)$

Thus, $C = 5{,}000$ and $N(t) = 5{,}000 - 1{,}000\,\ln(1+t^2)$

Now, $N(10) = 5{,}000 - 1{,}000\,\ln(1 + 10^2) = 5{,}000 - 1{,}000\,\ln(101) \approx 385$ bacteria/ml

(C) Set $N(t) = 1{,}000$ and solve for t:

$1{,}000 = 5{,}000 - 1{,}000\,\ln(1+t^2)$

$\ln(1+t^2) = 4$

$1+t^2 = e^4$

$t^2 = e^4 - 1$

$t = \sqrt{e^4 - 1} \approx 7.32$ days

89. $N'(t) = 6e^{-0.1t}, \; 0 \le t \le 15$

$N(t) = \int N'(t)\,dt = \int 6e^{-0.1t}\,dt = 6\int e^{-0.1t}\,dt = \dfrac{6}{-0.1}\int e^{-0.1t}\,(-0.1)\,dt = -60e^{-0.1t} + C$

Given $N(0) = 40$:

$40 = -60e^0 + C$

Hence, $C = 100$ and $N(t) = 100 - 60e^{-0.1t}$, $0 \le t \le 15$.

The number of words per minute after completing the course is:

$N(15) = 100 - 60e^{-0.1(15)} = 100 - 60e^{-1.5} \approx 87$ words per minute.

91. $\dfrac{dE}{dt} = 5000(t + 1)^{-3/2}$, $t \ge 0$

Let $u = t + 1$, then $du = dt$

$E = \displaystyle\int 5000(t+1)^{-3/2}\, dt = 5000 \int (t+1)^{-3/2}\, dt = 5000 \int u^{-3/2}\, du = 5000 \dfrac{u^{-1/2}}{-1/2} + C$

$$= -10{,}000(t + 1)^{-1/2} + C = \dfrac{-10{,}000}{\sqrt{t+1}} + C$$

Given $E(0) = 2000$:

$2000 = \dfrac{-10{,}000}{\sqrt{1}} + C$

Hence, $C = 12{,}000$ and $E(t) = 12{,}000 - \dfrac{10{,}000}{\sqrt{t+1}}$.

The projected enrollment 15 years from now is:

$E(15) = 12{,}000 - \dfrac{10{,}000}{\sqrt{15+1}} = 12{,}000 - \dfrac{10{,}000}{\sqrt{16}} = 12{,}000 - \dfrac{10{,}000}{4} = 9500$ students.

EXERCISE 13-3

Things to remember:

1. A DIFFERENTIAL EQUATION is an equation that involves an unknown function and one or more of its derivatives. The ORDER of a differential equation is the order of the highest derivative of the unknown function.

2. A SLOPE FIELD for a first-order differential equation is obtained by drawing tangent line segments determined by the equation at each point in a grid.

3. EXPONENTIAL GROWTH LAW

   If $\dfrac{dQ}{dt} = rQ$ and $Q(0) = Q_0$, then $Q(t) = Q_0 e^{rt}$, where

   Q_0 = Amount at $t = 0$

   r = Relative growth rate (expressed as a decimal)

   t = Time

   Q = Quantity at time t

4. COMPARISON OF EXPONENTIAL GROWTH PHENOMENA

DESCRIPTION	MODEL	SOLUTION	GRAPH	USES
Unlimited growth : Rate of growth is proportional to the amount present	$\dfrac{dy}{dt} = ky$ $k,\ t > 0$ $y(0) = c$	$y = ce^{kt}$		• Short-term population growth (people, bacteria, etc.) • Growth of money at continuous compound interest • Price-supply curves • Depletion of natural resources
Exponential decay: Rate of growth is proportional to the amount present	$\dfrac{dy}{dt} = -ky$ $k,\ t > 0$ $y(0) = c$	$y = ce^{-kt}$		• Radioactive decay • Light absorption in water • Price-demand curves • Atmospheric pressure (t is altitude)
Limited growth: Rate of growth is proportional to the difference between the amount present and a fixed limit	$\dfrac{dy}{dt} = k(M - y)$ $k,\ t > 0$ $y(0) = 0$	$y = M(1 - e^{-kt})$		• Sales fads (e.g., skateboards) • Depreciation of equipment • Company growth • Learning
Logistic growth: Rate of growth is proportional to the amount present and to the difference between the amount present and a fixed limit	$\dfrac{dy}{dt} = ky(M - y)$ $k,\ t > 0$ $y(0) = \dfrac{M}{1 + c}$	$y = \dfrac{M}{1 - ce^{-kMt}}$		• Long-term population growth • Epidemics • Sales of new products • Rumor spread • Company growth

---

1. The derivative of $f(x) = e^{5x}$ is 5 times f; $y' = 5y$.

3. The derivative of $f(x) = 10e^{-x}$ is $-f$; $y' = -y$.

5. The derivative of $f(x) = 3.2e^{x^2}$ is $2x$ times f; $y' = 2xy$.

7. The derivative of $f(x) = 1 - e^{-x}$ is 1 minus f; $y' = 1 - y$.

9. $\dfrac{dy}{dx} = 6x$

$\displaystyle\int \dfrac{dy}{dx}\,dx = \int 6x\,dx = 6\int x\,dx$

$\displaystyle\int dy \ = 6\int x\,dx$

$y = 6 \cdot \dfrac{x^2}{2} + C = 3x^2 + C$

General solution: $y = 3x^2 + C$

11. $\dfrac{dy}{dx} = \dfrac{7}{x}$

$\displaystyle\int \dfrac{dy}{dx}\,dx = 7\int \dfrac{1}{x}\,dx$

$\displaystyle\int dy = 7\int \dfrac{1}{x}\,dx$

General solution: $y = 7 \ln|x| + C$

13. $\dfrac{dy}{dx} = e^{0.02x}$

$\displaystyle\int \dfrac{dy}{dx}\,dx = \int e^{0.02x}\,dx$

$\displaystyle\int dy = \int e^{0.02x}\,dx \quad (u = 0.02x,\ du = 0.02\,dx)$

$y = \displaystyle\int e^{u}\,\dfrac{1}{0.02}\,du = \dfrac{1}{0.02}\int e^{u}\,du = \dfrac{1}{0.02}\,e^{u} + C = 50e^{0.02x} + C$

General solution: $y = 50e^{0.02x} + C$

15. $\dfrac{dy}{dx} = x^2 - x;\ y(0) = 0$

$\displaystyle\int \dfrac{dy}{dx}\,dx = \int (x^2 - x)\,dx$

$y = \dfrac{1}{3}x^3 - \dfrac{1}{2}x^2 + C$

Given $y(0) = 0$: $\dfrac{1}{3}(0)^3 - \dfrac{1}{2}(0)^2 + C = 0,\quad C = 0$

Particular solution: $y = \dfrac{1}{3}x^3 - \dfrac{1}{2}x^2$

17. $\dfrac{dy}{dx} = -2xe^{-x^2};\ y(0) = 3$

$\displaystyle\int \dfrac{dy}{dx}\,dx = \int -2x\,e^{-x^2}\,dx$

$y = \displaystyle\int -2xe^{-x^2}\,dx$

Let $u = -x^2$. Then $du = -2x\ dx$ and

$\displaystyle\int -2x\,e^{-x^2}\,dx = \int e^{u}\,du = e^{u} + C = e^{-x^2} + C$

Thus, $y = e^{-x^2} + C.$

Given $y(0) = 3$: $\ 3 = e^{0} + C$

$\qquad\qquad\qquad 3 = 1 + C,\quad C = 2$

Particular solution: $y = e^{-x^2} + 2$

19. $\dfrac{dy}{dx} = \dfrac{2}{1+x};\ y(0) = 5$

$\displaystyle\int \dfrac{dy}{dx}\,dx = \int \dfrac{2}{1+x}\,dx = 2\int \dfrac{1}{1+x}\,dx$

$\displaystyle\int dy = 2\int \dfrac{1}{1+x}\,dx \quad (u = 1 + x,\ du = dx)$

$y = 2\displaystyle\int \dfrac{1}{u}\,du = 2\,\ln|u| + C = 2\,\ln|1+x| + C$

Given $y(0) = 5$: $\ 5 = 2\ln 1 + C$

$\qquad\qquad\qquad 5 = C$

Particular solution: $y = 2\,\ln|1+x| + 5.$

21. Second order

23. Third order

25. $y = 5x$, $\dfrac{dy}{dx} = 5 = \dfrac{5x}{x} = \dfrac{y}{x}$; yes.

27. $y = \sqrt{9 + x^2}$, $y' = \dfrac{1}{2\sqrt{9 + x^2}}(2x) = \dfrac{x}{\sqrt{9 + x^2}} = \dfrac{x}{y}$; yes.

29. $y = e^{3x}$, $y' = 3e^{3x}$, $y'' = 9e^{3x}$; $9e^{3x} - 4(3e^{3x}) + 3(e^{3x}) = 12e^{3x} - 12e^{3x} = 0$; yes.

31. $y = 100e^{3x}$, $y' = 300e^{3x}$, $y'' = 900e^{3x}$; $900e^{3x} - 4(300e^{3x}) + 3(100e^{3x}) = 1200e^{3x} - 1200e^{3x} = 0$; yes.

33. Figure (B). When $x = 1$, $\dfrac{dy}{dx} = 1 - 1 = 0$ for any y. When $x = 0$,

$\dfrac{dy}{dx} = 0 - 1 = -1$ for any y. When $x = 2$, $\dfrac{dy}{dx} = 2 - 1 = 1$ for any y; and so on. These facts are

consistent with the slope-field in Figure (B); they are not consistent with the slope-field in Figure (A).

35. $\dfrac{dy}{dx} = x - 1$

$\displaystyle\int \dfrac{dy}{dx}\, dx = \int (x - 1)\, dx$

General solution: $y = \dfrac{1}{2}x^2 - x + C$

Given $y(0) = -2$: $\dfrac{1}{2}(0)^2 - 0 + C = -2$, $C = -2$

Particular solution: $y = \dfrac{1}{2}x^2 - x - 2$

37.

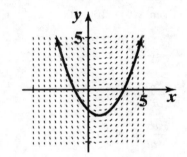

39. $\dfrac{dy}{dt} = 2y$

$\dfrac{1}{y}\dfrac{dy}{dt} = 2$

$\displaystyle\int \dfrac{1}{y}\dfrac{dy}{dt}\, dt = \int 2\, dt$

$\displaystyle\int \dfrac{1}{u}\, du = \int 2\, dt$ $\left[u = y,\ du = dy = \dfrac{dy}{dt}\cdot dt\right]$

$\ln|u| = 2t + K$ [K an arbitrary constant]

$|u| = e^{2t + K} = e^K e^{2t}$

$|u| = Ce^{2t}$ $[C = e^K, C > 0]$

so $|y| = Ce^{2t}$

Now, if we set $y(t) = Ce^{2t}$, C ANY constant, then

$y'(t) = 2Ce^{2t} = 2y(t)$,

So $y = Ce^{2t}$ satisfies the differential equation where C is any constant. This is the general solution. Note, the differential equation is the model for exponential growth with growth rate 2.

41. $\dfrac{dy}{dx} = -0.5y, \quad y(0) = 100$

$\dfrac{1}{y}\dfrac{dy}{dx} = -0.5$

$\displaystyle\int \dfrac{1}{y}\dfrac{dy}{dx}\,dx = \int -0.5\,dx$

$\displaystyle\int \dfrac{1}{u}\,du = \int -0.5\,dx \quad [u = y, \ du = dy = \dfrac{dy}{dx}\,dx]$

$\ln|u| = -0.5x + K$

$|u| = e^{-0.5x+K} = e^K e^{-0.5x}$

$|y| = Ce^{-0.5x}, \ C = e^K > 0.$

So, general solution: $y = Ce^{-0.5x}$, C any constant.

Given $y(0) = 100$: $100 = Ce^0 = C$; particular solution: $y = 100e^{-0.5x}$

43. $\dfrac{dx}{dt} = -5x$

$\dfrac{1}{x}\dfrac{dx}{dt} = -5$

$\displaystyle\int \dfrac{1}{x}\dfrac{dx}{dt}\,dt = \int -5\,dt$

$\displaystyle\int \dfrac{1}{x}\,dx = -5\int dt$

$\ln|x| = -5t + K$

$|x| = e^{-5t+K} = e^K e^{-5t} = Ce^{-5t}, \quad C = e^K > 0.$

General solution: $x = Ce^{-5t}$, C any constant.

45. $\dfrac{dx}{dt} = -5t$

$\displaystyle\int \dfrac{dx}{dt}\,dx = \int -5t\,dt = -5\int t\,dt$

General solution: $x = -\dfrac{5t^2}{2} + C$

47. $y' = 2.5y(300 - y);$ logistic growth   **49.** $y' = 0.43y;$ exponential growth

51. Figure (A). When $y = 1$, $\dfrac{dy}{dx} = 1 - 1 = 0$ for any x.

When $y = 2$, $\dfrac{dy}{dx} = 1 - 2 = -1$ for any x; and so on. This is consistent with the slope-field in Figure (A); it is not consistent with the slope-field in Figure (B).

53. $y = 1 - Ce^{-x}$

$\dfrac{dy}{dx} = \dfrac{d}{dx}[1 - Ce^{-x}] = Ce^{-x}$

From the original equation, $Ce^{-x} = 1 - y$
Thus, we have

$\dfrac{dy}{dx} = 1 - y$

and $y = 1 - Ce^{-x}$ is a solution of the differential equation for any number C.

Given $y(0) = 0$: $0 = 1 - Ce^0 = 1 - C$

$\qquad\qquad C = 1$

Particular solution: $y = 1 - e^{-x}$

55.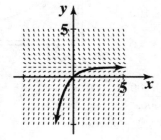

57.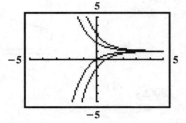

59. $y = \sqrt{C - x^2} = (C - x^2)^{1/2}$

$\dfrac{dy}{dx} = \dfrac{1}{2}(C - x^2)^{-1/2}(-2x) = \dfrac{-x}{\sqrt{C - x^2}} = -\dfrac{x}{y}$

Thus, $y = \sqrt{C - x^2}$ satisfies the differential equation $\dfrac{dy}{dx} = -\dfrac{x}{y}$.

Setting $x = 3$, $y = 4$, gives $4 = \sqrt{C - 9}$, $16 = C - 9$, $C = 25$.

The solution that passes through (3, 4) is: $y = \sqrt{25 - x^2}$.

61. $y = Cx; \ C = \dfrac{y}{x}$

$\dfrac{dy}{dx} = C = \dfrac{y}{x}$

Therefore $y = Cx$ satisfies the differential equation $\dfrac{dy}{dx} = \dfrac{y}{x}$.

Setting $x = -8, y = 24$ gives $24 = -8C, \ C = -3$
The solution that passes through (–8, 24) is $y = -3x$

63. $y = \dfrac{1}{1 + ce^{-t}} = \dfrac{e^t}{e^t + c}$ (multiply numerator and denominator by e^t)

$\dfrac{dy}{dt} = \dfrac{(e^t + c)e^t - e^t(e^t)}{(e^t + c)^2} = \dfrac{ce^t}{(e^t + c)^2}$

$$y(1-y) = \left(\frac{e^t}{e^t+c}\right)\left(1-\frac{e^t}{e^t+c}\right) = \frac{e^t}{e^t+c} \cdot \frac{c}{e^t+c} = \frac{ce^t}{(e^t+c)^2}$$

Thus $\dfrac{dy}{dt} = y(1-y)$ and $y = \dfrac{1}{1+ce^{-t}}$ satisfies the differential equation.

Setting $t = 0$, $y = -1$ gives

$$-1 = \frac{1}{1+c}, \; -1-c = 1, \; c = -2$$

The solution that passes through $(0, -1)$ is

$$y = \frac{1}{1-2e^{-t}}$$

65. $y = 1{,}000e^{0.08t}$
$0 \le t \le 15$,
$0 \le y \le 3{,}500$

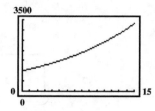

67. $p = 100e^{-0.05x}$
$0 \le x \le 30$,
$0 \le p \le 100$

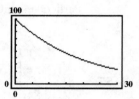

69. $N = 100(1 - e^{-0.05t})$
$0 \le t \le 100$, $0 \le N \le 100$

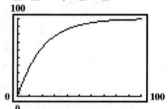

71. $N = \dfrac{1{,}000}{1+999e^{-0.4t}}$

$0 \le t \le 40$, $0 \le N \le 1{,}000$

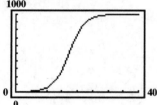

73. $\dfrac{dy}{dt} = ky(M - y)$, k, M positive constants. Set $f(y) = ky(M - y) = kMy - ky^2$.

This is a quadratic function which opens downward; it has a maximum value. Now
$f'(y) = kM - 2ky$

Critical value: $kM - 2ky = 0$ and $y = \dfrac{M}{2}$

$f''(y) = -2k < 0$. Thus, f has a maximum value at $y = \dfrac{M}{2}$.

75. In 1999: $\dfrac{dQ}{dt} = 6e^{0.013} \approx 6.079$

In 2009: $\dfrac{dQ}{dt} = 6.8e^{0.012} \approx 6.882$

The rate of growth in 2009 was greater than the rate of growth in 1999.

77. $\dfrac{dA}{dt} = 0.03A$ and $A(0) = 1{,}000$ is an unlimited growth model. From $\underline{4}$, the amount in the account after t years is: $A(t) = 1000e^{0.03t}$.

79. $\dfrac{dA}{dt} = rA,\ A(0) = 8{,}000$

is an unlimited growth model. From <u>4</u>, $A(t) = 8{,}000e^{rt}$.

Since $A(2) = 8{,}260.14,$ we solve $8{,}000e^{2r} = 8{,}260.14$ for r:

$$8000e^{2r} = 8{,}260.14$$
$$e^{2r} = \frac{8{,}260.14}{8{,}000}$$
$$2r = \ln(8{,}260.14/8{,}000)$$
$$r = \frac{\ln(8{,}260.14\,/\,8{,}000)}{2} \approx 0.016\ .$$

Thus, $A(t) = 8{,}000e^{0.016t}$.

81. (A) $\dfrac{dp}{dx} = rp,\ p(0) = 100$

This is an Unlimited Growth Model. From <u>4</u>, $p(x) = 100e^{rx}$.

Since $p(5) = 77.88,$ we have

$$77.88 = 100e^{5r}$$
$$e^{5r} = 0.7788$$
$$5r = \ln(0.7788)$$
$$r = \frac{\ln(0.7788)}{5} \approx -0.05$$

Thus, $p(x) = 100e^{-0.05x}$.

(B) $p(10) = 100e^{-0.05(10)} = 100e^{-0.5} \approx \$60.65$ per unit.

(C)

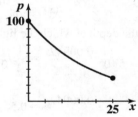

83. (A) $\dfrac{dN}{dt} = k(L - N);\ N(0) = 0$

This is a Limited Growth Model. From <u>4</u>, $N(t) = L(1 - e^{-kt})$.

Since $N(10) = 0.4L,$ we have

$$0.4L = L(1 - e^{-10k})$$
$$1 - e^{-10k} = 0.4$$
$$e^{-10k} = 0.6$$
$$-10k = \ln(0.6)$$
$$k = \frac{\ln(0.6)}{-10} \approx 0.051$$

Thus, $N(t) = L(1 - e^{-0.051t})$.

(B) $N(5) = L[1 - e^{-0.051(5)}] = L[1 - e^{-0.255}] \approx 0.225L$

Approximately 22.5% of the possible viewers will have been exposed after 5 days.

(C)   Solve $L(1 - e^{-0.051t}) = 0.8L$ for t:

$$1 - e^{-0.051t} = 0.8$$
$$e^{-0.051t} = 0.2$$
$$-0.051t = \ln(0.2)$$
$$t = \frac{\ln(0.2)}{-0.051} \approx 31.56$$

It will take 32 days for 80% of the possible viewers to be exposed.

(D)

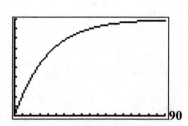

90

85. $\dfrac{dI}{dx} = -kI, \ I(0) = I_0$

This is an exponential decay model. From $\underline{4}$, $I(x) = I_0 e^{-kx}$ with $k = 0.00942$. We have

$$I(x) = I_0 e^{-0.00942x}$$

To find the depth at which the light is reduced to half of that at the surface, solve

$$I_0 e^{-0.00942x} = \frac{1}{2} I_0$$

for x:

$$e^{-0.00942x} = 0.5$$
$$-0.00942x = \ln(0.5)$$
$$x = \frac{\ln(0.5)}{-0.00942} \approx 74 \text{ feet}$$

87. $\dfrac{dQ}{dt} = -0.04Q, \ Q(0) = Q_0.$

(A)   This is a model for exponential decay. From 4,
$$Q(t) = Q_0 e^{-0.04t}$$
With $Q_0 = 3$, we have
$$Q(t) = 3e^{-0.04t}$$

(B) $Q(10) = 3e^{-0.04(10)} = 3e^{-0.4} \approx 2.01.$
There are approximately 2.01 milliliters in the body after 10 hours.

(C) $3e^{-0.04t} = 1$

$e^{-0.04t} = \dfrac{1}{3}$

$-0.04t = \ln(1/3)$

$t = \dfrac{\ln(1/3)}{-0.04} \approx 27.47$

(D)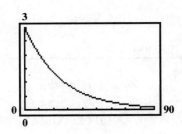

It will take approximately 27.47 hours for Q to decrease to 1 milliliter.

89. Using the exponential decay model, we have $\dfrac{dy}{dt} = -ky$, $y(0) = 100$,

$k > 0$ where $y = y(t)$ is the amount of cesium -137 present at time t. From 4,

$$y(t) = 100e^{-kt}$$

Since $y(3) = 93.3$, we solve $93.3 = 100e^{-3k}$ for k to find the continuous compound decay rate:

$$93.3 = 100e^{-3k}$$

$$e^{-3k} = 0.933$$

$$-3k = \ln(0.933)$$

$$k = \dfrac{\ln(0.933)}{-3} \approx 0.023117$$

91. From Example 3: $Q = Q_0 e^{-0.0001238t}$

Now, the amount of radioactive carbon-14 present is 5% of the original amount. Thus,

$0.05Q_0 = Q_0 e^{-0.0001238t}$ or $e^{-0.0001238t} = 0.05$.

Therefore, $-0.0001238t = \ln(0.05) \approx -2.9957$ and $t \approx 24{,}200$ years.

93. $N(k) = 180e^{-0.11(k-1)}$, $1 \le k \le 10$

Thus, $N(6) = 180e^{-0.11(6-1)} = 180e^{-0.55} \approx 104$ times

and $N(10) = 180e^{-0.11(10-1)} = 180e^{-0.99} \approx 67$ times.

95. (A) $x(t) = \dfrac{400}{1 + 399e^{-0.4t}}$

$x(5) = \dfrac{400}{1 + 399e^{(-0.4)5}} = \dfrac{400}{1 + 399e^{-2}} \approx \dfrac{400}{55} \approx 7$ people

$x(20) = \dfrac{400}{1 + 399e^{(-0.4)20}} = \dfrac{400}{1 + 399e^{-8}} \approx 353$ people

(B) $\displaystyle \lim_{t \to \infty} x(t) = 400$.

(C)

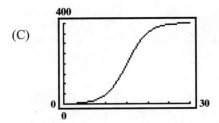

EXERCISE 13-4

Things to remember:

1.   APPROXIMATING AREAS BY LEFT AND RIGHT SUMS

   Let $f(x)$ be defined and positive on the interval $[a, b]$. Divide the interval into n subintervals of equal length

   $$\Delta x = \frac{b-a}{n},$$

   with endpoints $a = x_0 < x_1 < x_2 < \ldots < x_{n-1} < x_n = b$.

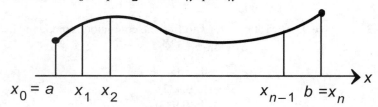

   Then

   $$L_n = f(x_0)\Delta x + f(x_1)\Delta x + f(x_2)\Delta x + \ldots + f(x_{n-1})\Delta x$$

   is called a LEFT SUM;

   $$R_n = f(x_1)\Delta x + f(x_2)\Delta x + \ldots + f(x_{n-1})\Delta x + f(x_n)\Delta x$$

   is called a RIGHT SUM.

   Left and right sums are approximations of the area between the graph of f and the x-axis from $x = a$ to $x = b$.

2.   ERROR IN AN APPROXIMATION

   The ERROR IN AN APPROXIMATION is the absolute value of the difference between the approximation and the actual value.

3.   ERROR BOUNDS FOR APPROXIMATIONS OF AREA BY LEFT AND RIGHT SUMS

   If $f(x) > 0$ and is either increasing on $[a, b]$ or decreasing on $[a, b]$, then

   $$|f(b) - f(a)| \cdot \frac{b-a}{n}$$

   is an error bound for the approximation of the area under the graph of f by L_n or R_n.

4.   LIMITS OF LEFT AND RIGHT SUMS

   If $f(x) > 0$ and is either increasing on $[a, b]$ or decreasing on $[a, b]$, then its left and right sums approach the same real number I as $n \to \infty$. This number is the area between the graph of f and the x-axis from $x = a$ to $x = b$.

<u>5.</u>   RIEMANN SUMS
Let f be defined on the interval $[a, b]$. Divide the interval into
n subintervals of equal length $\Delta x = \dfrac{b-a}{n}$ with endpoints
$a = x_0 < x_1 < x_2 < \ldots < x_{n-1} < x_n = b.$
Choose a point $c_1 \in [x_0, x_1]$, a point $c_2 \in [x_1, x_2]$, ..., and a point $c_n \in [x_{n-1}, x_n]$. Then

$$S_n = f(c_1)\Delta x + f(c_2)\Delta x + \ldots + f(c_n)\Delta x$$

is called a RIEMANN SUM. Note that left sums and right sums are special cases of Riemann Sums.

<u>6.</u>   LIMIT OF RIEMANN SUMS
If f is a continuous function on $[a, b]$ then the Riemann sums for f on $[a, b]$ approach a real number I
as $n \to \infty$.

<u>7.</u>   DEFINITE INTEGRAL
Let f be a continuous function on $[a, b]$. The limit I of Riemann sums for f on $[a, b]$ is called the
DEFINITE INTEGRAL of f from a to b, denoted

$$\int_a^b f(x)\,dx$$

The INTEGRAND is $f(x)$, the LOWER LIMIT OF INTEGRATION is a, and the UPPER LIMIT OF
INTEGRATION is b.

<u>8.</u>   GEOMETRIC INTERPRETATION OF THE DEFINITE INTEGRAL
If $f(x)$ is positive for some value of x on $[a, b]$ and negative for others, then the DEFINITE
INTEGRAL SYMBOL

$$\int_a^b f(x)\,dx$$

represents the cumulative sum of the signed areas between the curve $y = f(x)$ and the x-axis where
the areas above the x-axis are counted positively and the areas below the x-axis are counted
negatively (see the figure where A and B are actual areas of the indicated regions).

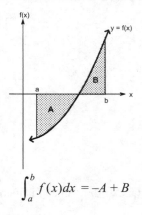

$$\int_a^b f(x)\,dx = -A + B$$

<u>9.</u>   PROPERTIES OF DEFINITE INTEGRALS

(a) $\displaystyle\int_a^a f(x)\,dx = 0$

(b) $\displaystyle\int_a^b f(x)\,dx = -\int_b^a f(x)\,dx$

(c) $\displaystyle\int_a^b Kf(x)\,dx = K\int_a^b f(x)\,dx$ K is a constant

(d) $\displaystyle\int_a^b [f(x) \pm g(x)]\,dx = \int_a^b f(x)\,dx \pm \int_a^b g(x)\,dx$

(e) $\displaystyle\int_a^b f(x)\,dx = \int_a^c f(x)\,dx + \int_c^b f(x)\,dx$

---

1. The area of each rectangle is $8 \times 2 = 16$ sq. in.; area of 5 such rectangles $5 \times 16 = 80$ sq. in.

3. $4[2(3+4+5+6)] = 4[2(18)] = 4(36) = 144$ sq. meters.

5. The square has side length $\sqrt{2}$ and area 2. The circle has area $\pi 1^2 = \pi$. No, the area inside the circle and outside the square is $\pi - 2 \approx 1.14 > 1$.

7. C, E   9. B   11. H, I   13. H

15.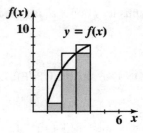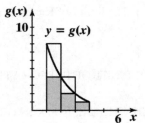

17. For Figure (A):

$$L_3 = f(1)\cdot 1 + f(2)\cdot 1 + f(3)\cdot 1$$
$$= 1 + 5 + 7 = 13$$
$$R_3 = f(2)\cdot 1 + f(3)\cdot 1 + f(4)\cdot 1$$
$$= 5 + 7 + 8 = 20$$

For Figure (B):

$$L_3 = g(1)\cdot 1 + g(2)\cdot 1 + g(3)\cdot 1$$
$$= 8 + 4 + 2 = 14$$
$$R_3 = g(2)\cdot 1 + g(3)\cdot 1 + g(4)\cdot 1$$
$$= 4 + 2 + 1 = 7$$

19. $L_3 \leq \displaystyle\int_1^4 f(x)\,dx \leq R_3,\ R_3 \leq \int_1^4 g(x)\,dx \leq L_3$; since f is increasing on [1, 4],

L_3 underestimates the area and R_3 overestimates the area; since g is decreasing on [1, 4], L_3 overestimates the area and R_3 underestimates the area.

21. For Figure (A).

Error bound for L_3 and R_3:

$$\text{Error} \le |f(4) - f(1)|\left(\frac{4-1}{3}\right) = |8 - 1| = 7$$

For Figure (B).

Error bound for L_3 and R_3:

$$\text{Error} \le |f(4) - f(1)|\left(\frac{4-1}{3}\right) = |1 - 8| = |-7| = 7$$

23. $f(x) = 25 - 3x^2$ on $[-2, 8]$

$\Delta x = \dfrac{8-(-2)}{5} = \dfrac{10}{5} = 2; x_0 = -2, x_1 = 0, x_2 = 2, \ldots, x_5 = 8$

$c_i = \dfrac{x_{i-1} + x_i}{2}; c_1 = -1, c_2 = 1, c_3 = 3, c_4 = 5, c_5 = 7$

$S_5 = f(-1)2 + f(1)2 + f(3)2 + f(5)2 + f(7)2$

$\quad = [22 + 22 - 2 - 50 - 122]2 = (-130)2 = -260$

25. $f(x) = 25 - 3x^2$ on $[0, 12]$

$\Delta x = \dfrac{12-0}{4} = 3; x_0 = 0, x_1 = 3, x_2 = 6, x_3 = 9, x_4 = 12$

$c_i = \dfrac{2x_{i-1} + x_i}{3}; c_1 = 1, c_2 = 4, c_3 = 7, c_4 = 10$

$S_4 = f(1)3 + f(4)3 + f(7)3 + f(10)3$

$\quad = [22 - 23 - 122 - 275]3 = (-398)3 = -1194$

27. $f(x) = x^2 - 5x - 6$

$\Delta x = \dfrac{3-0}{3} = 1; x_0 = 0, x_1 = 1, x_2 = 2, x_3 = 3$

$S_3 = f(0.7)1 + f(1.8)1 + f(2.4)1$

$\quad = -9.01 - 11.76 - 12.24 = -33.01$

29. $f(x) = x^2 - 5x - 6$

$\Delta x = \dfrac{7-1}{6} = 1; x_0 = 1, x_1 = 2, x_3 = 3, \ldots, x_6 = 7$

$S_6 = f(1)1 + f(3)1 + f(3)1 + f(5)1 + f(5)1 + f(7)1$

$\quad = -10 - 12 - 12 - 6 - 6 + 8 = -38$

31. $\displaystyle\int_b^0 f(x)dx = -\text{area } B = -2.475$

33. $\displaystyle\int_a^c f(x)dx = \text{area } A - \text{area } B + \text{area } C = 1.408 - 2.475 + 5.333 = 4.266$

35. $\displaystyle\int_a^d f(x)dx = \text{area } A - \text{area } B + \text{area } C - \text{area } D$

$\quad = 1.408 - 2.475 + 5.333 - 1.792 = 2.474$

37. $\int_c^0 f(x)dx = -\int_0^c f(x)dx = -\text{area } C = -5.333$

39. $\int_0^a f(x)dx = -\int_a^0 f(x)dx = -[\text{area } A - \text{area } B] = -[1.408 - 2.475] = 1.067$

41. $\int_d^b f(x)dx = -\int_b^d f(x)dx = -[\text{area } B + \text{area } C - \text{area } D]$
$$= -[-2.475 + 5.333 - 1.792] = -1.066$$

43. $\int_1^4 2x\,dx = 2\int_1^4 x\,dx = 2(7.5) = 15$

45. $\int_1^4 (5x + x^2)dx = 5\int_1^4 x\,dx + \int_1^4 x^2 dx = 5(7.5) + 21 = 58.5$

47. $\int_1^4 (x^2 - 10x)dx = \int_1^4 x^2 dx - 10\int_1^4 x\,dx = 21 - 10(7.5) = -54$

49. $\int_1^5 6x^2 dx = 6\int_1^5 x^2 dx = 6\left[\int_1^4 x^2 dx + \int_4^5 x^2 dx\right] = 6\left[21 + \dfrac{61}{3}\right] = 126 + 122 = 248$

51. $\int_4^4 (7x - 2)^2 dx = 0$

53. $\int_5^4 9x^2 dx = -\int_4^5 9x^2 dx = -9\int_4^5 x^2 dx = -9\left(\dfrac{61}{3}\right) = -183$

55.   False: Set $f(x) = x$ on $[-1, 1]$
$$\int_{-1}^1 f(x) = 0$$
by the Geometric Interpretation of the Definite Integral <u>8</u>.

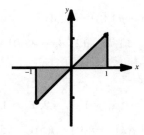

57.   False: $f(x) = 2x$ is increasing on $[0, 10]$ and so L_n is less than $\int_0^{10} f(x)\,dx$ for every n.

59.   False: Consider $f(x) = 1 - x^2$ on $[-1, 1]$ and let $n = 2$

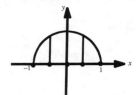

$L_2 = f(-1)1 + f(0)1 = 0 + 1 = 1$

$R_2 = f(0)1 + f(1)1 = 1 + 0 = 1$

(Note: In this case $L_n = R_n$ for every n.)

61. $h(x)$ is an increasing function; $\Delta x = 100$

$L_{10} = h(0)100 + h(100)100 + h(200)100 + \ldots + h(900)(100)$

$\qquad = [0 + 183 + 235 + 245 + 260 + 286 + 322 + 388 + 453 + 489]100$

$\qquad = (2{,}861)100 = 286{,}100$ sq ft

Error bound for L_{10}:

$$\text{Error} \leq |h(1{,}000) - h(0)| \left(\frac{1000 - 0}{10} \right) = 500(100) = 50{,}000 \text{ sq ft}$$

We want to find n such that $|I - L_n| \leq 2{,}500$:

$$|h(1000) - h(0)| \left(\frac{1000 - 0}{n} \right) \leq 2{,}500$$

$$500 \left(\frac{1000}{n} \right) \leq 2{,}500$$

$$500{,}000 \leq 2{,}500n$$

$$n \geq 200$$

63. $f(x) = 0.25x^2 - 4$ on $[2, 5]$

$L_6 = f(2)\Delta x + f(2.5)\Delta x + f(3)\Delta x + f(3.5)\Delta x + f(4)\Delta x + f(4.5)\Delta x$,   where $\Delta x = 0.5$

Thus,

$L_6 = [-3 - 2.44 - 1.75 - 0.94 + 0 + 1.06](0.5) = -3.53$

$R_6 = f(2.5)\Delta x + f(3)\Delta x + f(3.5)\Delta x + f(4)\Delta x + f(4.5)\Delta x + f(5)\Delta x$,   where $\Delta x = 0.5$

Thus,

$R_6 = [-2.44 - 1.75 - 0.94 + 0 + 1.06 + 2.25](0.5) = -0.91$

Error bound for L_6 and R_6: Since f is increasing on $[2, 5]$,

$$\text{Error} \leq |f(5) - f(2)| \left(\frac{5 - 2}{6} \right) = |2.25 - (-3)|(0.5) = 2.63$$

Geometrically, the definite integral over the interval $[2, 5]$ is the area of the region which lies above the x-axis minus the area of the region which lies below the x-axis. From the figure, if R_1 represents the region bounded by the graph of f and the x-axis for $2 \leq x \leq 4$ and R_2 represents the region bounded by the graph of f and the x-axis for $4 \leq x \leq 5$, then

$$\int_2^5 f(x)\,dx = \text{area}(R_2) - \text{area}(R_1)$$

65. $f(x) = e^{-x^2}$ 　　　　　　　　　　　　　**67.** $f(x) = x^4 - 2x^2 + 3$

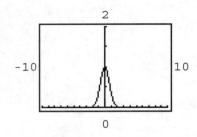

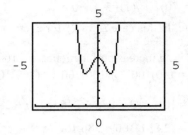

Thus, f is increasing on $(-\infty, 0]$ and decreasing on $[0, \infty)$.

Thus, f is decreasing on $(-\infty, -1]$ and on $[0, 1]$, and increasing on $[-1, 0]$ and on $[1, \infty)$.

69. $\displaystyle\int_1^3 \ln x \, dx$

$$|I - R_n| \le |\ln 3 - \ln 1| \frac{3-1}{n} \approx \frac{(1.0986)2}{n} = \frac{2.1972}{n}$$

Now $\dfrac{2.1972}{n} \le 0.1$ implies $n \ge \dfrac{2.1972}{0.1} = 21.972$

so $n \ge 22$.

71. $\displaystyle\int_1^3 x^x \, dx$

$$|I - L_n| \le |3^3 - 1^1| \frac{3-1}{n} = \frac{26 \cdot 2}{n} = \frac{52}{n}$$

Now $\dfrac{52}{n} \le 0.5$ implies $n \ge \dfrac{52}{0.5} = 104$

73. From $t = 0$ to $t = 60$

$$L_3 = N(0)20 + N(20)20 + N(40)20$$
$$= (10 + 51 + 68)20 = 2580$$

$$R_3 = N(20)20 + N(40)20 + N(60)20$$
$$= (51 + 68 + 76)20 = 3900$$

Error bound for L_3 and R_3: Since $N(t)$ is increasing,

$$\text{Error} \le |N(60) - N(0)|\left(\frac{60 - 0}{3}\right) = (76 - 10)20 = 1{,}320 \text{ units}$$

75. (A) $L_5 = A'(0)1 + A'(1)1 + A'(2)1 + A'(3)1 + A'(4)1$

$$= 0.90 + 0.81 + 0.74 + 0.67 + 0.60$$
$$= 3.72 \text{ sq cm}$$

$$R_5 = A'(1)1 + A'(2)1 + A'(3)1 + A'(4)1 + A'(5)1$$
$$= (0.81 + 0.74 + 0.67 + 0.60 + 0.55)$$
$$= 3.37 \text{ sq cm}$$

(B) Since $A'(t)$ is a decreasing function

$$R_5 = 3.37 \le \int_0^5 A'(t)dt \le 3.72 = L_5$$

77. $L_3 = N'(6)2 + N'(8)2 + N'(10)2$
$= (21 + 19 + 17)2 = 114$

$R_3 = N'(8)2 + N'(10)2 + N'(12)2$
$= (19 + 17 + 15)2 = 102$

Error bound for L_3 and R_3: Since $N'(x)$ is decreasing

$$\text{Error} \le |N'(12) - N'(6)|\left(\frac{12 - 6}{3}\right) = |15 - 21|(2) = 12 \text{ code symbols}$$

EXERCISE 13-5

1.   FUNDAMENTAL THEOREM OF CALCULUS

If f is a continuous function on the closed interval $[a, b]$ and F is any antiderivative of f, then

$$\int_a^b f(x)dx = F(x)\Big|_a^b = F(b) - F(a);$$
$$F'(x) = f(x)$$

2.   AVERAGE VALUE OF A CONTINUOUS FUNCTION OVER $[a, b]$

Let f be continuous on $[a, b]$. Then the AVERAGE VALUE of f over $[a, b]$ is:

$$\frac{1}{b - a}\int_a^b f(x)dx$$

---

1. $f(x) = 100$ on $[1,6]$. The region bounded by the graph of f and the x-axis is a rectangle of length 100 and width 5; area: $A = 500$.

3. $f(x) = x + 5$ on $[0,4]$. The region bounded by the graph of f and the x-axis is a trapezoid with bases of length 5 and 9, and height 4; area: $A = \frac{1}{2}(5 + 9)(4) = 28$.

5. $f(x) = 3x$ on $[-4,4]$. The region bounded by the graph of f and the x-axis consists of two congruent right triangles with base 4 and height 12. Area: $2\left(\frac{1}{2}\right)4 \cdot 12 = 48$.

7. $f(x) = \sqrt{9 - x^2}$ on $[-3,3]$. The region bounded by the graph of f and the x-axis is a semi-circle of radius 3; area $A = \frac{1}{2}\pi(3^2) = 4.5\pi \approx 14.14$.

9. $F(x) = 3x^2 + 160$
(A) $F(15) - F(10) = 3(15)^2 + 160 - [3(10)^2 + 160] = 675 - 300 = 375$

(B)

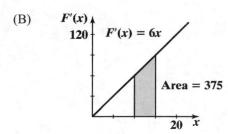

Area of trapezoid:

$$\frac{F'(15)+F'(10)}{2}\cdot 5 = \frac{90+60}{2}\cdot 5$$

$$= 75(5) = 375$$

(C)   By the Fundamental Theorem of Calculus:

$$\int_{10}^{15} 6x\, dx = 3x^2\Big|_{10}^{15} = 3(15)^2 - 3(10)^2 = 375$$

11. $F(x) = -x^2 + 42x + 240$

(A) $F(15) - F(10) = -(15)^2 + 42(15) + 240 - [-(10)^2 + 42(10) + 240]$

$\qquad = -225 + 630 + 240 - (-100 + 420 + 240) = -225 + 630 + 100 - 420 = 85$

(B)

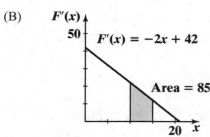

Area of trapezoid:

$$\frac{F'(15)+F'(10)}{2}\cdot 5$$

$$= \frac{(-30+42)+(-20+42)}{2}\cdot 5$$

$$= 17(5) = 85$$

(C)   By the Fundamental Theorem of Calculus:

$$\int_{10}^{15} (-2x + 42)dx = \Big[-x^2 + 42x\Big]_{10}^{15} = -(15)^2 + 42(15) - [-(10)^2 + 42(10)]$$

$$= -225 + 630 + 100 - 420 = 85$$

13. $\displaystyle\int_{0}^{10} 4\, dx = 4x\Big|_{0}^{10} = 4(10) - 4(0) = 40$

15. $\displaystyle\int_{0}^{6} x^2\, dx = \frac{1}{3}x^3\Big|_{0}^{6} = \frac{1}{3}(6)^3 - \frac{1}{3}(0)^3 = 72$

17. $\displaystyle\int_{1}^{4} (5x+3)\, dx = \left[\frac{5}{2}x^2 + 3x\right]_{1}^{4} = \left(\frac{5}{2}\cdot 4^2 + 3\cdot 4\right) - \left(\frac{5}{2}\cdot 1^2 + 3\cdot 1\right) = 52 - \frac{11}{2} = \frac{93}{2} = 46.5$

19. $\displaystyle\int_{0}^{1} e^x\, dx = e^x\Big|_{0}^{1} = e - 1 \approx 1.718$

21. $\displaystyle\int_{1}^{2} \frac{1}{x}\, dx = \ln|x|\big]_{1}^{2} = \ln 2 - \ln 1 = \ln 2 \approx 0.693$

23. $\displaystyle\int_{-2}^{2} (x^3 + 7x)\, dx = \left[\frac{1}{4}x^4 + \frac{7}{2}x^2\right]_{-2}^{2} = \left[\frac{1}{4}\cdot 2^4 + \frac{7}{2}\cdot 2^2\right] - \left[\frac{1}{2}(-2)^4 + \frac{7}{2}(-2)^2\right] = 18 - 18 = 0$

25. $\displaystyle\int_{2}^{5} (2x+9)\, dx = \left[x^2 + 9x\right]_{2}^{5} = (5^2 + 9\cdot 5) - (2^2 + 9\cdot 2) = 70 - 22 = 48$

27. $\displaystyle\int_{5}^{2} (2x+9)\, dx = -\int_{2}^{5} (2x+9)\, dx = -48$ (Property 2 and Problem 25)

29. $\int_2^3 (6-x^3)\,dx = \left[6x - \frac{1}{4}x^4\right]_2^3 = \left[6\cdot 3 - \frac{1}{4}(3)^4\right] - \left[6\cdot 2 - \frac{1}{4}(2)^4\right] = 18 - \frac{81}{4} - 8 = 10 - \frac{81}{4} = -\frac{41}{4} = -10.25$

31. $\int_6^6 (x^2 - 5x + 1)^{10}\,dx = 0$

33. $\int_1^2 (2x^{-2} - 3)\,dx = \left[-\frac{2}{x} - 3x\right]_1^2 = \left[-\frac{2}{2} - 3\cdot 2\right] - \left[-\frac{2}{1} - 3\cdot 1\right] = -7 + 5 = -2$

35. $\int_1^4 3\sqrt{x}\,dx = 3\int_1^4 x^{1/2}\,dx = 3\cdot\frac{2}{3}x^{3/2}\Big]_1^4 = 2x^{3/2}\Big]_1^4 = 2\cdot 8 - 2\cdot 1 = 14$

37. $\int_2^3 12(x^2 - 4)^5 x\,dx$. Consider the indefinite integral $\int 12(x^2 - 4)^5 x\,dx$.

Let $u = x^2 - 4$. Then $du = 2x\,dx$.

$\int 12(x^2 - 4)^5 x\,dx = 6\int (x^2 - 4)^5 2x\,dx = 6\int u^5\,du = 6\frac{u^6}{6} + C = u^6 + C = (x^2 - 4)^6 + C$

Thus,

$\int_2^3 12(x^2 - 4)^5 x\,dx = (x^2 - 4)^6\Big|_2^3 = (3^2 - 4)^6 - (2^2 - 4)^6 = 5^6 = 15{,}625.$

39. $\int_3^9 \frac{1}{x-1}\,dx$

Let $u = x - 1$. Then $du = dx$ and $u = 8$ when $x = 9$, $u = 2$ when $x = 3$.
Thus,

$\int_3^9 \frac{1}{x-1}\,dx = \int_2^8 \frac{1}{u}\,du = \ln u\Big|_2^8 = \ln 8 - \ln 2 = \ln 4 \approx 1.386.$

41. $\int_{-5}^{10} e^{-0.05x}\,dx$

Let $u = -0.05x$. Then $du = -0.05\,dx$ and $u = -0.5$ when $x = 10$, $u = 0.25$ when $x = -5$. Thus,

$\int_{-5}^{10} e^{-0.05x}\,dx = -\frac{1}{0.05}\int_{-5}^{10} e^{-0.05x}(-0.05)\,dx = -\frac{1}{0.05}\int_{0.25}^{-0.5} e^u\,du$

$= -\frac{1}{0.05}e^u\Big|_{0.25}^{-0.5} = -\frac{1}{0.05}\,[e^{-0.5} - e^{0.25}]$

$= 20(e^{0.25} - e^{-0.5}) \approx 13.550$

43. $\int_1^e \frac{\ln t}{t}\,dt = \int_0^1 u\,du = \frac{1}{2}u^2\Big|_0^1 = \frac{1}{2}$

Let $u = \ln t$. Then $du = \frac{1}{t}\,dt$

$t = 1$ implies $u = \ln 1 = 0$, $t = e$ implies $u = \ln e = 1$

45. $\int_0^1 xe^{-x^2}\,dx = \int_0^1 e^{-x^2}\left(\frac{-2}{-2}\right)x\,dx$

Let $u = -x^2$
Then $du = -2x\,dx$
$x = 0$ implies $u = 0$
$x = 1$ implies $u = -1$

$= -\dfrac{1}{2} \displaystyle\int_0^{-1} e^u \, du$

$= -\dfrac{1}{2} e^u \Big|_0^{-1} = -\dfrac{1}{2} e^{-1} + \dfrac{1}{2} e^0 = \dfrac{1}{2}(1 - e^{-1}) \approx 0.316$

47. $\displaystyle\int_1^1 e^{x^2} dx = 0$

49. $f(x) = 500 - 50x$ on $[0, 10]$

(A) Avg. $f(x) = \dfrac{1}{10-0} \displaystyle\int_0^{10} (500 - 50x)dx$

$\qquad\qquad = \dfrac{1}{10}\Big[500x - 25x^2\Big]_0^{10}$

$\qquad\qquad = \dfrac{1}{10}[5{,}000 - 2{,}500] = 250$

(B)

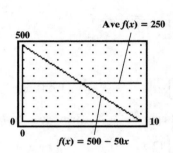

51. $f(t) = 3t^2 - 2t$ on $[-1, 2]$

(A) Avg. $f(t) = \dfrac{1}{2-(-1)} \displaystyle\int_{-1}^{2} (3t^2 - 2t)dt$

$\qquad\qquad = \dfrac{1}{3}(t^3 - t^2)\Big|_{-1}^{2}$

$\qquad\qquad = \dfrac{1}{3}[4 - (-2)] = 2$

(B)

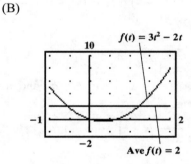

53. $f(x) = \sqrt[3]{x} = x^{1/3}$ on $[1, 8]$

(A) Avg. $f(x) = \dfrac{1}{8-1} \displaystyle\int_1^8 x^{1/3} \, dx$

$\qquad\qquad = \dfrac{1}{7}\left(\dfrac{3}{4} x^{4/3}\right)\Big|_1^8$

$\qquad\qquad = \dfrac{3}{28}(16-1) = \dfrac{45}{28} \approx 1.61$

(B)

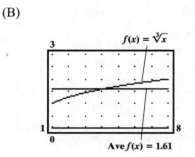

55. $f(x) = 4e^{-0.2x}$ on [0, 10]

(A) Avg. $f(x) = \dfrac{1}{10-0} \displaystyle\int_0^{10} 4e^{-0.2x}dx$

$= \dfrac{1}{10}\left[-20e^{-0.2x}\right]_0^{10}$

$= \dfrac{1}{10}(20 - 20e^{-2}) \approx 1.73$

(B)

Ave $f(x) = 1.73$

$f(x) = 4e^{-0.2x}$

57. $\displaystyle\int_2^3 x\sqrt{2x^2-3}\,dx = \int_2^3 x(2x^2-3)^{1/2}dx$

$= \dfrac{1}{4}\displaystyle\int_2^3 (2x^2-3)^{1/2}4x\,dx$

[Note: The integrand has the form $u^{1/2}du$;

the antiderivative is $\dfrac{2}{3}u^{3/2} = \dfrac{2}{3}(2x^2-3)^{3/2}$.]

$= \dfrac{1}{4}\left(\dfrac{2}{3}\right)(2x^2-3)^{3/2}\Big|_2^3$

$= \dfrac{1}{6}[2(3)^2-3]^{3/2} - \dfrac{1}{6}[2(2)^2-3]^{3/2}$

$= \dfrac{1}{6}(15)^{3/2} - \dfrac{1}{6}(5)^{3/2} = \dfrac{1}{6}[15^{3/2} - 5^{3/2}] \approx 7.819$

59. $\displaystyle\int_0^1 \dfrac{x-1}{x^2-2x+3}\,dx$

Consider the indefinite integral and let $u = x^2 - 2x + 3$.

Then $du = (2x-2)dx = 2(x-1)dx$.

$\displaystyle\int \dfrac{x-1}{x^2-2x+3}\,dx = \dfrac{1}{2}\int \dfrac{2(x-1)}{x^2-2x+3}\,dx = \dfrac{1}{2}\int \dfrac{1}{u}\,du = \dfrac{1}{2}\ln|u| + C$

Thus,

$\displaystyle\int_0^1 \dfrac{x-1}{x^2-2x+3}\,dx = \dfrac{1}{2}\ln|x^2-2x+3|\,\Big|_0^1 = \dfrac{1}{2}\ln 2 - \dfrac{1}{2}\ln 3 = \dfrac{1}{2}(\ln 2 - \ln 3) \approx -0.203$

61. $\displaystyle\int_{-1}^1 \dfrac{e^{-x}-e^x}{(e^{-x}+e^x)^2}\,dx$

Consider the indefinite integral and let $u = e^{-x} + e^x$.

Then $du = (-e^{-x} + e^x)\,dx = -(e^{-x} - e^x)\,dx$.

$\displaystyle\int \dfrac{e^{-x}-e^x}{(e^{-x}+e^x)^2}\,dx = -\int \dfrac{-(e^{-x}-e^x)}{(e^{-x}+e^x)^2}\,dx = -\int u^{-2}\,du = \dfrac{-u^{-1}}{-1} + C = \dfrac{1}{u} + C$

Thus,

$\displaystyle\int_{-1}^1 \dfrac{e^{-x}-e^x}{(e^{-x}+e^x)^2}\,dx = \dfrac{1}{e^{-x}+e^x}\,\Big|_{-1}^1 = \dfrac{1}{e^{-1}+e^1} - \dfrac{1}{e^{-(-1)}+e^{-1}} = \dfrac{1}{e^{-1}+e} - \dfrac{1}{e^{-1}+e} = 0$

63. $\displaystyle\int_{1.7}^{3.5} x \ln x\, dx \approx 4.566$

```
fnInt(X*ln X,X,1
.7,3.5)
      4.566415359
■
```

65. $\displaystyle\int_{-2}^{2} \frac{1}{1+x^2}\, dx \approx 2.214$

```
fnInt(1/(1+X²),X
,-2,2)
      2.214297436
■
```

67. If $F(t)$ denotes the position of the car at time t, then the average velocity over the time interval $t = a$ to $t = b$ is given by

$$\frac{F(b)-F(a)}{b-a}.$$

$F'(t)$ gives the instantaneous velocity of the car at time t. By the Mean Value Theorem, there exists at least one time $t = c$ at which

$$\frac{F(b)-F(a)}{b-a} = F'(c).$$

Thus, if $\dfrac{F(b)-F(a)}{b-a} = 60$, then the instantaneous velocity must equal 60 at least once during the 10 minute time interval.

69. $C'(x) = 500 - \dfrac{x}{3}$ on [300, 900]

The increase in cost from a production level of 300 bikes per month to a production level of 900 bikes per month is given by:

$$\int_{300}^{900} \left(500 - \frac{x}{3}\right) dx = \left[500x - \frac{1}{6}x^2\right]_{300}^{900} = 315{,}000 - (135{,}000) = \$180{,}000$$

71. Total loss in value in the first 5 years:

$$V(5) - V(0) = \int_{0}^{5} V'(t)\, dt = \int_{0}^{5} 500(t-12)\, dt = 500\left[\frac{t^2}{2} - 12t\right]_{0}^{5} = 500\left(\frac{25}{2} - 60\right) = -\$23{,}750$$

Total loss in value in the second 5 years:

$$V(10) - V(5) = \int_{5}^{10} V'(t)\, dt = \int_{5}^{10} 500(t-12)\, dt = 500\left[\frac{t^2}{2} - 12t\right]_{5}^{10}$$

$$= 500\left[(50-120) - \left(\frac{25}{2} - 60\right)\right] = -\$11{,}250$$

73. (A)

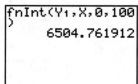
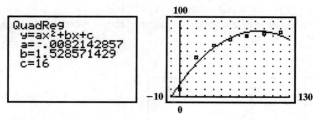

75. To find the useful life, set $C'(t) = R'(t)$ and solve for t.

$$\frac{1}{11}t = 5te^{-t^2}$$

$$e^{t^2} = 55$$

$$t^2 = \ln 55$$

$$t = \sqrt{\ln 55} \approx 2 \text{ years}$$

The total profit accumulated during the useful life is:

$$P(2) - P(0) = \int_0^2 [R'(t) - C'(t)]dt = \int_0^2 \left(5te^{-t^2} - \frac{1}{11}t\right)dt = \int_0^2 5te^{-t^2}dt - \int_0^2 \frac{1}{11}t\,dt$$

$$= -\frac{5}{2}\int_0^2 e^{-t^2}(-2t)dt - \frac{1}{11}\int_0^2 t\,dt$$

[Note: In the first integral, the integrand has the form $e^u du$, where $u = -t^2$; an antiderivative is $e^u = e^{-t^2}$.]

$$= -\frac{5}{2}e^{-t^2}\Big|_0^2 - \frac{1}{22}t^2\Big|_0^2$$

$$= -\frac{5}{2}e^{-4} + \frac{5}{2} - \frac{4}{22} = \frac{51}{22} - \frac{5}{2}e^{-4} \approx 2.272$$

Thus, the total profit is approximately \$2,272.

77. $C(x) = 60,000 + 300x$

(A)   Average cost per unit:

$$\overline{C}(x) = \frac{C(x)}{x} = \frac{60,000}{x} + 300$$

$$\overline{C}(500) = \frac{60,000}{500} + 300 = \$420$$

(B)   Avg. $C(x) = \dfrac{1}{500}\displaystyle\int_0^{500} (60,000 + 300x)dx$

$$= \frac{1}{500}(60,000x + 150x^2)\Big|_0^{500}$$

$$= \frac{1}{500}(30,000,000 + 37,500,000) = \$135,000$$

(C) $\overline{C}(500)$ is the average cost per unit at a production level of 500
units; Avg. $C(x)$ is the average value of the total cost as production increases from 0 units to 500
units.

79.   (A)

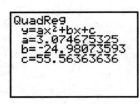

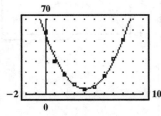

(B)   Let $q(x)$ be the quadratic regression model found in part (A).
The increase in cost in going from a production level of 2
thousand sunglasses per month to 8 thousand sunglasses per
month is given (approximately) by

$$\int_2^8 q(x)\,dx \approx 100.505$$

Therefore, the increase in cost is approximately $100,505.

81.   Average price:

$$\text{Avg. } S(x) = \frac{1}{30-20} \int_{20}^{30} 10(e^{0.02x} - 1)\,dx = \int_{20}^{30} (e^{0.02x} - 1)\,dx = \int_{20}^{30} e^{0.02x}\,dx - \int_{20}^{30} dx$$

$$= \frac{1}{0.02} \int_{20}^{30} e^{0.02x}(0.02)\,dx - x\Big|_{20}^{30} = 50 e^{0.02x}\Big|_{20}^{30} - (30 - 20)$$

$$= 50e^{0.6} - 50e^{0.4} - 10 \quad \approx 6.51 \text{ or } \$6.51$$

83. $g(x) = 2400x^{-1/2}$ and $L'(x) = g(x)$.
The number of labor hours to assemble the 17th through the 25th control units is:

$$L(25) - L(16) = \int_{16}^{25} g(x)\,dx = \int_{16}^{25} 2400x^{-1/2}\,dx = 2400(2)\,x^{1/2}\Big|_{16}^{25} = 4800x^{1/2}\Big|_{16}^{25}$$

$$= 4800[25^{1/2} - 16^{1/2}] = 4800 \text{ labor hours.}$$

85.   (A)   The inventory function is obtained by finding the equation of the line joining (0, 600) and (3, 0).

Slope: $m = \dfrac{0 - 600}{3 - 0} = -200$, y intercept: $b = 600$

Thus, the equation of the line is: $I = -200t + 600$

(B)   The average of I over [0, 3] is given by:

$$\text{Avg. } I(t) = \frac{1}{3-0} \int_0^3 I(t)\,dt = \frac{1}{3} \int_0^3 (-200t + 600)\,dt = \frac{1}{3}(-100t^2 + 600t)\Big|_0^3$$

$$= \frac{1}{3}[-100(3^2) + 600(3) - 0]$$

$$= \frac{900}{3} = 300 \text{ units}$$

87.   Rate of production: $R(t) = \dfrac{100}{t+1} + 5, \ 0 \le t \le 20$

Total production from year N to year M is given by:

$$P = \int_N^M R(t)\,dt = \int_N^M \left(\frac{100}{t+1} + 5\right) dt = 100 \int_N^M \frac{1}{t+1}\,dt + \int_N^M 5\,dt = 100\ln(t+1)\Big]_N^M + 5t\Big]_N^M$$

$$= 100\ln(M+1) - 100\ln(N+1) + 5(M - N)$$

Thus, for total production during the first 10 years, let $M = 10$ and $N = 0$.

$P = 100 \ln 11 - 100 \ln 1 + 5(10 - 0) = 100 \ln 11 + 50 \approx 290$ thousand barrels.

For the total production from the end of the 10th year to the end of the 20th year, let $M = 20$ and $N = 10$.

$P = 100 \ln 21 - 100 \ln 11 + 5(20 - 10) = 100 \ln 21 - 100 \ln 11 + 50 \approx 115$ thousand barrels.

89. $W'(t) = 0.2e^{0.1t}$

The weight increase during the first eight hours is given by:

$$W(8) - W(0) = \int_0^8 W'(t)\, dt = \int_0^8 0.2e^{0.1t} dt = 0.2 \int_0^8 e^{0.1t} dt$$

$$= \frac{0.2}{0.1} \int_0^8 e^{0.1t}(0.1)dt \quad \text{(Let } u = 0.1t, \text{ then } du = 0.1dt.)$$

$$= 2e^{0.1t} \Big|_0^8 = 2e^{0.8} - 2 \approx 2.45 \text{ grams}$$

The weight increase during the second eight hours, i.e., from the 8th hour through the 16th hour, is given by:

$$W(16) - W(8) = \int_8^{16} W'(t)d = \int_8^{16} 0.2e^{0.1t} dt = 2e^{0.1t} \Big|_8^{16} = 2e^{1.6} - 2e^{0.8} \approx 5.45 \text{ grams.}$$

91. $C(t) = t^3 - 2t + 10, \ 0 \le t \le 2.$

Average temperature over time period [0, 2] is given by:

$$\frac{1}{2-0} \int_0^2 C(t)dt = \frac{1}{2} \int_0^2 (t^3 - 2t + 10)dt = \frac{1}{2}\left(\frac{t^4}{4} - \frac{2t^2}{2} + 10t\right)\Bigg|_0^2 = \frac{1}{2}(4 - 4 + 20) = 10^\circ \text{ Celsius}$$

93. $P(t) = \dfrac{8.4t}{t^2 + 49} + 0.1, \ 0 \le t \le 24$

(A) Average fraction of people during the first seven months:

$$\frac{1}{7-0} \int_0^7 \left[\frac{8.4t}{t^2 + 49} + 0.1\right] dt = \frac{4.2}{7} \int_0^7 \frac{2t}{t^2 + 49} dt + \frac{1}{7} \int_0^7 0.1dt$$

$$= 0.6 \ln(t^2 + 49)\Big|_0^7 + \frac{0.1}{7}t\Big|_0^7$$

$$= 0.6[\ln 98 - \ln 49] + 0.1 = 0.6 \ln 2 + 0.1 \approx 0.516$$

(B) Average fraction of people during the first two years:

$$\frac{1}{24-0} \int_0^{24} \left[\frac{8.4t}{t^2 + 49} + 0.1\right] dt = \frac{4.2}{24} \int_0^{24} \frac{2t}{t^2 + 49} dt + \frac{1}{24} \int_0^{24} 0.1dt$$

$$= 0.175 \ln(t^2 + 49)\Big|_0^{24} + \frac{0.1}{24}t\Big|_0^{24}$$

$$= 0.175[\ln 625 - \ln 49] + 0.1 \approx 0.546$$

CHAPTER 13 REVIEW

1. $\displaystyle\int (6x + 3)\, dx = 6\int x\, dx + \int 3\, dx = 6 \cdot \frac{x^2}{2} + 3x + C = 3x^2 + 3x + C$ (5-1)

2. $\displaystyle\int_{10}^{20} 5dx = 5x\Big]_{10}^{20} = 5(20) - 5(10) = 50$ (5-5)

3. $\displaystyle\int_0^9 (4-t^2)\,dt = \int_0^9 4\,dt - \int_0^9 t^2\,dt = 4t\Big|_0^9 - \frac{t^3}{3}\Big|_0^9 = 36 - 243 = -207$ (5-5)

4. $\displaystyle\int (1-t^2)^3\, t\, dt = \int (1-t^2)^3 \left(\frac{-2}{-2}\right) t\, dt = -\frac{1}{2}\int (1-t^2)^3\,(-2t)\, dt = -\frac{1}{2}\int u^3\, du = -\frac{1}{2}\cdot\frac{u^4}{4} + C$

$$= -\frac{1}{8}(1-t^2)^4 + C \qquad (5\text{-}2)$$

5. $\displaystyle\int \frac{1+u^4}{u}\, du = \int \left(\frac{1}{u} + u^3\right) du = \int \frac{1}{u}\, du + \int u^3\, du = \ln|u| + \frac{1}{4}u^4 + C$ (5-1)

6. $\displaystyle\int_0^1 xe^{-2x^2}\, dx$

Let $u = -2x^2$. Then $du = -4x\, dx$.

$\displaystyle\int xe^{-2x^2}\, dx = \int e^{-2x^2}\left(\frac{-4}{-4}\right) x\, dx = -\frac{1}{4}\int e^u\, du = -\frac{1}{4}e^u + C = -\frac{1}{4}e^{-2x^2} + C$

$\displaystyle\int_0^1 xe^{-2x^2}\, dx = -\frac{1}{4}e^{-2x^2}\Big|_0^1 = -\frac{1}{4}e^{-2} + \frac{1}{4} \approx 0.216$ (5-5)

7. $F(x) = \ln x^2 = 2\ln x, \quad F'(x) = \dfrac{2}{x} \neq \ln 2x;$ no.   (5-1)

8. $F(x) = \ln x^2 = 2\ln x, \quad F'(x) = \dfrac{2}{x} = f(x);$ yes.   (5-1)

9. $F(x) = (\ln x)^2, \quad F'(x) = 2(\ln x)\left(\dfrac{1}{x}\right) = \dfrac{2\ln x}{x} \neq 2\ln x = f(x);$ no.   (5-1)

10. $F(x) = (\ln x)^2, \quad F'(x) = 2(\ln x)\left(\dfrac{1}{x}\right) = \dfrac{2\ln x}{x} = f(x);$ yes.   (5-1)

11. $y = 3x + 17, \quad y' = 3;$
$(x+5)y' = (x+5)(3) = 3x + 15 = 3x + 17 - 2 = y - 2;$ yes.   (5-3)

12. $y = 4x^3 + 7x^2 - 5x + 2, \quad y' = 12x^2 + 14x - 5, \quad y'' = 24x + 14, \quad y''' = 24;$
$(x+2)(24) - 24x = 24x + 48 - 24x = 48;$ yes.   (5-3)

13. $\dfrac{d}{dx}\left[\displaystyle\int e^{-x^2}\, dx\right] = e^{-x^2}$ (5-1)

14. $\displaystyle\int \frac{d}{dx}(\sqrt{4+5x})\, dx = \sqrt{4+5x} + C$ (5-1)

15. $\dfrac{dy}{dx} = 3x^2 - 2$

$y = f(x) = \displaystyle\int 3x^2 - 2)\, dx = f(x) = x^3 - 2x + C$

$f(0) = C = 4; \quad f(x) = x^3 - 2x + 4$ (5-3)

16. (A) $\int (8x^3 - 4x - 1)\,dx = 8\int x^3\,dx - 4\int x\,dx - \int dx = 8 \cdot \dfrac{1}{4}x^4 - 4\,\dfrac{1}{2}x^2 - x + C$

$$= 2x^4 - 2x^2 - x + C$$

(B) $\int \left(e^t - 4t^{-1}\right)dt = \int e^t\,dt - 4\int \dfrac{1}{t}\,dt = e^t - 4\ln|t| + C$ (5-1)

17. $f(x) = x^2 + 1,\ a = 1,\ b = 5,\ n = 2,\ \Delta x = \dfrac{5-1}{2} = 2;$

$R_2 = f(3)2 + f(5)2 = 10(2) + 26(2) = 72$

Error bound for R_2: f is increasing on [1, 5], so

$|I - R_2| \le [f(5) - f(1)]\dfrac{5-1}{2} = (26 - 2)(2) = 48$

Thus, $I = 72 \pm 48$.   (5-4)

18. $\displaystyle\int_1^5 (x^2 + 1)\,dx = \left[\dfrac{1}{3}x^3 + x\right]_1^5 = \dfrac{125}{3} + 5 - \left(\dfrac{1}{3} + 1\right) = \dfrac{136}{3} = 45\tfrac{1}{3}$

$|I - R_2| = \left|45\tfrac{1}{3} - 72\right| = 26\tfrac{2}{3} \approx 26.67$ (5-5)

19. Using the values of f in the table with $a = 1,\ b = 17,\ n = 4$

$\Delta x = \dfrac{17-1}{4} = 4$, we have

$L_4 = f(1)4 + f(5)4 + f(9)4 + f(13)4 = [1.2 + 3.4 + 2.6 + 0.5]4 = 30.8$ (5-4)

20. $f(x) = 6x^2 + 2x$ on $[-1, 2]$;

$\text{Ave } f(x) = \dfrac{1}{2-(-1)}\displaystyle\int_{-1}^2 (6x^2 + 2x)\,dx = \dfrac{1}{3}(2x^3 + x^2)\Big|_{-1}^2 = \dfrac{1}{3}[20 - (-1)] = 7$ (5-5)

21. width $= 2 - (-1) = 3$, height $=$ Avg. $f(x) = 7$ (5-5)

22. $f(x) = 100 - x^2$

$\Delta x = \dfrac{11-3}{4} = \dfrac{8}{4} = 2;\ c_i = \dfrac{x_{i-1} + x_i}{2}$ (= midpoint of interval)

$S_4 = f(4)2 + f(6)2 + f(8)\cdot 2 + f(10)\cdot 2 = [84 + 64 + 36 + 0]2 = (184)2 = 368$ (5-4)

23. $f(x) = 100 - x^2$

$\Delta x = \dfrac{5-(-5)}{5} = \dfrac{10}{5} = 2$

$S_5 = f(-4)2 + f(-1)2 + f(1)2 + f(2)2 + f(5)2 = [84 + 99 + 99 + 96 + 75]2 = (453)2 = 906$ (5-4)

24. $\displaystyle\int_a^b 5f(x)\,dx = 5\int_a^b f(x)\,dx = 5(-2) = -10$ (5-4, 5-5)

25. $\displaystyle\int_b^c \dfrac{f(x)}{5}\,dx = \dfrac{1}{5}\int_b^c f(x)\,dx = \dfrac{1}{5}(2) = \dfrac{2}{5} = 0.4$ (5-4, 5-5)

26. $\displaystyle\int_b^d f(x)\,dx = \int_b^c f(x)\,dx + \int_c^d f(x)\,dx = 2 - 0.6 = 1.4$ (5–4, 5-5)

27. $\displaystyle\int_a^c f(x)\,dx = \int_a^b f(x)\,dx + \int_b^c f(x)\,dx = -2 + 2 = 0$ (5-4, 5-5)

28. $\displaystyle\int_0^d f(x)\,dx = \int_0^a f(x)\,dx + \int_a^b f(x)\,dx + \int_b^c f(x)\,dx + \int_c^d f(x)\,dx = 1 - 2 + 2 - 0.6 = 0.4$ (5-4, 5-5)

29. $\displaystyle\int_b^a f(x)\,dx = -\int_a^b f(x)\,dx = -(-2) = 2$ (5-4, 5-5)

30. $\displaystyle\int_c^b f(x)\,dx = -\int_b^c f(x)\,dx = -2$ (5-4, 5-5)

31. $\displaystyle\int_d^0 f(x)\,dx = -\int_0^d f(x)\,dx = -0.4$ (from Problem 28)   (5-4, 5-5)

32. (A) $\dfrac{dy}{dx} = \dfrac{2y}{x}$; $\left.\dfrac{dy}{dx}\right|_{(2,1)} = \dfrac{2(1)}{2} = 1$, $\left.\dfrac{dy}{dx}\right|_{(-2,-1)} = \dfrac{2(-1)}{-2} = 1$

     (B) $\dfrac{dy}{dx} = \dfrac{2x}{y}$; $\left.\dfrac{dy}{dx}\right|_{(2,1)} = \dfrac{2(2)}{1} = 4$, $\left.\dfrac{dy}{dx}\right|_{(-2,-1)} = \dfrac{2(-2)}{-1} = 4$ (5-3)

33. $\dfrac{dy}{dx} = \dfrac{2y}{x}$; from the figure, the slopes at $(2, 1)$ and $(-2, -1)$ are approximately equal to 1 as computed in

     Problem 32(A), not 4 as computed in Problem 32(B).   (5-3)

34. Let $y = Cx^2$. Then $\dfrac{dy}{dx} = 2Cx$. From the original equation, $C = \dfrac{y}{x^2}$ so

$$\frac{dy}{dx} = 2x\left(\frac{y}{x^2}\right) = \frac{2y}{x} \qquad (5\text{-}3)$$

35. Letting $x = 2$ and $y = 1$ in $y = Cx^2$, we get

     $1 = 4C$ so $C = \dfrac{1}{4}$ and $y = \dfrac{1}{4}x^2$

     Letting $x = -2$ and $y = -1$ in $y = Cx^2$, we get

     $-1 = 4C$ so $C = -\dfrac{1}{4}$ and $y = -\dfrac{1}{4}x^2$ (5-3)

36.
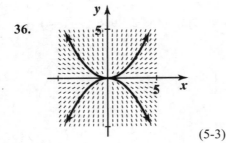
(5-3)

37.
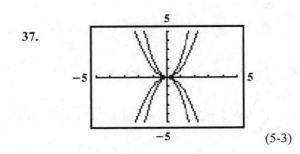
(5-3)

38. $\displaystyle\int_{-1}^{1}\sqrt{1+x}\,dx = \int_{0}^{2}u^{1/2}\,du = \left.\frac{u^{3/2}}{3/2}\right|_{0}^{2} = \frac{2}{3}(2)^{3/2} \approx 1.886$

Let $u = 1 + x,\quad du = dx.$
When $x = -1, u = 0,\quad$ when $x = 1, u = 2.$ $\qquad$ (5-5)

39. $\displaystyle\int_{-1}^{0}x^2(x^3+2)^{-2}\,dx = \int_{-1}^{0}(x^3+2)^{-2}\left(\frac{3}{3}\right)x^2\,dx = \frac{1}{3}\int_{-1}^{0}(x^3+2)^{-2}3x^2\,dx = \frac{1}{3}\int_{1}^{2}u^{-2}\,du$

Let $u = x^3 + 2.$ Then $du = 3x^2\,dx$
When $x = -1, u = 1,\quad$ when $x = 0, u = 2.$

$\displaystyle = \frac{1}{3}\cdot\left.\frac{u^{-1}}{-1}\right]_{1}^{2} = \left.-\frac{1}{3u}\right]_{1}^{2}$

$\displaystyle = -\frac{1}{6} + \frac{1}{3} = \frac{1}{6}\qquad$ (5-5)

40. $\displaystyle\int 5e^{-t}\,dt = -5\int e^{-t}(-dt) = -5\int e^{u}\,du = -5e^{u} + C = -5e^{-t} + C$
Let $u = -t.$ Then $du = -dt\qquad$ (5-2)

41. $\displaystyle\int_{1}^{e}\frac{1+t^2}{t}\,dt = \int_{1}^{e}\left(\frac{1}{t}+t\right)dt = \int_{1}^{e}\frac{1}{t}\,dt + \int_{1}^{e}t\,dt = \ln t\,\Big]_{1}^{e} + \left.\frac{1}{2}t^2\,\right]_{1}^{e} = \ln e - \ln 1 + \frac{1}{2}e^2 - \frac{1}{2}$

$\displaystyle = \frac{1}{2} + \frac{1}{2}e^2\qquad$ (5-5)

42. $\displaystyle\int xe^{3x^2}\,dx = \int e^{3x^2}\left(\frac{6}{6}\right)x\,dx = \frac{1}{6}\int e^{3x^2}6x\,dx = \frac{1}{6}\int e^{u}\,du = \frac{1}{6}e^{u} + C = \frac{1}{6}e^{3x^2} + C$

Let $u = 3x^2$ Then $du = 6x\,dx\qquad$ (5-2)

43. $\displaystyle\int_{-3}^{1}\frac{1}{\sqrt{2-x}}\,dx = -\int_{-3}^{1}\frac{1}{\sqrt{2-x}}(-dx) = -\int_{5}^{1}u^{-1/2}\,du = \int_{1}^{5}u^{-1/2}\,du = 2u^{1/2}\Big]_{1}^{5} = 2\sqrt{5} - 2 \approx 2.472$

Let $u = 2 - x.$ Then $du = -dx$
When $x = -3,\ u = 5,\quad$ when $x = 1,\ u = 1$ $\qquad$ (5-5)

44. Let $u = 1 + x^2.$ Then $du = 2x\,dx.$

$\displaystyle\int_{0}^{3}\frac{x}{1+x^2}\,dx = \int_{0}^{3}\frac{1}{1+x^2}\frac{2}{2}x\,dx = \frac{1}{2}\int_{0}^{3}\frac{1}{1+x^2}2x\,dx = \left.\frac{1}{2}\ln(1+x^2)\right|_{0}^{3}$

$\displaystyle = \frac{1}{2}\ln 10 - \frac{1}{2}\ln 1 = \frac{1}{2}\ln 10 \approx 1.151\qquad$ (5-5)

45. Let $u = 1 + x^2.$ Then $du = 2x\,dx.$

$\displaystyle\int_{0}^{3}\frac{x}{(1+x^2)^2}\,dx = \int_{0}^{3}(1+x^2)^{-2}\frac{2}{2}x\,dx = \frac{1}{2}\int_{0}^{3}(1+x^2)^{-2}2x\,dx$

$\displaystyle = \frac{1}{2}\cdot\left.\frac{(1+x^2)^{-1}}{-1}\right|_{0}^{3} = \left.\frac{-1}{2(1+x^2)}\right|_{0}^{3} = -\frac{1}{20} + \frac{1}{2} = \frac{9}{20} = 0.45\qquad$ (5-5)

46. $\int x^3 (2x^4 + 5)^5 \, dx$

Let $u = 2x^4 + 5$. Then $du = 8x^3 dx$.

$$\int x^3 (2x^4 + 5)^5 \, dx = \int (2x^4 + 5)^5 \left(\frac{8}{8}\right) x^3 \, dx = \frac{1}{8} \int (2x^4 + 5)^5 \, 8x^3 \, dx = \frac{1}{8} \int u^5 \, du = \frac{1}{8} \cdot \frac{u^6}{6} + C$$

$$= \frac{(2x^4 + 5)^6}{48} + C \qquad (5\text{-}2)$$

47. $\int \dfrac{e^{-x}}{e^{-x} + 3} \, dx = \int \dfrac{1}{e^{-x} + 3} \left(\dfrac{-1}{-1}\right) e^{-x} \, dx = -\int \dfrac{1}{u} \, du = -\ln|u| + C = -\ln|e^{-x} + 3| + C = -\ln(e^{-x} + 3) + C$

Let $u = e^{-x} + 3$.   [Note: Absolute value not needed since $e^{-x} + 3 > 0$.]

Then $du = -e^{-x} dx$.   (5-2)

48. $\int \dfrac{e^x}{(e^x + 2)^2} \, dx = \int (e^x + 2)^{-2} e^x \, dx = \int u^{-2} \, du \, dx = \dfrac{u^{-1}}{-1} + C = -(e^x + 2)^{-1} + C = \dfrac{-1}{(e^x + 2)} + C \qquad (5\text{-}2)$

Let $u = e^x + 2$. Then $du = e^x dx$

49. $\dfrac{dy}{dx} = 3x^{-1} - x^{-2}$

$$y = \int (3x^{-1} - x^{-2}) \, dx = 3 \int \frac{1}{x} \, dx - \int x^{-2} \, dx = 3 \ln|x| - \frac{x^{-1}}{-1} + C = 3 \ln|x| + x^{-1} + C$$

Given $y(1) = 5$:

$5 = 3 \ln 1 + 1 + C$ and $C = 4$

Thus, $y = 3 \ln|x| + x^{-1} + 4.$ (5-2, 5-3)

50. $\dfrac{dy}{dx} = 6x + 1$

$$f(x) = y = \int (6x + 1) \, dx = \frac{6x^2}{2} + x + C = 3x^2 + x + C$$

We have $y = 10$ when $x = 2$: $3(2)^2 + 2 + C = 10$, $C = 10 - 12 - 2 = -4$

Thus, the equation of the curve is $y = 3x^2 + x - 4.$ (5-3)

51.   (A) $f(x) = 3\sqrt{x} = 3x^{1/2}$ on $[1, 9]$

Avg. $f(x) = \dfrac{1}{9 - 1} \displaystyle\int_1^9 3x^{1/2} \, dx$

$= \dfrac{3}{8} \cdot \dfrac{x^{3/2}}{3/2} \Big|_1^9 = \dfrac{1}{4} x^{3/2} \Big|_1^9$

$= \dfrac{27}{4} - \dfrac{1}{4} = \dfrac{26}{4} = 6.5$

(B)

(5-5)

52. Let $u = \ln x$. Then $du = \dfrac{1}{x}dx$.

$$\int \frac{(\ln x)^2}{x}dx = \int (\ln x)^2\frac{1}{x}dx = \int u^2\,du = \frac{u^3}{3} + C = \frac{(\ln x)^3}{3} + C \qquad (5\text{-}2)$$

53. $\displaystyle\int x(x^3-1)^2\,dx = \int x(x^6 - 2x^3 + 1)\,dx$ (square $x^3 - 1$)

$$= \int (x^7 - 2x^4 + x)\,dx = \frac{x^8}{8} - \frac{2x^5}{5} + \frac{x^2}{2} + C \qquad (5\text{-}2)$$

54. $\displaystyle\int \frac{x}{\sqrt{6-x}}dx$

Let $u = 6 - x$. Then $x = 6 - u$ and $dx = -du$.

$$\int \frac{x}{\sqrt{6-x}}dx = -\int \frac{6-u}{u^{1/2}}du = \int (u^{1/2} - 6u^{-1/2})\,du = \frac{u^{3/2}}{3/2} - \frac{6u^{1/2}}{1/2} + C = \frac{2}{3}u^{3/2} - 12u^{1/2} + C$$

$$= \frac{2}{3}(6-x)^{3/2} - 12(6-x)^{1/2} + C \qquad (5\text{-}2)$$

55. $\displaystyle\int_0^7 x\sqrt{16-x}\,dx$. First consider the indefinite integral:

Let $u = 16 - x$. Then $x = 16 - u$ and $dx = -du$.

$$\int x\sqrt{16-x}\,dx = -\int (16-u)u^{1/2}\,du = \int (u^{3/2} - 16u^{1/2})\,du = \frac{u^{5/2}}{5/2} - \frac{16u^{3/2}}{3/2} + C = \frac{2}{5}u^{5/2} - \frac{32}{3}u^{3/2} + C$$

$$= \frac{2(16-x)^{5/2}}{5} - \frac{32(16-x)^{3/2}}{3} + C$$

$$\int_0^7 x\sqrt{16-x}\,dx = \left[\frac{2(16-x)^{5/2}}{5} - \frac{32(16-x)^{3/2}}{3}\right]_0^7 = \frac{2\cdot 9^{5/2}}{5} - \frac{32\cdot 9^{3/2}}{3} - \left(\frac{2\cdot 16^{5/2}}{5} - \frac{32\cdot 16^{3/2}}{3}\right)$$

$$= \frac{2\cdot 3^5}{5} - \frac{32\cdot 3^3}{3} - \left(\frac{2\cdot 4^5}{5} - \frac{32\cdot 4^3}{3}\right) = \frac{486}{5} - 288 - \left(\frac{2048}{5} - \frac{2048}{3}\right)$$

$$= \frac{1234}{15} \approx 82.267 \qquad (5\text{-}5)$$

56. $\displaystyle\int_1^1 (x+1)^9\,dx = 0$.   (Property 9a)   (5-4)

57. $\dfrac{dy}{dx} = 9x^2 e^{x^3}$, $f(0) = 2$

Let $u = x^3$. Then $du = 3x^2\,dx$.

$$y = \int 9x^2 e^{x^3}\,dx = 3\int e^{x^3}\cdot 3x^2\,dx = 3\int e^u\,du = 3e^u + C = 3e^{x^3} + C$$

Given $f(0) = 2$:

$2 = 3e^0 + C = 3 + C$

Hence, $C = -1$ and $y = f(x) = 3e^{x^3} - 1$.   (5-3)

58. $\dfrac{dN}{dt} = 0.06N$, $N(0) = 800$, $N > 0$

From the differential equation, $N(t) = Ce^{0.06t}$, where C is an arbitrary constant. Since $N(0) = 800$, we have

$800 = Ce^0 = C$.

Hence, $C = 800$ and $N(t) = 800e^{0.06t}$.   (5-3)

59. $N = 50(1 - e^{-0.07t})$,
$0 \le t \le 80$, $0 \le N \le 60$

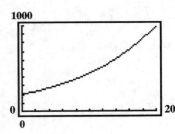

Limited growth

(5-3)

60. $p = 500e^{-0.03x}$,
$0 \le x \le 100$, $0 \le p \le 500$

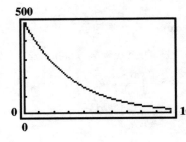

Exponential decay

(5-3)

61. $A = 200e^{0.08t}$,
$0 \le t \le 20$, $0 \le A \le 1{,}000$

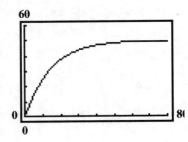

Unlimited growth

(5-3)

62. $N = \dfrac{100}{1 + 9e^{-0.3t}}$,
$0 \le t \le 25$, $0 \le N \le 100$

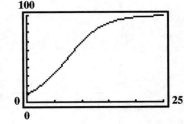

Logistic growth

(5-3)

63. $\displaystyle\int_{-0.5}^{0.6} \dfrac{1}{\sqrt{1 - x^2}}\, dx \approx 1.167$

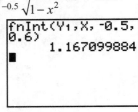

(5-5)

64. $\displaystyle\int_{-2}^{3} x^2 e^x\, dx \approx 99.074$

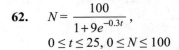

(5-5)

65. $\displaystyle\int_{0.5}^{2.5}\frac{\ln x}{x^2}\,dx \approx -0.153$

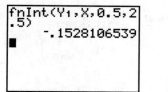

(5-5)

66. $a = 200$, $b = 600$, $n = 2$, $\Delta x = \dfrac{600-200}{2} = 200$

$L_2 = C'(200)\Delta x + C'(400)\Delta x$

$\quad = [500 + 400]200 = \$180,000$

$R_2 = C'(400)\Delta x + C'(600)\Delta x$

$\quad = [400 + 300]200 = \$140,000$

$\qquad 140,000 \le \displaystyle\int_{200}^{600} C'(x)\,dx \le 180,000$ (5-4)

67. The graph of $C'(x)$ is a straight line with y-intercept = 600 and slope = $\dfrac{300-600}{600-0} = -\dfrac{1}{2}$.

Thus, $C'(x) = -\dfrac{1}{2}x + 600$

Increase in costs:

$\displaystyle\int_{200}^{600}\left(600 - \frac{1}{2}x\right)dx = \left[600x - \frac{1}{4}x^2\right]_{200}^{600} = 270,000 - 110,000 = \$160,000$ (5-5)

68. The total change in profit for a production change from 10 units per week to 40 units per week is given by:

$\displaystyle\int_{10}^{40}\left(150 - \frac{x}{10}\right)dx = \left[150x - \frac{x^2}{20}\right]_{10}^{40} = \left(150(40) - \frac{40^2}{20}\right) - \left(150(10) - \frac{10^2}{20}\right) = 5920 - 1495 = \$4425$

(5-5)

69. $P'(x) = 100 - 0.02x$

$P(x) = \displaystyle\int (100 - 0.02x)\,dx = 100x - 0.02\frac{x^2}{2} + C = 100x - 0.01x^2 + C$

$P(0) = 0 - 0 + C = 0$

$\qquad\quad C = 0$

Thus, $P(x) = 100x - 0.01x^2$.

The profit on 10 units of production is given by: $P(10) = 100(10) - 0.01(10)^2 = \$999$ (5-3)

70. The required definite integral is:

$\displaystyle\int_0^{15}(60 - 4t)\,dt = \left[60t - 2t^2\right]_0^{15} = 60(15) - 2(15)^2 = 450$ or $450,000$ barrels

The total production in 15 years is 450,000 barrels.   (5-5)

71. Average inventory from $t = 3$ to $t = 6$:

Avg. $I(t) = \dfrac{1}{6-3}\displaystyle\int_3^6 (10 + 36t - 3t^2)\,dt = \frac{1}{3}\left[10t + 18t^2 - t^3\right]_3^6 = \frac{1}{3}[60 + 648 - 216 - (30 + 162 - 27)]$

$\qquad\quad = 109$ items   (5-5)

72. $S(x) = 8(e^{0.05x} - 1)$
Average price over the interval [40, 50]:

Avg. $S(x) = \dfrac{1}{50-40} \displaystyle\int_{40}^{50} 8(e^{0.05x} - 1)\,dx = \dfrac{8}{10}\int_{40}^{50}(e^{0.05x}-1)\,dx = \dfrac{4}{5}\left[\dfrac{e^{0.05x}}{0.05} - x\right]_{40}^{50}$

$= \dfrac{4}{5}[20e^{2.5} - 50 - (20e^2 - 40)] = 16e^{2.5} - 16e^2 - 8 \approx \$68.70$

(5-5)

73. To find the useful life, set $R'(t) = C'(t)$:
$20e^{-0.1t} = 3$

$e^{-0.1t} = \dfrac{3}{20}$

$-0.1t = \ln\left(\dfrac{3}{20}\right) \approx -1.897; \quad t = 18.97 \text{ or } 19 \text{ years}$

Total profit $= \displaystyle\int_0^{19}[R'(t)-C(t)]\,dt = \int_0^{19}(20e^{-0.1t}-3)\,dt$

$= 20\displaystyle\int_0^{19}e^{-0.1t}\,dt - \int_0^{19}3\,dt = \dfrac{20}{-0.1}\int_0^{19}e^{-0.1t}(-0.1)\,dt - \int_0^{19}3\,dt$

$= -200e^{-0.1t}\Big|_0^{19} - 3t\Big|_0^{19} = -200e^{-1.9} + 200 - 57 \approx 113.086 \text{ or } \$113,086$ (5-5)

74. $S'(t) = 4e^{-0.08t}$, $0 \le t \le 24$. Therefore,

$S(t) = \displaystyle\int 4e^{-0.08t}\,dt = \dfrac{4e^{-0.08t}}{-0.08} + C = -50e^{-0.08t} + C.$

Now, $S(0) = 0$, so
$0 = -50e^{-0.08(0)} + C = -50 + C.$

Thus, $C = 50$, and $S(t) = 50(1 - e^{-0.08t})$ gives the total sales after t months.
Estimated sales after 12 months:

$S(12) = 50(1 - e^{-0.08(12)}) = 50(1 - e^{-0.96}) \approx 31 \text{ or } \$31 \text{ million.}$

To find the time to reach \$40 million in sales, solve $40 = 50(1 - e^{-0.08t})$ for t.

$0.8 = 1 - e^{-0.08t}$

$e^{-0.08t} = 0.2$

$-0.08t = \ln(0.2)$

$t = \dfrac{\ln(0.2)}{-0.08} \approx 20 \text{ months}$ (5-3)

75. $\dfrac{dA}{dt} = -5t^{-2}$, $1 \le t \le 5$

$A = \displaystyle\int -5t^{-2}\,dt = -5\int t^{-2}\,dt = -5\cdot\dfrac{t^{-1}}{-1} + C = \dfrac{5}{t} + C$

Now $A(1) = \dfrac{5}{1} + C = 5$. Therefore, $C = 0$ and

$A(t) = \dfrac{5}{t}$

$A(5) = \dfrac{5}{5} = 1$

The area of the wound after 5 days is 1 cm^2.   (5-3)

76.   The total amount of seepage during the first four years is given by:

$$T = \int_0^4 R(t)dt = \int_0^4 \frac{1000}{(1+t)^2} dt = 1000 \int_0^4 (1+t)^{-2} dt = 1000 \frac{(1+t)^{-1}}{-1} \Big|_0^4$$

   [Let $u = 1 + t$. Then $du = dt$.] $\qquad\qquad = \frac{-1000}{1+t} \Big|_0^4 = \frac{-1000}{5} + 1000 = 800$ gallons

   (5-5)

77.   (A)   The exponential growth law applies and we have:

   $$\frac{dP}{dt} = 0.0107P, \quad P(0) = 116 \text{ (million)}$$

   Thus $P(t) = 116e^{0.0107t}$

   The year 2025: $t = 12$,   and $P(12) = 116e^{0.0107(12)} = 116e^{0.1284} \approx 132$

   Assuming that the population continues to grow at the rate 1.07% per year, the population in 2025 will be approximately 132 million.

   (B)   Time to double:
   $$116e^{0.0107t} = 232$$
   $$e^{0.0107t} = 2$$
   $$0.0107t = \ln 2$$
   $$t = \frac{\ln 2}{0.0107} \approx 65$$

   At the current growth rate it will take approximately 65 years for the population to double.

   (5-3)

78.   Let $Q = Q(t)$ be the amount of carbon-14 present in the bone at time t. Then,

   $$\frac{dQ}{dt} = -0.0001238Q \text{ and } Q(t) = Q_0 e^{-0.0001238t},$$

   where Q_0 is the amount present originally (i.e., at the time the animal died). We want to find t such that $Q(t) = 0.04Q_0$:

   $$0.04 \, Q_0 = Q_0 e^{-0.0001238t}$$
   $$e^{-0.0001238t} = 0.04$$
   $$-0.0001238t = \ln 0.04$$
   $$t = \frac{\ln 0.04}{-0.0001238} \approx 26{,}000 \text{ years} \qquad (5\text{-}3)$$

79. $N'(t) = 7e^{-0.1t}$ and $N(0) = 25$.

   $$N(t) = \int 7e^{-0.1t} \, dt = 7 \int e^{-0.1t} \, dt = \frac{7}{-0.1} \int e^{-0.1t} (-0.1) \, dt = -70e^{-0.1t} + C, \quad 0 \le t \le 15$$

   Given $N(0) = 25$: $25 = -70e^0 + C = -70 + C$

   Hence, $C = 95$ and $N(t) = 95 - 70e^{-0.1t}$. The student would be expected to type $N(15) = 95 - 70e^{-0.1(15)}$
   $= 95 - 70e^{-1.5} \approx 79$ words per minute after completing the course.   (5-3)

14   ADDITIONAL INTEGRATION TOPICS

EXERCISE 14-1

Things to remember:

1.   AREA BETWEEN TWO CURVES
     If f and g are continuous and $f(x) \geq g(x)$ over the interval
     $[a, b]$, then the area bounded by $y = f(x)$ and $y = g(x)$, for
     $a \leq x \leq b$, is given exactly by:

$$A = \int_a^b [f(x) - g(x)]dx.$$

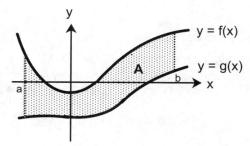

2.   GINI INDEX OF INCOME CONCENTRATION
     If $y = f(x)$ is the equation of a Lorenz curve, then the

$$\text{Gini Index} = 2\int_0^1 [x - f(x)]dx.$$

1.   The region bounded by the graphs of f and g is a rectangle with dimensions 15×10: area = 150.

3.   The region bounded by the graphs of f and g is a triangle with vertices $(0,6)$, $(5, 16)$, $(5,1)$. The base has length 15, the height is 5: area $\dfrac{1}{2}(15 \times 5) = 37.5$.

5.   The region bounded by the graphs of f and g is a trapezoid with vertices $(-1,2)$, $(2,8)$, $(2,-5)$, $(-1,-2)$. The bases have lengths 4 and 13, and height 3. Area:
     $$\frac{1}{2}(4+13)3 = 25.5.$$

7.   The region bounded by the graphs of f and g is one-eighth of the disk centered at the origin with radius 4: area: $\dfrac{1}{8}(\pi 4^2) = 2\pi$.

9. $A = \displaystyle\int_a^b g(x)dx$ 11. $A = \displaystyle\int_a^b [-h(x)]dx$

13.   Since the shaded region in Figure (c) is below the x-axis, $h(x) \leq 0$. Thus, $\displaystyle\int_a^b h(x)dx$ represents the negative of the area of the region.

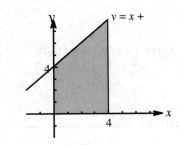

15. $y = x + 4$; $y = 0$ on $[0, 4]$

$$A = \int_0^4 (x + 4)dx = \left(\frac{1}{2}x^2 + 4x\right)\Big|_0^4 = (8 + 16) - 0 = 24$$

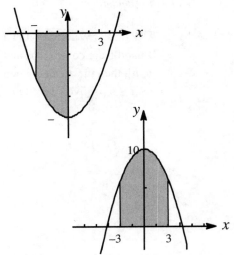

17. $y = x^2 - 20$; $y = 0$ on $[-3, 0]$

$$A = -\int_{-3}^0 (x^2 - 20)dx$$

$$= \int_{-3}^0 (20 - x^2)dx = \left(20x - \frac{1}{3}x^3\right)\Big|_{-3}^0$$

$$= 0 - (-60 + 9) = 51$$

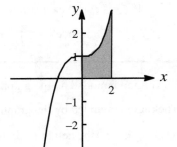

19. $y = -x^2 + 10$; $y = 0$ on $[-3, 3]$

$$A = \int_{-3}^3 (-x^2 + 10)\,dx = \left[-\frac{1}{3}x^3 + 10x\right]_{-3}^3$$

$$= (-9 + 30) - (9 - 30) = 42$$

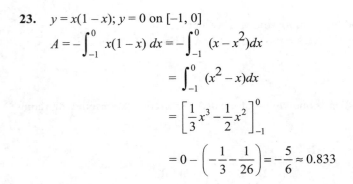

21. $y = x^3 + 1$; $y = 0$ on $[0, 2]$

$$A = \int_0^2 (x^3 + 1)dx = \left(-\frac{1}{4}x^4 + x\right)\Big|_0^2$$

$$= 4 + 2 = 6$$

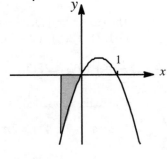

23. $y = x(1 - x)$; $y = 0$ on $[-1, 0]$

$$A = -\int_{-1}^0 x(1 - x)\,dx = -\int_{-1}^0 (x - x^2)dx$$

$$= \int_{-1}^0 (x^2 - x)dx$$

$$= \left[\frac{1}{3}x^3 - \frac{1}{2}x^2\right]_{-1}^0$$

$$= 0 - \left(-\frac{1}{3} - \frac{1}{26}\right) = -\frac{5}{6} \approx 0.833$$

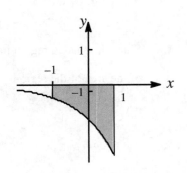

25. $y = -e^x$; $y = 0$ on $[-1, 1]$

$$A = -\int_{-1}^1 -e^x\,dx \qquad = \int_{-1}^1 e^x\,dx$$

$$= e^x\Big|_{-1}^1 = e - e^{-1}$$

$$\approx 2.350$$

27. $y = \dfrac{1}{x}$; $y = 0$ on $[1, e]$

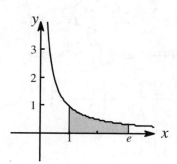

$A = \displaystyle\int_1^e \frac{1}{x}\,dx = \ln x \,\Big|_1^e$

$= \ln e - \ln 1 = 1$

29. Most equally distributed: Canada, Gini index 0.32; least equally distributed: Mexico, Gini index 0.48.

31. Most equally distributed: India, Gini index 0.37; least equally distributed: Brazil, Gini index 0.52.

33. $A = \displaystyle\int_a^b [-f(x)]\,dx$

35. $A = \displaystyle\int_b^c f(x)\,dx + \int_c^d [-f(x)]\,dx$

37. $A = \displaystyle\int_c^d [f(x) - g(x)]\,dx$

39. $A = \displaystyle\int_a^b [f(x) - g(x)]\,dx + \int_b^c [g(x) - f(x)]\,dx$

41. Find the x-coordinates of the points of intersection of the two curves on $[a, d]$ by solving the equation $f(x) = g(x)$, $a \le x \le d$. This gives $x = b$ and $x = c$. Then note that $f(x) \ge g(x)$ on $[a, b]$, $g(x) \ge f(x)$ on $[b, c]$ and $f(x) \ge g(x)$ on $[c, d]$.

Thus,

$$\text{Area} = \int_a^b [f(x) - g(x)]\,dx + \int_b^c [g(x) - f(x)]\,dx + \int_c^d [f(x) - g(x)]\,dx$$

43. $A = A_1 + A_2 = \displaystyle\int_{-2}^0 -x\,dx + \int_0^1 -(-x)\,dx$

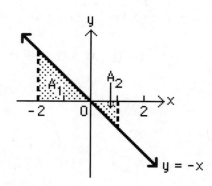

$= -\displaystyle\int_{-2}^0 x\,dx + \int_0^1 x\,dx$

$= -\dfrac{x^2}{2} \,\Big|_{-2}^0 + \dfrac{x^2}{2} \,\Big|_0^1$

$= -\left(0 - \dfrac{(-2)^2}{2}\right) + \left(\dfrac{1^2}{2} - 0\right) = 2 + \dfrac{1}{2} = \dfrac{5}{2} = 2.5$

45. $A = A_1 + A_2 = \int_0^2 -(x^2 - 4)dx + \int_2^3 (x^2 - 4)dx$

$$= \int_0^2 (4 - x^2)dx + \int_2^3 (x^2 - 4)dx$$

$$= \left(4x - \frac{x^3}{3}\right)\Big|_0^2 + \left(\frac{x^3}{3} - 4x\right)\Big|_2^3$$

$$= \left(8 - \frac{8}{3}\right) + \left(\frac{27}{3} - 12\right) - \left(\frac{8}{3} - 8\right)$$

$$= 13 - \frac{16}{3} = \frac{39}{3} - \frac{16}{3} = \frac{23}{3} \approx 7.667$$

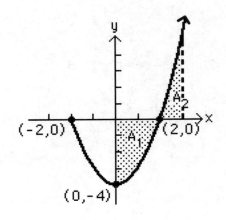

47. $A = A_1 + A_2 = \int_{-2}^0 (x^2 - 3x)dx + \int_0^2 -(x^2 - 3x)dx$

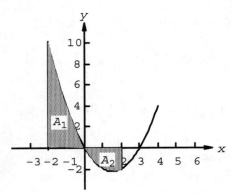

$$= \int_{-2}^0 (x^2 - 3x)dx + \int_0^2 (3x - x^2)dx$$

$$= \left(\frac{1}{3}x^3 - \frac{3}{2}x^2\right)\Big|_{-2}^0 + \left(\frac{3}{2}x^2 - \frac{1}{3}x^3\right)\Big|_0^2$$

$$= 0 - \left(-\frac{8}{3} - 6\right) + \left(6 - \frac{8}{3}\right) - 0 = 12$$

49. $A = \int_{-1}^2 [12 - (-2x + 8)]dx = \int_{-1}^2 (2x + 4)dx$

$$= \left(\frac{2x^2}{2} + 4x\right)\Big|_{-1}^2 = (x^2 + 4x)\Big|_{-1}^2$$

$$= (4 + 8) - (1 - 4) = 12 + 3 = 15$$

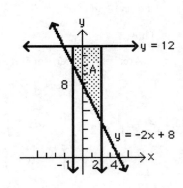

51. $A = \int_{-2}^2 (12 - 3x^2)dx = \left(12x - \frac{3x^3}{3}\right)\Big|_{-2}^2 = (12x - x^3)\Big|_{-2}^2$

$$= (12 \cdot 2 - 2^3) - [12 \cdot (-2) - (-2)^3]$$

$$= 16 - (-16) = 32$$

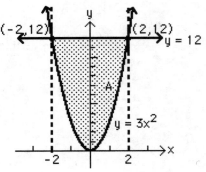

53. $(3, -5)$ and $(-3, -5)$ are the points of intersection.

$$A = \int_{-3}^{3} [4 - x^2 - (-5)]dx$$

$$= \int_{-3}^{3} (9 - x^2)dx = \left(9x - \frac{x^3}{3} \right) \Bigg|_{-3}^{3}$$

$$= \left(9 \cdot 3 - \frac{3^3}{3} \right) - \left(9(-3) - \frac{(-3)^3}{3} \right)$$

$$= 18 + 18 = 36$$

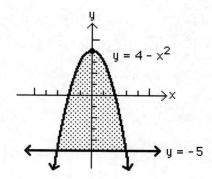

55. $A = \int_{-1}^{2} [(x^2 + 1) - (2x - 2)]dx$

$$= \int_{-1}^{2} (x^2 - 2x + 3)dx = \left(\frac{x^3}{3} - x^2 + 3x \right) \Bigg|_{-1}^{2}$$

$$= \left(\frac{8}{3} - 4 + 6 \right) - \left(-\frac{1}{3} - 1 - 3 \right)$$

$$= 3 - 4 + 6 + 1 + 3 = 9$$

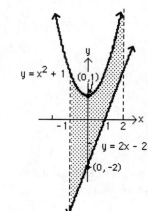

57. $A = \int_{1}^{2} \left[e^{0.5x} - \left(-\frac{1}{x} \right) \right] dx$

$$= \int_{1}^{2} \left(e^{0.5x} + \frac{1}{x} \right) dx$$

$$= \left(\frac{e^{0.5x}}{0.5} + \ln|x| \right) \Bigg|_{1}^{2}$$

$$= 2e + \ln 2 - 2e^{0.5} \approx 2.832$$

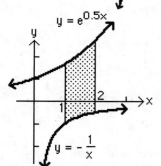

59. $y = \sqrt{9 - x^2}$; $y = 0$ on $[-3, 3]$

area $= \int_{-3}^{3} \sqrt{9 - x^2}\, dx =$ area of semicircle of radius 3

$$= \frac{9}{2}\pi \approx 14.137$$

61. $y = -\sqrt{16 - x^2}$; $y = 0$ on $[0, 4]$

area $= \int_{0}^{4} \sqrt{16 - x^2}\, dx =$ area of quarter circle of radius 4

$$= \frac{1}{4} \cdot 16\pi = 4\pi \approx 12.566$$

63. $y = -\sqrt{4 - x^2}$; $y = \sqrt{4 - x^2}$, on $[-2, 2]$

area $= \int_{-2}^{2} [\sqrt{4 - x^2} - (-\sqrt{4 - x^2})]dx = \int_{-2}^{2} 2\sqrt{4 - x^2} \, dx = 2\int_{-2}^{2} \sqrt{4 - x^2} \, dx$

$= 2$ area of semi- circle of radius 2 = area of circle of radius $2 = 4\pi \approx 12.566$

65. The graphs of $y = e^x$ and $y = e^{-x}$, $0 \le x \le 4$, are shown at the right.

$A = \int_{0}^{4} (e^x - e^{-x})dx = (e^x + e^{-x})\Big|_{0}^{4}$

$$= e^4 + e^{-4} - (1 + 1)$$

$$\approx 52.616$$

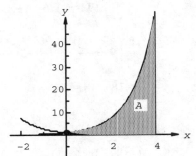

67. The graphs are given at the right. To find the points of intersection, solve:

$$x^3 = 4x$$
$$x^3 - 4x = 0$$
$$x(x^2 - 4) = 0$$
$$x(x + 2)(x - 2) = 0$$

Thus, the points of intersection are

$(-2, -8)$, $(0, 0)$, and $(2, 8)$.

$A = A_1 + A_2 = \int_{-2}^{0} (x^3 - 4x)dx + \int_{0}^{2} (4x - x^3)dx$

$$= \left(\frac{x^4}{4} - 2x^2\right)\Big|_{-2}^{0} + \left(2x^2 - \frac{x^4}{4}\right)\Big|_{0}^{2}$$

$$= 0 - \left[\frac{(-2)^4}{4} - 2(-2)^2\right] + \left[2(2^2) - \frac{2^4}{4}\right] - 0$$

$$= -4 + 8 + 8 - 4 = 8$$

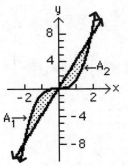

69. The graphs are given at the right.
To find the points of intersection, solve:

$$x^3 - 3x^2 - 9x + 12 = x + 12$$
$$x^3 - 3x^2 - 10x = 0$$
$$x(x^2 - 3x - 10) = 0$$
$$x(x - 5)(x + 2) = 0$$
$$x = -2, x = 0, x = 5$$

Thus, $(-2, 10)$, $(0, 12)$, and $(5, 17)$ are the points of intersection.

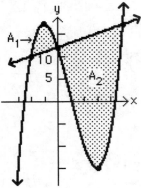

$A = A_1 + A_2$

$$= \int_{-2}^{0} [x^3 - 3x^2 - 9x + 12 - (x + 12)]dx + \int_{0}^{5} [x + 12 - (x^3 - 3x^2 - 9x + 12)]dx$$

$$= \int_{-2}^{0} (x^3 - 3x^2 - 10x)dx + \int_{0}^{5} (-x^3 + 3x^2 + 10x)dx$$

$$= \left(\frac{x^4}{4} - x^3 - 5x^2\right)\Bigg|_{-2}^{0} + \left(-\frac{x^4}{4} + x^3 + 5x^2\right)\Bigg|_{0}^{5} = -\left[\frac{(-2)^4}{4} - (-2)^3 - 5(-2)^2\right] + \left(\frac{-5^4}{4} + 5^3 + 5 \cdot 5^2\right)$$

$$= 8 + \frac{375}{4} = \frac{407}{4} = 101.75$$

71. The graphs are given below. The x-coordinates of the points of intersection are: $x_1 = -2$, $x_2 = 0.5$, $x_3 = 2$

$$A = A_1 + A_2$$

$$= \int_{-2}^{0.5} [(x^3 - x^2 + 2) - (-x^3 + 8x - 2)]\,dx + \int_{0.5}^{2} [(-x^3 + 8x - 2) - (x^3 - x^2 + 2)]dx$$

$$= \int_{-2}^{0.5} (2x^3 - x^2 - 8x + 4)dx + \int_{0.5}^{2} (-2x^3 + x^2 + 8x - 4)dx$$

$$= \left(\frac{1}{2}x^4 - \frac{1}{3}x^3 - 4x^2 + 4x\right)\Bigg|_{-2}^{0.5} + \left(-\frac{1}{2}x^4 + \frac{1}{3}x^3 + 4x^2 - 4x\right)\Bigg|_{0.5}^{2}$$

$$= \left(\frac{1}{32} - \frac{1}{24} - 1 + 2\right) - \left(8 + \frac{8}{3} - 16 - 8\right) + \left(-8 + \frac{8}{3} + 16 - 8\right) - \left(-\frac{1}{32} + \frac{1}{24} + 1 - 2\right)$$

$$= 18 + \frac{1}{16} - \frac{1}{12} \approx 17.979$$

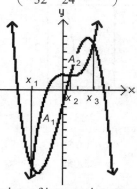

73. The graphs are given at the right. The x-coordinates of the points of intersection are: $x_1 \approx -1.924$, $x_2 \approx 1.373$

$$A = \int_{-1.924}^{1.373} [(3 - 2x) - e^{-x}]\,dx$$

$$= \left[3x - x^2 + e^{-x}\right]_{-1.924}^{1.373}$$

$$\approx 2.487 - (-2.626) = 5.113$$

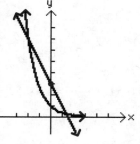

75. The graphs are given at the right. The x-coordinates of the points of intersection are: $x_1 \approx -2.247$, $x_2 \approx 0.264$, $x_3 \approx 1.439$

$$A = A_1 + A_2 = \int_{-2.247}^{0.264} [e^x - (5x - x^3)]dx + \int_{0.264}^{1.439} [5x - x^3 - e^x]dx$$

$$= \left(e^x - \frac{5}{2}x^2 + \frac{1}{4}x^4\right)\Bigg|_{-2.247}^{0.264} + \left(\frac{5}{2}x^2 - \frac{1}{4}x^4 - e^x\right)\Bigg|_{0.264}^{1.439}$$

$$\approx (1.129) - (-6.144) + (-0.112) - (-1.129) = 8.290$$

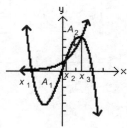

77. $y = e^{-x}; y = \sqrt{\ln x}$; $2 \le x \le 5$

The graphs of $y_1 = e^{-x}$ and $y_2 = \sqrt{\ln x}$ are

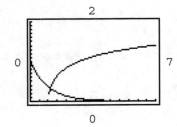   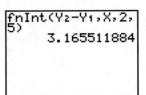

Thus, $A = \int_2^5 (\sqrt{\ln x} - e^{-x})dx \approx 3.166$

79. $y = e^{x^2}; y = x + 2$

The graphs of $y_1 = e^{x^2}$ and $y_2 = x + 2$ are

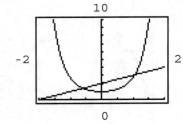   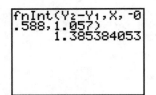

The curves intersect at

$\begin{array}{l} x \approx -0.588 \\ y \approx 1.412 \end{array}$ and $\begin{array}{l} x \approx 1.057 \\ y \approx 3.057 \end{array}$ $\int_{-0.588}^{1.057} (x + 2 - e^{x^2})dx \approx 1.385$

81. Solve $2\int_0^1 (x - x^c)\,dx = 0.52$ for c.

$$2\int_0^1 (x - x^c)\,dx = 2\left[\frac{x^2}{2} - \frac{x^{c+1}}{c+1}\right]_0^1 = 2\left[\frac{1}{2} - \frac{1}{c+1}\right] = 1 - \frac{2}{c+1}$$

$$1 - \frac{2}{c+1} = 0.52, \quad \frac{2}{c+1} = 1 - 0.52 = 0.48, \quad c+1 = \frac{2}{0.48} \approx 4.17, \quad c = 3.17$$

83. Solve $2\int_0^1 (x - x^c)\,dx = 0.29$ for c.

$$2\int_0^1 (x - x^c)\,dx = 2\left[\frac{x^2}{2} - \frac{x^{c+1}}{c+1}\right]_0^1 = 2\left[\frac{1}{2} - \frac{1}{c+1}\right] = 1 - \frac{2}{c+1}$$

$$1 - \frac{2}{c+1} = 0.29, \quad \frac{2}{c+1} = 1 - 0.29 = 0.71, \quad c+1 = \frac{2}{0.71} \approx 2.82, \quad c = 1.82$$

85. $\displaystyle\int_5^{10} R(t)dt = \int_5^{10}\left(\frac{100}{t+10}+10\right)dt = 100\int_5^{10}\frac{1}{t+10}\,dt + \int_5^{10}10dt$

$\qquad\qquad = 100\ln(t+10)\Big|_5^{10} + 10t\Big|_5^{10} = 100\ln 20 - 100\ln 15 + 10(10-5)$

$\qquad\qquad = 100\ln 20 - 100\ln 15 + 50 \approx 79$

The total production from the end of the fifth year to the end of the tenth year is approximately 79 thousand barrels.

87. To find the useful life, set $R'(t) = C'(t)$ and solve for t:

$9e^{-0.3t} = 2$

$e^{-0.3t} = \dfrac{2}{9}$

$-0.3t = \ln\dfrac{2}{9}$

$-0.3t \approx -1.5$

$\quad t \approx 5$ years

$\displaystyle\int_0^5 [R'(t) - C'(t)]\,dt = \int_0^5 [9e^{-0.3t} - 2]\,dt = 9\int_0^5 e^{-0.3t}dt - \int_0^5 2dt = \frac{9}{-0.3}e^{-0.3t}\Big|_0^5 - 2t\Big|_0^5$

$\qquad\qquad\qquad\qquad\qquad\qquad = -30e^{-1.5} + 30 - 10 = 20 - 30e^{-1.5} \approx 13.306$

The total profit over the useful life of the game is approximately \$13,306.

89. For 1935: $f(x) = x^{2.4}$

Gini Index $= 2\displaystyle\int_0^1 [x - f(x)]\,dx = 2\int_0^1 (x - x^{2.4})\,dx = 2\left(\frac{x^2}{2} - \frac{x^{3.4}}{3.4}\right)\Big|_0^1 = 2\left(\frac{1}{2} - \frac{1}{3.4}\right) \approx 0.412$

For 1947: $g(x) = x^{1.6}$

Gini Index $= 2\displaystyle\int_0^1 [x - g(x)]\,dx = 2\int_0^1 (x - x^{1.6})dx = 2\left(\frac{x^2}{2} - \frac{x^{2.6}}{2.6}\right)\Big|_0^1 = 2\left(\frac{1}{2} - \frac{1}{2.6}\right) \approx 0.231$

Interpretation: Income was more equally distributed in 1947.

91. For 1963: $f(x) = x^{10}$

Gini Index $= 2\displaystyle\int_0^1 [x - f(x)]\,dx = 2\int_0^1 (x - x^{10})\,dx = 2\left(\frac{x^2}{2} - \frac{x^{11}}{11}\right)\Big|_0^1 = 2\left(\frac{1}{2} - \frac{1}{11}\right) \approx 0.818$

For 1983: $g(x) = x^{12}$

Gini Index $= 2\displaystyle\int_0^1 [x - g(x)]\,dx = 2\int_0^1 (x - x^{12})\,dx = 2\left(\frac{x^2}{2} - \frac{x^{13}}{13}\right)\Big|_0^1 = 2\left(\frac{1}{2} - \frac{1}{13}\right) \approx 0.846$

Interpretation: Total assets were less equally distributed in 1983.

93.   (A)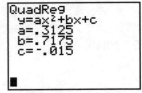

Lorenz curve:

$$f(x) = 0.3125x^2 + 0.7175x - 0.015.$$

(B)   Gini Index:

$$2\int_0^1 [x - f(x)]dx \approx 0.104$$

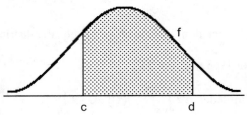

95. $W(t) = \int_0^{10} W'(t)\,dt = \int_0^{10} 0.3e^{0.1t}dt = 0.3\int_0^{10} e^{0.1t}\,dt = \dfrac{0.3}{0.1}e^{0.1t}\big|_0^{10} = 3e^{0.1t}\big|_0^{10} = 3e - 3 \approx 5.15$

Total weight gain during the first 10 hours is approximately 5.15 grams.

97. $V = \int_2^4 \dfrac{15}{t}\,dt = 15\int_2^4 \dfrac{1}{t}\,dt = 15\ln t\big|_2^4 = 15\ln 4 - 15\ln 2 = 15\ln\left(\dfrac{4}{2}\right) = 15\ln 2 \approx 10$

Average number of words learned from $t = 2$ to $t = 4$ is 10.

EXERCISE 14-2

Things to remember:

1.   PROBABILITY DENSITY FUNCTION

A function f which satisfies the following three conditions:

a. $f(x) \geq 0$ for all real x.

b.   The area under the graph of f over the interval $(-\infty, \infty)$ is exactly 1.

c.   If $[c, d]$ is a subinterval of $(-\infty, \infty)$, then the probability that the outcome x of an experiment will be in the interval $[c, d]$, denoted Probability $(c \leq x \leq d)$, is given by

$$\text{Probability } (c \leq x \leq d) = \int_c^d f(x)dx$$

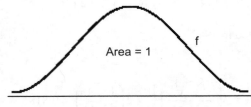

$$\int_c^d f(x)dx = \text{Probability } (c \leq x \leq d)$$

2. TOTAL INCOME FOR A CONTINUOUS INCOME STREAM

If $f(t)$ is the rate of flow of a continuous income stream, then the TOTAL INCOME produced during the time period from
$t = a$ to $t = b$ is:

$$\text{Total income} = \int_a^b f(t)dt$$

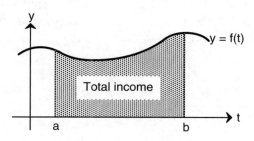

3. FUTURE VALUE OF A CONTINUOUS INCOME STREAM

If $f(t)$ is the rate of flow of a continuous income stream,
$0 \le t \le T$, and if the income is continuously invested at a rate r, compounded continuously, then the FUTURE VALUE, FV, at the end of T years is given by:

$$FV = \int_0^T f(t)e^{r(T-t)}dt = e^{rT}\int_0^T f(t)e^{-rt}dt$$

The future value of a continuous income stream is the total value of all money produced by the continuous income stream (income and interest) at the end of T years.

4. CONSUMERS' SURPLUS

If $(\overline{x}, \overline{p})$ is a point on the graph of the price-demand equation $p = D(x)$ for a particular product, then the CONSUMERS' SURPLUS, CS, at a price level of $\overline{p}$ is

$$CS = \int_0^{\overline{x}} [D(x) - \overline{p}]dx$$

which is the area between $p = \overline{p}$ and
$p = D(x)$ from $x = 0$ to $x = \overline{x}$.

Consumers' surplus represents the total savings to consumers who are willing to pay more than $\overline{p}$ for the product but are still able
to buy the product for $\overline{p}$.

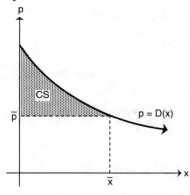

5. PRODUCERS' SURPLUS

If $(\overline{x}, \overline{p})$ is a point on the graph of the price-supply equation $p = S(x)$, then the PRODUCERS' SURPLUS, PS, at a price level of $\overline{p}$ is

$$PS = \int_0^{\overline{x}} [\overline{p} - S(x)]dx$$

which is the area between $p = \overline{p}$ and $p = S(x)$ from $x = 0$ to
$x = \overline{x}$

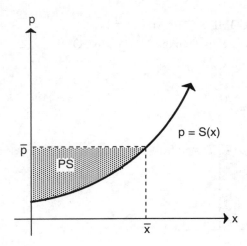

Producers' surplus represents the total gain to producers who are willing to supply units at a lower price than $\overline{p}$ but are still able to supply units at $\overline{p}$.

6.   EQUILIBRIUM PRICE AND EQUILIBRIUM QUANTITY

If $p = D(x)$ and $p = S(x)$ are the price-demand and the price-supply equations, respectively, for a product and if $(\overline{x}, \overline{p})$ is the point of intersection of these equations, then $\overline{p}$ is called the EQUILIBRIUM PRICE and $\overline{x}$ is called the EQUILIBRIUM QUANTITY.

---

1. $f(t) = e^{5(4-t)} = e^{20-5t} = e^{20} e^{-5t}, \quad b = 20, \quad c = -5$

3. $f(t) = e^{0.04(8-t)} = e^{0.32-0.04t} = e^{0.32} e^{-0.04t}, \quad b = 0.32, \quad c = -0.04$

5. $f(t) = e^{0.05t} e^{0.08(20-t)} = e^{0.05t} e^{1.6-0.08t} = e^{0.05t} e^{1.6} e^{-0.08t} = e^{1.6} e^{-0.03t}, \quad b = 1.6, \quad c = -0.03$

7. $f(t) = e^{0.09t} e^{0.07(25-t)} = e^{0.09t} e^{1.75-0.07t} = e^{0.09t} e^{1.75} e^{-0.07t} = e^{1.75} e^{0.02t}, \quad b = 1.75, \quad c = 0.02$

9. $\displaystyle\int_0^8 e^{0.06(8-t)} dt = \int_0^8 e^{0.48} \cdot e^{-0.06t} dt = e^{0.48} \int_0^8 e^{-0.06t} dt = e^{0.48} \left(-\frac{1}{0.06} e^{-0.06t} \right) \Big|_0^8$

$$= -\frac{e^{0.48}}{0.06} [e^{-0.48} - 1] = \frac{e^{0.48} - 1}{0.06} \approx 10.27$$

11. $\displaystyle\int_0^{20} e^{0.08t} e^{0.12(20-t)} dt = \int_0^{20} e^{0.08t} \cdot e^{2.4} \cdot e^{-0.12t} dt = e^{2.4} \int_0^{20} e^{-0.04t} dt$

$$= e^{2.4} \left(-\frac{1}{0.04} e^{-0.04t} \right) \Big|_0^{20} = -\frac{e^{2.4}}{0.04} (e^{-0.8} - 1) = \frac{-e^{1.6} + e^{2.4}}{0.04} \approx 151.75$$

13. $\displaystyle\int_0^{30} 500 e^{0.02t} e^{0.09(30-t)} dt = 500 \int_0^{30} e^{0.02t} \cdot e^{2.7} \cdot e^{-0.09t} dt = 500 e^{2.7} \int_0^{30} e^{-0.07t} dt$

$$= 500 e^{2.7} \left(-\frac{1}{0.07} e^{-0.07t} \right) \Big|_0^{30} = -\frac{500 e^{2.7}}{0.07} (e^{-2.1} - 1)$$

$$= \frac{500(e^{2.7} - e^{0.6})}{0.07} \approx 93{,}268.66$$

15. (A) $\displaystyle\int_0^8 e^{0.07(8-t)}dt = \int_0^8 e^{0.56-0.07t}\,dt = \int_0^8 e^{0.56}\cdot e^{-0.07t}\,dt$

$\displaystyle = e^{0.56}\int_0^8 e^{-0.07t}\,dt = \frac{e^{0.56}}{-0.07}e^{-0.07t}\,\Big|_0^8 = -\frac{e^{0.56}}{0.07}[e^{-0.56}-1] \approx 10.72$

(B) $\displaystyle\int_0^8 (e^{0.56}-e^{0.07t})\,dt = (e^{0.56})t\,\Big|_0^8 - \frac{e^{0.07t}}{0.07}\,\Big|_0^8 = 8e^{0.56} - \frac{1}{0.07}[e^{0.56}-1] \approx 3.28$

(C) $\displaystyle e^{0.56}\int_0^8 e^{-0.07t}\,dt \approx 10.72$ as in (A)

17.

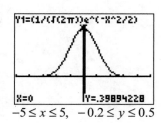

$-5 \le x \le 5, \quad -0.2 \le y \le 0.5$

19.

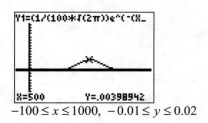

$-100 \le x \le 1000, \quad -0.01 \le y \le 0.02$

21. (A) $\displaystyle\int_{-1}^1 \frac{1}{\sqrt{2\pi}}e^{-x^2/2}\,dx \approx 0.6827$ (B) $\displaystyle\int_{-2}^2 \frac{1}{\sqrt{2\pi}}e^{-x^2/2}\,dx \approx 0.9545$

(C) $\displaystyle\int_{-3}^3 \frac{1}{\sqrt{2\pi}}e^{-x^2/2}\,dx \approx 0.9973$

23. (A) $\displaystyle\int_{400}^{600} \frac{1}{100\sqrt{2\pi}}e^{-(x-500)^2/20,000}\,dx \approx 0.6827$ (B) $\displaystyle\int_{300}^{700} \frac{1}{100\sqrt{2\pi}}e^{-(x-500)^2/20,000}\,dx \approx 0.9545$

(C) $\displaystyle\int_{200}^{800} \frac{1}{100\sqrt{2\pi}}e^{-(x-500)^2/20,000}\,dx \approx 0.9973$

25. $f(x) = \begin{cases} \dfrac{2}{(x+2)^2}, & x \ge 0 \\[2mm] 0 & x < 0 \end{cases}$

(A) Probability $(0 \le x \le 6) = \displaystyle\int_0^6 f(x)dx = \int_0^6 \frac{2}{(x+2)^2}\,dx = 2\frac{(x+2)^{-1}}{-1}\,\Big|_0^6 = \frac{-2}{(x+2)}\,\Big|_0^6$

$\displaystyle = -\frac{1}{4} + 1 = \frac{3}{4} = 0.75$

Thus, Probability $(0 \le x \le 6) = 0.75$

(B) Probability $(6 \le x \le 12) = \displaystyle\int_6^{12} f(x)dx = \int_6^{12} \frac{2}{(x+2)^2}\,dx = \frac{-2}{x+2}\,\Big|_6^{12} = -\frac{1}{7} + \frac{1}{4} = \frac{3}{28} \approx 0.11$

(C)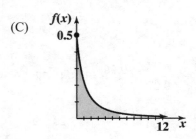

27. We want to find d such that

Probability $(0 \leq x \leq d) = \int_0^d f(x)\,dx = 0.8$:

$$\int_0^d f(x)\,dx = \int_0^d \frac{2}{(x+2)^2}\,dx = -\frac{2}{x+2}\bigg|_0^d = \frac{-2}{d+2} + 1 = \frac{d}{d+2}$$

Now, $\dfrac{d}{d+2} = 0.8$

$$d = 0.8d + 1.6$$
$$0.2d = 1.6$$
$$d = 8 \text{ years}$$

29. $f(t) = \begin{cases} 0.01e^{-0.01t} & \text{if } t \geq 0 \\ 0 & \text{otherwise} \end{cases}$

(A) Since t is in months, the probability of failure during the warranty period of the first year is

Probability $(0 \leq t \leq 12) = \int_0^{12} f(t)\,dt = \int_0^{12} 0.01e^{-0.01t}\,dt = \frac{0.01}{-0.01}e^{-0.01t}\bigg|_0^{12}$

$$= -1(e^{-0.12} - 1) \approx 0.11$$

(B) Probability $(12 \leq t \leq 24) = \int_{12}^{24} 0.01e^{-0.01t}\,dt = -1e^{-0.01t}\big|_{12}^{24} = -1(e^{-0.24} - e^{-0.12}) \approx 0.10$

31. Probability $(0 \leq t \leq \infty) = 1 = \int_0^\infty f(t)\,dt$. But, $\int_0^\infty f(t)\,dt = \int_0^{12} f(t)\,dt + \int_{12}^\infty f(t)\,dt$

Thus, Probability $(t \geq 12) = 1 - $ Probability $(0 \leq t \leq 12) \approx 1 - 0.11 = 0.89$

33. $f(t) = 2500$

Total income $= \int_0^5 2500\,dt = 2500t\big|_0^5 = \$12{,}500$

35.

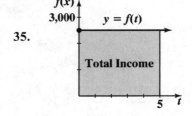

If $f(t)$ is the rate of flow of a continuous income stream, then the total income produced from 0 to 5 years is the area under the curve $y = f(t)$ from $t = 0$ to $t = 5$.

37. $f(t) = 400e^{0.05t}$

Total income $= \int_0^3 400e^{0.05t}\,dt = \dfrac{400}{0.05}e^{0.05t}\Big|_0^3 = 8000(e^{0.15} - 1) \approx \$1295$

39.

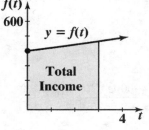

If $f(t)$ is the rate of flow of a continuous income stream, then the total income produced from 0 to 3 years is the area under the curve $y = f(t)$ from $t = 0$ to $t = 3$.

41. $f(t) = 2,000e^{0.05t}$

The amount in the account after 40 years is given by:

$\int_0^{40} 2,000e^{0.05t}\,dt = 40,000e^{0.05t}\Big|_0^{40} = 295,562.24 - 40,000 \approx \$255,562$

Since $\$2,000 \times 40 = \$80,000$ was deposited into the account, the interest earned is:
$\$255,562 - \$80,000 = \$175,562$

43. $f(t) = 1,650e^{-0.02t},\ r = 0.0325,\ T = 4.$

$FV = e^{0.0325(4)}\int_0^4 1,650e^{-0.02t}e^{-0.0325t}\,dt = 1,650e^{0.13}\int_0^4 e^{-0.0525t}\,dt$

$= 1,650e^{0.13}\left[\dfrac{e^{-0.0525t}}{-0.0525}\right]_0^4 = 31,428.57(e^{0.13} - e^{-0.08}) \approx \$6,779.52$.

45. Total Income $= \int_0^4 1,650e^{-0.02t}\,dt = \dfrac{1650}{-0.02}e^{-0.02t}\Big]_0^4 = -82,500(e^{-0.08} - 1) \approx \$6,342.90$

From Problem 43,
Interest earned $= \$6,779.52 - \$6,342.90 = \$436.62$

47. Clothing store: $f(t) = 12,000,\ r = 0.04,\ T = 5.$

$FV = e^{0.04(5)}\int_0^5 12,000e^{-0.04t}\,dt = 12,000e^{0.2}\int_0^5 e^{-0.04t}\,dt$

$= 12,000e^{0.2}\dfrac{e^{-0.04t}}{-0.04}\Big]_0^5 = -300,000e^{0.2}\left(e^{-0.2} - 1\right) \approx \$66,420.83$

Computer store: $g(t) = 10,000e^{0.05t},\ r = 0.04,\ T = 5.$

$FV = e^{0.04(5)}\int_0^5 10,000e^{0.05t}e^{-0.04t}\,dt = 10,000e^{0.2}\int_0^5 e^{0.01t}\,dt$

$= 10,000e^{0.2}\left[\dfrac{e^{0.01t}}{0.01}\right]_0^5 = 1,000,000e^{0.2}\left(e^{0.05} - 1\right) \approx \$62,622.66.$

The clothing store is the better investment.

49. Bond: $P = \$10,000,\ r = 0.0375,\ t = 5.$
$FV = 10,000e^{0.0375(5)} = 10,000e^{0.1875} \approx \$12,062.30$
Business: $f(t) = 2150,\ r = 0.0375,\ T = 5.$

$$FV = e^{0.0375(5)} \int_0^5 2150 e^{-0.0375t} dt = 2150 e^{0.1875} \int_0^5 e^{-0.0375t} dt$$

$$= 2150 e^{0.1875} \left[\frac{e^{-0.0375t}}{-0.0375} \right]_0^5 = -57,333.33 e^{0.1875} \left(e^{-0.1875} - 1 \right) = 57,333,33 \left(e^{0.1875} - 1 \right) \approx \$11,823.87.$$

The bond is the better investment.

51. $f(t) = 9,000, \ r = 0.0695, \ T = 8.$

$$FV = e^{0.0695(8)} \int_0^8 9000 e^{-0.0695t} \, dx = 9000 e^{0.556} \int_0^8 e^{-0.0695t} = \frac{9000 e^{0.556}}{-0.0695} e^{-0.0695t} \Bigg]_0^8$$

$$\approx -225,800.78 (e^{-0.556} - 1) \approx \$96,304.$$

The relationship between present value (PV) and future value (FV) at a continuously compounded interest rate r (expressed as a decimal) for t years is:

$$FV = PVe^{rt} \ \text{ or } \ PV = FVe^{-rt}$$

Thus, we have:

$$PV = 96,304 e^{-0.0695(8)} = 96,304 e^{-0.556} \approx \$55,230$$

53. $f(t) = k$, rate r (expressed as a decimal), years T:

$$FV = e^{rT} \int_0^T k e^{-rt} dt = k e^{rT} \int_0^T e^{-rt} dt = \frac{k e^{rT}}{-r} e^{-rt} \Big|_0^T \ = -\frac{k}{r} e^{rT} (e^{-rT} - 1) = \frac{k}{r} (e^{rT} - 1)$$

55. $D(x) = 400 - \dfrac{1}{20} x, \ \overline{p} = 150$

First, find $\overline{x}$: $150 = 400 - \dfrac{1}{20} \overline{x}$

$$\overline{x} = 5000$$

$$CS = \int_0^{5000} \left[400 - \frac{1}{20} x - 150 \right] dx = \int_0^{5000} \left(250 - \frac{1}{20} x \right) dx = \left(250x - \frac{1}{40} x^2 \right) \Bigg|_0^{5000} = \$625,000$$

57.

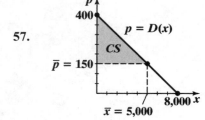

The shaded area is the consumers' surplus and represents the total savings to consumers who are willing to pay more than $150 for a product but are still able to buy the product for $150.

59. $p = S(x) = 10 + 0.1x + 0.0003x^2, \ \overline{p} = 67.$

First find $\overline{x}$: $67 = 10 + 0.1 \overline{x} + 0.0003 \overline{x}^2$

$$0.0003 \overline{x}^2 + 0.1 \overline{x} - 57 = 0$$

$$\overline{x} = \frac{-0.1 + \sqrt{0.01 + 0.0684}}{0.0006} = \frac{-0.1 + 0.28}{0.0006} = 300$$

$$PS = \int_0^{300} [67 - (10 + 0.1x + 0.0003x^2)] \, dx = \int_0^{300} (57 - 0.1x - 0.0003x^2) \, dx$$

$$= (57x - 0.05x^2 - 0.0001x^3) \Big|_0^{300} = \$9,900$$

61.

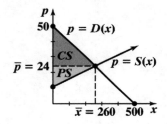

The area of the region PS is the producers' surplus and represents the total gain to producers who are willing to supply units at a lower price than \$67 but are still able to supply the product at \$67.

63. $p = D(x) = 50 - 0.1x;\ \ p = S(x) = 11 + 0.05x$

Equilibrium price: $D(x) = S(x)$

$$50 - 0.1x = 11 + 0.05x$$
$$39 = 0.15x$$
$$x = 260$$

Thus, $\bar{x} = 260$ and $\bar{p} = 50 - 0.1(260) = 24$.

$$CS = \int_0^{260} [(50 - 0.1x) - 24]\, dx = \int_0^{260} (26 - 0.1x)dx = (26x - 0.05x^2)\Big|_0^{260} = \$3{,}380$$

$$PS = \int_0^{260} [24 - (11 + 0.05x)]\, dx = \int_0^{260} [13 - 0.05x]\, dx = (13x - 0.025x^2)\Big|_0^{260} = \$1{,}690$$

65. $D(x) = 80e^{-0.001x}$ and $S(x) = 30e^{0.001x}$

Equilibrium price: $D(x) = S(x)$

$$80e^{-0.001x} = 30e^{0.001x}$$
$$e^{0.002x} = \frac{8}{3}$$
$$0.002x = \ln\left(\frac{8}{3}\right)$$
$$\bar{x} = \frac{\ln\left(\dfrac{8}{3}\right)}{0.002} \approx 490$$

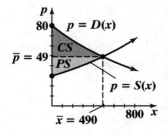

Thus, $\bar{p} = 30e^{0.001(490)} \approx 49$.

$$CS = \int_0^{490} [80e^{-0.001x} - 49]\, dx = \left(\frac{80e^{-0.001x}}{-0.001} - 49x\right)\Bigg|_0^{490} = -80{,}000e^{-0.49} + 80{,}000 - 24{,}010 \approx \$6{,}980$$

$$PS = \int_0^{490} [49 - 30e^{0.001x}]dx = \left(49x - \frac{30e^{0.001x}}{0.001}\right)\Bigg|_0^{490} = 24{,}010 - 30{,}000(e^{0.49} - 1) \approx \$5{,}041$$

67. $D(x) = 80 - 0.04x$; $S(x) = 30e^{0.001x}$

Equilibrium price: $D(x) = S(x)$

$$80 - 0.04x = 30e^{0.001x}$$

Using a graphing utility, we find that

$\overline{x} \approx 614$

Thus, $\overline{p} = 80 - (0.04)614 \approx 55$

$$CS = \int_0^{614} [80 - 0.04x - 55] \, dx = \int_0^{614} (25 - 0.04x) \, dx = (25x - 0.02x^2)\Big|_0^{614} \approx \$7,810$$

$$PS = \int_0^{614} [55 - (30e^{0.001x})] \, dx = \int_0^{614} (55 - 30e^{0.001x}) \, dx = (55x - 30,000e^{0.001x})\Big|_0^{614}$$

$$\approx \$8,336$$

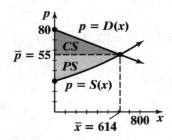

69. $D(x) = 80e^{-0.001x}$; $S(x) = 15 + 0.0001x^2$

Equilibrium price: $D(x) = S(x)$

Using a graphing utility, we find that $\overline{x} \approx 556$

Thus, $\overline{p} = 15 + 0.0001(556)^2 \approx 46$

$$CS = \int_0^{556} [80e^{-0.001x} - 46] \, dx = (-80,000e^{-0.001x} - 46x)\Big|_0^{556} \approx \$8,544$$

$$PS = \int_0^{556} [46 - (15 + 0.0001x^2)] \, dx = \int_0^{556} (31 - 0.0001x^2) \, dx = \left(31x - \frac{0.0001}{3}x^3\right)\Big|_0^{556} \approx \$11,507$$

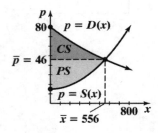

71. (A) Price-Demand   Price-Supply

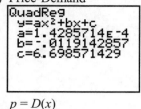

$p = D(x)$ $p = S(x)$

Graph the price-demand and price-supply models and find their point of intersection.

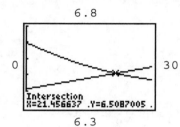

Equilibrium quantity $\overline{x} = 21.457$

Equilibrium price $\overline{p} = 6.51$

(B)   Let $D(x)$ be the quadratic regression model   in part (A).
Consumers' surplus:

$$CS = \int_0^{21.457} [D(x) - 6.51]dx \approx \$1,774.$$

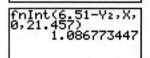

Let $S(x)$ be the linear regression model in   part (A).
Producers' surplus

$$PS = \int_0^{21.457} [6.51 - S(x)]dx \approx \$1,087$$

EXERCISE 14-3

Things to remember:

<u>1.</u>   INTEGRATION-BY-PARTS FORMULA

$$\int udv = uv - \int vdu$$

<u>2.</u>   INTEGRATION-BY-PARTS: SELECTION OF u AND dv

(a)   The product udv must equal the original integrand.

(b)   It must be possible to integrate dv (preferably by using standard formulas or simple substitutions.)

(c)   The new integral, $\int vdu$, should not be more complicated than

the original integral $\int udv$.

(d)   For integrals involving $x^p e^{ax}$, try

$u = x^p$; $dv = e^{ax}dx$.

(e)   For integrals involving $x^p (\ln x)^q$, try

$u = (\ln x)^q$; $dv = x^p dx$.

---

1. $f(x) = 5x, \quad f'(x) = 5; \quad g(x) = x^3, \quad \int x^3 dx = \dfrac{x^4}{4} + C$

3. $f(x) = x^3, \quad f'(x) = 3x^2; \quad g(x) = 5x, \quad \int 5x \, dx = \dfrac{5}{2}x^2 + C$

5 $f(x) = e^{4x}, \quad f'(x) = 4e^{4x}; \quad g(x) = \dfrac{1}{x}, \quad \int \dfrac{1}{x} dx = \ln|x| + C$

7. $f(x) = \dfrac{1}{x}, \quad f'(x) = -x^{-2} = \dfrac{-1}{x^2}; \quad g(x) = e^{4x}, \quad \int e^{4x} dx = \dfrac{1}{4}e^{4x} + C$

9. $\displaystyle\int xe^{3x} dx$

Let $u = x$ and $dv = e^{3x} dx$. Then $du = dx$ and $v = \dfrac{e^{3x}}{3}$.

$\displaystyle\int xe^{3x} dx = \dfrac{xe^{3x}}{3} - \int \dfrac{e^{3x}}{3} dx = \dfrac{1}{3}xe^{3x} - \dfrac{1}{3}\int e^{3x} dx = \dfrac{1}{3}xe^{3x} - \dfrac{1}{9}e^{3x} + C$

11. $\displaystyle\int x^2 \ln x \, dx$

Let $u = \ln x$ and $dv = x^2 dx$. Then $du = \dfrac{dx}{x}$ and $v = \dfrac{x^3}{3}$.

$\displaystyle\int x^2 \ln x \, dx = (\ln x)\left(\dfrac{x^3}{3}\right) - \int \dfrac{x^3}{3} \cdot \dfrac{dx}{x} = \dfrac{1}{3}x^3 \ln x - \dfrac{1}{3}\int x^2 dx$

$\qquad\qquad = \dfrac{x^3 \ln x}{3} - \dfrac{1}{3} \cdot \dfrac{x^3}{3} + C = \dfrac{x^3 \ln x}{3} - \dfrac{x^3}{9} + C$

13. $\displaystyle\int (x+1)^5 (x+2) \, dx$

The better choice is $u = x + 2, \; dv = (x+1)^5 dx$

The alternative is $u = (x+1)^5, \; dv = (x+2) \, dx$, which will lead to an integral of the form
$\displaystyle\int (x+1)^4 (x+2)^2 dx.$

Let $u = x + 2$ and $dv = (x+1)^5 dx$. Then $du = dx$ and $v = \dfrac{1}{6}(x+1)^6$.

Substitute into the integration by parts formula:

$\displaystyle\int (x+1)^5 (x+2) \, dx = \dfrac{1}{6}(x+1)^6 (x+2) - \int \dfrac{1}{6}(x+1)^6 dx = \dfrac{1}{6}(x+1)^6 (x+2) - \dfrac{1}{42}(x+1)^7 + C$

15. $\displaystyle\int xe^{-x} dx$

Let $u = x$ and $dv = e^{-x} dx$. Then $du = dx$ and $v = -e^{-x}$.

$\displaystyle\int xe^{-x} dx = x(-e^{-x}) - \int (-e^{-x}) dx = -xe^{-x} + \int e^{-x} dx = -xe^{-x} - e^{-x} + C$

17. $\int x e^{x^2} dx$

Let $u = x^2$. Then $du = 2x\, dx$.

$$\int x e^{x^2} dx = \frac{1}{2} \int e^{x^2} 2x\, dx = \frac{1}{2} \int e^u\, du = \frac{1}{2} e^u + C = \frac{1}{2} e^{x^2} + C$$

19. $\int_0^1 (x-3)e^x dx$

Let $u = (x-3)$ and $dv = e^x dx$. Then $du = dx$ and $v = e^x$.

$$\int (x-3)e^x dx = (x-3)e^x - \int e^x dx = (x-3)e^x - e^x + C = xe^x - 4e^x + C.$$

Thus, $\int_0^1 (x-3)e^x dx = (xe^x - 4e^x)\Big|_0^1 = (e - 4e) - (-4) = -3e + 4 \approx -4.1548.$

21. $\int_1^3 \ln 2x\, dx$

Let $u = \ln 2x$ and $dv = dx$. Then $du = \dfrac{dx}{x}$ and $v = x$.

$$\int \ln 2x\, dx = (\ln 2x)(x) - \int x \cdot \frac{dx}{x} = x \ln 2x - x + C$$

Thus, $\int_1^3 \ln 2x\, dx = (x \ln 2x - x)\Big|_1^3 = (3 \ln 6 - 3) \ - (\ln 2 - 1) \approx 2.6821.$

23. $\int \dfrac{2x}{x^2+1}\, dx = \int \dfrac{1}{u}\, du = \ln|u| + C = \ln(x^2 + 1) + C$

Substitution: $u = x^2 + 1, \quad du = 2x\, dx$

[Note: Absolute value not needed, since $x^2 + 1 \geq 0$.]

25. $\int \dfrac{\ln x}{x}\, dx = \int u\, du = \dfrac{u^2}{2} + C = \dfrac{(\ln x)^2}{2} + C$

Substitution: $u = \ln x, \quad du = \dfrac{1}{x}\, dx$

27. $\int \sqrt{x}\, \ln x\, dx = \int x^{1/2} \ln x\, dx$

Let $u = \ln x$ and $dv = x^{1/2} dx$. Then $du = \dfrac{dx}{x}$ and $v = \dfrac{2}{3} x^{3/2}$.

$$\int x^{1/2} \ln x\, dx = \frac{2}{3} x^{3/2} \ln x - \int \frac{2}{3} x^{3/2} \frac{dx}{x} = \frac{2}{3} x^{3/2} \ln x - \frac{2}{3} \int x^{1/2} dx = \frac{2}{3} x^{3/2} \ln x - \frac{4}{9} x^{3/2} + C$$

29. $\int (x-3)(x+1)^2 dx$

Let $u = x-3$ and $dv = (x+1)^2 dx$. Then $du = dx$ and $v = \dfrac{1}{3}(x+1)^3$.

$$\int (x-3)(x+1)^2 \, dx = \frac{1}{3}(x-3)(x+1)^3 - \int \frac{1}{3}(x+1)^3 \, dx = \frac{1}{3}(x-3)(x+1)^3 - \frac{1}{12}(x+1)^4 + C$$

OR

$$\int (x-3)(x+1)^2 \, dx = \int (x-3)(x^2+2x+1) \, dx = \int (x^3 - x^2 - 5x - 3) \, dx = \frac{1}{4}x^4 - \frac{1}{3}x^3 - \frac{5}{2}x^2 - 3x + C$$

31. $\int (2x+1)(x-2)^2 \, dx$

Let $u = 2x+1$ and $dv = (x-2)^2 \, dx$. Then $du = 2dx$ and $v = \frac{1}{3}(x-2)^3$.

$$\int (2x+1)(x-2)^2 \, dx = \frac{1}{3}(2x+1)(x-2)^3 - \frac{2}{3} \int (x-2)^3 \, dx = \frac{1}{3}(2x+1)(x-2)^3 - \frac{1}{6}(x-2)^4 + C$$

OR

$$\int (2x+1)(x-2)^2 \, dx = \int (2x+1)(x^2-4x+4) \, dx = \int (2x^3 - 7x^2 + 4x + 4) dx = \frac{1}{2}x^4 - \frac{7}{3}x^3 + 2x^2 + 4x + C$$

33.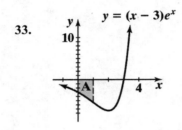

Since $f(x) = (x-3)e^x < 0$ on $[0, 1]$, the integral represents the negative of the area between the graph of f and the x-axis from $x = 0$ to $x = 1$.

35.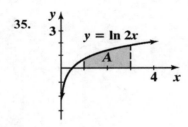

The integral represents the area between the curve $y = \ln 2x$ and the x-axis from $x = 1$ to $x = 3$.

37. $\int x^2 e^x \, dx$

Let $u = x^2$ and $dv = e^x dx$. Then $du = 2x \, dx$ and $v = e^x$.

$$\int x^2 e^x \, dx = x^2 e^x - \int e^x (2x) dx = x^2 e^x - 2 \int x e^x \, dx$$

$\int x e^x \, dx$ can be computed by using integration-by-parts again.

Let $u = x$ and $dv = e^x dx$. Then $du = dx$ and $v = e^x$.

$$\int x e^x \, dx = x e^x - \int e^x \, dx = x e^x - e^x + C$$

and

$$\int x^2 e^x \, dx = x^2 e^x - 2(x e^x - e^x) + C = x^2 e^x - 2x e^x + 2e^x + C \; = (x^2 - 2x + 2)e^x + C$$

39. $\int xe^{ax}dx$

Let $u = x$ and $dv = e^{ax}dx$. Then $du = dx$ and $v = \dfrac{e^{ax}}{a}$.

$\int xe^{ax}dx = \dfrac{xe^{ax}}{a} - \int \dfrac{e^{ax}}{a}dx = \dfrac{xe^{ax}}{a} - \dfrac{e^{ax}}{a^2} + C$

41. $\int_1^e \dfrac{\ln x}{x^2}dx$

Let $u = \ln x$ and $dv = \dfrac{dx}{x^2}$. Then $du = \dfrac{dx}{x}$ and $v = \dfrac{-1}{x}$.

$\int \dfrac{\ln x}{x^2}dx = (\ln x)\left(-\dfrac{1}{x}\right) - \int -\dfrac{1}{x} \cdot \dfrac{dx}{x} = -\dfrac{\ln x}{x} + \int \dfrac{dx}{x^2} = -\dfrac{\ln x}{x} - \dfrac{1}{x} + C$

Thus, $\int_1^e \dfrac{\ln x}{x^2}dx = \left[-\dfrac{\ln x}{x} - \dfrac{1}{x}\right]_1^e = -\dfrac{\ln e}{e} - \dfrac{1}{e} - \left(-\dfrac{\ln 1}{1} - \dfrac{1}{1}\right) = 1 - \dfrac{2}{e} \approx 0.2642$.

[Note: $\ln 1 = 0$, $\ln e = 1$.]

43. $\int_0^2 \ln(x + 4)\,dx$

Let $t = x + 4$. Then $dt = dx$ and

$\int \ln(x + 4)\,dx = \int \ln t\,dt$.

Now, let $u = \ln t$ and $dv = dt$. Then $du = \dfrac{dt}{t}$ and $v = t$.

$\int \ln t\,dt = t\ln t - \int t\left(\dfrac{1}{t}\right)dt = t\ln t - \int dt = t\ln t - t + C$

Thus, $\int \ln(x + 4)\,dx = (x + 4)\ln(x + 4) - (x + 4) + C$

and

$\int_0^2 \ln(x + 4)\,dx = [(x + 4)\ln(x + 4) - (x + 4)]\Big|_0^2 = 6\ln 6 - 6 - (4\ln 4 - 4) = 6\ln 6 - 4\ln 4 - 2$

≈ 3.205.

45. $\int xe^{x-2}dx$

Let $u = x$ and $dv = e^{x-2}dx$. Then $du = dx$ and $v = e^{x-2}$.

$\int xe^{x-2}dx = xe^{x-2} - \int e^{x-2}\,dx = xe^{x-2} - e^{x-2} + C$

47. $\int x\ln(1 + x^2)\,dx$

Let $t = 1 + x^2$. Then $dt = 2x\,dx$ and

$\int x\ln(1 + x^2)\,dx = \int \ln(1 + x^2)x\,dx = \int \ln t\dfrac{dt}{2} = \dfrac{1}{2}\int \ln t\,dt$.

Now, for $\int \ln t\,dt$, let $u = \ln t$, $dv = dt$. Then $du = \dfrac{dt}{t}$ and $v = t$.

$\int \ln t\,dt = t\ln t - \int t\left(\dfrac{1}{t}\right)dt = t\ln t - \int dt = t\ln t - t + C$

Therefore,

$\int x\ln(1 + x^2)\,dx = \dfrac{1}{2}(1 + x^2)\ln(1 + x^2) - \dfrac{1}{2}(1 + x^2) + C$.

49. $\displaystyle\int e^x \ln(1 + e^x)\, dx$

Let $t = 1 + e^x$. Then $dt = e^x dx$ and

$\displaystyle\int e^x \ln(1 + e^x)\, dx = \int \ln t\, dt.$

Now, as shown in Problems 43 and 47,

$\displaystyle\int \ln t\, dt = t \ln t - t + C.$

Thus, $\displaystyle\int e^x \ln(1 + e^x)\, dx = (1 + e^x) \ln(1 + e^x) - (1 + e^x) + C.$

51. $\displaystyle\int (\ln x)^2 dx$

Let $u = (\ln x)^2$ and $dv = dx$. Then $du = \dfrac{2 \ln x}{x} dx$ and $v = x$.

$\displaystyle\int (\ln x)^2 dx = x(\ln x)^2 - \int x \cdot \frac{2 \ln x}{x}\, dx = x(\ln x)^2 - 2\int \ln x\, dx$

$\displaystyle\int \ln x\, dx$ can be computed by using integration-by-parts again.

As shown in Problems 43 and 47,

$\displaystyle\int \ln x\, dx = x \ln x - x + C.$

Thus, $\displaystyle\int (\ln x)^2 dx = x(\ln x)^2 - 2(x \ln x - x) + C = x(\ln x)^2 - 2x \ln x + 2x + C.$

53. $\displaystyle\int (\ln x)^3 dx$

Let $u = (\ln x)^3$ and $dv = dx$. Then $du = 3(\ln x)^2\, \dfrac{1}{x} dx$ and $v = x$.

$\displaystyle\int (\ln x)^3 dx = x(\ln x)^3 - \int x \cdot 3(\ln x)^2 \cdot \frac{1}{x}\, dx = x(\ln x)^3 - 3\int (\ln x)^2 dx$

Now, using Problem 51,

$\displaystyle\int (\ln x)^2 dx = x(\ln x)^2 - 2x \ln x + 2x + C.$

Therefore, $\displaystyle\int (\ln x^3) dx = x(\ln x)^3 - 3[x(\ln x)^2 - 2x \ln x + 2x] + C$

$$= x(\ln x)^3 - 3x(\ln x)^2 + 6x \ln x - 6x + C.$$

55. $\displaystyle\int_1^e \ln(x^2)\, dx = \int_1^e 2 \ln x\, dx = 2\int_1^e \ln x\, dx$

By Example 4, $\displaystyle\int \ln x\, dx = x \ln x - x + C.$ Therefore

$\displaystyle 2\int_1^e \ln x\, dx = 2(x \ln x - x)\Big|_1^e = 2[e \ln e - e] - 2[\ln 1 - 1] = 2$

57. $\displaystyle\int_0^1 \ln(e^{x^2})\, dx = \int_0^1 x^2\, dx = \frac{1}{3} x^3 \Big|_0^1 = \frac{1}{3}$

(Note: $\ln(e^{x^2}) = x^2 \ln e = x^2.$)

59. $y = x - 2 - \ln x,\ 1 \le x \le 4$

$y = 0$ at $x \approx 3.146$

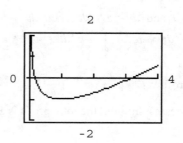

$$A = \int_1^{3.146} [-(x - 2 - \ln x)]\,dx + \int_{3.146}^4 (x - 2 - \ln x)\,dx$$

$$= \int_1^{3.146} (\ln x + 2 - x)\,dx + \int_{3.146}^4 (x - 2 - \ln x)\,dx$$

Now, $\displaystyle\int \ln x\,dx$ is found using integration-by-parts. Let $u = \ln x$ and

$dv = dx$. Then $du = \dfrac{1}{x}\,dx$ and $v = x$.

$$\int \ln x\,dx = x \ln x - \int x\left(\frac{1}{x}\right)dx = x \ln x - \int dx = x \ln x - x + C$$

Thus,

$$A = \left(x \ln x - x + 2x - \frac{1}{2}x^2\right)\Big|_1^{3.146} + \left(\frac{1}{2}x^2 - 2x - x \ln x + x\right)\Big|_{3.146}^4$$

$$= \left(x \ln x + x - \frac{1}{2}x^2\right)\Big|_1^{3.146} + \left(\frac{1}{2}x^2 - x - x \ln x\right)\Big|_{3.146}^4$$

$$\approx (1.803 - 0.5) + (-1.545 + 1.803) = 1.561$$

61. $y = 5 - xe^x,\ 0 \le x \le 3$

$y = 0$ at $x \approx 1.327$

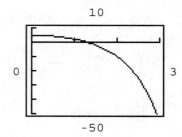

$$A = \int_0^{1.327} (5 - xe^x)\,dx + \int_{1.327}^3 [-(5 - xe^x)]\,dx$$

$$= \int_0^{1.327} (5 - xe^x)\,dx + \int_{1.327}^3 (xe^x - 5)\,dx$$

Now, $\displaystyle\int xe^x\,dx$ is found using integration-by-parts. Let $u = x$ and $dv = e^x\,dx$. Then, $du = dx$

and $v = e^x$.

$$\int xe^x\,dx = xe^x - \int e^x\,dx = xe^x - e^x + C$$

Thus,

$$A = (5x - [xe^x - e^x])\Big|_0^{1.327} + (xe^x - e^x - 5x)\Big|_{1.327}^3 \approx (5.402 - 1) + (25.171 - [-5.402]) \approx 34.98$$

63. Marginal profit: $P'(t) = 2t - te^{-t}$.

The total profit over the first 5 years is given by the definite integral:

$$\int_0^5 (2t - te^{-t})\,dt = \int_0^5 2t\,dt - \int_0^5 te^{-t}\,dt$$

We calculate the second integral using integration-by-parts.

Let $u = t$ and $dv = e^{-t}\,dt$. Then $du = dt$ and $v = -e^{-t}$

$$\int te^{-t}dt = -te^{-t} - \int -e^{-t}dt = -te^{-t} - e^{-t} + C = -e^{-t}[t+1] + C$$

Thus,

Total profit $= t^2\Big|_0^5 + (e^{-t}[t+1])\Big|_0^5 \approx 25 + (0.040 - 1) = 24.040$

To the nearest million, the total profit is $24 million.

65.

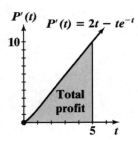

The total profit for the first five years (in millions of dollars) is the same as the area under the marginal profit function, $P'(t) = 2t - te^{-t}$, from $t = 0$ to $t = 5$.

67. From Section 6-2, Future Value $= e^{rT}\int_0^T f(t)e^{-rt}dt$. Now $r = 0.0395$, $T = 5$, $f(t) = 1000 - 200t$. Thus,

$$FV = e^{(0.0395)5}\int_0^5 (1000 - 200t)e^{-0.0395t}dt$$

$$= 1000e^{0.1975}\int_0^5 e^{-0.0395t}dt - 200e^{0.1975}\int_0^5 te^{-0.0395t}dt.$$

We calculate the second integral using integration-by-parts.

Let $u = t$, $dv = e^{-0.0395t}dt$. Then $du = dt$ and $v = \dfrac{e^{-0.0395t}}{-0.0395}$.

$$\int te^{-0.0395t}\,dt = -\frac{te^{-0.0395t}}{0.0395} - \int \frac{e^{-0.0395t}}{-0.0395}\,dt = -\frac{te^{-0.0395t}}{0.0395} - \frac{e^{-0.0395t}}{(0.0395)^2} + C$$

Thus, we have:

$$FV = 1000\,e^{0.1975}\left[\frac{e^{-0.0395t}}{-0.0395}\right]_0^5 - 200\,e^{0.1975}\left[-\frac{te^{-0.0395t}}{0.0395} - \frac{e^{-0.0395t}}{(0.0395)^2}\right]_0^5$$

$$= 1000\,e^{0.1975}\left[\frac{1}{0.0395} - \frac{e^{-0.1975}}{0.0395}\right] - 200\,e^{0.1975}\left[\frac{1}{(0.0395)^2} - \frac{5e^{-0.1975}}{0.0395} - \frac{e^{-0.1975}}{(0.0395)^2}\right] = \$2854.88$$

69. Gini Index $= 2\int_0^1 (x - xe^{x-1})dx = 2\int_0^1 x\,dx - 2\int_0^1 xe^{x-1}\,dx$

We calculate the second integral using integration-by-parts.

Let $u = x$, $dv = e^{x-1}dx$. Then $du = dx$, $v = e^{x-1}$.

$$\int xe^{x-1}dx = xe^{x-1} - \int e^{x-1}dx = xe^{x-1} - e^{x-1} + C$$

Therefore, $2\int_0^1 xdx - 2\int_0^1 xe^{x-1}dx = x^2\Big|_0^1 - 2[xe^{x-1} - e^{x-1}]\Big|_0^1 = 1 - 2[1 - 1 + (e^{-1})]$

$$= 1 - 2e^{-1} \approx 0.264.$$

71.

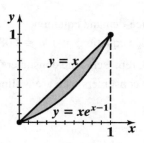

The area bounded by $y = x$ and the Lorenz curve $y = xe^{(x-1)}$ divided by the area under the curve $y = x$ from $x = 0$ to $x = 1$ is the index of income concentration, in this case 0.264. It is a measure of the concentration of income—the closer to zero, the closer to all the income being equally distributed; the closer to one, the closer to all the income being concentrated in a few hands.

73. $S'(t) = -4te^{0.1t}$, $S(0) = 2,000$

$$S(t) = \int -4te^{0.1t}\,dt = -4\int te^{0.1t}\,dt$$

Let $u = t$ and $dv = e^{0.1t}\,dt$. Then $du = dt$ and $v = \dfrac{e^{0.1t}}{0.1} = 10e^{0.1t}$

$$\int te^{0.1t}\,dt = 10te^{0.1t} - \int 10e^{0.1t}\,dt = 10te^{0.1t} - 100e^{0.1t} + C$$

Now, $S(t) = -40te^{0.1t} + 400e^{0.1t} + C$

Since $S(0) = 2,000$, we have
  $2,000 = 400 + C, C = 1,600$

Thus,
  $$S(t) = 1,600 + 400e^{0.1t} - 40te^{0.1t}$$

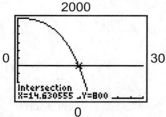

To find how long the company will continue to manufacture this computer, solve $S(t) = 800$ for t.

The company will manufacture the computer for 15 months.

75. $p = D(x) = 9 - \ln(x + 4)$; $\overline{p} = \$2.089$. To find $\overline{x}$, solve

$$9 - \ln(\overline{x} + 4) = 2.089$$
$$\ln(\overline{x} + 4) = 6.911$$
$$\overline{x} + 4 = e^{6.911} \quad \text{(take the exponential of both sides)}$$
$$\overline{x} \approx 1,000$$

Now,

$$CS = \int_0^{1,000}(D(x) - \overline{p})\,dx = \int_0^{1,000}[9 - \ln(x + 4) - 2.089]\,dx = \int_0^{1,000} 6.911\,dx - \int_0^{1,000}\ln(x + 4)\,dx$$

To calculate the second integral, we first let $z = x + 4$ and $dz = dx$ to get

$$\int \ln(x + 4)\,dx = \int \ln z\,dz$$

Then we use integration-by-parts. Let $u = \ln z$ and $dv = dz$. Then $du = \dfrac{1}{z}\,dz$ and $v = z$.

$$\int \ln z\,dz = z\ln z - \int z \cdot \frac{1}{z}\,dz = z\ln z - z + C$$

Therefore,

$$\int \ln(x+4)\, dx = (x+4)\ln(x+4) - (x+4) + C$$

and

$$CS = 6.911x \Big|_0^{1,000} - [(x+4)\ln(x+4) - (x+4)] \Big|_0^{1,000} \approx 6911 - (5935.39 - 1.55) \approx \$977$$

77.

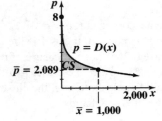

$\bar{p} = 2.089$

$\bar{x} = 1,000$

The area bounded by the price-demand equation, $p = 9 - \ln(x+4)$, and the price equation, $y = \bar{p} = 2.089$, from $x = 0$ to $x = \bar{x} = 1,000$, represents the consumers' surplus. This is the amount saved by consumers who are willing to pay more than \$2.089.

79. Average concentration: $= \dfrac{1}{5-0} \displaystyle\int_0^5 \dfrac{20\ln(t+1)}{(t+1)^2}\, dt = 4 \int_0^5 \dfrac{\ln(t+1)}{(t+1)^2}\, dt$

$\displaystyle\int \dfrac{\ln(t+1)}{(t+1)^2}\, dt$ is found using integration-by-parts.

Let $u = \ln(t+1)$ and $dv = (t+1)^{-2}dt$.

Then $du = \dfrac{1}{t+1}\, dt = (t+1)^{-1}dt$ and $v = -(t+1)^{-1}$.

$$\int \dfrac{\ln(t+1)}{(t+1)^2}\, dt = -\dfrac{\ln(t+1)}{t+1} - \int -(t+1)^{-1}(t+1)^{-1}dt$$

$$= -\dfrac{\ln(t+1)}{t+1} + \int (t+1)^{-2}dt = -\dfrac{\ln(t+1)}{t+1} - \dfrac{1}{t+1} + C$$

Therefore, the average concentration is:

$$\dfrac{1}{5}\int_0^5 \dfrac{20\ln(t+1)}{(t+1)^2}\, dt = 4\left[-\dfrac{\ln(t+1)}{t+1} - \dfrac{1}{t+1}\right]_0^5 = 4\left(-\dfrac{\ln 6}{6} - \dfrac{1}{6}\right) - 4(-\ln 1 - 1)$$

$$= 4 - \dfrac{2}{3}\ln 6 - \dfrac{2}{3} = \dfrac{1}{3}(10 - 2\ln 6) \approx 2.1388 \text{ ppm}$$

81. $N'(t) = (t+6)e^{-0.25t}, \ 0 \le t \le 15; \ N(0) = 40$

$N(t) - N(0) = \displaystyle\int_0^t N'(x)\, dx;$

$$N(t) = 40 + \int_0^t (x+6)e^{-0.25x}dx = 40 + 6\int_0^t e^{-0.25x}dx + \int_0^t xe^{-0.25x}dx$$

$$= 40 + 6(-4e^{-0.25x})\Big|_0^t + \int_0^t xe^{-0.25x}dx$$

$$= 64 - 24e^{-0.25t} + \int_0^t xe^{-0.25x}dx$$

Let $u = x$ and $dv = e^{-0.25x}dx$. Then $du = dx$ and $v = -4e^{-0.25x}$;

$$\int xe^{-0.25x}dx = -4xe^{-0.25x} - \int -4e^{-0.25x}dx = -4xe^{-0.25x} - 16e^{-0.25x} + C$$

Now, $\int_0^t xe^{-0.25x}dx = (-4xe^{-0.25x} - 16e^{-0.25x})\Big|_0^t = -4te^{-0.25t} - 16e^{-0.25t} + 16$

and

$N(t) = 80 - 40e^{-0.25t} - 4te^{-0.25t}$

To find how long it will take a student to achieve the 70 words per minute level, solve $N(t) = 70$:

It will take 8 weeks.

By the end of the course, a student should be able to type

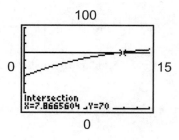

$N(15) = 80 - 40e^{-0.25(15)} - 60e^{-0.25(15)} \approx 78$ words per minute.

83. Average number of voters $= \dfrac{1}{5}\int_0^5 (20 + 4t - 5te^{-0.1t})dt$

$$= \dfrac{1}{5}\int_0^5 (20 + 4t)dt - \int_0^5 te^{-0.1t}dt$$

$\int te^{-0.1t}dt$ is found using integration-by-parts.

Let $u = t$ and $dv = e^{-0.1t}dt$. Then $du = dt$ and $v = \dfrac{e^{-0.1t}}{-0.1} = -10e^{-0.1t}$.

$\int te^{-0.1t}dt = -10te^{-0.1t} - \int -10e^{-0.1t}dt = -10te^{-0.1t} + 10\int e^{-0.1t}dt$

$$= -10te^{-0.1t} + \dfrac{10e^{-0.1t}}{-0.1} + C = -10te^{-0.1t} - 100e^{-0.1t} + C$$

Therefore, the average number of voters is:

$\dfrac{1}{5}\int_0^5 (20 + 4t)dt - \int_0^5 te^{-0.1t}dt = \dfrac{1}{5}(20t + 2t^2)\Big|_0^5 - (-10te^{-0.1t} - 100e^{-0.1t})\Big|_0^5$

$$= \dfrac{1}{5}(100 + 50) + (10te^{-0.1t} + 100e^{-0.1t})\Big|_0^5$$

$$= 30 + (50e^{-0.5} + 100e^{-0.5}) - 100 = 150e^{-0.5} - 70$$

$$\approx 20.98 \text{ (thousands) or } 20{,}980$$

EXERCISE 14-4

Things to remember:

1. **TRAPEZOIDAL RULE**

    Let f be a function defined on an interval $[a,b]$. Partition $[a,b]$ into n subintervals of equal length $\Delta x = (b-a)/n$ with endpoints $a = x_0 < x_1 < x_2 < \cdots < x_n = b$. Then

    $$T_n = \left[f(x_0) + 2f(x_1) + 2f(x_2) + \cdots + 2f(x_{n-1}) + f(x_n)\right]\dfrac{\Delta x}{2}$$

    is an approximation of $\displaystyle\int_a^b f(x)\,dx$.

2.   SIMPSON'S RULE

Let f be a function defined on an interval $[a,b]$. Partition $[a,b]$ into $2n$ subintervals of equal length $\Delta x = (b-a)/n$ with endpoints $a = x_0 < x_1 < x_2 < \cdots < x_n = b$. Then

$$S_n = \left[f(x_0) + 4f(x_1) + 2f(x_2) + 4f(x_3) + 2f(x_4) + \cdots + 4f(x_{2n-1}) + f(x_{2n}) \right]\frac{\Delta x}{3}$$

is an approximation of $\displaystyle\int_a^b f(x)\,dx$.

---

1. $\displaystyle\int_0^6 \sqrt{1+x^4}\,dx$, $f(x) = \sqrt{1+x^4}$.   Partition $[0,6]$ into three equal subintervals:

$x_0 = 0,\ x_1 = 2,\ x_2 = 4,\ x_3 = 6,\quad \Delta x = 6/3 = 2$.

x	$f(x)$
0	1
2	4.1231
4	16.0312
6	36.0139

By the trapezoidal rule:

$$T_3 = [f(0) + 2f(2) + 2f(4) + f(6)](2/2) = [1 + 8.2462 + 32.0624 + 36.0139] \approx 77.32$$

3. $\displaystyle\int_0^6 \sqrt{1+x^4}\,dx$, $f(x) = \sqrt{1+x^4}$.   Partition $[0,6]$ into six equal subintervals:

$x_0 = 0,\ x_1 = 1,\ x_2 = 2,\ x_3 = 3,\ x_4 = 4,\ x_5 = 5,\ x_6 = 6;\quad \Delta x = 6/6 = 1$.

x	$f(x)$
0	1
1	1.4142
2	4.1231
3	9.0554
4	16.0312
5	25.0200
6	36.0139

By the trapezoidal rule:

$$T_6 = [f(0) + 2f(1) + 2f(2) + 2f(3) + 2f(4) + 2f(5) + f(6)](1/2)$$
$$= [1 + 2.8284 + 8.4262 + 18.1108 + 32.0624 + 50.0400 + 36.0139](1/2) \approx 74.24$$

5. $\displaystyle\int_1^3 \frac{1}{1+x^2}\,dx, \ f(x) = \frac{1}{1+x^2}.$ Partition $[1,3]$ into two equal subintervals:

$x_0 = 1, \ x_1 = 2, \ x_2 = 3; \ \Delta x = 1.$

x	$f(x)$
1	0.5000
2	0.2000
3	0.1000

By Simpson's rule:

$S_2 = [f(1) + 4f(2) + f(3)](1/3) = [0.5000 + 0.8000 + 0.1000](1/3) \approx 0.47$

7. $\displaystyle\int_1^3 \frac{1}{1+x^2}\,dx, \ f(x) = \frac{1}{1+x^2}.$ Partition $[1,3]$ into four equal subintervals:

$x_0 = 1, \ x_1 = 1.5, \ x_2 = 2, \ x_3 = 2.5, \ x_4 = 3; \ \Delta x = 0.5.$

x	$f(x)$
1	0.5000
1.5	0.3077
2	0.2000
2.5	0.1379
3	0.1000

By Simpson's rule:

$S_4 = [f(1) + 4f(1.5) + 2f(2) + 4f(2.5) + f(3)](1/6)$
$= [0.5000 + 1.2308 + 0.4000 + 0.5516 + 0.1000](1/6)] \approx 0.46.$

9. Use Formula 9 with $a = b = 1.$ $\displaystyle\int \frac{1}{x(1+x)}\,dx = \frac{1}{1}\ln\left|\frac{x}{1+x}\right| + C = \ln\left|\frac{x}{x+1}\right| + C$

11. Use Formula 18 with $a = 3, \ b = 1, \ c = 5, \ d = 2$:

$\displaystyle\int \frac{1}{(3+x)^2(5+2x)}\,dx = \frac{1}{3\cdot2 - 5\cdot1}\cdot\frac{1}{3+x} + \frac{2}{(3\cdot2 - 5\cdot1)^2}\ln\left|\frac{5+2x}{3+x}\right| + C = \frac{1}{3+x} + 2\ln\left|\frac{5+2x}{3+x}\right| + C$

13. Use Formula 25 with $a = 16$ and $b = 1$:

$\displaystyle\int \frac{x}{\sqrt{16+x}}\,dx = \frac{2(x - 2\cdot16)}{3\cdot1^2}\sqrt{16+x} + C = \frac{2(x-32)}{3}\sqrt{16+x} + C$

15. Use Formula 29 with $a = 1$: $\displaystyle\int \frac{1}{x\sqrt{1-x^2}}\,dx = -\frac{1}{1}\ln\left|\frac{1+\sqrt{1-x^2}}{x}\right| + C = -\ln\left|\frac{1+\sqrt{1-x^2}}{x}\right| + C$

17. Use Formula 37 with $a = 2$ ($a^2 = 4$): $\displaystyle\int \frac{1}{x\sqrt{x^2+4}}\,dx = \frac{1}{2}\ln\left|\frac{x}{2+\sqrt{x^2+4}}\right| + C$

19. Use Formula 51 with $n = 2$: $\displaystyle\int x^2 \ln x\,dx = \frac{x^{2+1}}{2+1}\ln x - \frac{x^{2+1}}{(2+1)^2} + C = \frac{x^3}{3}\ln x - \frac{x^3}{9} + C$

21. Use Formula 48 with $a = c = d = 1$: $\displaystyle\int \frac{1}{1+e^x}\,dx = x - \ln|1 + e^x| + C.$

23. First use Formula 5 with $a = 3$ and $b = 1$ to find the indefinite integral.

$$\int \frac{x^2}{3+x}\,dx = \frac{(3+x)^2}{2\cdot 1^3} - \frac{2\cdot 3(3+x)}{1^3} + \frac{3^2}{1^3}\ln|3+x| + C = \frac{(3+x)^2}{2} - 6(3+x) + 9\ln|3+x| + C$$

Thus, $\displaystyle\int_1^3 \frac{x^2}{3+x}\,dx = \left[\frac{(3+x)^2}{2} - 6(3+x) + 9\ln|3+x|\right]_1^3$

$$= \frac{(3+3)^2}{2} - 6(3+3) + 9\ln|3+3| - \left[\frac{(3+1)^2}{2} - 6(3+1) + 9\ln|3+1|\right]$$

$$= 9\ln\frac{3}{2} - 2 \approx 1.6492.$$

25. First use Formula 15 with $a = 3$, $b = c = d = 1$ to find the indefinite integral.

$$\int \frac{1}{(3+x)(1+x)}\,dx = \frac{1}{3\cdot 1 - 1\cdot 1}\ln\left|\frac{1+x}{3+x}\right| + C = \frac{1}{2}\ln\left|\frac{1+x}{3+x}\right| + C$$

Thus, $\displaystyle\int_0^7 \frac{1}{(3+x)(1+x)}\,dx = \frac{1}{2}\ln\left|\frac{1+x}{3+x}\right|\Big|_0^7 = \frac{1}{2}\ln\left|\frac{1+7}{3+7}\right| - \frac{1}{2}\ln\left|\frac{1}{3}\right|$

$$= \frac{1}{2}\ln\left|\frac{4}{5}\right| - \frac{1}{2}\ln\left|\frac{1}{3}\right| = \frac{1}{2}\ln\frac{12}{5} \approx 0.4377.$$

27. First use Formula 36 with $a = 3$ ($a^2 = 9$) to find the indefinite integral:

$$\int \frac{1}{\sqrt{x^2+9}}\,dx = \ln\left|x + \sqrt{x^2+9}\right| + C$$

Thus, $\displaystyle\int_0^4 \frac{1}{\sqrt{x^2+9}}\,dx = \ln\left|x + \sqrt{x^2+9}\right|\Big|_0^4 = \ln\left|4 + \sqrt{16+9}\right| - \ln\left|\sqrt{9}\right| = \ln 9 - \ln 3 = \ln 3 \approx 1.0986.$

29. $\int\limits_{3}^{13} x^2 dx,\ f(x) = x^2.$ Partition $[3,13]$ into five equal subintervals:

$x_0 = 3,\ x_1 = 5,\ x_2 = 7,\ x_3 = 9,\ x_4 = 11,\ x_5 = 13;\ \Delta x = 2.$

x	$f(x)$
3	9
5	25
7	49
9	81
11	121
13	169

$T_5 = [f(3) + 2f(5) + 2f(7) + 2f(9) + 2f(11) + f(13)](2/2)$

$\quad = [9 + 50 + 98 + 162 + 242 + 169] = 730.$

Exact value: $\int\limits_{3}^{13} x^2\, dx = \dfrac{x^3}{3}\bigg|_{3}^{13} = \dfrac{13^3}{3} - 9 = 723.33.$

31. $\int\limits_{1}^{5} \dfrac{1}{x} dx,\ f(x) = \dfrac{1}{x}.$ Partition $[1,5]$ into eight equal subintervals:

$x_0 = 1,\ x_1 = 1.5,\ x_2 = 2,\ x_3 = 2.5,\ x_4 = 3,\ x_5 = 3.5,\ x_6 = 4,\ x_7 = 4.5,\ x_8 = 5,\ \Delta x = 0.5.$

x	$f(x)$
1	1
1.5	0.6667
2	0.5000
2.5	0.4000
3	0.3333
3.5	0.2857
4	0.2500
4.5	0.2222
5	0.2000

$S_8 = [f(1) + 4f(1.5) + 2f(2) + 4f(2.5) + 2f(3) + 4f(3.5) + 2f(4) + 4f(4.5) + f(5)](1/6)$

$\quad = [1 + 2.6667 + 1 + 1.6000 + 0.6667 + 1.1428 + 0.5000 + 0.8888 + .2000](1/6) \approx 1.61$

Exact value: $\int\limits_{1}^{5} \dfrac{1}{x} dx = \ln x\big|_{1}^{5} = \ln 5 \approx 1.61.$

33. $\displaystyle\int_5^8 (4x - 3)\,dx;\ f(x) = 4x - 3.$. Partition [5,8] into three equal subintervals:

$x_0 = 5,\ x_1 = 6,\ x_2 = 7,\ x_3 = 8,\ \Delta x = 1.$

x	$f(x)$
5	17
6	21
7	25
8	29

$T_3 = [f(5) + 2f(6) + 2f(7) + f(8)](1/2) = [17 + 42 + 50 + 29](1/2) = 69.$

Exact value: $\displaystyle\int_5^8 (4x - 3)\,dx = \left[2x^2 - 3x\right]_5^8 = (128 - 24) - (50 - 15) = 69.$

35. $\displaystyle\int_5^9 (3x^2 + 5x + 3)\,dx;\ f(x) = 3x^2 + 5x + 3.$ Partition [5,9] into four equal subintervals:

$x_0 = 5,\ x_1 = 6,\ x_2 = 7,\ x_3 = 8,\ x_4 = 9,\ \Delta x = 1.$

x	$f(x)$
5	103
6	141
7	185
8	235
9	291

$S_4 = [f(5) + 4f(6) + 2f(7) + 4f(8) + f(9)](1/3) = [103 + 564 + 370 + 940 + 291](1/3) = 756.$

Exact value: $\displaystyle\int_5^9 (3x^2 + 5x + 3)\,dx = \left[x^3 + \frac{5x^2}{2} + 3x\right]_5^9 = \left(729 + \frac{405}{2} + 27\right) - \left(125 + \frac{125}{2} + 15\right) = 756.$

37. Consider Formula 35. Let $u = 2x$. Then $u^2 = 4x^2$, $x = \dfrac{u}{2}$, and $dx = \dfrac{du}{2}$.

$$\int \frac{\sqrt{4x^2 + 1}}{x^2}\,dx = \int \frac{\sqrt{u^2 + 1}}{\dfrac{u^2}{4}}\frac{du}{2} = 2\int \frac{\sqrt{u^2 + 1}}{u^2}\,du = 2\left[-\frac{\sqrt{u^2 + 1}}{u} + \ln\left|u + \sqrt{u^2 + 1}\right|\right] + C$$

$$= 2\left[-\frac{\sqrt{4x^2 + 1}}{2x} + \ln\left|2x + \sqrt{4x^2 + 1}\right|\right] + C = -\frac{\sqrt{4x^2 + 1}}{x} + 2\ln\left|2x + \sqrt{4x^2 + 1}\right| + C$$

39. Let $u = x^2$. Then $du = 2x\,dx$.

$$\int \frac{x}{\sqrt{x^4 - 16}}\,dx = \frac{1}{2} \int \frac{1}{\sqrt{u^2 - 16}}\,du$$

Now use Formula 43 with $a = 4$ ($a^2 = 16$):

$$\frac{1}{2} \int \frac{1}{\sqrt{u^2 - 16}}\,du = \frac{1}{2}\,\ln\left|u + \sqrt{u^2 - 16}\right| + C = \frac{1}{2}\ln\left|x^2 + \sqrt{x^4 - 16}\right| + C$$

41. Let $u = x^3$. Then $du = 3x^2 dx$.

$$\int x^2 \sqrt{x^6 + 4}\,dx = \frac{1}{3} \int \sqrt{u^2 + 4}\,du$$

Now use Formula 32 with $a = 2$ ($a^2 = 4$):

$$\frac{1}{3} \int \sqrt{u^2 + 4}\,du = \frac{1}{3} \cdot \frac{1}{2}\left[u\sqrt{u^2 + 4} + 4\ln\left|u + \sqrt{u^2 + 4}\right|\right] + C$$

$$= \frac{1}{6}\left[x^3\sqrt{x^6 + 4} + 4\ln\left|x^3 + \sqrt{x^6 + 4}\right|\right] + C$$

43. $$\int \frac{1}{x^3\sqrt{4 - x^4}}\,dx = \int \frac{x}{x^4\sqrt{4 - x^4}}\,dx$$

Let $u = x^2$. Then $du = 2x\,dx$.

$$\int \frac{x}{x^4\sqrt{4 - x^4}}\,dx = \frac{1}{2} \int \frac{1}{u^2\sqrt{4 - u^2}}\,du$$

Now use Formula 30 with $a = 2$ ($a^2 = 4$):

$$\frac{1}{2} \int \frac{1}{u^2\sqrt{4 - u^2}}\,du = -\frac{1}{2} \cdot \frac{\sqrt{4 - u^2}}{4u} + C = \frac{-\sqrt{4 - x^4}}{8x^2} + C$$

45. $$\int \frac{e^x}{(2 + e^x)(3 + 4e^x)}\,dx = \int \frac{1}{(2 + u)(3 + 4u)}\,du$$

Substitution: $u = e^x$, $du = e^x\,dx$.
Now use Formula 15 with $a = 2$, $b = 1$, $c = 3$, $d = 4$:

$$\int \frac{1}{(2 + u)(3 + 4u)}\,du = \frac{1}{2 \cdot 4 - 3 \cdot 1}\,\ln\left|\frac{3 + 4u}{2 + u}\right| + C = \frac{1}{5}\,\ln\left|\frac{3 + 4e^x}{2 + e^x}\right| + C$$

47. $$\int \frac{\ln x}{x\sqrt{4 + \ln x}}\,dx = \int \frac{u}{\sqrt{4 + u}}\,du$$

Substitution: $u = \ln x$, $du = \frac{1}{x}\,dx$.
Use Formula 25 with $a = 4$, $b = 1$:

$$\int \frac{u}{\sqrt{4 + u}}\,du = \frac{2(u - 2 \cdot 4)}{3 \cdot 1^2}\sqrt{4 + u} + C = \frac{2(u - 8)}{3}\sqrt{4 + u} + C = \frac{2(\ln x - 8)}{3}\sqrt{4 + \ln x} + C$$

49. Use Formula 47 with $n = 2$ and $a = 5$:

$$\int x^2 e^{5x} dx = \frac{x^2 e^{5x}}{5} - \frac{2}{5} \int xe^{5x} dx$$

To find $\int xe^{5x} dx$, use Formula 47 with $n = 1$, $a = 5$:

$$\int xe^{5x} dx = \frac{xe^{5x}}{5} - \frac{1}{5} \int e^{5x} dx = \frac{xe^{5x}}{5} - \frac{1}{5} \cdot \frac{e^{5x}}{5}$$

Thus, $\displaystyle \int x^2 e^{5x} dx = \frac{x^2 e^{5x}}{5} - \frac{2}{5}\left[\frac{xe^{5x}}{5} - \frac{1}{25} e^{5x} \right] + C = \frac{x^2 e^{5x}}{5} - \frac{2xe^{5x}}{25} + \frac{2e^{5x}}{125} + C.$

51. Use Formula 47 with $n = 3$ and $a = -1$.

$$\int x^3 e^{-x} dx = \frac{x^3 e^{-x}}{-1} - \frac{3}{-1} \int x^2 e^{-x} dx = -x^3 e^{-x} + 3 \int x^2 e^{-x} dx$$

Now $\displaystyle \int x^2 e^{-x} dx = \frac{x^2 e^{-x}}{-1} - \frac{2}{-1} \int xe^{-x} dx = -x^2 e^{-x} + 2 \int xe^{-x} dx$

and $\displaystyle \int xe^{-x} dx = \frac{xe^{-x}}{-1} - \frac{1}{-1} \int e^{-x} dx = -xe^{-x} - e^{-x}$, using Formula 47.

Thus, $\displaystyle \int x^3 e^{-x} dx = -x^3 e^{-x} + 3[-x^2 e^{-x} + 2(-xe^{-x} - e^{-x})] + C = -x^3 e^{-x} - 3x^2 e^{-x} - 6xe^{-x} - 6e^{-x} + C.$

53. Use Formula 52 with $n = 3$:

$$\int (\ln x)^3 dx = x(\ln x)^3 - 3 \int (\ln x)^2 dx$$

Now $\displaystyle \int (\ln x)^2 dx = x(\ln x)^2 - 2 \int \ln x\, dx$ using Formula 52 again, and

$\displaystyle \int \ln x\, dx = x \ln x - x$ by Formula 49.

Thus, $\displaystyle \int (\ln x)^3 dx = x(\ln x)^3 - 3[x(\ln x)^2 - 2(x \ln x - x)] + C = x(\ln x)^3 - 3x(\ln x)^2 + 6x \ln x - 6x + C.$

55. $\displaystyle \int_3^5 x\sqrt{x^2 - 9}\ dx.$ First consider the indefinite integral.

Let $u = x^2 - 9$. Then $du = 2x\, dx$ or $x\, dx = \frac{1}{2} du$. Thus,

$$\int x\sqrt{x^2 - 9}\ dx = \frac{1}{2} \int u^{1/2} du = \frac{1}{2} \cdot \frac{u^{3/2}}{3/2} + C = \frac{1}{3}(x^2 - 9)^{3/2} + C.$$

Now, $\displaystyle \int_3^5 x\sqrt{x^2 - 9}\ dx = \frac{1}{3}(x^2 - 9)^{3/2} \Big|_3^5 = \frac{1}{3} \cdot 16^{3/2} = \frac{64}{3}.$

57. $\displaystyle \int_2^4 \frac{1}{x^2 - 1} dx.$ Consider the indefinite integral:

$$\int \frac{1}{x^2 - 1} dx = \frac{1}{2 \cdot 1} \ln \left| \frac{x-1}{x+1} \right| + C, \text{ using Formula 13 with } a = 1.$$

Thus,

$$\int_2^4 \frac{1}{x^2 - 1} dx = \frac{1}{2} \ln \left| \frac{x-1}{x+1} \right| \Big|_2^4 = \frac{1}{2} \ln \left| \frac{3}{5} \right| - \frac{1}{2} \ln \left| \frac{1}{3} \right| = \frac{1}{2} \ln \frac{9}{5} \approx 0.2939.$$

59. $\displaystyle \int \frac{\ln x}{x^2}\,dx \;=\; \int x^{-2}\ln x \,dx = \frac{x^{-1}}{-1}\ln x - \frac{x^{-1}}{(-1)^2} + C$ [Formula 51 with $n = -2$]

$$= -\frac{1}{x}\ln x - \frac{1}{x} + C = \frac{-1 - \ln x}{x} + C$$

61. $\displaystyle \int \frac{x}{\sqrt{x^2 - 1}}\,dx = \int \frac{1}{\sqrt{x^2-1}}\left(\frac{2}{2}\right)x\,dx = \frac{1}{2}\int u^{-1/2}\,du = u^{1/2} + C = \sqrt{x^2 - 1} + C$

Let $u = x^2 - 1$. Then $du = 2x\,dx$

63. $\displaystyle \int_{-1}^{1} (ax^2 + bx + c)\,dx;\; f(x) = ax^2 + bx + c..$ Partition $[-1,1]$ into two equal subintervals:

$x_0 = -1,\; x_1 = 0,\; x_2 = 1,\; \Delta x = 1.$

x	$f(x)$
-1	$a - b + c$
0	c
1	$a + b + c$

$$S_2 = [f(-1) + 4f(0) + f(1)](1/3) = [(a - b + c) + 4c + (a + b + c)](1/3) = \frac{2}{3}a + 2c.$$

Exact value:

$$\int_{-1}^{1} (ax^2 + bx + c)\,dx = \left[\frac{a}{3}x^3 + \frac{b}{2}x^2 + cx\right]_{-1}^{1} = \left(\frac{a}{3} + \frac{b}{2} + c\right) - \left(-\frac{a}{3} + \frac{b}{2} - c\right) = \frac{2a}{3} + 2c.$$

65. $\displaystyle f(x) = \frac{10}{\sqrt{x^2 + 1}},\; g(x) = x^2 + 3x$

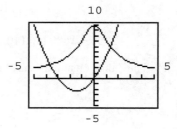

The graphs of f and g are shown at the right. The x-coordinates of the points of intersection are: $x_1 \approx -3.70,\, x_2 \approx 1.36$

$$A = \int_{-3.70}^{1.36} \left[\frac{10}{\sqrt{x^2+1}} - (x^2 + 3x)\right]dx$$

$$= 10\int_{-3.70}^{1.36} \frac{1}{\sqrt{x^2+1}}\,dx - \int_{-3.70}^{1.36} (x^2 + 3x)\,dx$$

For the first integral, use Formula 36 with $a = 1$:

$$A = (10\ln\;|x + \sqrt{x^2+1}\;|)\Big|_{-3.70}^{1.36} - \left(\frac{1}{3}x^3 + \frac{3}{2}x^2\right)\Big|_{-3.70}^{1.36}$$

$$\approx [11.15 - (-20.19)] - [3.61 - (3.65)] = 31.38$$

67. $f(x) = x\sqrt{x+4}$, $g(x) = 1 + x$

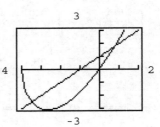

The graphs of f and g are shown at the right. The x-coordinates of the points of intersection are: $x_1 \approx -3.49$, $x_2 \approx 0.83$

$$A = \int_{-3.49}^{0.83} [1 + x - x\sqrt{x+4}\,]dx = \int_{-3.49}^{0.83} (1+x)dx - \int_{-3.49}^{0.83} x\sqrt{x+4}\ dx$$

For the second integral, use Formula 22 with $a = 4$ and $b = 1$:

$$A = \left(x + \frac{1}{2}x^2\right)\Big|_{-3.49}^{0.83} - \left(\frac{2[3x-8]}{15}\sqrt{(x+4)^3}\right)\Big|_{-3.49}^{0.83}$$

$$\approx (1.17445 - 2.60005) - (-7.79850 + 0.89693) \approx 5.48$$

69. Find $\overline{x}$, the demand when the price $\overline{p} = 15$:

$$15 = \frac{7500 - 30\overline{x}}{300 - \overline{x}}$$

$$4500 - 15\overline{x} = 7500 - 30\overline{x}$$

$$15\overline{x} = 3000$$

$$\overline{x} = 200$$

Consumers' surplus:

$$CS = \int_0^{\overline{x}} [D(x) - \overline{p}\,]dx = \int_0^{200} \left[\frac{7500 - 30x}{300 - x} - 15\right]dx = \int_0^{200} \left[\frac{3000 - 15x}{300 - x}\right]dx$$

Use Formula 20 with $a = 3000$, $b = -15$, $c = 300$, $d = -1$:

$$CS = \left[\frac{-15x}{-1} + \frac{3000(-1) - (-15)(300)}{(-1)^2}\ln|300 - x|\right]\Big|_0^{200} = [15x + 1500\ln|300 - x|]\Big|_0^{200}$$

$$= 3000 + 1500\ln(100) - 1500\ln(300) = 3000 + 1500\ln\left(\frac{1}{3}\right) \approx 1352$$

Thus, the consumers' surplus is \$1352.

71.

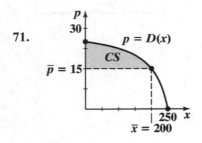

The shaded region represents the consumers' surplus.

73. $C'(x) = \dfrac{250 + 10x}{1 + 0.05x}$, $C(0) = 25,000$

$$C(x) = \int \frac{250 + 10x}{1 + 0.05x}\,dx = 250\int \frac{1}{1 + 0.05x}\,dx + 10\int \frac{x}{1 + 0.05x}\,dx$$

$$= 250\left(\frac{1}{0.05}\ln|1+0.05x|\right) + 10\left(\frac{x}{0.05} - \frac{1}{(0.05)^2}\ln|1+0.05x|\right) + K$$

(Formulas 3 and 4)

$$= 5{,}000\ \ln\ \ |1+0.05x| + 200x - 4{,}000\ \ln\ \ |1+0.05x| + K = 1{,}000\ \ln\ \ |1+0.05x| + 200x + K$$

Since $C(0) = 25{,}000,\ K = 25{,}000$ and $C(x) = 1{,}000\ \ln(1+0.05x) + 200x + 25{,}000,\ x \geq 0$

To find the production level that produces a cost of $150,000, solve $C(x) = 150{,}000$ for x:

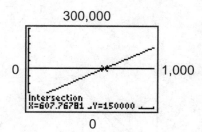

The production level is $x = 608$ pairs of skis.

At a production level of 850 pairs of skis,

$C(850) = 1{,}000\ \ln(1 + 0.05[850]) + 200(850) + 25{,}000 \approx \$198{,}773.$

75. $FV = e^{rT}\displaystyle\int_0^T f(t)e^{-rt}dt$

Now, $r = 0.044,\ T = 10,\ f(t) = 50t^2.$

$$FV = e^{(0.044)10}\int_0^{10} 50t^2 e^{-0.044t}\ dt = 50\,e^{0.44}\int_0^{10} t^2 e^{-0.044t}\ dt$$

To evaluate the integral, use Formula 47 with $n = 2$ and $a = -0.044$:

$$\int t^2 e^{-0.044t}\ dt = \frac{t^2 e^{-0.044t}}{-0.044} - \frac{2}{-0.044}\int t\,e^{-0.044t}\ dt = \frac{t^2 e^{-0.044t}}{-0.044} + \frac{2}{0.044}\int t\,e^{-0.044t}\ dt$$

Now, using Formula 47 again:

$$\int t\,e^{-0.044t}\ dt = \frac{te^{-0.044t}}{-0.044} - \frac{1}{-0.044}\int e^{-0.044t}\ dt = -\frac{te^{-0.044t}}{0.044} - \frac{e^{-0.044t}}{(0.044)^2}$$

Thus

$$\int t^2 e^{-0.044t}\ dt = -\frac{t^2 e^{-0.044t}}{0.044} - \frac{2t\,e^{-0.044t}}{(0.044)^2} - \frac{2e^{-0.044t}}{(0.044)^3} + C\ .$$

Now,

$$FV = 50\,e^{0.44}\left[-\frac{t^2 e^{-0.044t}}{0.044} - \frac{2t\,e^{-0.044t}}{(0.044)^2} - \frac{2\,e^{-0.044t}}{(0.044)^3}\right]_0^{10}$$

$$= 50\,e^{0.44}\left[-\frac{100\,e^{-0.44}}{0.044} - \frac{20\,e^{-0.44}}{(0.044)^2} - \frac{2\,e^{-0.44}}{(0.044)^3} + \frac{2}{(0.044)^3}\right] = \$18{,}673.95$$

77. Gini Index:

$$2\int_0^1 [x - f(x)]dx = 2\int_0^1 \left[x - \frac{1}{2}x\sqrt{1+3x} \right]dx = \int_0^1 [2x - x\sqrt{1+3x}]dx = \int_0^1 2xdx - \int_0^1 x\sqrt{1+3x}\,dx$$

For the second integral, use Formula 22 with $a = 1$ and $b = 3$:

$$= x^2\Big|_0^1 - \frac{2(3\cdot 3x - 2\cdot 1)}{15(3)^2}\sqrt{(1+3x)^3}\,\Big|_0^1 = 1 - \frac{2(9x-2)}{135}\sqrt{(1+3x)^3}\,\Big|_0^1$$

$$= 1 - \frac{14}{135}\sqrt{4^3} - \frac{4}{135}\sqrt{1^3} = 1 - \frac{112}{135} - \frac{4}{135} = \frac{19}{135} \approx 0.1407$$

79.

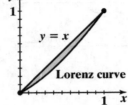

As the area bounded by the two curves gets smaller, the Lorenz curve approaches $y = x$ and the distribution of income approaches perfect equality — all individuals share equally in the income.

81. $S'(t) = \dfrac{t^2}{(1+t)^2}$; $S(t) = \displaystyle\int \dfrac{t^2}{(1+t)^2}\,dt$

Use Formula 7 with $a = 1$ and $b = 1$:

$$S(t) = \frac{1+t}{1^3} - \frac{1^2}{1^3(1+t)} - \frac{2(1)}{1^3}\ln|1+t| + C = 1 + t - \frac{1}{1+t} - 2\ln|1+t| + C$$

Since $S(0) = 0$, we have $0 = 1 - 1 - 2\ln 1 + C$ and $C = 0$. Thus,

$$S(t) = 1 + t - \frac{1}{1+t} - 2\ln|1+t|.$$

Now, the total sales during the first two years (= 24 months) is given by:

$$S(24) = 1 + 24 - \frac{1}{1+24} - 2\ln|1+24| = 24.96 - 2\ln 25 \approx 18.5$$

Thus, total sales during the first two years is approximately $18.5 million.

83.

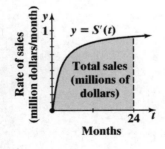

The total sales, in millions of dollars, over the first two years (24 months) is the area under the curve $y = S'(t)$ from $t = 0$ to $t = 24$.

85. $P'(x) = x\sqrt{2+3x}$, $P(1) = -\$2,000$

$$P(x) = \int x\sqrt{2+3x}\,dx = \frac{2(9x-4)}{135}(2+3x)^{3/2} + C$$

(Formula 22)

$$P(1) = \frac{2(5)}{135}5^{3/2} + C = -2,000$$

$$C = -2,000 - \frac{2}{27}5^{3/2} \approx -2,000.83$$

Thus, $P(x) = \frac{2(9x-4)}{135}(2+3x)^{3/2} - 2,000.83$.

The number of cars that must be sold to have a profit of \$13,000: 54
Profit if 80 cars are sold per week:

$$P(80) = \frac{2(716)}{135}(242)^{3/2} - 2,000.83 \approx \$37,932.20$$

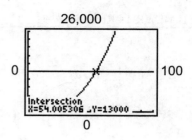

87. $\dfrac{dR}{dt} = \dfrac{100}{\sqrt{t^2+9}}$. Therefore,

$$R = \int \frac{100}{\sqrt{t^2+9}}\,dt = 100\int \frac{1}{\sqrt{t^2+9}}\,dt$$

Using Formula 36 with $a = 3$ ($a^2 = 9$), we have:

$$R = 100\ln\left|t+\sqrt{t^2+9}\right| + C$$

Now $R(0) = 0$, so $0 = 100\ln|3| + C$ or $C = -100\ln 3$. Thus,

$$R(t) = 100\ln\left|t+\sqrt{t^2+9}\right| - 100\ln 3$$

and

$$R(4) = 100\ln(4 + \sqrt{4^2+9}) - 100\ln 3 = 100\ln 9 - 100\ln 3 = 100\ln 3 \approx 110 \text{ feet}$$

89. $N'(t) = \dfrac{60}{\sqrt{t^2+25}}$

The number of items learned in the first twelve hours of study is given by:

$$N = \int_0^{12}\frac{60}{\sqrt{t^2+25}}\,dt = 60\int_0^{12}\frac{1}{\sqrt{t^2+25}}\,dt = 60\left(\ln\left|t+\sqrt{t^2+25}\right|\right)\Big|_0^{12},\text{ using Formula 36}$$

$$= 60\left[\ln\left|12+\sqrt{12^2+25}\right| - \ln\sqrt{25}\right] = 60(\ln 25 - \ln 5) = 60\ln 5 \approx 96.57 \text{ or } 97 \text{ items}$$

91.

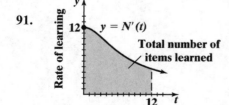

The area under the rate of learning curve, $y = N'(t)$, from $t = 0$ to $t = 12$ represents the total number of items learned in that time interval.

CHAPTER 14 REVIEW

1. $A = \int_a^b f(x)\,dx$ (6-1)

2. $A = \int_b^c [-f(x)]\,dx$ (6-1)

3. $A = \int_a^b f(x)\,dx + \int_b^c [-f(x)]\,dx$ (6-1)

4. $A = \int_{0.5}^1 [-\ln x]\,dx + \int_1^e \ln x\,dx$

We evaluate the integral using integration-by-parts.
Let $u = \ln x,\ dv = dx$.

Then $du = \dfrac{1}{x}\,dx,\ v = x,$ and $\int \ln x\,dx =$

$x \ln x - \int x\left(\dfrac{1}{x}\right) dx = x \ln x - x + C$

Thus,

$A = -\int_{0.5}^1 \ln x\,dx + \int_1^e \ln x\,dx$

$= (-x \ln x + x)\Big|_{0.5}^1 + (x \ln x - x)\Big|_1^e$

$\approx (1 - 0.847) + (1) = 1.153$

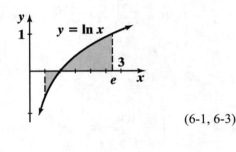

(6-1, 6-3)

5. $\int xe^{4x}\,dx.$ Use integration-by-parts:

Let $u = x$ and $dv = e^{4x}\,dx$. Then $du = dx$ and $v = \dfrac{e^{4x}}{4}$.

$\int xe^{4x}\,dx = \dfrac{xe^{4x}}{4} - \int \dfrac{e^{4x}}{4}\,dx = \dfrac{xe^{4x}}{4} - \dfrac{e^{4x}}{16} + C$ (6-3)

6. $\int x \ln x\,dx.$ Use integration-by-parts:

Let $u = \ln x$ and $dv = x\,dx$. Then $du = \dfrac{1}{x}\,dx$ and $v = \dfrac{x^2}{2}$.

$\int x \ln x\,dx = \dfrac{x^2 \ln x}{2} - \int \dfrac{1}{x} \cdot \dfrac{x^2}{2}\,dx = \dfrac{x^2 \ln x}{2} - \dfrac{1}{2} \int x\,dx = \dfrac{x^2 \ln x}{2} - \dfrac{x^2}{4} + C$ (6-3)

7. $\displaystyle\int \frac{\ln x}{x}\, dx$

Let $u = \ln x$. Then $du = \dfrac{1}{x}\, dx$ and

$$\int \frac{\ln x}{x}\, dx = \int u\, du = \frac{1}{2}u^2 + C = \frac{1}{2}[\ln x]^2 + C \qquad (6\text{-}2)$$

8. $\displaystyle\int \frac{x}{1+x^2}\, dx$

Let $u = 1 + x^2$. Then $du = 2x\, dx$ and

$$\int \frac{x}{1+x^2}\, dx = \int \frac{1/2\, du}{u} = \frac{1}{2}\int \frac{1}{u}\, du = \frac{1}{2}\ln|u| + C = \frac{1}{2}\ln(1+x^2) + C \qquad (6\text{-}2)$$

9. Use Formula 11 with $a = 1$ and $b = 1$.

$$\int \frac{1}{x(1+x)^2}\, dx = \frac{1}{1(1+x)} + \frac{1}{1^2}\ln\left|\frac{x}{1+x}\right| + C = \frac{1}{1+x} + \ln\left|\frac{x}{1+x}\right| + C \qquad (6\text{-}4)$$

10. Use Formula 28 with $a = 1$ and $b = 1$.

$$\int \frac{1}{x^2\sqrt{1+x}}\, dx = -\frac{\sqrt{1+x}}{1 \cdot x} - \frac{1}{2 \cdot 1\sqrt{1}}\ln\left|\frac{\sqrt{1+x} - \sqrt{1}}{\sqrt{1+x} + \sqrt{1}}\right| + C = -\frac{\sqrt{1+x}}{x} - \frac{1}{2}\ln\left|\frac{\sqrt{1+x} - 1}{\sqrt{1+x} + 1}\right| + C \qquad (6\text{-}4)$$

11. $y = 5 - 2x - 6x^2;\ y = 0$ on $[1, 2]$

$$A = -\int_1^2 (5 - 2x - 6x^2)\, dx$$

$$= \int_1^2 (6x^2 + 2x - 5)\, dx = (2x^3 + x^2 - 5x)\Big|_1^2$$

$$= 10 - (-2) = 12 \qquad (6\text{-}1)$$

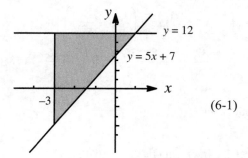

12. $y = 5x + 7;\ y = 12$ on $[-3, 1]$

$$A = \int_{-3}^1 [12 - (5x + 7)]\, dx$$

$$= \int_{-3}^1 (5 - 5x)\, dx = \left(5x - \frac{5}{2}x^2\right)\Big|_{-3}^1$$

$$= \left(5 - \frac{5}{2}\right) - \left(-15 - \frac{45}{2}\right) = 40$$

(6-1)

13. $y = -x + 2; y = x^2 + 3$ on $[-1, 4]$

$$A = \int_{-1}^{4} [(x^2 + 3) - (-x + 2)]\, dx$$

$$= \int_{-1}^{4} (x^2 + x + 1)\, dx$$

$$= \left(\frac{1}{3}x^3 + \frac{1}{2}x^2 + x \right) \Big|_{-1}^{4}$$

$$= \frac{64}{3} + 8 + 4 - \left(-\frac{1}{3} + \frac{1}{2} - 1 \right)$$

$$= \frac{205}{6} \approx 34.167$$

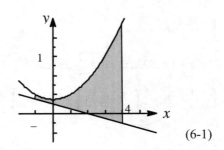

(6-1)

14. $y = \dfrac{1}{x}; y = -e^{-x}$ on $[1, 2]$

$$A = \int_{1}^{2} \left(\frac{1}{x} - e^{-x} \right) dx = \int_{1}^{2} \left(\frac{1}{x} + e^{-x} \right) dx$$

$$= (\ln x - e^{-x}) \Big|_{1}^{2} = (\ln 2 - e^{-2}) + e^{-1} \approx 0.926$$

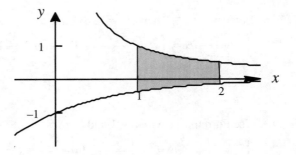

(6-1)

15. $y = x; y = -x^3$ on $[-2, 2]$

$$A = \int_{-2}^{0} (-x^3 - x)\,dx + \int_{0}^{2} x - (-x^3)\,dx$$

$$= -\int_{-2}^{0} (x^3 + x)\,dx + \int_{0}^{2} (x + x^3)\,dx$$

$$= -\left(\frac{1}{4}x^4 + \frac{1}{2}x^2 \right) \Big|_{-2}^{0} + \left(\frac{1}{2}x^2 + \frac{1}{4}x^4 \right) \Big|_{0}^{2}$$

$$= 0 + 6 + 6 - 0 = 12$$

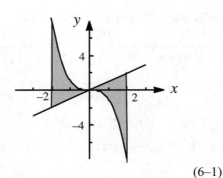

(6–1)

16. $y = x^2; y = -x^4$ on $[-2, 2]$

$$A = \int_{-2}^{2} [x^2 - (-x^4)]\,dx$$

$$= \int_{-2}^{2} (x^2 + x^4)\,dx = \left(\frac{1}{3}x^3 + \frac{1}{5}x^5 \right) \Big|_{-2}^{2}$$

$$= \frac{8}{3} + \frac{32}{5} - \left(-\frac{8}{3} - \frac{32}{5} \right) = \frac{16}{3} + \frac{64}{5} \approx 18.133$$

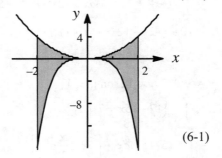

(6-1)

17.   Indonesia   (6-1)

18.   Vietnam   (6-1)

19. $A = \displaystyle\int_{a}^{b} [f(x) - g(x)]\,dx$ (6-1)

20. $A = \displaystyle\int_{b}^{c} [g(x) - f(x)]\,dx$ (6-1)

21. $A = \displaystyle\int_{b}^{c} [g(x) - f(x)]\,dx + \int_{c}^{d} [f(x) - g(x)]\,dx$ (6-1)

22. $A = \displaystyle\int_a^b [f(x) - g(x)]dx + \int_b^c [g(x) - f(x)]dx + \int_c^d [f(x) - g(x)] \, dx$ (6-1)

23. $A = \displaystyle\int_0^5 [(9 - x) - (x^2 - 6x + 9)]dx$

$\qquad = \displaystyle\int_0^5 (5x - x^2)dx$

$\qquad = \left(\dfrac{5}{2}x^2 - \dfrac{1}{3}x^3\right)\Big|_0^5$

$\qquad = \dfrac{125}{2} - \dfrac{125}{3} = \dfrac{125}{6} \approx 20.833$

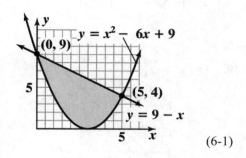

(6-1)

24. $\displaystyle\int_0^1 xe^x dx.$ Use integration-by-parts.

Let $u = x$ and $dv = e^x dx$. Then $du = dx$ and $v = e^x$.

$\displaystyle\int xe^x dx = xe^x - \int e^x dx = xe^x - e^x + C$

Therefore, $\displaystyle\int_0^1 xe^x dx = (xe^x - e^x)\Big|_0^1 = 1 \cdot e - e - (0 \cdot 1 - 1) = 1$ (6-3)

25. Use Formula 38 with $a = 4$

$\displaystyle\int_0^3 \dfrac{x^2}{\sqrt{x^2 + 16}} \, dx = \dfrac{1}{2}\left[x\sqrt{x^2 + 16} - 16\ln\left|x + \sqrt{x^2 + 16}\right|\right]\Big|_0^3$

$\qquad = \dfrac{1}{2}\left[3\sqrt{25} - 16\ln(3 + \sqrt{25})\right] - \dfrac{1}{2}(-16\ln\sqrt{16}) = \dfrac{1}{2}[15 - 16\ln 8] + 8\ln 4$

$\qquad = \dfrac{15}{2} - 8\ln 8 + 8\ln 4 \approx 1.955$ (6-4)

26. Let $u = 3x$, then $du = 3 \, dx$. Now, use Formula 40 with $a = 7$.

$\displaystyle\int \sqrt{9x^2 - 49} \, dx = \dfrac{1}{3}\int \sqrt{u^2 - 49} \, du = \dfrac{1}{3} \cdot \dfrac{1}{2}\left(u\sqrt{u^2 - 49} - 49\ln\left|u + \sqrt{u^2 - 49}\right|\right) + C$

$\qquad = \dfrac{1}{6}\left(3x\sqrt{9x^2 - 49} - 49\ln\left|3x + \sqrt{9x^2 - 49}\right|\right) + C$ (6-4)

27. $\displaystyle\int te^{-0.5t} \, dt.$ Use integration-by-parts.

Let $u = t$ and $dv = e^{-0.5t}dt$. Then $du = dt$ and $v = \dfrac{e^{-0.5t}}{-0.5}$.

$\displaystyle\int te^{-0.5t} \, dt = \dfrac{-te^{-0.5t}}{0.5} + \int \dfrac{e^{-0.5t}}{0.5} \, dt = \dfrac{-te^{-0.5t}}{0.5} + \dfrac{e^{-0.5t}}{-0.25} + C = -2te^{-0.5t} - 4e^{-0.5t} + C$ (6-3)

28. $\displaystyle\int x^2 \ln x \, dx.$ Use integration-by-parts.

Let $u = \ln x$ and $dv = x^2 dx$. Then $du = \dfrac{1}{x} \, dx$ and $v = \dfrac{x^3}{3}$.

$\displaystyle\int x^2 \ln x \, dx = \dfrac{x^3 \ln x}{3} - \int \dfrac{1}{x} \cdot \dfrac{x^3}{3} \, dx = \dfrac{x^3 \ln x}{3} - \dfrac{1}{3}\int x^2 dx = \dfrac{x^3 \ln x}{3} - \dfrac{x^3}{9} + C$ (6-3)

29. Use Formula 48 with $a = 1$, $c = 1$, and $d = 2$.

$$\int \frac{1}{1 + 2e^x}\, dx = \frac{x}{1} - \frac{1}{1 \cdot 1} \ln|1 + 2e^x| + C = x - \ln|1 + 2e^x| + C \qquad (6\text{-}4)$$

30. (A)

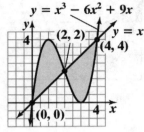

$$A = \int_0^2 [(x^3 - 6x^2 + 9x) - x]\, dx + \int_2^4 [x - (x^3 - 6x^2 + 9x)]\, dx$$

$$= \int_0^2 (x^3 - 6x^2 + 8x)\, dx + \int_2^4 (-x^3 + 6x^2 - 8x)\, dx$$

$$= \left(\frac{1}{4}x^4 - 2x^3 + 4x^2 \right) \Big|_0^2 + \left(-\frac{1}{4}x^4 + 2x^3 - 4x^2 \right) \Big|_2^4 = 4 + 4 = 8$$

 (B)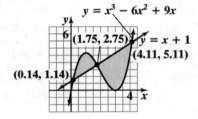

The x-coordinates of the points of intersection are: $x_1 \approx 0.14$,
$x_2 \approx 1.75$, $x_3 \approx 4.11$.

$$A = \int_{0.14}^{1.75} [(x^3 - 6x^2 + 9x) - (x + 1)]\, dx + \int_{1.75}^{4.11} [(x + 1) - (x^3 - 6x^2 + 9x)]\, dx$$

$$= \int_{0.14}^{1.75} (x^3 - 6x^2 + 8x - 1)\, dx + \int_{1.75}^{4.11} (1 - x^3 + 6x^2 - 8x)\, dx$$

$$= \left(\frac{1}{4}x^4 - 2x^3 + 4x^2 - x \right) \Big|_{0.14}^{1.75} + \left(x - \frac{1}{4}x^4 + 2x^3 - 4x^2 \right) \Big|_{1.75}^{4.11}$$

$$= [2.126 - (-0.066)] + [4.059 - (-2.126)] \approx 8.38 \qquad (6\text{-}1)$$

31. $\displaystyle\int_0^3 e^{x^2}\,dx;\ \ f(x)=e^{x^2}$.   Partition $[0,3]$ into three equal subintervals:

$x_0=0,\ \ x_1=1,\ \ x_2=2,\ \ x_3=3,\ \ \Delta x=1.$

x	$f(x)$
0	1
1	2.7183
2	54.5982
3	8103.0839

$T_3=[f(0)+2f(1)+2f(2)+f(3)](1/2)=[1+5.4366+109.1964+8103.0839](1/2)\approx 4109.36.$
(6.4)

32. $\displaystyle\int_0^3 e^{x^2}\,dx;\ \ f(x)=e^{x^2}$.   Partition $[0,3]$ into five equal subintervals:

$x_0=0,\ \ x_1=0.6,\ \ x_2=1.2,\ \ x_3=1.8,\ \ x_4=2.4,\ \ x_5=3,\ \ \Delta x=0.6.$

x	$f(x)$
0	1
0.6	1.4333
1.2	4.2207
1.8	25.5337
2.4	317.3483
3	8103.0839

$T_5=[f(0)+2f(0.6)+2f(1.2)+2f(1.8)+2f(2.4)+f(3)](0.6/2)$
$\quad=[1+2.8666+8.4414+51.0674+634.6966+8103.0839](0.3)\approx 2640.35.$
(6.4)

33. $\displaystyle\int_1^5 (\ln x)^2\, dx,$ $f(x) = (\ln x)^2.$ Partition $[1,5]$ into four equal subintervals:

$x_0 = 1,\ x_1 = 2,\ x_2 = 3,\ x_3 = 4,\ x_4 = 5,\ \Delta x = 1.$

x	$f(x)$
1	0
2	0.4805
3	1.2069
4	1.9218
5	2.5903

$S_4 = [f(1) + 4f(2) + 2f(3) + 4f(4) + f(5)](1/3) = [0 + 1.9220 + 2.4138 + 7.6872 + 2.5903](1/3) \approx 4.87$
(6.4)

34. $\displaystyle\int_1^5 (\ln x)^2\, dx,$ $f(x) = (\ln x)^2.$ Partition $[1,5]$ into eight equal subintervals:

$x_0 = 1,\ x_1 = 1.5,\ x_2 = 2,\ x_3 = 2.5,\ x_4 = 3,\ x_5 = 3.5,\ x_6 = 4,\ x_7 = 4.5,\ x_8 = 5,\ \Delta x = 0.5.$

x	$f(x)$
1	0
1.5	0.1644
2	0.4805
2.5	0.8396
3	1.2069
3.5	1.5694
4	1.9218
4.5	2.2622
5	2.5903

$S_8 = [f(1) + 4f(1.5) + 2f(2) + 4f(2.5) + 2f(3) + 4f(3.5) + 2f(4) + 4f(4.5) + f(5)](1/6)$

$= [0 + 0.6576 + 0.9610 + 3.3584 + 2.4138 + 6.2776 + 3.8436 + 9.0648 + 2.5903](1/6) \approx 4.86$
(6.4)

35. $\displaystyle\int \frac{(\ln x)^2}{x}\, dx = \int u^2\, du = \frac{u^3}{3} + C = \frac{(\ln x)^3}{3} + C$ (5-2)

Substitution: $u = \ln x, \quad du = \dfrac{1}{x}\, dx$

36. $\displaystyle\int x(\ln x)^2\, dx.$ Use integration-by-parts.

Let $u = (\ln x)^2$ and $dv = x\, dx$. Then $du = 2(\ln x)\dfrac{1}{x}\, dx$ and $v = \dfrac{x^2}{2}$.

$\displaystyle\int x(\ln x)^2\, dx = \frac{x^2(\ln x)^2}{2} - \int 2(\ln x)\frac{1}{x}\cdot\frac{x^2}{2}\, dx = \frac{x^2(\ln x)^2}{2} - \int x\ln x\, dx$

Let $u = \ln x$ and $dv = x\, dx$. Then $du = \dfrac{1}{x}\, dx$ and $v = \dfrac{x^2}{2}$.

$\displaystyle\int x\ln x\, dx = \frac{x^2\ln x}{2} - \frac{x^2}{4}$

Thus, $\displaystyle\int x(\ln x)^2\, dx = \frac{x^2(\ln x)^2}{2} - \left[\frac{x^2\ln x}{2} - \frac{x^2}{4}\right] + C = \frac{x^2(\ln x)^2}{2} - \frac{x^2\ln x}{2} + \frac{x^2}{4} + C.$ (6-3)

37. Let $u = x^2 - 36$. Then $du = 2x\, dx$.

$\displaystyle\int \frac{x}{\sqrt{x^2-36}}\, dx = \int \frac{x}{(x^2-36)^{1/2}}\, dx = \frac{1}{2}\int \frac{1}{u^{1/2}}\, du = \frac{1}{2}\int u^{-1/2}\, du$

$\displaystyle\qquad\qquad = \frac{1}{2}\cdot\frac{u^{1/2}}{1/2} + C = u^{1/2} + C = \sqrt{x^2-36} + C$ (6-2)

38. Let $u = x^2, \ du = 2x\, dx$.

Then use Formula 43 with $a = 6$.

$\displaystyle\int \frac{x}{\sqrt{x^4-36}}\, dx = \frac{1}{2}\int \frac{du}{\sqrt{u^2-36}} = \frac{1}{2}\ln\left|u + \sqrt{u^2-36}\right| + C = \frac{1}{2}\ln\left|x^2 + \sqrt{x^4-36}\right| + C$ (6-4)

39. $\displaystyle\int_0^4 x\ln(10-x)\, dx$

Consider

$\displaystyle\int x\ln(10-x)\, dx = \int (10-t)\ln t\,(-dt) = \int t\ln t\, dt - 10\int \ln t\, dt.$

Substitution: $t = 10 - x, \ dt = -dx, \ x = 10 - t$

Now use integration-by-parts on the two integrals.

Let $u = \ln t, dv = t\, dt$. Then $du = \dfrac{1}{t}\, dt, \ v = \dfrac{t^2}{2}$.

$\displaystyle\int \ln t\, dt = \frac{t^2}{2}\ln t - \int \frac{t^2}{2}\cdot\frac{1}{t}\, dt = \frac{t^2\ln t}{2} - \frac{t^2}{4} + C$

Let $u = \ln t, \ dv = dt$. Then $du = \dfrac{1}{t}\, dt, \ v = t$.

$\displaystyle\int t\ln t\, dt = t\ln t - \int t\cdot\frac{1}{t}\, dt = t\ln t - t + C$

Thus,

$$\int_0^4 x\ln(10-x)dx = \left[\frac{(10-x)^2\ln(10-x)}{2} - \frac{(10-x)^2}{4} - 10(10-x)\ln(10-x) + 10(10-x)\right]_0^4$$

$$= \frac{36\ln 6}{2} - \frac{36}{4} - 10(6)\ln 6 + 10(6) - \left[\frac{100\ln 10}{2} - \frac{100}{4} - 10(10)\ln 10 + 10(10)\right]$$

$$= 18\ln 6 - 9 - 60\ln 6 + 60 - 50\ln 10 + 25 + 100\ln 10 - 100$$

$$= 50\ln 10 - 42\ln 6 - 24 \approx 15.875. \qquad (6\text{-}3)$$

40. Use Formula 52 with $n = 2$.

$$\int (\ln x)^2 dx = x(\ln x)^2 - 2\int \ln x\, dx$$

Now use integration-by-parts to calculate $\int \ln x\, dx$.

Let $u = \ln x,\, dv = dx$. Then $du = \dfrac{1}{x}dx,\; v = x$.

$$\int \ln x\, dx = x\ln x - \int x\cdot\frac{1}{x}dx = x\ln x - x + C$$

Therefore, $\displaystyle\int (\ln x)^2 dx = x(\ln x)^2 - 2[x\ln x - x] + C = x(\ln x)^2 - 2x\ln x + 2x + C.$ $(6\text{-}3, 6\text{-}4)$

41. $\displaystyle\int xe^{-2x^2}\, dx$

Let $u = -2x^2$. Then $du = -4x\, dx$.

$$\int xe^{-2x^2}\, dx = -\frac{1}{4}\int e^u\, du = -\frac{1}{4}e^u + C = -\frac{1}{4}e^{-2x^2} + C \qquad (6\text{-}2)$$

42. $\displaystyle\int x^2 e^{-2x}\, dx.$

Use integration-by-parts. Let $u = x^2$ and $dv = e^{-2x}dx$. Then $du = 2x\, dx$ and $v = -\dfrac{1}{2}e^{-2x}$.

$$\int x^2 e^{-2x}\, dx = -\frac{1}{2}x^2 e^{-2x} + \int xe^{-2x}\, dx$$

Now use integration-by-parts again. Let $u = x$ and $dv = e^{-2x}dx$. Then $du = dx$ and $v = -\dfrac{1}{2}e^{-2x}$.

$$\int xe^{-2x}\, dx = -\frac{1}{2}xe^{-2x} + \frac{1}{2}\int e^{-2x}\, dx = -\frac{1}{2}xe^{-2x} - \frac{1}{4}e^{-2x} + C$$

Thus,

$$\int x^2 e^{-2x}\, dx = -\frac{1}{2}x^2 e^{-2x} + \left[-\frac{1}{2}xe^{-2x} - \frac{1}{4}e^{-2x}\right] + C = -\frac{1}{2}x^2 e^{-2x} - \frac{1}{2}xe^{-2x} - \frac{1}{4}e^{-2x} + C$$

$(6\text{-}3)$

43. First graph the two functions to find the points of intersection.

The curves intersect at the points where $x = 1.448$ and $x = 6.965$.

$$\text{Area } A = \int_{1.448}^{6.965} \left(\frac{6}{2 + 5e^{-x}} - [0.2x + 1.6] \right) dx$$
$$\approx 1.703$$

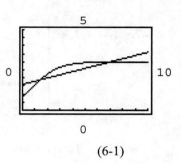

(6-1)

44. (A) Probability $(0 \le t \le 1) = \int_0^1 0.21e^{-0.21t} \, dt = -e^{-0.21t} \Big|_0^1 = -e^{-0.21} + 1 \approx 0.189$

(B) Probability $(1 \le t \le 2) = \int_1^2 0.21e^{-0.21t} \, dt = -e^{-0.21t} \Big|_1^2 = e^{-0.21} - e^{-0.42} \approx 0.154$ (6-2)

45.

The probability that the product will fail during the second year of warranty is the area under the probability density function $y = f(t)$ from $t = 1$ to $t = 2$.   (6-2)

46. $R'(x) = 65 - 6 \ln(x + 1)$, $R(0) = 0$

$$R(x) = \int [65 - 6 \ln(x + 1)] \, dx = 65x - 6 \int \ln(x + 1) \, dx$$

Let $z = x + 1$. Then $dz = dx$ and $\int \ln(x + 1) \, dx = \int \ln z \, dz$.

Now, let $u = \ln z$ and $dv = dz$. Then $du = \frac{1}{z} \, dz$ and $v = z$:

$$\int \ln z \, dz = z \ln z - \int z \left(\frac{1}{z} \right) dz = z \ln z - \int dz = z \ln z - z + C$$

Therefore, $\int \ln(x + 1) dx = (x + 1)\ln(x + 1) - (x + 1) + C$ and

$R(x) = 65x - 6[(x + 1)\ln(x + 1) - (x + 1)] + C.$

Since $R(0) = 0$, $C = -6$. Thus, $R(x) = 65x - 6[(x + 1)\ln(x + 1) - x]$

To find the production level for a revenue of $20,000 per week, solve $R(x) = 20,000$ for x.

The production level should be 618 hair dryers per week.

At a production level of 1,000 hair dryers per week, revenue

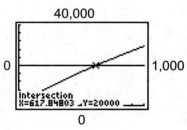

$R(1,000) = 65,000 - 6[(1,001)\ln(1,001) - 1,000] \approx \$29,506$ (6-3)

47. (A)

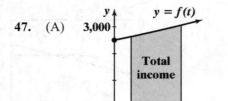

(B)   Total income = $\int_1^4 2{,}500e^{0.05t}\,dt = 50{,}000e^{0.05t}\Big|_1^4 = 50{,}000[e^{0.2} - e^{0.05}] \approx \$8{,}507$ (6-2)

48. $f(t) = 2{,}500e^{0.05t}, \ r = 0.04, \ T = 5$

(A) $FV = e^{(0.04)5}\int_0^5 2{,}500e^{0.05t}\,e^{-0.04t}\,dt = 2{,}500e^{0.2}\int_0^5 e^{0.01t}\,dt = 250{,}000e^{0.2}\ e^{0.01t}\Big|_0^5$

$$= 250{,}000[e^{0.25} - e^{0.2}] \approx \$15{,}655.66$$

(B)   Total income = $\int_0^5 2{,}500e^{0.05t}\,dt = 50{,}000e^{0.05t}\Big|_0^5 = 50{,}000[e^{0.25} - 1] \approx \$14{,}201.27$

Interest = FV – Total income = $\$15{,}655.66 - \$14{,}201.27 = \$1{,}454.39$ (6-2)

49. (A)

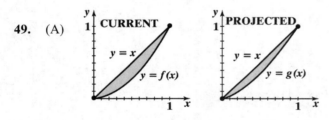

(B)   The income will be more equally distributed 10 years from now since the area between $y = x$ and the projected Lorenz curve is less than the area between $y = x$ and the current Lorenz curve.

(C)   Current:

Gini Index = $2\int_0^1 [x - (0.1x + 0.9x^2)]\,dx = 2\int_0^1 (0.9x - 0.9x^2)\,dx = 2(0.45x^2 - 0.3x^3)\Big|_0^1 = 0.30$

Projected:

Gini Index = $2\int_0^1 (x - x^{1.5})\,dx = 2\int_0^1 (x - x^{3/2})\,dx = 2\left(\frac{1}{2}x^2 - \frac{2}{5}x^{5/2}\right)\Big|_0^1 = 2\left(\frac{1}{10}\right) = 0.2$

Thus, income will be more equally distributed 10 years from now, as indicated in part (B).   (6-1)

50. (A) $p = D(x) = 70 - 0.2x, \ p = S(x) = 13 + 0.0012x^2$

Equilibrium price: $D(x) = S(x)$

$$70 - 0.2x = 13 + 0.0012x^2$$

$$0.0012x^2 + 0.2x - 57 = 0$$

$$x = \frac{-0.2 \pm \sqrt{0.04 + 0.2736}}{0.0024} = \frac{-0.2 \pm 0.56}{0.0024}$$

Therefore, $\bar{x} = \dfrac{-0.2 + 0.56}{0.0024} = 150$, and $\bar{p} = 70 - 0.2(150) = 40$.

$$CS = \int_0^{150} (70 - 0.2x - 40)\,dx = \int_0^{150} (30 - 0.2x)\,dx = (30x - 0.1x^2)\Big|_0^{150} = \$2{,}250$$

$$PS = \int_0^{150} [40 - (13 + 0.0012x^2)]\, dx = \int_0^{150} (27 - 0.0012x^2)\, dx = (27x - 0.0004x^3)\Big|_0^{150} = \$2,700$$

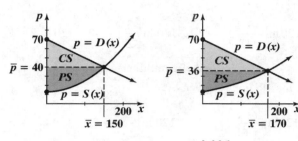

$$\bar{x} = 150 \qquad\qquad\qquad \bar{x} = 170$$

(B) $p = D(x) = 70 - 0.2x,\ p = S(x) = 13e^{0.006x}$

Equilibrium price: $\qquad\qquad\qquad D(x) = S(x)$

$$70 - 0.2x = 13e^{0.006x}$$

Using a graphing utility to solve for x, we get $\bar{x} \approx 170$
and $\bar{p} = 70 - 0.2(170) \approx 36$.

$$CS = \int_0^{170} (70 - 0.2x - 36)\, dx = \int_0^{170} (34 - 0.2x)\, dx = (34x - 0.1x^2)\Big|_0^{170} = \$2,890$$

$$PS = \int_0^{170} (36 - 13e^{0.006x})\, dx = (36x - 2{,}166.67e^{0.006x})\Big|_0^{170} \approx \$2,278 \qquad (6\text{-}2)$$

51. (A)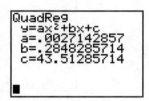

Graph the quadratic regression model and the
line $p = 52.50$ to find the point of intersection.

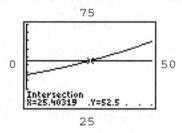

The demand at a price of 52.50 cents per pound is 25,403 lbs.

(B) Let $S(x)$ be the quadratic regression model found in part (A). Then the producers' surplus at the
price level of 52.5 cents per pound is given by

$$PS = \int_0^{25.403} [52.5 - S(x)]\, dx \approx \$1,216 \qquad (6\text{-}2)$$

52. $R(t) = \dfrac{60t}{(t+1)^2(t+2)}$

The amount of the drug eliminated during the first hour is given by

$$A = \int_0^1 \frac{60t}{(t+1)^2(t+2)}\, dt$$

First use Formula 19 with $a = 1,\ b = 1,\ c = 2,\ d = 1$ to find the indefinite integral:

$$\int \frac{60t}{(t+1)^2(t+2)}\,dt\,dt = 60 \int\; \int \frac{t}{(t+1)^2\cdot(t+2)}\,dt = 60\left[\frac{1}{t+1} - 2\ln\left|\frac{t+2}{t+1}\right|\right] + C = \frac{60}{t+1} - 120\ln\left|\frac{t+2}{t+1}\right| + C$$

Now,

$$A = \int_0^1 \frac{60t}{(t+1)^2(t+2)}\,dt = \left[\frac{60}{t+1} - 120\ln\left(\frac{t+2}{t+1}\right)\right]_0^1 = 30 - 120\ln\left(\frac{3}{2}\right) - 60 + 120\ln 2$$

$$\approx 4.522 \text{ milliliters}$$

The amount of drug eliminated during the 4th hour is given by:

$$A = \int_3^4 \frac{60t}{(t+1)^2(t+2)}\,dt = \left[\frac{60}{t+1} - 120\ln\left(\frac{t+2}{t+1}\right)\right]_3^4 = 12 - 120\ln\left(\frac{6}{5}\right) - 15 + 120\ln\left(\frac{5}{4}\right)$$

$$\approx 1.899 \text{ milliliters} \qquad (5\text{-}5,\ 6\text{-}4)$$

53.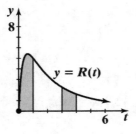

$$\qquad\qquad\qquad\qquad (5\text{-}5,\ 6\text{-}1)$$

54. $f(t) = \begin{cases} \dfrac{4/3}{(t+1)^2} & 0 \le t \le 3 \\[2mm] 0 & \text{otherwise} \end{cases}$

(A)   Probability $(0 \le t \le 1) = \displaystyle\int_0^1 \frac{4/3}{(t+1)^2}\,dt$

To calculate the integral, let $u = t + 1$, $du = dt$. Then,

$$\int \frac{4/3}{(t+1)^2}\,dt = \frac{4}{3}\int u^{-2}\,du = \frac{4}{3}\frac{u^{-1}}{-1} = -\frac{4}{3u} + C = \frac{-4}{3(t+1)} + C$$

Thus,

$$\int_0^1 \frac{4/3}{(t+1)^2}\,dt = \left[\frac{-4}{3(t+1)}\right]_0^1 = -\frac{2}{3} + \frac{4}{3} = \frac{2}{3} \approx 0.667$$

(B)   Probability $(t \ge 1) = \displaystyle\int_1^3 \frac{4/3}{(t+1)^2}\,dt = \left[\frac{-4}{3(t+1)}\right]_1^3 = -\frac{1}{3} + \frac{2}{3} = \frac{1}{3} \approx 0.333 \qquad (6\text{-}2)$

55.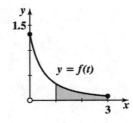

The probability that the doctor will spend more than an hour with a randomly selected patient is the area under the probability density function $y = f(t)$ from $t = 1$ to $t = 3$.   (6-2)

56. $N'(t) = \dfrac{100t}{(1+t^2)^2}$. To find $N(t)$, we calculate

$$\int \frac{100t}{(1+t^2)^2}\, dt$$

Let $u = 1 + t^2$. Then $du = 2t\, dt$, and

$$N(t) = \int \frac{100t}{(1+t^2)^2}\, dt = 50 \int \frac{1}{u^2}\, du = 50 \int u^{-2}\, du = -50\frac{1}{u} + C = \frac{-50}{1+t^2} + C$$

At $t = 0$, we have

$$N(0) = -50 + C$$

Therefore, $C = N(0) + 50$ and

$$N(t) = \frac{-50}{1+t^2} + 50 + N(0)$$

Now,

$$N(3) = \frac{-50}{1+3^2} + 50 + N(0) = 45 + N(0)$$

Thus, the population will increase by 45 thousand during the next 3 years.   (5-5, 6-1)

57. We want to find Probability $(t \geq 2) = \displaystyle\int_2^\infty f(t)\, dt$

Since

$$\int_{-\infty}^\infty f(t)\, dt = \int_{-\infty}^2 f(t)\, dt + \int_2^\infty f(t)\, dt = 1,$$

$$\int_2^\infty f(t)\, dt = 1 - \int_{-\infty}^2 f(t)\, dt = 1 - \int_0^2 f(t)\, dt \qquad (\text{since } f(t) = 0 \text{ for } t \leq 0)$$

$$= 1 - \text{Probability } (0 \leq t \leq 2)$$

Now, Probability $(0 \leq t \leq 2) = \displaystyle\int_0^2 0.5e^{-0.5t}\, dt = -e^{-0.5t}\Big]_0^2 = -e^{-1} + 1 \approx 0.632$

Therefore, Probability $(t \geq 2) = 1 - 0.632 = 0.368.$ (6-2)

15 MULTIVARIABLE CALCULUS

EXERCISE 15-1

Things to remember:

1. An equation of the form $z = f(x, y)$ describes a FUNCTION OF TWO INDEPENDENT VARIABLES if for each permissible ordered pair (x, y) there is one and only one value of z determined by $f(x, y)$. The variables x and y are INDEPENDENT VARIABLES, and the variable z is a DEPENDENT VARIABLE. The set of all ordered pairs of permissible values of x and y is the DOMAIN of the function, and the set of all corresponding values $f(x, y)$ is the RANGE of the function.

2. CONVENTION ON DOMAINS

    Unless otherwise stated, the domain of a function specified by an equation of the form $z = f(x, y)$ is the set of all ordered pairs of real numbers (x, y) such that $f(x, y)$ is also a real number.

3. Functions of three independent variables $w = f(x, y, z)$, four independent variables $u = f(x, y, z, w)$, and so on, are defined similarly.

---

1. $a = 8,\ b = 5,\ h = 3;\ A = \dfrac{1}{2}(a+b)h = \dfrac{1}{2}(8+5)3 = \dfrac{39}{2} = 19.5\ \text{ft}^2$

3. $V = lwh = 12 \times 5 \times 4 = 240\ \text{in}^3$

5. Radius $r = 2,\ h = 8;\ V = \pi r^2 h = \pi 4(8) = 32\pi \approx 100.5\ \text{m}^3$

7. Radius $r = 20,\ h = 48;\ T = \pi r\left(r + \sqrt{r^2 + h^2}\right) = \pi(20)\left(20 + \sqrt{20^2 + 48^2}\right) = \pi(20)(72) = 1,440\pi$

$\approx 4,523.9\ \text{cm}^2$

In Problems 9 -15 , $f(x, y) = 2x + 7y - 5$ and $g(x, y) = \dfrac{88}{x^2 + 3y}$.

9. $f(4, -1) = 2(4) + 7(-1) - 5 = -4$

11. $f(8, 0) = 2(8) + 7(0) - 5 = 11$

13. $g(1, 7) = \dfrac{88}{1^2 + 3(7)} = \dfrac{88}{22} = 4$

15. $g(3, -3)$ not defined; $3^2 + 3(-3) = 0$

In Problems 17 - 20, $f(x, y, z) = 2x - 3y^2 + 5z^3 - 1$.

17. $f(0, 0, 0) = 2(0) - 3(0) + 5(0) - 1 = -1$

19. $f(6, -5, 0) = 2(6) - 3(-5)^2 + 5(0) - 1 = 12 - 75 - 1 = -64$

21. $P(n, r) = \dfrac{n!}{(n-r)!};\ P(13, 5) = \dfrac{13!}{(13-5)!} = \dfrac{13!}{8!} = \dfrac{13 \cdot 12 \cdot 11 \cdot 10 \cdot 9 \cdot 8!}{8!} = 13 \cdot 12 \cdot 11 \cdot 10 \cdot 9 = 154,440$

23. $V(R,h) = \pi R^2 h;\ V(4,12) = \pi(4)^2 12 = \pi \cdot 16 \cdot 12 = 192\pi \approx 603.2$

25. $S(R,h) = \pi R \sqrt{R^2 + h^2}\ ;\ S(3,10) = \pi \cdot 3\sqrt{3^2 + 10^2} = 3\pi\sqrt{109} \approx 98.4$

27. $A(P, r, t) = P + Prt$
$A(100, 0.06, 3) = 100 + 100(0.06)3 = 118$
$(P = 100,\ r = 0.06,\ \text{and}\ t = 3)$

29. $P(r,T) = \displaystyle\int_0^T 4000 e^{-rt} dt,$

$P(0.05, 12) = \displaystyle\int_0^{12} 4000 e^{-0.05t} dt = \frac{4000}{-0.05} e^{-0.05t} \Big|_0^{12} = -80{,}000[e^{-0.6} - 1] \approx 36{,}095.07$

31. $G(x, y) = x^2 + 3xy + y^2 - 7;\ f(x) = G(x, 0) = x^2 + 3x(0) + 0^2 - 7 = x^2 - 7$

33. $K(x, y) = 10xy + 3x - 2y + 8;\ f(y) = K(4, y) = 10(4)y + 3(4) - 2y + 8 = 38y + 20$

35. $M(x, y) = x^2 y - 3xy^2 + 5;\ f(y) = M(y, y) = (y)^2 y - 3(y)y^2 + 5 = -2y^3 + 5$

37. $F(x, y) = 2x + 3y - 6;\ F(0, y) = 2(0) + 3y - 6 = 3y - 6$
$F(0, y) = 0:\ 3y - 6 = 0$
$\qquad\qquad\qquad y = 2$

39. $F(x, y) = 2xy + 3x - 4y - 1;\ F(x, x) = 2x(x) + 3x - 4x - 1 = 2x^2 - x - 1$

$F(x, x) = 0:\ 2x^2 - x - 1 = 0$
$\qquad\qquad (2x + 1)(x - 1) = 0$
$\qquad\qquad\qquad\quad x = 1, -\dfrac{1}{2}$

41. $F(x, y) = x^2 + e^x y - y^2;\ F(x, 2) = x^2 + 2e^x - 4.$

We use a graphing utility to solve $F(x, 2) = 0$. The graph of $u = F(x, 2)$ is shown at the right.

The solutions of $F(x, 2) = 0$ are: $x_1 \approx -1.926,\ x_2 \approx 0.599$

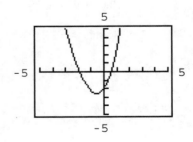

43. $f(x, y) = x^2 + 2y^2$
$$\frac{f(x+h, y) - f(x, y)}{h} = \frac{(x+h)^2 + 2y^2 - (x^2 + 2y^2)}{h} = \frac{x^2 + 2xh + h^2 + 2y^2 - x^2 - 2y^2}{h}$$
$$= \frac{2xh + h^2}{h} = \frac{h(2x + h)}{h} = 2x + h,\ h \neq 0$$

45. $f(x, y) = 2xy^2$

$$\frac{f(x+h, y) - f(x, y)}{h} = \frac{2(x+h)y^2 - 2xy^2}{h} = \frac{2xy^2 + 2hy^2 - 2xy^2}{h} = \frac{2hy^2}{h} = 2y^2, \quad h \neq 0$$

47. Coordinates of point $E = E(0, 0, 3)$.

Coordinates of point $F = F(2, 0, 3)$.

49. $f(x, y) = x^2$

(A) In the plane $y = c$, c any constant, the graph of $z = x^2$ is a parabola.

(B) Cross-section corresponding to $x = 0$: the y-axis

Cross-section corresponding to $x = 1$: the line passing through $(1, 0, 1)$ parallel to the y-axis.

Cross-section corresponding to $x = 2$: the line passing through $(2, 0, 4)$ parallel to the y-axis.

(C) The surface $z = x^2$ is a parabolic trough lying on the y-axis.

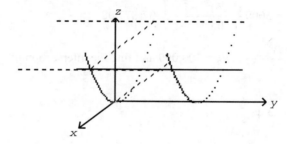

51. $f(x, y) = \sqrt{36 - x^2 - y^2}$

(A) Cross-sections corresponding to $y = 1$, $y = 2$, $y = 3$, $y = 4$, $y = 5$: Upper semicircles with centers at $(0, 1, 0)$, $(0, 2, 0)$, $(0, 3, 0)$, $(0, 4, 0)$, and $(0, 5, 0)$, respectively.

(B) Cross-sections corresponding to $x = 0$, $x = 1$, $x = 2$, $x = 3$, $x = 4$, $x = 5$: Upper semicircles with centers at $(0, 0, 0)$, $(1, 0, 0)$, $(2, 0, 0)$, $(3, 0, 0)$, $(4, 0, 0)$ and $(5, 0, 0)$, respectively.

(C) The upper hemisphere of radius 6 with center at the origin.

53. (A) If the points (a, b) and (c, d) both lie on the same circle centered at the origin, then $a^2 + b^2 = r^2 = c^2 + d^2$, where r is the radius of the circle.

(B) The cross-sections are:

(i) $x = 0$, $f(0, y) = e^{-y^2}$

(ii) $y = 0$, $f(x, 0) = e^{-x^2}$

(iii) $x = y$, $f(x, x) = e^{-2x^2}$

These are bell-shaped curves with maximum value 1 at $y = 0$ in (i) and $x = 0$ in (ii) and (iii).

(C) A "bell" with maximum value 1 at the origin, extending infinitely far in all directions, and approaching the x-y plane as $x, y \to \pm\infty$.

55. Monthly cost function $= C(x, y) = 6000 + 210x + 300y$

$$C(20, 10) = 6000 + 210 \cdot 20 + 300 \cdot 10 = \$13,200$$
$$C(50, 5) = 6000 + 210 \cdot 50 + 300 \cdot 5 = \$18,000$$
$$C(30, 30) = 6000 + 210 \cdot 30 + 300 \cdot 30 = \$21,3000$$

57. $R(p, q) = p \cdot x + q \cdot y = 200p - 5p^2 + 4pq + 300q + 2pq - 4q^2$ or

$R(p, q) = -5p^2 + 6pq - 4q^2 + 200p + 300q$

$R(2, 3) = -5 \cdot 2^2 + 6 \cdot 2 \cdot 3 - 4 \cdot 3^2 + 200 \cdot 2 + 300 \cdot 3 = 1280$ or $1280

$R(3, 2) = -5 \cdot 3^2 + 6 \cdot 3 \cdot 2 - 4 \cdot 2^2 + 200 \cdot 3 + 300 \cdot 2 = 1175$ or $1175

59. $f(x, y) = 20x^{0.4}y^{0.6}$

$f(1250, 1700) = 20(1250)^{0.4}(1700)^{0.6} \approx 20(17.3286)(86.7500) \approx 30,065$ units

61. $F(P, i, n) = P\dfrac{(1+i)^n - 1}{i}$

(A) $P = 5,000$, $i = 0.03$, $n = 30$

$$F(5000, 0.03, 30) = 5000\frac{(1.03)^{30} - 1}{0.03} \approx \$237,877.08$$

(B)   Set $y_1 = 5000\dfrac{(1+i)^{30} - 1}{i}$, $y_2 = 300,000$ and find the intersection of

the two curves.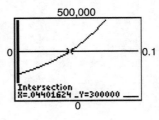

Rate of interest: $i = 4.4\%$

63. $T(V, x) = \dfrac{33V}{x + 33}$

$T(70, 47) = \dfrac{33 \cdot 70}{47 + 33} = \dfrac{33 \cdot 70}{80} = 28.875 \approx 29$ minutes

$T(60, 27) = \dfrac{33 \cdot 60}{27 + 33} = 33$ minutes

65. $C(W, L) = 100\dfrac{W}{L}$

$C(6, 8) = 100\dfrac{6}{8} = 75$

$C(8.1, 9) = 100\dfrac{8.1}{9} = 90$

67. $Q(M, C) = \dfrac{M}{C}100$

$Q(12, 10) = \dfrac{12}{10}100 = 120$

$Q(10, 12) = \dfrac{10}{12}100 = 83.33 \approx 83$

EXERCISE 15-2

Things to remember:

1.   Let $z = f(x, y)$ be a function of two independent variables. The PARTIAL DERIVATIVE
     OF f WITH RESPECT TO x, denoted by

$\dfrac{\partial z}{\partial x}$, f_x, or $f_x(x, y)$, is given by

$$\frac{\partial z}{\partial x} = \lim_{h \to 0} \frac{f(x+h, y) - f(x, y)}{h}$$

provided this limit exists. Similarly, the PARTIAL DERIVATIVE OF f WITH RESPECT TO y, denoted by $\dfrac{\partial z}{\partial y}$, f_y, or $f_y(x, y)$, is given by

$$\frac{\partial z}{\partial y} = \lim_{k \to 0} \frac{f(x, y+k) - f(x, y)}{k}$$

provided this limit exists.

<u>2.</u>   SECOND-ORDER PARTIAL DERIVATIVES

If $z = f(x, y)$, then:

$$\frac{\partial^2 z}{\partial x^2} = \frac{\partial\left(\frac{\partial z}{\partial x}\right)}{\partial x} = f_{xx}(x, y) = f_{xx}$$

$$\frac{\partial^2 z}{\partial y \partial x} = \frac{\partial\left(\frac{\partial z}{\partial x}\right)}{\partial y} = f_{xy}(x, y) = f_{xy}$$

$$\frac{\partial^2 z}{\partial x \partial y} = \frac{\partial\left(\frac{\partial z}{\partial y}\right)}{\partial x} = f_{yx}(x, y) = f_{yx}$$

$$\frac{\partial^2 z}{\partial y^2} = \frac{\partial\left(\frac{\partial z}{\partial y}\right)}{\partial y} = f_{yy}(x, y) = f_{yy}$$

[<u>Note:</u> For the functions being considered in this text, the mixed partial derivatives f_{xy} and f_{yx} are equal, i.e., $\dfrac{\partial^2 z}{\partial x \partial y} = \dfrac{\partial^2 z}{\partial y \partial x}$.]

---

1. $f(x) = \pi x^3 + x\pi^3,\quad f'(x) = 3\pi x^2 + \pi^3$

3. $f(x) = x^e + e^x;\quad f'(x) = ex^{e-1} + e^x$

5. $z = \dfrac{x}{e} + \dfrac{e}{x};\quad \dfrac{dz}{dx} = \dfrac{1}{e} - \dfrac{e}{x^2}$

7. $z = \ln(x^2 + e^2);\quad \dfrac{dz}{dx} = \dfrac{1}{x^2 + e^2}(2x) = \dfrac{2x}{x^2 + e^2}$

9. $f(x, y) = 4x - 3y + 6$

   $f_x(x, y) = 4$

11. $f(x, y) = x^2 - 3xy + 2y^2$

   $f_y(x, y) = 0 - 3x + 4y = -3x + 4y$

13. $z = x^3 + 4x^2 y + 2y^3$

   $\dfrac{\partial z}{\partial x} = 3x^2 + 8xy + 0 = 3x^2 + 8xy$

15. $z = (5x + 2y)^{10}$

   $\dfrac{\partial z}{\partial y} = 10(5x + 2y)^9 \dfrac{\partial(5x + 2y)}{\partial y}$

   $= 10(5x + 2y)^9(2) = 20(5x + 2y)^9$

17. $f(x, y) = 5x^3y - 4xy^2;$

$f_x(x, y) = 15x^2y - 4y^2$

$f_x(1, 3) = 15(1)^2 3 - 4(3)^2 = 45 - 36 = 9$

19. $f(x, y) = 3xe^y;$

$f_y(x, y) = 3xe^y$

$f_y(1, 0) = 3(1)e^0 = 3$

21. $f(x, y) = e^{x^2} - 4y;$ $f_y(x, y) = 0 - 4 = -4;$ $f_y(2, 1) = -4$

23. $f(x, y) = \dfrac{2xy}{1 + x^2y^2};$ $f_x(x, y) = \dfrac{(1 + x^2y^2)(2y) - 2xy(2xy^2)}{(1 + x^2y^2)^2} = \dfrac{2y - 2x^2y^3}{(1 + x^2y^2)^2} = \dfrac{2y(1 - x^2y^2)}{(1 + x^2y^2)^2}$

$f_x(1, -1) = \dfrac{2(-1)[1 - 1^2(-1)^2]}{[1 + 1^2(-1)^2]^2} = 0$

25. $M(x, y) = 68 + 0.3x - 0.8y;$ $M(32, 40) = 68 + 0.3(32) - 0.8(40) = 68 + 9.6 - 32 = 45.6$

Mileage is 45.6 mpg at a tire pressure of 32 psi and a speed of 40 mph.

27. $M(x, y) = 68 + 0.3x - 0.8y;$ $M(32, 50) = 68 + 0.3(32) - 0.8(50) = 68 + 9.6 - 40 = 37.6$

Mileage is 37.6 mpg at a tire pressure of 32 psi and a speed of 50 mph.

29. $M(x, y) = 68 + 0.3x - 0.8y;$ $M_x = 0.3$ mpg per psi; mileage increases at the rate of 0.3 mpg per psi of tire

pressure.

31. $f(x, y) = 6x - 5y + 3;$ $f_x(x, y) = 6;$ $f_{xx}(x, y) = 0$

33. $f(x, y) = 4x^2 + 6y^2 - 10;$ $f_x(x, y) = 8x;$ $f_{xy}(x, y) = 0$

35. $f(x, y) = e^{xy^2}$

$f_x(x, y) = e^{xy^2}y^2 = y^2e^{xy^2}$

$f_{xy}(x, y) = y^2e^{xy^2}(2xy) + 2ye^{xy^2} = 2xy^3e^{xy^2} + 2ye^{xy^2} = 2y(1 + xy^2)e^{xy^2}$

37. $f(x, y) = \dfrac{\ln x}{y};$ $f_y(x, y) = -\dfrac{\ln x}{y^2};$ $f_{yy}(x, y) = \dfrac{2\ln x}{y^3}$

39. $f(x, y) = (2x + y)^5;$ $f_x(x, y) = 5(2x + y)^4 \cdot 2 = 10(2x + y)^4$

$f_{xx}(x, y) = 40(2x + y)^3(2) = 80(2x + y)^3$

41. $f(x, y) = (x^2 + y^4)^{10};$ $f_x(x, y) = 10(x^2 + y^4)^9(2x) = 20x(x^2 + y^4)^9$

$f_{xy}(x, y) = 180x(x^2 + y^4)^8(4y^3) = 720xy^3(x^2 + y^4)^8$

In Problems 43 – 51, $C(x, y) = 3x^2 + 10xy - 8y^2 + 4x - 15y - 120$

43. $C_x(x, y) = 6x + 10y + 4$

45. From Problem 43, $C_x(3, -2) = 6(3) + 10(-2) + 4$

$= 2$

47. From Problem 43, $C_{xx}(x, y) = 6$

49. From Problem 43, $C_{xy}(x, y) = 10$

51. From Problem 47, $C_{xx}(3, -2) = 6$

In Problems 53 –57, S(T,r) =50(T − 40)(5 − r).

53. $S(T,r) = 50(T-40)(5-r);\ \ S(60,2) = 50(60-40)(5-2) = 50(20)(3) = 3{,}000$

Daily sales are $3,000 when the temperature is $60°$ and the rainfall is 2 inches.

55. $S(T,r) = 50(T-40)(5-r);\ \ S_r(T,r) = 50(T-40)(-1) = -50(T-40)$

$S_r(90,1) = -50(90-40) = -50(50) = -2{,}500\,.$ Daily sales decrease at a rate of $2,500 per inch of rain

when the temperature is $90°$ and the rainfall is 1 inch.

57. $S(T,r) = 50(T-40)(5-r);\ \ S_T(T,r) = 50(5-r);\ \ S_{Tr}(T,r) = -50$

S_r decreases at the rate of $50 per inch per degree of temperature.

59. (A) $f(x,\ y) = y^3 + 4y^2 - 5y + 3$

Since f is independent of x, $\dfrac{\partial f}{\partial x} = 0$

(B) If $g(x,\ y)$ depends on y only, that is, if $g(x,\ y) = G(y)$ is

independent of x, then

$$\frac{\partial g}{\partial x} = 0$$

Clearly there are an infinite number of such functions.

61. $f(x,y) = x^2 y^2 + x^3 + y$

$f_x(x,y) = 2xy^2 + 3x^2$ $\qquad\qquad\qquad$ $f_y(x,y) = 2x^2 y + 1$

$f_{xx}(x,y) = 2y^2 + 6x$ $\qquad\qquad\qquad$ $f_{yx}(x,y) = 4xy$

$f_{xy}(x,y) = 4xy$ $\qquad\qquad\qquad\qquad$ $f_{yy}(x,y) = 2x^2$

63. $f(x,y) = \dfrac{x}{y} - \dfrac{y}{x}$

$f_x(x,y) = \dfrac{1}{y} + \dfrac{y}{x^2}$ $\qquad\qquad\qquad$ $f_y(x,y) = -\dfrac{x}{y^2} - \dfrac{1}{x}$

$f_{xx}(x,y) = -\dfrac{2y}{x^3}$ $\qquad\qquad\qquad$ $f_{yx}(x,y) = -\dfrac{1}{y^2} + \dfrac{1}{x^2}$

$f_{xy}(x,y) = -\dfrac{1}{y^2} + \dfrac{1}{x^2}$ $\qquad\qquad$ $f_{yy}(x,y) = \dfrac{2x}{y^3}$

65. $f(x,y) = xe^{xy}$

$f_x(x,y) = xye^{xy} + e^{xy}$ $\qquad\qquad$ $f_y(x,y) = x^2 e^{xy}$

$f_{xx}(x,y) = xy^2 e^{xy} + 2ye^{xy}$ $\qquad\qquad$ $f_{yx}(x,y) = x^2 ye^{xy} + 2xe^{xy}$

$f_{xy}(x,y) = x^2 ye^{xy} + 2xe^{xy}$ $\qquad\qquad$ $f_{yy}(x,y) = x^3 e^{xy}$

67. $P(x, y) = -x^2 + 2xy - 2y^2 - 4x + 12y - 5$

$P_x(x, y) = -2x + 2y - 0 - 4 + 0 - 0 = -2x + 2y - 4$

$P_y(x, y) = 0 + 2x - 4y - 0 + 12 - 0 = 2x - 4y + 12$

$P_x(x, y) = 0$ and $P_y(x, y) = 0$ when

$-2x + 2y - 4 = 0$ (1)

$2x - 4y + 12 = 0$ (2)

Add equations (1) and (2): $-2y + 8 = 0$

$$y = 4$$

Substitute $y = 4$ into (1): $-2x + 2 \cdot 4 - 4 = 0$

$$-2x + 4 = 0$$

$$x = 2$$

Thus, $P_x(x, y) = 0$ and $P_y(x, y) = 0$ when $x = 2$ and $y = 4$.

69. $F(x, y) = x^3 - 2x^2y^2 - 2x - 4y + 10;$

$F_x(x, y) = 3x^2 - 4xy^2 - 2; \; F_y(x, y) = -4x^2y - 4.$

Set $F_x(x, y) = 0$ and $F_y(x, y) = 0$ and solve simultaneously:

$$3x^2 - 4xy^2 - 2 = 0 \qquad (1)$$

$$-4x^2y - 4 = 0 \qquad (2)$$

From (2), $y = -\dfrac{1}{x^2}$. Substituting this into (1),

$$3x^2 - 4x\left(-\frac{1}{x^2}\right)^2 - 2 = 0$$

$$3x^2 - 4x\left(\frac{1}{x^4}\right) - 2 = 0$$

$$3x^5 - 2x^3 - 4 = 0$$

Using a graphing utility, we find that $x \approx 1.1996$. Then, $y \approx -0.695$.

71. $f(x, y) = 3x^2 + y^2 - 4x - 6y + 2$

(A) $f(x, 1) = 3x^2 + 1 - 4x - 6 + 2$

$$= 3x^2 - 4x - 3$$

$\dfrac{d}{dx}[f(x, 1)] = 6x - 4$; critical values: $6x - 4 = 0, \; x = \dfrac{2}{3}$

$\dfrac{d^2}{dx^2}[f(x, 1)] = 6 > 0$

Therefore, $f\left(\dfrac{2}{3}, 1\right) = 3\left(\dfrac{2}{3}\right)^2 - 4\left(\dfrac{2}{3}\right) - 3 = -\dfrac{13}{3}$ is the minimum value of $f(x, 1)$.

(B) $-\dfrac{13}{3}$ is the minimum value of $f(x, y)$ on the curve $z = f(x, 1)$; $f(x, y)$ may have smaller values on other curves $z = f(x, k)$, k constant, or $z = f(h, y)$, h constant. For example, the minimum value of

$$f(x, 2) = 3x^2 - 4x - 6 \text{ is } f\left(\dfrac{2}{3}, 2\right) = -\dfrac{22}{3}; \text{ the minimum value of } f(0, y) = y^2 - 6y + 2 \text{ is } f(0,$$

73. $f(x, y) = 4 - x^4 y + 3xy^2 + y^5$

(A) Let $y = 2$ and find the maximum value of $f(x, 2) = 4 - 2x^4 + 12x + 32 = -2x^4 + 12x + 36$. Using a graphing utility we find that the maximum value of $f(x, 2)$ is 46.302 at $x = 1.1447152 \approx 1.145$.

(B) $f_x(x, y) = -4x^3 y + 3y^2$

$f_x(1.145, 2) = -0.008989 \approx 0$

$f_y(x, y) = -x^4 + 6xy + 5y^4$

$f_y(1.145, 2) = 92.021$

75. $f(x, y) = x^2 + 2y^2$

(A) $\displaystyle\lim_{h \to 0} \dfrac{f(x+h, y) - f(x, y)}{h} = \lim_{h \to 0} \dfrac{(x+h)^2 + 2y^2 - (x^2 + 2y^2)}{h}$

$\displaystyle = \lim_{h \to 0} \dfrac{x^2 + 2xh + h^2 + 2y^2 - x^2 - 2y^2}{h}$

$\displaystyle = \lim_{h \to 0} \dfrac{h(2x + h)}{h} = \lim_{h \to 0} (2x + h) = 2x$

(B) $\displaystyle\lim_{k \to 0} \dfrac{f(x, y+k) - f(x, y)}{k} = \lim_{k \to 0} \dfrac{x^2 + 2(y+k)^2 - (x^2 + 2y^2)}{k}$

$\displaystyle = \lim_{k \to 0} \dfrac{x^2 + 2(y^2 + 2yk + k^2) - x^2 - 2y^2}{k}$

$\displaystyle = \lim_{k \to 0} \dfrac{4yk + 2k^2}{k} = \lim_{k \to 0} (4y + 2k) = 4y$

77. $R(x, y) = 80x + 90y + 0.04xy - 0.05x^2 - 0.05y^2$
$C(x, y) = 8x + 6y + 20,000$
The profit $P(x, y)$ is given by:
$P(x, y) = R(x, y) - C(x, y)$

$\quad = 80x + 90y + 0.04xy - 0.05x^2 - 0.05y^2 - (8x + 6y + 20,000)$

$\quad = 72x + 84y + 0.04xy - 0.05x^2 - 0.05y^2 - 20,000$

Now
$P_x(x, y) = 72 + 0.04y - 0.1x$

and
$P_x(1200, 1800) = 72 + 0.04(1800) - 0.1(1200) = 72 + 72 - 120 = 24;$

$P_y(x, y) = 84 + 0.04x - 0.1y$

and
$P_y(1200, 1800) = 84 + 0.04(1200) - 0.1(1800) = 84 + 48 - 180 = -48.$

Thus, at the (1200, 1800) output level, profit will increase approximately $24 per unit increase in production of type A calculators; and profit will decrease $48 per unit increase in production of type B calculators.

79. $x = 200 - 5p + 4q$

$y = 300 - 4q + 2p$

$\dfrac{\partial x}{\partial p} = -5, \dfrac{\partial y}{\partial p} = 2$

A $1 increase in the price of brand A will decrease the demand for brand A by 5 pounds at any price level (p, q).

A $1 increase in the price of brand A will increase the demand for brand B by 2 pounds at any price level (p, q).

81. $f(x, y) = 10x^{0.75}y^{0.25}$

(A) $f_x(x, y) = 10(0.75)x^{-0.25}y^{0.25} = 7.5x^{-0.25}y^{0.25}$

$f_y(x, y) = 10(0.25)x^{0.75}y^{-0.75} = 2.5x^{0.75}y^{-0.75}$

(B) Marginal productivity of labor $= f_x(600, 100) = 7.5(600)^{-0.25}(100)^{0.25} \approx 4.79$

Marginal productivity of capital $= f_y(600, 100) = 2.5(600)^{0.75}(100)^{-0.75} \approx 9.58$

(C) The government should encourage the increased use of capital.

83. $x = f(p, q) = 8000 - 0.09p^2 + 0.08q^2$ (Butter)

$y = g(p, q) = 15,000 + 0.04p^2 - 0.3q^2$ (Margarine)

$f_q(p, q) = 0.08(2)q = 0.16q > 0$

$g_p(p, q) = 0.04(2)p = 0.08p > 0$

Thus, the products are competitive.

85. $x = f(p, q) = 800 - 0.004p^2 - 0.003q^2$ (Skis)

$y = g(p, q) = 600 - 0.003p^2 - 0.002q^2$ (Ski boots)

$f_q(p, q) = -0.003(2)q = -0.006q < 0$

$g_p(p, q) = -0.003(2)p = -0.006p < 0$

Thus, the products are complementary.

87. $A = f(w, h) = 15.64w^{0.425}h^{0.725}$

(A) $f_w(w, h) = 15.64(0.425)w^{-0.575}h^{0.725} \approx 6.65w^{-0.575}h^{0.725}$

$f_h(w, h) = 15.64(0.725)w^{0.425}h^{-0.275} \approx 11.34w^{0.425}h^{-0.275}$

(B) $f_w(65, 57) = 6.65(65)^{-0.575}(57)^{0.725} \approx 11.31$

For a 65 pound child 57 inches tall, the rate of change of surface area is approximately 11.31 square inches for a one-pound gain in weight, height held fixed.

$f_h(65, 57) = 11.34(65)^{0.425}(57)^{-0.275} \approx 21.99$

For a 65 pound child 57 inches tall, the rate of change of surface area is approximately 21.99 square inches for a one-inch gain in height, weight held fixed.

89. $C(W, L) = 100 \dfrac{W}{L}$ 　　　　　　　　　　　$C_L(W, L) = -\dfrac{100W}{L^2}$

$C_W(W, L) = \dfrac{100}{L}$ 　　　　　　　　　　　$C_L(6, 8) = -\dfrac{100 \times 6}{8^2}$

$C_W(6, 8) = \dfrac{100}{8} = 12.5$ 　　　　　　　　　　$= -\dfrac{600}{64} = -9.38$

The index increases 12.5 units
per 1-inch increase in the width
of the head (length held fixed)
when $W = 6$ and $L = 8$.

The index decreases 9.38 units
per 1-inch increase in length
(width held fixed) when $W = 6$
and $L = 8$.

EXERCISE 15-3

Things to remember:

1.　$f(a, b)$ is a LOCAL MAXIMUM if there exists a circular region in the domain of $f(x, y)$ with (a, b) as the center, such that $f(a, b) \geq f(x, y)$ for all (x, y) in the region. Similarly, $f(a, b)$ is a LOCAL MINIMUM if $f(a, b) \leq f(x, y)$ for all (x, y) in the region.

2.　LOCAL EXTREMA AND PARTIAL DERIVATIVES
Let $f(a, b)$ be a local extremum (a local maximum or a local minimum) for the function f. If both f_x and f_y exist at (a, b) then

$$f_x(a, b) = 0 \quad \text{and} \quad f_y(a, b) = 0$$

Points (a, b) such that $f_x(a, b) = f_y(a, b) = 0$ are called CRITICAL POINTS OF f.

3.　SECOND-DERIVATIVE TEST FOR LOCAL EXTREMA FOR $z = f(x, y)$
Given:

(a)　$f_x(a, b) = 0$ and $f_y(a, b) = 0$ [(a, b) is a critical point].

(b)　All second-order partial derivatives of f exist in some circular region containing (a, b) as center.

(c)　$A = f_{xx}(a, b)$, $B = f_{xy}(a, b)$, $C = f_{yy}(a, b)$.

Then:

i)　If $AC - B^2 > 0$ and $A < 0$, then $f(a, b)$ is a local maximum.

ii)　If $AC - B^2 > 0$ and $A > 0$, then $f(a, b)$ is a local minimum.

iii)　If $AC - B^2 < 0$, then f has a saddle point at (a, b).

iv)　If $AC - B^2 = 0$, then the test fails.

1.　$f(x) = 2x^3 - 9x^2 + 4$, 　$f'(x) = 6x^2 - 18x$, 　$f''(x) = 12x - 18$

$f'(0) = 6(0)^2 - 18(0) = 0$, 　$f''(0) = 12(0) - 18 = -18 < 0$; 　f has a local maximum at $x = 0$.

3. $f(x) = \dfrac{1}{1-x^2}$, $f'(x) = \dfrac{2x}{(1-x^2)^2}$,

   $f''(x) = \dfrac{2(1-x^2)^2 - 2x[2(1-x^2)(-2x)]}{(1-x^2)^4} = \dfrac{2(1-x^2)+8x^2}{(1-x^2)^3} = \dfrac{2+6x^2}{(1-x^2)^3}$

   $f'(0) = \dfrac{2(0)}{(1-0^2)^2} = 0$, $f''(0) = \dfrac{2+6(0)^2}{(1-0^2)^3} = 2 > 0$, f has a local minimum at $x = 0$.

5. $f(x) = e^{-x^2}$, $f'(x) = -2xe^{-x^2}$, $f''(x) = -2e^{-x^2} + 4x^2 e^{-x^2}$.

   $f'(0) = -2(0)e^{-0^2} = 0$, $f''(0) = -2e^{-0^2} + 4(0)^2 e^{-0^2} = -2 < 0$, f has a local maximum at $x = 0$.

7. $f(x) = x^3 - x^2 + x + 1$; $f'(x) = 3x^2 - 2x + 1$, $f''(x) = 6x - 2$

   $f'(0) = 3(0)^2 - 2(0) + 1 = 1$, $f''(0) = 6(0) - 2 = -2$

   $f'(0) = 1 \neq 0$. Therefore f has neither a maximum nor a minimum at $x = 0$.

9. $f(x, y) = 4x + 5y - 6$

   $f_x(x, y) = 4 \neq 0$; $f_y(x, y) = 5 \neq 0$; the functions $f_x(x, y)$ and $f_y(x, y)$ are nonzero for all (x, y).

11. $f(x, y) = 3.7 - 1.2x + 6.8y + 0.2y^3 + x^4$

    $f_x(x, y) = -1.2 + 4x^3$; $f_y = 6.8 + 0.6y^2$; the function $f_y(x, y)$ is nonzero for all (x, y).

13. $f(x, y) = 6 - x^2 - 4x - y^2$

    $f_x(x, y) = -2x - 4 = 0$ implies $x = -2$.

    $f_y(x, y) = -2y = 0$ implies $y = 0$.

    Thus, $(-2, 0)$ is a critical point.

    $f_{xx} = -2$, $f_{xy} = 0$, $f_{yy} = -2$,

    $f_{xx}(-2, 0) \cdot f_{yy}(-2, 0) - [f_{xy}(-2, 0)]^2 = (-2)(-2) - 0^2 = 4 > 0$ and $f_{xx}(-2, 0) = -2 < 0$.

    Thus, $f(-2, 0) = 6 - (-2)^2 - 4(-2) - 0^2 = 10$ is a local maximum (using 3).

15. $f(x, y) = x^2 + y^2 + 2x - 6y + 14$

    $f_x(x, y) = 2x + 2 = 0$ implies $x = -1$

    $f_y(x, y) = 2y - 6 = 0$ implies $y = 3$

    Thus, $(-1, 3)$ is a critical point.

    $f_{xx} = 2$, $f_{xy} = 0$, $f_{yy} = 2$

    $f_{xx}(-1, 3) = 2 > 0$, $f_{xy}(-1, 3) = 0$, $f_{yy}(-1, 3) = 2$

    $f_{xx}(-1, 3) \cdot f_{yy}(-1, 3) - [f_{xy}(-1, 3)]^2 = 2 \cdot 2 - 0^2 = 4 > 0$

    Thus, using 3, $f(-1, 3) = 4$ is a local minimum.

17. $f(x, y) = xy + 2x - 3y - 2$

$f_x = y + 2 = 0$ implies $y = -2$

$f_y = x - 3 = 0$ implies $x = 3$

Thus, $(3, -2)$ is a critical point.

$f_{xx} = 0$, $f_{xy} = 1$, $f_{yy} = 0$

$f_{xx}(3, -2) = 0$, $f_{xy}(3, -2) = 1$, $f_{yy}(3, -2) = 0$

$f_{xx}(3, -2) \cdot f_{yy}(3, -2) - [f_{xy}(3, -2)]^2 = 0 \cdot 0 - [1]^2 = -1 < 0$

Thus, using $\underline{3}$, f has a saddle point at $(3, -2)$.

19. $f(x, y) = -3x^2 + 2xy - 2y^2 + 14x + 2y + 10$

$f_x = -6x + 2y + 14 = 0$ (1)

$f_y = 2x - 4y + 2 = 0$ (2)

Solving (1) and (2) for x and y, we obtain $x = 3$ and $y = 2$.
Thus, $(3, 2)$ is a critical point.

$f_{xx} = -6$, $f_{xy} = 2$, $f_{yy} = -4$

$f_{xx}(3, 2) = -6 < 0$, $f_{xy}(3, 2) = 2$, $f_{yy}(3, 2) = -4$

$f_{xx}(3, 2) \cdot f_{yy}(3, 2) - [f_{xy}(3, 2)]^2 = (-6)(-4) - 2^2 = 20 > 0$

Thus, using $\underline{3}$, $f(3, 2)$ is a local maximum and $f(3, 2) = -3 \cdot 3^2 + 2 \cdot 3 \cdot 2 - 2 \cdot 2^2 + 14 \cdot 3 + 2 \cdot 2 + 10 = 33$.

21. $f(x, y) = 2x^2 - 2xy + 3y^2 - 4x - 8y + 20$

$f_x = 4x - 2y - 4 = 0$ (1)

$f_y = -2x + 6y - 8 = 0$ (2)

Solving (1) and (2) for x and y, we obtain $x = 2$ and $y = 2$.
Thus, $(2, 2)$ is a critical point.

$f_{xx} = 4$, $f_{xy} = -2$, $f_{yy} = 6$

$f_{xx}(2, 2) = 4 > 0$, $f_{xy}(2, 2) = -2$, $f_{yy}(2, 2) = 6$

$f_{xx}(2, 2) \cdot f_{yy}(2, 2) - [f_{xy}(2, 2)]^2 = 4 \cdot 6 - [-2]^2 = 20 > 0$

Thus, using $\underline{3}$, $f(2, 2)$ is a local minimum and $f(2, 2) = 2 \cdot 2^2 - 2 \cdot 2 \cdot 2 + 3 \cdot 2^2 - 4 \cdot 2 - 8 \cdot 2 + 20 = 8$.

23. $f(x, y) = e^{xy}$

$f_x = e^{xy} \dfrac{\partial(xy)}{\partial x}$ $\qquad\qquad$ $f_y = e^{xy} \dfrac{\partial(xy)}{\partial y}$

$\quad = e^{xy} y = 0$ $\qquad\qquad\qquad$ $= e^{xy} x = 0$

$\qquad y = 0$ $\;(e^{xy} \neq 0)$ $\qquad\qquad$ $x = 0$ $\;(e^{xy} \neq 0)$

Thus, $(0, 0)$ is a critical point.

$f_{xx} = ye^{xy} \dfrac{\partial(xy)}{\partial x}$, $\qquad$ $f_{xy} = e^{xy} \cdot 1 + ye^{xy} x$ $\qquad$ $f_{yy} = xe^{xy} \dfrac{\partial(xy)}{\partial y}$

$\quad = ye^{xy} y$ $\qquad\qquad\qquad$ $= e^{xy} + xye^{xy}$ $\qquad\qquad$ $= x^2 e^{xy}$

$\quad = y^2 e^{xy}$

$f_{xx}(0, 0) = 0$, $\qquad\qquad$ $f_{xy}(0, 0) = 1 + 0 = 1$, $\qquad$ $f_{yy}(0, 0) = 0$

$f_{xx}(0, 0) \cdot f_{yy}(0, 0) - [f_{xy}(0, 0)]^2 = 0 - [1]^2 = -1 < 0$

Thus, using $\underline{3}$, $f(x, y)$ has a saddle point at $(0, 0)$.

25. $f(x, y) = x^3 + y^3 - 3xy$

$f_x = 3x^2 - 3y = 3(x^2 - y) = 0$

Thus, $y = x^2$. (1)

$f_y = 3y^2 - 3x = 3(y^2 - x) = 0$

Thus, $y^2 = x$. (2)

Combining (1) and (2), we obtain $x = x^4$ or $x(x^3 - 1) = 0$. Therefore, $x = 0$ or $x = 1$, and the critical points are (0, 0) and (1, 1).

$f_{xx} = 6x$ $\qquad\qquad$ $f_{xy} = -3$ $\qquad\qquad$ $f_{yy} = 6y$

For the critical point (0, 0):

$f_{xx}(0, 0) = 0$ $\qquad\qquad$ $f_{xy}(0, 0) = -3$ $\qquad\qquad$ $f_{yy}(0, 0) = 0$

$f_{xx}(0, 0) \cdot f_{yy}(0, 0) - [f_{xy}(0, 0)]^2 = 0 - (-3)^2 = -9 < 0$

Thus, using $\underline{3}$, $f(x, y)$ has a saddle point at (0, 0).

For the critical point (1, 1):

$f_{xx}(1, 1) = 6$ $\qquad\qquad$ $f_{xy}(1, 1) = -3$ $\qquad\qquad$ $f_{yy}(1, 1) = 6$

$f_{xx}(1, 1) \cdot f_{yy}(1, 1) - [f_{xy}(1, 1)]^2 = 6 \cdot 6 - (-3)^2 = 27 > 0$

$f_{xx}(1, 1) > 0$

Thus, using $\underline{3}$, $f(1, 1)$ is a local minimum and $f(1, 1) = 1^3 + 1^3 - 3 \cdot 1 \cdot 1 = 2 - 3 = -1$.

27. $f(x, y) = 2x^4 + y^2 - 12xy$

$f_x = 8x^3 - 12y = 0$

Thus, $y = \frac{2}{3}x^3$.

$f_y = 2y - 12x = 0$

Thus, $y = 6x$

Therefore, $6x = \frac{2}{3}x^3$

$x^3 - 9x = 0$

$x(x^2 - 9) = 0$

$x = 0, \quad x = 3, \quad x = -3$

Thus, the critical points are (0, 0), (3, 18), (−3, −18). Now,

$f_{xx} = 24x^2$ $\qquad$ $f_{xy} = -12$ $\qquad$ $f_{yy} = 2$

For the critical point (0, 0):

$f_{xx}(0, 0) = 0$ $\qquad$ $f_{xy}(0, 0) = -12$ $\qquad$ $f_{yy}(0, 0) = 2$

and

$f_{xx}(0, 0) \cdot f_{yy}(0, 0) - [f_{xy}(0, 0)]^2 = 0 \cdot 2 - (-12)^2 = -144 < 0.$

Thus, $f(x, y)$ has a saddle point at (0, 0).

For the critical point (3, 18):

$f_{xx}(3, 18) = 24 \cdot 3^2 = 216 > 0$ $\qquad$ $f_{xy}(3, 18) = -12$ $\qquad$ $f_{yy}(3, 18) = 2$

and

$$f_{xx}(3, 18) \cdot f_{yy}(3, 18) - [f_{xy}(3, 18)]^2 = 216 \cdot 2 - (-12)^2 = 288 > 0$$

Thus, $f(3, 18) = -162$ is a local minimum.

For the critical point $(-3, -18)$:
$$f_{xx}(-3, -18) = 216 > 0 \qquad f_{xy}(-3, -18) = -12 \qquad f_{yy}(-3, -18) = 2$$

and

$$f_{xx}(-3, -18) \cdot f_{yy}(-3, -18) - [f_{xy}(-3, -18)]^2 = 288 > 0$$

Thus, $f(-3, -18) = -162$ is a local minimum.

29. $f(x, y) = x^3 - 3xy^2 + 6y^2$
$$f_x = 3x^2 - 3y^2 = 0$$

Thus, $y^2 = x^2$ or $y = \pm x$.
$$f_y = -6xy + 12y = 0 \quad \text{or} \quad -6y(x - 2) = 0$$

Thus, $y = 0$ or $x = 2$.

Therefore, the critical points are $(0, 0)$, $(2, 2)$, and $(2, -2)$.

Now,
$$f_{xx} = 6x, \qquad f_{xy} = -6y, \qquad f_{yy} = -6x + 12$$

For the critical point $(0, 0)$:
$$f_{xx}(0, 0) \cdot f_{yy}(0, 0) - [f_{xy}(0, 0]^2 = 0 \cdot 12 - 0^2 = 0$$

Thus, the second-derivative test fails.

For the critical point $(2, 2)$:
$$f_{xx}(2, 2) \cdot f_{yy}(2, 2) - [f_{xy}(2, 2)]^2 = 12 \cdot 0 - (-12)^2 = -144 < 0$$

Thus, $f(x, y)$ has a saddle point at $(2, 2)$.

For the critical point $(2, -2)$:
$$f_{xx}(2, -2) \cdot f_{yy}(2, -2) - [f_{xy}(2, -2)]^2 = 12 \cdot 0 - (12)^2 = -144 < 0$$

Thus, $f(x, y)$ has a saddle point at $(2, -2)$.

31. $f(x, y) = y^3 + 2x^2y^2 - 3x - 2y + 8;$
$$f_x = 4xy^2 - 3; \quad f_y = 3y^2 + 4x^2y - 2$$

Set $f_x = 0$ and $f_y = 0$ to find the critical points:

$$4xy^2 - 3 = 0 \qquad (1)$$
$$3y^2 + 4x^2y - 2 = 0 \qquad (2)$$

From (1) $x = \dfrac{3}{4y^2}$. Substituting this into (2), we have

$$3y^2 + 4\left(\frac{3}{4y^2}\right)^2 y - 2 = 0$$

$$3y^2 + 4\left(\frac{9}{16y^4}\right)y - 2 = 0$$

$$12y^5 - 8y^3 + 9 = 0$$

Using a graphing utility, we find that $y \approx -1.105$ and $x \approx 0.614$.

Now, $f_{xx} = 4y^2$ and $f_{xx}(0.614, -1.105) \approx 4.884$

$f_{xy} = 8xy$ and $f_{xy}(0.614, -1.105) \approx -5.428$

$f_{yy} = 6y + 4x^2$ and $f_{yy}(0.614, -1.105) \approx -5.122$

$f_{xx}(0.614, -1.105) \cdot f_{yy}(0.614, -1.105) - [f_{xy}(0.614, -1.105)]^2 \approx -54.479 < 0$

Thus, $f(x, y)$ has a saddle point at $(0.614, -1.105)$.

33. $f(x, y) = x^2 \geq 0$ for all (x, y) and $f(x, y) = 0$ when $x = 0$. Thus, f has a local minimum at each point $(0, y, 0)$ on the y–axis.

35. $f(x, y) = x^4 e^y + x^2 y^4 + 1$

(A) $f_x = 4x^3 e^y + 2xy^4 = 0$ \hfill (1)

$f_y = x^4 e^y + 4x^2 y^3 = 0$ \hfill (2)

The values $x = 0$, $y = 0$ satisfy (1) and (2) so $(0, 0)$ is a critical point.

$A = f_{xx} = 12x^2 e^y + 2y^4 = 0$ at $(0, 0)$,

$B = f_{xy} = 4x^3 e^y + 8xy^3 = 0$ at $(0, 0)$,

$C = f_{yy} = x^4 e^y + 12x^2 y^2 = 0$ at $(0, 0)$.

$AC - B^2 = 0$; the second derivative test fails.

(B) Cross–sections of f by the planes $y = 0$, $x = 0$, $y = x$ and $y = -x$ are shown at the right. The cross-sections indicate that f has a local minimum at $(0, 0)$.

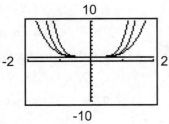

37. $P(x, y) = R(x, y) - C(x, y)$

$= 2x + 3y - (x^2 - 2xy + 2y^2 + 6x - 9y + 5) = -x^2 + 2xy - 2y^2 - 4x + 12y - 5$

$P_x = -2x + 2y - 4 = 0$ (1)

$P_y = 2x - 4y + 12 = 0$ (2)

Solving (1) and (2) for x and y, we obtain $x = 2$ and $y = 4$. Thus, $(2, 4)$ is a critical point.

$P_{xx} = -2$ and $P_{xx}(2, 4) = -2 < 0$

$P_{xy} = 2$ and $P_{xy}(2, 4) = 2$

$P_{yy} = -4$ and $P_{yy}(2, 4) = -4$

$P_{xx}(2, 4) \cdot P_{yy}(2, 4) - [P_{xy}(2, 4)]^2 = (-2)(-4) - [2]^2 = 4 > 0$

The maximum occurs when 2000 type A and 4000 type B earphones are produced. The maximum profit is given by $P(2, 4)$. Hence,

$\max P = P(2, 4) = -(2)^2 + 2 \cdot 2 \cdot 4 - 2 \cdot 4^2 - 4 \cdot 2 + 12 \cdot 4 - 5 = -4 + 16 - 32 - 8 + 48 - 5 = \$15$ million.

39. $x = 260 - 3p + q$ (Brand A)

$y = 180 + p - 2q$ (Brand B)

(A)

p	q	x	y
100	120	80	40
110	110	40	70

(B)   In terms of p and q, the cost function C is given by:

$C = 60x + 80y = 60(260 - 3p + q) + 80(180 + p - 2q) = 30{,}000 - 100p - 100q$

The revenue function R is given by:

$R = px + qy = p(260 - 3p + q) + q(180 + p - 2q)$

$\qquad = -3p^2 + 2pq - 2q^2 + 260p + 180q$

Thus, the profit $P = R - C$ is given by:

$P = -3p^2 + 2pq - 2q^2 + 260p + 180q - (30{,}000 - 100p - 100q)$

$\quad = -3p^2 + 2pq - 2q^2 + 360p + 280q - 30{,}000$

Now, calculating P_p and P_q and setting these equal to 0, we have:

$P_p = -6p + 2q + 360 = 0$ (1)

$P_q = 2p - 4q + 280 = 0$ (2)

Solving (1) and (2) for p and q, we get $p = 100$ and $q = 120$. Thus, (100, 120) is a critical point of the profit function P.

$P_{pp} = -6$ and $P_{pp}(100, 120) = -6$

$P_{pq} = 2$ and $P_{pq}(100, 120) = 2$

$P_{qq} = -4$ and $P_{qq}(100, 120) = -4$

$P_{pp} \cdot P_{qq} - [P_{pq}]^2 = (-6)(-4) - (2)^2 = 20 > 0$

Since $P_{pp}(100, 120) = -6 < 0$, we conclude that the maximum profit occurs when $p = \$100$ and $q = \$120$. The maximum profit is:

$P(100, 120) = -3(100)^2 + 2(100)(120) - 2(120)^2 + 360(100) + 280(120) - 30{,}000 = \$4{,}800$

41. The square of the distance from P to A is: $x^2 + y^2$

The square of the distance from P to B is:

$(x - 2)^2 + (y - 6)^2 = x^2 - 4x + y^2 - 12y + 40$

The square of the distance from P to C is:

$(x - 10)^2 + y^2 = x^2 - 20x + y^2 + 100$

Thus, we have:

$P(x, y) = 3x^2 - 24x + 3y^2 - 12y + 140$

$P_x = 6x - 24 = 0$ $\qquad\qquad\qquad\qquad$ $P_y = 6y - 12 = 0$

$\qquad x = 4$ $\qquad\qquad\qquad\qquad\qquad\qquad$ $y = 2$

Therefore, (4, 2) is a critical point.

$P_{xx} = 6$ and $P_{xx}(4, 2) = 6 > 0$

$P_{xy} = 0$ and $P_{xy}(4, 2) = 0$

$P_{yy} = 6$ and $P_{yy}(4, 2) = 6$

$P_{xx} \cdot P_{yy} - [P_{xy}]^2 = 6 \cdot 6 - 0 = 36 > 0$

Therefore, P has a minimum at the point (4, 2).

43. Let x = length, y = width, and z = height. Then $V = xyz = 64$ or $z = \dfrac{64}{xy}$. The surface area of the box is:

$$S = xy + 2xz + 4yz \qquad\qquad \text{or} \qquad\qquad S(x, y) = xy + \frac{128}{y} + \frac{256}{x}, \quad x > 0, \ y > 0$$

$$S_x = y - \frac{256}{x^2} = 0 \qquad \text{or} \qquad\qquad y = \frac{256}{x^2} \quad (1)$$

$$S_y = x - \frac{128}{y^2} = 0 \qquad \text{or} \qquad\qquad x = \frac{128}{y^2}$$

Thus, $y = \dfrac{256}{\dfrac{(128)^2}{y^4}}$ $\qquad\qquad$ or $\qquad y^4 - 64y = 0$

$\qquad\qquad\qquad\qquad\qquad\qquad\qquad\qquad\qquad y(y^3 - 64) = 0 \quad$ (Since $y > 0$, $y = 0$ does not yield a critical point.)

$\qquad\qquad\qquad\qquad\qquad\qquad\qquad\qquad\qquad\qquad y = 4$

Setting $y = 4$ in (1), we find $x = 8$. Therefore, the critical point is $(8, 4)$.

Now we have:

$$S_{xx} = \frac{512}{x^3} \quad \text{and} \ S_{xx}(8, 4) = 1 > 0$$

$$S_{xy} = 1$$

$$S_{yy} = \frac{256}{y^3} \quad \text{and} \ S_{yy}(8, 4) = 4$$

$$S_{xx}(8, 4) \cdot S_{yy}(8, 4) - [S_{xy}(8, 4)]^2 = 1 \cdot 4 - 1^2 = 3 > 0$$

Thus, the dimensions that will require the least amount of material are: length $x = 8$ inches; width $y = 4$ inches; height $z = \dfrac{64}{8(4)} = 2$ inches.

45. Let x = length of the package, y = width, and z = height. Then
$x + 2y + 2z = 120$ (1)
Volume $= V = xyz$.

From (1), $z = \dfrac{120 - x - 2y}{2}$. Thus, we have:

$$V(x, y) = xy\left(\frac{120 - x - 2y}{2}\right) = 60xy - \frac{x^2 y}{2} - xy^2, \quad x > 0, \ y > 0$$

$$V_x = 60y - xy - y^2 = 0$$

$$y(60 - x - y) = 0$$

$$60 - x - y = 0 \quad (2) \quad \text{(Since } y > 0, \ y = 0 \text{ does not yield a critical point.)}$$

$V_y = 60x - \dfrac{x^2}{2} - 2xy = 0$

$x\left(60 - \dfrac{x}{2} - 2y\right) = 0$

$120 - x - 4y = 0$ (3)   (Since $x > 0$, $x = 0$ does not yield a critical point.)

Solving (2) and (3) for x and y, we obtain $x = 40$ and $y = 20$. Thus, $(40, 20)$ is the critical point.

$V_{xx} = -y$ and $V_{xx}(40, 20) = -20 < 0$

$V_{xy} = 60 - x - 2y$ and $V_{xy}(40, 20) = 60 - 40 - 40 = -20$

$V_{yy} = -2x$ and $V_{yy}(40, 20) = -80$

$V_{xx}(40, 20) \cdot V_{yy}(40, 20) - [V_{xy}(40, 20)]^2 = (-20)(-80) - [-20]^2 = 1600 - 400 = 1200 > 0$

Thus, the maximum volume of the package is obtained when $x = 40$, $y = 20$, and

$z = \dfrac{120 - 40 - 2 \cdot 20}{2} = 20$ inches. The package has dimensions: length $x = 40$ inches; width $y = 20$ inches; height $z = 20$ inches.

EXERCISE 15-4

Things to remember:

1.   Any local maxima or minima of the function $z = f(x, y)$ subject to the constraint $g(x, y) = 0$ will be among those points (x_0, y_0) for which (x_0, y_0, λ_0) is a solution to the system:

$$F_x(x, y, \lambda) = 0$$
$$F_y(x, y, \lambda) = 0$$
$$F_\lambda(x, y, \lambda) = 0$$

where $F(x, y, \lambda) = f(x, y) + \lambda g(x, y)$, provided all the partial derivatives exist.

2.   METHOD OF LAGRANGE MULTIPLIERS FOR FUNCTIONS OF TWO INDEPENDENT VARIABLES

Step 1.   Formulate the problem in the form:
Maximize (or Minimize) $z = f(x, y)$
Subject to: $g(x, y) = 0$

Step 2.   Form the function F:
$F(x, y, \lambda) = f(x, y) + \lambda g(x, y)$

Step 3.   Find the critical points (x_0, y_0, λ_0) for F, that is, solve the system:

$$F_x(x, y, \lambda) = 0$$
$$F_y(x, y, \lambda) = 0$$
$$F_\lambda(x, y, \lambda) = 0$$

Step 4.   If (x_0, y_0, λ_0) is the only critical point of F, then assume that (x_0, y_0) is the solution to the problem. If F has more than one critical point, then evaluate $z = f(x, y)$ at (x_0, y_0) for each critical point (x_0, y_0, λ_0) of F. Assume that the largest of these values is the maximum value of $f(x, y)$ subject to the constraint $g(x, y) = 0$, and the smallest is the minimum value of $f(x, y)$ subject to the constraint $g(x, y) = 0$.

3. **METHOD OF LAGRANGE MULTIPLIERS FOR FUNCTIONS OF THREE VARIABLES**

   Any local maxima or minima of the function $w = f(x, y, z)$ subject to the constraint $g(x, y, z) = 0$ will be among the set of points (x_0, y_0, z_0) for which $(x_0, y_0, z_0, \lambda_0)$ is a solution to the system

   $$F_x(x, y, z, \lambda) = 0$$
   $$F_y(x, y, z, \lambda) = 0$$
   $$F_z(x, y, z, \lambda) = 0$$
   $$F_\lambda(x, y, z, \lambda) = 0$$

   where $F(x, y, z, \lambda) = f(x, y, z) + \lambda g(x, y, z)$, provided that all the partial derivatives exist.

---

1. Minimize $f(x, y) = x^2 + xy + y^2$, subject to: $y = 4$.

   $f(x, 4) = x^2 + 4x + 16$, $f'(x, 4) = 2x + 4$, $f''(x, 4) = 2$

   $f'(x, 4) = 0$: $2x + 4 = 0$, $x = -2$; $f''(-2, 4) = 2 > 0$.

   Therefore, f has a minimum at $(-2, 4)$; min $f(x, y) = f(-2, 4) = 4 - 8 + 16 = 12$.

3. Minimize $f(x, y) = 4xy$ subject to: $x - y = 2$ or $y = x - 2$

   $f(x, x - 2) = 4x(x - 2) = 4x^2 - 8x$, $f'(x, x - 2) = 8x - 8$, $f''(x, x - 2) = 8$

   $f'(x, x - 2) = 0$: $8x - 8 = 0$, $x = 1$; $f''(1, -1) = 8 > 0$.

   Therefore, f has a minimum at $(1, -1)$; min $f(x, y) = f(1, -1) = 4(1)(-1) = -4$.

5. Maximize $f(x, y) = 2x + y$, subject to: $x^2 + y = 1$ or $y = 1 - x^2$.

   $f(x, 1 - x^2) = F(x) = 2x + 1 - x^2 = -x^2 + 2x + 1$, $F'(x) = -2x + 2$, $F''(x) = -2$.

   $F'(x) = 0$: $-2x + 2 = 0$, $x = 1$, $y = 0$; $F''(1) = -2 < 0$.

   Therefore, f has a maximum at $(1, 0)$; max $f(x, y) = f(1, 0) = 2$.

7. Step 1.   Maximize $f(x, y) = 2xy$

   Subject to: $g(x, y) = x + y - 6 = 0$

   Step 2. $F(x, y, \lambda) = f(x, y) + \lambda g(x, y)$
   $$= 2xy + \lambda (x + y - 6)$$

   Step 3. $F_x = 2y + \lambda = 0$ (1)

   $F_y = 2x + \lambda = 0$ (2)

   $F_\lambda = x + y - 6 = 0$ (3)

   From (1) and (2), we obtain:

   $$x = -\frac{\lambda}{2}, y = -\frac{\lambda}{2}$$

   Substituting these into (3), we have:

   $$-\frac{\lambda}{2} - \frac{\lambda}{2} - 6 = 0$$
   $$\lambda = -6.$$

   Thus, the critical point is $(3, 3, -6)$.

Step 4.   Since $(3, 3, -6)$ is the only critical point for F,
we conclude that max $f(x, y) = f(3, 3) = 2 \cdot 3 \cdot 3 = 18$.

9.   Step 1.   Minimize $f(x, y) = x^2 + y^2$
Subject to: $g(x, y) = 3x + 4y - 25 = 0$

Step 2. $F(x, y, \lambda) = f(x, y) + \lambda\, g(x, y) = x^2 + y^2 + \lambda\,(3x + 4y - 25)$

Step 3. $F_x = 2x + 3\lambda = 0$ (1)

$F_y = 2y + 4\lambda = 0$ (2)

$F_\lambda = 3x + 4y - 25 = 0$ (3)

From (1) and (2), we obtain:

$x = -\dfrac{3\lambda}{2}$, $y = -2\lambda$

Substituting these into (3), we have:

$3\left(-\dfrac{3\lambda}{2}\right) + 4(-2\lambda) - 25 = 0$

$\dfrac{25}{2}\lambda = -25$

$\lambda = -2$

The critical point is $(3, 4, -2)$.

Step 4.   Since $(3, 4, -2)$ is the only critical point for F, we conclude that
min $f(x, y) = f(3, 4) = 3^2 + 4^2 = 25$.

11.   Step 1.   Maximize $f(x, y) = 4y - 3x$ subject to $2x + 5y - 3 = 0$

Step 2. $F(x, y, \lambda) = f(x, y) + \lambda\, g(x, y) = 4y - 3x + \lambda\,(2x + 5y - 3)$

Step 3. $F_x = -3 + 2\lambda = 0$ (1)

$F_y = 4 + 5\lambda = 0$ (2)

$F_\lambda = 2x + 5y - 3 = 0$ (3)

From (1), $\lambda = \dfrac{3}{2}$, from (2), $\lambda = -\dfrac{4}{5}$. Thus, the system (1), (2), (3) does not have a

solution; by Theorem 1 there are no maxima or minima.

13.   Step 1.   Maximize and minimize $f(x, y) = 2xy$
Subject to: $g(x, y) = x^2 + y^2 - 18 = 0$

Step 2. $F(x, y, \lambda) = f(x, y) + \lambda\, g(x, y) = 2xy + \lambda\,(x^2 + y^2 - 18)$

Step 3. $F_x = 2y + 2\lambda x = 0$ (1)

$F_y = 2x + 2\lambda y = 0$ (2)

$F_\lambda = x^2 + y^2 - 18 = 0$ (3)

From (1), (2), and (3), we obtain the critical points
$(3, 3, -1), (3, -3, 1), (-3, 3, 1)$ and $(-3, -3, -1)$.

Step 4. $f(3, 3) = 2 \cdot 3 \cdot 3 = 18$

$f(3, -3) = 2 \cdot 3(-3) = -18$

$f(-3, 3) = 2(-3) \cdot 3 = -18$

$f(-3, -3) = 2(-3)(-3) = 18$

Thus, max $f(x, y) = f(3, 3) = f(-3, -3) = 18;$ min $f(x, y) = f(3, -3) = f(-3, 3) = -18.$

15. Let x and y be the required numbers.

Step 1.   Maximize $f(x, y) = xy$

Subject to: $x + y = 10$ or $g(x, y) = x + y - 10 = 0$

Step 2. $F(x, y, \lambda) = xy + \lambda\ (x + y - 10)$

Step 3. $F_x = y + \lambda = 0$ (1)

$F_y = x + \lambda = 0$ (2)

$F_\lambda = x + y - 10 = 0$ (3)

From (1) and (2), we obtain:

$x = -\lambda, y = -\lambda$

Substituting these into (3), we have:

$\lambda = -5$

The critical point is $(5, 5, -5)$.

Step 4.   Since $(5, 5, -5)$ is the only critical point for F, we conclude that max $f(x, y) = f(5, 5) =$ $5 \cdot 5 = 25.$ Thus, the maximum product is 25 when $x = 5$ and $y = 5$.

17. Step 1.   Minimize $f(x, y, z) = x^2 + y^2 + z^2$

Subject to: $g(x, y) = 2x - y + 3z + 28 = 0$

Step 2. $F(x, y, z, \lambda) = x^2 + y^2 + z^2 + \lambda\ (2x - y + 3z + 28)$

Step 3. $F_x = 2x + 2\lambda = 0$ (1)

$F_y = 2y - \lambda = 0$ (2)

$F_z = 2z + 3\lambda = 0$ (3)

$F_\lambda = 2x - y + 3z + 28 = 0$ (4)

From (1), (2), and (3), we obtain:

$x = -\lambda, y = \dfrac{\lambda}{2}, z = -\dfrac{3}{2}\lambda$

Substituting these into (4), we have:

$2(-\lambda) - \dfrac{\lambda}{2} + 3\left(-\dfrac{3}{2}\lambda\right) + 28 = 0$

$-\dfrac{14}{2}\lambda + 28 = 0$

$\lambda = 4$

The critical point is $(-4, 2, -6, 4)$.

Step 4.   Since $(-4, 2, -6, 4)$ is the only critical point for F, we conclude that

min $f(x, y, z) = f(-4, 2, -6) = 56.$

19. <u>Step 1</u>.   Maximize and minimize $f(x, y, z) = x + y + z$

Subject to: $g(x, y, z) = x^2 + y^2 + z^2 - 12 = 0$

<u>Step 2</u>. $F(x, y, z, \lambda) = f(x, y, z) + \lambda\, g(x, y, z) = x + y + z + \lambda\,(x^2 + y^2 + z^2 - 12)$

<u>Step 3</u>. $F_x = 1 + 2x\lambda = 0$ (1)

$F_y = 1 + 2y\lambda = 0$ (2)

$F_z = 1 + 2z\lambda = 0$ (3)

$F_\lambda = x^2 + y^2 + z^2 - 12 = 0$ (4)

From (1), (2), and (3), we obtain:

$$x = -\frac{1}{2\lambda},\quad y = -\frac{1}{2\lambda},\quad z = -\frac{1}{2\lambda}$$

Substituting these into (4), we have:

$$\left(-\frac{1}{2\lambda}\right)^2 + \left(-\frac{1}{2\lambda}\right)^2 + \left(-\frac{1}{2\lambda}\right)^2 - 12 = 0$$

$$\frac{3}{4\lambda^2} - 12 = 0$$

$$1 - 16\lambda^2 = 0$$

$$\lambda = \pm\frac{1}{4}$$

Thus, the critical points are $\left(2, 2, 2, -\frac{1}{4}\right)$ and $\left(-2, -2, -2, \frac{1}{4}\right)$.

<u>Step 4</u>. $f(2, 2, 2) = 2 + 2 + 2 = 6$

$f(-2, -2, -2) = -2 - 2 - 2 = -6$

Thus, max $f(x, y, z) = f(2, 2, 2) = 6$; min $f(x, y, z) = f(-2, -2, -2) = -6$.

21. <u>Step 1</u>.   Maximize $f(x, y) = y + xy^2$

Subject to: $x + y^2 = 1$ or $g(x, y) = x + y^2 - 1 = 0$

<u>Step 2</u>. $F(x, y, \lambda) = y + xy^2 + \lambda\,(x + y^2 - 1)$

<u>Step 3</u>. $F_x = y^2 + \lambda = 0$ (1)

$F_y = 1 + 2xy + 2y\lambda = 0$ (2)

$F_\lambda = x + y^2 - 1 = 0$ (3)

From (1), $\lambda = -y^2$ and from (3), $x = 1 - y^2$. Substituting these values into (2), we have

$$1 + 2(1 - y^2)y - 2y^3 = 0$$

or $4y^3 - 2y - 1 = 0$

Using a graphing utility to solve this equation, we get $y \approx 0.885$. Then $x \approx 0.217$ and max $f(x, y) = f(0.217, 0.885) \approx 1.055$.

23. Step 1. Maximize $f(x, y) = e^x + 3e^y$ subject to $g(x, y) = x - 2y - 6 = 0$

Step 2. $F(x, y, \lambda) = f(x, y) + \lambda\, g(x, y) = e^x + 3e^y + \lambda\,(x - 2y - 6)$

Step 3. $F_x = e^x + \lambda = 0$ (1)

$F_y = 3e^y - 2\lambda = 0$ (2)

$F_\lambda = x - 2y - 6 = 0$ (3)

From (1), $\lambda = -e^x$, which implies λ is negative.

From (2), $\lambda = \dfrac{3}{2}e^y$ which implies λ is positive.

Thus, (1) and (2) have no simultaneous solution.

25. The constraint $g(x, y) = y - 5 = 0$ implies $y = 5$. Replacing y by 5 in the function f, the problem reduces to maximizing the function $h(x) = f(x, 5)$, a function of one independent variable.

27. Maximize $f(x, y) = e^{-(x^2+y^2)}$

Subject to: $g(x, y) = x^2 + y - 1 = 0$

(A) $x^2 + y - 1 = 0;\ y = 1 - x^2$

Substituting $y = 1 - x^2$ into $f(x, y)$, we get

$$h(x) = f(x, 1 - x^2) = e^{-(x^2 + [1 - x^2]^2)} = e^{-(x^4 - x^2 + 1)}$$

Now, $h'(x) = e^{-(x^4 - x^2 + 1)}(-4x^3 + 2x)$.

Critical numbers: $(2x - 4x^3)e^{-(x^4 - x^2 + 1)} = 0$

$$2x(1 - 2x^2) = 0$$

$$x = 0,\ \frac{\sqrt{2}}{2},\ -\frac{\sqrt{2}}{2}$$

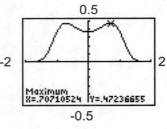

From the constraint equation, $y = \dfrac{1}{2}$ when $x = \pm\dfrac{\sqrt{2}}{2}$, and $y = 1$ when $x = 0$.

$f(0,1) = e^{-1} \approx 0.368$ and $f\left(\pm\dfrac{\sqrt{2}}{2}, \dfrac{1}{2}\right) \approx 0.472$

Thus, $\max f(x, y) = f\left(-\dfrac{\sqrt{2}}{2}, \dfrac{1}{2}\right) = f\left(\dfrac{\sqrt{2}}{2}, \dfrac{1}{2}\right) \approx 0.472.$

(B) $F(x, y, \lambda) = e^{-(x^2+y^2)} + \lambda\,(x^2 + y - 1)$

$F_x = -2xe^{-(x^2+y^2)} + 2x\lambda = 0$ (1)

$F_y = -2ye^{-(x^2+y^2)} + \lambda = 0$ (2)

$F_\lambda = x^2 + y - 1 = 0$ (3)

From (3), $y = 1 - x^2$ and from (2)

$$\lambda = 2ye^{-(x^2+y^2)} = 2(1-x^2)e^{-(x^4-x^2+1)}$$

Substituting these values into (1), we have

$$-2xe^{-(x^4-x^2+1)} + 2x[2(1-x^2)e^{-(x^4-x^2+1)}] = 0$$

$$2x[2-2x^2-1] = 0$$

$$2x(1-2x^2) = 0$$

$$x = 0, \frac{\sqrt{2}}{2}, -\frac{\sqrt{2}}{2}$$

Now, $y = 1$ when $x = 0$, and $y = \frac{1}{2}$ when $x = \pm\frac{\sqrt{2}}{2}$.

$$f(0, 1) = e^{-1} \approx 0.368$$

$$f\left(\frac{\sqrt{2}}{2}, \frac{1}{2}\right) = f\left(-\frac{\sqrt{2}}{2}, \frac{1}{2}\right) \approx 0.472$$

Thus, $\max f(x, y) \approx 0.472$.

29. Step 1.   Minimize cost function $C(x, y) = 6x^2 + 12y^2$
Subject to: $x + y = 90$ or $g(x, y) = x + y - 90 = 0$

Step 2. $F(x, y, \lambda) = 6x^2 + 12y^2 + \lambda(x + y - 90)$

Step 3. $F_x = 12x + \lambda = 0$ (1)

$F_y = 24y + \lambda = 0$ (2)

$F_\lambda = x + y - 90 = 0$ (3)

From (1) and (2), we obtain

$$x = -\frac{\lambda}{12}, \; y = -\frac{\lambda}{24}$$

Substituting these into (3), we have:

$$-\frac{\lambda}{12} - \frac{\lambda}{24} - 90 = 0$$

$$\frac{3\lambda}{24} = -90$$

$$\lambda = -720$$

The critical point is $(60, 30, -720)$.

Step 4.   Since $(60, 30, -720)$ is the only critical point for F, we conclude that:

$\min C(x, y) = C(60, 30) = 6 \cdot 60^2 + 12 \cdot 30^2 = 21,600 + 10,800 = \$32,400$

Thus, 60 of model A and 30 of model B will yield a minimum cost of \$32,400 per week.

31. (A)   Step 1.   Maximize the production function $N(x, y) = 50x^{0.8}y^{0.2}$
Subject to the constraint: $C(x, y) = 40x + 80y = 400,000$
i.e., $g(x, y) = 40x + 80y - 400,000 = 0$

Step 2. $F(x, y, \lambda) = 50x^{0.8}y^{0.2} + \lambda(40x + 80y - 400,000)$

<u>Step 3.</u> $F_x = 40x^{-0.2}y^{0.2} + 40\lambda = 0$ (1)

$F_y = 10x^{0.8}y^{-0.8} + 80\lambda = 0$ (2)

$F_\lambda = 40x + 80y - 400{,}000 = 0$ (3)

From (1), $\lambda = -\dfrac{y^{0.2}}{x^{0.2}}$. From (2), $\lambda = -\dfrac{x^{0.8}}{8y^{0.8}}$.

Thus, we obtain

$-\dfrac{y^{0.2}}{x^{0.2}} = -\dfrac{x^{0.8}}{8y^{0.8}}$ or $x = 8y$

Substituting into (3), we have:

$320y + 80y - 400{,}000 = 0$

$y = 1000$

Therefore, $x = 8000$, $\lambda \approx -0.6598$, and the critical point   is (8000, 1000, –0.6598).
Thus, we conclude that:

max $N(x, y) = N(8000, 1000) = 50(8000)^{0.8}(1000)^{0.2} \approx 263{,}902$ units
and production is maximized when 8000 labor units and 1000 capital units are
used.

(B)   The marginal productivity of money is $-\lambda \approx 0.6598$. The increase in production if an additional
$50,000 is budgeted for production is: $0.6598(50{,}000) = 32{,}990$ units

33.   Let $x =$ length, $y =$ width, and $z =$ height.

<u>Step 1.</u>   Maximize volume $V = xyz$
Subject to: $S(x, y, z) = xy + 3xz + 3yz - 192 = 0$

<u>Step 2.</u> $F(x, y, z, \lambda) = xyz + \lambda(xy + 3xz + 3yz - 192)$

<u>Step 3.</u> $F_x = yz + \lambda(y + 3z) = 0$ (1)
$F_y = xz + \lambda(x + 3z) = 0$ (2)

$F_z = xy + \lambda(3x + 3y) = 0$ (3)
$F_\lambda = xy + 3xz + 3yz - 192 = 0$ (4)

Solving this system of equations, (1)–(4), simultaneously, yields:

$x = 8$, $y = 8$, $z = \dfrac{8}{3}$, $\lambda = -\dfrac{4}{3}$

Thus, the critical point is $\left(8, 8, \dfrac{8}{3}, -\dfrac{4}{3}\right)$.

<u>Step 4.</u>   Since $\left(8, 8, \dfrac{8}{3}, -\dfrac{4}{3}\right)$ is the only critical point for F:

max $V(x, y, z) = V\left(8, 8, \dfrac{8}{3}\right) = \dfrac{512}{3} \approx 170.67$

Thus, the dimensions that will maximize the volume of the box are: length $x = 8$ inches;

width $y = 8$ inches; height $z = \dfrac{8}{3}$ inches.

35. <u>Step 1.</u>   Maximize $A = xy$

Subject to: $P(x, y) = y + 4x - 400 = 0$

<u>Step 2.</u> $F(x, y, \lambda) = xy + \lambda \ (y + 4x - 400)$

<u>Step 3.</u> $F_x = y + 4\lambda = 0$ (1)

$F_y = x + \lambda = 0$ (2)

$F_\lambda = y + 4x - 400 = 0$ (3)

From (1) and (2), we have:

$y = -4\lambda$ and $x = -\lambda$

Substituting these into (3), we obtain:

$-4\lambda - 4\lambda - 400 = 0$

Thus, $\lambda = -50$ and the critical point is (50, 200, –50).

<u>Step 4.</u>   Since (50, 200, -50) is the only critical point for F, max $A(x, y) = A(50, 200) = 10{,}000$.

Therefore, $x = 50$ feet, $y = 200$ feet will produce the maximum area $A(50, 200) = 10{,}000$ square feet.

EXERCISE 15-5

Things to remember:

<u>1.</u>   LEAST SQUARES APPROXIMATION FORMULAS

For a set of n points $(x_1, y_1), (x_2, y_2), \ldots, (x_n, y_n)$, the coefficients of the least squares line

$y = ax + b$ are the solutions of the system of NORMAL EQUATIONS

$$\left(\sum_{k=1}^{n} x_k^2 \right) a + \left(\sum_{k=1}^{n} x_k \right) b = \sum_{k=1}^{n} x_k y_k \tag{1}$$

$$\left(\sum_{k=1}^{n} x_k \right) a + nb = \sum_{k=1}^{n} y_k$$

and are given by the formulas

$$a = \frac{n \left(\sum_{k=1}^{n} x_k y_k \right) - \left(\sum_{k=1}^{n} x_k \right) \left(\sum_{k=1}^{n} y_k \right)}{n \left(\sum_{k=1}^{n} x_k^2 \right) - \left(\sum_{k=1}^{n} x_k \right)^2} \tag{2}$$

$$b = \frac{\sum_{k=1}^{n} y_k - a \left(\sum_{k=1}^{n} x_k \right)}{n} \tag{3}$$

[Note: To find a and b, either solve system (1) directly, or use formulas (2) and (3). If the formulas are used, the value of a must be calculated first since it is used in the formula for b.

1. $\displaystyle\sum_{k=1}^{5} x_k = 0+1+2+3+4 = 10$

3. $\displaystyle\sum_{k=1}^{5} x_k\, y_k = 0(4)+1(5)+2(7)+3(9)+4(13) = 98$

5. $\displaystyle\sum_{k=1}^{5} x_k \sum_{k=1}^{5} y_k = 10(4+5+7+9+13) = 380$

7.

	x_k	y_k	$x_k y_k$	x_k^2
	1	1	1	1
	2	3	6	4
	3	4	12	9
	4	3	12	16
Totals	10	11	31	30

Thus, $\displaystyle\sum_{k=1}^{4} x_k = 10, \ \sum_{k=1}^{4} y_k = 11, \ \sum_{k=1}^{4} x_k y_k = 31, \ \sum_{k=1}^{4} x_k^2 = 30$.

Substituting these values into formulas (2) and (3) for a and b, respectively, we have:

$$a = \frac{n\left(\displaystyle\sum_{k=1}^{n} x_k y_k\right) - \left(\displaystyle\sum_{k=1}^{n} x_k\right)\left(\displaystyle\sum_{k=1}^{n} y_k\right)}{n\left(\displaystyle\sum_{k=1}^{n} x_k^2\right) - \left(\displaystyle\sum_{k=1}^{n} x_k\right)^2} = \frac{4(31) - (10)(11)}{4(30) - (10)^2} = \frac{14}{20} = 0.7$$

$$b = \frac{\displaystyle\sum_{k=1}^{n} y_k - a\left(\displaystyle\sum_{k=1}^{n} x_k\right)}{n} = \frac{11 - 0.7(10)}{4} = 1$$

Thus, the least squares line is $y = ax + b = 0.7x + 1$. Refer to the graph at the right.

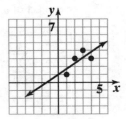

9.

	x_k	y_k	$x_k y_k$	x_k^2
	1	8	8	1
	2	5	10	4
	3	4	12	9
	4	0	0	16
Totals	10	17	30	30

Thus, $\displaystyle\sum_{k=1}^{4} x_k = 10, \ \sum_{k=1}^{4} y_k = 17, \ \sum_{k=1}^{4} x_k y_k = 30,$

$\displaystyle\sum_{k=1}^{4} x_k^2 = 30$.

Substituting these values into system (1), we have:

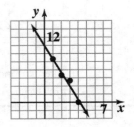

$$30a + 10b = 30$$
$$10a + \;\; 4b = 17$$

The solution of this system is $a = -2.5$, $b = 10.5$. Thus, the least squares line is $y = ax + b = -2.5x + 10.5$. Refer to the graph on the previous page.

11.

	x_k	y_k	$x_k y_k$	x_k^2
	1	3	3	1
	2	4	8	4
	3	5	15	9
	4	6	24	16
Totals	10	18	50	30

Thus, $\displaystyle\sum_{k=1}^{4} x_k = 10$, $\displaystyle\sum_{k=1}^{4} y_k = 18$, $\displaystyle\sum_{k=1}^{4} x_k y_k = 50$, $\displaystyle\sum_{k=1}^{4} x_k^2 = 30$.

Substituting these values into the formulas for a and b [formulas (2) and (3)], we have:

$$a = \frac{4(50) - (10)(18)}{4(30) - (10)^2} = \frac{20}{20} = 1$$

$$b = \frac{18 - 1(10)}{4} = \frac{8}{4} = 2$$

Thus, the least squares line is $y = ax + b = x + 2$. Refer to the graph at the right.

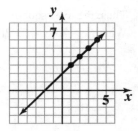

13.

	x_k	y_k	$x_k y_k$	x_k^2
	1	3	3	1
	2	1	2	4
	2	2	4	4
	3	0	0	9
Totals	8	6	9	18

Thus, $\displaystyle\sum_{k=1}^{4} x_k = 8$, $\displaystyle\sum_{k=1}^{4} y_k = 6$, $\displaystyle\sum_{k=1}^{4} x_k y_k = 9$, $\displaystyle\sum_{k=1}^{4} x_k^2 = 18$.

Substituting these values into formulas (2) and (3) for a and b, respectively, we have:

$$a = \frac{4(9) - 8(6)}{4(18) - 8^2} = \frac{36 - 48}{72 - 64} = \frac{-12}{8} = -\frac{3}{2} = -1.5$$

$$b = \frac{6 - (-3/2)(8)}{4} = \frac{6 + 12}{4} = \frac{9}{2} = 4.5$$

Thus, the least squares line is $y = -1.5x + 4.5$.
When $x = 2.5$, $y = -1.5(2.5) + 4.5 = 0.75$.

15.

	x_k	y_k	$x_k y_k$	x_k^2

	0	10	0	0
	5	22	110	25
	10	31	310	100
	15	46	690	225
	20	51	1020	400
Totals	50	160	2130	750

Thus, $\displaystyle\sum_{k=1}^{5} x_k = 50$, $\displaystyle\sum_{k=1}^{5} y_k = 160$, $\displaystyle\sum_{k=1}^{5} x_k y_k = 2130$, $\displaystyle\sum_{k=1}^{5} x_k^2 = 750$.

Substituting these values into formulas (2) and (3) for a and b, respectively, we have:

$$a = \frac{5(2130) - (50)(160)}{5(750) - (50)^2} = \frac{2650}{1250} = 2.12$$

$$b = \frac{160 - 2.12(50)}{5} = \frac{54}{5} = 10.8$$

Thus, the least squares line is $y = 2.12x + 10.8$.

When $x = 25$, $y = 2.12(25) + 10.8 = 63.8$.

17.

	x_k	y_k	$x_k y_k$	x_k^2
	−1	14	−14	1
	1	12	12	1
	3	8	24	9
	5	6	30	25
	7	5	35	49
Totals	15	45	87	85

Thus, $\displaystyle\sum_{k=1}^{5} x_k = 15$, $\displaystyle\sum_{k=1}^{5} y_k = 45$, $\displaystyle\sum_{k=1}^{5} x_k y_k = 87$, $\displaystyle\sum_{k=1}^{5} x_k^2 = 85$.

Substituting these values into formulas (2) and (3) for a and b, respectively, we have:

$$a = \frac{5(87) - (15)(45)}{5(85) - (15)^2} = \frac{-240}{200} = -1.2$$

$$b = \frac{45 - (-1.2)(15)}{5} = 12.6$$

Thus, the least squares line is

$y = -1.2x + 12.6$.

When $x = 2$, $y = -1.2(2) + 12.6 = 10.2$.

19.

	x_k	y_k	$x_k y_k$	x_k^2
	0.5	25	12.5	0.25

	2.0	22	44.0	4.00
	3.5	21	73.5	12.25
	5.0	21	105.0	25.00
	6.5	18	117.0	42.25
	9.5	12	114.0	90.25
	11.0	11	121.0	121.00
	12.5	8	100.0	156.25
	14.0	5	70.0	196.00
	15.5	1	15.5	240.25
Totals	80.0	144	772.5	887.50

Thus, $\sum_{k=1}^{10} x_k = 80$, $\sum_{k=1}^{10} y_k = 144$, $\sum_{k=1}^{10} x_k y_k = 772.5$, $\sum_{k=1}^{10} x_k^2 = 887.5$.

Substituting these values into formulas (2) and (3) for a and b, respectively, we have:

$$a = \frac{10(772.5) - (80)(144)}{10(887.5) - (80)^2} = \frac{-3795}{2475} \approx -1.53$$

$$b = \frac{144 - (-1.53)(80)}{10} = \frac{266.4}{10} = 26.64$$

Thus, the least squares line is
$y = -1.53x + 26.64$.
When $x = 8$, $y = -1.53(8) + 26.64 = 14.4$.

21. Minimize

$F(a, b, c) = (a + b + c - 2)^2 + (4a + 2b + c - 1)^2 + (9a + 3b + c - 1)^2 + (16a + 4b + c - 3)^2$

$F_a(a, b, c) = 2(a + b + c - 2) + 8(4a + 2b + c - 1) + 18(9a+3b+c-1) + 32(16a+4b+c-3)$

$\qquad = 708a + 200b + 60c - 126$

$F_b(a, b, c) = 2(a + b + c - 2) + 4(4a + 2b + c - 1) + 6(9a + 3b + c - 1)+8(16a + 4b + c - 3)$

$\qquad = 200a + 60b + 20c - 38$

$F_c(a, b, c) = 2(a + b + c - 2) + 2(4a + 2b + c - 1) + 2(9a + 3b + c - 1)+2(16a + 4b + c - 3)$

$\qquad = 60a + 20b + 8c - 14$

The system is:
$\qquad\qquad F_a(a, b, c) = 0$
$\qquad\qquad F_b(a, b, c) = 0$
$\qquad\qquad F_c(a, b, c) = 0$

or:
$\qquad\qquad 708a + 200b + 60c = 126$
$\qquad\qquad 200a + 60b + 20c = 38$
$\qquad\qquad 60a + 20b + 8c = 14$

The solution is $(a, b, c) = (0.75, -3.45, 4.75)$, which gives us the equation for the parabola shown at the right:

$$y = ax^2 + bx + c$$

or

$$y = 0.75x^2 - 3.45x + 4.75$$

The given points: $(1, 2), (2, 1), (3, 1), (4, 3)$ also shown.

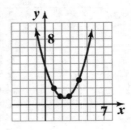

23.   System (1) is:

$$\left(\sum_{k=1}^{n} x_k\right) a + nb = \sum_{k=1}^{n} y_k \qquad\qquad \text{(a)}$$

$$\left(\sum_{k=1}^{n} x_k^2\right) a + \left(\sum_{k=1}^{n} x_k\right) b = \sum_{k=1}^{n} x_k y_k \qquad\qquad \text{(b)}$$

Multiply equation (a) by $-\left(\sum_{k=1}^{n} x_k\right)$, equation (b) by n, and add the resulting equations. This will eliminate b from the system.

$$\left[-\left(\sum_{k=1}^{n} x_k\right)^2 + n\sum_{k=1}^{n} x_k^2\right] a = -\left(\sum_{k=1}^{n} x_k\right)\left(\sum_{k=1}^{n} y_k\right) + n\sum_{k=1}^{n} x_k y_k$$

Thus, $a = \dfrac{n\left(\sum_{k=1}^{n} x_k y_k\right) - \left(\sum_{k=1}^{n} x_k\right)\left(\sum_{k=1}^{n} y_k\right)}{n\left(\sum_{k=1}^{n} x_k^2\right) - \left(\sum_{k=1}^{n} x_k\right)^2}$

which is equation (2). Solving equation (a) for b, we have

$$b = \frac{\sum_{k=1}^{n} y_k - m\left(\sum_{k=1}^{n} x_k\right)}{n}, \quad \text{which is equation (3)}.$$

25.   (A) Suppose that $n = 5$ and $x_1 = -2, x_2 = -1, x_3 = 0, x_4 = 1, x_5 = 2$. Then

$$\sum_{k=1}^{5} x_k = -2 - 1 + 0 + 1 + 2 = 0 .$$ Therefore, from formula (2),

$$a = \frac{5\sum_{k=1}^{5} x_k y_k}{5\sum_{k=1}^{5} x_k^2} = \frac{\sum x_k y_k}{\sum x_k^2} \quad \text{From formula (3),} \quad b = \frac{\sum_{k=1}^{5} y_k}{5},$$

which is the average of $y_1, y_2, y_3, y_4,$ and y_5.

(B)   If the average of the x-coordinates is 0, then

$$\frac{\sum\limits_{k=1}^{n} x_k}{n} = 0, \quad \text{i.e.,} \quad \sum\limits_{k=1}^{n} x_k = 0.$$

Then all calculations will be the same as in part (A) with "n" instead of 5.

27.   (A)

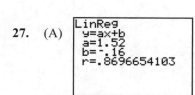   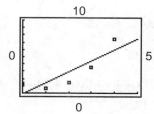   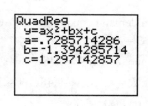   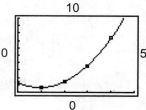

(B)   The quadratic regression function best fits the data.

29.   The cubic regression function has the form $y = ax^3 + bx^2 + cx + d$. The normal equations form a system of 4 linear equations in the 4 variables a, b, c and d. The system can be solved using Gauss-Jordan elimination.

31.   (A)

x_k	y_k	$x_k y_k$	x_k^2
1	3,658	3,658	1
3	3,591	10,773	9
5	3,431	17,155	25
7	3,276	22,932	49
9	3,041	27,369	81
11	2,908	31,988	121
Totals	36	19,905	113,875	286

Thus, $\sum\limits_{k=1}^{6} x_k = 36$, $\sum\limits_{k=1}^{6} y_k = 19,905$, $\sum\limits_{k=1}^{6} x_k y_k = 113,875$, $\sum\limits_{k=1}^{6} x_k^2 = 286$.

Substituting these values in the formulas for a and b, we have:

$$a = \frac{6(113,875) - (36)(19,905)}{6(286) - (36)^2} \approx -79.36 \quad \text{and} \quad b = \frac{19,905 - (-79.36)(36)}{6} \approx 3,793.6$$

Thus, the least squares line is $y = -79.36x + 3,793.6$

(B)   The property crime rate in 2024 will be (approximately) $-79.36(24) + 3,793.6 \approx 1,889$ crimes per 100,000 population.

33.   (A)

	x_k	y_k	$x_k y_k$	x_k^2
	5.0	2.0	10	25
	5.5	1.8	9.9	30.25
	6.0	1.4	8.4	36
	6.5	1.2	7.8	42.25
	7.0	1.1	7.7	49
Totals	30	7.5	43.8	182.5

Thus, $\displaystyle\sum_{k=1}^{5} x_k = 30$, $\displaystyle\sum_{k=1}^{5} y_k = 7.5$, $\displaystyle\sum_{k=1}^{5} x_k y_k = 43.8$, $\displaystyle\sum_{k=1}^{5} x_k^2 = 182.5$.

Substituting these values into the formulas for a and b, we have:

$$a = \frac{5(43.8) - (30)(7.5)}{5(182.5) - (30)^2} = \frac{-6}{12.5} = -0.48$$

$$b = \frac{7.5 - (-0.48)(30)}{5} = 4.38$$

Thus, a demand equation is $y = -0.48x + 4.38$.

(B) Cost: $C = 4y$

Revenue: $R = xy = -0.48x^2 + 4.38x$

Profit: $P = R - C = -0.48x^2 + 4.38x - 4(-0.48x + 4.38)$

or $P(x) = -0.48x^2 + 6.3x - 17.52$

Now, $P'(x) = -0.96x + 6.3$.

Critical value: $P'(x) = -0.96x + 6.3 = 0$

$$x = \frac{6.3}{0.96} \approx 6.56$$

$P''(x) = -0.96$ and $P''(6.56) = -0.96 < 0$

Thus, $P(x)$ has a maximum at $x = 6.56$; the price per bottle should be \$6.56 to maximize the monthly profit.

35. (A) We use the linear regression feature on a graphing utility. Thus, the least squares line is:
$y = 0.0222x + 18.94$.

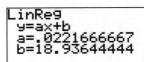

(B) The year 2024 corresponds to $x = 44$;
$y(44) = 0.0222(44) + 18.94 \approx 19.92$

An estimate for the winning height in the Olympic Games is 19.92 feet.

37. (A) We use the linear regression feature on a graphing utility with 1885 as $x = 0$, 1895 as $x = 10$, ..., etc.

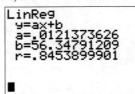

Thus, the least squares line is: $y = 0.00121x + 56.35$

(B) The year 2085 corresponds to $x = 200$;
$y(200) = 0.0121(200) + 56.35 \approx 58.77°$ F.

Things to remember:

GIVEN A FUNCTION $z = f(x, y)$:

<u>1.</u> $\displaystyle\int f(x, y)dx$ means antidifferentiate $f(x, y)$ with respect to x, holding y fixed.

$\displaystyle\int f(x, y)dy$ means antidifferentiate $f(x, y)$ with respect to y, holding x fixed.

<u>2.</u>   The DOUBLE INTEGRAL of $f(x, y)$ over the rectangle
$R = \{(x, y) \mid a \leq x \leq b, c \leq y \leq d\}$ is:

$$\iint\limits_{R} f(x, y)dA = \int_{a}^{b} \left[\int_{c}^{d} f(x, y)dy \right] dx$$

$$= \int_{c}^{d} \left[\int_{a}^{b} f(x, y)dx \right] dy$$

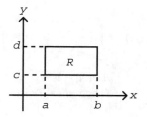

<u>3.</u>   The AVERAGE VALUE of $f(x, y)$ over the rectangle
$R = \{(x, y) | a \leq x \leq b, c \leq y \leq d\}$ is:

$$\frac{1}{(b-a)(d-c)} \iint\limits_{R} f(x, y)dA$$

<u>4.</u>   VOLUME UNDER A SURFACE

If $f(x, y) \geq 0$ over a rectangle $R = \{(x, y) | a \leq x \leq b, c \leq y \leq d\}$,
then the volume of the solid formed by graphing f over the rectangle R is given by:

$$V = \iint\limits_{R} f(x, y)dA$$

---

1. $\displaystyle\int (\pi + x)\,dx = \pi x + \frac{1}{2}x^2 + C$

3. $\displaystyle\int \left(1 + \frac{\pi}{x}\right)dx = x + \pi \ln x + C$

5. $\displaystyle\int e^{\pi x}\,dx$ (Let $u = \pi x$. Then $du = \pi\,dx$.) $\displaystyle\int e^{\pi x}\,dx = \frac{1}{\pi}\int e^{u}\,du = \frac{1}{\pi}e^{u} + C = \frac{e^{\pi x}}{\pi} + C$

7.   (A) $\displaystyle\int 12x^2 y^3\,dy = 12x^2 \int y^3\,dy$ (x is treated as a constant.)

$\displaystyle = 12x^2 \frac{y^4}{4} + C(x)$ (The "constant" of integration is a function of x.)

$= 3x^2 y^4 + C(x)$

(B) $\displaystyle\int_{0}^{1} 12x^2 y^3\,dy = 3x^2 y^4 \Big|_{0}^{1} = 3x^2$

9. (A) $\int (4x + 6y + 5)\, dx$

    $= \int 4x\, dx + \int (6y + 5)\, dx$ (y is treated as a constant.)

    $= 2x^2 + (6y + 5)x + E(y)$ (The "constant" of integration is a function of y.)

    $= 2x^2 + 6xy + 5x + E(y)$

    (B) $\int_{-2}^{3} (4x + 6y + 5)\, dx = (2x^2 + 6xy + 5x)\Big|_{-2}^{3} = 2\cdot 3^2 + 6\cdot 3y + 5\cdot 3 - [2(-2)^2 + 6(-2)y + 5(-2)]$

    $= 30y + 35$

11. (A) $\int \dfrac{x}{\sqrt{y + x^2}}\, dx = \int (y + x^2)^{-1/2}x\, dx = \dfrac{1}{2}\int (y + x^2)^{-1/2}2x\, dx$

    Let $u = y + x^2$.

    Then $du = 2x\, dx$.

    $= \dfrac{1}{2}\int u^{-1/2}\, du$

    $= u^{1/2} + E(y) = \sqrt{y + x^2} + E(y)$

    (B) $\int_{0}^{2} \dfrac{x}{\sqrt{y + x^2}}\, dx = \sqrt{y + x^2}\,\Big|_{0}^{2} = \sqrt{y + 4} - \sqrt{y}$

13. (A) $\int \dfrac{\ln x}{xy}\, dy = \dfrac{\ln x}{x}\int \dfrac{1}{y}\, dy = \dfrac{\ln x}{x}\cdot \ln y + C(x)$

    (B) $\int_{1}^{e^2} \dfrac{\ln x}{xy}\, dy = \dfrac{\ln x \ln y}{x}\,\Big]_{1}^{e^2} = \dfrac{\ln x \ln e^2}{x} - \dfrac{\ln x \ln 1}{x} = \dfrac{2\ln x}{x}$

15. $\int_{-1}^{2}\int_{0}^{1} 12x^2 y^3\, dy\, dx = \int_{-1}^{2}\left[\int_{0}^{1} 12x^2 y^3\, dy\right] dx = \int_{-1}^{2} 3x^2\, dx$ (see Problem 7)

    $= x^3\,\Big|_{-1}^{2} = 8 + 1 = 9$

17. $\int_{1}^{4}\int_{-2}^{3} (4x + 6y + 5)\, dx\, dy = \int_{1}^{4}\left[\int_{-2}^{3} (4x + 6y + 5)\, dx\right] dy = \int_{1}^{4} (30y + 35)\, dy$ (see Problem 9)

    $= (15y^2 + 35y)\,\Big|_{1}^{4} = 15\cdot 4^2 + 35\cdot 4 - (15 + 35) = 330$

19. $\int_{1}^{5}\int_{0}^{2} \dfrac{x}{\sqrt{y + x^2}}\, dx\, dy = \int_{1}^{5}\left[\int_{0}^{2} \dfrac{x}{\sqrt{y + x^2}}\, dx\right] dy = \int_{1}^{5} (\sqrt{4 + y} - \sqrt{y})\, dy$ (see Problem 11)

    $= \left[\dfrac{2}{3}(4 + y)^{3/2} - \dfrac{2}{3}y^{3/2}\right]_{1}^{5} = \dfrac{2}{3}(9)^{3/2} - \dfrac{2}{3}(5)^{3/2} - \left(\dfrac{2}{3}\cdot 5^{3/2} - \dfrac{2}{3}\cdot 1^{3/2}\right)$

    $= 18 - \dfrac{4}{3}(5)^{3/2} + \dfrac{2}{3} = \dfrac{56 - 20\sqrt{5}}{3}$

21. $\displaystyle\int_1^e \int_1^{e^2} \frac{\ln x}{xy}\,dy\,dx = \int_1^e \left[\int_1^{e^2} \frac{\ln x}{xy}\,dy\right] dx = \int_1^e \frac{2\ln x}{x}\,dx$ (see Problem 13)

$\displaystyle\qquad\qquad = 2\int_1^e \frac{\ln x}{x}\,dx = (\ln x)^2 \Big]_1^e = 1$ Substitution: $u = \ln x,\ du = \frac{1}{x}\,dx$

23. $\displaystyle\iint_R xy\,dA = \int_0^2 \int_0^4 xy\,dy\,dx = \int_0^2 \left[\int_0^4 xy\,dy\right] dx = \int_0^2 \left[\frac{xy^2}{2}\right]_0^4 dx = \int_0^2 8x\,dx = 4x^2 \Big|_0^2 = 16$

$\displaystyle\qquad\iint_R xy\,dA = \int_0^4 \int_0^2 xy\,dx\,dy = \int_0^4 \left[\int_0^2 xy\,dx\right] dy = \int_0^4 \left[\frac{x^2 y}{2}\right]_0^2 dy = \int_0^4 2y\,dy = y^2 \Big|_0^4 = 16$

25. $\displaystyle\iint_R (x+y)^5\,dA = \int_{-1}^1 \int_1^2 (x+y)^5\,dy\,dx = \int_{-1}^1 \left[\frac{(x+y)^6}{6}\right]_1^2 dx = \int_{-1}^1 \left[\frac{(x+2)^6}{6} - \frac{(x+1)^6}{6}\right] dx$

$\displaystyle\qquad\qquad = \left[\frac{(x+2)^7}{42} - \frac{(x+1)^7}{42}\right]_{-1}^1 = \frac{3^7}{42} - \frac{2^7}{42} - \frac{1}{42} = 49$

$\displaystyle\qquad\iint_R (x+y)^5\,dA = \int_1^2 \int_{-1}^1 (x+y)^5\,dx\,dy = \int_1^2 \left[\int_{-1}^1 (x+y)^5\,dx\right] dy = \int_1^2 \left[\frac{(x+y)^6}{6}\right]_{-1}^1 dy$

$\displaystyle\qquad\qquad = \int_1^2 \left[\frac{(y+1)^6}{6} - \frac{(y-1)^6}{6}\right] dy = \left[\frac{(y+1)^7}{42} - \frac{(y-1)^7}{42}\right]_1^2 = \frac{3^7}{42} - \frac{1}{42} - \frac{2^7}{42} = 49$

27. Average value $\displaystyle= \frac{1}{(5-1)[1-(-1)]} \iint_R (x+y)^2\,dA = \frac{1}{8}\int_{-1}^1 \int_1^5 (x+y)^2\,dx\,dy = \frac{1}{8}\int_{-1}^1 \left[\frac{(x+y)^3}{3}\right]_1^5 dy$

$\displaystyle\qquad\qquad = \frac{1}{8}\int_{-1}^1 \left[\frac{(5+y)^3}{3} - \frac{(1+y)^3}{3}\right] dy = \frac{1}{8}\left[\frac{(5+y)^4}{12} - \frac{(1+y)^4}{12}\right]_{-1}^1 = \frac{1}{96}[6^4 - 2^4 - 4^4] = \frac{32}{3}$

29. Average value $\displaystyle= \frac{1}{(4-1)(7-2)} \iint_R \frac{x}{y}\,dA = \frac{1}{15}\int_1^4 \int_2^7 \frac{x}{y}\,dy\,dx = \frac{1}{15}\int_1^4 \left[x\ln y\right]_2^7 dx$

$\displaystyle\qquad\qquad = \frac{1}{15}\int_1^4 [x\ln 7 - x\ln 2]\,dx = \frac{\ln 7 - \ln 2}{15}\int_1^4 x\,dx = \left[\frac{\ln 7 - \ln 2}{15} \cdot \frac{x^2}{2}\right]_1^4$

$\displaystyle\qquad\qquad = \frac{\ln 7 - \ln 2}{15}\left(\frac{4^2}{2} - \frac{1^2}{2}\right) = \frac{1}{2}(\ln 7 - \ln 2) = \frac{1}{2}\ln\left(\frac{7}{2}\right) \approx 0.626$

31. $\displaystyle V = \iint_R (2 - x^2 - y^2)\,dA = \int_0^1 \int_0^1 \iint_R (2 - x^2 - y^2)\,dy\,dx$

$\displaystyle\qquad = \int_0^1 \left[\int_0^1 (2 - x^2 - y^2)\,dy\right] dx = \int_0^1 \left[2y - x^2 y - \frac{y^3}{3}\right]_0^1 dx$

$\displaystyle\qquad = \int_0^1 \left(2 - x^2 - \frac{1}{3}\right) dx = \int_0^1 \left(\frac{5}{3} - x^2\right) dx = \left[\frac{5}{3}x - \frac{x^3}{3}\right]_0^1 = \frac{5}{3} - \frac{1}{3} = \frac{4}{3}$

33. $V = \displaystyle\iint_R (4 - y^2)dA = \int_0^2 \int_0^2 (4 - y^2)dx\,dy = \int_0^2 \left[\int_0^2 (4 - y^2)dx \right] dy$

$\qquad = \int_0^2 \left[(4x - xy^2)\big|_0^2 \right] dy = \int_0^2 (8 - 2y^2)dy = \left[8y - \dfrac{2}{3}y^3 \right]_0^2 = 16 - \dfrac{16}{3} = \dfrac{32}{3}$

35. $\displaystyle\iint_R xe^{xy}\,dA = \int_0^1 \int_1^2 xe^{xy}\,dy\,dx = \int_0^1 \left[\int_1^2 xe^{xy}\,dy \right] dx = \int_0^1 \left[x \int_1^2 e^{xy}\,dy \right] dx = \int_0^1 \left[x \cdot \dfrac{e^{xy}}{x} \right]_1^2 dx$

$\qquad = \int_0^1 (e^{2x} - e^x)\,dx = \left[\dfrac{e^{2x}}{2} - e^x \right]_0^1 = \dfrac{e^2}{2} - e - \left(\dfrac{1}{2} - 1 \right) = \dfrac{e^2}{2} - e + \dfrac{1}{2}$

37. $\displaystyle\iint_R \dfrac{2y + 3xy^2}{1 + x^2}\,dA = \int_0^1 \int_{-1}^1 \dfrac{2y + 3xy^2}{1 + x^2}\,dy\,dx = \int_0^1 \left[\int_{-1}^1 \dfrac{2y + 3xy^2}{1 + x^2}\,dy \right] dx = \int_0^1 \left[\dfrac{1}{1 + x^2}(y^2 + xy^3) \right]_{-1}^1 dx$

$\qquad = \int_0^1 \left[\dfrac{1}{1 + x^2}(1 + x - [1 - x]) \right] dx = \int_0^1 \dfrac{2x}{1 + x^2}\,dx = \ln(1 + x^2)\big|_0^1 = \ln 2$

$\qquad\qquad\qquad\qquad\qquad\qquad \text{Substitution: } u = 1 + x^2, \, du = 2x\,dx$

39. $\displaystyle\int_0^2 \int_0^2 (1 - y)dx\,dy = \int_0^2 \left[\int_0^2 (1 - y)dx \right] dy = \int_0^2 \left[(x - xy) \right]_0^2 dy = \int_0^2 (2 - 2y)\,dy = (2y - y^2)\big|_0^2 = 0$

Since $f(x, y) = 1 - y$ is NOT positive over the entire rectangle $R = \{(x, y) \mid 0 \le x \le 2, 0 \le y \le 2\}$, the double integral does not represent the volume of a solid.

41. $f(x, y) = x^3 + y^2 - e^{-x} - 1$ on $R = \{(x, y) \mid -2 \le x \le 2, -2 \le y \le 2\}$.

(A) Average value of f:

$\qquad \dfrac{1}{b - a} \cdot \dfrac{1}{d - c} \displaystyle\iint_R f(x, y)\,dA = \dfrac{1}{2 - (-2)} \cdot \dfrac{1}{2 - (-2)} \int_{-2}^2 \int_{-2}^2 (x^3 + y^2 - e^{-x} - 1)\,dx\,dy$

$\qquad = \dfrac{1}{16} \int_{-2}^2 \left[\dfrac{1}{4}x^4 + xy^2 + e^{-x} - x \right]_{-2}^2 dy = \dfrac{1}{16} \int_{-2}^2 [4y^2 + e^{-2} - e^2 - 4]\,dy$

$\qquad = \dfrac{1}{16} \left[\dfrac{4}{3}y^3 + e^{-2}y - e^2 y - 4y \right]_{-2}^2 = \dfrac{1}{16} \left[\dfrac{64}{3} + 4e^{-2} - 4e^2 - 16 \right] = \dfrac{1}{3} + \dfrac{1}{4}e^{-2} - \dfrac{1}{4}e^2$

(B)

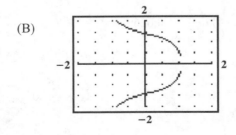

(C) $f(x, y) > 0$ at the points which lie to the right of the curve in part (B);

$f(x, y) < 0$ at the points which lie to the left of the curve in part (B).

43. $S(x, y) = \dfrac{y}{1-x}$, $0.6 \le x \le 0.8$, $5 \le y \le 7$.

The *average* total amount of spending is given by:

$$T = \frac{1}{(0.8 - 0.6)(7 - 5)} \iint_R \frac{y}{1-x} \, dA = \frac{1}{0.4} \int_{0.6}^{0.8} \int_5^7 \frac{y}{1-x} \, dy \, dx = \frac{1}{0.4} \int_{0.6}^{0.8} \left[\frac{1}{1-x} \cdot \frac{y^2}{2} \right]_5^7 dx$$

$$= \frac{1}{0.4} \int_{0.6}^{0.8} \frac{1}{1-x} \left(\frac{49}{2} - \frac{25}{2} \right) dx = \frac{12}{0.4} \int_{0.6}^{0.8} \frac{1}{1-x} \, dx = 30[-\ln(1-x)] \Big|_{0.6}^{0.8}$$

$$= 30[-\ln(0.2) + \ln(0.4)] = 30 \ln 2 \approx \$20.8 \text{ billion}$$

45. $N(x, y) = x^{0.75} y^{0.25}$, $10 \le x \le 20$, $1 \le y \le 2$

$$\text{Average value} = \frac{1}{(20 - 10)(2 - 1)} \int_{10}^{20} \int_1^2 x^{0.75} y^{0.25} \, dy \, dx = \frac{1}{10} \int_{10}^{20} \left[x^{0.75} \frac{y^{1.25}}{1.25} \right]_1^2 dx$$

$$= \frac{1}{10} \int_{10}^{20} \left[x^{0.75} \frac{2^{1.25} - 1}{1.25} \right] dx = \frac{1}{12.5} (2^{1.25} - 1) \int_{10}^{20} x^{0.75} \, dx = \left[\frac{1}{12.5} (2^{1.25} - 1) \frac{x^{1.75}}{1.75} \right]_{10}^{20}$$

$$= \frac{1}{21.875} (2^{1.25} - 1)(20^{1.75} - 10^{1.75}) \approx 8.375 \text{ or } 8,375 \text{ items}$$

47. $C = 10 - \dfrac{1}{10} d^2 = 10 - \dfrac{1}{10}(x^2 + y^2) = C(x, y)$, $-8 \le x \le 8$, $-6 \le y \le 6$

Average concentration

$$= \frac{1}{16(12)} \int_{-8}^{8} \int_{-6}^{6} \left[10 - \frac{1}{10}(x^2 + y^2) \right] dy \, dx = \frac{1}{192} \int_{-8}^{8} \left[10y - \frac{1}{10} \left(x^2 y + \frac{y^3}{3} \right) \right]_{-6}^{6} dx$$

$$= \frac{1}{192} \int_{-8}^{8} \left\{ 60 - \frac{1}{10} \left(6x^2 + \frac{216}{3} \right) - \left[-60 - \frac{1}{10} \left(-6x^2 - \frac{216}{3} \right) \right] \right\} dx$$

$$= \frac{1}{192} \int_{-8}^{8} \left[120 - \frac{1}{10} (12x^2 + 144) \right] dx = \frac{1}{192} \left[120x - \frac{1}{10} (4x^3 + 144x) \right]_{-8}^{8}$$

$$= \frac{1}{192} (1280) = \frac{20}{3} \approx 6.67 \text{ insects per square foot}$$

49. $C = 100 - 15d^2 = 100 - 15(x^2 + y^2) = C(x, y)$, $-2 \le x \le 2$, $-1 \le y \le 1$

$$\text{Average concentration} = \frac{1}{4(2)} \int_{-2}^{2} \int_{-1}^{1} [100 - 15(x^2 + y^2)] \, dy \, dx$$

$$= \frac{1}{8} \int_{-2}^{2} \left[100y - 15x^2 y - 5y^3 \right]_{-1}^{1} dx$$

$$= \frac{1}{8} \int_{-2}^{2} (190 - 30x^2) \, dx = \left[\frac{1}{8} \left(190x - 10x^3 \right) \right]_{-2}^{2} = \frac{1}{8} (600) = 75 \text{ parts per million}$$

51. $L = 0.0000133xy^2$, $2000 \leq x \leq 3000$, $50 \leq y \leq 60$

Average length $= \dfrac{1}{10,000} \displaystyle\int_{2000}^{3000} \int_{50}^{60} 0.0000133xy^2 \, dy \, dx$

$= \dfrac{0.0000133}{10,000} \displaystyle\int_{2000}^{3000} \left[\dfrac{xy^3}{3}\right]_{50}^{60} dx = \dfrac{0.0000133}{10,000} \displaystyle\int_{2000}^{3000} \dfrac{91,000}{3} x \, dx$

$= \left[\dfrac{1.2103}{30,000} \cdot \dfrac{x^2}{2}\right]_{2000}^{3000} = \dfrac{1.2103}{60,000}(5,000,000) \approx 100.86 \text{ feet}$

53. $Q(x, y) = 100\left(\dfrac{x}{y}\right)$, $8 \leq x \leq 16$, $10 \leq y \leq 12$

Average intelligence $= \dfrac{1}{16} \displaystyle\int_{8}^{16} \int_{10}^{12} 100\left(\dfrac{x}{y}\right) dy \, dx = \dfrac{100}{16} \displaystyle\int_{8}^{16} \left[x \ln y\right]_{10}^{12} dx$

$= \dfrac{100}{16} \displaystyle\int_{8}^{16} x(\ln 12 - \ln 10) \, dx = \left[\dfrac{100(\ln 12 - \ln 10)}{16} \cdot \dfrac{x^2}{2}\right]_{8}^{16}$

$= \dfrac{100(\ln 12 - \ln 10)}{32}(192) = 600 \ln(1.2) \approx 109.4$

EXERCISE 15-7

Things to remember:

1. REGULAR REGIONS:

    A region R in the xy plane is a REGULAR x REGION if there exist functions $f(x)$ and $g(x)$ and numbers a and b so that

    $R = \{(x, y) \mid g(x) \leq y \leq f(x), a \leq x \leq b\}$.

    A region R is a REGULAR y REGION if there exist functions $h(y)$ and $k(y)$ and numbers c and d so that

    $R = \{(x, y) \mid h(y) \leq x \leq k(y), c \leq y \leq d\}$.

    See Figure 3 in the text for a geometric interpretation.

2. DOUBLE INTEGRATION OVER REGULAR REGIONS:

    If $R = \{(x, y) \mid g(x) \leq y \leq f(x), a \leq x \leq b\}$, then

    $$\iint\limits_{R} F(x, y) \, dA = \int_{a}^{b}\left[\int_{g(x)}^{f(x)} F(x, y) dy\right] dx.$$

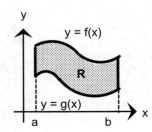

y = f(x)

R

y = g(x)

a   b

Regular x region

If $R = \{(x, y) \mid h(y) \leq x \leq k(y), c \leq y \leq d\}$, then

$$\iint\limits_R F(x, y)\, dA = \int_c^d \left[\int_{h(y)}^{k(y)} F(x, y)\, dx \right] dy$$

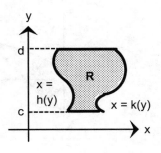

d

R

x = h(y)

x = k(y)

c

Regular y region

3.   If $f(x, y) \geq 0$ over a regular region R, then the double integral of f over R is the VOLUME of the solid formed by the graph f over R.

---

1. $y = 4 - x^2$, $y = 0, 0 \leq x \leq 2$

The region is

$R = \{(x, y) \mid 0 \leq y \leq 4 - x^2, 0 \leq x \leq 2\}$

which is a regular x region.

Also, on the interval $0 \leq x \leq 2$, we can write the equation $y = 4 - x^2$ as $x = \sqrt{4 - y}$, $0 \leq y \leq 4$. Thus,

$R = \{(x, y) \mid 0 \leq x \leq \sqrt{4 - y}, 0 \leq y \leq 4\}$

which is a regular y region. Hence R is both a regular x region and a regular y region.

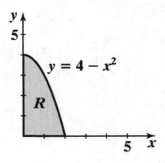

3. $y = x^3$, $y = 12 - 2x$, $x = 0$

or $g(x) = x^3$, $f(x) = 12 - 2x$, $x = 0$

We set $f(x) = g(x)$ to find the points of intersection of the two graphs:

$12 - 2x = x^3$

or $x^3 + 2x - 12 = 0$

$(x - 2)(x^2 + 2x + 6) = 0$

Therefore, $x = 2$ and $(2, 8)$ is the point of intersection. The region is:

$R = \{(x, y) \mid x^3 \leq y \leq 12 - 2x, 0 \leq x \leq 2\}$, which is a regular x region.

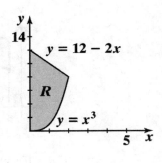

5. $y^2 = 2x$, $y = x - 4$ or $k(y) = \frac{1}{2}y^2$ and $x = h(y) = y + 4$.

To find the point(s) of intersection,

set $h(y) = k(y)$:

$y + 4 = \frac{1}{2}y^2$

or $y^2 - 2y - 8 = 0$

$(y - 4)(y + 2) = 0$

and $y = -2$, $y = 4$. The points of intersection are $(2, -2)$ and $(8, 4)$.

Therefore, the region R is

$R = \left\{ (x, y) \,\middle|\, \frac{1}{2}y^2 \leq x \leq y + 4, -2 \leq y \leq 4 \right\}$, which is a regular y region.

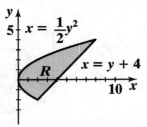

7. $\displaystyle\int_0^1 \int_0^x (x + y)\,dy\,dx = \int_0^1 \left[\int_0^x (x + y)\,dy \right] dx = \int_0^1 \left[xy + \frac{1}{2}y^2 \right]_0^x dx$

$\displaystyle = \int_0^1 \left[x^2 + \frac{1}{2}x^2 \right] dx = \int_0^1 \frac{3}{2}x^2\,dx = \left[\frac{1}{2}x^3 \right]_0^1 = \frac{1}{2}(1 - 0) = \frac{1}{2}$

9. $\displaystyle\int_0^1 \int_{y^3}^{\sqrt{y}} (2x + y)\,dx\,dy = \int_0^1 \left[\int_{y^3}^{\sqrt{y}} (2x + y)\,dx \right] dy = \int_0^1 \left[(x^2 + xy) \right]_{y^3}^{\sqrt{y}} dy$

$\displaystyle = \int_0^1 \{ [(\sqrt{y})^2 + y\sqrt{y}] - [(y^3)^2 + y(y^3)] \}\,dy = \int_0^1 (y + y^{3/2} - y^6 - y^4)\,dy$

$\displaystyle = \left[\frac{y^2}{2} + \frac{2}{5}y^{5/2} - \frac{1}{7}y^7 - \frac{1}{5}y^5 \right]_0^1 = \frac{1}{2} + \frac{2}{5} - \frac{1}{7} - \frac{1}{5} = \frac{39}{70}$

11. $R = \{(x, y) \mid |x| \leq 2, |y| \leq 3\}$.

R is the rectangle with vertices $(-2, 3)$, $(2, 3)$, $(2, -3)$, $(-2, -3)$;

R is a regular x region **and** a regular y region.

13. $R = \{(x, y) \mid x^2 + y^2 \geq 1, |x| \leq 2, 0 \leq y \leq 2\}$

R is the region that lies outside the circle of radius 1 centered at the origin and inside the rectangle with vertices $(-2, 2)$, $(2, 2)$, $(2, 0)$, $(-2, 0)$; R is a regular x region but not a regular y region.

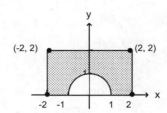

15. $R = \{(x, y) \mid 0 \leq y \leq 2x, 0 \leq x \leq 2\}$

$$\iint\limits_R (x^2 + y^2)\,dA = \int_0^2 \int_0^{2x} (x^2 + y^2)\,dy\,dx = \int_0^2 \left[\int_0^{2x}(x^2 + y^2)\,dy\right]dx = \int_0^2 \left[x^2 y + \frac{1}{3}y^3\right]_0^{2x} dx$$

$$= \int_0^2 \left[x^2(2x) + \frac{1}{3}(2x)^3\right]dx = \int_0^2 \left(2x^3 + \frac{8}{3}x^3\right)dx = \int_0^2 \frac{14}{3}x^3\,dx = \frac{7}{6}x^4\Big|_0^2 = \frac{56}{3}$$

17. $R = \{(x, y) \mid 0 \leq x \leq y + 2, 0 \leq y \leq 1\}$

$$\iint\limits_R (x + y - 2)^3\,dA = \int_0^1 \int_0^{y+2} (x + y - 2)^3\,dx\,dy = \int_0^1 \left[\int_0^{y+2}(x + y - 2)^3\,dx\right]dy$$

$$= \int_0^1 \left[\frac{1}{4}(x + y - 2)^4\right]_0^{y+2} dy = \int_0^1 \left[4y^4 - \frac{1}{4}(y - 2)^4\right]dy = \left[\frac{4}{5}y^5 - \frac{1}{20}(y - 2)^5\right]_0^1$$

$$= \frac{4}{5} - \frac{1}{20}(-1)^5 - \left[-\frac{1}{20}(-2)^5\right] = \frac{4}{5} + \frac{1}{20} - \frac{32}{20} = -\frac{3}{4}$$

19. $R = \{(x, y) \mid -x \leq y \leq x, 0 \leq x \leq 2\}$

$$\iint\limits_R e^{x+y}\,dA = \int_0^2 \int_{-x}^x e^{x+y}\,dy\,dx = \int_0^2 \left[\int_{-x}^x e^x e^y\,dy\right]dx = \int_0^2 e^x\left[e^y\right]_{-x}^x dx = \int_0^2 e^x(e^x - e^{-x})\,dx$$

$$= \int_0^2 (e^{2x} - 1)\,dx = \left[\frac{1}{2}e^{2x} - x\right]_0^2 = \frac{1}{2}e^4 - 2 - \left(\frac{1}{2}e^0\right) = \frac{1}{2}e^4 - \frac{5}{2}$$

21. $R = \{(x, y) \mid 0 \leq y \leq x + 1, 0 \leq x \leq 1\}$

$$\iint\limits_R \sqrt{1 + x + y}\,dA = \int_0^1 \int_0^{x+1} (1 + x + y)^{1/2}\,dy\,dx$$

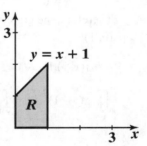

$$= \int_0^1 \left[\frac{2}{3}(1 + x + y)^{3/2}\right]_0^{x+1} dx$$

$$= \int_0^1 \left[\frac{2}{3}(2 + 2x)^{3/2} - \frac{2}{3}(1 + x)^{3/2}\right]dx$$

$$= \int_0^1 \left[\frac{4\sqrt{2}}{3}(1 + x)^{3/2} - \frac{2}{3}(1 + x)^{3/2}\right]dx$$

[Note: $(2 + 2x)^{3/2}$
$= 2\sqrt{2}\,(1 + x)^{3/2}$]

$$= \frac{(4\sqrt{2} - 2)}{3}\int_0^1 (1 + x)^{3/2}\,dx = \left[\frac{(4\sqrt{2} - 2)}{3}\cdot\frac{2}{5}(1 + x)^{5/2}\right]_0^1$$

$$= \frac{(8\sqrt{2} - 4)}{15}\cdot 2^{5/2} - \frac{(8\sqrt{2} - 4)}{15} = \frac{68 - 24\sqrt{2}}{15}$$

23. $y = f(x) = 4x - x^2, \ y = g(x) = 0$

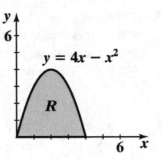

$4x - x^2 = 0$

$x(4 - x) = 0$

$x = 0, \ x = 4$

Therefore,

$R = \{(x, y)|0 \le y \le 4x - x^2, \ 0 \le x \le 4.\}$

$$\iint\limits_R \sqrt{y + x^2} \ dA = \int_0^4 \int_0^{4x - x^2} (y + x^2)^{1/2} \ dy \ dx$$

$$= \int_0^4 \left[\frac{2}{3}(y + x^2)^{3/2} \right]_0^{4x - x^2} dx$$

$$= \int_0^4 \frac{2}{3} [(4x - x^2 + x^2)^{3/2} - (0 + x^2)^{3/2}] \ dx$$

$$= \int_0^4 \left(\frac{16}{3} x^{3/2} - \frac{2}{3} x^3 \right) dx \qquad [\underline{\text{Note}}: (4x)^{3/2} = 4^{3/2}x^{3/2} = 8x^{3/2}.]$$

$$= \left[\frac{16}{3} \cdot \frac{2}{5} x^{5/2} - \frac{2}{3} \cdot \frac{1}{4} x^4 \right]_0^4 = \frac{32}{15}(4)^{5/2} - \frac{1}{6}(4)^4 \qquad [\underline{\text{Note}}: 4^{5/2} = 2^5 = 32.]$$

$$= \frac{32 \cdot 32}{15} - \frac{256}{6} = \frac{128}{5}$$

25. $y = g(x) = 1 - \sqrt{x} \ , \ y = f(x) = 1 + \sqrt{x} \ , \ x = 4$

$$f(x) = g(x)$$

$$1 + \sqrt{x} = 1 - \sqrt{x}$$

$$2\sqrt{x} = 0$$

$$x = 0$$

Therefore, the graphs intersect at the point
$(0, 1)$:

$R = \{(x, y)|1 - \sqrt{x} \ \le y \le 1 + \sqrt{x} \ , \ 0 \le x \le 4\}$

$$\iint\limits_R x(y - 1)^2 dA = \int_0^4 \int_{1 - \sqrt{x}}^{1 + \sqrt{x}} x(y - 1)^2 dy \ dx$$

$$= \int_0^4 \left[\frac{1}{3} x(y - 1)^3 \right]_{1 - \sqrt{x}}^{1 + \sqrt{x}} dx$$

$$= \int_0^4 \frac{1}{3} x[(1 + \sqrt{x} \ - 1)^3 - (1 - \sqrt{x} \ - 1)^3] \ dx = \int_0^4 \frac{2}{3} x^{5/2} \ dx$$

$$= \left[\frac{2}{3} \cdot \frac{2}{7} x^{7/2} \right]_0^4 = \frac{4}{21}(4)^{7/2} = \frac{4 \cdot 128}{21} = \frac{512}{21}$$

27. $\displaystyle\int_0^3 \int_0^{3-x} (x+2y)\, dy\, dx = \int_0^3 \left[\int_0^{3-x}(x+2y)\,dy\right] dx = \int_0^3 \left[(xy+y^2)\right]_0^{3-x} dx$

$$= \int_0^3 [x(3-x) + (3-x)^2]\, dx = \int_0^3 (9-3x)\, dx$$

$$= \left[9x - \frac{3}{2}x^2\right]_0^3 = 27 - \frac{27}{2} = \frac{27}{2}$$

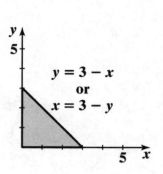

$R = \{(x,y) \mid 0 \le y \le 3-x,\ 0 \le x \le 3\}$

Now, $y = 3-x$ or $x = 3-y$; and if $x = 0$, then $y = 3$, if $x = 3$, then $y = 0$. Therefore,

$R = \{(x,y) \mid 0 \le x \le 3-y,\ 0 \le y \le 3\}$

and integration with the order reversed is:

$$\int_0^3 \int_0^{3-y} (x+2y)dx\, dy = \int_0^3 \left[\frac{1}{2}x^2 + 2xy\right]_0^{3-y} dy = \int_0^3 \left[\frac{1}{2}(3-y)^2 + 2y(3-y)\right] dy$$

$$= \int_0^3 \left(-\frac{3}{2}y^2 + 3y + \frac{9}{2}\right) dy = \left[-\frac{1}{2}y^3 + \frac{3}{2}y^2 + \frac{9}{2}y\right]_0^3 = -\frac{27}{2} + \frac{27}{2} + \frac{27}{2} = \frac{27}{2}$$

29. $\displaystyle\int_0^1 \int_0^{1-x^2} x\sqrt{y}\, dy\, dx = \int_0^1 \left[\int_0^{1-x^2} xy^{1/2}dy\right]dx = \int_0^1 \left[\frac{2}{3}xy^{3/2}\right]_0^{1-x^2} dx = \int_0^1 \frac{2}{3}x(1-x^2)^{3/2}dx$

$$= \frac{2}{3}\int_0^1 x(1-x^2)^{3/2}dx \qquad\qquad \text{Let } u = 1-x^2. \text{ Then } du = -2x\, dx \text{ or}$$

$$\qquad\qquad\qquad\qquad\qquad x\, dx = -\frac{1}{2}\, du$$

$$= -\frac{1}{3}\int_1^0 u^{3/2}du \qquad\qquad u = 1 \text{ when } x = 0,\ u = 0 \text{ when } x = 1.$$

$$= \frac{1}{3}\int_0^1 u^{3/2}du = \left[\frac{1}{3}\cdot\frac{2}{5}u^{5/2}\right]_0^1 = \frac{2}{15}$$

$R = \{(x,y) \mid 0 \le y \le 1-x^2,\ 0 \le x \le 1\}$

Now, $y = 1-x^2$; therefore, $x^2 = 1-y$ or

$x = \sqrt{1-y}$, and if $x = 0$, then $y = 1$, if $x = 1$, then $y = 0$.

Thus,

$R = \{(x,y) \mid 0 \le x \le \sqrt{1-y},\ 0 \le y \le 1\}$, and integration with the order reversed is:

$$\int_0^1 \int_0^{\sqrt{1-y}} x\sqrt{y}\, dx\, dy = \int_0^1 \left[\int_0^{\sqrt{1-y}} x\sqrt{y}dx\right] dy = \int_0^1 \left[\frac{1}{2}x^2\sqrt{y}\right]_0^{\sqrt{1-y}} dy = \int_0^1 \frac{1}{2}(1-y)\sqrt{y}\ dy$$

$$= \frac{1}{2}\int_0^1 (y^{1/2} - y^{3/2})\, dy = \frac{1}{2}\left[\frac{2}{3}y^{3/2} - \frac{2}{5}y^{5/2}\right]_0^1 = \frac{1}{2}\left(\frac{2}{3} - \frac{2}{5}\right) = \frac{2}{15}$$

31. $\int_0^4 \int_{x/4}^{\sqrt{x}/2} x \, dy \, dx = \int_0^4 \left[\int_{x/4}^{\sqrt{x}/2} x \, dy \right] dx = \int_0^4 \left[xy \right]_{x/4}^{\sqrt{x}/2} dx = \int_0^4 \left(\frac{1}{2} x^{3/2} - \frac{1}{4} x^2 \right) dx$

$\qquad\qquad = \left[\frac{1}{2} \cdot \frac{2}{5} x^{5/2} - \frac{1}{12} x^3 \right]_0^4 = \frac{32}{5} - \frac{16}{3} = \frac{16}{15}$

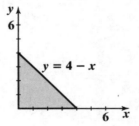

$R = \left\{ (x, y) \; \middle| \; \frac{x}{4} \le y \le \frac{\sqrt{x}}{2}, 0 \le x \le 4 \right\}$

Now, $y = \dfrac{\sqrt{x}}{2}$ or $x = 4y^2$, and $y = \dfrac{x}{4}$ or $x = 4y$.

Also, $y = 0$ when $x = 0$ and $y = 1$ when $x = 4$.

Therefore,

$R = \{ (x, y) \mid 4y^2 \le x \le 4y, 0 \le y \le 1 \}$

$\int_0^4 \int_{x/4}^{\sqrt{x}/2} x \, dy \, dx = \int_0^1 \int_{4y^2}^{4y} x \, dx \, dy = \int_0^1 \left[\frac{1}{2} x^2 \right]_{4y^2}^{4y} dy = \int_0^1 \frac{1}{2} [(4y)^2 - (4y^2)^2] \, dy$

$\qquad\qquad = \int_0^1 (8y^2 - 8y^4) \, dy = \left[\frac{8}{3} y^3 - \frac{8}{5} y^5 \right]_0^1 = \frac{8}{3} - \frac{8}{5} = \frac{16}{15}$

33. $v = \iint\limits_R (4 - x - y) \, dA$

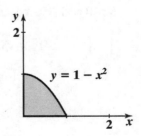

$x + y = 4$ or $y = 4 - x$, and $x = 0$, $y = 0$.

Therefore,
$R = \{ (x, y) \mid 0 \le y \le 4 - x, 0 \le x \le 4 \}$

$v = \int_0^4 \int_0^{4-x} (4 - x - y) \, dy \, dx = \int_0^4 \left[-\frac{1}{2} (4 - x - y)^2 \right]_0^{4-x} dx$

$\quad = \int_0^4 -\frac{1}{2} [(4 - x - 4 + x)^2 - (4 - x - 0)^2] \, dx = \int_0^4 \frac{1}{2} (4 - x)^2 \, dx$

$\quad = \left[-\frac{1}{6} (4 - x)^3 \right]_0^4 = -\frac{1}{6} [(4 - 4)^3 - (4 - 0)^3] = -\frac{1}{6} (-64) = \frac{32}{3}$

35. $v = \int_0^1 \int_0^{1-x^2} 4 \, dy \, dx = \int_0^1 \left[(4y) \right]_0^{1-x^2} dx = \int_0^1 4[(1 - x^2) - 0] \, dx$

$\qquad = \int_0^1 4(1 - x^2) \, dx = \left[4x - \frac{4}{3} x^3 \right]_0^1 = 4 - \frac{4}{3} = \frac{8}{3}$

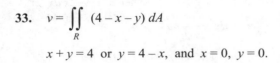

37. $R = \{(x, y) \mid x^2 \leq y \leq 4, 0 \leq x \leq 2\}$

Now, $y = x^2$ or $x = \sqrt{y}$; and $y = 0$ when $x = 0$, $y = 4$ when $x = 2$. Therefore,

$R = \{(x, y) \mid 0 \leq x \leq \sqrt{y}, 0 \leq y \leq 4\}$

and

$$\int_0^2 \int_{x^2}^4 \frac{4x}{1+y^2}\, dy\, dx = \int_0^4 \int_0^{\sqrt{y}} \frac{4x}{1+y^2}\, dx\, dy = \int_0^4 \left[\frac{2x^2}{1+y^2}\right]_0^{\sqrt{y}} dy = \int_0^4 \frac{2y}{1+y^2}\, dy$$

$$= \ln(1+y^2)\Big]_0^4 = \ln 17 - \ln 1 = \ln 17$$

39. $R = \{(x, y) \mid y^2 \leq x \leq 1, 0 \leq y \leq 1\}$

Now, $x = y^2$ or $y = \sqrt{x}$; and $y = 0$ when $x = 0$; $y = 1$ when $x = 1$.

Therefore, $R = \{(x, y) \mid 0 \leq y \leq \sqrt{x}, 0 \leq x \leq 1\}$ and

$$\int_0^1 \int_{y^2}^1 4ye^{x^2} dx\, dy = \int_0^1 \int_0^{\sqrt{x}} 4ye^{x^2} dy\, dx = \int_0^1 \left[2y^2 e^{x^2}\right]_0^{\sqrt{x}} dx = \int_0^1 2xe^{x^2}\, dx = e^{x^2}\Big]_0^1 = e - 1.$$

41. $y = 1 + \sqrt{x}$, $y = x^2$, $x = 0$

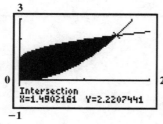

$R = \{(x, y) \mid x^2 \leq y \leq 1 + \sqrt{x}, 0 \leq x \leq 1.49\}$

$$\iint x\, dA = \int_0^{1.49} \left[\int_{x^2}^{1+\sqrt{x}} x\, dy\right] dx = \int_0^{1.49} \left[xy\right]_{x^2}^{1+\sqrt{x}} dx = \int_0^{1.49} (x + x^{3/2} - x^3)\, dx$$

$$= \left[\frac{x^2}{2} + \frac{2}{5}x^{5/2} - \frac{x^4}{4}\right]_0^{1.49} \approx 0.96$$

43. $y = \sqrt[3]{x}$, $y = 1 - x$, $y = 0$

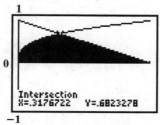

$R = \{(x, y) \mid y^3 \leq x \leq 1 - y, \ 0 \leq y \leq 0.68\}$

$$\iint_R 24xy\, dA = \int_0^{0.68} \left[\int_{y^3}^{1-y} 24xy\, dx\right] dy = \int_0^{0.68} \left[12x^2 y\right]_{y^3}^{1-y} dy = 12\int_0^{0.68} \left[y(1-y)^2 - y^7\right] dy$$

$$= 12\int_0^{0.68} (y^3 - 2y^2 + y - y^7)\, dy = 12\left[\frac{1}{4}y^4 - \frac{2}{3}y^3 + \frac{1}{2}y^2 - \frac{1}{8}y^8\right]_0^{0.68} \approx 0.83$$

45. $y = e^{-x}, y = 3 - x$

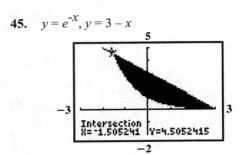

$R = \{(x, y) \mid e^{-x} \le y \le 3 - x, -1.51 \le x \le 2.95\}$ Regular x region

$\quad = \{(x, y) \mid -\ln y \le x \le 3 - y, 0.05 \le y \le 4.51\}$ Regular y region

$\displaystyle\iint\limits_{R} 4y \, dA = \int_{0.05}^{4.51} \left[\int_{-\ln y}^{3-y} 4y \, dx \right] dy = \int_{0.05}^{4.51} \left[4xy \right]_{-\ln y}^{3-y} dy = 4\int_{0.05}^{4.51} [y(3 - y) + y \ln y] \, dy$

$\qquad\qquad = 4 \left[\dfrac{3}{2} y^2 - \dfrac{1}{3} y^3 + \dfrac{1}{2} y^2 \ln y - \dfrac{1}{4} y^2 \right]_{0.05}^{4.51} \approx 40.67 \qquad \left(\displaystyle\int y \ln y \, dy = \dfrac{1}{2} y^2 \ln y - \dfrac{1}{4} y^2 \right)$

CHAPTER 15 REVIEW

1. $f(x, y) = 2000 + 40x + 70y$

$f(5, 10) = 2000 + 40 \cdot 5 + 70 \cdot 10 = 2900$

$f_x(x, y) = 40$

$f_y(x, y) = 70 \qquad$ (8-1, 8-2)

2. $z = x^3 y^2$

$\dfrac{\partial z}{\partial x} = 3x^2 y^2, \qquad\qquad \dfrac{\partial^2 z}{\partial x^2} = \dfrac{\partial\left(\dfrac{\partial z}{\partial x}\right)}{\partial x} = \dfrac{\partial(3x^2 y^2)}{\partial x} = 6xy^2$

$\dfrac{\partial z}{\partial y} = 2x^3 y, \qquad\qquad \dfrac{\partial^2 z}{\partial x \partial y} = \dfrac{\partial\left(\dfrac{\partial z}{\partial y}\right)}{\partial x} = \dfrac{\partial(2x^3 y)}{\partial x} = 6x^2 y \qquad$ (8-2)

3. $\displaystyle\int (6xy^2 + 4y) \, dy = 6x \int y^2 \, dy + 4 \int y \, dy = 6x \cdot \dfrac{y^3}{3} + 4 \cdot \dfrac{y^2}{2} + C(x) = 2xy^3 + 2y^2 + C(x) \qquad$ (8-6)

4. $\displaystyle\int (6xy^2 + 4y) \, dx = 6y^2 \int x \, dx + 4y \int dx = 6y^2 \cdot \dfrac{x^2}{2} + 4yx + E(y) = 3x^2 y^2 + 4xy + E(y) \qquad$ (8-6)

5. $\displaystyle\int_0^1 \int_0^1 4xy \, dy \, dx = \int_0^1 \left[\int_0^1 4xy \, dy \right] dx = \int_0^1 \left[2xy^2 \right]_0^1 dx = \int_0^1 2x \, dx = x^2 \Big]_0^1 = 1 \qquad$ (8-6)

6. $f(x, y) = 6 + 5x - 2y + 3x^2 + x^3$

$f_x(x, y) = 5 + 6x + 3x^2, \qquad\qquad f_y(x, y) = -2 \ne 0$

The function $f_y(x, y)$ is nonzero for all (x, y).   (8-3)

7. $f(x, y) = 3x^2 - 2xy + y^2 - 2x + 3y - 7$

$f(2, 3) = 3 \cdot 2^2 - 2 \cdot 2 \cdot 3 + 3^2 - 2 \cdot 2 + 3 \cdot 3 - 7 = 7$

$f_y(x, y) = -2x + 2y + 3$

$f_y(2, 3) = -2 \cdot 2 + 2 \cdot 3 + 3 = 5$ (8-1, 8-2)

8. $f(x, y) = -4x^2 + 4xy - 3y^2 + 4x + 10y + 81$

$f_x(x, y) = -8x + 4y + 4,$ $\qquad\qquad\qquad$ $f_y(x, y) = 4x - 6y + 10$

$f_{xx}(x, y) = -8,$ $\qquad\qquad\qquad\qquad$ $f_{yy}(x, y) = -6$

$f_{xy}(x, y) = 4$

Now, $f_{xx}(2, 3) \cdot f_{yy}(2, 3) - [f_{xy}(2, 3)]^2 = (-8)(-6) - 4^2 = 32.$ (8-2)

9. $f(x, y) = x + 3y$ and $g(x, y) = x^2 + y^2 - 10.$

Let $F(x, y, \lambda) = f(x, y) + \lambda g(x, y) = x + 3y + \lambda (x^2 + y^2 - 10)$. Then, we have:

$F_x = 1 + 2x\lambda$

$F_y = 3 + 2y\lambda$

$F_\lambda = x^2 + y^2 - 10$

Setting $F_x = F_y = F_\lambda = 0$, we obtain:

$1 + 2x\lambda = 0$ $\qquad\qquad\qquad$ (1)

$3 + 2y\lambda = 0$ $\qquad\qquad\qquad$ (2)

$x^2 + y^2 - 10 = 0$ $\qquad\qquad$ (3)

From the first equation, $x = -\dfrac{1}{2\lambda}$; from the second equation, $y = -\dfrac{3}{2\lambda}$. Substituting these into the third

equation gives:

$$\frac{1}{4\lambda^2} + \frac{9}{4\lambda^2} - 10 = 0$$

$$40\lambda^2 = 10$$

$$\lambda^2 = \frac{1}{4}$$

$$\lambda = \pm\frac{1}{2}$$

Thus, the critical points are $\left(-1, -3, \dfrac{1}{2}\right)$ and $\left(1, 3, -\dfrac{1}{2}\right)$.   (8-4)

10.

	x_k	y_k	$x_k y_k$	x_k^2
	2	12	24	4
	4	10	40	16
	6	7	42	36
	8	3	24	64
Totals	20	32	130	120

Thus, $\sum_{k=1}^{4} x_k = 20$, $\sum_{k=1}^{4} y_k = 32$, $\sum_{k=1}^{4} x_k y_k = 130$, $\sum_{k=1}^{4} x_k^2 = 120$.

Substituting these values into the formulas for a and b, we have:

$$a = \frac{4\left(\sum\limits_{k=1}^{4} x_k y_k\right) - \left(\sum\limits_{k=1}^{4} x_k\right)\left(\sum\limits_{k=1}^{4} y_k\right)}{4\left(\sum\limits_{k=1}^{4} x_k^2\right) - \left(\sum\limits_{k=1}^{4} x_k\right)^2} = \frac{4(130) - (20)(32)}{4(120) - (20)^2} = \frac{-120}{80} = -1.5$$

$$b = \frac{\sum\limits_{k=1}^{4} y_k - (-1.5)\sum\limits_{k=1}^{4} x_k}{4} = \frac{32 + (1.5)(20)}{4} = \frac{62}{4} = 15.5$$

Thus, the least squares line is: $y = ax + b = -1.5x + 15.5$
When $x = 10$, $y = -1.5(10) + 15.5 = 0.5$.   (8-5)

11.

$$\iint\limits_{R} (4x + 6y)\,dA = \int_{-1}^{1} \int_{1}^{2} (4x + 6y)\,dy\,dx = \int_{-1}^{1} \left[\int_{1}^{2}(4x+6y)\,dy\right]dx = \int_{-1}^{1} \left[(4xy + 3y^2)\right]_{1}^{2} dx$$

$$= \int_{-1}^{1} (8x + 12 - 4x - 3)\,dx = \int_{-1}^{1} (4x + 9)\,dx = (2x^2 + 9x)\Big|_{-1}^{1} = 2 + 9 - (2 - 9) = 18$$

$$\iint\limits_{R} (4x + 6y)\,dA = \int_{1}^{2} \int_{-1}^{1} (4x + 6y)\,dx\,dy = \int_{1}^{2} \left[\int_{-1}^{1}(4x+6y)\,dx\right]dy = \int_{1}^{2} \left[(2x^2 + 6xy)\right]_{-1}^{1} dy$$

$$= \int_{1}^{2} [2 + 6y - (2 - 6y)]\,dy = \int_{1}^{2} 12y\,dy = 6y^2\Big|_{1}^{2} = 24 - 6 = 18 \qquad (8\text{-}6)$$

12. $R = \{(x, y) \mid \sqrt{y} \leq x \leq 1, 0 \leq y \leq 1\}$

R is a regular y-region.
(R is also a regular x-region.)

$$\iint\limits_{R} (6x + y)\,dA = \int_{0}^{1} \int_{\sqrt{y}}^{1} (6x + y)\,dx\,dy$$

$$= \int_{0}^{1} \left[3x^2 + xy\right]_{\sqrt{y}}^{1} dy$$

$$= \int_{0}^{1} [(3 + y) - (3y + y^{3/2})]\,dy$$

$$= \int_{0}^{1} (3 - 2y - y^{3/2})\,dy = \left[3y - y^2 - \frac{2}{5}y^{5/2}\right]_{0}^{1} = 3 - 1 - \frac{2}{5} = \frac{8}{5} \qquad (8\text{-}7)$$

13. $f(x, y) = e^{x^2 + 2y}$

$f_x(x, y) = e^{x^2 + 2y} \cdot 2x = 2xe^{x^2 + 2y}$, $f_y(x, y) = e^{x^2 + 2y} \cdot 2 = 2e^{x^2 + 2y}$

$f_{xy}(x, y) = 2xe^{x^2 + 2y} \cdot 2 = 4xe^{x^2 + 2y}$ (8-2)

14. $f(x, y) = (x^2 + y^2)^5$

$f_x(x, y) = 5(x^2 + y^2)^4 \cdot 2x = 10x(x^2 + y^2)^4$

$f_{xy}(x, y) = 10x(4)(x^2 + y^2)^3 \cdot 2y = 80xy(x^2 + y^2)^3$ (8-2)

15. $f(x, y) = x^3 - 12x + y^2 - 6y$

$f_x(x, y) = 3x^2 - 12$ $\qquad\qquad\qquad$ $f_y(x, y) = 2y - 6$

$3x^2 - 12 = 0$ $\qquad\qquad\qquad\qquad$ $2y - 6 = 0$

$\qquad x^2 = 4$ $\qquad\qquad\qquad\qquad\qquad$ $y = 3$

$\qquad x = \pm 2$

Thus, the critical points are $(2, 3)$ and $(-2, 3)$.

$f_{xx}(x, y) = 6x,$ $\qquad\qquad$ $f_{xy}(x, y) = 0,$ $\qquad$ $f_{yy}(x, y) = 2$

For the critical point $(2, 3)$:

$f_{xx}(2, 3) = 12 > 0$

$f_{xy}(2, 3) = 0$

$f_{yy}(2, 3) = 2$

$f_{xx}(2, 3) \cdot f_{yy}(2, 3) - [f_{xy}(2, 3)]^2 = 12 \cdot 2 = 24 > 0$

Therefore, $f(2, 3) = 2^3 - 12 \cdot 2 + 3^2 - 6 \cdot 3 = -25$ is a local minimum.

For the critical point $(-2, 3)$:

$f_{xx}(-2, 3) = -12 < 0$

$f_{xy}(-2, 3) = 0$

$f_{yy}(-2, 3) = 2$

$f_{xx}(-2, 3) \cdot f_{yy}(-2, 3) - [f_{xy}(-2, 3)]^2 = -12 \cdot 2 - 0 = -24 < 0$

Thus, f has a saddle point at $(-2, 3)$.   (8-3)

16. Step 1. $\qquad$ Maximize $f(x, y) = xy$

$\qquad\qquad\qquad$ Subject to: $g(x, y) = 2x + 3y - 24 = 0$

$\quad$ Step 2. $\qquad$ $F(x, y, \lambda) = f(x, y) + \lambda g(x, y) = xy + \lambda (2x + 3y - 24)$

$\quad$ Step 3. $\qquad$ $F_x = y + 2\lambda = 0$ $\qquad\qquad\qquad$ (1)

$\qquad\qquad\qquad$ $F_y = x + 3\lambda = 0$ $\qquad\qquad\qquad$ (2)

$\qquad\qquad\qquad$ $F_\lambda = 2x + 3y - 24 = 0$ $\qquad\quad$ (3)

$\qquad\qquad\qquad$ From (1) and (2), we obtain:

$\qquad\qquad\qquad$ $y = -2\lambda$ $\quad$ and $\quad$ $x = -3\lambda$

$\qquad\qquad\qquad$ Substituting these into (3), we have:

$\qquad\qquad\qquad$ $-6\lambda - 6\lambda - 24 = 0$

$\qquad\qquad\qquad\qquad$ $\lambda = -2$

$\qquad\qquad\qquad$ Thus, the critical point is $(6, 4, -2)$.

$\quad$ Step 4. $\qquad$ Since $(6, 4, -2)$ is the only critical point for F, we conclude that

$\qquad\qquad\qquad$ max $f(x, y) = f(6, 4) = 6 \cdot 4 = 24$.   (8-4)

17.   <u>Step 1.</u>   Minimize $f(x, y, z) = x^2 + y^2 + z^2$

Subject to: $2x + y + 2z = 9$ or $g(x, y, z) = 2x + y + 2z - 9 = 0$

<u>Step 2.</u> $F(x, y, z, \lambda) = x^2 + y^2 + z^2 + \lambda(2x + y + 2z - 9)$

<u>Step 3.</u>

$$F_x = 2x + 2\lambda = 0 \tag{1}$$

$$F_y = 2y + \lambda = 0 \tag{2}$$

$$F_z = 2z + 2\lambda = 0 \tag{3}$$

$$F_\lambda = 2x + y + 2z - 9 = 0 \tag{4}$$

From equations (1), (2), and (3), we have:

$$x = -\lambda, \; y = -\frac{\lambda}{2}, \text{ and } z = -\lambda$$

Substituting these into (4), we obtain:

$$-2\lambda - \frac{\lambda}{2} - 2\lambda - 9 = 0$$

$$\frac{9}{2}\lambda = -9$$

$$\lambda = -2$$

The critical point is: (2, 1, 2, –2)

<u>Step 4.</u>   Since (2, 1, 2, –2) is the only critical point for F, we conclude that

$$\min f(x, y, z) = f(2, 1, 2) = 2^2 + 1^2 + 2^2 = 9. \tag{8-4}$$

18.

x_k	y_k	$x_k y_k$	x_k^2
10	50	500	100
20	45	900	400
30	50	1,500	900
40	55	2,200	1,600
50	65	3,250	2,500
60	80	4,800	3,600
70	85	5,950	4,900
80	90	7,200	6,400
90	90	8,100	8,100
100	110	11,000	10,000
Totals 550	720	45,400	38,500

Thus, $\displaystyle\sum_{k=1}^{10} x_k = 550$, $\displaystyle\sum_{k=1}^{10} y_k = 720$, $\displaystyle\sum_{k=1}^{10} x_k y_k = 45,400$, $\displaystyle\sum_{k=1}^{10} x_k^2 = 38,500$.

Substituting these values into the formulas for a and b, we have:

$$a = \frac{10(45,400) - (550)(720)}{10(38,500) - (550)^2} = \frac{58,000}{82,500} = \frac{116}{165}$$

$$b = \frac{720 - \left(\frac{116}{165}\right)550}{10} = \frac{100}{3}$$

Therefore, the least squares line is:

$$y = \frac{116}{165}x + \frac{100}{3} \approx 0.703x + 33.33 \qquad (8\text{-}5)$$

19. $$\frac{1}{(b-a)(d-c)} \iint\limits_{R} f(x, y)\, dA = \frac{1}{[8-(-8)](27-0)} \int_{-8}^{8} \int_{0}^{27} x^{2/3} y^{1/3}\, dy\, dx$$

$$= \frac{1}{16 \cdot 27} \int_{-8}^{8} \left[\frac{3}{4} x^{2/3} y^{4/3} \right]_{0}^{27} dx = \frac{1}{16 \cdot 27} \int_{-8}^{8} \frac{3^5}{4} x^{2/3}\, dx = \frac{9}{64} \int_{-8}^{8} x^{2/3}\, dx$$

$$= \frac{9}{64} \cdot \frac{3}{5} x^{5/3} \Big|_{-8}^{8} = \frac{9}{64} \cdot \frac{3}{5} [2^5 - (-2)^5] = \frac{9}{64} \cdot \frac{3}{5} \cdot 2^6 = \frac{27}{5} \qquad (8\text{-}6)$$

20. $$V = \iint\limits_{R} (3x^2 + 3y^2)\, dA = \int_{0}^{1} \int_{-1}^{1} (3x^2 + 3y^2)\, dy\, dx = \int_{0}^{1} \left[\int_{-1}^{1} (3x^2 + 3y^2)\, dy \right] dx$$

$$= \int_{0}^{1} \left[3x^2 y + y^3 \right]_{-1}^{1} dx = \int_{0}^{1} [3x^2 + 1 - (-3x^2 - 1)]\, dx = \int_{0}^{1} (6x^2 + 2)\, dx = (2x^3 + 2x) \Big|_{0}^{1} = 4 \text{ cubic units}$$

(8-6)

21. $f(x, y) = x + y;\ -10 \le x \le 10,\ -10 \le y \le 10$

Prediction: average value $= f(0, 0) = 0$.

Verification:

$$\text{average value} = \frac{1}{[10-(-10)][10-(-10)]} \int_{-10}^{10} \int_{-10}^{10} (x + y)\, dy\, dx$$

$$= \frac{1}{400} \int_{-10}^{10} \left[xy + \frac{1}{2} y^2 \right]_{-10}^{10} dx = \frac{1}{400} \int_{-10}^{10} 20x\, dx = \left[\frac{1}{400} (10x^2) \right]_{-10}^{10} = 0 \qquad (8\text{-}6)$$

22. $f(x, y) = \dfrac{e^x}{y+10}$

(A) $S = \{x, y)|-a \le x \le a, -a \le y \le a\}$

The average value of f over S is given by:

$$\frac{1}{[a-(-a)][a-(-a)]} \int_{-a}^{a} \int_{-a}^{a} \frac{e^x}{y+10}\, dx\ dy = \frac{1}{4a^2} \int_{-a}^{a} \left[\frac{e^x}{y+10} \right]_{-a}^{a} dy$$

$$= \frac{1}{4a^2} \int_{-a}^{a} \left(\frac{e^a}{y+10} - \frac{e^{-a}}{y+10} \right) dy = \frac{e^a - e^{-a}}{4a^2} \int_{-a}^{a} \frac{1}{y+10}\, dy$$

$$= \left[\frac{e^a - e^{-a}}{4a^2} \ln|y+10| \right]_{-a}^{a} = \frac{e^a - e^{-a}}{4a^2} [\ln(10+a) - \ln(10-a)] = \frac{e^a - e^{-a}}{4a^2} \ln\left(\frac{10+a}{10-a} \right)$$

Now, $\dfrac{e^a - e^{-a}}{4a^2} \ln\left(\dfrac{10+a}{10-a} \right) = 5$ is equivalent to $(e^a - e^{-a})\ln\left(\dfrac{10+a}{10-a} \right) - 20a^2 = 0.$

Using a graphing utility, the graph of

$$f(x) = (e^x - e^{-x})\ln\left(\frac{10+x}{10-x} \right) - 20x^2$$

is shown at the right and $f(x) = 0$ at $x \approx \pm 6.28.$

The dimensions of the square are:
$12.56 \times 12.56.$

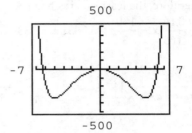

(B)   To determine whether there is a square centered at $(0, 0)$
such that

$$\frac{e^a - e^{-a}}{4a^2} \ln\left(\frac{10+a}{10-a} \right) = 0.05,$$

graph,

$$f(x) = (e^x - e^{-x})\ln\left(\frac{10+x}{10-x} \right) - 0.20x^2$$

The result is shown at the right and $f(x) = 0$ only at $x = 0.$
Thus, there does not exist a square centered at $(0, 0)$ such
that the average value of $f = 0.05.$

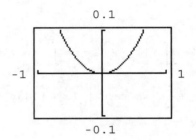

(8-6)

23.   <u>Step 1.</u>   Extremize $f(x, y) = 4x^3 - 5y^3$
        subject to $g(x, y) = 3x + 2y - 7 = 0.$

   <u>Step 2.</u> $F(x, y, \lambda) = 4x^3 - 5y^3 + \lambda\,(3x + 2y - 7)$

   <u>Step 3.</u> $F_x = 12x^2 + 3\lambda = 0$ $\qquad$ (1)

        $F_y = -15y^2 + 2\lambda = 0$ $\qquad$ (2)

        $F_\lambda = 3x + 2y - 7 = 0$ $\qquad$ (3)

From (1), $\lambda = -4x^2 \le 0$; from (2), $\lambda = \dfrac{15}{2}y^2 \ge 0.$ This implies

$x = y = \lambda = 0$ and $x = y = 0$ does not satisfy (3). The system
(1), (2), (3) does not have a simultaneous solution.   (8-4)

24. R is the region shown in the figure; it is a regular x-region and a regular y-region. $F(x, y) = 60x^2y \ge 0$ on R
so

$$V = \iint\limits_R 60x^2y \, dA = \int_0^1 \int_0^{1-x} 60x^2y \, dy \, dx \quad \text{(treating } R \text{ as an } x\text{-region)}$$

$$= \int_0^1 \left[30x^2y^2 \right]_0^{1-x} dx = \int_0^1 30x^2(1-x)^2 dx$$

$$= \int_0^1 30x^2(1 - 2x + x^2)\, dx = \int_0^1 (30x^2 - 60x^3 + 30x^4)\, dx$$

$$= (10x^3 - 15x^4 + 6x^5) \Big|_0^1 = 1$$

(8-7)

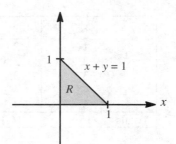

25. $P(x, y) = -4x^2 + 4xy - 3y^2 + 4x + 10y + 81$

(A) $P_x(x, y) = -8x + 4y + 4$
  $P_x(1, 3) = -8 \cdot 1 + 4 \cdot 3 + 4 = 8$

At the output level (1, 3), profit will increase by $8000 for 100 units increase in product A if the production of product B is held fixed.

(B) $P_x = -8x + 4y + 4 = 0$ (1)
  $P_y = 4x - 6y + 10 = 0$ (2)

Solving (1) and (2) for x and y, we obtain $x = 2$, $y = 3$.
Thus, (2, 3) is a critical point.

$P_{xx} = -8$	$P_{yy} = -6$	$P_{xy} = 4$
$P_{xx}(2, 3) = -8 < 0$	$P_{yy}(2, 3) = -6$	$P_{xy}(2, 3) = 4$

$$P_{xx}(2, 3) \cdot P_{yy}(2, 3) - [P_{xy}(2, 3)]^2 = (-8)(-6) - 4^2 = 32 > 0$$

Thus, $P(2, 3)$ is a maximum and
$$\max P(x, y) = P(2, 3) = -4 \cdot 2^2 + 4 \cdot 2 \cdot 3 - 3 \cdot 3^2 + 4 \cdot 2 + 10 \cdot 3 + 81 = -16 + 24 - 27 + 8 + 30 + 81$$
$$= 100.$$
The maximum profit is $100,000. This is obtained when 200 units of A and 300 units of B are produced per month.   (8-2, 8-3)

26. Minimize $S(x, y, z) = xy + 4xz + 3yz$
Subject to: $V(x, y, z) = xyz - 96 = 0$
Put $F(x, y, z, \lambda) = S(x, y, z) + \lambda V(x, y, z) = xy + 4xz + 3yz + \lambda(xyz - 96)$. Then, we have:

$F_x = y + 4z + \lambda yz = 0$ (1)

$F_y = x + 3z + \lambda xz = 0$ (2)

$F_z = 4x + 3y + \lambda xy = 0$ (3)

$F_\lambda = xyz - 96 = 0$ (4)

Solving the system of equations, (1) – (4), simultaneously, yields $x = 6$, $y = 8$, $z = 2$, and $\lambda = -1$. Thus, the critical point is (6, 8, 2, –1) and $S(6, 8, 2) = 6 \cdot 8 + 4 \cdot 6 \cdot 2 + 3 \cdot 8 \cdot 2 = 144$ is the minimum value of S subject to the constraint $V = xyz - 96 = 0$.

The dimensions of the box that will require the minimum amount of material are:
length $x = 6$ inches; width $y = 8$ inches; height $z = 2$ inches.   (8-3)

27.

	x_k	y_k	$x_k y_k$	x_k^2
	1	2.0	2.0	1
	2	2.5	5.0	4
	3	3.1	9.3	9
	4	4.2	16.8	16
	5	4.3	21.5	25
Totals	15	16.1	54.6	55

Thus, $\displaystyle\sum_{k=1}^{5} x_k = 15$, $\displaystyle\sum_{k=1}^{5} y_k = 16.1$, $\displaystyle\sum_{k=1}^{5} x_k y_k = 54.6$, $\displaystyle\sum_{k=1}^{5} x_k^2 = 55$.

Substituting these values into the formulas for a and b, we have:

$$a = \frac{5(54.6) - (15)(16.1)}{5(55) - (15)^2} = \frac{31.5}{50} \approx 0.63$$

$$b = \frac{16.1 - (0.63)(15)}{5} = 1.33$$

Therefore, the least squares line is: $y = 0.63x + 1.33$

When $x = 6$, $y = 0.63(6) + 1.33 = 5.11$, and the profit for the sixth year is estimated to be $5.11 million. (8-4)

28. $N(x, y) = 10x^{0.8}y^{0.2}$

(A) $N_x(x, y) = 8x^{-0.2}y^{0.2}$

$N_x(40, 50) = 8(40)^{-0.2}(50)^{0.2} \approx 8.37$

$N_y(x, y) = 2x^{0.8}y^{-0.8}$

$N_y(40, 50) = 2(40)^{0.8}(50)^{-0.8} \approx 1.67$

Thus, at the level of 40 units of labor and 50 units of capital, the marginal productivity of labor is approximately 8.36 and the marginal productivity of capital is approximately 1.67. Management should encourage increased use of labor.

(B) Step 1. Maximize the production function $N(x, y) = 10x^{0.8}y^{0.2}$

Subject to the constraint: $C(x, y) = 100x + 50y = 10,000$

i.e., $g(x, y) = 100x + 50y - 10,000 = 0$

Step 2. $F(x, y, \lambda) = 10x^{0.8}y^{0.2} + \lambda (100x + 50y - 10,000)$

Step 3. $F_x = 8x^{-0.2}y^{0.2} + 100\lambda = 0$ (1)

$F_y = 2x^{0.8}y^{-0.8} + 50\lambda = 0$ (2)

$F_\lambda = 100x + 50y - 10,000 = 0$ (3)

From equation (1), $\lambda = \dfrac{-0.08y^{0.2}}{x^{0.2}}$, and from (2), $\lambda = \dfrac{-0.04x^{0.8}}{y^{0.8}}$.

Thus, $\dfrac{0.08y^{0.2}}{x^{0.2}} = \dfrac{0.04x^{0.8}}{y^{0.8}}$ and $x = 2y$.

Substituting into (3) yields:

$200y + 50y = 10,000$

$250y = 10,000$

$y = 40$

Therefore, $x = 80$ and $\lambda \approx -0.0696$. The critical point is $(80, 40, -0.0696)$. Thus, we conclude that max $N(x, y) = N(80, 40) = 10(80)^{0.8}(40)^{0.2} \approx 696$ units.

Production is maximized when 80 units of labor and 40 units of capital are used.

The marginal productivity of money is $-\lambda \approx 0.0696$. The increase in production resulting from an increase of \$2000 in the budget is: $0.0696(2000) \approx 139$ units.

(C)   Average number of units

$$= \frac{1}{(100-50)(40-20)} \int_{50}^{100} \int_{20}^{40} 10x^{0.8} y^{0.2} \, dy \, dx = \frac{1}{(50)(20)} \int_{50}^{100} \left[\frac{10x^{0.8} y^{1.2}}{1.2} \right]_{20}^{40} dx$$

$$= \frac{1}{1000} \int_{50}^{100} \frac{10}{1.2} x^{0.8}(40^{1.2} - 20^{1.2}) \, dx = \frac{40^{1.2} - 20^{1.2}}{120} \int_{50}^{100} x^{0.8} dx = \frac{40^{1.2} - 20^{1.2}}{120} \cdot \left[\frac{x^{1.8}}{1.8} \right]_{50}^{100}$$

$$= \frac{(40^{1.2} - 20^{1.2})(100^{1.8} - 50^{1.8})}{216} \approx \frac{(47.24)(2837.81)}{216} \approx 621$$

Thus, the average number of units produced is approximately 621.   (8-4)

29. $T(V, x) = \dfrac{33V}{x+33} = 33V(x + 33)^{-1}$

$T_x(V, x) = -33V(x + 33)^{-2} = \dfrac{-33V}{(x+33)^2}$

$T_x(70, 17) = \dfrac{-33(70)}{(17+33)^2} = \dfrac{-33(70)}{2500} = -0.924$ minutes per unit increase in depth when $V = 70$ cubic feet

and $x = 17$ ft.   (8-2)

30. $C = 100 - 24d^2 = 100 - 24(x^2 + y^2)$

$C(x, y) = 100 - 24(x^2 + y^2), \ -2 \le x \le 2, \ -2 \le y \le 2$

Average concentration

$$= \frac{1}{4(4)} \int_{-2}^{2} \int_{-2}^{2} [100 - 24(x^2 + y^2)] \, dy \, dx = \frac{1}{16} \int_{-2}^{2} \int_{-2}^{2} \left[100y - 24x^2 y - 8y^3 \right]_{-2}^{2} dx$$

$$= \frac{1}{16} \int_{-2}^{2} [400 - 96x^2 - 128] \, dx = \frac{1}{16} \int_{-2}^{2} (272 - 96x^2) \, dx$$

$$= \frac{1}{16} \left[272x - 32x^3 \right]_{-2}^{2} = \frac{1}{16}(544 - 256) - \frac{1}{16}(-544 + 256) = 18 + 18 = 36 \text{ parts per million.}$$

(8-6)

31. $n(P_1, P_2, d) = 0.001 \dfrac{P_1 P_2}{d}$

$n(100{,}000, 50{,}000, 100) = 0.001 \dfrac{100{,}000 \times 50{,}000}{100} = 50{,}000$ (8-1)

32.

x_k	y_k	$x_k y_k$	x_k^2
30	60	1,800	900
50	75	3,750	2,500
60	80	4,800	3,600
70	85	5,950	4,900
90	90	8,100	8,100
Totals 300	390	24,400	20,000

Thus, $\displaystyle\sum_{k=1}^{5} x_k = 300,\ \sum_{k=1}^{5} y_k = 390,\ \sum_{k=1}^{5} x_k y_k = 24,400,\ \sum_{k=1}^{5} x_k^2 = 20,000.$

Substituting these values into the formulas for a and d, we have:

$$a = \frac{5(24,400) - (300)(390)}{5(20,000) - (300)^2} = \frac{5000}{10,000} = 0.5$$

$$d = \frac{390 - 0.5(300)}{5} = \frac{240}{5} = 48$$

Therefore, the least squares line is: $y = 0.5x + 48$
When $x = 40$, $y = 0.5(40) + 48 = 68$.　　(8-5)

33. (A)　We use the linear regression feature on a graphing utility with 1960 as $x = 0$.

Thus, the least squares line is: $y = 0.734x + 49.93$.

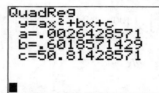

(B)　The year 2025 corresponds to $x = 65$; $y(65) \approx 97.64$ people/sq. mi.

(C)

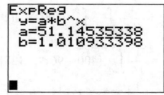

Quadratic regression:
$y(65) \approx 101.10$ people/sq. mi.

Exponential regression:
$y(65) \approx 103.70$ people/sq. mi.　　(8-5)

34.

(A)   The least squares line is $y \approx 1.069x + 0.522$.

(B)   Evaluate the result in (A) at $x = 60$: $y \approx 64.68$ yr

(C)

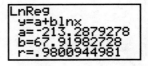

Evaluate at $x = 60$: $y \approx 64.78$ yr   Evaluate at $x = 60$: $y \approx 64.80$ yr   (8-5)

EXERCISE A–1

Things to remember:

1. THE SET OF REAL NUMBERS

SYMBOL	NAME	DESCRIPTION	EXAMPLES
N	Natural numbers	Counting numbers (also called positive integers)	1, 2, 3, ...
Z	Integers	Natural numbers, their negatives, and 0	... –2, –1, 0, 1, 2, ...
Q	Rational numbers	Any number that can be represented as $\frac{a}{b}$, where a and b are integers and $b \neq 0$. Decimal representations are repeating or terminating.	$-4;\ 0;\ 1;\ 25;\ \dfrac{-3}{5};\ \dfrac{2}{3};$ $3.67;\ -0.333\,\overline{3};\ 5.2727\,\overline{27}$
I	Irrational numbers	Any number with a decimal representation that is nonrepeating and non–terminating.	$\sqrt{2}\,;\ \pi;\ \sqrt[3]{7}\,;\ 1.414213...;$ $2.718281828...$
R	Real numbers	Rationals and irrationals	

2. BASIC PROPERTIES OF THE SET OF REAL NUMBERS

Let a, b, and c be arbitrary elements in the set of real numbers R.

ADDITION PROPERTIES

ASSOCIATIVE: $(a + b) + c = a + (b + c)$

COMMUTATIVE: $a + b = b + a$

IDENTITY: 0 is the additive identity; that is, $0 + a = a + 0$ for all a in R, and 0 is the only element in R with this property.

INVERSE: For each a in R, $-a$ is its unique additive inverse; that is, $a + (-a) = (-a) + a = 0$, and $-a$ is the only element in R relative to a with this property.

MULTIPLICATION PROPERTIES

ASSOCIATIVE: $(ab)c = a(bc)$

COMMUTATIVE: $ab = ba$

IDENTITY: 1 is the multiplicative identity; that is, $1a = a1 = a$ for all a in R, and 1 is the only element in R with this property.

INVERSE: For each a in R, $a \neq 0$, $\dfrac{1}{a}$ is its unique multiplicative inverse; that is, $a\left(\dfrac{1}{a}\right) = \left(\dfrac{1}{a}\right)a = 1$, and $\dfrac{1}{a}$ is the only element in R relative to a with this property.

DISTRIBUTIVE PROPERTIES

$$a(b + c) = ab + ac$$

$$(a + b)c = ac + bc$$

<u>3.</u>   **SUBTRACTION AND DIVISION**

For all real numbers a and b.

SUBTRACTION: $a - b = a + (-b)$
$7 - (-5) = 7 + [-(-5)] = 7 + 5 = 12$

DIVISION: $a \div b = a\left(\dfrac{1}{b}\right), b \neq 0$

$9 \div 4 = 9\left(\dfrac{1}{4}\right) = \dfrac{9}{4}$

NOTE: 0 can never be used as a divisor!

<u>4.</u>   **PROPERTIES OF NEGATIVES**

For all real numbers a and b.

a. $-(-a) = a$

b. $(-a)b = -(ab) = a(-b) = -ab$

c. $(-a)(-b) = ab$

d. $(-1)a = -a$

e. $\dfrac{-a}{b} = -\dfrac{a}{b} = \dfrac{a}{-b}, b \neq 0$

f. $\dfrac{-a}{-b} = -\dfrac{-a}{b} = -\dfrac{a}{-b} = \dfrac{a}{b}, b \neq 0$

<u>5.</u>   **ZERO PROPERTIES**

For all real numbers a and b.

a. $a \cdot 0 = 0$

b. $ab = 0$ if and only if $a = 0$ or $b = 0$ (or both)

6.   FRACTION PROPERTIES

For all real numbers a, b, c, d, and k (division by 0 excluded).

a. $\dfrac{a}{b} = \dfrac{c}{d}$ if and only if $ad = bc$

b. $\dfrac{ka}{kb} = \dfrac{a}{b}$

c. $\dfrac{a}{b} \cdot \dfrac{c}{d} = \dfrac{ac}{bd}$

d. $\dfrac{a}{b} \div \dfrac{c}{d} = \dfrac{a}{b} \cdot \dfrac{d}{c}$

e. $\dfrac{a}{b} + \dfrac{c}{b} = \dfrac{a+c}{b}$

f. $\dfrac{a}{b} - \dfrac{c}{b} = \dfrac{a-c}{b}$

g. $\dfrac{a}{b} + \dfrac{c}{d} = \dfrac{ad+bc}{bd}$

---

1. $uv = vu$

3. $3 + (7 + y) = (3 + 7) + y$

5. $1(u + v) = u + v$

7. T; Associative property of multiplication

9. T; Distributive property

11. F; $-2(-a)(2x - y) = 2a(2x - y)$

13. T; Commutative property of addition

15. T; Property of negatives

17. T; Multiplicative inverse property

19. T; Property of negatives

21. F; $\dfrac{a}{b} + \dfrac{c}{d} = \dfrac{ad+bc}{bd}$

23. T; Distributive property

25. T; Zero property

27. No. For example: $2\left(\dfrac{1}{2}\right) = 1$. In general $a\left(\dfrac{1}{a}\right) = 1$ whenever $a \neq 0$.

29. (A) False. For example, -3 is an integer but not a natural number.

(B) True

(C) True. For example, for any natural number n, $n = \dfrac{n}{1}$.

31. $\sqrt{2}$, $\sqrt{3}$, … ; in general, the square root of any rational number that is not a perfect square; π, e.

33. (A) $8 \in N, Z, Q, R$

(B) $\sqrt{2} \in R$

(C) $-1.414 = -\dfrac{1414}{1000} \in Q, R$

(D) $\dfrac{-5}{2} \in Q, R$

35. (A) F; $a(b - c) = ab - ac$; for example $2(3 - 1) = 2 \cdot 2 = 4 \neq 2 \cdot 3 - 1 = 5$

(B) F; for example, $(3 - 7) - 4 = -4 - 4 = -8 \neq 3 - (7 - 4) = 3 - 3 = 0$.

(C) T; this is the associative property of multiplication.

(D) F; for example, $(12 \div 4) \div 2 = 3 \div 2 = \dfrac{3}{2} \neq 12 \div (4 \div 2) = 12 \div 2 = 6$.

37.
$$C = 0.090909\ldots$$
$$100C = 9.090909\ldots$$
$$100C - C = (9.090909\ldots) - (0.090909\ldots)$$
$$99C = 9$$
$$C = \frac{9}{99} = \frac{1}{11}$$

39. (A) $\dfrac{13}{6} \approx 2.166\ 666\ 666$ (B) $\sqrt{21} \approx 4.582\ 575\ 695$

(C) $\dfrac{7}{16} = 0.4375$ (D) $\dfrac{29}{111} \approx 0.261\ 261\ 261$

41. (A) $\dfrac{43}{13} \approx 3$ (B) $\dfrac{37}{19} \approx 2$

43. (A) $\dfrac{7}{8} + \dfrac{11}{12} \approx 2$ (B) $\dfrac{55}{9} - \dfrac{7}{55} \approx 6$

45. Tax: $182.39(0.09) \approx 16.4151$;   \$16.42

47. Increase $4.37 - 4.25 = 0.12$; $\dfrac{0.12}{4.25} \approx 0.02824$;   2.8%

EXERCISE A–2

Things to remember:

<u>1.</u>   NATURAL NUMBER EXPONENT

For n a natural number and b any real number,

$b^n = b \cdot b \cdot \ldots \cdot b$, n factors of b.

For example, $2^3 = 2 \cdot 2 \cdot 2\ (= 8)$,

$3^5 = 3 \cdot 3 \cdot 3 \cdot 3 \cdot 3\ (= 243)$.

In the expression b^n, n is called the EXPONENT, and b is called the BASE.

<u>2.</u>   FIRST PROPERTY OF EXPONENTS

For any natural numbers m and n, and any real number b,

$b^m \cdot b^n = b^{m+n}$.

For example, $3^3 \cdot 3^4 = 3^{3+4} = 3^7$.

3.   POLYNOMIALS

    a.   A POLYNOMIAL IN ONE VARIABLE x is constructed by adding or subtracting constants and terms of the form ax^n, where a is a real number and n is a natural number.

    b.   A POLYNOMIAL IN TWO VARIABLES x AND y is constructed by adding or subtracting constants and terms of the form $ax^m y^n$, where a is a real number and m and n are natural numbers.

    c.   Polynomials in more than two variables are defined similarly.

    d.   A polynomial with only one term is called a MONOMIAL.
A polynomial with two terms is called a BINOMIAL.
A polynomial with three terms is called a TRINOMIAL.

4.   DEGREE OF A POLYNOMIAL

    a.   A term of the form ax^n, $a \neq 0$, has degree n. A term of the form $ax^m y^n$, $a \neq 0$, has degree $m + n$. A nonzero constant has degree 0.

    b.   The DEGREE OF A POLYNOMIAL is the degree of the nonzero term with the highest degree. For example, $3x^4 + \sqrt{2}\, x^3 - 2x + 7$ has degree 4; $2x^3 y^2 - 3x^2 y + 7x^4 - 5y^3 + 6$ has degree 5; the polynomial 4 has degree 0.

    c.   The constant 0 is a polynomial but it is not assigned a degree.

5.   Two terms in a polynomial are called LIKE TERMS if they have exactly the same variable factors raised to the same powers. For example, in
$$7x^5 y^2 - 3x^3 y + 2x + 4x^3 y - 1,$$
$-3x^3 y$ and $4x^3 y$ are like terms.

6.   To multiply two polynomials, multiply each term of one by each term of the other, and then combine like terms.

7.   SPECIAL PRODUCTS

    a. $(a - b)(a + b) = a^2 - b^2$

    b. $(a + b)^2 = a^2 + 2ab + b^2$

    c. $(a - b)^2 = a^2 - 2ab + b^2$

8.   ORDER OF OPERATIONS

Multiplication and division precede addition and subtraction, and taking powers precedes multiplication and division.

---

1.   The term of highest degree in $x^3 + 2x^2 - x + 3$ is x^3 and the degree of this term is 3.

3. $(2x^2 - x + 2) + (x^3 + 2x^2 - x + 3) = x^3 + 2x^2 + 2x^2 - x - x + 2 + 3 = x^3 + 4x^2 - 2x + 5$

5. $(x^3 + 2x^2 - x + 3) - (2x^2 - x + 2) = x^3 + 2x^2 - x + 3 - 2x^2 + x - 2 = x^3 + 1$

7. Using a vertical arrangement:

$$
\begin{array}{r}
x^3 + 2x^2 - x + 3 \\
2x^2 - x + 2 \\
\hline
2x^5 + 4x^4 - 2x^3 + 6x^2 \\
- x^4 - 2x^3 + x^2 - 3x \\
2x^3 + 4x^2 - 2x + 6 \\
\hline
2x^5 + 3x^4 - 2x^3 + 11x^2 - 5x + 6
\end{array}
$$

9. $2(u - 1) - (3u + 2) - 2(2u - 3) = 2u - 2 - 3u - 2 - 4u + 6 = -5u + 2$

11. $4a - 2a[5 - 3(a + 2)] = 4a - 2a[5 - 3a - 6] = 4a - 2a[-3a - 1] = 4a + 6a^2 + 2a = 6a^2 + 6a$

13. $(a + b)(a - b) = a^2 - b^2$ (Special product $\underline{7}$a)

15. $(3x - 5)(2x + 1) = 6x^2 + 3x - 10x - 5 = 6x^2 - 7x - 5$

17. $(2x - 3y)(x + 2y) = 2x^2 + 4xy - 3xy - 6y^2 = 2x^2 + xy - 6y^2$

19. $(3y + 2)(3y - 2) = (3y)^2 - 2^2 = 9y^2 - 4$ (Special product $\underline{7}$a)

21. $-(2x - 3)^2 = -[(2x)^2 - 2(2x)(3) + 3^2] = -[4x^2 - 12x + 9] = -4x^2 + 12x - 9$ (Special product $\underline{7}$c)

23. $(4m + 3n)(4m - 3n) = 16m^2 - 9n^2$ (Special product $\underline{7}$a)

25. $(3u + 4v)^2 = 9u^2 + 24uv + 16v^2$ (Special product $\underline{7}$b)

27. $(a - b)(a^2 + ab + b^2) = a(a^2 + ab + b^2) - b(a^2 + ab + b^2) = a^3 + a^2b + ab^2 - a^2b - ab^2 - b^3$
$$= a^3 - b^3$$

29. $[(x - y) + 3z][(x - y) - 3z] = (x - y)^2 - 9z^2 = x^2 - 2xy + y^2 - 9z^2$ (Special product $\underline{7}$a)

31. $m - \{m - [m - (m - 1)]\} = m - \{m - [m - m + 1]\} = m - \{m - 1\} = m - m + 1 = 1$

33. $(x^2 - 2xy + y^2)(x^2 + 2xy + y^2) = (x - y)^2(x + y)^2 = [(x - y)(x + y)]^2 = [x^2 - y^2]^2 = x^4 - 2x^2y^2 + y^4$

35. $(5a - 2b)^2 - (2b + 5a)^2 = 25a^2 - 20ab + 4b^2 - [4b^2 + 20ab + 25a^2] = -40ab$

37. $(m - 2)^2 - (m - 2)(m + 2) = m^2 - 4m + 4 - [m^2 - 4] = m^2 - 4m + 4 - m^2 + 4 = -4m + 8$

39. $(x - 2y)(2x + y) - (x + 2y)(2x - y) = 2x^2 - 4xy + xy - 2y^2 - [2x^2 + 4xy - xy - 2y^2]$
$$= 2x^2 - 3xy - 2y^2 - 2x^2 - 3xy + 2y^2 = -6xy$$

41. $(u + v)^3 = (u + v)(u + v)^2 = (u + v)(u^2 + 2uv + v^2) = u^3 + 3u^2v + 3uv^2 + v^3$

43. $(x - 2y)^3 = (x - 2y)(x - 2y)^2 = (x - 2y)(x^2 - 4xy + 4y^2) = x(x^2 - 4xy + 4y^2) - 2y(x^2 - 4xy + 4y^2)$
$$= x^3 - 4x^2y + 4xy^2 - 2x^2y + 8xy^2 - 8y^3$$
$$= x^3 - 6x^2y + 12xy^2 - 8y^3$$

45. $[(2x^2 - 4xy + y^2) + (3xy - y^2)] - [(x^2 - 2xy - y^2) + (-x^2 + 3xy - 2y^2)]$
$$= [2x^2 - xy] - [xy - 3y^2] = 2x^2 - 2xy + 3y^2$$

47. $[(2x-1)^2 - x(3x+1)]^2 = [4x^2 - 4x + 1 - 3x^2 - x]^2 = [x^2 - 5x + 1]^2 = (x^2 - 5x + 1)(x^2 - 5x + 1)$
$$= x^4 - 10x^3 + 27x^2 - 10x + 1$$

49. $2\{(x-3)(x^2 - 2x + 1) - x[3 - x(x-2)]\} = 2\{x^3 - 5x^2 + 7x - 3 - x[3 - x^2 + 2x]\}$
$$= 2\{x^3 - 5x^2 + 7x - 3 + x^3 - 2x^2 - 3x\} = 2\{2x^3 - 7x^2 + 4x - 3\} = 4x^3 - 14x^2 + 8x - 6$$

51. $m + n$

53. Given two polynomials, one with degree m and the other of degree n, their product will have degree $m + n$ regardless of the relationship between m and n.

55. Since $(a+b)^2 = a^2 + 2ab + b^2$, $(a+b)^2 = a^2 + b^2$ only when $2ab = 0$; that is, only when $a = 0$ or $b = 0$.

57. Let x = amount invested at 9%.
Then $10,000 - x$ = amount invested at 12%.
The total annual income I is:

$I = 0.09x + 0.12(10,000 - x)$
$\quad = 1,200 - 0.03x$

59. Let x = number of tickets at \$20.
Then $3x$ = number of tickets at \$30 and $4,000 - x - 3x = 4,000 - 4x$ = number of tickets at \$50.
The total receipts R are:

$R = 20x + 30(3x) + 50(4,000 - 4x)$
$\quad = 20x + 90x + 200,000 - 200x = 200,000 - 90x$

61. Let x = number of kilograms of food A.
Then $10 - x$ = number of kilograms of food B.
The total number of kilograms F of fat in the final food mix is:

$F = 0.02x + 0.06(10 - x) = 0.6 - 0.04x$

EXERCISE A-3

Things to remember:

<u>1.</u>   The discussion is limited to polynomials with integer coefficients.

<u>2.</u>   FACTORED FORMS

A polynomial is in FACTORED FORM if it is written as the product of two or more polynomials. A polynomial with integer coefficients is FACTORED COMPLETELY if each factor cannot be expressed as the product of two or more polynomials with integer coefficients, other than itself and 1.

<u>3.</u>   METHODS
   a.   Factor out all factors common to all terms, if they are present.
   b.   Try grouping terms.
   c.   *ac*–Test for polynomials of the form
   $$ax^2 + bx + c \quad \text{or} \quad ax^2 + bxy + cy^2$$
   If the product ac has two integer factors p and q whose sum is the coefficient b of the middle term, i.e., if integers p and q exist so that
   $$pq = ac \quad \text{and} \quad p + q = b$$
   then the polynomials have first–degree factors with integer coefficients. If no such integers exist then the polynomials will not have first–degree factors with integer coefficients; the polynomials are *not factorable*.

4. SPECIAL FACTORING FORMULAS

a. $u^2 + 2uv + v^2 = (u + v)^2$ Perfect square

b. $u^2 - 2uv + v^2 = (u - v)^2$ Perfect square

c. $u^2 - v^2 = (u - v)(u + v)$ Difference of squares

d. $u^3 - v^3 = (u - v)(u^2 + uv + v^2)$ Difference of cubes

e. $u^3 + v^3 = (u + v)(u^2 - uv + v^2)$ Sum of cubes

---

1. $3m^2$ is a common factor: $6m^4 - 9m^3 - 3m^2 = 3m^2(2m^2 - 3m - 1)$

3. $2uv$ is a common factor: $8u^3v - 6u^2v^2 + 4uv^3 = 2uv(4u^2 - 3uv + 2v^2)$

5. $(2m - 3)$ is a common factor: $7m(2m - 3) + 5(2m - 3) = (7m + 5)(2m - 3)$

7. $4ab(2c + d) - (2c + d) = (4ab - 1)(2c + d)$

9. $2x^2 - x + 4x - 2 = (2x^2 - x) + (4x - 2) = x(2x - 1) + 2(2x - 1) = (2x - 1)(x + 2)$

11. $3y^2 - 3y + 2y - 2 = (3y^2 - 3y) + (2y - 2) = 3y(y - 1) + 2(y - 1) = (y - 1)(3y + 2)$

13. $2x^2 + 8x - x - 4 = (2x^2 + 8x) - (x + 4) = 2x(x + 4) - (x + 4) = (x + 4)(2x - 1)$

15. $wy - wz + xy - xz = (wy - wz) + (xy - xz) = w(y - z) + x(y - z) = (y - z)(w + x)$

Or $wy - wz + xy - xz = (wy + xy) - (wz + xz) = y(w + x) - z(w + x) = (w + x)(y - z)$

17. $am - 3bm + 2na - 6bn = m(a - 3b) + 2n(a - 3b) = (a - 3b)(m + 2n)$

19. $3y^2 - y - 2$

$a = 3, b = -1, c = -2$

Step 1.   Use the ac-test to test for factorability

$ac = (3)(-2) = -6$

pq

$(1)(-6)$

$(-1)(6)$

$\boxed{(2)(-3)}$

$(-2)(3)$

Note that $2 + (-3) = -1 = b$. Thus, $3y^2 - y - 2$ has first-degree factors with integer coefficients.

Step 2.   Split the middle term using $b = p + q$ and factor by grouping.

$-1 = -3 + 2$

$3y^2 - y - 2 = 3y^2 - 3y + 2y - 2 = (3y^2 - 3y) + (2y - 2)$ $= 3y(y - 1) + 2(y - 1)$

$= (y - 1)(3y + 2)$

21. $u^2 - 2uv - 15v^2$

$a = 1, b = -2, c = -15$

Step 1.   Use the ac–test

$ac = 1(-15) = -15$

$\underline{pq}$

$(1)(-15)$

$(-1)(15)$

$\boxed{(3)(-5)}$

$(-3)(5)$

Note that $3 + (-5) = -2 = b$. Thus $u^2 - 2uv - 15v^2$ has first- degree factors with integer coefficients.

Step 2.   Factor by grouping

$-2 = 3 + (-5)$

$u^2 + 3uv - 5uv - 15v^2 = (u^2 + 3uv) - (5uv + 15v^2) = u(u + 3v) - 5v(u + 3v)$

$= (u + 3v)(u - 5v)$

23. $m^2 - 6m - 3$

$a = 1, b = -6, c = -3$

Step 1. Use the ac–test

$ac = (1)(-3) = -3$

$\underline{pq}$

$(1)(-3)$

$(-1)(3)$

None of the factors add up to $-6 = b$. Thus, this polynomial is *not factorable*.

25. $w^2x^2 - y^2 = (wx - y)(wx + y)$ (difference of squares)

27. $9m^2 - 6mn + n^2 = (3m - n)^2$ (perfect square)

29. $y^2 + 16$

$a = 1, b = 0, c = 16$

Step 1. Use the ac–test

$ac = (1)(16)$

$\underline{pq}$

$(1)(16)$

$(-1)(-16)$

$(2)(8)$

$(-2)(-8)$

$(4)(4)$

$(-4)(-4)$

None of the factors add up to $0 = b$. Thus this polynomial is *not factorable*.

31. $4z^2 - 28z + 48 = 4(z^2 - 7z + 12) = 4(z - 3)(z - 4)$

33. $2x^4 - 24x^3 + 40x^2 = 2x^2(x^2 - 12x + 20) = 2x^2(x - 2)(x - 10)$

35. $4xy^2 - 12xy + 9x = x(4y^2 - 12y + 9) = x(2y - 3)^2$

37. $6m^2 - mn - 12n^2 = (2m - 3n)(3m + 4n)$

39. $4u^3v - uv^3 = uv(4u^2 - v^2) = uv[(2u)^2 - v^2] = uv(2u - v)(2u + v)$

41. $2x^3 - 2x^2 + 8x = 2x(x^2 - x + 4)$ [Note: $x^2 - x + 4$ is *not factorable*.]

43. $8x^3 - 27y^3 = (2x)^3 - (3y)^3 = (2x - 3y)[(2x)^2 + (2x)(3y) + (3y)^2]$
$\qquad\qquad = (2x - 3y)[4x^2 + 6xy + 9y^2]$ (difference of cubes)

45. $x^4y + 8xy = xy[x^3 + 8] = xy(x + 2)(x^2 - 2x + 4)$

47. $(x + 2)^2 - 9y^2 = [(x + 2) - 3y][(x + 2) + 3y] = (x + 2 - 3y)(x + 2 + 3y)$

49. $5u^2 + 4uv - 2v^2$ is *not factorable*.

51. $6(x - y)^2 + 23(x - y) - 4 = [6(x - y) - 1][(x - y) + 4] = (6x - 6y - 1)(x - y + 4)$

53. $y^4 - 3y^2 - 4 = (y^2)^2 - 3y^2 - 4 = (y^2 - 4)(y^2 + 1) = (y - 2)(y + 2)(y^2 + 1)$

55. $15y(x - y)^3 + 12x(x - y)^2 = 3(x - y)^2[5y(x - y) + 4x] = 3(x - y)^2[5xy - 5y^2 + 4x]$

57. True: $u^n - v^n = (u - v)(u^{n-1} + u^{n-2}v + \ldots + uv^{n-2} + v^{n-1})$

59. False; For example, $u^2 + v^2$ cannot be factored.

EXERCISE A-4

Things to remember:

<u>1.</u>   FUNDAMENTAL PROPERTY OF FRACTIONS

If a, b, and k are real numbers with b, $k \neq 0$, then

$$\frac{ka}{kb} = \frac{a}{b}.$$

A fraction is in LOWEST TERMS if the numerator and denominator have no common factors other than 1 or –1.

<u>2.</u>   MULTIPLICATION AND DIVISION

For a, b, c, and d real numbers:

a. $\dfrac{a}{b} \cdot \dfrac{c}{d} = \dfrac{ac}{bd}$, $b, d \neq 0$

b. $\dfrac{a}{b} \div \dfrac{c}{d} = \dfrac{\dfrac{a}{b}}{\dfrac{c}{d}} = \dfrac{a}{b} \cdot \dfrac{d}{c}$, $b, c, d \neq 0$

The same procedures are used to multiply or divide two rational expressions.

<u>3.</u>   ADDITION AND SUBTRACTION

For a, b, and c real numbers:

a. $\dfrac{a}{b} + \dfrac{c}{b} = \dfrac{a+c}{b}$, $b \neq 0$

b. $\dfrac{a}{b} - \dfrac{c}{b} = \dfrac{a-c}{b}$, $b \neq 0$

The same procedures are used to add or subtract two rational expressions (with the same denominator).

4.   THE LEAST COMMON DENOMINATOR (LCD)

The LCD of two or more rational expressions is found as follows:

a.   Factor each denominator completely, including integer factors.

b.   Identify each different factor from all the denominators.

c.   Form a product using each different factor to the highest power that occurs in any one denominator. This product is the LCD.

The least common denominator is used to add or subtract rational expressions having different denominators.

---

1. $\dfrac{5\cdot 9\cdot 13}{3\cdot 5\cdot 7} = \dfrac{\cancel{5}\cdot \cancel{9}^3\cdot 13}{\cancel{3}\cdot \cancel{5}\cdot 7} = \dfrac{3\cdot 13}{7} = \dfrac{39}{7}$

3. $\dfrac{\cancel{12}\cdot 11\cdot \cancel{10}\cdot 9}{\cancel{4}\cdot \cancel{3}\cdot \cancel{2}\cdot 1} = 11\cdot 5\cdot 9 = 495$

5.

$\dfrac{d^5}{3a} \div \left(\dfrac{d^2}{6a^2}\cdot \dfrac{a}{4d^3} \right) = \dfrac{d^5}{3a} \div \left(\dfrac{\cancel{a}\,\cancel{d}^2}{24\,\cancel{a}^2\,\cancel{d}^3} \right) = \dfrac{d^5}{3a} \div \dfrac{1}{24ad} = \dfrac{d^5}{\cancel{3}\,\cancel{a}}\cdot \dfrac{\cancel{24}^{\,8}\,\cancel{a}d}{1} = 8d^6$

$\qquad\qquad\qquad\qquad\qquad\qquad a\ d$

7. $\dfrac{x^2}{12} + \dfrac{x}{18} - \dfrac{1}{30} = \dfrac{15x^2}{180} + \dfrac{10x}{180} - \dfrac{6}{180}$

$\qquad\qquad\qquad\quad = \dfrac{15x^2 + 10x - 6}{180}$

We find the LCD of 12, 18, 30:

$12 = 2^2\cdot 3,\ 18 = 2\cdot 3^2,\ 30 = 2\cdot 3\cdot 5$

Thus, LCD $= 2^2\cdot 3^2\cdot 5 = 180$.

9. $\dfrac{4m-3}{18m^3} + \dfrac{3}{4m} - \dfrac{2m-1}{6m^2}$

$= \dfrac{2(4m-3)}{36m^3} + \dfrac{3(9m^2)}{36m^3} - \dfrac{6m(2m-1)}{36m^3}$

$= \dfrac{8m-6+27m^2 - 6m(2m-1)}{36m^3}$

$= \dfrac{8m-6+27m^2 - 12m^2 + 6m}{36m^3} = \dfrac{15m^2 + 14m - 6}{36m^3}$

Find the LCD of $18m^3,\ 4m,\ 6m^2$:

$18m^3 = 2\cdot 3^2 m^3,\ 4m = 2^2 m,$

$6m^2 = 2\cdot 3m^2$

Thus, LCD $= 36m^3$.

11. $\dfrac{x^2-9}{x^2-3x} \div (x^2-x-12) = \dfrac{\cancel{(x-3)}\,(x+3)}{x\,\cancel{(x-3)}}\cdot \dfrac{1}{(x-4)\cancel{(x+3)}} = \dfrac{1}{x(x-4)}$

13. $\dfrac{2}{x} - \dfrac{1}{x-3} = \dfrac{2(x-3)}{x(x-3)} - \dfrac{x}{x(x-3)}$

$\qquad\qquad = \dfrac{2x-6-x}{x(x-3)} = \dfrac{x-6}{x(x-3)}$

LCD $= x(x-3)$

15. $\dfrac{2}{(x+1)^2} - \dfrac{5}{x^2-x-2} = \dfrac{2}{(x+1)^2} - \dfrac{5}{(x+1)(x-2)}$ $\qquad$ LCD $= (x+1)^2(x-2)$

$\qquad\qquad = \dfrac{2(x-2)}{(x+1)^2(x-2)} - \dfrac{5(x+1)}{(x+1)^2(x-2)} = \dfrac{2x-4-5x-5}{(x+1)^2(x-2)} = \dfrac{-3x-9}{(x+1)^2(x-2)}$

17. $\dfrac{x+1}{x-1} - 1 = \dfrac{x+1}{x-1} - \dfrac{x-1}{x-1} = \dfrac{x+1-(x-1)}{x-1} = \dfrac{2}{x-1}$

19. $\dfrac{3}{a-1} - \dfrac{2}{1-a} = \dfrac{3}{a-1} - \dfrac{-2}{-(1-a)} = \dfrac{3}{a-1} + \dfrac{2}{a-1} = \dfrac{5}{a-1}$

21. $\dfrac{2x}{x^2-16} - \dfrac{x-4}{x^2+4x} = \dfrac{2x}{(x-4)(x+4)} - \dfrac{x-4}{x(x+4)}$ $\qquad$ LCD $= x(x-4)(x+4)$

$\qquad\qquad = \dfrac{2x(x) - (x-4)(x-4)}{x(x-4)(x+4)} = \dfrac{2x^2 - (x^2-8x+16)}{x(x-4)(x+4)} = \dfrac{x^2+8x-16}{x(x-4)(x+4)}$

23. $\dfrac{x^2}{x^2+2x+1} + \dfrac{x-1}{3x+3} - \dfrac{1}{6} = \dfrac{x^2}{(x+1)^2} + \dfrac{x-1}{3(x+1)} - \dfrac{1}{6}$ $\qquad$ LCD $= 6(x+1)^2$

$\qquad\qquad = \dfrac{6x^2}{6(x+1)^2} + \dfrac{2(x+1)(x-1)}{6(x+1)^2} - \dfrac{(x+1)^2}{6(x+1)^2}$

$\qquad\qquad = \dfrac{6x^2 + 2(x^2-1) - (x^2+2x+1)}{6(x+1)^2} = \dfrac{7x^2-2x-3}{6(x+1)^2}$

25. $\dfrac{1-\dfrac{x}{y}}{2-\dfrac{y}{x}} = \dfrac{\dfrac{y-x}{y}}{\dfrac{2x-y}{x}} = \dfrac{y-x}{y} \cdot \dfrac{x}{2x-y} = \dfrac{x(y-x)}{y(2x-y)}$

27. $\dfrac{c+2}{5c-5} - \dfrac{c-2}{3c-3} + \dfrac{c}{1-c} = \dfrac{c+2}{5(c-1)} - \dfrac{c-2}{3(c-1)} - \dfrac{c}{c-1}$ $\qquad$ LCD $= 15(c-1)$

$\qquad\qquad = \dfrac{3(c+2)}{15(c-1)} - \dfrac{5(c-2)}{15(c-1)} - \dfrac{15c}{15(c-1)} = \dfrac{3c+6-5c+10-15c}{15(c-1)} = \dfrac{-17c+16}{15(c-1)}$

29. $\dfrac{1+\dfrac{3}{x}}{x-\dfrac{9}{x}} = \dfrac{\dfrac{x+3}{x}}{\dfrac{x^2-9}{x}} = \dfrac{x+3}{x} \cdot \dfrac{x}{x^2-9} = \dfrac{\cancel{x+3}}{\cancel{x}} \cdot \dfrac{\cancel{x}}{\cancel{(x+3)}(x-3)} = \dfrac{1}{x-3}$

31. $\dfrac{\dfrac{1}{2(x+h)} - \dfrac{1}{2x}}{h} = \left(\dfrac{1}{2(x+h)} - \dfrac{1}{2x}\right) \div \dfrac{h}{1} = \dfrac{x-x-h}{2x(x+h)} \cdot \dfrac{1}{h} = \dfrac{-\cancel{h}}{2x(x+h)\cancel{h}} = \dfrac{-1}{2x(x+h)}$

33. $\dfrac{\dfrac{x}{y} - 2 + \dfrac{y}{x}}{\dfrac{x}{y} - \dfrac{y}{x}} = \dfrac{\dfrac{x^2-2xy+y^2}{xy}}{\dfrac{x^2-y^2}{xy}} = \dfrac{(x-y)^2}{\cancel{xy}} \cdot \dfrac{\cancel{xy}}{(x-y)(x+y)} = \dfrac{x-y}{x+y}$

35. (A) $\dfrac{x^2+4x+3}{x+3} = x+4:$ Incorrect

   (B) $\dfrac{x^2+4x+3}{x+3} = \dfrac{\cancel{(x+3)}(x+1)}{\cancel{x+3}} = x+1 \quad (x \neq -3)$

37. (A) $\dfrac{(x+h)^2-x^2}{h} = 2x+1:$ Incorrect

   (B) $\dfrac{(x+h)^2-x^2}{h} = \dfrac{x^2+2xh+h^2-x^2}{h} = \dfrac{2xh+h^2}{h} = \dfrac{\cancel{h}(2x+h)}{\cancel{h}} = 2x+h \quad (h \neq 0)$

39. (A) $\dfrac{x^2-3x}{x^2-2x-3} + x - 3 = 1:$ Incorrect

   (B) $\dfrac{x^2-3x}{x^2-2x-3} + x - 3 = \dfrac{x\cancel{(x-3)}}{\cancel{(x-3)}(x+1)} + x - 3 = \dfrac{x}{x+1} + x - 3 = \dfrac{x+(x-3)(x+1)}{x+1} = \dfrac{x^2-x-3}{x+1}$

41. (A) $\dfrac{2x^2}{x^2-4} - \dfrac{x}{x-2} = \dfrac{x}{x+2}:$ Correct

   $\dfrac{2x^2}{x^2-4} - \dfrac{x}{x-2} = \dfrac{2x^2}{(x-2)(x+2)} - \dfrac{x}{x-2} = \dfrac{2x^2-x(x+2)}{(x-2)(x+2)}$

   $= \dfrac{x^2-2x}{(x-2)(x+2)} = \dfrac{x\cancel{(x-2)}}{\cancel{(x-2)}(x+2)} = \dfrac{x}{x+2}$

43. $\dfrac{\dfrac{1}{3(x+h)^2} - \dfrac{1}{3x^2}}{h} = \left[\dfrac{1}{3(x+h)^2} - \dfrac{1}{3x^2}\right] \div \dfrac{h}{1} = \dfrac{x^2-(x+h)^2}{3x^2(x+h)^2} \cdot \dfrac{1}{h} = \dfrac{x^2-(x^2+2xh+h^2)}{3x^2(x+h)^2 h}$

   $= \dfrac{-2xh-h^2}{3x^2(x+h)^2 h} = \dfrac{-\cancel{h}(2x+h)}{3x^2(x+h)^2 \cancel{h}} = -\dfrac{(2x+h)}{3x^2(x+h)^2} = \dfrac{-2x-h}{3x^2(x+h)^2}$

45. $x - \dfrac{2}{1-\dfrac{1}{x}} = x - \dfrac{2}{\dfrac{x-1}{x}} = x - \dfrac{2x}{x-1} \qquad (\text{LCD} = x-1)$

   $= \dfrac{x(x-1)}{x-1} - \dfrac{2x}{x-1} = \dfrac{x^2-x-2x}{x-1} = \dfrac{x(x-3)}{x-1}$

<u>**EXERCISE A-5**</u>

Things to remember:

<u>1.</u>   DEFINITION OF a^n, where n is an integer and a is a real number:

    a.   For n a positive integer,
$$a^n = a \cdot a \cdot \cdots \cdot a, \, n \text{ factors of } a.$$

    b.   For $n = 0$,
$$a^0 = 1, \, a \neq 0, \, 0^0 \text{ is not defined.}$$

    c.   For n a negative integer,
$$a^n = \frac{1}{a^{-n}}, \, a \neq 0.$$

    [<u>Note</u>: If n is negative, then $-n$ is positive.]

<u>2.</u>   PROPERTIES OF EXPONENTS
    GIVEN: n and m are integers and a and b are real numbers.

    a. $a^m a^n = a^{m+n}$ $a^8 a^{-3} = a^{8+(-3)} = a^5$

    b. $(a^n)^m = a^{mn}$ $(a^{-2})^3 = a^{3(-2)} = a^{-6}$

    c. $(ab)^m = a^m b^m$ $(ab)^{-2} = a^{-2} b^{-2}$

    d. $\left(\dfrac{a}{b}\right)^m = \dfrac{a^m}{b^m}, \, b \neq 0$ $\left(\dfrac{a}{b}\right)^5 = \dfrac{a^5}{b^5}$

    e. $\dfrac{a^m}{a^n} = a^{m-n} = \dfrac{1}{a^{n-m}}, \, a \neq 0$ $\dfrac{a^{-3}}{a^7} = \dfrac{1}{a^{7-(-3)}} = \dfrac{1}{a^{10}}$

<u>3.</u>   SCIENTIFIC NOTATION
    Let r be any finite decimal. Then r can be expressed as the product of a number between 1 and 10 and an integer power of 10; that is, r can be written $r = a \times 10^n$, $1 \leq a < 10$, a in decimal form, n an integer.   A number expressed in this form is said to be in SCIENTIFIC NOTATION.

    Examples:

    $7 = 7 \times 10^0$ $0.5 = 5 \times 10^{-1}$

    $67 = 6.7 \times 10$ $0.45 = 4.5 \times 10^{-1}$

    $580 = 5.8 \times 10^2$ $0.0032 = 3.2 \times 10^{-3}$

    $43{,}000 = 4.3 \times 10^4$ $0.000\,045 = 4.5 \times 10^{-5}$

---

1. $2x^{-9} = \dfrac{2}{x^9}$ **3.** $\dfrac{3}{2w^{-7}} = \dfrac{3w^7}{2}$

5. $2x^{-8}x^5 = 2x^{-8+5} = 2x^{-3} = \dfrac{2}{x^3}$ **7.** $\dfrac{w^{-8}}{w^{-3}} = \dfrac{1}{w^{-3+8}} = \dfrac{1}{w^5}$

9. $(2a^{-3})^2 = 2^2(a^{-3})^2 = 4a^{-6} = \dfrac{4}{a^6}$ **11.** $(a^{-3})^2 = a^{-6} = \dfrac{1}{a^6}$

13. $(2x^4)^{-3} = 2^{-3}(x^4)^{-3} = \dfrac{1}{8} \cdot x^{-12} = \dfrac{1}{8x^{12}}$

15. $82{,}300{,}000{,}000 = 8.23 \times 10^{10}$

17. $0.783 = 7.83 \times 10^{-1}$

19. $0.000\,034 = 3.4 \times 10^{-5}$

21. $4 \times 10^4 = 40{,}000$

23. $7 \times 10^{-3} = 0.007$

25. $6.171 \times 10^7 = 61{,}710{,}000$

27. $8.08 \times 10^{-4} = 0.000\,808$

29. $(22 + 31)^0 = (53)^0 = 1$

31. $\dfrac{10^{-3} \times 10^4}{10^{-11} \times 10^{-2}} = \dfrac{10^{-3+4}}{10^{-11-2}} = \dfrac{10^1}{10^{-13}} = 10^{1+13} = 10^{14}$

33. $(5x^2y^{-3})^{-2} = 5^{-2}x^{-4}y^6 = \dfrac{y^6}{5^2 x^4} = \dfrac{y^6}{25x^4}$

35. $\left(\dfrac{-5}{2x^3}\right)^{-2} = \dfrac{(-5)^{-2}}{(2x^3)^{-2}} = \dfrac{\dfrac{1}{(-5)^2}}{\dfrac{1}{(2x^3)^2}} = \dfrac{\dfrac{1}{25}}{\dfrac{1}{4x^6}} = \dfrac{4x^6}{25}$

37. $\dfrac{8x^{-3}y^{-1}}{6x^2 y^{-4}} = \dfrac{4y^{-1+4}}{3x^{2+3}} = \dfrac{4y^3}{3x^5}$

39. $\dfrac{7x^5 - x^2}{4x^5} = \dfrac{7x^5}{4x^5} - \dfrac{x^2}{4x^5} = \dfrac{7}{4} - \dfrac{1}{4x^3} = \dfrac{7}{4} - \dfrac{1}{4}x^{-3}$

41. $\dfrac{5x^4 - 3x^2 + 8}{2x^2} = \dfrac{5x^4}{2x^2} - \dfrac{3x^2}{2x^2} + \dfrac{8}{2x^2} = \dfrac{5}{2}x^2 - \dfrac{3}{2} + 4x^{-2}$

43. $\dfrac{3x^2(x-1)^2 - 2x^3(x-1)}{(x-1)^4} = \dfrac{x^2(x-1)[3(x-1) - 2x]}{(x-1)^4} = \dfrac{x^2(x-3)}{(x-1)^3}$

45. $2x^{-2}(x-1) - 2x^{-3}(x-1)^2 = \dfrac{2(x-1)}{x^2} - \dfrac{2(x-1)^2}{x^3} = \dfrac{2x(x-1) - 2(x-1)^2}{x^3}$

$\qquad = \dfrac{2(x-1)[x - (x-1)]}{x^3} = \dfrac{2(x-1)}{x^3}$

47. $\dfrac{9{,}600{,}000{,}000}{(1{,}600{,}000)(0.00000025)} = \dfrac{9.6 \times 10^9}{(1.6 \times 10^6)(2.5 \times 10^{-7})} = \dfrac{9.6 \times 10^9}{1.6(2.5) \times 10^{6-7}}$

$\qquad = \dfrac{9.6 \times 10^9}{4.0 \times 10^{-1}} = 2.4 \times 10^{9+1} = 2.4 \times 10^{10} = 24{,}000{,}000{,}000$

49. $\dfrac{(1{,}250{,}000)(0.00038)}{0.0152} = \dfrac{(1.25 \times 10^6)(3.8 \times 10^{-4})}{1.52 \times 10^{-2}} = \dfrac{1.25(3.8) \times 10^{6-4}}{1.52 \times 10^{-2}} = 3.125 \times 10^4 = 31{,}250$

51. On a calculator : $2^{3^2} = 64$. A calculator interprets 2^{3^2} as $(2^3)^2 = 8^2 = 64$.

53. $a^m a^0 = a^{m+0} = a^m$. Therefore, $a^m a^0 = a^m$ which implies $a^0 = 1$.

55. $\dfrac{u+v}{u^{-1}+v^{-1}} = \dfrac{u+v}{\dfrac{1}{u}+\dfrac{1}{v}} = \dfrac{u+v}{\dfrac{v+u}{uv}} = (u+v)\cdot\dfrac{uv}{v+u} = uv$

57. $\dfrac{b^{-2}-c^{-2}}{b^{-3}-c^{-3}} = \dfrac{\dfrac{1}{b^2}-\dfrac{1}{c^2}}{\dfrac{1}{b^3}-\dfrac{1}{c^3}} = \dfrac{\dfrac{c^2-b^2}{b^2c^2}}{\dfrac{c^3-b^3}{b^3c^3}} = \dfrac{(c-b)(c+b)}{b^2c^2}\cdot\dfrac{b^3c^3{}^{bc}}{(c-b)(c^2+cb+b^2)} = \dfrac{bc(c+b)}{c^2+cb+b^2}$

59. (A) Per capita debt: $\dfrac{1.606600\times10^{13}}{3.13\times10^8} \approx 0.51329\times10^5 = 51{,}329$ or $51{,}329$

    (B) Per capita interest: $\dfrac{3.60\times10^{11}}{3.13\times10^8} \approx 1.15016\times10^3 = 1{,}150$ or $1{,}150$

    (C) Percentage interest paid on debt: $\dfrac{3.60\times10^{11}}{1.606\times10^{13}} \approx 2.24159\times10^{-2} \approx 0.0224$ or 2.24%

61. (A) $9\text{ ppm} = \dfrac{9}{1{,}000{,}000} = \dfrac{9}{10^6} = 9\times10^{-6}$ (B) 0.000 009   (C) 0.0009%

63. $\dfrac{404}{100{,}000}\times309{,}000{,}000 = \dfrac{4.04\times10^2}{10^5}\times3.09\times10^8 = 12.4836\times10^5 \approx 1{,}248{,}000$

To the nearest thousand, there were 1,248,000 violent crimes committed in 2010.

EXERCISE A-6

Things to remember:

1.   *n*th ROOT

      Let *b* be a real number. For any natural number *n*,

         r is an *n*th ROOT of *b* if $r^n = b$

      If *n* is odd, then *b* has exactly one real *n*th root.

      If *n* is even, and $b < 0$, then *b* has NO real *n*th roots.

      If *n* is even, and $b > 0$, then *b* has two real *n*th roots;

        if *r* is an *n*th root, then −*r* is also an *n*th root.

      0 is an *n*th root of 0 for all *n*

2.   NOTATION

Let b be a real number and let $n > 1$ be a natural number. If n is odd, then the nth root of b is denoted

$$b^{1/n} \quad \text{or} \quad \sqrt[n]{b}$$

If n is even and $b > 0$, then the PRINCIPAL nth ROOT OF b is the positive nth root; the principal nth root is denoted

$$b^{1/n} \quad \text{or} \quad \sqrt[n]{b}$$

In the $\sqrt[n]{b}$ notation, the symbol $\sqrt{}$ is called a RADICAL, n is the INDEX of the radical and b is called the RADICAND.

3.   RATIONAL EXPONENTS

If m and n are natural numbers without common prime factors, b is a real number, and b is nonnegative when b is even, then

$$b^{m/n} = \begin{cases} \left(b^{1/n}\right)^m = \left(\sqrt[n]{b}\right)^m \\ \left(b^m\right)^{1/n} = \sqrt[n]{b^m} \end{cases} \quad \text{and} \quad b^{-m/n} = \frac{1}{b^{m/n}}, \ b \neq 0$$

The two definitions of $b^{m/n}$ are equivalent under the indicated restrictions on m, n, and b.

4.   PROPERTIES OF RADICALS

If m and n are natural numbers greater than or equal to 2 and x and y are positive real numbers, then

a. $\sqrt[n]{x^n} = x$ $\sqrt[3]{x^3} = x$

b. $\sqrt[n]{xy} = \sqrt[n]{x}\,\sqrt[n]{y}$ $\sqrt[5]{xy} = \sqrt[5]{x}\,\sqrt[5]{y}$

c. $\sqrt[n]{\dfrac{x}{y}} = \dfrac{\sqrt[n]{x}}{\sqrt[n]{y}}$ $\sqrt[4]{\dfrac{x}{y}} = \dfrac{\sqrt[4]{x}}{\sqrt[4]{y}}$

---

1. $6x^{3/5} = 6\sqrt[5]{x^3}$

3. $(32x^2 y^3)^{3/5} = \sqrt[5]{(32x^2 y^3)^3}$

5. $(x^2 + y^2)^{1/2} = \sqrt{x^2 + y^2}$

[Note: $\sqrt{x^2 + y^2} \neq x + y$.]

7. $5\sqrt[4]{x^3} = 5x^{3/4}$

9. $\sqrt[5]{(2x^2 y)^3} = (2x^2 y)^{3/5}$

11. $\sqrt[3]{x} + \sqrt[3]{y} = x^{1/3} + y^{1/3}$

13. $25^{1/2} = (5^2)^{1/2} = 5$

15. $16^{3/2} = (4^2)^{3/2} = 4^3 = 64$

17. $-49^{1/2} = -\sqrt{49} = -7$

19. $-64^{2/3} = -(\sqrt[3]{64})^2 = -16$

21. $\left(\dfrac{4}{25}\right)^{3/2} = \left(\left(\dfrac{2}{5}\right)^2\right)^{3/2} \left(\dfrac{2}{5}\right)^3 = \dfrac{2^3}{5^3} = \dfrac{8}{125}$

23. $9^{-3/2} = (3^2)^{-3/2} = 3^{-3} = \dfrac{1}{3^3} = \dfrac{1}{27}$

25. $x^{4/5} x^{-2/5} = x^{4/5 - 2/5} = x^{2/5}$

27. $\dfrac{m^{2/3}}{m^{-1/3}} = m^{2/3 - (-1/3)} = m^1 = m$

29. $(8x^3y^{-6})^{1/3} = (2^3x^3y^{-6})^{1/3} = 2^{3/3}x^{3/3}y^{-6/3} = 2xy^{-2} = \dfrac{2x}{y^2}$

31. $\left(\dfrac{4x^{-2}}{y^4}\right)^{-1/2} = \left(\dfrac{2^2x^{-2}}{y^4}\right)^{-1/2} = \dfrac{2^{2(-1/2)}x^{-2(-1/2)}}{y^{4(-1/2)}} = \dfrac{2^{-1}x^1}{y^{-2}} = \dfrac{xy^2}{2}$

33. $\dfrac{(8x)^{-1/3}}{12x^{1/4}} = \dfrac{\frac{1}{(8x)^{1/3}}}{12x^{1/4}} = \dfrac{\frac{1}{2x^{1/3}}}{12x^{1/4}} = \dfrac{1}{24x^{1/4+1/3}} = \dfrac{1}{24x^{7/12}}$

35. $\sqrt[5]{(2x+3)^5} = [(2x+3)^5]^{1/5} = 2x+3$

37. $\sqrt{6x}\,\sqrt{15x^3}\,\sqrt{30x^7} = \sqrt{6(15)(30)x^{11}} = \sqrt{3(30)^2x^{11}} = 30x^5\sqrt{3x}$

39. $\dfrac{\sqrt{6x}\sqrt{10}}{\sqrt{15x}} = \sqrt{\dfrac{60x}{15x}} = \sqrt{4} = 2$

41. $3x^{3/4}(4x^{1/4} - 2x^8) = 12x^{3/4+1/4} - 6x^{3/4+8} = 12x - 6x^{3/4+32/4} = 12x - 6x^{35/4}$

43. $(3u^{1/2} - v^{1/2})(u^{1/2} - 4v^{1/2}) = 3u - 12u^{1/2}v^{1/2} - u^{1/2}v^{1/2} + 4v = 3u - 13u^{1/2}v^{1/2} + 4v$

45. $(6m^{1/2} + n^{-1/2})(6m - n^{-1/2}) = 36m^{3/2} + 6mn^{-1/2} - 6m^{1/2}n^{-1/2} - n^{-1} = 36m^{3/2} + \dfrac{6m}{n^{1/2}} - \dfrac{6m^{1/2}}{n^{1/2}} - \dfrac{1}{n}$

47. $(3x^{1/2} - y^{1/2})^2 = (3x^{1/2})^2 - 6x^{1/2}y^{1/2} + (y^{1/2})^2 = 9x - 6x^{1/2}y^{1/2} + y$

49. $\dfrac{\sqrt[3]{x^2}+2}{2\sqrt[3]{x}\cdot} = \dfrac{x^{2/3}+2}{2x^{1/3}} = \dfrac{x^{2/3}}{2x^{1/3}} + \dfrac{2}{2x^{1/3}} = \dfrac{1}{2}x^{1/3} + \dfrac{1}{x^{1/3}} = \dfrac{1}{2}x^{1/3} + x^{-1/3}$

51. $\dfrac{2\sqrt[4]{x^3}+\sqrt[3]{x}}{3x} = \dfrac{2x^{3/4}+x^{1/3}}{3x} = \dfrac{2x^{3/4}}{3x} + \dfrac{x^{1/3}}{3x} = \dfrac{2}{3}x^{3/4-1} + \dfrac{1}{3}x^{1/3-1} = \dfrac{2}{3}x^{-1/4} + \dfrac{1}{3}x^{-2/3}$

53. $\dfrac{2\sqrt[3]{x}-\sqrt{x}}{4\sqrt{x}} = \dfrac{2x^{1/3}-x^{1/2}}{4x^{1/2}} = \dfrac{2x^{1/3}}{4x^{1/2}} - \dfrac{x^{1/2}}{4x^{1/2}} = \dfrac{1}{2}x^{1/3-1/2} - \dfrac{1}{4} = \dfrac{1}{2}x^{-1/6} - \dfrac{1}{4}$

55. $\dfrac{12mn^2}{\sqrt{3mn}} = \dfrac{12mn^2}{\sqrt{3mn}} \cdot \dfrac{\sqrt{3mn}}{\sqrt{3mn}} = \dfrac{12mn^2\,\sqrt{3mn}}{3mn} = 4n\sqrt{3mn}$

57. $\dfrac{2(x+3)}{\sqrt{x-2}} = \dfrac{2(x+3)}{\sqrt{x-2}} \cdot \dfrac{\sqrt{x-2}}{\sqrt{x-2}} = \dfrac{2(x+3)\sqrt{x-2}}{x-2}$

59. $\dfrac{7(x-y)^2}{\sqrt{x}-\sqrt{y}} = \dfrac{7(x-y)^2}{\sqrt{x}-\sqrt{y}} \cdot \dfrac{\sqrt{x}+\sqrt{y}}{\sqrt{x}+\sqrt{y}} = \dfrac{7(x-y)^2\,(\sqrt{x}+\sqrt{y})}{x-y} = 7(x-y)(\sqrt{x}+\sqrt{y})$

61. $\dfrac{\sqrt{5xy}}{5x^2y^2} = \dfrac{\sqrt{5xy}}{5x^2y^2} \cdot \dfrac{\sqrt{5xy}}{\sqrt{5xy}} = \dfrac{5xy}{5x^2y^2\,\sqrt{5xy}} = \dfrac{1}{xy\sqrt{5xy}}$

63. $\dfrac{\sqrt{x+h}-\sqrt{x}}{h} = \dfrac{\sqrt{x+h}-\sqrt{x}}{h} \cdot \dfrac{\sqrt{x+h}+\sqrt{x}}{\sqrt{x+h}+\sqrt{x}} = \dfrac{x+h-x}{h(\sqrt{x+h}+\sqrt{x})} = \dfrac{h}{h(\sqrt{x+h}+\sqrt{x})} = \dfrac{1}{\sqrt{x+h}+\sqrt{x}}$

65. $\dfrac{\sqrt{t}-\sqrt{x}}{t^2-x^2} = \dfrac{\sqrt{t}-\sqrt{x}}{(t-x)(t+x)} \cdot \dfrac{\sqrt{t}+\sqrt{x}}{\sqrt{t}+\sqrt{x}} = \dfrac{t-x}{(t-x)(t+x)(\sqrt{t}+\sqrt{x})} = \dfrac{1}{(t+x)(\sqrt{t}+\sqrt{x})}$

67. $(x+y)^{1/2} \overset{?}{=} x^{1/2}+y^{1/2}$

Let $x = y = 1$. Then

$(1+1)^{1/2} = 2^{1/2} = \sqrt{2} \approx 1.414$

$1^{1/2}+1^{1/2} = \sqrt{1}+\sqrt{1} = 1+1 = 2;\ \sqrt{2} \neq 2$

71. $\sqrt{x^2} = x$ for all real numbers x: False

$\sqrt{(-2)^2} = \sqrt{4} = 2 \neq -2$

73. $\sqrt[3]{x^3} = |x|$ for all real numbers x: False

$\sqrt[3]{(-1)^3} = \sqrt[3]{-1} = -1 \neq |-1| = 1$

75. False: $(-8)^{1/3} = -2$ since $(-2)^3 = -8$

77. True: $r^{1/2} = \sqrt{r}$ and $-r^{1/2} = -\sqrt{r}$ are each square roots of r.

79. True: $(\sqrt{10})^4 = (10^{1/2})^4 = 10^2 = 100$

$(-\sqrt{10})^4 = (-1)^4(\sqrt{10})^4 = (1)(100) = 100$

81. False: $5\sqrt{7}-6\sqrt{5} \approx -0.1877;\ \sqrt{a}$ is never negative.

83. $-\dfrac{1}{2}(x-2)(x+3)^{-3/2} + (x+3)^{-1/2} = \dfrac{-(x-2)}{2(x+3)^{3/2}} + \dfrac{1}{(x+3)^{1/2}} = \dfrac{-x+2+2(x+3)}{2(x+3)^{3/2}} = \dfrac{x+8}{2(x+3)^{3/2}}$

85. $\dfrac{(x-1)^{1/2}-x\left(\frac{1}{2}\right)(x-1)^{-1/2}}{x-1} = \dfrac{(x-1)^{1/2}-\dfrac{x}{2(x-1)^{1/2}}}{x-1} = \dfrac{\dfrac{2(x-1)^{1/2}(x-1)^{1/2}}{2(x-1)^{1/2}}-\dfrac{x}{2(x-1)^{1/2}}}{x-1}$

$= \dfrac{\dfrac{2(x-1)-x}{2(x-1)^{1/2}}}{x-1} = \dfrac{x-2}{2(x-1)^{3/2}}$

87. $\dfrac{(x+2)^{2/3}-x\left(\frac{2}{3}\right)(x+2)^{-1/3}}{(x+2)^{4/3}} = \dfrac{(x+2)^{2/3}-\dfrac{2x}{3(x+2)^{1/3}}}{(x+2)^{4/3}} = \dfrac{\dfrac{3(x+2)^{1/3}(x+2)^{2/3}-2x}{3(x+2)^{1/3}}}{(x+2)^{4/3}}$

$= \dfrac{3(x+2)-2x}{3(x+2)^{5/3}} = \dfrac{x+6}{3(x+2)^{5/3}}$

89. $22^{3/2} = 22^{1.5} \approx 103.2$ or $22^{3/2} = \sqrt{(22)^3} = \sqrt{10,648} \approx 103.2$

91. $827^{-3/8} = \dfrac{1}{827^{3/8}} = \dfrac{1}{827^{0.375}} \approx \dfrac{1}{12.42} \approx 0.0805$

93. $37.09^{7/3} \approx 37.09^{2.3333} \approx 4{,}588$

95. (A) $\sqrt{3} + \sqrt{5} \approx 1.732 + 2.236 = 3.968$

(B) $\sqrt{2+\sqrt{3}} + \sqrt{2-\sqrt{3}} \approx 2.449$

(C) $1 + \sqrt{3} \approx 2.732$

(D) $\sqrt[3]{10+6\sqrt{3}} \approx 2.732$

(E) $\sqrt{8+\sqrt{60}} \approx 3.968$

(F) $\sqrt{6} \approx 2.449$

(A) and (E) have the same value:

$$\left(\sqrt{3}+\sqrt{5}\right)^2 = 3 + 2\sqrt{3}\sqrt{5} + 5 = 8 + 2\sqrt{15}$$

$$\left[\sqrt{8+\sqrt{60}}\right]^2 = 8 + \sqrt{4 \cdot 15} = 8 + 2\sqrt{15}$$

(B) and (F) have the same value.

$$\left(\sqrt{2+\sqrt{3}} + \sqrt{2-\sqrt{3}}\right)^2 = 2 + \sqrt{3} + 2\sqrt{2+\sqrt{3}}\sqrt{2-\sqrt{3}} + 2 - \sqrt{3} = 4 + 2\sqrt{4-3} = 4 + 2 = 6$$

$$(\sqrt{6})^2 = 6.$$

(C) and (D) have the same value.

$$\left(1+\sqrt{3}\right)^3 = \left(1+\sqrt{3}\right)^2\left(1+\sqrt{3}\right) = (1 + 2\sqrt{3} + 3)(1 + \sqrt{3}) = (4 + 2\sqrt{3})(1 + \sqrt{3})$$
$$= 4 + 6\sqrt{3} + 6 = 10 + 6\sqrt{3}$$

$$\left(\sqrt[3]{10+6\sqrt{3}}\right)^3 = 10 + 6\sqrt{3}$$

EXERCISE A-7

Things to remember:

<u>1.</u>   A QUADRATIC EQUATION in one variable is any equation that can be written in the form

$$ax^2 + bx + c = 0, \, a \neq 0 \quad \text{STANDARD FORM}$$

where x is a variable and a, b, and c are constants.

<u>2.</u>   Quadratic equations of the form $ax^2 + c = 0$ can be solved by the SQUARE ROOT METHOD. The solutions are:

$$x = \pm\sqrt{\frac{-c}{a}} \quad \text{provided} \quad \frac{-c}{a} \geq 0;$$

otherwise, the equation has no real solutions.

<u>3.</u>   If the left side of the quadratic equation when written in standard form can be FACTORED,

$$ax^2 + bx + c = (px + q)(rx + s),$$

then the solutions are

$$x = \frac{-q}{p} \quad \text{or} \quad x = \frac{-s}{r}.$$

4.   The solutions of the quadratic equation written in standard form are given by the QUADRATIC FORMULA:

$$x = \frac{-b \pm \sqrt{b^2 - 4ac}}{2a}$$

The quantity $b^2 - 4ac$ under the radical is called the DISCRIMINANT and the equation:

(i) Has two real solutions if $b^2 - 4ac > 0$.

(ii) Has one real solution if $b^2 - 4ac = 0$.

(iii) Has no real solution if $b^2 - 4ac < 0$.

5.   FACTORABILITY THEOREM

The second-degree polynomial, $ax^2 + bx + c$, with integer coefficients, can be expressed as the product of two first-degree polynomials with integer coefficients if and only if $\sqrt{b^2 - 4ac}$ is an integer.

6.   FACTOR THEOREM

If r_1 and r_2 are solutions of $ax^2 + bx + c = 0$, then

$ax^2 + bx + c = a(x - r_1)(x - r_2)$.

---

1. $2x^2 - 22 = 0$

$\quad x^2 - 11 = 0$

$\qquad x^2 = 11$

$\qquad x = \pm\sqrt{11}$

3. $(3x - 1)^2 = 25$

$\quad 3x - 1 = \pm\sqrt{25} = \pm 5$

$\qquad 3x = 1 \pm 5 = -4 \text{ or } 6$

$\qquad x = -\dfrac{4}{3} \text{ or } 2$

5. $2u^2 - 8u - 24 = 0$

$\quad u^2 - 4u - 12 = 0$

$\quad (u - 6)(u + 2) = 0$

$\quad u - 6 = 0 \quad \text{or} \quad u + 2 = 0$

$\qquad u = 6 \quad \text{or} \qquad u = -2$

7. $\qquad x^2 = 2x$

$\quad x^2 - 2x = 0$

$\quad x(x - 2) = 0$

$\quad x = 0 \quad \text{or} \quad x - 2 = 0$

$\qquad\qquad\qquad x = 2$

9. $x^2 - 6x - 3 = 0$

$$x = \frac{-b \pm \sqrt{b^2 - 4ac}}{2a}, \qquad a = 1, \ b = -6, \ c = -3$$

$$= \frac{-(-6) \pm \sqrt{(-6)^2 - 4(1)(-3)}}{2(1)} = \frac{6 \pm \sqrt{48}}{2} = \frac{6 \pm 4\sqrt{3}}{2} = 3 \pm 2\sqrt{3}$$

11. $3u^2 + 12u + 6 = 0$

Since 3 is a factor of each coefficient, divide both sides by 3.

$u^2 + 4u + 2 = 0$

$$u = \frac{-b \pm \sqrt{b^2 - 4ac}}{2a}, \quad a = 1, \; b = 4, \; c = 2$$

$$= \frac{-4 \pm \sqrt{4^2 - 4(1)(2)}}{2(1)} = \frac{-4 \pm \sqrt{8}}{2} = \frac{-4 \pm 2\sqrt{2}}{2} = -2 \pm \sqrt{2}$$

13.
$$\frac{2x^2}{3} = 5x$$

$$2x^2 = 15x$$

$2x^2 - 15x = 0$

$x(2x - 15) = 0$

$x = 0 \quad \text{or} \quad 2x - 15 = 0$

$$x = \frac{15}{2}$$

15. $4u^2 - 9 = 0$

$4u^2 = 9$ (solve by square root method)

$$u^2 = \frac{9}{4}$$

$$u = \pm\sqrt{\frac{9}{4}} = \pm\frac{3}{2}$$

17.
$$8x^2 + 20x = 12$$

$8x^2 + 20x - 12 = 0$

$2x^2 + 5x - 3 = 0$

$(x + 3)(2x - 1) = 0$

$x + 3 = 0 \quad \text{or} \quad 2x - 1 = 0$

$x = -3 \quad \text{or} \quad 2x = 1$

$$x = \frac{1}{2}$$

19.
$$x^2 = 1 - x$$

$x^2 + x - 1 = 0$

$$x = \frac{-b \pm \sqrt{b^2 - 4ac}}{2a}, \quad a = 1, \; b = 1, \; c = -1$$

$$= \frac{-1 \pm \sqrt{(1)^2 - 4(1)(-1)}}{2(1)} = \frac{-1 \pm \sqrt{5}}{2}$$

21.
$$2x^2 = 6x - 3$$

$2x^2 - 6x + 3 = 0$

$$x = \frac{-b \pm \sqrt{b^2 - 4ac}}{2a}, \quad a = 2, \; b = -6, \; c = 3$$

$$= \frac{-(-6) \pm \sqrt{(-6)^2 - 4(2)(3)}}{2(2)} = \frac{6 \pm \sqrt{12}}{4} = \frac{6 \pm 2\sqrt{3}}{4} = \frac{3 \pm \sqrt{3}}{2}$$

23. $y^2 - 4y = -8$

$y^2 - 4y + 8 = 0$

$$y = \frac{-b \pm \sqrt{b^2 - 4ac}}{2a}, \quad a = 1, \ b = -4, \ c = 8$$

$$= \frac{-(-4) \pm \sqrt{(-4)^2 - 4(1)(8)}}{2(1)} = \frac{4 \pm \sqrt{-16}}{2}$$

Since $\sqrt{-16}$ is not a real number, there are no real solutions.

25. $(2x + 3)^2 = 11$

$2x + 3 = \pm\sqrt{11}$

$2x = -3 \pm \sqrt{11}$

$x = -\dfrac{3}{2} \pm \dfrac{1}{2}\sqrt{11}$

27. $\dfrac{3}{p} = p$

$p^2 = 3$

$p = \pm\sqrt{3}$

29. $2 - \dfrac{2}{m^2} = \dfrac{3}{m}$

$2m^2 - 2 = 3m$

$2m^2 - 3m - 2 = 0$

$(2m + 1)(m - 2) = 0$

$m = -\dfrac{1}{2}, \ 2$

31. $x^2 + 40x - 84$

Step 1. Test for factorability

$$\sqrt{b^2 - 4ac} = \sqrt{(40)^2 - 4(1)(-84)} = \sqrt{1936} = 44$$

Since the result is an integer, the polynomial has first-degree factors with integer coefficients.

Step 2. Use the factor theorem

$x^2 + 40x - 84 = 0$

$$x = \frac{-40 \pm 44}{2} = 2, -42 \ \text{(by the quadratic formula)}$$

Thus, $x^2 + 40x - 84 = (x - 2)(x - [-42]) = (x - 2)(x + 42)$

33. $x^2 - 32x + 144$

Step 1. Test for factorability

$$\sqrt{b^2 - 4ac} = \sqrt{(-32)^2 - 4(1)(144)} = \sqrt{448} \approx 21.166$$

Since this is not an integer, the polynomial is not factorable.

35. $2x^2 + 15x - 108$

Step 1. Test for factorability

$$\sqrt{b^2 - 4ac} = \sqrt{(15)^2 - 4(2)(-108)} = \sqrt{1089} = 33$$

Thus, the polynomial has first-degree factors with integer coefficients.

Step 2. Use the factor theorem

$2x^2 + 15x - 108 = 0$

$$x = \frac{-15 \pm 33}{4} = \frac{9}{2}, -12$$

Thus, $2x^2 + 15x - 108 = 2\left(x - \dfrac{9}{2}\right)(x - [-12]) = (2x - 9)(x + 12)$

37. $4x^2 + 241x - 434$

> Step 1.   Test for factorability
>
> $$\sqrt{b^2 - 4ac} = \sqrt{(241)^2 - 4(4)(-434)} = \sqrt{65025} = 255$$
>
> Thus, the polynomial has first-degree factors with integer coefficients.

> Step 2.   Use the factor theorem
>
> $$4x^2 + 241x - 434 = 0$$
>
> $$x = \frac{-241 \pm 255}{8} = \frac{14}{8}, \ -\frac{496}{8} \ \text{ or } \ \frac{7}{4}, \ -62$$
>
> Thus, $4x^2 + 241x - 434 = 4\left(x - \frac{7}{4}\right)(x + 62) = (4x - 7)(x + 62)$

39.
$$A = P(1 + r)^2$$
$$(1 + r)^2 = \frac{A}{P}$$
$$1 + r = \sqrt{\frac{A}{P}}$$
$$r = \sqrt{\frac{A}{P}} - 1$$

41. $x^2 + 4x + c = 0$

The discriminant is: $16 - 4c$

(A)   If $16 - 4c > 0$, i.e., if $c < 4$, then the equation has two distinct real roots.

(B)   If $16 - 4c = 0$, i.e., if $c = 4$, then the equation has one real double root.

(C)   If $16 - 4c < 0$, i.e., if $c > 4$, then there are no real roots.

43. $x^3 + 8 = (x + 2)(x^2 - 2x + 4) = 0; \ \ x = -2$

45. $5x^4 - 500 = 0, \ \ x^4 - 100 = 0, \ \ (x^2 - 10)(x^2 + 10) = 0, \ \ x = \pm\sqrt{10}$

47. $x^4 - 8x^2 + 15 = 0, \ \ (x^2 - 5)(x^2 - 3) = 0, \ \ x = \pm\sqrt{5}, \ x = \pm\sqrt{3}$

49. Setting the supply equation equal to the demand equation, we have

$$\frac{x}{450} + \frac{1}{2} = \frac{6,300}{x}$$

$$\frac{1}{450}x^2 + \frac{1}{2}x = 6,300$$

$$x^2 + 225x - 2,835,000 = 0$$

$$x = \frac{-225 \pm \sqrt{(225)^2 - 4(1)(-2,835,000)}}{2} \quad \text{(quadratic formula)}$$

$$= \frac{-225 \pm \sqrt{11,390,625}}{2} = \frac{-225 \pm 3375}{2} = 1,575 \text{ units}$$

Note, we discard the negative root since a negative number of units cannot be produced or sold.

Substituting $x = 1,575$ into either equation (we use the demand equation), we get

$$p = \frac{6,300}{1,575} = 4$$

Supply equals demand at $4 per unit.

51. $A = P(1 + r)^2 = P(1 + 2r + r^2) = Pr^2 + 2Pr + P$

Let $A = 625$ and $P = 484$. Then,

$484r^2 + 968r + 484 = 625$

$484r^2 + 968r - 141 = 0$

Using the quadratic formula,

$$r = \frac{-968 \pm \sqrt{(968)^2 - 4(484)(-141)}}{968} = \frac{-968 \pm \sqrt{1,210,000}}{968}$$

$$= \frac{-968 \pm 1100}{968} \approx 0.1364 \text{ or } -2.136$$

Since $r > 0$, we have $r = 0.1364$ or 13.64%.

53. $v^2 = 64h$

For $h = 1$, $v^2 = 64(1) = 64$. Therefore, $v = 8$ ft/sec.

For $h = 0.5$, $v^2 = 64(0.5) = 32$.

Therefore, $v = \sqrt{32} = 4\sqrt{2} \approx 5.66$ ft/sec.

APPENDIX B SPECIAL TOPICS

EXERCISE B-1

Things to remember:

1. SEQUENCES
    A SEQUENCE is a function whose domain is a set of successive integers. If the domain of a given sequence is a finite set, then the sequence is called a FINITE SEQUENCE; otherwise, the sequence is an INFINITE SEQUENCE. In general, unless stated to the contrary or the context specifies otherwise, the domain of a sequence will be understood to be the set N of natural numbers.

2. NOTATION FOR SEQUENCES

    Rather than function notation $f(n)$, n in the domain of a given sequence f, subscript notation a_n is normally used to denote the value in the range corresponding to n, and the sequence itself is denoted $\{a_n\}$ rather than f or $f(n)$. The elements in the range, a_n, are called the TERMS of the sequence; a_1 is the first term, a_2 is the second term, and a_n is the nth term or general term.

3. SERIES
    Given a sequence $\{a_n\}$. The sum of the terms of the sequence,
    $a_1 + a_2 + a_3 + \cdots$ is called a SERIES. If the sequence is finite,
    the corresponding series is a FINITE SERIES; if the sequence is infinite, then the corresponding series is an INFINITE SERIES.
    Only finite series are considered in this section.

4. NOTATION FOR SERIES
    Series are represented using SUMMATION NOTATION.
    If $\{a_k\}$, $k = 1, 2, \ldots, n$ is a finite sequence, then the series

    $$a_1 + a_2 + a_3 + \cdots + a_n$$

    is denoted

    $$\sum_{k=1}^{n} a_k.$$

    The symbol $\sum$ is called the SUMMATION SIGN and k is called the SUMMING INDEX.

5. ARITHMETIC MEAN
    If $\{a_k\}$, $k = 1, 2, \ldots, n$, is a finite sequence, then the ARITHMETIC MEAN $\bar{a}$ of the sequence is defined as

    $$\bar{a} = \frac{1}{n} \sum_{k=1}^{n} a_k.$$

1. $a_n = 2n + 3;$ $a_1 = 2\cdot1 + 3 = 5$
$a_2 = 2\cdot2 + 3 = 7$
$a_3 = 2\cdot3 + 3 = 9$
$a_4 = 2\cdot4 + 3 = 11$

3. $a_n = \dfrac{n+2}{n+1};$ $a_1 = \dfrac{1+2}{1+1} = \dfrac{3}{2}$

$a_2 = \dfrac{2+2}{2+1} = \dfrac{4}{3}$

$a_3 = \dfrac{3+2}{3+1} = \dfrac{5}{4}$

$a_4 = \dfrac{4+2}{4+1} = \dfrac{6}{5}$

5. $a_n = (-3)^{n+1};$ $a_1 = (-3)^{1+1} = (-3)^2 = 9$
$a_2 = (-3)^{2+1} = (-3)^3 = -27$
$a_3 = (-3)^{3+1} = (-3)^4 = 81$
$a_4 = (-3)^{4+1} = (-3)^5 = -243$

7. $a_n = 2n + 3; a_{10} = 2\cdot10 + 3 = 23$

9. $a_n = \dfrac{n+2}{n+1}; a_{99} = \dfrac{99+2}{99+1} = \dfrac{101}{100}$

11. $\displaystyle\sum_{k=1}^{6} k = 1 + 2 + 3 + 4 + 5 + 6 = 21$

13. $\displaystyle\sum_{k=4}^{7} (2k - 3) = (2\cdot4 - 3) + (2\cdot5 - 3) + (2\cdot6 - 3) + (2\cdot7 - 3) = 5 + 7 + 9 + 11 = 32$

15. $\displaystyle\sum_{k=0}^{3} \dfrac{1}{10^k} = \dfrac{1}{10^0} + \dfrac{1}{10^1} + \dfrac{1}{10^2} + \dfrac{1}{10^3} = 1 + \dfrac{1}{10} + \dfrac{1}{100} + \dfrac{1}{1000} = \dfrac{1111}{1000} = 1.111$

17. $a_1 = 5, a_2 = 4, a_3 = 2, a_4 = 1, a_5 = 6.$ Here $n = 5$ and the arithmetic mean is given by:
$$\bar{a} = \dfrac{1}{5}\sum_{k=1}^{5} a_k = \dfrac{1}{5}(5 + 4 + 2 + 1 + 6) = \dfrac{18}{5} = 3.6$$

19. $a_1 = 96, a_2 = 65, a_3 = 82, a_4 = 74, a_5 = 91, a_6 = 88, a_7 = 87, a_8 = 91, a_9 = 77,$ and $a_{10} = 74.$ Here $n = 10$ and the arithmetic mean is given by:
$$\bar{a} = \dfrac{1}{10}\sum_{k=1}^{10} a_k = \dfrac{1}{10}(96 + 65 + 82 + 74 + 91 + 88 + 87 + 91 + 77 + 74) = \dfrac{825}{10} = 82.5$$

21. $a_n = \dfrac{(-1)^{n+1}}{2^n};$ $a_1 = \dfrac{(-1)^2}{2^1} = \dfrac{1}{2}$

$a_2 = \dfrac{(-1)^3}{2^2} = -\dfrac{1}{4}$

$a_3 = \dfrac{(-1)^4}{2^3} = \dfrac{1}{8}$

$a_4 = \dfrac{(-1)^5}{2^4} = -\dfrac{1}{16}$

$a_5 = \dfrac{(-1)^6}{2^5} = \dfrac{1}{32}$

23. $a_n = n[1 + (-1)^n]$;

$a_1 = 1[1 + (-1)^1] = 0$

$a_2 = 2[1 + (-1)^2] = 4$

$a_3 = 3[1 + (-1)^3] = 0$

$a_4 = 4[1 + (-1)^4] = 8$

$a_5 = 5[1 + (-1)^5] = 0$

25. $a_n = \left(-\dfrac{3}{2}\right)^{n-1}$;

$a_1 = \left(-\dfrac{3}{2}\right)^0 = 1$

$a_2 = \left(-\dfrac{3}{2}\right)^1 = -\dfrac{3}{2}$

$a_3 = \left(-\dfrac{3}{2}\right)^2 = \dfrac{9}{4}$

$a_4 = \left(-\dfrac{3}{2}\right)^3 = -\dfrac{27}{8}$

$a_5 = \left(-\dfrac{3}{2}\right)^4 = \dfrac{81}{16}$

27. Given $-2, -1, 0, 1, \ldots$ The sequence is the set of successive integers beginning with -2. Thus, $a_n = n - 3$, $n = 1, 2, 3, \ldots$.

29. Given $4, 8, 12, 16, \ldots$ The sequence is the set of positive integer multiples of 4. Thus, $a_n = 4n$, $n = 1, 2, 3, \ldots$.

31. Given $\dfrac{1}{2}, \dfrac{3}{4}, \dfrac{5}{6}, \dfrac{7}{8}, \ldots$ The sequence is the set of fractions whose numerators are the odd positive integers and whose denominators are the even positive integers. Thus,

$a_n = \dfrac{2n-1}{2n}$, $n = 1, 2, 3, \ldots$.

33. Given $1, -2, 3, -4, \ldots$ The sequence consists of the positive integers with alternating signs. Thus,

$a_n = (-1)^{n+1}n$, $n = 1, 2, 3, \ldots$.

35. Given $1, -3, 5, -7, \ldots$ The sequence consists of the odd positive integers with alternating signs. Thus,

$a_n = (-1)^{n+1}(2n - 1)$, $n = 1, 2, 3, \ldots$.

37. Given $1, \dfrac{2}{5}, \dfrac{4}{25}, \dfrac{8}{125}, \ldots$ The sequence consists of the nonnegative integral powers of $\dfrac{2}{5}$. Thus,

$a_n = \left(\dfrac{2}{5}\right)^{n-1}$, $n = 1, 2, 3, \ldots$.

39. Given $x, x^2, x^3, x^4, \ldots$ The sequence is the set of positive integral powers of x. Thus, $a_n = x^n$, $n = 1, 2, 3, \ldots$.

41. Given $x, -x^3, x^5, -x^7, \ldots$ The sequence is the set of positive odd integral powers of x with alternating signs. Thus,

$$a_n = (-1)^{n+1}x^{2n-1}, \; n = 1, 2, 3, \ldots.$$

43. $\displaystyle\sum_{k=1}^{5}(-1)^{k+1}(2k-1)^2 = (-1)^2(2\cdot 1 - 1)^2 + (-1)^3(2\cdot 2 - 1)^2 + (-1)^4(2\cdot 3 - 1)^2 + (-1)^5(2\cdot 4 - 1)^2$

$$+ (-1)^6(2\cdot 5 - 1)^2 = 1 - 9 + 25 - 49 + 81$$

45. $\displaystyle\sum_{k=2}^{5}\frac{2^k}{2k+3} = \frac{2^2}{2\cdot 2 + 3} + \frac{2^3}{2\cdot 3 + 3} + \frac{2^4}{2\cdot 4 + 3} + \frac{2^5}{2\cdot 5 + 3} = \frac{4}{7} + \frac{8}{9} + \frac{16}{11} + \frac{32}{13}$

47. $\displaystyle\sum_{k=1}^{5}x^{k-1} = x^0 + x^1 + x^2 + x^3 + x^4 = 1 + x + x^2 + x^3 + x^4$

49. $\displaystyle\sum_{k=0}^{4}\frac{(-1)^k x^{2k+1}}{2k+1} = \frac{(-1)^0 x}{2\cdot 0 + 1} + \frac{(-1)^1 x^3}{2\cdot 1 + 1} + \frac{(-1)^2 x^5}{2\cdot 2 + 1} + \frac{(-1)^3 x^7}{2\cdot 3 + 1} + \frac{(-1)^4 x^9}{2\cdot 4 + 1} = x - \frac{x^3}{3} + \frac{x^5}{5} - \frac{x^7}{7} + \frac{x^9}{9}$

51. (A) $\; 2 + 3 + 4 + 5 + 6 = \displaystyle\sum_{k=1}^{5}(k+1)$ (B) $\; 2 + 3 + 4 + 5 + 6 = \displaystyle\sum_{j=0}^{4}(j+2)$

53. (A) $\; 1 - \dfrac{1}{2} + \dfrac{1}{3} - \dfrac{1}{4} = \displaystyle\sum_{k=1}^{4}\frac{(-1)^{k+1}}{k}$ (B) $\; 1 - \dfrac{1}{2} + \dfrac{1}{3} - \dfrac{1}{4} = \displaystyle\sum_{j=0}^{3}\frac{(-1)^j}{j+1}$

55. $2 + \dfrac{3}{2} + \dfrac{4}{3} + \ldots + \dfrac{n+1}{n} = \displaystyle\sum_{k=1}^{n}\frac{k+1}{k}$

57. $\dfrac{1}{2} - \dfrac{1}{4} + \dfrac{1}{8} - \ldots + \dfrac{(-1)^{n+1}}{2^n} = \displaystyle\sum_{k=1}^{n}\frac{(-1)^{k+1}}{2^k}$

59. False: $\; 1 + \dfrac{1}{2} + \dfrac{1}{3} + \dfrac{1}{4} + \dfrac{1}{5} + \dfrac{1}{6} + \ldots + \dfrac{1}{64}$

$$= 1 + \frac{1}{2} + \left(\frac{1}{3} + \frac{1}{4}\right) + \left(\frac{1}{5} + \frac{1}{6} + \frac{1}{7} + \frac{1}{8}\right) + \left(\frac{1}{9} + \cdots + \frac{1}{16}\right) + \left(\frac{1}{17} + \cdots + \frac{1}{32}\right) + \left(\frac{1}{33} + \cdots + \frac{1}{64}\right)$$

$$> 1 + \frac{1}{2} + \frac{1}{2} + \frac{1}{2} + \frac{1}{2} + \frac{1}{2} + \frac{1}{2} = 4$$

61. True: $\; \dfrac{1}{2} - \dfrac{1}{4} + \dfrac{1}{8} - \dfrac{1}{16} + \dfrac{1}{32} - \ldots$

$$= \left(\frac{1}{2} - \frac{1}{4}\right) + \left(\frac{1}{8} - \frac{1}{16}\right) + \left(\frac{1}{32} - \frac{1}{64}\right) + \text{(positive terms)}$$

$$= \frac{1}{4} + \frac{1}{16} + \frac{1}{64} + \ldots > \frac{1}{4}$$

63. $a_1 = 2$ and $a_n = 3a_{n-1} + 2$

for $n \geq 2$.

$a_1 = 2$

$a_2 = 3 \cdot a_1 + 2 = 3 \cdot 2 + 2 = 8$

$a_3 = 3 \cdot a_2 + 2 = 3 \cdot 8 + 2 = 26$

$a_4 = 3 \cdot a_3 + 2 = 3 \cdot 26 + 2 = 80$

$a_5 = 3 \cdot a_4 + 2 = 3 \cdot 80 + 2 = 242$

65. $a_1 = 1$ and $a_n = 2a_{n-1}$

for $n \geq 2$.

$a_1 = 1$

$a_2 = 2 \cdot a_1 = 2 \cdot 1 = 2$

$a_3 = 2 \cdot a_2 = 2 \cdot 2 = 4$

$a_4 = 2 \cdot a_3 = 2 \cdot 4 = 8$

$a_5 = 2 \cdot a_4 = 2 \cdot 8 = 16$

67. If $a_1 = \dfrac{A}{2}$, $a_n = \dfrac{1}{2}\left(a_{n-1} + \dfrac{A}{a_{n-1}}\right)$, $n \geq 2$, let $A = 2$. Then:

$a_1 = \dfrac{2}{2} = 1$

$a_2 = \dfrac{1}{2}\left(a_1 + \dfrac{A}{a_1}\right) = \dfrac{1}{2}(1+2) = \dfrac{3}{2}$

$a_3 = \dfrac{1}{2}\left(a_2 + \dfrac{A}{a_2}\right) = \dfrac{1}{2}\left(\dfrac{3}{2} + \dfrac{2}{3/2}\right) = \dfrac{1}{2}\left(\dfrac{3}{2} + \dfrac{4}{3}\right) = \dfrac{17}{12}$

$a_4 = \dfrac{1}{2}\left(a_3 + \dfrac{A}{a_3}\right) = \dfrac{1}{2}\left(\dfrac{17}{12} + \dfrac{2}{17/12}\right) = \dfrac{1}{2}\left(\dfrac{17}{12} + \dfrac{24}{17}\right) = \dfrac{577}{408} \approx 1.414216$

and $\sqrt{2} \approx 1.414214$

69. $a_1 = 1$, $a_2 = 1$, $a_n = a_{n-1} + a_{n-2}$, $n \geq 3$

$a_3 = a_2 + a_1 = 2$, $a_4 = a_3 + a_2 = 3$, $a_5 = a_4 + a_3 = 5$,

$a_6 = a_5 + a_4 = 8$, $a_7 = a_6 + a_5 = 13$, $a_8 = a_7 + a_6 = 21$

$a_9 = a_8 + a_7 = 34$, $a_{10} = a_9 + a_8 = 55$

EXERCISE B-2

Things to remember:

<u>1</u>.   A sequence of numbers $a_1, a_2, a_3, \ldots, a_n, \ldots,$ is called an ARITHMETIC SEQUENCE if

there is constant d, called the COMMON DIFFERENCE, such that

$a_n - a_{n-1} = d,$

that is,

$a_n = a_{n-1} + d$

for all $n > 1$.

<u>2</u>.   A sequence of numbers $a_1, a_2, a_3, \ldots, a_n, \ldots,$ is called a GEOMETRIC SEQUENCE if

there exists a nonzero constant r, called the COMMON RATIO, such that

$\dfrac{a_n}{a_{n-1}} = r,$

that is,

$a_n = ra_{n-1}$

for all $n > 1$.

3.   *n*TH TERM OF AN ARITHMETIC SEQUENCE

If $\{a_n\}$ is an arithmetic sequence with common difference d, then

$$a_n = a_1 + (n-1)d$$

for all $n > 1$.

4.   *n*TH TERM OF A GEOMETRIC SEQUENCE

If $\{a_n\}$ is a geometric sequence with common ratio r, then

$$a_n = a_1 r^{n-1}$$

for all $n > 1$.

5.   SUM FORMULAS FOR FINITE ARITHMETIC SERIES

The sum S_n of the first n terms of an arithmetic series
$a_1 + a_2 + a_3 + \ldots + a_n$ with common difference d, is given by

(a) $S_n = \dfrac{n}{2}[2a_1 + (n-1)d]$ (First Form)

or by

(b) $S_n = \dfrac{n}{2}(a_1 + a_n).$ (Second Form)

6.   SUM FORMULAS FOR FINITE GEOMETRIC SERIES

The sum S_n of the first n terms of a geometric series
$a_1 + a_2 + a_3 + a_n$ with common ratio r, is given by:

$$S_n = \frac{a_1(r^n - 1)}{r-1},\ r \neq 1, \quad \text{(First Form)}$$

or by

$$S_n = \frac{ra_n - a_1}{r-1},\ r \neq 1. \quad \text{(Second Form)}$$

7.   SUM OF AN INFINITE GEOMETRIC SERIES

If $a_1 + a_2 + a_3 + \ldots + a_n + \ldots$, is an infinite geometric series with common ratio r having
the property $-1 < r < 1$, then the sum S_∞ is defined to be:

$$S_\infty = \frac{a_1}{1-r}.$$

1. (A) $-11, -16, -21, \ldots$
        This is an arithmetic sequence with common difference $d = -5$;
        $a_4 = -26$, $a_5 = -31$.

    (B) $2, -4, 8, \ldots$
        This is a geometric sequence with common ratio $r = -2$;
        $a_4 = -16$, $a_5 = 32$.

    (C) $1, 4, 9, \ldots$
        This is neither an arithmetic sequence ($4 - 1 \neq 9 - 4$) nor a geometric sequence $\left(\dfrac{4}{1} \neq \dfrac{9}{4}\right)$.

    (D) $\dfrac{1}{2}, \dfrac{1}{6}, \dfrac{1}{18}, \ldots$
        This is a geometric sequence with common ratio $r = \dfrac{1}{3}$;
        $a_4 = \dfrac{1}{54}$, $a_5 = \dfrac{1}{162}$.

3. $\displaystyle\sum_{k=1}^{101} (-1)^{k+1} = 1 - 1 + 1 - 1 + \ldots + 1$

    This is a geometric series with $a_1 = 1$ and common ratio $r = -1$.

    $S_{101} = \dfrac{1[(-1)^{101} - 1]}{-1 - 1} = \dfrac{-2}{-2} = 1$

5. This series is neither arithmetic nor geometric.

7. $5 + 4.9 + 4.8 + \ldots + 0.1$ is an arithmetic series with $a_1 = 5$, $a_{50} = 0.1$ and common difference $d = -0.1$:

    $S_{50} = \dfrac{50}{2}[5 + 0.1] = 25(5.1) = 127.5$

9. $a_2 = a_1 + d = 7 + 4 = 11$
    $a_3 = a_2 + d = 11 + 4 = 15$ (using $\underline{1}$)

11. $a_{21} = a_1 + (21 - 1)d = 2 + 20\cdot4 = 82$ (using $\underline{3}$)

    $S_{31} = \dfrac{31}{2}[2a_1 + (31-1)d] = \dfrac{31}{2}[2\cdot2 + 30\cdot4] = \dfrac{31}{2} \cdot 124 = 1922$ [using $\underline{5}$(a)]

13. Using $\underline{5}$(b), $S_{20} = \dfrac{20}{2}(a_1 + a_{20}) = 10(18 + 75) = 930$

15. $a_2 = a_1 r = 3(-2) = -6$
    $a_3 = a_1 r^2 = 3(-2)^2 = 3\cdot4 = 12$
    $a_4 = a_1 r^3 = 3(-2)^3 = 3(-8) = -24$ (using $\underline{4}$)

17. Using $\underline{6}$, $S_7 = \dfrac{-3\cdot729 - 1}{-3 - 1} = \dfrac{-2188}{-4} = 547$.

19. Using $\underline{4}$, $a_{10} = 100(1.08)^9 = 199.90$.

21. Using $\underline{4}$, $200 = 100\,r^8$. Thus, $r^8 = 2$ and $r = \pm\sqrt[8]{2} \approx \pm 1.09$.

23. Using $\underline{6}$, $S_{10} = \dfrac{500[(0.6)^{10} - 1]}{0.6 - 1} \approx 1242$,

$$S_\infty = \frac{500}{1 - 0.6} = 1250.$$

25. $S_{41} = \displaystyle\sum_{k=1}^{41}(3k + 3)$. The sequence of terms is an arithmetic sequence. Therefore,

$$S_{41} = \frac{41}{2}(a_1 + a_{41}) = \frac{41}{2}(6 + 126) = \frac{41}{2}(132) = 41(66) = 2{,}706$$

27. $S_8 = \displaystyle\sum_{k=1}^{8}(-2)^{k-1}$. The sequence of terms is a geometric sequence with common ratio $r = -2$ and

$a_1 = (-2)^0 = 1$.

$$S_8 = \frac{1[(-2)^8 - 1]}{-2 - 1} = \frac{256 - 1}{-3} = -85$$

29. Let $a_1 = 13$, $d = 2$. Then, using $\underline{3}$, we can find n:

$$67 = 13 + (n - 1)2 \quad \text{or} \quad 2(n - 1) = 54$$
$$n - 1 = 27$$
$$n = 28$$

Therefore, using $\underline{5}$(b), $S_{28} = \dfrac{28}{2}[13 + 67] = 14 \cdot 80 = 1120$.

31. (A) $2 + 4 + 8 + \cdots$. Since $r = \dfrac{4}{2} = \dfrac{8}{4} = \cdots = 2$ and $|2| = 2 > 1$, the sum does not exist.

(B) $2, -\dfrac{1}{2}, \dfrac{1}{8}, \ldots$. In this case, $r = \dfrac{-1/2}{2} = \dfrac{1/8}{-1/2} = \cdots = -\dfrac{1}{4}$.

Since $|r| < 1$, $S_\infty = \dfrac{2}{1 - (-1/4)} = \dfrac{2}{5/4} = \dfrac{8}{5} = 1.6$.

33. $f(1) = -1$, $f(2) = 1$, $f(3) = 3$, $\ldots$. This is an arithmetic sequence $a_1 = -1$, $d = 2$. Thus, using $\underline{5}$(a),

$$f(1) + f(2) + f(3) + \cdots + f(50) = \frac{50}{2}[2(-1) + 49 \cdot 2] = 25 \cdot 96 = 2400.$$

35. $f(1) = \dfrac{1}{2}$, $f(2) = \left(\dfrac{1}{2}\right)^2 = \dfrac{1}{4}$, $f(3) = \left(\dfrac{1}{2}\right)^3 = \dfrac{1}{8}$, $\ldots$. This is a geometric sequence with $a_1 = \dfrac{1}{2}$ and $r = \dfrac{1}{2}$. Thus, using $\underline{6}$:

$$f(1) + f(2) + \cdots + f(10) = S_{10} = \frac{\dfrac{1}{2}\left[\left(\dfrac{1}{2}\right)^{10} - 1\right]}{\dfrac{1}{2} - 1} \approx 0.999$$

Copyright © 2015 Pearson Education, Inc.

37. Consider the arithmetic sequence with $a_1 = 1$, $d = 2$. This is the sequence of odd positive integers. Now, using $\underline{5}$(a), the sum of the first n odd positive integers is:

$$S_n = \frac{n}{2}[2 \cdot 1 + (n-1)2] = \frac{n}{2}(2 + 2n - 2) = \frac{n}{2} \cdot 2n = n^2$$

39. $S_n = a_1 + a_1 r + \ldots + a_1 r^{n-1}$. If $r = 1$, then $S_n = na_1$.

41. No: $\frac{n}{2}(1 + 1.1) = 100$ implies $(2.1)n = 200$ and this equation does not have an integer solution.

43. Yes: Solve the equation $6 = \frac{10}{1-r}$ for r. This yields $r = -\frac{2}{3}$.

The infinite geometric series: $10 - 10\left(\frac{2}{3}\right) + 10\left(\frac{2}{3}\right)^2 - 10\left(\frac{2}{3}\right)^3 + \ldots$

has sum $S_\infty = 6$.

45. Consider the time line:

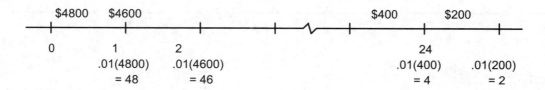

The total cost of the loan is $2 + 4 + 6 + \cdots + 46 + 48$. The terms form an arithmetic sequence with $n = 24$, $a_1 = 2$, and $a_{24} = 48$. Thus, using $\underline{5}$(b):

$$S_{24} = \frac{24}{2}(2 + 48) = 24 \cdot 25 = \$600$$

47. This is a geometric sequenceprogression with $a_1 = 3,500,000$ and $r = 0.7$. Thus, using $\underline{7}$:

$$S_\infty = \frac{3,500,000}{1 - 0.7} \approx \$11,666,666.67$$

49.

$\$1000$	$\$1050$	$\$1102.50$	$\$1157.63$
	1	2	3
	$(1.05)1000$	$(1.05)(1050)$	$1.05(1102.50)$
	$= 1050$	$= 1000(1.05)^2$	$= 1000(1.05)^3$
		$= 1102.50$	$= 1157.63$

In general, after n years, the amount A_n in the account is:

$$A_n = 1000(1.05)^n$$

Thus, $A_{10} = 1000(1.05)^{10} \approx \$1628.89$

and $A_{20} = 1000(1.05)^{20} \approx \$2653.30$

EXERCISE B-3

Things to remember:

1.   If n is a positive integer, then n FACTORIAL, denoted $n!$, is the product of the integers from 1 to n; that is,

$$n! = n \cdot (n-1) \cdot \ldots \cdot 3 \cdot 2 \cdot 1 = n(n-1)!$$

Also, $1! = 1$ and $0! = 1$.

2.   If n and r are nonnegative integers and $r \leq n$, then:

$$C_{n,r} = \frac{n!}{r!(n-r)!}$$

3.   BINOMIAL THEOREM

For all natural numbers n:

$$(a+b)^n = {}_n C_0 a^n + {}_n C_1 a^{n-1} b + {}_n C_2 a^{n-2} b^2 + \cdots + {}_n C_{n-1} ab^{n-1} + {}_n C_n b^n$$

---

1. $6! = 6 \cdot 5 \cdot 4 \cdot 3 \cdot 2 \cdot 1 = 720$

3. $\dfrac{10!}{9!} = \dfrac{10 \cdot 9!}{9!} = 10$

5. $\dfrac{12!}{9!} = \dfrac{12 \cdot 11 \cdot 10 \cdot 9!}{9!} = 1320$

7. $\dfrac{5!}{2!3!} = \dfrac{5 \cdot 4 \cdot 3!}{2 \cdot 1 \cdot 3!} = 10$

9. $\dfrac{6!}{5!(6-5)!} = \dfrac{6 \cdot 5!}{5!1!} = 6$

11. $\dfrac{20!}{3!17!} = \dfrac{20 \cdot 19 \cdot 18 \cdot 17!}{3!17!}$

$= \dfrac{20 \cdot 19 \cdot 18}{3 \cdot 2 \cdot 1} = 1140$

13. ${}_5 C_3 = \dfrac{5!}{3!(5-3)!} = \dfrac{5!}{3!2!} = 10$ (see Problem 7)

15. ${}_6 C_5 = \dfrac{6!}{5!(6-5)!} = 6$ (see Problem 9)

17. ${}_5 C_0 = \dfrac{5!}{0!(5-0)!} = \dfrac{5!}{1 \cdot 5!} = 1$

19. ${}_{18} C_{15} = \dfrac{18!}{15!(18-15)!} = \dfrac{18 \cdot 17 \cdot 16 \cdot 15!}{15!3!} = \dfrac{18 \cdot 17 \cdot 16}{3 \cdot 2 \cdot 1} = 816$

21. Using 3,

$(a+b)^4 = {}_4 C_0 a^4 + {}_4 C_1 a^3 b + {}_4 C_2 a^2 b^2 + {}_4 C_3 ab^3 + {}_4 C_4 b^4 = a^4 + 4a^3 b + 6a^2 b^2 + 4ab^4 + b^4$

23. Using 3,

$$(x-1)^6 = [x+(-1)]^6$$

$$= {}_6C_0\,x^6 + {}_6C_1\,x^5(-1) + {}_6C_2x^4(-1)^2 + {}_6C_3\,x^3(-1)^3 + {}_6C_4\,x^2(-1)^4 + {}_6C_5\,x(-1)^5 + {}_6C_6(-1)^6$$

$$= x^6 - 6x^5 + 15x^4 - 20x^3 + 15x^2 - 6x + 1$$

25. $(2a-b)^5 = [2a+(-b)]^5$

$$= {}_5C_0\,(2a)^5 + {}_5C_1\,(2a)^4(-b) + {}_5C_2\,(2a)^3(-b)^2 + {}_5C_3\,(2a)^2(-b)^3 + {}_5C_4\,(2a)(-b)^4 + {}_5C_5\,(-b)^5$$

$$= 32a^5 - 80a^4b + 80a^3b^2 - 40a^2b^3 + 10ab^4 - b^5$$

27. The fifth term in the expansion of $(x-1)^{18}$ is:

$$_{18}C_4\,x^{14}(-1)^4 = \frac{18\cdot17\cdot16\cdot15}{4\cdot3\cdot2\cdot1}x^{14} = 3060x^{14}$$

29. The seventh term in the expansion of $(p+q)^{15}$ is:

$$_{15}C_6\,p^9q^6 = \frac{15\cdot14\cdot13\cdot12\cdot11\cdot10}{6\cdot5\cdot4\cdot3\cdot2\cdot1}p^9q^6 = 5005p^9q^6$$

31. The eleventh term in the expansion of $(2x+y)^{12}$ is:

$$_{12}C_{10}\,(2x)^2y^{10} = \frac{12\cdot11}{2\cdot1}4x^2y^{10} = 264x^2y^{10}$$

33. $_nC_0 = \dfrac{n!}{0!(n-0)!} = \dfrac{n!}{1\cdot n!} = 1, \qquad _nC_n = \dfrac{n!}{n!(n-n)!} = \dfrac{n!}{n!0!} = 1$

35. The next two rows are:

1 5 10 10 5 1   and   1 6 15 20 15 6 1,

respectively. These are the coefficients in the binomial expansions of $(a+b)^5$ and $(a+b)^6$.

37. The nth row of Pascal's triangle gives the coefficients of $(a+b)^k$.
If we let $a=1$ and $b=-1$, we get

$$0 = (1-1)^n = {}_nC_0\,1^n + {}_nC_1\,(1)^{n-1}(-1) + {}_nC_2\,1^{n-2}(-1)^2 + \dots + {}_nC_n\,(-1)^n$$

$$= {}_nC_0 - {}_nC_1 + {}_nC_2 - {}_nC_3 + \dots + (-1)^n\,{}_nC_n.$$

39. $_nC_{r-1} + {}_nC_r = \dfrac{n!}{(r-1)!(n-[r-1])!} + \dfrac{n!}{r!(n-r)!} = \dfrac{n!}{(r-1)!(n-r+1)!} + \dfrac{n!}{r!(n-r)!}$

$$= \frac{r\cdot n! + (n-r+1)n!}{r!(n-r+1)!} = \frac{(n+1)n!}{r!(n-r+1)!} = \frac{(n+1)!}{r!(n+1-r)!} = {}_{n+1}C_r$$